# PRECALCULUS:

## FUNCTIONS

## AND GRAPHS

## DEDICATION

This Seventh Edition of *Precalculus: Functions and Graphs* is dedicated to the memory of Earl W. Swokowski.

I am grateful for having been given the opportunity to work with a great mentor.

## A NOTE FROM THE PUBLISHER

Earl W. Swokowski died on June 2, 1992. We at PWS Publishing are grateful to have been associated with one of the premier mathematics authors of the last twenty-five years, and we will miss him.

Earl has left us a rich legacy of extraordinary work and a tradition of excellence. Jeffery A. Cole has worked with Earl over the course of several editions of this precalculus series, and we are confident that he will carry on this tradition of uncompromising quality.

# SEVENTH EDITION

# PRECALCULUS:

# FUNCTIONS

# AND GRAPHS

**EARL W. SWOKOWSKI**

**JEFFERY A. COLE**
*Anoka-Ramsey Community College*

**PWS PUBLISHING COMPANY**
*Boston*

**PWS PUBLISHING COMPANY**
20 Park Plaza, Boston, MA 02116-4324

PWS Publishing Company is a division of Wadsworth, Inc.

International Thomson Publishing
The trademark ITP is used under license

**LIBRARY OF CONGRESS CATALOGING-IN-PUBLICATION DATA**

Swokowski, Earl William,
    Precalculus / Earl W. Swokowski, Jeffery A. Cole.—[7th ed.]
        p.    cm.
    Includes index.
    ISBN 0-534-93702-0
     1. Functions.  2. Algebra—Graphic methods.  3. Trigonometry.
    I. Cole, Jeffery A. (Jeffery Alan),        II. Title.
    QA331.3.S95   1993
    515'.1—dc20                      93-28896
                                                 CIP

Chapter opening photos appear courtesy of *The Stock Market*. Ch. 1, page 2 © David Lawrence; Ch. 2, page 112 © Otto Rogge; Ch. 3, page 182 © Pete Saloutos; Ch. 4, page 240 and Ch. 9, page 612 © Ted Horowitz; Ch. 5, page 300 © Peter Wilson; Ch. 6, page 402 © William E. Whitehurst; Ch. 7, page 464 © John Madere; Ch. 8, page 522 © Pina Oddo and Livio Sinibaldi; Ch. 10, page 682 © Ken Straiton.

Sponsoring Editor: *David Dietz*
Production Coordinator: *Susan M. C. Caffey*
Editorial Assistant: *Clark Benbow*
Assistant Editor: *Kelle Flannery*
Marketing Manager: *Marianne Rutter*
Manufacturing Coordinator: *Marcia A. Locke*
Production: *Lifland et al., Bookmakers/Sally Lifland*
Interior Designer: *Elise S. Kaiser*
Interior Illustrator: *Scientific Illustrators*
Cover Designer: *Susan M. C. Caffey*
Cover Art: *Steven Hunt © 1993*
Compositor: *York Graphic Services*
Cover Printer: *New England Book Components, Inc.*
Text Printer and Binder: *Arcata Graphics/Hawkins*

*Printed and bound in the United States of America.*
*93 94 95 96 97 98 — 10 9 8 7 6 5 4 3 2 1*

# CONTENTS

# 5

## The Trigonometric Functions                                301

# 6

## Analytic Trigonometry                                       403

*This book contains a Quick Reference Card. If this card is removed, the purchaser forfeits all returns, rights, and privileges to the seller.*

# PREFACE

One of our main goals in preparing the seventh edition of *Precalculus: Functions and Graphs* was to enhance the clarity of the discussions. We wanted to enable the student to more easily understand the concepts presented, but we did not want to sacrifice the mathematical soundness that has been paramount to the success of this text. By improving the explanations of the more difficult concepts and including step-by-step comments on the solutions of the examples, we have furthered our goal of helping the student while maintaining mathematical rigor.

Treatment of various topics has been updated to include the use of a graphics calculator. Graphics calculator examples and exercises are found throughout the text, wherever they are appropriate to the material being presented. The appendix on graphics calculators offers a great opportunity for both students and professors to familiarize themselves with examples and exercises (some from the main text, some designed specifically for the appendix) as well as with the use of a graphics calculator. The graphics calculator material in this text is versatile enough to satisfy the needs of every professor, from those who require minimal or no use of the graphics calculator to those who require its frequent use.

Suggestions provided by professors for improvements in the arrangement of topics and the writing have resulted in some reorganization and a great deal of rewriting for this edition. A main goal in the rewriting of the text was to include more intuition in the explanations. Hence, some explanations have been greatly expanded—making the material more accessible to average students while thoroughly preparing them for calculus. Throughout the text, there is an increased emphasis on approximating and interpreting results.

The principal changes made to each chapter are highlighted below.

## CHANGES FOR THE SEVENTH EDITION

**Chapter 1**  A comparison of the table of contents for this edition and the Sixth Edition would suggest that there is significantly more material in Chapter 1 of this edition. This conclusion is a bit misleading, however, because

Sections 1.2 and 1.3 are essentially Section 1.2 from the Sixth Edition, Sections 1.4 and 1.6 are more complete separate discussions of equations and inequalities from Section 1.3 of the Sixth Edition, and Section 1.5 (Complex Numbers) was previously Section 3.3. These changes were made to accommodate professors who requested a more thorough review chapter. Professors who are not interested in reviewing this material could easily begin their course with Section 1.7 or Section 2.1. We recommend beginning with Section 1.7 (where graphics calculator examples and exercises first appear) if you do not wish to review but do have students who are using graphics calculators for the first time. Many improvements have been made in the coverage of topics.

**Chapter 2**  Section 2.1 includes a greater emphasis on the difference quotient and linear functions, as well as a short discussion of variation. In Section 2.2, the improved charts on vertically or horizontally stretching and compressing graphs, along with their pedagogically sound color scheme, should help the student develop a true understanding of these topics. Section 2.3 includes a more intuitive method for finding the $x$-coordinate of the vertex of a parabola (by finding the average of the $x$-intercepts). The topic of composite function forms, fundamental to the study of the chain rule in calculus, is included in Section 2.4.

**Chapter 3**  Added emphasis has been placed on the graphical interpretation of the relationship of a zero of a polynomial and its multiplicity. The review material on complex numbers has been moved to Section 1.5 so as not to interrupt the flow of discussion on polynomials. The discussion of rational functions in Section 3.5 has been extensively rewritten with a more intuitive flavor. In general, the rewriting of material on graphing in this edition has deemphasized point plotting and emphasized understanding.

**Chapter 4**  Two new theorems highlight the fact that exponential and logarithmic functions are one-to-one. Greater emphasis has been placed on changing exponential forms to equivalent logarithmic forms (and vice versa) in the solving of exponential and logarithmic equations. The material on common and natural logarithms, a separate section in the Sixth Edition, is now incorporated throughout the chapter.

**Chapter 5**  The formula for the area of a circular sector has been added to Section 5.1. Verification of simple trigonometric identities has been added to Sections 5.2 and 5.3. In Section 5.4, the discussion of calculator solutions for equations requiring inverse trigonometric functions has been enhanced.

**Chapters 6 and 7**  The inclusion of calculator usage and exercises has strengthened these chapters, and a general rewriting has made all sections easier to understand.

**Chapter 8**  Systems of inequalities and linear programming have been moved forward to Sections 8.5 and 8.6, respectively. The concept of test points is emphasized in solving systems of inequalities. The inverse of a matrix appears

Examples, well structured and graded by difficulty, have been titled for this edition.

Examples requiring the use of a graphics calculator have been added to the text wherever appropriate. These are designated by an icon and illustrated with a figure reproduced from a graphics calculator screen.

**EXAMPLE 7    A graph containing a hole**
Sketch the graph of $f$ if

$$f(x) = \frac{x-4}{(x-2)(x-4)}.$$

FIGURE 33

SOLUTION    If $x \neq 4$, we may cancel the factor $x - 4$; that is,

$$f(x) = \frac{1}{x-2}, \quad \text{if} \quad x \neq 4.$$

Thus, the graph of $f$ is the same as that sketched in Figure 25, with one exception: There is a hole in the graph at $x = 4$. To determine the corresponding $y$ value, we use the reduced form of the function, obtaining $f(4) = 1/(4-2) = \frac{1}{2}$. The point $(4, \frac{1}{2})$ is indicated by the small circle in Figure 33. It should be noted that graphics calculators or computer programs may not be designed to show this feature of a graph.

**EXAMPLE 8    Sketching the graph of a rational function**
Sketch the graph of

$$f(x) = \frac{x^2 - x}{9x^3 - 9x^2 - 22x + 8}$$

FIGURE 34   $[-2, 3]$ by $[-1, 1]$

and find equations of the vertical asymptotes.

SOLUTION    Some specific keystrokes for the TI-81 graphics calculator are given in Appendix I. We begin by making the assignments

$$Y_1 = x^2 - x, \quad Y_2 = 9x^3 - 9x^2 - 22x + 8, \quad \text{and} \quad Y_3 = Y_1/Y_2.$$

Selecting only $Y_3$ to be graphed and using a standard viewing rectangle, we obtain a graph that gives us virtually no indication of the true shape of $f$. Changing to a viewing rectangle of $[-6, 6]$ by $[-4, 4]$ gives us a hint that the vertical asymptotes are confined to the interval $-2 < x < 3$.

FIGURE 35   $[-2, 3]$ by $[-1, 1]$

Using a viewing rectangle of $[-2, 3]$ by $[-1, 1]$ and changing to *dot mode* (so as not to graph the function across the vertical asymptotes) leads us to the sketch in Figure 34. Since the degree of the numerator, 2, is less than the degree of the denominator, 3, we know that the horizontal asymptote is the $x$-axis. The zeros of the numerator, 0 and 1, are the only $x$-intercepts.

To determine the equations of the vertical asymptotes, we will abandon the graph of $Y_3$ and examine the graph of $Y_2$—looking for its zeros. Graphing $Y_2$ with the same viewing rectangle, but using *connected mode*, gives us Figure 35.

*ILLUSTRATION*

**Exponential Functions Are One-to-One**

▪ If $7^{3x} = 7^{2x+5}$, then $3x = 2x + 5$, or $x = 5$.

In the following example we solve a simple *exponential equation*—that is, an equation in which the variable appears in an exponent.

**EXAMPLE 1    Solving an exponential equation**
Solve the equation $3^{5x-8} = 9^{x+2}$.

SOLUTION

$$\begin{aligned}
3^{5x-8} &= 9^{x+2} & &\text{given} \\
3^{5x-8} &= (3^2)^{x+2} & &\text{express both sides with the same base} \\
3^{5x-8} &= 3^{2x+4} & &\text{law of exponents} \\
5x - 8 &= 2x + 4 & &\text{exponential functions are one-to-one} \\
3x &= 12 & &\text{subtract } 2x \text{ and add 8} \\
x &= 4 & &\text{divide by 3}
\end{aligned}$$

In the next two examples we sketch the graphs of several different exponential functions.

FIGURE 3

**EXAMPLE 2    Sketching graphs of exponential functions**
If $f(x) = (\frac{3}{2})^x$ and $g(x) = 3^x$, sketch the graphs of $f$ and $g$ on the same coordinate plane.

SOLUTION    Since $\frac{3}{2} > 1$ and $3 > 1$, each graph *rises* as $x$ increases. The following table displays coordinates for several points on the graphs.

| $x$ | $-2$ | $-1$ | 0 | 1 | 2 | 3 | 4 |
|---|---|---|---|---|---|---|---|
| $y = (\frac{3}{2})^x$ | $\frac{4}{9} \approx 0.4$ | $\frac{2}{3} \approx 0.7$ | 1 | $\frac{3}{2}$ | $\frac{9}{4} \approx 2.3$ | $\frac{27}{8} \approx 3.4$ | $\frac{81}{16} \approx 5.1$ |
| $y = 3^x$ | $\frac{1}{9} \approx 0.1$ | $\frac{1}{3} \approx 0.3$ | 1 | 3 | 9 | 27 | 81 |

Plotting points and being familiar with the general graph of $y = a^x$ leads to the graphs in Figure 3.

Illustrations provide brief demonstrations of the use of definitions, laws, and theorems.

Many examples contain graphs, charts, or tables to help the student understand procedures and solutions.

Exercise sets begin with drill problems and progress to more challenging problems.

Graphics calculator exercises, designated by a , are included in many sections. Over 200 of these exercises have been added to this edition.

Exercise applications are identified to show students how the problems are related to current real-life situations.

Warnings are interspersed throughout the text to alert students to common mistakes.

Step-by-step explanations have been added to examples to help students more easily follow the solutions.

in a separate section, and Cramer's rule has been consolidated into Section 8.10.

**Chapter 9**  More attention has been paid to sequences defined recursively; their generation on a computational device is discussed. Mathematical induction now appears in Section 9.4, following the discussion of sequences. The topics of mutually exclusive events, independent events, and expected value have been included in Section 9.8.

**Chapter 10**  A formula for finding the distance from the vertex of a parabola to its focus is presented in Section 10.1. The graphing of half-parabolas, half-ellipses, and portions of a hyperbola has been added, as have exercises dealing with properties of paraboloids and ellipsoids. The orientation of parametrized curves is thoroughly discussed in Section 10.4.

**Appendixes**  Appendix I, ''Graphing with a TI Graphics Calculator,'' has specific keystrokes for the TI-81 and TI-82 graphics calculators for selected examples. Many of the twenty-two examples are examples or exercises from the text. This appendix can easily be worked through in two different ways: (1) before studying any of the material in the text or (2) as each appendix cross-reference is encountered in the text.

Appendix II, ''Common Graphs and Their Equations,'' is a pictorial summary of some of the most common graphs and their equations that a student encounters in precalculus mathematics.

Appendix III, ''A Summary of Graph Transformations,'' is an illustrative synopsis of the basic graph transformations discussed in the text: shifting, stretching, compressing, and reflecting.

Appendix IV, ''Graphs of Trigonometric Functions and Their Inverses,'' contains graphs, domains, and ranges of the six trigonometric functions and their inverses.

**FEATURES**

**Illustrations**  Brief demonstrations of the use of definitions, laws, and theorems are provided in the form of illustrations.

**Charts**  Many new charts have been added that summarize properties, laws, graphs, and definitions. These charts often contain simple illustrations of the concepts that are being introduced.

**Examples**  All examples have been titled for easy reference. They provide detailed solutions of problems similar to those that appear in exercise sets. Many examples contain graphs, charts, or tables to help the student understand procedures and solutions.

**Step-by-Step Explanations**  In order to help students follow them more easily, step-by-step explanations have been added to many of the solutions in examples.

**Checks**  The solutions to some examples are explicitly checked, to remind students to verify that their solutions satisfy the conditions of the problems.

**Graphics Calculator Examples**   Wherever appropriate, examples requiring the use of a graphing utility have been added to the text. These are designated by a gray calculator icon and illustrated with a figure reproduced from a graphics calculator screen.

**Graphics Calculator Exercises**   Exercises specifically designed to be solved with a graphing utility have been included in appropriate sections. Over 200 of these exercises, also designated by a gray calculator icon, have been added to this edition.

**Applications**   Applied exercises have been titled to arouse student interest and help students relate the exercises to current real-life situations. Titles of some of the new topical applications are "Computer chip reliability," "Ozone layer," "Radon gas," and "Robotic movement."

**Exercises**   Exercise sets begin with routine drill problems and gradually progress to more difficult problems. Many exercises containing graphs and data were added to this edition. Applied problems generally appear near the end of an exercise set to allow students to gain confidence in working with the new ideas that have been presented before attempting problems that require greater analysis and synthesis of these ideas. Review exercises at the end of each chapter may be used to prepare for examinations.

**Guidelines**   Guidelines for solving problems have been added or improved.

**Warnings**   Interspersed throughout the text are warnings to alert students to common mistakes.

**Text Art**   Every figure has been redrawn and relabeled, and graphs have been computer-generated for accuracy, using the latest technology. Additional colors are used to distinguish between different parts of figures. For example, the graph of one function may be shown in blue and that of a second function in red. Labels are the same color as the parts of the figure they identify. We believe that you will find the total art package in this text second to none.

**Text Design**   The text has been completely redesigned to ensure that discussions are easy to follow and important concepts are highlighted. Color is employed pedagogically to clarify complex graphs and to help students visualize applied problems.

**Endpapers**   The endpapers in the front and back of the text have been revised to include useful summaries from algebra, geometry, and trigonometry.

**Answer Section**   The answer section at the end of the text provides answers for most of the odd-numbered exercises as well as answers for *all* chapter review exercises. Considerable thought and effort were devoted to making this section a learning device for the student instead of merely a place to check answers. For instance, verifications are given for trigonometric identities and mathematical induction problems. Numerical answers for many exercises are stated in both an exact and an approximate form. Graphs, proofs, and hints are included whenever appropriate. To ensure a high degree of consistency among the text, the solutions, and the answers, the authors prepared their own solutions and answers.

**Flexibility** Syllabi from schools that used previous editions attest to the flexibility of this text—it has been successfully used for separate College Algebra and College Trigonometry courses besides Precalculus courses. Sections and chapters can be rearranged in many different ways, depending on objectives and the length of the course.

## TEACHING TOOLS FOR THE INSTRUCTOR

**Instructor's Solutions Manual** (*Cole*) This manual contains answers to all exercises and reasonably detailed solutions to most. The manual has been thoroughly reviewed for accuracy.

**EXPTest** This computerized test bank for IBM and compatibles contains hundreds of problems; 500 are new to this edition. Questions are multiple choice, true-false, and open-ended. Instructors can interact with the program, adding to existing questions and producing individual tests.

**EXPTest Sampler with Demonstration Disk** The sampler demonstrates the computerized test bank's capabilities and contains sample tests.

**ExamBuilder** This computerized test bank for the Macintosh has features and questions similar to those of EXPTest, described above. A demo is available.

**Test Bank with Chapter Tests** All the questions found in EXPTest (and ExamBuilder), along with their answers, appear in print form in the *Test Bank*. Two sample tests are also included for each chapter.

**Transparencies** Full-color acetates provide enlarged versions of illustrations found in the text. This teaching tool is available to adopters only.

## LEARNING TOOLS FOR THE STUDENT

**Quick Reference Card** Bound into this edition of the text is a new tool for solving exercises—a formula card. This perforated card will aid students in mastering the key formulas, equations, and graphs in the course. By serving as a quick reference and minimizing the need for page turning, the formula card will reduce the time spent on tedious tasks so that the student can focus on the central concepts and principles of the course.

**Student's Solutions Manual** (*Cole*) This manual contains detailed solutions for nearly all of the odd-numbered exercises, as well as strategies for solving additional exercises. Many helpful hints and warnings are also included.

**INVESTIGATE** (*Rockswold*) Tutorial software in IBM and Macintosh formats allows students to practice the skills taught in the textbook. For topics in the text, the tutorial provides multiple-choice problems and instruction. The student is presented with solutions to incorrectly answered problems. This diagnostic program keeps track of right and wrong responses and can report to the instructor on students' progress. Demo disks are available.

**Videotapes**  A series of videotapes, produced by Educational Video Resources, is available to qualified adopters. These tutorial videos can be viewed by your students in your math lab. They are designed to enhance comprehension of the key precalculus topics in the text.

**Precalculus in Context: Functioning in the Real World** *(Davis/Moran/Murphy)* This lab manual consists of 12 projects that encourage students to explore precalculus concepts. Graphics calculators and computer graphing software are used to solve each experiment and its corresponding exercises.

**Student Edition of Theorist**  This Macintosh-compatible software combines powerful, symbolic algebra and graphics capabilities with an intuitive, user-friendly interface.

**Theorist Notebook Software for the Macintosh**  Through an interactive, symbolic/graphic mathematics environment, students have the opportunity to study, graph, and solve selected exercises from the textbook. Exercises and examples ''come alive'' with the help of *Theorist*.

**Derive Notebook Software for the IBM**  Like *Theorist, Derive* provides students the opportunity to study, graph, and solve selected exercises from the textbook. Exercises and examples ''come alive'' when the *Notebook* is run using *Derive*.

## ACKNOWLEDGMENTS

Special thanks go to Gary Rockswold, of Mankato State University, for supplying most of the fine assortment of new applied problems and calculator exercises. I also wish to thank Thomas Vanden Eynden, of Thomas More College, and Teresa Bittner, of Laurel Tutoring Services, for doing a detailed accuracy check of the new exercises; Joan Cole, for proofreading the galleys; and George Morris, of Scientific Illustrators, for creating the mathematically precise art package.

Many changes for this edition are due to the following individuals, who reviewed the manuscript:

Daniel Anderson
*University of Iowa*

Judy Carter
*North Shore Community College*

Susan M. Dimick
*Spokane Community College*

John Edwards
*Anne Arundel Community College*

Elaine C. Ensign
*University of Alaska at Fairbanks*

Martin Forrest
*Louisiana State University*

M. D. Hendon
*University of Georgia*

Dennis T. Johnson
*SUNY College of Technology*

J. Robert Knott
*University of Evansville*

Emmett Larson
*Brevard Community College*

Marilyn McCollom
*North Carolina State University*

Michael J. Mears
*Manatee Community College*

James E. Moran
*Diablo Valley College*

William Radulovich
*Florida Community College*

Bruce Riley
*University of Wisconsin at La
Crosse*

Sue Ellen Skillings
*Miami University*

Gerald Thompson
*Augusta College*

Special credit goes to the mathematics educators who met with me and representatives from PWS Publishing Company for several days in the summer of 1992:

Daniel D. Anderson
*University of Iowa*

Martin Forrest
*Louisiana State University*

Emmett Larson
*Brevard Community College*

James E. Moran
*Diablo Valley College*

Lynda S. Morton
*University of Missouri at Columbia*

Gary Rockswold
*Mankato State University*

Gloria B. Shier
*North Hennepin Community
College*

Their comments about pedagogy and their specific recommendations about the content of a precalculus course helped me to improve the book.

I am thankful for the excellent cooperation of the staff of PWS Publishing Company. Several people in the company deserve special mention. Editors Tim Anderson and David Dietz supervised the project and were constant sources of information and advice. Elise Kaiser and Sue Caffey coordinated the production of the text and are responsible for the exquisite design. Sally Lifland, Denise Throckmorton, and Jeannie Yost, all of Lifland et al., Bookmakers, took exceptional care in seeing that no inconsistencies occurred and offered many helpful suggestions.

In addition to all the persons named here, I express my sincere gratitude to the many students and teachers who have helped shape my views on mathematics education. Please feel free to write to me about any aspect of this text—I value your opinion.

**JEFFERY A. COLE**

*Real numbers and equations may be used for estimating distances, speeds of objects in motion, and magnitudes of physical quantities.*

# CHAPTER 1

## TOPICS

## FROM

## ALGEBRA

■ The word algebra comes from ilm al-jabr w'al muqabala, the title of a book written in the ninth century by the Arabian mathematician al-Khworizimi. The title has been translated as the science of restoration and reduction, which means transposing and combining similar terms (of an equation). The Latin transliteration of al-jabr led to the name of the branch of mathematics we now call algebra.

We begin this chapter with a review of real numbers and their properties, which will be used throughout this text and further mathematics courses. A discussion of some fundamental algebraic techniques follows before we turn our attention to solving basic equations and inequalities—those whose solutions are subsets of the one-dimensional real number line. In Section 1.7, we introduce a two-dimensional coordinate system, graphs of equations, and the use of a graphics calculator. The chapter concludes with a study of lines and their applications. ■

## 1.1   REAL NUMBERS

Real numbers are used throughout mathematics, and you should be acquainted with symbols that represent them, such as

$$1, \quad 73, \quad -5, \quad \tfrac{49}{12}, \quad \sqrt{2}, \quad 0, \quad \sqrt[3]{-85}, \quad 0.33333\ldots, \quad 596.25,$$

and so on. The **positive integers,** or **natural numbers,** are

$$1, \quad 2, \quad 3, \quad 4, \quad \ldots.$$

The **whole numbers** (or *nonnegative integers*) are the natural numbers combined with the number 0. The **integers** are often listed as follows:

$$\ldots, \quad -4, \quad -3, \quad -2, \quad -1, \quad 0, \quad 1, \quad 2, \quad 3, \quad 4, \quad \ldots$$

Throughout this text lowercase letters $a, b, c, x, y, \ldots$ represent arbitrary real numbers. If $a$ and $b$ denote the same real number, we write $a = b$, which is read "*a* **is equal to** *b*." An expression of this type is called an **equality.** The notation $a \neq b$ is read "*a* **is *not* equal to** *b*."

If $a$, $b$, and $c$ are integers and $c = ab$, then $a$ and $b$ are **factors,** or **divisors,** of $c$. For example, since

$$6 = 2 \cdot 3 = (-2)(-3) = 1 \cdot 6 = (-1)(-6),$$

we know that $1, -1, 2, -2, 3, -3, 6$, and $-6$ are factors of 6.

A positive integer $p$ different from 1 is **prime** if its only positive factors are 1 and $p$. The first few primes are 2, 3, 5, 7, 11, 13, 17, and 19. The **Fundamental Theorem of Arithmetic** states that every positive integer different from 1 can be expressed as a product of primes in one and only one way (except for order of factors). Some examples are

$$12 = 2 \cdot 2 \cdot 3, \quad 126 = 2 \cdot 3 \cdot 3 \cdot 7, \quad 540 = 2 \cdot 2 \cdot 3 \cdot 3 \cdot 3 \cdot 5.$$

A **rational number** is a real number that can be expressed in the form $a/b$, where $a$ and $b$ are integers and $b \neq 0$. Note that every integer $a$ is a rational number, since it can be expressed in the form $a/1$. Every real number can be expressed as a decimal, and the decimal representations for rational numbers are either *terminating* or *nonterminating and repeating*. For example, we can show by using the arithmetic process of division that

$$\tfrac{5}{4} = 1.25 \quad \text{and} \quad \tfrac{177}{55} = 3.2181818\ldots,$$

where the digits 1 and 8 in the representation of $\tfrac{177}{55}$ repeat indefinitely (sometimes written $3.2\overline{18}$).

Real numbers that are not rational are **irrational numbers.** Decimal representations for irrational numbers are always *nonterminating and nonrepeating.* One common irrational number is the ratio of the circumference of a circle to its diameter, denoted by $\pi$. We sometimes use the notation $\pi \approx 3.1416$ to indicate that $\pi$ **is approximately equal to** 3.1416.

There is no rational number $b$ such that $b^2 = 2$, where $b^2$ denotes $b \cdot b$. However, there is an irrational number, denoted by $\sqrt{2}$ (the **square root** of 2), such that $(\sqrt{2})^2 = 2$.

The system of **real numbers** consists of all rational and irrational numbers. The diagram in Figure 1 indicates relationships among the types of numbers used in algebra, where a line connecting two rectangles means that the numbers named in the higher rectangle include those in the lower rectangle. The complex numbers, discussed in Section 1.5, contain all real numbers.

**FIGURE 1**   Types of numbers used in algebra

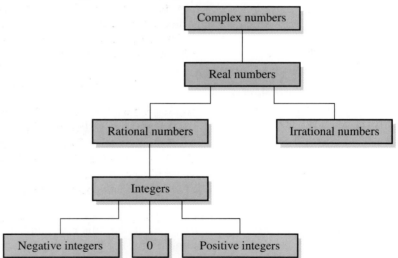

The real numbers are **closed relative to the operation of addition** (denoted by $+$); that is, to every pair $a, b$ of real numbers there corresponds exactly one real number $a + b$ called the **sum** of $a$ and $b$. The real numbers are also **closed relative to multiplication** (denoted by $\cdot$); that is, to every pair $a, b$ of real numbers there corresponds exactly one real number $a \cdot b$ (also denoted by $ab$) called the **product** of $a$ and $b$.

Important properties of addition and multiplication of real numbers are listed in the following chart.

**Properties of Real Numbers**

| Terminology | General case | Meaning |
|---|---|---|
| (1) Addition is **commutative.** | $a + b = b + a$ | Order is immaterial when adding two numbers. |
| (2) Addition is **associative.** | $a + (b + c) = (a + b) + c$ | Grouping is immaterial when adding three numbers. |
| (3) 0 is the **additive identity.** | $a + 0 = a$ | Adding 0 to any number yields the same number. |
| (4) $-a$ is the **additive inverse,** or **negative,** of $a$. | $a + (-a) = 0$ | Adding a number and its negative yields 0. |
| (5) Multiplication is **commutative.** | $ab = ba$ | Order is immaterial when multiplying two numbers. |
| (6) Multiplication is **associative.** | $a(bc) = (ab)c$ | Grouping is immaterial when multiplying three numbers. |
| (7) 1 is the **multiplicative identity.** | $a \cdot 1 = a$ | Multiplying any number by 1 yields the same number. |
| (8) If $a \neq 0$, $\dfrac{1}{a}$ is the **multiplicative inverse,** or **reciprocal,** of $a$. | $a\left(\dfrac{1}{a}\right) = 1$ | Multiplying a nonzero number by its reciprocal yields 1. |
| (9) Multiplication is **distributive** over addition. | $a(b + c) = ab + ac$ and $(a + b)c = ac + bc$ | Multiplying a number and a sum of two numbers is equivalent to multiplying each of the two numbers by the number and then adding the products. |

Since $a + (b + c)$ and $(a + b) + c$ are always equal, we may use $a + b + c$ to denote this real number. We use $abc$ for either $a(bc)$ or $(ab)c$. Similarly, if four or more real numbers $a, b, c, d$ are added or multiplied, we may write $a + b + c + d$ for their sum and $abcd$ for their product, regardless of how the numbers are grouped or interchanged.

The distributive properties are useful for finding products of many types of expressions involving sums. The next example provides one illustration.

**EXAMPLE I**     *Using distributive properties*

If $p$, $q$, $r$, and $s$ denote real numbers, show that

$$(p + q)(r + s) = pr + ps + qr + qs.$$

*SOLUTION*    We use both of the distributive properties listed in (9) of the preceding chart:

$$(p + q)(r + s) = p(r + s) + q(r + s) \qquad \text{second distributive property with } c = r + s$$

$$= (pr + ps) + (qr + qs) \qquad \text{first distributive property}$$

$$= pr + ps + qr + qs \qquad \text{remove parentheses}$$

The following are basic properties of equality.

**Properties of Equality**

If $a = b$ and $c$ is any real number, then

**(1)** $a + c = b + c$

**(2)** $ac = bc$

Properties 1 and 2 state that the same number may be added to both sides of an equality, and both sides of an equality may be multiplied by the same number. We will use these properties extensively throughout the text to help find solutions of equations.

The next result can be proved.

**Products Involving Zero**

**(1)** $a \cdot 0 = 0$ for every real number $a$.

**(2)** If $ab = 0$, then either $a = 0$ or $b = 0$.

When we use the word *or* as we do in (2), we mean that *at least* one of $a$ or $b$ is 0.

Some properties of negatives are listed in the following chart.

**Properties of Negatives**

| Property | Illustration |
|---|---|
| (1) $-(-a) = a$ | $-(-3) = 3$ |
| (2) $(-a)b = -(ab) = a(-b)$ | $(-2)3 = -(2 \cdot 3) = 2(-3)$ |
| (3) $(-a)(-b) = ab$ | $(-2)(-3) = 2 \cdot 3$ |
| (4) $(-1)a = -a$ | $(-1)3 = -3$ |

The reciprocal $\dfrac{1}{a}$ of a nonzero real number $a$ is often denoted by $a^{-1}$, as in the next chart.

**Notation for Reciprocals**

| Definition | Illustration |
|---|---|
| If $a \neq 0$, then $a^{-1} = \dfrac{1}{a}$. | $2^{-1} = \dfrac{1}{2}$ <br><br> $\left(\dfrac{3}{4}\right)^{-1} = \dfrac{1}{3/4} = \dfrac{4}{3}$ |

Note that if $a \neq 0$, then

$$a \cdot a^{-1} = a\left(\frac{1}{a}\right) = 1.$$

The operations of **subtraction** $(-)$ and **division** $(\div)$ are defined as follows.

**Subtraction and Division**

| Definition | Meaning | Illustration |
|---|---|---|
| $a - b = a + (-b)$ | To subtract one number from another, add the negative. | $3 - 7 = 3 + (-7)$ |
| $a \div b = a \cdot \left(\dfrac{1}{b}\right)$ <br> $= a \cdot b^{-1}; \ b \neq 0$ | To divide one number by a nonzero number, multiply by the reciprocal. | $3 \div 7 = 3 \cdot \left(\dfrac{1}{7}\right)$ <br> $= 3 \cdot 7^{-1}$ |

We use either $a/b$ or $\dfrac{a}{b}$ for $a \div b$ and refer to $a/b$ as the **quotient of $a$ and $b$** or the **fraction $a$ over $b$.** The numbers $a$ and $b$ are the **numerator** and **denominator,** respectively, of $a/b$. Since 0 has no multiplicative inverse, $a/b$ is not defined if $b = 0$; that is, *division by zero is not defined.* It is for this reason that the real numbers are not closed relative to division. Note that

$$1 \div b = \frac{1}{b} = b^{-1} \quad \text{if} \quad b \neq 0.$$

The following properties of quotients are true, provided all denominators are nonzero real numbers.

**Properties of Quotients**

| Property | Illustration |
|---|---|
| (1) $\dfrac{a}{b} = \dfrac{c}{d}$ if $ad = bc$ | $\dfrac{2}{5} = \dfrac{6}{15}$ because $2 \cdot 15 = 5 \cdot 6$ |
| (2) $\dfrac{ad}{bd} = \dfrac{a}{b}$ | $\dfrac{2 \cdot 3}{5 \cdot 3} = \dfrac{2}{5}$ |
| (3) $\dfrac{a}{-b} = \dfrac{-a}{b} = -\dfrac{a}{b}$ | $\dfrac{2}{-5} = \dfrac{-2}{5} = -\dfrac{2}{5}$ |
| (4) $\dfrac{a}{b} + \dfrac{c}{b} = \dfrac{a + c}{b}$ | $\dfrac{2}{5} + \dfrac{9}{5} = \dfrac{2 + 9}{5} = \dfrac{11}{5}$ |
| (5) $\dfrac{a}{b} + \dfrac{c}{d} = \dfrac{ad + bc}{bd}$ | $\dfrac{2}{5} + \dfrac{4}{3} = \dfrac{2 \cdot 3 + 5 \cdot 4}{5 \cdot 3} = \dfrac{26}{15}$ |
| (6) $\dfrac{a}{b} \cdot \dfrac{c}{d} = \dfrac{ac}{bd}$ | $\dfrac{2}{5} \cdot \dfrac{7}{3} = \dfrac{2 \cdot 7}{5 \cdot 3} = \dfrac{14}{15}$ |
| (7) $\dfrac{a}{b} \div \dfrac{c}{d} = \dfrac{a}{b} \cdot \dfrac{d}{c} = \dfrac{ad}{bc}$ | $\dfrac{2}{5} \div \dfrac{7}{3} = \dfrac{2}{5} \cdot \dfrac{3}{7} = \dfrac{6}{35}$ |

Real numbers may be represented by points on a line $l$ such that to each real number $a$ there corresponds exactly one point on $l$ and to each point $P$ on $l$ there corresponds one real number. This is called a **one-to-one correspondence.** We first choose an arbitrary point $O$, called the **origin,** and associate with it the real number 0. Points associated with the integers are then determined by laying off successive line segments of equal length on either side of $O$, as illustrated in Figure 2. The point corresponding to a rational number, such as $\frac{23}{5}$, is obtained by subdividing these line segments. Points associated with certain irrational numbers, such as $\sqrt{2}$, can be found by construction (see Exercise 41).

**FIGURE 2**

The number $a$ that is associated with a point $A$ on $l$ is the **coordinate** of $A$. We refer to these coordinates as a **coordinate system** and call $l$ a **coordinate line,** or a **real line.** A direction can be assigned to $l$ by taking the **positive direction** to the right and the **negative direction** to the left. The positive direction is noted by placing an arrowhead on $l$, as shown in Figure 2.

The numbers that correspond to points to the right of $O$ in Figure 2 are **positive real numbers.** Numbers that correspond to points to the left of $O$ are **negative real numbers.** *The real number $0$ is neither positive nor negative.*

Note the difference between a negative real number and the *negative of* a real number. In particular, the negative of a real number $a$ can be positive. For example, if $a$ is negative, say $a = -3$, then the negative of $a$ is $-a = -(-3) = 3$, which is positive. In general, we have the following relationships.

**Relationships Between $a$ and $-a$**

> **(1)** If $a$ is positive, then $-a$ is negative.
>
> **(2)** If $a$ is negative, then $-a$ is positive.

In the following chart we define the notions of **greater than** and **less than** for real numbers $a$ and $b$. The symbols $>$ and $<$ are **inequality signs,** and the expressions $a > b$ and $a < b$ are called **inequalities.**

**Greater Than or Less Than**

| Notation | Definition | Terminology |
|----------|------------|-------------|
| $a > b$ | $a - b$ is positive | $a$ is greater than $b$ |
| $a < b$ | $a - b$ is negative | $a$ is less than $b$ |

Referring to a coordinate line, if points $A$ and $B$ have coordinates $a$ and $b$, respectively, then $a > b$ is equivalent to the statement "$A$ is to the *right* of $B$," whereas $a < b$ is equivalent to "$A$ is to the *left* of $B$."

**ILLUSTRATION**

**Greater Than ($>$) and Less Than ($<$)**

- $5 > 3$, since $5 - 3 = 2$ is positive.
- $-6 < -2$, since $-6 - (-2) = -6 + 2 = -4$ is negative.
- $\frac{1}{3} > 0.33$, since $\frac{1}{3} - 0.33 = \frac{1}{3} - \frac{33}{100} = \frac{1}{300}$ is positive.
- $7 > 0$, since $7 - 0 = 7$ is positive.
- $-4 < 0$, since $-4 - 0 = -4$ is negative.

The next law enables us to compare, or *order,* any two real numbers.

**Trichotomy Law**

> If $a$ and $b$ are real numbers, then exactly one of the following is true:
> $$a = b, \quad a > b, \quad \text{or} \quad a < b$$

We refer to the **sign** of a real number as positive or negative if the number is positive or negative, respectively. Two real numbers have *the same sign* if both are positive or both are negative. The numbers have *opposite signs* if one is positive and the other is negative. The following results about the signs of products and quotients of two real numbers $a$ and $b$ can be proved using properties of negatives and quotients.

**Laws of Signs**

> **(1)** If $a$ and $b$ have the same sign, then $ab$ and $\dfrac{a}{b}$ are positive.
>
> **(2)** If $a$ and $b$ have opposite signs, then $ab$ and $\dfrac{a}{b}$ are negative.

The **converses\*** of the laws of signs are also true. For example, if a quotient is negative, then the numerator and denominator have opposite signs.

The notation $a \geq b$, read "*a* **is greater than or equal to** *b*," means that either $a > b$ or $a = b$ (but not both). For example, $a^2 \geq 0$ for every real number $a$. The symbol $a \leq b$, which is read "*a* **is less than or equal to** *b*," means that either $a < b$ or $a = b$. An expression of the form $a < b < c$ is called a **continued inequality** and means that both $a < b$ *and* $b < c$, and we say, "*b* **is between** *a* and *c*." Similarly, the expression $c > b > a$ means that both $c > b$ and $b > a$.

### ILLUSTRATION

#### Ordering Three Real Numbers

■ $1 < 5 < \frac{11}{2}$   ■ $-4 < \frac{2}{3} < \sqrt{2}$   ■ $3 > -6 > -10$

There are other types of inequalities. For example, $a < b \leq c$ means both $a < b$ and $b \leq c$. Similarly, $a \leq b < c$ means both $a \leq b$ and $b < c$. Finally, $a \leq b \leq c$ means both $a \leq b$ and $b \leq c$.

If $a$ is an integer, then it is the coordinate of some point $A$ on a coordinate line, and the symbol $|a|$ denotes the number of units between $A$ and

---

\*If a theorem is written in the form "if $P$, then $Q$," where $P$ and $Q$ are mathematical statements called the *hypothesis* and *conclusion,* respectively, then the *converse* of the theorem has the form "if $Q$, then $P$." If both the theorem and its converse are true, we often write "$P$ if and only if $Q$."

**FIGURE 3**

the origin, without regard to direction. The nonnegative number $|a|$ is called the *absolute value of a*. Referring to Figure 3, we see that for the point with coordinate $-4$ we have $|-4| = 4$. Similarly, $|4| = 4$. In general, *if a is negative, we change its sign to find* $|a|$; *if a is nonnegative, then* $|a| = a$. The next definition extends this concept to every real number.

**Definition of Absolute Value**

> The **absolute value** of a real number $a$, denoted by $|a|$, is defined as follows.
>
> **(1)** If $a \geq 0$, then $|a| = a$.
>
> **(2)** If $a < 0$, then $|a| = -a$.

Since $a$ is negative in part (2) of the definition, $-a$ represents a *positive* real number. Some special cases of this definition are given in the following illustration.

**ILLUSTRATION**

**The Absolute Value Notation $|a|$**

- $|3| = 3$, since $3 > 0$.
- $|-3| = -(-3)$, since $-3 < 0$. Thus, $|-3| = 3$.
- $|2 - \sqrt{2}| = 2 - \sqrt{2}$, since $2 - \sqrt{2} > 0$.
- $|\sqrt{2} - 2| = -(\sqrt{2} - 2)$, since $\sqrt{2} - 2 < 0$. Thus, $|\sqrt{2} - 2| = 2 - \sqrt{2}$.

In the preceding illustration, $|3| = |-3|$ and $|2 - \sqrt{2}| = |\sqrt{2} - 2|$. In general, we have the following:

$$|a| = |-a| \quad \text{for every real number } a$$

**EXAMPLE 2**    *Removing an absolute value symbol*

If $x < 1$, rewrite $|x - 1|$ without using the absolute value symbol.

**SOLUTION**    If $x < 1$, then $x - 1 < 0$; that is, $x - 1$ is negative. Hence, by part (2) of the definition of absolute value,

$$|x - 1| = -(x - 1) = -x + 1 = 1 - x.$$

**FIGURE 4**

$$5 = |7 - 2| = |2 - 7|$$

We shall use the concept of absolute value to define the distance between any two points on a coordinate line. First note that the distance between the points with coordinates 2 and 7, shown in Figure 4, equals 5 units. This distance is the difference obtained by subtracting the smaller coordinate from the larger $(7 - 2 = 5)$. If we employ absolute values, then, since $|7 - 2| = |2 - 7|$, it is unnecessary to be concerned about the order of subtraction. This fact motivates the next definition.

| | |
|---|---|
| **Definition of the Distance Between Points on a Coordinate Line** | Let $a$ and $b$ be the coordinates of two points $A$ and $B$, respectively, on a coordinate line. The **distance between $A$ and $B$,** denoted by $d(A, B)$, is defined by $$d(A, B) = |b - a|.$$ |

The number $d(A, B)$ is the length of the line segment $AB$.

Since $d(B, A) = |a - b|$ and $|b - a| = |a - b|$, we see that

$$d(A, B) = d(B, A).$$

Note that the distance between the origin $O$ and the point $A$ is

$$d(O, A) = |a - 0| = |a|,$$

which agrees with the geometric interpretation of absolute value illustrated in Figure 4. The formula $d(A, B) = |b - a|$ is true regardless of the signs of $a$ and $b$, as illustrated in the next example.

**EXAMPLE 3   _Finding distances between points_**

Let $A$, $B$, $C$, and $D$ have coordinates $-5$, $-3$, 1, and 6, respectively, on a coordinate line, as shown in Figure 5. Find $d(A, B)$, $d(C, B)$, $d(O, A)$, and $d(C, D)$.

**SOLUTION**   Using the definition of the distance between points, we obtain the distances:

$$d(A, B) = |-3 - (-5)| = |-3 + 5| = |2| = 2$$
$$d(C, B) = |-3 - 1| = |-4| = 4$$
$$d(O, A) = |-5 - 0| = |-5| = 5$$
$$d(C, D) = |6 - 1| = |5| = 5$$

**FIGURE 5**

The concept of absolute value has uses other than finding distances between points; it is employed whenever we are interested in the magnitude or numerical value of a real number without regard to its sign.

In the next section we shall discuss the _exponential notation_ $a^n$, where $a$ is a real number (called the _base_) and $n$ is an integer (called an _exponent_). In particular, for base 10 we have

$$10^0 = 1, \quad 10^1 = 10, \quad 10^2 = 10 \cdot 10 = 100, \quad 10^3 = 10 \cdot 10 \cdot 10 = 1000,$$

and so on. For negative exponents we use reciprocals of the corresponding positive exponent, as follows:

$$10^{-1} = \frac{1}{10^1} = \frac{1}{10}, \quad 10^{-2} = \frac{1}{10^2} = \frac{1}{100}, \quad 10^{-3} = \frac{1}{10^3} = \frac{1}{1000}$$

We can employ this notation to write any finite decimal representation of a real number as a sum of the following type:

$$437.56 = 4(100) + 3(10) + 7(1) + 5\left(\tfrac{1}{10}\right) + 6\left(\tfrac{1}{100}\right)$$
$$= 4(10^2) + 3(10^1) + 7(10^0) + 5(10^{-1}) + 6(10^{-2})$$

In the sciences it is often necessary to work with very large or very small numbers and to compare the relative magnitudes of very large or very small quantities. We usually represent a large or small positive number $a$ in *scientific form,* using $\times$ to denote multiplication.

**Scientific Form**

$$a = c \times 10^n, \quad \text{where} \quad 1 \leq c < 10 \text{ and } n \text{ is an integer}$$

The distance a ray of light travels in one year is approximately 5,900,000,000,000 miles. This number may be written in scientific form as $5.9 \times 10^{12}$. The positive exponent 12 indicates that the decimal point should be moved 12 places to the *right.* The notation works equally well for small numbers. To illustrate, the weight of an oxygen molecule is estimated to be

$$0.000\ 000\ 000\ 000\ 000\ 000\ 000\ 053 \text{ gram,}$$

or, in scientific form, $5.3 \times 10^{-23}$ gram. The negative exponent indicates that the decimal point should be moved 23 places to the *left.*

### ILLUSTRATION

**Scientific Form**

- $513 = 5.13 \times 10^2$
- $93,000,000 = 9.3 \times 10^7$
- $0.000\ 000\ 000\ 43 = 4.3 \times 10^{-10}$

- $7.3 = 7.3 \times 10^0$
- $20,700 = 2.07 \times 10^4$
- $0.000\ 648 = 6.48 \times 10^{-4}$

**FIGURE 6**

2.025 13

or  2.025E13

or  2.025 ¹³

Many calculators employ scientific form in their display panels: for the number $c \times 10^n$, the 10 is suppressed and the exponent is often shown preceded by the letter E. For example, to find $(4,500,000)^2$ on a scientific calculator, we could enter the integer 4,500,000 and press the $\boxed{x^2}$ (or squaring) key, obtaining a display similar to one of those in Figure 6. We would translate this as $2.025 \times 10^{13}$. Thus,

$$(4,500,000)^2 = 20,250,000,000,000.$$

Calculators may also use scientific form in the entry of numbers. The user's manual for your calculator should give specific details.

Before we conclude this section, we should make a few final remarks about rounding off results. Applied problems often include numbers that are obtained by various types of measurements and, hence, are *approximations* to exact values. Such answers should be rounded off, since the final result of a calculation cannot be more accurate than the data that have been used. For example, if the length and width of a rectangle are measured to two-decimal-place accuracy, we cannot expect more than two-decimal-place accuracy in the calculated value of the area of the rectangle. For purely *mathematical* work, if values of the length and width of a rectangle are given, we assume that the dimensions are *exact,* and no rounding off is required.

If a number $a$ is written in scientific form as $a = c \times 10^n$ for $1 \le c < 10$ and if $c$ is rounded off to $k$ decimal places, then we say that $a$ is accurate (or has been rounded off) to $k + 1$ **significant figures,** or **digits.** For example, 37.2638 rounded to 5 significant figures is $3.7264 \times 10^1$ or 37.264; to 3 significant figures, $3.73 \times 10^1$ or 37.3; and to 1 significant figure, $4 \times 10^1$ or 40.

## 1.1 EXERCISES

Exer. 1–2: If $x < 0$ and $y > 0$, determine the sign of the real number.

1  (a) $xy$    (b) $x^2y$    (c) $\dfrac{x}{y} + x$    (d) $y - x$

2  (a) $\dfrac{x}{y}$    (b) $xy^2$    (c) $\dfrac{x - y}{xy}$    (d) $y(y - x)$

Exer. 3–6: Replace the symbol □ with either <, >, or = to make the resulting statement true.

3  (a) $-7 \square -4$    (b) $\dfrac{\pi}{2} \square 1.57$    (c) $\sqrt{225} \square 15$

4  (a) $-3 \square -5$    (b) $\dfrac{\pi}{4} \square 0.8$    (c) $\sqrt{289} \square 17$

5  (a) $\tfrac{1}{11} \square 0.09$    (b) $\tfrac{2}{3} \square 0.6666$    (c) $\tfrac{22}{7} \square \pi$

6  (a) $\tfrac{1}{7} \square 0.143$    (b) $\tfrac{5}{6} \square 0.833$    (c) $\sqrt{2} \square 1.4$

Exer. 7–8: Express the statement as an inequality.

7  (a) $x$ is negative.

(b) $y$ is nonnegative.

(c) $q$ is less than or equal to $\pi$.

(d) $d$ is between 4 and 2.

(e) $t$ is not less than 5.

(f) The negative of $z$ is not greater than 3.

(g) The quotient of $p$ and $q$ is at most 7.

(h) The reciprocal of $w$ is at least 9.

(i) The absolute value of $x$ is greater than 7.

8  (a) $b$ is positive.

(b) $s$ is nonpositive.

(c) $w$ is greater than or equal to $-4$.

(d) $c$ is between $\tfrac{1}{5}$ and $\tfrac{1}{3}$.

(e) $p$ is not greater than $-2$.

(f) The negative of $m$ is not less than $-2$.

(g) The quotient of $r$ and $s$ is at least $\tfrac{1}{5}$.

(h) The reciprocal of $f$ is at most 14.

(i) The absolute value of $x$ is less than 4.

Exer. 9–14: Rewrite the number without using the absolute value symbol, and simplify the result.

9  (a) $|-3 - 2|$    (b) $|-5| - |2|$    (c) $|7| + |-4|$

10 (a) $|-11 + 1|$     (b) $|6| - |-3|$     (c) $|8| + |-9|$

11 (a) $(-5)|3 - 6|$     (b) $|-6|/(-2)$     (c) $|-7| + |4|$

12 (a) $(4)|6 - 7|$     (b) $5/|-2|$     (c) $|-1| + |-9|$

13 (a) $|4 - \pi|$     (b) $|\pi - 4|$     (c) $|\sqrt{2} - 1.5|$

14 (a) $|\sqrt{3} - 1.7|$     (b) $|1.7 - \sqrt{3}|$     (c) $|\frac{1}{5} - \frac{1}{3}|$

**Exer. 15–18:** The given numbers are coordinates of points $A$, $B$, and $C$, respectively, on a coordinate line. For the set of coordinates find the following distances:

(a) $d(A, B)$          (b) $d(B, C)$
(c) $d(C, B)$          (d) $d(A, C)$

15  $3, 7, -5$                16  $-6, -2, 4$

17  $-9, 1, 10$             18  $8, -4, -1$

**Exer. 19–24:** The two given numbers are coordinates of points $A$ and $B$, respectively, on a coordinate line. Express the indicated statement as an inequality involving the absolute value symbol.

19  $x,$       $7;$       $d(A, B)$ is less than 5

20  $x,$      $-\sqrt{2};$   $d(A, B)$ is greater than 1

21  $x,$      $-3;$      $d(A, B)$ is at least 8

22  $x,$       $4;$       $d(A, B)$ is at most 2

23  $4,$       $x;$       $d(A, B)$ is not greater than 3

24  $-2,$     $x;$       $d(A, B)$ is not less than 2

**Exer. 25–32:** Rewrite the expression without using the absolute value symbol, and simplify the result.

25  $|3 + x|$ if $x < -3$          26  $|5 - x|$ if $x > 5$

27  $|2 - x|$ if $x < 2$           28  $|7 + x|$ if $x \geq -7$

29  $|a - b|$ if $a < b$          30  $|a - b|$ if $a > b$

31  $|x^2 + 4|$                      32  $|-x^2 - 1|$

**Exer. 33–40:** Replace the symbol $\square$ with either $=$ or $\neq$ to make the resulting statement true for all real numbers $a$, $b$, $c$, and $d$, whenever the expressions are defined.

33  $\dfrac{ab + ac}{a} \ \square \ b + ac$          34  $\dfrac{ab + ac}{a} \ \square \ b + c$

35  $\dfrac{b + c}{a} \ \square \ \dfrac{b}{a} + \dfrac{c}{a}$          36  $\dfrac{a + c}{b + d} \ \square \ \dfrac{a}{b} + \dfrac{c}{d}$

37  $(a \div b) \div c \ \square \ a \div (b \div c)$   38  $(a - b) - c \ \square \ a - (b - c)$

39  $\dfrac{a - b}{b - a} \ \square \ -1$          40  $-(a + b) \ \square \ -a + b$

41 The point on a coordinate line corresponding to $\sqrt{2}$ may be determined by constructing a right triangle with sides of length 1, as shown in the figure. Determine the points that correspond to $\sqrt{3}$ and $\sqrt{5}$, respectively. (*Hint:* Use the Pythagorean theorem.)

**EXERCISE 41**

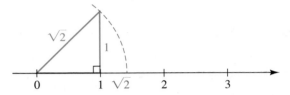

42 A circle of radius 1 rolls along a coordinate line in the positive direction, as shown in the figure. If point $P$ is initially at the origin, find the coordinate of $P$ after

(a) one,   (b) two, and   (c) ten complete revolutions.

**EXERCISE 42**

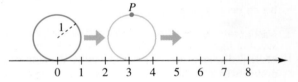

43 Geometric proofs of properties of real numbers were first given by the ancient Greeks. In order to establish the distributive property $a(b + c) = ab + ac$ for positive real numbers $a$, $b$, and $c$, find the area of the rectangle shown in the figure in two ways.

**EXERCISE 43**

44 Rational approximations to square roots can be found using a formula discovered by the ancient Babylonians. Let $x_1$ be the first rational approximation for $\sqrt{n}$. If we let

$$x_2 = \frac{1}{2}\left(x_1 + \frac{n}{x_1}\right),$$

then $x_2$ will be a better approximation for $\sqrt{n}$, and we can repeat the computation with $x_2$ replacing $x_1$. Starting with $x_1 = \frac{3}{2}$, find the next two rational approximations for $\sqrt{2}$.

**Exer. 45–46: Express the number in scientific form.**

45 (a) 427,000   (b) 0.000 000 098   (c) 810,000,000

46 (a) 85,200   (b) 0.000 005 5   (c) 24,900,000

**Exer. 47–48: Express the number in decimal form.**

47 (a) $8.3 \times 10^5$   (b) $2.9 \times 10^{-12}$   (c) $5.63 \times 10^8$

48 (a) $2.3 \times 10^7$   (b) $7.01 \times 10^{-9}$   (c) $1.23 \times 10^{10}$

49 *Mass of a hydrogen atom* The mass of a hydrogen atom is approximately

$\qquad$ 0.000 000 000 000 000 000 000 001 7 gram.

Express this number in scientific form.

50 *Mass of an electron* The mass of an electron is approximately $9.1 \times 10^{-31}$ kilogram. Express this number in decimal form.

51 *Light year* In astronomy, distances to stars are measured in light years. One light year is the distance a ray of light travels in one year. If the speed of light is approximately 186,000 miles per second, estimate the number of miles in one light year.

52 *Milky Way galaxy*

(a) Astronomers have estimated that the Milky Way galaxy contains 100 billion stars. Express this number in scientific form.

(b) The diameter $d$ of the Milky Way galaxy is estimated as 100,000 light years. Express $d$ in miles. (Refer to Exercise 51.)

53 *Avogadro's number* The number of hydrogen atoms in a mole is Avogadro's number, $6.02 \times 10^{23}$. If one mole of the gas has a mass of 1.01 grams, estimate the mass of a hydrogen atom.

54 *Fish population* The population dynamics of many fish are characterized by extremely high fertility rates among adults and very low survival rates among the young. A mature halibut may lay as many as 2.5 million eggs, but only 0.00035% of the offspring survive to the age of 3 years. Use scientific form to approximate the number of offspring that live to age 3.

55 *Frames in a movie film* The longest movie ever made is a 1970 British film that runs for 48 hours. Assuming that the film speed is 24 frames per second, approximate the total number of frames in this film. Express your answer in scientific form.

56 *Large prime numbers* The number $2^{44,497} - 1$ is prime. At the time that this number was determined to be prime, it took one of the world's fastest computers about 60 days to verify that it was prime. This computer was capable of performing $2 \times 10^{11}$ calculations per second. Use scientific form to estimate the number of calculations needed to perform this computation. (More recently, in 1992, $2^{756,839} - 1$, a number containing 227,832 digits, was shown to be prime.)

57 *Intelligence quotient* A person's intelligence quotient (IQ) is determined by multiplying the quotient of his or her mental age and chronological age by 100.

(a) Find the IQ of a 12-year-old child whose mental age is 15.

(b) Find the mental age of a person 15 years old whose IQ is 140.

58 *Earth's surface area* Water covers 70.8%, or $361 \times 10^6$ km$^2$, of the earth's surface. Approximate the total surface area of the earth.

# 1.2 EXPONENTS AND RADICALS

If $n$ is a positive integer, the exponential notation $a^n$, defined in the following chart, represents the product of the real number $a$ with itself $n$ times. We refer to $a^n$ as **$a$ to the $n$th power** or, simply, $a$ to the $n$. The positive integer $n$ is called the **exponent,** and the real number $a$ is called the **base.**

**Exponential Notation**

| General case ($n$ is any positive integer) | Special cases |
|---|---|
| $a^n = \underbrace{a \cdot a \cdot a \cdots \cdot a}_{n \text{ factors of } a}$ | $a^1 = a$ <br> $a^2 = a \cdot a$ <br> $a^3 = a \cdot a \cdot a$ <br> $a^6 = a \cdot a \cdot a \cdot a \cdot a \cdot a$ |

The next illustration contains several numerical examples of exponential notation.

**ILLUSTRATION**

**The Exponential Notation $a^n$**

- $5^4 = 5 \cdot 5 \cdot 5 \cdot 5 = 625$
- $\left(\frac{1}{2}\right)^5 = \frac{1}{2} \cdot \frac{1}{2} \cdot \frac{1}{2} \cdot \frac{1}{2} \cdot \frac{1}{2} = \frac{1}{32}$
- $(-3)^3 = (-3)(-3)(-3) = -27$
- $\left(-\frac{1}{3}\right)^4 = \left(-\frac{1}{3}\right)\left(-\frac{1}{3}\right)\left(-\frac{1}{3}\right)\left(-\frac{1}{3}\right) = \left(\frac{1}{9}\right)\left(\frac{1}{9}\right) = \frac{1}{81}$

It is important to note that if $n$ is a positive integer, then an expression such as $3a^n$ means $3(a^n)$, *not* $(3a)^n$. The real number 3 is the **coefficient** of $a^n$ in the expression $3a^n$. Similarly, $-3a^n$ means $(-3)a^n$, *not* $(-3a)^n$.

**ILLUSTRATION**

**The Notation $ca^n$**

- $5 \cdot 2^3 = 5 \cdot 8 = 40$
- $-5 \cdot 2^3 = -5 \cdot 8 = -40$
- $-2^4 = -(2^4) = -16$
- $3(-2)^3 = 3(-2)(-2)(-2) = 3(-8) = -24$

We next extend the definition of $a^n$ to nonpositive exponents.

**Zero and Negative Exponents**

| Definition ($a \neq 0$) | Illustration |
|---|---|
| $a^0 = 1$ | $3^0 = 1, \quad (-\sqrt{2})^0 = 1$ |
| $a^{-n} = \dfrac{1}{a^n}$ | $5^{-3} = \dfrac{1}{5^3}, \quad (-3)^{-5} = \dfrac{1}{(-3)^5}$ |

If $m$ and $n$ are positive integers, then

$$a^m a^n = \underbrace{a \cdot a \cdot a \cdots a}_{m \text{ factors of } a} \cdot \underbrace{a \cdot a \cdot a \cdots a}_{n \text{ factors of } a}.$$

Since the total number of factors of $a$ on the right is $m + n$, this expression is equal to $a^{m+n}$; that is,

$$a^m a^n = a^{m+n}.$$

We can extend this formula to $m \leq 0$ or $n \leq 0$ by using the definitions of zero exponent and negative exponents. This gives us law 1, stated in the next chart.

To prove law 2, we may write, for $m$ and $n$ positive,

$$(a^m)^n = \underbrace{a^m \cdot a^m \cdot a^m \cdots a^m}_{n \text{ factors of } a^m}$$

and count the number of times $a$ appears as a factor on the right-hand side. Since $a^m = a \cdot a \cdot a \cdots a$, with $a$ occurring as a factor $m$ times, and since the number of such groups of $m$ factors is $n$, the total number of factors of $a$ is $m \cdot n$. Thus,

$$(a^m)^n = a^{mn}.$$

The cases $m \leq 0$ or $n \leq 0$ can be proved using the definition of nonpositive exponents. The remaining three laws can be established in similar fashion by counting factors. In laws 4 and 5 we assume that denominators are not 0.

**Laws of Exponents for Real Numbers $a$ and $b$ and Integers $m$ and $n$**

| Law | Illustration |
|---|---|
| (1) $a^m a^n = a^{m+n}$ | $2^3 \cdot 2^4 = 2^{3+4} = 2^7 = 128$ |
| (2) $(a^m)^n = a^{mn}$ | $(2^3)^4 = 2^{3 \cdot 4} = 2^{12} = 4096$ |
| (3) $(ab)^n = a^n b^n$ | $(20)^3 = (2 \cdot 10)^3 = 2^3 \cdot 10^3 = 8 \cdot 1000 = 8000$ |
| (4) $\left(\dfrac{a}{b}\right)^n = \dfrac{a^n}{b^n}$ | $\left(\dfrac{2}{5}\right)^3 = \dfrac{2^3}{5^3} = \dfrac{8}{125}$ |
| (5) (a) $\dfrac{a^m}{a^n} = a^{m-n}$ | $\dfrac{2^5}{2^3} = 2^{5-3} = 2^2 = 4$ |
| (b) $\dfrac{a^m}{a^n} = \dfrac{1}{a^{n-m}}$ | $\dfrac{2^3}{2^5} = \dfrac{1}{2^{5-3}} = \dfrac{1}{2^2} = \dfrac{1}{4}$ |

We usually use 5(a) if $m > n$ and 5(b) if $m < n$.

We can extend laws of exponents to obtain rules such as $(abc)^n = a^n b^n c^n$ and $a^m a^n a^p = a^{m+n+p}$. Some other examples of the laws of exponents are given in the next illustration.

## ILLUSTRATION

### Laws of Exponents

$x^5x^6x^2 = x^{5+6+2} = x^{13}$

$(y^5)^7 = y^{5\cdot7} = y^{35}$

$(3st)^4 = 3^4s^4t^4 = 81s^4t^4$

$\left(\dfrac{p}{2}\right)^5 = \dfrac{p^5}{2^5} = \dfrac{p^5}{32}$

$\dfrac{c^8}{c^3} = c^{8-3} = c^5$

$\dfrac{u^3}{u^8} = \dfrac{1}{u^{8-3}} = \dfrac{1}{u^5}$

To **simplify** an expression involving powers of real numbers means to change it to an expression in which each real number appears only once and all exponents are positive. *We shall assume that denominators always represent nonzero real numbers.*

**EXAMPLE I**    *Simplifying expressions containing exponents*

Use laws of exponents to simplify the expression:

**(a)** $(3x^3y^4)(4xy^5)$    **(b)** $(2a^2b^3c)^4$    **(c)** $\left(\dfrac{2r^3}{s}\right)^2\left(\dfrac{s}{r^3}\right)^3$    **(d)** $(u^{-2}v^3)^{-3}$

### SOLUTION

**(a)** 
$$(3x^3y^4)(4xy^5) = (3)(4)x^3xy^4y^5 \qquad \text{rearrange factors}$$
$$= 12x^4y^9 \qquad \text{law 1}$$

**(b)** 
$$(2a^2b^3c)^4 = 2^4(a^2)^4(b^3)^4c^4 \qquad \text{law 3}$$
$$= 16a^8b^{12}c^4 \qquad \text{law 2}$$

**(c)** 
$$\left(\frac{2r^3}{s}\right)^2\left(\frac{s}{r^3}\right)^3 = \frac{(2r^3)^2}{s^2}\cdot\frac{s^3}{(r^3)^3} \qquad \text{law 4}$$

$$= \frac{2^2(r^3)^2}{s^2}\cdot\frac{s^3}{(r^3)^3} \qquad \text{law 3}$$

$$= \left(\frac{4r^6}{s^2}\right)\left(\frac{s^3}{r^9}\right) \qquad \text{law 2}$$

$$= 4\left(\frac{r^6}{r^9}\right)\left(\frac{s^3}{s^2}\right) \qquad \text{rearrange factors}$$

$$= 4\left(\frac{1}{r^3}\right)(s) \qquad \text{laws 5(b) and 5(a)}$$

$$= \frac{4s}{r^3} \qquad \text{rearrange factors}$$

**(d)**
$$(u^{-2}v^3)^{-3} = (u^{-2})^{-3}(v^3)^{-3} \quad \text{law 3}$$
$$= u^6 v^{-9} \quad \text{law 2}$$
$$= \frac{u^6}{v^9} \quad \text{definition of } a^{-n}$$

The following theorem is useful for problems that involve negative exponents.

**Theorem on Negative Exponents**

$$\textbf{(1)} \quad \frac{a^{-m}}{b^{-n}} = \frac{b^n}{a^m} \qquad \textbf{(2)} \quad \left(\frac{a}{b}\right)^{-n} = \left(\frac{b}{a}\right)^n$$

*PROOF* Using properties of negative exponents and quotients, we obtain

$$\textbf{(1)} \quad \frac{a^{-m}}{b^{-n}} = \frac{1/a^m}{1/b^n} = \frac{1}{a^m} \cdot \frac{b^n}{1} = \frac{b^n}{a^m}$$

$$\textbf{(2)} \quad \left(\frac{a}{b}\right)^{-n} = \frac{a^{-n}}{b^{-n}} = \frac{b^n}{a^n} = \left(\frac{b}{a}\right)^n \quad ■$$

**EXAMPLE 2** *Simplifying expressions containing negative exponents*

Simplify:

**(a)** $\dfrac{8x^3 y^{-5}}{4x^{-1} y^2}$ **(b)** $\left(\dfrac{u^2}{2v}\right)^{-3}$

*SOLUTION* We apply the theorem on negative exponents and the laws of exponents.

**(a)**
$$\frac{8x^3 y^{-5}}{4x^{-1} y^2} = \frac{8x^3}{4y^2} \cdot \frac{y^{-5}}{x^{-1}} \qquad \begin{array}{l}\text{rearrange quotients so that negative}\\ \text{exponents are in one fraction}\end{array}$$

$$= \frac{8x^3}{4y^2} \cdot \frac{x^1}{y^5} \qquad \text{theorem on negative exponents (1)}$$

$$= \frac{2x^4}{y^7} \qquad \text{law 1 of exponents}$$

*(continued)*

**(b)**

$$\left(\frac{u^2}{2v}\right)^{-3} = \left(\frac{2v}{u^2}\right)^3 \qquad \text{theorem on negative exponents (2)}$$

$$= \frac{2^3 v^3}{(u^2)^3} \qquad \text{laws 4 and 3 of exponents}$$

$$= \frac{8v^3}{u^6} \qquad \text{law 2 of exponents}$$

We next define the **principal $n$th root** $\sqrt[n]{a}$ of a real number $a$.

**Definition of $\sqrt[n]{a}$**

> Let $n$ be a positive integer greater than 1 and let $a$ be a real number.
>
> **(1)** If $a = 0$, then $\sqrt[n]{a} = 0$.
> **(2)** If $a > 0$, then $\sqrt[n]{a}$ is the *positive* real number $b$ such that $b^n = a$.
> **(3)** (a) If $a < 0$ and $n$ is odd, then $\sqrt[n]{a}$ is the *negative* real number $b$ such that $b^n = a$.
>   (b) If $a < 0$ and $n$ is even, then $\sqrt[n]{a}$ is not a real number.

Complex numbers, discussed in Section 1.5, are needed to define $\sqrt[n]{a}$ if $a < 0$ and $n$ is an *even* positive integer, since for all real numbers $b$, $b^n \geq 0$ whenever $n$ is even.

If $n = 2$, we write $\sqrt[2]{a}$ instead of $\sqrt[2]{a}$ and call $\sqrt{a}$ the **principal square root** of $a$ or, simply, the **square root** of $a$. The number $\sqrt[3]{a}$ is the (principal) **cube root** of $a$.

### ILLUSTRATION

#### The Principal $n$th Root $\sqrt[n]{a}$

- $\sqrt{16} = 4$, since $4^2 = 16$.
- $\sqrt[5]{\frac{1}{32}} = \frac{1}{2}$, since $\left(\frac{1}{2}\right)^5 = \frac{1}{32}$.
- $\sqrt[3]{-8} = -2$, since $(-2)^3 = -8$.
- $\sqrt[4]{-16}$ is not a real number.

Note that $\sqrt{16} \neq \pm 4$, since, by definition, roots of positive real numbers are positive.

To complete our terminology, the expression $\sqrt[n]{a}$ is a **radical,** the number $a$ is the **radicand,** and $n$ is the **index** of the radical. The symbol $\sqrt{\phantom{a}}$ is called a **radical sign.**

If $\sqrt{a} = b$, then $b^2 = a$; that is, $(\sqrt{a})^2 = a$. If $\sqrt[3]{a} = b$, then $b^3 = a$, or $(\sqrt[3]{a})^3 = a$. Generalizing this pattern gives us property 1 in the next chart.

**Properties of $\sqrt[n]{\phantom{x}}$ ($n$ is a positive integer)**

| Property | Illustration |
|---|---|
| (1) $(\sqrt[n]{a})^n = a$, if $\sqrt[n]{a}$ is a real number | $(\sqrt{5})^2 = 5, \quad (\sqrt[3]{-8})^3 = -8$ |
| (2) $\sqrt[n]{a^n} = a$, if $a \geq 0$ | $\sqrt{5^2} = 5, \quad \sqrt[3]{2^3} = 2$ |
| (3) $\sqrt[n]{a^n} = a$, if $a < 0$ and $n$ is odd | $\sqrt[3]{(-2)^3} = -2, \quad \sqrt[5]{(-2)^5} = -2$ |
| (4) $\sqrt[n]{a^n} = |a|$, if $a < 0$ and $n$ is even | $\sqrt{(-3)^2} = |-3| = 3, \quad \sqrt[4]{(-2)^4} = |-2| = 2$ |

If $a \geq 0$, then property 4 reduces to property 2. We also see from property 4 that

$$\sqrt{x^2} = |x|$$

for every real number $x$. In particular, if $x \geq 0$, then $\sqrt{x^2} = x$; however, if $x < 0$, then $\sqrt{x^2} = -x$, which is positive.

The three laws listed in the next chart are true for positive integers $m$ and $n$, *provided the indicated roots exist*—that is, provided the roots are real numbers.

**Laws of Radicals**

| Law | Illustration |
|---|---|
| (1) $\sqrt[n]{ab} = \sqrt[n]{a}\,\sqrt[n]{b}$ | $\sqrt{50} = \sqrt{25 \cdot 2} = \sqrt{25}\,\sqrt{2} = 5\sqrt{2}$ <br> $\sqrt[3]{-108} = \sqrt[3]{(-27)(4)} = \sqrt[3]{-27}\,\sqrt[3]{4} = -3\sqrt[3]{4}$ |
| (2) $\sqrt[n]{\dfrac{a}{b}} = \dfrac{\sqrt[n]{a}}{\sqrt[n]{b}}$ | $\sqrt[3]{\dfrac{5}{8}} = \dfrac{\sqrt[3]{5}}{\sqrt[3]{8}} = \dfrac{\sqrt[3]{5}}{2}$ |
| (3) $\sqrt[m]{\sqrt[n]{a}} = \sqrt[mn]{a}$ | $\sqrt{\sqrt[3]{64}} = \sqrt[2(3)]{64} = \sqrt[6]{2^6} = 2$ |

The radicands in laws 1 and 2 involve products and quotients. Care must be taken if sums or differences occur in the radicand. The following chart contains two particular warnings concerning mistakes that frequently occur in working with radicals containing sums.

**Warnings**

| If $a \neq 0$ and $b \neq 0$ | Illustration |
|---|---|
| (1) $\sqrt{a^2 + b^2} \neq a + b$ | $\sqrt{3^2 + 4^2} = \sqrt{25} = 5 \neq 3 + 4 = 7$ |
| (2) $\sqrt{a + b} \neq \sqrt{a} + \sqrt{b}$ | $\sqrt{4 + 9} = \sqrt{13} \neq \sqrt{4} + \sqrt{9} = 5$ |

If $c$ is a real number and $c^n$ occurs as a factor in a radical of index $n$, then we can remove $c$ from the radicand if the sign of $c$ is taken into account. For example, if $c > 0$, or if $c < 0$ and $n$ is *odd,* then

$$\sqrt[n]{c^n d} = \sqrt[n]{c^n}\,\sqrt[n]{d} = c\sqrt[n]{d},$$

provided $\sqrt[n]{d}$ exists. If $c < 0$ and $n$ is *even*, then

$$\sqrt[n]{c^n d} = \sqrt[n]{c^n}\,\sqrt[n]{d} = |c|\sqrt[n]{d},$$

provided $\sqrt[n]{d}$ exists.

### ILLUSTRATION

**Removing $n$th Powers from $\sqrt[n]{\ }$**

- $\sqrt[5]{x^7} = \sqrt[5]{x^5 \cdot x^2} = \sqrt[5]{x^5}\,\sqrt[5]{x^2} = x\sqrt[5]{x^2}$
- $\sqrt[3]{x^7} = \sqrt[3]{x^6 \cdot x} = \sqrt[3]{(x^2)^3 x} = \sqrt[3]{(x^2)^3}\,\sqrt[3]{x} = x^2\sqrt[3]{x}$
- $\sqrt{x^2 y} = \sqrt{x^2}\,\sqrt{y} = |x|\sqrt{y}$
- $\sqrt{x^6} = \sqrt{(x^3)^2} = |x^3|$
- $\sqrt[4]{x^6 y^3} = \sqrt[4]{x^4 \cdot x^2 y^3} = \sqrt[4]{x^4}\,\sqrt[4]{x^2 y^3} = |x|\sqrt[4]{x^2 y^3}$

**Note:** To avoid considering absolute values, *in examples and exercises involving radicals in this chapter, we shall assume that all letters—a, b, c, d, x, y, and so on—that appear in radicands represent positive real numbers, unless otherwise specified.*

As shown in the preceding illustration and in the following examples, if the index of a radical is $n$, then we rearrange the radicand, isolating a factor of the form $p^n$, where $p$ may consist of several letters. We then remove $\sqrt[n]{p^n} = p$ from the radical as previously indicated. Thus, in Example 3(b) the index of the radical is 3 and we rearrange the radicand into *cubes*, obtaining a factor $p^3$, with $p = 2xy^2 z$. In part (c) the index of the radical is 2 and we rearrange the radicand into squares, obtaining a factor $p^2$, with $p = 3a^3 b^2$.

To *simplify a radical* means to remove factors from the radical until no factor in the radicand has an exponent greater than or equal to the index of the radical and the index is as low as possible.

### EXAMPLE 3   *Removing factors from radicals*

Simplify the radical (all letters denote positive real numbers):

**(a)** $\sqrt[3]{320}$    **(b)** $\sqrt[3]{16x^3 y^8 z^4}$    **(c)** $\sqrt{3a^2 b^3}\,\sqrt{6a^5 b}$

**SOLUTION**

**(a)**
$$\sqrt[3]{320} = \sqrt[3]{64 \cdot 5} \qquad \text{factor out the largest cube in 320}$$
$$= \sqrt[3]{4^3}\,\sqrt[3]{5} \qquad \text{law 1 of radicals}$$
$$= 4\sqrt[3]{5} \qquad \text{property 2 of } \sqrt[n]{\ }$$

**(b)**
$$\sqrt[3]{16x^3 y^8 z^4} = \sqrt[3]{(2^3 x^3 y^6 z^3)(2y^2 z)} \qquad \text{rearrange radicand into cubes}$$
$$= \sqrt[3]{(2xy^2 z)^3 (2y^2 z)} \qquad \text{laws 2 and 3 of exponents}$$
$$= \sqrt[3]{(2xy^2 z)^3}\,\sqrt[3]{2y^2 z} \qquad \text{law 1 of radicals}$$
$$= 2xy^2 z\sqrt[3]{2y^2 z} \qquad \text{property 2 of } \sqrt[n]{\ }$$

**(c)** $\sqrt{3a^2b^3} \sqrt{6a^5b} = \sqrt{3a^2b^3 \cdot 2 \cdot 3a^5b}$    law 1 of radicals

$\qquad = \sqrt{(3^2a^6b^4)(2a)}$    rearrange radicand into squares

$\qquad = \sqrt{(3a^3b^2)^2(2a)}$    laws 2 and 3 of exponents

$\qquad = \sqrt{(3a^3b^2)^2} \sqrt{2a}$    law 1 of radicals

$\qquad = 3a^3b^2\sqrt{2a}$    property 2 of $\sqrt[n]{\phantom{x}}$

If the denominator of a quotient contains a factor of the form $\sqrt[n]{a^k}$, with $k < n$ and $a > 0$, then multiplying numerator and denominator by $\sqrt[n]{a^{n-k}}$ will eliminate the radical from the denominator, since

$$\sqrt[n]{a^k} \, \sqrt[n]{a^{n-k}} = \sqrt[n]{a^{k+n-k}} = \sqrt[n]{a^n} = a.$$

This process is called **rationalizing a denominator.** Some special cases are listed in the following chart.

**Rationalizing Denominators of Quotients** $(a > 0)$

| Factor in denominator | Multiply numerator and denominator by | Resulting factor |
|:---:|:---:|:---:|
| $\sqrt{a}$ | $\sqrt{a}$ | $\sqrt{a} \sqrt{a} = \sqrt{a^2} = a$ |
| $\sqrt[3]{a}$ | $\sqrt[3]{a^2}$ | $\sqrt[3]{a} \sqrt[3]{a^2} = \sqrt[3]{a^3} = a$ |
| $\sqrt[7]{a^3}$ | $\sqrt[7]{a^4}$ | $\sqrt[7]{a^3} \sqrt[7]{a^4} = \sqrt[7]{a^7} = a$ |

The next example illustrates this technique.

**EXAMPLE 4**    *Rationalizing denominators*

Rationalize the denominator:

**(a)** $\dfrac{1}{\sqrt{5}}$    **(b)** $\dfrac{1}{\sqrt[3]{x}}$    **(c)** $\sqrt{\dfrac{2}{3}}$    **(d)** $\sqrt[5]{\dfrac{x}{y^2}}$

**SOLUTION**

**(a)** $\quad \dfrac{1}{\sqrt{5}} = \dfrac{1}{\sqrt{5}} \dfrac{\sqrt{5}}{\sqrt{5}} = \dfrac{\sqrt{5}}{\sqrt{5^2}} = \dfrac{\sqrt{5}}{5}$

**(b)** $\quad \dfrac{1}{\sqrt[3]{x}} = \dfrac{1}{\sqrt[3]{x}} \dfrac{\sqrt[3]{x^2}}{\sqrt[3]{x^2}} = \dfrac{\sqrt[3]{x^2}}{\sqrt[3]{x^3}} = \dfrac{\sqrt[3]{x^2}}{x}$

**(c)** $\quad \sqrt{\dfrac{2}{3}} = \dfrac{\sqrt{2}}{\sqrt{3}} = \dfrac{\sqrt{2}}{\sqrt{3}} \dfrac{\sqrt{3}}{\sqrt{3}} = \dfrac{\sqrt{2 \cdot 3}}{\sqrt{3^2}} = \dfrac{\sqrt{6}}{3}$

**(d)** $\quad \sqrt[5]{\dfrac{x}{y^2}} = \dfrac{\sqrt[5]{x}}{\sqrt[5]{y^2}} = \dfrac{\sqrt[5]{x}}{\sqrt[5]{y^2}} \dfrac{\sqrt[5]{y^3}}{\sqrt[5]{y^3}} = \dfrac{\sqrt[5]{xy^3}}{\sqrt[5]{y^5}} = \dfrac{\sqrt[5]{xy^3}}{y}$

If we use a calculator to find decimal approximations of radicals, there is no advantage in rationalizing denominators, such as $1/\sqrt{5} = \sqrt{5}/5$ or $\sqrt{2/3} = \sqrt{6}/3$, as we did in Example 4(a) and (c). However, for *algebraic* simplifications, changing expressions to such forms is sometimes desirable. Similarly, in advanced mathematics courses such as calculus, changing $1/\sqrt[3]{x}$ to $\sqrt[3]{x^2}/x$, as in Example 4(b), could make a problem *more* complicated. In such courses it is simpler to work with the expression $1/\sqrt[3]{x}$ than with its rationalized form.

We next use radicals to define *rational exponents*.

**Definition of Rational Exponents**

Let $m/n$ be a rational number, where $n$ is a positive integer greater than 1. If $a$ is a real number such that $\sqrt[n]{a}$ exists, then

**(1)** $a^{1/n} = \sqrt[n]{a}$

**(2)** $a^{m/n} = (\sqrt[n]{a})^m = \sqrt[n]{a^m}$

**(3)** $a^{m/n} = (a^{1/n})^m = (a^m)^{1/n}$

When evaluating $a^{m/n}$ in (2), we usually use $(\sqrt[n]{a})^m$; that is, we take the $n$th root of $a$ first and then raise that result to the $m$th power, as shown in the following illustration.

## ILLUSTRATION

### The Exponential Notation $a^{m/n}$

■ $x^{1/3} = \sqrt[3]{x}$

■ $x^{3/5} = (\sqrt[5]{x})^3 = \sqrt[5]{x^3}$

■ $125^{2/3} = (\sqrt[3]{125})^2 = (\sqrt[3]{5^3})^2 = 5^2 = 25$

■ $\left(\frac{32}{243}\right)^{3/5} = \left(\sqrt[5]{\frac{32}{243}}\right)^3 = \left(\sqrt[5]{\left(\frac{2}{3}\right)^5}\right)^3 = \left(\frac{2}{3}\right)^3 = \frac{8}{27}$

The laws of exponents are true for rational exponents and also for *irrational* exponents, such as $3^{\sqrt{2}}$ or $5^{\pi}$, considered in Chapter 4.

To simplify an expression involving rational powers of letters that represent real numbers, we change it to an expression in which each letter appears only once and all exponents are positive. As we did with radicals, we shall assume that all letters represent positive real numbers unless otherwise specified.

### EXAMPLE 5    *Simplifying rational powers*

Simplify:

**(a)** $(-27)^{2/3}(4)^{-5/2}$     **(b)** $(r^2 s^6)^{1/3}$     **(c)** $\left(\frac{2x^{2/3}}{y^{1/2}}\right)^2 \left(\frac{3x^{-5/6}}{y^{1/3}}\right)$

**SOLUTION**

**(a)** $(-27)^{2/3}(4)^{-5/2} = (\sqrt[3]{-27})^2(\sqrt{4})^{-5}$     definition of rational exponents

$= (-3)^2(2)^{-5}$     take roots

$= \dfrac{(-3)^2}{2^5}$     definition of negative exponent

$= \frac{9}{32}$     take powers

**(b)** $(r^2 s^6)^{1/3} = (r^2)^{1/3}(s^6)^{1/3}$     law 3 of exponents

$= r^{2/3} s^2$     law 2 of exponents

**(c)** $\left(\dfrac{2x^{2/3}}{y^{1/2}}\right)^2\left(\dfrac{3x^{-5/6}}{y^{1/3}}\right) = \left(\dfrac{4x^{4/3}}{y}\right)\left(\dfrac{3x^{-5/6}}{y^{1/3}}\right)$     laws of exponents

$= \dfrac{(4 \cdot 3)x^{4/3-5/6}}{y^{1+(1/3)}}$     law 1 of exponents

$= \dfrac{12x^{8/6-5/6}}{y^{4/3}}$     common denominator

$= \dfrac{12x^{1/2}}{y^{4/3}}$     simplify

Rational exponents are useful for problems involving radicals that do not have the same index, as illustrated in the next example.

**EXAMPLE 6    *Combining radicals***

Change to an expression containing one radical of the form $\sqrt[n]{a^m}$:

**(a)** $\sqrt[3]{a}\sqrt{a}$     **(b)** $\dfrac{\sqrt[4]{a}}{\sqrt[3]{a^2}}$

**SOLUTION**     Introducing rational exponents, we obtain

**(a)** $\sqrt[3]{a}\sqrt{a} = a^{1/3}a^{1/2} = a^{(1/3)+(1/2)} = a^{5/6} = \sqrt[6]{a^5}$

**(b)** $\dfrac{\sqrt[4]{a}}{\sqrt[3]{a^2}} = \dfrac{a^{1/4}}{a^{2/3}} = a^{(1/4)-(2/3)} = a^{-5/12} = \dfrac{1}{a^{5/12}} = \dfrac{1}{\sqrt[12]{a^5}}$

In Exercises 1.2, whenever an index of a radical is even (or a rational exponent $m/n$ with $n$ even is employed), assume that the letters that appear in the radicand denote positive real numbers unless otherwise specified.

## 1.2 EXERCISES

**Exer. 1–10:** Express the number in the form $a/b$, where $a$ and $b$ are integers.

**1** $\left(-\frac{2}{3}\right)^4$

**2** $(-3)^3$

**3** $\dfrac{2^{-3}}{3^{-2}}$

**4** $\dfrac{2^0 + 0^2}{2 + 0}$

**5** $-2^4 + 3^{-1}$

**6** $\left(-\frac{3}{2}\right)^4 - 2^{-4}$

**7** $16^{-3/4}$

**8** $9^{5/2}$

**9** $(-0.008)^{2/3}$

**10** $(0.008)^{-2/3}$

**Exer. 11–38:** Simplify.

**11** $\left(\frac{1}{4}x^4\right)(16x^5)$

**12** $(-3x^{-2})(4x^4)$

**13** $\dfrac{(6x^3)^2}{(2x^2)^3}$

**14** $\dfrac{(3y^3)(2y^2)^2}{(y^4)^3}$

**15** $(8x^4y^{-3})\left(\frac{1}{2}x^{-5}y^2\right)$

**16** $\left(\dfrac{4a^2b}{a^3b^2}\right)\left(\dfrac{5a^2b}{2b^4}\right)$

**17** $\left(\frac{1}{3}x^4y^{-3}\right)^{-2}$

**18** $(-2xy^2)^5\left(\dfrac{x^7}{8y^3}\right)$

**19** $(3y^3)^4(4y^2)^{-3}$

**20** $(-3a^2b^{-5})^3$

**21** $(-2r^4s^{-3})^{-2}$

**22** $(2x^2y^{-5})(6x^{-3}y)(\frac{1}{3}x^{-1}y^3)$

**23** $(5x^2y^{-3})(4x^{-5}y^4)$

**24** $(-2r^2s)^5(3r^{-1}s^3)^2$

**25** $\left(\dfrac{3x^5y^4}{x^0y^{-3}}\right)^2$

**26** $(4a^2b)^4\left(\dfrac{-a^3}{2b}\right)^2$

**27** $(3x^{5/6})(8x^{2/3})$

**28** $(8r)^{1/3}(2r^{1/2})$

**29** $(27a^6)^{-2/3}$

**30** $(25z^4)^{-3/2}$

**31** $(8x^{-2/3})x^{1/6}$

**32** $(3x^{1/2})(-2x^{5/2})$

**33** $\left(\dfrac{-8x^3}{y^{-6}}\right)^{2/3}$

**34** $\left(\dfrac{-y^{3/2}}{y^{-1/3}}\right)^3$

**35** $\left(\dfrac{x^6}{9y^{-4}}\right)^{-1/2}$

**36** $\left(\dfrac{c^{-4}}{16d^8}\right)^{3/4}$

**37** $\dfrac{(x^6y^3)^{-1/3}}{(x^4y^2)^{-1/2}}$

**38** $a^{4/3}a^{-3/2}a^{1/6}$

**Exer. 39–44:** Rewrite the expression using rational exponents.

**39** $\sqrt[4]{x^3}$

**40** $\sqrt[3]{x^5}$

**41** $\sqrt[3]{(a + b)^2}$

**42** $\sqrt{a + \sqrt{b}}$

**43** $\sqrt{x^2 + y^2}$

**44** $\sqrt[3]{r^3 - s^3}$

**Exer. 45–48:** Rewrite the expression using a radical.

**45** (a) $4x^{3/2}$      (b) $(4x)^{3/2}$

**46** (a) $4 + x^{3/2}$      (b) $(4 + x)^{3/2}$

**47** (a) $8 - y^{1/3}$      (b) $(8 - y)^{1/3}$

**48** (a) $8y^{1/3}$      (b) $(8y)^{1/3}$

**Exer. 49–66:** Simplify the expression, and rationalize the denominator when appropriate.

**49** $\sqrt{81}$

**50** $\sqrt[3]{-125}$

**51** $\sqrt[5]{-64}$

**52** $\sqrt[4]{256}$

**53** $\dfrac{1}{\sqrt[3]{2}}$

**54** $\sqrt{\dfrac{1}{7}}$

**55** $\sqrt{9x^{-4}y^6}$

**56** $\sqrt{16a^8b^{-2}}$

**57** $\sqrt[3]{8a^6b^{-3}}$

**58** $\sqrt[4]{81r^5s^8}$

**59** $\sqrt{\dfrac{3x}{2y^3}}$

**60** $\sqrt[3]{\dfrac{2x^4y^4}{9x}}$

**61** $\sqrt[4]{\dfrac{5x^8y^3}{27x^2}}$

**62** $\sqrt[5]{\dfrac{8x^3}{y^4}}\sqrt[5]{\dfrac{4x^4}{y^2}}$

**63** $\sqrt[4]{(3x^5y^{-2})^4}$

**64** $\sqrt[6]{(2u^{-3}v^4)^6}$

**65** $\sqrt{5xy^7}\sqrt{10x^3y^3}$

**66** $\sqrt[3]{3t^4v^2}\sqrt[3]{-9t^{-1}v^4}$

**Exer. 67–70:** Simplify the expression, assuming $x$ and $y$ may be negative.

**67** $\sqrt{x^6y^4}$

**68** $\sqrt{x^4y^{10}}$

**69** $\sqrt[4]{x^8(y - 1)^{12}}$

**70** $\sqrt[4]{(x + 2)^{12}y^4}$

**Exer. 71–76:** Replace the symbol $\square$ with either $=$ or $\neq$ to make the resulting statement true, whenever the expression has meaning. Give a reason for your answer.

**71** $(a^r)^2 \square a^{(r^2)}$

**72** $(a^2 + 1)^{1/2} \square a + 1$

**73** $a^x b^y \square (ab)^{xy}$

**74** $\sqrt{a^r} \square (\sqrt{a})^r$

**75** $\sqrt[n]{\dfrac{1}{c}} \square \dfrac{1}{\sqrt[n]{c}}$

**76** $a^{1/k} \square \dfrac{1}{a^k}$

**77** *Savings account* One of the oldest banks in the United States is the Bank of America, founded in 1812. If \$200 had been deposited at that time into an account that paid 4% annual interest, then 180 years later the amount would have grown to $200(1.04)^{180}$ dollars. Approximate this amount to the nearest cent.

**78 *Viewing distance*** On a clear day, the distance $d$ (in miles) that can be seen from the top of a tall building of height $h$ (in feet) can be approximated by $d = 1.2\sqrt{h}$. Approximate the distance that can be seen from the top of the Chicago Sears Tower, which is 1454 feet tall.

**79 *Length of a halibut*** The length-weight relationship for Pacific halibut can be approximated by $L = 0.46\sqrt[3]{W}$, where $W$ is in kilograms and $L$ is in meters. The largest documented halibut weighed 230 kilograms. Estimate its length.

**80 *Weight of a whale*** The length-weight relationship for the sei whale can be approximated by $W = 0.0016L^{2.43}$, where $W$ is in tons and $L$ is in feet. Estimate the weight of a whale that is 25 feet long.

**81 *Weight lifters' handicaps*** O'Carroll's formula is used to handicap weight lifters. If a lifter who weighs $b$ kilo-grams lifts $w$ kilograms of weight, then the handicapped weight $W$ is given by

$$W = \frac{w}{\sqrt[3]{b - 35}}.$$

Suppose two lifters weighing 75 kilograms and 120 kilograms lift weights of 180 and 250 kilograms, respectively. Use O'Carroll's formula to determine the superior weight lifter.

**82 *Body surface area*** A person's body surface area $S$ (in square feet) can be approximated by

$$S = (0.1091)w^{0.425}h^{0.725},$$

where height $h$ is in inches and weight $w$ is in pounds.

**(a)** Estimate $S$ for a person 6 feet tall and weighing 175 pounds.

**(b)** If a person is 5 feet 6 inches tall, what effect does a 10% increase in weight have on $S$?

## *1.3*   ALGEBRAIC EXPRESSIONS

We sometimes use the notation and terminology of sets to describe mathematical relationships. A **set** is a collection of objects of some type, and the objects are called **elements** of the set. Capital letters $R$, $S$, $T$, . . . are often used to denote sets, and lowercase letters $a$, $b$, $x$, $y$, . . . usually represent elements of sets. Throughout this book, $\mathbb{R}$ denotes the set of real numbers and $\mathbb{Z}$ denotes the set of integers.

Two sets $S$ and $T$ are **equal,** denoted by $S = T$, if $S$ and $T$ contain exactly the same elements. We write $S \neq T$ if $S$ and $T$ are not equal. Additional notation and terminology are listed in the following chart.

| Notation or terminology | Meaning | Illustration |
|---|---|---|
| $a \in S$ | $a$ is an element of $S$. | $3 \in \mathbb{Z}$ |
| $a \notin S$ | $a$ is not an element of $S$. | $\frac{3}{5} \notin \mathbb{Z}$ |
| $S$ is a **subset** of $T$ | Every element of $S$ is an element of $T$. | $\mathbb{Z}$ is a subset of $\mathbb{R}$. |
| **Constant** | A letter or symbol that represents a *specific* element of a set. | $5, -\sqrt{2}, \pi$ |
| **Variable** | A letter or symbol that represents *any* element of a set | Let $x$ denote any real number. |

We usually use letters near the end of the alphabet such as $x$, $y$, and $z$ for variables, and letters near the beginning of the alphabet such as $a$, $b$, and $c$ for constants. Throughout this text, unless otherwise specified, variables represent real numbers.

If the elements of a set $S$ have a certain property, we sometimes write $S = \{x: \}$ and state the property describing the variable $x$ in the space after the colon. The expression involving the braces and colon is read ''the set of all $x$ such that . . . ,'' where we complete the phrase by stating the desired property. For example, $\{x: x > 3\}$ is read ''the set of all $x$ such that $x$ is greater than 3.''

For finite sets, we sometimes list all the elements of the set within braces. Thus, if the set $T$ consists of the first five positive integers, we may write $T = \{1, 2, 3, 4, 5\}$. When we describe sets in this way, the order used in listing the elements is irrelevant, so we could also write $T = \{1, 3, 2, 4, 5\}$, $T = \{4, 3, 2, 5, 1\}$, and so on.

If we begin with any collection of variables and real numbers, then an **algebraic expression** is the result obtained by applying additions, subtractions, multiplications, divisions, powers, or the taking of roots to this collection. If specific numbers are substituted for the variables in an algebraic expression, the resulting number is called the **value** of the expression for these numbers. The **domain** of an algebraic expression consists of all real numbers that may represent the variables. Thus, unless otherwise specified, *we assume that the domain consists of the real numbers that, when substituted for the variables, do not make the expression meaningless, in the sense that denominators cannot equal zero and roots always exist.* Two illustrations are given in the following chart.

## Algebraic Expressions

| Illustration | Domain | Typical value |
|---|---|---|
| $x^3 - 5x + \dfrac{6}{\sqrt{x}}$ | all $x > 0$ | At $x = 4$: $$4^3 - 5(4) + \frac{6}{\sqrt{4}} = 64 - 20 + 3 = 47$$ |
| $\dfrac{2xy + (3/x^2)}{\sqrt[3]{y} - 1}$ | all $x \neq 0$ and all $y \neq 1$ | At $x = 1$ and $y = 9$: $$\frac{2(1)(9) + (3/1^2)}{\sqrt[3]{9} - 1} = \frac{18 + 3}{\sqrt[3]{8}} = \frac{21}{2}$$ |

If $x$ is a variable, then a **monomial** in $x$ is an expression of the form $ax^n$, where $a$ is a real number and $n$ is a nonnegative integer. A **binomial** is a sum of two monomials, and a **trinomial** is a sum of three monomials. A *polynomial in x* is a sum of any number of monomials in $x$. Another way of stating this is as follows.

**Definition of Polynomial**

A **polynomial in $x$** is a sum of the form

$$a_n x^n + a_{n-1} x^{n-1} + \cdots + a_1 x + a_0,$$

where $n$ is a nonnegative integer and each coefficient $a_k$ is a real number. If $a_n \neq 0$, then the polynomial is said to have **degree $n$**.

Each expression $a_k x^k$ in the sum is a **term** of the polynomial. If a coefficient $a_k$ is zero, we usually delete the term $a_k x^k$. The coefficient $a_k$ of the highest power of $x$ is called the **leading coefficient** of the polynomial. The next chart contains illustrations of polynomials.

**Polynomials**

| Example | Leading coefficient | Degree |
|---|---|---|
| $3x^4 + 5x^3 + (-7)x + 4$ | 3 | 4 |
| $x^8 + 9x^2 + (-2)x$ | 1 | 8 |
| $-5x^2 + 1$ | $-5$ | 2 |
| $7x + 2$ | 7 | 1 |
| $8$ | 8 | 0 |

By definition, two polynomials are **equal** if and only if they have the same degree and the coefficients of like powers of $x$ are equal. If all the coefficients of a polynomial are zero, it is called the **zero polynomial** and is denoted by 0. However, by convention, the degree of the zero polynomial is *not* zero but, instead, is undefined. If $c$ is a *nonzero real number,* then $c$ is a polynomial of degree 0. Such polynomials (together with the zero polynomial) are **constant polynomials.**

If a coefficient of a polynomial is negative, we usually use a minus sign between appropriate terms. To illustrate,

$$3x^2 + (-5)x + (-7) = 3x^2 - 5x - 7.$$

We may also consider polynomials in variables other than $x$. For example, $\frac{2}{5}z^2 - 3z^7 + 8 - \sqrt{5}z^4$ is a polynomial in $z$ of degree 7. We often arrange the terms of a polynomial in order of decreasing powers of the variable; thus, we write

$$\tfrac{2}{5}z^2 - 3z^7 + 8 - \sqrt{5}z^4 = -3z^7 - \sqrt{5}z^4 + \tfrac{2}{5}z^2 + 8.$$

We may regard a polynomial in $x$ as an algebraic expression obtained by employing a finite number of additions, subtractions, and multiplications involving $x$. If an algebraic expression contains divisions or roots involving a variable $x$, then it is not a polynomial in $x$.

### ILLUSTRATION

**Nonpolynomials**

■ $\dfrac{1}{x} + 3x$         ■ $\dfrac{x - 5}{x^2 + 2}$         ■ $3x^2 + \sqrt{x} - 2$

Since polynomials represent real numbers, we may use the properties described in Section 1.1. In particular, if additions, subtractions, and multiplications are carried out with polynomials, we may simplify the results by using properties of real numbers, as demonstrated in the following example.

**EXAMPLE I    *Multiplying polynomials***

Find the product $(x^2 + 5x - 4)(2x^3 + 3x - 1)$.

**SOLUTION**    We begin by using a distributive property, treating the polynomial $2x^3 + 3x - 1$ as a single real number:

$$(x^2 + 5x - 4)(2x^3 + 3x - 1) = x^2(2x^3 + 3x - 1) + 5x(2x^3 + 3x - 1) - 4(2x^3 + 3x - 1)$$

We next use another distributive property three times and simplify the result, obtaining

$$(x^2 + 5x - 4)(2x^3 + 3x - 1) = 2x^5 + 3x^3 - x^2 + 10x^4 + 15x^2 - 5x - 8x^3 - 12x + 4$$
$$= 2x^5 + 10x^4 - 5x^3 + 14x^2 - 17x + 4.$$

Note that the three monomials in the first polynomial were multiplied by each of the three monomials in the second polynomial, giving a total of nine terms.

We may consider polynomials in more than one variable. For example, a polynomial in *two* variables, $x$ and $y$, is a finite sum of terms, each of the form $ax^m y^k$ for some real number $a$ and nonnegative integers $m$ and $k$. An example is

$$3x^4y + 2x^3y^5 + 7x^2 - 4xy + 8y - 5.$$

Other polynomials may involve three variables—such as $x$, $y$, $z$—or, for that matter, *any* number of variables. Addition, subtraction, and multiplication are performed using properties of real numbers, just as for polynomials in one variable.

The products listed in the next chart occur so frequently that they deserve special attention. You can check the validity of each formula by multiplication.

## Product Formulas

| Formula | Illustration |
|---|---|
| (1) $(x + y)(x - y) = x^2 - y^2$ | $(2a + 3)(2a - 3) = (2a)^2 - 3^2 = 4a^2 - 9$ |
| (2) $(x + y)^2 = x^2 + 2xy + y^2$ | $(a + 3\sqrt{x})^2 = (a)^2 + 2(a)(3\sqrt{x}) + (3\sqrt{x})^2$ |
| | $\qquad\qquad = a^2 + 6a\sqrt{x} + 9x$ |
| (3) $(x - y)^2 = x^2 - 2xy + y^2$ | $(2a - 3)^2 = (2a)^2 - 2(2a)(3) + (3)^2$ |
| | $\qquad\qquad = 4a^2 - 12a + 9$ |
| (4) $(x + y)^3 = x^3 + 3x^2y + 3xy^2 + y^3$ | $(2a + 3)^3 = (2a)^3 + 3(2a)^2(3) + 3(2a)(3)^2 + (3)^3$ |
| | $\qquad\qquad = 8a^3 + 36a^2 + 54a + 27$ |
| (5) $(x - y)^3 = x^3 - 3x^2y + 3xy^2 - y^3$ | $(2a - 5b)^3 = (2a)^3 - 3(2a)^2(5b) + 3(2a)(5b)^2 - (5b)^3$ |
| | $\qquad\qquad = 8a^3 - 60a^2b + 150ab^2 - 125b^3$ |

Note that the last expression in the illustration for product formula 2, $a^2 + 6a\sqrt{x} + 9x$, is *not* a polynomial.

If a polynomial is a product of other polynomials, then each polynomial in the product is a **factor** of the original polynomial. **Factoring** is the process of expressing a sum of terms as a product. For example, since $x^2 - 9 = (x + 3)(x - 3)$, the polynomials $x + 3$ and $x - 3$ are factors of $x^2 - 9$.

Factoring is an important process in mathematics, since it may be used to reduce the study of a complicated expression to the study of several simpler expressions. For example, properties of the polynomial $x^2 - 9$ can be determined by examining the factors $x + 3$ and $x - 3$. As we shall see later in this chapter, another important use for factoring is in finding solutions of equations and inequalities.

We shall be interested primarily in **nontrivial factors** of polynomials— that is, factors that contain polynomials of positive degree. However, if the coefficients are restricted to *integers,* then we usually remove a common integral factor from each term of the polynomial. For example,

$$4x^2y + 8z^3 = 4(x^2y + 2z^3).$$

A polynomial with coefficients in some set $S$ of numbers is **prime,** or **irreducible** over $S$, if it cannot be written as a product of two polynomials of positive degree with coefficients in $S$. A polynomial may be irreducible over one set $S$ but not over another. For example, $x^2 - 2$ is irreducible over the rational numbers, since it cannot be expressed as a product of two polynomials of positive degree that have *rational* coefficients. However, $x^2 - 2$ is *not* irreducible over the real numbers, since we can write

$$x^2 - 2 = (x + \sqrt{2})(x - \sqrt{2}).$$

Similarly, $x^2 + 1$ is irreducible over the real numbers, but, as we shall see in Section 1.5, not over the complex numbers.

Every polynomial $ax + b$ of degree 1 is irreducible.

Before we factor a polynomial, we must specify the number system (or set) from which the coefficients of the factors are to be chosen. In this chapter we shall use the rule that *if a polynomial has integral coefficients, then the factors should be polynomials with integral coefficients*. To **factor a polynomial** means to express it as a product of prime polynomials.

The **greatest common factor (gcf)** of an expression is the product of the factors that appear in each term, with each of these factors raised to the smallest nonzero exponent appearing in any term. When factoring polynomials, it is advisable to first factor out the gcf, as shown in the last two parts of the following illustration.

### ILLUSTRATION

**Factored Polynomials**

- $x^2 + x - 6 = (x + 3)(x - 2)$
- $6x^2 - 7x - 3 = (2x - 3)(3x + 1)$
- $12x^3 - 36x^2y + 27xy^2 = 3x(4x^2 - 12xy + 9y^2)$
  $$= 3x(2x - 3y)(2x - 3y) = 3x(2x - 3y)^2$$
- $4x^4y - 11x^3y^2 + 6x^2y^3 = x^2y(4x^2 - 11xy + 6y^2)$
  $$= x^2y(4x - 3y)(x - 2y)$$

It is usually difficult to factor polynomials of degree greater than 2. In simple cases the following factoring formulas may be useful. It can be shown that the factors $x^2 + xy + y^2$ and $x^2 - xy + y^2$ in the difference and sum of two cubes, respectively, are irreducible over the real numbers.

**Factoring Formulas**

| Formula | Illustration |
|---|---|
| (1) Difference of two squares: $x^2 - y^2 = (x + y)(x - y)$ | $9a^2 - 16 = (3a)^2 - (4)^2 = (3a + 4)(3a - 4)$ |
| (2) Difference of two cubes: $x^3 - y^3 = (x - y)(x^2 + xy + y^2)$ | $8a^3 - 27 = (2a)^3 - (3)^3$ $= (2a - 3)[(2a)^2 + (2a)(3) + (3)^2]$ $= (2a - 3)(4a^2 + 6a + 9)$ |
| (3) Sum of two cubes: $x^3 + y^3 = (x + y)(x^2 - xy + y^2)$ | $125a^3 + 1 = (5a)^3 + (1)^3$ $= (5a + 1)[(5a)^2 - (5a)(1) + (1)^2]$ $= (5a + 1)(25a^2 - 5a + 1)$ |

If a sum contains four or more terms, it may be possible to group the terms in a suitable manner and then find a factorization by using distributive properties. This technique, called **factoring by grouping**, is illustrated in the next example.

### EXAMPLE 2    *Factoring by grouping*

Factor:

**(a)** $4ac + 2bc - 2ad - bd$      **(b)** $3x^3 + 2x^2 - 12x - 8$

**(c)** $x^2 - 16y^2 + 10x + 25$

### SOLUTION

**(a)** We group the first two terms and the last two terms and then proceed as follows:

$$4ac + 2bc - 2ad - bd = (4ac + 2bc) - (2ad + bd)$$
$$= 2c(2a + b) - d(2a + b)$$

At this stage we have not factored the given expression, because the right-hand side has the form

$$2ck - dk \quad \text{with} \quad k = 2a + b.$$

However, if we factor out $k$, then

$$2ck - dk = (2c - d)k = (2c - d)(2a + b).$$

Hence,

$$4ac + 2bc - 2ad - bd = 2c(2a + b) - d(2a + b)$$
$$= (2c - d)(2a + b).$$

**(b)** We group the first two terms and the last two terms and then proceed as follows:

$$3x^3 + 2x^2 - 12x - 8 = (3x^3 + 2x^2) - (12x + 8)$$
$$= x^2(3x + 2) - 4(3x + 2)$$
$$= (x^2 - 4)(3x + 2)$$

Finally, using the difference of two squares formula for $x^2 - 4$, we obtain the factorization:

$$3x^3 + 2x^2 - 12x - 8 = (x + 2)(x - 2)(3x + 2)$$

**(c)** First we rearrange and group terms, and then we apply the difference of two squares formula, as follows:

$$x^2 - 16y^2 + 10x + 25 = (x^2 + 10x + 25) - 16y^2$$
$$= (x + 5)^2 - (4y)^2$$
$$= [(x + 5) + 4y][(x + 5) - 4y]$$
$$= (x + 4y + 5)(x - 4y + 5)$$

A **fractional expression** is a quotient of two algebraic expressions. As a special case, a **rational expression** is a quotient $p/q$ of two *polynomials* $p$ and $q$. Since division by zero is not allowed, the domain of $p/q$ consists of all real numbers except those that make the denominator zero. Two illustrations are given in the next chart.

**Rational Expressions**

| Example | Domain |
|---------|--------|
| $\dfrac{6x^2 - 5x + 4}{x + 7}$ | All $x \neq -7$ |
| $\dfrac{x^3 - 3x^2y + 4y^2}{y - x^3}$ | All $x$ and $y$ such that $y \neq x^3$ |

In most of our work we will be concerned with rational expressions in which both numerator and denominator are polynomials in only one variable.

Since the variables in a rational expression represent real numbers, we may use the properties of quotients in Section 1.1, replacing the letters $a$, $b$, $c$, and $d$ with polynomials. The following property is of particular importance, where $bd \neq 0$:

$$\frac{ad}{bd} = \frac{a}{b} \cdot \frac{d}{d} = \frac{a}{b} \cdot 1 = \frac{a}{b}$$

We sometimes describe this simplification process by saying that *a common nonzero factor in the numerator and denominator of a quotient may be canceled.*

A rational expression is *simplified,* or *reduced to lowest terms,* if the numerator and denominator have no common polynomial factors of positive degree and no common integral factors greater than 1. To simplify a rational expression, we factor both the numerator and the denominator into prime factors and then, assuming the factors in the denominator are not zero, cancel common factors, as in the following example.

**EXAMPLE 3**   *Simplifying products and quotients of rational expressions*

Perform the indicated operation and simplify:

**(a)** $\dfrac{x^2 - 6x + 9}{x^2 - 1} \cdot \dfrac{2x - 2}{x - 3}$     **(b)** $\dfrac{x + 2}{2x - 3} \div \dfrac{x^2 - 4}{2x^2 - 3x}$

**SOLUTION**

**(a)**  $\dfrac{x^2 - 6x + 9}{x^2 - 1} \cdot \dfrac{2x - 2}{x - 3} = \dfrac{(x^2 - 6x + 9)(2x - 2)}{(x^2 - 1)(x - 3)}$    property of quotients

$$= \dfrac{(x - 3)^{\cancel{2}^{1}} \cdot 2(\cancel{x - 1})}{(x + 1)(\cancel{x - 1})(\cancel{x - 3})}$$    factor all polynomials

if $x \neq 3$, $x \neq 1$
$$\overset{\downarrow}{=}\quad \dfrac{2(x - 3)}{x + 1}$$    cancel common factors

**(b)**  $\dfrac{x + 2}{2x - 3} \div \dfrac{x^2 - 4}{2x^2 - 3x} = \dfrac{x + 2}{2x - 3} \cdot \dfrac{2x^2 - 3x}{x^2 - 4}$    property of quotients

$$= \dfrac{(\cancel{x + 2})x(\cancel{2x - 3})}{(\cancel{2x - 3})(\cancel{x + 2})(x - 2)}$$    property of quotients; factor all polynomials

if $x \neq -2$, $x \neq \frac{3}{2}$
$$\overset{\downarrow}{=} \dfrac{x}{x - 2}$$    cancel common factors

To add or subtract two rational expressions, we usually find a *common denominator* and use the following properties of quotients:

$$\dfrac{a}{d} + \dfrac{c}{d} = \dfrac{a + c}{d} \quad \text{and} \quad \dfrac{a}{d} - \dfrac{c}{d} = \dfrac{a - c}{d}$$

If the denominators of the expressions are not the same, we may obtain a common denominator by multiplying the numerator and denominator of each fraction by a suitable expression. We usually use the ***least* common denominator (lcd)** of the two quotients. To find the lcd we factor each denominator into primes and then form the product of the different prime factors, using the *largest* exponent that appears with each prime factor.

**EXAMPLE 4**    ***Simplifying sums and differences of rational expressions***

Perform the operations and simplify:

$$\dfrac{6}{x(3x - 2)} + \dfrac{5}{3x - 2} - \dfrac{2}{x^2}$$

**SOLUTION**    The denominators are already in factored form. The lcd is $x^2(3x - 2)$. To obtain three quotients having the denominator $x^2(3x - 2)$,

*(continued)*

we multiply the numerator and denominator of the first quotient by $x$, those of the second by $x^2$, and those of the third by $3x - 2$:

$$\frac{6}{x(3x-2)} + \frac{5}{3x-2} - \frac{2}{x^2} = \frac{6}{x(3x-2)} \cdot \frac{x}{x} + \frac{5}{3x-2} \cdot \frac{x^2}{x^2} - \frac{2}{x^2} \cdot \frac{3x-2}{3x-2}$$

$$= \frac{6x}{x^2(3x-2)} + \frac{5x^2}{x^2(3x-2)} - \frac{2(3x-2)}{x^2(3x-2)}$$

$$= \frac{6x + 5x^2 - 2(3x-2)}{x^2(3x-2)}$$

$$= \frac{5x^2 + 4}{x^2(3x-2)}$$

A **complex fraction** is a quotient in which the numerator and/or the denominator is a fractional expression. Certain problems in calculus require simplifying complex fractions of the type given in the next example.

**EXAMPLE 5     *Simplifying a complex fraction***

Simplify the complex fraction

$$\frac{\dfrac{2}{x+3} - \dfrac{2}{a+3}}{x-a}.$$

**SOLUTION**     We change the numerator of the given expression into a single quotient and then use a property for simplifying quotients:

$$\frac{\dfrac{2}{x+3} - \dfrac{2}{a+3}}{x-a} = \frac{\dfrac{2(a+3) - 2(x+3)}{(x+3)(a+3)}}{x-a} \qquad \text{combine fractions in the numerator}$$

$$= \frac{2a - 2x}{(x+3)(a+3)} \cdot \frac{1}{x-a} \qquad \text{simplify; property of quotients}$$

$$= \frac{2(a-x)}{(x+3)(a+3)(x-a)} \qquad \begin{array}{l}\text{factor } 2a - 2x; \\ \text{property of quotients}\end{array}$$

$$\overset{\text{if } x \neq a}{\underset{\downarrow}{=}} -\frac{2}{(x+3)(a+3)} \qquad \text{replace } \frac{a-x}{x-a} \text{ with } -1$$

An alternative method is to multiply the numerator and denominator of the given expression by $(x+3)(a+3)$, the lcd of the numerator and denominator, and then simplify the result.

Some quotients that are not rational expressions contain denominators of the form $a + \sqrt{b}$ or $\sqrt{a} + \sqrt{b}$. As in the next example, these quotients can be simplified by multiplying the numerator and denominator by the **conjugate** $a - \sqrt{b}$ or $\sqrt{a} - \sqrt{b}$, respectively, Of course, if $a - \sqrt{b}$ appears, multiply by $a + \sqrt{b}$ instead.

In Example 4 of Section 1.2, we rationalized denominators. In calculus it is sometimes necessary to rationalize the *numerator* of a quotient, as shown in the following example.

**EXAMPLE 6**   *Rationalizing a numerator*

If $h \neq 0$, rationalize the numerator of

$$\frac{\sqrt{x+h} - \sqrt{x}}{h}.$$

*SOLUTION*

$$\frac{\sqrt{x+h} - \sqrt{x}}{h} = \frac{\sqrt{x+h} - \sqrt{x}}{h} \cdot \frac{\sqrt{x+h} + \sqrt{x}}{\sqrt{x+h} + \sqrt{x}}$$ multiply numerator and denominator by the conjugate of $\sqrt{x+h} - \sqrt{x}$

$$= \frac{(\sqrt{x+h})^2 - (\sqrt{x})^2}{h(\sqrt{x+h} + \sqrt{x})}$$ property of quotients and difference of two squares

$$= \frac{(x+h) - x}{h(\sqrt{x+h} + \sqrt{x})}$$ law of radicals

$$= \frac{\cancel{h}}{\cancel{h}(\sqrt{x+h} + \sqrt{x})}$$ simplify

$$= \frac{1}{\sqrt{x+h} + \sqrt{x}}$$ cancel $h \neq 0$

It may seem as though we have accomplished very little, since radicals occur in the denominator. In calculus, however, it is of interest to determine what is true if $h$ is very close to zero. Note that if we use the *given* expression we obtain the following:

If $\quad h \approx 0, \quad$ then $\quad \dfrac{\sqrt{x+h} - \sqrt{x}}{h} \approx \dfrac{\sqrt{x+0} - \sqrt{x}}{0} = \dfrac{0}{0},$

a meaningless expression. If we use the *rationalized* form, however, we obtain the following information:

If $\quad h \approx 0, \quad$ then $\quad \dfrac{1}{\sqrt{x+h} + \sqrt{x}} \approx \dfrac{1}{\sqrt{x} + \sqrt{x}} = \dfrac{1}{2\sqrt{x}}.$

Problems of the type given in the next example also occur in calculus.

**EXAMPLE 7    *Simplifying a fractional expression***
Simplify

$$\frac{3x^2(2x + 5)^{1/2} - x^3(\frac{1}{2})(2x + 5)^{-1/2}(2)}{[(2x + 5)^{1/2}]^2}.$$

**SOLUTION**    One way to simplify the expression is as follows:

$$\frac{3x^2(2x + 5)^{1/2} - x^3(\frac{1}{2})(2x + 5)^{-1/2}(2)}{[(2x + 5)^{1/2}]^2}$$

$$= \frac{3x^2(2x + 5)^{1/2} - \dfrac{x^3}{(2x + 5)^{1/2}}}{2x + 5} \qquad \text{definition of negative exponent}$$

$$= \frac{\dfrac{3x^2(2x + 5) - x^3}{(2x + 5)^{1/2}}}{2x + 5} \qquad \text{combine terms in numerator}$$

$$= \frac{6x^3 + 15x^2 - x^3}{(2x + 5)^{1/2}} \cdot \frac{1}{2x + 5} \qquad \text{property of quotients}$$

$$= \frac{5x^3 + 15x^2}{(2x + 5)^{3/2}} \qquad \text{simplify}$$

$$= \frac{5x^2(x + 3)}{(2x + 5)^{3/2}} \qquad \text{factor numerator}$$

An alternative simplification is to eliminate the negative power, $-\frac{1}{2}$, in the given expression, as follows:

$$\frac{3x^2(2x + 5)^{1/2} - x^3(\frac{1}{2})(2x + 5)^{-1/2}(2)}{[(2x + 5)^{1/2}]^2} \cdot \frac{(2x + 5)^{1/2}}{(2x + 5)^{1/2}} \qquad \begin{array}{l}\text{multiply numerator}\\ \text{and denominator by}\\ (2x + 5)^{1/2}\end{array}$$

$$= \frac{3x^2(2x + 5) - x^3}{(2x + 5)(2x + 5)^{1/2}} \qquad \begin{array}{l}\text{property of quotients}\\ \text{and law of exponents}\end{array}$$

The remainder of the simplification is similar.

A third method of simplification is to first factor out the gcf. In this case, the common factors are $x$ and $(2x + 5)$, and the smallest exponents are 2 and $-\frac{1}{2}$, respectively. Thus, the gcf is $x^2(2x + 5)^{-1/2}$, and we factor the numerator and simplify as follows:

$$\frac{x^2(2x + 5)^{-1/2}[3(2x + 5)^1 - x]}{(2x + 5)^1} = \frac{x^2(5x + 15)}{(2x + 5)^{3/2}} = \frac{5x^2(x + 3)}{(2x + 5)^{3/2}}$$

## 1.3 EXERCISES

**Exer. 1–12: Express as a polynomial.**

1  $(2u + 3)(u - 4) + 4u(u - 2)$

2  $(3u - 1)(u + 2) + 7u(u + 1)$

3  $\dfrac{8x^2y^3 - 10x^3y}{2x^2y}$

4  $\dfrac{6x^2yz^3 - xy^2z}{xyz}$

5  $(2x + 3y)(2x - 3y)$

6  $(5x + 4y)(5x - 4y)$

7  $(3x + 2y)^2$

8  $(5x - 4y)^2$

9  $(\sqrt{x} + \sqrt{y})(\sqrt{x} - \sqrt{y})$

10  $(\sqrt{x} + \sqrt{y})^2(\sqrt{x} - \sqrt{y})^2$

11  $(x - 2y)^3$

12  $(x + 3y)^3$

**Exer. 13–30: Factor the polynomial.**

13  $8x^2 - 53x - 21$

14  $7x^2 + 10x - 8$

15  $x^2 + 3x + 4$

16  $3x^2 - 4x + 2$

17  $4x^2 - 20x + 25$

18  $9x^2 + 24x + 16$

19  $x^4 - 4x^2$

20  $x^3 - 25x$

21  $64x^3 - y^6$

22  $x^6 - 27y^3$

23  $64x^3 + 27$

24  $125x^3 + 8$

25  $3x^3 + 3x^2 - 27x - 27$

26  $5x^3 + 10x^2 - 20x - 40$

27  $a^6 - b^6$

28  $x^8 - 16$

29  $x^2 + 4x + 4 - 9y^2$

30  $x^2 - 4y^2 - 6x + 9$

**Exer. 31–60: Simplify the expression.**

31  $\dfrac{y^2 - 25}{y^3 - 125}$

32  $\dfrac{12 + r - r^2}{r^3 + 3r^2}$

33  $\dfrac{9x^2 - 4}{3x^2 - 5x + 2} \cdot \dfrac{9x^4 - 6x^3 + 4x^2}{27x^4 + 8x}$

34  $\dfrac{5a^2 + 12a + 4}{a^4 - 16} \div \dfrac{25a^2 + 20a + 4}{a^2 - 2a}$

35  $\dfrac{2}{3s + 1} - \dfrac{9}{(3s + 1)^2}$

36  $\dfrac{4}{(5s - 2)^2} + \dfrac{s}{5s - 2}$

37  $\dfrac{2}{x} + \dfrac{3x + 1}{x^2} - \dfrac{x - 2}{x^3}$

38  $\dfrac{5}{x} - \dfrac{2x - 1}{x^2} + \dfrac{x + 5}{x^3}$

39  $\dfrac{3t}{t + 2} + \dfrac{5t}{t - 2} - \dfrac{40}{t^2 - 4}$

40  $\dfrac{t}{t + 3} + \dfrac{4t}{t - 3} - \dfrac{18}{t^2 - 9}$

41  $\dfrac{4x}{3x - 4} + \dfrac{8}{3x^2 - 4x} + \dfrac{2}{x}$

42  $\dfrac{12x}{2x + 1} - \dfrac{3}{2x^2 + x} + \dfrac{5}{x}$

43  $\dfrac{2x}{x + 2} - \dfrac{8}{x^2 + 2x} + \dfrac{3}{x}$

44  $\dfrac{5x}{2x + 3} - \dfrac{6}{2x^2 + 3x} + \dfrac{2}{x}$

45  $3 + \dfrac{5}{u} + \dfrac{2u}{3u + 1}$

46  $4 + \dfrac{2}{u} - \dfrac{3u}{u + 5}$

47  $\dfrac{2x + 1}{x^2 + 4x + 4} - \dfrac{6x}{x^2 - 4} + \dfrac{3}{x - 2}$

48  $\dfrac{2x + 6}{x^2 + 6x + 9} + \dfrac{5x}{x^2 - 9} + \dfrac{7}{x - 3}$

49  $\dfrac{\dfrac{b}{a} - \dfrac{a}{b}}{\dfrac{1}{a} - \dfrac{1}{b}}$

50  $\dfrac{\dfrac{x}{y^2} - \dfrac{y}{x^2}}{\dfrac{1}{y^2} - \dfrac{1}{x^2}}$

51  $\dfrac{y^{-1} + x^{-1}}{(xy)^{-1}}$

52  $\dfrac{y^{-2} - x^{-2}}{y^{-2} + x^{-2}}$

53  $\dfrac{\dfrac{r}{s} + \dfrac{s}{r}}{\dfrac{r^2}{s^2} - \dfrac{s^2}{r^2}}$

54  $\dfrac{\dfrac{3}{w} - \dfrac{6}{2w + 1}}{\dfrac{5}{w} + \dfrac{8}{2w + 1}}$

55  $\dfrac{(x + h)^2 - 3(x + h) - (x^2 - 3x)}{h}$

56  $\dfrac{(x + h)^3 + 5(x + h) - (x^3 + 5x)}{h}$

57  $\dfrac{\dfrac{3}{x - 1} - \dfrac{3}{a - 1}}{x - a}$

58  $\dfrac{\dfrac{x + 2}{x} - \dfrac{a + 2}{a}}{x - a}$

59  $\dfrac{\dfrac{1}{(x + h)^3} - \dfrac{1}{x^3}}{h}$

60  $\dfrac{\dfrac{1}{x + h} - \dfrac{1}{x}}{h}$

**Exer. 61–64: Rationalize the denominator.**

**61** $\dfrac{\sqrt{t}+5}{\sqrt{t}-5}$

**62** $\dfrac{16x^2-y^2}{2\sqrt{x}-\sqrt{y}}$

**63** $\dfrac{1}{\sqrt[3]{a}-\sqrt[3]{b}}$    (*Hint:* Multiply numerator and denominator by $\sqrt[3]{a^2}+\sqrt[3]{ab}+\sqrt[3]{b^2}$.)

**64** $\dfrac{1}{\sqrt[3]{x}+\sqrt[3]{y}}$

**Exer. 65–68: Rationalize the numerator.**

**65** $\dfrac{\sqrt{a}-\sqrt{b}}{a^2-b^2}$

**66** $\dfrac{\sqrt{b}+\sqrt{c}}{b^2-c^2}$

**67** $\dfrac{\sqrt{2(x+h)+1}-\sqrt{2x+1}}{h}$

**68** $\dfrac{\sqrt{x}-\sqrt{x+h}}{h\sqrt{x}\sqrt{x+h}}$

**Exer. 69–72: Express as a sum of terms of the form $ax^r$, where $r$ is a rational number.**

**69** $\dfrac{4x^2-x+5}{x^{2/3}}$

**70** $\dfrac{x^2+4x-6}{\sqrt{x}}$

**71** $\dfrac{(x^2+2)^2}{x^5}$

**72** $\dfrac{(\sqrt{x}-3)^2}{x^3}$

**Exer. 73–76: Express as a quotient.**

**73** $x^{-3}+x^2$

**74** $x^{-4}-x$

**75** $x^{-1/2}-x^{3/2}$

**76** $x^{-2/3}+x^{7/3}$

**Exer. 77–90: Simplify the expression.**

**77** $(2x^2-3x+1)(4)(3x+2)^3(3)+(3x+2)^4(4x-3)$

**78** $(6x-5)^3(2)(x^2+4)(2x)+(x^2+4)^2(3)(6x-5)^2(6)$

**79** $(x^2-4)^{1/2}(3)(2x+1)^2(2)+(2x+1)^3(\tfrac{1}{2})(x^2-4)^{-1/2}(2x)$

**80** $(3x+2)^{1/3}(2)(4x-5)(4)+(4x-5)^2(\tfrac{1}{3})(3x+2)^{-2/3}(3)$

**✳81** $(3x+1)^6(\tfrac{1}{2})(2x-5)^{-1/2}(2)+(2x-5)^{1/2}(6)(3x+1)^5(3)$

**82** $(x^2+9)^4(-\tfrac{1}{3})(x+6)^{-4/3}+(x+6)^{-1/3}(4)(x^2+9)^3(2x)$

**83** $\dfrac{(6x+1)^3(27x^2+2)-(9x^3+2x)(3)(6x+1)^2(6)}{(6x+1)^6}$

**84** $\dfrac{(x^2-1)^4(2x)-x^2(4)(x^2-1)^3(2x)}{(x^2-1)^8}$

**85** $\dfrac{(x^2+2)^3(2x)-x^2(3)(x^2+2)^2(2x)}{[(x^2+2)^3]^2}$

**86** $\dfrac{(x^2-5)^4(3x^2)-x^3(4)(x^2-5)^3(2x)}{[(x^2-5)^4]^2}$

**87** $\dfrac{(x^2+4)^{1/3}(3)-(3x)(\tfrac{1}{3})(x^2+4)^{-2/3}(2x)}{[(x^2+4)^{1/3}]^2}$

**88** $\dfrac{(1-x^2)^{1/2}(2x)-x^2(\tfrac{1}{2})(1-x^2)^{-1/2}(-2x)}{[(1-x^2)^{1/2}]^2}$

**89** $\dfrac{(4x^2+9)^{1/2}(2)-(2x+3)(\tfrac{1}{2})(4x^2+9)^{-1/2}(8x)}{[(4x^2+9)^{1/2}]^2}$

**✗90** $\dfrac{(3x+2)^{1/2}(\tfrac{1}{3})(2x+3)^{-2/3}(2)-(2x+3)^{1/3}(\tfrac{1}{2})(3x+2)^{-1/2}}{[(3x+2)^{1/2}]^2}$

**Exer. 91–92: The ancient Greeks gave geometric proofs of the factoring formulas for the difference of two squares and the difference of two cubes. Establish the formula for the special case described.**

**91** Find the areas of regions I and II in the figure to establish the difference of two squares formula for the special case $x>y$.

**EXERCISE 91**

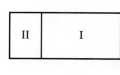

**92** Find the volumes of boxes I, II, and III in the figure to establish the difference of two cubes formula for the special case $x>y$.

**EXERCISE 92**

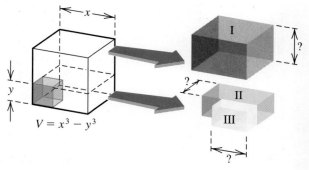

# 1.4 EQUATIONS

An **equation** (or **equality**) is a statement that two quantities or expressions are equal. Equations are employed in every field that uses real numbers. The following chart applies to a variable $x$, but any other variable may be considered. The abbreviations LS and RS in the second illustration stand for the equation's left side and right side.

| Terminology | Definition | Illustration |
|---|---|---|
| **Equation** in $x$ | A statement of equality involving one variable $x$ | $x^2 - 5 = 4x$ |
| **Solution,** or **root,** of an equation in $x$ | A number $b$ that yields a true statement when substituted for $x$ | 5 is a solution of $x^2 - 5 = 4x$, since substitution gives us LS: $5^2 - 5 = 25 - 5 = 20$ and RS: $4 \cdot 5 = 20$, and $20 = 20$ is a true statement. |
| A number $b$ **satisfies** an equation in $x$ | $b$ is a solution of the equation | 5 satisfies $x^2 - 5 = 4x$. |
| **Equivalent equations** | Equations that have exactly the same solutions | $2x + 1 = 7$ $2x = 7 - 1$ $2x = 6$ $x = 3$ |
| **Solve** an equation in $x$ | Find all solutions of the equation | To solve $(x + 3)(x - 5) = 0$, set each factor equal to 0: $x + 3 = 0$, $x - 5 = 0$, obtaining the solutions $-3$ and $5$. |

An **algebraic equation** in $x$ contains only algebraic expressions such as polynomials, rational expressions, radicals, and so on. An equation of this type is called a **conditional equation** if there are numbers in the domains of the expressions that are not solutions. For example, the equation $x^2 = 9$ is conditional, since the number $x = 4$ (and others) is not a solution. If *every* number in the domains of the expressions in an algebraic equation is a solution, the equation is called an **identity.**

We shall assume some experience in finding solutions of equations in one variable. To solve a **linear equation** $ax + b = 0$, where $a$ and $b$ are real numbers and $a \neq 0$, we subtract $b$ from both sides and divide by $a$ as follows:

$$ax + b = 0, \quad ax = -b, \quad x = -\frac{b}{a}$$

Thus, the linear equation $ax + b = 0$ has exactly one solution, $-b/a$.

In the following examples, the phrases in color indicate how an equivalent equation was obtained from the preceding equation. To shorten these phrases we have, as in Example 1, used "divide by 2" instead of the more accurate but lengthy *divide both sides by* 2. If an equation contains rational expressions, we often eliminate denominators by multiplying both sides by the lcd of these expressions. If we multiply both sides by an expression that equals zero for some value of $x$, then the resulting equation may not be equivalent to the original equation, as illustrated in the following example.

**EXAMPLE I    *An equation with no solutions***

Solve the equation $\dfrac{3x}{x-2} = 1 + \dfrac{6}{x-2}$.

**SOLUTION**

$$\frac{3x}{x-2} = 1 + \frac{6}{x-2} \qquad \text{given}$$

$$\left(\frac{3x}{x-2}\right)(x-2) = (1)(x-2) + \left(\frac{6}{x-2}\right)(x-2) \qquad \text{multiply by } x-2$$

$$3x = (x-2) + 6 \qquad \text{simplify}$$

$$2x = 4 \qquad \text{simplify; subtract } x$$

$$x = 2 \qquad \text{divide by 2}$$

☑ **Check $x = 2$     LS:** $\dfrac{3(2)}{(2)-2} = \dfrac{6}{0}$

Since division by 0 is not permissible, $x = 2$ is not a solution ($x = 2$ is called an **extraneous solution,** or **extraneous root,** of the given equation). Hence, *the given equation has no solutions.*

Formulas involving several variables occur in many applications of mathematics. Sometimes it is necessary to solve for a specific variable in terms of the remaining variables that appear in the formula, as the next example illustrates.

**EXAMPLE 2    *Resistors connected in parallel***

In electrical theory, the formula

$$\frac{1}{R} = \frac{1}{R_1} + \frac{1}{R_2}$$

is used to find the total resistance $R$ when two resistors $R_1$ and $R_2$ are connected in parallel. Solve the formula for $R_1$.

SOLUTION

$$\frac{1}{R} \cdot RR_1R_2 = \frac{1}{R_1} \cdot RR_1R_2 + \frac{1}{R_2} \cdot RR_1R_2 \qquad \text{multiply by the lcd, } RR_1R_2$$

$$R_1R_2 = RR_2 + RR_1 \qquad\qquad \text{cancel common factors}$$

$$R_1R_2 - RR_1 = RR_2 \qquad\qquad \text{collect terms with } R_1 \text{ on one side}$$

$$R_1(R_2 - R) = RR_2 \qquad\qquad \text{factor out } R_1$$

$$R_1 = \frac{RR_2}{R_2 - R} \qquad\qquad \text{divide by } R_2 - R$$

An alternative method of solution is to first solve the given formula for $1/R_1$, combine the fractions $1/R$ and $-1/R_2$, and then apply the property that if two nonzero numbers are equal, so are their reciprocals. The same result is obtained.

Equations are often used to solve *applied problems*—that is, problems that involve applications of mathematics to other fields. Because of the unlimited variety of applied problems, it is difficult to state specific rules for finding solutions. The following guidelines may be helpful, provided the problem can be formulated in terms of an equation in one variable. As you read the solution in Example 3, identify the specific guidelines used.

**Guidelines for Solving Applied Problems**

1   If the problem is stated in writing, read it carefully several times and think about the given facts, together with the unknown quantity that is to be found.

2   Introduce a letter to denote the unknown quantity. This is one of the most crucial steps in the solution. Phrases containing words such as *what, find, how much, how far,* or *when* should alert you to the unknown quantity.

3   If appropriate, draw a picture and label it.

4   List the known facts, together with any relationships that involve the unknown quantity. A relationship may be described by an equation in which written statements, instead of letters or numbers, appear on one or both sides of the equals sign.

5   After analyzing the list in guideline 4, formulate an equation that describes precisely what is stated in words.

6   Solve the equation formulated in guideline 5.

7   Check the solutions obtained in guideline 6 by referring to the original statement of the problem. Verify that the solution agrees with the stated conditions.

Banks and other financial institutions pay interest on investments. Usually this interest is *compounded* (as described in Section 4.1); however, if money is invested or loaned for a short period of time, *simple interest* may be paid using the following formula.

**Simple Interest Formula**

> If a sum of money $P$ (the **principal**) is invested at a simple interest rate $r$ (expressed as a decimal), then the **simple interest** $I$ at the end of $t$ years is
>
> $$I = Prt.$$

### EXAMPLE 3     *Investing money in two stocks*

An investment firm has $100,000 to invest for a client and decides to invest it in two stocks A and B. The expected annual rate of return, or simple interest, for stock A is 15%, but there is some risk involved, and the client does not wish to invest more than $50,000 in this stock. The annual rate of return on the more stable stock B is anticipated to be 10%. Determine whether there is a way of investing the money so that the annual interest is

**(a)** $12,000        **(b)** $13,000

**SOLUTION**     The annual interest is given by $I = Pr$, which comes from the simple interest formula $I = Prt$ with $t = 1$. If we let $x$ denote the amount invested in stock A, then $100,000 - x$ will be invested in stock B. This leads to the following equalities:

$$x = \text{amount invested in stock A at 15\%}$$
$$100,000 - x = \text{amount invested in stock B at 10\%}$$
$$0.15x = \text{annual interest from stock A}$$
$$0.10(100,000 - x) = \text{annual interest from stock B}$$

Adding the interest from both stocks, we obtain

$$\text{total annual interest} = 0.15x + 0.10(100,000 - x).$$

Simplifying the right side gives us

(*)                total annual interest $= 10,000 + 0.05x.$

**(a)** The total annual interest is $12,000 if

$$10,000 + 0.05x = 12,000 \qquad \text{from (*)}$$
$$0.05x = 2000 \qquad \text{subtract 10,000}$$
$$x = \frac{2000}{0.05} = 40,000 \qquad \text{divide by 0.05}$$

Thus, $40,000 should be invested in stock A, and the remaining $60,000 should be invested in stock B. Since the amount invested in stock A is not more than $50,000, this manner of investing the money meets the requirement of the client.

✔ **Check**  If $40,000 is invested in stock A and $60,000 in stock B, then the total annual interest is

$$40,000(0.15) + 60,000(0.10) = 6000 + 6000 = 12,000.$$

**(b)** The total annual interest is $13,000 if

$$10,000 + 0.05x = 13,000 \qquad \text{from } (*)$$
$$0.05x = 3000 \qquad \text{subtract } 10,000$$
$$x = \frac{3000}{0.05} = 60,000 \qquad \text{divide by } 0.05$$

Thus, $60,000 should be invested in stock A and the remaining $40,000 in stock B. This plan does *not* meet the client's requirement that no more than $50,000 be invested in stock A. Hence the firm cannot invest the client's money in stocks A and B such that the total annual interest is $13,000.

---

**EXAMPLE 4**  *Constructing a grain elevator hopper*

A grain elevator hopper is to be constructed as shown in Figure 7, with a right circular cylinder of radius 2 feet and altitude $h$ feet on top of a right circular cone whose altitude is one-half that of the cylinder. What value of $h$ will make the total volume $V$ of the hopper 500 ft$^3$?

**SOLUTION**  If $V_{\text{cylinder}}$ and $V_{\text{cone}}$ denote the volumes (in ft$^3$) and $h_{\text{cylinder}}$ and $h_{\text{cone}}$ denote the heights (in feet) of the cylinder and the cone, respectively, then, using the formulas for volume stated on the endpapers at the front of the text, we obtain the following:

$$V_{\text{cylinder}} = \pi r^2 \cdot h_{\text{cylinder}} = \pi(2)^2 h = 4\pi h$$
$$V_{\text{cone}} = \tfrac{1}{3}\pi r^2 \cdot h_{\text{cone}} = \tfrac{1}{3}\pi(2)^2(\tfrac{1}{2}h) = \tfrac{2}{3}\pi h$$

Since the total volume $V$ of the hopper is to be 500 ft$^3$, we must have

$$4\pi h + \tfrac{2}{3}\pi h = 500 \qquad V_{\text{cylinder}} + V_{\text{cone}} = V_{\text{total}}$$
$$12\pi h + 2\pi h = 1500 \qquad \text{multiply by } 3$$
$$h = \frac{1500}{14\pi} \approx 34.1 \text{ ft} \qquad \text{simplify; divide by } 14\pi$$

**FIGURE 7**

A **quadratic equation** in $x$ is an equation that can be written in the form $ax^2 + bx + c = 0$, where $a \neq 0$. This form is called the **standard form** of a quadratic equation in $x$. One method for solving a quadratic equation, called the **method of factoring,** is demonstrated in the next illustration. To use the method of factoring, *it is essential that only the number* $0$ *appear on one side of the equation.*

**ILLUSTRATION**

**Solving Quadratic Equations by Factoring**

$$3x^2 = 10 - x$$
$$3x^2 + x - 10 = 0$$
$$(3x - 5)(x + 2) = 0$$
$$x = \tfrac{5}{3} \quad \text{or} \quad x = -2$$

$$x^2 + 16 = 8x$$
$$x^2 - 8x + 16 = 0$$
$$(x - 4)(x - 4) = 0$$
$$x = 4$$

Since $x - 4$ appears as a factor twice in the second part of the previous illustration, we call 4 a **double root,** or **root of multiplicity 2,** of the equation $x^2 + 16 = 8x$.

If a quadratic equation has the form $x^2 = d$, for some $d > 0$, then $x^2 - d = 0$ or, equivalently,

$$(x + \sqrt{d})(x - \sqrt{d}) = 0.$$

Setting each factor equal to zero gives us the solutions $-\sqrt{d}$ and $\sqrt{d}$. We frequently use the symbol $\pm\sqrt{d}$ (*plus or minus* $\sqrt{d}$) to represent both $\sqrt{d}$ and $-\sqrt{d}$. Thus, for $d > 0$, we have proved the following result. (The case $d < 0$ requires the system of complex numbers discussed in Section 1.5.)

**A Special Quadratic Equation**

> If $x^2 = d$, then $x = \pm\sqrt{d}$.

The process of solving $x^2 = d$ as indicated in the preceding box is referred to as *taking the square root of both sides of the equation.* Note that if $d > 0$ we obtain both a positive square root and a negative square root, not just the principal square root defined in Section 1.2.

**ILLUSTRATION**

**Solving Equations of the Form $x^2 = d$**

$$x^2 = 169$$
$$x = \pm\sqrt{169}$$
$$= \pm 13$$

$$(x + 3)^2 = 5$$
$$x + 3 = \pm\sqrt{5}$$
$$x = -3 \pm \sqrt{5}$$

The solutions of a **quadratic equation** $ax^2 + bx + c = 0$, for $a \neq 0$, may be obtained by means of the following formula.

**Quadratic Formula**

If $a \neq 0$, the roots of $ax^2 + bx + c = 0$ are given by

$$x = \frac{-b \pm \sqrt{b^2 - 4ac}}{2a}.$$

**PROOF**   We shall assume that $b^2 - 4ac \geq 0$ so that $\sqrt{b^2 - 4ac}$ is a real number. (The case in which $b^2 - 4ac < 0$ will be discussed in the next section.) Let us proceed as follows:

$$ax^2 + bx + c = 0 \qquad \text{given}$$

$$ax^2 + bx = -c \qquad \text{subtract } c$$

$$x^2 + \frac{b}{a}x = -\frac{c}{a} \qquad \text{divide by } a$$

$$x^2 + \frac{b}{a}x + \left(\frac{b}{2a}\right)^2 = \left(\frac{b}{2a}\right)^2 - \frac{c}{a} \qquad \text{complete the square}$$

$$\left(x + \frac{b}{2a}\right)^2 = \frac{b^2 - 4ac}{4a^2} \qquad \text{equivalent equation}$$

$$x + \frac{b}{2a} = \pm\sqrt{\frac{b^2 - 4ac}{4a^2}} \qquad \text{take the square root}$$

$$x = -\frac{b}{2a} \pm \sqrt{\frac{b^2 - 4ac}{4a^2}} \qquad \text{subtract } \frac{b}{2a}$$

We may write the radical in the last equation as

$$\pm\sqrt{\frac{b^2 - 4ac}{4a^2}} = \pm\frac{\sqrt{b^2 - 4ac}}{\sqrt{(2a)^2}} = \pm\frac{\sqrt{b^2 - 4ac}}{|2a|}.$$

Since $|2a| = 2a$ if $a > 0$, or $|2a| = -2a$ if $a < 0$, we see that in all cases

$$x = -\frac{b}{2a} \pm \frac{\sqrt{b^2 - 4ac}}{2a} = \frac{-b \pm \sqrt{b^2 - 4ac}}{2a}. \quad \blacksquare$$

Note that if the quadratic formula is executed properly, it is unnecessary to check the solutions because no extraneous solutions are introduced. The number $b^2 - 4ac$ under the radical sign in the quadratic formula is called the **discriminant** of the quadratic equation. If the discriminant is positive, there are two real and unequal roots of the quadratic equation. If it is zero, there is one root of multiplicity 2; and if it is negative, there are no real roots of the quadratic equation.

### EXAMPLE 5   *Using the quadratic formula*

Solve the equation $2x(3 - x) = 3$.

**SOLUTION**   To use the quadratic formula we must write the equation in the standard form $ax^2 + bx + c = 0$. Doing so gives us the equation $-2x^2 + 6x = 3$ or, equivalently, $2x^2 - 6x + 3 = 0$. We now let $a = 2$, $b = -6$, and $c = 3$ in the quadratic formula, obtaining

$$x = \frac{-(-6) \pm \sqrt{(-6)^2 - 4(2)(3)}}{2(2)} = \frac{6 \pm \sqrt{12}}{4} = \frac{6 \pm 2\sqrt{3}}{4} = \frac{3 \pm \sqrt{3}}{2}.$$

Hence, the solutions are $\dfrac{3 + \sqrt{3}}{2} \approx 2.37$ and $\dfrac{3 - \sqrt{3}}{2} \approx 0.63$.

Many applied problems lead to quadratic equations. One is illustrated in the following example.

### EXAMPLE 6   *Constructing a rectangular box*

A box with a square base and no top is to be made from a square piece of tin by cutting out a 3-inch square from each corner and folding up the sides. If the box is to hold 48 in.$^3$, what size piece of tin should be used?

**FIGURE 8**

**SOLUTION**   We begin by drawing the picture in Figure 8, letting $x$ denote the unknown length of the side of the piece of tin. Subsequently, each side of the base of the box will have length $x - 3 - 3 = x - 6$.

Since the area of the base of the box is $(x - 6)^2$ and the height is 3, we obtain

$$\text{volume of box} = 3(x - 6)^2.$$

Since the box is to hold 48 in.$^3$,

$$3(x - 6)^2 = 48.$$

We now solve for $x$:

$$(x - 6)^2 = 16 \qquad \text{divide by 3}$$
$$x - 6 = \pm 4 \qquad \text{take the square root}$$
$$x = 6 \pm 4 \qquad \text{add 6}$$

Consequently,

$$x = 10 \quad \text{or} \quad x = 2.$$

**Check**   Referring to Figure 8, we see that $x = 2$ is unacceptable, since no box is possible in this case. However, if we begin with a 10-inch square of tin, cut out 3-inch corners, and fold, we obtain a box having dimensions 4 inches, 4 inches, and 3 inches. The box has the desired volume of 48 in.$^3$. Thus, 10 inches is the answer to the problem.

The equations considered so far are inadequate for many problems. For example, in applications it is often necessary to consider powers $x^k$ with $k > 2$. Some equations involve absolute values or radicals. We conclude this section by giving examples of equations of various types that can be solved using elementary methods.

**EXAMPLE 7**    *Solving an equation containing an absolute value*

Solve the equation $2|x - 5| + 3 = 9$.

*SOLUTION*    We first isolate the absolute value expression by subtracting 3 and dividing by 2, to obtain

$$|x - 5| = \frac{9 - 3}{2} = 3.$$

If $a$ and $b$ are real numbers with $b > 0$, then $|a| = b$ if and only if $a = b$ or $a = -b$. Hence if $|x - 5| = 3$, then either

$$x - 5 = 3 \quad \text{or} \quad x - 5 = -3.$$

Solving for $x$ gives us

$$x = 5 + 3 = 8 \quad \text{or} \quad x = 5 - 3 = 2.$$

Thus, the given equation has two solutions, 8 and 2.

**EXAMPLE 8**    *Solving an equation using grouping*

Solve the equation $x^3 + 2x^2 - x - 2 = 0$.

*SOLUTION*

| | |
|---|---|
| $x^3 + 2x^2 - x - 2 = 0$ | given |
| $x^2(x + 2) - 1(x + 2) = 0$ | group terms |
| $(x^2 - 1)(x + 2) = 0$ | factor out $x + 2$ |
| $(x + 1)(x - 1)(x + 2) = 0$ | factor $x^2 - 1$ |
| $x + 1 = 0, \quad x - 1 = 0, \quad x + 2 = 0$ | set each factor equal to 0 |
| $x = -1, \qquad x = 1, \qquad x = -2$ | solve for $x$ |

**EXAMPLE 9**    *Solving an equation containing rational exponents*
Solve the equation $x^{3/2} = x^{1/2}$.

*SOLUTION*

$$x^{3/2} = x^{1/2} \qquad \text{given}$$
$$x^{3/2} - x^{1/2} = 0 \qquad \text{subtract } x^{1/2}$$
$$x^{1/2}(x - 1) = 0 \qquad \text{factor out } x^{1/2}$$
$$x^{1/2} = 0, \quad x - 1 = 0 \qquad \text{set each factor equal to 0}$$
$$x = 0, \qquad x = 1 \qquad \text{solve for } x$$

In Example 9 it would have been incorrect to divide both sides of the equation $x^{3/2} = x^{1/2}$ by $x^{1/2}$, obtaining $x = 1$, since the solution $x = 0$ would be lost. In general, *avoid dividing both sides of an equation by an expression that contains variables;* always *factor* instead.

If an equation involves radicals or fractional exponents, we often raise both sides to a positive power. The solutions of the new equation always contain the solutions of the given equation. For example, the solutions of

$$2x - 3 = \sqrt{x + 6}$$

are also solutions of

$$(2x - 3)^2 = (\sqrt{x + 6})^2.$$

In some cases the new equation has *more* solutions than the given equation. To illustrate, if we are given the equation $x = 3$ and we square both sides, we obtain $x^2 = 9$. Note that the given equation $x = 3$ has only one solution, 3, but the new equation $x^2 = 9$ has two solutions, 3 and $-3$. Any solution of the new equation that is not a solution of the given equation is an extraneous solution. Since extraneous solutions may occur, *it is absolutely essential to check all solutions obtained after raising both sides of an equation to an even power.* Such checks are unnecessary if both sides are raised to an *odd* power, because in this case extraneous (real number) solutions are not introduced.

In general, for the equation $x^{m/n} = a$, we raise both sides to the power $n/m$ (the reciprocal of $m/n$) to solve for $x$. If $m$ is odd, we obtain $x = a^{n/m}$, but if $m$ is even, we have $x = \pm a^{n/m}$. If $n$ is even, extraneous solutions may occur.

**EXAMPLE 10**    *Solving an equation containing a radical*
Solve the equation $3 + \sqrt{3x + 1} = x$.

**SOLUTION**

$$\sqrt{3x + 1} = x - 3 \qquad \text{isolate the radical}$$

$$(\sqrt{3x + 1})^2 = (x - 3)^2 \qquad \text{square both sides}$$

$$3x + 1 = x^2 - 6x + 9 \qquad \text{simplify}$$

$$x^2 - 9x + 8 = 0 \qquad \text{subtract } 3x + 1$$

$$(x - 1)(x - 8) = 0 \qquad \text{factor}$$

$$x = 1, \quad x = 8 \qquad \text{solve for } x$$

We raised both sides to an even power, so checks are required.

✔ **Check** $x = 1$    LS: $3 + \sqrt{3(1) + 1} = 3 + \sqrt{4} = 3 + 2 = 5$
                     RS: 1

Since $5 \neq 1$, $x = 1$ is not a solution.

✔ **Check** $x = 8$    LS: $3 + \sqrt{3(8) + 1} = 3 + \sqrt{25} = 3 + 5 = 8$
                     RS: 8

Since $8 = 8$ is a true statement, $x = 8$ is a solution.
Hence, the given equation has one solution, $x = 8$.

---

An equation is of **quadratic type** if it can be written in the form

$$au^2 + bu + c = 0,$$

where $a \neq 0$ and $u$ is an expression in some variable. If we find the solutions in terms of $u$, then the solutions of the given equation can be obtained by referring to the specific form of $u$, as demonstrated in the next example.

**EXAMPLE 11**    *Solving an equation of quadratic type*

Solve the equation $x^6 + 7x^3 = 8$.

**SOLUTION**    Since $x^6 = (x^3)^2$, the form of the equation suggests that we let $u = x^3$, as in the second line below:

$$x^6 + 7x^3 - 8 = 0 \qquad \text{make one side } 0$$

$$u^2 + 7u - 8 = 0 \qquad \text{let } u = x^3$$

$$(u + 8)(u - 1) = 0 \qquad \text{factor}$$

$$u = -8, \quad u = 1 \qquad \text{solve for } u$$

$$x^3 = -8, \quad x^3 = 1 \qquad u = x^3$$

$$x = -2, \quad x = 1 \qquad \text{take the cube root}$$

## I.4 EXERCISES

Exer. 1–10: Solve the equation.

1  $4x - 3 = -5x + 6$

2  $5x - 4 = 2(x - 2)$

3  $(3x - 2)^2 = (x - 5)(9x + 4)$

4  $(x + 5)^2 + 3 = (x - 2)^2$

5  $\dfrac{3x + 1}{6x - 2} = \dfrac{2x + 5}{4x - 13}$

6  $\dfrac{5x + 2}{10x - 3} = \dfrac{x - 8}{2x + 3}$

7  $\dfrac{4}{x + 2} + \dfrac{1}{x - 2} = \dfrac{5x - 6}{x^2 - 4}$

8  $\dfrac{2}{2x + 5} + \dfrac{3}{2x - 5} = \dfrac{10x + 5}{4x^2 - 25}$

9  $\dfrac{5}{2x + 3} + \dfrac{4}{2x - 3} = \dfrac{14x + 3}{4x^2 - 9}$

10  $\dfrac{-3}{x + 4} + \dfrac{7}{x - 4} = \dfrac{-5x + 4}{x^2 - 16}$

Exer. 11–14: Solve the equation by factoring.

11  $75x^2 + 35x - 10 = 0$

12  $48x^2 + 12x - 90 = 0$

13  $\dfrac{2x}{x + 3} + \dfrac{5}{x} - 4 = \dfrac{18}{x^2 + 3x}$

14  $\dfrac{3x}{x - 2} + \dfrac{1}{x + 2} = \dfrac{-4}{x^2 - 4}$

Exer. 15–18: Solve the equation by using the special quadratic equation on page 48.

15  $25x^2 = 9$

16  $16x^2 = 49$

17  $(x - 3)^2 = 17$

18  $(x + 4)^2 = 31$

Exer. 19–20: Solve the equation by using the quadratic formula.

19  $x^2 + 4x + 2 = 0$

20  $x^2 - 6x - 3 = 0$

Exer. 21–38: Solve the equation.

21  $|3x - 2| + 3 = 7$

22  $2|5x + 2| - 1 = 5$

23  $3|x + 1| - 2 = -11$

24  $|x - 2| + 5 = 5$

25  $9x^3 - 18x^2 - 4x + 8 = 0$

26  $4x^4 + 10x^3 = 6x^2 + 15x$

27  $y^{3/2} = 5y$

28  $y^{4/3} = -3y$

29  $\sqrt{7 - x} = x - 5$

30  $\sqrt{3 - x} - x = 3$

31  $x = 3 + \sqrt{5x - 9}$

32  $x + \sqrt{5x + 19} = -1$

33  $5y^4 - 7y^2 + 1 = 0$

34  $3y^4 - 5y^2 + 1 = 0$

35  $36x^{-4} - 13x^{-2} + 1 = 0$

36  $x^{-2} - 2x^{-1} - 35 = 0$

37  $3x^{2/3} + 4x^{1/3} - 4 = 0$

38  $2y^{1/3} - 3y^{1/6} + 1 = 0$

Exer. 39–40: Find the real solutions of the equation.

39  (a) $x^{5/3} = 32$    (b) $x^{4/3} = 16$    (c) $x^{2/3} = -36$
    (d) $x^{3/4} = 125$    (e) $x^{3/2} = -27$

40  (a) $x^{3/5} = -27$    (b) $x^{2/3} = 25$    (c) $x^{4/3} = -49$
    (d) $x^{3/2} = 27$    (e) $x^{3/4} = -8$

Exer. 41–42: Use the quadratic formula to solve the equation for
(a) $x$ in terms of $y$   (b) $y$ in terms of $x$

41  $4x^2 - 4xy + 1 - y^2 = 0$

42  $2x^2 - xy = 3y^2 + 1$

Exer. 43–44: When computations are carried out on a calculator, the quadratic formula will not always give accurate results if $b^2$ is large in comparison to $ac$, because one of the roots will be close to zero and difficult to approximate.
(a) Use the quadratic formula to approximate the roots of the given equation.
(b) To obtain a better approximation for the root near zero, rationalize the numerator to change

$$x = \frac{-b \pm \sqrt{b^2 - 4ac}}{2a} \quad \text{to} \quad x = \frac{2c}{-b \mp \sqrt{b^2 - 4ac}},$$

and use the second formula.

43  $x^2 + 4,500,000x - 0.96 = 0$

44  $x^2 - 73,000,000x + 2.01 = 0$

Exer. 45–56: The formula occurs in the indicated application. Solve for the specified variable.

45  $A = P + Prt$ for $r$      (principal plus interest)

46  $s = \frac{1}{2}gt^2 + v_0 t$ for $v_0$      (distance an object falls)

47  $S = \dfrac{p}{q + p(1 - q)}$ for $q$      (Amdahl's law for supercomputers)

48 $S = 2(lw + hw + hl)$ for $h$    (surface area of a rectangular box)

49 $\dfrac{1}{f} = \dfrac{1}{p} + \dfrac{1}{q}$ for $q$    (lens equation)

50 $\dfrac{1}{R} = \dfrac{1}{R_1} + \dfrac{1}{R_2} + \dfrac{1}{R_3}$ for $R_2$    (three resistors connected in parallel)

51 $K = \frac{1}{2}mv^2$ for $v$    (kinetic energy)

52 $F = g\dfrac{mM}{d^2}$ for $d$    (Newton's law of gravitation)

53 $A = 2\pi r(r + h)$ for $r$    (surface area of a closed cylinder)

54 $s = \frac{1}{2}gt^2 + v_0 t$ for $t$    (distance an object falls)

55 $d = \frac{1}{2}\sqrt{4R^2 - C^2}$ for $C$    (segments of circles)

56 $S = \pi r\sqrt{r^2 + h^2}$ for $h$    (surface area of a cone)

57 **Cost of insulation** The cost of installing insulation in a particular two-bedroom home is $1080. Present monthly heating costs average $60, but the insulation is expected to reduce heating costs by 10%. How many months will it take to recover the cost of the insulation?

58 **Municipal funding** A city government has approved the construction of a $50 million sports arena. Up to $30 million will be raised by selling bonds that pay simple interest at a rate of 12% annually. The remaining amount (up to $40 million) will be obtained by borrowing money from an insurance company at a simple interest rate of 10%. Determine whether the arena can be financed so that the annual interest is $5.2 million.

59 **Walking rates** Two children, who are 224 meters apart, start walking toward each other at the same instant at rates of 1.5 m/sec and 2 m/sec, respectively (see the figure).

(a) When will they meet?

(b) How far will each have walked?

EXERCISE 59

1.5 m/s    2 m/s

224 m

60 **Fencing a region** A farmer plans to use 180 feet of fencing to enclose a rectangular region, using part of a straight river bank instead of fencing as one side of the rectangle, as shown in the figure. Find the area of the region if the length of the side parallel to the river bank is

(a) twice the length of an adjacent side

(b) one-half the length of an adjacent side

(c) the same as the length of an adjacent side

EXERCISE 60

61 **Distance to a target** A bullet is fired horizontally at a target, and the sound of its impact is heard 1.5 seconds later. If the speed of the bullet is 3300 ft/sec and the speed of sound is 1100 ft/sec, how far away is the target?

62 **Jogging rates** A woman begins jogging at 3:00 P.M., running due north at a 6-minute-mile pace. Later, she reverses direction and runs due south at a 7-minute-mile pace. If she returns to her starting point at 3:45 P.M., find the total number of miles run.

63 **Drainage ditch dimensions** Every cross section of a drainage ditch is an isosceles trapezoid with a small base of 3 feet and a height of 1 foot, as shown in the figure. Determine the width of the larger base that would give the ditch a cross-sectional area of 5 ft$^2$.

EXERCISE 63

3'

1'

**64 Constructing a silo**  A large grain silo is to be constructed in the shape of a circular cylinder with a hemisphere attached to the top (see the figure). The diameter of the silo is to be 30 feet, but the height is yet to be determined. Find the height $h$ of the silo that will result in a capacity of $11,250\pi$ ft$^3$.

EXERCISE 64

**65 Braking distance**  The distance that a car travels between the time the driver makes the decision to hit the brakes and the time the car actually stops is called the braking distance. For a certain car traveling $v$ mi/hr, the braking distance $d$ (in feet) is given by $d = v + (v^2/20)$.

(a) Find the braking distance when $v$ is 55 mi/hr.

(b) If a driver decides to brake 120 feet from a stop sign, how fast can the car be going and still stop by the time it reaches the sign?

**66 Temperature of boiling water**  The temperature $T$ (in °C) at which water boils is related to the elevation $h$ (in meters above sea level) by the formula

$$h = 1000(100 - T) + 580(100 - T)^2$$

for $95 \le T \le 100$.

(a) At what elevation does water boil at a temperature of 98°C?

(b) The elevation of Mt. Everest is approximately 8840 meters. Estimate the temperature at which water boils at the top of this mountain. (*Hint:* Use the quadratic formula with $x = 100 - T$.)

**67 Distance between airplanes**  An airplane flying north at 200 mi/hr passed over a point on the ground at 2:00 P.M. Another airplane at the same altitude passed over the point at 2:30 P.M., flying east at 400 mi/hr (see the figure).

(a) If $t$ denotes the time in hours after 2:30 P.M., express the distance $d$ between the airplanes in terms of $t$.

(b) At what time after 2:30 P.M. were the airplanes 500 miles apart?

EXERCISE 67

**68 Height of a cliff**  If a rock is dropped from a cliff into an ocean, it travels approximately $16t^2$ feet in $t$ seconds. If the splash is heard 4 seconds later and the speed of sound is 1100 ft/sec, approximate the height of the cliff.

**69 Price of a CD player**  When a popular brand of CD player is priced at $300 per unit, a store sells 15 units per week. Each time the price is reduced by $10, however, the sales increase by 2 per week. What selling price will result in weekly revenues of $7000?

**70 Dimensions of an oil drum**  A closed cylindrical oil drum of height 4 feet is to be constructed so that the total surface area is $10\pi$ ft$^2$. Find the diameter of the drum.

**71 Dimensions of a vitamin tablet**  The rate at which a tablet of vitamin C begins to dissolve depends on the surface area of the tablet. One brand of tablet is 2 centimeters long and is in the shape of a cylinder with hemispheres of diameter 0.5 centimeter attached to both ends, as shown in the figure. A second brand of tablet is to be manufactured in the shape of a right circular cylinder of altitude 0.5 centimeter.

(a) Find the diameter of the second tablet so that its surface area is equal to that of the first tablet.

(b) Find the volume of each tablet.

EXERCISE 71

72  **Withdrawal resistance of nails**  The *withdrawal resistance* of a nail indicates its holding strength in wood. A formula that is used for bright common nails is $P = 15{,}700S^{5/2}RD$, where $P$ is the maximum withdrawal resistance (in pounds), $S$ is the specific gravity of the wood at 12% moisture content, $R$ is the radius of the nail (in inches), and $D$ is the depth (in inches) that the nail has penetrated the wood. A 6d (six-penny) bright common nail of length 2 inches and diameter 0.113 inch is driven completely into a piece of Douglas fir. If it requires a maximum force of 380 pounds to remove the nail, approximate the specific gravity of Douglas fir.

73  **The effect of price on demand**  The demand for a commodity usually depends on its price. If other factors do not affect the demand, then the quantity $Q$ purchased at price $P$ (in cents) is given by $Q = kP^{-c}$, where $k$ and $c$ are positive constants. If $k = 10^5$ and $c = \frac{1}{2}$, find the price that will result in the purchase of 5000 items.

74  **The urban heat island**  Urban areas have higher average air temperatures than rural areas as a result of the presence of buildings, asphalt, and concrete. This phenomenon has become known as the *urban heat island.*

The temperature difference $T$ (in °C) between urban and rural areas near Montreal, with a population $P$ between 1000 and 1,000,000, can be described by the formula $T = 0.25P^{1/4}/\sqrt{v}$, where $v$ is the average wind speed (in mi/hr) and $v \geq 1$. If $T = 3$ and $v = 5$, find $P$.

75  **Installing a power line**  A power line is to be installed across a river that is 1 mile wide to a town that is 5 miles downstream (see the figure). It costs $7500 per mile to lay the cable underwater and $6000 per mile to lay it overland. Determine how the cable should be installed if $35,000 has been allocated for this project.

**EXERCISE 75**

# 1.5   COMPLEX NUMBERS

*Complex numbers* are needed to find solutions of equations that cannot be solved using only the set $\mathbb{R}$ of real numbers. The following chart illustrates several simple quadratic equations and the types of numbers required for solutions.

| Equation | Solutions | Type of numbers required |
|---|---|---|
| $x^2 = 9$ | $3, -3$ | Integers |
| $x^2 = \frac{9}{4}$ | $\frac{3}{2}, -\frac{3}{2}$ | Rational numbers |
| $x^2 = 5$ | $\sqrt{5}, -\sqrt{5}$ | Irrational numbers |
| $x^2 = -9$ | ? | Complex numbers |

The solutions of the first three equations in the chart are in $\mathbb{R}$; however, since squares of real numbers are never negative, $\mathbb{R}$ does not contain the solutions of $x^2 = -9$. To solve this equation we need the **complex number system** $\mathbb{C}$, which contains both $\mathbb{R}$ and numbers whose squares are negative.

We begin by introducing the **imaginary unit,** denoted by $i$, which has the following properties.

**Properties of $i$**

$$i = \sqrt{-1}, \quad i^2 = -1$$

Because its square is negative, the letter $i$ does not represent a real number. It is a new mathematical entity that will enable us to obtain $\mathbb{C}$. Since $i$, together with $\mathbb{R}$, is to be contained in $\mathbb{C}$, we must consider products of the form $bi$ for a real number $b$ and also expressions of the form $a + bi$ for real numbers $a$ and $b$. The next chart provides definitions we shall use.

| Terminology | Definition |
|---|---|
| **Complex number** | $a + bi$, where $a$ and $b$ are real numbers and $i^2 = -1$ (examples: 3, $2 + i$, $2i$) |
| **Imaginary number** | $a + bi$ with $b \neq 0$ (examples: $3 + 2i$, $-4i$) |
| **Pure imaginary number** | $bi$ with $b \neq 0$ (examples: $-4i$, $\sqrt{3}i$, $i$) |
| **Equality** | $a + bi = c + di$ if and only if $a = c$ and $b = d$ |
| **Sum** | $(a + bi) + (c + di) = (a + c) + (b + d)i$ |
| **Product** | $(a + bi)(c + di) = (ac - bd) + (ad + bc)i$ |

Note that the pure imaginary numbers are a subset of the imaginary numbers and the imaginary numbers are a subset of the complex numbers. We use the phrase *nonreal complex number* interchangeably with *imaginary number.*

It is unnecessary to memorize the definitions of addition and multiplication of complex numbers given in the preceding chart. Instead, *we may treat all symbols as having properties of real numbers, with exactly one exception: we replace $i^2$ by $-1$.* Thus, for the product $(a + bi)(c + di)$ we simply use the distributive laws and the fact that

$$(bi)(di) = bdi^2 = bd(-1) = -bd.$$

**EXAMPLE I    *Addition and multiplication of complex numbers***

Express in the form $a + bi$, where $a$ and $b$ are real numbers:

**(a)** $(3 + 4i) + (2 + 5i)$     **(b)** $(3 + 4i)(2 + 5i)$

*SOLUTION*

**(a)**         $(3 + 4i) + (2 + 5i) = (3 + 2) + (4 + 5)i = 5 + 9i$

**(b)**         $(3 + 4i)(2 + 5i) = (3 + 4i)2 + (3 + 4i)(5i)$

$$= 6 + 8i + 15i + 20i^2$$
$$= 6 + 23i + 20(-1)$$
$$= -14 + 23i$$

The set $\mathbb{R}$ of real numbers may be identified with the set of complex numbers of the form $a + 0i$. It is also convenient to denote the complex number $0 + bi$ by $bi$. Thus,

$$(a + 0i) + (0 + bi) = (a + 0) + (0 + b)i = a + bi.$$

Hence we may regard $a + bi$ as the sum of two complex numbers $a$ and $bi$ (that is, $a + 0i$ and $0 + bi$). For the complex number $a + bi$, we call $a$ the **real part** and $b$ the **imaginary part.**

We can now solve an equation such as $x^2 = -9$. Specifically, since

$$(3i)(3i) = 3^2 i^2 = 9(-1) = -9,$$

we see that one solution is $3i$ and another is $-3i$.

In the next chart we define the difference of complex numbers and multiplication of a complex number by a real number.

| Terminology | Definition |
|---|---|
| **Difference** | $(a + bi) - (c + di) = (a - c) + (b - d)i$ |
| **Multiplication by a real number $k$** | $k(a + bi) = ka + (kb)i$ |

If we are asked to write an expression in the form $a + bi$, we shall also accept the form $a - di$, since $a - di = a + (-d)i$.

### EXAMPLE 2 *Operations with complex numbers*

Express in the form $a + bi$, where $a$ and $b$ are real numbers:

**(a)** $4(2 + 5i) - (3 - 4i)$     **(b)** $(4 - 3i)(2 + i)$

**(c)** $i(3 - 2i)^2$              **(d)** $i^{51}$

### SOLUTION

**(a)**       $4(2 + 5i) - (3 - 4i) = 8 + 20i - 3 + 4i = 5 + 24i$

**(b)**        $(4 - 3i)(2 + i) = 8 - 6i + 4i - 3i^2 = 11 - 2i$

**(c)**   $i(3 - 2i)^2 = i(9 - 12i + 4i^2) = i(5 - 12i) = 5i - 12i^2 = 12 + 5i$

**(d)** Taking successive powers of $i$, we obtain

$$i^1 = i, \quad i^2 = -1, \quad i^3 = -i, \quad i^4 = 1,$$

and then the cycle starts over:

$$i^5 = i, \quad i^6 = i^2 = -1, \quad \text{and so on.}$$

In particular,

$$i^{51} = i^{48} i^3 = (i^4)^{12} i^3 = (1)^{12} i^3 = (1)(-i) = -i.$$

The following concept has important uses in working with complex numbers.

| | |
|---|---|
| **Definition of Conjugate of a Complex Number** | If $z = a + bi$ is a complex number, then its **conjugate,** denoted by $\bar{z}$, is $a - bi$. |

Since $a - bi = a + (-bi)$, it follows that the conjugate of $a - bi$ is

$$a - (-bi) = a + bi.$$

Therefore, $a + bi$ and $a - bi$ are conjugates of each other. Some properties of conjugates are given in Exercises 55–60.

## ILLUSTRATION

### Conjugates

| Complex number | Conjugate |
|---|---|
| $5 + 7i$ | $5 - 7i$ |
| $5 - 7i$ | $5 + 7i$ |
| $4i$ | $-4i$ |
| $3$ | $3$ |

The following two properties are consequences of the definitions of sum and product of complex numbers.

| Properties of conjugates | Illustration |
|---|---|
| $(a + bi) + (a - bi) = 2a$ | $(4 + 3i) + (4 - 3i) = 4 + 4 = 2 \cdot 4$ |
| $(a + bi)(a - bi) = a^2 + b^2$ | $(4 + 3i)(4 - 3i) = 4^2 - (3i)^2 = 4^2 - 3^2i^2 = 4^2 + 3^2$ |

Note that *the sum and the product of a complex number and its conjugate are real numbers.* Conjugates are useful for finding the **multiplicative inverse** of $a + bi$, $1/(a + bi)$, or for simplifying the quotient of two complex numbers. As illustrated in the next example, we may think of these types of simplifications as merely *rationalizing the denominator,* since we are multiplying the quotient by the conjugate of the denominator divided by itself.

**EXAMPLE 3**     *Quotients of complex numbers*

Express in the form $a + bi$, where $a$ and $b$ are real numbers:

**(a)** $\dfrac{1}{9 + 2i}$     **(b)** $\dfrac{7 - i}{3 - 5i}$

## SOLUTION

(a) $\dfrac{1}{9 + 2i} = \dfrac{1}{9 + 2i} \cdot \dfrac{9 - 2i}{9 - 2i} = \dfrac{9 - 2i}{81 + 4} = \dfrac{9}{85} - \dfrac{2}{85}i$

(b) $\dfrac{7 - i}{3 - 5i} = \dfrac{7 - i}{3 - 5i} \cdot \dfrac{3 + 5i}{3 + 5i} = \dfrac{21 + 35i - 3i - 5i^2}{9 + 25}$

$$= \dfrac{26 + 32i}{34} = \dfrac{13}{17} + \dfrac{16}{17}i$$

If $p$ is a positive real number, then the equation $x^2 = -p$ has solutions in $\mathbb{C}$. One solution is $\sqrt{p}\,i$, since

$$(\sqrt{p}\,i)^2 = (\sqrt{p})^2 i^2 = p(-1) = -p.$$

Similarly, $-\sqrt{p}\,i$ is also a solution.

The definition of $\sqrt{-r}$ in the next chart is motivated by $(\sqrt{r}\,i)^2 = -r$ for $r > 0$. When using this definition, take care *not* to write $\sqrt{ri}$ when $\sqrt{r}\,i$ is intended.

| Terminology | Definition | Illustrations |
|---|---|---|
| **Principal square root** $\sqrt{-r}$ for $r > 0$ | $\sqrt{-r} = \sqrt{r}\,i$ | $\sqrt{-9} = \sqrt{9}\,i = 3i$ $\sqrt{-5} = \sqrt{5}\,i$ $\sqrt{-1} = \sqrt{1}\,i = i$ |

The radical sign must be used with caution when the radicand is negative. For example, the formula $\sqrt{a}\sqrt{b} = \sqrt{ab}$, which holds for positive real numbers, is not true when $a$ and $b$ are both negative, as shown below:

$$\sqrt{-3}\sqrt{-3} = (\sqrt{3}\,i)(\sqrt{3}\,i) = (\sqrt{3})^2 i^2 = 3(-1) = -3,$$

but $\qquad\qquad\qquad \sqrt{(-3)(-3)} = \sqrt{9} = 3.$

Hence, $\qquad\qquad\qquad \sqrt{-3}\sqrt{-3} \neq \sqrt{(-3)(-3)}.$

If only *one* of $a$ or $b$ is negative, then $\sqrt{a}\sqrt{b} = \sqrt{ab}$. In general, we shall not apply laws of radicals if radicands are negative. Instead, we shall change the form of radicals before performing any operations, as illustrated in the next example.

**EXAMPLE 4**    *Working with square roots of negative numbers*

Express in the form $a + bi$, where $a$ and $b$ are real numbers:

$$(5 - \sqrt{-9})(-1 + \sqrt{-4})$$

**SOLUTION** First we use the definition $\sqrt{-r} = \sqrt{r}i$, and then we simplify:

$$
\begin{aligned}
(5 - \sqrt{-9})(-1 + \sqrt{-4}) &= (5 - \sqrt{9}i)(-1 + \sqrt{4}i) \\
&= (5 - 3i)(-1 + 2i) \\
&= -5 + 10i + 3i - 6i^2 \\
&= -5 + 13i + 6 = 1 + 13i
\end{aligned}
$$

In Section 1.4 we stated that if the discriminant $b^2 - 4ac$ of the quadratic equation $ax^2 + bx + c = 0$ is negative, then there are no real roots of the equation. In fact, the solutions of the equation are two *imaginary* numbers. Moreover, the solutions are conjugates of each other, as shown in the next example.

**EXAMPLE 5    *A quadratic equation with complex solutions***
Solve the equation $5x^2 + 2x + 1 = 0$.

**SOLUTION** Applying the quadratic formula with $a = 5$, $b = 2$, and $c = 1$, we see that

$$
\begin{aligned}
x &= \frac{-2 \pm \sqrt{2^2 - 4(5)(1)}}{2(5)} = \frac{-2 \pm \sqrt{-16}}{10} = \frac{-2 \pm 4i}{10} = \frac{-1 \pm 2i}{5} \\
&= -\frac{1}{5} \pm \frac{2}{5}i.
\end{aligned}
$$

Thus, the solutions of the equation are $-\frac{1}{5} + \frac{2}{5}i$ and $-\frac{1}{5} - \frac{2}{5}i$.

**EXAMPLE 6    *An equation with complex solutions***
Solve the equation $x^3 - 1 = 0$.

**SOLUTION** Using the difference of two cubes factoring formula in Section 1.3, we write $x^3 - 1 = 0$ as

$$(x - 1)(x^2 + x + 1) = 0.$$

Setting each factor equal to zero and solving the resulting equations, we obtain the solutions

$$
1, \quad \frac{-1 \pm \sqrt{1 - 4}}{2} = \frac{-1 \pm \sqrt{3}i}{2}
$$

or, equivalently,

$$
1, \quad -\frac{1}{2} + \frac{\sqrt{3}}{2}i, \quad -\frac{1}{2} - \frac{\sqrt{3}}{2}i.
$$

Since the number 1 is called the **unit real number** and the given equation may be written as $x^3 = 1$, we call these three solutions the **cube roots of unity.**

---

In Section 1.3 we mentioned that $x^2 + 1$ is irreducible over the *real* numbers. However, if we factor over the *complex* numbers, then $x^2 + 1$ may be factored as follows:

$$x^2 + 1 = (x + i)(x - i)$$

## 1.5 EXERCISES

**Exer. 1–34: Write the expression in the form $a + bi$, where $a$ and $b$ are real numbers.**

1  $(5 - 2i) + (-3 + 6i)$     2  $(-5 + 7i) + (4 + 9i)$

3  $(7 - 6i) - (-11 - 3i)$     4  $(-3 + 8i) - (2 + 3i)$

5  $(3 + 5i)(2 - 7i)$     6  $(-2 + 6i)(8 - i)$

7  $(1 - 3i)(2 + 5i)$     8  $(8 + 2i)(7 - 3i)$

9  $(5 - 2i)^2$     10  $(6 + 7i)^2$

11  $i(3 + 4i)^2$     12  $i(2 - 7i)^2$

13  $(3 + 4i)(3 - 4i)$     14  $(4 + 9i)(4 - 9i)$

15  $i^{43}$     16  $i^{92}$

17  $i^{73}$     18  $i^{66}$

19  $\dfrac{3}{2 + 4i}$     20  $\dfrac{5}{2 - 7i}$

21  $\dfrac{1 - 7i}{6 - 2i}$     22  $\dfrac{2 + 9i}{-3 - i}$

23  $\dfrac{-4 + 6i}{2 + 7i}$     24  $\dfrac{-3 - 2i}{5 + 2i}$

25  $\dfrac{4 - 2i}{-5i}$     26  $\dfrac{-2 + 6i}{3i}$

27  $(2 + 5i)^3$     28  $(3 - 2i)^3$

29  $(2 - \sqrt{-4})(3 - \sqrt{-16})$

30  $(-3 + \sqrt{-25})(8 - \sqrt{-36})$

31  $\dfrac{4 + \sqrt{-81}}{7 - \sqrt{-64}}$     32  $\dfrac{5 - \sqrt{-121}}{1 + \sqrt{-25}}$

33  $\dfrac{\sqrt{-36}\sqrt{-49}}{\sqrt{-16}}$     34  $\dfrac{\sqrt{-25}}{\sqrt{-16}\sqrt{-81}}$

**Exer. 35–38: Find the values of $x$ and $y$, where $x$ and $y$ are real numbers.**

35  $8 + (3x + y)i = 2x - 4i$     36  $(x - y) + 3i = 7 + yi$

37  $(3x + 2y) - y^3 i = 9 - 27i$

38  $x^3 + (2x - y)i = -8 - 3i$

**Exer. 39–54: Find the solutions of the equation.**

39  $x^2 - 6x + 13 = 0$     40  $x^2 - 2x + 26 = 0$

41  $x^2 + 4x + 13 = 0$     42  $x^2 + 8x + 17 = 0$

43  $x^2 - 5x + 20 = 0$     44  $x^2 + 3x + 6 = 0$

45  $4x^2 + x + 3 = 0$     46  $-3x^2 + x - 5 = 0$

47  $x^3 + 125 = 0$     48  $x^3 - 27 = 0$

49  $x^4 = 256$     50  $x^4 = 81$

51  $4x^4 + 25x^2 + 36 = 0$     52  $27x^4 + 21x^2 + 4 = 0$

53  $x^3 + 3x^2 + 4x = 0$

54  $8x^3 - 12x^2 + 2x - 3 = 0$

**Exer. 55–60: Verify the property.**

55  $\overline{z + w} = \bar{z} + \bar{w}$     56  $\overline{z - w} = \bar{z} - \bar{w}$

57  $\overline{z \cdot w} = \bar{z} \cdot \bar{w}$     58  $\overline{z/w} = \bar{z}/\bar{w}$

59  $\bar{z} = z$ if and only if $z$ is real.

60  $\overline{z^2} = (\bar{z})^2$

## *1.6* INEQUALITIES

An **inequality** is a statement that two quantities or expressions are not equal. It may be the case that one quantity is less than ($<$), less than or equal to ($\leq$), greater than ($>$), or greater than or equal to ($\geq$) another quantity. Consider the inequality

$$2x + 3 > 11,$$

where $x$ is a variable. If a true statement is obtained when a number $b$ is substituted for $x$, then $b$ is a **solution** of the inequality. Thus, $x = 5$ is a solution of $2x + 3 > 11$ since $13 > 11$ is true, but $x = 3$ is not a solution since $9 > 11$ is false. To **solve** an inequality means to find *all* solutions. Two inequalities are **equivalent** if they have exactly the same solutions.

Most inequalities have an infinite number of solutions. To illustrate, the solutions of the continued inequality

$$2 < x < 5$$

consist of *every* real number $x$ between 2 and 5. We call this set of numbers an **open interval** and denote it by $(2, 5)$. The **graph** of the open interval $(2, 5)$ is the set of all points on a coordinate line that lie between (but do not include) the points corresponding to $x = 2$ and $x = 5$. The graph is represented by shading an appropriate part of the axis, as shown in Figure 9. We refer to this process as **sketching the graph** of the interval. The numbers 2 and 5 are called the **end points** of the interval $(2, 5)$. The parentheses in the notation $(2, 5)$ and in Figure 9 are used to indicate that the end points of the interval are not included.

**FIGURE 9**

If we wish to include an end point, we use a bracket instead of a parenthesis. For example, the solutions of the inequality $2 \leq x \leq 5$ are denoted by $[2, 5]$ and referred to as a **closed interval.** The graph of $[2, 5]$ is sketched in Figure 10, where brackets indicate that the end points are included. We shall also consider **half-open intervals** $[a, b)$ and $(a, b]$ and **infinite intervals,** as described in the following chart. The symbol $\infty$ (read "infinity") used for infinite intervals is merely a notational device and does *not* represent a real number.

**FIGURE 10**

**Intervals**

| Notation | Inequality | Graph |
|----------|-----------|-------|
| (1) $(a, b)$ | $a < x < b$ | |
| (2) $[a, b]$ | $a \leq x \leq b$ | |
| (3) $[a, b)$ | $a \leq x < b$ | |
| (4) $(a, b]$ | $a < x \leq b$ | |
| (5) $(a, \infty)$ | $x > a$ | |
| (6) $[a, \infty)$ | $x \geq a$ | |
| (7) $(-\infty, b)$ | $x < b$ | |
| (8) $(-\infty, b]$ | $x \leq b$ | |
| (9) $(-\infty, \infty)$ | $-\infty < x < \infty$ | |

Methods for solving inequalities in $x$ are similar to those used for solving equations. In particular, we often use properties of inequalities to replace a given inequality with a list of equivalent inequalities, ending with an inequality from which solutions are easily obtained. The properties in the following chart can be proved for real numbers $a$, $b$, $c$, and $d$.

**Properties of Inequalities**

| Property | Illustration |
|----------|-------------|
| (1) If $a < b$ and $b < c$, then $a < c$. | $2 < 5$ and $5 < 9$, so $2 < 9$. |
| (2) If $a < b$, then $a + c < b + c$ and $a - c < b - c$. | $2 < 7$, so $2 + 3 < 7 + 3$ and $2 - 3 < 7 - 3$. |
| (3) If $a < b$ and $c > 0$, then $ac < bc$ and $\dfrac{a}{c} < \dfrac{b}{c}$. | $2 < 5$ and $3 > 0$, so $2 \cdot 3 < 5 \cdot 3$ and $\dfrac{2}{3} < \dfrac{5}{3}$. |
| (4) If $a < b$ and $c < 0$, then $ac > bc$ and $\dfrac{a}{c} > \dfrac{b}{c}$. | $2 < 5$ and $-3 < 0$, so $2(-3) > 5(-3)$ and $\dfrac{2}{-3} > \dfrac{5}{-3}$. |

It is important to remember that multiplying or dividing both sides of an inequality by a negative real number *reverses* the inequality sign (see property 4). Properties similar to those above are true for other inequalities and for $\leq$ or $\geq$. Thus, if $a > b$, then $a + c > b + c$; if $a \geq b$ and $c < 0$, then $ac \leq bc$; and so on.

If $x$ represents a real number, then, by property 2, adding or subtracting the same expression containing $x$ on both sides of an inequality yields an equivalent inequality. By property 3, we may multiply or divide both sides of an inequality by an expression containing $x$ if we are certain that the expression is positive for all values of $x$ under consideration. To illustrate, multiplication or division by $x^4 + 3x^2 + 5$ would be permissible, since this expression is always positive. If we multiply or divide both sides of an inequality by an expression that is always negative, such as $-7 - x^2$, then, by property 4, the inequality sign is reversed. We will demonstrate these properties in the upcoming examples.

In examples we shall describe solutions of inequalities by means of intervals and also represent some of them graphically.

**EXAMPLE 1    *Solving an inequality***

Solve the inequality $-4x - 3 > -2x - 11$.

**SOLUTION**

$$-4x - 3 > -2x - 11 \quad \text{given}$$
$$-2x > -8 \quad \text{add } 2x + 3$$
$$x < 4 \quad \text{divide by } -2; \text{ reverse the inequality sign}$$

**FIGURE 11**

Hence, the solutions of the given inequality consist of all real numbers $x$ such that $x < 4$. This is the interval $(-\infty, 4)$ sketched in Figure 11.

**EXAMPLE 2    *Solving a continued inequality***

Solve the inequality $-5 \leq \dfrac{4 - 3x}{2} < 1$.

**SOLUTION**    A number $x$ is a solution of the given inequality if and only if

$$-5 \leq \frac{4 - 3x}{2} \quad \text{and} \quad \frac{4 - 3x}{2} < 1.$$

We can either work with each inequality separately or solve both inequalities simultaneously, as follows (keep in mind that our goal is to isolate $x$):

$$-5 \leq \frac{4 - 3x}{2} < 1 \quad \text{given}$$
$$-10 \leq 4 - 3x < 2 \quad \text{multiply by 2}$$
$$-14 \leq -3x < -2 \quad \text{subtract 4}$$
$$\tfrac{14}{3} \geq x > \tfrac{2}{3} \quad \text{divide by } -3; \text{ reverse the inequality signs}$$
$$\tfrac{2}{3} < x \leq \tfrac{14}{3} \quad \text{equivalent inequality}$$

**FIGURE 12**

Thus, the solutions of the inequality are all numbers in the half-open interval $\left(\tfrac{2}{3}, \tfrac{14}{3}\right]$ sketched in Figure 12.

**EXAMPLE 3** *Solving a simple rational inequality*

Solve the inequality $\dfrac{1}{x-2} > 0$.

**FIGURE 13**

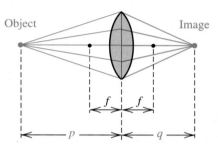

**SOLUTION** Since the numerator is positive, the fraction is positive if and only if the denominator, $x - 2$, is also positive. Thus, $x - 2 > 0$, or equivalently, $x > 2$, and the solutions are all numbers in the infinite interval $(2, \infty)$ sketched in Figure 13.

**EXAMPLE 4** *Using a lens formula*

**FIGURE 14**

As illustrated in Figure 14, if a convex lens has focal length $f$ centimeters and if an object is placed a distance $p$ centimeters from the lens with $p > f$, then the distance $q$ from the lens to the image is related to $p$ and $f$ by the formula

$$\frac{1}{p} + \frac{1}{q} = \frac{1}{f}.$$

If $f = 5$ cm, how close must the object be to the lens for the image to be more than 12 centimeters from the lens?

**SOLUTION** Since $f = 5$, the given formula may be written as

$$\frac{1}{p} + \frac{1}{q} = \frac{1}{5}.$$

We wish to determine the values of $q$ such that $q > 12$. Let us first solve the equation for $q$:

$5q + 5p = pq$          multiply by the lcd, $5pq$

$q(5 - p) = -5p$      collect $q$ terms on one side and factor

$q = -\dfrac{5p}{5 - p} = \dfrac{5p}{p - 5}$    divide by $5 - p$

To solve the inequality $q > 12$, we proceed as follows:

$\dfrac{5p}{p - 5} > 12$          $q = \dfrac{5p}{p - 5}$

$5p > 12(p - 5)$    allowable, since $p > f$ implies $p - 5 > 0$

$-7p > -60$     multiply factors and collect $p$ terms on one side

$p < \frac{60}{7}$      divide by $-7$; reverse the inequality

Combining the last inequality with the fact that $p$ is greater than 5, we obtain the solution

$$5 < p < \tfrac{60}{7}.$$

**FIGURE 15**

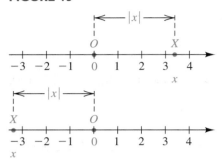

If a point $X$ on a coordinate line has coordinate $x$, as shown in Figure 15, then $X$ is to the right of the origin $O$ if $x > 0$ and to the left of $O$ if $x < 0$. From Section 1.1, the distance $d(O, X)$ between $O$ and $X$ is the *nonnegative* real number given by

$$d(O, X) = |x - 0| = |x|.$$

It follows that the solutions of an inequality such as $|x| < 3$ consist of the coordinates of all points whose distance from $O$ is less than 3. This is the open interval $(-3, 3)$ sketched in Figure 16. Thus,

$$|x| < 3 \quad \text{is equivalent to} \quad -3 < x < 3.$$

**FIGURE 16**

Similarly, for $|x| > 3$, the distance between $O$ and a point with coordinate $x$ is greater than 3; that is,

$$|x| > 3 \quad \text{is equivalent to} \quad x < -3 \text{ or } x > 3.$$

**FIGURE 17**

The graph of the solutions to $|x| > 3$ is sketched in Figure 17. We often use the union symbol $\cup$ and write

$$(-\infty, -3) \cup (3, \infty)$$

to denote all real numbers that are in either $(-\infty, -3)$ or $(3, \infty)$.

The intersection symbol $\cap$ is used to denote the elements that are *common* to two sets. For example,

$$(-\infty, 3) \cap (-3, \infty) = (-3, 3),$$

since the intersection of $(-\infty, 3)$ and $(-3, \infty)$ consists of all real numbers $x$ such that both $x < 3$ *and* $x > -3$.

The preceding discussion may be generalized to obtain the following.

**Properties of Absolute Values ($b > 0$)**

> **(1)** $|a| < b$   is equivalent to   $-b < a < b$.
> **(2)** $|a| > b$   is equivalent to   $a < -b \text{ or } a > b$.

In the next example we use property 1 with $a = x - 3$ and $b = 0.5$.

**EXAMPLE 5**   *Solving an inequality containing an absolute value*

Solve the inequality $|x - 3| - 0.1 < 0.4$.

**SOLUTION**

$$|x - 3| < 0.5 \qquad \text{isolate } |x - 3|$$
$$-0.5 < x - 3 < 0.5 \qquad \text{property 1}$$
$$2.5 < x < 3.5 \qquad \text{isolate } x \text{ by adding 3}$$

Thus, the solutions are the real numbers in the open interval $(2.5, 3.5)$. The graph is sketched in Figure 18.

**FIGURE 18**

In the next example we use property 2 with $a = 2x + 3$ and $b = 9$.

**EXAMPLE 6**    *Solving an inequality containing an absolute value*

Solve the inequality $|2x + 3| > 9$.

**SOLUTION**

$$|2x + 3| > 9 \qquad \text{given}$$
$$2x + 3 < -9 \quad \text{or} \quad 2x + 3 > 9 \qquad \text{property 2}$$
$$2x < -12 \quad \text{or} \qquad 2x > 6 \qquad \text{subtract 3}$$
$$x < -6 \quad \text{or} \qquad x > 3 \qquad \text{divide by 2}$$

**FIGURE 19**

Consequently, the solutions of the inequality $|2x + 3| > 9$ consist of the numbers in $(-\infty, -6) \cup (3, \infty)$. The graph is sketched in Figure 19.

The trichotomy law in Section 1.1 states that for any real numbers $a$ and $b$, exactly one of the following is true:

$$a > b, \quad a < b, \quad \text{or} \quad a = b$$

Thus, after solving $|2x + 3| > 9$ in Example 6, we readily obtain the solutions for $|2x + 3| < 9$ and $|2x + 3| = 9$—namely, the interval $(-6, 3)$ and the set $\{-6, 3\}$, respectively. Note that the union of these three sets of solutions is necessarily the set $\mathbb{R}$ of real numbers.

When using the notation $a < x < b$, we must have $a < b$. Thus, *it is incorrect to write the solution $x < -6$ or $x > 3$ (in Example 6) as $3 < x < -6$*. Another misuse of inequality notation is to write $a < x > b$, since when several inequality symbols are used in one expression, *they must point in the same direction*.

To solve an inequality involving polynomials of degree greater than 1, we shall express each polynomial as a product of linear factors $ax + b$ and/or irreducible quadratic factors $ax^2 + bx + c$. If any such factor is not zero in an interval, then it is either positive throughout the interval or negative throughout the interval. Hence, if we choose any $k$ in the interval and if the factor is positive (or negative) for $x = k$, then it is positive (or negative) throughout the interval. The value of the factor at $x = k$ is called a **test value** of the factor at the test number $k$. This concept is exhibited in the following example.

### EXAMPLE 7    *Solving a quadratic inequality*

Solve the inequality $2x^2 - x < 3$.

**SOLUTION**    To use test values *it is essential to have* 0 *on one side* of the inequality sign. Thus, we proceed as follows:

$$2x^2 - x - 3 < 0 \quad \text{make one side 0}$$
$$(x + 1)(2x - 3) < 0 \quad \text{factor}$$

The factors $x + 1$ and $2x - 3$ are zero at $-1$ and $\frac{3}{2}$, respectively. The corresponding points on a coordinate line (see Figure 20) determine the nonintersecting intervals

$$(-\infty, -1), \quad (-1, \tfrac{3}{2}), \quad \text{and} \quad (\tfrac{3}{2}, \infty).$$

**FIGURE 20**

We may find the signs of $x + 1$ and $2x - 3$ in each interval by using a test value. To illustrate, if we choose $k = -10$ in $(-\infty, -1)$, the values of both $x + 1$ and $2x - 3$ are negative, and hence they are negative throughout $(-\infty, -1)$. A similar procedure for the remaining two intervals gives us the following *sign chart,* where the term *resulting sign* in the last row refers to the sign obtained by applying laws of signs to the product of the factors. Note that the resulting sign is positive or negative according to whether the number of negative signs of factors is even or odd, respectively.

**FIGURE 21**

| Interval | $(-\infty, -1)$ | $(-1, \frac{3}{2})$ | $(\frac{3}{2}, \infty)$ |
|---|---|---|---|
| **Sign of $x + 1$** | $-$ | $+$ | $+$ |
| **Sign of $2x - 3$** | $-$ | $-$ | $+$ |
| Resulting sign | $+$ | $-$ | $+$ |

Sometimes it is convenient to represent the signs of $x + 1$ and $2x - 3$ by using a coordinate line and a *sign diagram,* of the type illustrated in Figure 21. The vertical lines indicate where the factors are zero, and signs of factors are shown above the coordinate line. The resulting signs are shown in red.

The solution of $(x + 1)(2x - 3) < 0$ are the values of $x$ for which the product of the factors is *negative*—that is, where the resulting sign is negative. This corresponds to the open interval $(-1, \frac{3}{2})$.

In future examples (here and in Chapter 3) we will use either a sign chart or a sign diagram, but not both. When working exercises, you should choose the method of solution with which you feel most comfortable.

**EXAMPLE 8** *Using a sign diagram to solve an inequality*

Solve the inequality $\dfrac{(x - 5)(x + 2)^2(3 - x)}{(x - 5)(x + 4)} \leq 0$.

**SOLUTION** We see that $x - 5$ is a factor of both the numerator and the denominator. Thus, *assuming that* $x - 5 \neq 0$ (that is, $x \neq 5$), we may cancel this factor and reduce our search for solutions to the case

$$\frac{(x + 2)^2(3 - x)}{x + 4} \leq 0 \quad \text{and} \quad x \neq 5.$$

**FIGURE 22**

We next observe that this quotient is 0 if $x + 2 = 0$ (that is, if $x = -2$). Hence, $-2$ is a solution. We do not include $(x + 2)^2$ in the sign diagram, since this expression is always positive if $x \neq -2$ and so has no effect on the sign of the quotient.

The signs of the factors $x + 4$ and $3 - x$ can be found using test values. Alternatively, we need only remember that as $x$ increases, the sign of a linear factor $ax + b$ changes from negative to positive if the coefficient $a$ of $x$ is positive, and the sign changes from positive to negative if $a$ is negative.

To determine where the quotient is less than or equal to 0, we first note from the sign diagram that it is *negative* for numbers in $(-\infty, -4) \cup (3, \infty)$. Since the quotient is *zero* at $x = -2$ and $x = 3$, the numbers $-2$ and 3 are also solutions and must be *included* in our solution. Lastly, the quotient is *undefined* at $x = -4$ and $x = 5$, so $-4$ and 5 must be *excluded* from our solution. Thus, the solutions of the given inequality are given by

$$(-\infty, -4) \cup \{-2\} \cup [3, 5) \cup (5, \infty).$$

**EXAMPLE 9** *Using a sign diagram to solve an inequality*

Solve the inequality $\dfrac{x + 1}{x + 3} \leq 2$.

**SOLUTION** A common mistake in solving such an inequality is to first multiply both sides by $x + 3$. If we did so, we would have to consider two cases, since $x + 3$ may be positive or negative (assuming $x + 3 \neq 0$), and we might have to reverse the inequality. A simpler method is to first

*(continued)*

obtain an equivalent inequality that has 0 on the right side and proceed from there:

$$\frac{x+1}{x+3} - 2 \leq 0 \quad \text{make one side 0}$$

$$\frac{x+1-2(x+3)}{x+3} \leq 0 \quad \text{combine into one fraction}$$

$$\frac{-x-5}{x+3} \leq 0 \quad \text{simplify}$$

$$\frac{x+5}{x+3} \geq 0 \quad \text{multiply by } -1$$

**FIGURE 23**

Note that the direction of the inequality is changed in the last step, since we multiplied by a negative number. This multiplication was performed for convenience, so that all factors would have positive coefficients of $x$.

The factors $x + 5$ and $x + 3$ are 0 at $x = -5$, and $x = -3$, respectively, leading to the sign diagram in Figure 23, where the signs are determined as in previous examples. We see from the diagram that the resulting sign, and hence the sign of the quotient, is positive in $(-\infty, -5) \cup (-3, \infty)$. The quotient is 0 at $x = -5$ (include $-5$) and undefined at $x = -3$ (exclude $-3$). Therefore, the solution of $(x + 5)/(x + 3) \geq 0$ is $(-\infty, -5] \cup (-3, \infty)$.

An alternative method of solution is to begin by multiplying both sides of the given inequality by $(x + 3)^2$, *assuming that $x \neq -3$.* In this case, $(x + 3)^2 > 0$, and the multiplication is permissible; however, after the resulting inequality is solved, the value $x = -3$ must be excluded.

---

**EXAMPLE 10**    *Determining minimum therapeutic levels*

For a drug to have a beneficial effect, its concentration in the bloodstream must exceed a certain value, which is called the *minimum therapeutic level*. Suppose that the concentration $c$ (in mg/L) of a particular drug $t$ hours after it is taken orally is given by

$$c = \frac{20t}{t^2 + 4}.$$

If the minimum therapeutic level is 4 mg/L, determine when this level is exceeded.

**SOLUTION**    The minimum therapeutic level, 4 mg/L, is exceeded if $c > 4$. Thus, we must solve the inequality

$$\frac{20t}{t^2 + 4} > 4.$$

Since $t^2 + 4 > 0$ for every $t$, we may multiply both sides by $t^2 + 4$ and proceed as follows:

$$20t > 4t^2 + 16 \qquad \text{allowable, since } t^2 + 4 > 0$$
$$-4t^2 + 20t - 16 > 0 \qquad \text{make one side } 0$$
$$t^2 - 5t + 4 < 0 \qquad \text{divide by the common factor } -4$$
$$(t - 1)(t - 4) < 0 \qquad \text{factor}$$

The factors in the last inequality are 0 when $t = 1$ and $t = 4$. These are the times at which $c$ is *equal* to 4. As in previous examples, we may use a sign chart or sign diagram (with $t \geq 0$) to show that $(t - 1)(t - 4) < 0$ for every $t$ in the interval $(1, 4)$. Hence, the minimum therapeutic level is exceeded if $1 < t < 4$.

Because graphs in a coordinate plane are introduced in the next section, it would be premature to demonstrate here the use of a graphics calculator or computer software to solve inequalities in $x$. Such methods will be considered later in the text.

Some basic properties of inequalities were stated at the beginning of this section. The following additional properties are sometimes helpful for solving certain inequalities. Proofs of the properties are given after the chart.

**Additional Properties of Inequalities**

| Property | Illustration |
|---|---|
| (1) If $0 < a < b$, then $\dfrac{1}{a} > \dfrac{1}{b}$. | If $0 < \dfrac{1}{x} < 4$, then $\dfrac{1}{1/x} > \dfrac{1}{4}$, or $x > \dfrac{1}{4}$. |
| (2) If $0 < a < b$, then $0 < a^2 < b^2$. | If $0 < \sqrt{x} < 4$, then $0 < (\sqrt{x})^2 < 4^2$, or $0 < x < 16$. |
| (3) If $0 < a < b$, then $0 < \sqrt{a} < \sqrt{b}$. | If $0 < x^2 < 4$, then $0 < \sqrt{x^2} < \sqrt{4}$, or $0 < |x| < 2$. |

*PROOFS*

**(1)** If $0 < a < b$, then (by multiplying by $1/(ab)$)

$$a \cdot \frac{1}{ab} < b \cdot \frac{1}{ab}, \quad \text{or} \quad \frac{1}{b} < \frac{1}{a}; \quad \text{that is,} \quad \frac{1}{a} > \frac{1}{b}.$$

**(2)** If $0 < a < b$, then (by multiplying by $a$) $a \cdot a < a \cdot b$ and (by multiplying by $b$) $b \cdot a < b \cdot b$, so $a^2 < ab < b^2$ and hence $a^2 < b^2$.

**(3)** If $0 < a < b$, then $b - a > 0$, or equivalently,

$$(\sqrt{b} + \sqrt{a})(\sqrt{b} - \sqrt{a}) > 0.$$

Dividing both sides of the last inequality by $\sqrt{b} + \sqrt{a}$, we obtain $\sqrt{b} - \sqrt{a} > 0$; that is, $\sqrt{b} > \sqrt{a}$.  ∎

## 1.6 EXERCISES

**Exer. 1–4:** Express the inequality as an interval, and sketch its graph.

1  $x < -2$

2  $x \geq 4$

3  $-3 \geq x > -5$

4  $3 \leq x \leq 7$

**Exer. 5–6:** Express the interval as an inequality in the variable $x$.

5  $(-5, 8]$

6  $(-3, \infty)$

**Exer. 7–50:** Solve the inequality, and express the solutions in terms of intervals whenever possible.

7  $2x + 5 < 3x - 7$

8  $x - 8 > 5x + 3$

9  $3 \leq \dfrac{2x - 3}{5} < 7$

10  $-2 < \dfrac{4x + 1}{3} \leq 0$

11  $\dfrac{4}{3x + 2} \geq 0$

12  $\dfrac{3}{2x + 5} \leq 0$

13  $\dfrac{-2}{4 - 3x} > 0$

14  $\dfrac{-3}{2 - x} < 0$

15  $\dfrac{2}{(1 - x)^2} > 0$

16  $\dfrac{4}{x^2 + 4} < 0$

17  $|x + 3| < 0.01$

18  $|x - 4| \leq 0.03$

19  $|3x - 7| \geq 5$

20  $-\frac{1}{3}|6 - 5x| + 2 \leq 1$

21  $|7x + 2| > -2$

22  $|6x - 5| \leq -2$

23  $|3x - 9| > 0$

24  $|5x + 2| \leq 0$

25  $-2 < |x| < 4$

26  $1 < |x| < 5$

27  $(3x + 1)(5 - 10x) > 0$

28  $(x + 2)(x - 1)(4 - x) \leq 0$

29  $x^2 - x - 6 < 0$

30  $x^2 + 4x + 3 \geq 0$

31  $x(2x + 3) \geq 5$

32  $6x - 8 > x^2$

33  $25x^2 - 9 < 0$

34  $25x^2 - 9x < 0$

35  $\dfrac{x^2(x + 2)}{(x + 2)(x + 1)} \leq 0$

36  $\dfrac{(x^2 + 1)(x - 3)}{x^2 - 9} \geq 0$

37  $\dfrac{x^2 - x}{x^2 + 2x} \leq 0$

38  $\dfrac{(x + 3)^2(2 - x)}{(x + 4)(x^2 - 4)} \leq 0$

39  $\dfrac{x - 2}{x^2 - 3x - 10} \geq 0$

40  $\dfrac{x + 5}{x^2 - 7x + 12} \leq 0$

41  $\dfrac{-3x}{x^2 - 9} > 0$

42  $\dfrac{2x}{16 - x^2} < 0$

43  $\dfrac{x + 1}{2x - 3} > 2$

44  $\dfrac{x - 2}{3x + 5} \leq 4$

45  $\dfrac{1}{x - 2} \geq \dfrac{3}{x + 1}$

46  $\dfrac{2}{2x + 3} \leq \dfrac{2}{x - 5}$

47  $\dfrac{x}{3x - 5} \leq \dfrac{2}{x - 1}$

48  $\dfrac{x}{2x - 1} \geq \dfrac{3}{x + 2}$

49  $x^3 > x$

50  $x^4 \geq x^2$

**Exer. 51–52:** Solve part (a), and use that answer to determine the answers to parts (b) and (c).

51  (a) $|x + 5| = 3$   (b) $|x + 5| < 3$   (c) $|x + 5| > 3$

52  (a) $|x - 3| < 2$   (b) $|x - 3| = 2$   (c) $|x - 3| > 2$

**Exer. 53–54:** Express the statement in terms of an inequality involving an absolute value.

53  The weight $w$ of a wrestler must be within 2 pounds of 148 pounds.

54  The radius $r$ of a ball bearing must be within 0.01 centimeter of 1 centimeter.

55  **Linear magnification**  Shown in the figure is a simple magnifier consisting of a convex lens. The object to be magnified is positioned so that the distance $p$ from the lens is less than the focal length $f$. The linear magnification $M$ is the ratio of the image size to the object size. It is shown in physics that $M = f/(f - p)$. If $f = 6$ cm, how far should the object be placed from the lens so that its image appears at least three times as large? (Compare with Example 4.)

**EXERCISE 55**

**56** *Drug concentration* To treat arrhythmia (irregular heartbeat), a drug is fed intravenously into the bloodstream. Suppose that the concentration $c$ of the drug after $t$ hours is given by $c = 3.5t/(t + 1)$ mg/L. If the minimum therapeutic level is 1.5 mg/L, determine when this level is exceeded.

**57** *Business expenditure* A construction firm is trying to decide which of two models of a crane to purchase. Model A costs $50,000 and requires $4000 per year to maintain. Model B has an initial cost of $40,000 and a maintenance cost of $5500 per year. For how many years must model A be used before it becomes more economical than B?

**58** *Buying a car* A consumer is trying to decide whether to purchase car A or car B. Car A costs $10,000 and has an mpg rating of 30, and insurance on it is $550 per year. Car B costs $12,000 and has an mpg rating of 50, and insurance on it is $600 per year. Assume that the consumer drives 15,000 miles per year and that the price of gas remains constant at $1.25 per gallon. Based only on these facts, determine how long it will take for the total cost of car B to become less than that of car A.

**59** *Vertical leap record* *Guinness Book of World Records* reports that German shepherds can make vertical leaps of over 10 feet when scaling walls. If the distance $s$ (in feet) off the ground after $t$ seconds is given by the equation $s = -16t^2 + 24t + 1$, for how many seconds is the dog more than 9 feet off the ground?

**60** *Height of a projected object* If an object is projected vertically upward from ground level with an initial velocity of 320 ft/sec, then its distance $s$ above the ground after $t$ seconds is given by $s = -16t^2 + 320t$. For what values of $t$ will the object be more than 1536 feet above the ground?

**61** *Braking distance* The braking distance $d$ (in feet) of a certain car traveling $v$ mi/hr is given by the equation $d = v + (v^2/20)$. Determine the velocities that result in braking distances of less than 75 feet.

**62** *Gas mileage* The number of miles $M$ that a certain compact car can travel on 1 gallon of gasoline is related to its speed $v$ (in mi/hr) by

$$M = -\tfrac{1}{30}v^2 + \tfrac{5}{2}v \quad \text{for} \quad 0 < v < 70.$$

For what speeds will $M$ be at least 45?

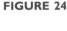

## RECTANGULAR COORDINATE SYSTEMS

**FIGURE 24**

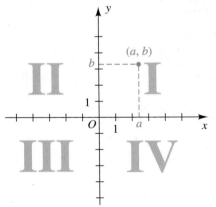

In Section 1.1 we discussed how to assign a real number (coordinate) to each point on a line. We shall now show how to assign an **ordered pair** $(a, b)$ of real numbers to each point in a plane. Although we have also used the notation $(a, b)$ to denote an open interval, there is little chance for confusion, since it should always be clear from our discussion whether $(a, b)$ represents a point or an interval.

We introduce a **rectangular,** or **Cartesian,*** **coordinate system** in a plane by means of two perpendicular coordinate lines, called **coordinate axes,** that intersect at the **origin** $O$, as shown in Figure 24. We often refer to the horizontal line as the **x-axis** and the vertical line as the **y-axis** and label them $x$ and $y$, respectively. The plane is then a **coordinate plane,** or an **xy-plane.** The coordinate axes divide the plane into four parts called the **first, second, third,** and **fourth quadrants,** labeled I, II, III, and IV, respectively (see Figure 24). Points on the axes do not belong to any quadrant.

Each point $P$ in an $xy$-plane may be assigned an ordered pair $(a, b)$, as shown in Figure 24. We call $a$ the **x-coordinate** (or **abscissa**) of $P$, and $b$

*The term *Cartesian* is used in honor of the French mathematician and philosopher René Descartes (1596–1650), who was one of the first to employ such coordinate systems.

the **y-coordinate** (or **ordinate**). We say that $P$ *has coordinates* $(a, b)$ and refer to the *point* $(a, b)$, or the *point* $P(a, b)$. Conversely, every ordered pair $(a, b)$ determines a point $P$ with coordinates $a$ and $b$. We **plot a point** by using a dot, as illustrated in Figure 25.

**FIGURE 25**

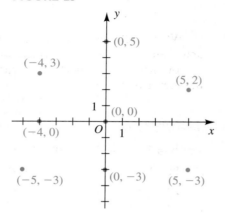

We may use the following formula to find the distance between two points in a coordinate plane.

**Distance Formula**

> The distance $d(P_1, P_2)$ between any two points $P_1(x_1, y_1)$ and $P_2(x_2, y_2)$ in a coordinate plane is
> $$d(P_1, P_2) = \sqrt{(x_2 - x_1)^2 + (y_2 - y_1)^2}.$$

**FIGURE 26**

**PROOF**   If $x_1 \neq x_2$ and $y_1 \neq y_2$, then, as illustrated in Figure 26, the points $P_1$, $P_2$, and $P_3(x_2, y_1)$ are vertices of a right triangle. By the Pythagorean theorem,

$$[d(P_1, P_2)]^2 = [d(P_1, P_3)]^2 + [d(P_3, P_2)]^2.$$

From the figure we see that

$$d(P_1, P_3) = |x_2 - x_1| \quad \text{and} \quad d(P_3, P_2) = |y_2 - y_1|.$$

Since $|a|^2 = a^2$ for every real number $a$, we may write

$$[d(P_1, P_2)]^2 = (x_2 - x_1)^2 + (y_2 - y_1)^2.$$

Taking the square root of each side of the last equation and using the fact that $d(P_1, P_2) \geq 0$ gives us the distance formula.

If $y_1 = y_2$, the points $P_1$ and $P_2$ lie on the same horizontal line, and

$$d(P_1, P_2) = |x_2 - x_1| = \sqrt{(x_2 - x_1)^2}.$$

Similarly, if $x_1 = x_2$, the points are on the same vertical line, and
$$d(P_1, P_2) = |y_2 - y_1| = \sqrt{(y_2 - y_1)^2}.$$
These are special cases of the distance formula.

Although we referred to the points shown in Figure 26, our proof is independent of the positions of $P_1$ and $P_2$. ▬

When applying the distance formula, note that $d(P_1, P_2) = d(P_2, P_1)$ and, hence, the order in which we subtract the $x$-coordinates and the $y$-coordinates of the points is immaterial. We may think of the distance between two points as the length of the hypotenuse of a right triangle.

## ILLUSTRATION

### Finding the Distance Between Two Points $A$ and $B$

▬ For $A(0, 0)$ and $B(4, 3)$, $d(A, B) = \sqrt{(4 - 0)^2 + (3 - 0)^2} = \sqrt{25} = 5$.

▬ For $A(-3, 6)$ and $B(5, 1)$, $d(A, B) = \sqrt{(5 + 3)^2 + (1 - 6)^2} = \sqrt{89}$
$$\approx 9.43.$$

### EXAMPLE I  *Applying the distance formula*

Given $A(1, 7)$, $B(-3, 2)$, and $C(4, \frac{1}{2})$, prove that $C$ is on the perpendicular bisector of segment $AB$.

**FIGURE 27**

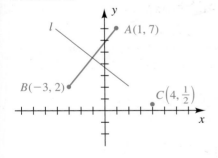

**SOLUTION**  The points $A$, $B$, $C$ and the *perpendicular bisector l* are illustrated in Figure 27. From plane geometry, $l$ can be characterized by either of the following conditions:

**(1)** $l$ is the line perpendicular to segment $AB$ at its midpoint.

**(2)** $l$ is the set of all points equidistant from the end points of segment $AB$.

We shall use condition (2) to show that $C$ is on $l$ by verifying that
$$d(A, C) = d(B, C).$$
We apply the distance formula:
$$d(A, C) = \sqrt{(4 - 1)^2 + (\tfrac{1}{2} - 7)^2} = \sqrt{9 + \tfrac{169}{4}} = \sqrt{\tfrac{205}{4}}$$
$$d(B, C) = \sqrt{[4 - (-3)]^2 + (\tfrac{1}{2} - 2)^2} = \sqrt{49 + \tfrac{9}{4}} = \sqrt{\tfrac{205}{4}}$$
Thus, $C$ is equidistance from $A$ and $B$, and the verification is complete.

In Example 9 of Section 1.8, we will find a formula for the perpendicular bisector of a segment using condition (1) of Example 1.

We can find the midpoint of a line segment by using the following formula.

**Midpoint Formula**

> The midpoint $M$ of the line segment from $P_1(x_1, y_1)$ to $P_2(x_2, y_2)$ is
> $$\left(\frac{x_1 + x_2}{2}, \frac{y_1 + y_2}{2}\right).$$

**PROOF**    The lines through $P_1$ and $P_2$ parallel to the $y$-axis intersect the $x$-axis at $A_1(x_1, 0)$ and $A_2(x_2, 0)$. From plane geometry, the line through the midpoint $M$ parallel to the $y$-axis bisects the segment $A_1A_2$ at point $M_1$ (see Figure 28). If $x_1 < x_2$, then $x_2 - x_1 > 0$, and hence $d(A_1, A_2) = x_2 - x_1$. Since $M_1$ is halfway from $A_1$ to $A_2$, the $x$-coordinate of $M_1$ is equal to the $x$-coordinate of $A_1$ plus one-half the distance from $A_1$ to $A_2$; that is,

$$x\text{-coordinate of } M_1 = x_1 + \tfrac{1}{2}(x_2 - x_1).$$

The expression on the right side of the last equation simplifies to

$$\frac{x_1 + x_2}{2}.$$

This quotient is the *average* of the numbers $x_1$ and $x_2$. It follows that the $x$-coordinate of $M$ is also $(x_1 + x_2)/2$. Similarly, the $y$-coordinate of $M$ is $(y_1 + y_2)/2$. These formulas hold for all positions of $P_1$ and $P_2$. ∎

**FIGURE 28**

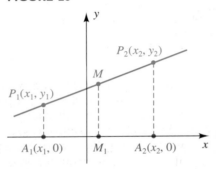

To apply the midpoint formula, it may suffice to remember that

the $x$-coordinate of the midpoint = the *average* of the $x$-coordinates

and that

the $y$-coordinate of the midpoint = the *average* of the $y$-coordinates.

## ILLUSTRATION

### Finding the Midpoint $M$ of Two Points $A$ and $B$

■ For $A(-2, 3)$ and $B(4, -2)$, $M = \left(\dfrac{-2 + 4}{2}, \dfrac{3 + (-2)}{2}\right) = \left(1, \dfrac{1}{2}\right).$

Graphs are often used to illustrate changes in quantities. A graph in the business section of a newspaper may show the fluctuation of the Dow-Jones average during a given month; a meteorologist might use a graph to indicate how the air temperature varied throughout a day; a cardiologist employs graphs (electrocardiograms) to analyze heart irregularities; an engineer or physicist may turn to a graph to illustrate the manner in which the pressure of a confined gas increases as the gas is heated. Such visual aids usually reveal the behavior of quantities more readily than a long table of numerical values.

Two quantities are sometimes related by means of an equation or formula that involves two variables. In this section we discuss how to represent such an equation geometrically, by a graph in a coordinate plane. The graph may then be used to discover properties of the quantities that are not evident from the equation alone. The following chart introduces the basic concept of the graph of an equation in two variables $x$ and $y$. Of course, other letters can also be used for the variables.

| Terminology | Definition | Illustration |
|---|---|---|
| **Solution** of an equation in $x$ and $y$ | An ordered pair $(a, b)$ that yields a true statement if $x = a$ and $y = b$ | $(2, 3)$ is a solution of $y^2 = 5x - 1$, since substituting $x = 2$ and $y = 3$ gives us LS: $3^2 = 9$ RS: $5(2) - 1 = 10 - 1 = 9$ |

For each solution $(a, b)$ of an equation in $x$ and $y$ there is a point $P(a, b)$ in a coordinate plane. The set of all such points is called the **graph** of the equation. To *sketch the graph of an equation,* we illustrate the significant features of the graph in a coordinate plane. In simple cases, a graph can be sketched by plotting few, if any, points. For a complicated equation, plotting points may give very little information about the graph. In such cases, methods of calculus or computer graphics are often employed. Let us begin with a simple example.

### EXAMPLE 2   *Sketching a simple graph by plotting points*

Sketch the graph of the equation $y = 2x - 1$.

**FIGURE 29**

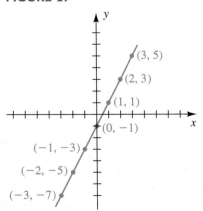

**SOLUTION**   We wish to find the points $(x, y)$ in a coordinate plane that correspond to the solutions of the equation. It is convenient to list coordinates of several such points in a table, where for each $x$ we obtain the value for $y$ from $y = 2x - 1$:

| $x$ | $-3$ | $-2$ | $-1$ | $0$ | $1$ | $2$ | $3$ |
|---|---|---|---|---|---|---|---|
| $y$ | $-7$ | $-5$ | $-3$ | $-1$ | $1$ | $3$ | $5$ |

The points with these coordinates appear to lie on a line, and we can sketch the graph in Figure 29. Ordinarily, the few points we have plotted would not be enough to illustrate the graph of an equation; however, in this elementary case we can be reasonably sure that the graph is a line. In the next section we will establish this fact.

It is impossible to sketch the entire graph in Example 2, because we can assign values to $x$ that are numerically as large as desired. Nevertheless, we call the drawing in Figure 29 *the graph of the equation* or *a sketch of the graph*. In general, the sketch of a graph should illustrate its essential features so that the remaining (unsketched) parts are self-evident. If a graph terminates at some point (as would be the case for a half-line or line segment), we place a dot at the appropriate *end point* of the graph. As a final general remark, *if ticks on the coordinate axes are not labeled* (as in Figure 29), *then each tick represents one unit*. We shall label ticks only when different units are used on the axes. For *arbitrary* graphs, where units of measurement are irrelevant, we omit ticks completely (see, for example, Figures 33 and 34).

**EXAMPLE 3**     *Sketching the graph of an equation*

Sketch the graph of the equation $y = x^2 - 3$.

**FIGURE 30**

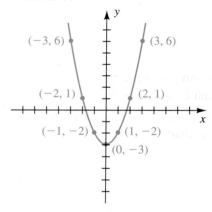

**SOLUTION**     Substituting values for $x$ and finding the corresponding values of $y$ using $y = x^2 - 3$, we obtain a table of coordinates for several points on the graph:

| $x$ | $-3$ | $-2$ | $-1$ | $0$ | $1$ | $2$ | $3$ |
|-----|------|------|------|-----|-----|-----|-----|
| $y$ | $6$ | $1$ | $-2$ | $-3$ | $-2$ | $1$ | $6$ |

Larger values of $|x|$ produce larger values of $y$. For example, the points $(4, 13)$, $(5, 22)$, and $(6, 33)$ are on the graph, as are $(-4, 13)$, $(-5, 22)$, and $(-6, 33)$. Plotting the points given by the table and drawing a smooth curve through these points (in the order of increasing values of $x$) gives us the sketch in Figure 30.

The graph in Figure 30 is a **parabola,** and the $y$-axis is the **axis of the parabola.** The lowest point $(0, -3)$ is the **vertex** of the parabola, and we say that the parabola **opens upward.** If we invert the graph, then the parabola **opens downward** and the vertex is the highest point on the graph. In general, the graph of *any* equation of the form $y = ax^2 + c$ with $a \neq 0$ is a parabola with vertex $(0, c)$, opening upward if $a > 0$ or downward if $a < 0$. If $c = 0$, the equation reduces to $y = ax^2$ and the vertex is at the origin $(0, 0)$. Parabolas may also open to the right or to the left (see Example 5) or in other directions.

We shall use the following terminology to describe where the graph of an equation in $x$ and $y$ intersects the $x$-axis or the $y$-axis.

**Intercepts of the Graph of an Equation in $x$ and $y$**

| Terminology | Definition | Graphical interpretation | How to find |
|---|---|---|---|
| $x$-intercepts | The $x$-coordinates of points where the graph intersects the $x$-axis | | Let $y = 0$, and solve for $x$. |
| $y$-intercepts | The $y$-coordinates of points where the graph intersects the $y$-axis | | Let $x = 0$, and solve for $y$. |

**EXAMPLE 4**    *Finding x-intercepts and y-intercepts*

Find the $x$- and $y$-intercepts of the graph of $y = x^2 - 3$.

**SOLUTION**    The graph is sketched in Figure 30 (Example 3). We find the intercepts as stated in the preceding chart.

**(1)** *x-intercepts:*

$$y = x^2 - 3 \qquad \text{given}$$
$$0 = x^2 - 3 \qquad \text{let } y = 0$$
$$x^2 = 3 \qquad \text{equivalent equation}$$
$$x = \pm\sqrt{3} \approx \pm 1.73 \quad \text{take the square root}$$

Thus, the $x$-intercepts are $-\sqrt{3}$ and $\sqrt{3}$. The points at which the graph crosses the $x$-axis are $(-\sqrt{3}, 0)$ and $(\sqrt{3}, 0)$.

**(2)** *y-intercepts:*

$$y = x^2 - 3 \qquad \text{given}$$
$$y = 0 - 3 = -3 \quad \text{let } x = 0$$

Thus, the $y$-intercept is $-3$, and the point at which the graph crosses the $y$-axis is $(0, -3)$.

If the coordinate plane in Figure 30 is folded along the $y$-axis, the graph that lies in the left half of the plane coincides with that in the right half, and we say that **the graph is symmetric with respect to the y-axis.** A graph is symmetric with respect to the $y$-axis provided that the point $(-x, y)$ is on the graph whenever $(x, y)$ is on the graph. The graph of $y = x^2 - 3$ in Example 3 has this property, since substitution of $-x$ for $x$ yields the same equation:

$$y = (-x)^2 - 3 = x^2 - 3$$

This substitution is an application of symmetry test 1 in the following chart. Two other types of symmetry and the appropriate tests are also listed. The graphs of $x = y^2$ and $4y = x^3$ in the illustration column are discussed in Examples 5 and 6, respectively.

## Symmetries of Graphs of Equations in $x$ and $y$

| Terminology | Graphical interpretation | Test for symmetry | Illustration |
|---|---|---|---|
| The graph is symmetric with respect to the $y$-axis. | | (1) Substitution of $-x$ for $x$ leads to the same equation. | |
| The graph is symmetric with respect to the $x$-axis. | | (2) Substitution of $-y$ for $y$ leads to the same equation. | |
| The graph is symmetric with respect to the origin. | | (3) Simultaneous substitution of $-x$ for $x$ and $-y$ for $y$ leads to the same equation. | |

If a graph is symmetric with respect to an axis, it is sufficient to determine the graph in half of the coordinate plane, since we can sketch the remainder of the graph by taking a *mirror image*, or *reflection*, through the appropriate axis.

**EXAMPLE 5** *A graph that is symmetric with respect to the x-axis*

Sketch the graph of the equation $y^2 = x$.

**SOLUTION** Since substitution of $-y$ for $y$ does not change the equation, the graph is symmetric with respect to the $x$-axis (see symmetry test 2). Hence, if the point $(x, y)$ is on the graph, then the point $(x, -y)$ is on the graph. Thus, it is sufficient to find points with nonnegative $y$-coordinates and then reflect through the $x$-axis. The equation $y^2 = x$ is equivalent to $y = \pm\sqrt{x}$. The $y$-coordinates of points *above* the $x$-axis ($y$ is *positive*) are given by $y = \sqrt{x}$, whereas the $y$-coordinates of points *below* the $x$-axis ($y$ is *negative*) are given by $y = -\sqrt{x}$. Coordinates of some points on the graph are listed below. The graph is sketched in Figure 31.

**FIGURE 31**

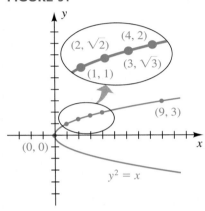

| $x$ | 0 | 1 | 2 | 3 | 4 | 9 |
|---|---|---|---|---|---|---|
| $y$ | 0 | 1 | $\sqrt{2} \approx 1.4$ | $\sqrt{3} \approx 1.7$ | 2 | 3 |

The graph is a parabola that opens to the right, with its vertex at the origin. In this case, the $x$-axis is the axis of the parabola.

**EXAMPLE 6** *A graph that is symmetric with respect to the origin*

Sketch the graph of the equation $4y = x^3$.

**SOLUTION** If we simultaneously substitute $-x$ for $x$ and $-y$ for $y$, then

$$4(-y) = (-x)^3 \quad \text{or, equivalently,} \quad -4y = -x^3.$$

Multiplying both sides by $-1$, we see that the last equation has the same solutions as the equation $4y = x^3$. Hence, from symmetry test 3, the graph is symmetric with respect to the origin, and if the point $(x, y)$ is on the graph, then the point $(-x, -y)$ is on the graph. The following table lists coordinates of some points on the graph.

**FIGURE 32**

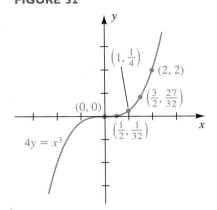

| $x$ | 0 | $\frac{1}{2}$ | 1 | $\frac{3}{2}$ | 2 | $\frac{5}{2}$ |
|---|---|---|---|---|---|---|
| $y$ | 0 | $\frac{1}{32}$ | $\frac{1}{4}$ | $\frac{27}{32}$ | 2 | $\frac{125}{32}$ |

Because of the symmetry, we can see that the points $(-1, -\frac{1}{4})$, $(-2, -2)$, and so on, are also on the graph. The graph is sketched in Figure 32.

If $C(h, k)$ is a point in a coordinate plane, then a circle with center $C$ and radius $r > 0$ consists of all points in the plane that are $r$ units from $C$. As shown in Figure 33, a point $P(x, y)$ is on the circle provided $d(C, P) = r$ or, by the distance formula,

$$\sqrt{(x - h)^2 + (y - k)^2} = r.$$

The above equation is equivalent to the following equation.

**Standard Equation of a Circle with Center $(h, k)$ and Radius $r$**

$$(x - h)^2 + (y - k)^2 = r^2$$

**FIGURE 33**

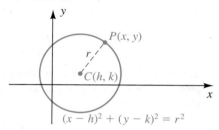

$$(x - h)^2 + (y - k)^2 = r^2$$

If $h = 0$ and $k = 0$, this equation reduces to $x^2 + y^2 = r^2$, which is an equation of a circle of radius $r$ with center at the origin (see Figure 34). If $r = 1$, we call the graph a **unit circle.**

**FIGURE 34**

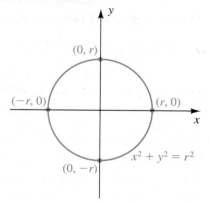

$$x^2 + y^2 = r^2$$

**EXAMPLE 7**    *Finding an equation of a circle*

Find an equation of the circle that has center $C(-2, 3)$ and contains the point $D(4, 5)$.

**FIGURE 35**

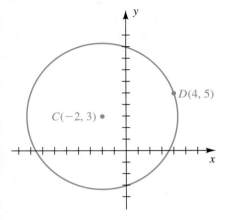

**SOLUTION** The circle is shown in Figure 35. Since $D$ is on the circle, the radius $r$ is $d(C, D)$. By the distance formula,

$$r = \sqrt{(4 + 2)^2 + (5 - 3)^2} = \sqrt{36 + 4} = \sqrt{40}.$$

Using the standard equation of a circle with $h = -2$, $k = 3$, and $r = \sqrt{40}$, we obtain

$$(x + 2)^2 + (y - 3)^2 = 40.$$

By squaring terms and simplifying the last equation, we may write it as

$$x^2 + y^2 + 4x - 6y - 27 = 0.$$

As in the solution to Example 7, squaring terms of an equation of the form $(x - h)^2 + (y - k)^2 = r^2$ and simplifying leads to an equation of the form

$$x^2 + y^2 + ax + by + c = 0,$$

where $a$, $b$, and $c$ are real numbers. Conversely, if we begin with this equation, it is always possible, by *completing squares,* to obtain an equation of the form

$$(x - h)^2 + (y - k)^2 = d.$$

This method will be illustrated in Example 8. If $d > 0$, the graph is a circle with center $(h, k)$ and radius $r = \sqrt{d}$. If $d = 0$, the graph consists of only one point $(h, k)$. Finally, if $d < 0$, the equation has no real solutions, and hence there is no graph.

**EXAMPLE 8** *Finding the center and radius of a circle*

Find the center and radius of the circle with equation

$$3x^2 + 3y^2 - 12x + 18y = 9.$$

**SOLUTION** Since it is easier to complete the square if the coefficients of $x^2$ and $y^2$ are 1, we begin by dividing the given equation by 3, obtaining

$$x^2 + y^2 - 4x + 6y = 3.$$

Next, we rewrite the equation as follows:

$$(x^2 - 4x + \_\_) + (y^2 + 6y + \_\_) = 3 + \_\_ + \_\_,$$

where the underscored spaces represent numbers to be determined. We then complete the squares for the expressions within parentheses, taking care to add the appropriate numbers to *both* sides of the equation. To complete the square for an expression of the form $x^2 + ax$, we add the square of half the coefficient of $x$ (that is, $(a/2)^2$) to both sides of the equation. Similarly, for $y^2 + by$, we add $(b/2)^2$ to both sides. In this example, $a = -4$, $b = 6$, $(a/2)^2 = (-2)^2 = 4$, and $(b/2)^2 = 3^2 = 9$. These additions lead to

$$(x^2 - 4x + \underline{4}) + (y^2 + 6y + \underline{9}) = 3 + \underline{4} + \underline{9} \qquad \text{completing the squares}$$
$$(x - 2)^2 + (y + 3)^2 = 16. \qquad \text{equivalent equation}$$

Comparing the last equation with the standard equation of a circle, we see that $h = 2$ and $k = -3$ and conclude that the circle has center $(2, -3)$ and radius $\sqrt{16} = 4$.

In some applications it is necessary to work with only one-half of a circle—that is, a **semicircle**. The next example indicates how to find equations for semicircles for circles with centers at the origin.

**EXAMPLE 9**   *Finding equations of semicircles*

Find equations for the upper half, lower half, right half, and left half of the circle $x^2 + y^2 = 81$.

**SOLUTION**   The graph of $x^2 + y^2 = 81$ is a circle of radius 9 with center at the origin (compare with Figure 34). To find equations for the upper and lower halves, we solve for $y$ in terms of $x$:

$$x^2 + y^2 = 81 \qquad \text{given}$$
$$y^2 = 81 - x^2 \qquad \text{subtract } x^2$$
$$y = \pm\sqrt{81 - x^2} \qquad \text{take the square root}$$

Since $\sqrt{81 - x^2} \geq 0$, it follows that the upper half of the circle has the equation $y = \sqrt{81 - x^2}$ ($y$ is positive) and the lower half is given by $y = -\sqrt{81 - x^2}$ ($y$ is negative), as illustrated in Figure 36(a) and (b).

**FIGURE 36**

(a) $y = \sqrt{81 - x^2}$

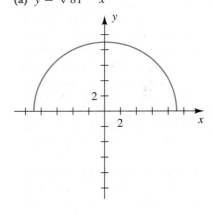

(b) $y = -\sqrt{81 - x^2}$

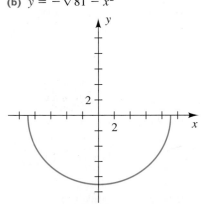

(c) $x = \sqrt{81 - y^2}$

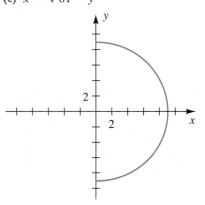

(d) $x = -\sqrt{81 - y^2}$

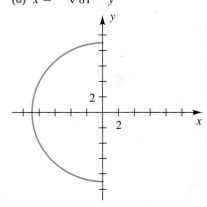

Similarly, to find equations for the right and left halves, we solve $x^2 + y^2 = 81$ for $x$ in terms of $y$, obtaining

$$x = \pm\sqrt{81 - y^2}.$$

Since $\sqrt{81 - y^2} \geq 0$, it follows that the right half of the circle has the equation $x = \sqrt{81 - y^2}$ ($x$ is positive), and the left half is given by $x = -\sqrt{81 - y^2}$ ($x$ is negative), as illustrated in Figure 36(c) and (d).

---

Engineers, scientists, and mathematicians often use graphics calculators or computer software to obtain sketches of graphs. These devices have the capability to *zoom in* (or *zoom out*) on any part of a graph, allowing the *user* to approximate $x$-intercepts, $y$-intercepts, high points, low points, and other important aspects of the graph. Because of the great variety of graphics calculators and computer software, we shall not discuss methods of using any specific type. User's manuals usually give adequate instructions.

The term **graphing utility** refers to either a graphics calculator or a computer equipped with appropriate software packages. The **viewing rectangle** of a graphing utility is the portion of the $xy$-plane shown on the screen. The boundaries (sides) of the viewing rectangle can be manually set by assigning a minimum $x$ value (Xmin), a maximum $x$ value (Xmax), a minimum $y$ value (Ymin), and a maximum $y$ value (Ymax). In examples, we often use the standard (or default) values for the viewing rectangle. These values depend on the dimensions (measured in pixels) of the graphing utility screen. If we want a different view of the graph, we use the phrase ''using [Xmin, Xmax] by [Ymin, Ymax]'' to indicate the change in the viewing rectangle.

In many applications it is essential to find the points at which the graphs of two equations in $x$ and $y$ intersect. To approximate such points of intersection with a graphing utility, it is often necessary to solve each equation for $y$ in terms of $x$. For example, suppose one equation is

$$4x^2 - 3x + 2y + 6 = 0.$$

Solving for $y$ gives us

$$y = \frac{-4x^2 + 3x - 6}{2} = -2x^2 + \frac{3}{2}x - 3.$$

The graph of the equation is then found by *making the assignment*

$$Y_1 = -2x^2 + \tfrac{3}{2}x - 3$$

in the graphing utility. (The symbol $Y_1$ indicates the *first* equation, or the first $y$ value.) We also solve the second equation for $y$ in terms of $x$ and make the assignment

$$Y_2 = \text{an expression in } x.$$

Pressing appropriate keys gives us sketches of the graphs, which we refer to as the graphs of $Y_1$ and $Y_2$. We then use the zoom-in and *tracing* features of the graphing utility to estimate the coordinates of the points of intersection. When estimating coordinates in examples, we usually find one-decimal-place approximations unless otherwise specified.

In the next example we demonstrate this technique for the graphs discussed in Examples 2 and 3.

**EXAMPLE 10**    *Estimating points of intersection of graphs*

Use a graphing utility to estimate the points of intersection of the graphs of $y = 2x - 1$ and $y = x^2 - 3$.

**SOLUTION**    Some specific keystrokes for the TI-81/82 graphics calculator are given in Appendix I. We first make the assignments

$$Y_1 = 2x - 1 \quad \text{and} \quad Y_2 = x^2 - 3.$$

**FIGURE 37**  $[-15, 15]$ by $[-10, 10]$

Using a standard viewing rectangle, $[-15, 15]$ by $[-10, 10]$, we see from the graphs of $Y_1$ and $Y_2$ in Figure 37 that there are two points of intersection: $P_1$ in quadrant I and $P_2$ in quadrant III.

We next either manually move the cursor or use a tracing feature (consult your user's manual for specific directions) to get close to $P_1$. Using the zoom-in feature, we estimate the coordinates of $P_1$ as $(2.7, 4.5)$.

We can now either zoom out or redraw the original graphs to view $P_2$. Using the tracing and zoom-in features, we obtain $(-0.7, -2.5)$ as approximate coordinates of $P_2$.

**EXAMPLE 11**    *Estimating points of intersection of graphs*

Use a graphing utility to estimate the points of intersection of the circles $x^2 + y^2 = 25$ and $x^2 + y^2 - 4y = 12$.

**SOLUTION**    Some specific keystrokes for the TI-81/82 graphics calculator are given in Appendix I. As in Example 9, we solve $x^2 + y^2 = 25$ for $y$ in terms of $x$, obtaining

$$y = \pm\sqrt{25 - x^2}.$$

To display the entire circle on the graphing utility, we make the following assignments:

$$Y_1 = \sqrt{25 - x^2} \quad \text{and} \quad Y_2 = -Y_1$$

(We often assign $Y_2$ in terms of $Y_1$ to avoid repetitive key stroking.)

At this point you should graph $Y_1$ and $Y_2$ using a standard viewing rectangle. If the circle has an oval shape, consult your user's manual to find out which dimensions for the viewing rectangle will yield a more circular shape.

We may regard the equation of the second circle as a quadratic equation $ay^2 + by + c = 0$ in $y$ by rearranging terms as follows:

$$y^2 - 4y + (x^2 - 12) = 0$$

Applying the quadratic formula with $a = 1$, $b = -4$, and $c = x^2 - 12$ ($x^2 - 12$ is considered to be the constant term, since it does not contain the variable $y$) gives us

$$y = \frac{-(-4) \pm \sqrt{(-4)^2 - 4(1)(x^2 - 12)}}{2(1)}$$

$$= \frac{4 \pm \sqrt{16 - 4(x^2 - 12)}}{2} = \frac{4 \pm 2\sqrt{4 - (x^2 - 12)}}{2}$$

$$= 2 \pm \sqrt{16 - x^2}.$$

(It is unnecessary to simplify the equation as much as we have, but the simplified form is easier to enter in a graphing utility.)

We now make the assignments

$$Y_3 = \sqrt{16 - x^2}, \quad Y_4 = 2 + Y_3, \quad \text{and} \quad Y_5 = 2 - Y_3.$$

(If a $Y_5$ is not available on your graphing utility, consult the user's manual for information about displaying additional graphs on the screen.) We then select $Y_1$, $Y_2$, $Y_4$, and $Y_5$ to be graphed, obtaining a display similar to Figure 38. There are two points of intersection. Zooming in on the point in the first quadrant, we estimate its coordinates as $(3.8, 3.25)$. Since both circles are symmetric with respect to the $y$-axis, the other point of intersection is approximately $(-3.8, 3.25)$.

**FIGURE 38** $[-15, 15]$ by $[-10, 10]$

It should be noted that the approximate solutions found in Examples 10 and 11 do not satisfy the given equations because of the inaccuracy of the estimates made from the graph. In Chapter 8 we will discuss how to find the *exact* values for the points of intersection.

## 1.7 EXERCISES

Exer. 1–2: Describe the set of all points $P(x, y)$ in a coordinate plane that satisfy the given condition.

1 (a) $x = -2$    (b) $y = 3$    (c) $x \geq 0$
   (d) $xy > 0$    (e) $y < 0$    (f) $x = 0$

2 (a) $y = -2$    (b) $x = -4$    (c) $x/y < 0$
   (d) $xy = 0$    (e) $y > 1$    (f) $y = 0$

**Exer. 3–4: (a)** Find the distance $d(A, B)$ between $A$ and $B$. **(b)** Find the midpoint of the segment $AB$.

3  $A(4, -3)$,   $B(6, 2)$        4  $A(-2, -5)$,   $B(4, 6)$

**Exer. 5–6:** Show that the triangle with vertices $A$, $B$, and $C$ is a right triangle, and find its area.

5

6

7  Show that $A(-4, 2)$, $B(1, 4)$, $C(3, -1)$, and $D(-2, -3)$ are vertices of a square.

8  Show that $A(-4, -1)$, $B(0, -2)$, $C(6, 1)$, and $D(2, 2)$ are vertices of a parallelogram.

9  Given $A(-3, 8)$, find the coordinates of the point $B$ such that $C(5, -10)$ is the midpoint of segment $AB$.

10  Given $A(5, -8)$ and $B(-6, 2)$, find the point on segment $AB$ that is three-fourths of the way from $A$ to $B$.

**Exer. 11–12:** Prove that $C$ is on the perpendicular bisector of segment $AB$.

11  $A(-4, -3)$,   $B(6, 1)$,       $C(5, -11)$

12  $A(-3, 2)$,      $B(5, -4)$,   $C(7, 7)$

13  Find all points on the $y$-axis that are a distance 6 from $P(5, 3)$.

14  Find all points on the $x$-axis that are a distance 5 from $P(-2, 4)$.

15  For what values of $a$ is the distance between $P(a, 3)$ and $Q(5, 2a)$ greater than $\sqrt{26}$?

16  Find all points with coordinates of the form $(a, a)$ that are a distance 3 from $P(-2, 1)$.

**Exer. 17–36:** Sketch the graph of the equation, and label the $x$- and $y$-intercepts.

17  $y = 2x - 3$                     18  $y = 3x + 2$

19  $y = -x + 1$                    20  $y = -2x - 3$

21  $y = -4x^2$                      22  $y = \frac{1}{3}x^2$

23  $y = 2x^2 - 1$                  24  $y = -x^2 + 2$

25  $x = \frac{1}{4}y^2$            26  $x = -2y^2$

27  $x = -y^2 + 3$                  28  $x = 2y^2 - 4$

29  $y = -\frac{1}{2}x^3$          30  $y = \frac{1}{2}x^3$

31  $y = x^3 - 8$                    32  $y = -x^3 + 1$

33  $y = \sqrt{x}$                    34  $y = \sqrt{-x}$

35  $y = \sqrt{x} - 4$              36  $y = \sqrt{x - 4}$

**Exer. 37–38:** Use tests for symmetry to determine which graphs in the indicated exercises are symmetric with respect to **(a)** the $y$-axis, **(b)** the $x$-axis, and **(c)** the origin.

37  The odd-numbered exercises in 17–36

38  The even-numbered exercises in 17–36

**Exer. 39–50:** Sketch the graph of the circle or semicircle.

39  $x^2 + y^2 = 11$                40  $x^2 + y^2 = 7$

41  $(x + 3)^2 + (y - 2)^2 = 9$

42  $(x - 4)^2 + (y + 2)^2 = 4$

43  $(x + 3)^2 + y^2 = 16$         44  $x^2 + (y - 2)^2 = 25$

45  $4x^2 + 4y^2 = 25$             46  $9x^2 + 9y^2 = 1$

47  $y = -\sqrt{16 - x^2}$         48  $y = \sqrt{4 - x^2}$

49  $x = \sqrt{9 - y^2}$            50  $x = -\sqrt{25 - y^2}$

**Exer. 51–62:** Find an equation of the circle that satisfies the stated conditions.

51  Center $C(2, -3)$, radius 5

52  Center $C(-4, 1)$, radius 3

53 Center $C(\frac{1}{4}, 0)$, radius $\sqrt{5}$

54 Center $C(\frac{3}{4}, -\frac{2}{3})$, radius $3\sqrt{2}$

55 Center $C(-4, 6)$, passing through $P(1, 2)$

56 Center at the origin, passing through $P(4, -7)$

57 Center $C(-3, 6)$, tangent to the $y$-axis

58 Center $C(4, -1)$, tangent to the $x$-axis

59 Tangent to both axes, center in the second quadrant, radius 4

60 Tangent to both axes, center in the fourth quadrant, radius 3

61 End points of a diameter $A(4, -3)$ and $B(-2, 7)$

62 End points of a diameter $A(-5, 2)$ and $B(3, 6)$

**Exer. 63–72: Find the center and radius of the circle with the given equation.**

63 $x^2 + y^2 - 4x + 6y = 36$

64 $x^2 + y^2 + 8x - 10y = -37$

65 $x^2 + y^2 + 4y - 117 = 0$

66 $x^2 + y^2 - 10x + 18 = 0$

67 $2x^2 + 2y^2 - 12x + 4y = 15$

68 $9x^2 + 9y^2 + 12x - 6y = -4$

69 $x^2 + y^2 + 4x - 2y + 5 = 0$

70 $x^2 + y^2 - 6x + 4y + 13 = 0$

71 $x^2 + y^2 - 2x - 8y + 19 = 0$

72 $x^2 + y^2 + 4x + 6y + 16 = 0$

**Exer. 73–76: Find equations for the upper half, lower half, right half, and left half of the circle.**

73 $x^2 + y^2 = 36$          74 $(x + 3)^2 + y^2 = 64$

75 $(x - 2)^2 + (y + 1)^2 = 49$

76 $(x - 3)^2 + (y - 5)^2 = 4$

**Exer. 77–78: Determine whether the point $P$ is inside, outside, or on the circle with center $C$ and radius $r$.**

77 (a) $P(2, 3)$,    $C(4, 6)$,    $r = 4$
   (b) $P(4, 2)$,    $C(1, -2)$,    $r = 5$
   (c) $P(-3, 5)$,    $C(2, 1)$,    $r = 6$

78 (a) $P(3, 8)$,    $C(-2, -4)$,    $r = 13$
   (b) $P(-2, 5)$,    $C(3, 7)$,    $r = 6$
   (c) $P(1, -2)$,    $C(6, -7)$,    $r = 7$

**Exer. 79–80: For the given circle, find (a) the $x$-intercepts and (b) the $y$-intercepts.**

79 $x^2 + y^2 - 4x - 6y + 4 = 0$

80 $x^2 + y^2 - 10x + 4y + 13 = 0$

81 Find an equation of the circle that is concentric with $x^2 + y^2 + 4x - 6y + 4 = 0$ and passes through the point $P(2, 6)$.

82 **Radio broadcasting ranges** The signal from a radio station has a circular range of 50 miles. A second radio station, located 100 miles east and 80 miles north of the first station, has a range of 80 miles. Are there locations where signals can be received from both radio stations? Explain your answer.

83 A circle $C_1$ of radius 5 has its center at the origin. Inside this circle there is a first-quadrant circle $C_2$ of radius 2 that is tangent to $C_1$. The $y$-coordinate of the center of $C_2$ is 2. Find the $x$-coordinate of the center of $C_2$.

84 A circle $C_1$ of radius 5 has its center at the origin. Outside this circle is a first-quadrant circle $C_2$ of radius 2 that is tangent to $C_1$. The $y$-coordinate of the center of $C_2$ is 3. Find the $x$-coordinate of the center of $C_2$.

**Exer. 85–88: The figure represents the screen of a graphing utility, where $Y_1$ and $Y_2$ are $y$-value assignments for two equations in $x$. Express, in interval form, the $x$ values such that $Y_1 < Y_2$ for the indicated viewing rectangle. Assume that the $x$ and $y$ values of each point of intersection are integers.**

85

86

87

88

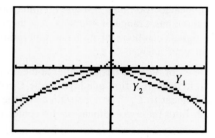

Exer. 89–90: Graph the equation, and estimate the *x*-intercepts.

**89** $y = x^3 - \frac{9}{10}x^2 - \frac{43}{25}x + \frac{24}{25}$

**90** $y = x^4 + 0.85x^3 - 2.46x^2 - 1.07x + 0.51$

Exer. 91–94: Graph the two equations on the same coordinate plane, and estimate the coordinates of their points of intersection.

**91** $y = x^3 + x;$         $x^2 + y^2 = 1$

**92** $y = 3x^4 - \frac{3}{2};$         $x^2 + y^2 = 1$

**93** $x^2 + (y - 1)^2 = 1;$         $(x - \frac{5}{4})^2 + y^2 = 1$

**94** $(x + 1)^2 + (y - 1)^2 = \frac{1}{4};$         $(x + \frac{1}{2})^2 + (y - \frac{1}{2})^2 = 1$

## *1.8*   LINES

One of the basic concepts in geometry is that of a *line*. In this section we will restrict our discussion to lines that lie in a coordinate plane. This will allow us to use algebraic methods to study their properties. Two of our principal objectives may be stated as follows:

**(1)** Given a line *l* in a coordinate plane, find an equation whose graph corresponds to *l*.

**(2)** Given an equation of a line *l* in a coordinate plane, sketch the graph of the equation.

The following concept is fundamental to the study of lines.

**Definition of Slope of a Line**

Let *l* be a line that is not parallel to the *y*-axis, and let $P_1(x_1, y_1)$ and $P_2(x_2, y_2)$ be distinct points on *l*. The **slope *m*** of *l* is

$$m = \frac{y_2 - y_1}{x_2 - x_1}.$$

If *l* is parallel to the *y*-axis, then the slope of *l* is not defined.

Typical points $P_1$ and $P_2$ on a line $l$ are shown in Figure 39. The numerator $y_2 - y_1$ in the formula for $m$ is the vertical change in direction from $P_1$ to $P_2$ and may be positive, negative, or zero. The denominator $x_2 - x_1$ is the horizontal change from $P_1$ to $P_2$, and it may be positive or negative, but never zero, because $l$ is not parallel to the $y$-axis if a slope exists. In Figure 39(a) the slope is positive, and we say that the line *rises.* In Figure 39(b) the slope is negative, and the line *falls.*

**FIGURE 39**

(a) Positive slope (line rises)

(b) Negative slope (line falls)

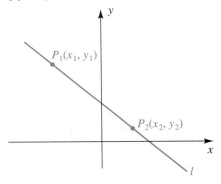

In finding the slope of a line it is immaterial which point we label as $P_1$ and which as $P_2$, since

$$\frac{y_2 - y_1}{x_2 - x_1} = \frac{y_2 - y_1}{x_2 - x_1} \cdot \frac{(-1)}{(-1)} = \frac{y_1 - y_2}{x_1 - x_2}.$$

If the points are labeled so that $x_1 < x_2$, as in Figure 39, then $x_2 - x_1 > 0$, and hence the slope is positive, negative, or zero, depending on whether $y_2 > y_1$, $y_2 < y_1$, or $y_2 = y_1$, respectively.

The definition of slope is independent of the two points that are chosen on $l$. If other points $P_1'(x_1', y_1')$ and $P_2'(x_2', y_2')$ are used, then, as in Figure 40, the triangle with vertices $P_1'$, $P_2'$, and $P_3'(x_2', y_1')$ is similar to the triangle

**FIGURE 40**

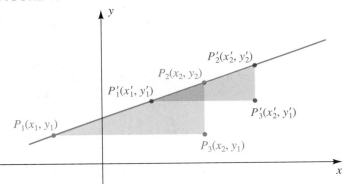

with vertices $P_1$, $P_2$, and $P_3(x_2, y_1)$. Since the ratios of corresponding sides of similar triangles are equal,

$$\frac{y_2 - y_1}{x_2 - x_1} = \frac{y_2' - y_1'}{x_2' - x_1'}.$$

### EXAMPLE I    *Finding slopes*

Sketch the line through each pair of points, and find its slope:

**(a)** $A(-1, 4)$ and $B(3, 2)$        **(b)** $A(2, 5)$ and $B(-2, -1)$

**(c)** $A(4, 3)$ and $B(-2, 3)$        **(d)** $A(4, -1)$ and $B(4, 4)$

*SOLUTION*    The lines are sketched in Figure 41. We use the definition of slope to find the slope of each line.

**(a)**
$$m = \frac{2 - 4}{3 - (-1)} = \frac{-2}{4} = -\frac{1}{2}$$

**(b)**
$$m = \frac{5 - (-1)}{2 - (-2)} = \frac{6}{4} = \frac{3}{2}$$

**(c)**
$$m = \frac{3 - 3}{-2 - 4} = \frac{0}{-6} = 0$$

**(d)** The slope is undefined because the line is parallel to the $y$-axis. Note that if the formula for $m$ is used, the denominator is zero.

**FIGURE 41**

**(a)** $m = -\frac{1}{2}$ 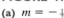            **(b)** $m = \frac{3}{2}$              **(c)** $m = 0$              **(d)** $m$ undefined

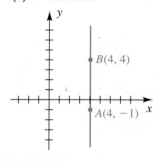

### EXAMPLE 2    *Sketching a line with a given slope*

Sketch a line through $P(2, 1)$ that has

**(a)** slope $\frac{5}{3}$        **(b)** slope $-\frac{5}{3}$

*SOLUTION*    If the slope of a line is $a/b$ and $b$ is positive, then for every change of $b$ units in the horizontal direction, the line rises or falls $|a|$ units, depending on whether $a$ is positive or negative.

**(a)** If $P(2, 1)$ is on the line and $m = \frac{5}{3}$, we can obtain another point on the line by starting at $P$ and moving 3 units to the right and 5 units upward. This gives us the point $Q(5, 6)$, and the line is determined as in Figure 42(a).

**(b)** If $P(2, 1)$ is on the line and $m = -\frac{5}{3}$, we move 3 units to the right and 5 units downward, obtaining the line through $Q(5, -4)$, as in Figure 42(b).

**FIGURE 42**

(a) $m = \frac{5}{3}$

(b) $m = -\frac{5}{3}$

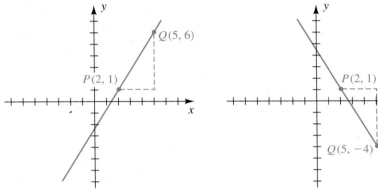

The diagram in Figure 43 indicates the slopes of several lines through the origin. The line that lies on the $x$-axis has slope $m = 0$. If this line is rotated about $O$ in the counterclockwise direction (as indicated by the blue

**FIGURE 43**

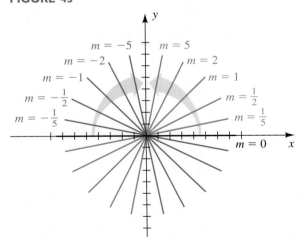

arrow), the slope is positive and increases, reaching the value 1 when the line bisects the first quadrant and continuing to increase as the line gets closer to the $y$-axis. If we rotate the line of slope $m = 0$ in the *clockwise* direction (as indicated by the red arrow), the slope is negative, reaching the value $-1$ when the line bisects the second quadrant and becoming large and negative as the line gets closer to the $y$-axis.

Lines that are horizontal or vertical have simple equations, as indicated in the following chart.

| Terminology | Definition | Graph | Equation |
|---|---|---|---|
| **Horizontal line** | A line parallel to the $x$-axis | $(0, b)$ | $y = a$ |
| **Vertical line** | A line parallel to the $y$-axis | $(a, 0)$ | $x = a$ |

A common error is to regard the graph of $y = b$ as consisting of only the one point $(0, b)$. If we express the equation in the form $0 \cdot x + y = b$, we see that the value of $x$ is immaterial; thus, the graph of $y = b$ consists of the points $(x, b)$ for *every* $x$ and hence is a horizontal line. Similarly, the graph of $x = a$ is the vertical line consisting of all points $(a, y)$, where $y$ is a real number.

**FIGURE 44**

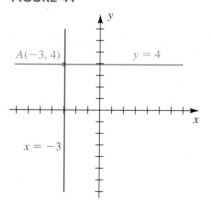

**EXAMPLE 3**   *Finding equations of horizontal and vertical lines*
Find an equation of the line through $A(-3, 4)$ that is parallel to

**(a)** the $x$-axis     **(b)** the $y$-axis

**SOLUTION**   The two lines are sketched in Figure 44. As indicated in the preceding chart, the equations are $y = 4$ for part (a) and $x = -3$ for part (b).

**FIGURE 45**

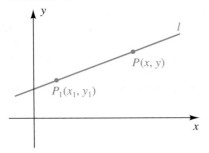

Let us next find an equation of a line $l$ through a point $P_1(x_1, y_1)$ with slope $m$. If $P(x, y)$ is any point with $x \neq x_1$ (see Figure 45), then $P$ is on $l$ if and only if the slope of the line through $P_1$ and $P$ is $m$—that is, if

$$\frac{y - y_1}{x - x_1} = m.$$

This equation may be written in the form

$$y - y_1 = m(x - x_1).$$

Note that $(x_1, y_1)$ is a solution of the last equation, and hence the points on $l$ are precisely the points that correspond to the solutions. This equation for $l$ is referred to as the **point-slope form.**

**Point-Slope Form for the Equation of a Line**

> An equation for the line through the point $(x_1, y_1)$ with slope $m$ is
>
> $$y - y_1 = m(x - x_1).$$

The point-slope form is only one possibility for an equation of a line. There are an unlimited number of equivalent equations. We sometimes simplify the equation obtained using the point-slope form to either

$$ax + by = c \quad \text{or} \quad ax + by + d = 0,$$

where $a$, $b$, and $c$ are integers with no common factor, $a > 0$, and $d = -c$.

**EXAMPLE 4** *Finding an equation of a line through two points*

Find an equation of the line through $A(1, 7)$ and $B(-3, 2)$.

**FIGURE 46**

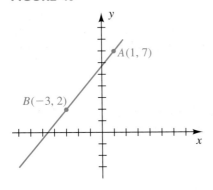

*SOLUTION* The line is sketched in Figure 46. The formula for the slope $m$ gives us

$$m = \frac{7 - 2}{1 - (-3)} = \frac{5}{4}.$$

We may use the coordinates of either $A$ or $B$ for $(x_1, y_1)$ in the point-slope form. Using $A(1, 7)$ gives us

| | |
|---|---|
| $y - 7 = \frac{5}{4}(x - 1)$ | point-slope form |
| $4(y - 7) = 5(x - 1)$ | multiply by 4 |
| $4y - 28 = 5x - 5$ | multiply factors |
| $-5x + 4y = 23$ | subtract $5x$ and add 28 |
| $5x - 4y = -23$ | multiply by $-1$ |

The last equation is one of the desired forms for an equation of a line. Another is $5x - 4y + 23 = 0$.

**FIGURE 47**

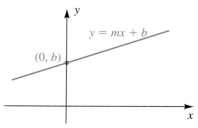

The point-slope form for the equation of a line may be rewritten as $y = mx - mx_1 + y_1$, which is of the form

$$y = mx + b$$

with $b = -mx_1 + y_1$. The real number $b$ is the $y$-intercept of the graph, as indicated in Figure 47. Since the equation $y = mx + b$ displays the slope $m$ and $y$-intercept $b$ of $l$, it is called the **slope-intercept form** for the equation of a line. Conversely, if we start with $y = mx + b$, we may write

$$y - b = m(x - 0).$$

Comparing this equation with the point-slope form, we see that the graph is a line with slope $m$ and passing through the point $(0, b)$. We have proved the following result.

**Slope-Intercept Form for the Equation of a Line**

> The graph of $y = mx + b$ is a line having slope $m$ and $y$-intercept $b$.

**EXAMPLE 5**    *Expressing an equation in slope-intercept form*

Express the equation $2x - 5y = 8$ in slope-intercept form.

**SOLUTION**    Our goal is to solve the given equation for $y$ to obtain the form $y = mx + b$. Doing so gives us the following equivalent equations:

$$2x - 5y = 8, \quad -5y = -2x + 8, \quad y = \tfrac{2}{5}x + \left(-\tfrac{8}{5}\right)$$

The last equation is the slope-intercept form $y = mx + b$ with slope $m = \tfrac{2}{5}$ and $y$-intercept $b = -\tfrac{8}{5}$.

It follows from the point-slope form that every line is a graph of an equation

$$ax + by = c,$$

where $a$, $b$, and $c$ are real numbers and $a$ and $b$ are not both zero. We call such an equation a **linear equation** in $x$ and $y$. Let us show, conversely, that the graph of $ax + by = c$, with $a$ and $b$ not both zero, is always a line. If $b \neq 0$, we may solve for $y$, obtaining

$$y = \left(-\frac{a}{b}\right)x + \frac{c}{b},$$

which, by the slope-intercept form, is an equation of a line with slope $-a/b$ and $y$-intercept $c/b$. If $b = 0$ but $a \neq 0$, we may solve for $x$, obtaining $x = c/a$, which is the equation of a vertical line with $x$-intercept $c/a$. This discussion establishes the following result.

**General Form for the Equation of a Line**

> The graph of a linear equation $ax + by = c$ is a line, and conversely, every line is the graph of a linear equation.

For simplicity, we use the terminology *the line $ax + by = c$* rather than *the line with equation $ax + by = c$*.

**EXAMPLE 6    *Sketching the graph of a linear equation***

Sketch the graph of $2x - 5y = 8$.

**FIGURE 48**

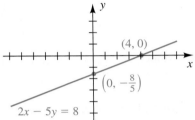

**SOLUTION**    We know from the preceding discussion that the graph is a line, so it is sufficient to find two points on the graph. Let us find the $x$- and $y$-intercepts by substituting $y = 0$ and $x = 0$, respectively, in the given equation, $2x - 5y = 8$.

$x$-intercept:    If $y = 0$, then $2x = 8$, or $x = 4$.

$y$-intercept:    If $x = 0$, then $-5y = 8$, or $y = -\frac{8}{5}$.

Plotting the intercepts $(4, 0)$ and $\left(0, -\frac{8}{5}\right)$ and drawing a line through them gives us the graph in Figure 48.

The following theorem specifies the relationship between **parallel lines** (lines in a plane that do not intersect) and slope.

**Theorem on Slopes of Parallel Lines**

> Two nonvertical lines are parallel if and only if they have the same slope.

**FIGURE 49**

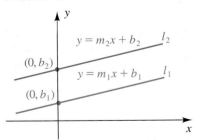

**PROOF**    Let $l_1$ and $l_2$ be distinct lines of slopes $m_1$ and $m_2$, respectively. If the $y$-intercepts are $b_1$ and $b_2$ (see Figure 49), then, by the slope-intercept form, the lines have equations

$$y = m_1x + b_1 \quad \text{and} \quad y = m_2x + b_2.$$

The lines intersect at some point $(x, y)$ if and only if the values of $y$ are equal for some $x$—that is, if

$$m_1x + b_1 = m_2x + b_2,$$

or

$$(m_1 - m_2)x = b_2 - b_1.$$

The last equation can be solved for $x$ if and only if $m_1 - m_2 \neq 0$. We have shown that the lines $l_1$ and $l_2$ intersect if and only if $m_1 \neq m_2$. Hence they do *not* intersect (are parallel) if and only if $m_1 = m_2$. ∎

**EXAMPLE 7**     *Finding an equation of a line parallel to a given line*

Find an equation of the line through $P(5, -7)$ that is parallel to the line $6x + 3y = 4$.

**SOLUTION**     We first express $6x + 3y = 4$ in slope-intercept form, obtaining $y = -2x + \frac{4}{3}$. Thus, the slope of the given line is $-2$. Since parallel lines have the same slope, the required line also has slope $-2$. Using the point $P(5, -7)$ in the point-slope form gives us

$$y - (-7) = -2(x - 5),$$

which can be written as

$$y = -2x + 3.$$

The last equation is in slope-intercept form and shows that the parallel line we have found has $y$-intercept 3. This line and the given line are sketched in Figure 50.

As an alternative solution, we might use the fact that lines of the form $6x + 3y = k$ have the same slope as the given line and hence are parallel to it. Substituting $x = 5$ and $y = -7$ into $6x + 3y = k$ gives us $6(5) + 3(-7) = k$ or, equivalently, $k = 9$. The equation $6x + 3y = 9$ is equivalent to $y = -2x + 3$.

**FIGURE 50**

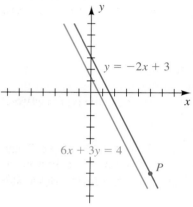

The next theorem gives us information about **perpendicular lines** (lines that intersect at a right angle).

**Theorem on Slopes of Perpendicular Lines**

> Two lines with slope $m_1$ and $m_2$ are perpendicular if and only if
> $$m_1 m_2 = -1.$$

**FIGURE 51**

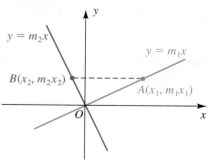

**PROOF**     For simplicity, let us consider the special case of two lines that intersect at the origin $O$, as illustrated in Figure 51. Equations of these lines are $y = m_1 x$ and $y = m_2 x$. If, as in the figure, we choose points $A(x_1, m_1 x_1)$ and $B(x_2, m_2 x_2)$ different from $O$ on the lines, then the lines are perpendicular if and only if angle $AOB$ is a right angle. Applying the Pythagorean theorem, we know that angle $AOB$ is a right angle if and only if

$$[d(A, B)]^2 = [d(O, B)]^2 + [d(O, A)]^2$$

or, by the distance formula,

$$(x_2 - x_1)^2 + (m_2 x_2 - m_1 x_1)^2 = x_2^2 + (m_2 x_2)^2 + x_1^2 + (m_1 x_1)^2.$$

Squaring terms, simplifying, and factoring gives us

$$-2m_1 m_2 x_1 x_2 - 2x_1 x_2 = 0$$
$$-2x_1 x_2 (m_1 m_2 + 1) = 0.$$

Since both $x_1$ and $x_2$ are not zero, we may divide both sides by $-2x_1x_2$, obtaining $m_1m_2 + 1 = 0$. Thus, the lines are perpendicular if and only if $m_1m_2 = -1$.

The same type of proof may be given if the lines intersect at *any* point $(a, b)$. ∎

A convenient way to remember the conditions on slopes of perpendicular lines is to note that $m_1$ and $m_2$ must be *negative reciprocals* of each other—that is, $m_1 = -1/m_2$ and $m_2 = -1/m_1$.

**EXAMPLE 8**   *Finding an equation of a line perpendicular to a given line*

**FIGURE 52**

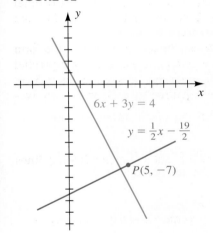

Find the slope-intercept form for the line through $P(5, -7)$ that is perpendicular to the line $6x + 3y = 4$.

**SOLUTION**   We considered the line $6x + 3y = 4$ in Example 7 and found that its slope is $-2$. Hence, the slope of the required line is the negative reciprocal $-[1/(-2)]$, or $\frac{1}{2}$. Using $P(5, -7)$ gives us the following:

$$
\begin{aligned}
y - (-7) &= \tfrac{1}{2}(x - 5) & \text{point-slope form} \\
y + 7 &= \tfrac{1}{2}x - \tfrac{5}{2} & \text{simplify} \\
y &= \tfrac{1}{2}x - \tfrac{19}{2} & \text{put in slope-intercept form}
\end{aligned}
$$

The last equation is in slope-intercept form and shows that the perpendicular line has $y$-intercept $-\frac{19}{2}$. This line and the given line are sketched in Figure 52.

**EXAMPLE 9**   *Finding an equation of a perpendicular bisector*

Given $A(-3, 1)$ and $B(5, 4)$, find the general form of the perpendicular bisector $l$ of the line segment $AB$.

**FIGURE 53**

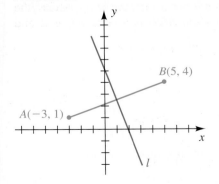

**SOLUTION**   The line segment $AB$ and its perpendicular bisector $l$ are shown in Figure 53. We calculate the following, where $M$ is the midpoint of $AB$:

Coordinates of $M$:   $\left(\dfrac{-3 + 5}{2}, \dfrac{1 + 4}{2}\right) = \left(1, \dfrac{5}{2}\right)$   midpoint formula

Slope of $AB$:   $\dfrac{4 - 1}{5 - (-3)} = \dfrac{3}{8}$   slope formula

Slope of $l$:   $-\dfrac{1}{\frac{3}{8}} = -\dfrac{8}{3}$   negative reciprocal of $\frac{3}{8}$

Using the point $M(1, \frac{5}{2})$ and slope $-\frac{8}{3}$ gives us the following equivalent equations for $l$:

$$y - \tfrac{5}{2} = -\tfrac{8}{3}(x - 1) \qquad \text{point-slope form}$$
$$6y - 15 = -16(x - 1) \qquad \text{multiply by the lcd, 6}$$
$$6y - 15 = -16x + 16 \qquad \text{multiply}$$
$$16x + 6y = 31 \qquad \text{put in general form}$$

Two variables $x$ and $y$ are **linearly related** if $y = ax + b$, where $a$ and $b$ are real numbers and $a \neq 0$. Linear relationships between variables occur frequently in applied problems. The following example gives one illustration.

**EXAMPLE 10**    *Relating air temperature to altitude*

The relationship between the air temperature $T$ (in °F) and the altitude $h$ (in feet above sea level) is approximately linear for $0 \leq h \leq 20{,}000$. If the temperature at sea level is 60°, an increase of 5000 feet in altitude lowers the air temperature about 18°.

**(a)** Express $T$ in terms of $h$, and sketch the graph on an $hT$-coordinate system.

**(b)** Approximate the air temperature at an altitude of 15,000 feet.

**(c)** Approximate the altitude at which the temperature is 0°.

**SOLUTION**

**(a)** If $T$ is linearly related to $h$, then

$$T = ah + b$$

for some constants $a$ and $b$ ($a$ represents the slope and $b$ the $T$-intercept). Since $T = 60°$ when $h = 0$ ft (sea level), the $T$-intercept is 60, and the temperature $T$ for $0 \leq h \leq 20{,}000$ is given by

$$T = ah + 60.$$

From the given data, we note that when $h = 5000$ ft, the temperature $T = 60° - 18° = 42°$. Hence, we may find $a$ as follows:

$$42 = a(5000) + 60 \qquad \text{let } T = 42 \text{ and } h = 5000$$
$$a = \frac{42 - 60}{5000} = -\frac{9}{2500} \qquad \text{solve for } a$$

Substituting for $a$ in $T = ah + 60$ gives us the following formula for $T$:

$$T = -\tfrac{9}{2500}h + 60$$

The graph is sketched in Figure 54, with different scales on the axes.

**FIGURE 54**

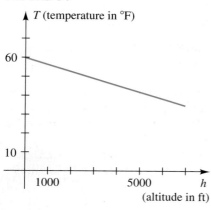

$T$ (temperature in °F)

60

10

1000        5000        $h$
(altitude in ft)

(b) Using the last formula for $T$ obtained in part (a), we find that the temperature (in °F) when $h = 15{,}000$ is

$$T = -\tfrac{9}{2500}(15{,}000) + 60 = -54 + 60 = 6.$$

(c) To find the altitude $h$ that corresponds to $T = 0°$, we proceed as follows:

| | |
|---|---|
| $T = -\tfrac{9}{2500}h + 60$ | from part (a) |
| $0 = -\tfrac{9}{2500}h + 60$ | let $T = 0$ |
| $\tfrac{9}{2500}h = 60$ | add $\tfrac{9}{2500}h$ |
| $h = 60 \cdot \tfrac{2500}{9}$ | multiply by $\tfrac{2500}{9}$ |
| $h = \dfrac{50{,}000}{3} \approx 16{,}667 \text{ ft}$ | simplify and approximate |

In Chapter 8 we will discuss algebraic methods for finding the point of intersection of two nonparallel lines. In some applications, numbers that occur in equations of lines are irrational or involve approximate data. In such cases it may be sufficient to estimate the coordinates of the point using a graphing utility, as in the next example.

**EXAMPLE 11**  *Estimating the point of intersection of two lines*

Suppose the following lines were obtained using approximate data:

$$1.018x + 0.230y = 0.447$$
$$1.847x + 4.538y = 1.414$$

Use a graphing utility to estimate the coordinates of the point of intersection to two decimal places.

*SOLUTION*    Solving each equation for $y$, we obtain

$$y = (0.447 - 1.018x)/0.230,$$
$$y = (1.414 - 1.847x)/4.538.$$

We next assign the preceding expressions for $y$ to $Y_1$ and $Y_2$, respectively. A display of the graph (similar to Figure 55) shows that there is a point of intersection near the origin. Using the tracing and zoom-in features yields the approximate solution $(0.41, 0.15)$.

Problems similar to this one are often encountered in statistical analysis when the *method of least squares* is employed to find a *linear regression line*.

**FIGURE 55** $[-15, 15]$ by $[-10, 10]$

## 1.8 EXERCISES

Exer. 1–6: Sketch the line through $A$ and $B$, and find its slope $m$.

1  $A(-3, 2)$,   $B(5, -4)$      2  $A(4, -1)$,   $B(-6, -3)$

3  $A(2, 5)$,      $B(-7, 5)$      4  $A(5, -1)$,   $B(5, 6)$

5  $A(-3, 2)$,   $B(-3, 5)$      6  $A(4, -2)$,   $B(-3, -2)$

Exer. 7–10: Use slopes to show that the points are vertices of the specified polygon.

7  $A(-3, 1)$, $B(5, 3)$, $C(3, 0)$, $D(-5, -2)$;   parallelogram

8  $A(2, 3)$, $B(5, -1)$, $C(0, -6)$, $D(-6, 2)$;   trapezoid

9  $A(6, 15)$, $B(11, 12)$, $C(-1, -8)$, $D(-6, -5)$;   rectangle

10  $A(1, 4)$, $B(6, -4)$, $C(-15, -6)$;   right triangle

11  If three consecutive vertices of a parallelogram are $A(-1, -3)$, $B(4, 2)$, and $C(-7, 5)$, find the fourth vertex.

12  Let $A(x_1, y_1)$, $B(x_2, y_2)$, $C(x_3, y_3)$, and $D(x_4, y_4)$ denote the vertices of an arbitrary quadrilateral. Show that the line segments joining midpoints of adjacent sides form a parallelogram.

Exer. 13–14: Sketch the graph of $y = mx$ for the given values of $m$.

13  $m = 3, -2, \frac{2}{3}, -\frac{1}{4}$      14  $m = 5, -3, \frac{1}{2}, -\frac{1}{3}$

Exer. 15–16: Sketch the graph of the line through $P$ for each value of $m$.

15  $P(3, 1)$;      $m = \frac{1}{2}, -1, -\frac{1}{5}$

16  $P(-2, 4)$;   $m = 1, -2, -\frac{1}{2}$

Exer. 17–18: Sketch the graphs of the lines on the same coordinate plane.

17  $y = x + 3$,      $y = x + 1$,      $y = -x + 1$

18  $y = -2x - 1$,   $y = -2x + 3$,   $y = \frac{1}{2}x + 3$

Exer. 19–30: Find a general form of an equation of the line through the point $A$ that satisfies the given condition.

19  $A(5, -2)$
    (a) parallel to the $y$-axis
    (b) perpendicular to the $y$-axis

20  $A(-4, 2)$
    (a) parallel to the $x$-axis
    (b) perpendicular to the $x$-axis

21  $A(5, -3)$;   slope $-4$      22  $A(-1, 4)$;   slope $\frac{2}{3}$

23  $A(4, 0)$;      slope $-3$      24  $A(0, -2)$;   slope $5$

25  $A(4, -5)$;   through $B(-3, 6)$

26  $A(-1, 6)$;   $x$-intercept $5$

27  $A(2, -4)$;   parallel to the line $5x - 2y = 4$

28  $A(-3, 5)$;   parallel to the line $x + 3y = 1$

29  $A(7, -3)$;   perpendicular to the line $2x - 5y = 8$

30  $A(4, 5)$;      perpendicular to the line $3x + 2y = 7$

Exer. 31–34: Find the slope-intercept form of the line that satisfies the given conditions.

31  $x$-intercept 4,      $y$-intercept $-3$

32  $x$-intercept $-5$,   $y$-intercept $-1$

33  Through $A(5, 2)$ and $B(-1, 4)$

34  Through $A(-2, 1)$ and $B(3, 7)$

Exer. 35–36: Find a general form of an equation for the perpendicular bisector of the segment $AB$.

35  $A(3, -1)$,   $B(-2, 6)$      36  $A(4, 2)$,   $B(-2, 10)$

Exer. 37–38: Find an equation for the line that bisects the given quadrants.

37  II and IV      38  I and III

Exer. 39–42: Use the slope-intercept form to find the slope and $y$-intercept of the given line, and sketch its graph.

39  $2x = 15 - 3y$      40  $7x = -4y - 8$

41  $4x - 3y = 9$      42  $x - 5y = -15$

**Exer. 43–44: Find an equation of the line shown in the figure.**

43 (a)

(b)

(c)

(d)

44 (a)

(b)

(c)

(d)

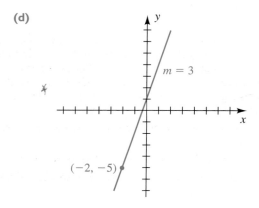

**Exer. 45–46:** If a line *l* has nonzero *x*- and *y*-intercepts *a* and *b*, respectively, then its *intercept form* is

$$\frac{x}{a} + \frac{y}{b} = 1.$$

**Find the intercept form for the given line.**

45  $4x - 2y = 6$          46  $x - 3y = -2$

47  Find an equation of the circle that has center $C(3, -2)$ and is tangent to the line $y = 5$.

48  Find an equation of the line that is tangent to the circle $x^2 + y^2 = 25$ at the point $P(3, 4)$.

49  **Fetal growth**  The growth of a fetus more than 12 weeks old can be approximated by the formula $L = 1.53t - 6.7$, where $L$ is the length in centimeters and $t$ is the age in weeks. Prenatal length can be determined by ultrasound. Approximate the age of a fetus whose length is 28 centimeters.

50  **Estimating salinity**  Salinity of the ocean refers to the amount of dissolved material found in a sample of seawater. Salinity $S$ can be estimated from the amount $C$ of chlorine in seawater using $S = 0.03 + 1.805C$, where $S$ and $C$ are measured by weight in parts per thousand. Approximate $C$ if $S$ is 0.35.

51  **Weight of a humpback whale**  The expected weight $W$ (in tons) of a humpback whale can be approximated from its length $L$ (in feet) by using $W = 1.70L - 42.8$ for $30 \le L \le 50$.

(a) Estimate the weight of a 40-foot humpback whale.

(b) If the error in estimating the length could be as large as 2 feet, what is the corresponding error for the weight estimate?

52  **Growth of a blue whale**  Newborn blue whales are approximately 24 feet long and weigh 3 tons. Young whales are nursed for 7 months, and by the time of weaning they often are 53 feet long and weigh 23 tons. Let $L$ and $W$ denote the length (in feet) and the weight (in tons), respectively, of a whale that is $t$ months of age.

(a) If $L$ and $t$ are linearly related, express $L$ in terms of $t$.

(b) What is the daily increase in the length of a young whale? (Use 1 month = 30 days.)

(c) If $W$ and $t$ are linearly related, express $W$ in terms of $t$.

(d) What is the daily increase in the weight of a young whale?

53  **Baseball stats**  Suppose a major league baseball player has hit 5 home runs in the first 14 games, and he keeps up this pace throughout the 162-game season.

(a) Express the number $y$ of home runs in terms of the number $x$ of games played.

(b) How many home runs will the player hit for the season?

54  **Cheese production**  A cheese manufacturer produces 18,000 pounds of cheese from January 1 through March 24. Suppose that this rate of production continues for the remainder of the year.

(a) Express the number $y$ of pounds of cheese produced in terms of the number $x$ of the day in a 365-day year.

(b) Predict, to the nearest pound, the number of pounds produced for the year.

**55** *Childhood weight*  A baby weighs 10 pounds at birth, and three years later the child's weight is 30 pounds. Assume that childhood weight $W$ (in pounds) is linearly related to age $t$ (in years).

(a) Express $W$ in terms of $t$.

(b) What is $W$ on the child's sixth birthday?

(c) At what age will the child weigh 70 pounds?

(d) Sketch, on a $tW$-plane, a graph that shows the relationship between $W$ and $t$ for $0 \le t \le 12$.

**56** *Loan repayment*  A college student receives an interest-free loan of $8250 from a relative. The student will repay $125 per month until the loan is paid off.

(a) Express the amount $P$ (in dollars) remaining to be paid in terms of time $t$ (in months).

(b) After how many months will the student owe $5000?

(c) Sketch, on a $tP$-plane, a graph that shows the relationship between $P$ and $t$ for the duration of the loan.

**57** *Vaporizing water*  The amount of heat $H$ (in joules) required to convert one gram of water into vapor is linearly related to the temperature $T$ (in °C) of the atmosphere. At 10°C this conversion requires 2480 joules, and each increase in temperature of 15°C lowers the amount of heat needed by 40 joules. Express $H$ in terms of $T$.

**58** *Aerobic power*  In exercise physiology, aerobic power $P$ is defined in terms of maximum oxygen intake. For altitudes up to 1800 meters, aerobic power is optimal—that is, 100%. Beyond 1800 meters, $P$ decreases linearly from the maximum of 100% to a value near 40% at 5000 meters.

(a) Express aerobic power $P$ in terms of altitude $h$ (in meters) for $1800 \le h \le 5000$.

(b) Estimate aerobic power in Mexico City (altitude: 2400 meters), the site of the 1968 Summer Olympic Games.

**59** *Urban heat island*  The urban heat island phenomenon has been observed in Tokyo. The average temperature was 13.5°C in 1915, and since then has risen 0.032°C per year.

(a) Assuming that temperature $T$ (in °C) is linearly related to time $t$ (in years) and that $t = 0$ corresponds to 1915, express $T$ in terms of $t$.

(b) Predict the average temperature in the year 2000.

**60** *Rising ground temperature*  In 1870 the average ground temperature in Paris was 11.8°C. Since then it has risen at a nearly constant rate, reaching 13.5°C in 1969.

(a) Express the temperature $T$ (in °C) in terms of time $t$ (in years), where $t = 0$ corresponds to the year 1870 and $0 \le t \le 99$.

(b) During what year was the average ground temperature 12.5°C?

**61** *Business expenses*  The owner of an ice cream franchise must pay the parent company $1000 per month plus 5% of the monthly revenue $R$. Operating cost of the franchise includes a fixed cost of $2600 per month for items such as utilities and labor. The cost of ice cream and supplies is 50% of the revenue.

(a) Express the owner's monthly expense $E$ in terms of $R$.

(b) Express the monthly profit $P$ in terms of $R$.

(c) Determine the monthly revenue needed to break even.

**62** *Drug dosage*  Pharmacological products must specify recommended dosages for adults and children. Two formulas for modification of adult dosage levels for young children are

$$\text{Cowling's rule:}\quad y = \tfrac{1}{24}(t + 1)a$$

and

$$\text{Friend's rule:}\quad y = \tfrac{2}{25}ta,$$

where $a$ denotes adult dose (in milligrams) and $t$ denotes the age of the child (in years).

(a) If $a = 100$, graph the two linear equations on the same coordinate plane for $0 \le t \le 12$.

(b) For what age do the two formulas specify the same dosage?

**63** *Video game*  In the video game shown in the figure, an airplane flies from left to right along the path given by $y = 1 + (1/x)$ and shoots bullets in the tangent direction at creatures placed along the $x$-axis at $x = 1, 2, 3, 4$.

EXERCISE 63

From calculus, the slope of the tangent line to the path at $P(1, 2)$ is $m = -1$ and at $Q(\frac{3}{2}, \frac{5}{3})$ is $m = -\frac{4}{9}$. Determine whether a creature will be hit if bullets are shot when the airplane is at

(a) $P$     (b) $Q$

64 **Temperature scales**   The relationship between the temperature reading $F$ on the Fahrenheit scale and the temperature reading $C$ on the Celsius scale is given by $C = \frac{5}{9}(F - 32)$.

(a) Find the temperature at which the reading is the same on both scales.

(b) When is the Fahrenheit reading twice the Celsius reading?

65 **Vertical wind shear**   Vertical wind shear occurs when wind speed varies at different heights above the ground. Wind shear is of great importance to pilots during take-offs and landings. If the wind speed is $v_1$ at height $h_1$ and $v_2$ at height $h_2$, then the average wind shear $s$ is given by the slope formula

$$s = \frac{v_2 - v_1}{h_2 - h_1}.$$

If the wind speed at ground level is 22 mi/hr and $s$ has been determined to be 0.07, find the wind speed 185 feet above the ground.

66 **Vertical wind shear**   In the study of vertical wind shear, the formula

$$\frac{v_1}{v_2} = \left(\frac{h_1}{h_2}\right)^P$$

is sometimes used, where $P$ is a variable that depends on the terrain and structures near ground level. In Montreal, average daytime values for $P$ with north winds over 29 mi/hr were determined to be 0.13. If a 32 mi/hr north wind is measured 20 feet above the ground, approximate the average wind shear (see Exercise 65) between 20 feet and 200 feet.

Exer. 67–68: The given points were found using empirical methods. Determine whether they lie on the same line $y = ax + b$, and if so, find the values of $a$ and $b$.

67  $A(-1.3, -1.3598)$,   $B(-0.55, -1.11905)$,
    $C(1.2, -0.5573)$,    $D(3.25, 0.10075)$

68  $A(-0.22, 1.6968)$,   $B(-0.12, 1.6528)$,
    $C(1.3, 1.028)$,      $D(1.45, 0.862)$

Exer. 69–70: Graph the lines on the same coordinate plane, and find the coordinates of the points of intersection (the coordinates are integers).

69  $x - 3y = -58$;   $3x - y = -70$

70  $x + 10y = 123$;   $2x - y = -6$

Exer. 71–72: Graph the lines on the same coordinate plane, and estimate the coordinates of the points of intersection. Identify the polygon determined by the lines.

71  $2x - y = -1$;   $x + 2y = -2$;   $3x + y = 11$

72  $10x - 42y = -7.14$;   $8.4x + 2y = -3.8$;
    $0.5x - 2.1y = 2.73$;   $16.8x + 4y = 14$

# CHAPTER 1 REVIEW EXERCISES

Exer. 1–2: Rewrite the expression without using the absolute value symbol, and simplify the result.

1  $|x + 3|$ if $x \le -3$

2  $|(x - 2)(x - 3)|$ if $2 < x < 3$

Exer. 3–14: Simplify the expression, and rationalize the denominator when appropriate.

3  $\left(\dfrac{a^{2/3}b^{3/2}}{a^2b}\right)^6$

4  $(-2p^2q)^3\left(\dfrac{p}{4q^2}\right)^2$

5 $\left(\dfrac{xy^{-1}}{\sqrt{z}}\right)^4 \div \left(\dfrac{x^{1/3}y^2}{z}\right)^3$

6 $\left(\dfrac{-64x^3}{z^6y^9}\right)^{2/3}$

7 $[(a^{2/3}b^{-2})^3]^{-1}$

8 $x^{-2} - y^{-1}$

9 $\sqrt[3]{8x^5y^3z^4}$

10 $\sqrt[4]{(-4a^3b^2c)^2}$

11 $\dfrac{1}{\sqrt{t}}\left(\dfrac{1}{\sqrt{t}} - 1\right)$

12 $\sqrt{\sqrt[3]{(c^3d^6)^4}}$

13 $\dfrac{\sqrt{12x^4y}}{\sqrt{3x^2y^5}}$

14 $\dfrac{3 + \sqrt{x}}{3 - \sqrt{x}}$

**Exer. 15–22: Express as a polynomial.**

15 $(3x^3 - 4x^2 + x - 7) + (x^4 - 2x^3 + 3x^2 + 5)$

16 $(x + 4)(x + 3) - (2x - 1)(x - 5)$

17 $(3a - 5b)(2a + 7b)$

18 $(4r^2 - 3s)^2$

19 $(13a^2 + 4b)(13a^2 - 4b)$

20 $(2a + b)^3$

21 $(3x + 2y)^2(3x - 2y)^2$

22 $(a + b + c + d)^2$

**Exer. 23–30: Factor the polynomial.**

23 $28x^2 + 4x - 9$

24 $16a^4 + 24a^2b^2 + 9b^4$

25 $8x^3 + 64y^3$

26 $u^3v^4 - u^6v$

27 $p^8 - q^8$

28 $x^4 - 8x^3 + 16x^2$

29 $x^2 - 49y^2 - 14x + 49$

30 $x^5 - 4x^3 + 8x^2 - 32$

**Exer. 31–36: Simplify the expression.**

31 $\dfrac{2}{4x - 5} - \dfrac{5}{10x + 1}$

32 $\dfrac{7}{x + 2} + \dfrac{3x}{(x + 2)^2} - \dfrac{5}{x}$

33 $\dfrac{x + x^{-2}}{1 + x^{-2}}$

34 $(a^{-1} + b^{-1})^{-1}$

35 $\dfrac{\dfrac{x}{x + 2} - \dfrac{4}{x + 2}}{x - 3 - \dfrac{6}{x + 2}}$

36 $\dfrac{(4 - x^2)(\frac{1}{3})(6x + 1)^{-2/3}(6) - (6x + 1)^{1/3}(-2x)}{(4 - x^2)^2}$

**Exer. 37–60: Solve the equation or inequality. Express the solutions in terms of intervals whenever possible.**

37 $\dfrac{3x + 1}{5x + 7} = \dfrac{6x + 11}{10x - 3}$

38 $2x^2 + 5x - 12 = 0$

39 $x(3x + 4) = 5$

40 $4x^4 - 33x^2 + 50 = 0$

41 $20x^3 + 8x^2 - 35x - 14 = 0$

42 $|4x - 1| = 7$

43 $2|2x + 1| + 1 = 19$

44 $\dfrac{1}{x} + 6 = \dfrac{5}{\sqrt{x}}$

45 $\sqrt{7x + 2} + x = 6$

46 $\sqrt{3x + 1} - \sqrt{x + 4} = 1$

47 $10 - 7x < 4 + 2x$

48 $-\dfrac{1}{2} < \dfrac{2x + 3}{5} < \dfrac{3}{2}$

49 $\dfrac{6}{10x + 3} < 0$

50 $|4x + 7| < 21$

51 $2|3 - x| + 1 > 5$

52 $|16 - 3x| \geq 5$

53 $10x^2 + 11x > 6$

54 $x(x - 3) \leq 10$

55 $\dfrac{x^2(3 - x)}{x + 2} \leq 0$

56 $\dfrac{x^2 - x - 2}{x^2 + 4x + 3} \leq 0$

57 $\dfrac{3}{2x + 3} < \dfrac{1}{x - 2}$

58 $\dfrac{x + 1}{x^2 - 25} \leq 0$

59 $x^3 > x^2$

60 $(x^2 - x)(x^2 - 5x + 6) < 0$

**Exer. 61–62: Solve for the specified variable.**

61 $F = \dfrac{\pi P R^4}{8VL}$ for $R$ (Poiseuille's law for fluids)

62 $V = \frac{1}{3}\pi h(r^2 + R^2 + rR)$ for $r$ (volume of a frustum of a cone)

**Exer. 63–66: Express in the form $a + bi$, where $a$ and $b$ are real numbers.**

63 $(3 + 8i)^2$

64 $\dfrac{1}{9 - \sqrt{-4}}$

65 $\dfrac{6 - 3i}{2 + 7i}$

66 $\dfrac{20 - 8i}{4i}$

**67** Show that the triangle with vertices $A(3, 1)$, $B(-5, -3)$, and $C(4, -1)$ is a right triangle, and find its area.

**68** Given $P(-5, 9)$ and $Q(-8, -7)$, find

(a) the distance $d(P, Q)$

(b) the midpoint of the segment $PQ$

(c) a point $R$ such that $Q$ is the midpoint of $PR$

**69** Find all points on the $y$-axis that are a distance 13 from $P(12, 6)$.

**70** For what values of $a$ is the distance between $P(a, 1)$ and $Q(-2, a)$ less than 3?

**71** Find an equation of the circle that has center $C(7, -4)$ and passes through $P(-3, 3)$.

**72** Find an equation of the circle that has end points of a diameter $A(8, 10)$ and $B(-2, -14)$.

**73** Find an equation for the left half of the circle given by $(x + 2)^2 + y^2 = 9$.

**74** Show that $A(-3, 1)$, $B(1, -1)$, $C(4, 1)$, and $D(3, 5)$ are vertices of a trapezoid.

**75** Find an equation of the line through $A(\frac{1}{2}, -\frac{1}{3})$ that is

(a) parallel to the line $6x + 2y + 5 = 0$

(b) perpendicular to the line $6x + 2y + 5 = 0$

**76** Find an equation of the circle that has center $C(-5, -1)$ and is tangent to the line $x = 4$.

**77** Find an equation of the line that has $x$-intercept $-3$ and passes through the center of the circle that has equation $x^2 + y^2 - 4x + 10y + 26 = 0$.

**78** Find a general form of an equation of the line through $P(4, -3)$ with slope 5.

**79** Given $A(-1, 2)$ and $B(3, -4)$, find a general form of an equation for the perpendicular bisector of segment $AB$.

**80** Find the center and radius of the circle that has equation $x^2 + y^2 - 12y + 31 = 0$.

**Exer. 81–88: Sketch the graph of the equation and label the $x$- and $y$-intercepts.**

**81** $2y + 5x - 8 = 0$       **82** $x = 3y + 4$

**83** $9y + 2x^2 = 0$       **84** $3x - 7y^2 = 0$

**85** $y = \sqrt{1 - x}$       **86** $y^2 = 16 - x^2$

**87** $x^2 + y^2 - 8x = 0$       **88** $x = -\sqrt{9 - y^2}$

**Exer. 89–90: Find an equation for the graph shown in the figure.**

**89**

**90**

**91 *Investment income*** An investor has a choice of two investments: a bond fund and a stock fund. The bond fund yields 7.186% interest annually, which is nontaxable at both the federal and state levels. Suppose the investor pays federal income tax at a rate of 28% and state income tax at a rate of 7%. Determine what the annual yield must be on the taxable stock fund so that the two funds pay the same amount of net interest income to the investor.

**92 *Preparing hospital food*** A hospital dietitian wishes to prepare a 10-ounce meat-vegetable dish that will provide 7 grams of protein. If an ounce of the vegetable portion supplies $\frac{1}{2}$ gram of protein and an ounce of meat supplies 1 gram of protein, how much of each should be used?

**93 *Solar heating*** A large solar heating panel requires 120 gallons of a fluid that is 30% antifreeze. The fluid comes in either a 50% solution or a 20% solution. How many gallons of each should be used to prepare the 120-gallon solution?

**94** *Filling a bin*  An extruder can fill an empty bin in 2 hours, and a packaging crew can empty a full bin in 5 hours. If a bin is half full when an extruder begins to fill it and a crew begins to empty it, how long will it take to fill the bin?

**95** *Highway travel*  A north-south highway intersects an east-west highway at a point $P$. An automobile crosses $P$ at 10:00 A.M., traveling east at a constant rate of 20 mi/hr. At that same instant another automobile is 2 miles north of $P$, traveling south at 50 mi/hr.

(a) Find a formula for the distance $d$ between the automobiles $t$ hours after 10:00 A.M.

(b) At approximately what time will the automobiles be 104 miles apart?

**96** *Fencing a kennel*  A kennel owner has 270 feet of fencing material to be used to divide a rectangular area into 10 equal pens, as shown in the figure. Find dimensions that would allow 100 ft² for each pen.

EXERCISE 96

**97** *Dimensions of an aquarium*  An open-topped aquarium is to be constructed with 6-foot-long sides and square ends, as shown in the figure.

EXERCISE 97

$\longleftarrow$ 6' $\longrightarrow$

(a) Find the height of the aquarium if the volume is to be 48 ft³.

(b) Find the height if 44 ft² of glass is to be used.

**98** *Sales commission*  A recent college graduate has job offers for a sales position in two computer firms. Job A pays $25,000 per year plus a 5% commission. Job B pays only $20,000 per year, but the commission rate is 10%. How much yearly business must the salesman do for the second job to be more lucrative?

**99** *Speed of sound*  The speed of sound in air at 0°C (or 273 K) is 1087 ft/sec, but this speed increases as the temperature rises. The speed $v$ of sound at temperature $T$ in K is given by $v = 1087\sqrt{T/273}$. At what temperatures does the speed of sound exceed 1100 ft/sec?

**100** *Planting an apple orchard*  The owner of an apple orchard estimates that if 24 trees are planted per acre, then each mature tree will yield 600 apples per year. For each additional tree planted per acre, the number of apples produced by each tree decreases by 12 per year. How many trees should be planted per acre to obtain at least 16,416 apples per year?

**101** *Discus throw*  Based on Olympic records, the winning distance for the discus throw can be approximated by $d = 175 + 1.75t$, where $d$ is in feet and $t = 0$ corresponds to the year 1948.

(a) Predict the winning distance for the 1996 Summer Olympics in Atlanta.

(b) Estimate the year in which the winning distance will be 270 feet.

**102** *House appreciation*  Six years ago a house was purchased for $89,000. This year it was appraised at $125,000. Assume that the value $V$ of the house after its purchase is a linear function of time $t$ (in years).

(a) Express $V$ in terms of $t$.

(b) How many years after the purchase date was the house worth $103,000?

Flights of leaping animals typically have
parabolic paths.

The mathematical term function (or its Latin equivalent) dates back to the late seventeenth century, when calculus was in the early stages of development. This important concept is now the backbone of advanced courses in mathematics and is indispensable in every field of science.

In this chapter we study properties of functions using algebraic and graphical methods that include plotting points, determining symmetries, and making horizontal and vertical shifts. These techniques are adequate for obtaining rough sketches of graphs that help us understand properties of functions; modern-day methods, however, employ sophisticated computer software and advanced mathematics to generate extremely accurate graphical representations of functions.

# CHAPTER 2

# FUNCTIONS

## 2.1   DEFINITION OF FUNCTION

The notion of **correspondence** occurs frequently in everyday life. Some examples are given in the following illustration.

### ILLUSTRATION

#### Correspondence

- To each book in a library there corresponds the number of pages in the book.
- To each human being there corresponds a birth date.
- If the temperature of the air is recorded throughout a day, then to each instant of time there corresponds a temperature.

Each correspondence in the previous illustration involves two sets, $D$ and $E$. In the first illustration, $D$ denotes the set of books in a library and $E$ the set of positive integers. To each book $x$ in $D$ there corresponds a positive integer $y$ in $E$—namely, the number of pages in the book.

We sometimes depict correspondences by diagrams of the type shown in Figure 1, where the sets $D$ and $E$ are represented by points within regions in a plane. The curved arrow indicates that the element $y$ of $E$ corresponds to the element $x$ of $D$. The two sets may have elements in common. As a matter of fact, we often have $D = E$. It is important to note that *to each $x$ in $D$ there corresponds exactly one $y$ in $E$.* However, the same element of $E$ may correspond to different elements of $D$. For example, two books may have the same number of pages, two people may have the same birthday, and the temperature may be the same at different times.

In most of our work, $D$ and $E$ will be sets of numbers. To illustrate, let both $D$ and $E$ denote the set $\mathbb{R}$ of real numbers, and to each real number $x$ let us assign its square $x^2$. This gives us a correspondence from $\mathbb{R}$ to $\mathbb{R}$.

Each of our illustrations of a correspondence is a *function,* which we define as follows.

**FIGURE I**

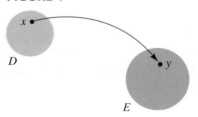

**Definition of Function**

> A **function** $f$ from a set $D$ to a set $E$ is a correspondence that assigns to each element $x$ of $D$ exactly one element $y$ of $E$.

The element $y$ of $E$ is the **value** of $f$ at $x$ (or the **image** of $x$ under $f$) and is denoted by $f(x)$, read "$f$ of $x$." The set $D$ is the **domain** of the function. The **range** of $f$ is the subset $R$ of $E$ consisting of all possible values $f(x)$ for $x$ in $D$. Note that there may be elements in the set $E$ that are not in the range $R$ of $f$.

**FIGURE 2**

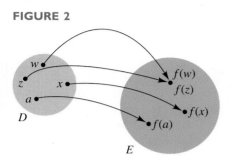

Consider the diagram in Figure 2. The curved arrows indicate that the elements $f(w)$, $f(z)$, $f(x)$, and $f(a)$ of $E$ correspond to the elements $w$, $z$, $x$, and $a$ of $D$. *To each element in $D$ there is assigned exactly one function value in $E$*; however, different elements of $D$, such as $w$ and $z$ in Figure 2, may have the same value in $E$.

The symbols

$$D \xrightarrow{f} E, \quad f : D \to E, \quad \text{and}$$

signify that $f$ is a function from $D$ to $E$, and we say that $f$ **maps** $D$ *into* $E$. Initially the notations $f$ and $f(x)$ may be confusing. Remember that $f$ is used to represent the function. It is neither in $D$ nor in $E$. However, $f(x)$ is an element of the range $R$—the element that the function $f$ assigns to the element $x$, which is in the domain $D$.

Two functions $f$ and $g$ from $D$ to $E$ are **equal**, and we write

$$f = g \quad \text{provided} \quad f(x) = g(x) \quad \text{for every } x \text{ in } D.$$

For example, if $g(x) = \frac{1}{2}(2x^2 - 6) + 3$ and $f(x) = x^2$ for every $x$ in $\mathbb{R}$, then $g = f$.

**EXAMPLE I    *Finding function values***

Let $f$ be the function with domain $\mathbb{R}$ such that $f(x) = x^2$ for every $x$ in $\mathbb{R}$.
**(a)** Find $f(-6)$, $f(\sqrt{3})$, $f(a + b)$, and $f(a) + f(b)$, where $a$ and $b$ are real numbers.
**(b)** What is the range of $f$?

*SOLUTION*

**(a)** We find values of $f$ by substituting for $x$ in the equation $f(x) = x^2$:

$$f(-6) = (-6)^2 = 36$$
$$f(\sqrt{3}) = (\sqrt{3})^2 = 3$$
$$f(a + b) = (a + b)^2 = a^2 + 2ab + b^2$$
$$f(a) + f(b) = a^2 + b^2$$

**(b)** By definition, the range of $f$ consists of all numbers of the form $f(x) = x^2$ for $x$ in $\mathbb{R}$. Since the square of every real number is nonnegative, the range is contained in the set of all nonnegative real numbers. Moreover, every nonnegative real number $c$ is a value of $f$, since $f(\sqrt{c}) = (\sqrt{c})^2 = c$. Hence, the range of $f$ is the set of all nonnegative real numbers.

*If a function is defined as in Example 1, the symbols used for the function and variable are immaterial; that is, expressions such as $f(x) = x^2$, $f(s) = s^2$, $g(t) = t^2$, and $k(r) = r^2$ all define the same function. This is true because if $a$ is any number in the domain, then the same value $a^2$ is obtained regardless of which expression is employed.*

In the remainder of our work, the phrase *f is a function* will mean that the domain and range are sets of real numbers. If a function is defined by means of an expression, as in Example 1, and the domain $D$ is not stated, then we will consider $D$ to be the totality of real numbers $x$ such that $f(x)$ is real. This is sometimes called the **implied domain** of $f$. To illustrate, if $f(x) = \sqrt{x - 2}$, then the implied domain is the set of real numbers $x$ such that $\sqrt{x - 2}$ is real—that is, $x - 2 \geq 0$, or $x \geq 2$. Thus, the domain is the infinite interval $[2, \infty)$. If $x$ is in the domain, we say that *f is defined at x* or that *f(x) exists*. If a set $S$ is contained in the domain, *f is defined on S*. The terminology *f is undefined at x* means that $x$ is not in the domain of $f$.

**EXAMPLE 2**     *Finding function values*

Let $g(x) = \dfrac{\sqrt{4 + x}}{1 - x}$.

**(a)** Find the domain of $g$.

**(b)** Find $g(5)$, $g(-2)$, $g(-a)$, and $-g(a)$.

*SOLUTION*

**(a)** The expression $\sqrt{4 + x}/(1 - x)$ is a real number if and only if the radicand $4 + x$ is nonnegative and the denominator $1 - x$ is not equal to 0. Thus, $g(x)$ exists if and only if

$$4 + x \geq 0 \qquad \text{and} \qquad 1 - x \neq 0$$

or, equivalently, $\qquad x \geq -4 \qquad \text{and} \qquad x \neq 1.$

We may express the domain in terms of intervals as $[-4, 1) \cup (1, \infty)$.

**(b)** To find values of $g$, we substitute for $x$:

$$g(5) = \frac{\sqrt{4 + 5}}{1 - 5} = \frac{\sqrt{9}}{-4} = -\frac{3}{4}$$

$$g(-2) = \frac{\sqrt{4 + (-2)}}{1 - (-2)} = \frac{\sqrt{2}}{3}$$

$$g(-a) = \frac{\sqrt{4 + (-a)}}{1 - (-a)} = \frac{\sqrt{4 - a}}{1 + a}$$

$$-g(a) = -\frac{\sqrt{4 + a}}{1 - a} = \frac{\sqrt{4 + a}}{a - 1}$$

Of special interest in calculus is the expression

$$\frac{f(x + h) - f(x)}{h}$$

(with $h \neq 0$), which is commonly called a **difference quotient.**

### EXAMPLE 3  *Simplifying a difference quotient*

Simplify the difference quotient

$$\frac{f(x + h) - f(x)}{h}$$

using the function $f(x) = x^2 + 6x - 4$.

**SOLUTION**

$$\frac{f(x + h) - f(x)}{h} = \frac{[(x + h)^2 + 6(x + h) - 4] - [x^2 + 6x - 4]}{h}$$

definition of $f$

$$= \frac{(x^2 + 2xh + h^2 + 6x + 6h - 4) - (x^2 + 6x - 4)}{h}$$

expand numerator

$$= \frac{(x^2 + 2xh + h^2 + 6\!\!\!/x + 6h - 4\!\!\!/) - (x^2 + 6\!\!\!/x - 4\!\!\!/)}{h}$$

subtract terms

$$= \frac{2xh + h^2 + 6h}{h}$$

simplify

$$= \frac{h(2x + h + 6)}{h}$$

factor out $h$

$$= 2x + h + 6$$

cancel $h \neq 0$

Graphs are often used to describe the variation of physical quantities. For example, a scientist may use the graph in Figure 3 to indicate the temperature $T$ of a certain solution at various times $t$ during an experiment. The sketch shows that the temperature increased gradually from time $t = 0$ to time $t = 5$, did not change between $t = 5$ and $t = 8$, and then decreased rapidly from $t = 8$ to $t = 9$.

Similarly, if $f$ is a function, we may use a graph to indicate the change in $f(x)$ as $x$ varies through the domain of $f$. Specifically, we have the following definition.

**FIGURE 3**

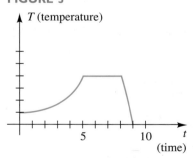

$T$ (temperature)

5    10    $t$
(time)

**Definition of Graph of a Function**

The **graph of a function** $f$ is the graph of the equation $y = f(x)$ for $x$ in the domain of $f$.

We often attach the label $y = f(x)$ to a sketch of the graph. If $P(a, b)$ is a point on the graph, then the $y$-coordinate $b$ is the function value $f(a)$, as illustrated in Figure 4. The figure displays the domain of $f$ (the set of possible values of $x$) and the range of $f$ (the corresponding values of $y$). Although we have pictured the domain and range as closed intervals, they may be infinite intervals or other sets of real numbers.

**FIGURE 4**

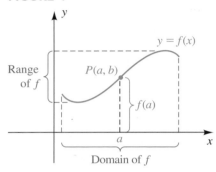

Since there is exactly one value $f(a)$ for each $a$ in the domain of $f$, only *one* point on the graph of $f$ has $x$-coordinate $a$. In general, we may use the following graphical test to determine whether a graph is the graph of a function.

**Vertical Line Test**

> The graph of a set of points in a coordinate plane is the graph of a function if every vertical line intersects the graph in at most one point.

Thus, *every vertical line intersects the graph of a function in at most one point.* Consequently, the graph of a function cannot be a figure such as a circle, in which a vertical line may intersect the graph in more than one point.

The $x$-intercepts of the graph of a function $f$ are the solutions of the equation $f(x) = 0$. These numbers are called the **zeros** of the function. The $y$-intercept of the graph is $f(0)$, if it exists.

**EXAMPLE 4     *Sketching the graph of a function***

Let $f(x) = \sqrt{x - 1}$.

**(a)** Sketch the graph of $f$.

**(b)** Find the domain and range of $f$.

SOLUTION

**(a)** By definition, the graph of $f$ is the graph of the equation $y = \sqrt{x - 1}$. The following table lists coordinates of several points on the graph.

| $x$ | 1 | 2 | 3 | 4 | 5 | 6 |
|-----|---|---|---|---|---|---|
| $y = f(x)$ | 0 | 1 | $\sqrt{2} \approx 1.4$ | $\sqrt{3} \approx 1.7$ | 2 | $\sqrt{5} \approx 2.2$ |

Plotting points, we obtain the sketch shown in Figure 5. Note that the $x$-intercept is 1 and there is no $y$-intercept.

**FIGURE 5**

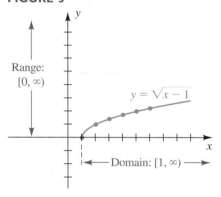

**(b)** Referring to Figure 5, note that the domain of $f$ consists of all real numbers $x$ such that $x \geq 1$ or, equivalently, the interval $[1, \infty)$. The range of $f$ is the set of all real numbers $y$ such that $y \geq 0$ or, equivalently, $[0, \infty)$.

---

The **square root function,** defined by $f(x) = \sqrt{x}$, has a graph similar to the one in Figure 5, but the end point is at $(0, 0)$. The $y$-value of a point on this graph is the number displayed on a calculator when a square root is requested. This graphical relationship may help you remember that $\sqrt{9}$ is 3 and that $\sqrt{9}$ is *not* $\pm 3$. Similarly, $f(x) = x^2$, $f(x) = x^3$, and $f(x) = \sqrt[3]{x}$ are often referred to as the **squaring function,** the **cubing function,** and the **cube root function,** respectively.

In Example 4, as $x$ increases, the function value $f(x)$ also increases, and we say that the graph of $f$ *rises* (see Figure 5). A function of this type is said to be *increasing*. For certain functions, $f(x)$ decreases as $x$ increases. In this case the graph *falls*, and $f$ is a *decreasing* function. In general, we shall consider functions that increase or decrease on an interval $I$, as described in the following chart, where $x_1$ and $x_2$ denote numbers in $I$.

## Increasing, Decreasing, and Constant Functions

| Terminology | Definition | Graphical interpretation |
|---|---|---|
| $f$ is **increasing** on an interval $I$ | $f(x_1) < f(x_2)$ whenever $x_1 < x_2$ | |
| $f$ is **decreasing** on an interval $I$ | $f(x_1) > f(x_2)$ whenever $x_1 < x_2$ | |
| $f$ is **constant** on an interval $I$ | $f(x_1) = f(x_2)$ for every $x_1$ and $x_2$ | |

An example of an *increasing function* is the **identity function,** whose equation is $f(x) = x$ and whose graph is the line through the origin with slope 1. An example of a *decreasing function* is $f(x) = -x$, an equation of the line through the origin with slope $-1$. If $f(x) = c$ for every real number $x$, then $f$ is called a *constant function.*

We shall use the phrases *f is increasing* and *f(x) is increasing* interchangeably. We shall do the same with the terms *decreasing* and *constant.*

**EXAMPLE 5**    *Using a graph to find domain, range, and where a function increases or decreases*

Let $f(x) = \sqrt{9 - x^2}$.

**(a)** Sketch the graph of $f$.

**(b)** Find the domain and range of $f$.

**(c)** Find the intervals on which $f$ is increasing or is decreasing.

*SOLUTION*

**(a)** By definition, the graph of $f$ is the graph of the equation $y = \sqrt{9 - x^2}$. We know from our work with circles in Section 1.7 that the graph of $x^2 + y^2 = 9$ is a circle of radius 3 with center at the origin. Solving the equation $x^2 + y^2 = 9$ for $y$ gives us $y = \pm\sqrt{9 - x^2}$. It follows that the graph of $f$ is the *upper half* of the circle, as illustrated in Figure 6.

**(b)** Referring to Figure 6, we see that the domain of $f$ is the closed interval $[-3, 3]$, and the range of $f$ is the interval $[0, 3]$.

**(c)** The graph rises as $x$ increases from $-3$ to $0$, so $f$ is increasing on the closed interval $[-3, 0]$. Thus, as shown in the preceding chart, if $x_1 < x_2$ in $[-3, 0]$, then $f(x_1) < f(x_2)$ (note that *possibly* $x_1 = -3$ or $x_2 = 0$).

The graph falls as $x$ increases from $0$ to $3$, so $f$ is decreasing on the closed interval $[0, 3]$. In this case, the chart indicates that if $x_1 < x_2$ in $[0, 3]$, then $f(x_1) > f(x_2)$ (note that *possibly* $x_1 = 0$ or $x_2 = 3$).

**FIGURE 6**

The following type of function is one of the most basic in algebra.

**Definition of Linear Function**

A function $f$ is a **linear function** if

$$f(x) = ax + b,$$

where $x$ is any real number and $a$ and $b$ are constants.

The graph of $f$ in the preceding definition is the graph of $y = ax + b$, which, by the slope-intercept form, is a line with slope $a$ and $y$-intercept $b$.

Thus, *the graph of a linear function is a line.* Since $f(x)$ exists for every $x$, the domain of $f$ is $\mathbb{R}$. As illustrated in the next example, if $a \neq 0$, then the range of $f$ is also $\mathbb{R}$.

**EXAMPLE 6**   *Sketching the graph of a linear function*

Let $f(x) = 2x + 3$.

**(a)** Sketch the graph of $f$.

**(b)** Find the domain and range of $f$.

**(c)** Determine where $f$ is increasing or is decreasing.

**FIGURE 7**

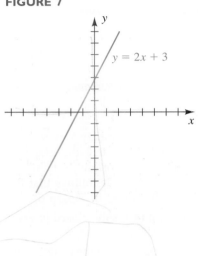

**SOLUTION**

**(a)** Since $f(x)$ has the form $ax + b$, with $a = 2$ and $b = 3$, $f$ is a linear function. The graph of $y = 2x + 3$ is the line with slope 2 and $y$-intercept 3, illustrated in Figure 7.

**(b)** We see from the graph that $x$ and $y$ may be any real numbers, so both the domain and the range of $f$ are $\mathbb{R}$.

**(c)** Since the slope $a$ is positive, the graph of $f$ rises as $x$ increases; that is, $f(x_1) < f(x_2)$ whenever $x_1 < x_2$. Thus, $f$ is increasing throughout its domain.

In applications it is sometimes necessary to determine a specific linear function from given data, as in the next example.

**EXAMPLE 7**   *Finding a linear function*

If $f$ is a linear function such that $f(-2) = 5$ and $f(6) = 3$, find $f(x)$, where $x$ is any real number.

**FIGURE 8**

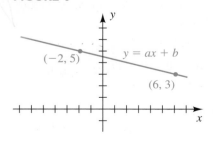

**SOLUTION**   By the definition of linear function, $f(x) = ax + b$, where $a$ and $b$ are constants. Moreover, the given function values tell us that the points $(-2, 5)$ and $(6, 3)$ are on the graph of $f$—that is, on the line $y = ax + b$ illustrated in Figure 8. The slope $a$ of this line is

$$a = \frac{5 - 3}{-2 - 6} = \frac{2}{-8} = -\frac{1}{4},$$

and hence $f(x)$ has the form

$$f(x) = -\tfrac{1}{4}x + b.$$

To find the value of $b$, we may use the fact that $f(6) = 3$, as follows:

$$f(6) = -\tfrac{1}{4}(6) + b \qquad \text{let } x = 6 \text{ in } f(x) = -\tfrac{1}{4}x + b$$
$$3 = -\tfrac{3}{2} + b \qquad f(6) = 3$$
$$b = 3 + \tfrac{3}{2} = \tfrac{9}{2} \qquad \text{solve for } b$$

Thus, the linear function satisfying $f(-2) = 5$ and $f(6) = 3$ is

$$f(x) = -\tfrac{1}{4}x + \tfrac{9}{2}.$$

Many formulas that occur in mathematics and the sciences determine functions. For instance, the formula $A = \pi r^2$ for the area $A$ of a circle of radius $r$ assigns to each positive real number $r$ exactly one value of $A$. This determines a function $f$ such that $f(r) = \pi r^2$, and we may write $A = f(r)$. The letter $r$, which represents an arbitrary number from the domain of $f$, is called an **independent variable.** The letter $A$, which represents a number from the range of $f$, is a **dependent variable,** since its value depends on the number assigned to $r$. If two variables $r$ and $A$ are related in this manner, we say that *A is a function of r.* In applications, the independent variable and dependent variable are sometimes referred to as the **input variable** and **output variable,** respectively. As another example, if an automobile travels at a uniform rate of 50 mi/hr, then the distance $d$ (miles) traveled in time $t$ (hours) is given by $d = 50t$, and hence *the distance d is a function of time t.*

**EXAMPLE 8**   *Expressing the volume of a tank as a function of its radius*

A steel storage tank for propane gas is to be constructed in the shape of a right circular cylinder of altitude 10 feet with a hemisphere attached to each end. The radius $r$ is yet to be determined. Express the volume $V$ (in ft$^3$) of the tank as a function of $r$ (in ft).

**FIGURE 9**

**SOLUTION**   The tank is illustrated in Figure 9. We may find the volume of the cylindrical part of the tank by multiplying the altitude 10 by the area $\pi r^2$ of the base of the cylinder. This gives us

$$\text{volume of cylinder} = 10(\pi r^2) = 10\pi r^2.$$

The two hemispherical ends, taken together, form a sphere of radius $r$. Using the formula for the volume of a sphere, we obtain

$$\text{volume of the two ends} = \tfrac{4}{3}\pi r^3.$$

*(continued)*

Thus, the volume $V$ of the tank is

$$V = \tfrac{4}{3}\pi r^3 + 10\pi r^2.$$

This formula expresses $V$ as a function of $r$. In factored form,

$$V = \tfrac{1}{3}\pi r^2 (4r + 30) = \tfrac{2}{3}\pi r^2 (2r + 15).$$

### EXAMPLE 9    *Expressing a distance as a function of time*

Two ships leave port at the same time, one sailing west at a rate of 17 mi/hr and the other sailing south at 12 mi/hr. If $t$ is the time (in hours) after their departure, express the distance $d$ between the ships as a function of $t$.

**FIGURE 10**

*SOLUTION*    To help visualize the problem, we begin by drawing a picture and labeling it, as in Figure 10. By the Pythagorean theorem,

$$d^2 = a^2 + b^2, \quad \text{or} \quad d = \sqrt{a^2 + b^2}.$$

Since distance = (rate)(time) and the rates are 17 and 12, respectively,

$$a = 17t \quad \text{and} \quad b = 12t.$$

Substitution in $d = \sqrt{a^2 + b^2}$ gives us

$$d = \sqrt{(17t)^2 + (12t)^2} = \sqrt{289t^2 + 144t^2} = \sqrt{433t^2} \approx (20.8)t.$$

Ordered pairs can be used to obtain an alternative approach to functions. We first observe that a function $f$ from $D$ to $E$ determines the following set $W$ of ordered pairs:

$$W = \{(x, f(x)): x \text{ is in } D\}.$$

Thus, $W$ consists of all ordered pairs such that the first number $x$ is in $D$ and the second number is the function value $f(x)$. In Example 1, where $f(x) = x^2$, $W$ is the set of all ordered pairs of the form $(x, x^2)$. It is important to note that, *for each $x$, there is exactly one ordered pair $(x, y)$ in W having $x$ in the first position.*

Conversely, if we begin with a set $W$ of ordered pairs such that each $x$ in $D$ appears exactly once in the first position of an ordered pair, then $W$ determines a function. Specifically, for each $x$ in $D$ there is exactly one pair $(x, y)$ in $W$, and by letting $y$ correspond to $x$, we obtain a function with domain $D$. The range consists of all real numbers $y$ that appear in the second position of the ordered pairs.

It follows from the preceding discussion that the next statement could also be used as a definition of function.

**Alternative Definition of Function**

> A **function** with domain $D$ is a set $W$ of ordered pairs such that, for each $x$ in $D$, there is exactly one ordered pair $(x, y)$ in $W$ having $x$ in the first position.

In terms of the preceding definition, the ordered pairs $(x, \sqrt{x - 1})$ determine the function of Example 4 given by $f(x) = \sqrt{x - 1}$. Note, however, that if

$$W = \{(x, y)\colon x^2 = y^2\},$$

then $W$ is *not* a function, since for a given $x$, there may be more than one pair in $W$ with $x$ in the first position. For example, if $x = 2$, then both $(2, 2)$ and $(2, -2)$ are in $W$.

In some scientific investigations, the terminology of *variation* or *proportion* is used to describe relationships between variable quantities. In the following chart, $k$ is a nonzero real number called a **constant of variation** or a **constant of proportionality.**

**Directly and Inversely Proportional**

| Terminology | General formula | Illustration |
|---|---|---|
| $y$ **varies directly** as $x$, or $y$ is **directly proportional** to $x$ | $y = kx$ | $C = 2\pi r$, where $C$ is the circumference of a circle, $r$ is the radius, and $k = 2\pi$ |
| $y$ **varies inversely** as $x$, or $y$ is **inversely proportional** to $x$ | $y = \dfrac{k}{x}$ | $I = \dfrac{110}{R}$, where $I$ is the current in an electrical circuit, $R$ is the resistance, and $k = 110$ is the voltage |

The variable $x$ in the chart can also represent a power. For example, the formula $A = \pi r^2$ states that the area $A$ of a circle varies directly as the *square* of the radius $r$, where $\pi$ is the constant of variation. Similarly, the formula $V = \frac{4}{3}\pi r^3$ states that the volume $V$ of a sphere is directly proportional to the *cube* of the radius. In this case the constant of proportionality is $\frac{4}{3}\pi$.

**EXAMPLE 10** *Finding the support load of a rectangular beam*

The weight $W$ that can be safely supported by a beam with a rectangular cross section varies directly as the product of the width $w$ and square of the depth $d$ of the cross section and inversely as the length $l$ of the beam. If a 2-inch by 4-inch beam that is 8 feet long safely supports a load of 500 pounds, what weight can be safely supported by a 2-inch by 8-inch beam that is 10 feet long? (Assume that the width is the *shorter* dimension of the cross section.)

**SOLUTION**    A general formula for the weight $W$ is

$$W = k\frac{wd^2}{l},$$

where $k$ is a constant of variation. Using the given data—that is, $w = 2$, $d = 4$, $l = 8$, and $W = 500$—we substitute and determine the value of $k$ as follows:

$$500 = k\frac{2(4^2)}{8} \quad \text{or, equivalently,} \quad k = 125$$

To answer the question, we substitute $k = 125$, $w = 2$, $d = 8$, and $l = 10$ into the general formula for $W$, obtaining

$$W = 125\frac{2 \cdot 8^2}{10} = 1600 \text{ lb.}$$

In the next example we illustrate how some of the concepts presented in this section may be studied with the aid of a graphing utility. Hereafter, when making assignments on a graphing utility, we will frequently refer to variables such as $Y_1$ and $Y_2$ as the *functions* $Y_1$ and $Y_2$.

**EXAMPLE 11**    *Analyzing the graph of a function*

Let $f(x) = x^{2/3} - 3$.

**(a)** Find $f(-2)$.

**(b)** Sketch the graph of $f$.

**(c)** State the domain and range of $f$.

**(d)** State the intervals on which $f$ is increasing or is decreasing.

**(e)** Estimate the $x$-intercepts of the graph to one-decimal-place accuracy.

**SOLUTION**    Some specific keystrokes for parts (a) and (b) using the TI-81/82 graphics calculator are given in Appendix I.

**(a)** A representation of $f$ on a computational device may have the form

X^(2/3) − 3    or    (X^(1/3))^2 − 3    or    (X^2)^(1/3) − 3.

We assign one of these expressions to the function $Y_1$. To find the value of $Y_1$ at $x = -2$, we first assign the value $-2$ to a memory location identified as X. This is usually done with a "store" or "assign" operation on a computational device. We next determine the value of $Y_1$ by requesting the computational device to indicate the contents of the memory location that contains the value of $Y_1$. We often refer to this process of finding a value of $Y_1$ as "querying $Y_1$." Upon querying $Y_1$, we see that its value is approximately $-1.41$ (that is, $f(-2) \approx -1.41$). It should be noted that not all computational devices process rational exponents in the same fashion. If you did not obtain this answer for $f(-2)$, change your representation of $Y_1$ before proceeding.

**(b)** Using the viewing rectangle $[-15, 15]$ by $[-10, 10]$ to graph $Y_1$ gives us a display similar to that of Figure 11. The v-shaped part of the graph of $f$ at $x = 0$ is called a **cusp.**

**(c)** The domain of $f$ is $\mathbb{R}$, since we may input any value for $x$. The figure indicates that $y \geq -3$, so we conclude that the range of $f$ is $[-3, \infty)$.

**(d)** From the figure, we see that $f$ is decreasing on $(-\infty, 0]$ and is increasing on $[0, \infty)$.

**(e)** Using the tracing and zoom-in features, we find that the positive $x$-intercept in Figure 11 is approximately 5.2. Since $f$ is symmetric with respect to the $y$-axis, the negative $x$-intercept is about $-5.2$.

**FIGURE 11**   $[-15, 15]$ by $[-10, 10]$

As a reference aid, some common graphs and their equations are listed in Appendix II. Many of these graphs are graphs of functions.

## 2.1 EXERCISES

1 If $f(x) = -x^2 - x - 4$, find $f(-2)$, $f(0)$, and $f(4)$.

2 If $f(x) = -x^3 - x^2 + 3$, find $f(-3)$, $f(0)$, and $f(2)$.

3 If $f(x) = \sqrt{x - 4} - 3x$, find $f(4)$, $f(8)$, and $f(13)$.

4 If $f(x) = \dfrac{x}{x - 3}$, find $f(-2)$, $f(0)$, and $f(3)$.

**Exer. 5–8: If $a$ and $h$ are real numbers, find**

(a) $f(a)$     (b) $f(-a)$     (c) $-f(a)$     (d) $f(a + h)$

(e) $f(a) + f(h)$          (f) $\dfrac{f(a + h) - f(a)}{h}$, if $h \neq 0$

5 $f(x) = 5x - 2$                         6 $f(x) = 3 - 4x$

7 $f(x) = x^2 - x + 3$                   8 $f(x) = 2x^2 + 3x - 7$

**Exer. 9–10: Simplify the difference quotient**
$$\dfrac{f(x + h) - f(x)}{h} \text{ if } h \neq 0.$$

9 $f(x) = x^2 + 5$                        10 $f(x) = 1/x^2$

**Exer. 11–12: Simplify the difference quotient** $\dfrac{f(x) - f(a)}{x - a}$

**if $x \neq a$.**

11 $f(x) = \sqrt{x - 3}$   (*Hint:* Rationalize the numerator.)

12 $f(x) = x^3 - 2$

**Exer. 13–16: If $a$ is a positive real number, find**

(a) $g\left(\dfrac{1}{a}\right)$     (b) $\dfrac{1}{g(a)}$     (c) $g(\sqrt{a})$     (d) $\sqrt{g(a)}$

13 $g(x) = 4x^2$                           14 $g(x) = 2x - 5$

15 $g(x) = \dfrac{2x}{x^2 + 1}$    16 $g(x) = \dfrac{x^2}{x + 1}$

Exer. 17–18: For the graph of the function $f$ sketched in the figure, determine

(a) the domain    (b) the range    (c) $f(1)$

(d) all $x$ such that $f(x) = 1$

(e) all $x$ such that $f(x) > 1$

17

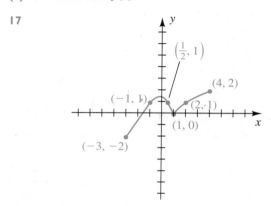

18

y

$(-1, 1)$
$(-2, 2)$    $(4, 2)$
$(-3, 1)$    $(3, 1)$    $(5, 1)$

x

$(-5, -1)$    $(1, -1)$    $(7, -1)$

Exer. 19–30: Find the domain of $f$.

19 $f(x) = \sqrt{2x + 7}$    20 $f(x) = \sqrt{8 - 3x}$

21 $f(x) = \sqrt{9 - x^2}$    22 $f(x) = \sqrt{x^2 - 25}$

23 $f(x) = \dfrac{x + 1}{x^3 - 4x}$    24 $f(x) = \dfrac{4x}{6x^2 + 13x - 5}$

25 $f(x) = \dfrac{\sqrt{2x - 3}}{x^2 - 5x + 4}$    26 $f(x) = \dfrac{\sqrt{4x - 3}}{x^2 - 4}$

27 $f(x) = \dfrac{x - 4}{\sqrt{x - 2}}$    28 $f(x) = \dfrac{1}{(x - 3)\sqrt{x + 3}}$

29 $f(x) = \sqrt{x + 2} + \sqrt{2 - x}$

30 $f(x) = \sqrt{(x - 2)(x - 6)}$

Exer. 31–40: (a) Sketch the graph of $f$. (b) Find the domain $D$ and range $R$ of $f$. (c) Find the intervals on which $f$ is increasing, is decreasing, or is constant.

31 $f(x) = 3x - 2$    32 $f(x) = -2x + 3$

33 $f(x) = 4 - x^2$    34 $f(x) = x^2 - 1$

35 $f(x) = \sqrt{x + 4}$    36 $f(x) = \sqrt{4 - x}$

37 $f(x) = -2$    38 $f(x) = 3$

39 $f(x) = -\sqrt{36 - x^2}$    40 $f(x) = \sqrt{16 - x^2}$

Exer. 41–42: If a linear function $f$ satisfies the given conditions, find $f(x)$.

41 $f(-3) = 1$ and $f(3) = 2$

42 $f(-2) = 7$ and $f(4) = -2$

Exer. 43–52: Determine whether the set $W$ of ordered pairs is a function in the sense of the alternative definition of function on page 125.

43 $W = \{(x, y): 2y = x^2 + 5\}$

44 $W = \{(x, y): x = 3y + 2\}$

45 $W = \{(x, y): x^2 + y^2 = 4\}$

46 $W = \{(x, y): y^2 - x^2 = 1\}$

47 $W = \{(x, y): y = 3\}$    48 $W = \{(x, y): x = 3\}$

49 $W = \{(x, y): xy = 0\}$

50 $W = \{(x, y): x + y = 0\}$

51 $W = \{(x, y): |y| = |x|\}$    52 $W = \{(x, y): y < x\}$

53 **Constructing a box** From a rectangular piece of cardboard having dimensions 20 inches × 30 inches, an open box is to be made by cutting out identical squares of area $x^2$ from each corner and turning up the sides (see the figure). Express the volume $V$ of the box as a function of $x$.

EXERCISE 53

**54** *Constructing a storage tank* Refer to Example 8. A steel storage tank for propane gas is to be constructed in the shape of a right circular cylinder of altitude 10 feet with a hemisphere attached to each end. The radius $r$ is yet to be determined. Express the surface area $S$ of the tank as a function of $r$.

**55** *Dimensions of a building* A small office unit is to contain 500 ft$^2$ of floor space. A simplified model is shown in the figure.

(a) Express the length $y$ of the building as a function of the width $x$.

(b) If the walls cost $100 per running foot, express the cost $C$ of the walls as a function of the width $x$. (Disregard the wall space above the doors and the thickness of the walls.)

EXERCISE 55

**56** *Dimensions of an aquarium* An aquarium of height 1.5 feet is to have a volume of 6 ft$^3$. Let $x$ denote the length of the base and $y$ the width (see the figure).

(a) Express $y$ as a function of $x$.

(b) Express the total number $S$ of square feet of glass needed as a function of $x$.

EXERCISE 56

**57** *Skyline ordinance* A city council is proposing a new skyline ordinance. It would require the setback $S$ for any

building from a residence to be a minimum of 100 feet, plus an additional 6 feet for each foot of height above 25 feet. Find a linear function for $S$ in terms of $h$.

EXERCISE 57

Setback

**58** *Energy tax* A proposed energy tax $T$ on gasoline, which would affect the cost of driving a vehicle, is to be computed by multiplying the number $x$ of gallons of gasoline that you buy by 125,000 (the number of BTUs per gallon of gasoline) and then multiplying the total BTUs by the tax—34.2 cents per million BTUs. Find a linear function for $T$ in terms of $x$.

**59** *Childhood growth* For children between ages 6 and 10, height $y$ (in inches) is frequently a linear function of age $t$ (in years). The height of a certain child is 48 inches at age 6 and 50.5 inches at age 7.

(a) Express $y$ as a function of $t$.

(b) Sketch the line in part (a), and interpret the slope.

(c) Predict the height of the child at age 10.

**60** *Radioactive contamination* It has been estimated that 1000 curies of a radioactive substance introduced at a point on the surface of the open sea would spread over an area of 40,000 km$^2$ in 40 days. Assuming that the area covered by the radioactive substance is a linear function of time $t$ and is always circular in shape, express the radius $r$ of the contamination as a function of $t$.

**61** *Distance to a hot-air balloon* A hot-air balloon is released at 1:00 P.M. and rises vertically at a rate of 2 m/sec. An observation point is situated 100 meters from a point on the ground directly below the balloon (see the figure on page 130). If $t$ denotes the time (in seconds) after 1:00 P.M., express the distance $d$ between the balloon and the observation point as a function of $t$.

**EXERCISE 61**

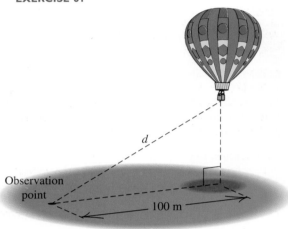

$d$

Observation point

100 m

62 Triangle $ABC$ is inscribed in a semicircle of diameter 15 (see the figure).

(a) If $x$ denotes the length of side $AC$, express the length $y$ of side $BC$ as a function of $x$. (*Hint:* Angle $ACB$ is a right angle.)

(b) Express the area $\mathcal{A}$ of triangle $ABC$ as a function of $x$, and state the domain of this function.

**EXERCISE 62**

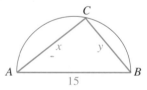

63 **Distance to the earth**   From an exterior point $P$ that is $h$ units from a circle of radius $r$, a tangent line is drawn to the circle (see the figure). Let $y$ denote the distance from the point $P$ to the point of tangency $T$.

(a) Express $y$ as a function of $h$. (*Hint:* If $C$ is the center of the circle, then $PT$ is perpendicular to $CT$.)

(b) If $r$ is the radius of the earth and $h$ is the altitude of a space shuttle, then $y$ is the maximum distance to the earth that an astronaut can see from the shuttle. In particular, if $h = 200$ mi and $r \approx 4000$ mi, approximate $y$.

**EXERCISE 63**

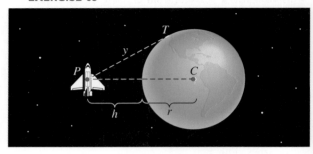

64 **Length of a tightrope**   The figure illustrates the apparatus for a tightrope walker. Two poles are set 50 feet apart, but the point of attachment $P$ for the rope is yet to be determined.

(a) Express the length $L$ of the rope as a function of the distance $x$ from $P$ to the ground.

(b) If the total walk is to be 75 feet, determine the distance from $P$ to the ground.

**EXERCISE 64**

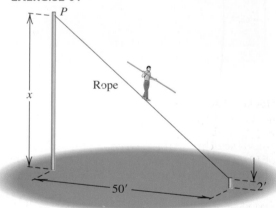

65 **Airport runway**   The relative positions of an aircraft runway and a 20-foot-tall control tower are shown in the figure. The beginning of the runway is at a perpendicu-

**EXERCISE 65**

lar distance of 300 feet from the base of the tower. If $x$ denotes the distance an airplane has moved down the runway, express the distance $d$ between the airplane and the control booth as a function of $x$.

66 **Destination time** A man in a rowboat that is 2 miles from the nearest point $A$ on a straight shoreline wishes to reach a house located at a point $B$ that is 6 miles farther down the shoreline (see the figure). He plans to row to a point $P$ that is between $A$ and $B$ and is $x$ miles from the house, and then he will walk the remainder of the distance. Suppose he can row at a rate of 3 mi/hr and can walk at a rate of 5 mi/hr. If $T$ is the total time required to reach the house, express $T$ as a function of $x$.

**EXERCISE 66**

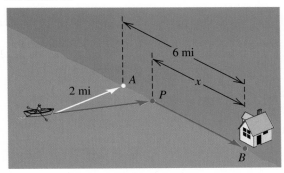

67 **Hooke's law** Hooke's law states that the force $F$ required to stretch a spring $x$ units beyond its natural length is directly proportional to $x$. If a weight of 4 pounds stretches a certain spring from its natural length of 10 inches to a length of 10.3 inches, what weight will stretch the spring to a length of 11.5 inches?

68 **Period of a planet** Kepler's third law states that the period $T$ of a planet (the time needed to make one complete revolution about the sun) is directly proportional to the $\frac{3}{2}$ power of its average distance $d$ (in millions of miles) from the sun. For the planet Earth, $T = 365$ days and $d = 93$. Estimate the period of Venus if its average distance from the sun is 67 million miles.

69 **Electrical resistance** The electrical resistance $R$ of a wire varies directly as its length $l$ and inversely as the square of its diameter $d$. A wire 100 feet long of diameter 0.01 inch has a resistance of 25 ohms. Find the resistance of a wire made of the same material that has a diameter of 0.015 inch and is 50 feet long.

70 **Coulomb's law** Coulomb's law in electrical theory states that the force $F$ of attraction between two oppositely charged particles varies directly as the product of the magnitudes $Q_1$ and $Q_2$ of the charges and inversely as the square of the distance $d$ between the particles. What is the effect of reducing the distance between the particles by a factor of one-fourth?

71 **Density at a point** A thin flat plate is situated in an $xy$-plane such that the density $d$ (in lb/ft$^2$) at the point $P(x, y)$ is inversely proportional to the square of the distance from the origin. What is the effect on the density at $P$ if the $x$- and $y$-coordinates are each multiplied by $\frac{1}{3}$?

72 **Temperature at a point** A flat metal plate is positioned in an $xy$-plane such that the temperature $T$ (in °C) at the point $(x, y)$ is inversely proportional to the distance from the origin. If the temperature at the point $P(3, 4)$ is 20 °C, find the temperature at the point $Q(24, 7)$.

Exer. 73–76: (a) Sketch the graph of $f$ on the given interval $[a, b]$. (b) Estimate the range of $f$ on $[a, b]$. (c) Estimate the intervals on which $f$ is increasing or is decreasing.

73 $f(x) = \dfrac{x^{1/3}}{1 + x^4}$;  $\quad [-2, 2]$

74 $f(x) = x^4 - 0.4x^3 - 0.8x^2 + 0.2x + 0.1$;  $\quad [-1, 1]$

75 $f(x) = x^5 - 3x^2 + 1$;  $\quad [-0.7, 1.4]$

76 $f(x) = \dfrac{1 - x^3}{1 + x^4}$;  $\quad [-4, 4]$

Exer. 77–78: In Exercises 39–40 of Section 1.4, algebraic methods were used to find solutions to each of the following equations. Now solve the equation graphically by assigning the expression on the left side to $Y_1$ and the number on the right side to $Y_2$ and then finding the $x$-coordinates of all points of intersection of the two graphs.

77 (a) $x^{5/3} = 32$  (b) $x^{4/3} = 16$  (c) $x^{2/3} = -36$
   (d) $x^{3/4} = 125$  (e) $x^{3/2} = -27$

78 (a) $x^{3/5} = -27$  (b) $x^{2/3} = 25$  (c) $x^{4/3} = -49$
   (d) $x^{3/2} = 27$  (e) $x^{3/4} = -8$

## 2.2   GRAPHS OF FUNCTIONS

In this section we discuss aids for sketching graphs of certain types of functions. In particular, a function $f$ is called **even** if $f(-x) = f(x)$ for every $x$ in its domain. In this case, the equation $y = f(x)$ is not changed if $-x$ is substituted for $x$, and hence, from symmetry test (1) of Section 1.7, the graph of an even function is symmetric with respect to the $y$-axis.

A function $f$ is called **odd** if $f(-x) = -f(x)$ for every $x$ in its domain. If we apply symmetry test (3) of Section 1.7 to the equation $y = f(x)$, we see that the graph of an odd function is symmetric with respect to the origin.

These facts are summarized in the first two columns of the next chart.

### Even and Odd Functions

| Terminology | Definition | Illustration | Symmetry of graph |
|---|---|---|---|
| $f$ is an **even function.** | $f(-x) = f(x)$ for every $x$ in the domain. | $y = f(x) = x^2$ | $y$-axis |
| $f$ is an **odd function.** | $f(-x) = -f(x)$ for every $x$ in the domain. | $y = f(x) = x^3$ | the origin |

**EXAMPLE I**   *Determining whether a function is even or odd*

Determine whether $f$ is even, odd, or neither even nor odd.

(a) $f(x) = 3x^4 - 2x^2 + 5$    (b) $f(x) = 2x^5 - 7x^3 + 4x$

(c) $f(x) = x^3 + x^2$

**SOLUTION**   In each case the domain of $f$ is $\mathbb{R}$. To determine whether $f$ is even or odd, we begin by examining $f(-x)$, where $x$ is any real number.

(a) $f(-x) = 3(-x)^4 - 2(-x)^2 + 5$   substitute $-x$ for $x$ in $f(x)$

$\quad\quad\quad = 3x^4 - 2x^2 + 5$    simplify

$\quad\quad\quad = f(x)$          definition of $f$

Since $f(-x) = f(x)$, $f$ is an even function.

(b) $f(-x) = 2(-x)^5 - 7(-x)^3 + 4(-x)$   substitute $-x$ for $x$ in $f(x)$

$\quad\quad\quad = -2x^5 + 7x^3 - 4x$     simplify

$\quad\quad\quad = -(2x^5 - 7x^3 + 4x)$    factor out $-1$

$\quad\quad\quad = -f(x)$           definition of $f$

Since $f(-x) = -f(x)$, $f$ is an odd function.

**(c)** $f(-x) = (-x)^3 + (-x)^2$        substitute $-x$ for $x$ in $f(x)$
$\qquad\quad = -x^3 + x^2$                simplify

Since $f(-x) \neq f(x)$ and $f(-x) \neq -f(x)$ (note that $-f(x) = -x^3 - x^2$), the function $f$ is neither even nor odd.

In the next example we consider the **absolute value function** $f$, defined by $f(x) = |x|$.

**EXAMPLE 2**    *Sketching the graph of the absolute value function*
Let $f(x) = |x|$.
**(a)** Determine whether $f$ is even or odd.
**(b)** Sketch the graph of $f$.
**(c)** Find the intervals on which $f$ is increasing or is decreasing.

**SOLUTION**
**(a)** The domain of $f$ is $\mathbb{R}$, because the absolute value of $x$ exists for every real number $x$. If $x$ is in $\mathbb{R}$, then

$$f(-x) = |-x| = |x| = f(x).$$

Thus, $f$ is an even function, since $f(-x) = f(x)$.
**(b)** Since $f$ is even, its graph is symmetric with respect to the $y$-axis. If $x \geq 0$, then $|x| = x$, and therefore the first quadrant part of the graph coincides with the line $y = x$. Sketching this half-line and using symmetry gives us Figure 12.
**(c)** Referring to the graph, we see that $f$ is decreasing on $(-\infty, 0]$ and is increasing on $[0, \infty)$.

**FIGURE 12**

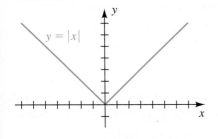

If we know the graph of $y = f(x)$, it is easy to sketch the graphs of

$$y = f(x) + c \quad \text{and} \quad y = f(x) - c$$

for any positive real number $c$. As in the next chart, for $y = f(x) + c$, we add $c$ to the $y$-coordinate of each point on the graph of $y = f(x)$. This *shifts* the graph of $f$ *upward* a distance $c$. For $y = f(x) - c$ with $c > 0$, we subtract $c$ from each $y$-coordinate, thereby shifting the graph of $f$ a distance $c$ *downward*. These are called **vertical shifts** of graphs.

**Vertically Shifting the Graph of** $y = f(x)$

| Equation | $y = f(x) + c$ with $c > 0$ | $y = f(x) - c$ with $c > 0$ |
|---|---|---|
| Effect on graph | The graph of $f$ is shifted vertically upward a distance $c$. | The graph of $f$ is shifted vertically downward a distance $c$. |
| Graphical interpretation | 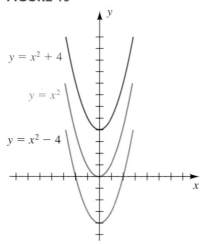 | |

**EXAMPLE 3    *Vertically shifting a graph***

Sketch the graph of $f$:

**(a)** $f(x) = x^2$    **(b)** $f(x) = x^2 + 4$    **(c)** $f(x) = x^2 - 4$

**FIGURE 13**

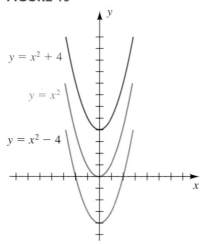

*SOLUTION*    We shall sketch all graphs on the same coordinate plane.

**(a)** Since

$$f(-x) = (-x)^2 = x^2 = f(x),$$

the function $f$ is even, and hence its graph is symmetric with respect to the $y$-axis. Several points on the graph of $y = x^2$ are $(0, 0)$, $(1, 1)$, $(2, 4)$, and $(3, 9)$. Drawing a smooth curve through these points and reflecting through the $y$-axis gives us the sketch in Figure 13. The graph is a parabola with vertex at the origin and opening upward.

**(b)** To sketch the graph of $y = x^2 + 4$, we add 4 to the $y$-coordinate of each point on the graph of $y = x^2$; that is, we shift the graph in part (a) upward 4 units, as shown in the figure.

**(c)** To sketch the graph of $y = x^2 - 4$, we decrease $y$-coordinates of $y = x^2$ by 4; that is, we shift the graph in part (a) downward 4 units.

We can also consider **horizontal shifts** of graphs. Specifically, if $c > 0$, consider the graphs of $y = f(x)$ and $y = f(x - c)$ sketched on the same coordinate plane, as illustrated in the next chart. Since

$$f(a) = f([a + c] - c),$$

we see that the point with $x$-coordinate $a$ on the graph of $y = f(x)$ has the same $y$-coordinate as the point with $x$-coordinate $a + c$ on the graph of $y = f(x - c)$. This implies that the graph of $y = f(x - c)$ can be obtained by shifting the graph of $y = f(x)$ *to the right* a distance $c$. Similarly, the graph of $y = f(x + c)$ can be obtained by shifting the graph of $f$ *to the left* a distance $c$, as shown in the chart.

**Horizontally Shifting the Graph of $y = f(x)$**

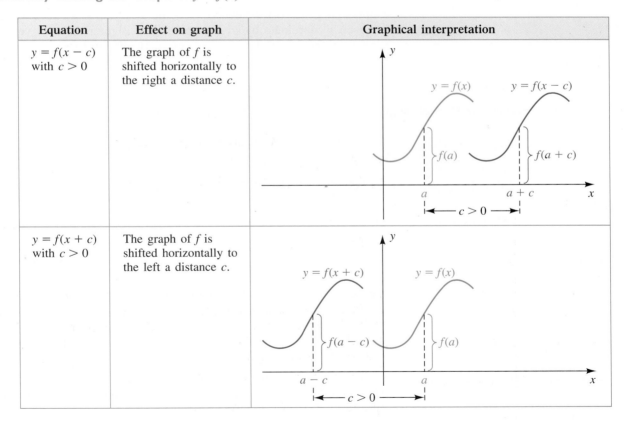

| Equation | Effect on graph | Graphical interpretation |
|---|---|---|
| $y = f(x - c)$ with $c > 0$ | The graph of $f$ is shifted horizontally to the right a distance $c$. | |
| $y = f(x + c)$ with $c > 0$ | The graph of $f$ is shifted horizontally to the left a distance $c$. | |

Horizontal and vertical shifts are also referred to as *translations*.

**FIGURE 14**

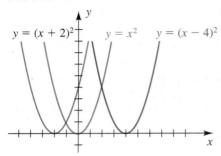

$y = (x + 2)^2$    $y = x^2$    $y = (x - 4)^2$

**EXAMPLE 4**    *Horizontally shifting a graph*

Sketch the graph of $f$:

**(a)** $f(x) = (x - 4)^2$    **(b)** $f(x) = (x + 2)^2$

**SOLUTION**    The graph of $y = x^2$ is sketched in Figure 14.

**(a)** Shifting the graph of $y = x^2$ to the right 4 units gives us the graph of $y = (x - 4)^2$, shown in the figure.

**(b)** Shifting the graph of $y = x^2$ to the left 2 units leads to the graph of $y = (x + 2)^2$, shown in the figure.

To obtain the graph of $y = cf(x)$ for some real number $c$, we may *multiply* the $y$-coordinates of points on the graph of $y = f(x)$ by $c$. For example, if $y = 2f(x)$, we double $y$-coordinates; or if $y = \frac{1}{2}f(x)$, we multiply each $y$-coordinate by $\frac{1}{2}$. This procedure is referred to as **vertically stretching** the graph of $f$ (if $c > 1$) or **vertically compressing** the graph (if $0 < c < 1$) and is summarized in the following chart.

**Vertically Stretching or Compressing the Graph of $y = f(x)$**

| Equation | $y = cf(x)$ with $c > 1$ | $y = cf(x)$ with $0 < c < 1$ |
|---|---|---|
| Effect on graph | The graph of $f$ is stretched vertically by a factor $c$. | The graph of $f$ is compressed vertically by a factor $1/c$. |
| Graphical interpretation | | |

**EXAMPLE 5**    *Vertically stretching or compressing a graph*

Sketch the graph of the equation:

**(a)** $y = 4x^2$    **(b)** $y = \frac{1}{4}x^2$

**FIGURE 15**

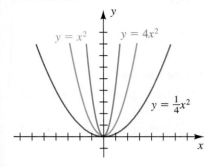

**SOLUTION**

**(a)** To sketch the graph of $y = 4x^2$, we may refer to the graph of $y = x^2$ in Figure 15 and multiply the $y$-coordinate of each point by 4. This stretches the graph of $y = x^2$ vertically by a factor 4 and gives us a narrower parabola that is sharper at the vertex, as illustrated in the figure.

**(b)** The graph of $y = \frac{1}{4}x^2$ may be sketched by multiplying $y$-coordinates of points on the graph of $y = x^2$ by $\frac{1}{4}$. This compresses the graph of $y = x^2$ vertically by a factor $1/\frac{1}{4} = 4$ and gives us a wider parabola that is flatter at the vertex, as shown in Figure 15.

**FIGURE 16**

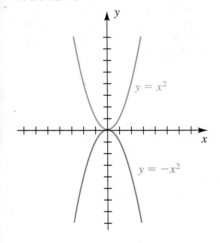

We may obtain the graph of $y = -f(x)$ by multiplying the $y$-coordinate of each point on the graph of $y = f(x)$ by $-1$. Thus, every point $(a, b)$ on the graph of $y = f(x)$ that lies above the $x$-axis determines a point $(a, -b)$ on the graph of $y = -f(x)$ that lies below the $x$-axis. Similarly, if $(c, d)$ lies below the $x$-axis (that is, $d < 0$), then $(c, -d)$ lies above the $x$-axis. The graph of $y = -f(x)$ is a **reflection** of the graph of $y = f(x)$ through the $x$-axis.

**EXAMPLE 6**   *Reflecting a graph through the x-axis*

Sketch the graph of $y = -x^2$.

**SOLUTION**   The graph may be found by plotting points; however, since the graph of $y = x^2$ is familiar to us, we sketch it as in Figure 16 and then multiply $y$-coordinates of points by $-1$. This procedure gives us the reflection through the $x$-axis indicated in the figure.

Sometimes it is useful to compare the graphs of $y = f(x)$ and $y = f(cx)$ if $c \neq 0$. In this case the function values $f(x)$ for

$$a \leq x \leq b$$

are the same as the function values $f(cx)$ for

$$a \leq cx \leq b \quad \text{or, equivalently,} \quad \frac{a}{c} \leq x \leq \frac{b}{c}.$$

This implies that the graph of $f$ is **horizontally compressed** (if $c > 1$) or **horizontally stretched** (if $0 < c < 1$), as summarized in the following chart.

### Horizontally Compressing or Stretching the Graph of $y = f(x)$

| Equation | Effect on graph | Graphical interpretation |
|---|---|---|
| $y = f(cx)$ with $c > 1$ | The graph of $f$ is compressed horizontally by a factor $c$. | |
| $y = f(cx)$ with $0 < c < 1$ | The graph of $f$ is stretched horizontally by a factor $1/c$. | |

If $c < 0$, then the graph of $y = f(cx)$ may be obtained by reflecting the graph of $y = f(|c|x)$ through the $y$-axis. For example, to sketch the graph of $y = f(-2x)$, we reflect the graph of $y = f(2x)$ through the $y$-axis. As a special case, the graph of $y = f(-x)$ is a **reflection** of the graph of $y = f(x)$ through the $y$-axis.

**EXAMPLE 7**   *Horizontally stretching or compressing a graph*

If $f(x) = x^3 - 4x^2$, sketch the graphs of $y = f(x)$, $y = f(2x)$, and $y = f(\frac{1}{2}x)$.

**SOLUTION**   We have the following:

$$y = f(x) = x^3 - 4x^2 = x^2(x - 4)$$
$$y = f(2x) = (2x)^3 - 4(2x)^2 = 8x^3 - 16x^2 = 8x^2(x - 2)$$
$$y = f(\tfrac{1}{2}x) = (\tfrac{1}{2}x)^3 - 4(\tfrac{1}{2}x)^2 = \tfrac{1}{8}x^3 - x^2 = \tfrac{1}{8}x^2(x - 8)$$

Note that the $x$-intercepts of the graph of $y = f(2x)$ are 0 and 2, which are $\frac{1}{2}$ the $x$-intercepts of 0 and 4 for $y = f(x)$. This indicates a horizontal compression by a factor 2.

**FIGURE 17** $[-6, 15]$ by $[-10, 4]$

The $x$-intercepts of the graph of $y = f(\frac{1}{2}x)$ are 0 and 8, which are 2 times the $x$-intercepts for $y = f(x)$. This indicates a horizontal stretching by a factor $1/\frac{1}{2} = 2$.

The graphs, obtained by using a graphing utility with viewing rectangle $[-6, 15]$ by $[-10, 4]$, are shown in Figure 17.

Functions are sometimes described by more than one expression, as in the next examples. We call such functions **piecewise-defined functions.**

**EXAMPLE 8**   *Sketching the graph of a piecewise-defined function*

Sketch the graph of the function $f$ if

$$f(x) = \begin{cases} 2x + 3 & \text{if } x < 0 \\ x^2 & \text{if } 0 \le x < 2 \\ 1 & \text{if } x \ge 2 \end{cases}$$

**FIGURE 18**

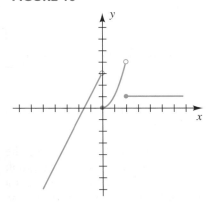

**SOLUTION**   If $x < 0$, then $f(x) = 2x + 3$ and the graph of $f$ coincides with the line $y = 2x + 3$. This gives that portion of the graph to the left of the $y$-axis, sketched in Figure 18. The small circle indicates that the point $(0, 3)$ is *not* on the graph.

If $0 \le x < 2$, we use $x^2$ to find values of $f$, and therefore this part of the graph of $f$ coincides with the parabola $y = x^2$, as indicated in the figure. Note that the point $(2, 4)$ is not on the graph.

Finally, if $x \ge 2$, the values of $f$ are always 1. Thus, the graph of $f$ for $x \ge 2$ is the horizontal half-line illustrated in Figure 18.

If $x$ is a real number, we define the symbol $[\![x]\!]$ as follows:

$$[\![x]\!] = n, \quad \text{where } n \text{ is the greatest integer such that } n \le x$$

If we identify $\mathbb{R}$ with points on a coordinate line, then $n$ is the first integer to the *left* of (or *equal* to) $x$.

**ILLUSTRATION**

**The Symbol** $[\![x]\!]$

| | | |
|---|---|---|
| ▬ $[\![0.5]\!] = 0$ | ▬ $[\![1.8]\!] = 1$ | ▬ $[\![\sqrt{5}]\!] = 2$ |
| ▬ $[\![3]\!] = 3$ | ▬ $[\![-3]\!] = -3$ | ▬ $[\![-2.7]\!] = -3$ |
| ▬ $[\![-\sqrt{3}]\!] = -2$ | ▬ $[\![-0.5]\!] = -1$ | |

The **greatest integer function** $f$ is defined by $f(x) = [\![x]\!]$.

**FIGURE 19**

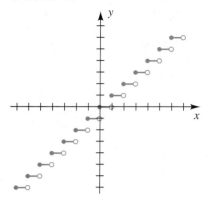

**EXAMPLE 9** *Sketching the graph of the greatest integer function*

Sketch the graph of the greatest integer function.

*SOLUTION*    The $x$- and $y$-coordinates of some points on the graph may be listed as follows:

| Values of $x$ | $f(x) = [\![x]\!]$ |
|:---:|:---:|
| $\vdots$ | $\vdots$ |
| $-2 \le x < -1$ | $-2$ |
| $-1 \le x < 0$ | $-1$ |
| $0 \le x < 1$ | $0$ |
| $1 \le x < 2$ | $1$ |
| $2 \le x < 3$ | $2$ |
| $\vdots$ | $\vdots$ |

Whenever $x$ is between successive integers, the corresponding part of the graph is a segment of a horizontal line. Part of the graph is sketched in Figure 19. The graph continues indefinitely to the right and to the left.

**FIGURE 20**

**(a)**

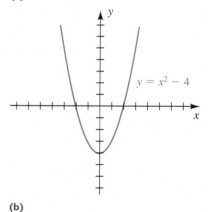

The next example involves absolute values.

**EXAMPLE 10** *Sketching the graph of an equation containing an absolute value*

Sketch the graph of $y = |x^2 - 4|$.

*SOLUTION*    The graph of $y = x^2 - 4$ was sketched in Figure 13 and is resketched in Figure 20(a). We note the following facts:

**(b)**

**(1)** If $x \le -2$ or $x \ge 2$, then $x^2 - 4 \ge 0$, and hence $|x^2 - 4| = x^2 - 4$.

**(2)** If $-2 < x < 2$, then $x^2 - 4 < 0$, and hence $|x^2 - 4| = -(x^2 - 4)$.

It follows from (1) that the graphs of $y = |x^2 - 4|$ and $y = x^2 - 4$ coincide for $|x| \ge 2$. We see from (2) that if $|x| < 2$, then the graph of $y = |x^2 - 4|$ is the reflection of the graph of $y = x^2 - 4$ through the $x$-axis. This gives us the sketch in Figure 20(b).

In general, if the graph of $y = f(x)$ contains a point $P(c, -d)$ with $d$ positive, then the graph of $y = |f(x)|$ contains the point $Q(c, d)$—that is, $Q$ is the reflection of $P$ through the $x$-axis. Points with nonnegative $y$-values are the same for the graphs of $y = f(x)$ and $y = |f(x)|$.

In Chapter 1 we used algebraic methods to solve inequalities involving absolute values of polynomials of degree 1, such as

$$|2x - 5| < 7 \quad \text{and} \quad |5x + 2| \geq 3.$$

Much more complicated inequalities can be investigated using a graphing utility, as illustrated in the next example.

**EXAMPLE 11**  *Solving an absolute value inequality graphically*

Estimate the solutions of

$$|0.14x^2 - 13.72| > |0.58x| + 11.$$

**SOLUTION**  To solve the inequality, we make the assignments

$$Y_1 = \text{ABS}(0.14x^2 - 13.72) \quad \text{and} \quad Y_2 = \text{ABS}(0.58x) + 11$$

**FIGURE 21**  $[-30, 30]$ by $[0, 40]$

and estimate the values of $x$ for which the graph of $Y_1$ is *above* the graph of $Y_2$. After perhaps several trials, we choose the viewing rectangle $[-30, 30]$ by $[0, 40]$, obtaining graphs similar to those in Figure 21, where each tick represents 5 units. Since there is symmetry with respect to the $y$-axis, it is sufficient to find the $x$-coordinates of the points of intersection of the graphs for $x > 0$. Using tracing and zoom-in features, we obtain $x \approx 2.80$ and $x \approx 15.52$. Referring to Figure 21, we obtain the (approximate) solution

$$(-\infty, -15.52) \cup (-2.80, 2.80) \cup (15.52, \infty).$$

Later in this text and in calculus, you will encounter functions such as

$$g(x) = \ln |x| \quad \text{and} \quad h(x) = \sin |x|.$$

Both functions are of the form $y = f(|x|)$. The effect of substituting $|x|$ for $x$ can be described as follows: If the graph of $y = f(x)$ contains a point $P(c, d)$ with $c$ positive, then the graph of $y = f(|x|)$ contains the point $Q(-c, d)$—that is, $Q$ is the reflection of $P$ through the $y$-axis. Points on the $y$-axis ($x = 0$) are the same for the graphs of $y = f(x)$ and $y = f(|x|)$. Points with negative $x$-values on the graph of $y = f(x)$ are not on the graph of $y = f(|x|)$, since the result of the absolute value is always nonnegative.

The processes of shifting, stretching, compressing, and reflecting a graph may be collectively termed **transforming** a graph, and the resulting graph is called a **transformation** of the original graph. A graphical summary of the types of transformations encountered in this section appears in Appendix III.

## 2.2 EXERCISES

Exer. 1–10: Determine whether $f$ is even, odd, or neither even nor odd.

1  $f(x) = 5x^3 + 2x$

2  $f(x) = |x| - 3$

3  $f(x) = 3x^4 + 2x^2 - 5$

4  $f(x) = 7x^5 - 4x^3$

5  $f(x) = 8x^3 - 3x^2$

6  $f(x) = 12$

7  $f(x) = \sqrt{x^2 + 4}$

8  $f(x) = 3x^2 - 5x + 1$

9  $f(x) = \sqrt[3]{x^3 - x}$

10  $f(x) = x^3 - \dfrac{1}{x}$

Exer. 11–24: Sketch, on the same coordinate plane, the graphs of $f$ for the given values of $c$. (Make use of symmetry, shifting, stretching, compressing, or reflecting.)

11  $f(x) = |x| + c;$           $c = -3, 1, 3$

12  $f(x) = |x - c|;$           $c = -3, 1, 3$

13  $f(x) = -x^2 + c;$           $c = -4, 2, 4$

14  $f(x) = 2x^2 - c;$           $c = -4, 2, 4$

15  $f(x) = 2\sqrt{x} + c;$           $c = -3, 0, 2$

16  $f(x) = \sqrt{9 - x^2} + c;$           $c = -3, 0, 2$

17  $f(x) = \frac{1}{2}\sqrt{x - c};$           $c = -2, 0, 3$

18  $f(x) = -\frac{1}{2}(x - c)^2;$           $c = -2, 0, 3$

19  $f(x) = c\sqrt{4 - x^2};$           $c = -2, 1, 3$

20  $f(x) = (x + c)^3;$           $c = -2, 1, 2$

21  $f(x) = cx^3;$           $c = -\frac{1}{3}, 1, 2$

22  $f(x) = (cx)^3 + 1;$           $c = -1, 1, 4$

23  $f(x) = \sqrt{cx} - 1;$           $c = -1, \frac{1}{9}, 4$

24  $f(x) = -\sqrt{16 - (cx)^2};$   $c = 1, \frac{1}{2}, 4$

Exer. 25–26: If the point $P$ is on the graph of a function $f$, find the corresponding point on the graph of the given function.

25  $P(3, -2);$   $y = 2f(x - 4) + 1$

26  $P(-2, 1);$   $y = -3f(2x) - 5$

Exer. 27–28: The graph of a function $f$ with domain $[0, 4]$ is shown in the figure. Sketch the graph of the given equation.

27

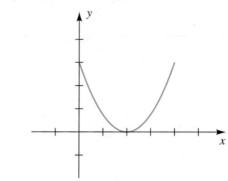

(a) $y = f(x + 3)$           (b) $y = f(x - 3)$
(c) $y = f(x) + 3$           (d) $y = f(x) - 3$
(e) $y = -3f(x)$           (f) $y = -\frac{1}{3}f(x)$
(g) $y = f(-\frac{1}{3}x)$           (h) $y = f(2x)$
(i) $y = -f(x + 2) - 3$           (j) $y = f(x - 2) + 3$
(k) $y = |f(x)|$           (l) $y = f(|x|)$

28

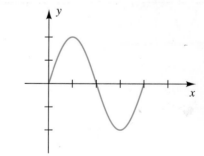

(a) $y = f(x - 2)$           (b) $y = f(x + 2)$
(c) $y = f(x) - 2$           (d) $y = f(x) + 2$
(e) $y = -2f(x)$           (f) $y = -\frac{1}{2}f(x)$
(g) $y = f(-2x)$           (h) $y = f(\frac{1}{2}x)$
(i) $y = -f(x + 4) - 2$           (j) $y = f(x - 4) + 2$
(k) $y = |f(x)|$           (l) $y = f(|x|)$

Exer. 29–32: The graph of a function $f$ is shown, together with graphs of three other functions (a), (b), and (c). Use properties of symmetry, shifts, and reflecting to find equations for graphs (a), (b), and (c) in terms of $f$.

29

30

31

32

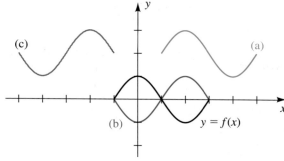

**Exer. 33–38: Sketch the graph of** $f$.

33  $f(x) = \begin{cases} 3 & \text{if } x \le -1 \\ -2 & \text{if } x > -1 \end{cases}$

34  $f(x) = \begin{cases} -1 & \text{if } x \text{ is an integer} \\ -2 & \text{if } x \text{ is not an integer} \end{cases}$

35  $f(x) = \begin{cases} 3 & \text{if } x < -2 \\ -x + 1 & \text{if } |x| \le 2 \\ -3 & \text{if } x > 2 \end{cases}$

36  $f(x) = \begin{cases} -2x & \text{if } x < -1 \\ x^2 & \text{if } -1 \le x < 1 \\ -2 & \text{if } x \ge 1 \end{cases}$

37  $f(x) = \begin{cases} x + 2 & \text{if } x \le -1 \\ x^3 & \text{if } |x| < 1 \\ -x + 3 & \text{if } x \ge 1 \end{cases}$

38  $f(x) = \begin{cases} x - 3 & \text{if } x \le -2 \\ -x^2 & \text{if } -2 < x < 1 \\ -x + 4 & \text{if } x \ge 1 \end{cases}$

**Exer. 39–40: The symbol** $[\![x]\!]$ **denotes values of the greatest integer function. Sketch the graph of** $f$.

39  (a) $f(x) = [\![x - 3]\!]$           (b) $f(x) = [\![x]\!] - 3$
    (c) $f(x) = 2[\![x]\!]$              (d) $f(x) = [\![2x]\!]$
    (e) $f(x) = [\![-x]\!]$

40  (a) $f(x) = [\![x + 2]\!]$           (b) $f(x) = [\![x]\!] + 2$
    (c) $f(x) = \frac{1}{2}[\![x]\!]$   (d) $f(x) = [\![\frac{1}{2}x]\!]$
    (e) $f(x) = -[\![-x]\!]$

**Exer. 41–42: Explain why the graph of the equation is not the graph of a function.**

41  $x = y^2$                           42  $x = -|y|$

**Exer. 43–44:** For the graph of $y = f(x)$ shown in the figure, sketch the graph of $y = |f(x)|$.

**43**

**44**

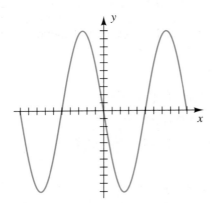

**Exer. 45–48:** Sketch the graph of the equation.

**45** $y = |9 - x^2|$          **46** $y = |x^3 - 1|$

**47** $y = |\sqrt{x} - 1|$          **48** $y = ||x| - 1|$

**49** Let $y = f(x)$ be a function with domain $D = [-2, 6]$ and range $R = [-4, 8]$. Find the domain $D$ and range $R$ for each function. Assume $f(2) = 8$ and $f(6) = -4$.

    (a) $y = -2f(x)$          (b) $y = f(\frac{1}{2}x)$

    (c) $y = f(x - 3) + 1$          (d) $y = f(x + 2) - 3$

    (e) $y = f(-x)$          (f) $y = -f(x)$

    (g) $y = f(|x|)$          (h) $y = |f(x)|$

**50** Let $y = f(x)$ be a function with domain $D = [-6, -2]$ and range $R = [-10, -4]$. Find the domain $D$ and range $R$ for each function.

    (a) $y = \frac{1}{2}f(x)$          (b) $y = f(2x)$

    (c) $y = f(x - 2) + 5$          (d) $y = f(x + 4) - 1$

    (e) $y = f(-x)$          (f) $y = -f(x)$

    (g) $y = f(|x|)$          (h) $y = |f(x)|$

**51** *Income tax rates*   A certain country taxes the first $20,000 of an individual's income at a rate of 15%, and all income over $20,000 is taxed at 20%. Find a piecewise-defined function $T$ that specifies the total tax on an income of $x$ dollars.

**52** *Telephone rates*   A telephone company charges 25 cents for a long-distance call that does not exceed one minute; for longer calls it charges 15 cents for each additional minute. Find a piecewise-defined function $C$ that specifies the total cost of a long-distance call of $x$ minutes.

**53** *Royalty rates*   A certain paperback sells for $12. The author is paid royalties of 10% on the first 10,000 copies sold, 12.5% on the next 5000 copies, and 15% on any additional copies. Find a piecewise-defined function $R$ that specifies the total royalties if $x$ copies are sold.

**54** *Electricity rates*   An electric company charges its customers $0.0577 per kilowatt-hour (kWh) for the first 1000 kWh used, $0.0532 for the next 4000 kWh, and $0.0511 for any kWh over 5000. Find a piecewise-defined function $C$ for a customer's bill of $x$ kWh.

**Exer. 55–58:** Estimate the solutions of the inequality.

**55** $|1.3x + 2.8| < 1.2x + 5$

**56** $|0.3x| - 2 > 2.2 - 0.63x^2$

**57** $|1.2x^2 - 10.8| > 1.36x + 4.08$

**58** $|\sqrt{16 - x^2} - 3| < 0.12x^2 - 0.3$

## 2.3   QUADRATIC FUNCTIONS

If $a \neq 0$, then the graph of $y = ax^2$ is a parabola with vertex at the origin $(0, 0)$, opening upward if $a > 0$ or downward if $a < 0$ (see, for example, Figures 15 and 16). In this section we show that the graph of an equation of the form

$$y = ax^2 + bx + c$$

**FIGURE 22**

can be obtained by vertical and/or horizontal shifts of the graph of $y = ax^2$ and hence is also a parabola. An important application of such equations is to describe the trajectory, or path, of an object near the surface of the earth when the only force acting on the object is gravitational attraction. To illustrate, if an outfielder on a baseball team throws a ball into the infield, as illustrated in Figure 22, and if air resistance and other outside forces are negligible, then the path of the ball is a parabola. If suitable coordinate axes are introduced, then the path coincides with the graph of the equation $y = ax^2 + bx + c$ for some $a$, $b$, and $c$. We call the function determined by this equation a *quadratic function*.

**Definition of Quadratic Function**

A function $f$ is a **quadratic function** if

$$f(x) = ax^2 + bx + c,$$

where $a$, $b$, and $c$ are real numbers with $a \neq 0$.

If $b = c = 0$ in the preceding definition, then $f(x) = ax^2$, and the graph is a parabola with vertex at the origin. If $b = 0$ and $c \neq 0$, then

$$f(x) = ax^2 + c,$$

and, from our discussion of vertical shifts in Section 2.2, the graph is a parabola with vertex at the point $(0, c)$ on the $y$-axis. The following example contains additional illustrations.

**EXAMPLE 1**   *Sketching the graph of a quadratic function*

Sketch the graph of $f$ if

**(a)** $f(x) = -\frac{1}{2}x^2$      **(b)** $f(x) = -\frac{1}{2}x^2 + 4$

*SOLUTION*

**(a)** Since $f$ is even, the graph of $f$ (that is, of $y = -\frac{1}{2}x^2$) is symmetric with respect to the $y$-axis. It is similar in shape to but wider than the parabola $y = -x^2$, sketched in Figure 16 of Section 2.2. Several points on the graph
*(continued)*

are $(0, 0)$, $\left(1, -\frac{1}{2}\right)$, $(2, -2)$, and $\left(3, -\frac{9}{2}\right)$. Plotting and using symmetry, we obtain the sketch in Figure 23.

**(b)** To find the graph of $y = -\frac{1}{2}x^2 + 4$, we shift the graph of $y = -\frac{1}{2}x^2$ upward a distance 4, obtaining the sketch in Figure 24.

**FIGURE 23**

**FIGURE 24**

If $f(x) = ax^2 + bx + c$ and $b \neq 0$, then, by completing the square, we can change the form to

$$f(x) = a(x - h)^2 + k$$

for some real numbers $h$ and $k$. This technique is illustrated in the next example.

**EXAMPLE 2**　　***Expressing a quadratic function as $f(x) = a(x - h)^2 + k$***

If $f(x) = 3x^2 + 24x + 50$, express $f(x)$ in the form $a(x - h)^2 + k$.

*SOLUTION*　　Before completing the square, *it is essential that we factor out the coefficient of $x^2$ from the first two terms of $f(x)$*, as follows:

$$f(x) = 3x^2 + 24x + 50 \qquad \text{given}$$
$$= 3(x^2 + 8x + \quad) + 50 \qquad \text{factor out 3 from } 3x^2 + 24x$$

We now complete the square for the expression $x^2 + 8x$ within the parentheses by adding the square of half the coefficient of $x$—that is, $\left(\frac{8}{2}\right)^2$, or 16. However, if we add 16 to the expression within parentheses, then, because of the factor 3, we are actually adding 48 to $f(x)$. Hence, we must compensate by subtracting 48:

$$f(x) = 3(x^2 + 8x + \quad) + 50 \qquad \text{given}$$
$$= 3(x^2 + 8x + 16) + (50 - 48) \qquad \text{complete the square for } x^2 + 8x$$
$$= 3(x + 4)^2 + 2 \qquad \text{equivalent equation}$$

The last expression has the form $a(x - h)^2 + k$ with $a = 3$, $h = -4$, and $k = 2$.

If $f(x) = ax^2 + bx + c$, then, by completing the square as in Example 2, we see that the graph of $f$ is the same as the graph of an equation of the form

$$y = a(x - h)^2 + k.$$

The graph of this equation can be obtained from the graph of $y = ax^2$ shown in Figure 25(a) by means of a horizontal and a vertical shift, as follows. First, as in Figure 25(b), we obtain the graph of $y = a(x - h)^2$ by shifting the graph of $y = ax^2$ either to the left or to the right, depending on the sign of $h$ (the figure illustrates the case with $h > 0$). Next, as in Figure 25(c), we shift the graph in (b) vertically a distance $|k|$ (the figure illustrates the case with $k > 0$). It follows that *the graph of a quadratic function is a parabola with a vertical axis.*

**FIGURE 25**

(a)

(b)

(c)

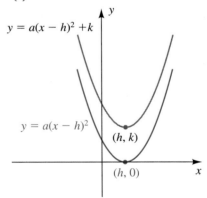

The sketch in Figure 25(c) illustrates one possible graph of the equation $y = ax^2 + bx + c$. If $a > 0$, the point $(h, k)$ is the lowest point on the parabola, and the function $f$ has a **minimum value** $f(h) = k$. If $a < 0$, the parabola opens downward, and the point $(h, k)$ is the highest point on the parabola. In this case, the function $f$ has a **maximum value** $f(h) = k$.

We have obtained the following result.

**Standard Equation of a Parabola with Vertical Axis**

> The graph of the equation
>
> $$y = a(x - h)^2 + k$$
>
> for $a \neq 0$ is a parabola that has vertex $V(h, k)$ and a vertical axis. The parabola opens upward if $a > 0$ or downward if $a < 0$.

For convenience, we often refer to the *parabola* $y = ax^2 + bx + c$ when considering the graph of this equation.

### EXAMPLE 3   *Finding a standard equation of a parabola*

Express $y = 2x^2 - 6x + 4$ as a standard equation of a parabola with a vertical axis. Find the vertex and sketch the graph.

**FIGURE 26**

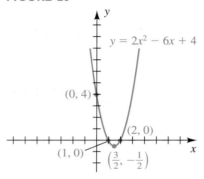

### SOLUTION

$$
\begin{aligned}
y &= 2x^2 - 6x + 4 && \text{given} \\
&= 2(x^2 - 3x + \phantom{xx}) + 4 && \text{factor out 2 from } 2x^2 - 6x \\
&= 2\left(x^2 - 3x + \tfrac{9}{4}\right) + \left(4 - \tfrac{9}{2}\right) && \text{complete the square for } x^2 - 3x \\
&= 2\left(x - \tfrac{3}{2}\right)^2 - \tfrac{1}{2} && \text{equivalent equation}
\end{aligned}
$$

The last equation has the form of the standard equation of a parabola with $a = 2$, $h = \frac{3}{2}$, and $k = -\frac{1}{2}$. Hence, the vertex $V(h, k)$ of the parabola is $V\left(\frac{3}{2}, -\frac{1}{2}\right)$. Since $a = 2 > 0$, the parabola opens upward.

   To find the $y$-intercept of the graph of $y = 2x^2 - 6x + 4$, we let $x = 0$, obtaining $y = 4$. To find the $x$-intercepts, we let $y = 0$ and solve the equation $2x^2 - 6x + 4 = 0$ or the equivalent equation $2(x - 1)(x - 2) = 0$, obtaining $x = 1$ and $x = 2$. Plotting the vertex and using the $x$- and $y$-intercepts provides enough points for a reasonably accurate sketch (see Figure 26).

### EXAMPLE 4   *Finding a standard equation of a parabola*

Express $y = -x^2 - 2x + 8$ as a standard equation of a parabola with a vertical axis. Find the vertex and sketch the graph.

**FIGURE 27**

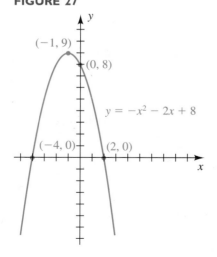

### SOLUTION

$$
\begin{aligned}
y &= -x^2 - 2x + 8 && \text{given} \\
&= -(x^2 + 2x + \phantom{xx}) + 8 && \text{factor out } -1 \text{ from } -x^2 - 2x \\
&= -(x^2 + 2x + 1) + (8 + 1) && \text{complete the square for } x^2 + 2x \\
&= -(x + 1)^2 + 9 && \text{equivalent equation}
\end{aligned}
$$

This is the standard equation of a parabola with $h = -1$, $k = 9$, and hence the vertex is $(-1, 9)$. Since $a = -1 < 0$, the parabola opens downward.

   The $y$-intercept of the graph of $y = -x^2 - 2x + 8$ is the constant term, 8. To find the $x$-intercepts, we solve $-x^2 - 2x + 8 = 0$ or, equivalently, $x^2 + 2x - 8 = 0$. Factoring gives us $(x + 4)(x - 2) = 0$, and hence the intercepts are $x = -4$ and $x = 2$. Using this information gives us the sketch in Figure 27.

**FIGURE 28**

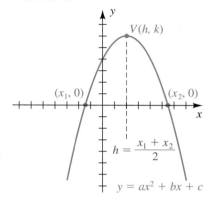

If a parabola $y = ax^2 + bx + c$ has $x$-intercepts $x_1$ and $x_2$, as illustrated in Figure 28 for the case $a < 0$, then the axis of the parabola is the vertical line $x = (x_1 + x_2)/2$ through the midpoint of $(x_1, 0)$ and $(x_2, 0)$. Therefore, the $x$-coordinate $h$ of the vertex $(h, k)$ is $h = (x_1 + x_2)/2$. Some special cases are illustrated in Figures 26 and 27.

In the following example we find an equation of a parabola from given data.

**EXAMPLE 5**   *Finding an equation of a parabola*
                *with a given vertex*

Find an equation of a parabola that has vertex $V(2, 3)$ and a vertical axis and passes through the point $(5, 1)$.

**FIGURE 29**

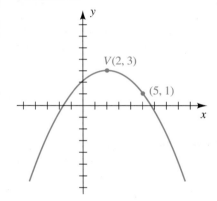

**SOLUTION**   Figure 29 shows the vertex $V$, the point $(5, 1)$, and a possible position of the parabola. Using the standard equation

$$y = a(x - h)^2 + k$$

with $h = 2$ and $k = 3$ gives us

$$y = a(x - 2)^2 + 3.$$

To find $a$, we use the fact that $(5, 1)$ is on the parabola and so is a solution of the last equation. Thus,

$$1 = a(5 - 2)^2 + 3, \quad \text{or} \quad a = -\tfrac{2}{9}.$$

Hence, an equation for the parabola is

$$y = -\tfrac{2}{9}(x - 2)^2 + 3.$$

The next theorem gives us a simple formula for locating the vertex of a parabola.

**Theorem for Locating the**
**Vertex of a Parabola**

The vertex of the parabola $y = ax^2 + bx + c$ has $x$-coordinate

$$-\frac{b}{2a}.$$

**PROOF**   Let us begin by writing $y = ax^2 + bx + c$ as

$$y = a\left(x^2 + \frac{b}{a}x + \quad\right) + c.$$

Next we complete the square by adding $\left(\dfrac{1}{2}\dfrac{b}{a}\right)^2$ to the expression within parentheses:

$$y = a\left(x^2 + \frac{b}{a}x + \frac{b^2}{4a^2}\right) + \left(c - \frac{b^2}{4a}\right)$$

Note that if $b^2/(4a^2)$ is added *inside* the parentheses, then, because of the factor $a$ on the *outside,* we have actually added $b^2/(4a)$ to $y$. Therefore, we must compensate by subtracting $b^2/(4a)$. The last equation may be written

$$y = a\left(x + \frac{b}{2a}\right)^2 + \left(c - \frac{b^2}{4a}\right).$$

This is the equation of a parabola that has vertex $(h, k)$ with $h = -b/(2a)$ and $k = c - b^2/(4a)$. ■

It is unnecessary to remember the formula for the $y$-coordinate of the vertex of the parabola in the preceding result. Once the $x$-coordinate has been found, we can calculate the $y$-coordinate by substituting $-b/(2a)$ for $x$ in the equation of the parabola.

### EXAMPLE 6   *Finding the vertex of a parabola*

Find the vertex of the parabola $y = 2x^2 - 6x + 4$.

*SOLUTION*   We considered this parabola in Example 3 and found the vertex by completing the square. We shall now use the vertex formula with $a = 2$ and $b = -6$, obtaining the $x$-coordinate

$$\frac{-b}{2a} = \frac{-(-6)}{2(2)} = \frac{6}{4} = \frac{3}{2}.$$

We next find the $y$-coordinate by substituting $\frac{3}{2}$ for $x$ in the given equation:

$$y = 2\left(\tfrac{3}{2}\right)^2 - 6\left(\tfrac{3}{2}\right) + 4 = -\tfrac{1}{2}$$

Thus, the vertex is $\left(\tfrac{3}{2}, -\tfrac{1}{2}\right)$ (see Figure 26).

Since the graph of $f(x) = ax^2 + bx + c$ for $a \neq 0$ is a parabola, we can use the vertex formula to help find the maximum or minimum value of a quadratic function. Specifically, since the $x$-coordinate of the vertex $V$ is $-b/(2a)$, the $y$-coordinate of $V$ is the function value $f(-b/(2a))$. Moreover, since the parabola opens downward if $a < 0$ and upward if $a > 0$, this function value is the maximum or minimum value, respectively, of $f$. We may summarize these facts as follows.

**Theorem on the Maximum or Minimum Value of a Quadratic Function**

If $f(x) = ax^2 + bx + c$, where $a \neq 0$, then $f\left(-\dfrac{b}{2a}\right)$ is

    **(1)** the maximum value of $f$ if $a < 0$

    **(2)** the minimum value of $f$ if $a > 0$

We shall use this theorem in the next example.

**EXAMPLE 7**    *Finding the maximum value of a quadratic function*

A long rectangular sheet of metal, 12 inches wide, is to be made into a rain gutter by turning up two sides so that they are perpendicular to the sheet. How many inches should be turned up to give the gutter its greatest capacity?

**FIGURE 30**

*SOLUTION*    The gutter is illustrated in Figure 30. If $x$ denotes the number of inches turned up on each side, the width of the base of the gutter is $12 - 2x$ inches. The capacity will be greatest when the cross-sectional area of the rectangle with sides of lengths $x$ and $12 - 2x$ has its greatest value. Letting $f(x)$ denote this area, we have

$$
\begin{aligned}
f(x) &= x(12 - 2x) \\
&= 12x - 2x^2 \\
&= -2x^2 + 12x,
\end{aligned}
$$

which has the form $f(x) = ax^2 + bx + c$ with $a = -2$, $b = 12$, and $c = 0$. Since $f$ is a quadratic function and $a = -2 < 0$, it follows from the preceding theorem that the maximum value of $f$ occurs at

$$
x = -\frac{b}{2a} = -\frac{12}{2(-2)} = 3.
$$

Thus, 3 inches should be turned up on each side to achieve maximum capacity.

As an alternative solution, we may note that the graph of the function $f(x) = x(12 - 2x)$ has $x$-intercepts at $x = 0$ and $x = 6$. Hence, the average of the intercepts,

$$
x = \frac{0 + 6}{2} = 3,
$$

is the $x$-coordinate of the vertex of the parabola and the value that yields the maximum capacity.

In Chapter 1 we solved quadratic equations and inequalities algebraically. The next example indicates how they can be solved with the aid of a graphing utility.

**EXAMPLE 8**    *Analyzing the flight of a projectile*

A projectile is fired vertically upward from a height of 600 feet above the ground. Its height $h(t)$ in feet, after $t$ seconds, is given by

$$h(t) = -16t^2 + 803t + 600.$$

**(a)** Determine a reasonable viewing rectangle that includes all pertinent features of the graph of $h$.

**(b)** Use a graphing utility to estimate when the height is 5000 feet.

**(c)** Determine when the projectile will be more than 5000 feet above the ground.

**(d)** How long will the projectile be in flight?

**SOLUTION**

**(a)** The graph of $h$ is a parabola that opens downward. To estimate Ymax (note that we use $x$ and $y$ interchangeably with $t$ and $h$), let us approximate the maximum value of $h$. Using

$$t = -\frac{b}{2a} = -\frac{803}{2(-16)} \approx 25.1,$$

we see that the maximum height is approximately $h(25) = 10{,}675$.

The projectile rises for approximately the first 25 seconds, and because its height at $t = 0$, 600 feet, is small in comparison to 10,675, it will take only slightly more than an additional 25 seconds to fall to the ground. Since $h$ and $t$ are positive, a reasonable viewing rectangle is

$$[0, 60] \quad \text{by} \quad [0, 11{,}000].$$

**(b)** We wish to estimate where the graph of $h$ intersects the horizontal line $h(t) = 5000$, so we make the assignments

$$Y_1 = -16x^2 + 803x + 600 \quad \text{and} \quad Y_2 = 5000$$

and obtain a display similar to Figure 31. It is important to remember that the graph of $Y_1$ shows only the height at time $t$—it is *not* the path of the projectile, which is vertical. Using the tracing and zoom-in features, we find that the smallest value of $t$ for which $h(t) = 5000$ is about 6.3 seconds.

Since the vertex is on the axis of the parabola, the other time at which $h(t)$ is 5000 is approximately $25.1 - 6.3$, or 18.8, seconds *after* $t = 25.1$—that is, at $t \approx 25.1 + 18.8 = 43.9$ sec.

**FIGURE 31**    [0, 60] by [0, 11,000]

**(c)** The projectile is more than 5000 feet above the ground when the graph of the parabola in Figure 31 is above the horizontal line—that is, when

$$6.3 < t < 43.9.$$

**(d)** The projectile will be in flight until $h(t) = 0$. This corresponds to the $x$-intercept in Figure 31. Using the tracing and zoom-in features, we obtain $t \approx 50.9$ sec. (Note that since the $y$-intercept is not zero, it is incorrect to merely double the $t$ value of the vertex to find the total time of the flight; however, this *would* be acceptable for problems in which $h(0) = 0$.)

When working with quadratic functions, we are often most interested in finding the vertex and the $x$-intercepts. Typically, a given quadratic function closely resembles one of the three forms listed in the following chart.

**Relationship Between Quadratic Function Forms and Their Vertex and $x$-intercepts**

| Form | Vertex $(h, k)$ | $x$-intercepts (if there are any) |
|---|---|---|
| (1) $y = f(x) = a(x - h)^2 + k$ | $h$ and $k$ as in the form | $x = h \pm \sqrt{-k/a}$ (see below) |
| (2) $y = f(x) = a(x - x_1)(x - x_2)$ | $h = \dfrac{x_1 + x_2}{2}, \quad k = f(h)$ | $x = x_1, x_2$ |
| (3) $y = f(x) = ax^2 + bx + c$ | $h = -\dfrac{b}{2a}, \quad k = f(h)$ | $x = -\dfrac{b}{2a} \pm \dfrac{\sqrt{b^2 - 4ac}}{2a}$ (see below) |

If the radicands in (1) or (3) are negative, then there are no $x$-intercepts. To find the $x$-intercepts with form (1), use the special quadratic equation on page 48. If you have a quadratic function in form (3) and want to find the vertex and the $x$-intercepts, it may be best to first find the $x$-intercepts by using the quadratic formula. Then you can easily obtain the $x$-coordinate of the vertex, $h$, since

$$-\frac{b}{2a} \pm \frac{\sqrt{b^2 - 4ac}}{2a} = h \pm \frac{\sqrt{b^2 - 4ac}}{2a}.$$

Of course, if the function in form (3) is easily factorable, it is not necessary to use the quadratic formula.

We will discuss parabolas further in Section 10.1.

## 2.3 EXERCISES

**Exer. 1–4:** Find the standard equation of any parabola that has vertex $V$.

1  $V(-3, 1)$      2  $V(4, -2)$

3  $V(0, -3)$      4  $V(-2, 0)$

**Exer. 5–12:** Express $f(x)$ in the form $a(x - h)^2 + k$.

5  $f(x) = -x^2 - 4x - 8$      6  $f(x) = x^2 - 6x + 11$

7  $f(x) = 2x^2 - 12x + 22$      8  $f(x) = 5x^2 + 20x + 17$

9  $f(x) = -3x^2 - 6x - 5$

10  $f(x) = -4x^2 + 16x - 13$

11  $f(x) = -\frac{3}{4}x^2 + 9x - 34$      12  $f(x) = \frac{2}{3}x^2 - \frac{12}{5}x + \frac{23}{5}$

**Exer. 13–22:** (a) Use the quadratic formula to find the zeros of $f$. (b) Find the maximum or minimum value of $f(x)$. (c) Sketch the graph of $f$.

13  $f(x) = x^2 - 4x$      14  $f(x) = -x^2 - 6x$

15  $f(x) = -12x^2 + 11x + 15$

16  $f(x) = 6x^2 + 7x - 24$

17  $f(x) = 9x^2 + 24x + 16$      18  $f(x) = -4x^2 + 4x - 1$

19  $f(x) = x^2 + 4x + 9$      20  $f(x) = -3x^2 - 6x - 6$

21  $f(x) = -2x^2 + 20x - 43$

22  $f(x) = 2x^2 - 4x - 11$

**Exer. 23–26:** Find the standard equation of the parabola shown in the figure.

23

26

24

25

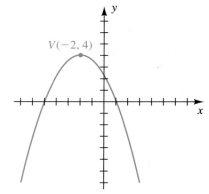

**Exer. 27–32:** Find the standard equation of a parabola that has a vertical axis and satisfies the given conditions.

27  Vertex $(0, -2)$, passing through $(3, 25)$

28 Vertex $(0, 5)$, passing through $(2, -3)$

29 Vertex $(3, 5)$, $x$-intercept 0

30 Vertex $(4, -7)$, $x$-intercept $-4$

31 $x$-intercepts $-3$ and 5, highest point has $y$-coordinate 4

32 $x$-intercepts 8 and 0, lowest point has $y$-coordinate $-48$

Exer. 33–34: Find the maximum vertical distance $d$ between the parabola and the line for the green region.

33

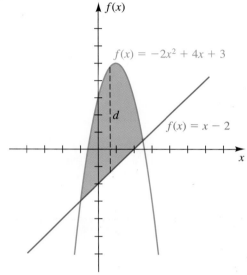

$f(x)$

$f(x) = -2x^2 + 4x + 3$

$d$

$f(x) = x - 2$

$x$

34

$f(x)$

$f(x) = 2x^2 + 8x + 4$

$d$

$f(x) = -x + 3$

$x$

Exer. 35–36: Ozone occurs at all levels of the earth's atmosphere. The density of ozone varies both seasonally and latitudinally. At Edmonton, Canada, the density $D(h)$ of ozone (in $10^{-3}$ cm/km) for altitudes $h$ between 20 kilometers and 35 kilometers was determined experimentally. For each $D(h)$ and season, approximate the altitude at which the density of ozone is greatest.

35 $D(h) = -0.058h^2 + 2.867h - 24.239$ (autumn)

36 $D(h) = -0.078h^2 + 3.811h - 32.433$ (spring)

37 **Infant growth rate** The growth rate $y$ (in pounds per month) of an infant is related to present weight $x$ (in pounds) by the formula $y = cx(21 - x)$, where $c$ is a positive constant and $0 < x < 21$. At what weight does the maximum growth rate occur?

38 **Gasoline mileage** The number of miles $M$ that a certain automobile can travel on one gallon of gasoline at a speed of $v$ mi/hr is given by

$$M = -\tfrac{1}{30}v^2 + \tfrac{5}{2}v \quad \text{for} \quad 0 < v < 70.$$

(a) Find the most economical speed for a trip.

(b) Find the largest value of $M$.

39 **Height of a projectile** An object is projected vertically upward from the top of a building with an initial velocity of 144 ft/sec. Its distance $s(t)$ in feet above the ground after $t$ seconds is given by the equation $s(t) = -16t^2 + 144t + 100$.

(a) Find its maximum distance above the ground.

(b) Find the height of the building.

40 **Flight of a projectile** An object is projected vertically upward with an initial velocity of $v_0$ ft/sec, and its distance $s(t)$ in feet above the ground after $t$ seconds is given by $s(t) = -16t^2 + v_0 t$.

(a) If the object hits the ground after 12 seconds, find its initial velocity $v_0$.

(b) Find its maximum distance above the ground.

41 Find two positive real numbers whose sum is 40 and whose product is a maximum.

42 Find two real numbers whose difference is 40 and whose product is a minimum.

**43 Constructing cages** One thousand feet of chain-link fence is to be used to construct six animal cages, as shown in the figure.

(a) Express the width $y$ as a function of the length $x$.

(b) Express the total enclosed area $A$ of the exhibit as a function of $x$.

(c) Find the dimensions that maximize the enclosed area.

EXERCISE 43

**44 Fencing a field** A farmer wishes to put a fence around a rectangular field and then divide the field into three rectangular plots by placing two fences parallel to one of the sides. If the farmer can afford only 1000 yards of fencing, what dimensions will give the maximum rectangular area?

**45 Leaping animals** Flights of leaping animals typically have parabolic paths. The figure illustrates a frog jump superimposed on a coordinate plane. The length of the leap is 9 feet, and the maximum height off the ground is 3 feet. Find a standard equation for the path of the frog.

EXERCISE 45

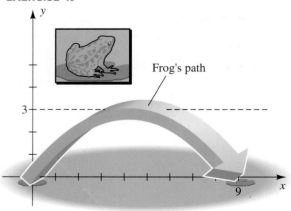

**46 The human cannonball** In the 1940s, the human cannonball stunt was performed regularly by Emmanuel Zacchini for The Ringling Brothers and Barnum & Bailey Circus. The tip of the cannon rose 15 feet off the ground, and the total horizontal distance traveled was 175 feet. When the cannon is aimed at an angle of 45°, an equation of the parabolic flight (see the figure) has the form $y = ax^2 + x + c$.

(a) Use the given information to find an equation of the flight.

(b) Find the maximum height attained by the human cannonball.

EXERCISE 46

**47 Shape of a suspension bridge** One section of a suspension bridge has its weight uniformly distributed between twin towers that are 400 feet apart and rise 90 feet above the horizontal roadway (see the figure). A cable strung between the tops of the towers has the shape of a parabola, and its center point is 10 feet above the roadway. Suppose coordinate axes are introduced, as shown in the figure.

EXERCISE 47

(a) Find an equation for the parabola.

(b) Nine equally spaced vertical cables are used to support the bridge (see the figure). Find the total length of these supports.

48 *Designing a highway* Traffic engineers are designing a stretch of highway that will connect a horizontal highway with one having a 20% grade (that is, slope $\frac{1}{5}$), as illustrated in the figure. The smooth transition is to take place over a horizontal distance of 800 feet, with a parabolic piece of highway used to connect points $A$ and $B$. If the equation of the parabolic segment is of the form $y = ax^2 + bx + c$, it can be shown that the slope of the tangent line at the point $P(x, y)$ on the parabola is given by $m = 2ax + b$.

(a) Find an equation of the parabola that has a tangent line of slope 0 at $A$ and $\frac{1}{5}$ at $B$.

(b) Find the coordinates of $B$.

**EXERCISE 48**

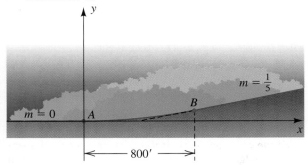

49 *Parabolic doorway* A doorway has the shape of a parabolic arch and is 9 feet high at the center and 6 feet wide at the base. If a rectangular box 8 feet high must fit through the doorway, what is the maximum width the box can have?

50 *Wire rectangle* A piece of wire 24 inches long is bent into the shape of a rectangle having width $x$ and length $y$.

(a) Express $y$ as a function of $x$.

(b) Express the area $A$ of the rectangle as a function of $x$.

(c) Show that the area $A$ is greatest if the rectangle is a square.

51 *Quantity discount* A company sells running shoes to dealers at a rate of $40 per pair if less than 50 pairs are ordered. If a dealer orders 50 or more pairs (up to 600), the price per pair is reduced at a rate of 4 cents times the number ordered. What size order will produce the maximum amount of money for the company?

52 *Group discount* A travel agency offers group tours at a rate of $60 per person for the first 30 participants. For larger groups—up to 90—each person receives a $0.50 discount for every participant in excess of 30. For example, if 31 people participate, then the cost per person is $59.50. Determine the size of the group that will produce the maximum amount of money for the agency.

53 *Cable TV fee* A cable television firm presently serves 5000 households and charges $20 per month. A marketing survey indicates that each decrease of $1 in the monthly charge will result in 500 new customers. Let $R(x)$ denote the total monthly revenue when the monthly charge is $x$ dollars.

(a) Determine the revenue function $R$.

(b) Sketch the graph of $R$ and find the value of $x$ that results in maximum monthly revenue.

54 *Apartment rentals* A real estate company owns 180 efficiency apartments, which are fully occupied when the rent is $300 per month. The company estimates that for each $10 increase in rent, 5 apartments will become unoccupied. What rent should be charged so that the company will receive the maximum monthly income?

Exer. 55–56: Graph $y = x^3 - x^{1/3}$ and $f$ on the same coordinate plane, and estimate the points of intersection.

55 $f(x) = x^2 - x - \frac{1}{4}$

56 $f(x) = -x^2 + 0.5x + 0.4$

57 Graph, on the same coordinate plane, $y = ax^2 + x + 1$ for $a = \frac{1}{4}, \frac{1}{2}, 1, 2,$ and 4, and describe how the value of $a$ affects the graph.

58 Graph, on the same coordinate plane, $y = x^2 + bx + 1$ for $b = 0, \pm1, \pm2,$ and $\pm3$, and describe how the value of $b$ affects the graph.

## 2.4   OPERATIONS ON FUNCTIONS

Functions are often defined using sums, differences, products, and quotients of various expressions. For example, if

$$h(x) = x^2 + \sqrt{5x + 1},$$

we may regard $h(x)$ as a sum of values of the functions $f$ and $g$ given by

$$f(x) = x^2 \quad \text{and} \quad g(x) = \sqrt{5x + 1}.$$

We call $h$ the *sum* of $f$ and $g$ and denote it by $f + g$. Thus,

$$h(x) = (f + g)(x) = x^2 + \sqrt{5x + 1}.$$

In general, if $f$ and $g$ are *any* functions, we use the terminology and notation given in the following chart.

**Sum, Difference, Product, and Quotient of Functions**

| Terminology | Function value |
|---|---|
| **sum** $f + g$ | $(f + g)(x) = f(x) + g(x)$ |
| **difference** $f - g$ | $(f - g)(x) = f(x) - g(x)$ |
| **product** $fg$ | $(fg)(x) = f(x)g(x)$ |
| **quotient** $\dfrac{f}{g}$ | $\left(\dfrac{f}{g}\right)(x) = \dfrac{f(x)}{g(x)}$ |

The domains of $f + g$, $f - g$, and $fg$ are the intersection $I$ of the domains of $f$ and $g$—that is, the numbers that are *common* to both domains. The domain of $f/g$ is the subset of $I$ consisting of all $x$ in $I$ such that $g(x) \neq 0$.

**EXAMPLE 1**   *Finding function values of $f + g$, $f - g$, $fg$, and $f/g$*

If $f(x) = 3x - 2$ and $g(x) = x^3$, find $(f + g)(2)$, $(f - g)(2)$, $(fg)(2)$, and $(f/g)(2)$.

**SOLUTION**   Since $f(2) = 3(2) - 2 = 4$ and $g(2) = 2^3 = 8$, we have

$$(f + g)(2) = f(2) + g(2) = 4 + 8 = 12$$
$$(f - g)(2) = f(2) - g(2) = 4 - 8 = -4$$
$$(fg)(2) = f(2)g(2) = (4)(8) = 32$$
$$\left(\frac{f}{g}\right)(2) = \frac{f(2)}{g(2)} = \frac{4}{8} = \frac{1}{2}$$

**EXAMPLE 2** *Finding* $(f + g)(x)$, $(f - g)(x)$, $(fg)(x)$, *and* $(f/g)(x)$

If $f(x) = \sqrt{4 - x^2}$ and $g(x) = 3x + 1$, find $(f + g)(x)$, $(f - g)(x)$, $(fg)(x)$, and $(f/g)(x)$, and state the domains of the respective functions.

*SOLUTION* The domain of $f$ is the closed interval $[-2, 2]$, and the domain of $g$ is $\mathbb{R}$. The intersection of these domains is $[-2, 2]$, which is the domain of $f + g, f - g$, and $fg$. For the domain of $f/g$ we exclude each number $x$ in $[-2, 2]$ such that $g(x) = 3x + 1 = 0$ (namely, $x = -\frac{1}{3}$). Thus, we have the following:

$$(f + g)(x) = \sqrt{4 - x^2} + (3x + 1), \quad -2 \le x \le 2$$
$$(f - g)(x) = \sqrt{4 - x^2} - (3x + 1), \quad -2 \le x \le 2$$
$$(fg)(x) = \sqrt{4 - x^2}(3x + 1), \quad -2 \le x \le 2$$
$$\left(\frac{f}{g}\right)(x) = \frac{\sqrt{4 - x^2}}{3x + 1}, \quad -2 \le x \le 2 \text{ and } x \ne -\frac{1}{3}$$

A function $f$ is a **polynomial function** if $f(x)$ is a polynomial—that is, if

$$f(x) = a_n x^n + a_{n-1} x^{n-1} + \cdots + a_1 x + a_0,$$

where the coefficients $a_0, a_1, \ldots, a_n$ are real numbers and the exponents are nonnegative integers. A polynomial function may be regarded as a sum of functions whose values are of the form $cx^k$, where $c$ is a real number and $k$ is a nonnegative integer. Note that the quadratic functions considered in the previous section are polynomial functions.

An **algebraic function** is a function that can be expressed in terms of finite sums, differences, products, quotients, or roots of polynomial functions.

*ILLUSTRATION*

**Algebraic Function**

■ $f(x) = 5x^4 - 2\sqrt[3]{x} + \dfrac{x(x^2 + 5)}{\sqrt{x^3 + \sqrt{x}}}$

Functions that are not algebraic are **transcendental.** The exponential and logarithmic functions considered in Chapter 4 are examples of transcendental functions.

In the remainder of this section we shall discuss how two functions $f$ and $g$ may be used to obtain the *composite functions* $f \circ g$ and $g \circ f$ (read "$f$ circle $g$" and "$g$ circle $f$," respectively). Functions of this type are very important in calculus. The function $f \circ g$ is defined as follows.

**Definition of Composite Function**

The **composite function** $f \circ g$ of two functions $f$ and $g$ is defined by

$$(f \circ g)(x) = f(g(x)).$$

The domain of $f \circ g$ is the set of all $x$ in the domain of $g$ such that $g(x)$ is in the domain of $f$.

**FIGURE 32**

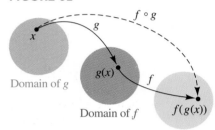

Figure 32 is a schematic diagram that illustrates relationships among $f$, $g$, and $f \circ g$. Note that for $x$ in the domain of $g$, *first we find* $g(x)$ (which must be in the domain of $f$) and then, *second, we find* $f(g(x))$.

For the composite function $g \circ f$, we reverse this order, first finding $f(x)$ and second finding $g(f(x))$. The domain of $g \circ f$ is the set of all $x$ in the domain of $f$ such that $f(x)$ is in the domain of $g$.

Since the notation $g(x)$ is read "$g$ of $x$," we sometimes say that $g$ *is a function of $x$.* For the composite function $f \circ g$, the notation $f(g(x))$ is read "$f$ of $g$ of $x$," and we could regard $f$ as a function of $g(x)$. In this sense, *a composite function is a function of a function* or, more precisely, a function of another function's values.

**EXAMPLE 3**   *Finding composite functions*

Let $f(x) = x^2 - 1$ and $g(x) = 3x + 5$.

**(a)** Find $(f \circ g)(x)$ and the domain of $f \circ g$.

**(b)** Find $(g \circ f)(x)$ and the domain of $g \circ f$.

**(c)** Find $f(g(2))$ in two different ways: first using the functions $f$ and $g$ separately and second using the composite function $f \circ g$.

*SOLUTION*

**(a)**
$$\begin{aligned}
(f \circ g)(x) &= f(g(x)) && \text{definition of } f \circ g \\
&= f(3x + 5) && \text{definition of } g \\
&= (3x + 5)^2 - 1 && \text{definition of } f \\
&= 9x^2 + 30x + 24 && \text{simplify}
\end{aligned}$$

The domain of both $f$ and $g$ is $\mathbb{R}$. Since for each $x$ in $\mathbb{R}$ (the domain of $g$), the function value $g(x)$ is in $\mathbb{R}$ (the domain of $f$), the domain of $f \circ g$ is also $\mathbb{R}$.

**(b)**
$$\begin{aligned}
(g \circ f)(x) &= g(f(x)) && \text{definition of } g \circ f \\
&= g(x^2 - 1) && \text{definition of } f \\
&= 3(x^2 - 1) + 5 && \text{definition of } g \\
&= 3x^2 + 2 && \text{simplify}
\end{aligned}$$

Since for each $x$ in $\mathbb{R}$ (the domain of $f$), the function value $f(x)$ is in $\mathbb{R}$ (the domain of $g$), the domain of $g \circ f$ is $\mathbb{R}$.

**(c)** To find $f(g(2))$ using $f(x) = x^2 - 1$ and $g(x) = 3x + 5$ separately, we may proceed as follows:

$$g(2) = 3(2) + 5 = 11$$
$$f(g(2)) = f(11) = 11^2 - 1 = 120$$

To find $f(g(2))$ using $f \circ g$, we refer to part (a), where we found

$$(f \circ g)(x) = f(g(x)) = 9x^2 + 30x + 24.$$

Hence,                  $$f(g(2)) = 9(2)^2 + 30(2) + 24$$
$$= 36 + 60 + 24 = 120.$$

Note that in Example 3, $f(g(x))$ and $g(f(x))$ are not always the same; that is, $f \circ g \neq g \circ f$.

If two functions $f$ and $g$ both have domain $\mathbb{R}$, then the domain of $f \circ g$ and $g \circ f$ is also $\mathbb{R}$. This was illustrated in Example 3. The next example shows that the domain of a composite function may differ from those of the two given functions.

**EXAMPLE 4**   *Finding composite functions*

Let $f(x) = x^2 - 16$ and $g(x) = \sqrt{x}$.

**(a)** Find $(f \circ g)(x)$ and the domain of $f \circ g$.

**(b)** Find $(g \circ f)(x)$ and the domain of $g \circ f$.

**SOLUTION**   We first note that the domain of $f$ is $\mathbb{R}$ and the domain of $g$ is the set of all nonnegative real numbers—that is, the interval $[0, \infty)$. We may proceed as follows.

**(a)**
$$
\begin{aligned}
(f \circ g)(x) &= f(g(x)) & &\text{definition of } f \circ g \\
&= f(\sqrt{x}) & &\text{definition of } g \\
&= (\sqrt{x})^2 - 16 & &\text{definition of } f \\
&= x - 16 & &\text{simplify}
\end{aligned}
$$

If we consider only the final expression $x - 16$, we might be led to believe that the domain of $f \circ g$ is $\mathbb{R}$, since $x - 16$ is defined for every real number $x$. However, this is not the case. By definition, the domain of $f \circ g$ is the set of all $x$ in $[0, \infty)$ (the domain of $g$) such that $g(x)$ is in $\mathbb{R}$ (the domain of $f$). Since $g(x) = \sqrt{x}$ is in $\mathbb{R}$ for every $x$ in $[0, \infty)$, it follows that the domain of $f \circ g$ is $[0, \infty)$.

**(b)**
$$
\begin{aligned}
(g \circ f)(x) &= g(f(x)) & &\text{definition of } g \circ f \\
&= g(x^2 - 16) & &\text{definition of } f \\
&= \sqrt{x^2 - 16} & &\text{definition of } g
\end{aligned}
$$

*(continued)*

By definition, the domain of $g \circ f$ is the set of all $x$ in $\mathbb{R}$ (the domain of $f$) such that $f(x) = x^2 - 16$ is in $[0, \infty)$ (the domain of $g$). The statement "$x^2 - 16$ is in $[0, \infty)$" is equivalent to each of the inequalities

$$x^2 - 16 \geq 0, \quad x^2 \geq 16, \quad |x| \geq 4.$$

Thus, the domain of $g \circ f$ is the union $(-\infty, -4] \cup [4, \infty)$. Note that this domain is different from the domains of both $f$ and $g$.

The next example illustrates how special values of composite functions may sometimes be obtained from tables.

**EXAMPLE 5    Finding composite function values from tables**

Several values of two functions $f$ and $g$ are listed in the following tables.

| $x$ | 1 | 2, | 3 | 4 |
|------|---|----|---|---|
| $f(x)$ | 3 | 4 | 2 | 1 |

| $x$ | 1 | 2 | 3 | 4 |
|------|---|---|---|---|
| $g(x)$ | 4 | 1 | 3 | 2 |

Find $(f \circ g)(2)$, $(g \circ f)(2)$, $(f \circ f)(2)$, and $(g \circ g)(2)$.

**SOLUTION**    Using the definition of composite function and referring to the tables above, we obtain

$$(f \circ g)(2) = f(g(2)) = f(1) = 3$$
$$(g \circ f)(2) = g(f(2)) = g(4) = 2$$
$$(f \circ f)(2) = f(f(2)) = f(4) = 1$$
$$(g \circ g)(2) = g(g(2)) = g(1) = 4.$$

In some applied problems it is necessary to express a quantity $y$ as a function of time $t$. The following example illustrates that it is often easier to introduce a third variable $x$, express $x$ as a function of $t$ (that is, $x = g(t)$), express $y$ as a function of $x$ (that is, $y = f(x)$), and finally form the composite function given by $y = f(x) = f(g(t))$.

**EXAMPLE 6    Using a composite function
to find the volume of a balloon**

A meteorologist is inflating a spherical balloon with helium gas. If the radius of the balloon is changing at a rate of 1.5 cm/sec, express the volume $V$ of the balloon as a function of time $t$ (in seconds).

**SOLUTION** Let $x$ denote the radius of the balloon. If we assume that the radius is 0 initially, then after $t$ seconds

$$x = 1.5t. \quad \text{radius of balloon after } t \text{ seconds}$$

To illustrate, after 1 second, the radius is 1.5 centimeters; after 2 seconds, it is 3.0 centimeters; after 3 seconds, it is 4.5 centimeters; and so on.

Next we write

$$V = \tfrac{4}{3}\pi x^3. \quad \text{volume of a sphere of radius } x$$

This gives us a composite function relationship in which $V$ is a function of $x$ and $x$ is a function of $t$. By substitution, we obtain

$$V = \tfrac{4}{3}\pi x^3 = \tfrac{4}{3}\pi (1.5t)^3 = \tfrac{4}{3}\pi \left(\tfrac{3}{2}t\right)^3 = \tfrac{4}{3}\pi \left(\tfrac{27}{8}t^3\right).$$

Simplifying, we obtain the following formula for $V$ as a function of $t$:

$$V = \tfrac{9}{2}\pi t^3$$

If $f$ and $g$ are functions such that

$$y = f(u) \quad \text{and} \quad u = g(x),$$

then substituting for $u$ in $y = f(u)$ yields

$$y = f(g(x)).$$

For certain problems in calculus we *reverse* this procedure; that is, given $y = h(x)$ for some function $h$, we find a *composite function form* $y = f(u)$ and $u = g(x)$ such that $h(x) = f(g(x))$.

## EXAMPLE 7 *Finding a composite function form*

Express $y = (2x + 5)^8$ as a composite function form.

**SOLUTION** Suppose, for a real number $x$, we wanted to evaluate the expression $(2x + 5)^8$ by using a calculator. We would first calculate the value of $2x + 5$ and then raise the result to the eighth power. This suggests that we let

$$u = 2x + 5 \quad \text{and} \quad y = u^8,$$

which is a composite function form for $y = (2x + 5)^8$.

The method used in the preceding example can be extended to other functions. In general, suppose we are given $y = h(x)$. To choose the *inside* expression $u = g(x)$ in a composite function form, ask the following question: If a calculator were being used, which part of the expression $h(x)$

would be evaluated first? This often leads to a suitable choice for $u = g(x)$. After choosing $u$, refer to $h(x)$ to determine $y = f(u)$. The following illustration contains typical problems.

### ILLUSTRATION

#### Composite Function Forms

| Function value | Choice for $u = g(x)$ | Choice for $y = f(u)$ |
|---|---|---|
| $y = (x^3 - 5x + 1)^4$ | $u = x^3 - 5x + 1$ | $y = u^4$ |
| $y = \sqrt{x^2 - 4}$ | $u = x^2 - 4$ | $y = \sqrt{u}$ |
| $y = \dfrac{2}{3x + 7}$ | $u = 3x + 7$ | $y = \dfrac{2}{u}$ |

The composite function form is never unique. For example, consider the first expression in the preceding illustration:

$$y = (x^3 - 5x + 1)^4$$

If $n$ is any nonzero integer, we could choose

$$u = (x^3 - 5x + 1)^n \quad \text{and} \quad y = u^{4/n}.$$

Thus, there are an *unlimited* number of composite function forms. Generally, our goal is to choose a form such that the expression for $y$ is simple, as we did in the illustration.

The next example illustrates how a graphing utility can help determine the domain of a composite function. We use the same functions that appeared in Example 4.

**EXAMPLE 8**    *Graphically analyzing a composite function*

Let $f(x) = x^2 - 16$ and $g(x) = \sqrt{x}$.

**(a)** Find $f(g(3))$.

**(b)** Sketch $y = (f \circ g)(x)$ and use the graph to find the domain of $f \circ g$.

#### SOLUTION

**(a)** We begin by making the assignments

$$Y_1 = \sqrt{x} \quad \text{and} \quad Y_2 = (Y_1)^2 - 16.$$

Note that we have substituted $Y_1$ for $x$ in $f(x)$ and assigned this expression to $Y_2$, much the same way as we substituted $g(x)$ for $x$ in Example 4.

Next we store the value 3 in the memory location for $x$ and then query the value of $Y_2$. We see that the value of $Y_2$ at 3 is $-13$; that is, $f(g(3)) = -13$.

**(b)** To determine a viewing rectangle for the graph of $f \circ g$, we first note that $f(x) \geq -16$ for all $x$ and therefore choose Ymin *less* than $-16$; say, Ymin $= -20$. If we want the rectangle to have a vertical dimension of 40, we must choose Ymax $= 20$.

If your screen is in 1:1 proportion (horizontal:vertical), then a reasonable choice for [Xmin, Xmax] would be $[-10, 30]$, a horizontal dimension of 40. If your screen is in 3:2 proportion, choose [Xmin, Xmax] to be $[-10, 50]$, a horizontal dimension of 60.

Selecting $Y_2$ and then displaying the graph of $Y_2$ using the viewing rectangle $[-10, 50]$ by $[-20, 20]$ gives us a graph similar to Figure 33. We see that the graph is a half-line with end point $(0, -16)$. Thus, the domain of $Y_2$ is all $x \geq 0$.

**FIGURE 33**   $[-10, 50]$ by $[-20, 20]$

The next example demonstrates how to use a graphing utility to graph composite functions of the form $af(bx)$. We will use the function from Example 7 of Section 2.2.

**EXAMPLE 9**    *Graphing composite functions*

If $f(x) = x^3 - 4x^2$, sketch the graph of $y = -\frac{1}{2}f(\frac{1}{3}x)$.

**SOLUTION**    Some specific keystrokes for the TI-81/82 graphics calculator are given in Appendix I.

From our discussion on compressing and stretching graphs in Section 2.2, we recognize that the graph of $f$ will be vertically compressed by a factor 2 and horizontally stretched by a factor 3. To relate this problem to composite functions, we may think of

$$y = -\tfrac{1}{2}f(\tfrac{1}{3}x) \quad \text{as} \quad y = -\tfrac{1}{2}f(g(x)), \quad \text{where} \quad g(x) = \tfrac{1}{3}x.$$

The last equation for $y$ suggests the assignments

$$Y_1 = \tfrac{1}{3}x, \quad Y_2 = (Y_1)^3 - 4(Y_1)^2, \quad \text{and} \quad Y_3 = -\tfrac{1}{2}Y_2.$$

Note that $Y_2 = f(Y_1) = f(g(x))$. We select only $Y_3$ to be graphed and choose the viewing rectangle $[-7, 14]$ by $[-3, 11]$ to obtain Figure 34.

There are two advantages of assigning the functions in the fashion above:

**FIGURE 34**   $[-7, 14]$ by $[-3, 11]$

**(1)** We did not actually have to compute the polynomial function to be graphed, as we did in Example 7 of Section 2.2.

**(2)** By merely changing the coefficients in $Y_1$ and $Y_3$, we can easily examine their effect on the graph of $Y_3$.

As an illustration of item (2), you should try graphing $y = \frac{1}{2}f(3x)$ by changing $Y_1$ to $3x$, $Y_3$ to $\frac{1}{2}Y_2$, and the viewing rectangle to $[-1, 3]$ by $[-5, 1]$ and then graphing $Y_3$ to obtain Figure 35.

**FIGURE 35**   $[-1, 3]$ by $[-5, 1]$

## 2.4 EXERCISES

**Exer. 1–2: Find**

(a) $(f + g)(3)$    (b) $(f - g)(3)$

(c) $(fg)(3)$    (d) $(f/g)(3)$

1 $f(x) = x + 3,$        $g(x) = x^2$

2 $f(x) = -x^2,$        $g(x) = 2x - 1$

**Exer. 3–8: Find**

(a) $(f + g)(x)$, $(f - g)(x)$, $(fg)(x)$, and $(f/g)(x)$

(b) the domain of $f + g$, $f - g$, and $fg$

(c) the domain of $f/g$

3 $f(x) = x^2 + 2,$        $g(x) = 2x^2 - 1$

4 $f(x) = x^2 + x,$        $g(x) = x^2 - 3$

5 $f(x) = \sqrt{x + 5},$        $g(x) = \sqrt{x + 5}$

6 $f(x) = \sqrt{3 - 2x},$        $g(x) = \sqrt{x + 4}$

7 $f(x) = \dfrac{2x}{x - 4},$        $g(x) = \dfrac{x}{x + 5}$

8 $f(x) = \dfrac{x}{x - 2},$        $g(x) = \dfrac{3x}{x + 4}$

**Exer. 9–10: Find**

(a) $(f \circ g)(x)$    (b) $(g \circ f)(x)$

(c) $(f \circ f)(x)$    (d) $(g \circ g)(x)$

9 $f(x) = 2x - 1,$        $g(x) = -x^2$

10 $f(x) = 3x^2,$        $g(x) = x - 1$

**Exer. 11–20: Find**

(a) $(f \circ g)(x)$    (b) $(g \circ f)(x)$

(c) $f(g(-2))$    (d) $g(f(3))$

11 $f(x) = 2x - 5,$        $g(x) = 3x + 7$

12 $f(x) = 5x + 2,$        $g(x) = 6x - 1$

13 $f(x) = 3x^2 + 4,$        $g(x) = 5x$

14 $f(x) = 3x - 1,$        $g(x) = 4x^2$

15 $f(x) = 2x^2 + 3x - 4,$    $g(x) = 2x - 1$

16 $f(x) = 5x - 7,$        $g(x) = 3x^2 - x + 2$

17 $f(x) = 4x,$        $g(x) = 2x^3 - 5x$

18 $f(x) = x^3 + 2x^2,$        $g(x) = 3x$

19 $f(x) = |x|,$        $g(x) = -7$

20 $f(x) = 5,$        $g(x) = x^2$

**Exer. 21–34: Find (a) $(f \circ g)(x)$ and the domain of $f \circ g$ and (b) $(g \circ f)(x)$ and the domain of $g \circ f$.**

21 $f(x) = x^2 - 3x,$        $g(x) = \sqrt{x + 2}$

22 $f(x) = \sqrt{x - 15},$        $g(x) = x^2 + 2x$

23 $f(x) = x^2 - 4,$        $g(x) = \sqrt{3x}$

24 $f(x) = -x^2 + 1,$        $g(x) = \sqrt{x}$

25 $f(x) = \sqrt{x - 2},$        $g(x) = \sqrt{x + 5}$

26 $f(x) = \sqrt{3 - x},$        $g(x) = \sqrt{x + 2}$

27 $f(x) = \sqrt{3 - x},$        $g(x) = \sqrt{x^2 - 16}$

28 $f(x) = x^3 + 5,$        $g(x) = \sqrt[3]{x - 5}$

29 $f(x) = \dfrac{3x + 5}{2},$        $g(x) = \dfrac{2x - 5}{3}$

30 $f(x) = \dfrac{1}{x - 1},$        $g(x) = x - 1$

31 $f(x) = x^2,$        $g(x) = \dfrac{1}{x^3}$

32 $f(x) = \dfrac{x}{x - 2},$        $g(x) = \dfrac{3}{x}$

33 $f(x) = \dfrac{x - 1}{x - 2},$        $g(x) = \dfrac{x - 3}{x - 4}$

34 $f(x) = \dfrac{x + 2}{x - 1},$        $g(x) = \dfrac{x - 5}{x + 4}$

**Exer. 35–36: Solve the equation $(f \circ g)(x) = 0$.**

35 $f(x) = x^2 - 2,$        $g(x) = x + 3$

36 $f(x) = x^2 - x - 2,$        $g(x) = 2x - 1$

**37** Several values of two functions $f$ and $g$ are listed in the following tables:

| $x$ | 5 | 6 | 7 | 8 |
|---|---|---|---|---|
| $f(x)$ | 8 | 7 | 6 | 5 |

| $x$ | 5 | 6 | 7 | 8 |
|---|---|---|---|---|
| $g(x)$ | 7 | 8 | 6 | 5 |

Find

(a) $(f \circ g)(6)$    (b) $(g \circ f)(6) = g(f(6)) = g(7) = 6$
(c) $(f \circ f)(6)$    (d) $(g \circ g)(6)$  $g(g(6)) = g(8) = 5$
    $= f(7) = 6$

**38** Several values of two functions $T$ and $S$ are listed in the following tables:

| $t$ | 0 | 1 | 2 | 3 |
|---|---|---|---|---|
| $T(t)$ | 2 | 3 | 1 | 0 |

| $x$ | 0 | 1 | 2 | 3 |
|---|---|---|---|---|
| $S(x)$ | 1 | 0 | 3 | 2 |

Find $T(S(1)) = T(0) = 2$

(a) $(T \circ S)(1)$    (b) $(S \circ T)(1)$  $S(T(1)) = S(3) = 2$
(c) $(T \circ T)(1)$    (d) $(S \circ S)(1)$

**39** If $D(t) = \sqrt{400 + t^2}$ and $R(x) = 20x$, find $(D \circ R)(x)$.

**40** If $S(r) = 4\pi r^2$ and $D(t) = 2t + 5$, find $(S \circ D)(t)$.

**41** If $f$ is an odd function and $g$ is an even function, is $fg$ even, odd, or neither even nor odd?

**42** There is one function with domain $\mathbb{R}$ that is both even and odd. Find that function.

**43** *Payroll functions* Let the social security tax function SSTAX be defined as SSTAX$(x) = 0.0715x$, where $x \geq 0$ is the weekly income. Let ROUND2 be the function that rounds a number to two decimal places. Find (ROUND2 $\circ$ SSTAX)(437.21).

**44** *Computer science functions* Let the function CHR be defined by CHR(65) = "A", CHR(66) = "B", . . . , CHR(91) = "Z". Then let the function ORD be defined by ORD("A") = 65, ORD("B") = 66, . . . , ORD("Z") = 91. Find

(a) (CHR $\circ$ ORD)("C")    (b) CHR(ORD("A") + 3)

**45** *Spreading fire* A fire has started in a dry open field and is spreading in the form of a circle. If the radius of this circle increases at the rate of 6 ft/min, express the total fire area $A$ as a function of time $t$ (in minutes).

**46** *Dimensions of a balloon* A spherical balloon is being inflated at a rate of $\frac{9}{2}\pi$ ft³/min. Express its radius $r$ as a function of time $t$ (in minutes), assuming that $r = 0$ when $t = 0$.

**47** *Dimensions of a sand pile* The volume of a conical pile of sand is increasing at a rate of $243\pi$ ft³/min, and the height of the pile always equals the radius $r$ of the base. Express $r$ as a function of time $t$ (in minutes), assuming that $r = 0$ when $t = 0$.

**48** *Diameter of a cube* The diameter $d$ of a cube is the distance between two opposite vertices. Express $d$ as a function of the edge $x$ of the cube. (*Hint:* First express the diagonal $y$ of a face as a function of $x$.)

**49** *Altitude of a balloon* A hot-air balloon rises vertically from ground level as a rope attached to the base of the balloon is released at the rate of 5 ft/sec (see the figure). The pulley that releases the rope is 20 feet from a platform where passengers board the balloon. Express the altitude $h$ of the balloon as a function of time $t$.

**EXERCISE 49**

20'

**50** *Tightrope walker* Refer to Exercise 64 of Section 2.1. Starting at the lowest point, the tightrope walker moves up the rope at a steady rate of 2 ft/sec. If the rope is attached 30 feet up the pole, express the height $h$ of the walker above the ground as a function of time $t$. (*Hint:* Let $d$ denote the total distance traveled along the wire. First express $d$ as a function of $t$, and then $h$ as a function of $d$.)

**51** *Airplane take-off*   Refer to Exercise 65 of Section 2.1. When the airplane is 500 feet down the runway, it has reached a speed of 150 ft/sec (or about 102 mi/hr), which it will maintain until take-off. Express the distance $d$ of the plane from the control tower as a function of time $t$ (in seconds). (*Hint:* In the figure, first write $x$ as a function of $t$.)

**52** *Cable corrosion*   A 100-foot-long cable of diameter 4 inches is submerged in seawater. Because of corrosion, the surface area of the cable decreases at the rate of 750 in.$^2$ per year. Express the diameter $d$ of the cable as a function of time $t$ (in years). (Disregard corrosion at the ends of the cable.)

**Exer. 53–60: Find a composite function form for $y$.**

**53** $y = (x^2 + 3x)^{1/3}$      **54** $y = \sqrt[4]{x^4 - 16}$

**55** $y = \dfrac{1}{(x - 3)^4}$      **56** $y = 4 + \sqrt{x^2 + 1}$

**57** $y = (x^4 - 2x^2 + 5)^5$      **58** $y = \dfrac{1}{(x^2 + 3x - 5)^3}$

**59** $y = \dfrac{\sqrt{x + 4} - 2}{\sqrt{x + 4} + 2}$      **60** $y = \dfrac{\sqrt[3]{x}}{1 + \sqrt[3]{x}}$

**61** If $f(x) = \sqrt{x} - 1$ and $g(x) = x^3 + 1$, approximate $(f \circ g)(0.0001)$. In order to avoid calculating a zero value

for $(f \circ g)(0.0001)$, rewrite the formula for $f \circ g$ as
$$\frac{x^3}{\sqrt{x^3 + 1} + 1}.$$

**62** If $f(x) = \dfrac{x^3}{x^2 + x + 2}$ and $g(x) = (\sqrt{3x} - x^3)^{3/2}$, approximate
$$\frac{(f + g)(1.12) - (f/g)(1.12)}{[(f \circ f)(5.2)]^2}.$$

**63** Refer to Exercise 49 of Section 2.2. Make the assignments $Y_1 = x$ and $Y_2 = 3\sqrt{(Y_1 + 2)(6 - Y_1)} - 4$. Determine assignments for $Y_1$ and $Y_3$ that will enable you to graph each function in (a)–(h), and then graph the function. (Check the domain and range with the previously listed answer.)

(a) $y = -2f(x)$      (b) $y = f(\tfrac{1}{2}x)$
(c) $y = f(x - 3) + 1$      (d) $y = f(x + 2) - 3$
(e) $y = f(-x)$      (f) $y = -f(x)$
(g) $y = f(|x|)$      (h) $y = |f(x)|$

**64** Refer to Exercise 50 of Section 2.2. Make the assignments $Y_1 = x$ and $Y_2 = 3\sqrt{(-Y_1 - 6)(Y_1 + 2)} - 10$. Determine assignments for $Y_1$ and $Y_3$ that will enable you to graph each function, and then graph the function.

(a) $y = \tfrac{1}{2}f(x)$      (b) $y = f(2x)$
(c) $y = f(x - 2) + 5$      (d) $y = f(x + 4) - 1$
(e) $y = f(-x)$      (f) $y = -f(x)$
(g) $y = f(|x|)$      (h) $y = |f(x)|$

## 2.5   INVERSE FUNCTIONS

A function $f$ may have the same value for different numbers in its domain. For example, if $f(x) = x^2$, then $f(2) = 4$ and $f(-2) = 4$, but $2 \neq -2$. For *the inverse of a function* to be defined, it is essential that different numbers in the domain *always* give different values of $f$. Such functions are called *one-to-one functions.*

**Definition of One-to-One Function**

A function $f$ with domain $D$ and range $R$ is a **one-to-one function** if either of the following equivalent conditions is satisfied:

**(1)** Whenever $a \neq b$ in $D$, then $f(a) \neq f(b)$ in $R$.

**(2)** Whenever $f(a) = f(b)$ in $R$, then $a = b$ in $D$.

**FIGURE 36**

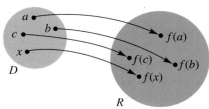

The arrow diagram in Figure 36 illustrates a one-to-one function. Note that each function value in the range $R$ corresponds to *exactly one* element in the domain $D$. The function illustrated in Figure 2 of Section 2.1 is not one-to-one, since $f(w) = f(z)$, but $w \neq z$.

**EXAMPLE 1    *Determining whether a function is one-to-one***

**(a)** If $f(x) = 3x + 2$, prove that $f$ is one-to-one.

**(b)** If $g(x) = x^4 + 2x^2$, prove that $g$ is not one-to-one.

**SOLUTION**

**(a)** We shall use condition (2) of the preceding definition. Thus, suppose that $f(a) = f(b)$ for some numbers $a$ and $b$ in the domain of $f$. This gives us

$$3a + 2 = 3b + 2 \qquad \text{definition of } f(x)$$
$$3a = 3b \qquad \text{subtract 2}$$
$$a = b \qquad \text{divide by 3}$$

Hence, $f$ is one-to-one.

**(b)** Showing that a function *is* one-to-one requires a *general* proof, as in part (a). To show that $g$ is *not* one-to-one we need only find two distinct real numbers in the domain that produce the same function value. For example, $-1 \neq 1$, but $g(-1) = g(1)$. In fact, since $g$ is an even function, $f(-a) = f(a)$ for every real number $a$.

**FIGURE 37**

If we know the graph of a function $f$, it is easy to determine whether $f$ is one-to-one. For example, the function whose graph is sketched in Figure 37 is not one-to-one, since $a \neq b$, but $f(a) = f(b)$. Note that the horizontal line $y = f(a)$ (or $y = f(b)$) intersects the graph in more than one point. In general, we may use the following graphical test to determine whether a function is one-to-one.

**Horizontal Line Test**

A function $f$ is one-to-one if and only if every horizontal line intersects the graph of $f$ in at most one point.

Since every increasing function or decreasing function passes the horizontal line test, we obtain the following result.

**Theorem: Increasing or Decreasing Functions Are One-to-One**

**(1)** A function that is increasing throughout its domain is one-to-one.

**(2)** A function that is decreasing throughout its domain is one-to-one.

Let $f$ be a one-to-one function with domain $D$ and range $R$. Thus, for each number $y$ in $R$, there is *exactly one* number $x$ in $D$ such that $y = f(x)$, as illustrated by the arrow in Figure 38(a). We may, therefore, define a function $g$ from $R$ to $D$ by means of the following rule:

$$x = g(y)$$

As in Figure 38(b), $g$ *reverses the correspondence given by* $f$. We call $g$ the *inverse function* of $f$, as in the next definition.

**FIGURE 38**

(a) $y = f(x)$

(b) $x = g(y)$

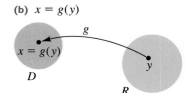

**Definition of Inverse Function**

Let $f$ be a one-to-one function with domain $D$ and range $R$. A function $g$ with domain $R$ and range $D$ is the **inverse function** of $f$, provided the following condition is true for every $x$ in $D$ and every $y$ in $R$:

$$y = f(x) \quad \text{if and only if} \quad x = g(y)$$

Remember that for the inverse of a function $f$ to be defined, *it is absolutely essential that $f$ be one-to-one.* The following theorem, stated without proof, is useful to verify that a function $g$ is the inverse of $f$.

**Theorem on Inverse Functions**

Let $f$ be a one-to-one function with domain $D$ and range $R$. If $g$ is a function with domain $R$ and range $D$, then $g$ is the inverse function of $f$ if and only if both of the following conditions are true:

**(1)** $g(f(x)) = x$ for every $x$ in $D$.

**(2)** $f(g(y)) = y$ for every $y$ in $R$.

Conditions (1) and (2) of the preceding theorem are illustrated in Figure 39(a) and (b), respectively, where the blue arrow indicates that $f$ is a function from $D$ to $R$ and the red arrow indicates that $g$ is a function from $R$ to $D$.

**FIGURE 39**

(a) First $f$, then $g$

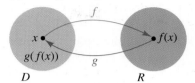

(b) First $g$, then $f$

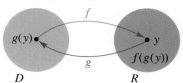

Note that in Figure 39(a) we first apply $f$ to the number $x$ in $D$, obtaining the function value $f(x)$ in $R$, and then apply $g$ to $f(x)$, obtaining the number $g(f(x))$ in $D$. Condition (1) of the theorem states that $g(f(x)) = x$ for every $x$; that is, $g$ *reverses* the correspondence given by $f$.

In Figure 39(b) we use the opposite order for the functions. We first apply $g$ to the number $y$ in $R$, obtaining the function value $g(y)$ in $D$, and then apply $f$ to $g(y)$, obtaining the number $f(g(y))$ in $R$. Condition (2) of the theorem states that $f(g(y)) = y$ for every $y$; that is, $f$ *reverses* the correspondence given by $g$.

If a function $f$ has an inverse function $g$, we often denote $g$ by $f^{-1}$. The $-1$ used in this notation should not be mistaken for an exponent; that is, $f^{-1}(y)$ *does not mean* $1/[f(y)]$. The reciprocal $1/[f(y)]$ may be denoted by $[f(y)]^{-1}$. It is important to remember the following facts about the domain and range of $f$ and $f^{-1}$.

**Domain and Range of $f$ and $f^{-1}$**

$$\text{domain of } f^{-1} = \text{range of } f$$
$$\text{range of } f^{-1} = \text{domain of } f$$

When we discuss functions, we often let $x$ denote an arbitrary number in the domain. Thus, for the inverse function $f^{-1}$, we may wish to consider $f^{-1}(x)$, *where $x$ is in the domain $R$ of $f^{-1}$*. In this event, the two conditions in the theorem on inverse functions are written as follows:

**(1)** $f^{-1}(f(x)) = x$ for every $x$ in the domain of $f$

**(2)** $f(f^{-1}(x)) = x$ for every $x$ in the domain of $f^{-1}$

Figure 39 contains a hint for finding the inverse of a one-to-one function in certain cases: If possible, *we solve the equation $y = f(x)$ for $x$ in terms of $y$*, obtaining an equation of the form $x = g(y)$. If the two conditions $g(f(x)) = x$ and $f(g(x)) = x$ are true for every $x$ in the domains of $f$ and $g$, respectively, then $g$ is the required inverse function $f^{-1}$. The following guidelines summarize this procedure; in guideline 2, in anticipation of finding $f^{-1}$, we write $x = f^{-1}(y)$ instead of $x = g(y)$.

**Guidelines for Finding $f^{-1}$ in Simple Cases**

> **1** Verify that $f$ is a one-to-one function throughout its domain.
> **2** Solve the equation $y = f(x)$ for $x$ in terms of $y$, obtaining an equation of the form $x = f^{-1}(y)$.
> **3** Verify the following two conditions:
>   **a** $f^{-1}(f(x)) = x$ for every $x$ in the domain of $f$
>   **b** $f(f^{-1}(x)) = x$ for every $x$ in the domain of $f^{-1}$

The success of this method depends on the nature of the equation $y = f(x)$, since we must be able to solve for $x$ in terms of $y$. For this reason, we include the phrase *in simple cases* in the title of the guidelines. We shall follow these guidelines in the next three examples.

**EXAMPLE 2**   *Finding the inverse of a function*

Let $f(x) = 3x - 5$. Find the inverse function of $f$.

**SOLUTION**

**Guideline 1**   The graph of the linear function $f$ is a line of slope 3, and hence $f$ is increasing throughout $\mathbb{R}$. Thus, $f$ is one-to-one and the inverse function $f^{-1}$ exists. Moreover, since the domain and range of $f$ is $\mathbb{R}$, the same is true for $f^{-1}$.

**Guideline 2**   Solve the equation $y = f(x)$ for $x$:

$$y = 3x - 5 \qquad \text{let } y = f(x)$$

$$x = \frac{y + 5}{3} \qquad \text{solve for } x \text{ in terms of } y$$

We now formally let $x = f^{-1}(y)$; that is,

$$f^{-1}(y) = \frac{y + 5}{3}.$$

Since the symbol used for the variable is immaterial, we may also write

$$f^{-1}(x) = \frac{x + 5}{3},$$

where $x$ is in the domain of $f^{-1}$.

**Guideline 3**   Since the domain and range of both $f$ and $f^{-1}$ is $\mathbb{R}$, we must verify conditions (a) and (b) for every real number $x$. We proceed as follows:

**(a)**
$$f^{-1}(f(x)) = f^{-1}(3x - 5) \qquad \text{definition of } f$$
$$= \frac{(3x - 5) + 5}{3} \qquad \text{definition of } f^{-1}$$
$$= x \qquad \text{simplify}$$

(b)      $$f(f^{-1}(x)) = f\left(\dfrac{x+5}{3}\right) \qquad \text{definition of } f^{-1}$$

$$= 3\left(\dfrac{x+5}{3}\right) - 5 \quad \text{definition of } f$$

$$= x \qquad\qquad\quad \text{simplify}$$

These verifications prove that the inverse function of $f$ is given by

$$f^{-1}(x) = \dfrac{x+5}{3}.$$

---

## EXAMPLE 3    *Finding the inverse of a function*

Let $f(x) = x^2 - 3$ for $x \geq 0$. Find the inverse function of $f$.

### SOLUTION

**FIGURE 40**

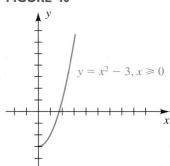

$y = x^2 - 3, x \geq 0$

**Guideline 1**   The graph of $f$ is sketched in Figure 40. The domain of $f$ is $[0, \infty)$, and the range is $[-3, \infty)$. Since $f$ is increasing, it is one-to-one and hence has an inverse function $f^{-1}$ with domain $[-3, \infty)$ and range $[0, \infty)$.

**Guideline 2**   We consider the equation

$$y = x^2 - 3$$

and solve for $x$, obtaining

$$x = \pm\sqrt{y+3}.$$

Since $x$ is nonnegative, we reject $x = -\sqrt{y+3}$ and let

$$f^{-1}(y) = \sqrt{y+3} \quad \text{or, equivalently,} \quad f^{-1}(x) = \sqrt{x+3}.$$

(Note that if the function $f$ had domain $x \leq 0$, we would choose the function $f^{-1}(x) = -\sqrt{x+3}$.)

**Guideline 3**   We verify conditions (a) and (b) for $x$ in the domains of $f$ and $f^{-1}$, respectively.

**(a)** $f^{-1}(f(x)) = f^{-1}(x^2 - 3)$
$$= \sqrt{(x^2 - 3) + 3} = \sqrt{x^2} = x \text{ for } x \geq 0$$

**(b)** $f(f^{-1}(x)) = f(\sqrt{x+3})$
$$= (\sqrt{x+3})^2 - 3 = (x+3) - 3 = x \text{ for } x \geq -3$$

Thus, the inverse function is given by

$$f^{-1}(x) = \sqrt{x+3} \quad \text{for} \quad x \geq -3.$$

---

There is an interesting relationship between the graph of a function $f$ and the graph of its inverse function $f^{-1}$. We first note that $b = f(a)$ is equivalent to $a = f^{-1}(b)$. These equations imply that *the point $(a, b)$ is on the graph of $f$ if and only if the point $(b, a)$ is on the graph of $f^{-1}$.*

**FIGURE 41**

**FIGURE 42**

**FIGURE 43**

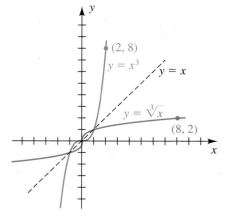

As an illustration, in Example 3 we found that the functions $f$ and $f^{-1}$ given by

$$f(x) = x^2 - 3 \quad \text{and} \quad f^{-1}(x) = \sqrt{x + 3}$$

are inverse functions of each other, provided that $x$ is suitably restricted. Some points on the graph of $f$ are $(0, -3)$, $(1, -2)$, $(2, 1)$, and $(3, 6)$. Corresponding points on the graph of $f^{-1}$ are $(-3, 0)$, $(-2, 1)$, $(1, 2)$, and $(6, 3)$. The graphs of $f$ and $f^{-1}$ are sketched on the same coordinate plane in Figure 41. If the page is folded along the line $y = x$ that bisects quadrants I and III (as indicated by the dashes in the figure), then the graphs of $f$ and $f^{-1}$ coincide. The two graphs are *reflections* of each other through the line $y = x$, or are *symmetric* with respect to this line. This is typical of the graph of every function $f$ that has an inverse function $f^{-1}$ (see Exercise 40).

**EXAMPLE 4**    *The relationship between the graphs of $f$ and $f^{-1}$*

Let $f(x) = x^3$. Find the inverse function $f^{-1}$ of $f$, and sketch the graphs of $f$ and $f^{-1}$ on the same coordinate plane.

**SOLUTION**    The graph of $f$ is sketched in Figure 42. Note that $f$ is an odd function, and hence the graph is symmetric with respect to the origin.

**Guideline 1**  Since $f$ is increasing throughout its domain $\mathbb{R}$, it is one-to-one and hence has an inverse function $f^{-1}$.

**Guideline 2**  We consider the equation

$$y = x^3$$

and solve for $x$ by taking the cube root of each side, obtaining

$$x = y^{1/3} = \sqrt[3]{y}.$$

We now let

$$f^{-1}(y) = \sqrt[3]{y} \quad \text{or, equivalently,} \quad f^{-1}(x) = \sqrt[3]{x}.$$

**Guideline 3**  We verify conditions (a) and (b):

**(a)**    $f^{-1}(f(x)) = f^{-1}(x^3) = \sqrt[3]{x^3} = x$   for every $x$ in $\mathbb{R}$
**(b)**    $f(f^{-1}(x)) = f(\sqrt[3]{x}) = (\sqrt[3]{x})^3 = x$   for every $x$ in $\mathbb{R}$

The graph of $f^{-1}$ (that is, the graph of the equation $y = \sqrt[3]{x}$) may be obtained by reflecting the graph in Figure 42 through the line $y = x$, as shown in Figure 43. Three points on the graph of $f^{-1}$ are $(0, 0)$, $(1, 1)$, and $(8, 2)$.

In the next example we illustrate how some of the concepts presented in this section are helpful in determining the sketch of a graph with the aid of a graphing utility.

**EXAMPLE 5    *Sketching the graph of the inverse of a function***

**(a)** Sketch the graph of $f(x) = x/\sqrt{x^2 + 1}$.

**(b)** Explain why $f$ is one-to-one, and sketch the graph of $f^{-1}$.

**SOLUTION**

**FIGURE 44**  $[-15, 15]$ by $[-10, 10]$

**(a)** We assign $x/\sqrt{x^2 + 1}$ to $Y_1$ and use a standard viewing rectangle to obtain a display similar to Figure 44. The graph is close to 1 if $x$ is large positive and close to $-1$ if $x$ is large negative. To improve the graph, we change the viewing rectangle to $[-15, 15]$ by $[-1, 1]$, obtaining a display similar to Figure 45.

**(b)** By the horizontal line test, the function $f$ appears to be one-to-one, with domain $\mathbb{R}$ and range $(-1, 1)$. (These facts can be proved algebraically; however, we shall not do so here.) Hence, $f$ has an inverse function $f^{-1}$ that has domain $(-1, 1)$ and range $\mathbb{R}$.

To find $f^{-1}$, we solve the equation $y = f(x)$ for $x$ in terms of $y$, as follows:

**FIGURE 45**  $[-15, 15]$ by $[-1, 1]$

$$y = \frac{x}{\sqrt{x^2 + 1}} \qquad \text{given}$$

$$y\sqrt{x^2 + 1} = x \qquad \text{multiply by } \sqrt{x^2 + 1}$$

$$y^2(x^2 + 1) = x^2 \qquad \text{square both sides}$$

$$y^2x^2 + y^2 = x^2 \qquad \text{multiply}$$

$$(y^2 - 1)x^2 = -y^2 \qquad \text{combine } x^2 \text{ terms}$$

$$x^2 = \frac{y^2}{1 - y^2} \qquad \text{divide by } 1 - y^2 \text{ and change signs}$$

$$x = \pm\sqrt{\frac{y^2}{1 - y^2}} = \pm\frac{\sqrt{y^2}}{\sqrt{1 - y^2}} \qquad \text{take the square root}$$

$$x = \pm\frac{|y|}{\sqrt{1 - y^2}} = \pm\frac{y}{\sqrt{1 - y^2}} \qquad \sqrt{y^2} = |y|$$

Referring to Figure 45, we see that *x and y are always both positive or both negative,* so the last equation can be simplified to

**FIGURE 46**  $[-1, 1]$ by $[-15, 15]$

$$x = \frac{y}{\sqrt{1 - y^2}}.$$

Hence,

$$f^{-1}(y) = \frac{y}{\sqrt{1 - y^2}} \quad \text{or, equivalently,} \quad f^{-1}(x) = \frac{x}{\sqrt{1 - x^2}}.$$

Using the viewing rectangle $[-1, 1]$ by $[-15, 15]$ gives us a display similar to Figure 46.

## 2.5 EXERCISES

**Exer. 1–12:** Determine whether the function $f$ is one-to-one.

**1** $f(x) = 3x - 7$

**2** $f(x) = \dfrac{1}{x - 2}$

**3** $f(x) = x^2 - 9$

**4** $f(x) = x^2 + 4$

**5** $f(x) = \sqrt{x}$

**6** $f(x) = \sqrt[3]{x}$

**7** $f(x) = |x|$

**8** $f(x) = 3$

**9** $f(x) = \sqrt{4 - x^2}$

**10** $f(x) = 2x^3 - 4$

**11** $f(x) = \dfrac{1}{x}$

**12** $f(x) = \dfrac{1}{x^2}$

**Exer. 13–16:** Use the theorem on inverse functions to prove that $f$ and $g$ are inverse functions of each other, and sketch the graphs of $f$ and $g$ on the same coordinate plane.

**13** $f(x) = 3x - 2;$      $g(x) = \dfrac{x + 2}{3}$

**14** $f(x) = x^2 + 5,\ x \le 0;$    $g(x) = -\sqrt{x - 5},\ x \ge 5$

**15** $f(x) = -x^2 + 3,\ x \ge 0;$    $g(x) = \sqrt{3 - x},\ x \le 3$

**16** $f(x) = x^3 - 4;$      $g(x) = \sqrt[3]{x + 4}$

**Exer. 17–32:** Find the inverse function of $f$.

**17** $f(x) = 3x + 5$

**18** $f(x) = 7 - 2x$

**19** $f(x) = \dfrac{1}{3x - 2}$

**20** $f(x) = \dfrac{1}{x + 3}$

**21** $f(x) = \dfrac{3x + 2}{2x - 5}$

**22** $f(x) = \dfrac{4x}{x - 2}$

**23** $f(x) = 2 - 3x^2,\ x \le 0$

**24** $f(x) = 5x^2 + 2,\ x \ge 0$

**25** $f(x) = 2x^3 - 5$

**26** $f(x) = -x^3 + 2$

**27** $f(x) = \sqrt{3 - x}$

**28** $f(x) = \sqrt{4 - x^2},\ 0 \le x \le 2$

**29** $f(x) = \sqrt[3]{x} + 1$

**30** $f(x) = (x^3 + 1)^5$

**31** $f(x) = x$

**32** $f(x) = -x$

**Exer. 33–34:** Find $f^{-1}(x)$ for the given function and condition.

**33** $f(x) = x^2 - 4;$      $f^{-1}(5) = -3$

**34** $f(x) = x^2 - 4x + 3;$   $f^{-1}(3) = 0$

**Exer. 35–38:** The graph of a one-to-one function $f$ is shown. **(a)** Use the reflection property to sketch the graph of $f^{-1}$. **(b)** Find the domain $D$ and range $R$ of the function $f$. **(c)** Find the domain $D_1$ and range $R_1$ of the inverse function $f^{-1}$.

**35**

**36**

**37**

**38**

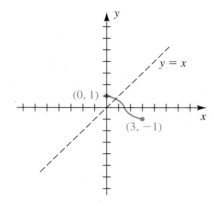

**39 (a)** Prove that the function defined by $f(x) = ax + b$ (a linear function) for $a \neq 0$ has an inverse function, and find $f^{-1}(x)$.

**(b)** Does a constant function have an inverse? Explain.

**40** Show that the graph of $f^{-1}$ is the reflection of the graph of $f$ through the line $y = x$ by verifying the following conditions:

(1) If $P(a, b)$ is on the graph of $f$, then $Q(b, a)$ is on the graph of $f^{-1}$.

(2) The midpoint of line segment $PQ$ is on the line $y = x$.

(3) The line $PQ$ is perpendicular to the line $y = x$.

**41** Verify that $f(x) = f^{-1}(x)$ if

**(a)** $f(x) = -x + b$        **(b)** $f(x) = \dfrac{ax + b}{cx - a}$ for $c \neq 0$

**(c)** $f(x)$ has the following graph:

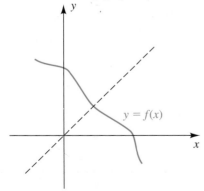

**42** Let $n$ be any positive integer. Find the inverse function of $f$ if

**(a)** $f(x) = x^n$ for $x \geq 0$

**(b)** $f(x) = x^{m/n}$ for $x \geq 0$ and $m$ any positive integer

**Exer. 43–44: Use the graph of $f$ to determine whether $f$ is one-to-one.**

**43** $f(x) = 0.4x^5 - 0.4x^4 + 1.2x^3 - 1.2x^2 + 0.8x - 0.8$

**44** $f(x) = \dfrac{x - 8}{x^{2/3} + 4}$

**Exer. 45–46: Graph $f$ on the given interval. (a) Estimate the largest interval $[a, b]$ with $a < 0 < b$ on which $f$ is one-to-one. (b) If $g$ is the function with domain $[a, b]$ such that $g(x) = f(x)$ for $a \leq x \leq b$, estimate the domain and range of $g^{-1}$.**

**45** $f(x) = 2.1x^3 - 2.98x^2 - 2.11x + 3$;        $[-1, 2]$

**46** $f(x) = 0.05x^4 - 0.24x^3 - 0.15x^2 + 1.18x + 0.24$;        $[-2, 2]$

**47** *Density of the ozone layer*  The density $D$ (in $10^{-3}$ cm/km) of the ozone layer at altitudes $x$ between 3 and 15 kilometers during winter at Edmonton, Canada, was determined experimentally to be

$$D = 0.0833x^2 - 0.4996x + 3.5491.$$

Express $x$ as a function of $D$.

# CHAPTER 2 REVIEW EXERCISES

1 If $f(x) = \dfrac{x}{\sqrt{x+3}}$, find

(a) $f(1)$    (b) $f(-1)$    (c) $f(0)$    (d) $f(-x)$

(e) $-f(x)$    (f) $f(x^2)$    (g) $[f(x)]^2$

**Exer. 2–3:** Find *the sign* of $f(4)$ without actually finding $f(4)$.

2 $f(x) = \dfrac{-32(x^2 - 4)}{(9 - x^2)^{5/3}}$

3 $f(x) = \dfrac{-2(x^2 - 20)(5 - x)}{(6 - x^2)^{4/3}}$

4 Find the domain and range of $f$ if

(a) $f(x) = \sqrt{3x - 4}$    (b) $f(x) = \dfrac{1}{(x+3)^2}$

**Exer. 5–6:** Find $\dfrac{f(a+h) - f(a)}{h}$ if $h \neq 0$.

5 $f(x) = -x^2 + x + 5$    6 $f(x) = \dfrac{1}{x+2}$

7 Find a linear function $f$ such that $f(1) = 2$ and $f(3) = 7$.

8 Determine whether $f$ is even, odd, or neither even nor odd.

(a) $f(x) = \sqrt[3]{x^3 + 4x}$    (b) $f(x) = \sqrt[3]{3x^2 - x^3}$

(c) $f(x) = \sqrt[3]{x^4 + 3x^2 + 5}$

**Exer. 9–10:** Sketch the graph of the equation and label the $x$- and $y$-intercepts.

9 $y = (x - 3)^2 - 2$    10 $y = -x^2 - 2x + 3$

**Exer. 11–20:** (a) Sketch the graph of $f$. (b) Find the domain $D$ and range $R$ of $f$. (c) Find the intervals on which $f$ is increasing, is decreasing, or is constant.

11 $f(x) = \dfrac{1 - 3x}{2}$    12 $f(x) = 1000$

13 $f(x) = |x + 3|$    14 $f(x) = -\sqrt{10 - x^2}$

15 $f(x) = 1 - \sqrt{x + 1}$    16 $f(x) = \sqrt{2 - x}$

17 $f(x) = 9 - x^2$    18 $f(x) = x^2 + 6x + 16$

19 $f(x) = \begin{cases} x^2 & \text{if } x < 0 \\ 3x & \text{if } 0 \leq x < 2 \\ 6 & \text{if } x \geq 2 \end{cases}$    20 $f(x) = 1 + 2[\![x]\!]$

21 Sketch the graphs of the following equations, making use of shifting, stretching, or reflecting:

(a) $y = \sqrt{x}$    (b) $y = \sqrt{x + 4}$    (c) $y = \sqrt{x} + 4$

(d) $y = 4\sqrt{x}$    (e) $y = \frac{1}{4}\sqrt{x}$    (f) $y = -\sqrt{x}$

22 The graph of a function $f$ with domain $[-3, 3]$ is shown in the figure. Sketch the graph of the given equation.

(a) $y = f(x - 2)$    (b) $y = f(x) - 2$

(c) $y = f(-x)$    (d) $y = f(2x)$

(e) $y = f(\frac{1}{2}x)$    (f) $y = f^{-1}(x)$

(g) $y = |f(x)|$    (h) $y = f(|x|)$

EXERCISE 22

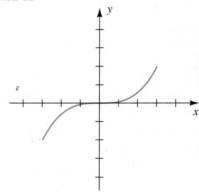

**Exer. 23–24:** Find an equation for the graph shown in the figure.

23

24

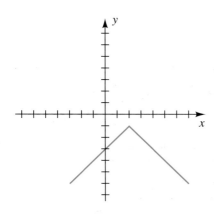

**Exer. 25–28: Find the maximum or minimum value of** $f(x)$.

25  $f(x) = 5x^2 + 30x + 49$

26  $f(x) = -3x^2 + 30x - 82$

27  $f(x) = -12(x + 1)^2 - 37$

28  $f(x) = 3(x + 2)(x - 10)$

29  Express the function $f(x) = -2x^2 + 12x - 14$ in the form $a(x - h)^2 + k$.

30  Find the standard equation of a parabola with a vertical axis that has vertex $V(3, -2)$ and passes through $(5, 4)$.

31  If $f(x) = \sqrt{4 - x^2}$ and $g(x) = \sqrt{x}$, find the domain of
  (a) $fg$    (b) $f/g$

32  If $f(x) = 8x - 1$ and $g(x) = \sqrt{x - 2}$, find
  (a) $(f \circ g)(2)$    (b) $(g \circ f)(2)$

**Exer. 33–34: Find (a)** $(f \circ g)(x)$ **and (b)** $(g \circ f)(x)$.

33  $f(x) = 2x^2 - 5x + 1$,   $g(x) = 3x + 2$

34  $f(x) = \sqrt{3x + 2}$,      $g(x) = 1/x^2$

**Exer. 35–36: Find (a)** $(f \circ g)(x)$ **and the domain of** $f \circ g$ **and (b)** $(g \circ f)(x)$ **and the domain of** $g \circ f$.

35  $f(x) = \sqrt{25 - x^2}$,  $g(x) = \sqrt{x - 3}$

36  $f(x) = \dfrac{x}{3x + 2}$,   $g(x) = \dfrac{2}{x}$

37  Find a composite function form for $y = \sqrt[3]{x^2 - 5x}$.

38  Is $f(x) = 2x^3 - 5$ a one-to-one function?

**Exer. 39–40: (a) Find** $f^{-1}(x)$. **(b) Sketch the graphs of** $f$ **and** $f^{-1}$ **on the same coordinate plane.**

39  $f(x) = 10 - 15x$          40  $f(x) = 9 - 2x^2$, $x \le 0$

41  Refer to the figure to determine each of the following.
  (a) $f(1)$    (b) $(f \circ f)(1)$    (c) $f^{-1}(4)$
  (d) all $x$ such that $f(x) = 4$
  (e) all $x$ such that $f(x) > 4$

**EXERCISE 41**

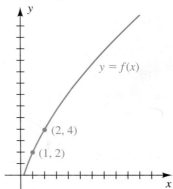

42  Suppose $f$ and $g$ are one-to-one functions such that $f(2) = 7, f(4) = 2$, and $g(2) = 5$. Find the value, if possible.
  (a) $(g \circ f^{-1})(7)$    (b) $(f \circ g^{-1})(5)$
  (c) $(f^{-1} \circ g^{-1})(5)$    (d) $(g^{-1} \circ f^{-1})(2)$

43  **House appreciation**  Six years ago a house was purchased for \$89,000. This year it was appraised at \$125,000. Assume that the value $V$ of the house after its purchase is a linear function of time $t$ (in years).
  (a) Express $V$ in terms of $t$.
  (b) How many years after the purchase date was the house worth \$103,000?

44  **Temperature scales**  The freezing point of water is 0 °C, or 32 °F, and the boiling point is 100 °C, or 212 °F.
  (a) Express the Fahrenheit temperature $F$ as a linear function of the Celsius temperature $C$.
  (b) What temperature increase in °F corresponds to an increase in temperature of 1 °C?

45  **Constructing a storage shelter**  An open rectangular storage shelter consisting of two vertical sides, 4 feet wide, and a flat roof is to be attached to an existing

structure, as illustrated in the figure. The flat roof is made of tin and costs $5 per square foot, and the two sides are made of plywood costing $2 per square foot.

(a) If $400 is available for construction, express the length $y$ as a function of the height $x$.

(b) Express the volume $V$ inside the shelter as a function of $x$.

**EXERCISE 45**

**46 Constructing a cylindrical container** A company plans to manufacture a container having the shape of a right circular cylinder, open at the top, and having a capacity of $24\pi$ in.³ If the cost of the material for the bottom is 30 cents per in.² and that for the curved sides is 10 cents per in.², express the total cost $C$ of the material as a function of the radius $r$ of the base of the container.

**47 Filling a pool** A cross section of a rectangular pool of dimensions 80 feet by 40 feet is shown in the figure. The pool is being filled with water at a rate of 10 ft³ per minute.

(a) Express the volume $V$ of the water in the pool as a function of time $t$.

(b) Express $V$ as a function of the depth $h$ at the deep end for $0 \le h \le 6$ and then for $6 < h \le 9$.

(c) Express $h$ as a function of $t$ for $0 \le h \le 6$ and then for $6 < h \le 9$.

**EXERCISE 47**

**48 Filtering water** Suppose 5 in.³ of water is poured into a conical filter and subsequently drips into a cup, as shown in the figure. Let $x$ denote the height of the water in the filter, and let $y$ denote the height of the water in the cup.

(a) Express the radius $r$ shown in the figure as a function of $x$. (*Hint:* Use similar triangles.)

(b) Express the height $y$ of the water in the cup as a function of $x$. (*Hint:* What is the sum of the two volumes shown in the figure?)

**EXERCISE 48**

**49 Frustum of a cone** The shape of the first spacecraft in the Apollo program was a frustum of a right circular cone—a solid formed by truncating a cone by a plane parallel to its base. For the frustum shown in the figure, the radii $a$ and $b$ have already been determined.

(a) Use similar triangles to express $y$ as a function of $h$.

(b) Derive a formula for the volume of the frustum as a function of $h$.

(c) If $a = 6$ ft and $b = 3$ ft, for what value of $h$ is the volume of the frustum 600 ft³?

**EXERCISE 49**

**50** *Radioactive decay of radon gas* When uranium disintegrates into lead, one step in the process is the radioactive decay of radium into radon gas. Radon enters through the soil into home basements, where it presents a health hazard if inhaled. In the simplest case of radon detection, a sample of air with volume $V$ is taken. After equilibrium has been established, the radioactive decay $D$ of the radon gas is counted with efficiency $E$ over time $t$. The radon concentration $C$ present in the sample of air varies directly as the product of $D$ and $E$ and inversely as the product of $V$ and $t$. For a fixed radon concentration $C$ and time $t$, find the change in the radioactive decay count $D$ if $V$ is doubled and $E$ is reduced by 20%.

**51** *Telephone calls* In a certain county, the average number of telephone calls per day between any two cities is directly proportional to the product of their populations and inversely proportional to the square of the distance between them. Cities A and B are 25 miles apart and have populations of 10,000 and 5000, respectively. Telephone records indicate an average of 2000 calls per day between the two cities. Estimate the average number of calls per day between city A and another city of 15,000 people that is 100 miles from A.

**52** *Power of a wind rotor* The power $P$ generated by a wind rotor is directly proportional to the product of the square of the area $A$ swept out by the blades and the third power of the wind velocity $v$. Suppose the diameter of the circular area swept out by the blades is 10 feet, and $P = 3000$ watts when $v = 20$ mi/hr. Find the power generated when the wind velocity is 30 mi/hr.

**53** *Distance between ships* At 1:00 P.M. ship A is 30 miles due south of ship B and is sailing north at a rate of 15 mi/hr. If ship B is sailing west at a rate of 10 mi/hr, find the time at which the distance $d$ between the ships is minimal (see the figure).

**EXERCISE 53**

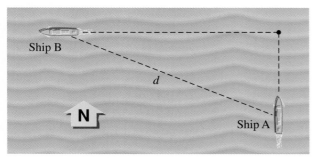

**54** *Dimensions of a race track* The interior of a half-mile race track consists of a rectangle with semicircles at two opposite ends. Find the dimensions that will maximize the area of the rectangle.

**55** *Vertical leaps* When a particular basketball player leaps straight up for a dunk, the player's distance $f(t)$ (in feet) off the floor after $t$ seconds is given by the formula $f(t) = -\frac{1}{2}gt^2 + 16t$, where $g$ is a gravitational constant.

(a) If $g = 32$, find the player's hang time—that is, the total number of seconds that the player is in the air.

(b) Find the player's vertical leap—that is, the maximum distance of the player's feet from the floor.

(c) On the moon, $g = \frac{32}{6}$. Rework parts (a) and (b) for the player on the moon.

**56** *Trajectory of a rocket* A rocket is fired up a hillside, following a path given by $y = -0.016x^2 + 1.6x$. The hillside has slope $\frac{1}{5}$, as illustrated in the figure.

(a) Where does the rocket land?

(b) Find the maximum height of the rocket *above the ground.*

**EXERCISE 56**

*Many geometric figures that occur in everyday life can be represented by graphs of polynomial and rational functions.*

■ *Polynomial functions are the most basic functions in mathematics, because they are defined only in terms of addition, subtraction, and multiplication. In applications it is often necessary to sketch their graphs and to find (or approximate) their zeros. In the first part of this chapter we discuss results that are useful in obtaining this information. We then turn our attention to quotients of polynomial functions—that is, rational functions.* ■

# CHAPTER 3

# POLYNOMIAL

# AND

# RATIONAL FUNCTIONS

## 3.1   GRAPHS OF POLYNOMIAL FUNCTIONS OF DEGREE GREATER THAN 2

If $f$ is a polynomial function with real coefficients of degree $n$, then

$$f(x) = a_n x^n + a_{n-1} x^{n-1} + \cdots + a_1 x + a_0,$$

with $a_n \neq 0$. The special cases listed in the following chart were discussed in Chapter 2.

| Degree of $f$ | Form of $f(x)$ | Graph of $f$ |
|---|---|---|
| 0 | $f(x) = a_0$ | A horizontal line with $y$-intercept $a_0$ |
| 1 | $f(x) = a_1 x + a_0$ | A line of slope $a_1$ and $y$-intercept $a_0$ |
| 2 | $f(x) = a_2 x^2 + a_1 x + a_0$ | A parabola with a vertical axis |

In this section we shall discuss graphs of polynomial functions of degree greater than 2.

If $f$ has degree $n$ *and all the coefficients except* $a_n$ *are zero,* then

$$f(x) = ax^n \quad \text{for some } a = a_n \neq 0.$$

In this case, if $n = 1$, the graph of $f$ is a line through the origin. If $n = 2$, the graph is a parabola with vertex at the origin. Two illustrations with $n = 3$ are given in the next example.

**FIGURE 1**

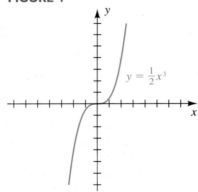

$y = \frac{1}{2} x^3$

**EXAMPLE 1   *Sketching graphs of $y = ax^3$***

Sketch the graph of $f$ if

**(a)** $f(x) = \frac{1}{2} x^3$     **(b)** $f(x) = -\frac{1}{2} x^3$

*SOLUTION*

**(a)** The following table lists several points on the graph of $y = \frac{1}{2} x^3$.

**FIGURE 2**

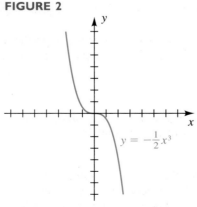

$y = -\frac{1}{2} x^3$

| $x$ | 0 | $\frac{1}{2}$ | 1 | $\frac{3}{2}$ | 2 | $\frac{5}{2}$ |
|---|---|---|---|---|---|---|
| $y$ | 0 | $\frac{1}{16} \approx 0.06$ | $\frac{1}{2}$ | $\frac{27}{16} \approx 1.7$ | 4 | $\frac{125}{16} \approx 7.8$ |

Since $f$ is an odd function, the graph of $f$ is symmetric with respect to the origin, and hence points such as $(-\frac{1}{2}, -\frac{1}{16})$ and $(-1, -\frac{1}{2})$ are also on the graph. The graph is sketched in Figure 1.

**(b)** If $y = -\frac{1}{2} x^3$, the graph can be obtained from that in part (a) by multiplying all $y$-coordinates by $-1$ (that is, by reflecting the graph in part (a) through the $x$-axis). This gives us the sketch in Figure 2.

If $f(x) = ax^n$ and $n$ is an *odd* positive integer, then $f$ is an odd function and the graph of $f$ is symmetric with respect to the origin, as illustrated in Figures 1 and 2. For $a > 0$, the graph is similar in shape to that in Figure 1; however, as either $n$ or $a$ increases, the graph rises more rapidly for $x > 1$. If $a < 0$, we reflect the graph through the $x$-axis, as in Figure 2.

If $f(x) = ax^n$ and $n$ is an *even* positive integer, then $f$ is an even function and the graph of $f$ is symmetric with respect to the $y$-axis, as illustrated in Figure 3 for the case $a = 1$. Note that as the exponent increases, the graph becomes flatter at the origin. It also rises more rapidly for $x > 1$. If $a < 0$, the graph lies below the $x$-axis.

**FIGURE 3**

$y = x^4$

$y = x^6$

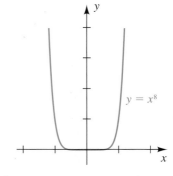

$y = x^8$

**FIGURE 4**

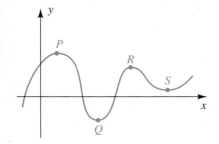

A complete analysis of graphs of polynomial functions of degree greater than 2 requires methods that are used in calculus. As the degree increases, the graphs usually become more complicated. They always have a smooth appearance, however, with a number of high points and low points, such as $P$, $Q$, $R$, and $S$ in Figure 4. Such points are sometimes called **turning points** for the graph. Each function value ($y$-coordinate) corresponding to a high or low point is called an **extremum** of the function $f$. At an extremum, $f$ changes from an increasing function to a decreasing function, or vice versa.

The intermediate value theorem specifies another important property of polynomial functions. The proof is omitted, since it requires advanced mathematics.

**Intermediate Value Theorem
for Polynomial Functions**

> If $f$ is a polynomial function and $f(a) \neq f(b)$ for $a < b$, then $f$ takes on every value between $f(a)$ and $f(b)$ in the interval $[a, b]$.

The intermediate value theorem states that if $w$ is any number between $f(a)$ and $f(b)$, there is at least one number $c$ between $a$ and $b$ such that $f(c) = w$. If we regard the graph of $f$ as extending continuously from the

**FIGURE 5**

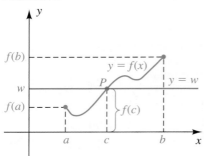

point $(a, f(a))$ to the point $(b, f(b))$, as illustrated in Figure 5, then for any number $w$ between $f(a)$ and $f(b)$, the horizontal line $y = w$ intersects the graph in at least one point $P$. The $x$-coordinate $c$ of $P$ is a number such that $f(c) = w$.

A consequence of the intermediate value theorem is that if $f(a)$ and $f(b)$ have opposite signs, there is at least one number $c$ between $a$ and $b$ such that $f(c) = 0$; that is, $f$ *has a zero at c.* Thus, if the point $(a, f(a))$ lies below the $x$-axis and the point $(b, f(b))$ lies above the $x$-axis, or vice versa, the graph crosses the $x$-axis at least once between $x = a$ and $x = b$, as illustrated in Figure 6.

**FIGURE 6**

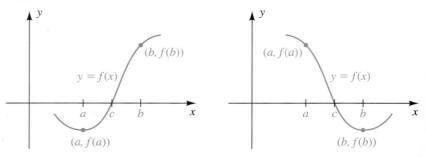

**EXAMPLE 2    *Using the intermediate value theorem***

Show that $f(x) = x^5 + 2x^4 - 6x^3 + 2x - 3$ has a zero between 1 and 2.

**SOLUTION**    Substituting 1 and 2 for $x$ gives us the following function values:

$$f(1) = 1 + 2 - 6 + 2 - 3 = -4$$
$$f(2) = 32 + 32 - 48 + 4 - 3 = 17$$

Since $f(1)$ and $f(2)$ have opposite signs, we see that $f(c) = 0$ for at least one real number $c$ between 1 and 2.

Example 2 illustrates a method for locating real zeros of polynomials. By using *successive approximations,* we can approximate each zero to any degree of accuracy by locating it in smaller and smaller intervals.

If $c$ and $d$ are *successive* real zeros of $f(x)$—that is, there are no other zeros between $c$ and $d$—then $f(x)$ *does not change sign on the interval* $(c, d)$. Thus, if we choose any number $k$ such that $c < k < d$ and if $f(k)$ is positive, then $f(x)$ is positive throughout $(c, d)$. Similarly, if $f(k)$ is negative, then $f(x)$ is negative throughout $(c, d)$. We shall call $f(k)$ a **test value** for $f(x)$ on the interval $(c, d)$. Test values may also be used on infinite intervals of the form $(-\infty, a)$ or $(a, \infty)$, provided that $f(x)$ has no zeros on these intervals. The use of test values in graphing is similar to the technique used for inequalities in Section 1.6.

**EXAMPLE 3**   *Sketching the graph of a polynomial function of degree 3*

Let $f(x) = x^3 + x^2 - 4x - 4$. Find all values of $x$ such that $f(x) > 0$ and all $x$ such that $f(x) < 0$, and then sketch the graph of $f$.

*SOLUTION*   The graph of $f$ lies above the $x$-axis for values of $x$ such that $f(x) > 0$, and it lies below the $x$-axis for all $x$ such that $f(x) < 0$. We may factor $f(x)$ as follows:

$$
\begin{aligned}
f(x) &= x^3 + x^2 - 4x - 4 && \text{given} \\
&= (x^3 + x^2) - (4x + 4) && \text{group terms} \\
&= x^2(x + 1) - 4(x + 1) && \text{factor } x^3 + x^2 \text{ and } 4x + 4 \\
&= (x^2 - 4)(x + 1) && \text{factor out } (x + 1) \\
&= (x + 2)(x - 2)(x + 1) && \text{factor } x^2 - 4
\end{aligned}
$$

**FIGURE 7**

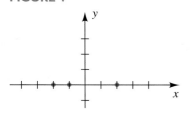

We see from the last equation that the zeros of $f(x)$ (the $x$-intercepts of the graph) are $-2$, $-1$, and $2$. The corresponding points on the graph (see Figure 7) divide the $x$-axis into four parts, and we consider the open intervals

$$(-\infty, -2), \ (-2, -1), \ (-1, 2), \ (2, \infty).$$

As in our work with inequalities in Section 1.6, the sign of $f(x)$ in each of these intervals can be determined by using a sign chart, as follows.

**FIGURE 8**

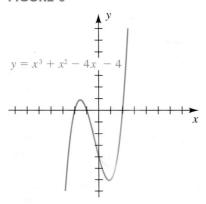

$y = x^3 + x^2 - 4x - 4$

| Interval | $(-\infty, -2)$ | $(-2, -1)$ | $(-1, 2)$ | $(2, \infty)$ |
|---|---|---|---|---|
| Sign of $x + 2$ | − | + | + | + |
| Sign of $x + 1$ | − | − | + | + |
| Sign of $x - 2$ | − | − | − | + |
| Sign of $f(x)$ | − | + | − | + |
| Position of graph | Below $x$-axis | Above $x$-axis | Below $x$-axis | Above $x$-axis |

Using the information from the sign chart leads to the sketch in Figure 8. To find the turning points of the graph, it would be necessary to use a computational device or methods developed in calculus.

The graph of every polynomial function of degree 3 has an appearance similar to that of Figure 8, or it has an inverted version of that graph if the coefficient of $x^3$ is negative. Sometimes, however, the graph may have only one $x$-intercept or the shape may be elongated, as in Figures 1 and 2.

**EXAMPLE 4**   *Sketching the graph of a polynomial function of degree 4*

Let $f(x) = x^4 - 4x^3 + 3x^2$. Find all values of $x$ such that $f(x) > 0$ and all $x$ such that $f(x) < 0$, and then sketch the graph of $f$.

*SOLUTION*   We shall follow the same steps used in the solution of Example 3. Thus, we begin by factoring $f(x)$:

$$\begin{aligned} f(x) &= x^4 - 4x^3 + 3x^2 &&\text{given} \\ &= x^2(x^2 - 4x + 3) &&\text{factor out } x^2 \\ &= x^2(x - 1)(x - 3) &&\text{factor} \end{aligned}$$

Next, we construct the sign diagram in Figure 9, where the vertical lines indicate the zeros 0, 1, and 3 of the factors. Since the factor $x^2$ is always positive if $x \neq 0$, it has no effect on the sign of the product and hence may be omitted from the diagram.

**FIGURE 9**

| Sign of $f(x)$ | + | + | − | + |
| Sign of $x - 3$ | − | − | − | + |
| Sign of $x - 1$ | − | − | + | + |

Referring to the sign of $f(x)$ in the diagram, we see that

$$f(x) > 0 \quad \text{in } (-\infty, 0) \cup (0, 1) \cup (3, \infty) \quad \text{and} \quad f(x) < 0 \quad \text{in } (1, 3).$$

Making use of these facts leads to the sketch in Figure 10.

**FIGURE 10**

$$y = x^4 - 4x^3 + 3x^2$$

In the following example we use a graphing utility to estimate coordinates of important points on a graph.

**EXAMPLE 5**   *Estimating high and low points on a graph*

**(a)** Estimate the real zeros of $f(x) = x^3 - 4.6x^2 + 5.72x - 0.656$ to three decimal places.

**(b)** Estimate the coordinates of the highest and lowest points on the graph for $0.5 \leq x \leq 2.5$.

*SOLUTION*

**(a)** We assign $f(x)$ to $Y_1$ and use a standard viewing rectangle to obtain a display similar to Figure 11(a). Since all the real roots appear to lie between 0 and 3, let us regraph, using the viewing rectangle $[-1, 3]$ by $[-1, 3]$. This gives us a display similar to Figure 11(b), which shows that

**FIGURE 11**

(a) $[-15, 15]$ by $[-10, 10]$          (b) $[-1, 3]$ by $[-1, 3]$

there is only one $x$-intercept and hence one real root. Using the tracing and zoom-in features, we estimate the zero as 0.127.

**(b)** We see from Figure 11(b) that if $0.5 \le x \le 2.5$, then $0 < y < 2$. This suggests that we use the viewing rectangle $[0.5, 2.5]$ by $[0, 2]$. Again using the tracing and zoom-in features, we estimate the highest point to be $(0.866, 1.497)$ and the lowest point to be $(2.200, 0.312)$.

---

In Section 1.6 we solved inequalities similar to the one in the next example, but we relied heavily on the fact that we could somehow factor the expression. We now use a graphing utility to solve an inequality involving an expression (a polynomial of degree 3) that is not easily factored.

**EXAMPLE 6**  *Solving an inequality graphically*

Estimate the solutions of the inequality

$$6x^2 - 3x^3 < 2.$$

**SOLUTION**    Let us subtract 2 from both sides and consider the equivalent inequality

$$6x^2 - 3x^3 - 2 < 0.$$

**FIGURE 12**  $[-15, 15]$ by $[-10, 10]$

We assign $6x^2 - 3x^3 - 2$ to $Y_1$ and use a standard viewing rectangle to obtain a display similar to Figure 12. We see that there are three $x$-intercepts. If we denote them by $x_1$, $x_2$, and $x_3$ (with $x_1 < x_2 < x_3$), then the solutions to the inequality are given by

$$(x_1, x_2) \cup (x_3, \infty),$$

since these are the intervals on which $Y_1$ is less than 0. Using the tracing and zoom-in features for each $x$-intercept, we find that

$$x_1 \approx -0.515, \quad x_2 \approx 0.722, \quad x_3 \approx 1.793.$$

## 3.1 EXERCISES

**Exer. 1–4: Sketch the graph of $f$ for the indicated value of $c$ or $a$.**

1 $f(x) = 2x^3 + c$

(a) $c = 3$ 　　　　(b) $c = -3$

2 $f(x) = -2x^3 + c$

(a) $c = -2$ 　　　(b) $c = 2$

3 $f(x) = ax^3 + 2$

(a) $a = 2$ 　　　　(b) $a = -\frac{1}{3}$

4 $f(x) = ax^3 - 3$

(a) $a = -2$ 　　　(b) $a = \frac{1}{4}$

**Exer. 5–10: Use the intermediate value theorem to show that $f$ has a zero between $a$ and $b$.**

5 $f(x) = x^3 - 4x^2 + 3x - 2$; 　$a = 3$, 　$b = 4$

6 $f(x) = 2x^3 + 5x^2 - 3$; 　$a = -3$, 　$b = -2$

7 $f(x) = -x^4 + 3x^3 - 2x + 1$; 　$a = 2$, 　$b = 3$

8 $f(x) = 2x^4 + 3x - 2$; 　$a = \frac{1}{2}$, 　$b = \frac{3}{4}$

9 $f(x) = x^5 + x^3 + x^2 + x + 1$; 　$a = -\frac{1}{2}$, 　$b = -1$

10 $f(x) = x^5 - 3x^4 - 2x^3 + 3x^2 - 9x - 6$; 　$a = 3$, $b = 4$

**Exer. 11–26: Find all values of $x$ such that $f(x) > 0$ and all $x$ such that $f(x) < 0$, and then sketch the graph of $f$.**

11 $f(x) = \frac{1}{4}x^3 - 2$ 　　　12 $f(x) = -\frac{1}{9}x^3 - 3$

13 $f(x) = -\frac{1}{16}x^4 + 1$ 　　14 $f(x) = x^5 + 1$

15 $f(x) = x^4 - 4x^2$ 　　　　16 $f(x) = 9x - x^3$

17 $f(x) = -x^3 + 3x^2 + 10x$

18 $f(x) = x^4 + 3x^3 - 4x^2$

19 $f(x) = \frac{1}{6}(x + 2)(x - 3)(x - 4)$

20 $f(x) = -\frac{1}{8}(x + 4)(x - 2)(x - 6)$

21 $f(x) = x^3 + 2x^2 - 4x - 8$

22 $f(x) = x^3 - 3x^2 - 9x + 27$

23 $f(x) = x^4 - 6x^2 + 8$

24 $f(x) = -x^4 + 12x^2 - 27$

25 $f(x) = x^2(x + 2)(x - 1)^2(x - 2)$

26 $f(x) = x^3(x + 1)^2(x - 2)(x - 4)$

27 Let $f(x)$ be a polynomial such that the coefficient of every odd power of $x$ is 0. Show that $f$ is an even function.

28 Let $f(x)$ be a polynomial such that the coefficient of every even power of $x$ is 0. Show that $f$ is an odd function.

29 If $f(x) = 3x^3 - kx^2 + x - 5k$, find a number $k$ such that the graph of $f$ contains the point $(-1, 4)$.

30 If one zero of $f(x) = x^3 - 2x^2 - 16x + 16k$ is 2, find two other zeros.

31 **A Legendre polynomial** The third-degree Legendre polynomial $P(x) = \frac{1}{2}(5x^3 - 3x)$ occurs in the solution of heat transfer problems in physics and engineering. Find all values of $x$ such that $P(x) > 0$ and all $x$ such that $P(x) < 0$, and sketch the graph of $P$.

32 **A Chebyshev polynomial** The fourth-degree Chebyshev polynomial $f(x) = 8x^4 - 8x^2 + 1$ occurs in statistical studies. Find all values of $x$ such that $f(x) > 0$. (*Hint:* Let $z = x^2$, and use the quadratic formula.)

33 **Constructing a box** From a rectangular piece of cardboard having dimensions 20 inches × 30 inches, an open box is to be made by cutting out identical squares of area $x^2$ from each corner and turning up the sides (see Exercise 53 of Section 2.1).

(a) Show that the volume of the box is given by the function $V(x) = x(20 - 2x)(30 - 2x)$.

(b) Find all positive values of $x$ such that $V(x) > 0$, and sketch the graph of $V$ for $x > 0$.

34 **Constructing a crate** The frame for a shipping crate is to be constructed from 24 feet of 2 × 2 lumber (see the figure).

**EXERCISE 34**

(a) If the crate is to have square ends of side $x$ feet, express the outer volume $V$ of the crate as a function of $x$ (disregard the thickness of the lumber).

(b) Sketch the graph of $V$ for $x > 0$.

**35** *Determining temperatures* A meteorologist determines that the temperature $T$ (in °F) for a certain 24-hour period in winter was given by the formula $T = \frac{1}{20}t(t - 12)(t - 24)$ for $0 \leq t \leq 24$, where $t$ is time in hours and $t = 0$ corresponds to 6 A.M.

(a) When was $T > 0$, and when was $T < 0$?

(b) Sketch the graph of $T$.

(c) Show that the temperature was 32 °F sometime between 12 noon and 1 P.M. (*Hint:* Use the intermediate value theorem.)

**36** *Deflections of diving boards* A diver stands at the very end of a diving board before beginning a dive. The deflection $d$ of the board at a position $s$ feet from the stationary end is given by $d = cs^2(3L - s)$ for $0 \leq s \leq L$, where $L$ is the length of the board and $c$ is a positive constant that depends on the weight of the diver and on the physical properties of the board (see the figure). Suppose the board is 10 feet long.

(a) If the deflection at the end of the board is 1 foot, find $c$.

(b) Show that the deflection is $\frac{1}{2}$ foot somewhere between $s = 6.5$ and $s = 6.6$.

**EXERCISE 36**

**37** *Deer population* A herd of 100 deer is introduced onto a small island. At first the herd increases rapidly, but eventually food resources dwindle and the population declines. Suppose that the number $N(t)$ of deer after $t$ years is given by $N(t) = -t^4 + 21t^2 + 100$, where $t > 0$.

(a) Determine the values of $t$ for which $N(t) > 0$, and sketch the graph of $N$.

(b) Does the population become extinct? If so, when?

**38** *Deer population* Refer to Exercise 37. It can be shown by means of calculus that the rate $R$ (in deer per year) at which the deer population changes at time $t$ is given by $R = -4t^3 + 42t$.

(a) When does the population cease to grow?

(b) Determine the positive values of $t$ for which $R > 0$.

**Exer. 39–42: Graph $f$ and estimate its zeros.**

**39** $f(x) = x^3 + 0.2x^2 - 2.6x + 1.1$

**40** $f(x) = -x^4 + 0.1x^3 + 4x^2 - 0.5x - 3$

**41** $f(x) = x^3 - 3x + 1$

**42** $f(x) = 2x^3 - 4x^2 - 3x + 1$

**Exer. 43–46: Graph $f$ and estimate all values of $x$ such that $f(x) > k$.**

**43** $f(x) = x^3 + 5x - 2;$        $k = 1$

**44** $f(x) = x^4 - 4x^3 + 3x^2 - 8x + 5;$   $k = 3$

**45** $f(x) = x^5 - 2x^2 + 2;$        $k = -2$

**46** $f(x) = x^4 - 2x^3 + 10x - 26;$    $k = -1$

**Exer. 47–48: Graph $f$ and $g$ on the same coordinate plane, and estimate the points of intersection.**

**47** $f(x) = x^3 - 2x^2 - 1.5x + 2.8;$
$g(x) = -x^3 - 1.7x^2 + 2x + 2.5$

**48** $f(x) = x^4 - 5x^2 + 4;$
$g(x) = x^4 - 3x^3 - 0.25x^2 + 3.75x$

## 3.2 PROPERTIES OF DIVISION

In this section we use $f(x)$, $g(x)$, and so on, to denote polynomials in $x$. If $g(x)$ is a factor of $f(x)$, then $f(x)$ is **divisible** by $g(x)$. For example, $x^4 - 16$ is divisible by $x^2 - 4$, by $x^2 + 4$, by $x + 2$, and by $x - 2$.

The polynomial $x^4 - 16$ is not divisible by $x^2 + 3x + 1$; however, we can use the process called **long division** to find a *quotient* and a *remainder,* as in the following illustrations, where we have inserted terms with zero coefficients.

### ILLUSTRATION

### Long Division of Polynomials

$$
\begin{array}{r}
\overset{\text{quotient}}{\overbrace{x^2 - 3x + 8}} \\
x^2 + 3x + 1 \,\overline{\smash{\big)}\, x^4 + 0x^3 + 0x^2 + 0x - 16} \\
\underline{x^4 + 3x^3 + x^2} \qquad \qquad x^2(x^2 + 3x + 1) \\
-3x^3 - x^2 \qquad \qquad \text{subtract} \\
\underline{-3x^3 - 9x^2 - 3x} \qquad -3x(x^2 + 3x + 1) \\
8x^2 + 3x - 16 \quad \text{subtract} \\
\underline{8x^2 + 24x + 8} \quad 8(x^2 + 3x + 1) \\
\underbrace{-21x - 24}_{\text{remainder}} \quad \text{subtract}
\end{array}
$$

The long division process ends when we arrive at a polynomial (the remainder) that either is 0 or has smaller degree than the divisor. The result of the long division in the preceding illustration can be written

$$
\frac{x^4 - 16}{x^2 + 3x + 1} = (x^2 - 3x + 8) + \left( \frac{-21x - 24}{x^2 + 3x + 1} \right).
$$

Multiplying both sides of this equation by $x^2 + 3x + 1$, we obtain

$$
x^4 - 16 = (x^2 + 3x + 1)(x^2 - 3x + 8) + (-21x - 24).
$$

This example illustrates the following theorem, which we state without proof.

**Division Algorithm**
**for Polynomials**

> If $f(x)$ and $p(x)$ are polynomials and if $p(x) \neq 0$, then there exist unique polynomials $q(x)$ and $r(x)$ such that
>
> $$f(x) = p(x)q(x) + r(x),$$
>
> where either $r(x) = 0$ or the degree of $r(x)$ is less than the degree of $p(x)$. The polynomial $q(x)$ is the **quotient,** and $r(x)$ is the **remainder** in the division of $f(x)$ by $p(x)$.

A useful special case of the division algorithm occurs if $f(x)$ is divided by $x - c$, where $c$ is a real number. If $x - c$ is a factor of $f(x)$, then

$$f(x) = (x - c)q(x)$$

for some quotient $q(x)$; that is, the remainder $r(x)$ is 0. If $x - c$ is not a factor of $f(x)$, then the degree of the remainder $r(x)$ is less than the degree of $x - c$, and hence $r(x)$ must have degree 0. This means that the remainder is a nonzero number. Consequently, for every $x - c$ we have

$$f(x) = (x - c)q(x) + d,$$

where the remainder $d$ is a real number (possibly $d = 0$). If we substitute $c$ for $x$, we obtain

$$
\begin{aligned}
f(c) &= (c - c)q(c) + d \\
&= 0 \cdot q(c) + d \\
&= d.
\end{aligned}
$$

This proves the following theorem.

**Remainder Theorem**

> If a polynomial $f(x)$ is divided by $x - c$, then the remainder is $f(c)$.

**EXAMPLE 1**    *Using the remainder theorem*

If $f(x) = x^3 - 3x^2 + x + 5$, use the remainder theorem to find $f(2)$.

*SOLUTION*    According to the remainder theorem, $f(2)$ is the remainder when $f(x)$ is divided by $x - 2$. By long division,

$$
\begin{array}{r}
x^2 - x - 1 \\
x - 2 \enclose{longdiv}{x^3 - 3x^2 + x + 5} \\
\end{array}
$$

| | |
|---|---|
| $x^3 - 2x^2$ | $x^2(x - 2)$ |
| $-x^2 + x$ | subtract |
| $-x^2 + 2x$ | $-x(x - 2)$ |
| $-x + 5$ | subtract |
| $-x + 2$ | $(-1)(x - 2)$ |
| $3$ | subtract |

Hence $f(2) = 3$. We may check this fact by direct substitution:

$$f(2) = 2^3 - 3(2)^2 + 2 + 5 = 3.$$

We shall use the remainder theorem to prove the following important result.

**Factor Theorem**

A polynomial $f(x)$ has a factor $x - c$ if and only if $f(c) = 0$.

*PROOF*    By the remainder theorem,

$$f(x) = (x - c)q(x) + f(c)$$

for some quotient $q(x)$.

If $f(c) = 0$, then $f(x) = (x - c)q(x)$; that is, $x - c$ is a factor of $f(x)$. Conversely, if $x - c$ is a factor of $f(x)$, then the remainder upon division of $f(x)$ by $x - c$ must be 0, and hence, by the remainder theorem, $f(c) = 0$.
∎

The factor theorem is useful for finding factors of polynomials, as illustrated in the next example.

**EXAMPLE 2    *Using the factor theorem***

Show that $x - 2$ is a factor of $f(x) = x^3 - 4x^2 + 3x + 2$.

*SOLUTION*    Since $f(2) = 8 - 16 + 6 + 2 = 0$, we see from the factor theorem that $x - 2$ is a factor of $f(x)$. Another method of solution would be to divide $f(x)$ by $x - 2$ and show that the remainder is 0. The quotient in the division would be another factor of $f(x)$.

**EXAMPLE 3    *Finding a polynomial with prescribed zeros***

Find a polynomial $f(x)$ of degree 3 that has zeros 2, $-1$, and 3.

*SOLUTION*    By the factor theorem, $f(x)$ has factors $x - 2$, $x + 1$, and $x - 3$. Thus,

$$f(x) = a(x - 2)(x + 1)(x - 3),$$

where any nonzero value may be assigned to $a$. If we let $a = 1$ and multiply, we obtain

$$f(x) = x^3 - 4x^2 + x + 6.$$

To apply the remainder theorem it is necessary to divide a polynomial $f(x)$ by $x - c$. The method of **synthetic division** may be used to simplify this work. The following guidelines state how to proceed. The method can be justified by a careful (and lengthy) comparison with the method of long division.

**Guidelines for Synthetic Division of** $a_n x^n + a_{n-1} x^{n-1} + \cdots + a_1 x + a_0$ **by** $x - c$

1  Begin with the following display, supplying zeros for any missing coefficients in the given polynomial.

$$\underline{c}\ \begin{array}{cccccc} a_n & a_{n-1} & a_{n-2} & \cdots & a_1 & a_0 \end{array}$$

$$a_n$$

2  Multiply $a_n$ by $c$, and place the product $ca_n$ underneath $a_{n-1}$, as indicated by the arrow in the following display. (This arrow, and others, is used only to clarify these guidelines and will not appear in *specific* synthetic divisions.) Next find $b_1 = a_{n-1} + ca_n$, and place it below the line as shown.

$$\underline{c}\ \begin{array}{ccccccc} a_n & a_{n-1} & a_{n-2} & \cdots & & a_1 & a_0 \\ & ca_n & cb_1 & cb_2 & \cdots & cb_{n-2} & cb_{n-1} \\ a_n & b_1 & b_2 & \cdots & & b_{n-1} & r \end{array}$$

3  Multiply $b_1$ by $c$, and place the product $cb_1$ underneath $a_{n-2}$, as indicated by the second arrow. Proceeding, we next find the sum $b_2 = a_{n-2} + cb_1$ and place it below the line as shown.

4  Continue this process, as indicated by the arrows, until the final sum $r = a_0 + cb_{n-1}$ is obtained. The numbers

$$a_n, \quad b_1, \quad b_2, \quad \ldots, \quad b_{n-2}, \quad b_{n-1}$$

are the coefficients of the quotient $q(x)$; that is,

$$q(x) = a_n x^{n-1} + b_1 x^{n-2} + \cdots + b_{n-2} x + b_{n-1},$$

and $r$ is the remainder.

The following examples illustrate synthetic division for some special cases.

**EXAMPLE 4**  *Using synthetic division to find a quotient and remainder*

Use synthetic division to find the quotient and remainder if the polynomial $2x^4 + 5x^3 - 2x - 8$ is divided by $x + 3$.

**SOLUTION**  Since the divisor is $x + 3$, the $c$ in the expression $x - c$ is $-3$. Hence, the synthetic division takes this form:

$$\underline{-3}\ \begin{array}{rrrrr} 2 & 5 & 0 & -2 & -8 \\ & -6 & 3 & -9 & 33 \\ \hline 2 & -1 & 3 & -11 & 25 \end{array}$$

$$\underbrace{\phantom{2 \quad -1 \quad 3 \quad -11}}_{\substack{\text{coefficients} \\ \text{of quotient}}} \quad \underbrace{\phantom{25}}_{\text{remainder}}$$

(continued)

As we have indicated, the first four numbers in the third row are the coefficients of the quotient $q(x)$, and the last number is the remainder $r$. Thus,

$$q(x) = 2x^3 - x^2 + 3x - 11 \quad \text{and} \quad r = 25.$$

Synthetic division can be used to find values of polynomial functions, as illustrated in the next example.

**EXAMPLE 5**   *Using synthetic division to find values of a polynomial*

If $f(x) = 3x^5 - 38x^3 + 5x^2 - 1$, use synthetic division to find $f(4)$.

**SOLUTION**   By the remainder theorem, $f(4)$ is the remainder when $f(x)$ is divided by $x - 4$. Dividing synthetically, we obtain

$$
\begin{array}{r|rrrrrr}
4 & 3 & 0 & -38 & 5 & 0 & -1 \\
  &   & 12 & 48 & 40 & 180 & 720 \\
\hline
  & 3 & 12 & 10 & 45 & 180 & 719 \\
\end{array}
$$

$$\underbrace{\hphantom{3 \quad 12 \quad 10 \quad 45 \quad 180}}_{\text{coefficients of quotient}} \quad \underbrace{\hphantom{719}}_{\text{remainder}}$$

Consequently, $f(4) = 719$.

Synthetic division may be employed to help find zeros of polynomials. By the method illustrated in the preceding example, $f(c) = 0$ if and only if the remainder in the synthetic division by $x - c$ is 0.

**EXAMPLE 6**   *Using synthetic division to find zeros of a polynomial*

Show that $-11$ is a zero of the polynomial

$$f(x) = x^3 + 8x^2 - 29x + 44.$$

**SOLUTION**   Dividing synthetically by $x - (-11) = x + 11$ gives us

$$
\begin{array}{r|rrrr}
-11 & 1 & 8 & -29 & 44 \\
    &   & -11 & 33 & -44 \\
\hline
    & 1 & -3 & 4 & 0 \\
\end{array}
$$

$$\underbrace{\hphantom{1 \quad -3 \quad 4}}_{\text{coefficients of quotient}} \quad \underbrace{\hphantom{0}}_{\text{remainder}}$$

Thus, $f(-11) = 0$.

Example 6 shows that the number $-11$ is a solution of the equation $x^3 + 8x^2 - 29x + 44 = 0$. In Section 3.4 we shall use synthetic division to find rational solutions of equations.

At this stage, you should recognize that the following three statements are equivalent for a polynomial function $f$ whose graph is the graph of the equation $y = f(x)$.

equivalent statements

**(1)** The point $(a, b)$ is on the graph of $f$.

**(2)** The value of $f$ at $x = a$ equals $b$—that is, $f(a) = b$.

**(3)** If $f(x)$ is divided by $x - a$, then the remainder is $b$.

Furthermore, if $b$ is equal to 0, then the next four statements are also equivalent.

additional equivalent statements

**(1)** The number $a$ is a zero of the function $f$.

**(2)** The point $(a, 0)$ is an $x$-intercept on the graph of $f$.

**(3)** The number $a$ is a solution of the equation $f(x) = 0$.

**(4)** The binomial $x - a$ is a factor of the polynomial $f(x)$.

You should become familiar with these statements—so familiar that if you know one of them is true, you can easily recall and apply any appropriate equivalent statement.

**EXAMPLE 7**   *Relating a graph to division*

Use the graph of

$$f(x) = 0.5x^5 + 3.5x^4 - 5.5x^3 - 7.5x^2 + 2x + 2$$

to approximate (to two decimal places) the remainder if $f(x)$ is divided by $x + 1.37$.

**SOLUTION**   We assign $f(x)$ to $Y_1$ and graph $f$ with a standard viewing rectangle as shown in Figure 13. From the preceding discussion, we know that to find a remainder $b$ by utilizing a graph, we should find the point $(a, b)$ that corresponds to dividing $f(x)$ by $x - a$. In this case, $a = -1.37$ and the point on the graph with $x$-coordinate $-1.37$ is approximately $(-1.37, 9.24)$. Hence, the remainder $b$ is approximately 9.24.

The easiest way to find the remainder using a graphing utility is to simply query the function value $Y_1$ when $x = -1.37$. However, the purpose of this example was to point out the graphical relationship to the division process.

**FIGURE 13**   $[-10, 10]$ by $[-10, 10]$

## 3.2 EXERCISES

**Exer. 1–8: Find the quotient and remainder if $f(x)$ is divided by $p(x)$.**

1 $f(x) = 2x^4 - x^3 - 3x^2 + 7x - 12$;  $p(x) = x^2 - 3$

2 $f(x) = 3x^4 + 2x^3 - x^2 - x - 6$;  $p(x) = x^2 + 1$

3 $f(x) = 3x^3 + 2x - 4$;  $p(x) = 2x^2 + 1$

4 $f(x) = 3x^3 - 5x^2 - 4x - 8$;  $p(x) = 2x^2 + x$

5 $f(x) = 7x + 2$;  $p(x) = 2x^2 - x - 4$

6 $f(x) = -5x^2 + 3$;  $p(x) = x^3 - 3x + 9$

7 $f(x) = 9x + 4$;                    $p(x) = 2x - 5$

8 $f(x) = 7x^2 + 3x - 10$;           $p(x) = x^2 - x + 10$

**Exer. 9–12: Use the remainder theorem to find $f(c)$.**

9 $f(x) = 3x^3 - x^2 + 5x - 4$; $=26$    $c = 2$

10 $f(x) = 2x^3 + 4x^2 - 3x - 1$;    $c = 3$

11 $f(x) = x^4 - 6x^2 + 4x - 8$;    $c = -3$

12 $f(x) = x^4 + 3x^2 - 12$;    $c = -2$

**Exer. 13–16: Use the factor theorem to show that $x - c$ is a factor of $f(x)$.**

13 $f(x) = x^3 + x^2 - 2x + 12$;    $c = -3$

14 $f(x) = x^3 + x^2 - 11x + 10$;    $c = 2$

15 $f(x) = x^{12} - 4096$;    $c = -2$

16 $f(x) = x^4 - 3x^3 - 2x^2 + 5x + 6$;  $c = 2$

**Exer. 17–20: Find a polynomial $f(x)$ with leading coefficient 1 and having the given degree and zeros.**

17 degree 3;   zeros $-2, 0, 5$

18 degree 3;   zeros $\pm 2, 3$

19 degree 4;   zeros $-2, \pm 1, 4$

20 degree 4;   zeros $-3, 0, 1, 5$

**Exer. 21–28: Use synthetic division to find the quotient and remainder if the first polynomial is divided by the second.**

21 $2x^3 - 3x^2 + 4x - 5$;    $x - 2$

22 $3x^3 - 4x^2 - x + 8$;    $x + 4$

23 $x^3 - 8x - 5$;    $x + 3$

24 $5x^3 - 6x^2 + 15$;    $x - 4$

25 $3x^5 + 6x^2 + 7$;    $x + 2$

26 $-2x^4 + 10x - 3$;    $x - 3$

27 $4x^4 - 5x^2 + 1$;    $x - \frac{1}{2}$

28 $9x^3 - 6x^2 + 3x - 4$;    $x - \frac{1}{3}$

**Exer. 29–34: Use synthetic division to find $f(c)$.**

29 $f(x) = 2x^3 + 3x^2 - 4x + 4$;    $c = 3$

30 $f(x) = -x^3 + 4x^2 + x$;    $c = -2$

31 $f(x) = 0.3x^3 + 0.04x - 0.034$;    $c = -0.2$

32 $f(x) = 8x^5 - 3x^2 + 7$;    $c = \frac{1}{2}$

33 $f(x) = x^2 + 3x - 5$;    $c = 2 + \sqrt{3}$

34 $f(x) = x^3 - 3x^2 - 8$;    $c = 1 + \sqrt{2}$

**Exer. 35–38: Use synthetic division to show that $c$ is a zero of $f(x)$.**

35 $f(x) = 3x^4 + 8x^3 - 2x^2 - 10x + 4$;   $c = -2$

36 $f(x) = 4x^3 - 9x^2 - 8x - 3$;    $c = 3$

37 $f(x) = 4x^3 - 6x^2 + 8x - 3$;    $c = \frac{1}{2}$

38 $f(x) = 27x^4 - 9x^3 + 3x^2 + 6x + 1$;   $c = -\frac{1}{3}$

**Exer. 39–40: Find all values of $k$ such that $f(x)$ is divisible by the given linear polynomial.**

39 $f(x) = kx^3 + x^2 + k^2x + 3k^2 + 11$;    $x + 2$

40 $f(x) = k^2x^3 - 4kx + 3$;    $x - 1$

**Exer. 41–42: Show that $x - c$ is not a factor of $f(x)$ for any real number $c$.**

41 $f(x) = 3x^4 + x^2 + 5$        42 $f(x) = -x^4 - 3x^2 - 2$

43 Find the remainder if the polynomial
$$3x^{100} + 5x^{85} - 4x^{38} + 2x^{17} - 6$$
is divided by $x + 1$. *plug in $(-1)$ to find remainder*

**Exer. 44–46: Use the factor theorem to verify the statement.**

44 $x - y$ is a factor of $x^n - y^n$ for every positive integer $n$.

45 $x + y$ is a factor of $x^n - y^n$ for every positive even integer $n$.

46 $x + y$ is a factor of $x^n + y^n$ for every positive odd integer $n$.

47 Let $P(x, y)$ be a first-quadrant point on $y = 6 - x$, and consider the vertical line segment $PQ$ shown in the figure.

**EXERCISE 47**

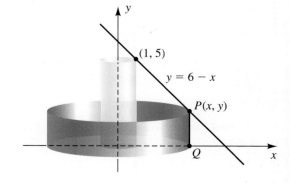

(a) If $PQ$ is rotated about the $y$-axis, determine the volume $V$ of the resulting cylinder.

(b) For what point $P(x, y)$ with $x \neq 1$ is the volume $V$ in part (a) the same as the volume of the cylinder of radius 1 and altitude 5 shown in the figure?

48 **Strength of a beam**  The strength of a rectangular beam is directly proportional to the product of its width and the square of the depth of a cross section (see the figure). A beam of width 1.5 feet has been cut from a cylindrical log of radius 1 foot. Find the width of a second rectangular beam of equal strength that could have been cut from the log.

EXERCISE 48

49 **Parabolic arch**  An arch has the shape of the parabola $y = 4 - x^2$. A rectangle is fit under the arch by selecting a point $(x, y)$ on the parabola (see the figure).

EXERCISE 49

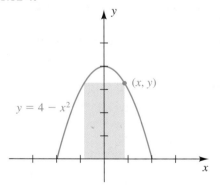

(a) Express the area $A$ of the rectangle in terms of $x$.

(b) If $x = 1$, the rectangle has base 2 and height 3. Find the base of a second rectangle that has the same area.

50 **Dimensions of a capsule**  An aspirin tablet in the shape of a right circular cylinder has height $\frac{1}{3}$ centimeter and radius $\frac{1}{2}$ centimeter. The manufacturer also wishes to market the aspirin in capsule form. The capsule is to be $\frac{3}{2}$ centimeters long, in the shape of a right circular cylinder with hemispheres attached at both ends (see the figure).

(a) If $r$ denotes the radius of a hemisphere, find a formula for the volume of the capsule.

(b) Find the radius of the capsule so that its volume is equal to that of the tablet.

EXERCISE 50

Exer. 51–52: Use the graph of $f$ to approximate the remainder if $f$ is divided by $x - 0.21$.

51 $f(x) = x^8 - 7.9x^5 - 0.8x^4 + x^3 + 1.2x - 9.81$

52 $f(x) = 3.33x^6 - 2.5x^5 + 6.9x^3 - 4.1x^2 + 1.22x - 6.78$

Exer. 53–54: Use the graph of $f$ to approximate all values of $k$ such that $f(x)$ is divisible by the given linear polynomial.

53 $f(x) = x^3 + k^3x^2 + 2kx - 2k^4;$    $x - 1.6$

54 $f(x) = k^5x^3 - 2.1x^2 + k^3x - 1.2k^2;$  $x + 0.4$

## 3.3  ZEROS OF POLYNOMIALS

The **zeros of a polynomial** $f(x)$ are the solutions of the equation $f(x) = 0$. Each real zero is an $x$-intercept of the graph of $f$. In applied fields, calculators and computers are usually employed to find or approximate zeros. Before using a calculator, however, it is worth knowing what type of zeros to expect. Some questions we could ask are

**(1)** How many zeros of $f(x)$ are real? imaginary?

**(2)** How many real zeros of $f(x)$ are positive? negative?

**(3)** How many real zeros of $f(x)$ are rational? irrational?

**(4)** Are the real zeros of $f(x)$ large or small in value?

In this and the following section we shall discuss results that help answer some of these questions. These results form the basis of the *theory of equations*.

The factor and remainder theorems can be extended to the system of complex numbers. Thus, a complex number $c = a + bi$ is a zero of a polynomial $f(x)$ if and only if $x - c$ is a factor of $f(x)$. Except in special cases, zeros of polynomials are very difficult to find. For example, there are no obvious zeros of $f(x) = x^5 - 3x^4 + 4x^3 - 4x - 10$. Although we have no formula that can be used to find the zeros, the next theorem states that there is at *least* one zero $c$, and hence, by the factor theorem, $f(x)$ has a factor of the form $x - c$.

**Fundamental Theorem of Algebra**

> If a polynomial $f(x)$ has positive degree and complex coefficients, then $f(x)$ has at least one complex zero.

The standard proof of this theorem requires results from an advanced field of mathematics called *functions of a complex variable*. A prerequisite for studying this field is a strong background in calculus. The first proof of the fundamental theorem of algebra was given by the German mathematician Carl Friedrich Gauss (1777–1855), who is considered by many to be the greatest mathematician of all time.

As a special case of the fundamental theorem, if all the coefficients of $f(x)$ are real, then $f(x)$ has at least one complex zero. If $a + bi$ is a complex zero, it may happen that $b = 0$, in which case the number $a$ is a real zero.

The fundamental theorem of algebra enables us, at least in theory, to express every polynomial $f(x)$ of positive degree as a product of polynomials of degree 1, as in the next theorem.

**Complete Factorization Theorem for Polynomials**

> If $f(x)$ is a polynomial of degree $n > 0$, then there exist $n$ complex numbers $c_1, c_2, \ldots, c_n$ such that
>
> $$f(x) = a(x - c_1)(x - c_2)\cdots(x - c_n),$$
>
> where $a$ is the leading coefficient of $f(x)$. Each number $c_k$ is a zero of $f(x)$.

**PROOF**   If $f(x)$ has degree $n > 0$, then, by the fundamental theorem of algebra, $f(x)$ has a complex zero $c_1$. Hence, by the factor theorem, $f(x)$ has a factor $x - c_1$; that is,

$$f(x) = (x - c_1)f_1(x),$$

where $f_1(x)$ is a polynomial of degree $n - 1$. If $n - 1 > 0$, then, by the same argument, $f_1(x)$ has a complex zero $c_2$ and therefore a factor $x - c_2$. Thus,

$$f_1(x) = (x - c_2)f_2(x),$$

where $f_2(x)$ is a polynomial of degree $n - 2$. Hence,

$$f(x) = (x - c_1)(x - c_2)f_2(x).$$

Continuing this process, after $n$ steps we arrive at a polynomial $f_n(x)$ of degree 0. Thus, $f_n(x) = a$ for some nonzero number $a$, and we may write

$$f(x) = a(x - c_1)(x - c_2)\cdots(x - c_n),$$

where each complex number $c_k$ is a zero of $f(x)$. The leading coefficient of the polynomial on the right-hand side in the last equation is $a$, and therefore $a$ is the leading coefficient of $f(x)$. ■

**ILLUSTRATION**

**Complete Factorization Theorem for Polynomials**

| A Polynomial $f(x)$ | A Factored Form of $f(x)$ | Zeros of $f(x)$ |
|---|---|---|
| $3x^2 - (12 + 6i)x + 24i$ | $3(x - 4)(x - 2i)$ | $4, 2i$ |
| $-6x^3 - 2x^2 - 6x - 2$ | $-6(x + \frac{1}{3})(x + i)(x - i)$ | $-\frac{1}{3}, \pm i$ |
| $5x^3 - 30x^2 + 65x$ | $5(x - 0)[x - (3 + 2i)][x - (3 - 2i)]$ | $0, 3 \pm 2i$ |
| $\frac{2}{3}x^3 + 8x^2 - \frac{2}{3}x - 8$ | $\frac{2}{3}(x + 12)(x + 1)(x - 1)$ | $-12, \pm 1$ |

We may now prove the following.

**Theorem on the Maximum Number of Zeros of a Polynomial**

> A polynomial of degree $n > 0$ has at most $n$ different complex zeros.

**PROOF**     We will give an indirect proof; that is, we will suppose $f(x)$ has *more* than $n$ different complex zeros and show that this supposition leads to a contradiction. Let us choose $n + 1$ of the zeros and label them $c_1$, $c_2$, . . . , $c_n$, and $c$. We may use the $c_k$ to obtain the factorization indicated in the statement of the complete factorization theorem. Substituting $c$ for $x$ and using the fact that $f(c) = 0$, we obtain

$$0 = a(c - c_1)(c - c_2) \cdots (c - c_n).$$

However, each factor on the right-hand side is different from zero because $c \neq c_k$ for every $k$. Since the product of nonzero numbers cannot equal zero, we have a contradiction.   ∎

**EXAMPLE I**     *Finding a polynomial with prescribed zeros*

Find a polynomial $f(x)$ in factored form that has degree 3; has zeros 2, $-1$, and 3; and satisfies $f(1) = 5$.

**SOLUTION**     By the factor theorem, $f(x)$ has factors $x - 2$, $x + 1$, and $x - 3$. No other factors of degree 1 exist, since, by the factor theorem, another linear factor $x - c$ would produce a fourth zero of $f(x)$, contrary to the preceding theorem. Hence, $f(x)$ has the form

$$f(x) = a(x - 2)(x + 1)(x - 3)$$

for some number $a$. Since $f(1) = 5$, we see that

$$
\begin{aligned}
5 &= a(1 - 2)(1 + 1)(1 - 3) &&\text{let } x = 1 \text{ in } f(x)\\
5 &= 4a &&\text{simplify}\\
a &= \tfrac{5}{4} &&\text{solve for } a
\end{aligned}
$$

Consequently,

$$f(x) = \tfrac{5}{4}(x - 2)(x + 1)(x - 3).$$

If we multiply the factors, we obtain the polynomial

$$f(x) = \tfrac{5}{4}x^3 - 5x^2 + \tfrac{5}{4}x + \tfrac{15}{2}.$$

---

The numbers $c_1$, $c_2$, . . . , $c_n$ in the complete factorization theorem are not necessarily all different. To illustrate, $f(x) = x^3 + x^2 - 5x + 3$ has the factorization

$$f(x) = (x + 3)(x - 1)(x - 1).$$

If a factor $x - c$ occurs $m$ times in the factorization, then $c$ is a **zero of multiplicity** $m$ of $f(x)$, or a **root of multiplicity** $m$ of the equation $f(x) = 0$. In the preceding display, 1 is a zero of multiplicity 2, and $-3$ is a zero of multiplicity 1.

If $c$ is a real zero of $f(x)$ of multiplicity $m$, then $f(x)$ has the factor $(x - c)^m$, and the graph of $f$ has an $x$-intercept $c$. The general shape of the graph at $(c, 0)$ depends on whether $m$ is an odd integer or an even integer. If $m$ is odd, then $(x - c)^m$ changes sign as $x$ increases through $c$, and hence the graph of $f$ crosses the $x$-axis at $(c, 0)$, as indicated in the first row of the following chart. The figures in the chart do not show the complete graph of $f$, but only its general shape near $(c, 0)$. If $m$ is even, then $(x - c)^m$ does not change sign at $c$, and the graph of $f$ near $(c, 0)$ has the appearance of one of the two figures in the second row.

| Factor of $f(x)$ | General shape of the graph of $f$ near $(c, 0)$ | |
|---|---|---|
| $(x - c)^m$, with $m$ odd and $m \neq 1$ | | |
| $(x - c)^m$, with $m$ even | | |

**FIGURE 14**

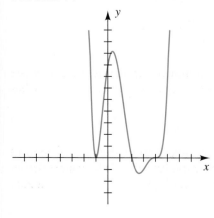

**EXAMPLE 2** *Finding multiplicities of zeros*

Find the zeros of the polynomial $f(x) = \frac{1}{16}(x - 2)(x - 4)^3(x + 1)^2$, state the multiplicity of each, and then sketch the graph of $f$.

**SOLUTION** We see from the factored form that $f(x)$ has three distinct zeros, 2, 4, and $-1$. The zero 2 has multiplicity 1, the zero 4 has multiplicity 3, and the zero $-1$ has multiplicity 2. Note that $f(x)$ has degree 6.

The $x$-intercepts of the graph of $f$ are the real zeros $-1$, 2, and 4. Since the multiplicity of $-1$ is an even integer, the graph intersects, but does not cross, the $x$-axis at $(-1, 0)$. Since the multiplicities of 2 and 4 are odd, the graph crosses the $x$-axis at $(2, 0)$ and $(4, 0)$. The $y$-intercept is $f(0) = 8$. The graph is shown in Figure 14.

If $f(x) = a(x - c_1)(x - c_2) \cdots (x - c_n)$ is a polynomial of degree $n$, then the $n$ complex numbers $c_1, c_2, \ldots, c_n$ are zeros of $f(x)$. Counting a zero of multiplicity $m$ as $m$ zeros tell us that $f(x)$ has at least $n$ zeros (not necessarily all different). Combining this fact with the fact that $f(x)$ has at most $n$ zeros gives us the next result.

**Theorem on the Exact Number of Zeros of a Polynomial**

> If $f(x)$ is a polynomial of degree $n > 0$ and if a zero of multiplicity $m$ is counted $m$ times, then $f(x)$ has precisely $n$ zeros.

**EXAMPLE 3**     *Finding the zeros of a polynomial*

Express $f(x) = x^5 - 4x^4 + 13x^3$ as a product of linear factors, and find the five zeros of $f(x)$.

**SOLUTION**     We begin by factoring out $x^3$:

$$f(x) = x^3(x^2 - 4x + 13)$$

By the quadratic formula, the zeros of the polynomial $x^2 - 4x + 13$ are

$$\frac{-(-4) \pm \sqrt{(-4)^2 - 4(1)(13)}}{2(1)} = \frac{4 \pm \sqrt{-36}}{2} = \frac{4 \pm 6i}{2} = 2 \pm 3i.$$

Hence, by the factor theorem, $x^2 - 4x + 13$ has factors $x - (2 + 3i)$ and $x - (2 - 3i)$, and we obtain the factorization

$$f(x) = x \cdot x \cdot x \cdot (x - 2 - 3i)(x - 2 + 3i).$$

Since $x - 0$ occurs as a factor three times, the number 0 is a zero of multiplicity 3, and the five zeros of $f(x)$ are 0, 0, 0, 2 + 3i, and 2 − 3i.

We next show how to use *Descartes' rule of signs* to obtain information about the zeros of a polynomial $f(x)$ with real coefficients. In the statement of the rule we assume that the terms of $f(x)$ are arranged in order of decreasing powers of $x$ and that terms with zero coefficients are deleted. We also assume that the **constant term**—that is, the term that does not contain $x$—is different from 0. We say there is a **variation of sign** in $f(x)$ if two consecutive coefficients have opposite signs. To illustrate, the polynomial $f(x)$ in the following illustration has three variations in sign, as indicated by the braces—one variation from $2x^5$ to $-7x^4$, a second from $-7x^4$ to $3x^2$, and a third from $6x$ to $-5$.

### ILLUSTRATION

**Variations of Sign in $f(x) = 2x^5 - 7x^4 + 3x^2 + 6x - 5$**

$$f(x) = 2x^5 \overbrace{\phantom{xx}}^{+ \text{ to } -} 7x^4 \overbrace{\phantom{xx}}^{- \text{ to } +} 3x^2 \overbrace{\phantom{xx}}^{\substack{\text{no}\\ \text{variation}}} 6x \overbrace{\phantom{xx}}^{+ \text{ to } -} 5$$

Descartes' rule also refers to the variations of sign in $f(-x)$. Using the previous illustration, note that

$$f(-x) = 2(-x)^5 - 7(-x)^4 + 3(-x)^2 + 6(-x) - 5$$
$$= -2x^5 - 7x^4 + 3x^2 - 6x - 5.$$

Hence, as indicated in the next illustration, there are two variations of sign in $f(-x)$—one from $-7x^4$ to $3x^2$ and a second from $3x^2$ to $-6x$.

### ILLUSTRATION

**Variations of Sign in $f(-x)$ if $f(x) = 2x^5 - 7x^4 + 3x^2 + 6x - 5$**

$$f(-x) = -2x^5 \overbrace{\phantom{xx}}^{\substack{\text{no}\\ \text{variation}}} 7x^4 \overbrace{\phantom{xx}}^{- \text{ to } +} 3x^2 \overbrace{\phantom{xx}}^{+ \text{ to } -} 6x \overbrace{\phantom{xx}}^{\substack{\text{no}\\ \text{variation}}} 5$$

We may state Descartes' rule as follows.

**Descartes' Rule of Signs**

> Let $f(x)$ be a polynomial with real coefficients and nonzero constant term.
>
> **(1)** The number of positive real zeros of $f(x)$ either is equal to the number of variations of sign in $f(x)$ or is less than that number by an even integer.
>
> **(2)** The number of negative real zeros of $f(x)$ either is equal to the number of variations of sign in $f(-x)$ or is less than that number by an even integer.

A proof of Descartes' rule will not be given.

### EXAMPLE 4   *Using Descartes' rule of signs*

Discuss the number of possible positive and negative real solutions and nonreal complex solutions of the equation

$$2x^5 - 7x^4 + 3x^2 + 6x - 5 = 0.$$

SOLUTION    The polynomial $f(x)$ on the left-hand side of the equation is the same as the one given in the two previous illustrations. Since there are three variations of sign in $f(x)$, the equation has either three positive real solutions or one positive real solution.

Since $f(-x) = -2x^5 - 7x^4 + 3x^2 - 6x - 5$ has two variations of sign, the given equation has either two negative solutions or no negative solution. Because $f(x)$ has degree 5, there are a total of 5 solutions. The solutions that are not positive or negative real numbers are imaginary numbers. The following table summarizes the various possibilities that can occur for solutions of the equation.

| Number of positive real solutions | 3 | 3 | 1 | 1 |
|---|---|---|---|---|
| Number of negative real solutions | 2 | 0 | 2 | 0 |
| Number of imaginary solutions | 0 | 2 | 2 | 4 |
| Total number of solutions | 5 | 5 | 5 | 5 |

Descartes' rule stipulates that the constant term of the polynomial $f(x)$ is different from 0. If the constant term is 0, as in the equation

$$x^4 - 3x^3 + 2x^2 - 5x = 0,$$

we factor out the lowest power of $x$, obtaining

$$x(x^3 - 3x^2 + 2x - 5) = 0.$$

Thus, one solution is $x = 0$, and we apply Descartes' rule to the polynomial $x^3 - 3x^2 + 2x - 5$ to determine the nature of the remaining three solutions.

### EXAMPLE 5    Using Descartes' rule of signs

Discuss the nature of the roots of the equation $3x^5 + 4x^3 + 2x - 5 = 0$.

SOLUTION    The polynomial $f(x)$ on the left-hand side of the equation has one variation of sign (from $2x$ to $-5$), and hence, by part (1) of Descartes' rule, the equation has precisely one positive real root. Since

$$f(-x) = 3(-x)^5 + 4(-x)^3 + 2(-x) - 5$$
$$= -3x^5 - 4x^3 - 2x - 5,$$

$f(-x)$ has no variation of sign, and hence, by part (2) of Descartes' rule, there is no negative real root. Thus, the equation has one real root and four nonreal, complex roots.

When applying Descartes' rule, we count roots of multiplicity $k$ as $k$ roots. For example, given $x^2 - 2x + 1 = 0$, the polynomial $x^2 - 2x + 1$ has two variations of sign, and hence the equation has either two positive real roots or none. The factored form of the equation is $(x - 1)^2 = 0$, and hence 1 is a root of multiplicity 2.

We conclude this section with a discussion of *bounds* for the real zeros of a polynomial $f(x)$ with real coefficients. By definition, a real number $b$ is an **upper bound** for the zeros if no zero is greater than $b$. A real number $a$ is a **lower bound** for the zeros if no zero is less than $a$. Thus, if $r$ is any real zero of $f(x)$, than $a \le r \le b$; that is, $r$ is in the closed interval $[a, b]$, as illustrated in Figure 15. Note that upper and lower bounds are not unique, since any number greater than $b$ is also an upper bound and any number less than $a$ is a lower bound.

**FIGURE 15**

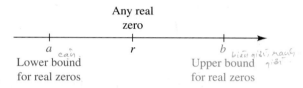

We may use synthetic division to find upper and lower bounds for the zeros of $f(x)$. Recall that if we divide $f(x)$ synthetically by $x - c$, the third row in the division process contains the coefficients of the quotient $q(x)$ together with the remainder $f(c)$. The following theorem indicates how this third row may be used to find upper and lower bounds for the real solutions.

**Bounds for Real Zeros of Polynomials**

Suppose that $f(x)$ is a polynomial with real coefficients and a positive leading coefficient and that $f(x)$ is divided synthetically by $x - c$.

**(1)** If $c > 0$ and if all numbers in the third row of the division process are either positive or zero, then $c$ is an upper bound for the real zeros of $f(x)$.

**(2)** If $c < 0$ and if the numbers in the third row of the division process are alternatively positive and negative (and a 0 in the third row is considered to be either positive or negative), then $c$ is a lower bound for the real zeros of $f(x)$.

A general proof of this result can be patterned after the solution given in the next example.

### EXAMPLE 6    *Finding bounds for the solutions of an equation*

Find upper and lower bounds for the real solutions of the equation $2x^3 + 5x^2 - 8x - 7 = 0$.

*SOLUTION*    We divide the polynomial $2x^3 + 5x^2 - 8x - 7$ synthetically by $x - 1$ and $x - 2$:

$$
\begin{array}{r|rrrr}
1 & 2 & 5 & -8 & -7 \\
  &   & 2 & 7  & -1 \\
\hline
  & 2 & 7 & -1 & -8
\end{array}
\qquad
\begin{array}{r|rrrr}
2 & 2 & 5 & -8 & -7 \\
  &   & 4 & 18 & 20 \\
\hline
  & 2 & 9 & 10 & 13
\end{array}
$$

The third row of the synthetic division by $x - 1$ contains negative numbers, and hence part (1) of the result on bounds for real zeros does not apply. However, since all numbers in the third row of the synthetic division by $x - 2$ are positive, it follows from part (1) that 2 is an upper bound for the real solutions of the equation. This fact is also evident if we express the division by $x - 2$ in the division algorithm form

$$2x^3 + 5x^2 - 8x - 7 = (x - 2)(2x^2 + 9x + 10) + 13,$$

for if $x > 2$, then the right-hand side of the equation is positive and hence is not zero. Consequently, $2x^3 + 5x^2 - 8x - 7$ is not zero if $x > 2$.

We now find the lower bound. After some trial-and-error attempts using $x - (-1)$, $x - (-2)$, and $x - (-3)$, we see that synthetic division by $x - (-4)$ gives us

$$
\begin{array}{r|rrrr}
-4 & 2 & 5 & -8 & -7 \\
   &   & -8 & 12 & -16 \\
\hline
   & 2 & -3 & 4 & -23
\end{array}
$$

Since the numbers in the third row are alternately positive and negative, it follows from part (2) of the preceding theorem that $-4$ is a lower bound for the real solutions. This can also be proved by expressing the division by $x + 4$ in the form

$$2x^3 + 5x^2 - 8x - 7 = (x + 4)(2x^2 - 3x + 4) - 23,$$

for if $x < -4$, then the right-hand side of this equation is negative and therefore is not zero. Hence, $2x^3 + 5x^2 - 8x - 7$ is not zero if $x < -4$.

Since lower and upper bounds for the real solutions are $-4$ and 2, respectively, it follows that all real solutions are in the closed interval $[-4, 2]$.

The graph of $f(x) = 2x^3 + 5x^2 - 8x - 7$ is sketched in Figure 16. The graph shows that the three zeros of $f$ are in the intervals $[-4, -3]$, $[-1, 0]$, and $[1, 2]$, respectively.

**FIGURE 16**

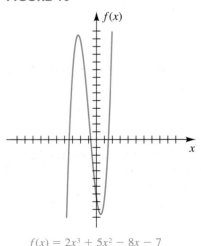

$f(x) = 2x^3 + 5x^2 - 8x - 7$

When a graphing utility is used *only* to find the zeros of a polynomial $f(x)$, it is not necessary to see the extrema of the polynomial. Hence, you might begin looking for the zeros of $f(x)$ by using the viewing rectangle dimensions

$$[-M, M] \text{ by } [-1, 1].$$

The value of M is equal to the absolute value of the ratio of the largest coefficient (in magnitude) of $f(x)$ and the leading coefficient of $f(x)$, plus 1. For example, using the polynomial $f(x) = 2x^3 + 5x^2 - 8x - 7$ in Example 6, we have

$$M = \left| \frac{-8}{2} \right| + 1 = 4 + 1 = 5.$$

By graphing $Y_1 = 2x^3 + 5x^2 - 8x - 7$ with a viewing rectangle of $[-5, 5]$ by $[-1, 1]$, you can almost "eyeball" the approximate solutions $-3.4$, $-0.7$, and $1.5$. Try it!

**EXAMPLE 7**   *Finding a polynomial from a graph*

Shown in Figure 17 is a polynomial function $f$ with viewing rectangle $[-4, 4]$ by $[-35, 35]$, Xscl = 1, and Yscl = 5.

**(a)** Find a factored form for $f$ that has minimal degree.

**(b)** Assuming the leading coefficient of $f$ is 1, find the $y$-intercept.

**FIGURE 17**   $[-4, 4]$ by $[-35, 35]$

*SOLUTION*

**(a)** The zero at $x = -2$ must have a multiplicity that is an even number, since $f$ does not change sign at $x = -2$. The zero at $x = 1$ must have an odd multiplicity of 3 or greater, since $f$ changes sign at $x = 1$ and levels off. The zero at $x = 3$ is of multiplicity 1, since $f$ changes sign and does not level off. Thus, a factored form of $f$ is

$$f(x) = a(x + 2)^m(x - 1)^n(x - 3)^1.$$

Because we desire the function having minimal degree, we let $m = 2$ and $n = 3$, obtaining

$$f(x) = a(x + 2)^2(x - 1)^3(x - 3),$$

which is a sixth-degree polynomial.

**(b)** If the leading coefficient of $f$ is to be 1, then from the complete factorization theorem for polynomials we know that the value of $a$ is 1. To find the $y$-intercept, we let $x = 0$ and compute $f(0)$:

$$f(0) = 1(0 + 2)^2(0 - 1)^3(0 - 3) = (4)(-1)(-3) = 12$$

Hence, the $y$-intercept is the point $(0, 12)$.

**EXAMPLE 8     *Exploring the graph of a polynomial***

Find the zeros of $f(x) = x^3 - 1000x^2 - x + 1000$.

**FIGURE 18**   $[-15, 15]$ by $[-10, 10]$

**SOLUTION**     Some specific keystrokes for the TI-81/82 graphics calculator are given in Appendix I. We assign $x^3 - 1000x^2 - x + 1000$ to $Y_1$
and use a standard viewing rectangle to obtain Figure 18. It appears that 1
is a root of $f$, and we can prove this fact by using synthetic division.

$$\underline{1|}\;\begin{array}{rrrr} 1 & -1000 & -1 & 1000 \\ & 1 & -999 & -1000 \\ \hline 1 & -999 & -1000 & 0 \end{array}$$

Using the **depressed equation**, $x^2 - 999x - 1000 = 0$, we can also
show that $-1$ is a root of $f$.

$$\underline{-1|}\;\begin{array}{rrr} 1 & -999 & -1000 \\ & -1 & 1000 \\ \hline 1 & -1000 & 0 \end{array}$$

**FIGURE 19**   $[-200, 1100]$ by $[-20, 10]$

From the last synthetic division, we see that $x - 1000$ is a factor of $f$, and
hence the third root is 1000.

Because of the relative sizes of the roots 1 and 1000, it is very difficult
to obtain a viewing rectangle that shows all three zeros. However, by
vertically compressing the graph of $f$, we will obtain a sketch of $f$ showing
its zeros and turning points. After some trial and error, we arrive at Figure
19, where we have used $10^{-7}$ as a multiplier of $Y_1$, a viewing rectangle of
$[-200, 1100]$ by $[-20, 10]$, $Xscl = 100$, and $Yscl = 1$.

## 3.3 EXERCISES

**Exer. 1–6: Find a polynomial $f(x)$ of degree 3 that has
the indicated zeros and satisfies the given condition.**

1  $-1, 2, 3;$     $f(-2) = 80$

2  $-5, 2, 4;$     $f(3) = -24$

3  $-4, 3, 0;$     $f(2) = -36$

4  $-3, -2, 0;$    $f(-4) = 16$

5  $-2i, 2i, 3;$    $f(1) = 20$

6  $-3i, 3i, 4;$    $f(-1) = 50$

7  Find a polynomial $f(x)$ of degree 4 with leading coefficient 1 such that both $-4$ and 3 are zeros of multiplicity
2, and sketch the graph of $f$.

8  Find a polynomial $f(x)$ of degree 4 with leading coefficient 1 such that both $-5$ and 2 are zeros of multiplicity
2, and sketch the graph of $f$.

9  Find a polynomial $f(x)$ of degree 6 such that 0 and 3 are
both zeros of multiplicity 3 and $f(2) = -24$. Sketch the
graph of $f$.

10  Find a polynomial $f(x)$ of degree 7 such that $-2$ and 2
are both zeros of multiplicity 2, 0 is a zero of multiplicity 3, and $f(-1) = 27$. Sketch the graph of $f$.

11 Find the third-degree polynomial function whose graph is shown in the figure.

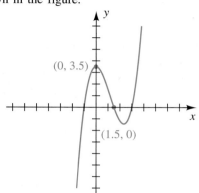

(0, 3.5)

(1.5, 0)

12 Find the fourth-degree polynomial function whose graph is shown in the figure.

(−1, 4)

**Exer. 13–14: Find the polynomial function of degree 3 whose graph is shown in the figure.**

13

14

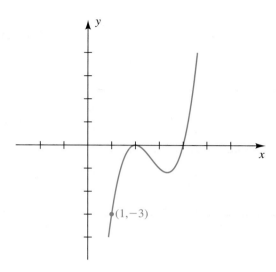

(1, −3)

**Exer. 15–22: Find the zeros of $f(x)$, and state the multiplicity of each zero.**

15 $f(x) = x^2(3x + 2)(2x - 5)^3$

16 $f(x) = x(x + 1)^4(3x - 7)^2$

17 $f(x) = 4x^5 + 12x^4 + 9x^3$

18 $f(x) = (4x^2 - 5)^2$

19 $f(x) = (x^2 + x - 12)^3(x^2 - 9)^2$

20 $f(x) = (6x^2 + 7x - 5)^4(4x^2 - 1)^2$

21 $f(x) = x^4 + 7x^2 - 144$

22 $f(x) = x^4 + 21x^2 - 100$

**Exer. 23–26: Show that the number is a zero of $f(x)$ of the given multiplicity, and express $f(x)$ as a product of linear factors.**

23 $f(x) = x^4 + 7x^3 + 13x^2 - 3x - 18$;
　　　　　　　　　　　−3 (multiplicity 2)

24 $f(x) = x^4 - 9x^3 + 22x^2 - 32$;　　4 (multiplicity 2)

25 $f(x) = x^6 - 4x^5 + 5x^4 - 5x^2 + 4x - 1$;
　　　　　　　　　　　1 (multiplicity 5)

26 $f(x) = x^5 + x^4 - 6x^3 - 14x^2 - 11x - 3$;
　　　　　　　　　　　−1 (multiplicity 4)

**Exer. 27–34: Use Descartes' rule of signs to determine the number of possible positive, negative, and nonreal complex solutions of the equation.**

27 $4x^3 - 6x^2 + x - 3 = 0$　　28 $5x^3 - 6x - 4 = 0$

29  $4x^3 + 2x^2 + 1 = 0$

30  $3x^3 - 4x^2 + 3x + 7 = 0$

31  $3x^4 + 2x^3 - 4x + 2 = 0$

32  $2x^4 - x^3 + x^2 - 3x + 4 = 0$

33  $x^5 + 4x^4 + 3x^3 - 4x + 2 = 0$

34  $2x^6 + 5x^5 + 2x^2 - 3x + 4 = 0$

**Exer. 35–40:** Find the smallest and largest integers that are upper and lower bounds, respectively, for the real solutions of the equation.

35  $x^3 - 4x^2 - 5x + 7 = 0$

36  $2x^3 - 5x^2 + 4x - 8 = 0$

37  $x^4 - x^3 - 2x^2 + 3x + 6 = 0$

38  $2x^4 - 9x^3 - 8x - 10 = 0$

39  $2x^5 - 13x^3 + 2x - 5 = 0$

40  $3x^5 + 2x^4 - x^3 - 8x^2 - 7 = 0$

**Exer. 41–42:** Find a factored form for a polynomial function $f$ that has minimal degree. Assume that the intercept values are integers and that Xscl = Yscl = 1.

41

42

**Exer. 43–44:** (a) Find a factored form for a polynomial function $f$ that has minimal degree. Assume that the intercept values are integers, Xscl = 1, and Yscl = 5. (b) If the leading coefficient of $f$ is $a$, find the $y$-intercept.

43  $a = 1$

44  $a = -1$

**Exer. 45–48:** Is there a polynomial of the given degree $n$ whose graph contains the indicated points?

45  $n = 4$;
    $(-2, 0), (0, -24), (1, 0), (3, 0), (2, 0), (-1, -52)$

46  $n = 5$;
    $(0, 0), (-3, 0), (-1, 0), (2, 0), (3, 0), (-2, 5), (1, 2)$

47  $n = 3$;
    $(1.1, -49.815), \quad (2, 0), \quad (3.5, 25.245), \quad (5.2, 0),$
    $(6.4, -29.304), (10.1, 0)$

48  $n = 4$;
    $(1.25, 0), \quad (2, 0), \quad (2.5, 56.25), \quad (3, 128.625), \quad (6.5, 0),$
    $(9, -307.75), (10, 0)$

49  **Using limited data**   A scientist has limited data on the temperature $T$ (in °C) during a 24-hour period. If $t$ denotes time in hours and $t = 0$ corresponds to midnight, find the fourth-degree polynomial that fits the information in the following table.

| $t$ (hours) | 0 | 5 | 12 | 19 | 24 |
|---|---|---|---|---|---|
| $T$ (°C) | 0 | 0 | 10 | 0 | 0 |

**50** *Lagrange interpolation polynomial* A polynomial $f(x)$ of degree 3 with zeros at $c_1$, $c_2$, and $c_3$ and with $f(c) = 1$ for $c_2 < c < c_3$ is a third-degree *Lagrange interpolation polynomial*. Find an explicit formula for $f(x)$ in terms of $c_1$, $c_2$, $c_3$, and $c$.

**Exer. 51–52: Graph $f$ for each value of $n$ on the same coordinate plane, and describe how the multiplicity of a zero affects the graph of $f$.**

**51** $f(x) = (x - 0.5)^n(x^2 + 1)$; $n = 1, 2, 3, 4$

**52** $f(x) = (x - 1)^n(x + 1)^n$; $n = 1, 2, 3, 4$

**Exer. 53–54: Graph $f$, estimate all real zeros, and determine the multiplicity of each zero.**

**53** $f(x) = x^3 + 1.3x^2 - 1.2x - 1.584$

**54** $f(x) = x^5 - \frac{1}{4}x^4 - \frac{19}{8}x^3 - \frac{9}{32}x^2 + \frac{405}{256}x + \frac{675}{1024}$

**55** *Greenhouse effect* Because of the combustion of fossil fuels, the concentration of carbon dioxide in the atmosphere is increasing. Research indicates that this will result in a *greenhouse effect* that will change the average global surface temperature. Assuming a vigorous expansion of coal use, the future amount $A(t)$ of atmospheric carbon dioxide concentration can be approximated (in parts per million) by

$$A(t) = -\tfrac{1}{2400}t^3 + \tfrac{1}{20}t^2 + \tfrac{7}{6}t + 340,$$

where $t$ is in years, $t = 0$ corresponds to 1980, and $0 \le t \le 60$. Use the graph of $A$ to estimate the year when the carbon dioxide concentration will be 400.

**56** *Greenhouse effect* The average increase in global surface temperature due to the greenhouse effect can be approximated by

$$T(t) = \frac{21}{5,000,000}t^3 - \frac{127}{1,000,000}t^2 + \frac{1293}{50,000}t,$$

where $0 \le t \le 60$ and $t = 0$ corresponds to 1980. Use the graph of $T$ to estimate the year when the average temperature will have risen by 1 °C.

## 3.4 COMPLEX AND RATIONAL ZEROS OF POLYNOMIALS

Example 3 of the preceding section illustrates an important fact about polynomials with real coefficients: The two complex zeros $2 + 3i$ and $2 - 3i$ of $x^5 - 4x^4 + 13x^3$ are conjugates of each other. The relationship is not accidental, since the following general result is true.

**Theorem on Conjugate Pair Zeros of a Polynomial**

If a polynomial $f(x)$ of degree $n > 1$ has real coefficients and if $z = a + bi$ with $b \ne 0$ is a complex zero of $f(x)$, then the conjugate $\bar{z} = a - bi$ is also a zero of $f(x)$.

*PROOF* Let us consider the polynomial

$$f(x) = a_n x^n + a_{n-1}x^{n-1} + \cdots + a_1 x + a_0,$$

where each coefficient $a_k$ is a real number and $a_n \ne 0$. If $f(z) = 0$, then

$$a_n z^n + a_{n-1}z^{n-1} + \cdots + a_1 z + a_0 = 0.$$

If two complex numbers are equal, then so are their conjugates. Hence, the conjugate of the left-hand side of the last equation equals the conjugate of the right-hand side; that is,

$$\overline{a_n z^n + a_{n-1} z^{n-1} + \cdots + a_1 z + a_0} = \overline{0} = 0.$$

The fact that $\overline{0} = 0$ follows from $\overline{0} = \overline{0 + 0i} = 0 - 0i = 0.$

If $z$ and $w$ are arbitrary complex numbers, then it can be shown that $\overline{z + w} = \overline{z} + \overline{w}$. More generally, the conjugate of *any* sum of complex numbers is the sum of the conjugates. Consequently,

$$\overline{a_n z^n} + \overline{a_{n-1} z^{n-1}} + \cdots + \overline{a_1 z} + \overline{a_0} = 0.$$

It can also be shown that $\overline{z \cdot w} = \overline{z} \cdot \overline{w}$, $\overline{z^n} = \overline{z}^n$ for every positive integer $n$, and $\overline{z} = z$ if and only if $z$ is real. Thus, for every $k$,

$$\overline{a_k z^k} = \overline{a_k} \cdot \overline{z^k} = \overline{a_k} \cdot \overline{z}^k = a_k \overline{z}^k,$$

and therefore

$$a_n \overline{z}^n + a_{n-1} \overline{z}^{n-1} + \cdots + a_1 \overline{z} + a_0 = 0.$$

The last equation states that $f(\overline{z}) = 0$, which completes the proof. ■

### EXAMPLE 1    *Finding a polynomial with prescribed zeros*

Find a polynomial $f(x)$ of degree 4 that has real coefficients and zeros $2 + i$ and $-3i$.

**SOLUTION**    By the theorem on conjugate pair zeros, $f(x)$ must also have zeros $2 - i$ and $3i$. Applying the factor theorem, we find that $f(x)$ has the following factors:

$$x - (2 + i), \quad x - (2 - i), \quad x - (-3i), \quad x - (3i)$$

Multiplying these four factors gives us

$$\begin{aligned} f(x) &= [x - (2 + i)][x - (2 - i)](x + 3i)(x - 3i) \\ &= (x^2 - 4x + 5)(x^2 + 9) \\ &= x^4 - 4x^3 + 14x^2 - 36x + 45. \end{aligned}$$

---

If a polynomial with real coefficients is factored as in the complete factorization theorem (see page 201), some of the factors $x - c_k$ may contain a complex number $c_k$. However, it is always possible to obtain a factorization into polynomials with real coefficients, as stated in the next theorem.

**Theorem on Expressing a Polynomial as a Product of Linear and Quadratic Factors**

Every polynomial with real coefficients and positive degree $n$ can be expressed as a product of linear and quadratic polynomials with real coefficients such that the quadratic factors are irreducible over $\mathbb{R}$.

**PROOF**   Since $f(x)$ has precisely $n$ complex zeros $c_1, c_2, \ldots, c_n$, we may write

$$f(x) = a(x - c_1)(x - c_2) \cdots (x - c_n),$$

where $a$ is the leading coefficient of $f(x)$. Of course, some of the zeros may be real. In such cases we obtain the linear factors referred to in the statement of the theorem.

If a zero $c_k$ is not real, then, by the theorem on conjugate pair zeros, the conjugate $\overline{c_k}$ is also a zero of $f(x)$ and hence must be one of the numbers $c_1, c_2, \ldots, c_n$. This implies that both $x - c_k$ and $x - \overline{c_k}$ appear in the factorization of $f(x)$. If those factors are multiplied, we obtain

$$(x - c_k)(x - \overline{c_k}) = x^2 - (c_k + \overline{c_k})x + c_k\overline{c_k},$$

which has *real* coefficients, since $c_k + \overline{c_k}$ and $c_k\overline{c_k}$ are real numbers. Thus, if $c_k$ is a complex zero, then the product $(x - c_k)(x - \overline{c_k})$ is a quadratic polynomial that is irreducible over $\mathbb{R}$. This completes the proof. ■

**EXAMPLE 2**   *Expressing a polynomial as a product of linear and quadratic factors*

Express $x^5 - 4x^3 + x^2 - 4$ as a product of

**(a)** linear and quadratic polynomials with real coefficients that are irreducible over $\mathbb{R}$

**(b)** linear polynomials

**SOLUTION**

**(a)** $x^5 - 4x^3 + x^2 - 4$

$$
\begin{aligned}
&= (x^5 - 4x^3) + (x^2 - 4) && \text{group terms} \\
&= x^3(x^2 - 4) + (x^2 - 4) && \text{factor out } x^3 \\
&= (x^3 + 1)(x^2 - 4) && \text{factor out } x^2 - 4 \\
&= (x + 1)(x^2 - x + 1)(x + 2)(x - 2) && \text{factor as the sum of cubes} \\
& && \text{and the difference of squares}
\end{aligned}
$$

Using the quadratic formula, we see that the polynomial $x^2 - x + 1$ has the complex zeros

$$\frac{-(-1) \pm \sqrt{(-1)^2 - 4(1)(1)}}{2(1)} = \frac{1 \pm \sqrt{3}i}{2} = \frac{1}{2} \pm \frac{\sqrt{3}}{2}i$$

and hence is irreducible over $\mathbb{R}$. Thus, the desired factorization is

$$(x + 1)(x^2 - x + 1)(x + 2)(x - 2).$$

*(continued)*

**(b)** Since the polynomial $x^2 - x + 1$ in part (a) has zeros $\frac{1}{2} \pm (\sqrt{3}/2)i$, it follows from the factor theorem that the polynomial has factors

$$x - \left(\frac{1}{2} + \frac{\sqrt{3}}{2}i\right) \quad \text{and} \quad x - \left(\frac{1}{2} - \frac{\sqrt{3}}{2}i\right).$$

Substituting in the factorization found in part (a), we obtain the following complete factorization into linear polynomials:

$$(x + 1)\left(x - \frac{1}{2} - \frac{\sqrt{3}}{2}i\right)\left(x - \frac{1}{2} + \frac{\sqrt{3}}{2}i\right)(x + 2)(x - 2)$$

---

We previously pointed out that it is generally very difficult to find the zeros of a polynomial of high degree. If all the coefficients are rational numbers, however, there is a method for finding the *rational* zeros, if they exist. The method is a consequence of the following result.

**Theorem on Rational Zeros of a Polynomial**

If the polynomial

$$f(x) = a_n x^n + a_{n-1}x^{n-1} + \cdots + a_1 x + a_0$$

has *integer* coefficients and if $c/d$ is a rational zero of $f(x)$ such that $c$ and $d$ have no common prime factor, then

**(1)** the numerator $c$ of the zero is a factor of the constant term $a_0$

**(2)** the denominator $d$ of the zero is a factor of the leading coefficient $a_n$

**PROOF**   Assume that $c > 0$. (The proof for $c < 0$ is similar.) Let us show that $c$ is a factor of $a_0$. The case $c = 1$ is trivial, since 1 is a factor of *any* number. Thus, suppose $c \neq 1$. In this case $c/d \neq 1$, for if $c/d = 1$, we obtain $c = d$, and since $c$ and $d$ have no prime factor in common, this implies that $c = d = 1$, a contradiction. Hence, in the following discussion we have $c \neq 1$ and $c \neq d$.

Since $f(c/d) = 0$,

$$a_n \frac{c^n}{d^n} + a_{n-1}\frac{c^{n-1}}{d^{n-1}} + \cdots + a_1 \frac{c}{d} + a_0 = 0.$$

We multiply by $d^n$ and then add $-a_0 d^n$ to both sides:

$$a_n c^n + a_{n-1}c^{n-1}d + \cdots + a_1 cd^{n-1} = -a_0 d^n$$
$$c(a_n c^{n-1} + a_{n-1}c^{n-2}d + \cdots + a_1 d^{n-1}) = -a_0 d^n$$

The last equation shows that $c$ is a factor of the integer $a_0 d^n$. Since $c$ and $d$ have no common factor, $c$ is a factor of $a_0$. A similar argument may be used to prove that $d$ is a factor of $a_n$. ■

The technique of using the theorem on rational zeros to find rational solutions of equations with integral coefficients is illustrated in the following examples.

**EXAMPLE 3   *A polynomial with no rational zeros***

Show that $f(x) = x^3 - 4x - 2$ has no rational zeros.

*SOLUTION*   If $f(x)$ has a rational zero $c/d$ such that $c$ and $d$ have no common prime factor, then, by the theorem on rational zeros, $c$ is a factor of the constant term $-2$ and hence is either 2 or $-2$ (which we write as $\pm 2$). The denominator $d$ is a factor of the leading coefficient 1 and hence is $\pm 1$. Thus, the only possibilities for $c/d$ are

$$\frac{\pm 1}{\pm 1} \quad \text{and} \quad \frac{\pm 2}{\pm 1} \quad \text{or, equivalently,} \quad \pm 1 \quad \text{and} \quad \pm 2.$$

Substituting each of these numbers for $x$, we obtain

$$f(1) = -5, \quad f(-1) = 1, \quad f(2) = -2, \quad \text{and} \quad f(-2) = -2.$$

Since $f(\pm 1) \neq 0$ and $f(\pm 2) \neq 0$, it follows that $f(x)$ has no rational zeros.

---

In the solution of the following example we shall assume that a graphing utility is not available. In Example 5 we will rework the problem to demonstrate the advantage of using a graphing utility.

**EXAMPLE 4   *Finding the rational solutions of an equation***

Find all rational solutions of

$$3x^4 + 14x^3 + 14x^2 - 8x - 8 = 0.$$

*SOLUTION*   The problem is equivalent to finding the rational zeros of the polynomial on the left-hand side of the equation. If $c/d$ is a rational zero and $c$ and $d$ have no common factor, then $c$ is a factor of the constant term $-8$ and $d$ is a factor of the leading coefficient 3. All possible choices are listed in the following table.

| Choices for the numerator $c$ | $\pm 1$, $\pm 2$, $\pm 4$, $\pm 8$ |
|---|---|
| Choices for the denominator $d$ | $\pm 1$, $\pm 3$ |
| Choices for $c/d$ | $\pm 1$, $\pm 2$, $\pm 4$, $\pm 8$, $\pm \frac{1}{3}$, $\pm \frac{2}{3}$, $\pm \frac{4}{3}$, $\pm \frac{8}{3}$ |

We can reduce the number of choices by finding upper and lower bounds for the real solutions; however, we shall not do so here. It is necessary to determine which of the choices for $c/d$, if any, are zeros. We see by substitution that neither 1 nor −1 is a solution. If we divide synthetically by $x + 2$, we obtain

$$\begin{array}{r|rrrrr} -2 & 3 & 14 & 14 & -8 & -8 \\ & & -6 & -16 & 4 & 8 \\ \hline & 3 & 8 & -2 & -4 & 0 \end{array}$$

This result shows that −2 is a zero. Moreover, the synthetic division provides the coefficients of the quotient in the division of the polynomial by $x + 2$. Hence, we have the following factorization of the given polynomial:

$$(x + 2)(3x^3 + 8x^2 - 2x - 4)$$

The remaining solutions of the equation must be zeros of the second factor, so we use that polynomial to check for solutions. *Do not* use the polynomial in the given equation. (Note that $\pm\frac{8}{3}$ are no longer candidates, since the numerator must be a factor of 4.) Again proceeding by trial and error, we ultimately find that synthetic division by $x + \frac{2}{3}$ gives us the following result:

$$\begin{array}{r|rrrr} -\frac{2}{3} & 3 & 8 & -2 & -4 \\ & & -2 & -4 & 4 \\ \hline & 3 & 6 & -6 & 0 \end{array}$$

Therefore, $-\frac{2}{3}$ is also a zero.

Using the coefficients of the quotient, we know that the remaining zeros are solutions of the equation $3x^2 + 6x - 6 = 0$. Dividing both sides by 3 gives us the equivalent equation $x^2 + 2x - 2 = 0$. By the quadratic formula, this equation has solutions

$$\frac{-2 \pm \sqrt{2^2 - 4(1)(-2)}}{2(1)} = \frac{-2 \pm \sqrt{12}}{2} = \frac{-2 \pm 2\sqrt{3}}{2} = -1 \pm \sqrt{3}.$$

Hence, the given polynomial has two rational roots, −2 and $-\frac{2}{3}$, and two irrational roots, $-1 + \sqrt{3} \approx 0.732$ and $-1 - \sqrt{3} \approx -2.732$.

---

**EXAMPLE 5    *Finding the rational solutions of an equation***

Find all rational solutions of the equation

$$3x^4 + 14x^3 + 14x^2 - 8x - 8 = 0.$$

**SOLUTION**    Assigning the indicated polynomial to $Y_1$ and choosing the viewing rectangle $[-7.5, 7.5]$ by $[-5, 5]$, we obtain a sketch similar to Figure 20. The graph indicates that −2 is a solution and that there is one solution in each of the intervals $(-3, -2)$, $(-1, 0)$, and $(0, 1)$. Referring to the choices for a rational zero $c/d$ in the preceding example, we see that the only possibilities are $-\frac{8}{3}$ in $(-3, -2)$, $-\frac{2}{3}$ or $-\frac{1}{3}$ in $(-1, 0)$, and $\frac{1}{3}$ or $\frac{2}{3}$ in $(0, 1)$. Thus, by referring to the graph, we have reduced the number of choices for zeros from sixteen, as in Example 4, to five.

FIGURE 20 $[-7.5, 7.5]$ by $[-5, 5]$

If we zoom in on the graph of $Y_1$, we see that $-\frac{8}{3}$, $-\frac{2}{3}$, and $\frac{2}{3}$ are the most reasonable choices for zeros. As in Example 4, synthetic division can be used to determine that the only rational solutions are $-2$ and $-\frac{2}{3}$.

The theorem on rational zeros may be applied to equations with rational coefficients. We merely multiply both sides of the equation by the lcd of all the coefficients to obtain an equation with integral coefficients, and then we proceed as in Example 4.

### EXAMPLE 6    *Finding the radius of a grain silo*

A grain silo has the shape of a right circular cylinder with a hemisphere attached to the top. If the total height of the structure is 30 feet, find the radius of the cylinder that results in a total volume of $1008\pi$ ft$^3$.

FIGURE 21

*SOLUTION*    Let $x$ denote the radius of the cylinder. A sketch of the silo, labeled appropriately, is shown in Figure 21. Since the volume of the cylinder is $\pi x^2(30 - x)$ and the volume of the hemisphere is $\frac{2}{3}\pi x^3$, we must solve the following:

$$\pi x^2(30 - x) + \tfrac{2}{3}\pi x^3 = 1008\pi \quad \text{total volume is } 1008\pi$$

$$3x^2(30 - x) + 2x^3 = 3024 \quad \text{multiply by } \frac{3}{\pi}$$

$$90x^2 - x^3 = 3024 \quad \text{simplify}$$

$$x^3 - 90x^2 + 3024 = 0 \quad \text{equivalent equation}$$

Since the leading coefficient of the polynomial on the left side of the last equation is 1, any rational root has the form $c/1 = c$, where $c$ is a factor of 3024. If we factor 3024 into primes, we find that $3024 = 2^4 \cdot 3^3 \cdot 7$. It follows that some of the positive factors of 3024 are

$$1, \quad 2, \quad 3, \quad 4, \quad 6, \quad 8, \quad 9, \quad 12, \quad \ldots .$$

To help us decide which of these numbers to test first, let us make a rough estimate of the radius by assuming that the silo has the shape of a right circular cylinder of height 30 feet. In that case, the volume would be $\pi r^2 h = 30\pi r^2$. Since this volume should be close to $1008\pi$, we see that

$$30r^2 = 1008, \quad \text{or} \quad r^2 = 1008/30 \approx 33.6.$$

This suggests that we use 6 in our first synthetic division, as follows:

$$
\begin{array}{r|rrrr}
6 & 1 & -90 & 0 & 3024 \\
  &   & 6 & -504 & -3024 \\
\hline
  & 1 & -84 & -504 & 0
\end{array}
$$

Thus, 6 is a solution of the equation $x^3 - 90x^2 + 3024 = 0$.

*(continued)*

**FIGURE 22**  [0, 20] by [0, 4000]

The remaining two solutions of the equation can be found by solving $x^2 - 84x - 504 = 0$, which we obtain from the quotient in the synthetic division. These zeros are approximately $-5.62$ and $89.62$—neither of which satisfies the conditions of the problem. Hence, the desired radius is 6 feet.

Using a graphing utility with viewing rectangle [0, 20] by [0, 4000], we obtain the graph of $f(x) = x^3 - 90x^2 + 3024$ in Figure 22, which shows the zero $x = 6$. An extended graph would also indicate the other two zeros.

## 3.4 EXERCISES

**Exer. 1–10:** A polynomial $f(x)$ with real coefficients and leading coefficient 1 has the given zero(s) and degree. Express $f(x)$ as a product of linear and quadratic polynomials with real coefficients that are irreducible over $\mathbb{R}$.

| | | |
|---|---|---|
| 1 | $3 + 2i$; | degree 2 |
| 2 | $-4 + 3i$; | degree 2 |
| 3 | $2, -2 - 5i$; | degree 3 |
| 4 | $-3, 1 - 7i$; | degree 3 |
| 5 | $-1, 0, 3 + i$; | degree 4 |
| 6 | $0, 2, -2 - i$; | degree 4 |
| 7 | $4 + 3i, -2 + i$; | degree 4 |
| 8 | $3 + 5i, -1 - i$; | degree 4 |
| 9 | $0, -2i, 1 - i$; | degree 5 |
| 10 | $0, 3i, 4 + i$; | degree 5 |

**Exer. 11–14:** Show that the equation has no rational root.

11  $x^3 + 3x^2 - 4x + 6 = 0$

12  $3x^3 - 4x^2 + 7x + 5 = 0$

13  $x^5 - 3x^3 + 4x^2 + x - 2 = 0$

14  $2x^5 + 3x^3 + 7 = 0$

**Exer. 15–24:** Find all solutions of the equation.

15  $x^3 - x^2 - 10x - 8 = 0$

16  $x^3 + x^2 - 14x - 24 = 0$

17  $2x^3 - 3x^2 - 17x + 30 = 0$

18  $12x^3 + 8x^2 - 3x - 2 = 0$

19  $x^4 + 3x^3 - 30x^2 - 6x + 56 = 0$

20  $3x^5 - 10x^4 - 6x^3 + 24x^2 + 11x - 6 = 0$

21  $6x^5 + 19x^4 + x^3 - 6x^2 = 0$

22  $6x^4 + 5x^3 - 17x^2 - 6x = 0$

23  $8x^3 + 18x^2 + 45x + 27 = 0$

24  $3x^3 - x^2 + 11x - 20 = 0$

**Exer. 25–26:** Find a factored form with integer coefficients of the polynomial $f$ shown in the figure. Assume that Xscl = Yscl = 1.

25  $f(x) = 6x^5 - 23x^4 + 24x^3 + x^2 - 12x + 4$

26  $f(x) = -6x^5 + 5x^4 + 14x^3 - 8x^2 - 8x + 3$

27 Does there exist a polynomial of degree 3 with real coefficients that has zeros 1, −1, and $i$? Justify your answer.

28 The polynomial $f(x) = x^3 - ix^2 + 2ix + 2$ has the complex number $i$ as a zero; however, the conjugate $-i$ of $i$ is not a zero. Why doesn't this result contradict the theorem on conjugate pair zeros of a polynomial?

29 If $n$ is an odd positive integer, prove that a polynomial of degree $n$ with real coefficients has at least one real zero.

30 If a polynomial of the form
$$x^n + a_{n-1}x^{n-1} + \cdots + a_1x + a_0,$$
where each $a_k$ is an integer, has a rational root $r$, show that $r$ is an integer and is a factor of $a_0$.

31 **Constructing a box** From a rectangular piece of cardboard having dimensions 20 inches × 30 inches, an open box is to be made by removing squares of area $x^2$ from each corner and turning up the sides. (See Exercise 33 of Section 3.1.)

(a) Show that there are two boxes that have a volume of 1000 in.$^3$.

(b) Which box has the smaller surface area?

32 **Constructing a crate** The frame for a shipping crate is to be constructed from 24 feet of 2 × 2 lumber. Assuming the crate is to have square ends of length $x$ feet, determine the value(s) of $x$ that result(s) in a volume of 4 ft$^3$. (See Exercise 34 of Section 3.1.)

33 A right triangle has area 30 ft$^2$ and a hypotenuse that is 1 foot longer than one of its sides.

(a) If $x$ denotes the length of this side, then show that $2x^3 + x^2 - 3600 = 0$.

(b) Show that there is a positive root of the equation in part (a) and that this root is less than 13.

(c) Find the lengths of the sides of the triangle.

34 **Constructing a storage tank** A storage tank for propane gas is to be constructed in the shape of a right circular cylinder of altitude 10 feet with a hemisphere attached to each end. Determine the radius $x$ so that the resulting volume is $27\pi$ ft$^3$. (See Example 8 of Section 2.1.)

35 **Constructing a storage shelter** A storage shelter is to be constructed in the shape of a cube with a triangular prism forming the roof (see the figure). The length $x$ of a side of the cube is yet to be determined.

(a) If the total height of the structure is 6 feet, show that its volume $V$ is given by $V = x^3 + \frac{1}{2}x^2(6 - x)$.

(b) Determine $x$ so that the volume is 80 ft$^3$.

EXERCISE 35

36 **Designing a tent** A canvas camping tent is to be constructed in the shape of a pyramid with a square base. An 8-foot pole will form the center support, as illustrated in the figure. Find the length $x$ of a side of the base so that the total amount of canvas needed for the sides and bottom is 384 ft$^2$.

EXERCISE 36

Exer. 37–38: Use a graph to determine the number of nonreal solutions of the equation.

37 $x^5 + 1.1x^4 - 3.21x^3 - 2.835x^2 + 2.7x + 0.62 = -1$

38 $x^4 - 0.4x^3 - 2.6x^2 + 1.1x + 3.5 = 2$

Exer. 39–40: Use a graph and synthetic division to find all solutions of the equation.

39 $x^4 + 1.4x^3 + 0.44x^2 - 0.56x - 0.96 = 0$

40 $x^5 + 1.1x^4 - 2.62x^3 - 4.72x^2 - 0.2x + 5.44 = 0$

41 **Atmospheric density**   The density $D(h)$ (in kg/m$^3$) of the earth's atmosphere at an altitude of $h$ meters can be approximated by

$$D(h) = 1.2 - ah + bh^2 - ch^3,$$

where

$a = 1.096 \times 10^{-4}$, $b = 3.42 \times 10^{-9}$, $c = 3.6 \times 10^{-14}$,

and $0 \le h \le 30{,}000$. Use the graph of $D$ to approximate the altitude $h$ at which the density is 0.4.

42 **The earth's density**   The earth's density $D(h)$ (in g/cm$^3$) $h$ meters underneath the surface can be approximated by

$$D(h) = 2.84 + ah + bh^2 - ch^3,$$

where

$a = 1.4 \times 10^{-3}$, $b = 2.49 \times 10^{-6}$, $c = 2.19 \times 10^{-9}$,

and $0 \le h \le 1000$. Use the graph of $D$ to approximate the depth $h$ at which the density of the earth is 3.7.

## 3.5   RATIONAL FUNCTIONS

A function $f$ is a **rational function** if

$$f(x) = \frac{g(x)}{h(x)},$$

where $g(x)$ and $h(x)$ are polynomials. The domain of $f$ consists of all real numbers *except* the zeros of the denominator $h(x)$.

**ILLUSTRATION**

**Rational Functions and Their Domains**

- $f(x) = \dfrac{1}{x - 2}$;   *domain:* all $x$ *except* $x = 2$

- $f(x) = \dfrac{3x}{x^2 - 9}$;   *domain:* all $x$ *except* $x = \pm 3$

- $f(x) = \dfrac{x^3 - 8}{x^2 + 4}$;   *domain:* all real numbers $x$

When sketching the graph of a rational function $f$, it is important to answer the following two questions.

**Question 1**   What can be said of the function values $f(x)$ when $x$ is close to (but not equal to) a zero of the denominator?

**Question 2**   What can be said of the function values $f(x)$ when $x$ is large positive or when $x$ is large negative?

As we shall see, if $a$ is a zero of the denominator, one of several situations often occurs. These are shown in Figure 23, where we have used notations from the following chart.

| Notation | Terminology |
|---|---|
| $x \to a^-$ | $x$ approaches $a$ from the left (through values *less* than $a$). |
| $x \to a^+$ | $x$ approaches $a$ from the right (through values *greater* than $a$). |
| $f(x) \to \infty$ | $f(x)$ increases without bound (can be made as large positive as desired). |
| $f(x) \to -\infty$ | $f(x)$ decreases without bound (can be made as large negative as desired). |

**FIGURE 23**

$f(x) \to \infty$ as $x \to a^-$       $f(x) \to \infty$ as $x \to a^+$       $f(x) \to -\infty$ as $x \to a^-$       $f(x) \to -\infty$ as $x \to a^+$

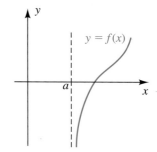

The symbols $\infty$ **(infinity)** and $-\infty$ **(minus infinity)** do not represent real numbers; they simply specify certain types of behavior of functions and variables.

The dashed line $x = a$ in Figure 23 is called a *vertical asymptote,* as in the following definition.

**Definition of Vertical Asymptote**

The line $x = a$ is a **vertical asymptote** for the graph of a function $f$ if

$$f(x) \to \infty \qquad \text{or} \qquad f(x) \to -\infty$$

as $x$ approaches $a$ from either the left or the right.

Thus, the answer to Question 1 is that if $a$ is a zero of the denominator of a rational function $f$, then the graph of $f$ *may* have a vertical asymptote $x = a$. There are rational functions where this is *not* the case (see Example 7). If the numerator and denominator have no common factor, then $f$ *must* have a vertical asymptote $x = a$.

Let us next consider Question 2. For *x large positive* or *large negative,* the graph of a rational function may look like one of those in Figure 24, where the notation

$$f(x) \to c \quad \text{as} \quad x \to \infty$$

is read "$f(x)$ approaches $c$ as $x$ increases without bound" and the notation

$$f(x) \to c \quad \text{as} \quad x \to -\infty$$

is read "$f(x)$ approaches $c$ as $x$ decreases without bound."

**FIGURE 24**

$$f(x) \to c \text{ as } x \to \infty$$

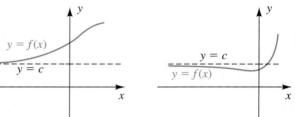

$$f(x) \to c \text{ as } x \to -\infty$$

We call the dashed line in Figure 24 a *horizontal asymptote,* as in the next definition.

**Definition of Horizontal Asymptote**

The line $y = c$ is a **horizontal asymptote** for the graph of a function $f$ if

$$f(x) \to c \quad \text{as} \quad x \to \infty \quad \text{or as} \quad x \to -\infty.$$

Thus, the answer to Question 2 is that $f(x)$ *may* be very close to some number $c$ when $x$ is large positive or large negative; that is, the graph of $f$ may have a horizontal asymptote $y = c$. There are rational functions where this is *not* the case (see Examples 2(c) and 6).

Note that as in the second and fourth sketches in Figure 24, the graph of $f$ may cross a horizontal asymptote.

In the next example we find the asymptotes for the graph of a simple rational function.

**EXAMPLE I**    *Sketching the graph of a rational function*

Sketch the graph of $f$ if

$$f(x) = \frac{1}{x - 2}.$$

**SOLUTION** Let us begin by considering Question 1, stated at the beginning of this section. The denominator $x - 2$ is zero at $x = 2$. If $x$ is close to 2 and $x > 2$, then $f(x)$ is large positive, as indicated in the following table.

| $x$ | 2.1 | 2.01 | 2.001 | 2.0001 | 2.00001 |
|---|---|---|---|---|---|
| $\dfrac{1}{x-2}$ | 10 | 100 | 1000 | 10,000 | 100,000 |

Since we can make $1/(x - 2)$ as large as desired by taking $x$ close to 2 (and $x > 2$), we see that

$$f(x) \to \infty \quad \text{as} \quad x \to 2^+.$$

If $f(x)$ is close to 2 and $x < 2$, then $f(x)$ is large negative; for example, $f(1.9999) = -10,000$ and $f(1.99999) = -100,000$. Thus,

$$f(x) \to -\infty \quad \text{as} \quad x \to 2^-.$$

The line $x = 2$ is a vertical asymptote for the graph of $f$, as illustrated in Figure 25.

We next consider Question 2. The following table lists some approximate values for $f(x)$ when $x$ is large and positive.

| $x$ | 100 | 1000 | 10,000 | 100,000 | 1,000,000 |
|---|---|---|---|---|---|
| $\dfrac{1}{x-2}$ (approx.) | 0.01 | 0.001 | 0.0001 | 0.00001 | 0.000001 |

We may describe this behavior of $f(x)$ by writing

$$f(x) \to 0 \quad \text{as} \quad x \to \infty.$$

Similarly, $f(x)$ is close to 0 when $x$ is large negative; for example, $f(-100,000) = -0.00001$. Thus,

$$f(x) \to 0 \quad \text{as} \quad x \to -\infty.$$

The line $y = 0$ (the $x$-axis) is a horizontal asymptote, as shown in Figure 25.

Plotting the points $(1, -1)$ and $(3, 1)$ gives us a rough sketch of the graph.

---

**FIGURE 25**

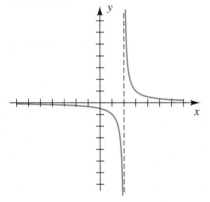

The function considered in Example 1, $f(x) = 1/(x - 2)$, closely resembles one of the simplest rational functions, the **reciprocal function.**

The reciprocal function has equation $f(x) = 1/x$, vertical asymptote $x = 0$ (the $y$-axis), and horizontal asymptote $y = 0$ (the $x$-axis). The graph of the reciprocal function (shown in Appendix II) is the graph of a *hyperbola* (discussed in Section 10.3). Note that we can obtain the graph of $y = 1/(x - 2)$ by shifting the graph of $y = 1/x$ to the right 2 units.

The following theorem is useful for finding the horizontal asymptote for the graph of a rational function.

**Theorem on Horizontal Asymptotes**

Let $f(x) = \dfrac{a_n x^n + a_{n-1}x^{n-1} + \cdots + a_1 x + a_0}{b_k x^k + b_{k-1}x^{k-1} + \cdots + b_1 x + b_0}$, where $a_n \neq 0$ and $b_k \neq 0$.

(1) If $n < k$, then the $x$-axis (the line $y = 0$) is a horizontal asymptote for the graph of $f$.

(2) If $n = k$, then the line $y = a_n/b_k$ (the ratio of leading coefficients) is a horizontal asymptote for the graph of $f$.

(3) If $n > k$, the graph of $f$ has no horizontal asymptote. Instead, either $f(x) \to \infty$ or $f(x) \to -\infty$ as $x \to \infty$ or as $x \to -\infty$.

Proofs for each part of this theorem may be patterned after the solutions in the next example. Concerning part (3), if $q(x)$ is the quotient obtained by dividing the numerator by the denominator, then $f(x) \to \infty$ if $q(x) \to \infty$ or $f(x) \to -\infty$ if $q(x) \to -\infty$.

**EXAMPLE 2** *Finding horizontal asymptotes*

Find the horizontal asymptote for the graph of $f$, if it exists.

**(a)** $f(x) = \dfrac{3x - 1}{x^2 - x - 6}$ **(b)** $f(x) = \dfrac{5x^2 + 1}{3x^2 - 4}$ **(c)** $f(x) = \dfrac{2x^4 - 3x^2 + 5}{x^2 + 1}$

*SOLUTION*

**(a)** The degree of the numerator $3x - 1$ is less than the degree of the denominator $x^2 - x - 6$, so, by part (1) of the theorem on horizontal asymptotes, the $x$-axis is a horizontal asymptote. To verify this directly, we divide the numerator and denominator of the quotient by $x^2$ (since 2 is the highest power on $x$ in the denominator), obtaining

$$f(x) = \frac{\dfrac{3x - 1}{x^2}}{\dfrac{x^2 - x - 6}{x^2}} = \frac{\dfrac{3}{x} - \dfrac{1}{x^2}}{1 - \dfrac{1}{x} - \dfrac{6}{x^2}} \quad \text{for} \quad x \neq 0.$$

If $x$ is large positive or large negative, then $3/x$, $1/x^2$, $1/x$, and $6/x^2$ are close to 0, and hence

$$f(x) \approx \frac{0 - 0}{1 - 0 - 0} = \frac{0}{1} = 0.$$

Thus,

$$f(x) \to 0 \quad \text{as} \quad x \to \infty \quad \text{or as} \quad x \to -\infty.$$

Since $f(x)$ is the $y$-coordinate of a point on the graph, the last statement means that the line $y = 0$ (that is, the $x$-axis) is a horizontal asymptote.

**(b)** If $f(x) = (5x^2 + 1)/(3x^2 - 4)$, then the numerator and denominator have the same degree and the leading coefficients are 5 and 3, respectively. Hence, by part (2) of the theorem on horizontal asymptotes, the line $y = \frac{5}{3}$ is a horizontal asymptote. We can also show that $y = \frac{5}{3}$ is a horizontal asymptote by dividing the numerator and denominator of $f(x)$ by $x^2$, obtaining

$$f(x) = \frac{\dfrac{5x^2 + 1}{x^2}}{\dfrac{3x^2 - 4}{x^2}} = \frac{5 + \dfrac{1}{x^2}}{3 - \dfrac{4}{x^2}} \quad \text{for} \quad x \neq 0.$$

Since $1/x^2 \to 0$ and $4/x^2 \to 0$ as $x \to \infty$ or as $x \to -\infty$, we see that

$$f(x) \to \frac{5 + 0}{3 - 0} = \frac{5}{3} \quad \text{as} \quad x \to \infty \quad \text{or as} \quad x \to -\infty.$$

**(c)** The degree of the numerator $2x^4 - 3x^2 + 5$ is greater than the degree of the denominator $x^2 + 1$, so, by part (3) of the theorem on horizontal asymptotes, the graph has no horizontal asymptote. If we use long division, we obtain

$$f(x) = 2x^2 - 5 + \frac{10}{x^2 + 1}.$$

As either $x \to \infty$ or $x \to -\infty$, the quotient $2x^2 - 5$ increases without bound and $10/(x^2 + 1) \to 0$. Hence, $f(x) \to \infty$ as $x \to \infty$ or as $x \to -\infty$.

We next list some guidelines for sketching the graph of a rational function. Their use will be illustrated in Examples 3 and 4.

**Guidelines for Sketching the Graph of a Rational Function**

Assume that $f(x) = \dfrac{g(x)}{h(x)}$, where $g(x)$ and $h(x)$ are polynomials that have no common factor.

1   Find the $x$-intercepts—that is, the real zeros of the numerator $g(x)$—and plot the corresponding points on the $x$-axis.

2   Find the real zeros of the denominator $h(x)$. For each real zero $a$, sketch the vertical asymptote $x = a$ with dashes.

3   Find the $y$-intercept $f(0)$, if it exists, and plot the point $(0, f(0))$ on the $y$-axis.

4   Apply the theorem on horizontal asymptotes. If there is a horizontal asymptote $y = c$, sketch it with dashes.

5   If there is a horizontal asymptote $y = c$, determine whether it intersects the graph. The $x$-coordinates of the points of intersection are the solutions of the equation $f(x) = c$. Plot these points, if they exist.

6   Sketch the graph of $f$ in each of the regions in the $xy$-plane determined by the vertical asymptotes in guideline 2. If necessary, use the sign of specific function values to tell whether the graph is above or below the $x$-axis or the horizontal asymptote. Use guideline 5 to decide whether the graph approaches the horizontal asymptote from above or below.

In the following examples our main objective is to determine the general shape of the graph, paying particular attention to how the graph approaches the asymptotes. We will plot only a few points, such as those corresponding to the $x$-intercepts and $y$-intercept or the intersection of the graph with a horizontal asymptote.

**EXAMPLE 3   *Sketching the graph of a rational function***

Sketch the graph of $f$ if

$$f(x) = \frac{x - 1}{x^2 - x - 6}.$$

**SOLUTION**   It is useful to express both numerator and denominator in factored form. Thus, we begin by writing

$$f(x) = \frac{x - 1}{x^2 - x - 6} = \frac{x - 1}{(x + 2)(x - 3)}.$$

We next follow the guidelines.

**Guideline 1**   To find the $x$-intercepts we find the zeros of the numerator. Letting $x - 1 = 0$ gives us $x = 1$, and we plot the point $(1, 0)$ on the $x$-axis, as shown in Figure 26.

FIGURE 26

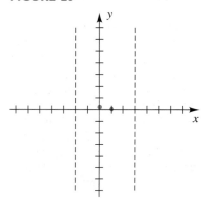

**Guideline 2**   The denominator has zeros $-2$ and $3$. Hence, the lines $x = -2$ and $x = 3$ are vertical asymptotes; we sketch them with dashes, as in Figure 26.

**Guideline 3**   The $y$-intercept is $f(0) = \frac{1}{6}$, and we plot the point $(0, \frac{1}{6})$ in Figure 26.

**Guideline 4**   The degree of the numerator of $f(x)$ is less than the degree of the denominator, so, by part (1) of the theorem on horizontal asymptotes, the $x$-axis is a horizontal asymptote.

**Guideline 5**   The points where the graph intersects the horizontal asymptote (the $x$-axis) found in guideline 4 correspond to the $x$-intercepts. We already plotted the point $(1, 0)$ in guideline 1.

**Guideline 6**   The vertical asymptotes in Figure 26 divide the $xy$-plane into three regions:

$$R_1: \quad \text{the region to the left of } x = -2$$

$$R_2: \quad \text{the region between } x = -2 \text{ and } x = 3$$

$$R_3: \quad \text{the region to the right of } x = 3$$

For $R_1$, we have $x < -2$. There are only two choices for the shape of the graph of $f$ in $R_1$: as $x \rightarrow -\infty$, the graph approaches the $x$-axis either from above or from below. To determine which choice is correct, we will examine the *sign* of a typical function value in $R_1$. Choosing $-10$ for $x$, we use the factored form of $f(x)$ to find the sign of $f(-10)$ (this process is similar to the one used in Section 1.6):

$$f(-10) = \frac{(-)}{(-)(-)} = -$$

The negative value of $f(-10)$ indicates that the graph approaches the horizontal asymptote $y = 0$ from *below* as $x \rightarrow -\infty$. Moreover, as $x \rightarrow -2^-$, the graph extends *downward;* that is, $f(x) \rightarrow -\infty$. A sketch of $f$ on $R_1$ is shown in Figure 27(a).

In $R_2$, we have $-2 < x < 3$, and the graph crosses the $x$-axis at $x = 1$. Since, for example, $f(0)$ is positive, it follows that the graph lies *above* the $x$-axis if $-2 < x < 1$. Thus, as $x \rightarrow -2^+$, the graph extends *upward;* that is, $f(x) \rightarrow \infty$. Since $f(2)$ is negative, the graph lies *below* the $x$-axis if $1 < x < 3$. Hence, as $x \rightarrow 3^-$, the graph extends *downward;* that is, $f(x) \rightarrow -\infty$. A sketch of $f$ on $R_2$ is shown in Figure 27(b).

Finally, in $R_3$, $x > 3$, and the graph does not cross the $x$-axis. Since, for example, $f(10)$ is positive, the graph lies *above* the $x$-axis. It follows that $f(x) \rightarrow \infty$ as $x \rightarrow 3^-$ and that the graph approaches the horizontal asymptote $y = 0$ from *above* as $x \rightarrow \infty$. The graph of $f$ is sketched in Figure 27(c).

**FIGURE 27**

**(a)**

**(b)**

**(c)**

**EXAMPLE 4    *Sketching the graph of a rational function***

Sketch the graph of $f$ if

$$f(x) = \frac{x^2}{x^2 - x - 2}.$$

*SOLUTION*    Factoring the denominator gives us

$$f(x) = \frac{x^2}{x^2 - x - 2} = \frac{x^2}{(x + 1)(x - 2)}.$$

We again follow the guidelines.

**FIGURE 28**

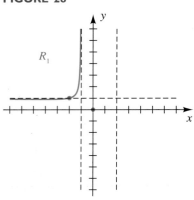

**Guideline 1**   To find the $x$-intercepts we find the zeros of the numerator. Letting $x^2 = 0$ gives us $x = 0$, and we plot the point $(0, 0)$ on the $x$-axis, as shown in Figure 28.

**Guideline 2**   The denominator has zeros $-1$ and $2$. Hence, the lines $x = -1$ and $x = 2$ are vertical asymptotes, and we sketch them with dashes, as in Figure 28.

**Guideline 3**   The $y$-intercept is $f(0) = 0$. This gives us the same point $(0, 0)$ found in guideline 1.

**Guideline 4**   The numerator and denominator of $f(x)$ have the same degree, and the leading coefficients are both 1. Hence, by part (2) of the theorem on horizontal asymptotes, the line $y = \frac{1}{1} = 1$ is a horizontal asymptote. We sketch the line with dashes in Figure 28.

**Guideline 5**   The $x$-coordinates of the points where the graph intersects the horizontal asymptote $y = 1$ are solutions of the equation $f(x) = 1$. We solve this equation as follows:

$$\frac{x^2}{x^2 - x - 2} = 1 \qquad \text{let } f(x) = 1$$

$$x^2 = x^2 - x - 2 \quad \text{multiply by } x^2 - x - 2$$

$$x = -2 \qquad\qquad \text{subtract } x^2 \text{ and add } x$$

The $y$-coordinate of the point of intersection is $f(-2) = 1$, and we plot the point $(-2, 1)$ on the horizontal asymptote in Figure 28.

**Guideline 6**   The vertical asymptotes in Figure 28 divide the $xy$-plane into three regions:

$R_1$:   the region to the left of $x = -1$

$R_2$:   the region between $x = -1$ and $x = 2$

$R_3$:   the region to the right of $x = 2$

For $R_1$, let us first consider the portion of the graph that corresponds to $-2 < x < -1$. From the point $(-2, 1)$ on the horizontal asymptote, the graph must extend *upward* as $x \to -1^-$ (it cannot extend downward, since there is no $x$-intercept between $x = -2$ and $x = -1$). As $x \to -\infty$, there will be a low point on the graph between $y = 0$ and $y = 1$, and then the graph will approach the horizontal asymptote $y = 1$ from *below*. It is difficult to see where the low point occurs in Figure 28, because the function values are very close to one another. Using calculus, it can be shown that the low point is $\left(-4, \frac{8}{9}\right)$.

In $R_2$, we have $-1 < x < 2$, and the graph intersects the $x$-axis at $x = 0$. Since the function does not cross the horizontal asymptote in this region, we know that the graph extends *downward* as $x \to -1^+$ and as $x \to 2^-$, as shown in Figure 29(a).

In $R_3$, the graph approaches the horizontal asymptote $y = 1$ (from either above or below) as $x \to \infty$. Furthermore, the graph must extend *upward* as $x \to 2^+$, because there are no $x$-intercepts in $R_3$. This implies that as $x \to \infty$, the graph approaches the horizontal asymptote from *above*, as in Figure 29(b).

The graph of $f$ is sketched in Figure 29(c).

**FIGURE 29**

(a)

(b)                                   (c)

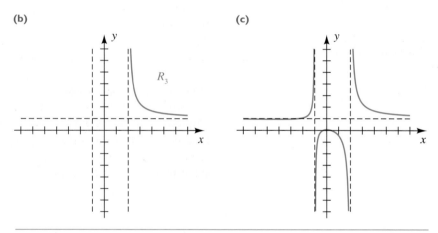

**EXAMPLE 5**    *Sketching the graph of a rational function*

Sketch the graph of $f$ if

$$f(x) = \frac{2x^4}{x^4 + 1}.$$

**SOLUTION**    In this solution we shall not formally write down each guideline. Note that since $f(-x) = f(x)$, the function is even, and hence the graph is symmetric with respect to the $y$-axis.

The graph intersects the $x$-axis at $(0, 0)$. Since the denominator of $f(x)$ has no real zero, the graph has no vertical asymptote.

The numerator and denominator of $f(x)$ have the same degree. Since the leading coefficients are 2 and 1, respectively, we see from part (2) of

*(continued)*

**FIGURE 30**

$$y = \frac{2x^4}{x^4 + 1}$$

the theorem on horizontal asymptotes that the line $y = \frac{2}{1} = 2$ is a horizontal asymptote. We represent this line with dashes in Figure 30. The graph does not cross the horizontal asymptote $y = 2$, since the equation $f(x) = 2$ has no real solution.

Plotting the points $(1, 1)$ and $(2, \frac{32}{17})$ and making use of symmetry leads to the sketch in Figure 30.

---

An **oblique asymptote** for a graph is a line $y = ax + b$, with $a \neq 0$, such that the graph approaches this line as $x \to \infty$ or as $x \to -\infty$. (If the graph is a line, we consider it to be its own asymptote.) If the rational function $f(x) = g(x)/h(x)$ for polynomials $g(x)$ and $h(x)$ and *if the degree of $g(x)$ is one greater than the degree of $h(x)$*, then the graph of $f$ has an oblique asymptote. To find this oblique asymptote we may use long division to express $f(x)$ in the form

$$f(x) = \frac{g(x)}{h(x)} = (ax + b) + \frac{r(x)}{h(x)},$$

where either $r(x) = 0$ or the degree of $r(x)$ is less than the degree of $h(x)$. From part (1) of the theorem on horizontal asymptotes,

$$\frac{r(x)}{h(x)} \to 0 \quad \text{as} \quad x \to \infty \quad \text{or as} \quad x \to -\infty.$$

Consequently, $f(x)$ approaches the line $y = ax + b$ as $x$ increases or decreases without bound; that is, $y = ax + b$ is an oblique asymptote.

### EXAMPLE 6    *Finding an oblique asymptote*

Find all the asymptotes and sketch the graph of $f$ if

$$f(x) = \frac{x^2 - 9}{2x - 4}.$$

**SOLUTION**    A vertical asymptote occurs if $2x - 4 = 0$ (that is, if $x = 2$).

The degree of the numerator of $f(x)$ is greater than the degree of the denominator. Hence, by part (3) of the theorem on horizontal asymptotes, there is no horizontal asymptote. Since the degree of the numerator $x^2 - 9$ is *one* greater than the degree of the denominator $2x - 4$, however, the graph has an oblique asymptote. By long division, we obtain

$$
\begin{array}{r}
\frac{1}{2}x + 1 \\
2x - 4 \overline{) \phantom{x^2}x^2 \phantom{-2x} - 9} \\
\underline{x^2 - 2x \phantom{xxxx}} \\
2x - 9 \\
\underline{2x - 4} \\
-5
\end{array}
$$

Therefore,    $\dfrac{x^2 - 9}{2x - 4} = \left(\dfrac{1}{2}x + 1\right) - \dfrac{5}{2x - 4}.$

As we indicated in the discussion preceding this example, the line $y = \frac{1}{2}x + 1$ is an oblique asymptote. This line and the vertical asymptote $x = 2$ are sketched with dashes in Figure 31.

The $x$-intercepts of the graph are the solutions of $x^2 - 9 = 0$ and hence are 3 and $-3$. The $y$-intercept is $f(0) = \frac{9}{4}$. The corresponding points are plotted in Figure 31. We may now show that the graph has the shape indicated in Figure 32.

FIGURE 31                                    FIGURE 32

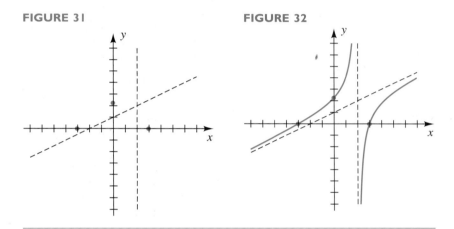

In Example 6, the graph of $f$ approaches the line $y = \frac{1}{2}x + 1$ *asymptotically* as $x \to \infty$ or as $x \to -\infty$. Graphs of rational functions may approach different types of curves asymptotically. For example, if

$$f(x) = \frac{x^4 - x}{x^2} = x^2 - \frac{1}{x},$$

then for large values of $|x|$, $1/x \approx 0$ and hence $f(x) \approx x^2$. Thus, the graph of $f$ approaches the parabola $y = x^2$ asymptotically as $x \to \infty$ or as $x \to -\infty$. In general, if $f(x) = g(x)/h(x)$ and if $q(x)$ is the quotient obtained by dividing $g(x)$ by $h(x)$, then the graph of $f$ approaches the graph of $y = q(x)$ asymptotically as $x \to \infty$ or as $x \to -\infty$.

In the guidelines for sketching the graph of a rational function, we assumed that the numerator and denominator have no common prime factor. If there *is* a common factor, then there may be a *hole* in the graph—that is, a single point may be missing—as illustrated in the following example.

**EXAMPLE 7    *A graph containing a hole***

Sketch the graph of $f$ if

$$f(x) = \frac{x - 4}{(x - 2)(x - 4)}.$$

**FIGURE 33**

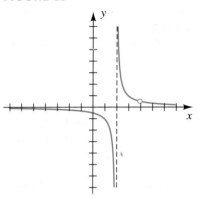

**SOLUTION**    If $x \neq 4$, we may cancel the factor $x - 4$; that is,

$$f(x) = \frac{1}{x - 2}, \quad \text{if} \quad x \neq 4.$$

Thus, the graph of $f$ is the same as that sketched in Figure 25, with one exception: There is a hole in the graph at $x = 4$. To determine the corresponding $y$ value, we use the reduced form of the function, obtaining $f(4) = 1/(4 - 2) = \frac{1}{2}$. The point $(4, \frac{1}{2})$ is indicated by the small circle in Figure 33. It should be noted that graphics calculators or computer programs may not be designed to show this feature of a graph.

**EXAMPLE 8    *Sketching the graph of a rational function***

Sketch the graph of

$$f(x) = \frac{x^2 - x}{9x^3 - 9x^2 - 22x + 8}$$

and find equations of the vertical asymptotes.

**FIGURE 34** $[-2, 3]$ by $[-1, 1]$

**SOLUTION**    Some specific keystrokes for the TI-81/82 graphics calculator are given in Appendix I. We begin by making the assignments

$$Y_1 = x^2 - x, \quad Y_2 = 9x^3 - 9x^2 - 22x + 8, \quad \text{and} \quad Y_3 = Y_1/Y_2.$$

Selecting only $Y_3$ to be graphed and using a standard viewing rectangle, we obtain a graph that gives us virtually no indication of the true shape of $f$. Changing to a viewing rectangle of $[-6, 6]$ by $[-4, 4]$ gives us a hint that the vertical asymptotes are confined to the interval $-2 < x < 3$.

**FIGURE 35** $[-2, 3]$ by $[-1, 1]$

Using a viewing rectangle of $[-2, 3]$ by $[-1, 1]$ and changing to *dot mode* (so as not to graph the function across the vertical asymptotes) leads us to the sketch in Figure 34. Since the degree of the numerator, 2, is less than the degree of the denominator, 3, we know that the horizontal asymptote is the $x$-axis. The zeros of the numerator, 0 and 1, are the only $x$-intercepts.

To determine the equations of the vertical asymptotes, we will abandon the graph of $Y_3$ and examine the graph of $Y_2$—looking for its zeros. Graphing $Y_2$ with the same viewing rectangle, but using *connected mode*, gives us Figure 35.

By the theorem on rational zeros of a polynomial, we know that the possible rational roots of $9x^3 - 9x^2 - 22x + 8 = 0$ are

$$\pm 1, \pm 2, \pm 4, \pm 8, \pm \tfrac{1}{3}, \pm \tfrac{2}{3}, \pm \tfrac{4}{3}, \pm \tfrac{8}{3}, \pm \tfrac{1}{9}, \pm \tfrac{2}{9}, \pm \tfrac{4}{9}, \pm \tfrac{8}{9}.$$

From the graph, we see that the only reasonable choice for the zero in the interval $(-2, -1)$ is $-\tfrac{4}{3}$. The number 2 appears to be a zero, and using the tracing feature makes it apparent that $\tfrac{1}{3}$ is also a zero. These facts can be proved by using synthetic division. Thus, the equations of the vertical asymptotes are

$$x = -\tfrac{4}{3}, \quad x = \tfrac{1}{3}, \quad \text{and} \quad x = 2.$$

Graphs of rational functions may become increasingly complicated as the degrees of the polynomials in the numerator and denominator increase. Techniques developed in calculus must be employed for a thorough treatment of such graphs.

Formulas that represent physical quantities may determine rational functions. For example, consider Ohm's law in electrical theory, which states that $I = V/R$, where $R$ is the resistance (in ohms) of a conductor, $V$ is the potential difference (in volts) across the conductor, and $I$ is the current (in amperes) that flows through the conductor. The resistance of certain alloys approaches zero as the temperature approaches absolute zero (approximately $-273\,°C$), and the alloy becomes a *superconductor* of electricity. If the voltage $V$ is fixed, then, for such a superconductor,

$$I = \frac{V}{R} \to \infty \quad \text{as} \quad R \to 0^+;$$

that is, as $R$ approaches 0, the current increases without bound. Superconductors allow very large currents to be used in generating plants and motors. They also have applications in experimental high-speed ground transportation, where the strong magnetic fields produced by superconducting magnets enable trains to levitate so that there is essentially no friction between the wheels and the track. Perhaps the most important use for superconductors is in circuits for computers, because such circuits produce very little heat.

## 3.5 EXERCISES

**Exer. 1–2:** (a) Sketch the graph of $f$. (b) Find the domain $D$ and range $R$ of $f$. (c) Find the intervals on which $f$ is increasing or is decreasing.

1  $f(x) = \dfrac{4}{x}$

2  $f(x) = \dfrac{1}{x^2}$

**Exer. 3–24:** Sketch the graph of $f$.

3  $f(x) = \dfrac{3}{x - 4}$

4  $f(x) = \dfrac{-3}{x + 3}$

5  $f(x) = \dfrac{-3x}{x + 2}$

6  $f(x) = \dfrac{4x}{2x - 5}$

**7** $f(x) = \dfrac{x - 2}{x^2 - x - 6}$

**8** $f(x) = \dfrac{x + 1}{x^2 + 2x - 3}$

**9** $f(x) = \dfrac{-4}{(x - 2)^2}$

**10** $f(x) = \dfrac{2}{(x + 1)^2}$

**11** $f(x) = \dfrac{x - 3}{x^2 - 1}$

**12** $f(x) = \dfrac{x + 4}{x^2 - 4}$

**13** $f(x) = \dfrac{2x^2 - 2x - 4}{x^2 + x - 12}$

**14** $f(x) = \dfrac{-3x^2 - 3x + 6}{x^2 - 9}$

**15** $f(x) = \dfrac{-x^2 - x + 6}{x^2 + 3x - 4}$

**16** $f(x) = \dfrac{x^2 - 3x - 4}{x^2 + x - 6}$

**17** $f(x) = \dfrac{3x^2 - 3x - 36}{x^2 + x - 2}$

**18** $f(x) = \dfrac{2x^2 + 4x - 48}{x^2 + 3x - 10}$

**19** $f(x) = \dfrac{-2x^2 + 10x - 12}{x^2 + x}$

**20** $f(x) = \dfrac{2x^2 + 8x + 6}{x^2 - 2x}$

**21** $f(x) = \dfrac{x - 1}{x^3 - 4x}$

**22** $f(x) = \dfrac{x^2 - 2x + 1}{x^3 - 9x}$

**23** $f(x) = \dfrac{-3x^2}{x^2 + 1}$

**24** $f(x) = \dfrac{x^2 - 4}{x^2 + 1}$

**Exer. 25–28: Find the oblique asymptote, and sketch the graph of $f$.**

**25** $f(x) = \dfrac{x^2 - x - 6}{x + 1}$

**26** $f(x) = \dfrac{2x^2 - x - 3}{x - 2}$

**27** $f(x) = \dfrac{8 - x^3}{2x^2}$

**28** $f(x) = \dfrac{x^3 + 1}{x^2 - 9}$

**Exer. 29–36: Simplify $f(x)$, and sketch the graph of $f$.**

**29** $f(x) = \dfrac{2x^2 + x - 6}{x^2 + 3x + 2}$

**30** $f(x) = \dfrac{x^2 - x - 6}{x^2 - 2x - 3}$

**31** $f(x) = \dfrac{x - 1}{1 - x^2}$

**32** $f(x) = \dfrac{x + 2}{x^2 - 4}$

**33** $f(x) = \dfrac{x^2 + x - 2}{x + 2}$

**34** $f(x) = \dfrac{x^3 - 2x^2 - 4x + 8}{x - 2}$

**35** $f(x) = \dfrac{x^2 + 4x + 4}{x^2 + 3x + 2}$

**36** $f(x) = \dfrac{(x^2 + x)(2x - 1)}{(x^2 - 3x + 2)(2x - 1)}$

**Exer. 37–40: Find an equation of a rational function $f$ that satisfies the given conditions.**

**37** vertical asymptote: $x = 4$
horizontal asymptote: $y = -1$
$x$-intercept: 3

**38** vertical asymptotes: $x = -2$, $x = 0$
horizontal asymptote: $y = 0$
$x$-intercept: 2; $f(3) = 1$

**39** vertical asymptotes: $x = -3$, $x = 1$
horizontal asymptote: $y = 0$
$x$-intercept: $-1$; $f(0) = -2$
hole at $x = 2$

**40** vertical asymptotes: $x = -1$, $x = 3$
horizontal asymptote: $y = 2$
$x$-intercepts: $-2$, 1; hole at $x = 0$

**41** *A container for radioactive waste* A cylindrical container for storing radioactive waste is to be constructed from lead. This container must be 6 inches thick. The volume of the outside cylinder shown in the figure is to be $16\pi$ ft$^3$.

(a) Express the height $h$ of the inside cylinder as a function of the inside radius $r$.

(b) Show that the inside volume $V(r)$ is given by

$$V(r) = \pi r^2 \left[ \dfrac{16}{(r + 0.5)^2} - 1 \right].$$

(c) What values of $r$ must be excluded in part (b)?

**EXERCISE 41**

**42 Drug dosage** Young's rule is a formula that is used to modify adult drug dosage levels for young children. If $a$ denotes the adult dosage (in milligrams) and if $t$ is the age of the child (in years), then the child's dose $y$ is given by $y = ta/(t + 12)$. Sketch the graph of this equation for $t > 0$ and $a = 100$.

**43 Salt concentration** Salt water of concentration 0.1 pound of salt per gallon flows into a large tank that initially contains 50 gallons of pure water.

(a) If the flow rate of salt water into the tank is 5 gallons per minute, find the volume $V(t)$ of water and the amount $A(t)$ of salt in the tank after $t$ minutes.

(b) Find a formula for the salt concentration $c(t)$ (in lb/gal) after $t$ minutes.

(c) Discuss the variation of $c(t)$ as $t \to \infty$.

**44 Amount of rainfall** The total number of inches $R(t)$ of rain during a storm of length $t$ hours can be approximated by

$$R(t) = \frac{at}{t + b},$$

where $a$ and $b$ are positive constants that depend on the geographical locale.

(a) Discuss the variation of $R(t)$ as $t \to \infty$.

(b) The intensity $I$ of the rainfall (in in./hr) is defined by $I = R(t)/t$. If $a = 2$ and $b = 8$, sketch the graph of $R$ and $I$ on the same coordinate plane for $t > 0$.

**45 Salmon propagation** For a particular salmon population, the relationship between the number $S$ of spawners and the number $R$ of offspring that survive to maturity is given by the formula

$$R = \frac{4500S}{S + 500}.$$

(a) Under what conditions is $R > S$?

(b) Find the number of spawners that would yield 90%

of the greatest possible number of offspring that survive to maturity.

(c) Work part (b) with 80% replacing 90%.

(d) Compare the results for $S$ and $R$ (in terms of percentage increases) from parts (b) and (c).

**46 Population density** The population density $D$ (in people/mi$^2$) in a large city is related to the distance $x$ (in miles) from the center of the city by

$$D = \frac{5000x}{x^2 + 36}.$$

(a) What happens to the density as the distance from the center of the city changes from 20 miles to 25 miles?

(b) What eventually happens to the density?

(c) In what areas of the city does the population density exceed 400 people/mi$^2$?

**Exer. 47–50: Graph $f$ and find equations of the vertical asymptotes.**

**47** $f(x) = \dfrac{20x^2 + 80x + 72}{10x^2 + 40x + 41}$

**48** $f(x) = \dfrac{15x^2 - 60x + 68}{3x^2 - 12x + 13}$

**49** $f(x) = \dfrac{(x - 1)^2}{(x - 0.999)^2}$   **50** $f(x) = \dfrac{x^2 - 9.01}{x - 3}$

**51** Let $f(x)$ be the polynomial
$(x + 3)(x + 2)(x + 1)(x)(x - 1)(x - 2)(x - 3)$.

(a) Describe the graph of $g(x) = f(x)/f(x)$.

(b) Describe the graph of $h(x) = g(x)p(x)$, where $p(x)$ is a polynomial function.

**52** Refer to Exercise 51.

(a) Describe the graph of $y = f(x)$.

(b) Describe the graph of $k(x) = 1/f(x)$.

## CHAPTER 3 REVIEW EXERCISES

**Exer. 1–6: Find all values of $x$ such that $f(x) > 0$ and all $x$ such that $f(x) < 0$, and then sketch the graph of $f$.**

1 $f(x) = (x + 2)^3$

2 $f(x) = x^6 - 32$

3 $f(x) = -\frac{1}{4}(x + 2)(x - 1)^2(x - 3)$

4 $f(x) = 2x^2 + x^3 - x^4$

5 $f(x) = x^3 + 2x^2 - 8x$

6 $f(x) = \frac{1}{15}(x^5 - 20x^3 + 64x)$

**7** If $f(x) = x^3 - 5x^2 + 7x - 9$, use the intermediate value theorem for polynomial functions to prove that there is a real number $a$ such that $f(a) = 100$.

**8** Prove that the equation $x^5 - 3x^4 - 2x^3 - x + 1 = 0$ has a solution between 0 and 1.

**Exer. 9–10: Find the quotient and remainder if $f(x)$ is divided by $p(x)$.**

**9** $f(x) = 3x^5 - 4x^3 + x + 5$;   $p(x) = x^3 - 2x + 7$

**10** $f(x) = 4x^3 - x^2 + 2x - 1$;   $p(x) = x^2$

**11** If $f(x) = -4x^4 + 3x^3 - 5x^2 + 7x - 10$, use the remainder theorem to find $f(-2)$.

**12** Use the factor theorem to show that $x - 3$ is a factor of $f(x) = 2x^4 - 5x^3 - 4x^2 + 9$.

**Exer. 13–14: Use synthetic division to find the quotient and remainder if $f(x)$ is divided by $p(x)$.**

**13** $f(x) = 6x^5 - 4x^2 + 8$;        $p(x) = x + 2$

**14** $f(x) = 2x^3 + 5x^2 - 2x + 1$;   $p(x) = x - \sqrt{2}$

**Exer. 15–16: A polynomial $f(x)$ with real coefficients has the indicated zero(s) and degree and satisfies the given condition. Express $f(x)$ as a product of linear and quadratic polynomials with real coefficients that are irreducible over $\mathbb{R}$.**

**15** $-3 + 5i$, $-1$;   degree 3;   $f(1) = 4$

**16** $1 - i$, 3, 0;     degree 4;   $f(2) = -1$

**17** Find a polynomial $f(x)$ of degree 7 with leading coefficient 1 such that $-3$ is a zero of multiplicity 2 and 0 is a zero of multiplicity 5, and sketch the graph of $f$.

**18** Show that 2 is a zero of multiplicity 3 of the polynomial $f(x) = x^5 - 4x^4 - 3x^3 + 34x^2 - 52x + 24$, and express $f(x)$ as a product of linear factors.

**Exer. 19–20: Find the zeros of $f(x)$, and state the multiplicity of each zero.**

**19** $f(x) = (x^2 - 2x + 1)^2(x^2 + 2x - 3)$

**20** $f(x) = x^6 + 2x^4 + x^2$

**Exer. 21–22: (a) Use Descartes' rule of signs to determine the number of possible positive, negative, and non-real complex solutions of the equation. (b) Find the smallest and largest integers that are upper and lower bounds, respectively, for the real solutions of the equation.**

**21** $2x^4 - 4x^3 + 2x^2 - 5x - 7 = 0$

**22** $x^5 - 4x^3 + 6x^2 + x + 4 = 0$

**23** Show that $7x^6 + 2x^4 + 3x^2 + 10$ has no real zero.

**Exer. 24–26: Find all solutions of the equation.**

**24** $x^4 + 9x^3 + 31x^2 + 49x + 30 = 0$

**25** $16x^3 - 20x^2 - 8x + 3 = 0$

**26** $x^4 - 7x^2 + 6 = 0$

**Exer. 27–36: Sketch the graph of $f$.**

**27** $f(x) = \dfrac{-2}{(x+1)^2}$         **28** $f(x) = \dfrac{1}{(x-1)^3}$

**29** $f(x) = \dfrac{3x^2}{16 - x^2}$

**30** $f(x) = \dfrac{x}{(x+5)(x^2 - 5x + 4)}$

**31** $f(x) = \dfrac{x^3 - 2x^2 - 8x}{-x^2 + 2x}$

**32** $f(x) = \dfrac{x^2 - 2x + 1}{x^3 - x^2 + x - 1}$

**33** $f(x) = \dfrac{3x^2 + x - 10}{x^2 + 2x}$     **34** $f(x) = \dfrac{-2x^2 - 8x - 6}{x^2 - 6x + 8}$

**35** $f(x) = \dfrac{x^2 + 2x - 8}{x + 3}$     **36** $f(x) = \dfrac{x^4 - 16}{x^3}$

**37** *Deflection of a beam*   A horizontal beam $l$ feet long is supported at one end and unsupported at the other end (see the figure). If the beam is subjected to a uniform load and if $y$ denotes the deflection of the beam at a position $x$ feet from the supported end, then it can be shown that $y = cx^2(x^2 - 4lx + 6l^2)$, where $c$ is a positive constant that depends on the weight of the load and the physical properties of the beam.

(a) If the beam is 10 feet long and the deflection at the unsupported end of the board is 2 feet, find $c$.

(b) Show that the deflection is 1 foot somewhere between $x = 6.1$ and $x = 6.2$.

**EXERCISE 37**

38 *Elastic cylinder*   A rectangle made of elastic material is to be made into a cylinder by joining edge $AD$ to edge $BC$, as shown in the figure. A wire of fixed length $l$ is placed along the diagonal of the rectangle to support the structure. Let $x$ denote the height of the cylinder.

(a) Express the volume $V$ of the cylinder in terms of $x$.

(b) For what positive values of $x$ is $V > 0$?

EXERCISE 38

39 *Determining temperatures*   A meteorologist determines that the temperature $T$ (in °F) for a certain 24-hour period in winter was given by the formula $T = \frac{1}{20}t(t - 12)(t - 24)$ for $0 \le t \le 24$, where $t$ is time in hours and $t = 0$ corresponds to 6 A.M. At what time(s) was the temperature 32 °F?

40 *Deer propagation*   A herd of 100 deer is introduced onto a small island. Assuming the number $N(t)$ of deer after $t$ years is given by $N(t) = -t^4 + 21t^2 + 100$ (for $t > 0$), determine when the herd size exceeds 180.

41 *Threshold response curve*   In biochemistry, the general threshold response curve is the graph of an equation

$$R = \frac{kS^n}{S^n + a^n},$$

where $R$ is the chemical response when the level of the substance being acted on is $S$ and $a$, $k$, and $n$ are positive constants. An example is the removal rate $R$ of alcohol from the bloodstream by the liver when the blood alcohol concentration is $S$.

(a) Find an equation of the horizontal asymptote for the graph.

(b) In the case of alcohol removal, $n = 1$ and a typical value of $k$ is 0.22 gram per liter per minute. What is the interpretation of $k$ in this setting?

42 *Oil spill clean-up*   The cost $C(x)$ of cleaning up $x$ percent of an oil spill that has washed ashore increases greatly as $x$ approaches 100. Suppose that

$$C(x) = \frac{20x}{101 - x} \text{ (thousand dollars)}.$$

(a) Compare $C(100)$ to $C(90)$.

(b) Sketch the graph of $C$ for $0 < x < 100$.

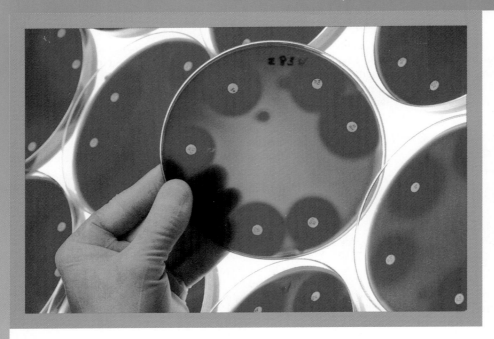

*The number of bacteria in a culture often increases exponentially.*

■ *Exponential and logarithmic func-*
*tions are transcendental functions, since*
*they cannot be defined in terms of only*
*addition, subtraction, multiplication, divi-*
*sion, and rational powers of a variable*
*x, as is the case for the algebraic func-*
*tions considered in previous chapters.*
*Such functions are of major importance*
*in mathematics and have applications in*
*almost every field of human endeavor.*
*They are especially useful in the fields*
*of chemistry, biology, physics, and engi-*
*neering, where they help describe the*
*manner in which quantities in nature*
*grow or decay. As we shall see in this*
*chapter, there is a close relationship*
*between specific exponential and loga-*
*rithmic functions—they are inverse*
*functions of each other.* ■

# CHAPTER 4

# EXPONENTIAL

# AND LOGARITHMIC

# FUNCTIONS

## 4.1   EXPONENTIAL FUNCTIONS

Let us consider the function $f$ defined by

$$f(x) = 2^x,$$

where $x$ is restricted to *rational* numbers. (Recall that if $x = m/n$ for integers $m$ and $n$ with $n > 0$, then $2^x = 2^{m/n} = (\sqrt[n]{2})^m$.) Coordinates of several points on the graph of $y = 2^x$ are listed in the following table.

| $x$ | $-10$ | $-3$ | $-2$ | $-1$ | 0 | 1 | 2 | 3 | 10 |
|---|---|---|---|---|---|---|---|---|---|
| $y = 2^x$ | $\frac{1}{1024}$ | $\frac{1}{8}$ | $\frac{1}{4}$ | $\frac{1}{2}$ | 1 | 2 | 4 | 8 | 1024 |

Other values of $y$ for $x$ rational, such as $2^{1/3}$, $2^{-9/7}$, and $2^{5.143}$, can be approximated with a calculator. We can show algebraically that if $x_1$ and $x_2$ are rational numbers such that $x_1 < x_2$, then $2^{x_1} < 2^{x_2}$. Thus, $f$ is an increasing function, and its graph rises. Plotting points leads to the sketch in Figure 1, where the small dots indicate that only the points with *rational* $x$-coordinates are on the graph. There is a *hole* in the graph whenever the $x$-coordinate of a point is irrational.

To extend the domain of $f$ to all real numbers, it is necessary to define $2^x$ for every *irrational* exponent $x$. To illustrate, if we wish to define $2^\pi$, we could use the nonterminating decimal representing 3.1415926 . . . for $\pi$ and consider the following *rational* powers of 2:

$$2^3, \quad 2^{3.1}, \quad 2^{3.14}, \quad 2^{3.141}, \quad 2^{3.1415}, \quad 2^{3.14159}, \quad \ldots$$

It can be shown, using calculus, that each successive power gets closer to a unique real number, denoted by $2^\pi$. Thus,

$$2^x \to 2^\pi \quad \text{as} \quad x \to \pi, \quad \text{with } x \text{ rational.}$$

The same technique can be used for any other irrational power of 2. To sketch the graph of $y = 2^x$ with $x$ *real,* we replace the holes in the graph in Figure 1 with points, and we obtain the graph in Figure 2. The function $f$ defined by $f(x) = 2^x$ for every real number $x$ is called the *exponential function with base* 2.

Let us next consider *any* base $a$, where $a$ is a positive real number different from 1. As in the preceding discussion, to each real number $x$ there corresponds exactly one positive number $a^x$ such that the laws of exponents are true. Thus, as in the following chart, we may define a function $f$ whose domain is $\mathbb{R}$ and range is the set of positive real numbers.

**FIGURE 1**

**FIGURE 2**

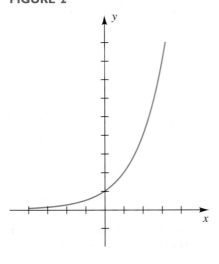

| Terminology | Definition | Graph of $f$ for $a > 1$ | Graph of $f$ for $0 < a < 1$ |
|---|---|---|---|
| Exponential function $f$ with base $a$ | $f(x) = a^x$ for every $x$ in $\mathbb{R}$, where $a > 0$ and $a \neq 1$ | | |

The graphs in the chart show that if $a > 1$, then $f$ is increasing on $\mathbb{R}$, and if $0 < a < 1$, then $f$ is decreasing on $\mathbb{R}$. (These facts can be proved using calculus.) The graphs merely indicate the *general* appearance—the *exact* shape depends on the value of $a$. Note, however, that since $a^0 = 1$, the $y$-intercept is 1 for every $a$.

If $a > 1$, then as $x$ *decreases* through negative values, the graph of $f$ approaches the $x$-axis (see the third column in the chart). Thus, the $x$-axis is a *horizontal asymptote.* As $x$ increases through positive values, the graph rises rapidly. This type of variation is characteristic of the **exponential law of growth,** and $f$ is sometimes called a **growth function.**

If $0 < a < 1$, then as $x$ *increases,* the graph of $f$ approaches the $x$-axis asymptotically (see the last column of the chart). This type of variation is known as **exponential decay.**

When considering $a^x$ we exclude the cases $a \leq 0$ and $a = 1$. Note that if $a < 0$, then $a^x$ is not a real number for many values of $x$, such as $\frac{1}{2}, \frac{3}{4}$, and $\frac{11}{6}$. If $a = 0$, then $a^0 = 0^0$ is undefined. Finally, if $a = 1$, then $a^x = 1$ for every $x$, and the graph of $y = a^x$ is a horizontal line.

The graph of an exponential function $f$ is either increasing throughout its domain or decreasing throughout its domain. Thus, $f$ is one-to-one by the theorem on page 169. Combining this result with the definition of a one-to-one function (see page 168) gives us parts (1) and (2) of the following theorem.

**Theorem: Exponential Functions Are One-to-One**

The exponential function $f$ given by

$$f(x) = a^x \quad \text{for} \quad 0 < a < 1 \quad \text{or} \quad a > 1$$

is one-to-one. Thus, the following equivalent conditions are satisfied for real numbers $x_1$ and $x_2$:

**(1)** If $x_1 \neq x_2$, then $a^{x_1} \neq a^{x_2}$.

**(2)** If $a^{x_1} = a^{x_2}$, then $x_1 = x_2$.

When using this theorem as a reason for a step in the solution to an example, we will state that *exponential functions are one-to-one.*

**ILLUSTRATION**

**Exponential Functions Are One-to-One**

■ If $7^{3x} = 7^{2x+5}$, then $3x = 2x + 5$, or $x = 5$.

In the following example we solve a simple *exponential equation*—that is, an equation in which the variable appears in an exponent.

**EXAMPLE I**   *Solving an exponential equation*

Solve the equation $3^{5x-8} = 9^{x+2}$.

**SOLUTION**

$$
\begin{array}{ll}
3^{5x-8} = 9^{x+2} & \text{given} \\
3^{5x-8} = (3^2)^{x+2} & \text{express both sides with the same base} \\
3^{5x-8} = 3^{2x+4} & \text{law of exponents} \\
5x - 8 = 2x + 4 & \text{exponential functions are one-to-one} \\
3x = 12 & \text{subtract } 2x \text{ and add } 8 \\
x = 4 & \text{divide by } 3
\end{array}
$$

In the next two examples we sketch the graphs of several different exponential functions.

**FIGURE 3**

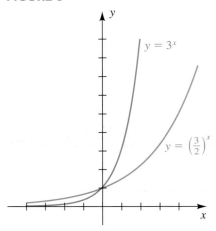

$y = 3^x$

$y = \left(\frac{3}{2}\right)^x$

**EXAMPLE 2**   *Sketching graphs of exponential functions*

If $f(x) = \left(\frac{3}{2}\right)^x$ and $g(x) = 3^x$, sketch the graphs of $f$ and $g$ on the same coordinate plane.

**SOLUTION**   Since $\frac{3}{2} > 1$ and $3 > 1$, each graph *rises* as $x$ increases. The following table displays coordinates for several points on the graphs.

| $x$ | $-2$ | $-1$ | $0$ | $1$ | $2$ | $3$ | $4$ |
|---|---|---|---|---|---|---|---|
| $y = \left(\frac{3}{2}\right)^x$ | $\frac{4}{9} \approx 0.4$ | $\frac{2}{3} \approx 0.7$ | $1$ | $\frac{3}{2}$ | $\frac{9}{4} \approx 2.3$ | $\frac{27}{8} \approx 3.4$ | $\frac{81}{16} \approx 5.1$ |
| $y = 3^x$ | $\frac{1}{9} \approx 0.1$ | $\frac{1}{3} \approx 0.3$ | $1$ | $3$ | $9$ | $27$ | $81$ |

Plotting points and being familiar with the general graph of $y = a^x$ leads to the graphs in Figure 3.

Example 2 illustrates the fact that if $1 < a < b$, then $a^x < b^x$ for positive values of $x$ and $b^x < a^x$ for negative values of $x$. In particular, since $\frac{3}{2} < 2 < 3$, the graph of $y = 2^x$ in Figure 2 lies between the graphs of $f$ and $g$ in Figure 3.

**EXAMPLE 3    *Sketching the graph of an exponential function***

Sketch the graph of the equation $y = \left(\frac{1}{2}\right)^x$.

**SOLUTION**    Since $0 < \frac{1}{2} < 1$, the graph *falls* as $x$ increases. Coordinates of some points on the graph are listed in the following table.

| $x$ | −3 | −2 | −1 | 0 | 1 | 2 | 3 |
|---|---|---|---|---|---|---|---|
| $y = \left(\frac{1}{2}\right)^x$ | 8 | 4 | 2 | 1 | $\frac{1}{2}$ | $\frac{1}{4}$ | $\frac{1}{8}$ |

The graph is sketched in Figure 4. Since $\left(\frac{1}{2}\right)^x = 2^{-x}$, the graph is the same as the graph of the equation $y = 2^{-x}$. Note that the graph is a reflection through the $y$-axis of the graph of $y = 2^x$ in Figure 2.

**FIGURE 4**

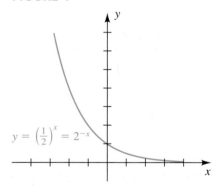

$y = \left(\frac{1}{2}\right)^x = 2^{-x}$

Equations of the form $y = a^u$, where $u$ is some expression in $x$, occur in applications. The next two examples illustrate equations of this form.

**EXAMPLE 4    *Shifting graphs of exponential functions***

Sketch the graph of the equation:

**(a)** $y = 3^{x-2}$        **(b)** $y = 3^x - 2$

**SOLUTION**

**(a)** The graph of $y = 3^x$ was sketched in Figure 3 and is resketched in Figure 5. From the discussion of horizontal shifts in Section 2.2 we can obtain the graph of $y = 3^{x-2}$ by shifting the graph of $y = 3^x$ two units to the right, as shown in Figure 5.

The graph of $y = 3^{x-2}$ can also be obtained by plotting several points and using them as a guide to sketch an exponential-type curve.

**(b)** From the discussion of vertical shifts in Section 2.2 we can obtain the graph of $y = 3^x - 2$ by shifting the graph of $y = 3^x$ two units downward, as shown in Figure 6. Note that the $y$-intercept is $-1$ and the line $y = -2$ is a horizontal asymptote for the graph.

**FIGURE 5**        **FIGURE 6**

$y = 3^x$    $y = 3^{x-2}$      $y = 3^x$    $y = 3^x - 2$    $y = -2$

The bell-shaped graph of the function in the next example is similar to a *normal probability curve* used in statistical studies.

**EXAMPLE 5**   *Sketching a bell-shaped graph*

If $f(x) = 2^{-x^2}$, sketch the graph of $f$.

**FIGURE 7**

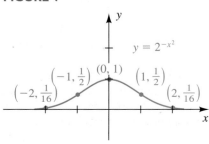

$y = 2^{-x^2}$, $(-1, \frac{1}{2})$, $(0, 1)$, $(1, \frac{1}{2})$, $(-2, \frac{1}{16})$, $(2, \frac{1}{16})$

**SOLUTION**   If we rewrite $f(x)$ as

$$f(x) = \frac{1}{2^{(x^2)}}$$

we see that as $x$ increases through positive values, $f(x)$ decreases rapidly, and hence the graph approaches the $x$-axis asymptotically. The maximum value of $f$ is $f(0) = 1$. Since $f$ is an even function, the graph is symmetric with respect to the $y$-axis. Some points on the graph are $(0, 1)$, $(1, \frac{1}{2})$, and $(2, \frac{1}{16})$. Plotting and using symmetry gives us the sketch in Figure 7.

## APPLICATION

### Bacterial Growth

Exponential functions may be used to describe the growth of certain populations. As an illustration, suppose it is observed experimentally that the number of bacteria in a culture doubles every day. If 1000 bacteria are present at the start, then we obtain the following table, where $t$ is the time in days and $f(t)$ is the bacteria count at time $t$.

**FIGURE 8**

$f(t)$ (bacteria count)

| $t$ (time in days) | 0 | 1 | 2 | 3 | 4 |
|---|---|---|---|---|---|
| $f(t)$ (bacteria count) | 1000 | 2000 | 4000 | 8000 | 16,000 |

It appears that $f(t) = (1000)2^t$. With this formula we can predict the number of bacteria present at any time $t$. For example, at $t = 1.5 = \frac{3}{2}$,

$$f(t) = (1000)2^{3/2} \approx 2828.$$

The graph of $f$ is sketched in Figure 8.

## APPLICATION

### Radioactive Decay

Certain physical quantities *decrease* exponentially. In such cases, if $a$ is the base of the exponential function, then $0 < a < 1$. One of the most common examples of exponential decrease is the decay of a radioactive substance, or isotope. The **half-life** of an isotope is the time it takes for one-half the original amount in a given sample to decay. The half-life is the principal characteristic used to distinguish one radioactive substance from another. The polonium isotope $^{210}$Po has a half-life of approximately 140 days; that is, given any amount, one-half of it will disintegrate in 140 days. If 20 milligrams of $^{210}$Po is present initially, then the following table indicates the amount remaining after various intervals of time.

**FIGURE 9**

$f(t)$ (mg remaining)

| $t$ (time in days) | 0 | 140 | 280 | 420 | 560 |
|---|---|---|---|---|---|
| $f(t)$ (mg remaining) | 20 | 10 | 5 | 2.5 | 1.25 |

The sketch in Figure 9 illustrates the exponential nature of the disintegration.

Other radioactive substances have much longer half-lives. In particular, a by-product of nuclear reactors is the radioactive plutonium isotope $^{239}$Pu, which has a half-life of approximately 24,000 years. It is for this reason that the disposal of radioactive waste is a major problem in modern society.

### APPLICATION

**Compound Interest**

*Compound interest* provides a good illustration of exponential growth. If a sum of money $P$, the *principal*, is invested at a *simple* interest rate $r$, then the interest at the end of one interest period is the product $Pr$ when $r$ is expressed as a decimal. For example, if $P = \$1000$ and the interest rate is 9% per year, then $r = 0.09$, and the interest at the end of one year is $\$1000(0.09)$, or $\$90$.

If the interest is reinvested with the principal at the end of the interest period, then the new principal is

$$P + Pr, \quad \text{or} \quad P(1 + r).$$

Note that to find the new principal we may multiply the original principal by $(1 + r)$. In the preceding example, the new principal is $\$1000(1.09)$, or $\$1090$.

After another interest period has elapsed, the new principal may be found by multiplying $P(1 + r)$ by $(1 + r)$. Thus, the principal after two interest periods is $P(1 + r)^2$. If we continue to reinvest, the principal after three periods is $P(1 + r)^3$; after four it is $P(1 + r)^4$; and, in general, the amount $A$ accumulated after $k$ interest periods is

$$A = P(1 + r)^k.$$

Interest accumulated by means of this formula is **compound interest.** Note that $A$ is expressed in terms of an exponential function with base $1 + r$. The interest period may be measured in years, months, weeks, days, or any other suitable unit of time. When applying the formula for $A$, remember that *$r$ is the interest rate per interest period expressed as a decimal.* For example, if the rate is stated as 6% *per year compounded monthly,* then the rate per month is $\frac{6}{12}\%$ or, equivalently, 0.5%. Thus, $r = 0.005$ and $k$ is the number of months. If $\$100$ is invested at this rate, then the formula for $A$ is

$$A = 100(1 + 0.005)^k = 100(1.005)^k.$$

In general, we have the following formula.

**Compound Interest Formula**

$$A = P\left(1 + \frac{r}{n}\right)^{nt},$$

where   $P = $ principal
        $r = $ interest rate expressed as a decimal
        $n = $ number of interest periods per year
        $t = $ number of years $P$ is invested
        $A = $ amount after $t$ years.

The next example illustrates a special case of the compound interest formula.

### EXAMPLE 6   *Using the compound interest formula*

Suppose that $1000 is invested at an interest rate of 9% compounded monthly. Find the new amount of principal after 5 years, after 10 years, and after 15 years. Illustrate graphically the growth of the investment.

**SOLUTION**   Applying the compound interest formula with $r = 0.09$, $n = 12$, and $P = \$1000$, we find that the amount after $t$ years is

$$A = 1000\left(1 + \frac{0.09}{12}\right)^{12t} = 1000(1.0075)^{12t}.$$

Substituting $t = 5$, 10, and 15 and using a calculator, we obtain the following table.

**FIGURE 10**
Compound interest: $A = 1000(1.0075)^{12t}$

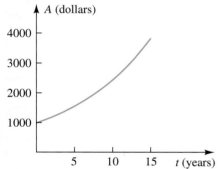

| Number of years | Amount |
|---|---|
| 5 | $A = \$1000(1.0075)^{60} = \$1565.68$ |
| 10 | $A = \$1000(1.0075)^{120} = \$2451.36$ |
| 15 | $A = \$1000(1.0075)^{180} = \$3838.04$ |

The exponential nature of the increase is indicated by the fact that during the first five years, the growth in the investment is $565.68; during the second five-year period, the growth is $885.68; and during the last five-year period, it is $1368.68.

The sketch in Figure 10 illustrates the growth of $1000 invested over a period of 15 years.

We conclude this section with an example involving a graphing utility.

### EXAMPLE 7   *Estimating amounts of a drug in the bloodstream*

If an adult takes a 100-milligram tablet of a certain prescription drug orally, the rate $R$ at which the drug enters the bloodstream $t$ minutes later is predicted to be

$$R = 5(0.95)^t \text{ mg/min.}$$

It can be shown using calculus that the amount $A$ of the drug in the bloodstream at time $t$ can be approximated by

$$A = 97.4786[1 - (0.95)^t] \text{ mg.}$$

**(a)** Estimate how long it takes for 50 milligrams of the drug to enter the bloodstream.

**(b)** Estimate the number of milligrams of the drug in the bloodstream when the drug is entering at a rate of 3 mg/min.

**SOLUTION**

**(a)** We wish to determine $t$ when $A$ is equal to 50. Since the value of $A$ cannot exceed 97.4786, we choose the viewing rectangle to be [0, 100] by [0, 100].

We next assign $97.4786[1 - (0.95)^x]$ to $Y_1$ and graph $Y_1$, obtaining a display similar to that in Figure 11 (note that $x = t$). Using the tracing and zoom-in features, we estimate that $A = 50$ mg when $x \approx 14$ min.

**(b)** We wish to determine $t$ when $R$ is equal to 3. Let us first assign $5(0.95)^x$ to $Y_2$. Since the maximum value of $Y_2$ is 5 (at $t = 0$), we use a viewing rectangle of dimensions [0, 15] by [0, 10] and obtain a display similar to that in Figure 12. Tracing $Y_2$ until $y = 3$ gives us $x \approx 9.96$. Thus, after almost 10 minutes, the drug will be entering the bloodstream at a rate of 3 mg/min. (Note that the initial rate, at $t = 0$, is 5 mg/min.) Finding the value of $Y_1$ at $x = 10$, we see that there is almost 39 milligrams of the drug in the bloodstream after 10 minutes.

**FIGURE 11**    [0, 100] by [0, 100]        **FIGURE 12**    [0, 15] by [0, 10]

## 4.1 EXERCISES

**Exer. 1–8: Solve the equation.**

**1** $7^{x+6} = 7^{3x-4}$                **2** $6^{7-x} = 6^{2x+1}$

**3** $3^{2x+3} = 3^{(x^2)}$              **4** $9^{(x^2)} = 3^{3x+2}$

**5** $2^{-100x} = (0.5)^{x-4}$            **6** $(\frac{1}{2})^{6-x} = 2$

**7** $4^{x-3} = 8^{4-x}$                  **8** $27^{x-1} = 9^{2x-3}$

**9** Sketch the graph of $f$ if $a = 2$.

   (a) $f(x) = a^x$                (b) $f(x) = -a^x$
   (c) $f(x) = 3a^x$               (d) $f(x) = a^{x+3}$
   (e) $f(x) = a^x + 3$            (f) $f(x) = a^{x-3}$
   (g) $f(x) = a^x - 3$            (h) $f(x) = a^{-x}$
   (i) $f(x) = \left(\dfrac{1}{a}\right)^x$        (j) $f(x) = a^{3-x}$

**10** Work Exercise 9 if $a = \frac{1}{2}$.

**Exer. 11–20: Sketch the graph of $f$.**

**11** $f(x) = (\frac{2}{3})^{-x}$          **12** $f(x) = (\frac{2}{3})^x$

**13** $f(x) = -(\frac{1}{2})^x + 4$        **14** $f(x) = -3^x + 9$

**15** $f(x) = 2^{|x|}$                     **16** $f(x) = 2^{-|x|}$

**17** $f(x) = 3^{1-x^2}$                   **18** $f(x) = 2^{-(x+1)^2}$

**19** $f(x) = 3^x + 3^{-x}$                **20** $f(x) = 3^x - 3^{-x}$

**21** **Elk population**   One hundred elk, each 1 year old, are introduced into a game preserve. The number $N(t)$ alive after $t$ years is predicted to be $N(t) = 100(0.9)^t$. Estimate the number alive after

   (a) 1 year   (b) 5 years   (c) 10 years

22 **Drug dosage** *lieu lieong* A drug is eliminated from the body through urine. Suppose that for an initial dose of 10 milligrams, the amount $A(t)$ in the body $t$ hours later is given by $A(t) = 10(0.8)^t$.

(a) Estimate the amount of the drug in the body 8 hours after the initial dose.

(b) What percentage of the drug still in the body is eliminated each hour?

23 **Bacterial growth**   The number of bacteria in a certain culture increased from 600 to 1800 between 7:00 A.M. and 9:00 A.M. Assuming growth is exponential, the number $f(t)$ of bacteria $t$ hours after 7:00 A.M. is given by $f(t) = 600(3)^{t/2}$.

(a) Estimate the number of bacteria in the culture at 8:00 A.M., 10:00 A.M., and 11:00 A.M.

(b) Sketch the graph of $f$ for $0 \le t \le 4$.

24 **Newton's law of cooling**   According to Newton's law of cooling, the rate at which an object cools is directly proportional to the difference in temperature between the object and the surrounding medium. The face of a household iron cools from 125° to 100° in 30 minutes in a room that remains at a constant temperature of 75°. From calculus, the temperature $f(t)$ of the face after $t$ hours of cooling is given by $f(t) = 50(2)^{-2t} + 75$.

(a) Assuming $t = 0$ corresponds to 1:00 P.M., approximate to the nearest tenth of a degree the temperature at 2:00 P.M., 3:30 P.M., and 4:00 P.M.

(b) Sketch the graph of $f$ for $0 \le t \le 4$.

25 **Radioactive decay**   The radioactive bismuth isotope $^{210}$Bi has a half-life of 5 days. If there is 100 milligrams of $^{210}$Bi present at $t = 0$, then the amount $f(t)$ remaining after $t$ days is given by $f(t) = 100(2)^{-t/5}$.

(a) How much $^{210}$Bi remains after 5 days? 10 days? 12.5 days?

(b) Sketch the graph of $f$ for $0 \le t \le 30$.

26 **Light penetration in an ocean**   An important problem in oceanography is to determine the amount of light that can penetrate to various ocean depths. The Beer-Lambert law asserts that the exponential function given by $I(x) = I_0 c^x$ is a model for this phenomenon (see the figure). For a certain location, $I(x) = 10(0.4)^x$ is the amount of light (in calories/cm²/sec) reaching a depth of $x$ meters.

(a) Find the amount of light at a depth of 2 meters.

(b) Sketch the graph of $I$ for $0 \le x \le 5$.

EXERCISE 26

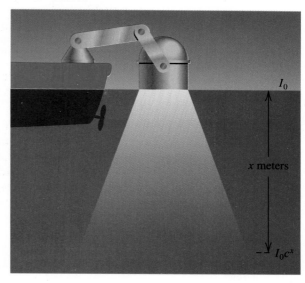

27 **Decay of radium**   The half-life of radium is 1600 years. If the initial amount is $q_0$ milligrams, then the quantity $q(t)$ remaining after $t$ years is given by $q(t) = q_0 2^{kt}$. Find $k$.

28 **Dissolving salt in water**   If 10 grams of salt is added to a quantity of water, then the amount $q(t)$ that is undissolved after $t$ minutes is given by $q(t) = 10(\frac{4}{5})^t$. Sketch a graph that shows the value $q(t)$ at any time from $t = 0$ to $t = 10$.

29 **Compound interest**   If $1000 is invested at a rate of 12% per year compounded monthly, find the principal after

(a) 1 month                     (b) 2 months

(c) 6 months                    (d) 1 year

30 **Compound interest**   If a savings fund pays interest at a rate of 10% compounded semiannually, how much money invested now will amount to $5000 after 1 year?

31 **Automobile trade-in value**   If a certain make of automobile is purchased for $C$ dollars, its trade-in value $V(t)$ at the end of $t$ years is given by $V(t) = 0.78C(0.85)^{t-1}$. If the original cost is $10,000, calculate, to the nearest dollar, the value after

(a) 1 year      (b) 4 years      (c) 7 years

**32 Real estate appreciation** If the value of real estate increases at a rate of 5% per year, after $t$ years the value $V$ of a house purchased for $P$ dollars is $V = P(1.05)^t$. A graph for the value of a house purchased for $80,000 in 1986 is shown in the figure. Approximate the value of the house, to the nearest $1000, in the year 2000.

**EXERCISE 32**

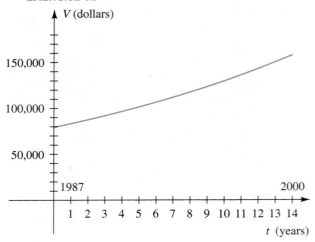

**33 Compound interest** If $1000 is invested at an interest rate of 6% per year compounded quarterly, find the principal at the end of

(a) 1 year     (b) 2 years

(c) 5 years    (d) 10 years

**34 Credit-card interest** A certain department store requires its credit-card customers to pay interest on unpaid bills at the rate of 18% per year compounded monthly. If a customer buys a television set for $500 on credit and makes no payments for one year, how much is owed at the end of the year?

**35 Depreciation** The declining balance method is an accounting method in which the amount of depreciation taken each year is a fixed percentage of the present value of the item. If $y$ is the value of the item in a given year, the depreciation taken is $ay$ for some depreciation rate $a$ with $0 < a < 1$, and the new value is $(1 - a)y$.

(a) If the initial value of the item is $y_0$, show that the value after $n$ years of depreciation is $(1 - a)^n y_0$.

(b) At the end of $T$ years, the item has a salvage value of $s$ dollars. The taxpayer wishes to choose a depreciation rate such that the value of the item after $T$ years will equal the salvage value (see the figure). Show that $a = 1 - \sqrt[T]{s/y_0}$.

**EXERCISE 35**

**36 Language dating** Glottochronology is a method of dating a language at a particular stage, based on the theory that over a long period of time linguistic changes take place at a fairly constant rate. Suppose that a language originally had $N_0$ basic words and that at time $t$, measured in millennia (1 millennium = 1000 years), the number $N(t)$ of basic words that remain in common use is given by $N(t) = N_0(0.805)^t$.

(a) Approximate the percentage of basic words lost every 100 years.

(b) If $N_0 = 200$, sketch the graph of $N$ for $0 \le t \le 5$.

Exer. 37–40: Some lending institutions calculate the monthly payment $M$ on a loan of $L$ dollars at an interest rate $r$ (expressed as a decimal) by using the formula

$$M = \frac{Lrk}{12(k - 1)},$$

where $k = [1 + (r/12)]^{12t}$ and $t$ is the number of years that the loan is in effect.

**37 Home mortgage**

(a) Find the monthly payment on a 30-year $90,000 home mortgage if the interest rate is 12%.

(b) Find the total interest paid on the loan in part (a).

**38 Home mortgage** Find the largest 25-year home mortgage that can be obtained at an interest rate of 10% if the monthly payment is to be $800.

**39 Car loan** An automobile dealer offers customers no-down-payment 3-year loans at an interest rate of 15%. If a customer can afford to pay $220 per month, find the price of the most expensive car that can be purchased.

**40 Business loan** The owner of a small business decides to finance a new computer by borrowing $3000 for 2 years at an interest rate of 12.5%.

(a) Find the monthly payment.

(b) Find the total interest paid on the loan.

**41 Trout population** One thousand trout, each 1 year old, are introduced into a large pond. The number $N(t)$ still alive after $t$ years is predicted to be given by $N(t) = 1000(0.9)^t$. Use the graph of $N$ to approximate when 500 trout will be alive.

**42 Buying power** An economist predicts that the buying power $B(t)$ of a dollar $t$ years from now will be given by $B(t) = (0.95)^t$. Use the graph of $B$ to approximate when the buying power will be half of what it is today.

**Exer. 43–44:** Sketch the graph of the equation. (a) Estimate $y$ if $x = 40$. (b) Estimate $x$ if $y = 2$.

**43** $y = (1.085)^x$

**44** $y = (1.0525)^x$

**Exer. 45–46:** Use a graph to estimate the roots of the equation.

**45** $1.4x^2 - 2.2^x = 1$

**46** $1.21^{3x} + 1.4^{-1.1x} - 2x = 0.5$

**Exer. 47–48:** Graph $f$ on the given interval. (a) Determine whether $f$ is one-to-one. (b) Estimate the zeros of $f$.

**47** $f(x) = \dfrac{3.1^x - 2.5^{-x}}{2.7^x + 4.5^{-x}}$; $\quad [-3, 3]$

**48** $f(x) = \pi^{0.6x} - 1.3^{(x^{1.8})}$; $\quad [-4, 4]$
(*Hint:* Change $x^{1.8}$ to an equivalent form that is defined for $x < 0$.)

**Exer. 49–50:** Graph $f$ on the given interval. (a) Estimate where $f$ is increasing or is decreasing. (b) Estimate the range of $f$.

**49** $f(x) = 0.7x^3 + 1.7^{(-1.8x)}$; $\quad [-4, 1]$

**50** $f(x) = \dfrac{3.1^{-x} - 4.1^x}{4.4^{-x} + 5.3^x}$; $\quad [-3, 3]$

**51 Gompertz function** The **Gompertz function**,

$y = ka^{(b^x)}$ with $k > 0$, $0 < a < 1$, and $0 < b < 1$,

is sometimes used to describe the sales of a new product whose sales are initially large but then level off toward a maximum saturation level. Graph, on the same coordinate plane, the line $y = k$ and the Gompertz function with $k = 4$, $a = \frac{1}{8}$, and $b = \frac{1}{4}$. What is the significance of the constant $k$?

**52 Logistic function** The **logistic function**,

$y = \dfrac{1}{k + ab^x}$ with $k > 0$, $a > 0$, and $0 < b < 1$,

is sometimes used to describe the sales of a new product that experiences slower sales initially, followed by growth toward a maximum saturation level. Graph, on the same coordinate plane, the line $y = 1/k$ and the logistic function with $k = \frac{1}{4}$, $a = \frac{1}{8}$, and $b = \frac{5}{8}$. What is the significance of the value $1/k$?

**Exer. 53–54:** If monthly payments $p$ are deposited in a savings account paying an annual interest rate $r$, then the amount $A$ in the account after $n$ years is given by

$$A = \frac{p\left(1 + \dfrac{r}{12}\right)\left[\left(1 + \dfrac{r}{12}\right)^{12n} - 1\right]}{\dfrac{r}{12}}.$$

Graph $A$ for each value of $p$ and $r$, and estimate $n$ for $A = \$100,000$.

**53** $p = 100$, $r = 0.05$

**54** $p = 250$, $r = 0.09$

## 4.2 THE NATURAL EXPONENTIAL FUNCTION

The *compound interest formula* discussed in the preceding section is

$$A = P\left(1 + \frac{r}{n}\right)^{nt},$$

where $P$ is the principal invested, $r$ is the interest rate (expressed as a decimal), $n$ is the number of interest periods per year, and $t$ is the number of years that the principal is invested. The next example illustrates what happens if the rate and total time invested are fixed, but the *interest period* is varied.

**EXAMPLE I**    *Using the compound interest formula*

Suppose $1000 is invested at a compound interest rate of 9%. Find the new amount of principal after one year if the interest is compounded quarterly, monthly, weekly, daily, hourly, and each minute.

*SOLUTION*    If we let $P = \$1000$, $t = 1$, and $r = 0.09$ in the compound interest formula, then

$$A = \$1000\left(1 + \frac{0.09}{n}\right)^n$$

for $n$ interest periods per year. The values of $n$ we wish to consider are listed in the following table, where we have assumed that there are 365 days in a year and hence $(365)(24) = 8760$ hours and $(8760)(60) = 525{,}600$ minutes. (In actual business transactions an investment year is considered to be 360 days.)

| Interest period | Quarter | Month | Week | Day | Hour | Minute |
|---|---|---|---|---|---|---|
| $n$ | 4 | 12 | 52 | 365 | 8760 | 525,600 |

Using the compound interest formula (and a calculator), we obtain the amounts given in the following table.

| Interest period | Amount after one year |
|---|---|
| Quarter | $\$1000\left(1 + \dfrac{0.09}{4}\right)^4 = \$1093.08$ |
| Month | $\$1000\left(1 + \dfrac{0.09}{12}\right)^{12} = \$1093.81$ |
| Week | $\$1000\left(1 + \dfrac{0.09}{52}\right)^{52} = \$1094.09$ |
| Day | $\$1000\left(1 + \dfrac{0.09}{365}\right)^{365} = \$1094.16$ |
| Hour | $\$1000\left(1 + \dfrac{0.09}{8760}\right)^{8760} = \$1094.17$ |
| Minute | $\$1000\left(1 + \dfrac{0.09}{525{,}600}\right)^{525{,}600} = \$1094.17$ |

Note that, in the preceding example, after we reach an interest period of one hour, the number of interest periods per year has no effect on the final amount. If interest had been compounded each *second,* the result would still be $1094.17, provided we truncate $A$ to the nearest cent. (Some decimal places *beyond* the first two *do* change.) Thus, the amount approaches a fixed value as $n$ increases. Interest is said to be **compounded continuously** if the number $n$ of time periods per year increases without bound.

If we let $P = 1$, $r = 1$, and $t = 1$ in the compound interest formula, we obtain

$$A = \left(1 + \frac{1}{n}\right)^n.$$

The expression on the right-hand side of the equation is important in calculus. In Example 1 we considered a similar situation: as $n$ increased, $A$ approached a limiting value. The same phenomenon occurs for this formula, as illustrated by the following table, which was obtained using a calculator.

| $n$ | Approximation to $\left(1 + \dfrac{1}{n}\right)^n$ |
|---|---|
| 1 | 2.00000000 |
| 10 | 2.59374246 |
| 100 | 2.70481383 |
| 1000 | 2.71692393 |
| 10,000 | 2.71814593 |
| 100,000 | 2.71826824 |
| 1,000,000 | 2.71828047 |
| 10,000,000 | 2.71828169 |
| 100,000,000 | 2.71828181 |
| 1,000,000,000 | 2.71828183 |

In calculus it is shown that as $n$ increases without bound, the value of $[1 + (1/n)]^n$ approaches a certain irrational number, denoted by $e$. The number $e$ arises in the investigation of many physical phenomena. An approximation is $e \approx 2.71828$. Using the notation we developed for rational functions in Section 3.5, we denote this fact as follows.

**The Number** *e*

If $n$ is a positive integer, then

$$\left(1 + \frac{1}{n}\right)^n \to e \approx 2.71828 \quad \text{as} \quad n \to \infty.$$

In the following definition we use $e$ as a base for an important exponential function.

**Definition of the Natural
Exponential Function**

The **natural exponential function** $f$ is defined by

$$f(x) = e^x$$

for every real number $x$.

The natural exponential function is one of the most useful functions in advanced mathematics and applications. Since $2 < e < 3$, the graph of $y = e^x$ lies between the graphs of $y = 2^x$ and $y = 3^x$, as shown in Figure 13. Scientific calculators have an $\boxed{e^x}$ key for approximating values of the natural exponential function.

**FIGURE 13**

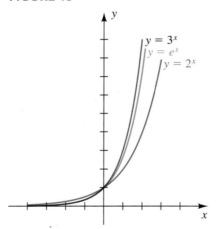

**APPLICATION**

**Continuously Compounded Interest**

The compound interest formula is

$$A = P\left(1 + \frac{r}{n}\right)^{nt}.$$

If we let $k = n/r$, then $n = kr$ and $nt = krt$, and we may rewrite the formula as

$$A = P\left(1 + \frac{1}{k}\right)^{krt} = P\left[\left(1 + \frac{1}{k}\right)^{k}\right]^{rt}.$$

For continuously compounded interest we let $n$ (the number of interest periods per year) increase without bound, denoted by $n \to \infty$ or, equivalently, by $k \to \infty$. Using the definition of $e$, we see that

$$P\left[\left(1 + \frac{1}{k}\right)^k\right]^{rt} \to P[e]^{rt} = Pe^{rt} \quad \text{as} \quad k \to \infty.$$

This result gives us the following formula.

**Continuously Compounded Interest Formula**

$$A = Pe^{rt},$$

where   $P = $ principal   *chints, usu*.
          *so phập phảis .*
        $r = $ interest rate expressed as a decimal
        $t = $ number of years $P$ is invested *nglúch đẫm*
        $A = $ amount after $t$ years.

The next example illustrates the use of this formula.

**EXAMPLE 2**   *Using the continuously compounded interest formula*

Suppose \$20,000 is deposited in a money market account that pays interest at a rate of 8% per year compounded continuously. Determine the balance in the account after 5 years.

**SOLUTION**   Applying the formula for continuously compounded interest with $P = 20{,}000$, $r = 0.08$, and $t = 5$, we have

$$A = Pe^{rt} = 20{,}000e^{0.08(5)} = 20{,}000e^{0.4}.$$

Using a calculator, we find that $A = \$29{,}836.49$.

The continuously compounded interest formula is just one specific case of the following law.

**Law of Growth (or Decay) Formula**

Let $q_0$ be the value of a quantity $q$ at time $t = 0$ (that is, $q_0$ is the initial amount of $q$). If $q$ changes instantaneously at a rate proportional to its current value, then

$$q = q_0 e^{rt},$$

where $r > 0$ is the rate of growth (or $r < 0$ is the rate of decay) of $q$.

### EXAMPLE 3    *The population of a city*

The population of a city in 1970 was 153,800. Assuming that the population increases continuously at a rate of 5% per year, predict the population of the city in the year 2000.

**SOLUTION**    We apply the growth formula $q = q_0 e^{rt}$ with initial population $q_0 = 153{,}800$, rate of growth $r = 0.05$, and time $t = 2000 - 1970 = 30$ years. Thus, a prediction for the population of the city in the year 2000 is

$$153{,}800 e^{(0.05)(30)} = 153{,}800 e^{1.5} \approx 689{,}284.$$

The function $f$ in the next example is important in advanced applications of mathematics.

### EXAMPLE 4    *A graph involving two exponential functions*

Sketch the graph of $f$ if

$$f(x) = \frac{e^x + e^{-x}}{2}.$$

**FIGURE 14**

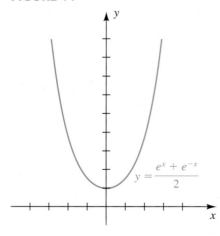

$$y = \frac{e^x + e^{-x}}{2}$$

**SOLUTION**    Note that $f$ is an even function, because

$$f(-x) = \frac{e^{-x} + e^{-(-x)}}{2} = \frac{e^{-x} + e^x}{2} = f(x).$$

Thus, the graph is symmetric with respect to the $y$-axis. Using a calculator, we obtain the following approximations of $f(x)$.

| $x$ | 0 | 0.5 | 1.0 | 1.5 | 2.0 |
|---|---|---|---|---|---|
| $f(x)$ (approx.) | 1 | 1.13 | 1.54 | 2.35 | 3.76 |

Plotting points and using symmetry with respect to the $y$-axis gives us the sketch in Figure 14. The graph *appears* to be a parabola; however, this is not actually the case.

### APPLICATION

#### Flexible Cables

The function $f$ of Example 4 occurs in applied mathematics and engineering, where it is called the **hyperbolic cosine function.** This function can

FIGURE 15

be used to describe the shape of a uniform flexible cable or chain whose ends are supported from the same height, such as a telephone or power line (see Figure 15). If we introduce a coordinate system, as indicated in the figure, then it can be shown that an equation that corresponds to the shape of the cable is

$$y = \frac{a}{2}(e^{x/a} + e^{-x/a}),$$

where $a$ is a real number. The graph is called a **catenary,** after the Latin word for *chain*. The function in Example 4 is the special case in which $a = 1$.

### APPLICATION

#### Radiotherapy

Exponential functions play an important role in the field of *radiotherapy,* the treatment of tumors by radiation. The fraction of cells in a tumor that survive a treatment, called the *surviving fraction,* depends not only on the energy and nature of the radiation, but also on the depth, size, and characteristics of the tumor itself. The exposure to radiation may be thought of as a number of potentially damaging events, where at least one *hit* is required to kill a tumor cell. For instance, suppose that each cell has exactly one *target* that must be hit. If $k$ denotes the average target size of a tumor cell and if $x$ is the number of damaging events (the *dose*), then the surviving fraction $f(x)$ is given by

$$f(x) = e^{-kx}.$$

This is called the *one target–one hit surviving fraction.*

Suppose next that each cell has $n$ targets and that each target must be hit once for the cell to die. In this case, the *n target–one hit surviving fraction* is given by

$$f(x) = 1 - (1 - e^{-kx})^n.$$

FIGURE 16 Surviving fraction of tumor cells after a radiation treatment

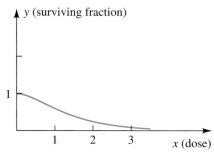

The graph of $f$ may be analyzed to determine what effect increasing the dosage $x$ will have on decreasing the surviving fraction of tumor cells. Note that $f(0) = 1$; that is, if there is no dose, then all cells survive. As an example, if $k = 1$ and $n = 2$, then

$$f(x) = 1 - (1 - e^{-x})^2$$
$$= 1 - (1 - 2e^{-x} + e^{-2x})$$
$$= 2e^{-x} - e^{-2x}.$$

A complete analysis of the graph of $f$ requires calculus. The graph is sketched in Figure 16. The *shoulder* on the curve near the point $(0, 1)$ represents the threshold nature of the treatment—that is, a small dose results in very little tumor cell elimination. Note that for a large $x$, an increase in dosage has little effect on the surviving fraction. To determine the ideal dose to administer to a patient, specialists in radiation therapy must also take into account the number of healthy cells that are killed during a treatment.

Problems of the type illustrated in the next example occur in the study of calculus.

**EXAMPLE 5     *Finding zeros of a function involving exponentials***
If $f(x) = x^2(-2e^{-2x}) + 2xe^{-2x}$, find the zeros of $f$.

*SOLUTION*     We may factor $f(x)$ as follows:

$$f(x) = 2xe^{-2x} - 2x^2e^{-2x}$$
$$= 2xe^{-2x}(1 - x)$$

To find the zeros of $f$, we solve the equation $f(x) = 0$. Since $e^{-2x} > 0$ for every $x$, we see that $f(x) = 0$ if and only if $x = 0$ or $1 - x = 0$. Thus, the zeros of $f$ are 0 and 1.

**EXAMPLE 6     *Sketching a Gompertz growth curve***
In biology, the **Gompertz growth function** $G$ given by

$$G(t) = ke^{(-Ae^{-Bt})},$$

where $k$, $A$, and $B$ are positive constants, is used to estimate the size of certain quantities at time $t$. The graph of $G$ is called a **Gompertz growth curve.** The function is always positive and increasing, and as $t$ increases without bound, $G(t)$ levels off and approaches the value $k$. Graph $G$ on the interval $[0, 5]$ for $k = 1.1$, $A = 3.2$, and $B = 1.1$, and estimate the time $t$ at which $G(t) = 1$.

*SOLUTION*     Some specific keystrokes for the TI-81/82 graphics calculator are given in Appendix I. We begin by assigning

$$1.1e^{(-3.2e^{-1.1t})}$$

**FIGURE 17**   $[0, 5]$ by $[0, 2]$

to $Y_1$. Since we wish to graph $G$ on the interval $[0, 5]$, we choose Xmin = 0 and Xmax = 5. Because $G(t)$ is always positive and does not exceed the value $k = 1.1$, we choose Ymin = 0 and Ymax = 2. Hence, the viewing rectangle dimensions are $[0, 5]$ by $[0, 2]$. Graphing $G$ gives us a display similar to Figure 17. The end point values of the graph are approximately $(0, 0.045)$ and $(5, 1.086)$.

To determine the time when $y = G(t) = 1$, we use the tracing and zoom-in features and obtain $x = t \approx 3.194$.

## 4.2 EXERCISES

**Exer. 1–4:** Use the graph of $y = e^x$ to help sketch the graph of $f$.

1 (a) $f(x) = e^{-x}$      (b) $f(x) = -e^x$

2 (a) $f(x) = e^{2x}$      (b) $f(x) = 2e^x$

3 (a) $f(x) = e^{x+4}$      (b) $f(x) = e^x + 4$

4 (a) $f(x) = e^{-2x}$      (b) $f(x) = -2e^x$

**Exer. 5–6:** If $P$ dollars is deposited in a savings account that pays interest at a rate of $r\%$ per year compounded continuously, find the balance after $t$ years.

5 $P = 1000$,    $r = 8\frac{1}{4}$,    $t = 5$

6 $P = 100$,    $r = 12\frac{1}{2}$,    $t = 10$

**Exer. 7–8:** How much money, invested at an interest rate of $r\%$ per year compounded continuously, will amount to $A$ dollars after $t$ years?

7 $A = 100,000$,    $r = 11$,    $t = 18$

8 $A = 15,000$,    $r = 9.5$,    $t = 4$

**Exer. 9–10:** An investment of $P$ dollars increased to $A$ dollars in $t$ years. If interest was compounded continuously, find the interest rate.

9 $A = 13,464$,    $P = 1000$,    $t = 20$

10 $A = 890.20$,    $P = 400$,    $t = 16$

**Exer. 11–12:** Solve the equation.

11 $e^{(x^2)} = e^{7x-12}$      12 $e^{3x} = e^{2x-1}$

**Exer. 13–16:** Find the zeros of $f$.

13 $f(x) = xe^x + e^x$

14 $f(x) = -x^2e^{-x} + 2xe^{-x}$

15 $f(x) = x^3(4e^{4x}) + 3x^2e^{4x}$

16 $f(x) = x^2(2e^{2x}) + 2xe^{2x} + e^{2x} + 2xe^{2x}$

**Exer. 17–18:** Simplify the expression.

17 $\dfrac{(e^x + e^{-x})(e^x + e^{-x}) - (e^x - e^{-x})(e^x - e^{-x})}{(e^x + e^{-x})^2}$

18 $\dfrac{(e^x - e^{-x})^2 - (e^x + e^{-x})^2}{(e^x + e^{-x})^2}$

19 **Crop growth** An exponential function $W$ such that $W(t) = W_0 e^{kt}$ for $k > 0$ describes the first month of growth for crops such as maize, cotton, and soybeans. The function value $W(t)$ is the total weight in milligrams, $W_0$ is the weight on the day of emergence, and $t$ is the time in days. If, for a species of soybean, $k = 0.2$ and $W_0 = 68$ mg, predict the weight at the end of 30 days.

20 **Crop growth** Refer to Exercise 19. It is often difficult to measure the weight $W_0$ of a plant when it first emerges from the soil. If, for a species of cotton, $k = 0.21$ and the weight after 10 days is 575 milligrams, estimate $W_0$.

21 **U.S. population growth** The 1980 population of the United States was approximately 227 million, and the population has been growing continuously at a rate of 0.7% per year. Predict the population $N(t)$ in the year 2000 if this growth trend continues.

22 **Population growth in India** The 1985 population estimate for India was 762 million, and the population has been growing continuously at a rate of about 2.2% per year. Assuming that this rapid growth rate continues, estimate the population $N(t)$ of India in the year 2000.

23 **Longevity of halibut** In fishery science, a cohort is the collection of fish that results from one annual reproduction. It is usually assumed that the number of fish $N(t)$ still alive after $t$ years is given by an exponential function. For Pacific halibut, $N(t) = N_0 e^{-0.2t}$, where $N_0$ is the initial size of the cohort. Approximate the percentage of the original number still alive after 10 years.

24 **Radioactive tracer** The radioactive tracer $^{51}$Cr can be used to locate the position of the placenta in a pregnant woman. Often the tracer must be ordered from a medical laboratory. If $A_0$ units (microcuries) are shipped, then because of radioactive decay, the number of units $A(t)$ present after $t$ days is given by $A(t) = A_0 e^{-0.0249t}$.

(a) If 35 units are shipped and it takes 2 days for the tracer to arrive, approximately how many units will be available for the test?

(b) If 35 units are needed for the test, approximately how many units should be shipped?

**25** *Blue whale population growth* In 1978, the population of blue whales in the southern hemisphere was thought to number 5000. Since whaling has been outlawed and an abundant food supply is available, the population $N(t)$ is expected to grow exponentially according to the formula $N(t) = 5000e^{0.0036t}$, where $t$ is in years and $t = 0$ corresponds to 1978. Predict the population in the year 2000.

**26** *Halibut growth* The length (in centimeters) of many common commercial fish $t$ years old can be approximated by a von Bertalanffy growth function of the form $f(t) = a(1 - be^{-kt})$, where $a$, $b$, and $k$ are constants.

   **(a)** For Pacific halibut, $a = 200$, $b = 0.956$, and $k = 0.18$. Estimate the length of a 10-year-old halibut.

   **(b)** Use the graph of $f$ to estimate the maximum attainable length of the Pacific halibut.

**27** *Atmospheric pressure* Under certain conditions the atmospheric pressure $p$ (in inches) at altitude $h$ feet is given by $p = 29e^{-0.000034h}$. What is the pressure at an altitude of 40,000 feet?

**28** *Polonium isotope decay* If we start with $c$ milligrams of the polonium isotope $^{210}$Po, the amount remaining after $t$ days may be approximated by $A = ce^{-0.00495t}$. If the initial amount is 50 milligrams, approximate, to the nearest hundredth, the amount remaining after

   **(a)** 30 days     **(b)** 180 days     **(c)** 365 days

**29** *Growth of children* The Jenss model is generally regarded as the most accurate formula for predicting the height of preschool children. If $y$ is height (in centimeters) and $x$ is age (in years), then

$$y = 79.041 + 6.39x - e^{3.261 - 0.993x}$$

for $\frac{1}{4} \le x \le 6$. From calculus, the rate of growth $R$ (in cm/year) is given by $R = 6.39 + 0.993e^{3.261 - 0.993x}$. Find the height and rate of growth of a typical 1-year-old child.

**30** *Particle velocity* A very small spherical particle (on the order of 5 microns in diameter) is projected into still air with an initial velocity of $v_0$ m/sec, but its velocity decreases because of drag forces. Its velocity $t$ seconds later is given by $v(t) = v_0e^{-at}$ for some $a > 0$, and the distance $s(t)$ the particle travels is given by

$$s(t) = \frac{v_0}{a}(1 - e^{-at}).$$

The stopping distance is the distance traveled by the particle before it comes to rest.

   **(a)** Express the stopping distance in terms of $v_0$ and $a$.

   **(b)** Use the formula in part (a) to estimate the stopping distance if $v_0 = 10$ m/sec and $a = 8 \times 10^5$.

**31** *Minimum wage* In 1971 the minimum wage in the United States was \$1.60 per hour. Assuming that the rate of inflation increases continuously at 5% per year, find the equivalent minimum wage in the year 2000.

**32** *Land value* In 1867 the United States purchased Alaska from Russia for \$7,200,000. There is 586,400 square miles of land in Alaska. Assuming that the value of the land increases continuously at 3% per year and that land can be purchased at an equivalent price, determine the price of 1 acre in the year 2000. (One square mile is equivalent to 640 acres.)

Exer. 33–34: The *effective yield* (or effective annual interest rate) for an investment is the simple interest rate that would yield at the end of one year the same amount as is yielded by the compounded rate that is actually applied. Approximate, to the nearest 0.01%, the effective yield corresponding to an interest rate of $r$% per year compounded **(a)** quarterly and **(b)** continuously.

**33** $r = 7$                      **34** $r = 12$

Exer. 35–36: Sketch the graph of the equation. **(a)** Estimate $y$ if $x = 40$. **(b)** Estimate $x$ if $y = 2$.

**35** $y = e^{0.085x}$               **36** $y = e^{0.0525x}$

Exer. 37–39: **(a)** Graph $f$ using a graphing utility. **(b)** Sketch the graph of $g$ by taking the reciprocals of $y$-coordinates in (a), *without* using a graphing utility.

**37** $f(x) = \dfrac{e^x - e^{-x}}{2}$;   $g(x) = \dfrac{2}{e^x - e^{-x}}$

**38** $f(x) = \dfrac{e^x + e^{-x}}{2}$;   $g(x) = \dfrac{2}{e^x + e^{-x}}$

**39** $f(x) = \dfrac{e^x - e^{-x}}{e^x + e^{-x}}$;   $g(x) = \dfrac{e^x + e^{-x}}{e^x - e^{-x}}$

**40** *Probability density function* In statistics, the probability density function for the normal distribution is defined by

$$f(x) = \frac{1}{\sigma\sqrt{2\pi}}e^{-z^2/2} \quad \text{with} \quad z = \frac{x - \mu}{\sigma},$$

where $\mu$ and $\sigma$ are real numbers ($\mu$ is the *mean* and $\sigma^2$ is the *variance* of the distribution). Sketch the graph of $f$ for the case $\sigma = 1$ and $\mu = 0$.

**Exer. 41–42:** Graph $f$ and $g$ on the same coordinate plane, and estimate the solutions of the equation $f(x) = g(x)$.

**41** $f(x) = e^{0.5x} - e^{-0.4x}$;   $g(x) = x^2 - 2$

**42** $f(x) = 0.3e^x$;             $g(x) = x^3 - x$

**Exer. 43–44:** The functions $f$ and $g$ can be used to approximate $e^x$ on the interval $[0, 1]$. Graph $f$, $g$, and $y = e^x$ on the same coordinate plane, and compare the accuracy of $f(x)$ and $g(x)$ as an approximation to $e^x$.

**43** $f(x) = x + 1$;           $g(x) = 1.72x + 1$

**44** $f(x) = \frac{1}{2}x^2 + x + 1$;   $g(x) = 0.84x^2 + 0.878x + 1$

**Exer. 45–46:** Graph $f$ and estimate its zeros.

**45** $f(x) = x^2 e^x - xe^{(x^2)} + 0.1$

**46** $f(x) = x^3 e^x - x^2 e^{2x} + 1$

**Exer. 47–48:** Graph $f$ on the interval $(0, 200]$. Find an approximate equation for the horizontal asymptote.

**47** $f(x) = \left(1 + \dfrac{1}{x}\right)^x$     **48** $f(x) = \left(1 + \dfrac{2}{x}\right)^x$

**Exer. 49–50:** Approximate the real root of the equation.

**49** $e^{-x} = x$           **50** $e^{3x} = 5 - 2x$

**Exer. 51–52:** Graph $f$ and determine where $f$ is increasing or is decreasing.

**51** $f(x) = xe^x$           **52** $f(x) = x^2 e^{-2x}$

**53** *Pollution from a smokestack*   The concentration $C$ (in units/m³) of pollution near a ground-level point that is downwind from a smokestack source of height $h$ is

sometimes given by

$$C = \frac{Q}{\pi v a b} e^{-y^2/(2a^2)} [e^{-(z-h)^2/(2b^2)} + e^{-(z+h)^2/(2b^2)}],$$

where $Q$ is the source strength (in units/sec), $v$ is the average wind velocity (in m/sec), $z$ is the height (in meters) above the downwind point, $y$ is the distance from the downwind point in the direction that is perpendicular to the wind (the cross-wind direction), and $a$ and $b$ are constants that depend on the downwind distance (see the figure).

**(a)** How does the concentration of pollution change at the ground-level, downwind position ($y = 0$ and $z = 0$) if the height of the smokestack is increased?

**(b)** How does the concentration of pollution change at ground level ($z = 0$) for a smokestack of fixed height $h$ if a person moves in the cross-wind direction by increasing $y$?

**EXERCISE 53**

**54** *Pollution concentration*   Refer to Exercise 53. If the smokestack height is 100 meters and $b = 12$, use a graph to estimate the height $z$ above the downwind point ($y = 0$) where the maximum pollution concentration occurs. (*Hint:* Let $h = 100$, $b = 12$, and graph the equation $C = e^{-(z-h)^2/(2b^2)} + e^{-(z+h)^2/(2b^2)}$.)

# 4.3   LOGARITHMIC FUNCTIONS

In Section 4.1 we observed that the exponential function given by $f(x) = a^x$ for $0 < a < 1$ or $a > 1$ is one-to-one. Hence, $f$ has an inverse function $f^{-1}$ (see Section 2.5). This inverse of the exponential function with base $a$ is called the **logarithmic function with base $a$** and is denoted by $\log_a$. Its values are written $\log_a (x)$ or $\log_a x$, read "the logarithm of $x$ with base $a$." Since, by the definition of an inverse function $f^{-1}$,

$$y = f^{-1}(x) \quad \text{if and only if} \quad x = f(y),$$

the definition of $\log_a$ may be expressed as follows.

**Definition of $\log_a$**

Let $a$ be a positive real number different from 1. The **logarithm of $x$ with base $a$** is defined by

$$y = \log_a x \quad \text{if and only if} \quad x = a^y$$

for every $x > 0$ and every real number $y$.

Note that the two equations in the definition are equivalent. We call the first equation the **logarithmic form** and the second the **exponential form.** You should strive to become an expert in changing each form into the other. The following diagram may help you achieve this goal.

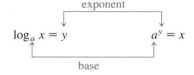

|  Logarithmic form  |  Exponential form  |
|:---:|:---:|

exponent

$$\log_a x = y \qquad\qquad\qquad a^y = x$$

base

Observe that when forms are changed, *the bases of the logarithmic and exponential forms are the same.* The number $y$ (that is, $\log_a x$) corresponds to the exponent in the exponential form. In words, $\log_a x$ *is the exponent to which the base $a$ must be raised to obtain $x$.*

The following illustration contains examples of equivalent forms.

**ILLUSTRATION**

**Equivalent Forms**

| Logarithmic form | Exponential form |
|---|---|
| $\log_5 u = 2$ | $5^2 = u$ |
| $\log_b 8 = 3$ | $b^3 = 8$ |
| $r = \log_p q$ | $p^r = q$ |
| $w = \log_4 (2t + 3)$ | $4^w = 2t + 3$ |
| $\log_3 x = 5 + 2z$ | $3^{5+2z} = x$ |

The next example contains an application that involves changing from an exponential form to a logarithmic form.

**EXAMPLE 1    *Changing exponential form to logarithmic form***

The number $N$ of bacteria in a certain culture after $t$ hours is given by $N = (1000)2^t$. Express $t$ as a logarithmic function of $N$ with base 2.

**SOLUTION**    If $N = (1000)2^t$, then

$$2^t = \frac{N}{1000}.$$

Changing to logarithmic form, we obtain

$$t = \log_2 \frac{N}{1000}.$$

Some special cases of logarithms are given in the next example.

**EXAMPLE 2    *Finding logarithms***

Find the number.

**(a)** $\log_{10} 100$      **(b)** $\log_2 \frac{1}{32}$      **(c)** $\log_9 3$      **(d)** $\log_7 1$

**SOLUTION**    In each case we are given $\log_a x$ and must find the exponent $y$ such that $a^y = x$. We obtain the following:

**(a)** $\log_{10} 100 = 2$    because    $10^2 = 100$.
**(b)** $\log_2 \frac{1}{32} = -5$    because    $2^{-5} = \frac{1}{32}$.
**(c)** $\log_9 3 = \frac{1}{2}$    because    $9^{1/2} = 3$.
**(d)** $\log_7 1 = 0$    because    $7^0 = 1$.

The following general properties follow from the interpretation of $\log_a x$ as an exponent.

| Property of $\log_a x$ | Reason | Illustration |
|---|---|---|
| (1) $\log_a 1 = 0$ | $a^0 = 1$ | $\log_3 1 = 0$ |
| (2) $\log_a a = 1$ | $a^1 = a$ | $\log_{10} 10 = 1$ |
| (3) $\log_a a^x = x$ | $a^x = a^x$ | $\log_2 8 = \log_2 2^3 = 3$ |
| (4) $a^{\log_a x} = x$ | as follows | $5^{\log_5 7} = 7$ |

**FIGURE 18**

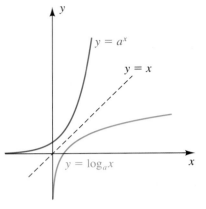

The reason for property 4 follows directly from the definition of $\log_a$, since

$$\text{if} \quad y = \log_a x, \quad \text{then} \quad x = a^y, \quad \text{or} \quad x = a^{\log_a x}.$$

The logarithmic function with base $a$ is the inverse of the exponential function with base $a$, so the graph of $y = \log_a x$ can be obtained by reflecting the graph of $y = a^x$ through the line $y = x$ (see Section 2.5). This procedure is illustrated in Figure 18 for the case $a > 1$. Note that the $x$-intercept of the graph is 1, the domain is the set of positive real numbers, the range is $\mathbb{R}$, and the $y$-axis is a vertical asymptote. Logarithms with base $a < 1$ are seldom used, and hence we will not emphasize their graphs.

We see from Figure 18 that if $a > 1$, then $\log_a x$ is increasing on $(0, \infty)$ and hence is one-to-one by the theorem on page 169. Combining this result with parts (1) and (2) of the definition of one-to-one function on page 168 gives us the following theorem, which can also be proved if $0 < a < 1$.

**Theorem: Logarithmic Functions Are One-to-One**

> The logarithmic function with base $a$ is one-to-one. Thus, the following equivalent conditions are satisfied for real numbers $x_1$ and $x_2$:
>
> **(1)** If $x_1 \neq x_2$, then $\log_a x_1 \neq \log_a x_2$.
> **(2)** If $\log_a x_1 = \log_a x_2$, then $x_1 = x_2$.

When using this theorem as a reason for a step in the solution to an example, we will state that *logarithmic functions are one-to-one.*

In the following example we solve a simple *logarithmic equation*—that is, an equation involving a logarithm of an expression that contains a variable. Extraneous solutions may be introduced when logarithmic equations are solved. Hence, we must check solutions of logarithmic equations to make sure that we are taking logarithms of *only positive real numbers;* otherwise, a logarithmic function is not defined.

**EXAMPLE 3     *Solving a logarithmic equation***

Solve the equation $\log_3 (4x - 5) = \log_3 (2x + 1)$.

*SOLUTION*

$$\begin{aligned}
\log_3 (4x - 5) &= \log_3 (2x + 1) && \text{given} \\
4x - 5 &= 2x + 1 && \text{logarithmic functions are one-to-one} \\
x &= 3 && \text{solve for } x
\end{aligned}$$

✓ *Check x = 3*     LS: $\log_3 (4 \cdot 3 - 5) = \log_3 7$

RS: $\log_3 (2 \cdot 3 + 1) = \log_3 7$

Since $\log_3 7 = \log_3 7$ is a true statement, $x = 3$ is a solution.

In the next example we use the definition of logarithm to solve a logarithmic equation.

**EXAMPLE 4**    *Solving a logarithmic equation*

Solve the equation $\log_4 (5 + x) = 3$.

**SOLUTION**

$$\log_4 (5 + x) = 3 \qquad \text{given}$$
$$5 + x = 4^3 \qquad \text{definition of logarithm}$$
$$x = 59 \qquad \text{solve for } x$$

✓ **Check** $x = 59$      LS: $\log_4 (5 + 59) = \log_4 64 = \log_4 4^3 = 3$

RS: 3

Since $3 = 3$ is a true statement, $x = 59$ is a solution.

We next sketch the graph of a specific logarithmic function.

**EXAMPLE 5**    *Sketching the graph of a logarithmic function*

Sketch the graph of $f$ if $f(x) = \log_3 x$.

**SOLUTION**    We will describe three methods for sketching the graph.

**Method 1**    Since the functions given by $\log_3 x$ and $3^x$ are inverses of each other, we proceed as we did for $y = \log_a x$ in Figure 18; that is, we first sketch the graph of $y = 3^x$ and then reflect it through the line $y = x$. This gives us the sketch in Figure 19. Note that points $(-1, 3^{-1})$, $(0, 1)$, and $(1, 3)$ on the graph of $y = 3^x$ reflect into the points $(3^{-1}, -1)$, $(1, 0)$, and $(3, 1)$ on the graph of $y = \log_3 x$.

**FIGURE 19**

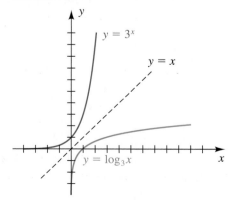

*(continued)*

**Method 2**  We can find points on the graph of $y = \log_3 x$ by letting $x = 3^k$, where $k$ is a real number, and then applying property (3) of logarithms on page 265, as follows:

$$y = \log_3 x = \log_3 3^k = k$$

Using this formula, we obtain the points on the graph listed in the following table.

| $x = 3^k$ | $3^{-3}$ | $3^{-2}$ | $3^{-1}$ | $3^0$ | $3^1$ | $3^2$ | $3^3$ |
|---|---|---|---|---|---|---|---|
| $y = \log_3 x = k$ | $-3$ | $-2$ | $-1$ | $0$ | $1$ | $2$ | $3$ |

This gives us the same points obtained using the first method.

**Method 3**  We can sketch the graph of $y = \log_3 x$ by sketching the graph of the exponential form $x = 3^y$.

As in the following examples, we often wish to sketch the graph of $f(x) = \log_a u$, where $u$ is some expression involving $x$.

**EXAMPLE 6**   *Sketching the graph of a logarithmic function*
Sketch the graph of $f$ if $f(x) = \log_3 |x|$ for $x \neq 0$.

**SOLUTION**   The graph is symmetric with respect to the $y$-axis, since

$$f(-x) = \log_3 |-x| = \log_3 |x| = f(x).$$

If $x > 0$, then $|x| = x$ and the graph coincides with the graph of $y = \log_3 x$ sketched in Figure 19. Using symmetry, we reflect that part of the graph through the $y$-axis, obtaining the sketch in Figure 20.

Alternatively, we may think of this function as $g(x) = \log_3 x$ with $|x|$ substituted for $x$ (refer to the discussion on page 141). Since all points on the graph of $g$ have positive $x$-coordinates, we can obtain the graph of $f$ by combining $g$ with the reflection of $g$ through the $y$-axis.

**FIGURE 20**

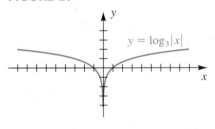

**EXAMPLE 7**   *Reflecting the graph of a logarithmic function*
Sketch the graph of $f$ if $f(x) = \log_3 (-x)$.

**SOLUTION**   The domain of $f$ is the set of negative real numbers, since $\log_3 (-x)$ exists only if $-x > 0$ or, equivalently, $x < 0$. We can obtain the graph of $f$ from the graph of $y = \log_3 x$ by replacing each point $(x, y)$ in Figure 19 by $(-x, y)$. This is equivalent to reflecting the graph of $y = \log_3 x$ through the $y$-axis. The graph is sketched in Figure 21.

Another method is to change $y = \log_3 (-x)$ to the exponential form $3^y = -x$ and then sketch the graph of $x = -3^y$.

**FIGURE 21**

**FIGURE 22**

**FIGURE 23**

**FIGURE 24**

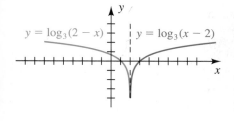

**EXAMPLE 8** *Shifting graphs of logarithmic equations*

Sketch the graph of the equation:

**(a)** $y = \log_3 (x - 2)$ **(b)** $y = \log_3 x - 2$

**SOLUTION**

**(a)** The graph of $y = \log_3 x$ was sketched in Figure 19 and is resketched in Figure 22. From the discussion of horizontal shifts in Section 2.2, we can obtain the graph of $y = \log_3 (x - 2)$ by shifting the graph of $y = \log_3 x$ two units to the right, as shown in Figure 22.

**(b)** From the discussion of vertical shifts in Section 2.2, the graph of $y = \log_3 x - 2$ can be obtained by shifting the graph of $y = \log_3 x$ two units downward, as shown in Figure 23. Note that the x-intercept is given by $\log_3 x = 2$, or $x = 3^2 = 9$.

**EXAMPLE 9** *Reflecting the graph of a logarithmic function*

Sketch the graph of $f$ if $f(x) = \log_3 (2 - x)$.

**SOLUTION** If we write

$$f(x) = \log_3 (2 - x) = \log_3 [-(x - 2)],$$

then, by applying the same technique used to obtain the graph of the equation $y = \log_3 (-x)$ in Example 7 (with $x$ replaced by $x - 2$), we see that the graph of $f$ is the reflection of the graph of $y = \log_3 (x - 2)$ through the vertical line $x = 2$. This gives us the sketch in Figure 24.

Another method is to change $y = \log_3 (2 - x)$ to the exponential form $3^y = 2 - x$ and then sketch the graph of $x = 2 - 3^y$.

Before electronic calculators were invented, logarithms with base 10 were used for complicated numerical computations involving products, quotients, and powers of real numbers. Base 10 was employed because it is well suited for numbers that are expressed in scientific form. Logarithms with base 10 are called **common logarithms.** The symbol **log $x$** is used as an abbreviation for $\log_{10} x$.

**Definition of Common Logarithm**

$$\log x = \log_{10} x \quad \text{for every} \quad x > 0$$

Since inexpensive calculators are now available, there is no need for common logarithms as a tool for computational work. Base 10 does occur in applications, however, and hence many calculators have a $\boxed{\text{LOG}}$ key, which can be used to approximate common logarithms.

The natural exponential function is given by $f(x) = e^x$ (see Section 4.2). The logarithmic function with base $e$ is called the **natural logarithmic function.** The symbol **ln $x$** (read ''ell-en of $x$'') is an abbreviation for $\log_e x$, and we refer to it as the **natural logarithm of $x$.** Thus, *the natural logarithmic function and the natural exponential function are inverse functions of each other.*

**Definition of**
**Natural Logarithm**

$$\ln x = \log_e x \quad \text{for every} \quad x > 0$$

Many calculators have a key labeled $\boxed{\text{LN}}$, which can be used to approximate natural logarithms. The next illustration gives several examples of equivalent forms involving common and natural logarithms.

**ILLUSTRATION**

**Equivalent Forms**

| Logarithmic form | Exponential form |
|---|---|
| ▬ $\log x = 2$ | $10^2 = x$ |
| ▬ $\log z = y + 3$ | $10^{y+3} = z$ |
| ▬ $\ln x = 2$ | $e^2 = x$ |
| ▬ $\ln z = y + 3$ | $e^{y+3} = z$ |

To find $x$ when given $\log x$ or $\ln x$, we may use the $\boxed{10^x}$ key or the $\boxed{e^x}$ key, respectively, on a calculator, as in the next example. If your calculator has an $\boxed{\text{INV}}$ key (for inverse), you may enter $x$ and successively press $\boxed{\text{INV}}$ $\boxed{\text{LOG}}$ or $\boxed{\text{INV}}$ $\boxed{\text{LN}}$.

**EXAMPLE 10**    *Solving a logarithmic equation*

Find $x$ if

**(a)** $\log x = 1.7959$      **(b)** $\ln x = 4.7$

*SOLUTION*

**(a)** Changing $\log x = 1.7959$ to its equivalent exponential form gives us

$$x = 10^{1.7959}.$$

Evaluating the last expression to three-decimal-place accuracy yields

$$x \approx 62.503.$$

**(b)** Changing $\ln x = 4.7$ to its equivalent exponential form gives us

$$x = e^{4.7} \approx 109.95.$$

The following chart lists common and natural logarithmic forms for some of the properties discussed earlier.

| Logarithms with base $a$ | Common logarithms | Natural logarithms |
|---|---|---|
| $\log_a 1 = 0$ | $\log 1 = 0$ | $\ln 1 = 0$ |
| $\log_a a = 1$ | $\log 10 = 1$ | $\ln e = 1$ |
| $\log_a a^x = x$ | $\log 10^x = x$ | $\ln e^x = x$ |
| $a^{\log_a x} = x$ | $10^{\log x} = x$ | $e^{\ln x} = x$ |

The next four examples illustrate applications of common and natural logarithms.

**EXAMPLE 11** *The Richter scale*

On the Richter scale, the magnitude $R$ of an earthquake of intensity $I$ is given by

$$R = \log \frac{I}{I_0},$$

where $I_0$ is a certain minimum intensity.

**(a)** If the intensity of an earthquake is $1000 I_0$, find $R$.

**(b)** Express $I$ in terms of $R$ and $I_0$.

**SOLUTION**

**(a)**

$$R = \log \frac{I}{I_0} \qquad \text{given}$$

$$= \log \frac{1000 I_0}{I_0} \qquad \text{let } I = 1000 I_0$$

$$= \log 1000 \qquad \text{cancel } I_0$$

$$= \log 10^3 \qquad 1000 = 10^3$$

$$= 3 \qquad \log 10^x = x \text{ for every } x$$

**(b)**

$$R = \log \frac{I}{I_0} \qquad \text{given}$$

$$\frac{I}{I_0} = 10^R \qquad \text{definition of } \log_{10}$$

$$I = I_0 \cdot 10^R \qquad \text{multiply by } I_0$$

### EXAMPLE 12    *Newton's law of cooling*

Newton's law of cooling states that the rate at which an object cools is directly proportional to the difference in temperature between the object and its surrounding medium. Newton's law can be used to show that under certain conditions the temperature $T$ (in °C) of an object at time $t$ (in hours) is given by $T = 75e^{-2t}$. Express $t$ as a function of $T$.

SOLUTION

$$T = 75e^{-2t} \qquad \text{given}$$

$$e^{-2t} = \frac{T}{75} \qquad \text{isolate the exponential expression}$$

$$-2t = \ln \frac{T}{75} \qquad \text{change to logarithmic form}$$

$$t = -\frac{1}{2} \ln \frac{T}{75} \qquad \text{divide by } -2$$

### EXAMPLE 13    *Approximating a doubling time*

Assume that a population is growing continuously at a rate of 4% per year. Approximate the amount of time it takes for the population to double its size—that is, its **doubling time.**

SOLUTION    Note that an initial population size is not given. This fact does not present a problem, since we wish only to determine the time needed to obtain a population size *relative* to the initial population size. Using the growth formula $q = q_0 e^{rt}$ with $r = 0.04$ gives us

$$2q_0 = q_0 e^{0.04t} \qquad \text{let } q = 2q_0$$

$$2 = e^{0.04t} \qquad \text{divide by } q_0 \ (q_0 \neq 0)$$

$$0.04t = \ln 2 \qquad \text{change to logarithmic form}$$

$$t = 25 \ln 2 \approx 17.3 \text{ yr.} \qquad \text{multiply by } \frac{1}{0.04} = 25$$

The fact that $q_0$ did not have any effect on the answer indicates that the doubling time for a population of 1000 is the same as the doubling time for a population of 1,000,000, or any other reasonable initial population.

From the last example we may obtain a general formula for the doubling time of a population—namely,

$$rt = \ln 2 \quad \text{or, equivalently,} \quad t = \frac{\ln 2}{r}.$$

Since $\ln 2 \approx 0.69$, we see that the doubling time $t$ for a growth of this type is approximately $0.69/r$. Because the number 72 is close to 69, but has more divisors, some resources refer to this doubling relationship as the **rule of 72.** As an illustration of the rule of 72, if the growth rate of a population is 8%, then it takes about 72/8 = 9 years for the population to double. More precisely, this value is

$$\frac{\ln 2}{8} \cdot 100 \approx 8.7 \text{ yr.}$$

**EXAMPLE 14**    *The half-life of a radioactive substance*

A physicist finds that an unknown radioactive substance registers 2000 counts per minute on a Geiger counter. Ten days later the substance registers 1500 counts per minute. Using calculus, it can be shown that after $t$ days the amount of radioactive material, and hence the number of counts per minute $N(t)$, is directly proportional to $e^{ct}$ for some constant $c$. Determine the half-life of the substance.

**SOLUTION**    Since $N(t)$ is directly proportional to $e^{ct}$,

$$N(t) = ke^{ct},$$

where $k$ is a constant. Letting $t = 0$ and using $N(0) = 2000$, we obtain

$$2000 = ke^{c0} = k \cdot 1 = k.$$

Hence, the formula for $N(t)$ may be written

$$N(t) = 2000e^{ct}.$$

Since $N(10) = 1500$, we may determine $c$ as follows:

$$1500 = 2000e^{c \cdot 10} \qquad \text{let } t = 10 \text{ in } N(t)$$
$$\tfrac{3}{4} = e^{10c} \qquad \text{divide by 2000 and simplify}$$
$$10c = \ln \tfrac{3}{4} \qquad \text{change to logarithmic form}$$
$$c = \tfrac{1}{10} \ln \tfrac{3}{4} \qquad \text{divide by 10}$$

Finally, since the half-life corresponds to the time $t$ at which $N(t)$ is equal to 1000, we have the following:

$$1000 = 2000e^{ct} \qquad \text{let } N(t) = 1000$$
$$\tfrac{1}{2} = e^{ct} \qquad \text{divide by 2000}$$
$$ct = \ln \tfrac{1}{2} \qquad \text{change to logarithmic form}$$
$$t = \frac{1}{c} \ln \frac{1}{2} \qquad \text{divide by } c$$
$$= \frac{1}{\tfrac{1}{10} \ln \tfrac{3}{4}} \ln \frac{1}{2} \qquad c = \frac{1}{10} \ln \frac{3}{4}$$
$$\approx 24 \text{ days} \qquad \text{approximate}$$

The following example is a good illustration of the power of a graphing utility, since it is impossible to find the exact solution using only algebraic methods.

**EXAMPLE 15    *Approximating a solution to an inequality***

Graph $f(x) = \log(x + 1)$ and $g(x) = \ln(3 - x)$, and estimate the solution of the inequality $f(x) \geq g(x)$.

**FIGURE 25**  $[-1, 3]$ by $[-2, 2]$

**SOLUTION**    Some specific keystrokes for the TI-81/82 graphics calculator are given in Appendix I. We begin by making the assignments

$$Y_1 = \log(x + 1) \quad \text{and} \quad Y_2 = \ln(3 - x).$$

Since the domain of $f$ is $(-1, \infty)$ and the domain of $g$ is $(-\infty, 3)$, we choose the viewing rectangle $[-1, 3]$ by $[-2, 2]$ and obtain the graph in Figure 25. Using the tracing and zoom-in features, we find that the point of intersection is approximately $(1.51, 0.40)$. Thus, the approximate solution of $f(x) \geq g(x)$ is the interval

$$1.51 < x < 3.$$

## 4.3 EXERCISES

**Exer. 1–2: Change to logarithmic form.**

1  (a) $4^3 = 64$    (b) $4^{-3} = \frac{1}{64}$    (c) $t^r = s$

(d) $3^x = 4 - t$    (e) $5^{7t} = \dfrac{a + b}{a}$    (f) $(0.7)^t = 5.3$

2  (a) $3^5 = 243$    (b) $3^{-4} = \frac{1}{81}$    (c) $c^p = d$

(d) $7^x = 100p$    (e) $3^{-2x} = \dfrac{P}{F}$    (f) $(0.9)^t = \frac{1}{2}$

**Exer. 3–4: Change to exponential form.**

3  (a) $\log_2 32 = 5$    (b) $\log_3 \frac{1}{243} = -5$

(c) $\log_t r = p$    (d) $\log_3 (x + 2) = 5$

(e) $\log_2 m = 3x + 4$    (f) $\log_b 512 = \frac{3}{2}$

4  (a) $\log_3 81 = 4$    (b) $\log_4 \frac{1}{256} = -4$

(c) $\log_v w = q$    (d) $\log_6 (2x - 1) = 3$

(e) $\log_4 p = 5 - x$    (f) $\log_a 343 = \frac{3}{4}$

**Exer. 5–8: Solve for $t$ using logarithms with base $a$.**

5  $2a^{t/3} = 5$    6  $3a^{4t} = 10$

7  $A = Ba^{Ct} + D$    8  $L = Ma^{t/N} - P$

**Exer. 9–10: Change to logarithmic form.**

9  (a) $10^5 = 100{,}000$    (b) $10^{-3} = 0.001$

(c) $10^x = y + 1$    (d) $e^7 = p$

(e) $e^{2t} = 3 - x$

10  (a) $10^4 = 10{,}000$    (b) $10^{-2} = 0.01$

(c) $10^x = 38z$    (d) $e^4 = D$

(e) $e^{0.1t} = x + 2$

**Exer. 11–12: Change to exponential form.**

11  (a) $\log x = 50$    (b) $\log x = 20t$

(c) $\ln x = 0.1$    (d) $\ln w = 4 + 3x$

(e) $\ln(z - 2) = \frac{1}{6}$

12  (a) $\log x = -8$    (b) $\log x = y - 2$

(c) $\ln x = \frac{1}{2}$    (d) $\ln z = 7 + x$

(e) $\ln(t - 5) = 1.2$

**Exer. 13–14: Find the number, if possible.**

13 (a) $\log_5 1$    (b) $\log_3 3$    (c) $\log_4 (-2)$

   (d) $\log_7 7^2$    (e) $3^{\log_3 8}$    (f) $\log_5 125$

   (g) $\log_4 \frac{1}{16}$

14 (a) $\log_8 1$    (b) $\log_9 9$    (c) $\log_5 0$

   (d) $\log_6 6^7$    (e) $5^{\log_5 4}$    (f) $\log_3 243$

   (g) $\log_2 128$

**Exer. 15–16: Find the number.**

15 (a) $10^{\log 3}$    (b) $\log 10^5$    (c) $\log 100$

   (d) $\log 0.0001$    (e) $e^{\ln 2}$    (f) $\ln e^{-3}$

   (g) $e^{2+\ln 3}$

16 (a) $10^{\log 7}$    (b) $\log 10^{-6}$    (c) $\log 100{,}000$

   (d) $\log 0.001$    (e) $e^{\ln 8}$    (f) $\ln e^{2/3}$

   (g) $e^{1+\ln 5}$

**Exer. 17–30: Solve the equation.**

17 $\log_4 x = \log_4 (8 - x)$

18 $\log_3 (x + 4) = \log_3 (1 - x)$

19 $\log_5 (x - 2) = \log_5 (3x + 7)$

20 $\log_7 (x - 5) = \log_7 (6x)$

21 $\log x^2 = \log (-3x - 2)$    22 $\ln x^2 = \ln (12 - x)$

23 $\log_3 (x - 4) = 2$    24 $\log_2 (x - 5) = 4$

25 $\log_9 x = \frac{3}{2}$    26 $\log_4 x = -\frac{3}{2}$

27 $\ln x^2 = -2$    28 $\log x^2 = -4$

29 $e^{2\ln x} = 9$    30 $e^{-\ln x} = 0.2$

31 Sketch the graph of $f$ if $a = 4$:

   (a) $f(x) = \log_a x$      (b) $f(x) = -\log_a x$

   (c) $f(x) = 2 \log_a x$      (d) $f(x) = \log_a (x + 2)$

   (e) $f(x) = (\log_a x) + 2$    (f) $f(x) = \log_a (x - 2)$

   (g) $f(x) = (\log_a x) - 2$    (h) $f(x) = \log_a |x|$

   (i) $f(x) = \log_a (-x)$    (j) $f(x) = \log_a (3 - x)$

   (k) $f(x) = |\log_a x|$

32 Work Exercise 31 if $a = 5$.

**Exer. 33–36: Sketch the graph of $f$.**

33 $f(x) = \log x$      34 $f(x) = \ln x$

35 $f(x) = \log_2 |x - 5|$    36 $f(x) = \log_3 |x + 1|$

**Exer. 37–44: Shown in the figure is the graph of a function $f$. Express $f(x)$ in terms of logarithms with base 2.**

37

38

39

40

43

41

44

42

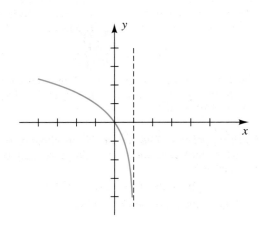

**Exer. 45–46: Approximate $x$ to three significant figures.**

45 (a) $\log x = 3.6274$     (b) $\log x = 0.9469$

   (c) $\log x = -1.6253$     (d) $\ln x = 2.3$

   (e) $\ln x = 0.05$     (f) $\ln x = -1.6$

46 (a) $\log x = 1.8965$     (b) $\log x = 4.9680$

   (c) $\log x = -2.2118$     (d) $\ln x = 3.7$

   (e) $\ln x = 0.95$     (f) $\ln x = -5$

47 **Radium decay**  If we start with $q_0$ milligrams of radium, the amount $q$ remaining after $t$ years is given by $q = q_0(2)^{-t/1600}$. Express $t$ in terms of $q$ and $q_0$.

48 **Bismuth isotope decay**  The radioactive bismuth isotope $^{210}$Bi disintegrates according to $Q = k(2)^{-t/5}$, where $k$ is a constant and $t$ is the time in days. Express $t$ in terms of $Q$ and $k$.

49 **Electrical circuit** A schematic of a simple electrical circuit consisting of a resistor and an inductor is shown in the figure. The current $I$ at time $t$ is given by the formula $I = 20e^{-Rt/L}$, where $R$ is the resistance and $L$ is the inductance. Solve this equation for $t$.

EXERCISE 49

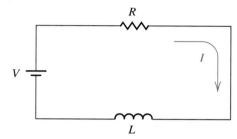

50 **Electrical condenser** An electrical condenser with initial charge $Q_0$ is allowed to discharge. After $t$ seconds the charge $Q$ is $Q = Q_0 e^{kt}$, where $k$ is a constant. Solve this equation for $t$.

51 **Richter scale** Use the Richter scale formula $R = \log (I/I_0)$ to find the magnitude of an earthquake that has an intensity

(a) 100 times that of $I_0$

(b) 10,000 times that of $I_0$

(c) 100,000 times that of $I_0$

52 **Richter scale** Refer to Exercise 51. The largest recorded magnitudes of earthquakes have been between 8 and 9 on the Richter scale. Find the corresponding intensities in terms of $I_0$.

53 **Sound intensity** The loudness of a sound, as experienced by the human ear, is based on its intensity level. A formula used for finding the intensity level $\alpha$ (in decibels) that corresponds to a sound intensity $I$ is $\alpha = 10 \log (I/I_0)$, where $I_0$ is a special value of $I$ agreed to be the weakest sound that can be detected by the ear under certain conditions. Find $\alpha$ if

(a) $I$ is 10 times as great as $I_0$

(b) $I$ is 1000 times as great as $I_0$

(c) $I$ is 10,000 times as great as $I_0$ (This is the intensity level of the average voice.)

54 **Sound intensity** Refer to Exercise 53. A sound intensity level of 140 decibels produces pain in the average human ear. Approximately how many times greater than $I_0$ must $I$ be in order for $\alpha$ to reach this level?

55 **U.S. population growth** The population $N(t)$ (in millions) of the United States $t$ years after 1980 may be approximated by the formula $N(t) = 227e^{0.007t}$. When will the population be twice what it was in 1980?

56 **Population growth in India** The population $N(t)$ (in millions) of India $t$ years after 1985 may be approximated by the formula $N(t) = 762e^{0.022t}$. When will the population reach 1 billion?

57 **Children's weight** The Ehrenberg relation
$$\ln W = \ln 2.4 + (1.84)h$$
is an empirically based formula relating the height $h$ (in meters) to the average weight $W$ (in kilograms) for children 5 through 13 years old.

(a) Express $W$ as a function of $h$ that does not contain ln.

(b) Estimate the average weight of an 8-year-old child who is 1.5 meters tall.

58 **Continuously compounded interest** If interest is compounded continuously at the rate of 10% per year, approximate the number of years it will take an initial deposit of $6000 to grow to $25,000.

59 **Air pressure** The air pressure $p(h)$ (in lb/in.$^2$) at an altitude of $h$ feet above sea level may be approximated by the formula $p(h) = 14.7e^{-0.0000385h}$. At approximately what altitude $h$ is the air pressure

(a) 10 lb/in.$^2$?

(b) one-half its value at sea level?

60 **Vapor pressure** A liquid's vapor pressure $P$ (in lb/in.$^2$), a measure of its volatility, is related to its temperature $T$ (in °F) by the Antoine equation
$$\ln P = a + \frac{b}{c + T},$$
where $a$, $b$, and $c$ are constants. Vapor pressure increases rapidly with an increase in temperature. Express $P$ as a function of $T$.

61 **Elephant growth** The weight $W$ (in kilograms) of a female African elephant at age $t$ (in years) may be approximated by
$$W = 2600(1 - 0.51e^{-0.075t})^3.$$

(a) Approximate the weight at birth.

(b) Estimate the age of a female African elephant weighing 1800 kilograms by using (1) the accompanying graph and (2) the formula for $W$.

EXERCISE 61

62 **Coal consumption**  A country presently has coal reserves of 50 million tons. Last year 6.5 million tons of coal was consumed. Past years' data and population projections suggest that the rate of consumption $R$ (in million tons/year) will increase according to the formula $R = 6.5e^{0.02t}$, and the total amount $T$ (in million tons) of coal that will be used in $t$ years is given by the formula $T = 325(e^{0.02t} - 1)$. If the country uses only its own resources, when will the coal reserves be depleted?

63 **Urban population density**  An urban density model is a formula that relates the population density $D$ (in thousands/mi$^2$) to the distance $x$ (in miles) from the center of the city. The formula $D = ae^{-bx}$ for the central density $a$ and coefficient of decay $b$ has been found to be appropriate for many large U.S. cities. For the city of Atlanta in 1970, $a = 5.5$ and $b = 0.10$. At approximately what distance was the population density 2000 per square mile?

64 **Brightness of stars**  Stars are classified into categories of brightness called magnitudes. The faintest stars, with light flux $L_0$, are assigned a magnitude 6. Brighter stars of light flux $L$ are assigned a magnitude $m$ by means of the formula

$$m = 6 - (2.5) \log \frac{L}{L_0}.$$

(a) Find $m$ if $L = 10^{0.4}L_0$.

(b) Solve the formula for $L$ in terms of $m$ and $L_0$.

65 **Radioactive iodine decay**  Radioactive iodine $^{131}$I is frequently used in tracer studies involving the thyroid gland. The substance decays according to the formula $A(t) = A_0 a^{-t}$, where $A_0$ is the initial dose and $t$ is the time in days. Find $a$, assuming the half-life of $^{131}$I is 8 days.

66 **Radioactive contamination**  Radioactive strontium $^{90}$Sr has been deposited in a large field by acid rain. If sufficient amounts make their way through the food chain to humans, bone cancer can result. It has been determined that the radioactivity level in the field is 2.5 times the safe level $S$. $^{90}$Sr decays according to the formula

$$A(t) = A_0 e^{-0.0239t},$$

where $A_0$ is the amount currently in the field and $t$ is the time in years. For how many years will the field be contaminated?

67 **Walking speed**  In a survey of 15 cities ranging in population $P$ from 300 to 3,000,000, it was found that the average walking speed $S$ (in ft/sec) of a pedestrian could be approximated by $S = 0.05 + 0.86 \log P$.

(a) How does the population affect the average walking speed?

(b) For what population is the average walking speed 5 ft/sec?

68 **Computer chips**  For manufacturers of computer chips, it is important to consider the fraction $F$ of chips that will fail after $t$ years of service. This fraction can sometimes be approximated by the formula $F = 1 - e^{-ct}$, where $c$ is a positive constant.

(a) How does the value of $c$ affect the reliability of a chip?

(b) If $c = 0.125$, after how many years will 35% of the chips have failed?

**Exer. 69–70: Approximate the real root of the equation.**

69 $x \ln x = 1$                70 $\ln x + x = 0$

**Exer. 71–72: Graph $f$ and $g$ on the same coordinate plane, and estimate the solution of the inequality $f(x) \geq g(x)$.**

71 $f(x) = 2.2 \log (x + 2)$;    $g(x) = \ln x$

72 $f(x) = x \ln |x|$;           $g(x) = 0.15e^x$

## 4.4 PROPERTIES OF LOGARITHMS

In the preceding section we observed that $\log_a x$ can be interpreted as an exponent. Thus, it seems reasonable to expect that the laws of exponents can be used to obtain corresponding laws of logarithms. This is demonstrated in the proofs of the following laws, which are fundamental for all work with logarithms.

**Laws of Logarithms**

> If $u$ and $w$ denote positive real numbers, then
>
> (1) $\log_a (uw) = \log_a u + \log_a w$
>
> (2) $\log_a \dfrac{u}{w} = \log_a u - \log_a w$
>
> (3) $\log_a (u^c) = c \log_a u$   for every real number $c$

**PROOF**   Throughout the proof, let

$$r = \log_a u \quad \text{and} \quad s = \log_a w.$$

Applying the definition of logarithm gives us

$$u = a^r \quad \text{and} \quad w = a^s.$$

We now proceed as follows:

(1)
| | |
|---|---|
| $uw = a^r a^s$ | $u = a^r$ and $w = a^s$ |
| $uw = a^{r+s}$ | law of exponents |
| $\log_a uw = r + s$ | definition of $\log_a$ |
| $\log_a uw = \log_a u + \log_a w$ | $r = \log_a u, s = \log_a w$ |

(2)
| | |
|---|---|
| $\dfrac{u}{w} = \dfrac{a^r}{a^s}$ | $u = a^r$ and $w = a^s$ |
| $\dfrac{u}{w} = a^{r-s}$ | law of exponents |
| $\log_a \dfrac{u}{w} = r - s$ | definition of $\log_a$ |
| $\log_a \dfrac{u}{w} = \log_a u - \log_a w$ | $r = \log_a u, s = \log_a w$ |

(3)
| | |
|---|---|
| $u^c = (a^r)^c$ | $u = a^r$ |
| $u^c = a^{cr}$ | law of exponents |
| $\log_a u^c = cr$ | definition of $\log_a$ |
| $\log_a u^c = c \log_a u$ | $r = \log_a u$ ∎ |

The laws of logarithms for the special cases $a = 10$ (common logs) and $a = e$ (natural logs) are written as in the following chart.

| Common logarithms | Natural logarithms |
|---|---|
| $\log(uw) = \log u + \log w$ | $\ln(uw) = \ln u + \ln w$ |
| $\log \dfrac{u}{w} = \log u - \log w$ | $\ln \dfrac{u}{w} = \ln u - \ln w$ |
| $\log(u^c) = c \log u$ | $\ln(u^c) = c \ln u$ |

As indicated by the following warnings, there are no general laws for expressing $\log_a (u + w)$ or $\log_a (u - w)$ in terms of simpler logarithms. The expressions on the right side equal $\log_a uw$ and $\log_a \dfrac{u}{w}$, respectively.

**Warnings**

$$\log_a (u + w) \neq \log_a u + \log_a w; \quad \log_a (u - w) \neq \log_a u - \log_a w$$

The following examples illustrate uses of the laws of logarithms.

**EXAMPLE 1    *Using laws of logarithms***

Express $\log_a \dfrac{x^3 \sqrt{y}}{z^2}$ in terms of logarithms of $x$, $y$, and $z$.

**SOLUTION**    We write $\sqrt{y}$ as $y^{1/2}$ and use laws of logarithms:

$$
\begin{aligned}
\log_a \frac{x^3 \sqrt{y}}{z^2} &= \log_a (x^3 y^{1/2}) - \log_a z^2 && \text{law 2} \\
&= \log_a x^3 + \log_a y^{1/2} - \log_a z^2 && \text{law 1} \\
&= 3 \log_a x + \tfrac{1}{2} \log_a y - 2 \log_a z && \text{law 3}
\end{aligned}
$$

**EXAMPLE 2    *Using laws of logarithms***

Express as one logarithm:

$$\tfrac{1}{3} \log_a (x^2 - 1) - \log_a y - 4 \log_a z$$

**SOLUTION**    We apply the laws of logarithms as follows:

$$
\begin{aligned}
\tfrac{1}{3} \log_a (x^2 - 1) &- \log_a y - 4 \log_a z \\
&= \log_a (x^2 - 1)^{1/3} - \log_a y - \log_a z^4 && \text{law 3} \\
&= \log_a \sqrt[3]{x^2 - 1} - (\log_a y + \log_a z^4) && \text{algebra} \\
&= \log_a \sqrt[3]{x^2 - 1} - \log_a (yz^4) && \text{law 1} \\
&= \log_a \frac{\sqrt[3]{x^2 - 1}}{yz^4} && \text{law 2}
\end{aligned}
$$

**EXAMPLE 3**  *Solving a logarithmic equation*

Solve the equation $\log_5 (2x + 3) = \log_5 11 + \log_5 3$.

**SOLUTION**

$$\log_5 (2x + 3) = \log_5 11 + \log_5 3 \qquad \text{given}$$
$$\log_5 (2x + 3) = \log_5 (11 \cdot 3) \qquad \text{law 1}$$
$$2x + 3 = 33 \qquad \text{logarithmic functions are one-to-one}$$
$$x = 15 \qquad \text{solve for } x$$

✔ **Check** $x = 15$  LS: $\log_5 (2 \cdot 15 + 3) = \log_5 33$

RS: $\log_5 11 + \log_5 3 = \log_5 (11 \cdot 3) = \log_5 33$

Since $\log_5 33 = \log_5 33$ is a true statement, $x = 15$ is a solution.

---

The laws of logarithms were proved for logarithms of *positive* real numbers $u$ and $w$. If we apply these laws to equations in which $u$ and $w$ are expressions involving a variable, then extraneous solutions may occur. Answers should therefore be substituted for the variable in $u$ and $w$ to determine whether these expressions are defined.

**EXAMPLE 4**  *Solving a logarithmic equation*

Solve the equation $\log_2 x + \log_2 (x + 2) = 3$.

**SOLUTION**

$$\log_2 x + \log_2 (x + 2) = 3 \qquad \text{given}$$
$$\log_2 [x(x + 2)] = 3 \qquad \text{law 1}$$
$$x(x + 2) = 2^3 \qquad \text{definition of logarithm}$$
$$x^2 + 2x - 8 = 0 \qquad \text{simplify}$$
$$(x - 2)(x + 4) = 0 \qquad \text{factor}$$
$$x - 2 = 0, \quad x + 4 = 0 \qquad \text{set each factor equal to 0}$$
$$x = 2, \qquad x = -4 \qquad \text{solve for } x$$

✔ **Check** $x = 2$  LS: $\log_2 2 + \log_2 (2 + 2) = 1 + \log_2 4$
$$= 1 + \log_2 2^2 = 1 + 2 = 3$$

RS: 3

Since $3 = 3$ is a true statement, $x = 2$ is a solution.

✔ **Check** $x = -4$  LS: $\log_2 (-4) + \log_2 (-4 + 2)$

Since logarithms of negative numbers are undefined, $x = -4$ is not a solution.

**EXAMPLE 5**   *Solving a logarithmic equation*

Solve the equation $\ln (x + 6) - \ln 10 = \ln (x - 1) - \ln 2$.

**SOLUTION**

$$\ln (x + 6) - \ln (x - 1) = \ln 10 - \ln 2 \quad \text{rearrange terms}$$

$$\ln \left( \frac{x + 6}{x - 1} \right) = \ln \frac{10}{2} \quad \text{law 2}$$

$$\frac{x + 6}{x - 1} = 5 \quad \text{ln is one-to-one}$$

$$x + 6 = 5x - 5 \quad \text{multiply by } x - 1$$

$$x = \tfrac{11}{4} \quad \text{solve for } x$$

✔ **Check**   Since both $\ln (x + 6)$ and $\ln (x - 1)$ are defined at $x = \tfrac{11}{4}$ (they are logarithms of positive real numbers) and since our algebraic steps are correct, it follows that $\tfrac{11}{4}$ is a solution of the given equation.

**EXAMPLE 6**   *Shifting the graph of a logarithmic equation*

Sketch the graph of $y = \log_3 (81x)$.

**FIGURE 26**

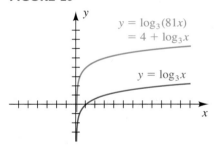

**SOLUTION**   We may rewrite the equation as follows:

$$y = \log_3 (81x) \quad \text{given}$$
$$= \log_3 81 + \log_3 x \quad \text{law 1}$$
$$= \log_3 3^4 + \log_3 x \quad 81 = 3^4$$
$$= 4 + \log_3 x \quad \log_a a^x = x$$

Thus, we can obtain the graph of $y = \log_3 (81x)$ by vertically shifting the graph of $y = \log_3 x$ in Figure 19 upward four units. This gives us the sketch in Figure 26.

**EXAMPLE 7**   *Sketching graphs of logarithmic equations*

Sketch the graph of the equation:

**(a)** $y = \log_3 (x^2)$   **(b)** $y = 2 \log_3 x$

**SOLUTION**

**(a)** Since $x^2 = |x|^2$, we may rewrite the given equation as

$$y = \log_3 |x|^2.$$

Using a law of logarithms, we have

$$y = 2 \log_3 |x|.$$

**FIGURE 27**

**(a)**

**(b)**

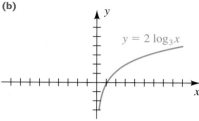

We can obtain the graph of $y = 2 \log_3 |x|$ by multiplying the $y$-coordinates of points on the graph of $y = \log_3 |x|$ in Figure 20 by 2. This gives us the graph in Figure 27(a).

**(b)** If $y = 2 \log_3 x$, then $x$ must be positive. Hence, the graph is identical to that part of the graph of $y = 2 \log_3 |x|$ in Figure 27(a) that lies to the right of the $y$-axis. This gives us Figure 27(b).

---

**EXAMPLE 8    *A relationship between selling price and demand***

In the study of economics, the demand $D$ for a product is often related to its selling price $p$ by an equation of the form

$$\log_a D = \log_a c - k \log_a p,$$

where $a$, $c$, and $k$ are positive constants.

**(a)** Solve the equation for $D$.

**(b)** How does increasing or decreasing the selling price affect the demand?

**SOLUTION**

**(a)**
$$\log_a D = \log_a c - k \log_a p \qquad \text{given}$$
$$\log_a D = \log_a c - \log_a p^k \qquad \text{law 3}$$
$$\log_a D = \log_a \frac{c}{p^k} \qquad \text{law 2}$$
$$D = \frac{c}{p^k} \qquad \log_a \text{ is one-to-one}$$

**(b)** If the price $p$ is increased, the denominator $p^k$ in $D = c/p^k$ will also increase and hence the demand $D$ for the product will decrease. If the price is decreased, then $p^k$ will decrease and the demand $D$ will increase.

---

## 4.4 EXERCISES

Exer. 1–8: Express in terms of logarithms of $x$, $y$, $z$, or $w$.

1 (a) $\log_4 (xz)$　(b) $\log_4 (y/x)$　(c) $\log_4 \sqrt[3]{z}$

2 (a) $\log_3 (xyz)$　(b) $\log_3 (xz/y)$　(c) $\log_3 \sqrt[5]{y}$

3 $\log_a \dfrac{x^3 w}{y^2 z^4}$

4 $\log_a \dfrac{y^5 w^2}{x^4 z^3}$

5 $\log \dfrac{\sqrt[3]{z}}{x\sqrt{y}}$

6 $\log \dfrac{\sqrt{y}}{x^4 \sqrt[3]{z}}$

7 $\ln \sqrt[4]{\dfrac{x^7}{y^5 z}}$

8 $\ln x \sqrt[3]{\dfrac{y^4}{z^5}}$

Exer. 9–16: Write the expression as one logarithm.

9 (a) $\log_3 x + \log_3 (5y)$　(b) $\log_3 (2z) - \log_3 x$
　(c) $5 \log_3 y$

10 (a) $\log_4 (3z) + \log_4 x$　(b) $\log_4 x - \log_4 (7y)$
　(c) $\tfrac{1}{3} \log_4 w$

11 $2 \log_a x + \tfrac{1}{3} \log_a (x - 2) - 5 \log_a (2x + 3)$

12 $5 \log_a x - \tfrac{1}{2} \log_a (3x - 4) - 3 \log_a (5x + 1)$

13 $\log (x^3 y^2) - 2 \log x \sqrt[3]{y} - 3 \log \left(\dfrac{x}{y}\right)$

14 $2 \log \dfrac{y^3}{x} - 3 \log y + \dfrac{1}{2} \log x^4 y^2$

15 $\ln y^3 + \tfrac{1}{3} \ln (x^3 y^6) - 5 \ln y$

16 $2 \ln x - 4 \ln (1/y) - 3 \ln (xy)$

Exer. 17–32: Solve the equation.

17 $\log_6 (2x - 3) = \log_6 12 - \log_6 3$

18 $\log_4 (3x + 2) = \log_4 5 + \log_4 3$

19 $2 \log_3 x = 3 \log_3 5$

20 $3 \log_2 x = 2 \log_2 3$

21 $\log x - \log (x + 1) = 3 \log 4$

22 $\log (x + 2) - \log x = 2 \log 4$

23 $\ln (-4 - x) + \ln 3 = \ln (2 - x)$

24 $\ln x + \ln (x + 6) = \tfrac{1}{2} \ln 9$

25 $\log_2 (x + 7) + \log_2 x = 3$

26 $\log_6 (x + 5) + \log_6 x = 2$

27 $\log_3 (x + 3) + \log_3 (x + 5) = 1$

28 $\log_3 (x - 2) + \log_3 (x - 4) = 2$

29 $\log (x + 3) = 1 - \log (x - 2)$

30 $\log (57x) = 2 + \log (x - 2)$

31 $\ln x = 1 - \ln (x + 2)$

32 $\ln x = 1 + \ln (x + 1)$

Exer. 33–44: Sketch the graph of $f$.

33 $f(x) = \log_3 (3x)$　　　34 $f(x) = \log_4 (16x)$

35 $f(x) = 3 \log_3 x$　　　36 $f(x) = \tfrac{1}{3} \log_3 x$

37 $f(x) = \log_3 (x^2)$　　38 $f(x) = \log_2 (x^2)$

39 $f(x) = \log_2 (x^3)$　　40 $f(x) = \log_3 (x^3)$

41 $f(x) = \log_2 \sqrt{x}$　　42 $f(x) = \log_2 \sqrt[3]{x}$

43 $f(x) = \log_3 \left(\dfrac{1}{x}\right)$　44 $f(x) = \log_2 \left(\dfrac{1}{x}\right)$

Exer. 45–48: Shown in the figure is the graph of a function $f$. Express $f(x)$ as one logarithm with base 2.

45

**46**

**47**

**48**

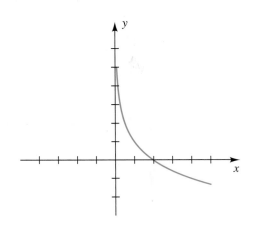

**49 Pareto's law** Pareto's law for capitalist countries states that the relationship between annual income $x$ and the number $y$ of individuals whose income exceeds $x$ is

$$\log y = \log b - k \log x,$$

where $b$ and $k$ are positive constants. Solve this equation for $y$.

**50 Price and demand** If $p$ denotes the selling price (in dollars) of a commodity and $x$ is the corresponding demand (in number sold per day), then the relationship between $p$ and $x$ is sometimes given by $p = p_0 e^{-ax}$, where $p_0$ and $a$ are positive constants. Express $x$ as a function of $p$.

**51 Wind velocity** If $v$ denotes the wind velocity (in m/sec) at a height of $z$ meters above the ground, then under certain conditions $v = c \ln (z/z_0)$, where $c$ is a positive constant and $z_0$ is the height at which the velocity is zero. Sketch the graph of this equation on a $zv$-plane for $c = 0.5$ and $z_0 = 0.1$ m.

**52 Eliminating pollution** If the pollution of Lake Erie were stopped suddenly, it has been estimated that the level $y$ of pollutants would decrease according to the formula $y = y_0 e^{-0.3821t}$, where $t$ is the time in years and $y_0$ is the pollutant level at which further pollution ceased. How many years would it take to clear 50% of the pollutants?

**53 Reaction to a stimulus** Let $R$ denote the reaction of a subject to a stimulus of strength $x$. There are many possibilities for $R$ and $x$. If the stimulus $x$ is saltiness (in grams of salt per liter), $R$ may be the subject's estimate of how salty the solution tasted, based on a scale from 0 to 10. One relationship between $R$ and $x$ is given by the Weber-Fechner formula $R(x) = a \log (x/x_0)$, where $a$ is a positive constant and $x_0$ is called the threshold stimulus.

(a) Find $R(x_0)$.

(b) Find a relationship between $R(x)$ and $R(2x)$.

**54 Electron energy** The energy $E(x)$ of an electron after passing through material of thickness $x$ is given by $E(x) = E_0 e^{-x/x_0}$, where $E_0$ is the initial energy and $x_0$ is the radiation length.

(a) Express, in terms of $E_0$, the energy of an electron after it passes through material of thickness $x_0$.

(b) Express, in terms of $x_0$, the thickness at which the electron loses 99% of its initial energy.

**55  *Ozone layer*** One method of estimating the thickness of the ozone layer is to use the formula

$$\ln I_0 - \ln I = kx,$$

where $I_0$ is the intensity of a particular wavelength of light from the sun before it reaches the atmosphere, $I$ is the intensity of the same wavelength after passing through a layer of ozone $x$ centimeters thick, and $k$ is the absorption constant of ozone for that wavelength. Suppose for a wavelength of $3176 \times 10^{-8}$ cm with $k \approx 0.39$, $I_0/I$ is measured as 1.12. Approximate the thickness of the ozone layer to the nearest 0.01 centimeter.

**56  *Ozone layer*** Refer to Exercise 55. Approximate the percentage decrease in the intensity of light with a wavelength of $3176 \times 10^{-8}$ cm if the ozone layer is 0.24 centimeter thick.

**Exer. 57–58: Graph $f$ and $g$ on the same coordinate plane, and estimate the solution of the inequality $f(x) \geq g(x)$.**

**57** $f(x) = x^3 - 3.5x^2 + 3x$;    $g(x) = \log 3x$

**58** $f(x) = 3^{-0.5x}$;               $g(x) = \log x$

**Exer. 59–60: Use a graph to estimate the roots of the equation on the given interval.**

**59** $e^{-x} - 2 \log (1 + x^2) + 0.5x = 0$;          $[0, 8]$

**60** $0.3 \ln x + x^3 - 3.1x^2 + 1.3x + 0.8 = 0$;   $(0, 3)$

**Exer. 61–62: Graph $f$ on the interval [0.2, 16]. (a) Estimate the intervals where $f$ is increasing or is decreasing. (b) Estimate the maximum and minimum values of $f$ on [0.2, 16].**

**61** $f(x) = 2 \log 2x - 1.5x + 0.1x^2$

**62** $f(x) = 1.1^{3x} + x - 1.35^x - \log x + 5$

---

## 4.5  EXPONENTIAL AND LOGARITHMIC EQUATIONS

In this section we shall consider various types of exponential and logarithmic equations and their applications.

**EXAMPLE 1     *Solving an exponential equation***
Solve the equation $3^x = 21$.

*SOLUTION*

$$3^x = 21 \qquad \text{given}$$
$$\log (3^x) = \log 21 \qquad \text{take log of both sides}$$
$$x \log 3 = \log 21 \qquad \text{law 3 of logarithms}$$
$$x = \frac{\log 21}{\log 3} \qquad \text{divide by } \log 3$$

We could also have used natural logarithms to obtain

$$x = \frac{\ln 21}{\ln 3}.$$

Using a calculator gives us the approximate solution $x \approx 2.77$. A partial check is to note that since $3^2 = 9$ and $3^3 = 27$, the number $x$ such that $3^x = 21$ should lie between 2 and 3, somewhat closer to 3 than to 2.

We could also have solved the equation in Example 1 by changing the exponential form $3^x = 21$ to logarithmic form, as we did in Section 4.3, obtaining

$$x = \log_3 21.$$

This is, in fact, the solution of the equation; however, since calculators typically have keys only for log and ln, we cannot approximate $\log_3 21$ directly. The next theorem gives us a simple *change of base formula* for finding $\log_b u$ if $u > 0$ and $b$ is *any* logarithmic base.

**Theorem: Change of Base Formula**

If $u > 0$ and if $a$ and $b$ are positive real numbers different from 1, then

$$\log_b u = \frac{\log_a u}{\log_a b}.$$

*PROOF*  We begin with the equivalent equations

$$w = \log_b u \quad \text{and} \quad b^w = u$$

and proceed as follows:

$$
\begin{aligned}
b^w &= u && \text{given} \\
\log_a b^w &= \log_a u && \text{take } \log_a \text{ of both sides} \\
w \log_a b &= \log_a u && \text{law 3 of logarithms} \\
w &= \frac{\log_a u}{\log_a b} && \text{divide by } \log_a b
\end{aligned}
$$

Since $w = \log_b u$, we obtain the formula.  ■

The following special case of the change of base formula is obtained by letting $u = a$ and using the fact that $\log_a a = 1$:

$$\log_b a = \frac{1}{\log_a b}$$

The change of base formula is sometimes confused with law 2 of logarithms. The following warning could be remembered with the phrase ''a quotient of logs is *not* the log of the quotient.''

**Warning**

$$\frac{\log_a u}{\log_a b} \neq \log_a \frac{u}{b}$$

The most frequently used special cases of the change of base formula are those for $a = 10$ (common logarithms) and $a = e$ (natural logarithms), as stated in the next box.

**Special Change of Base Formulas**

$$\textbf{(1)} \ \log_b u = \frac{\log u}{\log b} \qquad \textbf{(2)} \ \log_b u = \frac{\ln u}{\ln b}$$

We shall next rework Example 1 using a change of base formula.

**EXAMPLE 2**    *Using a change of base formula*

Solve the equation $3^x = 21$.

**SOLUTION**    We proceed as follows:

$$\begin{aligned}
3^x &= 21 && \text{given} \\
x &= \log_3 21 && \text{change to logarithmic form} \\
&= \frac{\log 21}{\log 3} && \text{special change of base formula 1}
\end{aligned}$$

Another method is to use special change of base formula 2, obtaining

$$x = \frac{\ln 21}{\ln 3}.$$

Logarithms with base 2 are used in computer science. The next example indicates how to find logarithms with base 2 using change of base formulas.

**EXAMPLE 3**    *Approximating a logarithm with base 2*

Approximate $\log_2 5$ using

**(a)** common logarithms    **(b)** natural logarithms

**SOLUTION**    Using special change of base formulas 1 and 2, we obtain the following:

**(a)**
$$\log_2 5 = \frac{\log 5}{\log 2} \approx 2.322$$

**(b)**
$$\log_2 5 = \frac{\ln 5}{\ln 2} \approx 2.322$$

**EXAMPLE 4**    *Solving an exponential equation*

Solve the equation $5^{2x+1} = 6^{x-2}$.

*SOLUTION*   We can use either common or natural logarithms. Using base 10 gives us the following:

$$5^{2x+1} = 6^{x-2} \qquad \text{given}$$

$$\log(5^{2x+1}) = \log(6^{x-2}) \qquad \text{take log of both sides}$$

$$(2x+1)\log 5 = (x-2)\log 6 \qquad \text{law 3 of logarithms}$$

$$2x \log 5 + \log 5 = x \log 6 - 2 \log 6 \qquad \text{multiply}$$

$$2x \log 5 - x \log 6 = -\log 5 - 2 \log 6 \qquad \text{subtract log 5 and } x \log 6$$

$$x(\log 5^2 - \log 6) = -(\log 5 + \log 6^2) \qquad \begin{array}{l}\text{factor and use law 3}\\ \text{of logarithms}\end{array}$$

$$x = -\frac{\log(5 \cdot 36)}{\log \frac{25}{6}} \qquad \begin{array}{l}\text{solve for } x \text{ and use laws of}\\ \text{logarithms}\end{array}$$

An approximation is $x \approx -3.64$.

**EXAMPLE 5**   *Solving an exponential equation*

Solve the equation $\dfrac{5^x - 5^{-x}}{2} = 3$.

*SOLUTION*

$$\frac{5^x - 5^{-x}}{2} = 3 \qquad \text{given}$$

$$5^x - 5^{-x} = 6 \qquad \text{multiply by 2}$$

$$5^x - \frac{1}{5^x} = 6 \qquad \text{definition of negative exponent}$$

$$5^x(5^x) - \frac{1}{5^x}(5^x) = 6(5^x) \qquad \text{multiply by the lcd, } 5^x$$

$$5^{2x} - 6(5^x) - 1 = 0 \qquad \text{simplify and subtract } 6(5^x)$$

We recognize this form of the equation as a quadratic in $5^x$ and proceed as follows:

$$(5^x)^2 - 6(5^x) - 1 = 0 \qquad \text{law of exponents}$$

$$u^2 - 6u - 1 = 0 \qquad \text{let } u = 5^x$$

$$u = \frac{6 \pm \sqrt{36 + 4}}{2} \qquad \text{quadratic formula}$$

$$5^x = 3 \pm \sqrt{10} \qquad u = 5^x$$

$$5^x = 3 + \sqrt{10} \qquad 5^x > 0, \text{ but } 3 - \sqrt{10} < 0$$

$$\log 5^x = \log(3 + \sqrt{10}) \qquad \text{take log of both sides}$$

$$x \log 5 = \log(3 + \sqrt{10}) \qquad \text{law 3 of logarithms}$$

$$x = \frac{\log(3 + \sqrt{10})}{\log 5} \qquad \text{divide by log 5}$$

*(continued)*

Natural logarithms could have also been used to obtain

$$x = \frac{\ln (3 + \sqrt{10})}{\ln 5}.$$

An approximation is $x \approx 1.13$.

---

**EXAMPLE 6** *Light penetration in an ocean*

The Beer-Lambert law states that the amount of light $I$ that penetrates to a depth of $x$ meters in an ocean is given by $I = I_0 c^x$, where $0 < c < 1$ and $I_0$ is the amount of light at the surface.

(a) Solve for $x$ using common logarithms.

(b) Solve for $x$ using natural logarithms.

(c) If $c = \frac{1}{4}$, approximate the depth at which $I = 0.01 I_0$. (This determines the zone where photosynthesis can take place.)

*SOLUTION*

(a) $\qquad\qquad I = I_0 c^x \qquad\qquad$ given

$\qquad\qquad\qquad \dfrac{I}{I_0} = c^x \qquad\qquad$ isolate the exponential expression

$\qquad\qquad\qquad x = \log_c \dfrac{I}{I_0} \qquad$ change to logarithmic form

$\qquad\qquad\qquad\quad = \dfrac{\log (I/I_0)}{\log c} \qquad$ special change of base formula 1

(b) Replacing log with ln throughout part (a), we obtain

$$x = \frac{\ln (I/I_0)}{\ln c}.$$

(c) Letting $I = 0.01 I_0$ and $c = \frac{1}{4}$ in the formula for $x$ obtained in part (a), we have

$$x = \frac{\log (0.01 I_0 / I_0)}{\log \frac{1}{4}} = \frac{\log (0.01)}{\log 1 - \log 4} = \frac{\log 10^{-2}}{0 - \log 4} = \frac{-2}{-\log 4} = \frac{2}{\log 4}.$$

An approximation is $x \approx 3.32$ m.

---

**EXAMPLE 7** *Comparing light intensities*

If a beam of light that has intensity $I_0$ is projected vertically downward into water, then its intensity $I(x)$ at a depth of $x$ meters is $I(x) = I_0 e^{-1.4x}$ (see Figure 28). At what depth is the intensity one-half its value at the surface?

**FIGURE 28**

**SOLUTION**   At the surface, $x = 0$, and the intensity is

$$I(0) = I_0 e^0 = I_0.$$

We wish to find the value of $x$ such that $I(x) = \frac{1}{2} I_0$. This leads to the following:

$$I(x) = \tfrac{1}{2} I_0 \qquad \text{desired intensity}$$

$$I_0 e^{-1.4x} = \tfrac{1}{2} I_0 \qquad \text{formula for } I(x)$$

$$e^{-1.4x} = \tfrac{1}{2} \qquad \text{divide by } I_0$$

$$-1.4x = \ln \tfrac{1}{2} \qquad \text{change to logarithmic form}$$

$$x = \frac{\ln \tfrac{1}{2}}{-1.4} \qquad \text{divide by } -1.4$$

An approximation is $x \approx 0.495$ m.

**EXAMPLE 8**   *A logistic curve*

A **logistic curve** is the graph of an equation of the form

$$y = \frac{k}{1 + be^{-cx}},$$

where $k$, $b$, and $c$ are positive constants. Such curves are useful for describing a population $y$ that grows rapidly initially, but whose growth rate decreases after $x$ reaches a certain value. In a famous study of the growth of protozoa by Gause, a population of *Paramecium caudata* was found to be described by a logistic equation with $c = 1.1244$, $k = 105$, and $x$ the time in days.

**(a)** Find $b$ if the initial population was 3 protozoa.

**(b)** In the study, the maximum growth rate took place at $y = 52$. At what time $x$ did this occur?

**(c)** Show that after a long period of time, the population described by any logistic curve approaches the constant $k$.

**SOLUTION**

**(a)** Letting $c = 1.1244$ and $k = 105$ in the logistic equation, we obtain

$$y = \frac{105}{1 + be^{-1.1244x}}.$$

We now proceed as follows:

$$3 = \frac{105}{1 + be^0} = \frac{105}{1 + b} \qquad y = 3 \text{ when } x = 0$$

$$1 + b = 35 \qquad \text{multiply by } \frac{1 + b}{3}$$

$$b = 34 \qquad \text{solve for } b$$

**(b)** Using the fact that $b = 34$ leads to the following:

$$52 = \frac{105}{1 + 34e^{-1.1244x}} \qquad \text{let } y = 52 \text{ in part (a)}$$

$$1 + 34e^{-1.1244x} = \frac{105}{52} \qquad \text{multiply by } \frac{1 + 34e^{-1.1244x}}{52}$$

$$e^{-1.1244x} = \left(\frac{105}{52} - 1\right) \cdot \frac{1}{34} = \frac{53}{1768} \qquad \text{isolate } e^{-1.1244x}$$

$$-1.1244x = \ln \frac{53}{1768} \qquad \text{change to logarithmic form}$$

$$x = \frac{\ln \frac{53}{1768}}{-1.1244} \approx 3.12 \text{ days} \qquad \text{divide by } -1.1244$$

**(c)** As $x \to \infty$, $e^{-cx} \to 0$. Hence,

$$y = \frac{k}{1 + be^{-cx}} \quad \to \quad \frac{k}{1 + b \cdot 0} = k.$$

In the next example we graph the equation obtained in part (a) of the preceding example.

**EXAMPLE 9**    *Sketching the graph of a logistic curve*

Graph the logistic curve given by

$$y = \frac{105}{1 + 34e^{-1.1244x}},$$

and estimate the value of $x$ for $y = 52$.

**SOLUTION**    We begin by assigning

$$\frac{105}{1 + 34e^{-1.1244x}}$$

**FIGURE 29**    [0, 10] by [0, 105]

to $Y_1$. Since the time $x$ is nonnegative, we choose Xmin = 0. We select Xmax = 10 in order to include the value of $x$ found in part (b) of Example 8. By part (c), we know that the value of $y$ cannot exceed 105. Thus, we choose Ymin = 0 and Ymax = 105 and obtain a display similar to Figure 29.

Using the tracing and zoom-in features, we see that for $y = 52$, the value of $x$ is approximately 3.12, which agrees with the approximation found in (b) of Example 8.

---

The following example shows how a change of base formula may be used to enable us to graph logarithmic functions with bases other than 10 and $e$ on a graphing utility.

**EXAMPLE 10**    *Estimating points of intersection of logarithmic graphs*

Estimate the point of intersection of the graphs of

$$f(x) = \log_3 x \quad \text{and} \quad g(x) = \log_6 (x + 2).$$

**SOLUTION**    Most graphing utilities are equipped to work with only common and natural logarithmic functions. Thus, we first use a change of base formula to rewrite $f$ and $g$ as

**FIGURE 30**    [−2, 4] by [−2, 2]

$$f(x) = \frac{\ln x}{\ln 3} \quad \text{and} \quad g(x) = \frac{\ln (x + 2)}{\ln 6}.$$

We next assign $(\ln x)/\ln 3$ and $(\ln (x + 2))/\ln 6$ to $Y_1$ and $Y_2$, respectively. After graphing $Y_1$ and $Y_2$ using a standard viewing rectangle, we see that there is a point of intersection in the first quadrant with $2 < x < 3$. Using the tracing and zoom-in features, we find that the point of intersection is approximately (2.52, 0.84).

Figure 30 was obtained using viewing rectangle dimensions [−2, 4] by [−2, 2]. There are no other points of intersection, since $f$ increases more rapidly than $g$ for $x > 3$.

## 4.5 EXERCISES

Exer. 1–4: Find the exact solution and a two-decimal-place approximation for it by using (a) the method of Example 1 and (b) the method of Example 2.

1  $5^x = 8$

2  $4^x = 3$

3  $3^{4-x} = 5$

4  $(\frac{1}{3})^x = 100$

*ước lượng, tautgia*

Exer. 5–10: Evaluate using the change of base formula.

5  $\log_5 6$

6  $\log_2 20$

7  $\log_9 0.2$

8  $\log_6 \frac{1}{2}$

9  $\dfrac{\log_5 16}{\log_5 4}$

10  $\dfrac{\log_7 243}{\log_7 3}$

Exer. 11–24: Find the exact solution, using common logarithms, and a two-decimal-place approximation of each solution, when appropriate.

11  $3^{x+4} = 2^{1-3x}$

12  $4^{2x+3} = 5^{x-2}$

13  $2^{2x-3} = 5^{x-2}$

14  $3^{2-3x} = 4^{2x+1}$

15  $2^{-x} = 8$

16  $2^{-x^2} = 5$

17  $\log x = 1 - \log(x - 3)$

18  $\log(5x + 1) = 2 + \log(2x - 3)$

19  $\log(x^2 + 4) - \log(x + 2) = 2 + \log(x - 2)$

20  $\log(x - 4) - \log(3x - 10) = \log(1/x)$

21  $5^x + 125(5^{-x}) = 30$

22  $3(3^x) + 9(3^{-x}) = 28$

23  $4^x - 3(4^{-x}) = 8$

24  $2^x - 6(2^{-x}) = 6$

Exer. 25–30: Solve the equation without using a calculator or a table.

25  $\log(x^2) = (\log x)^2$

26  $\log \sqrt{x} = \sqrt{\log x}$

27  $\log(\log x) = 2$

28  $\log \sqrt{x^3 - 9} = 2$

29  $x^{\sqrt{\log x}} = 10^8$

30  $\log(x^3) = (\log x)^3$

Exer. 31–34: Use common logarithms to solve for $x$ in terms of $y$.

31  $y = \dfrac{10^x + 10^{-x}}{2}$

32  $y = \dfrac{10^x - 10^{-x}}{2}$

33  $y = \dfrac{10^x - 10^{-x}}{10^x + 10^{-x}}$

34  $y = \dfrac{10^x + 10^{-x}}{10^x - 10^{-x}}$

Exer. 35–38: Use natural logarithms to solve for $x$ in terms of $y$.

35  $y = \dfrac{e^x - e^{-x}}{2}$

36  $y = \dfrac{e^x + e^{-x}}{2}$

37  $y = \dfrac{e^x + e^{-x}}{e^x - e^{-x}}$

38  $y = \dfrac{e^x - e^{-x}}{e^x + e^{-x}}$

Exer. 39–40: Sketch the graph of $f$, and use the change of base formula to approximate the $y$-intercept.

39  $f(x) = \log_2(x + 3)$

40  $f(x) = \log_3(x + 5)$

Exer. 41–42: Sketch the graph of $f$, and use the change of base formula to approximate the $x$-intercept.

41  $f(x) = 4^x - 3$

42  $f(x) = 3^x - 6$

Exer. 43–46: Chemists use a number denoted by pH to describe quantitatively the acidity or basicity of solutions. By definition, $\text{pH} = -\log[\text{H}^+]$, where $[\text{H}^+]$ is the hydrogen ion concentration in moles per liter.

43  Approximate the pH of each substance:

  (a) vinegar:    $[\text{H}^+] \approx 6.3 \times 10^{-3}$

  (b) carrots:    $[\text{H}^+] \approx 1.0 \times 10^{-5}$

  (c) sea water:  $[\text{H}^+] \approx 5.0 \times 10^{-9}$

44  Approximate the hydrogen ion concentration $[\text{H}^+]$ of each substance:

  (a) apples:  $\text{pH} \approx 3.0$

  (b) beer:    $\text{pH} \approx 4.2$

  (c) milk:    $\text{pH} \approx 6.6$

45  A solution is considered basic if $[\text{H}^+] < 10^{-7}$ or acidic if $[\text{H}^+] > 10^{-7}$. Find the corresponding inequalities involving pH.

46  Many solutions have a pH between 1 and 14. Find the corresponding range of $[\text{H}^+]$.

47  *Compound interest*  Use the compound interest formula to determine how long it will take for a sum of money to double if it is invested at a rate of 6% per year compounded monthly.

**48 Compound interest** Solve the compound interest formula

$$A = P\left(1 + \frac{r}{n}\right)^{nt}$$

for $t$ by using natural logarithms.

**49 Photic zone** Refer to Example 6. The most important zone in the sea from the viewpoint of marine biology is the photic zone, in which photosynthesis takes place. The photic zone ends at the depth where about 1% of the surface light penetrates. In very clear waters in the Caribbean, 50% of the light at the surface reaches a depth of about 13 meters. Estimate the depth of the photic zone.

**50 Photic zone** In contrast to the situation described in the previous exercise, in parts of New York harbor, 50% of the surface light does not reach a depth of 10 centimeters. Estimate the depth of the photic zone.

**51** Shown in the figure is a graph of $f(x) = (\ln x)/x$ for $x > 0$. The maximum value of $f(x)$ occurs at $x = e$.

(a) The integers 2 and 4 have the unusual property that $2^4 = 4^2$. Show that if $x^y = y^x$ for positive real numbers $x$ and $y$, then $(\ln x)/x = (\ln y)/y$.

(b) Use the graph of $f$ to explain why many pairs of real numbers satisfy the equation $x^y = y^x$.

**EXERCISE 51**

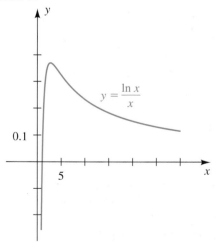

**52 Language history** Refer to Exercise 36 of Section 4.1. If a language originally had $N_0$ basic words of which $N(t)$ are still in use, then $N(t) = N_0(0.805)^t$, where time $t$ is measured in millennia. After how many years are one-half the basic words still in use?

**53 Drug absorption** If a 100-milligram tablet of an asthma drug is taken orally and if none of the drug is present in the body when the tablet is first taken, the total amount $A$ in the bloodstream after $t$ minutes is predicted to be

$$A = 100[1 - (0.9)^t] \quad \text{for} \quad 0 \le t \le 10.$$

(a) Sketch the graph of the equation.

(b) Determine the number of minutes needed for 50 milligrams of the drug to have entered the bloodstream.

**54 Drug dosage** A drug is eliminated from the body through urine. Suppose that for a dose of 10 milligrams, the amount $A(t)$ remaining in the body $t$ hours later is given by $A(t) = 10(0.8)^t$ and that in order for the drug to be effective, at least 2 milligrams must be in the body.

(a) Determine when 2 milligrams is left in the body.

(b) What is the half-life of the drug?

**55 Genetic mutation** The basic source of genetic diversity is mutation, or changes in the chemical structure of genes. If a gene mutates at a constant rate $m$ and if other evolutionary forces are negligible, then the frequency $F$ of the original gene after $t$ generations is given by $F = F_0(1 - m)^t$, where $F_0$ is the frequency at $t = 0$.

(a) Solve the equation for $t$ using common logarithms.

(b) If $m = 5 \times 10^{-5}$, after how many generations does $F = \frac{1}{2}F_0$?

**56 Employee productivity** Certain learning processes may be illustrated by the graph of an equation of the form $f(x) = a + b(1 - e^{-cx})$, where $a$, $b$, and $c$ are positive constants. Suppose a manufacturer estimates that a new employee can produce five items the first day on the job. As the employee becomes more proficient, the daily production increases until a certain maximum production is reached. Suppose that on the $n$th day on the job, the number $f(n)$ of items produced is approximated by

$$f(n) = 3 + 20(1 - e^{-0.1n}).$$

(a) Estimate the number of items produced on the fifth day, the ninth day, the twenty-fourth day, and the thirtieth day.

(b) Sketch the graph of $f$ from $n = 0$ to $n = 30$. (Graphs of this type are called *learning curves* and are used frequently in education and psychology.)

(c) What happens as $n$ increases without bound?

**57** *Height of trees*  The growth in height of trees is frequently described by a logistic equation. Suppose the height $h$ (in feet) of a tree at age $t$ (in years) is

$$h = \frac{120}{1 + 200e^{-0.2t}},$$

as illustrated by the graph in the figure.

(a) What is the height of the tree at age 10?

(b) At what age is the height 50 feet?

**EXERCISE 57**

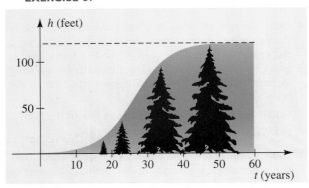

**58** *Employee productivity*  Manufacturers sometimes use empirically based formulas to predict the time required to produce the $n$th item on an assembly line for an integer $n$. If $T(n)$ denotes the time required to assemble the $n$th item and $T_1$ denotes the time required for the first, or prototype, item, then typically $T(n) = T_1 n^{-k}$ for some positive constant $k$.

(a) For many airplanes, the time required to assemble the second airplane, $T(2)$, is equal to $(0.80)T_1$. Find the value of $k$.

(b) Express, in terms of $T_1$, the time required to assemble the fourth airplane.

(c) Express, in terms of $T(n)$, the time $T(2n)$ required to assemble the $(2n)$th airplane.

**59** *Vertical wind shear*  Refer to Exercises 65–66 in Section 1.8. If $v_0$ is the wind speed at height $h_0$ and if $v_1$ is the wind speed at height $h_1$, then the vertical wind shear can be described by the equation

$$\frac{v_0}{v_1} = \left(\frac{h_0}{h_1}\right)^P,$$

where $P$ is a constant. During a one-year period in Montreal, the maximum vertical wind shear occurred when

the winds at the 200-foot level were 25 mi/hr while the winds at the 35-foot level were 6 mi/hr. Find $P$ for these conditions.

**60** *Vertical wind shear*  Refer to Exercise 59. The average vertical wind shear is given by the equation

$$s = \frac{v_1 - v_0}{h_1 - h_0}.$$

Suppose that the velocity of the wind increases with increasing altitude and that all values for wind speeds taken at the 35-foot and 200-foot altitudes are greater than 1 mi/hr. Does increasing the value of $P$ produce larger or smaller values of $s$?

**Exer. 61–62: An economist suspects that the following data points lie on the graph of $y = c2^{kx}$, where $c$ and $k$ are constants. If the data points have three-decimal-place accuracy, is this suspicion correct?**

**61**  $(0, 4)$, $(1, 3.249)$, $(2, 2.639)$, $(3, 2.144)$

**62**  $(0, -0.3)$, $(0.5, -0.345)$, $(1, -0.397)$, $(1.5, -0.551)$, $(2, -0.727)$

**Exer. 63–64: It is suspected that the following data points lie on the graph of $y = c \log{(kx + 10)}$, where $c$ and $k$ are constants. If the data points have three-decimal-place accuracy, is this suspicion correct?**

**63**  $(0, 1.5)$, $(1, 1.619)$, $(2, 1.720)$, $(3, 1.997)$

**64**  $(0, 0.7)$, $(1, 0.782)$, $(2, 0.847)$, $(3, 0.900)$, $(4, 0.945)$

**Exer. 65–66: Use a graph to estimate the roots of the equation on the given interval.**

**65**  $x - \ln{(0.3x)} - 3 \log_3 x = 0;$   $(0, 9)$

**66**  $2 \log 2x - \log_3 x^2 = 0;$     $(0, 3)$

**Exer. 67–68: Graph $f$ and $g$ on the same coordinate plane, and estimate the solution of the equation $f(x) = g(x)$.**

**67**  $f(x) = x;$   $g(x) = 3 \log_2 x$

**68**  $f(x) = x;$   $g(x) = -x^2 - \log_5 x$

**Exer. 69–70: Graph $f$ and $g$ on the same coordinate plane, and estimate the solution of the inequality $f(x) > g(x)$.**

**69**  $f(x) = 3^{-x} - 4^{0.2x};$     $g(x) = \ln{(1.2)} - x$

**70**  $f(x) = 3 \log_4 x - \log x;$   $g(x) = e^x - 0.25x^4$

# CHAPTER 4 REVIEW EXERCISES

**Exer. 1–16: Sketch the graph of f.**

1  $f(x) = 3^{x+2}$

2  $f(x) = (\frac{3}{5})^x$

3  $f(x) = (\frac{3}{2})^{-x}$

4  $f(x) = 3^{-2x}$

5  $f(x) = 3^{-x^2}$

6  $f(x) = 1 - 3^{-x}$

7  $f(x) = e^{x/2}$

8  $f(x) = \frac{1}{2}e^x$

9  $f(x) = e^{x-2}$

10  $f(x) = e^{2-x}$

11  $f(x) = \log_6 x$

12  $f(x) = \log_6 (36x)$

13  $f(x) = \log_4 (x^2)$

14  $f(x) = \log_4 \sqrt[3]{x}$

15  $f(x) = \log_2 (x + 4)$

16  $f(x) = \log_2 (4 - x)$

**Exer. 17–18: Evaluate without using a calculator or table.**

17  (a) $\log_2 \frac{1}{16}$   (b) $\log_\pi 1$   (c) $\ln e$
    (d) $6^{\log_6 4}$   (e) $\log 1{,}000{,}000$   (f) $10^{3 \log 2}$
    (g) $\log_4 2$

18  (a) $\log_5 \sqrt[3]{5}$   (b) $\log_5 1$   (c) $\log 10$
    (d) $e^{\ln 5}$   (e) $\log \log 10^{10}$   (f) $e^{2 \ln 5}$
    (g) $\log_{27} 3$

**Exer. 19–36: Solve the equation without using a calculator or table.**

19  $2^{3x-1} = \frac{1}{2}$

20  $\log \sqrt{x} = \log (x - 6)$

21  $\log_8 (x - 5) = \frac{2}{3}$

22  $\log_4 (x + 1) = 2 + \log_4 (3x - 2)$

23  $2 \ln (x + 3) - \ln (x + 1) = 3 \ln 2$

24  $\log \sqrt[4]{x + 1} = \frac{1}{2}$

25  $2^{5-x} = 6$

26  $3^{(x^2)} = 7$

27  $2^{5x+3} = 3^{2x+1}$

28  $\log_3 (3x) = \log_3 x + \log_3 (4 - x)$

29  $\log_4 x = \sqrt[3]{\log_4 x}$

30  $e^{x+\ln 4} = 3e^x$

31  $10^{2 \log x} = 5$

32  $e^{\ln (x+1)} = 3$

33  $x^2(-2xe^{-x^2}) + 2xe^{-x^2} = 0$

34  $e^x + 2 = 8e^{-x}$

35  (a) $\log x^2 = \log (6 - x)$   (b) $2 \log x = \log (6 - x)$

36  (a) $\ln (e^x)^2 = 16$   (b) $\ln e^{(x^2)} = 16$

37  Express $\log x^4 \sqrt[3]{y^2/z}$ in terms of logarithms of $x$, $y$, and $z$.

38  Express $\log (x^2/y^3) + 4 \log y - 6 \log \sqrt{xy}$ as one logarithm.

**Exer. 39–40: Use common logarithms to solve the equation for x in terms of y.**

39  $y = \dfrac{1}{10^x + 10^{-x}}$

40  $y = \dfrac{1}{10^x - 10^{-x}}$

**Exer. 41–42: Approximate x to three significant figures.**

41  (a) $x = \ln 6.6$   (b) $\log x = 1.8938$
    (c) $\ln x = -0.75$

42  (a) $x = \log 8.4$   (b) $\log x = -2.4260$
    (c) $\ln x = 1.8$

43  **Bacteria growth**   The number of bacteria in a certain culture at time $t$ (in hours) is given by $Q(t) = 2(3^t)$, where $Q(t)$ is measured in thousands.
    (a) What is the number of bacteria at $t = 0$?
    (b) Find the number after 10 minutes, 30 minutes, and 1 hour.

44  **Compound interest**   If $1000 is invested at a rate of 12% compounded quarterly, what is the principal after one year?

45  **Radioactive iodine decay**   Radioactive iodine $^{131}$I, which is frequently used in tracer studies involving the thyroid gland, decays according to $N = N_0(0.5)^{t/8}$, where $N_0$ is the initial dose and $t$ is the time in days.
    (a) Sketch the graph of the equation if $N_0 = 64$.
    (b) Find the half-life of $^{131}$I.

46  **Trout population**   A pond is stocked with 1000 trout. Three months later, it is estimated that 600 remain. Find a formula of the form $N = N_0 a^{ct}$ that can be used to estimate the number of trout remaining after $t$ months.

**47 Continuously compounded interest** Ten thousand dollars is invested in a savings fund in which interest is compounded continuously at the rate of 11% per year.

(a) When will the account contain $35,000?

(b) How long does it take for money to double in the account?

**48 Electrical current** The current $I(t)$ in a certain electrical circuit at time $t$ is given by $I(t) = I_0 e^{-Rt/L}$, where $R$ is the resistance, $L$ is the inductance, and $I_0$ is the initial current at $t = 0$. Find the value of $t$, in terms of $L$ and $R$, for which $I(t)$ is 1% of $I_0$.

**49 Sound intensity** The sound intensity level formula is $\alpha = 10 \log (I/I_0)$.

(a) Solve for $I$ in terms of $\alpha$ and $I_0$.

(b) Show that a one-decibel rise in the intensity level $\alpha$ corresponds to a 26% increase in the intensity $I$.

**50 Fish growth** The length $L$ of a fish is related to its age by means of the von Bertalanffy growth formula

$$L = a(1 - be^{-kt}),$$

where $a$, $b$, and $k$ are positive constants that depend on the type of fish. Solve this equation for $t$ to obtain a formula that can be used to estimate the age of a fish from a length measurement.

**51 Earthquake area in the West** In the western United States, the area $A$ (in mi$^2$) affected by an earthquake is related to the magnitude $R$ of the quake by the formula

$$R = 2.3 \log (A + 3000) - 5.1.$$

Solve for $A$ in terms of $R$.

**52 Earthquake area in the East** Refer to Exercise 51. For the eastern United States, the area-magnitude formula has the form

$$R = 2.3 \log (A + 34{,}000) - 7.5.$$

If $A_1$ is the area affected by an earthquake of magnitude $R$ in the West and $A_2$ is the area affected by a similar quake in the East, find a formula for $A_1/A_2$ in terms of $R$.

**53 Earthquake area in the Central states** Refer to Exercise 51. For the Rocky Mountain and Central states, the area-magnitude formula has the form

$$R = 2.3 \log (A + 14{,}000) - 6.6.$$

If an earthquake has magnitude 4 on the Richter scale, estimate the area $A$ of the region that will feel the quake.

**54 Atmospheric pressure** Under certain conditions, the atmospheric pressure $p$ at altitude $h$ is given by the formula $p = 29e^{-0.000034h}$. Express $h$ as a function of $p$.

**55 Rocket velocity** A rocket of mass $m_1$ is filled with fuel of initial mass $m_2$. If frictional forces are disregarded, the total mass $m$ of the rocket at time $t$ after ignition is related to its upward velocity $v$ by $v = -a \ln m + b$, where $a$ and $b$ are constants. At ignition time $t = 0$, $v = 0$ and $m = m_1 + m_2$. At burnout, $m = m_1$. Use this information to find a formula, in terms of one logarithm, for the velocity of the rocket at burnout.

**56 Earthquake frequency** Let $n$ be the average number of earthquakes per year that have magnitudes between $R$ and $R + 1$ on the Richter scale. A formula that approximates the relationship between $n$ and $R$ is

$$\log n = 7.7 - (0.9)R.$$

(a) Solve the equation for $n$ in terms of $R$.

(b) Find $n$ if $R = 4$, 5, and 6.

**57 Earthquake energy** The energy $E$ (in ergs) released during an earthquake of magnitude $R$ may be approximated by using the formula

$$\log E = 11.4 + (1.5)R.$$

(a) Solve for $E$ in terms of $R$.

(b) Find the energy released during the famous Alaskan quake of 1964, which measured 8.4 on the Richter scale.

**58 Radioactive decay** A certain radioactive substance decays according to the formula $q(t) = q_0 e^{-0.0063t}$, where $q_0$ is the initial amount of the substance and $t$ is the time in days. Approximate the half-life of the substance.

**59 Children's growth** The Count Model is a formula that can be used to predict the height of preschool children. If $h$ is height (in centimeters) and $t$ is age (in years), then

$$h = 70.228 + 5.104t + 9.222 \ln t$$

for $\frac{1}{4} \le t \le 6$. From calculus, the rate of growth $R$ (in cm/year) is given by $R = 5.104 + (9.222/t)$. Predict the height and rate of growth of a typical 2-year-old.

**60 Electrical circuit** The current $I$ in a certain electrical circuit at time $t$ is given by

$$I = \frac{V}{R}(1 - e^{-Rt/L}),$$

where $V$ is the electromotive force, $R$ is the resistance, and $L$ is the inductance. Solve the equation for $t$.

61  **Carbon 14 dating**   The technique of carbon 14 ($^{14}$C) dating is used to determine the age of archaeological and geological specimens. The formula $T = -8310 \ln x$ is sometimes used to predict the age $T$ (in years) of a bone fossil, where $x$ is the percentage (expressed as a decimal) of $^{14}$C still present in the fossil.

(a) Estimate the age of a bone fossil that contains 4% of the $^{14}$C found in an equal amount of carbon in present-day bone.

(b) Approximate the percentage of $^{14}$C present in a fossil that is 10,000 years old.

62  **Population of Kenya**   Based on present birth and death rates, the population of Kenya is expected to increase according to the formula $N = 20.2e^{0.041t}$, with $N$ in millions and $t = 0$ corresponding to 1985. How many years will it take for the population to double?

*The motion of water waves is an illustration of a periodic function.*

Trigonometry was invented over 2000 years ago by the Greeks, who needed precise methods for measuring angles and sides of triangles. In fact, the word trigonometry was derived from the two Greek words trigonon (triangle) and metria (measurement). The chapter begins with a discussion of angles and how they are measured. We next introduce the trigonometric functions by using ratios of sides of a right triangle. After extending the domains of the trigonometric functions to arbitrary angles and real numbers, we consider their graphs and graphing techniques that make use of amplitudes, periods, and phase shifts. We then examine a large variety of applications and conclude the chapter with a brief description of harmonic motion.

# CHAPTER 5

# THE

# TRIGONOMETRIC

# FUNCTIONS

# 5.1   ANGLES

**FIGURE 1**

**FIGURE 2**   Coterminal angles

In geometry an **angle** is defined as the set of points determined by two rays, or half-lines, $l_1$ and $l_2$, having the same end point $O$. If $A$ and $B$ are points on $l_1$ and $l_2$, as in Figure 1, we refer to **angle AOB,** or $\angle AOB$. An angle may also be considered as two finite line segments with a common end point.

In trigonometry we often interpret angles as rotations of rays. Start with a fixed ray $l_1$ having end point $O$, and rotate it about $O$, in a plane, to a position specified by ray $l_2$. We call $l_1$ the **initial side,** $l_2$ the **terminal side,** and $O$ the **vertex** of $\angle AOB$. The amount or direction of rotation is not restricted in any way. We might let $l_1$ make several revolutions in either direction about $O$ before coming to position $l_2$, as illustrated by the curved arrows in Figure 2. Thus, many different angles have the same initial and terminal sides. Any two such angles are called **coterminal angles.**

If we introduce a rectangular coordinate system, then the **standard position** of an angle is obtained by taking the vertex at the origin and letting the initial side $l_1$ coincide with the positive $x$-axis. If $l_1$ is rotated in a *counterclockwise* direction to the terminal position $l_2$, then the angle is considered **positive.** If $l_1$ is rotated in a *clockwise* direction, the angle is **negative.** We often denote angles by lowercase Greek letters such as $\alpha$ (*alpha*), $\beta$ (*beta*), $\gamma$ (*gamma*), $\theta$ (*theta*), $\phi$ (*phi*), and so on. Figure 3 contains sketches of two positive angles, $\alpha$ and $\beta$, and a negative angle, $\gamma$. If the terminal side of an angle in standard position is in a certain quadrant, we say that the *angle* is in that quadrant. In Figure 3, $\alpha$ is in quadrant III, $\beta$ is in quadrant I, and $\gamma$ is in quadrant II. An angle is called a **quadrantal angle** if its terminal side lies on a coordinate axis.

**FIGURE 3**   Standard position of an angle

Positive angle                     Positive angle                     Negative angle

One unit of measurement for angles is the **degree.** The angle in standard position obtained by one complete revolution in the counterclockwise direction has measure 360 degrees, written 360°. Thus, an angle of measure 1 degree (1°) is obtained by $\frac{1}{360}$ of one complete counterclockwise revolution. In Figure 4, several angles measured in degrees are shown in

**FIGURE 4**

standard position on rectangular coordinate systems. Note that the first three are quadrantal angles.

Throughout our work, a notation such as $\theta = 60°$ specifies an angle $\theta$ whose measure is 60°. We also refer to *an angle of* 60° or *a* 60° *angle*, instead of using the more precise (but cumbersome) phrase *an angle having measure* 60°.

**EXAMPLE 1**   *Finding coterminal angles*

If $\theta = 60°$ is in standard position, find two positive angles and two negative angles that are coterminal with $\theta$.

**SOLUTION**   The angle $\theta$ is shown in standard position in the first sketch in Figure 5. To find positive coterminal angles, we may add 360° or 720° (or any other positive multiple of 360°) to $\theta$, obtaining

$$60° + 360° = 420° \quad \text{and} \quad 60° + 720° = 780°.$$

These coterminal angles are also shown in Figure 5.

To find negative coterminal angles, we may add −360° or −720° (or any other negative multiple of 360°), obtaining

$$60° + (-360°) = -300° \quad \text{and} \quad 60° + (-720°) = -660°,$$

as shown in Figure 5.

**FIGURE 5**

An angle $\theta$ is a **right angle** if $\theta = 90°$. The following chart contains definitions of other special types of angles.

| Terminology | Definition | Illustration |
|---|---|---|
| acute angle $\theta$ | $0° < \theta < 90°$ | $12°$; $37°$ |
| obtuse angle $\theta$ | $90° < \theta < 180°$ | $95°$; $157°$ |
| complementary angles $\alpha$, $\beta$ | $\alpha + \beta = 90°$ | $20°$, $70°$; $7°$, $83°$ |
| supplementary angles $\alpha$, $\beta$ | $\alpha + \beta = 180°$ | $115°$, $65°$; $18°$, $162°$ |

If smaller measurements than the degree are required, we can use tenths, hundredths, or thousandths of degrees. Alternatively, we can divide the degree into 60 equal parts, called **minutes** (denoted by ′), and each minute into 60 equal parts, called **seconds** (denoted by ″). Thus, $1° = 60′$, and $1′ = 60″$. The notation $\theta = 73°56′18″$ refers to an angle $\theta$ that has measure 73 degrees, 56 minutes, 18 seconds.

**EXAMPLE 2**    *Finding complementary angles*

Find the angle that is complementary to $\theta$:

**(a)** $\theta = 25°43′37″$    **(b)** $\theta = 73.26°$

*SOLUTION*    We wish to find $90° - \theta$. Let us arrange our work as follows:

**(a)**
$$\begin{array}{rl} 90° & = 89°59′60″ \\ \theta & = 25°43′37″ \\ \hline 90° - \theta & = 64°16′23″ \end{array}$$

**(b)**
$$\begin{array}{rl} 90° & = 90.00° \\ \theta & = 73.26° \\ \hline 90° - \theta & = 16.74° \end{array}$$

Degree measure for angles is used in applied areas such as surveying, navigation, and the design of mechanical equipment. In scientific applications that require calculus, it is customary to employ *radian measure*. To define an angle of radian measure 1, we consider a circle of any radius $r$. A **central angle** of a circle is an angle whose vertex is at the center of the circle. If $\theta$ is the central angle shown in Figure 6, we say that the arc $AP$ (denoted $\overset{\frown}{AP}$) of the circle **subtends** $\theta$ or that $\theta$ **is subtended by** $\overset{\frown}{AP}$. If the length of $\overset{\frown}{AP}$ is equal to the radius $r$ of the circle, then $\theta$ has a measure of one radian, as in the next definition.

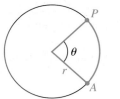

**FIGURE 6**   Central angle $\theta$

**Definition of Radian Measure**

**One radian** is the measure of the central angle of a circle subtended by an arc equal in length to the radius of the circle.

If we consider a circle of radius $r$, then an angle $\alpha$ whose measure is 1 radian intercepts an arc $AP$ of length $r$, as illustrated in Figure 7(a). The angle $\beta$ in Figure 7(b) has radian measure 2, since it is subtended by an arc of length $2r$. Similarly, $\gamma$ in (c) of the figure has radian measure 3, since it is subtended by an arc of length $3r$.

**FIGURE 7**

**(a)** $\alpha = 1$ radian

**(b)** $\beta = 2$ radians

**(c)** $\gamma = 3$ radians

**(d)** $360° = 2\pi \approx 6.28$ radians

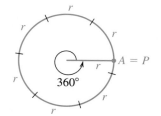

To find the radian measure corresponding to 360°, we must find the number of times that a circular arc of length $r$ can be laid off along the circumference (see Figure 7(d)). This number is not an integer or even a rational number. Since the circumference of the circle is $2\pi r$, the number of times $r$ units can be laid off is $2\pi$. Thus, an angle of measure $2\pi$ radians corresponds to the degree measure 360°, and we write $360° = 2\pi$ radians. This result gives us the following relationships.

**Relationships Between Degrees and Radians**

**(1)**    $180° = \pi$ radians

**(2)**    $1° = \dfrac{\pi}{180}$ radian $\approx 0.0175$ radian

**(3)** 1 radian $= \left(\dfrac{180}{\pi}\right)° \approx 57.2958°$

*When radian measure of an angle is used, no units will be indicated.* Thus, if an angle has radian measure 5, we write $\theta = 5$ instead of $\theta = 5$ *radians.* There should be no confusion as to whether radian or degree measure is being used, since if $\theta$ has *degree* measure 5°, we write $\theta = 5°$, and *not* $\theta = 5$.

The next chart illustrates how to change from one angular measure to another.

**Changing Angular Measures**

| To change | Multiply by | Illustration |
|---|---|---|
| degrees to radians | $\dfrac{\pi}{180}$ | $150° = 150\left(\dfrac{\pi}{180}\right) = \dfrac{5\pi}{6}$ |
| | | $225° = 225\left(\dfrac{\pi}{180}\right) = \dfrac{5\pi}{4}$ |
| radians to degrees | $\dfrac{180}{\pi}$ | $\dfrac{7\pi}{4} = \dfrac{7\pi}{4}\left(\dfrac{180}{\pi}\right)° = 315°$ |
| | | $\dfrac{\pi}{3} = \dfrac{\pi}{3}\left(\dfrac{180}{\pi}\right)° = 60°$ |

We may use the technique illustrated in the chart to obtain the following table, which displays the corresponding radian and degree measures of special angles.

| Radians | $0$ | $\dfrac{\pi}{6}$ | $\dfrac{\pi}{4}$ | $\dfrac{\pi}{3}$ | $\dfrac{\pi}{2}$ | $\dfrac{2\pi}{3}$ | $\dfrac{3\pi}{4}$ | $\dfrac{5\pi}{6}$ | $\pi$ | $\dfrac{7\pi}{6}$ | $\dfrac{5\pi}{4}$ | $\dfrac{4\pi}{3}$ | $\dfrac{3\pi}{2}$ | $\dfrac{5\pi}{3}$ | $\dfrac{7\pi}{4}$ | $\dfrac{11\pi}{6}$ | $2\pi$ |
|---|---|---|---|---|---|---|---|---|---|---|---|---|---|---|---|---|---|
| Degrees | $0°$ | $30°$ | $45°$ | $60°$ | $90°$ | $120°$ | $135°$ | $150°$ | $180°$ | $210°$ | $225°$ | $240°$ | $270°$ | $300°$ | $315°$ | $330°$ | $360°$ |

Several of these special angles, in radian measure, are shown in standard position in Figure 8.

**FIGURE 8**

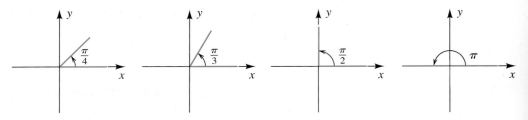

**EXAMPLE 3**   *Changing radians to degrees, minutes, and seconds*

If $\theta = 3$, approximate $\theta$ in terms of degrees, minutes, and seconds.

*SOLUTION*

$$3 \text{ radians} = 3\left(\frac{180}{\pi}\right)^{\circ} \qquad \text{multiply by } \frac{180}{\pi}$$

$$\approx 171.8873° \qquad\qquad \text{approximate}$$

$$= 171° + (0.8873)(60') \qquad 1° = 60'$$

$$= 171° + 53.238' \qquad\qquad \text{multiply}$$

$$= 171° + 53' + (0.238)(60'') \qquad 1' = 60''$$

$$= 171°53' + 14.28'' \qquad\qquad \text{multiply}$$

$$\approx 171°53'14'' \qquad\qquad \text{approximate}$$

Some calculators have a key that can be used to convert radian measure to degrees, and vice versa. Calculators may also have a $\boxed{\text{DMS}}$ key for converting decimal degree measure to degrees, minutes, and seconds, and vice versa. Since entries are made in terms of decimals, angles that are expressed in terms of degrees, minutes, and seconds must be changed to decimal form. If a $\boxed{\text{DMS}}$ key is not available, the procedure given in the next example may be followed.

**EXAMPLE 4** *Expressing minutes and seconds as decimal degrees*

Express $19°47'23''$ as a decimal, to the nearest ten-thousandth of a degree.

**SOLUTION** Since $1' = \left(\frac{1}{60}\right)°$ and $1'' = \left(\frac{1}{60}\right)' = \left(\frac{1}{3600}\right)°$,

$$19°47'23'' = 19° + \left(\frac{47}{60}\right)° + \left(\frac{23}{3600}\right)°$$
$$\approx 19° + 0.7833° + 0.0064°$$
$$= 19.7897°.$$

The next result specifies the relationship between the length of a circular arc and the central angle that it subtends.

**Formula for the Length of a Circular Arc**

If an arc of length $s$ on a circle of radius $r$ subtends a central angle of radian measure $\theta$, then

$$s = r\theta.$$

**FIGURE 9**

**(a)**

**PROOF** A typical arc of length $s$ and the corresponding central angle $\theta$ are shown in Figure 9(a). Figure 9(b) shows an arc of length $s_1$ and central angle $\theta_1$. If radian measure is used, then, from plane geometry, the ratio of the lengths of the arcs is the same as the ratio of the angular measures; that is,

$$\frac{s}{s_1} = \frac{\theta}{\theta_1}.$$

If we consider the special case in which $\theta_1$ has radian measure 1, then, from the definition of radian, $s_1 = r$ and

$$\frac{s}{r} = \frac{\theta}{1}, \quad \text{or} \quad s = r\theta. \quad \blacksquare$$

**(b)**

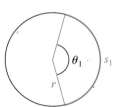

The next formula is proved in a similar manner.

**Formula for the Area of a Circular Sector**

If $\theta$ is the radian measure of a central angle of a circle of radius $r$ and if $A$ is the area of the circular sector determined by $\theta$, then

$$A = \tfrac{1}{2}r^2\theta.$$

**PROOF** If $A$ and $A_1$ are the areas of the sectors in Figure 9(a) and 9(b), respectively, then, from plane geometry,

$$\frac{A}{A_1} = \frac{\theta}{\theta_1}, \quad \text{or} \quad A = \frac{A_1}{\theta_1}\theta.$$

If we consider the special case $\theta_1 = 2\pi$, then $A_1 = \pi r^2$ and

$$A = \frac{\pi r^2}{2\pi}\theta = \frac{1}{2}r^2\theta. \quad \blacksquare$$

When using the preceding formulas, it is important to remember to use the radian measure of $\theta$ rather than the degree measure, as illustrated in the next example.

### EXAMPLE 5    *Using the circular arc and sector formulas*

A central angle $\theta$ is subtended by an arc 10 centimeters long on a circle of radius 4 centimeters.

**(a)** Approximate the measure of $\theta$ in degrees.

**(b)** Find the area of the circular sector determined by $\theta$.

*SOLUTION*    We proceed as follows:

**(a)**              $s = r\theta$           length of a circular arc formula

$\qquad\qquad \theta = \dfrac{s}{r}$           solve for $\theta$

$\qquad\qquad\quad = \frac{10}{4} = 2.5$    let $s = 10$, $r = 4$

This is the radian measure of $\theta$. Changing to degrees, we have

$$\theta = 2.5\left(\frac{180}{\pi}\right)^\circ = \left(\frac{450}{\pi}\right)^\circ \approx 143.24^\circ.$$

**(b)**              $A = \frac{1}{2}r^2\theta$           area of a circular sector formula

$\qquad\qquad = \frac{1}{2}(4)^2(2.5)$    let $r = 4$, $\theta = 2.5$

$\qquad\qquad = 20 \text{ cm}^2$        multiply

**FIGURE 10**

The **angular speed** of a wheel that is rotating at a constant rate is the angle generated in one unit of time by a line segment from the center of the wheel to a point $P$ on the circumference (see Figure 10). The **linear speed** of a point $P$ on the circumference is the distance that $P$ travels per unit of time.

### EXAMPLE 6    *Finding angular and linear speeds*

Suppose that a machine contains a wheel of diameter 3 feet, rotating at a rate of 1600 rpm (revolutions per minute).

**(a)** Find the angular speed of the wheel.

**(b)** Find the linear speed of a point $P$ on the circumference of the wheel.

## SOLUTION

**(a)** Let $O$ denote the center of the wheel, and let $P$ be a point on the circumference. Since the number of revolutions per minute is 1600 and since each revolution generates an angle of $2\pi$ radians, the angle generated by the line segment $OP$ in one minute has radian measure $(1600)(2\pi)$; that is,

$$\text{angular speed} = (1600)(2\pi) = 3200\pi \text{ radians per minute.}$$

Note that the diameter of the wheel is irrelevant in finding the angular speed.

**(b)** The linear speed of $P$ is the distance it travels per minute. We may find this distance by using the formula $s = r\theta$, with $r = \frac{3}{2}$ ft and $\theta = 3200\pi$. Thus,

$$s = \tfrac{3}{2}(3200\pi) = 4800\pi \text{ ft,}$$

and consequently the linear speed of $P$ is $4800\pi$ ft/min. To the nearest integer, this is approximately 15,080 ft/min, or 171.36 mi/hr. Unlike the angular speed, the linear speed *is* dependent on the diameter of the wheel.

## 5.1 EXERCISES

**Exer. 1–4: If the given angle is in standard position, find two positive coterminal angles and two negative coterminal angles.**

1. (a) 120°  (b) 135°  (c) −30°

2. (a) 240°  (b) 315°  (c) −150°

3. (a) 620°  (b) $\dfrac{5\pi}{6}$  (c) $-\dfrac{\pi}{4}$

4. (a) 570°  (b) $\dfrac{2\pi}{3}$  (c) $-\dfrac{5\pi}{4}$

**Exer. 5–6: Find the angle that is complementary to $\theta$.**

5. (a) $\theta = 5°17'34''$  (b) $\theta = 32.5°$

6. (a) $\theta = 63°4'15''$  (b) $\theta = 82.73°$

**Exer. 7–8: Find the angle that is supplementary to $\theta$.**

7. (a) $\theta = 48°51'37''$  (b) $\theta = 136.42°$

8. (a) $\theta = 152°12'4''$  (b) $\theta = 15.9°$

**Exer. 9–12: Find the exact radian measure of the angle.**

9. (a) 150°  (b) −60°  (c) 225°

10. (a) 120°  (b) −135°  (c) 210°

11. (a) 450°  (b) 72°  (c) 100°

12. (a) 630°  (b) 54°  (c) 95°

**Exer. 13–16: Find the exact degree measure of the angle.**

13. (a) $\dfrac{2\pi}{3}$  (b) $\dfrac{11\pi}{6}$  (c) $\dfrac{3\pi}{4}$

14. (a) $\dfrac{5\pi}{6}$  (b) $\dfrac{4\pi}{3}$  (c) $\dfrac{11\pi}{4}$

15. (a) $-\dfrac{7\pi}{2}$  (b) $7\pi$  (c) $\dfrac{\pi}{9}$

16. (a) $-\dfrac{5\pi}{2}$  (b) $9\pi$  (c) $\dfrac{\pi}{16}$

**Exer. 17–20: Express $\theta$ in terms of degrees, minutes, and seconds, to the nearest second.**

17. $\theta = 2$        18. $\theta = 1.5$

19. $\theta = 5$        20. $\theta = 4$

**Exer. 21–24:** Express the angle as a decimal, to the nearest ten-thousandth of a degree.

21  $37°41'$

22  $83°17'$

23  $115°26'27''$

24  $258°39'52''$

**Exer. 25–28:** Express the angle in terms of degrees, minutes, and seconds, to the nearest second.

25  $63.169°$

26  $12.864°$

27  $310.6215°$

28  $81.7238°$

**Exer. 29–30:** If a circular arc of the given length $s$ subtends the central angle $\theta$ on a circle, find the radius of the circle.

29  $s = 10$ cm,   $\theta = 4$

30  $s = 3$ km,   $\theta = 20°$

**Exer. 31–32: (a)** Find the length of the arc of the colored sector in the figure. **(b)** Find the area of the sector.

31

45°

8 cm

32

120°

9 cm

**Exer. 33–34: (a)** Find the radian and degree measures of the central angle $\theta$ subtended by the given arc of length $s$ on a circle of radius $r$. **(b)** Find the area of the sector determined by $\theta$.

33  $s = 7$ cm,   $r = 4$ cm

34  $s = 3$ ft,   $r = 20$ in.

**Exer. 35–36: (a)** Find the length of the arc that subtends the given central angle $\theta$ on a circle of diameter $d$. **(b)** Find the area of the sector determined by $\theta$.

35  $\theta = 50°$,   $d = 16$ m

36  $\theta = 2.2$,   $d = 120$ cm

37  **Measuring distances on the earth**   The distance between two points $A$ and $B$ on the earth is measured along a circle having center $C$ at the center of the earth and radius equal to the distance from $C$ to the surface (see the figure). If the diameter of the earth is approximately 8000 miles, approximate the distance between $A$ and $B$ if angle $ACB$ has the indicated measure:

(a) $60°$   (b) $45°$   (c) $30°$   (d) $10°$   (e) $1°$

**EXERCISE 37**

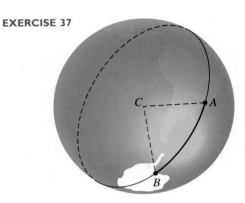

38  **Nautical miles**   Refer to Exercise 37. If angle $ACB$ has measure $1'$, then the distance between $A$ and $B$ is a nautical mile. Approximate the number of land (statute) miles in a nautical mile.

39  **Measuring angles using distance**   Refer to Exercise 37. If two points $A$ and $B$ are 500 miles apart, express angle $ACB$ in radians and in degrees.

40  **A tornado's core**   A simple model of the core of a tornado is a right circular cylinder that rotates about its axis. If a tornado has a core diameter of 200 feet and maximum wind speed of 180 mi/hr (or 264 ft/sec) at the perimeter of the core, approximate the number of revolutions the core makes each minute.

41  **The earth's rotation**   The earth rotates about its axis once every 23 hours, 56 minutes, and 4 seconds. Approximate the number of radians the earth rotates in one second.

42  **The earth's rotation**   Refer to Exercise 41. The equatorial radius of the earth is approximately 3963.3 miles. Find the linear speed of a point on the equator as a result of the earth's rotation.

**Exer. 43–44:** A wheel of the given radius is rotating at the indicated rate.
**(a)** Find the angular speed (in radians per minute).
**(b)** Find the linear speed of a point on the circumference (in ft/min).

43  radius 5 in.,   40 rpm

44  radius 9 in.,   2400 rpm

45  **Rotation of phonograph records**   The two common types of phonograph records, LP albums and singles, have diameters of 12 inches and 7 inches, respectively. The album rotates at a rate of $33\frac{1}{3}$ rpm, and the single rotates at 45 rpm.

(a) Find the angular speed (in radians per minute) of the album and of the single.

(b) Find the linear speed (in ft/min) of a point on the circumference of the album and of the single.

**46 Tire revolutions** A typical tire for a compact car is 22 inches in diameter. If the car is traveling at a speed of 60 mi/hr, find the number of revolutions the tire makes per minute.

**47 Cargo winch** A large winch of diameter 3 feet is used to hoist cargo, as shown in the figure.

(a) Find the distance the cargo is lifted if the winch rotates through an angle of radian measure $7\pi/4$.

(b) Find the angle (in radians) through which the winch must rotate in order to lift the cargo $d$ feet.

**EXERCISE 47**

**48 Pendulum's swing** A pendulum in a grandfather clock is 4 feet long and swings back and forth along a 6-inch arc. Approximate the angle (in degrees) through which the pendulum passes during one swing.

**49 Pizza values** A vendor sells two sizes of pizza by the slice. The *small* slice is $\frac{1}{6}$ of a circular 18-inch-diameter pizza, and it sells for $2.00. The *large* slice is $\frac{1}{8}$ of a circular 26-inch-diameter pizza, and it sells for $3.00. Which slice provides the most pizza per dollar?

**50 Bicycle mechanics** The sprocket assembly for a bicycle is shown in the figure. If the sprocket of radius $r_1$ rotates through an angle of $\theta_1$ radians, find the corresponding angle of rotation for the sprocket of radius $r_2$.

**EXERCISE 50**

**51 Bicycle mechanics** An expert cyclist can attain a speed of 40 mi/hr. If in Exercise 50 the sprocket assembly has $r_1 = 5$ in., $r_2 = 2$ in. and the wheel has a diameter of 28 inches, approximately how many revolutions per minute of the front sprocket wheel will produce a speed of 40 mi/hr? (*Hint:* First change 40 mi/hr to in./sec.)

**52 Magnetic pole drift** The geographic and magnetic north poles have different locations. Currently, the magnetic north pole is drifting westward through 0.0017 radian per year, where the angle of drift has its vertex at the center of the earth. If this movement continues, approximately how many years will it take for the magnetic north pole to drift a total of 5°?

## 5.2 TRIGONOMETRIC FUNCTIONS OF ANGLES

**FIGURE 11**

We shall introduce the trigonometric functions in the manner in which they originated historically—as ratios of sides of a right triangle. A triangle is a **right triangle** if one of its angles is a right angle. If $\theta$ is any acute angle, we may consider a right triangle having $\theta$ as one of its angles, as in Figure 11, where the symbol $\sqcap$ specifies the 90° angle. Six ratios can be obtained using the lengths $a$, $b$, and $c$ of the sides of the triangle:

$$\frac{b}{c}, \quad \frac{a}{c}, \quad \frac{b}{a}, \quad \frac{a}{b}, \quad \frac{c}{a}, \quad \frac{c}{b}$$

FIGURE 12

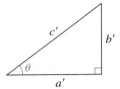

We can show that these ratios depend only on $\theta$, and not on the size of the triangle, as indicated in Figure 12. Since the two triangles have equal angles, they are similar, and therefore ratios of corresponding sides are proportional. For example,

$$\frac{b}{c} = \frac{b'}{c'}, \quad \frac{a}{c} = \frac{a'}{c'}, \quad \frac{b}{a} = \frac{b'}{a'}.$$

Thus, for each $\theta$, the six ratios are uniquely determined and hence are functions of $\theta$. They are called the **trigonometric functions** * and are designated as the **sine, cosine, tangent, cotangent, secant,** and **cosecant** functions, abbreviated **sin, cos, tan, cot, sec,** and **csc,** respectively. The symbol sin ($\theta$), or sin $\theta$, is used for the ratio $b/c$, which the sine function associates with $\theta$. Values of the other five functions are denoted in similar fashion. To summarize, if $\theta$ is the acute angle of the right triangle in Figure 11, then, by definition,

$$\sin \theta = \frac{b}{c} \quad \cos \theta = \frac{a}{c} \quad \tan \theta = \frac{b}{a}$$

$$\csc \theta = \frac{c}{b} \quad \sec \theta = \frac{c}{a} \quad \cot \theta = \frac{a}{b}$$

The domain of each of the six trigonometric functions is the set of all acute angles. Later in this section we will extend the domains to larger sets of angles, and in the next section, to real numbers.

If $\theta$ is the angle in Figure 11, we refer to the sides of the triangle of lengths $a$, $b$, and $c$ as the **adjacent side, opposite side,** and **hypotenuse,** respectively. We shall use **adj, opp,** and **hyp** to denote the lengths of the sides. We may then represent the triangle as in Figure 13. With this notation, the trigonometric functions may be expressed as follows.

FIGURE 13

**Definition of the Trigonometric Functions of an Acute Angle of a Right Triangle**

$$\sin \theta = \frac{\text{opp}}{\text{hyp}} \quad \cos \theta = \frac{\text{adj}}{\text{hyp}} \quad \tan \theta = \frac{\text{opp}}{\text{adj}}$$

$$\csc \theta = \frac{\text{hyp}}{\text{opp}} \quad \sec \theta = \frac{\text{hyp}}{\text{adj}} \quad \cot \theta = \frac{\text{adj}}{\text{opp}}$$

The formulas in the preceding definition can be applied to any right triangle without attaching the labels $a$, $b$, $c$ to the sides. Since the lengths of the sides of a triangle are positive real numbers, *the values of the six trigonometric functions are positive for every acute angle $\theta$*. Moreover, the hypotenuse is always greater than the adjacent or opposite side, and hence sin $\theta < 1$, cos $\theta < 1$, csc $\theta > 1$, and sec $\theta > 1$ for every acute angle $\theta$.

*There are other, less common trigonometric functions that we will not use in this text. Hence, we will refer to these six trigonometric functions as *the* trigonometric functions.

Note that since

$$\sin \theta = \frac{\text{opp}}{\text{hyp}} \quad \text{and} \quad \csc \theta = \frac{\text{hyp}}{\text{opp}},$$

$\sin \theta$ and $\csc \theta$ are reciprocals of each other, which gives us the two identities in the left-hand column of the next box. Similarly, $\cos \theta$ and $\sec \theta$ are reciprocals of each other, as are $\tan \theta$ and $\cot \theta$.

**Reciprocal Identities**

$$\sin \theta = \frac{1}{\csc \theta} \qquad \cos \theta = \frac{1}{\sec \theta} \qquad \tan \theta = \frac{1}{\cot \theta}$$

$$\csc \theta = \frac{1}{\sin \theta} \qquad \sec \theta = \frac{1}{\cos \theta} \qquad \cot \theta = \frac{1}{\tan \theta}$$

Several other important identities involving the trigonometric functions will be discussed at the end of this section.

**EXAMPLE 1**    *Finding trigonometric function values*

If $\theta$ is an acute angle and $\cos \theta = \frac{3}{4}$, find the values of the trigonometric functions of $\theta$.

**FIGURE 14**

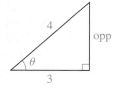

*SOLUTION*    We begin by sketching a right triangle having an acute angle $\theta$ with adj = 3 and hyp = 4, as shown in Figure 14, and proceed as follows:

$$3^2 + (\text{opp})^2 = 4^2 \qquad \text{Pythagorean theorem}$$
$$(\text{opp})^2 = 16 - 9 = 7 \qquad \text{isolate } (\text{opp})^2$$
$$\text{opp} = \sqrt{7} \qquad \text{take the square root}$$

Applying the definition of trigonometric functions of an acute angle of a right triangle, we obtain the following:

$$\sin \theta = \frac{\sqrt{7}}{4} \quad \cos \theta = \frac{3}{4} \quad \tan \theta = \frac{\sqrt{7}}{3}$$

$$\csc \theta = \frac{4}{\sqrt{7}} \quad \sec \theta = \frac{4}{3} \quad \cot \theta = \frac{3}{\sqrt{7}}$$

In Example 1 we could have rationalized the denominators for $\csc \theta$ and $\cot \theta$, writing

$$\csc \theta = \frac{4\sqrt{7}}{7} \quad \text{and} \quad \cot \theta = \frac{3\sqrt{7}}{7}.$$

However, in most examples and exercises we will leave expressions in unrationalized form. An exception to this practice is the special trigonometric function values corresponding to 60°, 30°, and 45°, which are obtained in the following example.

### EXAMPLE 2   Finding trigonometric function values of 60°, 30°, and 45°

Find the values of the trigonometric functions that correspond to $\theta$:

**(a)** $\theta = 60°$     **(b)** $\theta = 30°$     **(c)** $\theta = 45°$

**FIGURE 15**

**SOLUTION**   Consider an equilateral triangle with sides of length 2. The median from one vertex to the opposite side bisects the angle at that vertex, as illustrated by the dashes in Figure 15. By the Pythagorean theorem, the side opposite 60° in the shaded right triangle has length $\sqrt{3}$. Using the formulas for the trigonometric functions of an acute angle of a right triangle, we obtain the values corresponding to 60° and 30° as follows:

**(a)** $\sin 60° = \dfrac{\sqrt{3}}{2}$          $\cos 60° = \dfrac{1}{2}$          $\tan 60° = \dfrac{\sqrt{3}}{1} = \sqrt{3}$

$\csc 60° = \dfrac{2}{\sqrt{3}} = \dfrac{2\sqrt{3}}{3}$     $\sec 60° = \dfrac{2}{1} = 2$     $\cot 60° = \dfrac{1}{\sqrt{3}} = \dfrac{\sqrt{3}}{3}$

**(b)** $\sin 30° = \dfrac{1}{2}$          $\cos 30° = \dfrac{\sqrt{3}}{2}$          $\tan 30° = \dfrac{1}{\sqrt{3}} = \dfrac{\sqrt{3}}{3}$

$\csc 30° = \dfrac{2}{1} = 2$     $\sec 30° = \dfrac{2}{\sqrt{3}} = \dfrac{2\sqrt{3}}{3}$     $\cot 30° = \dfrac{\sqrt{3}}{1} = \sqrt{3}$

**FIGURE 16**

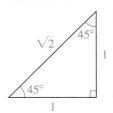

**(c)** To find the values for $\theta = 45°$, we may consider an isosceles right triangle whose two equal sides have length 1, as illustrated in Figure 16. By the Pythagorean theorem, the length of the hypotenuse is $\sqrt{2}$. Hence, the values corresponding to 45° are as follows:

$$\sin 45° = \dfrac{1}{\sqrt{2}} = \dfrac{\sqrt{2}}{2} = \cos 45° \quad \tan 45° = \dfrac{1}{1} = 1$$

$$\csc 45° = \dfrac{\sqrt{2}}{1} = \sqrt{2} = \sec 45° \quad \cot 45° = \dfrac{1}{1} = 1$$

For reference, we list the values found in Example 2, together with the radian measures of the angles, in the following table. Two reasons for stressing these values are that they are exact and that they occur frequently in work involving trigonometry. Because of the importance of these special values, it is a good idea either to memorize the table or to learn to find the values quickly by using triangles, as in Example 2.

## Special Values of the Trigonometric Functions

| $\theta$ (radians) | $\theta$ (degrees) | $\sin \theta$ | $\cos \theta$ | $\tan \theta$ | $\cot \theta$ | $\sec \theta$ | $\csc \theta$ |
|---|---|---|---|---|---|---|---|
| $\dfrac{\pi}{6}$ | $30°$ | $\dfrac{1}{2}$ | $\dfrac{\sqrt{3}}{2}$ | $\dfrac{\sqrt{3}}{3}$ | $\sqrt{3}$ | $\dfrac{2\sqrt{3}}{3}$ | $2$ |
| $\dfrac{\pi}{4}$ | $45°$ | $\dfrac{\sqrt{2}}{2}$ | $\dfrac{\sqrt{2}}{2}$ | $1$ | $1$ | $\sqrt{2}$ | $\sqrt{2}$ |
| $\dfrac{\pi}{3}$ | $60°$ | $\dfrac{\sqrt{3}}{2}$ | $\dfrac{1}{2}$ | $\sqrt{3}$ | $\dfrac{\sqrt{3}}{3}$ | $2$ | $\dfrac{2\sqrt{3}}{3}$ |

The next example illustrates a practical use for trigonometric functions of acute angles. Additional applications involving right triangles will be considered in Section 5.7.

**EXAMPLE 3    *Finding the height of a flagpole***

A surveyor observes that at a point A, located on level ground a distance 25.0 feet from the base B of a flagpole, the angle between the ground and the top of the pole is 30°. Approximate the height of the pole to the nearest tenth of a foot.

**FIGURE 17**

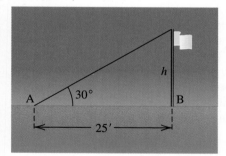

**SOLUTION**    If the height of the pole is $h$, then referring to Figure 17 we see that

$$\tan 30° = \frac{h}{25}, \quad \text{or} \quad h = 25 \tan 30°.$$

We use the value of $\tan 30°$ from Example 2 to find $h$:

$$h = 25\left(\frac{\sqrt{3}}{3}\right) \approx 14.4 \text{ ft}$$

It is possible to approximate, to any degree of accuracy, the values of the trigonometric functions for any acute angle. Scientific calculators have keys labeled $\boxed{\text{SIN}}$, $\boxed{\text{COS}}$, and $\boxed{\text{TAN}}$ that can be used to approximate values of these functions. The values of csc, sec, and cot may then be found by means of the reciprocal key. *Before using a calculator to find function values that correspond to the radian measure of an acute angle, be sure that the calculator is in radian mode. For values corresponding to degree measure, select degree mode.*
As an illustration, to find $\sin 30°$ on a typical calculator, place the calculator in degree mode and use the $\boxed{\text{SIN}}$ key to obtain $\sin 30° = 0.5$,

which is the exact value. Using the same procedure for 60°, we obtain a decimal approximation to $\sqrt{3}/2$, such as

$$\sin 60° \approx 0.8660.$$

Most calculators give eight- to ten-decimal-place accuracy for such function values; throughout the text, however, we will usually round off values to four decimal places.

To find a value such as cos 1.3, where 1.3 is the radian measure of an acute angle, we place the calculator in radian mode and use the $\boxed{\text{COS}}$ key, obtaining

$$\cos 1.3 \approx 0.2675.$$

For sec 1.3, we could find cos 1.3 and then use the reciprocal key, usually labeled $\boxed{1/x}$ or $\boxed{x^{-1}}$, to obtain

$$\sec 1.3 = \frac{1}{\cos 1.3} \approx 3.7383.$$

The formulas listed in the next box are, without doubt, the most important identities in trigonometry, because they can be used to simplify and unify many different aspects of the subject. Since the formulas are part of the foundation for work in trigonometry, they are called the *fundamental identities*.

Three of the fundamental identities involve squares, such as $(\sin \theta)^2$ and $(\cos \theta)^2$. In general, if $n$ is an integer different from $-1$, then a power such as $(\cos \theta)^n$ is written $\cos^n \theta$. The symbols $\sin^{-1} \theta$ and $\cos^{-1} \theta$ are reserved for inverse trigonometric functions, which we will discuss in the next chapter. With this agreement on notation, we have, for example,

$$\cos^2 \theta = (\cos \theta)^2 = (\cos \theta)(\cos \theta)$$

$$\tan^3 \theta = (\tan \theta)^3 = (\tan \theta)(\tan \theta)(\tan \theta)$$

$$\sec^4 \theta = (\sec \theta)^4 = (\sec \theta)(\sec \theta)(\sec \theta)(\sec \theta).$$

Let us first list all the fundamental identities and then discuss the proofs. These formulas are true for every acute angle $\theta$, and $\theta$ may take on various forms. For example, using the first Pythagorean identity with $\theta = 4\alpha$, we know that

$$\sin^2 4\alpha + \cos^2 4\alpha = 1.$$

We shall see later that they also are true for other angles and for real numbers.

**The Fundamental Identities**

*cô bản.*

**(1) The reciprocal identities:**   *Đảo nghịch.*

$$\csc \theta = \frac{1}{\sin \theta} \qquad \sec \theta = \frac{1}{\cos \theta} \qquad \cot \theta = \frac{1}{\tan \theta}$$

**(2) The tangent and cotangent identities:**

$$\tan \theta = \frac{\sin \theta}{\cos \theta} \qquad \cot \theta = \frac{\cos \theta}{\sin \theta}$$

**(3) The Pythagorean identities:**

$$\sin^2 \theta + \cos^2 \theta = 1 \quad 1 + \tan^2 \theta = \sec^2 \theta \quad 1 + \cot^2 \theta = \csc^2 \theta$$

*PROOF*

**(1)** The reciprocal identities were developed earlier in this section.

**(2)** To prove the tangent identity, we refer to the right triangle in Figure 11 and use definitions of trigonometric functions as follows:

$$\tan \theta = \frac{b}{a} = \frac{b/c}{a/c} = \frac{\sin \theta}{\cos \theta}$$

To verify the cotangent identity, we use a reciprocal identity and the tangent identity:

$$\cot \theta = \frac{1}{\tan \theta} = \frac{1}{\sin \theta / \cos \theta} = \frac{\cos \theta}{\sin \theta}$$

**(3)** The Pythagorean identities are so named because of the first step in the following proof. Referring to Figure 11, we obtain

$$b^2 + a^2 = c^2 \qquad \text{Pythagorean theorem}$$

$$\left(\frac{b}{c}\right)^2 + \left(\frac{a}{c}\right)^2 = \left(\frac{c}{c}\right)^2 \qquad \text{divide by } c^2$$

$$\sin^2 \theta + \cos^2 \theta = 1. \qquad \text{definitions of } \sin \theta \text{ and } \cos \theta$$

We may use this identity to verify the second Pythagorean identity, as follows:

$$\frac{\sin^2 \theta + \cos^2 \theta}{\cos^2 \theta} = \frac{1}{\cos^2 \theta} \qquad \text{divide by } \cos^2 \theta$$

$$\frac{\sin^2 \theta}{\cos^2 \theta} + \frac{\cos^2 \theta}{\cos^2 \theta} = \frac{1}{\cos^2 \theta} \qquad \text{equivalent equation}$$

$$\left(\frac{\sin \theta}{\cos \theta}\right)^2 + \left(\frac{\cos \theta}{\cos \theta}\right)^2 = \left(\frac{1}{\cos \theta}\right)^2 \qquad \text{law of exponents}$$

$$\tan^2 \theta + 1 = \sec^2 \theta \qquad \text{tangent and reciprocal identities}$$

To prove that $1 + \cot^2 \theta = \csc^2 \theta$, we could divide both sides of $\sin^2 \theta + \cos^2 \theta = 1$ by $\sin^2 \theta$. ∎

We can use the fundamental identities to express each trigonometric function in terms of any other trigonometric function. Two illustrations are given in the next example.

### EXAMPLE 4     *Using fundamental identities*

Let $\theta$ be an acute angle.

**(a)** Express $\sin \theta$ in terms of $\cos \theta$.

**(b)** Express $\tan \theta$ in terms of $\sin \theta$.

### SOLUTION

**(a)** We proceed as follows:

$$\sin^2 \theta + \cos^2 \theta = 1 \qquad \text{fundamental identity}$$
$$\sin^2 \theta = 1 - \cos^2 \theta \qquad \text{subtract } \cos^2 \theta$$
$$\sin \theta = \pm\sqrt{1 - \cos^2 \theta} \qquad \text{take the square root}$$
$$\sin \theta = \sqrt{1 - \cos^2 \theta} \qquad \sin \theta > 0 \text{ for acute angles}$$

Later in this section (Example 12) we will consider a simplification involving a *non*-acute angle $\theta$.

**(b)** If we begin with the fundamental identity

$$\tan \theta = \frac{\sin \theta}{\cos \theta},$$

then all that remains is to express $\cos \theta$ in terms of $\sin \theta$. We can do this by solving $\sin^2 \theta + \cos^2 \theta = 1$ for $\cos \theta$, obtaining

$$\cos \theta = \sqrt{1 - \sin^2 \theta} \quad \text{for} \quad 0 < \theta < \frac{\pi}{2}.$$

Hence,

$$\tan \theta = \frac{\sin \theta}{\cos \theta} = \frac{\sin \theta}{\sqrt{1 - \sin^2 \theta}} \quad \text{for} \quad 0 < \theta < \frac{\pi}{2}.$$

Fundamental identities are often used to simplify expressions involving trigonometric functions, as illustrated in the next example.

### EXAMPLE 5     *Showing that an equation is an identity*

Show that the following equation is an identity by transforming the left side into the right side:

$$(\sec \theta + \tan \theta)(1 - \sin \theta) = \cos \theta$$

*SOLUTION*   We begin with the left side and proceed as follows:

$$(\sec \theta + \tan \theta)(1 - \sin \theta)$$

$$= \left(\frac{1}{\cos \theta} + \frac{\sin \theta}{\cos \theta}\right)(1 - \sin \theta) \qquad \text{reciprocal and tangent identities}$$

$$= \left(\frac{1 + \sin \theta}{\cos \theta}\right)(1 - \sin \theta) \qquad \text{add fractions}$$

$$= \frac{1 - \sin^2 \theta}{\cos \theta} \qquad \text{multiply}$$

$$= \frac{\cos^2 \theta}{\cos \theta} \qquad \sin^2 \theta + \cos^2 \theta = 1$$

$$= \cos \theta \qquad \text{cancel } \cos \theta$$

There are other ways to simplify the expression on the left side in Example 5. We could first multiply the two factors and then simplify and combine terms. The method we employed—changing all expressions to expressions that involve only sines and cosines—is often useful. However, that technique does not always lead to the shortest possible simplification.

Hereafter, we shall use the phrase *verify an identity* instead of *show that an equation is an identity*. When verifying an identity, we often use fundamental identities and algebraic manipulations to simplify expressions, as we did in the preceding example.

**Warning**

> If an (assumed) identity contains fractions, do not multiply both sides by the lcd.

One reason for the preceding warning is that carelessness about angle restrictions may lead to erroneous conclusions. To illustrate, consider the equation

$$\frac{1 - \sin \theta}{1 - \sin \theta} = \frac{1 + \sin \theta}{1 + \sin \theta}.$$

If we multiply both sides by the lcd $(1 - \sin \theta)(1 + \sin \theta)$ and simplify, we obtain

$$(1 - \sin \theta)(1 + \sin \theta) = (1 + \sin \theta)(1 - \sin \theta)$$
$$1 - \sin^2 \theta = 1 - \sin^2 \theta.$$

The last equation is true for every $\theta$. This is not the case for the *given* equation, however, since if $\sin \theta = 1$, substitution gives us the following:

$$\text{LS: } \tfrac{0}{0} \text{ (undefined)}; \quad \text{RS: } \tfrac{2}{2} = 1$$

Similarly, the given equation is false for every $\theta$ such that $\sin \theta = -1$. If we had added the restriction $\sin \theta \neq \pm 1$, the given equation would be an identity, since it reduces to $1 = 1$ for every other $\theta$. Our warning allows us to avoid stating such restrictions each time we encounter an (assumed) identity.

**EXAMPLE 6**  *Verifying an identity*

Verify the following identity by transforming the left side into the right side:

$$\frac{\tan \theta + \cos \theta}{\sin \theta} = \sec \theta + \cot \theta$$

*SOLUTION*     By the warning preceding this example, *multiplication of both sides by* $\sin \theta$ *is not allowed.* We may transform the left side into the right side as follows:

$$\frac{\tan \theta + \cos \theta}{\sin \theta} = \frac{\tan \theta}{\sin \theta} + \frac{\cos \theta}{\sin \theta} \qquad \text{divide numerator by } \sin \theta$$

$$= \frac{\left(\dfrac{\sin \theta}{\cos \theta}\right)}{\sin \theta} + \cot \theta \qquad \text{tangent and cotangent identities}$$

$$= \frac{\sin \theta}{\cos \theta} \cdot \frac{1}{\sin \theta} + \cot \theta \qquad \text{rule for quotients}$$

$$= \frac{1}{\cos \theta} + \cot \theta \qquad \text{cancel } \sin \theta$$

$$= \sec \theta + \cot \theta \qquad \text{reciprocal identity}$$

In Section 6.1 we will verify many other identities using methods similar to those used in Examples 5 and 6.

Since many applied problems involve angles that are not acute, it is necessary to extend the definitions of the trigonometric functions. We make this extension by using the standard position of an angle $\theta$ on a rectangular coordinate system. If $\theta$ is acute, we have the situation illustrated in Figure 18, where we have chosen a point $P(x, y)$ on the terminal side of $\theta$ and where $d(O, P) = r = \sqrt{x^2 + y^2}$. Referring to triangle $OQP$, we have

$$\sin \theta = \frac{\text{opp}}{\text{hyp}} = \frac{y}{r}, \quad \cos \theta = \frac{\text{adj}}{\text{hyp}} = \frac{x}{r}, \quad \text{and} \quad \tan \theta = \frac{\text{opp}}{\text{adj}} = \frac{y}{x}.$$

**FIGURE 18**

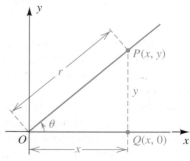

We now wish to consider angles of the types illustrated in Figure 19 (or any *other* angle, either positive, negative, or zero). Note that in Figure 19 the value of $x$ or $y$ may be negative. In each case, side $QP$ (opp in Figure 19) has length $|y|$, side $OQ$ (adj in Figure 19) has length $|x|$, and the hypotenuse $OP$ has length $r$. We shall define the six trigonometric functions so that their values agree with those given previously whenever the angle is acute. It is understood that if a zero denominator occurs, then the corresponding function value is undefined.

**FIGURE 19**

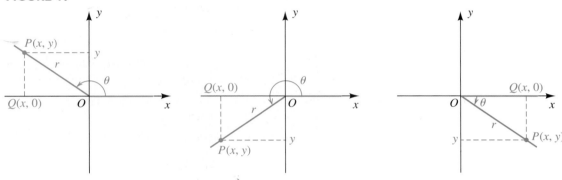

**Definition of Trigonometric Functions of Any Angle**

Let $\theta$ be an angle in standard position on a rectangular coordinate system, and let $P(x, y)$ be any point other than the origin $O$ on the terminal side of $\theta$.
    If $d(O, P) = r = \sqrt{x^2 + y^2}$, then

$$\sin \theta = \frac{y}{r} \qquad \cos \theta = \frac{x}{r} \qquad \tan \theta = \frac{y}{x}$$

$$\csc \theta = \frac{r}{y} \qquad \sec \theta = \frac{r}{x} \qquad \cot \theta = \frac{x}{y}$$

We can show, using similar triangles, that the formulas in this definition do not depend on the point $P(x, y)$ that is chosen on the terminal side of $\theta$. The fundamental identities, which were established for acute angles, are also true for trigonometric functions of any angle.
    The domains of the sine and cosine functions consist of all angles $\theta$. However, $\tan \theta$ and $\sec \theta$ are undefined if $x = 0$ (that is, if the terminal side of $\theta$ is on the $y$-axis). Thus, the domains of the tangent and the secant functions consist of all angles *except* those of radian measure $(\pi/2) + \pi n$

for any integer $n$. Some special cases are $\pm\pi/2$, $\pm3\pi/2$, and $\pm5\pi/2$. The corresponding degree measures are $\pm90°$, $\pm270°$, and $\pm450°$.

The domains of the cotangent and cosecant functions consist of all angles except those that lead to $y = 0$ (that is, all angles except those having terminal sides on the $x$-axis). These are the angles of radian measure $\pi n$ (or degree measure $180° \cdot n$) for any integer $n$.

Our discussion of domains is summarized in the following table, where $n$ denotes any integer.

| Function | Domain |
|---|---|
| sin, cos | every angle $\theta$ |
| tan, sec | every angle $\theta$ except $\theta = \dfrac{\pi}{2} + \pi n = 90° + 180° \cdot n$ |
| cot, csc | every angle $\theta$ except $\theta = \pi n = 180° \cdot n$ |

For any point $P(x, y)$ in the preceding definition, $|x| \le r$ and $|y| \le r$ or, equivalently, $|x/r| \le 1$ and $|y/r| \le 1$. Thus,

$$|\sin\theta| \le 1, \quad |\cos\theta| \le 1, \quad |\csc\theta| \ge 1, \quad \text{and} \quad |\sec\theta| \ge 1$$

for every $\theta$ in the domains of these functions.

**EXAMPLE 7**   *Finding trigonometric function values of an angle in standard position*

If $\theta$ is an angle in standard position on a rectangular coordinate system and if $P(-15, 8)$ is on the terminal side of $\theta$, find the values of the six trigonometric functions of $\theta$.

**FIGURE 20**

**SOLUTION**   The point $P(-15, 8)$ is shown in Figure 20. Applying the definition of trigonometric functions of any angle with $x = -15$, $y = 8$, and

$$r = \sqrt{x^2 + y^2} = \sqrt{(-15)^2 + 8^2} = \sqrt{289} = 17,$$

we obtain the following:

$$\sin\theta = \tfrac{8}{17} \quad \cos\theta = -\tfrac{15}{17} \quad \tan\theta = -\tfrac{8}{15}$$
$$\csc\theta = \tfrac{17}{8} \quad \sec\theta = -\tfrac{17}{15} \quad \cot\theta = -\tfrac{15}{8}$$

**EXAMPLE 8**   *Finding trigonometric function values of an angle in standard position*

An angle $\theta$ is in standard position, and its terminal side lies in quadrant III on the line $y = 3x$. Find the values of the trigonometric functions of $\theta$.

**FIGURE 21**

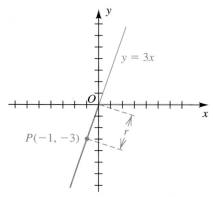

**SOLUTION**   The graph of $y = 3x$ is sketched in Figure 21, together with the initial and terminal sides of $\theta$. Since the terminal side of $\theta$ is in quadrant III, we begin by choosing a convenient negative value for $x$, say $x = -1$. Substituting in $y = 3x$ gives us $y = 3(-1) = -3$, and hence $P(-1, -3)$ is on the terminal side. Applying the definition of trigonometric functions of any angle with

$$x = -1, \quad y = -3, \quad \text{and} \quad r = \sqrt{x^2 + y^2} = \sqrt{(-1)^2 + (-3)^2} = \sqrt{10}$$

gives us

$$\sin\theta = -\frac{3}{\sqrt{10}} \quad \cos\theta = -\frac{1}{\sqrt{10}} \quad \tan\theta = \frac{-3}{-1} = 3$$

$$\csc\theta = -\frac{\sqrt{10}}{3} \quad \sec\theta = -\frac{\sqrt{10}}{1} \quad \cot\theta = \frac{-1}{-3} = \frac{1}{3}$$

The definitions of the trigonometric functions may be applied if $\theta$ is a quadrantal angle. The procedure is illustrated in the next example.

**EXAMPLE 9**   *Finding trigonometric function values of a quadrantal angle*

If $\theta = 3\pi/2$, find the values of the trigonometric functions of $\theta$.

**FIGURE 22**

**SOLUTION**   Note that $3\pi/2 = 270°$. If $\theta$ is placed in standard position, the terminal side of $\theta$ coincides with the negative $y$-axis, as shown in Figure 22. To apply the definition of trigonometric functions of any angle, we may choose *any* point $P$ on the terminal side of $\theta$. For simplicity, we use $P(0, -1)$. In this case, $x = 0$, $y = -1$, $r = 1$, and hence

$$\sin\frac{3\pi}{2} = \frac{-1}{1} = -1 \quad \cos\frac{3\pi}{2} = \frac{0}{1} = 0$$

$$\csc\frac{3\pi}{2} = \frac{1}{-1} = -1 \quad \cot\frac{3\pi}{2} = \frac{0}{-1} = 0$$

The tangent and secant functions are undefined, since the meaningless expressions $\tan\theta = (-1)/0$ and $\sec\theta = 1/0$ occur when we substitute in the appropriate formulas.

Let us determine the signs associated with values of the trigonometric functions. If $\theta$ is in quadrant II and $P(x, y)$ is a point on the terminal side, then $x$ is negative and $y$ is positive. Hence, $\sin\theta = y/r$ and $\csc\theta = r/y$ are positive, and the other four trigonometric functions are negative. Checking the remaining quadrants in a similar fashion, we obtain the following table.

**Signs of the Trigonometric Functions**

| Quadrant containing $\theta$ | Positive functions | Negative functions |
|:---:|:---:|:---:|
| I | all | none |
| II | sin, csc | cos, sec, tan, cot |
| III | tan, cot | sin, csc, cos, sec |
| IV | cos, sec | sin, csc, tan, cot |

**FIGURE 23** Positive trigonometric functions

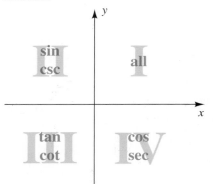

The diagram in Figure 23 may be useful for remembering quadrants in which trigonometric functions are *positive*. If a function is not listed (such as cos in quadrant II), then that function is negative. We finish this section with three examples that require using the information in the preceding table.

**EXAMPLE 10**   *Finding the quadrant containing an angle*

Find the quadrant containing $\theta$ if both $\cos \theta > 0$ and $\sin \theta < 0$.

**SOLUTION**   Referring to the table of signs or Figure 23, we see that $\cos \theta > 0$ if $\theta$ is in quadrant I or IV and that $\sin \theta < 0$ if $\theta$ is in quadrant III or IV. Hence, for both conditions to be satisfied, $\theta$ must be in quadrant IV.

**EXAMPLE 11**   *Finding values of trigonometric functions from prescribed conditions*

If $\sin \theta = \frac{3}{5}$ and $\tan \theta < 0$, use fundamental identities to find the values of the other five trigonometric functions.

**SOLUTION**   Since $\sin \theta = \frac{3}{5} > 0$ (positive) and $\tan \theta < 0$ (negative), $\theta$ is in quadrant II. Using the relationship $\sin^2 \theta + \cos^2 \theta = 1$ and the fact that $\cos \theta$ is negative in quadrant II, we have

$$\cos \theta = -\sqrt{1 - \sin^2 \theta} = -\sqrt{1 - (\tfrac{3}{5})^2} = -\sqrt{\tfrac{16}{25}} = -\tfrac{4}{5}.$$

Next we use the tangent identity to obtain

$$\tan \theta = \frac{\sin \theta}{\cos \theta} = \frac{3/5}{-4/5} = -\frac{3}{4}.$$

Finally, using the reciprocal identities gives us

$$\csc \theta = \frac{1}{\sin \theta} = \frac{1}{3/5} = \frac{5}{3}$$

$$\sec \theta = \frac{1}{\cos \theta} = \frac{1}{-4/5} = -\frac{5}{4}$$

$$\cot \theta = \frac{1}{\tan \theta} = \frac{1}{-3/4} = -\frac{4}{3}$$

### EXAMPLE 12    *Using fundamental identities*

Rewrite $\sqrt{\cos^2 \theta + \sin^2 \theta + \cot^2 \theta}$ in nonradical form without using absolute values for $\pi < \theta < 2\pi$.

### SOLUTION

$$\sqrt{\cos^2 \theta + \sin^2 \theta + \cot^2 \theta} = \sqrt{1 + \cot^2 \theta} \quad \cos^2 \theta + \sin^2 \theta = 1$$
$$= \sqrt{\csc^2 \theta} \quad\quad 1 + \cot^2 \theta = \csc^2 \theta$$
$$= |\csc \theta| \quad\quad\quad \sqrt{x^2} = |x|$$

Since $\pi < \theta < 2\pi$, we know that $\theta$ is in quadrant III or IV. Thus, $\csc \theta$ is *negative,* and by the definition of absolute value, we have

$$|\csc \theta| = -\csc \theta.$$

## 5.2 EXERCISES

**Exer. 1–8: Find the values of the six trigonometric functions for the angle $\theta$.**

1

2

3

4

5

6

7

8

**Exer. 9–14:** Find the exact values of *x* and *y*.

9

10

11

12

13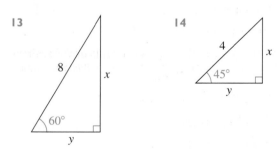

14

**Exer. 15–20:** Find the exact values of the trigonometric functions for the acute angle *θ*.

15  $\sin\theta = \frac{3}{5}$

16  $\cos\theta = \frac{8}{17}$

17  $\tan\theta = \frac{5}{12}$

18  $\cot\theta = \frac{7}{24}$

19  $\sec\theta = \frac{6}{5}$

20  $\csc\theta = 4$

21  **Height of a tree**   A forester, 200 feet from the base of a redwood tree, observes that the angle between the ground and the top of the tree is 60°. Estimate the height of the tree.

22  **Distance to Mt. Fuji**   The peak of Mt. Fuji in Japan is approximately 12,400 feet high. A trigonometry stu-

dent, several miles away, notes that the angle between level ground and the peak is 30°. Estimate the distance from the student to the point on level ground directly beneath the peak.

**Exer. 23–26:** Use the Pythagorean identities to write the expression as an integer.

23  (a) $\tan^2 4\beta - \sec^2 4\beta$    (b) $4\tan^2\beta - 4\sec^2\beta$

24  (a) $\csc^2 3\alpha - \cot^2 3\alpha$    (b) $3\csc^2\alpha - 3\cot^2\alpha$

25  (a) $5\sin^2\theta + 5\cos^2\theta$
    (b) $5\sin^2(\theta/4) + 5\cos^2(\theta/4)$

26  (a) $7\sec^2\gamma - 7\tan^2\gamma$
    (b) $7\sec^2(\gamma/3) - 7\tan^2(\gamma/3)$

**Exer. 27–30:** Simplify the expression.

27  $\dfrac{\sin^3\theta + \cos^3\theta}{\sin\theta + \cos\theta}$

28  $\dfrac{\cot^2\alpha - 4}{\cot^2\alpha - \cot\alpha - 6}$

29  $\dfrac{2 - \tan\theta}{2\csc\theta - \sec\theta}$

30  $\dfrac{\csc\theta + 1}{(1/\sin^2\theta) + \csc\theta}$

**Exer. 31–36:** Use fundamental identities to write the first expression in terms of the second, for any acute angle *θ*.

31  $\cot\theta$,   $\sin\theta$

32  $\tan\theta$,   $\cos\theta$

33  $\sec\theta$,   $\sin\theta$

34  $\csc\theta$,   $\cos\theta$

35  $\sin\theta$,   $\sec\theta$

36  $\cos\theta$,   $\cot\theta$

**Exer. 37–58:** Verify the identity by transforming the left side into the right side.

37  $\cos\theta\sec\theta = 1$

38  $\tan\theta\cot\theta = 1$

39  $\sin\theta\sec\theta = \tan\theta$

40  $\sin\theta\cot\theta = \cos\theta$

41  $\dfrac{\csc\theta}{\sec\theta} = \cot\theta$

42  $\cot\theta\sec\theta = \csc\theta$

43  $(1 + \cos\theta)(1 - \cos\theta) = \sin^2\theta$

44  $\cos^2\theta - \sin^2\theta = 2\cos^2\theta - 1$

45  $\cos^2\theta(\sec^2\theta - 1) = \sin^2\theta$

46  $(\tan\theta + \cot\theta)\tan\theta = \sec^2\theta$

47  $\dfrac{\sin\theta}{\csc\theta} + \dfrac{\cos\theta}{\sec\theta} = 1$

48  $1 - 2\sin^2\theta = 2\cos^2\theta - 1$

49  $(1 + \sin\theta)(1 - \sin\theta) = \dfrac{1}{\sec^2\theta}$

50 $(1 - \sin^2 \theta)(1 + \tan^2 \theta) = 1$

51 $\sec \theta - \cos \theta = \tan \theta \sin \theta$

52 $\dfrac{\sin \theta + \cos \theta}{\cos \theta} = 1 + \tan \theta$

53 $(\cot \theta + \csc \theta)(\tan \theta - \sin \theta) = \sec \theta - \cos \theta$

54 $\cot \theta + \tan \theta = \csc \theta \sec \theta$

55 $\sec^2 \theta \csc^2 \theta = \sec^2 \theta + \csc^2 \theta$

56 $\dfrac{1 + \cos^2 \theta}{\sin^2 \theta} = 2 \csc^2 \theta - 1$

57 $\log \csc \theta = -\log \sin \theta$

58 $\log \tan \theta = \log \sin \theta - \log \cos \theta$

**Exer. 59–62: Find the exact values of the six trigonometric functions of $\theta$ if $\theta$ is in standard position and $P$ is on the terminal side.**

59 $P(4, -3)$          60 $P(-8, -15)$

61 $P(-2, -5)$         62 $P(-1, 2)$

**Exer. 63–68: Find the exact values of the six trigonometric functions of $\theta$ if $\theta$ is in standard position and the terminal side of $\theta$ is in the specified quadrant and satisfies the given condition.**

63 II;     on the line $y = -4x$

64 IV;    on the line $3y + 5x = 0$

65 I;      on a line having slope $\frac{4}{3}$

66 III;    bisects the quadrant

67 III;    parallel to the line $2y - 7x + 2 = 0$

68 II;     parallel to the line through $A(1, 4)$ and $B(3, -2)$

**Exer. 69–70: Find the exact values of the six trigonometric functions of each angle, whenever possible.**

69 (a) $90°$    (b) $0°$    (c) $7\pi/2$    (d) $3\pi$

70 (a) $180°$   (b) $-90°$   (c) $2\pi$    (d) $5\pi/2$

**Exer. 71–72: Find the quadrant containing $\theta$ if the given conditions are true.**

71 (a) $\cos \theta > 0$ and $\sin \theta < 0$    IV

    (b) $\sin \theta < 0$ and $\cot \theta > 0$    III

    (c) $\csc \theta > 0$ and $\sec \theta < 0$    II

    (d) $\sec \theta < 0$ and $\tan \theta > 0$    III

72 (a) $\tan \theta < 0$ and $\cos \theta > 0$    IV

    (b) $\sec \theta > 0$ and $\tan \theta < 0$    IV

    (c) $\csc \theta > 0$ and $\cot \theta < 0$    II

    (d) $\cos \theta < 0$ and $\csc \theta < 0$    III

**Exer. 73–80: Use fundamental identities to find the values of the trigonometric functions for the given conditions.**

73 $\tan \theta = -\frac{3}{4}$ and $\sin \theta > 0$

74 $\cot \theta = \frac{3}{4}$ and $\cos \theta < 0$

75 $\sin \theta = -\frac{5}{13}$ and $\sec \theta > 0$

76 $\cos \theta = \frac{1}{2}$ and $\sin \theta < 0$

77 $\cos \theta = -\frac{1}{3}$ and $\sin \theta < 0$

78 $\csc \theta = 5$ and $\cot \theta < 0$

79 $\sec \theta = -4$ and $\csc \theta > 0$

80 $\sin \theta = \frac{2}{3}$ and $\cos \theta < 0$

**Exer. 81–86: Rewrite the expression in nonradical form without using absolute values for the indicated values of $\theta$.**

81 $\sqrt{\sec^2 \theta - 1}$;    $\pi/2 < \theta < \pi$

82 $\sqrt{1 + \cot^2 \theta}$;    $0 < \theta < \pi$

83 $\sqrt{1 + \tan^2 \theta}$;    $3\pi/2 < \theta < 2\pi$

84 $\sqrt{\csc^2 \theta - 1}$;    $3\pi/2 < \theta < 2\pi$

85 $\sqrt{\sin^2 (\theta/2)}$;    $2\pi < \theta < 4\pi$

86 $\sqrt{\cos^2 (\theta/2)}$;    $0 < \theta < \pi$

## 5.3 TRIGONOMETRIC FUNCTIONS OF REAL NUMBERS

The domain of each trigonometric function we have discussed is a set of angles. In calculus and in many applications, domains of functions consist of real numbers. To regard the domain of a trigonometric function as a subset of $\mathbb{R}$, we may use the following definition.

**Definition of Trigonometric Functions of Real Numbers**

The **value of a trigonometric function at a real number** $t$ is its value at an angle of $t$ radians, provided that value exists.

**FIGURE 24**

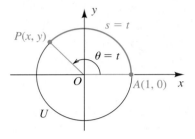

**FIGURE 25**   $\theta = t,\ t > 0$

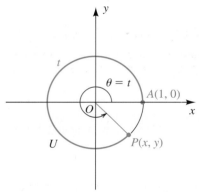

**FIGURE 26**   $\theta = t,\ t < 0$

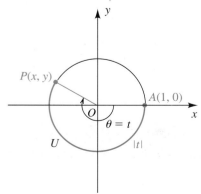

Using this definition, we may interpret a notation such as $\sin 2$ as *either* the sine of the real number 2 *or* the sine of an angle of 2 radians. As in Section 5.2, if degree measure is used, we shall write $\sin 2°$. With this understanding,

$$\sin 2 \neq \sin 2°.$$

To find the values of trigonometric functions of real numbers with a calculator, we use the radian mode.

We may interpret trigonometric functions of real numbers geometrically by using a unit circle $U$—that is, a circle of radius 1, with center at the origin $O$ of a rectangular coordinate plane. The circle $U$ is the graph of the equation $x^2 + y^2 = 1$. Let $t$ be a real number such that $0 < t < 2\pi$, and let $\theta$ denote the angle (in standard position) of radian measure $t$. One possibility is illustrated in Figure 24, where $P(x, y)$ is the point of intersection of the terminal side of $\theta$ and the unit circle $U$ and where $s$ is the length of the circular arc from $A(1, 0)$ to $P(x, y)$. Using the formula $s = r\theta$ for the length of a circular arc (see Section 5.1), with $r = 1$ and $\theta = t$, we see that

$$s = 1(t) = t.$$

Thus, *$t$ may be regarded either as the radian measure of the angle $\theta$ or as the length of the circular arc AP on U.*

Next consider *any* nonnegative real number $t$. If we regard the angle $\theta$ of radian measure $t$ as having been generated by rotating the line segment $OA$ about $O$ in the counterclockwise direction, then $t$ is the distance along $U$ that $A$ travels before reaching its final position $P(x, y)$. In Figure 25 we have illustrated a case for $t < 2\pi$; however, if $t > 2\pi$, then $A$ may travel around $U$ several times in a counterclockwise direction before reaching $P(x, y)$.

If $t < 0$, then the rotation of $OA$ is in the *clockwise* direction, and the distance $A$ travels before reaching $P(x, y)$ is $|t|$, as illustrated in Figure 26.

The preceding discussion indicates how *we may associate with each real number t a unique point P(x, y) on U.* We shall call $P(x, y)$ the **point on the unit circle $U$ that corresponds to $t$.** The coordinates $(x, y)$ of $P$ may be used to find the six trigonometric functions of $t$. Thus, by the definition of trigonometric functions of real numbers together with the definition of trigonometric functions of any angle (given in Section 5.2), we see that

$$\sin t = \sin \theta = \frac{y}{1} = y.$$

Using the same procedure for the remaining five trigonometric functions gives us the following formulas.

**Definition of Trigonometric Functions in Terms of a Unit Circle**

If $t$ is a real number and $P(x, y)$ is the point on the unit circle $U$ that corresponds to $t$, then

$$\sin t = y \qquad\qquad \cos t = x \qquad\qquad \tan t = \frac{y}{x} \text{ (if } x \neq 0)$$

$$\csc t = \frac{1}{y} \text{ (if } y \neq 0) \quad \sec t = \frac{1}{x} \text{ (if } x \neq 0) \quad \cot t = \frac{x}{y} \text{ (if } y \neq 0)$$

The formulas in this definition express function values in terms of coordinates of a point $P$ on a unit circle. For this reason, the trigonometric functions are sometimes referred to as the **circular functions.**

**FIGURE 27**

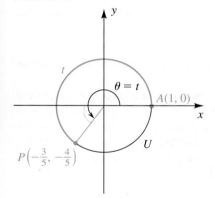

$P\left(-\frac{3}{5}, -\frac{4}{5}\right)$

**EXAMPLE I    *Finding values of the trigonometric functions***

A point $P(x, y)$ on the unit circle $U$ corresponding to a real number $t$ is shown in Figure 27, for $\pi < t < 3\pi/2$. Find the values of the trigonometric functions at $t$.

**SOLUTION**    Referring to Figure 27, we see that the coordinates of $P(x, y)$ are

$$x = -\tfrac{3}{5}, \quad y = -\tfrac{4}{5}.$$

Using the definition of trigonometric functions in terms of a unit circle gives us

$$\sin t = -\frac{4}{5} \qquad\qquad \cos t = -\frac{3}{5} \qquad\qquad \tan t = \frac{-4/5}{-3/5} = \frac{4}{3}$$

$$\csc t = \frac{1}{-4/5} = -\frac{5}{4} \quad \sec t = \frac{1}{-3/5} = -\frac{5}{3} \quad \cot t = \frac{-3/5}{-4/5} = \frac{3}{4}$$

**EXAMPLE 2    *Finding a point on $U$ relative to a given point***

Let $P(t)$ denote the point on the unit circle $U$ that corresponds to $t$ for $0 \leq t < 2\pi$. If $P(t) = (\tfrac{4}{5}, \tfrac{3}{5})$, find

**(a)** $P(t + \pi)$    **(b)** $P(t - \pi)$    **(c)** $P(-t)$

**FIGURE 28**

**(a)**

$$P(t) = \left(\tfrac{4}{5}, \tfrac{3}{5}\right)$$

$$A(1, 0) \quad x$$

$$P(t + \pi) = \left(-\tfrac{4}{5}, -\tfrac{3}{5}\right)$$

**(b)**

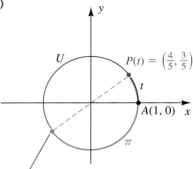

$$P(t) = \left(\tfrac{4}{5}, \tfrac{3}{5}\right)$$

$$A(1, 0) \quad x$$

$$P(t - \pi) = \left(-\tfrac{4}{5}, -\tfrac{3}{5}\right)$$

**(c)**

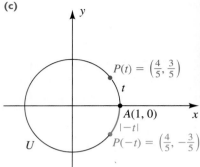

$$P(t) = \left(\tfrac{4}{5}, \tfrac{3}{5}\right)$$

$$A(1, 0) \quad x$$

$$|-t|$$

$$P(-t) = \left(\tfrac{4}{5}, -\tfrac{3}{5}\right)$$

**SOLUTION**

**(a)** The point $P(t)$ on $U$ is plotted in Figure 28(a), where we have also shown the arc $AP$ of length $t$. To find $P(t + \pi)$, we travel a distance $\pi$ in the *counterclockwise* direction along $U$ from $P(t)$, as indicated by the blue arc in the figure. Since $\pi$ is one-half the circumference of $U$, this gives us the point $P(t + \pi) = \left(-\tfrac{4}{5}, -\tfrac{3}{5}\right)$ diametrically opposite $P(t)$.

**(b)** To find $P(t - \pi)$, we travel a distance $\pi$ in the *clockwise* direction along $U$ from $P(t)$, as indicated in Figure 28(b). This gives us $P(t - \pi) = \left(-\tfrac{4}{5}, -\tfrac{3}{5}\right)$. Note that $P(t + \pi) = P(t - \pi)$.

**(c)** To find $P(-t)$, we travel along $U$ a distance $|-t|$ in the *clockwise* direction from $A(1, 0)$, as indicated in Figure 28(c). This is equivalent to reflecting $P(t)$ through the $x$-axis. Thus, we merely change the sign of the $y$-coordinate of $P(t)$ to obtain $P(-t) = \left(\tfrac{4}{5}, -\tfrac{3}{5}\right)$.

---

**EXAMPLE 3**    *Finding special values of the trigonometric functions*

Find the values of the trigonometric functions at $t$:

**(a)** $t = 0$    **(b)** $t = \dfrac{\pi}{4}$    **(c)** $t = \dfrac{\pi}{2}$

**SOLUTION**

**(a)** The point $P$ on the unit circle $U$ that corresponds to $t = 0$ has coordinates $(1, 0)$, as shown in Figure 29(a). Thus, we let $x = 1$ and $y = 0$ in the definition of trigonometric functions, obtaining

$$\sin 0 = 0, \quad \cos 0 = 1, \quad \tan 0 = \tfrac{0}{1} = 0, \quad \sec 0 = \tfrac{1}{1} = 1.$$

Note that $\csc 0$ and $\cot 0$ are undefined, since $y = 0$ is a denominator.

**FIGURE 29**

(a)

(b)

(c)

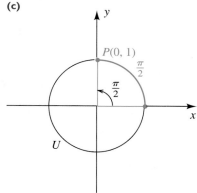

**(b)** If $t = \pi/4$, then the angle of radian measure $\pi/4$ shown in Figure 29(b) bisects the first quadrant and the point $P(x, y)$ lies on the line $y = x$. Since $P(x, y)$ is on the unit circle $x^2 + y^2 = 1$ and since $y = x$, we obtain

$$x^2 + x^2 = 1, \quad \text{or} \quad 2x^2 = 1.$$

Solving for $x$ and noting that $x > 0$ gives us

$$x = \frac{1}{\sqrt{2}} = \frac{\sqrt{2}}{2}.$$

Thus, $P$ is the point $(\sqrt{2}/2, \sqrt{2}/2)$. Letting $x = \sqrt{2}/2$ and $y = \sqrt{2}/2$ in the definition of the trigonometric functions gives us

$$\sin \frac{\pi}{4} = \frac{\sqrt{2}}{2} \qquad \cos \frac{\pi}{4} = \frac{\sqrt{2}}{2} \qquad \tan \frac{\pi}{4} = \frac{\sqrt{2}/2}{\sqrt{2}/2} = 1$$

$$\csc \frac{\pi}{4} = \frac{2}{\sqrt{2}} = \sqrt{2} \quad \sec \frac{\pi}{4} = \frac{2}{\sqrt{2}} = \sqrt{2} \quad \cot \frac{\pi}{4} = \frac{\sqrt{2}/2}{\sqrt{2}/2} = 1$$

**(c)** The point $P$ on $U$ that corresponds to $t = \pi/2$ has coordinates $(0, 1)$, as shown in Figure 29(c). Thus, we let $x = 0$ and $y = 1$ in the definition of trigonometric functions, obtaining

$$\sin \frac{\pi}{2} = 1, \quad \cos \frac{\pi}{2} = 0, \quad \csc \frac{\pi}{2} = \frac{1}{1} = 1, \quad \cot \frac{\pi}{2} = \frac{0}{1} = 0.$$

The tangent and secant functions are undefined, since $x = 0$ is a denominator in each case.

We shall use the unit circle formulation of the trigonometric functions to help obtain their graphs. If $t$ is a real number and $P(x, y)$ is the point on the unit circle $U$ that corresponds to $t$, then by the definitions of the trigonometric functions in terms of a unit circle,

$$x = \cos t \quad \text{and} \quad y = \sin t.$$

**FIGURE 30**

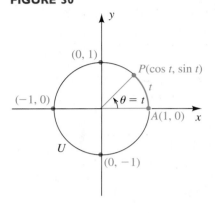

Thus, as shown in Figure 30, we may denote $P(x, y)$ by

$$P(\cos t, \sin t).$$

If $t > 0$, the real number $t$ may be interpreted either as the radian measure of the angle $t$ or as the length of arc $AP$.

If we let $t$ increase from 0 to $2\pi$ radians, the point $P(\cos t, \sin t)$ travels around the unit circle $U$ one time in the counterclockwise direction. By observing the variation of the $x$- and $y$-coordinates of $P$, we obtain the next table. The notation $0 \to \pi/2$ in the first row of the table means that $t$ increases from 0 to $\pi/2$, and the notation $(1, 0) \to (0, 1)$ denotes the corresponding variation of $P(\cos t, \sin t)$ as it travels along $U$ from $(1, 0)$ to $(0, 1)$. If $t$ increases from 0 to $\pi/2$, then $\sin t$ increases from 0 to 1, which we denote by $0 \to 1$. Moreover, $\sin t$ takes on every value between 0 and 1. If $t$ increases from $\pi/2$ to $\pi$, then $\sin t$ decreases from 1 to 0, which is denoted by $1 \to 0$. Other entries in the table may be interpreted in similar fashion.

| $t$ | $P(\cos t, \sin t)$ | $\cos t$ | $\sin t$ |
|---|---|---|---|
| $0 \to \dfrac{\pi}{2}$ | $(1, 0) \to (0, 1)$ | $1 \to 0$ | $0 \to 1$ |
| $\dfrac{\pi}{2} \to \pi$ | $(0, 1) \to (-1, 0)$ | $0 \to -1$ | $1 \to 0$ |
| $\pi \to \dfrac{3\pi}{2}$ | $(-1, 0) \to (0, -1)$ | $-1 \to 0$ | $0 \to -1$ |
| $\dfrac{3\pi}{2} \to 2\pi$ | $(0, -1) \to (1, 0)$ | $0 \to 1$ | $-1 \to 0$ |

If $t$ increases from $2\pi$ to $4\pi$, the point $P(\cos t, \sin t)$ in Figure 30 traces the unit circle $U$ again and the identical patterns for $\sin t$ and $\cos t$ are repeated—that is,

$$\sin (t + 2\pi) = \sin t \quad \text{and} \quad \cos (t + 2\pi) = \cos t$$

for every $t$ in the interval $[0, 2\pi]$. The same is true if $t$ increases from $4\pi$ to $6\pi$, from $6\pi$ to $8\pi$, and so on. In general, we have the following theorem.

**Theorem on Repeated Function Values for sin and cos**

If $n$ is any integer, then

$$\sin (t + 2\pi n) = \sin t \quad \text{and} \quad \cos (t + 2\pi n) = \cos t.$$

The repetitive variation of the sine and cosine functions is *periodic* in the sense of the following definition.

**Definition of Periodic Function**

A function $f$ is **periodic** if there exists a positive real number $k$ such that

$$f(t + k) = f(t)$$

for every $t$ in the domain of $f$. The least such positive real number $k$, if it exists, is the **period** of $f$.

The period of the sine and cosine functions is $2\pi$.

We may now readily obtain the graphs of the sine and cosine functions. Since we wish to sketch these graphs on an $xy$-plane, let us replace the variable $t$ by $x$ and consider the equations

$$y = \sin x \quad \text{and} \quad y = \cos x.$$

We may think of $x$ as the radian measure of any angle; however, in calculus, $x$ is usually regarded as a real number. These are equivalent points of view, since the sine (or cosine) of an angle of $x$ radians is the same as the sine (or cosine) of the real number $x$. The variable $y$ denotes the function value that corresponds to $x$.

The table in the margin lists coordinates of several points on the graph of $y = \sin x$ for $0 \le x \le 2\pi$. Additional points can be determined using results on special angles, such as $\sin(\pi/6) = 1/2$ and $\sin(\pi/3) = \sqrt{3}/2 \approx 0.87$.

To sketch the graph for $0 \le x \le 2\pi$, we plot the points given by the table and remember that $\sin x$ increases on $[0, \pi/2]$, decreases on $[\pi/2, \pi]$ and $[\pi, 3\pi/2]$, and increases on $[3\pi/2, 2\pi]$. This gives us the sketch in Figure 31. Since the sine function is periodic, the pattern shown in Figure 31 is repeated to the right and to the left, in intervals of length $2\pi$. This gives us the sketch in Figure 32.

| $x$ | $y = \sin x$ |
|---|---|
| $0$ | $0$ |
| $\dfrac{\pi}{4}$ | $\dfrac{\sqrt{2}}{2} \approx 0.7$ |
| $\dfrac{\pi}{2}$ | $1$ |
| $\dfrac{3\pi}{4}$ | $\dfrac{\sqrt{2}}{2} \approx 0.7$ |
| $\pi$ | $0$ |
| $\dfrac{5\pi}{4}$ | $-\dfrac{\sqrt{2}}{2} \approx -0.7$ |
| $\dfrac{3\pi}{2}$ | $-1$ |
| $\dfrac{7\pi}{4}$ | $-\dfrac{\sqrt{2}}{2} \approx -0.7$ |
| $2\pi$ | $0$ |

**FIGURE 31** $y = \sin x; \ 0 \le x \le 2\pi$

**FIGURE 32** $y = \sin x$

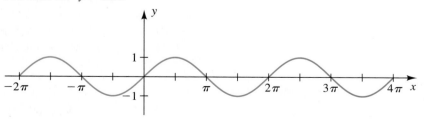

| $x$ | $y = \cos x$ |
|---|---|
| 0 | 1 |
| $\dfrac{\pi}{4}$ | $\dfrac{\sqrt{2}}{2} \approx 0.7$ |
| $\dfrac{\pi}{2}$ | 0 |
| $\dfrac{3\pi}{4}$ | $-\dfrac{\sqrt{2}}{2} \approx -0.7$ |
| $\pi$ | $-1$ |
| $\dfrac{5\pi}{4}$ | $-\dfrac{\sqrt{2}}{2} \approx -0.7$ |
| $\dfrac{3\pi}{2}$ | 0 |
| $\dfrac{7\pi}{4}$ | $\dfrac{\sqrt{2}}{2} \approx 0.7$ |
| $2\pi$ | 1 |

We can use the same procedure to sketch the graph of $y = \cos x$. The table in the margin lists coordinates of several points on the graph for $0 \leq x \leq 2\pi$. Plotting these points leads to the part of the graph shown in Figure 33. Repeating this pattern to the right and to the left, in intervals of length $2\pi$, we obtain the sketch in Figure 34.

FIGURE 33   $y = \cos x$; $0 \leq x \leq 2\pi$

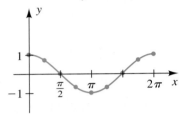

FIGURE 34   $y = \cos x$

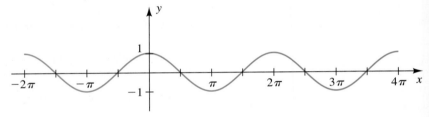

The part of the graph of the sine or cosine function corresponding to $0 \leq x \leq 2\pi$ is one **cycle.** We sometimes refer to a cycle as a **sine wave** or a **cosine wave.**

The range of the sine and cosine functions consists of all real numbers in the closed interval $[-1, 1]$. Since $\csc x = 1/\sin x$ and $\sec x = 1/\cos x$, it follows that the range of the cosecant and secant functions consists of all real numbers having absolute value greater than or equal to 1.

As we shall see, the range of the tangent and cotangent functions consists of all real numbers.

Before discussing graphs of the other trigonometric functions, let us establish formulas that involve functions of $-t$ for any $t$. Since a minus sign is involved, we call them *formulas for negatives.*

**Formulas for Negatives**

$$\sin (-t) = -\sin t \quad \cos (-t) = \cos t \quad \tan (-t) = -\tan t$$
$$\csc (-t) = -\csc t \quad \sec (-t) = \sec t \quad \cot (-t) = -\cot t$$

**FIGURE 35**

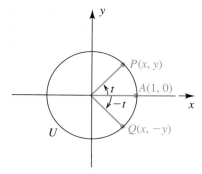

**PROOF**    Consider the unit circle $U$ in Figure 35. As $t$ increases from 0 to $2\pi$, the point $P(x, y)$ traces the unit circle $U$ once in the counterclockwise direction and the point $Q(x, -y)$, corresponding to $-t$, traces $U$ once in the clockwise direction. Applying the definition of trigonometric functions of any angle (with $r = 1$), we have

$$\sin(-t) = -y = -\sin t$$
$$\cos(-t) = x = \cos t$$
$$\tan(-t) = \frac{-y}{x} = -\frac{y}{x} = -\tan t.$$

The proofs of the remaining three formulas are similar. ■

In the following illustration, formulas for negatives are used to find an exact value for each trigonometric function.

■ **ILLUSTRATION**

**Use of Formulas for Negatives**

■  $\sin(-45°) = -\sin 45° = -\dfrac{\sqrt{2}}{2}$

■  $\cos(-30°) = \cos 30° = \dfrac{\sqrt{3}}{2}$

■  $\tan\left(-\dfrac{\pi}{3}\right) = -\tan\left(\dfrac{\pi}{3}\right) = -\sqrt{3}$

■  $\csc(-30°) = -\csc 30° = -2$

■  $\sec(-60°) = \sec 60° = 2$

■  $\cot\left(-\dfrac{\pi}{4}\right) = -\cot\left(\dfrac{\pi}{4}\right) = -1$

We shall next use formulas for negatives to verify a trigonometric identity.

**EXAMPLE 4**    *Using formulas for negatives to verify an identity*

Verify the following identity by transforming the left side into the right side:

$$\sin(-x)\tan(-x) + \cos(-x) = \sec x$$

**SOLUTION** We may proceed as follows:

$$\sin(-x)\tan(-x) + \cos(-x)$$

$$= (-\sin x)(-\tan x) + \cos x \qquad \text{formulas for negatives}$$

$$= \sin x \frac{\sin x}{\cos x} + \cos x \qquad \text{tangent identity}$$

$$= \frac{\sin^2 x}{\cos x} + \cos x \qquad \text{multiply}$$

$$= \frac{\sin^2 x + \cos^2 x}{\cos x} \qquad \text{add terms}$$

$$= \frac{1}{\cos x} \qquad \text{Pythagorean identity}$$

$$= \sec x \qquad \text{reciprocal identity}$$

We may use the formulas for negatives to prove the following theorem.

**Theorem on Even and Odd Trigonometric Functions**

(1) The cosine and secant functions are even.

(2) The sine, tangent, cotangent, and cosecant functions are odd.

**PROOF** We shall prove the theorem for the cosine and sine functions. If $f(x) = \cos x$, then

$$f(-x) = \cos(-x) = \cos x = f(x),$$

which means that the cosine function is even.
If $f(x) = \sin x$, then

$$f(-x) = \sin(-x) = -\sin x = -f(x).$$

Thus, the sine function is odd. ■

| $x$ | $y = \tan x$ |
|---|---|
| $-\dfrac{\pi}{3}$ | $-\sqrt{3} \approx -1.7$ |
| $-\dfrac{\pi}{4}$ | $-1$ |
| $-\dfrac{\pi}{6}$ | $-\dfrac{\sqrt{3}}{3} \approx -0.6$ |
| $0$ | $0$ |
| $\dfrac{\pi}{6}$ | $\dfrac{\sqrt{3}}{3} \approx 0.6$ |
| $\dfrac{\pi}{4}$ | $1$ |
| $\dfrac{\pi}{3}$ | $\sqrt{3} \approx 1.7$ |

Since the sine function is odd, its graph is symmetric with respect to the origin (see Figure 32). Since the cosine function is even, its graph is symmetric with respect to the $y$-axis (see Figure 34).

By the preceding theorem, the tangent function is odd, and hence the graph of $y = \tan x$ is symmetric with respect to the origin. The table in the margin lists some points on the graph if $-\pi/2 < x < \pi/2$. The corresponding points are plotted in Figure 36. The values of $\tan x$ near $x = \pi/2$ require special attention. If we consider $\tan x = \sin x/\cos x$, then as $x$ increases toward $\pi/2$, the numerator $\sin x$ approaches 1 and the denominator $\cos x$ approaches 0. Consequently, $\tan x$ takes on large positive values.

**FIGURE 36**

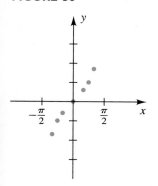

Following are some approximations of tan $x$ for $x$ close to $\pi/2 \approx 1.5708$:

$$\tan 1.57000 \approx \quad 1{,}255.8$$
$$\tan 1.57030 \approx \quad 2{,}014.8$$
$$\tan 1.57060 \approx \quad 5{,}093.5$$
$$\tan 1.57070 \approx \quad 10{,}381.3$$
$$\tan 1.57079 \approx 158{,}057.9$$

Notice how rapidly tan $x$ increases as $x$ approaches $\pi/2$. We say that tan $x$ *increases without bound as $x$ approaches $\pi/2$ through values less than $\pi/2$.* Similarly, if $x$ approaches $-\pi/2$ through values *greater* than $-\pi/2$, then tan $x$ *decreases without bound.* We may denote this variation using the notation introduced for rational functions in Section 3.5:

$$\text{as} \quad x \to \frac{\pi}{2}^{-}, \quad \tan x \to \infty$$

$$\text{as} \quad x \to -\frac{\pi}{2}^{+}, \quad \tan x \to -\infty$$

This variation of tan $x$ in the open interval $(-\pi/2, \pi/2)$ is illustrated in Figure 37. The lines $x = -\pi/2$ and $x = \pi/2$ are vertical asymptotes for the graph. The same pattern is repeated in the open intervals $(-3\pi/2, -\pi/2)$, $(\pi/2, 3\pi/2)$, and $(3\pi/2, 5\pi/2)$ and in similar intervals of length $\pi$, as shown in the figure. Thus, *the tangent function is periodic with period $\pi$.*

**FIGURE 37**  $y = \tan x$

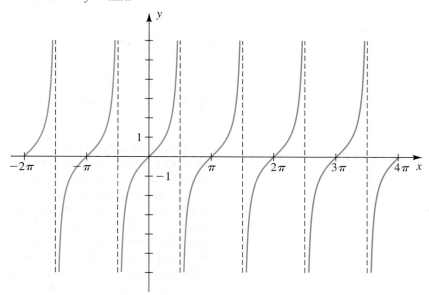

We may use the graphs of $y = \sin x$, $y = \cos x$, and $y = \tan x$ to help sketch the graphs of the remaining three trigonometric functions. For example, since $\csc x = 1/\sin x$, we may find the $y$-coordinate of a point on the graph of the cosecant function by taking the reciprocal of the corresponding $y$-coordinate on the sine graph for every value of $x$ except $x = \pi n$ for any integer $n$. (If $x = \pi n$, $\sin x = 0$, and hence $1/\sin x$ is undefined.) As an aid to sketching the graph of the cosecant function, it is convenient to sketch the graph of the sine function (shown in red in Figure 38) and then take reciprocals to obtain points on the cosecant graph.

**FIGURE 38**   $y = \csc x$

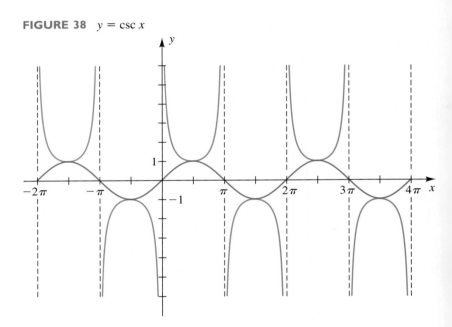

Notice the manner in which the cosecant function increases or decreases without bound as $x$ approaches $\pi n$ for any integer $n$. The graph has vertical asymptotes $x = \pi n$, as indicated in the figure.

Since $\sec x = 1/\cos x$ and $\cot x = 1/\tan x$, we may obtain the graphs of the secant and cotangent functions by taking reciprocals of $y$-coordinates of points on the graphs of the cosine and tangent functions, as illustrated in Figures 39 and 40.

A graphical summary of the six trigonometric functions and their inverses (discussed in Section 6.6) appears in Appendix IV.

**FIGURE 39** $y = \sec x$

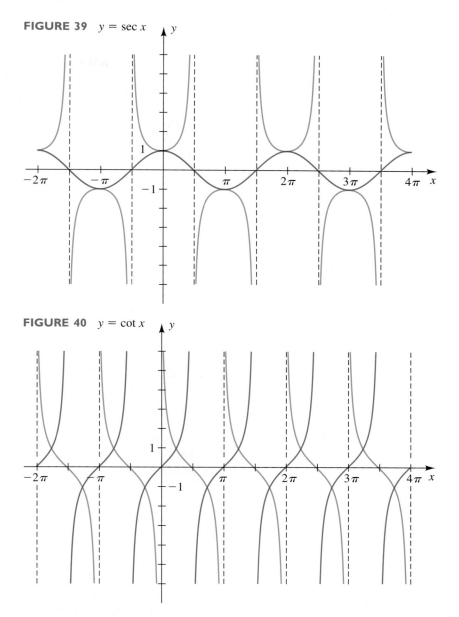

**FIGURE 40** $y = \cot x$

We have considered many properties of the six trigonometric functions of $x$, where $x$ is a real number or the radian measure of an angle. The following chart contains a summary of important features of these functions. As usual, $x$ and $y$ are coordinates of a point $P$ on the terminal side of the angle $x$ (in standard position), $r = \sqrt{x^2 + y^2}$, and $n$ denotes an arbitrary integer.

## Summary of Features of the Trigonometric Functions and Their Graphs

| Feature | $\sin x$ | $\cos x$ | $\tan x$ | $\cot x$ | $\sec x$ | $\csc x$ |
|---|---|---|---|---|---|---|
| Definition | $y/r$ | $x/r$ | $y/x$ | $x/y$ | $r/x$ | $r/y$ |
| Domain | $\mathbb{R}$ | $\mathbb{R}$ | $x \neq \frac{\pi}{2} + \pi n$ | $x \neq \pi n$ | $x \neq \frac{\pi}{2} + \pi n$ | $x \neq \pi n$ |
| Vertical asymptotes | none | none | $x = \frac{\pi}{2} + \pi n$ | $x = \pi n$ | $x = \frac{\pi}{2} + \pi n$ | $x = \pi n$ |
| Range | $[-1, 1]$ | $[-1, 1]$ | $\mathbb{R}$ | $\mathbb{R}$ | $(-\infty, -1] \cup [1, \infty)$ | $(-\infty, -1] \cup [1, \infty)$ |
| $x$-intercepts | $\pi n$ | $\frac{\pi}{2} + \pi n$ | $\pi n$ | $\frac{\pi}{2} + \pi n$ | none | none |
| $y$-intercept | 0 | 1 | 0 | none | 1 | none |
| Period | $2\pi$ | $2\pi$ | $\pi$ | $\pi$ | $2\pi$ | $2\pi$ |
| Even or odd | odd | even | odd | odd | even | odd |
| Symmetry | origin | $y$-axis | origin | origin | $y$-axis | origin |

**EXAMPLE 5    *Investigating the variation of* $\csc x$**

Investigate the variation of $\csc x$ as

$$x \to \pi^-, \quad x \to \pi^+, \quad x \to \frac{\pi^-}{2}, \quad \text{and} \quad x \to \frac{\pi^+}{6}.$$

**SOLUTION**    Referring to the graph of $y = \csc x$ in Figure 38 and using our knowledge of the special values of the cosecant function, we obtain the following:

$$\text{as} \quad x \to \pi^-, \quad \csc x \to \infty$$
$$\text{as} \quad x \to \pi^+, \quad \csc x \to -\infty$$
$$\text{as} \quad x \to \frac{\pi^-}{2}, \quad \csc x \to 1$$
$$\text{as} \quad x \to \frac{\pi^+}{6}, \quad \csc x \to 2$$

**EXAMPLE 6    *Solving equations and inequalities that involve a trigonometric function***

Find all values of $x$ in the interval $[-2\pi, 2\pi]$ such that

**(a)** $\cos x = \frac{1}{2}$     **(b)** $\cos x > \frac{1}{2}$     **(c)** $\cos x < \frac{1}{2}$

**SOLUTION**    This problem can be easily solved by referring to the graphs of $y = \cos x$ and $y = \frac{1}{2}$, sketched on the same $xy$-plane in Figure 41 for $-2\pi \le x \le 2\pi$.

**FIGURE 41**

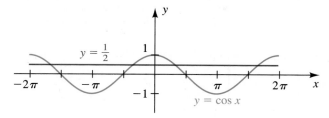

**(a)** The values of $x$ such that $\cos x = \frac{1}{2}$ are the $x$-coordinates of the points at which the graphs intersect. Recall that $x = \pi/3$ satisfies the equation. By symmetry, $x = -\pi/3$ is another solution of $\cos x = \frac{1}{2}$. Since the cosine function has period $2\pi$, the other values of $x$ in $[-2\pi, 2\pi]$ such that $\cos x = \frac{1}{2}$ are

$$-\frac{\pi}{3} + 2\pi = \frac{5\pi}{3} \quad \text{and} \quad \frac{\pi}{3} - 2\pi = -\frac{5\pi}{3}.$$

**(b)** The values of $x$ such that $\cos x > \frac{1}{2}$ can be found by determining where the graph of $y = \cos x$ in Figure 41 lies *above* the line $y = \frac{1}{2}$. This gives us the $x$-intervals

$$\left[-2\pi, -\frac{5\pi}{3}\right), \quad \left(-\frac{\pi}{3}, \frac{\pi}{3}\right), \quad \left(\frac{5\pi}{3}, 2\pi\right].$$

**(c)** To solve $\cos x < \frac{1}{2}$, we again refer to Figure 41 and note where the graph of $y = \cos x$ lies *below* the line $y = \frac{1}{2}$. This gives us the $x$-intervals

$$\left(-\frac{5\pi}{3}, -\frac{\pi}{3}\right), \quad \left(\frac{\pi}{3}, \frac{5\pi}{3}\right).$$

Another method of solving $\cos x < \frac{1}{2}$ is to note that the solutions are the open subintervals of $[-2\pi, 2\pi]$ that are *not* included in the intervals obtained in part (b).

The result discussed in the next example plays an important role in advanced mathematics.

**EXAMPLE 7**   *Sketching the graph of $f(x) = (\sin x)/x$*

If $f(x) = (\sin x)/x$, sketch the graph of $f$ on $[-\pi, \pi]$, and investigate the behavior of $f(x)$ as $x \to 0^-$ and as $x \to 0^+$.

**SOLUTION**   Some specific keystrokes for the TI-81/82 graphics calculator are given in Appendix I. Note that $f$ is undefined at $x = 0$, because substitution yields the meaningless expression 0/0.

We assign $(\sin x)/x$ to $Y_1$. Since $\pi \approx 3.1416$, we use the viewing rectangle $[-3.15, 3.15]$ by $[-2.1, 2.1]$, obtaining a sketch similar to Figure 42. Using tracing and zoom features, it appears that

$$\text{as} \quad x \to 0^-, \quad f(x) \to 1 \quad \text{and as} \quad x \to 0^+, \quad f(x) \to 1.$$

There is a hole in the graph at the point (0, 1); however, most graphing utilities are not capable of showing this fact.

Our graphical technique does not *prove* that $f(x) \to 1$ as $x \to 0$, but it does make it appear highly probable. A rigorous proof, based on the definition of $\sin x$ and geometric considerations, can be found in calculus texts.

**FIGURE 42**
$[-3.15, 3.15]$ by $[-2.1, 2.1]$

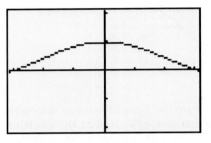

An interesting result obtained from Example 7 is that *if $x$ is in radians* and

$$\text{if} \quad x \approx 0, \quad \text{then} \quad \frac{\sin x}{x} \approx 1, \quad \text{or} \quad \sin x \approx x.$$

The last statement gives us an *approximation formula* for $\sin x$ if $x$ is close to 0. To illustrate, using a calculator we find that

$$\sin (0.06) \approx 0.0599640$$
$$\sin (0.05) \approx 0.0499792$$
$$\sin (0.04) \approx 0.0399893$$
$$\sin (0.03) \approx 0.0299955$$
$$\sin (0.02) \approx 0.0199987$$
$$\sin (0.01) \approx 0.0099998.$$

We have now discussed two different approaches to the trigonometric functions. The development in terms of angles and ratios, introduced in Section 5.2, has many applications in the sciences and engineering. The definition in terms of a unit circle, considered in this section, emphasizes the fact that the trigonometric functions have domains consisting of real numbers. Such functions are the building blocks for calculus. In addition, the unit circle approach is useful for discussing graphs and deriving trigonometric identities. You should work to become proficient in the use of both formulations of the trigonometric functions, since each will reinforce the other and thus facilitate your mastery of more advanced aspects of trigonometry.

## 5.3 EXERCISES

Exer. 1–4: A point $P(x, y)$ is shown on the unit circle $U$ corresponding to a real number $t$. Find the values of the trigonometric functions at $t$.

1

2

3

4

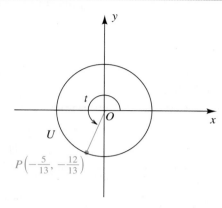

Exer. 5–8: Let $P(t)$ be the point on the unit circle $U$ that corresponds to $t$. If $P(t)$ has the given rectangular coordinates, find

(a) $P(t + \pi)$   (b) $P(t - \pi)$   (c) $P(-t)$   (d) $P(-t - \pi)$

5 $\left(\frac{3}{5}, \frac{4}{5}\right)$                           6 $\left(-\frac{8}{17}, \frac{15}{17}\right)$

7 $\left(-\frac{12}{13}, -\frac{5}{13}\right)$                      8 $\left(\frac{7}{25}, -\frac{24}{25}\right)$

Exer. 9–16: Let $P$ be the point on the unit circle $U$ that corresponds to $t$. Find the coordinates of $P$ and the exact values of the trigonometric functions at $t$, whenever possible.

9  (a) $2\pi$                          (b) $-3\pi$

10  (a) $-\pi$                         (b) $6\pi$

11  (a) $3\pi/2$                       (b) $-7\pi/2$

12  (a) $5\pi/2$                       (b) $-\pi/2$

13  (a) $9\pi/4$                       (b) $-5\pi/4$

14  (a) $3\pi/4$                       (b) $-7\pi/4$

15  (a) $5\pi/4$                       (b) $-\pi/4$

16  (a) $7\pi/4$                       (b) $-3\pi/4$

Exer. 17–20: Use a formula for negatives to find the exact value.

17  (a) $\sin(-90°)$     (b) $\cos\left(-\dfrac{3\pi}{4}\right)$     (c) $\tan(-45°)$

18  (a) $\sin\left(-\dfrac{3\pi}{2}\right)$     (b) $\cos(-225°)$     (c) $\tan(-\pi)$

**19** (a) $\cot\left(-\dfrac{3\pi}{4}\right)$  (b) $\sec(-180°)$  (c) $\csc\left(-\dfrac{3\pi}{2}\right)$

**20** (a) $\cot(-225°)$  (b) $\sec\left(-\dfrac{\pi}{4}\right)$  (c) $\csc(-45°)$

**Exer. 21–26:** Verify the identity by transforming the left side into the right side.

**21** $\sin(-x)\sec(-x) = -\tan x$

**22** $\csc(-x)\cos(-x) = -\cot x$

**23** $\dfrac{\cot(-x)}{\csc(-x)} = \cos x$  **24** $\dfrac{\sec(-x)}{\tan(-x)} = -\csc x$

**25** $\dfrac{1}{\cos(-x)} - \tan(-x)\sin(-x) = \cos x$

**26** $\cot(-x)\cos(-x) + \sin(-x) = -\csc x$

**Exer. 27–38:** Complete the statement by referring to a graph of a trigonometric function.

**27** (a) As $x \to 0^+$, $\sin x \to$ _____

(b) As $x \to -\dfrac{\pi}{2}^-$, $\sin x \to$ _____

**28** (a) As $x \to \pi^+$, $\sin x \to$ _____

(b) As $x \to \dfrac{\pi}{6}^-$, $\sin x \to$ _____

**29** (a) As $x \to \dfrac{\pi}{4}^+$, $\cos x \to$ _____

(b) As $x \to \pi^-$, $\cos x \to$ _____

**30** (a) As $x \to 0^+$, $\cos x \to$ _____

(b) As $x \to -\dfrac{\pi}{3}^-$, $\cos x \to$ _____

**31** (a) As $x \to \dfrac{\pi}{4}^+$, $\tan x \to$ _____

(b) As $x \to \dfrac{\pi}{2}^+$, $\tan x \to$ _cont._

**32** (a) As $x \to 0^+$, $\tan x \to$ _____

(b) As $x \to -\dfrac{\pi}{2}^-$, $\tan x \to$ _____

**33** (a) As $x \to -\dfrac{\pi}{4}^-$, $\cot x \to$ _____

(b) As $x \to 0^+$, $\cot x \to$ _____

**34** (a) As $x \to \dfrac{\pi}{6}^+$, $\cot x \to$ _____

(b) As $x \to \pi^-$, $\cot x \to$ _____

**35** (a) As $x \to \dfrac{\pi}{2}^-$, $\sec x \to$ _____

(b) As $x \to \dfrac{\pi}{4}^+$, $\sec x \to$ _____

**36** (a) As $x \to \dfrac{\pi}{2}^+$, $\sec x \to$ _____

(b) As $x \to 0^-$, $\sec x \to$ _____

**37** (a) As $x \to 0^-$, $\csc x \to$ _____

(b) As $x \to \dfrac{\pi}{2}^+$, $\csc x \to$ _____

**38** (a) As $x \to \pi^+$, $\csc x \to$ _____

(b) As $x \to \dfrac{\pi}{4}^-$, $\csc x \to$ _____

**Exer. 39–46:** Refer to the graph of $y = \sin x$ or $y = \cos x$ to find the exact values of $x$ in the interval $[0, 4\pi]$ that satisfy the equation.

**39** $\sin x = -1$  **40** $\sin x = 1$

**41** $\sin x = \tfrac{1}{2}$  **42** $\sin x = -\sqrt{2}/2$

**43** $\cos x = 1$  **44** $\cos x = -1$

**45** $\cos x = \sqrt{2}/2$  **46** $\cos x = -\tfrac{1}{2}$

**Exer. 47–50:** Refer to the graph of $y = \tan x$ to find the exact values of $x$ in the interval $(-\pi/2, 3\pi/2)$ that satisfy the equation.

**47** $\tan x = 1$  **48** $\tan x = \sqrt{3}$

**49** $\tan x = 0$  **50** $\tan x = -1/\sqrt{3}$

**Exer. 51–54:** Refer to the graph of the equation on the specified interval. Find all values of $x$ such that for the real number $a$, (a) $y = a$, (b) $y > a$, and (c) $y < a$.

**51** $y = \sin x$;  $[-2\pi, 2\pi]$;  $a = \tfrac{1}{2}$

**52** $y = \cos x$;  $[0, 4\pi]$;  $a = \sqrt{3}/2$

**53** $y = \cos x$;  $[-2\pi, 2\pi]$;  $a = -\tfrac{1}{2}$

**54** $y = \sin x$;  $[0, 4\pi]$;  $a = -\sqrt{2}/2$

**Exer. 55–62:** Use the graph of a trigonometric function to sketch the graph of the equation without plotting points.

**55** $y = 2 + \sin x$  **56** $y = 3 + \cos x$

57 $y = \cos x - 2$   58 $y = \sin x - 1$

59 $y = 1 + \tan x$   60 $y = \cot x - 1$

61 $y = \sec x - 2$   62 $y = 1 + \csc x$

**Exer. 63–66: Find the intervals between $-2\pi$ and $2\pi$ on which the given function is (a) increasing and (b) decreasing.**

63 secant   64 cosecant

65 tangent   66 cotangent

67 Practice sketching the graph of the sine function, taking different units of length on the horizontal and vertical axes. Practice sketching graphs of the cosine and tangent functions in the same manner. Continue this practice until you reach the stage at which, if you were awakened from a sound sleep in the middle of the night and asked to sketch one of these graphs, you could do so in less than thirty seconds.

68 Work Exercise 67 for the cosecant, secant, and cotangent functions.

**Exer. 69–72: Use the figure to approximate the following to one decimal place.**

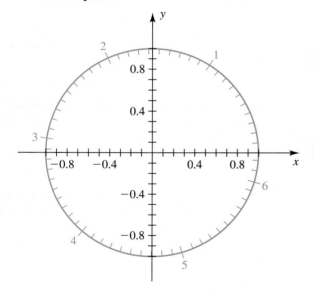

69 (a) sin 4   (b) sin (−1.2)
   (c) All numbers $t$ between 0 and $2\pi$ such that $\sin t = 0.5$

70 (a) sin 2   (b) sin (−2.3)
   (c) All numbers $t$ between 0 and $2\pi$ such that $\sin t = -0.2$

71 (a) cos 4   (b) cos (−1.2)
   (c) All numbers $t$ between 0 and $2\pi$ such that $\cos t = -0.6$

72 (a) cos 2   (b) cos (−2.3)
   (c) All numbers $t$ between 0 and $2\pi$ such that $\cos t = 0.2$

**Exer. 73–74: Graph the equation, and estimate the values of $x$ in the specified interval that correspond to the given value of $y$.**

73 $y = \sin(x^2)$,   $[-\pi, \pi]$;   $y = 0.5$

74 $y = \tan(\sqrt{x})$,   $[0, 25]$;   $y = 5$

**Exer. 75–76: Graph $f$ on the interval $[-2\pi, 2\pi]$, and estimate the coordinates of the high and low points.**

75 $f(x) = x \sin x$   76 $f(x) = \sin^2 x \cos x$

**Exer. 77–82: As $x \to 0^+$, $f(x) \to L$ for some real number $L$. Use a graph to predict $L$.**

77 $f(x) = \dfrac{1 - \cos x}{x}$   78 $f(x) = \dfrac{6x - 6 \sin x}{x^3}$

79 $f(x) = x \cot x$   80 $f(x) = \dfrac{x + \tan x}{\sin x}$

81 $f(x) = \dfrac{\tan x}{x}$   82 $f(x) = \dfrac{\cos(x + \frac{1}{2}\pi)}{x}$

# 5.4   VALUES OF THE TRIGONOMETRIC FUNCTIONS

In previous sections we calculated special values of the trigonometric functions by using either the unit circle definition or the theorem on trigonometric functions as ratios. In practice we often use a calculator to approximate function values.

We will next show how the values of any trigonometric function at a real number $t$ or at an angle of $\theta$ degrees can be found from its values in the $t$-interval $(0, \pi/2)$ or the $\theta$-interval $(0°, 90°)$, respectively. This technique is sometimes necessary when a calculator is used to find all angles or real numbers that correspond to a given function value.

We shall make use of the following concept.

**Definition of Reference Angle**

Let $\theta$ be a nonquadrantal angle in standard position. The **reference angle** for $\theta$ is the acute angle $\theta_R$ that the terminal side of $\theta$ makes with the $x$-axis.

Figure 43 illustrates the reference angle $\theta_R$ for a nonquadrantal angle $\theta$, with $0° < \theta < 360°$ or $0 < \theta < 2\pi$, in each of the four quadrants.

**FIGURE 43**   Reference angles

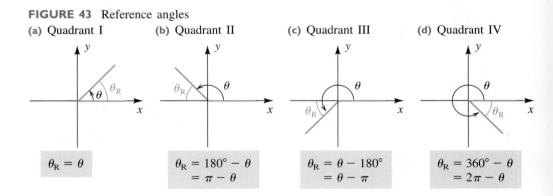

(a) Quadrant I    (b) Quadrant II    (c) Quadrant III    (d) Quadrant IV

$\theta_R = \theta$

$\theta_R = 180° - \theta$
$= \pi - \theta$

$\theta_R = \theta - 180°$
$= \theta - \pi$

$\theta_R = 360° - \theta$
$= 2\pi - \theta$

The formulas below the axes in Figure 43 may be used to find the degree or radian measure of $\theta_R$ when $\theta$ is in degrees or radians, respectively. *For a nonquadrantal angle greater than 360° or less than 0°, first find the coterminal angle $\theta$ with $0° < \theta < 360°$ or $0 < \theta < 2\pi$, and then use the formulas in Figure 43.*

**FIGURE 44**

(a)

(b)

(c)

(d)

**EXAMPLE I**   *Finding reference angles*

Find the reference angle $\theta_R$ for $\theta$ and sketch $\theta$ and $\theta_R$ in standard position on the same coordinate plane.

**(a)** $\theta = 315°$  **(b)** $\theta = -240°$  **(c)** $\theta = \dfrac{5\pi}{6}$  **(d)** $\theta = 4$

*SOLUTION*

**(a)** The angle $\theta = 315°$ is in quadrant IV, and hence, as in Figure 43(d),

$$\theta_R = 360° - 315° = 45°.$$

The angles $\theta$ and $\theta_R$ are sketched in Figure 44(a).

**(b)** The angle between $0°$ and $360°$ that is coterminal with $\theta = -240°$ is

$$-240° + 360° = 120°,$$

which is in quadrant II. Using the formula in Figure 43(b) gives

$$\theta_R = 180° - 120° = 60°.$$

The angles $\theta$ and $\theta_R$ are sketched in Figure 44(b).

**(c)** Since the angle $\theta = 5\pi/6$ is in quadrant II, we have

$$\theta_R = \pi - \frac{5\pi}{6} = \frac{\pi}{6},$$

as shown in Figure 44(c).

**(d)** Since $\pi < 4 < 3\pi/2$, the angle $\theta = 4$ is in quadrant III. Using the formula in Figure 43(c), we obtain

$$\theta_R = 4 - \pi.$$

The angles are sketched in Figure 44(d).

We shall next show how reference angles can be used to find values of the trigonometric functions.

If $\theta$ is a nonquadrantal angle with reference angle $\theta_R$, then we have $0° < \theta_R < 90°$ or $0 < \theta_R < \pi/2$. Let $P(x, y)$ be a point on the terminal side of $\theta$, and consider the point $Q(x, 0)$ on the $x$-axis. Figure 45 illustrates

**FIGURE 45**

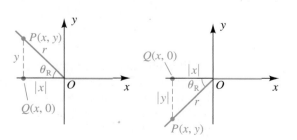

a typical situation for $\theta$ in each quadrant. In each case, the lengths of the sides of triangle $OPQ$ are

$$d(O, Q) = |x|, \quad d(Q, P) = |y|, \quad \text{and} \quad d(O, P) = \sqrt{x^2 + y^2} = r.$$

We may apply the definitions of trigonometric functions of any angle and also use triangle $OPQ$ to obtain the following formulas:

$$|\sin \theta| = \left|\frac{y}{r}\right| = \frac{|y|}{|r|} = \frac{|y|}{r} = \sin \theta_R$$

$$|\cos \theta| = \left|\frac{x}{r}\right| = \frac{|x|}{|r|} = \frac{|x|}{r} = \cos \theta_R$$

$$|\tan \theta| = \left|\frac{y}{x}\right| = \frac{|y|}{|x|} = \tan \theta_R$$

These formulas lead to the next theorem. If $\theta$ is a quadrantal angle, the definition of trigonometric functions should be used to find values.

**Theorem on Reference Angles**

> If $\theta$ is a nonquadrantal angle in standard position, then to find the value of a trigonometric function at $\theta$, find its value for the reference angle $\theta_R$ and prefix the appropriate sign.

The ''appropriate sign'' referred to in the theorem can be determined from the table of signs of the trigonometric functions given in Section 5.2.

### EXAMPLE 2    *Using reference angles*

Use reference angles to find the exact values of $\sin \theta$, $\cos \theta$, and $\tan \theta$ if

**(a)** $\theta = \dfrac{5\pi}{2}$    **(b)** $\theta = 315°$

*SOLUTION*

**(a)** The angle $\theta = 5\pi/6$ and its reference angle $\theta_R = \pi/6$ are sketched in Figure 44(c). Since $\theta$ is in quadrant II, $\sin \theta$ is positive and both $\cos \theta$ and $\tan \theta$ are negative. Hence, by the theorem on reference angles and known results about special angles, we obtain the following values:

$$\sin \frac{5\pi}{6} = \sin \frac{\pi}{6} = \frac{1}{2}$$

$$\cos \frac{5\pi}{6} = -\cos \frac{\pi}{6} = -\frac{\sqrt{3}}{2}$$

$$\tan \frac{5\pi}{6} = -\tan \frac{\pi}{6} = -\frac{\sqrt{3}}{3}$$

**(b)** The angle $\theta = 315°$ and its reference angle $\theta_R = 45°$ are sketched in Figure 44(a). Since $\theta$ is in quadrant IV, $\sin \theta < 0$, $\cos \theta > 0$, and $\tan \theta < 0$. Hence, by the theorem on reference angles, we obtain

$$\sin 315° = -\sin 45° = -\frac{\sqrt{2}}{2}$$

$$\cos 315° = \cos 45° = \frac{\sqrt{2}}{2}$$

$$\tan 315° = -\tan 45° = -1.$$

If we use a calculator to approximate function values, reference angles are unnecessary. As an illustration, to find $\sin 210°$, we place the calculator in degree mode and obtain $\sin 210° = -0.5$, which is the exact value. Using the same procedure for $240°$, we obtain a decimal representation:

$$\sin 240° \approx -0.8660$$

A calculator should not be used to find the *exact* value of $\sin 240°$. In this case, we find the reference angle $60°$ of $240°$ and use the theorem on reference angles, together with known results about special angles, to obtain

$$\sin 240° = -\sin 60° = -\frac{\sqrt{3}}{2}.$$

Let us next consider the problem of solving an equation of the following type:

**Problem:** If $\theta$ is an acute angle and $\sin \theta = 0.6635$, approximate $\theta$.

One method of approximating $\theta$ is to use a table. Some calculators have a key labeled $\boxed{\text{SIN}^{-1}}$ that can be used to help solve the equation. With other calculators, it may be necessary to use another key or a keystroke sequence such as $\boxed{\text{INV}}\boxed{\text{SIN}}$ (refer to the user manual for your calculator). We shall use the following notation when finding $\theta$, where $0 \leq k \leq 1$:

$$\text{if} \quad \sin \theta = k, \quad \text{then} \quad \theta = \sin^{-1} k$$

This notation is similar to that used for the inverse function $f^{-1}$ of a function $f$ in Section 2.5, where we saw that under certain conditions,

$$\text{if} \quad f(x) = y, \quad \text{then} \quad x = f^{-1}(y).$$

For the problem $\sin \theta = 0.6635$, $f$ is the sine function, $x = \theta$, and $y = 0.6635$. The notation $\sin^{-1}$ is based on the *inverse trigonometric functions* discussed in Section 6.6. At this stage of our work, *we shall regard* $\sin^{-1}$ *simply as an entry made on a calculator using a* $\boxed{\text{SIN}^{-1}}$ *key.* Thus, for the stated problem, we obtain

$$\theta = \sin^{-1}(0.6635) \approx 41.57° \approx 0.7255.$$

As indicated, when finding an angle we will usually round off degree measure to the nearest 0.01° and radian measure to four decimal places.

Similarly, given $\cos \theta = k$ or $\tan \theta = k$, where $\theta$ is acute, we write

$$\theta = \cos^{-1} k \quad \text{or} \quad \theta = \tan^{-1} k$$

to indicate the use of a $\boxed{\text{COS}^{-1}}$ or $\boxed{\text{TAN}^{-1}}$ key on a calculator.

Given $\csc \theta$, $\sec \theta$, or $\cot \theta$, we use a reciprocal relationship to find $\theta$, as indicated in the following illustration.

### ILLUSTRATION

**Finding Acute Angle Solutions of Equations with a Calculator**

| Equation | Calculator solution (degree and radian) |
|---|---|
| ▦ $\sin \theta = 0.5$ | $\theta = \sin^{-1}(0.5) = 30° \approx 0.5236$ |
| ▦ $\cos \theta = 0.5$ | $\theta = \cos^{-1}(0.5) = 60° \approx 1.0472$ |
| ▦ $\tan \theta = 0.5$ | $\theta = \tan^{-1}(0.5) \approx 26.57° \approx 0.4636$ |
| ▦ $\csc \theta = 2$ | $\theta = \sin^{-1}\left(\frac{1}{2}\right) = 30° \approx 0.5236$ |
| ▦ $\sec \theta = 2$ | $\theta = \cos^{-1}\left(\frac{1}{2}\right) = 60° \approx 1.0472$ |
| ▦ $\cot \theta = 2$ | $\theta = \tan^{-1}\left(\frac{1}{2}\right) \approx 26.57° \approx 0.4636$ |

The same technique may be employed if $\theta$ is *any* angle or real number. Thus, using the $\boxed{\text{SIN}^{-1}}$ key, we obtain, in degree or radian mode,

$$\theta = \sin^{-1}(0.6635) \approx 41.57° \approx 0.7255,$$

which is the reference angle for $\theta$. If $\sin \theta$ is *negative,* then a calculator gives us the *negative* of the reference angle. For example,

$$\sin^{-1}(-0.6635) \approx -41.57° \approx -0.7255.$$

Similarly, given $\cos \theta$ or $\tan \theta$, we find $\theta$ with a calculator by using $\boxed{\text{COS}^{-1}}$ or $\boxed{\text{TAN}^{-1}}$, respectively. The interval containing $\theta$ is listed in the next chart. It is important to note that if $\cos \theta$ is negative, then $\theta$ is *not* the negative of the reference angle, but instead is in the interval $\pi/2 < \theta \le \pi$, or $90° < \theta \le 180°$. The reasons for using these intervals are explained in Section 6.6. We may use reciprocal relationships to solve similar equations involving $\csc \theta$, $\sec \theta$, and $\cot \theta$.

| Equation | Values of $k$ | Calculator solution | Interval containing $\theta$ if a calculator is used |
|---|---|---|---|
| $\sin \theta = k$ | $-1 \le k \le 1$ | $\theta = \sin^{-1} k$ | $-\dfrac{\pi}{2} \le \theta \le \dfrac{\pi}{2}$ or $\quad -90° \le \theta \le 90°$ |
| $\cos \theta = k$ | $-1 \le k \le 1$ | $\theta = \cos^{-1} k$ | $0 \le \theta \le \pi$ or $\quad 0° \le \theta \le 180°$ |
| $\tan \theta = k$ | any $k$ | $\theta = \tan^{-1} k$ | $-\dfrac{\pi}{2} < \theta < \dfrac{\pi}{2}$ or $\quad -90° < \theta < 90°$ |

The following illustration contains some specific examples for both degree and radian modes.

## ILLUSTRATION

### Finding Angles with a Calculator

| Equation | Calculator solution (degree and radian) |
|---|---|
| $\sin \theta = -0.5$ | $\theta = \sin^{-1}(-0.5) = -30° \approx -0.5236$ |
| $\cos \theta = -0.5$ | $\theta = \cos^{-1}(-0.5) = 120° \approx 2.0944$ |
| $\tan \theta = -0.5$ | $\theta = \tan^{-1}(-0.5) \approx -26.57° \approx -0.4636$ |

When using a calculator to find $\theta$, be sure to keep the restrictions on $\theta$ in mind. If other values are desired, then reference angles or other methods may be employed, as illustrated in the next example.

**EXAMPLE 3** *Approximating an angle with a calculator*

If $\tan \theta = -0.4623$ and $0° \leq \theta < 360°$, find $\theta$ to the nearest $0.1°$.

**FIGURE 46**

$\theta = 180° - \theta_R$

**FIGURE 47**

$\theta = 360° - \theta_R$

*SOLUTION* As pointed out in the preceding discussion, if we use a calculator (in degree mode) to find $\theta$ when $\tan \theta$ is negative, then the degree measure is in the interval $(-90°, 0°)$. In particular, we obtain the following:

$$\theta = \tan^{-1}(-0.4623) \approx -24.8°$$

Since we wish to find values of $\theta$ between $0°$ and $360°$, we use the (approximate) reference angle $\theta_R \approx 24.8°$. There are two possible values of $\theta$ such that $\tan \theta$ is negative—one in quadrant II, the other in quadrant IV. If $\theta$ is in quadrant II and $0° \leq \theta < 360°$, we have the situation shown in Figure 46, and

$$\theta = 180° - \theta_R \approx 180° - 24.8° = 155.2°.$$

If $\theta$ is in quadrant IV and $0° \leq \theta < 360°$, then, as in Figure 47,

$$\theta = 360° - \theta_R \approx 360° - 24.8° = 335.2°.$$

## 5.4 EXERCISES

**Exer. 1–6:** Find the reference angle $\theta_R$ if $\theta$ has the given measure.

1  (a) $240°$  (b) $340°$  (c) $-202°$  (d) $-660°$

2  (a) $165°$  (b) $275°$  (c) $-110°$  (d) $400°$

3  (a) $\dfrac{3\pi}{4}$  (b) $\dfrac{4\pi}{3}$  (c) $-\dfrac{\pi}{6}$  (d) $\dfrac{9\pi}{4}$

4  (a) $\dfrac{7\pi}{4}$  (b) $\dfrac{2\pi}{3}$  (c) $-\dfrac{3\pi}{4}$  (d) $-\dfrac{23\pi}{6}$

5 (a) 3          (b) −2          (c) 5.5          (d) 100

6 (a) 6          (b) −4          (c) 4.5          (d) 80

## Exer. 7–18: Find the exact value.

7 (a) $\sin \dfrac{2\pi}{3}$                   (b) $\sin \left(-\dfrac{5\pi}{4}\right)$

8 (a) $\sin 210°$                              (b) $\sin (-315°)$

9 (a) $\cos 150°$                              (b) $\cos (-60°)$

10 (a) $\cos \dfrac{5\pi}{4}$                   (b) $\cos \left(-\dfrac{11\pi}{6}\right)$

11 (a) $\tan \dfrac{5\pi}{6}$                   (b) $\tan \left(-\dfrac{\pi}{3}\right)$

12 (a) $\tan 330°$                             (b) $\tan (-225°)$

13 (a) $\cot 120°$                             (b) $\cot (-150°)$

14 (a) $\cot \dfrac{3\pi}{4}$                   (b) $\cot \left(-\dfrac{2\pi}{3}\right)$

15 (a) $\sec \dfrac{2\pi}{3}$                   (b) $\sec \left(-\dfrac{\pi}{6}\right)$

16 (a) $\sec 135°$                             (b) $\sec (-210°)$

17 (a) $\csc 240°$                             (b) $\csc (-330°)$

18 (a) $\csc \dfrac{3\pi}{4}$                   (b) $\csc \left(-\dfrac{2\pi}{3}\right)$

## Exer. 19–24: Approximate to three decimal places.

19 (a) $\sin 73°20'$                            (b) $\cos 0.68$

20 (a) $\cos 38°30'$                            (b) $\sin 1.48$

21 (a) $\tan 21°10'$                            (b) $\cot 1.13$

22 (a) $\cot 9°10'$                             (b) $\tan 0.75$

23 (a) $\sec 67°50'$                            (b) $\csc 0.32$

24 (a) $\csc 43°40'$                            (b) $\sec 0.26$

## Exer. 25–32: Approximate the acute angle $\theta$ to the nearest

(a) $(0.01)°$   (b) $1'$

25 $\cos \theta = 0.8620$                       26 $\sin \theta = 0.6612$

27 $\tan \theta = 3.7$                          28 $\cos \theta = 0.8$

29 $\sin \theta = 0.4217$                       30 $\tan \theta = 4.91$

31 $\sec \theta = 4.246$                        32 $\csc \theta = 11$

## Exer. 33–34: Approximate to four decimal places.

33 (a) $\sin 98°10'$          (b) $\cos 623.7°$          (c) $\tan 3$
   (d) $\cot 231°40'$         (e) $\sec 1175.1°$         (f) $\csc 0.82$

34 (a) $\sin 496.4°$          (b) $\cos 0.65$            (c) $\tan 105°40'$
   (d) $\cot 1030.2°$         (e) $\sec 1.46$            (f) $\csc 320°50'$

## Exer. 35–36: Approximate, to the nearest $(0.1)°$, all angles $\theta$ in the interval $[0°, 360°)$ that satisfy the equation.

35 (a) $\sin \theta = -0.5640$                  (b) $\cos \theta = 0.7490$
   (c) $\tan \theta = 2.798$                     (d) $\cot \theta = -0.9601$
   (e) $\sec \theta = -1.116$                    (f) $\csc \theta = 1.485$

36 (a) $\sin \theta = 0.8225$                   (b) $\cos \theta = -0.6604$
   (c) $\tan \theta = -1.5214$                   (d) $\cot \theta = 1.3752$
   (e) $\sec \theta = 1.4291$                    (f) $\csc \theta = -2.3179$

## Exer. 37–38: Approximate, to the nearest 0.01 radian, all angles $\theta$ in the interval $[0, 2\pi)$ that satisfy the equation.

37 (a) $\sin \theta = 0.4195$                   (b) $\cos \theta = -0.1207$
   (c) $\tan \theta = -3.2504$                   (d) $\cot \theta = 2.6815$
   (e) $\sec \theta = 1.7452$                    (f) $\csc \theta = -4.8521$

38 (a) $\sin \theta = -0.0135$                  (b) $\cos \theta = 0.9235$
   (c) $\tan \theta = 0.42$                      (d) $\cot \theta = -2.731$
   (e) $\sec \theta = -3.51$                     (f) $\csc \theta = 1.258$

39 **Thickness of the ozone layer** The thickness of the ozone layer can be estimated using the formula

$$\ln I_0 - \ln I = kx \sec \theta,$$

where $I_0$ is the intensity of a particular wavelength of light from the sun before it reaches the atmosphere, $I$ is the intensity of the same wavelength after it passes through a layer of ozone $x$ centimeters thick, $k$ is the absorption constant of ozone for that wavelength, and $\theta$ is the acute angle that the sunlight makes with the vertical. Suppose that for a wavelength of $3055 \times 10^{-8}$ centimeter with $k \approx 1.88$, $I_0/I$ is measured as 1.72 and $\theta = 12°$. Approximate the thickness of the ozone layer to the nearest 0.01 centimeter.

40 **Ozone calculations** Refer to Exercise 39. If the ozone layer is estimated to be 0.31 centimeter thick and, for a wavelength of $3055 \times 10^{-8}$ centimeter, $I_0/I$ is measured as 2.05, approximate the angle the sun made with the vertical at the time of the measurement.

41 **Solar radiation** The amount of sunshine illuminating a wall of a building can greatly affect the energy efficiency of the building. The solar radiation striking a

vertical wall that faces east is given by the formula $R = R_0 \cos \theta \sin \phi$, where $R_0$ is the maximum solar radiation possible, $\theta$ is the angle that the sun makes with the horizontal, and $\phi$ is the direction of the sun in the sky, with $\phi = 90°$ when the sun is in the east and $\phi = 0°$ when the sun is in the south.

(a) When does the maximum solar radiation $R_0$ strike the wall?

(b) What percentage of $R_0$ is striking the wall when $\theta = 60°$ and the sun is in the southeast?

42 **Meteorological calculations** In the mid-latitudes it is sometimes possible to estimate the distance between consecutive regions of low pressure. If $\phi$ is the latitude (in degrees), $R$ is the earth's radius (in kilometers), and $v$ is the horizontal wind velocity (in km/hr), then the distance $d$ (in kilometers) from one low pressure area to the next can be estimated using the formula

$$d = 2\pi \left( \frac{vR}{0.52 \cos \phi} \right)^{1/3}.$$

(a) At a latitude of 48°, the earth's radius is approximately 6369 kilometers. Approximate $d$ if the wind speed is 45 km/hr.

(b) If $v$ and $R$ are constant, how does $d$ vary as the latitude increases?

43 **Robot's arm** Points on the terminal sides of angles play an important part in the design of arms for robots. Suppose a robot has a straight arm 18 inches long that can rotate about the origin in a coordinate plane. If the robot's hand is located at $(18, 0)$ and then rotates through an angle of 60°, what is the new location of the hand?

44 **Robot's arm** Suppose the robot's arm in Exercise 43 can change its length in addition to rotating about the origin. If the hand is initially at $(12, 12)$, approximately how many degrees should the arm be rotated and how

much should its length be changed to move the hand to $(-16, 10)$?

45 **Race track coordinates** Shown in the figure is a circular race track of diameter 2 kilometers. All races begin at $S$ and proceed in a counterclockwise direction. Approximate, to four decimal places, the coordinates of the point at which the following races end relative to a rectangular coordinate system with origin at the center of the track and $S$ on the positive $x$-axis.

(a) A drag race of length 2 kilometers

(b) An endurance race of length 500 kilometers

EXERCISE 45

46 **Race track coordinates** Work Exercise 46 for the track shown in the figure, if the origin of a rectangular coordinate system is at the center of the track and $S$ is on the negative $y$-axis.

EXERCISE 46

## 5.5 TRIGONOMETRIC GRAPHS

In this section we consider graphs of the equations

$$y = a \sin (bx + c) \quad \text{and} \quad y = a \cos (bx + c)$$

for real numbers $a$, $b$, and $c$. Our goal is to sketch such graphs without plotting many points. To do so we shall use facts about the graphs of the sine and cosine functions discussed in Section 5.3.

Let us begin by considering the special case $c = 0$ and $b = 1$—that is,

$$y = a \sin x \quad \text{and} \quad y = a \cos x.$$

We can find $y$-coordinates of points on the graphs by multiplying $y$-coordinates of points on the graphs of $y = \sin x$ and $y = \cos x$ by $a$. To illustrate, if $y = 2 \sin x$, we multiply the $y$-coordinate of each point on the graph of $y = \sin x$ by 2. This gives us Figure 48, where for comparison we have shown the graph of $y = \sin x$. The procedure is the same as that for stretching the graph of a function, discussed in Section 2.2.

**FIGURE 48**

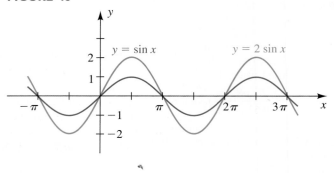

As another illustration, if $y = \frac{1}{2} \sin x$, we multiply $y$-coordinates of points on the graph of $y = \sin x$ by $\frac{1}{2}$. This multiplication vertically compresses the graph of $y = \sin x$ by a factor of 2, as illustrated in Figure 49.

**FIGURE 49**

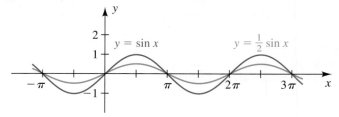

The following example illustrates a graph of $y = a \sin x$ with $a$ negative.

**EXAMPLE 1**   *Sketching the graph of an equation involving* $\sin x$

Sketch the graph of the equation $y = -2 \sin x$.

**SOLUTION**   The graph of $y = -2 \sin x$ sketched in Figure 50 can be obtained by first sketching the graph of $y = \sin x$ (shown in the figure) and then multiplying $y$-coordinates by $-2$. An alternative method is to reflect the graph of $y = 2 \sin x$ (see Figure 48) through the $x$-axis.

**FIGURE 50**

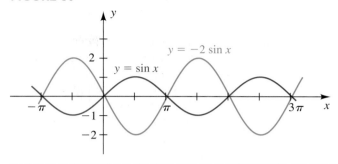

For any $a \neq 0$, the graph of $y = a \sin x$ has the general appearance of one of the graphs illustrated in Figures 48, 49, and 50. The amount of stretching of the graph of $y = \sin x$ and whether the graph is reflected are determined by the absolute value and sign of $a$. The largest $y$-coordinate $|a|$ is the **amplitude of the graph** or, equivalently, the **amplitude of the function** $f$ given by $f(x) = a \sin x$. In Figures 48 and 50 the amplitude is 2. In Figure 49 the amplitude is $\frac{1}{2}$. Similar remarks and techniques apply if $y = a \cos x$.

**EXAMPLE 2    *Sketching the graph of an equation involving* cos $x$**

Find the amplitude and sketch the graph of $y = 3 \cos x$.

**SOLUTION**    By the preceding discussion, the amplitude is 3. As indicated in Figure 51, we first sketch the graph of $y = \cos x$ and then multiply $y$-coordinates by 3.

**FIGURE 51**

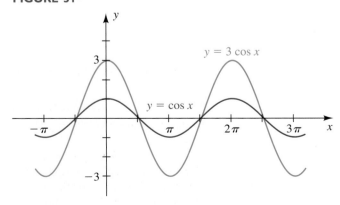

Let us next consider $y = a \sin bx$ and $y = a \cos bx$ for nonzero real numbers $a$ and $b$. As before, the amplitude is $|a|$. If $b > 0$, then exactly one cycle occurs as $bx$ increases from 0 to $2\pi$ or, equivalently, as $x$ increases from 0 to $2\pi/b$. If $b < 0$, then $-b > 0$ and one cycle occurs as $x$ increases from 0 to $2\pi/(-b)$. Thus, the period of the function $f$ given by $f(x) = a \sin bx$ or $f(x) = a \cos bx$ is $2\pi/|b|$. For convenience, we shall also refer to $2\pi/|b|$ as the period of the *graph of $f$*. The next theorem summarizes our discussion.

**Theorem on Amplitudes and Periods**

> If $y = a \sin bx$ or $y = a \cos bx$ for nonzero real numbers $a$ and $b$, then the graph has amplitude $|a|$ and period $\dfrac{2\pi}{|b|}$.

We can also relate the role of $b$ to the discussion of horizontal compressing and stretching of a graph in Section 2.2. If $|b| > 1$, the graph of $y = \sin bx$ or $y = \cos bx$ can be considered to be compressed horizontally by a factor $b$. If $0 < |b| < 1$, the graphs are stretched horizontally by a factor $1/b$. This concept is illustrated in the next two examples.

**EXAMPLE 3     *Finding an amplitude and a period***

Find the amplitude and the period and sketch the graph of $y = 3 \sin 2x$.

*SOLUTION*     Using the theorem on amplitudes and periods with $a = 3$ and $b = 2$, we obtain the following:

$$\text{amplitude:} \quad |a| = |3| = 3$$

$$\text{period:} \quad \frac{2\pi}{|b|} = \frac{2\pi}{2} = \pi$$

Thus, there is exactly one sine wave of amplitude 3 on the $x$-interval $[0, \pi]$. Sketching this wave and then extending the graph to the right and left gives us Figure 52.

**FIGURE 52**

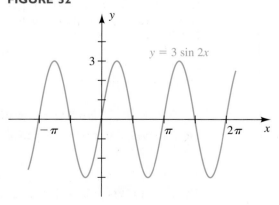

### EXAMPLE 4   Finding an amplitude and a period

Find the amplitude and the period and sketch the graph of $y = 2 \sin \frac{1}{2}x$.

SOLUTION   Using the theorem on amplitudes and periods with $a = 2$ and $b = \frac{1}{2}$, we obtain the following:

$$\text{amplitude:} \quad |a| = |2| = 2$$

$$\text{period:} \quad \frac{2\pi}{|b|} = \frac{2\pi}{\frac{1}{2}} = 4\pi$$

Thus, there is one sine wave of amplitude 2 on the interval $[0, 4\pi]$. Sketching this wave and extending it left and right gives us the graph in Figure 53.

**FIGURE 53**

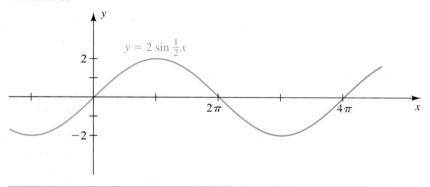

If $y = a \sin bx$ and if $b$ is a large positive number, then the period $2\pi/b$ is small and the sine waves are close together, with $b$ sine waves on the interval $[0, 2\pi]$. For example, in Figure 52, $b = 2$ and we have two sine waves on $[0, 2\pi]$. If $b$ is a small positive number, then the period $2\pi/b$ is large and the waves are far apart. To illustrate, if $y = \sin \frac{1}{10}x$, then one-tenth of a sine wave occurs on $[0, 2\pi]$ and an interval $20\pi$ units long is required for one complete cycle. (See also Figure 53—for $y = 2 \sin \frac{1}{2}x$, one-half of a sine wave occurs on $[0, 2\pi]$.)

If $b < 0$, we can use the fact that $\sin(-x) = -\sin x$ to obtain the graph of $y = a \sin bx$. To illustrate, the graph of $y = \sin(-2x)$ is the same as the graph of $y = -\sin 2x$.

### EXAMPLE 5   Finding an amplitude and a period

Find the amplitude and the period and sketch the graph of the equation $y = 2 \sin(-3x)$.

SOLUTION   Since $\sin(-3x) = -\sin 3x$, we may write $y = -2 \sin 3x$. The amplitude is $|-2| = 2$, and the period is $2\pi/3$. Thus, there is one cycle

*(continued)*

on an interval of length $2\pi/3$. The negative sign indicates a reflection through the $x$-axis. If we consider the interval $[0, 2\pi/3]$ and sketch a sine wave of amplitude 2 (reflected through the $x$-axis), the shape of the graph is apparent. The part of the graph in the interval $[0, 2\pi/3]$ is repeated periodically, as illustrated in Figure 54.

**FIGURE 54**

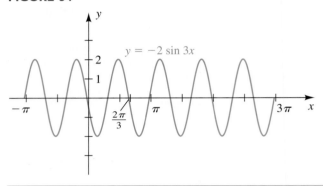

$y = -2 \sin 3x$

**FIGURE 55**

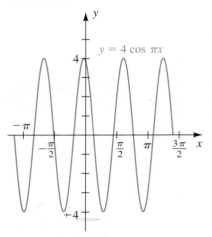

$y = 4 \cos \pi x$

**EXAMPLE 6**    *Finding an amplitude and a period*

Find the amplitude and the period and sketch the graph of $y = 4 \cos \pi x$.

**SOLUTION**    The amplitude is $|4| = 4$, and the period is $2\pi/\pi = 2$. Thus, there is exactly one cosine wave of amplitude 4 on the interval $[0, 2]$. Sketching this wave and extending it left and right gives us the graph in Figure 55.

As in Section 2.2, if $f$ is a function and $c$ is a positive real number, then the graph of $y = f(x) + c$ can be obtained by shifting the graph of $y = f(x)$ vertically upward a distance $c$. For the graph of $y = f(x) - c$, we shift the graph of $y = f(x)$ vertically downward a distance $c$. In the next example we use this technique for a trigonometric graph.

**EXAMPLE 7**    *Vertically shifting a trigonometric graph*

Sketch the graph of $y = 2 \sin x + 3$.

**SOLUTION**    It is important to note that $y \neq 2 \sin (x + 3)$. The graph of $y = 2 \sin x$ is sketched in Figure 56. If we shift this graph vertically upward a distance 3, we obtain the graph of $y = 2 \sin x + 3$.

**FIGURE 56**

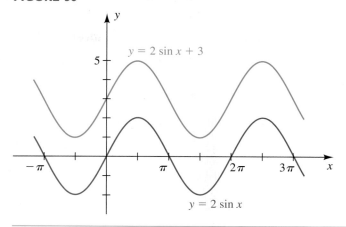

Let us next consider the graph of

$$y = a \sin (bx + c).$$

As before, the amplitude is $|a|$, and the period is $2\pi/|b|$. One cycle occurs if $bx + c$ increases from 0 to $2\pi$. Hence, we can find an interval containing exactly one sine wave by solving the equations

$$bx + c = 0 \quad \text{and} \quad bx + c = 2\pi.$$

The solutions are

$$x = -\frac{c}{b} \quad \text{and} \quad x = -\frac{c}{b} + \frac{2\pi}{b}.$$

The number $-c/b$ is the **phase shift** associated with the graph. The graph of $y = a \sin (bx + c)$ may be obtained by shifting the graph of $y = a \sin bx$ to the left if the phase shift is negative or to the right if the phase shift is positive.

Analogous results are true for $y = a \cos (bx + c)$. The next theorem summarizes our discussion.

**Theorem on Amplitudes, Periods, and Phase Shifts**

If $y = a \sin (bx + c)$ or $y = a \cos (bx + c)$ for nonzero real numbers $a$ and $b$, then

**(1)** the amplitude is $|a|$ and the period is $\dfrac{2\pi}{|b|}$;

**(2)** the phase shift and an interval containing exactly one cycle can be found by solving the two equations

$$bx + c = 0 \quad \text{and} \quad bx + c = 2\pi.$$

**EXAMPLE 8**    *Finding an amplitude, a period, and a phase shift*

Find the amplitude, the period, and the phase shift and sketch the graph of

$$y = 3 \sin \left(2x + \frac{\pi}{2}\right).$$

**SOLUTION**    The equation is of the form $y = a \sin (bx + c)$ with $a = 3$, $b = 2$, and $c = \pi/2$. Thus, the amplitude is $|a| = 3$, and the period is $2\pi/|b| = 2\pi/2 = \pi$.

By part (2) of the theorem on amplitudes, periods, and phase shifts, the phase shift and an interval containing one sine wave can be found by solving the two equations

$$2x + \frac{\pi}{2} = 0 \quad \text{and} \quad 2x + \frac{\pi}{2} = 2\pi.$$

This gives us

$$x = -\frac{\pi}{4} \quad \text{and} \quad x = \frac{3\pi}{4}.$$

Thus, the phase shift is $-\pi/4$, and one sine wave of amplitude 3 occurs on the interval $[-\pi/4, 3\pi/4]$. Sketching that wave and then repeating it to the right and left gives us the graph in Figure 57.

**FIGURE 57**

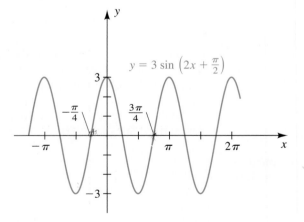

**EXAMPLE 9**    *Finding an amplitude, a period, and a phase shift*

Find the amplitude, the period, and the phase shift and sketch the graph of $y = 2 \cos (3x - \pi)$.

**SOLUTION** The equation has the form $y = a \cos(bx + c)$ with $a = 2$, $b = 3$, and $c = -\pi$. Thus, the amplitude is $|a| = 2$, and the period is $2\pi/|b| = 2\pi/3$.

By part (2) of the theorem on amplitudes, periods, and phase shifts, the phase shift and an interval containing one cycle can be found by solving the two equations

$$3x - \pi = 0 \quad \text{and} \quad 3x - \pi = 2\pi.$$

This gives us

$$x = \frac{\pi}{3} \quad \text{and} \quad x = \frac{3\pi}{3} = \pi.$$

Hence, the phase shift is $\pi/3$, and one cosine-type cycle of amplitude 2 occurs on the interval $[\pi/3, \pi]$. Sketching that part of the graph and then repeating it to the right and left gives us the sketch in Figure 58.

**FIGURE 58**

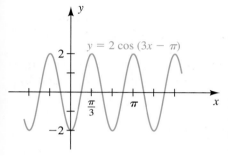

$y = 2 \cos(3x - \pi)$

**EXAMPLE 10** *Finding an equation for a sine wave*

Express the equation for the sine wave shown in Figure 59 in the form

$$y = a \sin(bx + c)$$

for $a > 0$, $b > 0$, and the least positive real number $c$.

**FIGURE 59**

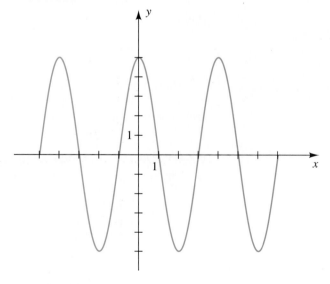

**SOLUTION** The largest and smallest $y$-coordinates of points on the graph are 5 and $-5$, respectively. Hence, the amplitude is $a = 5$.

*(continued)*

Since one wave occurs on the interval $[-1, 3]$, the period has value $3 - (-1) = 4$. Hence, by the theorem on amplitudes, periods, and phase shifts (with $b > 0$),

$$\frac{2\pi}{b} = 4, \quad \text{or} \quad b = \frac{\pi}{2}.$$

The phase shift is $-c/b = -c/(\pi/2)$. Since $c$ is to be positive, the phase shift must be *negative;* that is, the graph in Figure 59 must be obtained by shifting the graph of $y = 5 \sin [(\pi/2)x]$ to the *left*. Since we want $c$ to be as small as possible, we choose the phase shift $-1$. Hence,

$$-\frac{c}{\pi/2} = -1, \quad \text{or} \quad c = \frac{\pi}{2}.$$

Thus, the desired equation is

$$y = 5 \sin \left(\frac{\pi}{2}x + \frac{\pi}{2}\right).$$

There are many other equations for the graph. For example, we could use the phase shifts $-5$, $-9$, $-13$, and so on, but these would not give us the *least* positive value for $c$. Two other equations for the graph are

$$y = 5 \sin \left(\frac{\pi}{2}x - \frac{3\pi}{2}\right) \quad \text{and} \quad y = -5 \sin \left(\frac{\pi}{2}x + \frac{3\pi}{2}\right).$$

However, neither of these equations satisfies the given criteria for $a$, $b$, and $c$, since in the first, $c < 0$, and in the second, $a < 0$ and $c$ does not have its least positive value.

Many phenomena that occur in nature vary in a cyclic or rhythmic manner. It is sometimes possible to represent such behavior by means of trigonometric functions, as illustrated in the next two examples.

**EXAMPLE 11**   *Analyzing the process of breathing*

The rhythmic process of breathing consists of alternating periods of inhaling and exhaling. One complete cycle normally takes place every 5 seconds. If $F(t)$ denotes the air flow rate at time $t$ (in liters per second) and if the maximum flow rate is 0.6 liter per second, find a formula of the form $F(t) = a \sin bt$ that fits this information.

**SOLUTION**     If $F(t) = a \sin bt$ for some $b > 0$, then the period of $F$ is $2\pi/b$. In this application the period is 5 seconds, and hence

$$\frac{2\pi}{b} = 5, \quad \text{or} \quad b = \frac{2\pi}{5}.$$

Since the maximum flow rate corresponds to the amplitude $a$ of $F$, we let $a = 0.6$. This gives us the formula

$$F(t) = 0.6 \sin\left(\frac{2\pi}{5}t\right).$$

---

**EXAMPLE 12**   *Approximating the number of hours of daylight in a day*

The number of hours of daylight $D(t)$ at a particular time of the year can be approximated by

$$D(t) = \frac{K}{2}\sin\frac{2\pi}{365}(t - 79) + 12$$

for $t$ in days and $t = 0$ corresponding to January 1. The constant $K$ determines the total variation in day length and depends on the latitude of the locale.

**(a)** For Boston, $K \approx 6$. Sketch the graph of $D$ for $0 \le t \le 365$.

**(b)** When is the day length the longest? the shortest?

*SOLUTION*

**(a)** If $K = 6$, then $K/2 = 3$, and we may write $D(t)$ in the form

$$D(t) = f(t) + 12,$$

where

$$f(t) = 3\sin\frac{2\pi}{365}(t - 79).$$

We shall sketch the graph of $f$ and then apply a vertical shift through a distance 12. Let us rewrite $f(t)$ as

$$f(t) = 3\sin\left(\frac{2\pi}{365}t - \frac{158\pi}{365}\right),$$

which has the form $f(t) = a\sin(bt + c)$ with $a = 3$, $b = 2\pi/365$, and $c = -158\pi/365$. Thus, the amplitude is 3, and the period of $f$ is

$$\frac{2\pi}{b} = \frac{2\pi}{2\pi/365} = 365 \text{ days.}$$

As in part (2) of the theorem on amplitudes, periods, and phase shifts, we can obtain a $t$-interval containing exactly one cycle by solving the two equations

$$\frac{2\pi}{365}(t - 79) = 0 \quad \text{and} \quad \frac{2\pi}{365}(t - 79) = 2\pi.$$

*(continued)*

The solutions are $t = 79$ and $t = 444$. Hence, one sine wave occurs on the interval $[79, 444]$. Dividing this interval into four equal parts, we obtain the following table of values, which indicates the familiar sine wave pattern of amplitude 3.

**FIGURE 60**

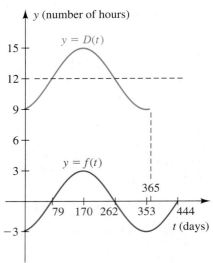

| $t$ | 79 | 170.25 | 261.5 | 352.75 | 444 |
|------|-----|--------|-------|--------|-----|
| $f(t)$ | 0 | 3 | 0 | −3 | 0 |

If $t = 0$,

$$f(0) = 3 \sin \frac{2\pi}{365}(-79) \approx 3 \sin(-1.36) \approx -2.9.$$

Since the period of $f$ is 365, this implies that $f(365) \approx -2.9$.

The graph of $f$ for the interval $[0, 444]$ is sketched in Figure 60, with different scales on the axes and $t$ rounded off to the nearest day.

Applying a vertical shift of 12 units gives us the graph of $D$ for $0 \leq t \leq 365$ shown in Figure 60.

**(b)** The longest day—that is, the largest value of $D(t)$—occurs 170 days after January 1. Except for leap year, this corresponds to June 20. The shortest day occurs 353 days after January 1, or December 20.

In the next example we use a graphing utility to approximate the solution of an inequality that involves trigonometric expressions.

**EXAMPLE 13**   *Approximating solutions of a trigonometric inequality*

Approximate the solution of the inequality

$$\sin 3x < x + \sin x.$$

**SOLUTION**   The given inequality is equivalent to

$$\sin 3x - x - \sin x < 0.$$

If we assign $\sin 3x - x - \sin x$ to $Y_1$, then the given problem is equivalent to finding where the graph of $Y_1$ is below the $x$-axis. Using the standard viewing rectangle gives us a sketch similar to Figure 61(a), where we see that the graph of $Y_1$ has an $x$-intercept $c$ between $-1$ and 0. It appears that the graph is below the $x$-axis on the interval $(c, \infty)$; however, this fact is not perfectly clear because of the small scale on the axes.

**FIGURE 61**

(a) $[-15, 15]$ by $[-10, 10]$          (b) $[-1.5, 1.5]$ by $[-1, 1]$

Using the viewing rectangle $[-1.5, 1.5]$ by $[-1, 1]$ with Xscl and Yscl both equal to 0.25, we obtain Figure 61(b), where we see that the $x$-intercepts are approximately $-0.5$, 0, and 0.5. Using tracing and zoom features yields the more accurate positive value 0.51. Since the function involved is odd, the negative value is approximately $-0.51$. Hence, the solutions of the inequality are in the (approximate) intervals

$$(-0.51, 0) \cup (0.51, \infty).$$

**EXAMPLE 14**   *Investigating alternating current in an electrical circuit*

The current $I$ (in amperes) in an alternating current circuit at time $t$ (in seconds) is given by

$$I = 30 \sin \left(50\pi t - \frac{7\pi}{3}\right).$$

Approximate the smallest value of $t$ for which $I = 15$.

**SOLUTION**   Letting $I = 15$ in the given formula, we obtain

$$15 = 30 \sin \left(50\pi t - \frac{7\pi}{3}\right)$$

or, equivalently,

$$\sin \left(50\pi t - \frac{7\pi}{3}\right) - \frac{1}{2} = 0.$$

If we assign $\sin (50\pi x - 7\pi/3) - \frac{1}{2}$ to $Y_1$, then the given problem is equivalent to approximating the smallest $x$-intercept of the graph.

*(continued)*

**FIGURE 62**  $[0, 0.04]$ by $[-1.5, 0.5]$

Since the period of $Y_1$ is

$$\frac{2\pi}{b} = \frac{2\pi}{50\pi} = \frac{1}{25} = 0.04$$

and since $-\frac{3}{2} \le Y_1 \le \frac{1}{2}$, we select the viewing rectangle $[0, 0.04]$ by $[-1.5, 0.5]$ with Xscl $= 0.01$ and Yscl $= 0.25$, obtaining a sketch similar to Figure 62. Using tracing and zoom features gives us $t \approx 0.01$ sec.

We will rework the preceding example in Section 6.2 and show how to find the exact value of $t$ without the aid of a graphing utility.

## 5.5 EXERCISES

1 Find the amplitude and the period and sketch the graph of the equation:

(a) $y = 4 \sin x$    (b) $y = \sin 4x$

(c) $y = \frac{1}{4} \sin x$    (d) $y = \sin \frac{1}{4}x$

(e) $y = 2 \sin \frac{1}{4}x$    (f) $y = \frac{1}{4} \sin 4x$

(g) $y = -4 \sin x$    (h) $y = \sin (-4x)$

2 Sketch the graphs of the equations that involve the cosine and are analogous to those in (a)–(h) of Exercise 1.

3 Find the amplitude and the period and sketch the graph of the equation:

(a) $y = 3 \cos x$    (b) $y = \cos 3x$

(c) $y = \frac{1}{3} \cos x$    (d) $y = \cos \frac{1}{3}x$

(e) $y = 2 \cos \frac{1}{3}x$    (f) $y = \frac{1}{3} \cos 3x$

(g) $y = -3 \cos x$    (h) $y = \cos (-3x)$

4 Sketch the graphs of the equations that involve the sine and are analogous to those in (a)–(h) of Exercise 3.

**Exer. 5–40: Find the amplitude, the period, and the phase shift and sketch the graph of the equation.**

5 $y = \sin \left(x - \frac{\pi}{2}\right)$    6 $y = \sin \left(x + \frac{\pi}{4}\right)$

7 $y = 3 \sin \left(x + \frac{\pi}{6}\right)$    8 $y = 2 \sin \left(x - \frac{\pi}{3}\right)$

9 $y = \cos \left(x + \frac{\pi}{2}\right)$    10 $y = \cos \left(x - \frac{\pi}{3}\right)$

11 $y = 4 \cos \left(x - \frac{\pi}{4}\right)$    12 $y = 3 \cos \left(x + \frac{\pi}{6}\right)$

13 $y = \sin (2x - \pi) + 1$    14 $y = -\sin (3x + \pi) - 1$

15 $y = -\cos (3x + \pi) - 2$    16 $y = \cos (2x - \pi) + 2$

17 $y = -2 \sin (3x - \pi)$    18 $y = 3 \cos (3x - \pi)$

19 $y = \sin \left(\frac{1}{2}x - \frac{\pi}{3}\right)$    20 $y = \sin \left(\frac{1}{2}x + \frac{\pi}{4}\right)$

21 $y = 6 \sin \pi x$    22 $y = 3 \cos \frac{\pi}{2}x$

23 $y = 2 \cos \frac{\pi}{2}x$    24 $y = 4 \sin 3\pi x$

25 $y = \frac{1}{2} \sin 2\pi x$    26 $y = \frac{1}{2} \cos \frac{\pi}{2}x$

27 $y = 5 \sin \left(3x - \frac{\pi}{2}\right)$    28 $y = -4 \cos \left(2x + \frac{\pi}{3}\right)$

**29**  $y = 3 \cos \left( \dfrac{1}{2} x - \dfrac{\pi}{4} \right)$      **30**  $y = -2 \sin \left( \dfrac{1}{2} x + \dfrac{\pi}{2} \right)$

**31**  $y = -5 \cos \left( \dfrac{1}{3} x + \dfrac{\pi}{6} \right)$      **32**  $y = 4 \sin \left( \dfrac{1}{3} x - \dfrac{\pi}{3} \right)$

**33**  $y = 3 \cos (\pi x + 4 \pi)$

**34**  $y = -2 \sin (2 \pi x + \pi)$

**35**  $y = -\sqrt{2} \sin \left( \dfrac{\pi}{2} x - \dfrac{\pi}{4} \right)$

**36**  $y = \sqrt{3} \cos \left( \dfrac{\pi}{4} x - \dfrac{\pi}{2} \right)$

**37**  $y = -2 \sin (2x - \pi) + 3$

**38**  $y = 3 \cos (x + 3\pi) - 2$

**39**  $y = 5 \cos (2x + 2\pi) + 2$

**40**  $y = -4 \sin (3x - \pi) - 3$

**Exer. 41–44: The graph of an equation is shown in the figure.**

**(a)** Find the amplitude, period, and phase shift.

**(b)** Write the equation in the form $y = a \sin (bx + c)$ for $a > 0$, $b > 0$, and the least positive real number $c$.

**41**

**42**

**43**

**44**

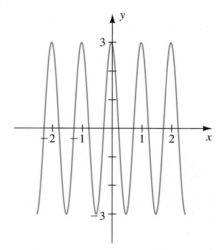

**45**  *Electroencephalography*  Shown in the figure is an electroencephalogram of human brain waves during deep sleep. If we use $W = a \sin (bt + c)$ to represent these waves, what is the value of $b$?

**EXERCISE 45**

0                1                2 (sec)

**46**  *Intensity of daylight*  On a certain spring day with 12 hours of daylight, the light intensity $I$ takes on its largest value of 510 calories/cm$^2$ at midday. If $t = 0$ corresponds to sunrise, find a formula $I = a \sin bt$ that fits this information.

**47**  *Heart action*  The pumping action of the heart consists of the systolic phase, in which blood rushes from the left ventricle into the aorta, and the diastolic phase, during which the heart muscle relaxes. The function whose graph is shown in the figure is sometimes used to model one complete cycle of this process. For a particular individual, the systolic phase lasts $\frac{1}{4}$ second and has a maximum flow rate of 8 liters per minute. Find $a$ and $b$.

48 **Biorhythms** The popular biorhythm theory uses the graphs of three simple sine functions to make predictions about an individual's physical, emotional, and intellectual potential for a particular day. The graphs are given by $y = a \sin bt$ for $t$ in days, with $t = 0$ corresponding to birth and $a = 1$ denoting 100% potential.

(a) Find the value of $b$ for the physical cycle, which has a period of 23 days; for the emotional cycle (period 28 days); and for the intellectual cycle (period 33 days).

(b) Evaluate the biorhythm cycles for a person who has just become 21 years of age and is exactly 7670 days old.

49 **Tidal components** The height of the tide at a particular point on shore can be predicted by using seven trigonometric functions (called tidal components) of the form $f(t) = a \cos (bt + c)$. The principal lunar component may be approximated by

$$f(t) = a \cos \left( \frac{\pi}{6}t - \frac{11\pi}{12} \right),$$

where $t$ is in hours and $t = 0$ corresponds to midnight. Sketch the graph of $f$ if $a = 0.5$ m.

50 **Tidal components** Refer to Exercise 49. The principal solar diurnal component may be approximated by

$$f(t) = a \cos \left( \frac{\pi}{12}t - \frac{7\pi}{12} \right).$$

Sketch the graph of $f$ if $a = 0.2$ m.

51 **Hours of daylight in Fairbanks** If the formula for $D(t)$ in Example 12 is used for Fairbanks, Alaska, then $K \approx 12$. Sketch the graph of $D$ in this case for $0 \le t \le 365$.

52 **Low temperature in Fairbanks** Based on years of weather data, the expected low temperature $T$ (in °F) in Fairbanks, Alaska, can be approximated by

$$T = 36 \sin \frac{2\pi}{365}(t - 101) + 14,$$

where $t$ is in days and $t = 0$ corresponds to January 1.

(a) Sketch the graph of $T$ for $0 \le t \le 365$.

(b) Predict when the coldest day of the year will occur.

**Exer. 53–56:** Scientists sometimes use the formula
$$f(t) = a \sin (bt + c) + d$$
to simulate temperature variations during the day, with time $t$ in hours, temperature $f(t)$ in °C, and $t = 0$ corresponding to midnight. Assume that $f(t)$ is decreasing at midnight.

(a) Determine values of $a$, $b$, $c$, and $d$ that fit the information.

(b) Sketch the graph of $f$ for $0 \le t \le 24$.

53 The high temperature is 10 °C, and the low temperature of $-10$ °C occurs at 4 A.M.

54 The temperature at midnight is 15 °C, and the high and low temperatures are 20 °C and 10 °C.

55 The temperature varies between 10 °C and 30 °C, and the average temperature of 20 °C first occurs at 9 A.M.

56 The high temperature of 28 °C occurs at 2 P.M., and the average temperature of 20 °C occurs 6 hours later.

**Exer. 57–60:** Graph the equation on the interval $[-2, 2]$, and describe the behavior of $y$ as $x \to 0^-$ and as $x \to 0^+$.

57 $y = \sin \dfrac{1}{x}$          58 $y = |x| \sin \dfrac{1}{x}$

59 $y = \dfrac{\sin 2x}{x}$          60 $y = \dfrac{1 - \cos 3x}{x}$

**Exer. 61–62:** Graph the equation on the interval $[-20, 20]$, and estimate the horizontal asymptote.

61 $y = x^2 \sin^2 \left( \dfrac{2}{x} \right)$          62 $y = \dfrac{1 - \cos^2 (2/x)}{\sin (1/x)}$

**Exer. 63–64:** Use a graph to solve the inequality on the interval $[-\pi, \pi]$.

63 $\cos 3x \ge \frac{1}{2}x - \sin x$

64 $\frac{1}{4} \tan (\frac{1}{3}x^2) < \frac{1}{2} \cos 2x + \frac{1}{3}x^2$

# 5.6   ADDITIONAL TRIGONOMETRIC GRAPHS

Methods we developed in Section 5.5 for the sine and cosine can be applied to the other four trigonometric functions; however, there are several differences. Since the tangent, cotangent, secant, and cosecant functions have no largest values, the notion of amplitude has no meaning. Moreover, we do not refer to cycles. For some tangent and cotangent graphs, we begin by sketching the portion between successive vertical asymptotes and then repeat that pattern to the right and to the left.

The graph of $y = a \tan x$ for $a > 0$ can be obtained by stretching or compressing the graph of $y = \tan x$. If $a < 0$, then we also use a reflection about the $x$-axis. Since the tangent function has period $\pi$, it is sufficient to sketch the graph between the two successive vertical asymptotes $x = -\pi/2$ and $x = \pi/2$. The same pattern occurs to the right and to the left, as in the next example.

**EXAMPLE 1    *Sketching the graph of an equation involving* tan $x$**

Sketch the graph of the equation:

**(a)** $y = 2 \tan x$     **(b)** $y = \frac{1}{2} \tan x$

**SOLUTION**    Begin by sketching the graph of $y = \tan x$, as shown in red in Figures 63 and 64, between the vertical asymptotes $x = -\pi/2$ and $x = \pi/2$.

**FIGURE 63**   $y = 2 \tan x$

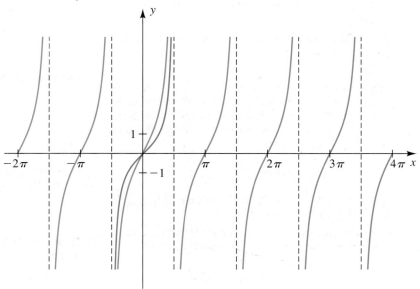

*(continued)*

**FIGURE 64**  $y = \frac{1}{2} \tan x$

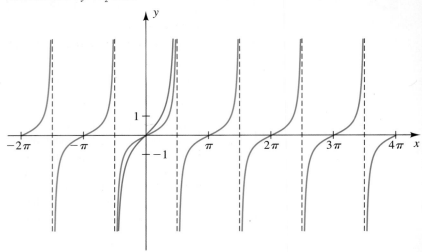

**(a)** For $y = 2 \tan x$, we multiply the $y$-coordinate of each point by 2 and then extend the resulting graph to the right and left, as shown in Figure 63.

**(b)** For $y = \frac{1}{2} \tan x$, we multiply the $y$-coordinates by $\frac{1}{2}$, obtaining the sketch in Figure 64.

The method used in Example 1 can be applied to other functions. Thus, to sketch the graph of $y = 3 \sec x$, we could first sketch the graph of $y = \sec x$ and then multiply the $y$-coordinate of each point by 3.

The next theorem is an analogue of the theorem stated in Section 5.5 for the sine and cosine functions.

**Theorem on the Graph of $y = a \tan (bx + c)$**

If $y = a \tan (bx + c)$ for nonzero real numbers $a$ and $b$, then

**(1)** the period is $\dfrac{\pi}{|b|}$;

**(2)** successive vertical asymptotes for the graph may be found by solving the equations

$$bx + c = -\frac{\pi}{2} \quad \text{and} \quad bx + c = \frac{\pi}{2};$$

**(3)** the phase shift is $-\dfrac{c}{b}$.

**EXAMPLE 2**    *Sketching the graph of an equation of the form $y = a \tan (bx + c)$*

Find the period and sketch the graph of $y = \dfrac{1}{2} \tan \left( x + \dfrac{\pi}{4} \right)$.

**SOLUTION**    The equation has the form given in the preceding theorem with $a = \frac{1}{2}$, $b = 1$, and $c = \pi/4$. Hence, by part (1), the period is given by $\pi/|b| = \pi/1 = \pi$.

As in part (2), to find successive vertical asymptotes we solve the two equations

$$x + \frac{\pi}{4} = -\frac{\pi}{2} \quad \text{and} \quad x + \frac{\pi}{4} = \frac{\pi}{2},$$

obtaining

$$x = -\frac{3\pi}{4} \quad \text{and} \quad x = \frac{\pi}{4}.$$

Since $a = \frac{1}{2}$, the graph of the equation on the interval $[-3\pi/4, \pi/4]$ has the shape of the graph of $y = \frac{1}{2} \tan x$ (see Figure 64). Sketching that part of the graph and extending it to the right and left gives us Figure 65.

**FIGURE 65**   $y = \dfrac{1}{2} \tan \left( x + \dfrac{\pi}{4} \right)$

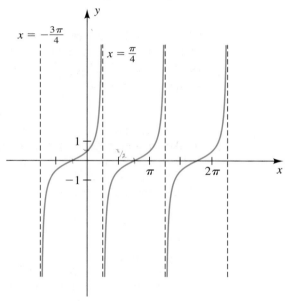

Note that since $c = \pi/4$ and $b = 1$, the phase shift is $-c/b = -\pi/4$. Hence, the graph can also be obtained by shifting the graph of $y = \frac{1}{2} \tan x$ in Figure 64 to the left a distance $\pi/4$.

---

If $y = a \cot (bx + c)$, we have a situation similar to that stated in the previous theorem. The only difference is part (2). Since successive vertical asymptotes for the graph of $y = \cot x$ are $x = 0$ and $x = \pi$ (see Figure 40), we obtain successive vertical asymptotes for the graph of $y = a \cot (bx + c)$ by solving the equations

$$bx + c = 0 \quad \text{and} \quad bx + c = \pi.$$

**EXAMPLE 3**    *Sketching the graph of an equation of the form $y = a \cot (bx + c)$*

Find the period and sketch the graph of $y = \cot \left( 2x - \dfrac{\pi}{2} \right)$.

**SOLUTION**    Using the usual notation, we see that $a = 1$, $b = 2$, and $c = -\pi/2$. The period is $\pi/|b| = \pi/2$. Hence, the graph repeats itself in intervals of length $\pi/2$.

As in the discussion preceding this example, to find two successive vertical asymptotes for the graph we solve the equations

$$2x - \frac{\pi}{2} = 0 \quad \text{and} \quad 2x - \frac{\pi}{2} = \pi,$$

obtaining

$$x = \frac{\pi}{4} \quad \text{and} \quad x = \frac{3\pi}{4}.$$

Since $a$ is positive, we sketch a cotangent-shaped graph on the interval $[\pi/4, 3\pi/4]$ and then repeat it to the right and left in intervals of length $\pi/2$, as shown in Figure 66.

**FIGURE 66**    $y = \cot \left( 2x - \dfrac{\pi}{2} \right)$

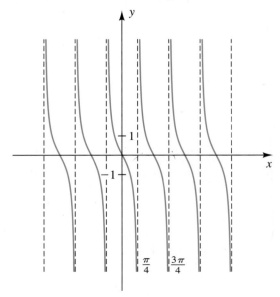

Graphs involving the secant and cosecant functions can be obtained by using methods similar to those for the tangent and cotangent or by taking reciprocals of corresponding graphs of the cosine and sine functions.

**EXAMPLE 4**    *Sketching the graph of an equation of the form $y = a \sec (bx + c)$*

Sketch the graph of the equation:

**(a)** $y = \sec \left( x - \dfrac{\pi}{4} \right)$    **(b)** $y = 2 \sec \left( x - \dfrac{\pi}{4} \right)$

*SOLUTION*

**(a)** The graph of $y = \sec x$ is sketched in red in Figure 67. We can obtain the graph of $y = \sec \left( x - \dfrac{\pi}{4} \right)$ by shifting this graph to the right a distance $\pi/4$, as shown in blue in Figure 67.

**FIGURE 67**   $y = \sec \left( x - \dfrac{\pi}{4} \right)$

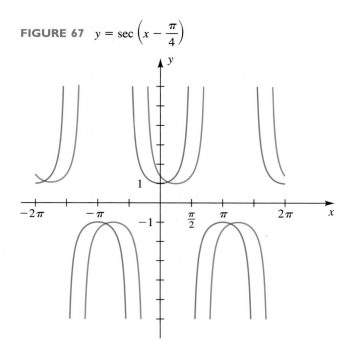

*(continued)*

**(b)** We can sketch the graph by multiplying the *y*-coordinates of the graph in part (a) by 2. This gives us Figure 68.

**FIGURE 68**   $y = 2 \sec\left(x - \dfrac{\pi}{4}\right)$

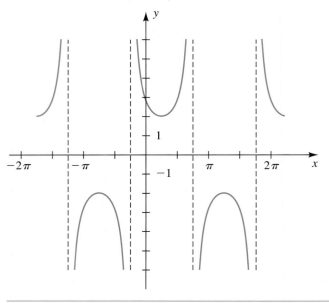

**EXAMPLE 5**     ***Sketching the graph of an equation of the form*** $y = a \csc(bx + c)$

Sketch the graph of $y = \csc(2x + \pi)$.

**SOLUTION**     Since $\csc\theta = 1/\sin\theta$,

$$y = \frac{1}{\sin(2x + \pi)}.$$

Thus, we may obtain the graph of $y = \csc(2x + \pi)$ by finding the graph of $y = \sin(2x + \pi)$ and then taking the reciprocal of the *y*-coordinate of each point. Using $a = 1$, $b = 2$, and $c = \pi$, we see that the amplitude of $y = \sin(2x + \pi)$ is 1 and the period is $2\pi/|b| = 2\pi/2 = \pi$. To find an interval containing one cycle we solve the equations

$$2x + \pi = 0 \quad \text{and} \quad 2x + \pi = 2\pi,$$

obtaining

$$x = -\frac{\pi}{2} \quad \text{and} \quad x = \frac{\pi}{2}.$$

This leads to the graph in red in Figure 69. Taking reciprocals gives us the graph of $y = \csc(2x + \pi)$ shown in the figure.

**FIGURE 69** $y = \csc(2x + \pi)$

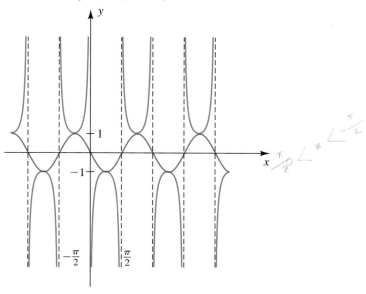

FIGURE 70

(a)

$y = \cos x$

(b)

$y = |\cos x|$

(c)

$y = |\cos x| + 1$

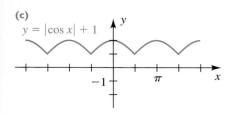

The next example involves the absolute value of a trigonometric function.

**EXAMPLE 6** *Sketching the graph of an equation involving an absolute value*

Sketch the graph of $y = |\cos x| + 1$.

**SOLUTION** We shall sketch the graph in three stages. First, we sketch the graph of $y = \cos x$, as in Figure 70(a).

Next, we obtain the graph of $y = |\cos x|$ by reflecting the negative $y$-coordinates in Figure 70(a) through the $y$-axis. This gives us Figure 70(b).

Finally, we vertically shift the graph in (b) upward 1 unit to obtain Figure 70(c).

We have used three graphs for clarity. In practice, we could sketch the graphs successively on one coordinate plane.

Mathematical applications often involve a function $f$ that is a sum of two or more other functions. To illustrate, suppose

$$f(x) = g(x) + h(x),$$

where $f$, $g$, and $h$ have the same domain $D$. Before electronic graphing devices were invented, a technique known as **addition of y-coordinates** was sometimes used to sketch the graph of $f$. The method is illustrated in Figure 71, where for each $x_1$, the $y$-coordinate $f(x_1)$ of a point on the graph of $f$ is the *sum* $g(x_1) + h(x_1)$ of $y$-coordinates of points on the graphs of $g$ and $f$. The graph of $f$ is obtained by *graphically adding* a sufficient number of such $y$-coordinates using a ruler and compass.

**FIGURE 71**

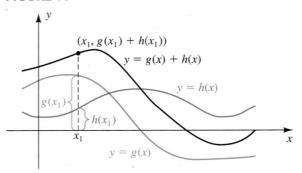

This cumbersome method is no longer necessary, since graphs can be readily sketched with the aid of a graphing utility. However, it is sometimes useful to compare the graph of a sum of functions with the individual functions, as illustrated in the next example.

**EXAMPLE 7**     *Sketching the graph of a sum of two trigonometric functions*

Sketch the graph of $y = \cos x$, $y = \sin x$, and $y = \cos x + \sin x$ on the same coordinate plane for $0 \le x \le 3\pi$.

**SOLUTION**     Some specific keystrokes for the TI-81/82 graphics calculator are given in Appendix I. We make the following assignments:

$$Y_1 = \cos x, \quad Y_2 = \sin x, \quad \text{and} \quad Y_3 = Y_1 + Y_2$$

Since $3\pi \approx 9.42$ and we desire a $3:2$ horizontal:vertical screen proportion, we choose the viewing rectangle $[0, 9.42]$ by $[-3.14, 3.14]$, obtaining Figure 72(a). The clarity of the graph can be enhanced by changing the viewing rectangle to $[0, 9.42]$ by $[-1.5, 1.5]$, as in Figure 72(b).

**FIGURE 72**

(a) [0, 9.42] by [−3.14, 3.14]

(b) [0, 9.42] by [−1.5, 1.5]

Note that the graph of $Y_3$ intersects the graph of $Y_1$ when $Y_2 = 0$, and the graph of $Y_2$ when $Y_1 = 0$. The $x$-intercepts for $Y_3$ correspond to the solutions of $Y_2 = -Y_1$. Finally, we see that the maximum and minimum values of $Y_3$ occur when $Y_1 = Y_2$ (that is, when $x = \pi/4$, $5\pi/4$, and $9\pi/4$). These values are

$$\sqrt{2}/2 + \sqrt{2}/2 = \sqrt{2} \quad \text{and} \quad -\sqrt{2}/2 + (-\sqrt{2}/2) = -\sqrt{2}.$$

The graph of an equation of the form

$$y = f(x) \sin (ax + b) \quad \text{or} \quad y = f(x) \cos (ax + b),$$

where $f$ is a function and $a$ and $b$ are real numbers, is called a **damped sine wave** or **damped cosine wave,** respectively, and $f(x)$ is called the **damping factor.** The next example illustrates a method for graphing such equations.

**EXAMPLE 8**    *Sketching the graph of a damped sine wave*

Sketch the graph of $f$ if $f(x) = 2^{-x} \sin x$.

**SOLUTION**    Using properties of absolute values, we see that

$$|f(x)| = |2^{-x}| \, |\sin x|.$$

Since $|\sin x| \leq 1$ and $2^{-x} > 0$, it follows that $|f(x)| < 2^{-x}$, and hence

$$-2^{-x} \leq f(x) \leq 2^{-x}.$$

**FIGURE 73**

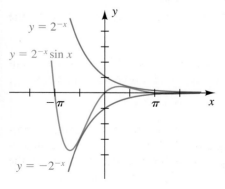

The last inequality implies that the graph of $f$ lies between the graphs of the equations $y = -2^{-x}$ and $y = 2^{-x}$. The graph of $f$ will coincide with one of these graphs if $|\sin x| = 1$—that is, if $x = (\pi/2) + \pi n$ for some integer $n$.

Since $2^{-x} > 0$, the $x$-intercepts on the graph of $f$ occur at $\sin x = 0$—that is, at $x = \pi n$. Because there are an infinite number of $x$-intercepts, this is an example of a function that intersects its horizontal asymptote an infinite number of times. With this information, we obtain the sketch shown in Figure 73.

The damping factor in Example 8 is $2^{-x}$. By using different damping factors, we can obtain other compressed or expanded variations of sine waves. The analysis of such graphs is important in physics and engineering.

## 5.6 EXERCISES

**Exer. 1–52: Find the period and sketch the graph of the equation. Show the asymptotes.**

1  $y = 4 \tan x$

2  $y = \frac{1}{4} \tan x$

3  $y = 3 \cot x$

4  $y = \frac{1}{3} \cot x$

5  $y = 2 \csc x$

6  $y = \frac{1}{2} \csc x$

7  $y = 3 \sec x$

8  $y = \frac{1}{4} \sec x$

9  $y = \tan \left( x - \frac{\pi}{4} \right)$

10  $y = \tan \left( x + \frac{3\pi}{4} \right)$

11  $y = \tan 2x$

12  $y = \tan \frac{1}{2} x$

13  $y = \tan \frac{1}{4} x$

14  $y = \tan 4x$

15  $y = 2 \tan \left( 2x + \frac{\pi}{2} \right)$

16  $y = \frac{1}{3} \tan \left( 2x - \frac{\pi}{4} \right)$

17  $y = -\frac{1}{4} \tan \left( \frac{1}{2} x + \frac{\pi}{3} \right)$

18  $y = -3 \tan \left( \frac{1}{3} x - \frac{\pi}{3} \right)$

19  $y = \cot \left( x - \frac{\pi}{2} \right)$

20  $y = \cot \left( x + \frac{\pi}{4} \right)$

21  $y = \cot 2x$

22  $y = \cot \frac{1}{2} x$

23  $y = \cot \frac{1}{3} x$

24  $y = \cot 3x$

25  $y = 2 \cot \left( 2x + \frac{\pi}{2} \right)$

26  $y = -\frac{1}{3} \cot (3x - \pi)$

27  $y = -\frac{1}{2} \cot \left( \frac{1}{2} x + \frac{\pi}{4} \right)$

28  $y = 4 \cot \left( \frac{1}{3} x - \frac{\pi}{6} \right)$

29  $y = \sec \left( x - \frac{\pi}{2} \right)$

30  $y = \sec \left( x - \frac{3\pi}{4} \right)$

31  $y = \sec 2x$

32  $y = \sec \frac{1}{2} x$

33  $y = \sec \frac{1}{3} x$

34  $y = \sec 3x$

35  $y = 2 \sec \left( 2x - \frac{\pi}{2} \right)$

36  $y = \frac{1}{2} \sec \left( 2x - \frac{\pi}{2} \right)$

37  $y = -\frac{1}{3} \sec \left( \frac{1}{2} x + \frac{\pi}{4} \right)$

38  $y = -3 \sec \left( \frac{1}{3} x + \frac{\pi}{3} \right)$

39  $y = \csc \left( x - \frac{\pi}{2} \right)$

40  $y = \csc \left( x + \frac{3\pi}{4} \right)$

41  $y = \csc 2x$

42  $y = \csc \frac{1}{2} x$

43  $y = \csc \frac{1}{3} x$

44  $y = \csc 3x$

45  $y = 2 \csc \left( 2x + \frac{\pi}{2} \right)$

46  $y = -\frac{1}{2} \csc (2x - \pi)$

47  $y = -\frac{1}{4} \csc \left( \frac{1}{2} x + \frac{\pi}{2} \right)$

48  $y = 4 \csc \left( \frac{1}{2} x - \frac{\pi}{4} \right)$

49  $y = \tan \frac{\pi}{2} x$

50  $y = \cot \pi x$

51  $y = \csc 2\pi x$

52  $y = \sec \frac{\pi}{8} x$

53  Find an equation using the cotangent function that has the same graph as $y = \tan x$.

54  Find an equation using the cosecant function that has the same graph as $y = \sec x$.

**Exer. 55–60: Use the graph of a trigonometric function to aid in sketching the graph of the equation without plotting points.**

55  $y = |\sin x|$

56  $y = |\cos x|$

57  $y = |\sin x| + 2$

58  $y = |\cos x| - 3$

59  $y = -|\cos x| + 1$

60  $y = -|\sin x| - 2$

**Exer. 61–66: Sketch the graph of the equation.**

**61** $y = x + \cos x$

**62** $y = x - \sin x$

**63** $y = 2^{-x} \cos x$

**64** $y = e^x \sin x$

**65** $y = |x| \sin x$

**66** $y = |x| \cos x$

**Exer. 67–68: Identify the damping factor $f(x)$ for the damped wave. Sketch graphs of $y = \pm f(x)$ and the equation on the same coordinate plane for $-2\pi \le x \le 2\pi$.**

**67** $y = e^{-x/4} \sin 4x$

**68** $y = 3^{-x/5} \cos 2x$

**Exer. 69–70: Graph the function $f$ on $[-\pi, \pi]$, and estimate the high and low points.**

**69** $f(x) = \cos 2x + 2 \sin 4x - \sin x$

**70** $f(x) = \tan \frac{1}{4}x - 2 \sin 2x$

**Exer. 71–72: Use a graph to estimate the largest interval $[a, b]$, with $a < 0$ and $b > 0$, on which $f$ is one-to-one.**

**71** $f(x) = \sin (2x + 2) \cos (1.5x - 1)$

**72** $f(x) = 1.5 \cos (\frac{1}{2}x - 0.3) + \sin (1.5x + 0.5)$

**Exer. 73–74: Use a graph to solve the inequality on the interval $[-\pi, \pi]$.**

**73** $\cos (2x - 1) + \sin 3x \ge \sin \frac{1}{3}x + \cos x$

**74** $\frac{1}{2} \cos 2x + 2 \cos (x - 2) <$
$$2 \cos (1.5x + 1) + \sin (x - 1)$$

**75** *The earth's magnetic field* The strength of the earth's magnetic field varies with the depth below the surface. The strength at depth $z$ and time $t$ can sometimes be approximated using the damped sine wave

$$S = A_0 e^{-\alpha z} \sin (kt - \alpha z),$$

where $A_0$, $\alpha$, and $k$ are constants.

**(a)** What is the damping factor?

**(b)** Find the phase shift at depth $z_0$.

**(c)** At what depth is the amplitude of the wave one-half the amplitude of the surface strength?

## 5.7 APPLICATIONS INVOLVING RIGHT TRIANGLES

Trigonometry was developed to help solve problems involving angles and lengths of sides of triangles. Problems of that type are no longer the most important applications; however, questions about triangles still arise in physical situations. We shall restrict our discussion in this section to right triangles. Triangles that do not contain a right angle will be considered in Chapter 7.

We shall often use the following notation. The vertices of a triangle will be denoted by $A$, $B$, and $C$; the angles at $A$, $B$, and $C$ will be denoted by $\alpha$, $\beta$, and $\gamma$, respectively; and the lengths of the sides opposite these angles by $a$, $b$, and $c$, respectively. The triangle itself will be referred to as *triangle ABC*. If a triangle is a right triangle and if one of the acute angles and a side are known or if two sides are given, then we may find the remaining parts by using the formulas in Section 5.2 that express the trigonometric functions as ratios of sides of a triangle.

In all examples *it is assumed that you know how to find trigonometric function values and angles by using either a calculator or results about special angles.*

**EXAMPLE I**   *Approximating parts of a right triangle*

In triangle $ABC$, $\gamma = 90°$, $\alpha = 34°$, and $b = 10.5$. Approximate the remaining parts.

**SOLUTION**    Since the sum of the three interior angles in a triangle is $180°$ and $\alpha + \gamma = 90° + 34° = 124°$,

$$\beta = 180° - (\alpha + \gamma)$$
$$= 180° - 124°$$
$$= 56°.$$

**FIGURE 74**

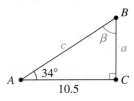

Referring to Figure 74, we obtain

$$\tan 34° = \frac{a}{10.5} \qquad\qquad \tan \alpha = \frac{\text{opp}}{\text{adj}}$$
$$a = (10.5) \tan 34° \qquad \text{solve for } a$$
$$\approx (10.5)(0.6745) \approx 7.1. \quad \text{approximate}$$

To find side $c$, we can use either the cosine or the secant function, as in (1) or (2), respectively:

**(1)**     $\cos 34° = \dfrac{10.5}{c} \qquad\qquad \cos \alpha = \dfrac{\text{adj}}{\text{hyp}}$

$$c = \frac{10.5}{\cos 34°} \qquad \text{solve for } c$$

$$\approx \frac{10.5}{0.8290} \approx 12.7 \qquad \text{approximate}$$

**(2)**     $\sec 34° = \dfrac{c}{10.5} \qquad\qquad \sec \alpha = \dfrac{\text{hyp}}{\text{adj}}$

$$c = (10.5) \sec 34° \qquad \text{solve for } c$$
$$\approx (10.5)(1.2062) \approx 12.7 \quad \text{approximate}$$

As illustrated in Example 1, when working with triangles, we usually round off answers. One reason for doing so is that in most applications, the lengths of sides of triangles and measures of angles are found by mechanical devices and hence are only approximations to the exact values. Consequently, a number such as 10.5 in Example 1 is assumed to have been rounded off to the nearest tenth. We cannot expect more accuracy in the calculated values for the remaining sides, and therefore they should also be rounded off to the nearest tenth.

In finding angles, answers should be rounded off as indicated in the following table.

| Number of significant figures for sides | Round off degree measure of angles to the nearest |
|---|---|
| 2 | 1° |
| 3 | 10′ or 0.1° |
| 4 | 1′ or 0.01° |

Justification of this table requires a careful analysis of problems that involve approximate data.

### EXAMPLE 2    *Approximating parts of a right triangle*

In triangle $ABC$, $\gamma = 90°$, $a = 12.3$, and $b = 31.6$. Approximate the remaining parts.

**FIGURE 75**

**SOLUTION**    Referring to the triangle illustrated in Figure 75 gives us

$$\tan \alpha = \frac{12.3}{31.6}.$$

Since the sides are given with three significant figures, the rule stated in the preceding table tells us that $\alpha$ should be rounded off to the nearest 0.1°, or the nearest multiple of 10′. Using the degree mode on a calculator, we have

$$\alpha = \tan^{-1} \frac{12.3}{31.6} \approx 21.3° \quad \text{or, equivalently,} \quad \alpha \approx 21°20'.$$

Since $\alpha$ and $\beta$ are complementary angles,

$$\beta = 90° - \alpha \approx 90° - 21.3° = 68.7°.$$

The only remaining part to find is $c$. We could use several relationships involving $c$ to determine its value. Among these are

$$\sec \alpha = \frac{c}{31.6}, \quad \sec \beta = \frac{c}{12.3}, \quad \text{and} \quad a^2 + b^2 = c^2.$$

Whenever possible, it is best to use a relationship that involves only given information, since it doesn't depend on any previously calculated value. Hence, with $a = 12.3$ and $b = 31.6$, we have

$$c = \sqrt{a^2 + b^2} = \sqrt{(12.3)^2 + (31.6)^2} = \sqrt{1149.85} \approx 33.9.$$

As illustrated in Figure 76, if an observer at point $X$ sights an object, then the angle that the line of sight makes with the horizontal line $l$ is the **angle of elevation** of the object, if the object is above the horizontal line, or the **angle of depression** of the object, if the object is below the horizontal line. We use this terminology in the next two examples.

**FIGURE 76**

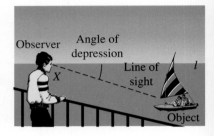

**EXAMPLE 3**    *Using an angle of elevation*

From a point on level ground 135 feet from the base of a tower, the angle of elevation of the top of the tower is $57°20'$. Approximate the height of the tower.

*SOLUTION*    If we let $d$ denote the height of the tower, then the given facts are represented by the triangle in Figure 77. Referring to the figure, we obtain

$$\tan 57°20' = \frac{d}{135} \qquad\qquad \tan 57°20' = \frac{\text{opp}}{\text{adj}}$$

$$d = 135 \tan 57°20' \qquad \text{solve for } d$$

$$\approx 135(1.560) \approx 211. \qquad \text{approximate}$$

The tower is approximately 211 feet high.

**FIGURE 77**

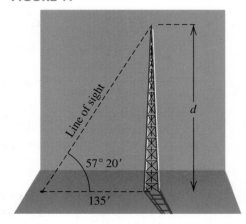

**EXAMPLE 4    *Using angles of depression***

From the top of a building that overlooks an ocean, an observer watches a boat sailing directly toward the building. If the observer is 100 feet above sea level and if the angle of depression of the boat changes from 25° to 40° during the period of observation, approximate the distance that the boat travels.

*SOLUTION*    As in Figure 78, let $A$ and $B$ be the positions of the boat that correspond to the 25° and 40° angles, respectively. Suppose that the observer is at point $D$ and that $C$ is the point 100 feet directly below. Let $d$ denote the distance the boat travels, and let $k$ denote the distance from $B$ to $C$. If $\alpha$ and $\beta$ denote angles $DAC$ and $DBC$, respectively, then it follows from geometry that $\alpha = 25°$ and $\beta = 40°$.

**FIGURE 78**

From triangle $BCD$:

$$\cot \beta = \cot 40° = \frac{k}{100} \qquad \cot \beta = \frac{\text{adj}}{\text{opp}}$$

$$k = 100 \cot 40° \qquad \text{solve for } k$$

From triangle $DAC$:

$$\cot \alpha = \cot 25° = \frac{d + k}{100} \qquad \cot \alpha = \frac{\text{adj}}{\text{opp}}$$

$$d + k = 100 \cot 25° \qquad \text{multiply by } 100$$

$$d = 100 \cot 25° - k \qquad \text{subtract } k$$

$$= 100 \cot 25° - 100 \cot 40° \qquad k = 100 \cot 40°$$

$$= 100(\cot 25° - \cot 40°) \qquad \text{factor out } 100$$

$$\approx 100(2.145 - 1.192) \approx 95 \qquad \text{approximate}$$

Hence, the boat travels approximately 95 feet.

In certain navigation or surveying problems, the **direction,** or **bearing,** from a point $P$ to a point $Q$ is specified by stating the acute angle that segment $PQ$ makes with the north-south line through $P$. We also state whether $Q$ is north or south and east or west of $P$. Figure 79 illustrates four possibilities. The bearing from $P$ to $Q_1$ is 25° east of north and is denoted by N25°E. We also refer to the **direction** N25°E, meaning the direction from $P$ to $Q_1$. The bearings from $P$ to $Q_2$, to $Q_3$, and to $Q_4$ are represented in a similar manner in the figure. Note that when this notation is used for bearings or directions, N or S always appears to the *left* of the angle and W or E to the *right*.

**FIGURE 79**

**FIGURE 80**

In air navigation, directions and bearings are specified by measuring from the north in a *clockwise* direction. In this case, a positive measure is assigned to the angle instead of the negative measure to which we are accustomed for clockwise rotations. Referring to Figure 80, we see that the direction of $PQ$ is 40° and the direction of $PR$ is 300°.

**EXAMPLE 5    Using bearings**

Two ships leave port at the same time, one ship sailing in the direction N23°E at a speed of 11 mi/hr and the second ship sailing in the direction S67°E at 15 mi/hr. Approximate the bearing from the second ship to the first, one hour later.

**FIGURE 81**

**SOLUTION**    The sketch in Figure 81 indicates the positions of the first and second ships at points $A$ and $B$, respectively, after one hour. Point $C$ represents the port. We wish to find the bearing from $B$ to $A$. Note that

$$\angle ACB = 180° - (23° + 67°) = 90°,$$

and hence triangle $ACB$ is a right triangle. Thus,

$$\tan \beta = \frac{11}{15} \qquad \tan \beta = \frac{\text{opp}}{\text{adj}}$$

$$\beta = \tan^{-1} \tfrac{11}{15} \approx 36°. \qquad \text{approximate}$$

**FIGURE 82**

We have rounded $\beta$ to the nearest degree, since the sides of the triangles are given with two significant figures.

Referring to Figure 82, we obtain the following:

$$\angle CBD = 90° - \angle BCD = 90° - 67° = 23°$$
$$\angle ABD = \angle ABC + \angle CBD \approx 36° + 23° = 59°$$
$$\theta = 90° - \angle ABD \approx 90° - 59° = 31°$$

Thus, the bearing from $B$ to $A$ is approximately N31°W.

## 5.7 EXERCISES

**Exer. 1–8:** Given the indicated parts of triangle $ABC$ with $\gamma = 90°$, find the exact values of the remaining parts.

1  $\alpha = 30°$,   $b = 20$       2  $\beta = 45°$,   $b = 35$

3  $\beta = 45°$,   $c = 30$       4  $\alpha = 60°$,   $c = 6$

5  $a = 5$,       $b = 5$         6  $a = 4\sqrt{3}$,   $c = 8$

7  $b = 5\sqrt{3}$,   $c = 10\sqrt{3}$    8  $b = 7\sqrt{2}$,   $c = 14$

**Exer. 9–16:** Given the indicated parts of triangle $ABC$ with $\gamma = 90°$, approximate the remaining parts.

9  $\alpha = 37°$,       $b = 24$      10  $\beta = 64°20'$,   $a = 20.1$

11  $\beta = 71°51'$,   $b = 240.0$   12  $\alpha = 31°10'$,   $a = 510$

13  $a = 25$,           $b = 45$      14  $a = 31$,         $b = 9.0$

15  $c = 5.8$,          $b = 2.1$     16  $a = 0.42$,       $c = 0.68$

**Exer. 17–24:** Given the indicated parts of triangle $ABC$ with $\gamma = 90°$, express the third part in terms of the first two.

17  $\alpha$, $c$;  $b$              18  $\beta$, $c$;  $b$

19  $\beta$, $b$;  $a$              20  $\alpha$, $b$;  $a$

21  $\alpha$, $a$;  $c$              22  $\beta$, $a$;  $c$

23  $a$, $c$;  $b$                  24  $a$, $b$;  $c$

25  **Height of a kite**  A person flying a kite holds the string 4 feet above ground level. The string of the kite is taut and makes an angle of 60° with the horizontal (see the figure). Approximate the height of the kite above level ground if 500 feet of string is payed out.

**EXERCISE 25**

26  **Surveying**  From a point 15 meters above level ground, a surveyor measures the angle of depression of an object on the ground at 68°. Approximate the distance from the object to the point on the ground directly beneath the surveyor.

27  **Airplane landing**  A pilot, flying at an altitude of 5000 feet, wishes to approach the numbers on a runway at an angle of 10°. Approximate, to the nearest 100 feet, the distance from the airplane to the numbers at the beginning of the descent.

28  **Radio antenna**  A guy wire is attached to the top of a radio antenna and to a point on horizontal ground that is 40.0 meters from the base of the antenna. If the wire makes an angle of 58°20' with the ground, approximate the length of the wire.

**29 Surveying** To find the distance $d$ between two points $P$ and $Q$ on opposite shores of a lake, a surveyor locates a point $R$ that is 50.0 meters from $P$ such that $RP$ is perpendicular to $PQ$, as shown in the figure. Next, using a transit, the surveyor measures angle $PRQ$ as $72°40'$. Find $d$.

EXERCISE 29

**30 Meteorological calculations** To measure the height $h$ of a cloud cover, a meteorology student directs a spotlight vertically upward from the ground. From a point $P$ on level ground that is $d$ meters from the spotlight, the angle of elevation $\theta$ of the light image on the clouds is then measured (see the figure).

(a) Express $h$ in terms of $d$ and $\theta$.

(b) Approximate $h$ if $d = 1000$ m and $\theta = 59°$.

EXERCISE 30

**31 Altitude of a rocket** A rocket is fired at sea level and climbs at a constant angle of $75°$ through a distance of 10,000 feet. Approximate its altitude to the nearest foot.

**32 Airplane takeoff** An airplane takes off at a $10°$ angle and travels at the rate of 250 ft/sec. Approximately how long does it take the airplane to reach an altitude of 15,000 feet?

**33 Designing a drawbridge** A drawbridge is 150 feet long when stretched across a river. As shown in the figure,

the two sections of the bridge can be rotated upward through an angle of $35°$.

(a) If the water level is 15 feet below the closed bridge, find the distance $d$ between the end of a section and the water level when the bridge is fully open.

(b) Approximately how far apart are the ends of the two sections when the bridge is fully opened, as shown in the figure?

EXERCISE 33

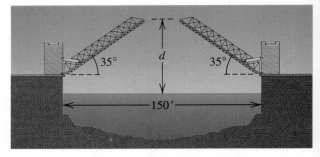

**34 Designing a water slide** Shown in the figure is part of a design for a water slide. Find the total length of the slide to the nearest foot.

EXERCISE 34

**35 Sun's elevation** Approximate the angle of elevation $\alpha$ of the sun if a person 5.0 feet tall casts a shadow 4.0 feet long on level ground (see the figure).

EXERCISE 35

**36 *Constructing a ramp*** A builder wishes to construct a ramp 24 feet long that rises to a height of 5.0 feet above level ground. Approximate the angle that the ramp should make with the horizontal.

**37 *Video game*** Shown in the figure is the screen for a simple video arcade game in which ducks move from $A$ to $B$ at the rate of 7 cm/sec. Bullets fired from point $O$ travel 25 cm/sec. If a player shoots as soon as a duck appears at $A$, at which angle $\varphi$ should the gun be aimed to score a direct hit?

EXERCISE 37

**38 *Conveyor belt*** A conveyor belt 9 meters long can be hydraulically rotated up to an angle of 40° to unload cargo from airplanes (see the figure).

(a) Find, to the nearest degree, the angle through which the conveyor belt should be rotated to reach a door that is 4 meters above the platform supporting the belt.

(b) Approximate the maximum height above the platform that the belt can reach.

EXERCISE 38

**39 *The Pentagon's ground area*** The Pentagon is the largest office building in the world in terms of ground area. The perimeter of the building has the shape of a regular pentagon with each side of length 921 feet. Find the area enclosed by the perimeter of the building.

**40 *Perimeter of an octagon*** A regular octagon is inscribed in a circle of radius 12.0 centimeters. Approximate the perimeter of the octagon.

**41 *Diagonals of a box*** A rectangular box has dimensions 8 inches × 6 inches × 4 inches. Approximate, to the nearest tenth of a degree, the angle $\theta$ formed by a diagonal of the base and the diagonal of the box, as shown in the figure.

EXERCISE 41

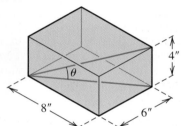

**42 *Volume of a conical cup*** A conical paper cup has a radius of 2 inches. Approximate, to the nearest degree, the angle $\beta$ (see the figure) so that the cone will have a volume of 20 in.$^3$.

EXERCISE 42

**43 *Height of a tower*** From a point $P$ on level ground, the angle of elevation of the top of a tower is 26°50′. From a point 25.0 meters closer to the tower and on the same line with $P$ and the base of the tower, the angle of elevation of the top is 53°30′. Approximate the height of the tower.

**44 Ladder calculations**  A ladder 20 feet long leans against the side of a building, and the angle between the ladder and the building is 22°.

(a) Approximate the distance from the bottom of the ladder to the building.

(b) If the distance from the bottom of the ladder to the building is increased by 3.0 feet, approximately how far does the top of the ladder move down the building?

**45 Ascent of a hot-air balloon**  As a hot-air balloon rises vertically, its angle of elevation from a point $P$ on level ground 110 kilometers from the point $Q$ directly underneath the balloon changes from $19°20'$ to $31°50'$ (see the figure). Approximately how far does the balloon rise during this period?

EXERCISE 45

**46 Height of a building**  From a point $A$ that is 8.20 meters above level ground, the angle of elevation of the top of the building is $31°20'$ and the angle of depression of the base of the building is $12°50'$. Approximate the height of the building.

**47 Radius of the earth**  A spacelab circles the earth at an altitude of 380 miles. When an astronaut views the horizon of the earth, the angle $\theta$ shown in the figure is 65.8°. Use this information to estimate the radius of the earth.

EXERCISE 47

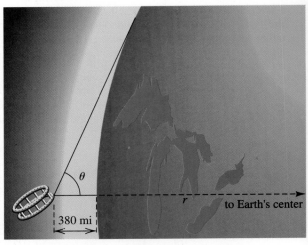

**48 Length of an antenna**  A CB antenna is located on the top of a garage that is 16 feet tall. From a point on level ground that is 100 feet from a point directly below the antenna, the antenna subtends an angle of 12°, as shown in the figure. Approximate the length of the antenna.

EXERCISE 48

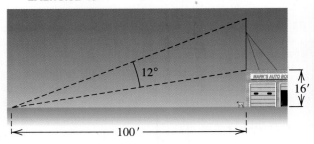

**49 Speed of an airplane**  An airplane flying at an altitude of 10,000 feet passes directly over a fixed object on the ground. One minute later, the angle of depression of the object is 42°. Approximate the speed of the airplane to the nearest mile per hour.

**50 Height of a mountain**  A motorist, traveling along a level highway at a speed of 60 km/hr directly toward a mountain, observes that between 1:00 P.M. and 1:10 P.M. the angle of elevation of the top of the mountain changes from 10° to 70°. Approximate the height of the mountain.

**51  Communications satellite**  Shown in the first figure is a communications satellite with an equatorial orbit—that is, a nearly circular orbit in the plane determined by the earth's equator. If the satellite circles the earth at an altitude of $a = 22{,}300$ mi, its speed is the same as the rotational speed of earth; to an observer on the equator, the satellite appears to be stationary—that is, its orbit is synchronous.

(a) Using $R = 4000$ mi for the radius of the earth, determine the percentage of the equator that is within signal range of such a satellite.

(b) As shown in the second figure, three satellites are equally spaced in equatorial synchronous orbits. Use the value of $\theta$ obtained in part (a) to explain why all points on the equator are within signal range of at least one of the three satellites.

EXERCISE 51

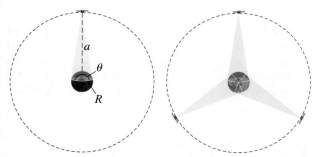

**52  Communications satellite**  Refer to Exercise 51. Shown in the figure is the area served by a communications satellite circling a planet of radius $R$ at an altitude $a$. The portion of the planet's surface within range of the satellite is a spherical cap of depth $d$ and surface area $A = 2\pi R d$.

EXERCISE 52

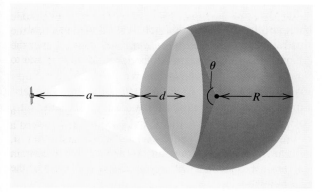

(a) Express $d$ in terms of $R$ and $\theta$.

(b) Estimate the percentage of the planet's surface that is within signal range of a single satellite in equatorial synchronous orbit.

**53  Height of a kite**  Generalize Exercise 25 to the case where the angle is $\alpha$, the number of feet of string payed out is $d$, and the end of the string is held $c$ feet above the ground. Express the height $h$ of the kite in terms of $\alpha$, $d$, and $c$.

**54  Surveying**  Generalize Exercise 26 to the case where the point is $d$ meters above level ground and the angle of depression is $\alpha$. Express the distance $x$ in terms of $d$ and $\alpha$.

**55  Height of a tower**  Generalize Exercise 43 to the case where the first angle is $\alpha$, the second angle is $\beta$, and the distance between the two points is $d$. Express the height $h$ of the tower in terms of $d$, $\alpha$, and $\beta$.

**56  Perimeter of an octagon**  Generalize Exercise 40 to the case of an $n$-sided polygon inscribed in a circle of radius $r$. Express the perimeter $P$ in terms of $n$ and $r$.

**57  Ascent of a hot-air balloon**  Generalize Exercise 45 to the case where the distance from $P$ to $Q$ is $d$ kilometers and the angle of elevation changes from $\alpha$ to $\beta$.

**58  Height of a building**  Generalize Exercise 46 to the case where point $A$ is $d$ meters above ground and the angles of elevation and depression are $\alpha$ and $\beta$, respectively. Express the height $h$ of the building in terms of $d$, $\alpha$, and $\beta$.

**Exer. 59–60: Find the bearing from $P$ to each of the points $A$, $B$, $C$, and $D$.**

59

60

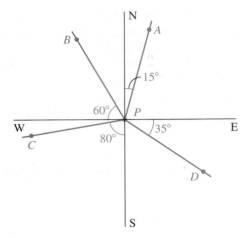

another ranger sights the same fire in the direction
S54°10′E. Approximate, to the nearest tenth of a mile,
the distance of the fire from *A*.

**EXERCISE 62**

**61  Ships' bearings**  A ship leaves port at 1:00 P.M. and
sails in the direction N34°W at a rate of 24 mi/hr. Another ship leaves port at 1:30 P.M. and sails in the direction N56°E at a rate of 18 mi/hr.

   **(a)** Approximately how far apart are the ships at
   3:00 P.M.?

   **(b)** What is the bearing, to the nearest degree, from the
   first ship to the second?

**62  Pinpointing a forest fire**  From an observation point *A*,
a forest ranger sights a fire in the direction S35°50′W
(see the figure). From a point *B*, 5 miles due west of *A*,

**63  Airplane flight**  An airplane flying at a speed of
360 mi/hr flies from a point *A* in the direction 137° for
30 minutes and then flies in the direction 227° for 45
minutes. Approximate, to the nearest mile, the distance
from the airplane to *A*.

**64  Airplane flight plan**  An airplane flying at a speed of
400 mi/hr flies from a point *A* in the direction 153° for 1
hour and then flies in the direction 63° for 1 hour.

   **(a)** In what direction does the plane need to fly in order
   to get back to point *A*?

   **(b)** How long will it take to get back to point *A*?

## 5.8   HARMONIC MOTION

**FIGURE 83**

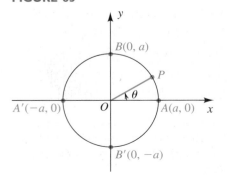

Trigonometric functions are useful in the investigation of vibratory or
oscillatory motion, such as the motion of a particle in a vibrating guitar
string or in a spring that has been compressed or elongated and then
released to oscillate back and forth. The fundamental type of particle displacement in these illustrations is *harmonic motion.*

To help introduce harmonic motion, let us consider a point *P* moving
at a constant rate around the circle of radius *a* shown in Figure 83. Suppose that the initial position of *P* is $A(a, 0)$ and that $\theta$ is the angle generated by ray *OP* after *t* units of time. The *angular speed* $\omega$ of *OP* is the rate
at which the measure of $\theta$ changes per unit time (see Example 6 of Section
5.1). To state that *P* moves around the circle at a constant rate is equivalent
to stating that the angular speed $\omega$ is constant. If $\omega$ is constant, then
$\theta = \omega t$. To illustrate, if $\omega = \pi/6$ radian per second, then $\theta = (\pi/6)t$. In
this case, at $t = 1$ sec, $\theta = (\pi/6)(1) = \pi/6$, and *P* is one-third of the way

from $A(a, 0)$ to $B(0, a)$. At $t = 2$ sec, $\theta = (\pi/6)(2) = \pi/3$, and $P$ is two-thirds of the way from $A$ to $B$. At $t = 6$ sec, $\theta = (\pi/6)(6) = \pi$, and $P$ is at $A'(-a, 0)$, and so on.

If the coordinates of $P$ are $(x, y)$, as shown in Figure 84, then we have $\cos \theta = x/a$ and $\sin \theta = y/a$, and therefore

$$x = a \cos \theta, \quad y = a \sin \theta.$$

Using the fact that $\theta = \omega t$ gives us

$$x = a \cos \omega t, \quad y = a \sin \omega t.$$

The last two equations specify the position $(x, y)$ of $P$ at any time $t$.

**FIGURE 84**

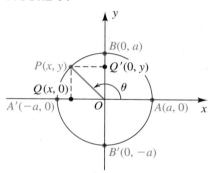

Next consider the point $Q(x, 0)$—that is, the **projection of $P$ on the x-axis.** This point is shown in Figure 84. The position of $Q$ is given by $x = a \cos \omega t$. As $P$ moves around the circle several times, the point $Q$ oscillates back and forth between $A(a, 0)$ and $A'(-a, 0)$. Similarly, the point $Q'(0, y)$ is the **projection of $P$ on the y-axis,** and its position is given by $y = a \sin \omega t$. As $P$ moves around the circle, $Q'$ oscillates between $B'(0, -a)$ and $B(0, a)$. The motions of $Q$ and $Q'$ are of the type described in the next definition.

**Definition of Simple Harmonic Motion**

A point moving on a coordinate line is in **simple harmonic motion** if its distance $d$ from the origin at time $t$ is given by either

$$d = a \cos \omega t \quad \text{or} \quad d = a \sin \omega t,$$

where $a$ and $\omega$ are constants, with $\omega > 0$.

In the preceding definition, the **amplitude** of the motion is the maximum displacement $|a|$ of the point from the origin. The **period** is the time $2\pi/\omega$ required for one complete oscillation. The **frequency** $\omega/(2\pi)$ is the number of oscillations per unit of time.

FIGURE 85

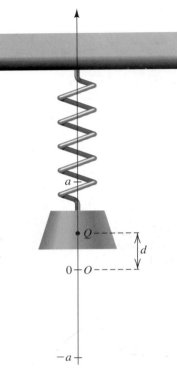

A physical interpretation of simple harmonic motion can be obtained by considering a spring with an attached weight that is oscillating vertically relative to a coordinate line, as illustrated in Figure 85. The number $d$ represents the coordinate of a fixed point $Q$ in the weight, and we assume that the amplitude $a$ of the motion is constant. In this case no frictional force is retarding the motion. If friction is present, then the amplitude decreases with time, and the motion is said to be *damped.*

### EXAMPLE   *Describing harmonic motion*

Suppose that the oscillation of the weight shown in Figure 85 is given by

$$d = 10 \cos\left(\frac{\pi}{6}t\right),$$

with $t$ measured in seconds and $d$ in centimeters. Discuss the motion of the weight.

**SOLUTION**   By definition, the motion is simple harmonic with amplitude $a = 10$ cm. Since $\omega = \pi/6$, we obtain the following:

$$\text{period} = \frac{2\pi}{\omega} = \frac{2\pi}{\pi/6} = 12$$

Thus, in 12 seconds the weight makes one complete oscillation. We also see that

$$\text{frequency} = \frac{\omega}{2\pi} = \frac{\pi/6}{2\pi} = \frac{1}{12},$$

which means that one-twelfth of an oscillation takes place each second. The following table indicates the position of $Q$ at various times.

| $t$ | 0 | 1 | 2 | 3 | 4 | 5 | 6 |
|---|---|---|---|---|---|---|---|
| $\frac{\pi}{6}t$ | 0 | $\frac{\pi}{6}$ | $\frac{\pi}{3}$ | $\frac{\pi}{2}$ | $\frac{2\pi}{3}$ | $\frac{5\pi}{6}$ | $\pi$ |
| $d$ | 10 | $5\sqrt{3} \approx 8.7$ | 5 | 0 | $-5$ | $-5\sqrt{3} \approx -8.7$ | $-10$ |

The initial position of $Q$ is 10 centimeters above the origin $O$. It moves downward, gaining speed until it reaches $O$. Note that $Q$ travels approximately $10 - 8.7 = 1.3$ cm during the first second, $8.7 - 5 = 3.7$ cm during the next second, and $5 - 0 = 5$ cm during the third second. It then slows down until it reaches a point 10 centimeters below $O$ at the end of 6 seconds. The direction of motion is then reversed, and the weight moves upward, gaining speed until it reaches $O$. Once it reaches $O$, it slows down until it returns to its original position at the end of 12 seconds. The direction of motion is then reversed again, and the same pattern is repeated indefinitely.

The preceding example is typical of simple harmonic motion. If the initial angle $AOP$ in Figure 84 is $\phi$ at $t = 0$, then the position $(x, y)$ of $P$ on the circle is given by

$$x = a \cos (\omega t + \phi), \quad y = a \sin (\omega t + \phi).$$

Points that oscillate on a coordinate line according to either of these formulas are also said to be in simple harmonic motion.

Simple harmonic motion takes place in many different types of wave motion, such as water waves, sound waves, radio waves, light waves, and distortional waves that are present in vibrating bodies. As a specific example, consider the waves made by holding one end of a long rope, as in Figure 86, and causing it to vibrate by raising and lowering the hand in simple harmonic motion. Waves appear to move along the rope, traveling away from the hand. A ruler, positioned vertically a fixed distance $c$ from the end, as shown in the figure, may be used to study the upward and downward motion of a specific particle $P$ in the rope. If a suitable horizontal axis is chosen, it can be shown that $P$ is in simple harmonic motion, as if it were attached to a spring. A similar motion takes place in water ripples on a large pond.

FIGURE 86

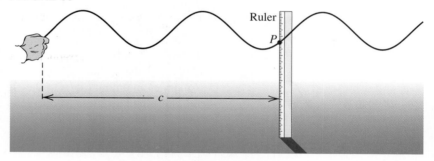

The definition of simple harmonic motion is usually ex‍nded to include situations where $d$ is *any* mathematical or physical quantity (not necessarily a distance). In the following discussion, $d$ is an angle.

Historically, the first type of harmonic motion that was scientifically investigated involved pendulums. Figure 87 illustrates a *simple pendulum* consisting of a bob of mass $m$ attached to one end of a string, with the other end of the string attached to a fixed point $P$. In the ideal case it is assumed that the string is weightless, that there is no air resistance, and that the only force acting on the bob is gravity. If the bob is displaced sideways and released, the pendulum oscillates back and forth in a vertical plane. Let $\alpha$ denote the *angular displacement* at time $t$ (see Figure 87). If the bob moves through a small arc (say, $|\alpha| < 5°$), then it can be shown, by using physical laws, that

$$\alpha = \beta \cos (\omega t + \alpha_0),$$

FIGURE 87

with $\alpha_0$ the initial displacement, $\omega$ the frequency of oscillation of the angle $\alpha$, and $\beta$ the (angular) amplitude of oscillation. This means that the *angle* $\alpha$ is in simple harmonic motion, and hence we refer to the motion as *angular* simple harmonic motion.

As a final illustration, in electrical circuits an alternating electromotive force (emf) and the current may vary harmonically. For example, the emf $V$ (measured in volts) at time $t$ may be given by

$$V = V_M \sin \omega t,$$

where $V_M$ is the maximum voltage. If an emf of this type is impressed on a circuit containing only a resistance of $R$ ohms, then, by Ohm's law, the current $I$ at time $t$ is

$$I = \frac{V}{R} = \frac{V_M \sin \omega t}{R} = I_M \sin \omega t,$$

where $I_M = V_M/R$ is the maximum current. A schematic drawing of an electrical circuit of this type is shown in Figure 88.

The maximum current $I_M$ occurs at the same time as the maximum voltage $V_M$. In other situations these maximum values may occur at different times, in which case we say there is a **phase difference** between $V$ and $I$. For example, if $V = V_M \sin \omega t$, we could have either

(a) $I = I_M \sin(\omega t - \phi)$    or    (b) $I = I_M \sin(\omega t + \phi)$

for some $\phi > 0$. In case (a), the current is said to *lag* the emf by an amount $\phi/\omega$, and the graph of $I$ can be obtained by shifting the graph of $I = I_M \sin \omega t$ an amount $\phi/\omega$ to the *right*. In case (b), we shift the graph to the *left*, and the current is said to *lead* the emf by an amount $\phi/\omega$. The situation is analogous to that described in our discussion of phase shifts in Section 5.5.

Other physical illustrations involving simple harmonic motion may be found in books on physics and engineering.

**FIGURE 88**

## 5.8 EXERCISES

1  A wheel of diameter 40 centimeters is rotating about an axle at a rate of 100 revolutions per minute. If a coordinate system is introduced, as in Figure 84, such that $P$ is a point on the rim of the wheel, find

(a) the angular speed of $OP$

(b) the position $(x, y)$ of $P$ after $t$ minutes

2  A wheel of radius 2 feet is rotating about an axle, and the angular speed of a ray from the center of the wheel to a point $P$ on the rim is $5\pi/6$ radians per second.

(a) How many revolutions does the wheel make in 10 minutes?

(b) If a coordinate system is introduced, as in Figure 84, find the position $(x, y)$ of $P$ after $t$ seconds.

**Exer. 3–6:** The formula specifies the position of a point $P$ that is moving harmonically on a vertical axis, where $t$ is in seconds and $d$ is in centimeters. Determine the amplitude, period, and frequency, and describe the motion of the point during one complete oscillation (starting at $t = 0$).

3 $d = 10 \sin 6\pi t$

4 $d = \dfrac{1}{3} \cos \dfrac{\pi}{4} t$

5 $d = 4 \cos \dfrac{3\pi}{2} t$

6 $d = 6 \sin \dfrac{2\pi}{3} t$

7 A point $P$ in simple harmonic motion has a period of 3 seconds and an amplitude of 5 centimeters. Express the motion of $P$ by means of an equation of the form $d = a \cos \omega t$.

8 A point $P$ in simple harmonic motion has a frequency of $\frac{1}{2}$ oscillation per minute and an amplitude of 4 feet. Express the motion of $P$ by means of an equation of the form $d = a \sin \omega t$.

**Exer. 9–10:** The equations state the electromotive force $V$ and the alternating current $I$ in an electrical circuit at time $t$. Sketch the graphs of both equations on the same coordinate plane, and determine the lag or lead.

9 $V = 220 \sin 360\pi t$; $\quad I = 20 \sin \left(360\pi t - \frac{1}{4}\pi\right)$

10 $V = 110 \sin 120\pi t$; $\quad I = 15 \sin \left(120\pi t + \frac{1}{3}\pi\right)$

11 **Revolution of a Ferris wheel** A large Ferris wheel is 100 feet in diameter and rises 110 feet off the ground, as illustrated in the figure. Each revolution of the wheel takes 30 seconds.

(a) Find the angular speed of the wheel.

(b) Express the height $h$ of a passenger off the ground as a function of time $t$ (in seconds) if $t = 0$ corresponds to a time when a passenger is at the bottom.

**EXERCISE 11**

12 **Tsunamis** A tsunami is a tidal wave caused by an earthquake beneath the sea. These waves can be more than 100 feet in height and can travel at great speeds. Engineers sometimes represent such waves by trigonometric expressions of the form $y = a \cos bt$ and use these representations to estimate the effectiveness of sea walls. Suppose that a wave has height $h = 50$ ft and period 30 minutes and is traveling at the rate of 180 feet per second.

(a) Let $(x, y)$ be a point on the wave represented in the figure. Express $y$ as a function of $t$ if $y = 25$ ft when $t = 0$.

(b) The wave length $L$ is the distance between two successive crests of the wave. Approximate $L$ in feet.

**EXERCISE 12**

13 **Some Hawaiian tsunamis** For an interval of 45 minutes, the tsunamis near Hawaii caused by the Chilean earthquake of 1960 could be modeled by the equation $y = 8 \sin \dfrac{\pi}{6} t$, where $y$ is in feet and $t$ is in minutes.

(a) Find the amplitude and period of the waves.

(b) If the distance from one crest of the wave to the next was 21 kilometers, what was the velocity of the wave? (Tidal waves can have velocities of more than 700 km/hr in deep sea water.)

14 **Crankshaft calculations** A piston is attached to a crankshaft, as shown in the figure on page 396. The connecting rod $AB$ has length 6 inches, and the radius of the crankshaft is 2 inches.

(a) If the crankshaft rotates counterclockwise twice per second, find the coordinates of point $A$ at $t$ seconds after $A$ has coordinates $(2, 0)$.

(b) Find the coordinates of point $B$ at time $t$.

**EXERCISE 14**

**15  Boston Harbor**  The graph in the figure shows the rise and fall of the water level in Boston Harbor during a particular 24-hour period. Approximate the water level $y$ during a 24-hour period by means of an equation of the form $y = a \sin (bt + c) + d$, with $t = 0$ corresponding to midnight.

**EXERCISE 15**

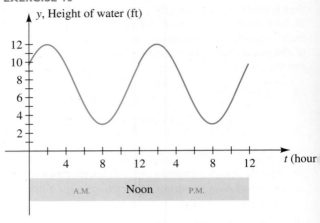

## CHAPTER 5 REVIEW EXERCISES

1  Find the radian measure that corresponds to each degree measure: $330°$, $405°$, $-150°$, $240°$, $36°$.

2  Find the degree measure that corresponds to each radian measure: $\dfrac{9\pi}{2}$, $-\dfrac{2\pi}{3}$, $\dfrac{7\pi}{4}$, $5\pi$, $\dfrac{\pi}{5}$.

3  A central angle $\theta$ is subtended by an arc 20 centimeters long on a circle of radius 2 meters.

   **(a)** Find the radian measure of $\theta$.

   **(b)** Find the area of the sector determined by $\theta$.

4  **(a)** Find the length of arc that subtends an angle of measure $70°$ on a circle of diameter 15 centimeters.

   **(b)** Find the area of the sector in part (a).

**Exer. 5–6: Find the exact values of $x$ and $y$.**

5

6

**Exer. 7–8: Use fundamental identities to write the first expression in terms of the second, for any acute angle $\theta$.**

7  $\tan \theta$,   $\sec \theta$

8  $\cot \theta$,   $\csc \theta$

**Exer. 9–18: Verify the identity by transforming the left side into the right side.**

9  $\sin \theta (\csc \theta - \sin \theta) = \cos^2 \theta$

10  $\cos \theta (\tan \theta + \cot \theta) = \csc \theta$

11  $(\cos^2 \theta - 1)(\tan^2 \theta + 1) = 1 - \sec^2 \theta$

12  $\dfrac{\sec \theta - \cos \theta}{\tan \theta} = \dfrac{\tan \theta}{\sec \theta}$

13  $\dfrac{1 + \tan^2 \theta}{\tan^2 \theta} = \csc^2 \theta$

14  $\dfrac{\sec \theta + \csc \theta}{\sec \theta - \csc \theta} = \dfrac{\sin \theta + \cos \theta}{\sin \theta - \cos \theta}$

15  $\dfrac{\cot \theta - 1}{1 - \tan \theta} = \cot \theta$

16  $\dfrac{1 + \sec \theta}{\tan \theta + \sin \theta} = \csc \theta$

17  $\dfrac{\tan (-\theta) + \cot (-\theta)}{\tan \theta} = -\csc^2 \theta$

18  $-\dfrac{1}{\csc (-\theta)} - \dfrac{\cot (-\theta)}{\sec (-\theta)} = \csc \theta$

19  If $\theta$ is an acute angle of a right triangle and if the adjacent side and hypotenuse have lengths 4 and 7, respectively, find the values of the trigonometric functions of $\theta$.

20  Whenever possible, find the exact values of the trigonometric functions of $\theta$ if $\theta$ is in standard position and satisfies the stated condition.

(a) The point $(30, -40)$ is on the terminal side of $\theta$.

(b) The terminal side of $\theta$ is in quadrant II and is parallel to the line $2x + 3y + 6 = 0$.

(c) The terminal side of $\theta$ is on the negative $y$-axis.

21  Find the quadrant containing $\theta$ if $\theta$ is in standard position.

(a) $\sec \theta < 0$ and $\sin \theta > 0$     II

(b) $\cot \theta > 0$ and $\csc \theta < 0$     III

(c) $\cos \theta > 0$ and $\tan \theta < 0$     IV

22  Find the exact values of the remaining trigonometric functions if

(a) $\sin \theta = -\frac{4}{5}$ and $\cos \theta = \frac{3}{5}$

(b) $\csc \theta = \dfrac{\sqrt{13}}{2}$ and $\cot \theta = -\dfrac{3}{2}$

**Exer. 23–24: $P(t)$ denotes the point on the unit circle $U$ that corresponds to the real number $t$.**

23  Find the rectangular coordinates of $P(7\pi)$, $P(-5\pi/2)$, $P(9\pi/2)$, $P(-3\pi/4)$, $P(18\pi)$, and $P(\pi/6)$.

24  If $P(t)$ has coordinates $\left(-\frac{3}{5}, -\frac{4}{5}\right)$, find the coordinates of $P(t + 3\pi)$, $P(t - \pi)$, $P(-t)$, and $P(2\pi - t)$.

25  (a) Find the reference angle for each radian measure:
$\dfrac{5\pi}{4}$, $-\dfrac{5\pi}{6}$, $-\dfrac{9\pi}{8}$

(b) Find the reference angle for each degree measure: $245°$, $137°$, $892°$

26  Without the use of tables or a calculator, find the exact values of the trigonometric functions corresponding to each real number, whenever possible.

(a) $\dfrac{9\pi}{2}$   (b) $-\dfrac{5\pi}{4}$   (c) $0$   (d) $\dfrac{11\pi}{6}$

27  Find the exact value.

(a) $\cos 225°$   (b) $\tan 150°$   (c) $\sin\left(-\dfrac{\pi}{6}\right)$

(d) $\sec \dfrac{4\pi}{3}$   (e) $\cot \dfrac{7\pi}{4}$   (f) $\csc 300°$

28  ask someone.
If $\sin \theta = -0.7604$ and $\sec \theta$ is positive, approximate $\theta$ to the nearest $0.1°$ for $0° \le \theta < 360°$.

**Exer. 29–36: Find the amplitude and period and sketch the graph of the equation.**

29  $y = 5 \cos x$

30  $y = \frac{2}{3} \sin x$

31  $y = \frac{1}{3} \sin 3x$

32  $y = -\frac{1}{2} \cos \frac{1}{3}x$

33  $y = -3 \cos \frac{1}{2}x$

34  $y = 4 \sin 2x$

35  $y = 2 \sin \pi x$

36  $y = 4 \cos \dfrac{\pi}{2}x - 2$

**Exer. 37–40: The graph of an equation is shown in the figure.**

(a) Find the amplitude and period.

(b) Express the equation in the form $y = a \sin bx$ or in the form $y = a \cos bx$.

37

$(1.5, -1.43)$

38

$\left(\dfrac{3\pi}{4}, -3.27\right)$

39

40

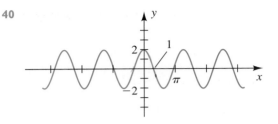

**Exer. 41–52: Sketch the graph of the equation.**

41  $y = 2 \sin \left( x - \dfrac{2\pi}{3} \right)$

42  $y = -3 \sin \left( \dfrac{1}{2}x - \dfrac{\pi}{4} \right)$

43  $y = -4 \cos \left( x + \dfrac{\pi}{6} \right)$    44  $y = 5 \cos \left( 2x + \dfrac{\pi}{2} \right)$

45  $y = 2 \tan \left( \dfrac{1}{2}x - \pi \right)$    46  $y = -3 \tan \left( 2x + \dfrac{\pi}{3} \right)$

47  $y = -4 \cot \left( 2x - \dfrac{\pi}{2} \right)$    48  $y = 2 \cot \left( \dfrac{1}{2}x + \dfrac{\pi}{4} \right)$

49  $y = \sec \left( \dfrac{1}{2}x + \pi \right)$    50  $y = \sec \left( 2x - \dfrac{\pi}{2} \right)$

51  $y = \csc \left( 2x - \dfrac{\pi}{4} \right)$    52  $y = \csc \left( \dfrac{1}{2}x + \dfrac{\pi}{4} \right)$

**Exer. 53–56: Given the indicated parts of triangle ABC with $\gamma = 90°$, approximate the remaining parts.**

53  $\beta = 60°$,   $b = 40$         54  $\alpha = 54°40'$,   $b = 220$

55  $a = 62$,    $b = 25$         56  $a = 9.0$,    $c = 41$

57  **Airplane propeller**  The length of the largest airplane propeller ever used was 22 feet 7.5 inches. The plane was powered by four engines that turned the propeller at 545 revolutions per minute.

   (a) What was the angular speed of the propeller in radians per second?

   (b) Approximately how fast (in mi/hr) did the tip of the propeller travel along the circle it generated?

58  **The Eiffel Tower**  When the top of the Eiffel Tower is viewed at a distance of 200 feet from the base, the angle of elevation is 79.2°. Estimate the height of the tower.

59  **The Great Pyramid**  The Great Pyramid of Egypt is 147 meters high, with a square base of side 230 meters (see the figure). Approximate, to the nearest degree, the

angle $\varphi$ formed when an observer stands at the midpoint of one of the sides and views the apex of the pyramid.

**EXERCISE 59**

60  **Venus**  When viewed from the earth over a period of time, the planet Venus appears to move back and forth along a line segment with the sun at its midpoint (see the figure). At the maximum apparent distance from the sun, angle SEV is approximately 47°. If ES is approximately 92,900,000 miles, estimate the distance of Venus from the sun.

**EXERCISE 60**

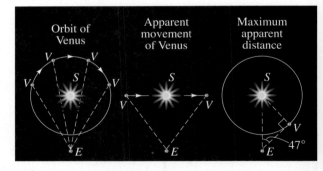

61  **Constructing a conical cup**  A conical paper cup is constructed by removing a sector from a circle of radius 5 inches and attaching edge OA to OB (see the figure). Find angle AOB so that the cup has a depth of 4 inches.

**EXERCISE 61**

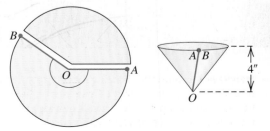

**62** *Length of a tunnel*  A tunnel for a new highway is to be cut through a mountain that is 260 feet high. At a distance of 200 feet from the base of the mountain, the angle of elevation is 36° (see the figure). From a distance of 150 feet on the other side, the angle of elevation is 47°. Approximate the length of the tunnel to the nearest foot.

EXERCISE 62

**63** *Height of a skyscraper*  When a certain skyscraper is viewed from the top of a building 50 feet tall, the angle of elevation is 59° (see the figure). When viewed from the street next to the shorter building, the angle of elevation is 62°.

(a) Approximately how far apart are the two structures?

(b) Approximate the height of the skyscraper to the nearest tenth of a foot.

EXERCISE 63

**64** *Height of a mountain*  When a mountaintop is viewed from the point $P$ shown in the figure, the angle of elevation is $\alpha$. From a point $Q$, which is $d$ miles closer to the mountain, the angle of elevation increases to $\beta$.

(a) Show that the height $h$ of the mountain is given by

$$h = \frac{d}{\cot \alpha - \cot \beta}.$$

(b) Approximate the height of the mountain if $d = 2$ mi, $\alpha = 15°$, and $\beta = 20°$.

EXERCISE 64

**65** *Height of a building*  An observer of height $h$ stands on an incline at a distance $d$ from the base of a building of height $T$, as shown in the figure. The angle of elevation from the observer to the top of the building is $\theta$, and the incline makes an angle of $\alpha$ with the horizontal.

(a) Express $T$ in terms of $h$, $d$, $\alpha$, and $\theta$.

(b) Estimate the height of the building if $h = 6$ ft, $d = 50$ ft, $\alpha = 15°$, and $\theta = 31.4°$.

EXERCISE 65

**66** *Illuminance*  A spotlight with intensity 5000 candles is located 15 feet above a stage. If the spotlight is rotated through an angle $\theta$ as shown in the figure on page 400, the illuminance $E$ (in foot-candles) in the lighted area of the stage is given by

$$E = \frac{5000 \cos \theta}{s^2},$$

where $s$ is the distance (in feet) that the light must travel.

(a) Find the illuminance if the spotlight is rotated through an angle of 30°.

(b) The maximum illuminance occurs when $\theta = 0°$. For what value of $\theta$ is the illuminance one-half the maximum value?

EXERCISE 66

67 **Height of a mountain** If a mountaintop is viewed from a point $P$ due south of the mountain, the angle of elevation is $\alpha$ (see the figure). If viewed from a point $Q$ that is $d$ miles east of $P$, the angle of elevation is $\beta$.

(a) Show that the height $h$ of the mountain is given by

$$h = \frac{d \sin \alpha \sin \beta}{\sqrt{\sin^2 \alpha - \sin^2 \beta}}.$$

(b) If $\alpha = 30°$, $\beta = 20°$, and $d = 10$ mi, approximate $h$ to the nearest hundredth of a mile.

EXERCISE 67

68 **Mounting a projection unit** The manufacturer of a computerized projection system recommends that a projection unit be mounted on the ceiling as shown in the figure. The distance from the end of the mounting bracket to the center of the screen is 85.5 inches, and the angle of depression is 30°.

(a) If the thickness of the screen is disregarded, how far from the wall should the bracket be mounted?

(b) If the bracket is 18 inches long and the screen is 6 feet high, determine the distance from the ceiling to the top edge of the screen.

EXERCISE 68

69 **Pyramid relationships** A pyramid has a square base and congruent triangular faces. Let $\theta$ be the angle that the altitude $a$ of a triangular face makes with the altitude $y$ of the pyramid, and let $x$ be the length of a side (see the figure).

(a) Express the total surface area $S$ of the four faces in terms of $a$ and $\theta$.

(b) The volume $V$ of the pyramid equals one-third the area of the base times the altitude. Express $V$ in terms of $a$ and $\theta$.

EXERCISE 69

70 **Surveying a bluff** A surveyor, using a transit, sights the edge $B$ of a bluff, as shown in the first figure (not drawn to scale). Because of the curvature of the earth, the true elevation $h$ of the bluff is larger than that measured by the surveyor. A cross-sectional schematic view of the earth is shown in the second figure.

(a) If $s$ is the length of arc $PQ$ and $R$ is the distance from $P$ to the center $C$ of the earth, express $h$ in terms of $R$ and $s$.

(b) If $R = 4000$ mi and $s = 50$ mi, estimate the elevation of the bluff in feet.

**EXERCISE 70**

71 **Earthquake response**   To simulate the response of a structure to an earthquake, an engineer must choose a shape for the initial displacement of the beams in the building. When the beam has length $L$ feet and the maximum displacement is $a$ feet, the equation

$$y = a - a \cos \frac{\pi}{2L} x$$

has been used by engineers to estimate the displacement $y$ (see the figure). If $a = 1$ and $L = 10$, sketch the graph of the equation for $0 \le x \le 10$.

**EXERCISE 71**

72 **Circadian rhythms**   The variation in body temperature is an example of a circadian rhythm, a biological process that repeats itself approximately every 24 hours. Body temperature is highest about 5 P.M. and lowest at 5 A.M. Let $y$ denote the body temperature (in °F), and let $t = 0$ correspond to midnight. If the low and high body temperatures are 98.3° and 98.9°, respectively, find an equation of the form $y = 98.6 + a \sin (bt + c)$ that fits this information.

73 **Temperature variation in Ottawa**   The annual variation in temperature $T$ (in °C) in Ottawa, Canada, may be approximated by

$$T(t) = 15.8 \sin \left[ \frac{\pi}{6}(t - 3) \right] + 5,$$

where $t$ is the time in months and $t = 0$ corresponds to January 1.

(a) Sketch the graph of $T$ for $0 \le t \le 12$.

(b) Find the highest temperature of the year and the date on which it occurs.

74 **Water demand**   A reservoir supplies water to a community. During the summer months, the demand $D(t)$ for water (in ft³/day) is given by

$$D(t) = 2000 \sin \frac{\pi}{90} t + 4000,$$

where $t$ is time in days and $t = 0$ corresponds to the beginning of summer.

(a) Sketch the graph of $D$ for $0 \le t \le 90$.

(b) When is the demand for water the greatest?

75 **Bobbing cork**   A cork bobs up and down in a lake. The distance from the bottom of the lake to the center of the cork at time $t \ge 0$ is given by $s(t) = 12 + \cos \pi t$, where $s(t)$ is in feet and $t$ is in seconds.

(a) Describe the motion of the cork for $0 \le t \le 2$.

(b) During what time intervals is the cork rising?

76 **Scotch yoke**   The figure shows a schematic diagram of a scotch yoke, a device that converts circular motion into linear motion. Point $P$ moves in a counterclockwise direction around a circle of radius 1 and up and down in the slider, causing point $Q$ to move back and forth along the $x$-axis. Suppose that $OP$ has a constant angular speed $\omega$ and that $d(O, Q) = 3$ when $t = 0$.

(a) Express the coordinates of $P$ in terms of $\omega$ and $t$.

(b) Express the $x$-coordinate of $Q$ in terms of $\omega$ and $t$.

(c) Describe the motion of $Q$.

**EXERCISE 76**

*Trigonometric equations are indispensable in the analysis of electrical circuits.*

In advanced mathematics, the natural sciences, and engineering, it is sometimes necessary to simplify complicated trigonometric expressions and to solve equations that involve trigonometric functions. These topics are discussed in the first two sections of this chapter. We then derive many useful formulas with respect to sums, differences, and multiples; for reference they are listed on the inside back cover of the text. In addition to formal manipulations, we also consider numerous applications of these formulas. The last section contains the definitions and properties of the inverse trigonometric functions. ■

# CHAPTER 6

# ANALYTIC

# TRIGONOMETRY

## 6.1    VERIFYING TRIGONOMETRIC IDENTITIES

A **trigonometric expression** contains symbols involving trigonometric functions.

### ILLUSTRATION

**Trigonometric Expressions**

- $x + \sin x$

- $\dfrac{\sqrt{\theta} + 2^{\sin \theta}}{\cot \theta}$

- $\dfrac{\cos (3t + 1)}{t^2 + \tan^2 (2 - t^2)}$

We assume that the domain of each variable in a trigonometric expression is the set of real numbers or angles for which the expression is meaningful. To provide manipulative practice in simplifying complicated trigonometric expressions, we shall use the fundamental identities (see page 317) and algebraic manipulations, as we did in Examples 5 and 6 of Section 5.2. In the first three examples our method consists of transforming the left side of a given identity into the right side, or vice versa. Let us repeat a warning issued in Section 5.2.

**Warning**

If an (assumed) identity contains fractions, do not multiply both sides by the lcd.

**EXAMPLE 1    *Verifying an identity***

Verify the identity $\sec \alpha - \cos \alpha = \sin \alpha \tan \alpha$.

*SOLUTION*    We transform the left side into the right side:

$$\sec \alpha - \cos \alpha = \frac{1}{\cos \alpha} - \cos \alpha \qquad \text{reciprocal identity}$$

$$= \frac{1 - \cos^2 \alpha}{\cos \alpha} \qquad \text{add expressions}$$

$$= \frac{\sin^2 \alpha}{\cos \alpha} \qquad \sin^2 \alpha + \cos^2 \alpha = 1$$

$$= \sin \alpha \left( \frac{\sin \alpha}{\cos \alpha} \right) \qquad \text{equivalent expression}$$

$$= \sin \alpha \tan \alpha \qquad \text{tangent identity}$$

**EXAMPLE 2    Verifying an identity**

Verify the identity $\sec \theta = \sin \theta \, (\tan \theta + \cot \theta)$.

**SOLUTION**    Since the expression on the right side is more complicated than that on the left side, we shall transform the right side into the left side:

$$\sin \theta \, (\tan \theta + \cot \theta) = \sin \theta \left( \frac{\sin \theta}{\cos \theta} + \frac{\cos \theta}{\sin \theta} \right) \qquad \text{tangent and cotangent identities}$$

$$= \sin \theta \left( \frac{\sin^2 \theta + \cos^2 \theta}{\cos \theta \, \sin \theta} \right) \qquad \text{add fractions}$$

$$= \sin \theta \left( \frac{1}{\cos \theta \, \sin \theta} \right) \qquad \text{Pythagorean identity}$$

$$= \frac{1}{\cos \theta} \qquad \text{cancel } \sin \theta$$

$$= \sec \theta \qquad \text{reciprocal identity}$$

**EXAMPLE 3    Verifying an identity**

Verify the identity $\dfrac{\cos x}{1 - \sin x} = \dfrac{1 + \sin x}{\cos x}$.

**SOLUTION**    By the previous warning, multiplication of both sides by the lcd $(1 - \sin x) \cos x$ is not allowed. Instead, we change the form of the fraction on the left side by multiplying the numerator and denominator by the conjugate of the denominator.

$$\frac{\cos x}{1 - \sin x} = \frac{\cos x}{1 - \sin x} \cdot \frac{1 + \sin x}{1 + \sin x} \qquad \begin{array}{l} \text{multiply numerator and} \\ \text{denominator by } 1 + \sin x \end{array}$$

$$= \frac{\cos x \, (1 + \sin x)}{1 - \sin^2 x} \qquad \text{property of quotients}$$

$$= \frac{\cos x \, (1 + \sin x)}{\cos^2 x} \qquad \sin^2 x + \cos^2 x = 1$$

$$= \frac{1 + \sin x}{\cos x} \qquad \text{cancel } \cos x$$

Another technique for showing that an equation $p = q$ is an identity is to begin by transforming the left side $p$ into another expression $s$, making sure that each step is *reversible*—that is, making sure it is possible to transform $s$ back into $p$ by reversing the procedure used in each step. In this case, the equation $p = s$ is an identity. Next, as a *separate* exercise, we show that the right side $q$ can also be transformed into the expression $s$ by means of reversible steps and, therefore, that $q = s$ is an identity. It then follows that $p = q$ is an identity. This method is illustrated in the next example.

**EXAMPLE 4**    *Verifying an identity*

Verify the identity $(\tan \theta - \sec \theta)^2 = \dfrac{1 - \sin \theta}{1 + \sin \theta}$.

**SOLUTION**    We shall verify the identity by showing that each side of the equation can be transformed into the same expression. First we work only with the left side:

$$(\tan \theta - \sec \theta)^2 = \tan^2 \theta - 2 \tan \theta \sec \theta + \sec^2 \theta \qquad \text{square expression}$$

$$= \left(\frac{\sin \theta}{\cos \theta}\right)^2 - 2\left(\frac{\sin \theta}{\cos \theta}\right)\left(\frac{1}{\cos \theta}\right) + \left(\frac{1}{\cos \theta}\right)^2$$

tangent and reciprocal identities

$$= \frac{\sin^2 \theta}{\cos^2 \theta} - \frac{2 \sin \theta}{\cos^2 \theta} + \frac{1}{\cos^2 \theta} \qquad \text{equivalent expression}$$

$$= \frac{\sin^2 \theta - 2 \sin \theta + 1}{\cos^2 \theta} \qquad \text{add fractions}$$

At this point it may not be obvious how we can obtain the right side of the given equation from the last expression. Thus, we next work with only the right side and try to obtain the last expression. Multiplying numerator and denominator by the conjugate of the denominator gives us

$$\frac{1 - \sin \theta}{1 + \sin \theta} = \frac{1 - \sin \theta}{1 + \sin \theta} \cdot \frac{1 - \sin \theta}{1 - \sin \theta} \qquad \begin{array}{l}\text{multiply numerator and} \\ \text{denominator by } 1 - \sin \theta\end{array}$$

$$= \frac{1 - 2 \sin \theta + \sin^2 \theta}{1 - \sin^2 \theta} \qquad \text{property of quotients}$$

$$= \frac{1 - 2 \sin \theta + \sin^2 \theta}{\cos^2 \theta}. \qquad \sin^2 \theta + \cos^2 \theta = 1$$

The last expression is the same as that obtained from $(\tan \theta - \sec \theta)^2$. Since all steps are reversible, the given equation is an identity.

In calculus it is sometimes convenient to change the forms of certain algebraic expressions by making a **trigonometric substitution,** as illustrated in the following example.

**EXAMPLE 5**    *Making a trigonometric substitution*

Express $\sqrt{a^2 - x^2}$ in terms of a trigonometric function of $\theta$, without radicals, by making the substitution $x = a \sin \theta$ for $-\pi/2 \le \theta \le \pi/2$ and $a > 0$.

*SOLUTION*    We proceed as follows:

$$\sqrt{a^2 - x^2} = \sqrt{a^2 - (a \sin \theta)^2} \qquad \text{let } x = a \sin \theta$$
$$= \sqrt{a^2 - a^2 \sin^2 \theta} \qquad \text{law of exponents}$$
$$= \sqrt{a^2(1 - \sin^2 \theta)} \qquad \text{factor out } a^2$$
$$= \sqrt{a^2 \cos^2 \theta} \qquad \sin^2 \theta + \cos^2 \theta = 1$$
$$= a \cos \theta \qquad \text{see below}$$

The last equality is true because (1) if $a > 0$, then $\sqrt{a^2} = a$, and (2) if $-\pi/2 \le \theta \le \pi/2$, then $\cos \theta \ge 0$ and hence $\sqrt{\cos^2 \theta} = \cos \theta$.

## 6.1 EXERCISES

Exer. 1–50: Verify the identity.

1  $\csc \theta - \sin \theta = \cot \theta \cos \theta$

2  $\sin x + \cos x \cot x = \csc x$

3  $\dfrac{\sec^2 u - 1}{\sec^2 u} = \sin^2 u$

4  $\tan t + 2 \cos t \csc t = \sec t \csc t + \cot t$

5  $\dfrac{\csc^2 \theta}{1 + \tan^2 \theta} = \cot^2 \theta$

6  $(\tan u + \cot u)(\cos u + \sin u) = \csc u + \sec u$

7  $\dfrac{1 + \cos t}{\sin t} + \dfrac{\sin t}{1 + \cos t} = 2 \csc t$

8  $\tan^2 \alpha - \sin^2 \alpha = \tan^2 \alpha \sin^2 \alpha$

9  $\dfrac{1}{1 - \cos \gamma} + \dfrac{1}{1 + \cos \gamma} = 2 \csc^2 \gamma$

10  $\dfrac{1 + \csc \beta}{\sec \beta} - \cot \beta = \cos \beta$

11  $(\sec u - \tan u)(\csc u + 1) = \cot u$

12  $\dfrac{\cot \theta - \tan \theta}{\sin \theta + \cos \theta} = \csc \theta - \sec \theta$

13  $\csc^4 t - \cot^4 t = \csc^2 t + \cot^2 t$

14  $\cos^4 \theta + \sin^2 \theta = \cos^2 \theta + \sin^4 \theta$

15  $\dfrac{\cos \beta}{1 - \sin \beta} = \sec \beta + \tan \beta$

16  $\dfrac{1}{\csc y - \cot y} = \csc y + \cot y$

17  $\dfrac{\tan^2 x}{\sec x + 1} = \dfrac{1 - \cos x}{\cos x}$

18  $\dfrac{\cot x}{\csc x + 1} = \dfrac{\csc x - 1}{\cot x}$

19  $\dfrac{\cot u - 1}{\cot u + 1} = \dfrac{1 - \tan u}{1 + \tan u}$

20  $\dfrac{1 + \sec x}{\sin x + \tan x} = \csc x$

21  $\sin^4 r - \cos^4 r = \sin^2 r - \cos^2 r$

22  $\sin^4 \theta + 2 \sin^2 \theta \cos^2 \theta + \cos^4 \theta = 1$

23  $\tan^4 k - \sec^4 k = 1 - 2 \sec^2 k$

24  $\sec^4 u - \sec^2 u = \tan^2 u + \tan^4 u$

25 $(\sec t + \tan t)^2 = \dfrac{1 + \sin t}{1 - \sin t}$

26 $\sec^2 \gamma + \tan^2 \gamma = (1 - \sin^4 \gamma) \sec^4 \gamma$

27 $(\sin^2 \theta + \cos^2 \theta)^3 = 1$

28 $\dfrac{\sin t}{1 - \cos t} = \csc t + \cot t$

29 $\dfrac{1 + \csc \beta}{\cot \beta + \cos \beta} = \sec \beta$

30 $\dfrac{\cos^3 x - \sin^3 x}{\cos x - \sin x} = 1 + \sin x \cos x$

31 $(\csc t - \cot t)^4 (\csc t + \cot t)^4 = 1$

32 $(a \cos t - b \sin t)^2 + (a \sin t + b \cos t)^2 = a^2 + b^2$

33 $\dfrac{\sin \alpha \cos \beta + \cos \alpha \sin \beta}{\cos \alpha \cos \beta - \sin \alpha \sin \beta} = \dfrac{\tan \alpha + \tan \beta}{1 - \tan \alpha \tan \beta}$

34 $\dfrac{\tan u - \tan v}{1 + \tan u \tan v} = \dfrac{\cot v - \cot u}{\cot u \cot v + 1}$

35 $\dfrac{\tan \alpha}{1 + \sec \alpha} + \dfrac{1 + \sec \alpha}{\tan \alpha} = 2 \csc \alpha$

36 $\dfrac{\csc x}{1 + \csc x} - \dfrac{\csc x}{1 - \csc x} = 2 \sec^2 x$

37 $\dfrac{1}{\tan \beta + \cot \beta} = \sin \beta \cos \beta$

38 $\dfrac{\cot y - \tan y}{\sin y \cos y} = \csc^2 y - \sec^2 y$

39 $\sec \theta + \csc \theta - \cos \theta - \sin \theta = \sin \theta \tan \theta + \cos \theta \cot \theta$

40 $\sin^3 t + \cos^3 t = (1 - \sin t \cos t)(\sin t + \cos t)$

41 $(1 - \tan^2 \phi)^2 = \sec^4 \phi - 4 \tan^2 \phi$

42 $\cos^4 w + 1 - \sin^4 w = 2 \cos^2 w$

43 $\dfrac{\cot (-t) + \tan (-t)}{\cot t} = -\sec^2 t$

44 $\dfrac{\csc (-t) - \sin (-t)}{\sin (-t)} = \cot^2 t$

45 $\log 10^{\tan t} = \tan t$

46 $10^{\log |\sin t|} = |\sin t|$

47 $\ln \cot x = -\ln \tan x$

48 $\ln \sec \theta = -\ln \cos \theta$

49 $\ln |\sec \theta + \tan \theta| = -\ln |\sec \theta - \tan \theta|$

50 $\ln |\csc x - \cot x| = -\ln |\csc x + \cot x|$

**Exer. 51–62:** Show that the equation is *not* an identity. (*Hint:* Find one number in the domain of $t$ or $\theta$ for which the equation is false.)

51 $\cos t = \sqrt{1 - \sin^2 t}$

52 $\sqrt{\sin^2 t + \cos^2 t} = \sin t + \cos t$

53 $\sqrt{\sin^2 t} = \sin t$    54 $\sec t = \sqrt{\tan^2 t + 1}$

55 $(\sin \theta + \cos \theta)^2 = \sin^2 \theta + \cos^2 \theta$

56 $\log \left( \dfrac{1}{\sin t} \right) = \dfrac{1}{\log \sin t}$

57 $\cos (-t) = -\cos t$    58 $\sin (t + \pi) = \sin t$

59 $\cos (\sec t) = 1$    60 $\cot (\tan \theta) = 1$

61 $\sin^2 t - 4 \sin t - 5 = 0$

62 $3 \cos^2 \theta + \cos \theta - 2 = 0$

**Exer. 63–66:** Refer to Example 5. Make the trigonometric substitution $x = a \sin \theta$ for $-\pi/2 < \theta < \pi/2$ and $a > 0$. Use fundamental identities to simplify the resulting expression.

63 $(a^2 - x^2)^{3/2}$    64 $\dfrac{\sqrt{a^2 - x^2}}{x}$

65 $\dfrac{x^2}{\sqrt{a^2 - x^2}}$    66 $\dfrac{1}{x \sqrt{a^2 - x^2}}$

**Exer. 67–70:** Make the trigonometric substitution $x = a \tan \theta$ for $-\pi/2 < \theta < \pi/2$ and $a > 0$. Simplify the resulting expression.

67 $\sqrt{a^2 + x^2}$    68 $\dfrac{1}{\sqrt{a^2 + x^2}}$

69 $\dfrac{1}{x^2 + a^2}$    70 $\dfrac{(x^2 + a^2)^{3/2}}{x}$

**Exer. 71–74:** Make the trigonometric substitution $x = a \sec \theta$ for $0 < \theta < \pi/2$ and $a > 0$. Simplify the resulting expression.

71 $\sqrt{x^2 - a^2}$    72 $\dfrac{1}{x^2 \sqrt{x^2 - a^2}}$

73 $x^3 \sqrt{x^2 - a^2}$    74 $\dfrac{\sqrt{x^2 - a^2}}{x^2}$

**Exer. 75–78:** Use the graph of $f$ to find the simplest expression $g(x)$ such that the equation $f(x) = g(x)$ is an identity. Verify this identity.

75 $f(x) = \dfrac{\sin^2 x - \sin^4 x}{(1 - \sec^2 x)\cos^4 x}$

76 $f(x) = \dfrac{\sin x - \sin^3 x}{\cos^4 x + \cos^2 x \sin^2 x}$

77 $f(x) = \sec x\,(\sin x \cos x + \cos^2 x) - \sin x$

78 $f(x) = \dfrac{\sin^3 x + \sin x \cos^2 x}{\csc x} + \dfrac{\cos^3 x + \cos x \sin^2 x}{\sec x}$

## 6.2    TRIGONOMETRIC EQUATIONS

A **trigonometric equation** is an equation that contains trigonometric expressions. Each identity considered in the preceding section is an example of a trigonometric equation with every number (or angle) in the domain of the variable a solution of the equation. If a trigonometric equation is not an identity, we often find solutions by using techniques similar to those used for algebraic equations. The main difference is that we first solve the trigonometric equation for $\sin x$, $\cos \theta$, and so on, and then find values of $x$ or $\theta$ that satisfy the equation. Solutions may be expressed either as real numbers or as angles. Throughout our work we shall use the following rule: *If degree measure is not specified, then solutions of a trigonometric equation should be expressed in radian measure (or as real numbers).* If solutions in degree measure are desired, an appropriate statement will be included in the example or exercise.

**EXAMPLE 1**    *Solving a trigonometric equation involving the sine function*

Find the solutions of the equation $\sin \theta = \frac{1}{2}$ if

**(a)** $\theta$ is in the interval $[0, 2\pi)$

**(b)** $\theta$ is any real number

*SOLUTION*

**(a)** If $\sin \theta = \frac{1}{2}$, then the reference angle for $\theta$ is $\theta_R = \pi/6$. If we regard $\theta$ as an angle in standard position, then, since $\sin \theta > 0$, the terminal side is in either quadrant I or quadrant II, as illustrated in Figure 1. Thus, there are two solutions for $0 \le \theta < 2\pi$:

$$\theta = \frac{\pi}{6} \quad \text{and} \quad \theta = \pi - \frac{\pi}{6} = \frac{5\pi}{6}$$

**(b)** Since the sine function has period $2\pi$, we may obtain all solutions by adding multiples of $2\pi$ to $\pi/6$ and $5\pi/6$. This gives us

$$\theta = \frac{\pi}{6} + 2\pi n \quad \text{and} \quad \theta = \frac{5\pi}{6} + 2\pi n \quad \text{for every integer } n.$$

*(continued)*

FIGURE 1

An alternative (graphical) solution involves determining where the graph of $y = \sin \theta$ intersects the horizontal line $y = \frac{1}{2}$, as illustrated in Figure 2.

**FIGURE 2**

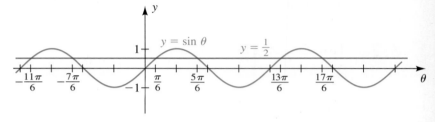

**EXAMPLE 2**     *Solving a trigonometric equation involving the tangent function*

Find the solutions of the equation $\tan u = -1$.

**SOLUTION**     Since the tangent function has period $\pi$, it is sufficient to find one real number $u$ such that $\tan u = -1$ and then add multiples of $\pi$.

A portion of the graph of $y = \tan u$ is sketched in Figure 3. Since $\tan (3\pi/4) = -1$, one solution is $3\pi/4$; hence,

$$\text{if}\quad \tan u = -1, \quad \text{then}\quad u = \frac{3\pi}{4} + \pi n \quad \text{for every integer } n.$$

**FIGURE 3**   $y = \tan u$

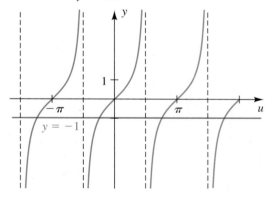

We could also have chosen $-\pi/4$ (or some other number $u$ such that $\tan u = -1$) for the initial solution and written

$$u = -\frac{\pi}{4} + \pi n \quad \text{for every integer } n.$$

**EXAMPLE 3** *Solving a trigonometric equation involving multiple angles*

**(a)** Solve the equation $\cos 2x = 0$, and express the solutions both in radians and in degrees.

**(b)** Find the solutions that are in the intervals $[0, 2\pi)$ and $[0°, 360°)$.

*SOLUTION*

**(a)** We proceed as follows, where $n$ denotes any integer:

$$\cos 2x = 0 \qquad \text{given}$$

$$\cos \theta = 0 \qquad \text{let } \theta = 2x$$

$$\theta = \frac{\pi}{2} + \pi n \qquad \text{refer to graph of } y = \cos x \text{ (page 334)}$$

$$2x = \frac{\pi}{2} + \pi n \qquad \theta = 2x$$

$$x = \frac{\pi}{4} + \frac{\pi}{2}n \qquad \text{divide by 2}$$

In degrees, we have $x = 45° + 90°n$.

**(b)** We may find particular solutions of the equation by substituting integers for $n$ in either of the formulas for $x$ obtained in part (a). Several such solutions are listed in the following table.

| $n$ | $\dfrac{\pi}{4} + \dfrac{\pi}{2}n$ | $45° + 90°n$ |
|---|---|---|
| $-1$ | $\dfrac{\pi}{4} + \dfrac{\pi}{2}(-1) = -\dfrac{\pi}{4}$ | $45° + 90°(-1) = -45°$ |
| $0$ | $\dfrac{\pi}{4} + \dfrac{\pi}{2}(0) = \dfrac{\pi}{4}$ | $45° + 90°(0) = 45°$ |
| $1$ | $\dfrac{\pi}{4} + \dfrac{\pi}{2}(1) = \dfrac{3\pi}{4}$ | $45° + 90°(1) = 135°$ |
| $2$ | $\dfrac{\pi}{4} + \dfrac{\pi}{2}(2) = \dfrac{5\pi}{4}$ | $45° + 90°(2) = 225°$ |
| $3$ | $\dfrac{\pi}{4} + \dfrac{\pi}{2}(3) = \dfrac{7\pi}{4}$ | $45° + 90°(3) = 315°$ |
| $4$ | $\dfrac{\pi}{4} + \dfrac{\pi}{2}(4) = \dfrac{9\pi}{4}$ | $45° + 90°(4) = 405°$ |

*(continued)*

Note that the solutions in the intervals $[0, 2\pi)$ and $[0°, 360°)$ are given by $n = 0$, $n = 1$, $n = 2$, and $n = 3$. These solutions are

$$\frac{\pi}{4}, \frac{3\pi}{4}, \frac{5\pi}{4}, \frac{7\pi}{4}, \quad \text{or} \quad 45°, 135°, 225°, 315°.$$

**EXAMPLE 4**   *Solving a trigonometric equation by factoring*

Solve the equation $\sin \theta \tan \theta = \sin \theta$.

*SOLUTION*

$$\sin \theta \tan \theta = \sin \theta \qquad \text{given}$$
$$\sin \theta \tan \theta - \sin \theta = 0 \qquad \text{subtract } \sin \theta$$
$$\sin \theta (\tan \theta - 1) = 0 \qquad \text{factor out } \sin \theta$$
$$\sin \theta = 0, \tan \theta = 1 \qquad \text{set each factor equal to 0}$$

The solutions of the equation $\sin \theta = 0$ are $0, \pm\pi, \pm 2\pi, \ldots$. Thus,

if   $\sin \theta = 0$,   then   $\theta = \pi n$   for every integer $n$.

The tangent function has period $\pi$, and hence we find the solutions of the equation $\tan \theta = 1$ that are in the interval $(-\pi/2, \pi/2)$ and then add multiples of $\pi$. Since the only solution of $\tan \theta = 1$ in $(-\pi/2, \pi/2)$ is $\pi/4$, we see that

if   $\tan \theta = 0$,   then   $\theta = \dfrac{\pi}{4} + \pi n$   for every integer $n$.

Thus, the solutions of $\sin \theta \tan \theta = \sin \theta$ are

$$\pi n \quad \text{and} \quad \frac{\pi}{4} + \pi n \quad \text{for every integer } n.$$

Some *particular* solutions, obtained by letting $n = 0$, $n = 1$, $n = 2$, and $n = -1$, are

$$0, \quad \frac{\pi}{4}, \quad \pi, \quad \frac{5\pi}{4}, \quad 2\pi, \quad \frac{9\pi}{4}, \quad -\pi, \quad -\frac{3\pi}{4}.$$

Other values of $n$ give other solutions.

In Example 4 it would have been incorrect to begin by dividing both sides by $\sin \theta$, since we would have lost the solutions of $\sin \theta = 0$.

**EXAMPLE 5    *Solving a trigonometric equation by factoring***

Solve the equation $2 \sin^2 t - \cos t - 1 = 0$, and express the solutions both in radians and in degrees.

**SOLUTION**    We shall first express the equation in terms of $\cos t$ alone and then solve by factoring.

$$2 \sin^2 t - \cos t - 1 = 0 \qquad \text{given}$$
$$2(1 - \cos^2 t) - \cos t - 1 = 0 \qquad \sin^2 t + \cos^2 t = 1$$
$$-2 \cos^2 t - \cos t + 1 = 0 \qquad \text{simplify}$$
$$2 \cos^2 t + \cos t - 1 = 0 \qquad \text{multiply by } -1$$
$$(2 \cos t - 1)(\cos t + 1) = 0 \qquad \text{factor}$$
$$2 \cos t - 1 = 0, \quad \cos t + 1 = 0 \qquad \text{set each factor equal to 0}$$
$$\cos t = \tfrac{1}{2}, \qquad \cos t = -1 \qquad \text{solve for } \cos t$$

Since the cosine function has period $2\pi$, we may find all solutions of these equations by adding multiples of $2\pi$ to the solutions that are in the interval $[0, 2\pi)$.

If $\cos t = \tfrac{1}{2}$, the reference angle is $\pi/3$ (or $60°$). Since $\cos t$ is positive, the angle of radian measure $t$ is in either quadrant I or quadrant IV. Hence, in the interval $[0, 2\pi)$, we see that

$$\text{if} \quad \cos t = \frac{1}{2}, \quad \text{then} \quad t = \frac{\pi}{3} \quad \text{or} \quad t = 2\pi - \frac{\pi}{3} = \frac{5\pi}{3}.$$

Referring to the graph of the cosine function, we see that

$$\text{if} \quad \cos t = -1, \quad \text{then} \quad t = \pi.$$

Thus, the solutions of the given equation are the following, where $n$ is any integer:

$$\frac{\pi}{3} + 2\pi n, \quad \frac{5\pi}{3} + 2\pi n, \quad \pi + 2\pi n$$

In degree measure, we have

$$60° + 360°n, \quad 300° + 360°n, \quad 180° + 360°n.$$

---

**EXAMPLE 6    *Solving a trigonometric equation by factoring***

Find the solutions of $4 \sin^2 x \tan x - \tan x = 0$ that are in the interval $[0, 2\pi)$.

*SOLUTION*

$$4 \sin^2 x \tan x - \tan x = 0 \qquad \text{given}$$

$$\tan x \, (4 \sin^2 x - 1) = 0 \qquad \text{factor out } \tan x$$

$$\tan x = 0, \quad 4 \sin^2 x - 1 = 0 \qquad \text{set each factor equal to zero}$$

$$\tan x = 0, \qquad \sin^2 x = \tfrac{1}{4} \qquad \text{solve for } \tan x, \, \sin^2 x$$

$$\tan x = 0, \qquad \sin x = \pm \tfrac{1}{2} \qquad \text{solve for } \sin x$$

**FIGURE 4**

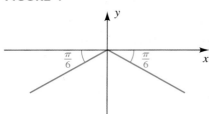

The reference angle $\pi/6$ for the third and fourth quadrants is shown in Figure 4. These angles, $7\pi/6$ and $11\pi/6$, are the solutions of the equation $\sin x = -\tfrac{1}{2}$ for $0 \le x < 2\pi$. The solutions of all three equations are listed in the following table.

| Equation | Solutions in $[0, 2\pi)$ | Refer to |
|---|---|---|
| $\tan x = 0$ | $0, \pi$ | Figure 3 |
| $\sin x = \dfrac{1}{2}$ | $\dfrac{\pi}{6}, \dfrac{5\pi}{6}$ | Example 1 |
| $\sin x = -\dfrac{1}{2}$ | $\dfrac{7\pi}{6}, \dfrac{11\pi}{6}$ | Figure 4 (use reference angle) |

Thus, the given equation has the six solutions listed in the second column of the table.

---

**EXAMPLE 7    *Solving a trigonometric equation involving multiple angles***

Find the solutions of $\csc^4 2u - 4 = 0$.

*SOLUTION*

$$\csc^4 2u - 4 = 0 \qquad \text{given}$$

$$(\csc^2 2u - 2)(\csc^2 2u + 2) = 0 \qquad \text{factor}$$

$$\csc^2 2u - 2 = 0, \qquad \csc^2 2u + 2 = 0 \qquad \text{set each factor equal to 0}$$

$$\csc^2 2u = 2, \qquad \csc^2 2u = -2 \qquad \text{solve for } \csc^2 2u$$

$$\csc 2u = \pm\sqrt{2}, \qquad \csc 2u = \pm\sqrt{-2} \qquad \text{take square roots}$$

The second equation has no solution, since $\sqrt{-2}$ is not a real number. The first equation is equivalent to

$$\sin 2u = \pm\frac{1}{\sqrt{2}} = \pm\frac{\sqrt{2}}{2}.$$

Since the reference angle for $2u$ is $\pi/4$, we obtain the following table, in which $n$ denotes any integer.

| Equation | Solution for $2u$ | Solution for $u$ |
|---|---|---|
| $\sin 2u = \dfrac{\sqrt{2}}{2}$ | $2u = \dfrac{\pi}{4} + 2\pi n$ | $u = \dfrac{\pi}{8} + \pi n$ |
| | $2u = \dfrac{3\pi}{4} + 2\pi n$ | $u = \dfrac{3\pi}{8} + \pi n$ |
| $\sin 2u = -\dfrac{\sqrt{2}}{2}$ | $2u = \dfrac{5\pi}{4} + 2\pi n$ | $u = \dfrac{5\pi}{8} + \pi n$ |
| | $2u = \dfrac{7\pi}{4} + 2\pi n$ | $u = \dfrac{7\pi}{8} + \pi n$ |

The solutions of the given equation are listed in the last column. Note that *all* of these solutions can be written in the one form

$$u = \frac{\pi}{8} + \frac{\pi}{4}n.$$

The next example illustrates the use of a calculator in solving a trigonometric equation.

**EXAMPLE 8** *Approximating the solutions of a trigonometric equation*

Approximate, to the nearest degree, the solutions of the following equation in the interval $[0, 360°)$:

$$5 \sin \theta \tan \theta - 10 \tan \theta + 3 \sin \theta - 6 = 0$$

**SOLUTION**

$5 \sin \theta \tan \theta - 10 \tan \theta + 3 \sin \theta - 6 = 0$     given

$(5 \sin \theta \tan \theta - 10 \tan \theta) + (3 \sin \theta - 6) = 0$     group terms

$5 \tan \theta (\sin \theta - 2) + 3(\sin \theta - 2) = 0$     factor each group

$(5 \tan \theta + 3)(\sin \theta - 2) = 0$     factor out $(\sin \theta - 2)$

$5 \tan \theta + 3 = 0, \quad \sin \theta - 2 = 0$     set each factor equal to 0

$\tan \theta = -\tfrac{3}{5}, \quad \sin \theta = 2$     solve for $\tan \theta$ and $\sin \theta$

*(continued)*

The equation $\sin \theta = 2$ has no solution, since $\sin \theta \leq 1$ for every $\theta$. For $\tan \theta = -\frac{3}{5}$, we use a calculator in degree mode, obtaining

$$\theta = \tan^{-1}\left(-\tfrac{3}{5}\right) \approx -31°.$$

Hence, the reference angle is $\theta_R \approx 31°$. Since $\theta$ is in either quadrant II or quadrant IV, we obtain the following solutions:

$$\theta = 180° - \theta_R \approx 180° - 31° = 149°$$
$$\theta = 360° - \theta_R \approx 360° - 31° = 329°$$

---

### EXAMPLE 9    *Investigating the number of hours of daylight*

In Boston, the number of hours of daylight $D(t)$ at a particular time of the year may be approximated by

$$D(t) = 3 \sin \frac{2\pi}{365}(t - 79) + 12,$$

with $t$ in days and $t = 0$ corresponding to January 1. How many days of the year have more than 10.5 hours of daylight?

**FIGURE 5**

**SOLUTION**    The graph of $D$ was discussed in Example 12 of Section 5.5 and is resketched in Figure 5. As illustrated in the figure, if we can find two numbers $a$ and $b$ with $D(a) = 10.5$, $D(b) = 10.5$, and $0 < a < b < 365$, then there will be more than 10.5 hours of daylight in the $t$th day of the year if $a < t < b$.

Let us solve the equation $D(t) = 10.5$ as follows:

$$3 \sin \frac{2\pi}{365}(t - 79) + 12 = 10.5 \qquad \text{let } D(t) = 10.5$$

$$3 \sin \frac{2\pi}{365}(t - 79) = -1.5 \qquad \text{subtract 12}$$

$$\sin \frac{2\pi}{365}(t - 79) = -0.5 = -\frac{1}{2} \qquad \text{divide by 3}$$

If $\sin \theta = -\frac{1}{2}$, then the reference angle is $\pi/6$ and the angle $\theta$ is in either quadrant III or quadrant IV. Thus, we can find the numbers $a$ and $b$ by solving the equations

$$\frac{2\pi}{365}(t - 79) = \frac{7\pi}{6} \quad \text{and} \quad \frac{2\pi}{365}(t - 79) = \frac{11\pi}{6}.$$

From the first of these equations we obtain

$$t - 79 = \frac{7\pi}{6} \cdot \frac{365}{2\pi} = \frac{2555}{12} \approx 213,$$

and hence        $t \approx 213 + 79, \quad \text{or} \quad t \approx 292.$

Similarly, the second equation gives us $t \approx 414$. Since the period of the function $D$ is 365 days (see Figure 5), we obtain

$$t \approx 414 - 365, \quad \text{or} \quad t \approx 49.$$

Thus, there will be at least 10.5 hours of daylight from $t = 49$ to $t = 292$—that is, for 243 days of the year.

A graphical solution of the next example was given in Example 14 of Section 5.5.

EXAMPLE 10     *Finding the minimum current in an electrical circuit*

The current $I$ (in amperes) in an alternating current circuit at time $t$ (in seconds) is given by

$$I = 30 \sin \left( 50\pi t - \frac{7\pi}{3} \right).$$

Find the smallest exact value of $t$ for which $I = 15$.

SOLUTION     Letting $I = 15$ in the given formula, we obtain

$$15 = 30 \sin \left( 50\pi t - \frac{7\pi}{3} \right),$$

or

$$\sin \left( 50\pi t - \frac{7\pi}{3} \right) = \frac{1}{2}.$$

Thus, the reference angle is $\pi/6$, and consequently

$$50\pi t - \frac{7\pi}{3} = \frac{\pi}{6} + 2\pi n \quad \text{or} \quad 50\pi t - \frac{7\pi}{3} = \frac{5\pi}{6} + 2\pi n,$$

where $n$ is any integer. Solving for $t$ gives us

$$t = \frac{(15/6) + 2n}{50} \quad \text{or} \quad t = \frac{(19/6) + 2n}{50}.$$

The smallest positive value of $t$ will occur when one of the numerators of these two fractions has its least positive value. Since $\frac{15}{6} = 2.5$, $\frac{19}{6} \approx 3.17$, and $2(-1) = -2$, we see that the smallest positive value of $t$ occurs when $n = -1$ in the first fraction—that is, when

$$t = \frac{(15/6) + 2(-1)}{50} = \frac{1}{100}.$$

The next example illustrates how a graphing utility can aid in solving a complicated trigonometric equation.

**EXAMPLE 11** *Using a graph to determine solutions of a trigonometric equation*

Find the solutions of the following equation that are in the interval $[0, 2\pi)$:

$$\sin x + \sin 2x + \sin 3x = 0$$

**FIGURE 6** $[0, 6.3]$ by $[-3, 3]$

**SOLUTION**    We assign $\sin x + \sin 2x + \sin 3x$ to $Y_1$. Since $2\pi \approx 6.3$ and $|\sin \theta| \leq 1$ for $\theta = x$, $2x$, and $3x$, we choose the viewing rectangle $[0, 6.3]$ by $[-3, 3]$ and obtain a sketch similar to Figure 6. Using the zoom and tracing features, we obtain the following approximations for the $x$-intercepts—that is, the approximate solutions of the given equation in $[0, 2\pi)$:

$$0, \quad 1.57, \quad 2.09, \quad 3.14, \quad 4.19, \quad 4.71$$

Changing to degree measure and rounding off to the nearest degree, we obtain

$$0°, \quad 90°, \quad 120°, \quad 180°, \quad 240°, \quad 270°.$$

Converting these degree measures to radian measures gives us

$$0, \quad \frac{\pi}{2}, \quad \frac{2\pi}{3}, \quad \pi, \quad \frac{4\pi}{3}, \quad \frac{3\pi}{2}.$$

Checking these values in the given equation, we see that all six are solutions. Figure 6 suggests that the graph has period $2\pi$. After studying Section 6.4, you will be able to change the form of $Y_1$ and *prove* that the period is $2\pi$ and, therefore, that *all* solutions of the given equation can be obtained by adding integer multiples of $2\pi$.

In the preceding example we were able to use a graphing utility to help us find the *exact* solutions of the equation. For many equations that occur in applications, however, it is only possible to approximate the solutions.

## 6.2 EXERCISES

**Exer. 1–36: Find all solutions of the equation.**

1  $\sin x = -\dfrac{\sqrt{2}}{2}$

2  $\cos t = -1$

3  $\tan \theta = \sqrt{3}$

4  $\cot \alpha = -\dfrac{1}{\sqrt{3}}$

5  $\sec \beta = 2$

6  $\csc \gamma = \sqrt{2}$

7  $\sin x = \dfrac{\pi}{2}$

8  $\cos x = -\dfrac{\pi}{3}$

9  $\cos \theta = \dfrac{1}{\sec \theta}$

10  $\csc \theta \sin \theta = 1$

11 $2 \cos 2\theta - \sqrt{3} = 0$

12 $2 \sin 3\theta + \sqrt{2} = 0$

13 $\sqrt{3} \tan \dfrac{1}{3} t = 1$

14 $\cos \dfrac{1}{4} x = -\dfrac{\sqrt{2}}{2}$

15 $\sin \left( \theta + \dfrac{\pi}{4} \right) = \dfrac{1}{2}$

16 $\cos \left( x - \dfrac{\pi}{3} \right) = -1$

17 $\sin \left( 2x - \dfrac{\pi}{3} \right) = \dfrac{1}{2}$

18 $\cos \left( 4x - \dfrac{\pi}{4} \right) = \dfrac{\sqrt{2}}{2}$

19 $2 \cos t + 1 = 0$

20 $\cot \theta + 1 = 0$

21 $\tan^2 x = 1$

22 $4 \cos \theta - 2 = 0$

23 $(\cos \theta - 1)(\sin \theta + 1) = 0$

24 $2 \cos x = \sqrt{3}$

25 $\sec^2 \alpha - 4 = 0$

26 $3 - \tan^2 \beta = 0$

27 $\sqrt{3} + 2 \sin \beta = 0$

28 $4 \sin^2 x - 3 = 0$

29 $\cot^2 x - 3 = 0$

30 $(\sin t - 1) \cos t = 0$

31 $(2 \sin \theta + 1)(2 \cos \theta + 3) = 0$

32 $(2 \sin u - 1)(\cos u - \sqrt{2}) = 0$

33 $\sin 2x (\csc 2x - 2) = 0$

34 $\tan \alpha + \tan^2 \alpha = 0$

35 $\cos (\ln x) = 0$

36 $\ln (\sin x) = 0$

**Exer. 37–60: Find the solutions of the equation that are in the interval $[0, 2\pi)$.**

37 $\cos \left( 2x - \dfrac{\pi}{4} \right) = 0$

38 $\sin \left( 3x - \dfrac{\pi}{4} \right) = 1$

39 $2 - 8 \cos^2 t = 0$

40 $\cot^2 \theta - \cot \theta = 0$

41 $2 \sin^2 u = 1 - \sin u$

42 $2 \cos^2 t + 3 \cos t + 1 = 0$

43 $\tan^2 x \sin x = \sin x$

44 $\sec \beta \csc \beta = 2 \csc \beta$

45 $2 \cos^2 \gamma + \cos \gamma = 0$

46 $\sin x - \cos x = 0$

47 $\sin^2 \theta + \sin \theta - 6 = 0$

48 $2 \sin^2 u + \sin u - 6 = 0$

49 $1 - \sin t = \sqrt{3} \cos t$

50 $\cos \theta - \sin \theta = 1$

51 $\cos \alpha + \sin \alpha = 1$

52 $\sqrt{3} \sin t + \cos t = 1$

53 $2 \tan t - \sec^2 t = 0$

54 $\tan \theta + \sec \theta = 1$

55 $\cot \alpha + \tan \alpha = \csc \alpha \sec \alpha$

56 $\sin x + \cos x \cot x = \csc x$

57 $2 \sin^3 x + \sin^2 x - 2 \sin x - 1 = 0$

58 $\sec^5 \theta = 4 \sec \theta$

59 $2 \tan t \csc t + 2 \csc t + \tan t + 1 = 0$

60 $2 \sin v \csc v - \csc v = 4 \sin v - 2$

**Exer. 61–66: Approximate, to the nearest 10′, the solutions of the equation in the interval $[0°, 360°)$.**

61 $\sin^2 t - 4 \sin t + 1 = 0$

62 $\cos^2 t - 4 \cos t + 2 = 0$

63 $\tan^2 \theta + 3 \tan \theta + 2 = 0$

64 $2 \tan^2 x - 3 \tan x - 1 = 0$

65 $12 \sin^2 u - 5 \sin u - 2 = 0$

66 $5 \cos^2 \alpha + 3 \cos \alpha - 2 = 0$

67 **Tidal waves** A tidal wave of height 50 feet and period 30 minutes is approaching a sea wall that is 12.5 feet above sea level (see the figure). From a particular point on shore, the distance $y$ from sea level to the top of the wave is given by

$$y = 25 \cos \dfrac{\pi}{15} t,$$

with $t$ in minutes. For approximately how many minutes of each 30-minute period is the top of the wave above the level of the top of the sea wall?

EXERCISE 67

68 **Temperature in Fairbanks** The expected low temperature $T$ (in °F) in Fairbanks, Alaska, may be approximated by

$$T = 36 \sin \left[ \dfrac{2\pi}{365} (t - 101) \right] + 14,$$

where $t$ is in days, with $t = 0$ corresponding to January 1. For how many days during the year is the low temperature expected to be below $-4$ °F?

**69** *Intensity of sunlight* On a clear day with $D$ hours of daylight, the intensity of sunlight $I$ (in calories/cm²) may be approximated by

$$I = I_M \sin^3 \frac{\pi t}{D} \quad \text{for} \quad 0 \le t \le D,$$

where $t = 0$ corresponds to sunrise and $I_M$ is the maximum intensity. If $D = 12$, approximately how many hours after sunrise is $I = \frac{1}{2} I_M$?

**70** *Intensity of sunlight* Refer to Exercise 69. On cloudy days, a better approximation of the sun intensity $I$ is given by

$$I = I_M \sin^2 \frac{\pi t}{D}.$$

If $D = 12$, approximately how many hours after sunrise is $I = \frac{1}{2} I_M$?

**71** *Protection from sunlight* Refer to Exercises 69 and 70. A dermatologist recommends protection from the sun when the intensity $I$ exceeds 75% of the maximum intensity. If $D = 12$ hours, approximate the number of hours for which protection is required on

(a) a clear day      (b) a cloudy day

**72** *Highway engineering* In the study of frost penetration problems in highway engineering, the temperature $T$ at time $t$ hours and depth $x$ feet is given by

$$T = T_0 e^{-\lambda x} \sin (\omega t - \lambda x),$$

where $T_0$, $\omega$, and $\lambda$ are constants and the period of $T$ is 24 hours.

(a) Find a formula for the temperature at the surface.

(b) At what times is the surface temperature a minimum?

(c) If $\lambda = 2.5$, find the times when the temperature is a minimum at a depth of 1 foot.

**73** *Rabbit population* Many animal populations, such as that of rabbits, fluctuate over ten-year cycles. Suppose that the number of rabbits at time $t$ (in years) is given by

$$N(t) = 1000 \cos \frac{\pi}{5} t + 4000.$$

(a) Sketch the graph of $N$ for $0 \le t \le 10$.

(b) For what values of $t$ in part (a) does the rabbit population exceed 4500?

**74** *River flow rate* The flow rate (or water discharge rate) at the mouth of the Orinoco River in South America may be approximated by

$$F(t) = 26,000 \sin \left[ \frac{\pi}{6} (t - 5.5) \right] + 34,000,$$

where $t$ is the time in months and $F(t)$ is the flow rate in m³/sec. For approximately how many months each year does the flow rate exceed 55,000 m³/sec?

**75** Shown in the figure is a graph of $y = \frac{1}{2} x + \sin x$ for $-2\pi \le x \le 2\pi$. Using calculus, it can be shown that the $x$-coordinates of the turning points $A$, $B$, $C$, and $D$ on the graph are solutions of the equation $\frac{1}{2} + \cos x = 0$. Determine the coordinates of these points.

EXERCISE 75

**76** Shown in the figure is the graph of the equation

$$y = e^{-x/2} \sin 2x.$$

The $x$-coordinates of the turning points on the graph are solutions of $4 \cos 2x - \sin 2x = 0$. Approximate the $x$-coordinates of these points for $x > 0$.

EXERCISE 76

Exer. 77–78: If $I(t)$ is the current (in amperes) in an alternating current circuit at time $t$ (in seconds), find the smallest exact value of $t$ for which $I(t) = k$.

**77** $I(t) = 20 \sin (60\pi t - 6\pi); \quad k = -10$

**78** $I(t) = 40 \sin (100\pi t - 4\pi); \quad k = 20$

Exer. 79–82: Approximate the solution to each inequality on the interval $[0, 2\pi]$.

**79** $\cos x \ge 0.3$      **80** $\sin x < -0.6$

**81** $\cos 3x < \sin x$      **82** $\tan x \le \sin 2x$

**Exer. 83–88: Estimate the solutions of the equation in the interval $[-\pi, \pi]$.**

**83** $\sin 2x = 2 - x^2$

**84** $\cos^3 x + \cos 3x - \sin^3 x = 0$

**85** $\ln (1 + \sin^2 x) = \cos x$

**86** $e^{\sin x} = \sec \left(\frac{1}{3}x - \frac{1}{2}\right)$

**87** $3 \cos^4 x - 2 \cos^3 x + \cos x - 1 = 0$

**88** $\cos 2x + \sin 3x - \tan \frac{1}{3}x = 0$

**89** *Weight at various latitudes* The weight $W$ of a person on the earth's surface is directly proportional to the force of gravity $g$ (in m/sec$^2$). Because of rotation, the earth is flattened at the poles, and as a result, weight will vary at different latitudes. If $\theta$ is the latitude, then $g$ can be approximated by $g = 9.8066(1 - 0.00264 \cos 2\theta)$.

**(a)** At what latitude is $g = 9.8$?

**(b)** If a person weighs 150 pounds at the equator ($\theta = 0°$), at what latitude will the person weigh 150.5 pounds?

# 6.3  THE ADDITION AND SUBTRACTION FORMULAS

In this section we derive formulas that involve trigonometric functions of $u + v$ or $u - v$ for any real numbers or angles $u$ and $v$. These formulas are known as *addition* and *subtraction formulas,* respectively, or as *sum* and *difference identities.* The first formula that we will consider may be stated as follows.

**Subtraction Formula for Cosine**

$$\cos (u - v) = \cos u \cos v + \sin u \sin v$$

FIGURE 7

FIGURE 8

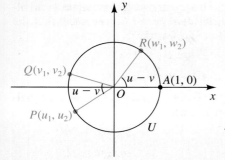

**PROOF**    Let $u$ and $v$ be any real numbers, and consider angles of radian measure $u$ and $v$. Let $w = u - v$. Figure 7 illustrates one possibility with the angles in standard position. For convenience we have assumed that both $u$ and $v$ are positive and that $0 \le u - v < v$.

As in Figure 8, let $P(u_1, u_2)$, $Q(v_1, v_2)$, and $R(w_1, w_2)$ be the points on the terminal sides of the indicated angles that are each a distance 1 from the origin. In this case $P$, $Q$, and $R$ are on the unit circle $U$ with center at the origin. From the definition of trigonometric functions of any angle (with $r = 1$),

(*) 
$$\cos u = u_1 \quad \cos v = v_1 \quad \cos (u - v) = w_1$$
$$\sin u = u_2 \quad \sin v = v_2 \quad \sin (u - v) = w_2.$$

We next observe that the distance between $A(1, 0)$ and $R$ must equal the distance between $Q$ and $P$, because angles $AOR$ and $QOP$ have the same measure, $u - v$. Using the distance formula yields

$$d(A, R) = d(Q, P)$$
$$\sqrt{(w_1 - 1)^2 + (w_2 - 0)^2} = \sqrt{(u_1 - v_1)^2 + (u_2 - v_2)^2}.$$

Squaring both sides and simplifying the expressions under the radicals gives us

$$w_1^2 - 2w_1 + 1 + w_2^2 = u_1^2 - 2u_1v_1 + v_1^2 + u_2^2 - 2u_2v_2 + v_2^2.$$

Since the points $(u_1, u_2)$, $(v_1, v_2)$, and $(w_1, w_2)$ are on the unit circle $U$ and since an equation for $U$ is $x^2 + y^2 = 1$, we may substitute 1 for each of $u_1^2 + u_2^2$, $v_1^2 + v_2^2$, and $w_1^2 + w_2^2$. Doing this and simplifying, we obtain

$$2 - 2w_1 = 2 - 2u_1v_1 - 2u_2v_2,$$

which reduces to

$$w_1 = u_1v_1 + u_2v_2.$$

Substituting from the formulas stated in $(*)$ gives us

$$\cos (u - v) = \cos u \cos v + \sin u \sin v,$$

which is what we wished to prove. It is possible to extend our discussion to all values of $u$ and $v$. ■

The next example demonstrates the use of the subtraction formula in finding the exact value of $\cos 15°$. Of course, if only an approximation were desired, we could use a calculator.

**EXAMPLE I**    *Using a subtraction formula*

Find the exact value of $\cos 15°$ by using the fact that $15° = 60° - 45°$.

**SOLUTION**    We use the subtraction formula with $u = 60°$ and $v = 45°$:

$$\cos 15° = \cos (60° - 45°)$$
$$= \cos 60° \cos 45° + \sin 60° \sin 45°$$
$$= \frac{1}{2} \frac{\sqrt{2}}{2} + \frac{\sqrt{3}}{2} \frac{\sqrt{2}}{2}$$
$$= \frac{\sqrt{2} + \sqrt{6}}{4}$$

It is relatively easy to obtain a formula for $\cos (u + v)$. We begin by writing $u + v$ as $u - (-v)$ and then employ the subtraction formula for cosine:

$$\cos (u + v) = \cos [u - (-v)]$$
$$= \cos u \cos (-v) + \sin u \sin (-v)$$

Using the formulas for negatives, $\cos (-v) = \cos v$ and $\sin (-v) = -\sin v$, gives us the following addition formula.

**Addition Formula for Cosine**

$$\cos (u + v) = \cos u \cos v - \sin u \sin v$$

**EXAMPLE 2**    *Using an addition formula*

Find the exact value of $\cos \dfrac{7\pi}{12}$ by using the fact that $\dfrac{7\pi}{12} = \dfrac{\pi}{3} + \dfrac{\pi}{4}$.

**SOLUTION**   We apply the formula for $\cos(u + v)$:

$$\cos \frac{7\pi}{12} = \cos \left( \frac{\pi}{3} + \frac{\pi}{4} \right)$$

$$= \cos \frac{\pi}{3} \cos \frac{\pi}{4} - \sin \frac{\pi}{3} \sin \frac{\pi}{4}$$

$$= \frac{1}{2} \frac{\sqrt{2}}{2} - \frac{\sqrt{3}}{2} \frac{\sqrt{2}}{2}$$

$$= \frac{\sqrt{2} - \sqrt{6}}{4}$$

**FIGURE 9**

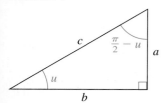

We refer to the sine and cosine functions as **cofunctions** of each other. Similarly, the tangent and cotangent functions are cofunctions, as are the secant and cosecant. If $u$ is the radian measure of an acute angle, then the angle with radian measure $\pi/2 - u$ is complementary to $u$, and we may consider the right triangle shown in Figure 9. Using ratios, we see that

$$\sin u = \frac{a}{c} = \cos \left( \frac{\pi}{2} - u \right)$$

$$\cos u = \frac{b}{c} = \sin \left( \frac{\pi}{2} - u \right)$$

$$\tan u = \frac{a}{b} = \cot \left( \frac{\pi}{2} - u \right).$$

These three formulas and their analogues for $\sec u$, $\csc u$, and $\cot u$ state that *the function value of $u$ equals the cofunction of the complementary angle $\pi/2 - u$.*

In the following formulas we use subtraction formulas to extend these relationships to any real number $u$, provided the function values are defined.

**Cofunction Formulas**

> If $u$ is a real number or the radian measure of an angle, then
>
> $$\cos \left( \frac{\pi}{2} - u \right) = \sin u \qquad \sin \left( \frac{\pi}{2} - u \right) = \cos u$$
>
> $$\tan \left( \frac{\pi}{2} - u \right) = \cot u \qquad \cot \left( \frac{\pi}{2} - u \right) = \tan u$$
>
> $$\sec \left( \frac{\pi}{2} - u \right) = \csc u \qquad \csc \left( \frac{\pi}{2} - u \right) = \sec u.$$

*PROOF*    Using the subtraction formula for cosine, we have

$$\cos\left(\frac{\pi}{2} - u\right) = \cos\frac{\pi}{2}\cos u + \sin\frac{\pi}{2}\sin u$$

$$= (0)\cos u + (1)\sin u = \sin u.$$

This gives us the first formula.

If we substitute $\pi/2 - v$ for $u$ in the first formula, we obtain

$$\cos\left[\frac{\pi}{2} - \left(\frac{\pi}{2} - v\right)\right] = \sin\left(\frac{\pi}{2} - v\right),$$

or
$$\cos v = \sin\left(\frac{\pi}{2} - v\right).$$

Since the symbol $v$ is arbitrary, this equation is equivalent to the next cofunction formula:

$$\sin\left(\frac{\pi}{2} - u\right) = \cos u$$

Using the tangent identity, the first two cofunction formulas, and the cotangent identity, we obtain

$$\tan\left(\frac{\pi}{2} - u\right) = \frac{\sin\left(\frac{\pi}{2} - u\right)}{\cos\left(\frac{\pi}{2} - u\right)} = \frac{\cos u}{\sin u} = \cot u.$$

The proofs of the remaining three formulas are similar. ∎

An easy way to remember the cofunction formulas is to refer to the triangle in Figure 9.

We may now prove the following identities.

**Addition and Subtraction Formulas for Sine and Tangent**

(1)  $\sin(u + v) = \sin u \cos v + \cos u \sin v$

(2)  $\sin(u - v) = \sin u \cos v - \cos u \sin v$

(3)  $\tan(u + v) = \dfrac{\tan u + \tan v}{1 - \tan u \tan v}$

(4)  $\tan(u - v) = \dfrac{\tan u - \tan v}{1 + \tan u \tan v}$

**PROOF**   We shall prove formulas (1) and (3). Using the cofunction formulas and the subtraction formula for cosine, we can verify formula (1):

$$\sin (u + v) = \cos \left[ \frac{\pi}{2} - (u + v) \right]$$

$$= \cos \left[ \left( \frac{\pi}{2} - u \right) - v \right]$$

$$= \cos \left( \frac{\pi}{2} - u \right) \cos v + \sin \left( \frac{\pi}{2} - u \right) \sin v$$

$$= \sin u \cos v + \cos u \sin v$$

To verify formula (3), we begin as follows:

$$\tan (u + v) = \frac{\sin (u + v)}{\cos (u + v)}$$

$$= \frac{\sin u \cos v + \cos u \sin v}{\cos u \cos v - \sin u \sin v}$$

If $\cos u \cos v \neq 0$, then we may divide the numerator and the denominator by $\cos u \cos v$, obtaining

$$\tan (u + v) = \frac{\left( \dfrac{\sin u}{\cos u} \right) \left( \dfrac{\cos v}{\cos v} \right) + \left( \dfrac{\cos u}{\cos u} \right) \left( \dfrac{\sin v}{\cos v} \right)}{\left( \dfrac{\cos u}{\cos u} \right) \left( \dfrac{\cos v}{\cos v} \right) - \left( \dfrac{\sin u}{\cos u} \right) \left( \dfrac{\sin v}{\cos v} \right)}$$

$$= \frac{\tan u + \tan v}{1 - \tan u \tan v}.$$

If $\cos u \cos v = 0$, then either $\cos u = 0$ or $\cos v = 0$. In this case, either $\tan u$ or $\tan v$ is undefined and the formula is invalid. ∎

**EXAMPLE 3**   *Using addition formulas to find the quadrant containing an angle*

Suppose $\sin \alpha = \frac{4}{5}$ and $\cos \beta = -\frac{12}{13}$, where $\alpha$ is in quadrant I and $\beta$ is in quadrant II.

**(a)** Find the exact values of $\sin (\alpha + \beta)$ and $\tan (\alpha + \beta)$.

**(b)** Find the quadrant containing $\alpha + \beta$.

**SOLUTION**   Angles $\alpha$ and $\beta$ are illustrated in Figure 10. There is no loss of generality in regarding $\alpha$ and $\beta$ as positive angles between 0 and $2\pi$, as we have done in the figure. Since $\sin \alpha = \frac{4}{5}$, we may choose the point $(3, 4)$ on the terminal side of $\alpha$. Similarly, since $\cos \beta = -\frac{12}{13}$, the
*(continued)*

**FIGURE 10**

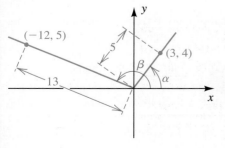

point $(-12, 5)$ is on the terminal side of $\beta$. Referring to Figure 10 and using the definition of trigonometric functions of any angle, we have

$$\cos \alpha = \tfrac{3}{5}, \quad \tan \alpha = \tfrac{4}{3}, \quad \sin \beta = \tfrac{5}{13}, \quad \tan \beta = -\tfrac{5}{12}.$$

**(a)** Addition formulas give us

$$\sin (\alpha + \beta) = \sin \alpha \cos \beta + \cos \alpha \sin \beta = (\tfrac{4}{5})(-\tfrac{12}{13}) + (\tfrac{3}{5})(\tfrac{5}{13}) = -\tfrac{33}{65}$$

$$\tan (\alpha + \beta) = \frac{\tan \alpha + \tan \beta}{1 - \tan \alpha \tan \beta} = \frac{\tfrac{4}{3} + (-\tfrac{5}{12})}{1 - (\tfrac{4}{3})(-\tfrac{5}{12})} = \frac{33}{56}.$$

**(b)** Since $\sin (\alpha + \beta)$ is negative and $\tan (\alpha + \beta)$ is positive, the angle $\alpha + \beta$ must be in quadrant III.

The type of simplification illustrated in the next example occurs in calculus.

**EXAMPLE 4    *A formula used in calculus***

If $f(x) = \sin x$ and $h \neq 0$, show that

$$\frac{f(x + h) - f(x)}{h} = \sin x \left( \frac{\cos h - 1}{h} \right) + \cos x \left( \frac{\sin h}{h} \right).$$

**SOLUTION**    We use the definition of $f$ and the addition formula for sine:

$$\frac{f(x + h) - f(x)}{h} = \frac{\sin (x + h) - \sin x}{h}$$

$$= \frac{\sin x \cos h + \cos x \sin h - \sin x}{h}$$

$$= \frac{\sin x (\cos h - 1) + \cos x \sin h}{h}$$

$$= \sin x \left( \frac{\cos h - 1}{h} \right) + \cos x \left( \frac{\sin h}{h} \right)$$

Addition formulas may also be employed to derive **reduction formulas.** Reduction formulas may be used to change expressions such as

$$\sin \left( \theta + \frac{\pi}{2} n \right) \quad \text{and} \quad \cos \left( \theta + \frac{\pi}{2} n \right) \quad \text{for any integer } n$$

to expressions involving only $\sin \theta$ or $\cos \theta$. Similar formulas are true for the other trigonometric functions. Instead of deriving general reduction formulas, we shall illustrate two special cases in the next example.

**EXAMPLE 5    Obtaining reduction formulas**

Express in terms of a trigonometric function of $\theta$ alone:

**(a)** $\sin\left(\theta - \dfrac{3\pi}{2}\right)$    **(b)** $\cos(\theta + \pi)$

**SOLUTION**    Using subtraction and addition formulas, we obtain the following:

**(a)**      $\sin\left(\theta - \dfrac{3\pi}{2}\right) = \sin\theta\cos\dfrac{3\pi}{2} - \cos\theta\sin\dfrac{3\pi}{2}$

$$= \sin\theta \cdot (0) - \cos\theta \cdot (-1) = \cos\theta$$

**(b)**      $\cos(\theta + \pi) = \cos\theta\cos\pi - \sin\theta\sin\pi$

$$= \cos\theta \cdot (-1) - \sin\theta \cdot (0) = -\cos\theta$$

---

skip

**EXAMPLE 6    Combining a sum involving the sine and cosine functions**

Let $a$ and $b$ be real numbers with $a > 0$. Show that for every $x$,

$$a\cos Bx + b\sin Bx = A\cos(Bx - C),$$

where $A = \sqrt{a^2 + b^2}$ and $\tan C = \dfrac{b}{a}$ with $-\dfrac{\pi}{2} < C < \dfrac{\pi}{2}$.

**SOLUTION**    Given $a\cos Bx + b\sin Bx$, let us consider $\tan C = b/a$ with $-\pi/2 < C < \pi/2$. Thus, $b = a\tan C$, and we may write

$$a\cos Bx + b\sin Bx = a\cos Bx + (a\tan C)\sin Bx$$

$$= a\cos Bx + a\,\frac{\sin C}{\cos C}\sin Bx$$

$$= \frac{a}{\cos C}(\cos C\cos Bx + \sin C\sin Bx)$$

$$= (a\sec C)\cos(Bx - C).$$

We shall complete the proof by showing that $a\sec C = \sqrt{a^2 + b^2}$. Since $-\pi/2 < C < \pi/2$, it follows that $\sec C$ is positive, and hence

$$a\sec C = a\sqrt{1 + \tan^2 C}.$$

Using $\tan C = b/a$ and $a > 0$, we obtain

$$a\sec C = a\sqrt{1 + \frac{b^2}{a^2}} = \sqrt{a^2\left(1 + \frac{b^2}{a^2}\right)} = \sqrt{a^2 + b^2}.$$

### EXAMPLE 7    *An application of Example 6*

If $f(x) = \cos x + \sin x$, use the formulas given in Example 6 to express $f(x)$ in the form $A \cos (Bx - C)$, and then sketch the graph of $f$.

**SOLUTION**    Letting $a = 1$, $b = 1$, and $B = 1$ in the formulas from Example 6, we have

$$A = \sqrt{a^2 + b^2} = \sqrt{1 + 1} = \sqrt{2} \quad \text{and} \quad \tan C = \frac{b}{a} = \frac{1}{1} = 1.$$

Since $\tan C = 1$ and $-\pi/2 < C < \pi/2$, we obtain $C = \pi/4$. Substituting for $a$, $b$, $A$, $B$, and $C$ in the formula

$$a \cos Bx + b \sin Bx = A \cos (Bx - C)$$

gives us

$$f(x) = \cos x + \sin x = \sqrt{2} \cos \left(x - \frac{\pi}{4}\right).$$

Comparing the last formula with the equation $y = a \cos (bx + c)$, which we discussed in Section 5.5, we see that the amplitude of the graph is $\sqrt{2}$, the period is $2\pi$, and the phase shift is $\pi/4$. The graph of $f$ is sketched in Figure 11, where we have also shown the graphs of $y = \sin x$ and $y = \cos x$. Our sketch agrees with that obtained in Section 5.6 using a graphing utility. (See Figure 72 in Chapter 5.)

**FIGURE 11**

## 6.3 EXERCISES

**Exer. 1–4: Express as a cofunction of a complementary angle.**

1  (a) $\sin 46°37'$      (b) $\cos 73°12'$

   (c) $\tan \dfrac{\pi}{6}$      (d) $\sec 17.28°$

2  (a) $\tan 24°12'$      (b) $\sin 89°41'$

   (c) $\cos \dfrac{\pi}{3}$      (d) $\cot 61.87°$

3  (a) $\cos \dfrac{7\pi}{20}$      (b) $\sin \dfrac{1}{4}$

   (c) $\tan 1$      (d) $\csc 0.53$

4  (a) $\sin \dfrac{\pi}{12}$      (b) $\cos 0.64$

   (c) $\tan \sqrt{2}$      (d) $\sec 1.2$

**Exer. 5–10: Find the exact values.**

5  (a) $\cos \dfrac{\pi}{4} + \cos \dfrac{\pi}{6}$

   (b) $\cos \dfrac{5\pi}{12} \quad \left(\text{use } \dfrac{5\pi}{12} = \dfrac{\pi}{4} + \dfrac{\pi}{6}\right)$

6  (a) $\sin \dfrac{2\pi}{3} + \sin \dfrac{\pi}{4}$

   (b) $\sin \dfrac{11\pi}{12} \quad \left(\text{use } \dfrac{11\pi}{12} = \dfrac{2\pi}{3} + \dfrac{\pi}{4}\right)$

7  (a) $\tan 60° + \tan 225°$

   (b) $\tan 285°$   (use $285° = 60° + 225°$)

8  (a) $\cos 135° - \cos 60°$

   (b) $\cos 75°$   (use $75° = 135° - 60°$)

9  (a) $\sin \dfrac{3\pi}{4} - \sin \dfrac{\pi}{6}$

   (b) $\sin \dfrac{7\pi}{12} \quad \left(\text{use } \dfrac{7\pi}{12} = \dfrac{3\pi}{4} - \dfrac{\pi}{6}\right)$

10  (a) $\tan \dfrac{3\pi}{4} - \tan \dfrac{\pi}{6}$

   (b) $\tan \dfrac{7\pi}{12} \quad \left(\text{use } \dfrac{7\pi}{12} = \dfrac{3\pi}{4} - \dfrac{\pi}{6}\right)$

**Exer. 11–16: Express as a trigonometric function of one angle.**

11  $\cos 48° \cos 23° + \sin 48° \sin 23°$

12  $\cos 13° \cos 50° - \sin 13° \sin 50°$

13  $\cos 10° \sin 5° - \sin 10° \cos 5°$

14  $\sin 57° \cos 4° + \cos 57° \sin 4°$

15  $\cos 3 \sin (-2) - \cos 2 \sin 3$

16  $\sin (-5) \cos 2 + \cos 5 \sin (-2)$

17  If $\alpha$ and $\beta$ are acute angles such that $\cos \alpha = \frac{4}{5}$ and $\tan \beta = \frac{8}{15}$, find
   (a) $\sin (\alpha + \beta)$   (b) $\cos (\alpha + \beta)$
   (c) the quadrant containing $\alpha + \beta$   I

18  If $\alpha$ and $\beta$ are acute angles such that $\csc \alpha = \frac{13}{12}$ and $\cot \beta = \frac{4}{3}$, find
   (a) $\sin (\alpha + \beta)$   (b) $\tan (\alpha + \beta)$
   (c) the quadrant containing $\alpha + \beta$

19  If $\sin \alpha = -\frac{4}{5}$ and $\sec \beta = \frac{5}{3}$ for a third-quadrant angle $\alpha$ and a first-quadrant angle $\beta$, find
   (a) $\sin (\alpha + \beta)$   (b) $\tan (\alpha + \beta)$
   (c) the quadrant containing $\alpha + \beta$

20  If $\tan \alpha = -\frac{7}{24}$ and $\cot \beta = \frac{3}{4}$ for a second-quadrant angle $\alpha$ and a third-quadrant angle $\beta$, find
   (a) $\sin (\alpha + \beta)$   (b) $\cos (\alpha + \beta)$   (c) $\tan (\alpha + \beta)$
   (d) $\sin (\alpha - \beta)$   (e) $\cos (\alpha - \beta)$   (f) $\tan (\alpha - \beta)$

21  If $\alpha$ and $\beta$ are third-quadrant angles such that $\cos \alpha = -\frac{2}{3}$ and $\cos \beta = -\frac{1}{3}$, find
   (a) $\sin (\alpha - \beta)$   (b) $\cos (\alpha - \beta)$
   (c) the quadrant containing $\alpha - \beta$

22  If $\alpha$ and $\beta$ are second-quadrant angles such that $\sin \alpha = \frac{2}{3}$ and $\cos \beta = -\frac{1}{3}$, find
   (a) $\sin (\alpha + \beta)$   (b) $\tan (\alpha + \beta)$
   (c) the quadrant containing $\alpha + \beta$

**Exer. 23–34: Verify the reduction formula.**

23  $\sin (\theta + \pi) = -\sin \theta$

24  $\sin \left(x + \dfrac{\pi}{2}\right) = \cos x$

25  $\sin \left(x - \dfrac{5\pi}{2}\right) = -\cos x$

26  $\sin \left(\theta - \dfrac{3\pi}{2}\right) = \cos \theta$

27  $\cos (\theta - \pi) = -\cos \theta$

28  $\cos \left(x + \dfrac{\pi}{2}\right) = -\sin x$

29  $\cos \left(x + \dfrac{3\pi}{2}\right) = \sin x$

30  $\cos \left(\theta - \dfrac{5\pi}{2}\right) = \sin \theta$

31  $\tan \left(x - \dfrac{\pi}{2}\right) = -\cot x$

32  $\tan (\pi - \theta) = -\tan \theta$

33  $\tan \left(\theta + \dfrac{\pi}{2}\right) = -\cot \theta$

34  $\tan (x + \pi) = \tan x$

**Exer. 35–44: Verify the identity.**

35  $\sin \left(\theta + \dfrac{\pi}{4}\right) = \dfrac{\sqrt{2}}{2} (\sin \theta + \cos \theta)$

36  $\cos \left(\theta + \dfrac{\pi}{4}\right) = \dfrac{\sqrt{2}}{2} (\cos \theta - \sin \theta)$

37  $\tan \left(u + \dfrac{\pi}{4}\right) = \dfrac{1 + \tan u}{1 - \tan u}$

38  $\tan \left(x - \dfrac{\pi}{4}\right) = \dfrac{\tan x - 1}{\tan x + 1}$

39  $\cos (u + v) + \cos (u - v) = 2 \cos u \cos v$

40  $\sin (u + v) + \sin (u - v) = 2 \sin u \cos v$

41  $\sin (u + v) \cdot \sin (u - v) = \sin^2 u - \sin^2 v$

42  $\cos (u + v) \cdot \cos (u - v) = \cos^2 u - \sin^2 v$

43  $\dfrac{1}{\cot \alpha - \cot \beta} = \dfrac{\sin \alpha \sin \beta}{\sin (\beta - \alpha)}$

44  $\dfrac{1}{\tan \alpha + \tan \beta} = \dfrac{\cos \alpha \cos \beta}{\sin (\alpha + \beta)}$

45  Express $\sin (u + v + w)$ in terms of trigonometric functions of $u$, $v$, and $w$. (*Hint:* Write
   $$\sin (u + v + w) \quad \text{as} \quad \sin [(u + v) + w]$$
   and use addition formulas.)

46  Express $\tan (u + v + w)$ in terms of trigonometric functions of $u$, $v$, and $w$.

47  Derive the formula $\cot (u + v) = \dfrac{\cot u \cot v - 1}{\cot u + \cot v}$.

**48** If $\alpha$ and $\beta$ are complementary angles, show that
$$\sin^2 \alpha + \sin^2 \beta = 1.$$

**49** Derive the subtraction formula for the sine function.

**50** Derive the subtraction formula for the tangent function.

**51** If $f(x) = \cos x$, show that
$$\frac{f(x+h) - f(x)}{h} = \cos x \left(\frac{\cos h - 1}{h}\right) - \sin x \left(\frac{\sin h}{h}\right).$$

**52** If $f(x) = \tan x$, show that
$$\frac{f(x+h) - f(x)}{h} = \sec^2 x \left(\frac{\sin h}{h}\right) \frac{1}{\cos h - \sin h \tan x}.$$

**Exer. 53–58:** Use an addition or subtraction formula to find the solutions of the equation that are in the interval $[0, \pi)$.

**53** $\sin 4t \cos t = \sin t \cos 4t$

**54** $\cos 5t \cos 3t = \frac{1}{2} + \sin(-5t) \sin 3t$

**55** $\cos 5t \cos 2t = -\sin 5t \sin 2t$

**56** $\sin 3t \cos t + \cos 3t \sin t = -\frac{1}{2}$

**57** $\tan 2t + \tan t = 1 - \tan 2t \tan t$

**58** $\tan t - \tan 4t = 1 + \tan 4t \tan t$

**Exer. 59–62:**

**(a)** Use the formula from Example 6 to express $f$ in terms of the cosine function.

**(b)** Determine the amplitude, period, and phase shift of $f$.

**(c)** Sketch the graph of $f$.

**59** $f(x) = \sqrt{3} \cos 2x + \sin 2x$

**60** $f(x) = \cos 4x + \sqrt{3} \sin 4x$

**61** $f(x) = 2 \cos 3x - 2 \sin 3x$

**62** $f(x) = 5 \cos 10x - 5 \sin 10x$

**Exer. 63–64:** For certain applications in electrical engineering, the sum of several voltage signals or radio waves of the same frequency is expressed in the compact form $y = A \cos(Bt - C)$. Express the given signal in this form.

**63** $y = 50 \sin 60\pi t + 40 \cos 60\pi t$

**64** $y = 10 \sin\left(120\pi t - \frac{\pi}{2}\right) + 5 \sin 120\pi t$

**65** **Motion of a mass** If a mass that is attached to a spring is raised $y_0$ feet and released with an initial vertical velocity of $v_0$ ft/sec, then the subsequent position $y$ of the mass is given by
$$y = y_0 \cos \omega t + \frac{v_0}{\omega} \sin \omega t,$$
where $t$ is time in seconds and $\omega$ is a positive constant.

**(a)** If $\omega = 1$, $y_0 = 2$ ft, and $v_0 = 3$ ft/sec, express $y$ in the form $A \cos(Bt - C)$, and find the amplitude and period of the resulting motion.

**(b)** Determine the times when $y = 0$—that is, the times when the mass passes through the equilibrium position.

**66** **Motion of a mass** Refer to Exercise 65. If $y_0 = 1$ and $\omega = 2$, find the initial velocities that result in an amplitude of 4 feet.

**67** **Pressure on the eardrum** If a tuning fork is struck and then held a certain distance from the eardrum, the pressure $p_1(t)$ on the outside of the eardrum at time $t$ may be represented by $p_1(t) = A \sin \omega t$, where $A$ and $\omega$ are positive constants. If a second identical tuning fork is struck with a possibly different force and held a different distance from the eardrum (see the figure), its effect may be represented by $p_2(t) = B \sin(\omega t + \tau)$, where $B$ is a positive constant and $0 \leq \tau \leq 2\pi$. The total pressure $p(t)$ on the eardrum is given by
$$p(t) = A \sin \omega t + B \sin(\omega t + \tau).$$

**(a)** Show that $p(t) = a \cos \omega t + b \sin \omega t$, where
$$a = B \sin \tau \quad \text{and} \quad b = A + B \cos \tau.$$

**(b)** Show that the amplitude $C$ of $p$ is given by
$$C^2 = A^2 + B^2 + 2AB \cos \tau.$$

**EXERCISE 67**

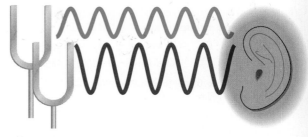

**68** **Destructive interference** Refer to Exercise 67. Destructive interference occurs if the amplitude of the resulting sound wave is less than $A$. Suppose that the two tuning forks are struck with the same force—that is, $A = B$.

(a) When total destructive interference occurs, the amplitude of $p$ is zero and no sound is heard. Find the least positive value of $\tau$ for which this occurs.

(b) Determine the $\tau$-interval $(a, b)$ for which destructive interference occurs and $a$ has its least positive value.

**69 Constructive interference** Refer to Exercise 67. When two tuning forks are struck, constructive interference occurs if the amplitude $C$ of the resulting sound wave is larger than either $A$ or $B$ (see the figure).

(a) Show that $C \leq A + B$.

**EXERCISE 69**

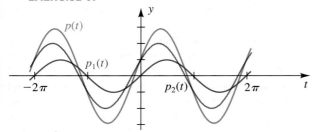

(b) Find the values of $\tau$ such that $C = A + B$.

(c) If $A \geq B$, determine a condition under which constructive interference will occur.

**70 Pressure on the eardrum** Refer to Exercise 67. If two tuning forks with different pitches are struck simultaneously with different forces, then the total pressure $p(t)$ on the eardrum at time $t$ is given by

$$p(t) = p_1(t) + p_2(t) = A \sin \omega_1 t + B \sin (\omega_2 t + \tau),$$

where $A$, $B$, $\omega_1$, $\omega_2$, and $\tau$ are constants.

(a) Graph $p$ for $-2\pi \leq t \leq 2\pi$ if $A = B = 2$, $\omega_1 = 1$, $\omega_2 = 20$, and $\tau = 3$.

(b) Use the graph to describe the variation of the tone that is produced.

**Exer. 71–72:** Refer to Exercise 69. Graph the equation for $-\pi \leq t \leq \pi$, and estimate the intervals on which constructive interference occurs.

**71** $y = 3 \sin 2t + 2 \sin (4t + 1)$

**72** $y = 2 \sin t + 2 \sin (3t + 3)$

## 6.4 MULTIPLE-ANGLE FORMULAS

We refer to the formulas considered in this section as **multiple-angle formulas.** In particular, the following identities are **double-angle formulas,** because they contain the expression $2u$.

**Double-Angle Formulas**

> **(1)** $\sin 2u = 2 \sin u \cos u$
>
> **(2)** $\cos 2u = \cos^2 u - \sin^2 u$
> $$= 1 - 2 \sin^2 u$$
> $$= 2 \cos^2 u - 1$$
>
> **(3)** $\tan 2u = \dfrac{2 \tan u}{1 - \tan^2 u}$

**PROOF** Each of these formulas may be proved by letting $v = u$ in the appropriate addition formulas. If we use the formula for $\sin (u + v)$, then

$$\sin 2u = \sin (u + u)$$
$$= \sin u \cos u + \cos u \sin u$$
$$= 2 \sin u \cos u.$$

Using the formula for cos $(u + v)$, we have

$$\cos 2u = \cos (u + u)$$
$$= \cos u \cos u - \sin u \sin u$$
$$= \cos^2 u - \sin^2 u.$$

To obtain the other two forms for cos $2u$, we use the fundamental identity $\sin^2 u + \cos^2 u = 1$. Thus,

$$\cos 2u = \cos^2 u - \sin^2 u$$
$$= (1 - \sin^2 u) - \sin^2 u$$
$$= 1 - 2 \sin^2 u.$$

Similarly, if we substitute for $\sin^2 u$ instead of $\cos^2 u$, we obtain

$$\cos 2u = \cos^2 u - (1 - \cos^2 u)$$
$$= 2 \cos^2 u - 1.$$

Formula 3 for tan $2u$ may be obtained by letting $v = u$ in the formula for tan $(u + v)$. ■

**EXAMPLE I    *Using double-angle formulas***

If $\sin \alpha = \frac{4}{5}$ and $\alpha$ is an acute angle, find the exact values of $\sin 2\alpha$ and $\cos 2\alpha$.

**FIGURE 12**

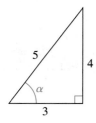

**SOLUTION**    If we regard $\alpha$ as an acute angle of a right triangle, as shown in Figure 12, we obtain $\cos \alpha = \frac{3}{5}$. We next substitute in double-angle formulas:

$$\sin 2\alpha = 2 \sin \alpha \cos \alpha = 2(\tfrac{4}{5})(\tfrac{3}{5}) = \tfrac{24}{25}$$
$$\cos 2\alpha = \cos^2 \alpha - \sin^2 \alpha = \tfrac{9}{25} - \tfrac{16}{25} = -\tfrac{7}{25}$$

**EXAMPLE 2    *Changing the form of* $\cos 3\theta$**

Express $\cos 3\theta$ in terms of $\cos \theta$.

**SOLUTION**

$$\cos 3\theta = \cos (2\theta + \theta) \qquad\qquad 3\theta = 2\theta + \theta$$
$$= \cos 2\theta \cos \theta - \sin 2\theta \sin \theta \qquad \text{addition formula}$$
$$= (2 \cos^2 \theta - 1) \cos \theta - (2 \sin \theta \cos \theta) \sin \theta \qquad \text{double-angle formulas}$$
$$= 2 \cos^3 \theta - \cos \theta - 2 \cos \theta \sin^2 \theta \qquad \text{multiply}$$
$$= 2 \cos^3 \theta - \cos \theta - 2 \cos \theta (1 - \cos^2 \theta) \qquad \sin^2 \theta + \cos^2 \theta = 1$$
$$= 4 \cos^3 \theta - 3 \cos \theta \qquad \text{simplify}$$

We call each of the next three formulas a **half-angle identity,** because the number $u$ is one-half the number $2u$.

**Half-Angle Identities**

$$\sin^2 u = \frac{1 - \cos 2u}{2} \qquad \cos^2 u = \frac{1 + \cos 2u}{2} \qquad \tan^2 u = \frac{1 - \cos 2u}{1 + \cos 2u}$$

*PROOF*    The first identity may be verified as follows:

$$\cos 2u = 1 - 2\sin^2 u \qquad \text{double-angle formula}$$

$$2\sin^2 u = 1 - \cos 2u \qquad \text{isolate } 2\sin^2 u$$

$$\sin^2 u = \frac{1 - \cos 2u}{2} \qquad \text{divide by 2}$$

The second identity can be derived in similar fashion by using

$$\cos 2u = 2\cos^2 u - 1.$$

The third identity is obtained from the first two by using the fact that

$$\tan^2 u = \left(\frac{\sin u}{\cos u}\right)^2 = \frac{\sin^2 u}{\cos^2 u}. \quad \blacksquare$$

Half-angle identities may be used to express even powers of trigonometric functions in terms of functions with exponent 1, as illustrated in the next two examples.

**EXAMPLE 3    *Using half-angle identities to verify an identity***
Verify the identity $\sin^2 x \cos^2 x = \frac{1}{8}(1 - \cos 4x)$.

*SOLUTION*

$$\sin^2 x \cos^2 x = \left(\frac{1 - \cos 2x}{2}\right)\left(\frac{1 + \cos 2x}{2}\right) \qquad \text{half-angle identities}$$

$$= \tfrac{1}{4}(1 - \cos^2 2x) \qquad \text{multiply}$$

$$= \tfrac{1}{4}\sin^2 2x \qquad \sin^2 \theta + \cos^2 \theta = 1$$

$$= \frac{1}{4}\left(\frac{1 - \cos 4x}{2}\right) \qquad \text{half-angle identity with } u = 2x$$

$$= \tfrac{1}{8}(1 - \cos 4x) \qquad \text{multiply}$$

**EXAMPLE 4**    *Using half-angle identities to reduce a power of* cos *t*

Express $\cos^4 t$ in terms of values of the cosine function with exponent 1.

**SOLUTION**

$$
\begin{aligned}
\cos^4 t &= (\cos^2 t)^2 & \text{law of exponents} \\
&= \left(\frac{1 + \cos 2t}{2}\right)^2 & \text{half-angle identity} \\
&= \tfrac{1}{4}(1 + 2\cos 2t + \cos^2 2t) & \text{square} \\
&= \frac{1}{4}\left(1 + 2\cos 2t + \frac{1 + \cos 4t}{2}\right) & \text{half-angle identity with } u = 2t \\
&= \tfrac{3}{8} + \tfrac{1}{2}\cos 2t + \tfrac{1}{8}\cos 4t & \text{simplify}
\end{aligned}
$$

Substituting $v/2$ for $u$ in the three half-angle identities gives us

$$
\sin^2 \frac{v}{2} = \frac{1 - \cos v}{2} \qquad \cos^2 \frac{v}{2} = \frac{1 + \cos v}{2} \qquad \tan^2 \frac{v}{2} = \frac{1 - \cos v}{1 + \cos v}.
$$

Taking the square roots of both sides of each of these equations, we obtain the following, which we call the *half-angle formulas* in order to distinguish them from the half-angle identities.

**Half-Angle Formulas**

$$
\sin \frac{v}{2} = \pm\sqrt{\frac{1 - \cos v}{2}} \qquad \cos \frac{v}{2} = \pm\sqrt{\frac{1 + \cos v}{2}}
$$

$$
\tan \frac{v}{2} = \pm\sqrt{\frac{1 - \cos v}{1 + \cos v}}
$$

*When using a half-angle formula, we choose either the* + *or the* −, *depending on the quadrant containing the angle of radian measure* $v/2$. Thus, for sin $(v/2)$, we use + if $v/2$ is an angle in quadrant I or II and − if $v/2$ is in quadrant III or IV. For cos $(v/2)$, we use + if $v/2$ is in quadrant I or IV, and so on.

**EXAMPLE 5**    *Using half-angle formulas for the sine and cosine*

Find the exact values of sin 22.5° and cos 22.5°.

***SOLUTION***    We use the formula for sin $(v/2)$ and the fact that 22.5° is
in quadrant I:

$$\sin 22.5° = \sin \frac{45°}{2} \qquad 22.5° = \frac{45°}{2}$$

$$= \sqrt{\frac{1 - \cos 45°}{2}} \qquad \text{half-angle formula for sine}$$

$$= \sqrt{\frac{1 - \sqrt{2}/2}{2}} \qquad \cos 45° = \frac{\sqrt{2}}{2}$$

$$= \frac{\sqrt{2 - \sqrt{2}}}{2} \qquad \text{multiply radicand by } \frac{2}{2} \text{ and simplify}$$

Similarly,

$$\cos 22.5° = \sqrt{\frac{1 + \cos 45°}{2}} \qquad \text{half-angle formula for cosine}$$

$$= \sqrt{\frac{1 + \sqrt{2}/2}{2}} \qquad \cos 45° = \frac{\sqrt{2}}{2}$$

$$= \frac{\sqrt{2 + \sqrt{2}}}{2}. \qquad \text{multiply radicand by } \frac{2}{2} \text{ and simplify}$$

---

We can obtain an alternative form for tan $(v/2)$. Multiplying the numerator and denominator of the radicand in the third half-angle formula by
$1 - \cos v$ gives us

$$\tan \frac{v}{2} = \pm \sqrt{\frac{1 - \cos v}{1 + \cos v} \cdot \frac{1 - \cos v}{1 - \cos v}}$$

$$= \pm \sqrt{\frac{(1 - \cos v)^2}{\sin^2 v}}$$

$$= \pm \frac{1 - \cos v}{\sin v}.$$

We can eliminate the $\pm$ sign in the preceding formula. First note that
the numerator $1 - \cos v$ is never negative. We can show that tan $(v/2)$ and
sin $v$ always have the same sign. For example, if $0 < v < \pi$, then $0 <
v/2 < \pi/2$, and consequently both sin $v$ and tan $(v/2)$ are positive. If $\pi <
v < 2\pi$, then $\pi/2 < v/2 < \pi$, and hence both sin $v$ and tan $(v/2)$ are nega-
tive, which gives us the first of the next two identities. The second identity
for tan $(v/2)$ may be obtained by multiplying the numerator and denominator of the radicand in the third half-angle formula by $1 + \cos v$.

**Half-Angle Formulas
for the Tangent**

$$\tan\frac{v}{2} = \frac{1 - \cos v}{\sin v} \qquad \tan\frac{v}{2} = \frac{\sin v}{1 + \cos v}$$

**FIGURE 13**

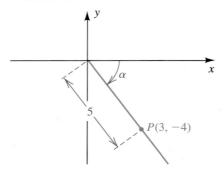

**EXAMPLE 6    Using a half-angle formula for the tangent**

If $\tan\alpha = -\dfrac{4}{3}$ and $\alpha$ is in quadrant IV, find $\tan\dfrac{\alpha}{2}$.

**SOLUTION**    If we choose the point $(3, -4)$ on the terminal side of $\alpha$, as illustrated in Figure 13, then $\sin\alpha = -\frac{4}{5}$ and $\cos\alpha = \frac{3}{5}$. Applying a half-angle formula, we obtain

$$\tan\frac{\alpha}{2} = \frac{1 - \cos\alpha}{\sin\alpha} = \frac{1 - \frac{3}{5}}{-\frac{4}{5}} = -\frac{1}{2}.$$

**FIGURE 14**

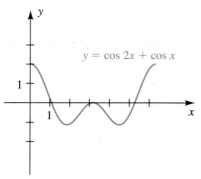

$y = \cos 2x + \cos x$

**EXAMPLE 7    Finding the x-intercepts of a graph**

A graph of the equation $y = \cos 2x + \cos x$ for $0 \le x \le 2\pi$ is sketched in Figure 14. The $x$-intercepts appear to be approximately 1.1, 3.1, and 5.2. Find their exact values and three-decimal-place approximations.

**SOLUTION**    To find the $x$-intercepts, we proceed as follows:

$$
\begin{aligned}
\cos 2x + \cos x &= 0 && \text{let } y = 0 \\
(2\cos^2 x - 1) + \cos x &= 0 && \text{double-angle formula} \\
2\cos^2 x + \cos x - 1 &= 0 && \text{equivalent equation} \\
(2\cos x - 1)(\cos x + 1) &= 0 && \text{factor} \\
2\cos x - 1 = 0, \quad \cos x + 1 &= 0 && \text{set each factor equal to 0} \\
\cos x = \tfrac{1}{2}, \qquad \cos x &= -1 && \text{solve for } \cos x
\end{aligned}
$$

The solutions of the last two equations in the interval $[0, 2\pi]$ give us the following $x$-intercepts:

$$\frac{\pi}{3} \approx 1.047, \quad \frac{5\pi}{3} \approx 5.236, \quad \pi \approx 3.142$$

**FIGURE 15**

**EXAMPLE 8    Deriving a formula for the area
of an isosceles triangle**

An isosceles triangle has two equal sides of length $a$, and the angle between them is $\theta$ (see Figure 15). Express the area $A$ of the triangle in terms of $a$ and $\theta$.

**FIGURE 16**

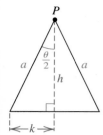

**SOLUTION**   From Figure 16 we see that the altitude from point $P$ bisects $\theta$ and that $A = \frac{1}{2}(2k)h = kh$. Thus, we have the following, where $\theta/2$ is an acute angle:

$$\sin \frac{\theta}{2} = \frac{k}{a} \qquad \cos \frac{\theta}{2} = \frac{h}{a} \qquad \text{Figure 16}$$

$$k = a \sin \frac{\theta}{2} \qquad h = a \cos \frac{\theta}{2} \quad \text{multiply by } a$$

We next find the area:

$$(*) \quad A = a^2 \sin \frac{\theta}{2} \cos \frac{\theta}{2} \qquad \text{substitute in } A = kh$$

$$= a^2 \sqrt{\frac{1 - \cos \theta}{2}} \sqrt{\frac{1 + \cos \theta}{2}} \qquad \begin{array}{l} \text{half-angle formulas with } \theta/2 \\ \text{in quadrant I} \end{array}$$

$$= a^2 \sqrt{\frac{1 - \cos^2 \theta}{4}} \qquad \text{law of radicals}$$

$$= a^2 \sqrt{\frac{\sin^2 \theta}{4}} \qquad \sin^2 \theta + \cos^2 \theta = 1$$

$$= \tfrac{1}{2} a^2 \, |\sin \theta| \qquad \text{take the square root}$$

$$= \tfrac{1}{2} a^2 \sin \theta \qquad \sin \theta > 0 \text{ for } 0° < \theta < 180°$$

Another method for simplifying $(*)$ is to write the double-angle formula for the sine, $\sin 2u = 2 \sin u \cos u$, as

$$(**) \qquad \sin u \cos u = \tfrac{1}{2} \sin 2u$$

and proceed as follows:

$$A = a^2 \sin \frac{\theta}{2} \cos \frac{\theta}{2} \qquad \text{substitute in } A = kh$$

$$= a^2 \cdot \frac{1}{2} \sin \left( 2 \cdot \frac{\theta}{2} \right) \qquad \text{let } u = \frac{\theta}{2} \text{ in } (**)$$

$$= \tfrac{1}{2} a^2 \sin \theta \qquad \text{simplify}$$

## 6.4 EXERCISES

**Exer. 1–4: Find the exact values of $\sin 2\theta$, $\cos 2\theta$, and $\tan 2\theta$ for the given values of $\theta$.**

1  $\cos \theta = \frac{3}{5}$;    $0° < \theta < 90°$

2  $\cot \theta = \frac{4}{3}$;    $180° < \theta < 270°$

3  $\sec \theta = -3$;    $90° < \theta < 180°$

4  $\sin \theta = -\frac{4}{5}$;    $270° < \theta < 360°$

**Exer. 5–8: Find the exact values of $\sin (\theta/2)$, $\cos (\theta/2)$, and $\tan (\theta/2)$ for the given conditions.**

5  $\sec \theta = \frac{5}{4}$;    $0° < \theta < 90°$

6  $\csc \theta = -\frac{5}{3}$;    $-90° < \theta < 0°$

**7** $\tan \theta = 1$;    $-180° < \theta < -90°$

**8** $\sec \theta = -4$;    $180° < \theta < 270°$

**Exer. 9–10: Use half-angle formulas to find the exact values.**

**9 (a)** $\cos 67°30'$    **(b)** $\sin 15°$    **(c)** $\tan \dfrac{3\pi}{8}$

**10 (a)** $\cos 165°$    **(b)** $\sin 157°30'$    **(c)** $\tan \dfrac{\pi}{8}$

**Exer. 11–28: Verify the identity.**

**11** $\sin 10\theta = 2 \sin 5\theta \cos 5\theta$

**12** $\cos^2 3x - \sin^2 3x = \cos 6x$

**13** $4 \sin \dfrac{x}{2} \cos \dfrac{x}{2} = 2 \sin x$

**14** $\dfrac{\sin^2 2\alpha}{\sin^2 \alpha} = 4 - 4 \sin^2 \alpha$

**15** $(\sin t + \cos t)^2 = 1 + \sin 2t$

**16** $\csc 2u = \tfrac{1}{2} \csc u \sec u$

**17** $\sin 3u = \sin u (3 - 4 \sin^2 u)$

**18** $\sin 4t = 4 \sin t \cos t (1 - 2 \sin^2 t)$

**19** $\cos 4\theta = 8 \cos^4 \theta - 8 \cos^2 \theta + 1$

**20** $\cos 6t = 32 \cos^6 t - 48 \cos^4 t + 18 \cos^2 t - 1$

**21** $\sin^4 t = \tfrac{3}{8} - \tfrac{1}{2} \cos 2t + \tfrac{1}{8} \cos 4t$

**22** $\cos^4 x - \sin^4 x = \cos 2x$

**23** $\sec 2\theta = \dfrac{\sec^2 \theta}{2 - \sec^2 \theta}$    **24** $\cot 2u = \dfrac{\cot^2 u - 1}{2 \cot u}$

**25** $2 \sin^2 2t + \cos 4t = 1$

**26** $\tan \theta + \cot \theta = 2 \csc 2\theta$

**27** $\tan 3u = \dfrac{\tan u (3 - \tan^2 u)}{1 - 3 \tan^2 u}$

**28** $\dfrac{1 + \sin 2v + \cos 2v}{1 + \sin 2v - \cos 2v} = \cot v$

**Exer. 29–32: Express in terms of the cosine function with exponent 1.**

**29** $\cos^4 \dfrac{\theta}{2}$    **30** $\cos^4 2x$

**31** $\sin^4 2x$    **32** $\sin^4 \dfrac{\theta}{2}$

**Exer. 33–40: Find the solutions of the equation that are in the interval $[0, 2\pi)$.**

**33** $\sin 2t + \sin t = 0$    **34** $\cos t - \sin 2t = 0$

**35** $\cos u + \cos 2u = 0$    **36** $\cos 2\theta - \tan \theta = 1$

**37** $\tan 2x = \tan x$    **38** $\tan 2t - 2 \cos t = 0$

**39** $\sin \tfrac{1}{2}u + \cos u = 1$    **40** $2 - \cos^2 x = 4 \sin^2 \tfrac{1}{2}x$

**41** If $a > 0$, $b > 0$, and $0 < u < \pi/2$, show that

$$a \sin u + b \cos u = \sqrt{a^2 + b^2} \sin (u + v)$$

for $0 < v < \dfrac{\pi}{2}$,

$$\sin v = \dfrac{b}{\sqrt{a^2 + b^2}}, \quad \text{and} \quad \cos v = \dfrac{a}{\sqrt{a^2 + b^2}}.$$

**42** Use Exercise 41 to express $8 \sin u + 15 \cos u$ in the form $c \sin (u + v)$.

**43** A graph of $y = \cos 2x + 2 \cos x$ for $0 \le x \le 2\pi$ is shown in the figure.

**(a)** Approximate the $x$-intercepts to two decimal places.

**(b)** The $x$-coordinates of the turning points $P$, $Q$, and $R$ on the graph are solutions of the equation $\sin 2x + \sin x = 0$. Find the coordinates of these points.

**EXERCISE 43**

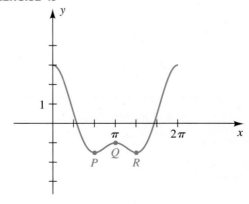

**44** A graph of $y = \cos x - \sin 2x$ for $-2\pi \le x \le 2\pi$ is shown in the figure.

(a) Find the $x$-intercepts.

(b) The $x$-coordinates of the eight turning points on the graph are solutions of $\sin x + 2 \cos 2x = 0$. Approximate these $x$-coordinates to two decimal places.

**EXERCISE 44**

**45** A graph of $y = \cos 3x - 3 \cos x$ for $-2\pi \le x \le 2\pi$ is shown in the figure.

(a) Find the $x$-intercepts. (*Hint:* Use the formula for $\cos 3\theta$ given in Example 2.)

(b) The $x$-coordinates of the 13 turning points on the graph are solutions of $\sin 3x - \sin x = 0$. Find these $x$-coordinates. (*Hint:* Use the formula for $\sin 3u$ in Exercise 17.)

**EXERCISE 45**

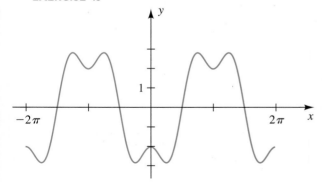

**46** A graph of $y = \sin 4x - 4 \sin x$ for $-2\pi \le x \le 2\pi$ is shown in the figure. Find the $x$-intercepts. (*Hint:* Use the formula for $\sin 4t$ in Exercise 18.)

**EXERCISE 46**

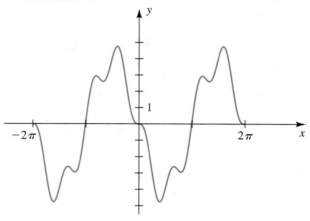

**47** *Planning a railroad route*   Shown in the figure is a proposed railroad route through three towns located at points $A$, $B$, and $C$. The track will branch out from $B$ toward $C$ at an angle $\theta$.

(a) Show that the total distance $d$ from $A$ to $C$ is given by $d = 20 \tan \frac{1}{2}\theta + 40$.

(b) Because of mountains between $A$ and $C$, the branching point $B$ must be at least 20 miles from $A$. Is there a route that avoids the mountains and measures exactly 50 miles?

**EXERCISE 47**

**48** *Projectile's range*   If a projectile is fired from ground level with an initial velocity of $v$ ft/sec and at an angle of $\theta$ degrees with the horizontal, the range $R$ of the projectile is given by

$$R = \frac{v^2}{16} \sin \theta \cos \theta.$$

If $v = 80$ ft/sec, approximate the angles that result in a range of 150 feet.

**49** *Constructing a rain gutter*   Shown in the figure is a design for a rain gutter.

  (a) Express the volume $V$ as a function of $\theta$. (*Hint:* See Example 8.)

  (b) Approximate the acute angle $\theta$ that results in a volume of $2\ \text{ft}^3$.

**EXERCISE 49**

**50** *Designing curbing*   A highway engineer is designing curbing for a street at an intersection where two highways meet at an angle $\phi$, as shown in the figure. The curbing between points $A$ and $B$ is to be constructed using a circle that is tangent to the highway at these two points.

  (a) Show that the relationship between the radius $R$ of the circle and the distance $d$ in the figure is given by $d = R\tan(\phi/2)$.

  (b) If $\phi = 45°$ and $d = 20$ ft, approximate $R$ and the length of the curbing.

**EXERCISE 50**

**51** *Arterial bifurcation*   A common form of cardiovascular branching is bifurcation, in which an artery splits into two smaller blood vessels. The bifurcation angle $\theta$ is the angle formed by the two smaller arteries. In the figure, the line through $A$ and $D$ bisects $\theta$ and is perpendicular to the line through $B$ and $C$.

  (a) Show that the length $l$ of the artery from $A$ to $B$ is given by $l = a + \dfrac{b}{2}\tan\dfrac{\theta}{4}$.

  (b) Estimate $l$ from the measurements $a = 10$ mm, $b = 6$ mm, and $\theta = 156°$.

**EXERCISE 51**

**52** *Heat production in an AC circuit*   By definition, the average value of $f(t) = c + a\cos bt$ for one or more complete cycles is $c$ (see the figure).

  (a) Use a double-angle formula to find the average value of $f(t) = \sin^2 \omega t$ for $0 \le t \le 2\pi/\omega$, with $t$ in seconds.

  (b) In an electrical circuit with an alternating current $I = I_0 \sin \omega t$, the rate $r$ (in calories/sec) at which heat is produced in an $R$-ohm resistor is given by $r = RI^2$. Find the average rate at which heat is produced for one complete cycle.

**EXERCISE 52**

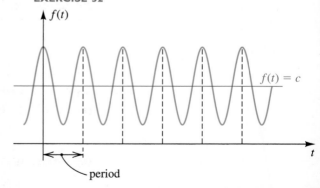

period

**Exer. 53–54:** Use the graph of $f$ to find the simplest expression $g(x)$ such that the equation $f(x) = g(x)$ is an identity. Verify this identity.

**53** $f(x) = \dfrac{\sin 2x + \sin x}{\cos 2x + \cos x + 1}$

**54** $f(x) = \dfrac{\sin x\,(1 + \cos 2x)}{\sin 2x}$

## 6.5    PRODUCT-TO-SUM AND SUM-TO-PRODUCT FORMULAS

The following formulas may be used to change the form of certain trigonometric expressions from products to sums. We refer to these as **product-to-sum formulas** even though two of the formulas express a product as a difference, because any difference $x - y$ of two real numbers is also a sum $x + (-y)$.

**Product-to-Sum Formulas**

> **(1)** $\sin u \cos v = \frac{1}{2}[\sin(u + v) + \sin(u - v)]$
>
> **(2)** $\cos u \sin v = \frac{1}{2}[\sin(u + v) - \sin(u - v)]$
>
> **(3)** $\cos u \cos v = \frac{1}{2}[\cos(u + v) + \cos(u - v)]$
>
> **(4)** $\sin u \sin v = \frac{1}{2}[\cos(u - v) - \cos(u + v)]$

*PROOF*    Let us add the left and right sides of the addition and subtraction formulas for the sine function, as follows:

$$\sin(u + v) = \sin u \cos v + \cos u \sin v$$
$$\underline{\sin(u - v) = \sin u \cos v - \cos u \sin v}$$
$$\sin(u + v) + \sin(u - v) = 2 \sin u \cos v$$

Dividing both sides of the last equation by 2 gives us formula 1.

Formula 2 is obtained by *subtracting* the left and right sides of the addition and subtraction formulas for the sine function. Formulas 3 and 4 are developed in a similar fashion, using the addition and subtraction formulas for the cosine function. ■

**EXAMPLE 1**    *Using product-to-sum formulas*

Express as a sum:

**(a)** $\sin 4\theta \cos 3\theta$      **(b)** $\sin 3x \sin x$

*SOLUTION*

**(a)** We use product-to-sum formula 1 with $u = 4\theta$ and $v = 3\theta$:

$$\sin 4\theta \cos 3\theta = \frac{1}{2}[\sin(4\theta + 3\theta) + \sin(4\theta - 3\theta)]$$
$$= \frac{1}{2}(\sin 7\theta + \sin \theta)$$

We can also obtain this relationship by using product-to-sum formula 2.

**(b)** We use product-to-sum formula 4 with $u = 3x$ and $v = x$:

$$\sin 3x \sin x = \frac{1}{2}[\cos(3x - x) - \cos(3x + x)]$$
$$= \frac{1}{2}(\cos 2x - \cos 4x)$$

We may use product-to-sum formulas to express a sum or difference as a product. To obtain forms that can be applied more easily, we shall change the notation as follows. If we let

$$u + v = a \quad \text{and} \quad u - v = b,$$

then $(u + v) + (u - v) = a + b$, which simplifies to

$$u = \frac{a + b}{2}.$$

Similarly, since $(u + v) - (u - v) = a - b$,

$$v = \frac{a - b}{2}.$$

We now substitute for $u + v$ and $u - v$ on the right-hand sides of the product-to-sum formulas and for $u$ and $v$ on the left-hand sides. If we then multiply by 2, we obtain the following formulas.

**Sum-to-Product Formulas**

> **(1)** $\sin a + \sin b = 2 \sin \dfrac{a + b}{2} \cos \dfrac{a - b}{2}$
>
> **(2)** $\sin a - \sin b = 2 \cos \dfrac{a + b}{2} \sin \dfrac{a - b}{2}$
>
> **(3)** $\cos a + \cos b = 2 \cos \dfrac{a + b}{2} \cos \dfrac{a - b}{2}$
>
> **(4)** $\cos a - \cos b = -2 \sin \dfrac{a + b}{2} \sin \dfrac{a - b}{2}$

**EXAMPLE 2**    *Using a sum-to-product formula*

Express $\sin 5x - \sin 3x$ as a product.

**SOLUTION**    We use sum-to-product formula 2 with $a = 5x$ and $b = 3x$:

$$\sin 5x - \sin 3x = 2 \cos \frac{5x + 3x}{2} \sin \frac{5x - 3x}{2}$$

$$= 2 \cos 4x \sin x$$

**EXAMPLE 3**    *Using sum-to-product formulas to verify an identity*

Verify the identity $\dfrac{\sin 3t + \sin 5t}{\cos 3t - \cos 5t} = \cot t$.

*SOLUTION*    We first use a sum-to-product formula for the numerator and one for the denominator:

$$\frac{\sin 3t + \sin 5t}{\cos 3t - \cos 5t} = \frac{2 \sin \dfrac{3t + 5t}{2} \cos \dfrac{3t - 5t}{2}}{-2 \sin \dfrac{3t + 5t}{2} \sin \dfrac{3t - 5t}{2}} \qquad \text{sum-to-product formulas 1 and 4}$$

$$= \frac{2 \sin 4t \cos (-t)}{-2 \sin 4t \sin (-t)} \qquad \text{simplify}$$

$$= \frac{\cos (-t)}{-\sin (-t)} \qquad \text{cancel } 2 \sin 4t$$

$$= \frac{\cos t}{\sin t} \qquad \text{formulas for negatives}$$

$$= \cot t \qquad \text{cotangent identity}$$

**EXAMPLE 4**    *Using a sum-to-product formula to solve an equation*

Find the solutions of $\sin 5x + \sin x = 0$.

*SOLUTION*

$$\sin 5x + \sin x = 0 \qquad \text{given}$$

$$2 \sin \frac{5x + x}{2} \cos \frac{5x - x}{2} = 0 \qquad \text{sum-to-product formula 1}$$

$$\sin 3x \cos 2x = 0 \qquad \text{simplify and divide by 2}$$

$$\sin 3x = 0, \quad \cos 2x = 0 \qquad \text{set each factor equal to 0}$$

The solutions of the last two equations are

$$3x = \pi n \quad \text{and} \quad 2x = \frac{\pi}{2} + \pi n \quad \text{for every integer } n.$$

Dividing by 3 and 2, respectively, we obtain

$$\frac{\pi}{3} n \quad \text{and} \quad \frac{\pi}{4} + \frac{\pi}{2} n \quad \text{for every integer } n.$$

**EXAMPLE 5**    *Finding the x-intercepts of a graph*

A graph of the equation $y = \cos x - \cos 3x - \sin 2x$ is shown in Figure 17. Find the 13 $x$-intercepts that are in the interval $[-2\pi, 2\pi]$.

**FIGURE 17**

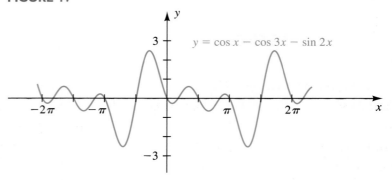

$y = \cos x - \cos 3x - \sin 2x$

*SOLUTION*    To find the $x$-intercepts, we proceed as follows:

$$\cos x - \cos 3x - \sin 2x = 0 \qquad \text{let } y = 0$$

$$(\cos x - \cos 3x) - \sin 2x = 0 \qquad \text{group the first two terms}$$

$$-2 \sin \frac{x + 3x}{2} \sin \frac{x - 3x}{2} - \sin 2x = 0 \qquad \text{sum-to-product formula 4}$$

$$-2 \sin 2x \sin (-x) - \sin 2x = 0 \qquad \text{simplify}$$

$$2 \sin 2x \sin x - \sin 2x = 0 \qquad \text{formula for negatives}$$

$$\sin 2x (2 \sin x - 1) = 0 \qquad \text{factor out } \sin 2x$$

$$\sin 2x = 0, \quad 2 \sin x - 1 = 0 \qquad \text{set each factor equal to 0}$$

$$\sin 2x = 0, \qquad \sin x = \tfrac{1}{2} \qquad \text{solve for } \sin x$$

The equation $\sin 2x = 0$ has solutions $2x = \pi n$, or, dividing by 2,

$$x = \frac{\pi}{2}n \quad \text{for every integer } n.$$

If we let $n = 0, \pm 1, \pm 2, \pm 3,$ and $\pm 4$, we obtain nine $x$-intercepts in $[-2\pi, 2\pi]$:

$$0, \quad \pm\frac{\pi}{2}, \quad \pm\pi, \quad \pm\frac{3\pi}{2}, \quad \pm 2\pi$$

The solutions of the equation $\sin x = \tfrac{1}{2}$ are

$$\frac{\pi}{6} + 2\pi n \quad \text{and} \quad \frac{5\pi}{6} + 2\pi n \quad \text{for every integer } n.$$

The four solutions in $[-2\pi, \ 2\pi]$ are obtained by letting $n = 0$ and $n = -1$:

$$\frac{\pi}{6}, \quad \frac{5\pi}{6}, \quad -\frac{11\pi}{6}, \quad -\frac{7\pi}{6}$$

## 6.5 EXERCISES

**Exer. 1–8: Express as a sum or difference.**

**1** $\sin 7t \sin 3t$

**2** $\sin (-4x) \cos 8x$

**3** $\cos 6u \cos (-4u)$

**4** $\cos 4t \sin 6t$

**5** $2 \sin 9\theta \cos 3\theta$

**6** $2 \sin 7\theta \sin 5\theta$

**7** $3 \cos x \sin 2x$

**8** $5 \cos u \cos 5u$

**Exer. 9–16: Express as a product.**

**9** $\sin 6\theta + \sin 2\theta$

**10** $\sin 4\theta - \sin 8\theta$

**11** $\cos 5x - \cos 3x$

**12** $\cos 5t + \cos 6t$

**13** $\sin 3t - \sin 7t$

**14** $\cos \theta - \cos 5\theta$

**15** $\cos x + \cos 2x$

**16** $\sin 8t + \sin 2t$

**Exer. 17–24: Verify the identity.**

**17** $\dfrac{\sin 4t + \sin 6t}{\cos 4t - \cos 6t} = \cot t$

**18** $\dfrac{\sin \theta + \sin 3\theta}{\cos \theta + \cos 3\theta} = \tan 2\theta$

**19** $\dfrac{\sin u + \sin v}{\cos u + \cos v} = \tan \dfrac{1}{2}(u + v)$

**20** $\dfrac{\sin u - \sin v}{\cos u - \cos v} = -\cot \dfrac{1}{2}(u + v)$

**21** $\dfrac{\sin u - \sin v}{\sin u + \sin v} = \dfrac{\tan \frac{1}{2}(u - v)}{\tan \frac{1}{2}(u + v)}$

**22** $\dfrac{\cos u - \cos v}{\cos u + \cos v} = -\tan \dfrac{1}{2}(u + v) \tan \dfrac{1}{2}(u - v)$

**23** $4 \cos x \cos 2x \sin 3x = \sin 2x + \sin 4x + \sin 6x$

**24** $\dfrac{\cos t + \cos 4t + \cos 7t}{\sin t + \sin 4t + \sin 7t} = \cot 4t$

**Exer. 25–26: Express as a sum.**

**25** $(\sin ax)(\cos bx)$

**26** $(\cos au)(\cos bu)$

**Exer. 27–34: Use sum-to-product formulas to find the solutions of the equation.**

**27** $\sin 5t + \sin 3t = 0$

**28** $\sin t + \sin 3t = \sin 2t$

**29** $\cos x = \cos 3x$

**30** $\cos 4x - \cos 3x = 0$

**31** $\cos 3x + \cos 5x = \cos x$

**32** $\cos 3x = -\cos 6x$

**33** $\sin 2x - \sin 5x = 0$

**34** $\sin 5x - \sin x = 2 \cos 3x$

**Exer. 35–36: Shown in the figure is a graph of the function $f$ for $0 \le x \le 2\pi$. Use a sum-to-product formula to help find the $x$-intercepts.**

**35** $f(x) = \cos x + \cos 3x$

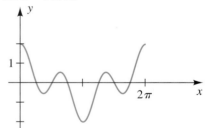

**36** $f(x) = \sin 4x - \sin x$

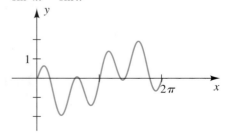

**37** Refer to Exercise 45 of Section 6.4. The graph of the equation $y = \cos 3x - 3 \cos x$ has 13 turning points for $-2\pi \le x \le 2\pi$. The $x$-coordinates of these points are solutions of the equation $\sin 3x - \sin x = 0$. Use a sum-to-product formula to find these $x$-coordinates.

**38** Refer to Exercise 46 of Section 6.4. The $x$-coordinates of the turning points on the graph of the equation $y = \sin 4x - 4 \sin x$ are solutions of $\cos 4x - \cos x = 0$. Use a sum-to-product formula to find these $x$-coordinates for $-2\pi \le x \le 2\pi$.

**39** *Vibration of a violin string* Mathematical analysis of a vibrating violin string of length $l$ involves functions such that

$$f(x) = \sin\left(\frac{\pi n}{l} x\right) \cos\left(\frac{k\pi n}{l} t\right),$$

where $n$ is an integer, $k$ is a constant, and $t$ is time. Express $f$ as a sum of two sine functions.

**40** *Pressure on the eardrum* If two tuning forks are struck simultaneously with the same force and are then held at the same distance from the eardrum, the pressure on the outside of the eardrum at time $t$ is given by

$$p(t) = a \cos \omega_1 t + a \cos \omega_2 t,$$

where $a$, $\omega_1$, and $\omega_2$ are constants. If $\omega_1$ and $\omega_2$ are almost equal, a tone is produced that alternates between loudness and virtual silence. This phenomenon is known as beats.

**(a)** Use a sum-to-product formula to express $p(t)$ as a product.

**(b)** Show that $p(t)$ may be considered as a cosine wave with approximate period $2\pi/\omega_1$ and variable amplitude $f(t) = 2a \cos \frac{1}{2}(\omega_1 - \omega_2)t$. Find the maximum amplitude.

**(c)** Shown in the figure is a graph of the equation

$$p(t) = \cos 4.5t + \cos 3.5t.$$

Near-silence occurs at points $A$ and $B$, where the variable amplitude $f(t)$ in part (b) is zero. Find the coordinates of these points, and determine how frequently near-silence occurs.

**(d)** Use the graph to show that the function $p$ in part (c) has period $4\pi$. Conclude that the maximum amplitude of 2 occurs every $4\pi$ units of time.

**EXERCISE 40**

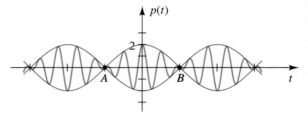

Exer. 41–42: Graph $f$ on the interval $[-\pi, \pi]$.

**(a)** Estimate the $x$-intercepts.

**(b)** Use sum-to-product formulas to find the exact values of the $x$-intercepts.

**41** $f(x) = \sin 4x + \sin 2x$     **42** $f(x) = \cos 3x - \cos 2x$

Exer. 43–44: Use the graph of $f$ to find the simplest expression $g(x)$ such that the equation $f(x) = g(x)$ is an identity. Verify this identity.

**43** $f(x) = \dfrac{\sin x + \sin 2x + \sin 3x}{\cos x + \cos 2x + \cos 3x}$

**44** $f(x) = \dfrac{\cos x - \cos 2x + \cos 3x}{\sin x - \sin 2x + \sin 3x}$

---

## 6.6    THE INVERSE TRIGONOMETRIC FUNCTIONS

Recall from Section 2.5 that to define the inverse function $f^{-1}$ of a function $f$, it is essential that $f$ be one-to-one; that is, if $a \ne b$ in the domain of $f$, then $f(a) \ne f(b)$. The inverse function $f^{-1}$ *reverses* the correspondence given by $f$; that is,

$$u = f(v) \quad \text{if and only if} \quad v = f^{-1}(u).$$

The following general relationships involving $f$ and $f^{-1}$ were discussed in Section 2.5.

**Relationships Between $f^{-1}$ and $f$**

> **(1)** $y = f^{-1}(x)$ if and only if $x = f(y)$, where $x$ is in the domain of $f^{-1}$ and $y$ is in the domain of $f$.
> **(2)** domain of $f^{-1}$ = range of $f$
> **(3)** range of $f^{-1}$ = domain of $f$
> **(4)** $f(f^{-1}(x)) = x$   for every $x$ in the domain of $f^{-1}$
> **(5)** $f^{-1}(f(y)) = y$   for every $y$ in the domain of $f$
> **(6)** The graphs of $f^{-1}$ and $f$ are reflections of each other through the line $y = x$.

We shall use relationship 1 to define each of the inverse trigonometric functions.

The sine function is not one-to-one, since different numbers, such as $\pi/6$, $5\pi/6$, and $-7\pi/6$, yield the same function value $(\frac{1}{2})$. If we restrict the domain to $[-\pi/2, \pi/2]$, then, as illustrated by the blue portion of the graph of $y = \sin x$ in Figure 18, we obtain a one-to-one (increasing) function that takes on every value of the sine function once and only once. We use this *new* function with domain $[-\pi/2, \pi/2]$ and range $[-1, 1]$ to define the *inverse sine function*.

**FIGURE 18**

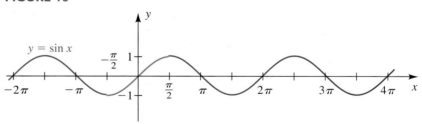

**Definition of the Inverse Sine Function**

The **inverse sine function,** denoted by $\sin^{-1}$, is defined by

$$y = \sin^{-1} x \quad \text{if and only if} \quad x = \sin y$$

for $-1 \le x \le 1$ and $-\dfrac{\pi}{2} \le y \le \dfrac{\pi}{2}$.

The domain of the inverse sine function is $[-1, 1]$, and the range is $[-\pi/2, \pi/2]$.

The notation $y = \sin^{-1} x$ is sometimes read "$y$ is the inverse sine of $x$." The equation $x = \sin y$ in the definition allows us to regard $y$ as an angle, so $y = \sin^{-1} x$ may also be read "$y$ is the angle whose sine is $x$" (with $-\pi/2 \le y \le \pi/2$).

The inverse sine function is also called the **arcsine function,** and arcsin $x$ may be used in place of $\sin^{-1} x$. If $t = \arcsin x$, then $\sin t = x$, and $t$ may be interpreted as an *arc length* on the unit circle $U$ with center at the origin. We will use both notations—$\sin^{-1}$ and arcsin—throughout our work.

Several values of the inverse sine function are listed in the next chart.

**Warning**

> It is *essential* to choose the value $y$ in the range $[-\pi/2, \pi/2]$ of $\sin^{-1}$. Thus, even though $\sin (5\pi/6) = \frac{1}{2}$, the number $y = 5\pi/6$ is not the inverse function value $\sin^{-1} \frac{1}{2}$.

| Equation | Equivalent statement | | Solution |
|---|---|---|---|
| $y = \sin^{-1}\left(\dfrac{1}{2}\right)$ | $\sin y = \dfrac{1}{2}$ and | $-\dfrac{\pi}{2} \le y \le \dfrac{\pi}{2}$ | $y = \dfrac{\pi}{6}$ |
| $y = \sin^{-1}\left(-\dfrac{1}{2}\right)$ | $\sin y = -\dfrac{1}{2}$ and | $-\dfrac{\pi}{2} \le y \le \dfrac{\pi}{2}$ | $y = -\dfrac{\pi}{6}$ |
| $y = \sin^{-1}(1)$ | $\sin y = 1$ and | $-\dfrac{\pi}{2} \le y \le \dfrac{\pi}{2}$ | $y = \dfrac{\pi}{2}$ |
| $y = \arcsin(0)$ | $\sin y = 0$ and | $-\dfrac{\pi}{2} \le y \le \dfrac{\pi}{2}$ | $y = 0$ |
| $y = \arcsin\left(-\dfrac{\sqrt{3}}{2}\right)$ | $\sin y = -\dfrac{\sqrt{3}}{2}$ and | $-\dfrac{\pi}{2} \le y \le \dfrac{\pi}{2}$ | $y = -\dfrac{\pi}{3}$ |

**FIGURE 19**

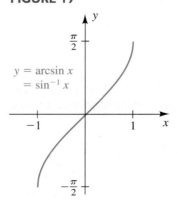

$y = \arcsin x$
$= \sin^{-1} x$

We have now justified the method of solving an equation of the form $\sin \theta = k$ as discussed in Chapter 5. We see that the calculator key $\boxed{\text{SIN}^{-1}}$ used to obtain $\theta = \sin^{-1} k$ gives us the value of the inverse sine function.

Relationship 6 for the graphs of $f$ and $f^{-1}$ tells us that we can sketch the graph of $y = \sin^{-1} x$ by reflecting the blue portion of Figure 18 through the line $y = x$. We can also use the equation $x = \sin y$ with the restriction $-\pi/2 \le y \le \pi/2$ to find points on the graph. This gives us Figure 19.

Relationship 4, $f(f^{-1}(x)) = x$, and relationship 5, $f^{-1}(f(y)) = y$, which hold for any inverse function $f^{-1}$, give us the following properties.

**Properties of $\sin^{-1}$**

> **(1)**  $\sin(\sin^{-1} x) = \sin(\arcsin x) = x$  if  $-1 \le x \le 1$
>
> **(2)**  $\sin^{-1}(\sin y) = \arcsin(\sin y) = y$  if  $-\dfrac{\pi}{2} \le y \le \dfrac{\pi}{2}$

**EXAMPLE 1    Using properties of $\sin^{-1}$**

Find the exact value:

**(a)** $\sin\left(\sin^{-1} \dfrac{1}{2}\right)$    **(b)** $\sin^{-1}\left(\sin \dfrac{\pi}{4}\right)$    **(c)** $\sin^{-1}\left(\sin \dfrac{2\pi}{3}\right)$

*SOLUTION*

**(a)** The *difficult* way to find the value of this expression is to first find the angle $\sin^{-1} \frac{1}{2}$, namely $\pi/6$, and then evaluate $\sin(\pi/6)$, obtaining $\frac{1}{2}$. The *easy* way is to use property 1 of $\sin^{-1}$, as follows:

$$\sin\left(\sin^{-1} \tfrac{1}{2}\right) = \tfrac{1}{2} \quad \text{since} \quad -1 < \tfrac{1}{2} < 1$$

**(b)** Since $-\pi/2 < \pi/4 < \pi/2$, we can use property 2 of $\sin^{-1}$ to obtain

$$\sin^{-1}\left(\sin \frac{\pi}{4}\right) = \frac{\pi}{4}.$$

**(c)** Be careful! Since $2\pi/3$ is *not* between $-\pi/2$ and $\pi/2$, we cannot use property 2 of $\sin^{-1}$. Instead, we first evaluate the inner expression, $\sin(2\pi/3)$, and then use the definition of $\sin^{-1}$, as follows:

$$\sin^{-1}\left(\sin \frac{2\pi}{3}\right) = \sin^{-1}\left(\frac{\sqrt{3}}{2}\right) = \frac{\pi}{3}$$

---

**EXAMPLE 2    Finding a value of $\sin^{-1}$**

Find the exact value of $\sin^{-1}\left(\tan \dfrac{3\pi}{4}\right)$.

*SOLUTION*    We cannot use property 2 of $\sin^{-1}$, since the given expression does not have the form $\sin^{-1}(\sin y)$. Instead, we first evaluate the inner expression—$\tan(3\pi/4)$—and then find the inverse sine of that number. If we let

$$y = \sin^{-1}\left(\tan \frac{3\pi}{4}\right),$$

then
$$y = \sin^{-1}(-1).$$

*(continued)*

By definition,

$$\sin y = -1 \quad \text{and} \quad -\frac{\pi}{2} \le y \le \frac{\pi}{2}.$$

It follows that $y = -\pi/2$, and hence

$$\sin^{-1}\left(\tan\frac{3\pi}{4}\right) = -\frac{\pi}{2}.$$

The other trigonometric functions may also be used to introduce inverse trigonometric functions. The procedure is first to determine a convenient subset of the domain in order to obtain a one-to-one function. If the domain of the cosine function is restricted to the interval $[0, \pi]$, as illustrated by the blue portion of the graph of $y = \cos x$ in Figure 20, we obtain a one-to-one (decreasing) function that takes on every value of the cosine function once and only once. Then, we use this *new* function with domain $[0, \pi]$ and range $[-1, 1]$ to define the *inverse cosine function.*

**FIGURE 20**

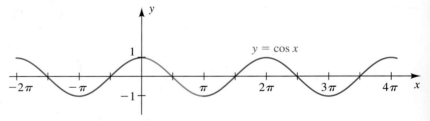

| | |
|---|---|
| **Definition of the Inverse Cosine Function** | The **inverse cosine function,** denoted by $\cos^{-1}$, is defined by $$y = \cos^{-1} x \quad \text{if and only if} \quad x = \cos y$$ for $-1 \le x \le 1$ and $0 \le y \le \pi$. |

The domain of the inverse cosine function is $[-1, 1]$, and the range is $[0, \pi]$. Note that the range of $\cos^{-1}$ is *not* the same as the range of $\sin^{-1}$ but their domains are equal.

The notation $y = \cos^{-1} x$ may be read "$y$ is the inverse cosine of $x$" or "$y$ is the angle whose cosine is $x$" (with $0 \le y \le \pi$).

The inverse cosine function is also called the **arccosine function,** and the notation arccos $x$ is used interchangeably with $\cos^{-1} x$.

Several values of the inverse cosine function are listed in the next chart.

**Warning**

It is *essential* to choose the value $y$ in the range $[0, \pi]$ of $\cos^{-1}$.

**FIGURE 21**

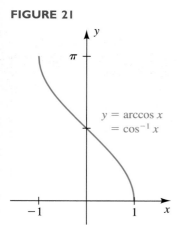

$y = \arccos x$
$= \cos^{-1} x$

| Equation | Equivalent statement | | Solution |
|---|---|---|---|
| $y = \cos^{-1}\left(\dfrac{1}{2}\right)$ | $\cos y = \dfrac{1}{2}$ | and $0 \le y \le \pi$ | $y = \dfrac{\pi}{3}$ |
| $y = \cos^{-1}\left(-\dfrac{1}{2}\right)$ | $\cos y = -\dfrac{1}{2}$ | and $0 \le y \le \pi$ | $y = \dfrac{2\pi}{3}$ |
| $y = \cos^{-1}(1)$ | $\cos y = 1$ | and $0 \le y \le \pi$ | $y = 0$ |
| $y = \arccos(0)$ | $\cos y = 0$ | and $0 \le y \le \pi$ | $y = \dfrac{\pi}{2}$ |
| $y = \arccos\left(-\dfrac{\sqrt{3}}{2}\right)$ | $\cos y = -\dfrac{\sqrt{3}}{2}$ | and $0 \le y \le \pi$ | $y = \dfrac{5\pi}{6}$ |

We can sketch the graph of $y = \cos^{-1} x$ by reflecting the blue portion of Figure 20 through the line $y = x$. This gives us the sketch in Figure 21. We could also use the equation $x = \cos y$, with $0 \le y \le \pi$, to find points on the graph. As indicated by the graph, *the values of the inverse cosine function are never negative.*

Using relationships 4 and 5 for general inverse functions $f$ and $f^{-1}$, we obtain the following properties.

**Properties of $\cos^{-1}$**

> **(1)** $\cos(\cos^{-1} x) = \cos(\arccos x) = x$   if   $-1 \le x \le 1$
>
> **(2)** $\cos^{-1}(\cos y) = \arccos(\cos y) = y$   if   $0 \le y \le \pi$

**EXAMPLE 3**   *Using properties of $\cos^{-1}$*

Find the exact value:

**(a)** $\cos[\cos^{-1}(-0.5)]$   **(b)** $\cos^{-1}(\cos 3.14)$   **(c)** $\cos^{-1}\left[\sin\left(-\dfrac{\pi}{6}\right)\right]$

**SOLUTION**   For parts (a) and (b), we may use the properties of $\cos^{-1}$.

**(a)** $\cos[\cos^{-1}(-0.5)] = -0.5$, since $-1 < -0.5 < 1$.

**(b)** $\cos^{-1}(\cos 3.14) = 3.14$, since $0 < 3.14 < \pi$.

**(c)** We first find $\sin(-\pi/6)$ and then use the definition of $\cos^{-1}$, as follows:

$$\cos^{-1}\left[\sin\left(-\frac{\pi}{6}\right)\right] = \cos^{-1}\left(-\frac{1}{2}\right) = \frac{2\pi}{3}$$

**EXAMPLE 4**    *Finding a trigonometric function value*

Find the exact value of $\sin\left[\arccos\left(-\frac{2}{3}\right)\right]$.

*SOLUTION*    We cannot use property 1 of $\cos^{-1}$, since the given expression does not have the form $\cos(\cos^{-1}x)$. However, if we let

$$\theta = \arccos\left(-\tfrac{2}{3}\right),$$

then, using the definition of the inverse cosine function, we have

$$\cos\theta = -\tfrac{2}{3} \quad \text{and} \quad 0 \le \theta \le \pi.$$

**FIGURE 22**

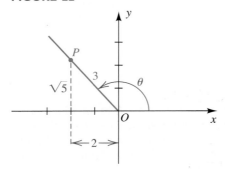

Hence, $\theta$ is in quadrant II, as illustrated in Figure 22. If we choose the point $P$ on the terminal side with $x$-coordinate $-2$, the hypotenuse of the triangle in the figure must have length 3, since $\cos\theta = -\frac{2}{3}$. Thus, by the Pythagorean theorem, the $y$-coordinate of $P$ is

$$\sqrt{3^2 - 2^2} = \sqrt{9 - 4} = \sqrt{5},$$

and therefore

$$\sin\left[\arccos\left(-\frac{2}{3}\right)\right] = \sin\theta = \frac{\sqrt{5}}{3}.$$

If we restrict the domain of the tangent function to the open interval $(-\pi/2, \pi/2)$, we obtain a one-to-one (increasing) function (see Figure 3 on page 410). We use this *new* function to define the *inverse tangent function.*

**Definition of the Inverse Tangent Function**

The **inverse tangent function,** or **arctangent function,** denoted by $\tan^{-1}$ or arctan, is defined by

$$y = \tan^{-1}x = \arctan x \quad \text{if and only if} \quad x = \tan y$$

for any real number $x$ and for $-\dfrac{\pi}{2} < y < \dfrac{\pi}{2}$.

**FIGURE 23**

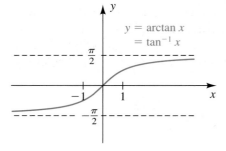

The domain of the arctangent function is $\mathbb{R}$, and the range is the open interval $(-\pi/2, \pi/2)$.

We can obtain the graph of $y = \tan^{-1}x$ in Figure 23 by sketching the graph of $x = \tan y$ for $-\pi/2 < y < \pi/2$.

As with $\sin^{-1}$ and $\cos^{-1}$, we have the following properties for $\tan^{-1}$.

**Properties of tan$^{-1}$**

> **(1)**  $\tan (\tan^{-1} x) = \tan (\arctan x) = x$   for every $x$
>
> **(2)**  $\tan^{-1} (\tan y) = \arctan (\tan y) = y$   if   $-\dfrac{\pi}{2} < y < \dfrac{\pi}{2}$

**EXAMPLE 5**   *Using properties of* tan$^{-1}$
Find the exact value:

**(a)** $\tan (\tan^{-1} 1000)$   **(b)** $\tan^{-1} \left( \tan \dfrac{\pi}{4} \right)$   **(c)** $\arctan (\tan \pi)$

*SOLUTION*

**(a)** By property 1 of tan$^{-1}$,

$$\tan (\tan^{-1} 1000) = 1000.$$

**(b)** Since $-\pi/2 < \pi/4 < \pi/2$, we have, by property 2 of tan$^{-1}$,

$$\tan^{-1} \left( \tan \dfrac{\pi}{4} \right) = \dfrac{\pi}{4}.$$

**(c)** Since $\pi > \pi/2$, we cannot use the second property of tan$^{-1}$. Thus, we first find tan $\pi$ and then evaluate as follows:

$$\arctan (\tan \pi) = \arctan 0 = 0$$

**EXAMPLE 6**   *Finding a trigonometric function value*
Find the exact value of $\sec \left( \arctan \frac{2}{3} \right)$.

*SOLUTION*   If we let $y = \arctan \frac{2}{3}$, then $\tan y = \frac{2}{3}$. We wish to find sec $y$.

Since $-\pi/2 < \arctan x < \pi/2$ for every $x$ and $\tan y > 0$, it follows that $0 < y < \pi/2$. Thus, we may regard $y$ as the radian measure of an angle of a right triangle such that $\tan y = \frac{2}{3}$, as illustrated in Figure 24. By the Pythagorean theorem, the hypotenuse is $\sqrt{3^2 + 2^2} = \sqrt{13}$. Referring to the triangle, we obtain

**FIGURE 24**

$$\sec \left( \arctan \dfrac{2}{3} \right) = \sec y = \dfrac{\sqrt{13}}{3}.$$

**EXAMPLE 7**    *Finding a trigonometric function value*

Find the exact value of $\sin \left( \arctan \frac{1}{2} - \arccos \frac{4}{5} \right)$.

**SOLUTION**    If we let

$$u = \arctan \tfrac{1}{2} \quad \text{and} \quad v = \arccos \tfrac{4}{5},$$

then

$$\tan u = \tfrac{1}{2} \quad \text{and} \quad \cos v = \tfrac{4}{5}.$$

We wish to find $\sin (u - v)$. Since $u$ and $v$ are in the interval $(0, \pi/2)$, they can be considered as the radian measures of positive acute angles, and we may refer to the right triangles in Figure 25. This gives us

$$\sin u = \frac{1}{\sqrt{5}}, \quad \cos u = \frac{2}{\sqrt{5}}, \quad \sin v = \frac{3}{5}, \quad \cos v = \frac{4}{5}.$$

Consequently,

$$\sin (u - v) = \sin u \cos v - \cos u \sin v$$

$$= \frac{1}{\sqrt{5}} \frac{4}{5} - \frac{2}{\sqrt{5}} \frac{3}{5}$$

$$= \frac{-2}{5\sqrt{5}} = \frac{-2\sqrt{5}}{25}.$$

**FIGURE 25**

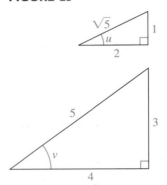

---

**EXAMPLE 8**    *Changing an expression involving $\sin^{-1} x$ to an algebraic expression*

If $-1 \leq x \leq 1$, rewrite $\cos (\sin^{-1} x)$ as an algebraic expression in $x$.

**SOLUTION**    Let

$$y = \sin^{-1} x \quad \text{or, equivalently,} \quad \sin y = x.$$

We wish to express $\cos y$ in terms of $x$. Since $-\pi/2 \leq y \leq \pi/2$, it follows that $\cos y \geq 0$, and hence

$$\cos y = \sqrt{1 - \sin^2 y} = \sqrt{1 - x^2}.$$

Consequently,    $\cos (\sin^{-1} x) = \sqrt{1 - x^2}.$

The last identity is also evident geometrically if $0 < x < 1$. In this case $0 < y < \pi/2$, and we may regard $y$ as the radian measure of an angle of a right triangle such that $\sin y = x$, as illustrated in Figure 26. (The side of length $\sqrt{1 - x^2}$ is found by the Pythagorean theorem.) Referring to the triangle, we have

$$\cos (\sin^{-1} x) = \cos y = \frac{\sqrt{1 - x^2}}{1} = \sqrt{1 - x^2}.$$

**FIGURE 26**

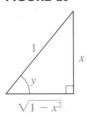

Most of the trigonometric equations we considered in Section 6.2 had solutions that were rational multiples of $\pi$, such as $\pi/3$, $3\pi/4$, $\pi$, and so on. If solutions of trigonometric equations are not of that type, we can sometimes use inverse functions to express them in exact form, as illustrated in the next example.

**EXAMPLE 9**   *Using inverse trigonometric functions to solve an equation*

Find the solutions of the equation $5 \sin^2 t + 3 \sin t - 1 = 0$ that are in the interval $[-\pi/2, \pi/2]$.

**SOLUTION**   The equation may be regarded as a quadratic equation in $\sin t$. We may apply the quadratic formula:

$$\sin t = \frac{-3 \pm \sqrt{3^2 - 4(5)(-1)}}{2(5)} = \frac{-3 \pm \sqrt{29}}{10}$$

Using the definition of the inverse sine function, we obtain the following solutions:

$$t = \sin^{-1} \tfrac{1}{10}(-3 + \sqrt{29}) \approx 0.2408$$
$$t = \sin^{-1} \tfrac{1}{10}(-3 - \sqrt{29}) \approx -0.9946$$

The next example illustrates one of many identities that are true for inverse trigonometric functions.

**EXAMPLE 10**   *Verifying an identity involving inverse trigonometric functions*

Verify the identity $\sin^{-1} x + \cos^{-1} x = \dfrac{\pi}{2}$ for $-1 \le x \le 1$.

**SOLUTION**   Let

$$\alpha = \sin^{-1} x \quad \text{and} \quad \beta = \cos^{-1} x.$$

We wish to show that $\alpha + \beta = \pi/2$. From the definitions of $\sin^{-1}$ and $\cos^{-1}$,

$$\sin \alpha = x \quad \text{for} \quad -\frac{\pi}{2} \le \alpha \le \frac{\pi}{2}$$

and

$$\cos \beta = x \quad \text{for} \quad 0 \le \beta \le \pi.$$

*(continued)*

Adding the two inequalities on the right, we see that

$$-\frac{\pi}{2} \le \alpha + \beta \le \frac{3\pi}{2}.$$

Note also that

$$\cos \alpha = \sqrt{1 - \sin^2 \alpha} = \sqrt{1 - x^2}$$

and

$$\sin \beta = \sqrt{1 - \cos^2 \beta} = \sqrt{1 - x^2}.$$

Using an addition formula, we obtain

$$\begin{aligned} \sin (\alpha + \beta) &= \sin \alpha \cos \beta + \cos \alpha \sin \beta \\ &= x \cdot x + \sqrt{1 - x^2}\sqrt{1 - x^2} \\ &= x^2 + (1 - x^2) = 1. \end{aligned}$$

Since $\alpha + \beta$ is in the interval $[-\pi/2, 3\pi/2]$, the equation $\sin (\alpha + \beta) = 1$ has only one solution, $\alpha + \beta = \pi/2$, which is what we wished to show.

We may interpret the identity geometrically if $0 < x < 1$. If we construct a right triangle with one side of length $x$ and hypotenuse of length 1, as illustrated in Figure 27, then angle $\beta$ at $B$ is an angle whose cosine is $x$; that is, $\beta = \cos^{-1} x$. Similarly, angle $\alpha$ at $A$ is an angle whose sine is $x$; that is, $\alpha = \sin^{-1} x$. Since the acute angles of a right triangle are complementary, $\alpha + \beta = \pi/2$ or, equivalently,

$$\sin^{-1} x + \cos^{-1} x = \frac{\pi}{2}.$$

**FIGURE 27**

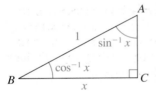

Each of the remaining inverse trigonometric functions is defined in the same manner as the first three—by choosing a domain $D$ in which the corresponding trigonometric function is one-to-one and then using the usual technique (where $y$ is in $D$):

$$\begin{aligned} y &= \cot^{-1} x \quad \text{if and only if} \quad x = \cot y \\ y &= \sec^{-1} x \quad \text{if and only if} \quad x = \sec y \\ y &= \csc^{-1} x \quad \text{if and only if} \quad x = \csc y \end{aligned}$$

The function $\sec^{-1}$ is used in calculus; however, $\cot^{-1}$ and $\csc^{-1}$ are seldom employed. Because of their limited use in applications, we shall not consider examples or exercises pertaining to these functions. We shall merely summarize typical domains, ranges, and graphs in the following chart. A similar summary for the six trigonometric functions and their inverses appears in Appendix IV.

**Summary of Features of $\cot^{-1}$, $\sec^{-1}$, and $\csc^{-1}$**

| Feature | $y = \cot^{-1} x$ | $y = \sec^{-1} x$ | $y = \csc^{-1} x$ |
|---|---|---|---|
| Domain | $\mathbb{R}$ | $\lvert x \rvert \geq 1$ | $\lvert x \rvert \geq 1$ |
| Range | $(0, \pi)$ | $\left[ 0, \dfrac{\pi}{2} \right) \cup \left[ \pi, \dfrac{3\pi}{2} \right)$ | $\left( -\pi, -\dfrac{\pi}{2} \right] \cup \left( 0, \dfrac{\pi}{2} \right]$ |
| Graph | | | |

It is often difficult to verify an identity involving inverse trigonometric functions, as we saw in Example 10. A graphing utility can be extremely helpful in determining whether an equation involving inverse trigonometric functions is an identity and, if it is not an identity, finding any solutions of the equation. The next example illustrates this process.

**EXAMPLE 11**   *Investigating an equation*

We know that $\tan x = (\sin x)/\cos x$ is an identity. Determine whether the equation

$$\arctan x = \frac{\arcsin x}{\arccos x}$$

is an identity. If it is not an identity, approximate the values of $x$ for which the equation is true—that is, solve the equation.

**SOLUTION**   Some specific keystrokes for the TI-81/82 graphics calculator are given in Appendix I. We begin by making the assignments

$$Y_1 = \tan^{-1} x \quad \text{and} \quad Y_2 = \sin^{-1} x / \cos^{-1} x.$$

Since the domain of $\sin^{-1}$ and $\cos^{-1}$ is $[-1, 1]$ and the range of $\tan^{-1}$ is $(-\pi/2, \pi/2)$, we choose the viewing rectangle dimensions $[-1, 1]$ by $[-1.57, 1.57]$, with $\text{Xscl} = 0.1$ and $\text{Yscl} = 0.2$, as shown in Figure 28.

*(continued)*

**FIGURE 28**   $[-1, 1]$ by $[-1.57, 1.57]$

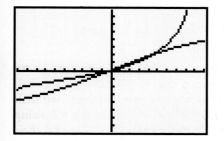

Since the graphs representing $Y_1$ and $Y_2$ are not the same, we know that *the given equation is not an identity.* Because the graphs intersect twice, however, we know that the equation has two solutions. It appears that $x = 0$ is a solution, and a quick check in the given equation verifies that this is true. To estimate the point of intersection in the first quadrant, we use the tracing and zoom features to determine that the point has the approximate coordinates $(0.450, 0.423)$. Hence,

$$x = 0 \quad \text{and} \quad x \approx 0.450$$

are the values of $x$ for which the given equation is true.

## 6.6 EXERCISES

**Exer. 1–22: Find the exact value of the expression whenever it is defined.**

1  (a) $\sin^{-1}\left(-\dfrac{\sqrt{2}}{2}\right)$    (b) $\cos^{-1}\left(-\frac{1}{2}\right)$

   (c) $\tan^{-1}(-\sqrt{3})$

2  (a) $\sin^{-1}(-\frac{1}{2})$    (b) $\cos^{-1}\left(-\dfrac{\sqrt{2}}{2}\right)$

   (c) $\tan^{-1}(-1)$

3  (a) $\arcsin\dfrac{\sqrt{3}}{2}$  (b) $\arccos\dfrac{\sqrt{2}}{2}$  (c) $\arctan\dfrac{1}{\sqrt{3}}$

4  (a) $\arcsin 0$    (b) $\arccos(-1)$    (c) $\arctan 0$

5  (a) $\sin^{-1}\dfrac{\pi}{3}$    (b) $\cos^{-1}\dfrac{\pi}{2}$    (c) $\tan^{-1} 1$

6  (a) $\arcsin\dfrac{\pi}{2}$    (b) $\arccos\dfrac{\pi}{3}$    (c) $\arctan\left(-\dfrac{\sqrt{3}}{3}\right)$

7  (a) $\sin\left[\arcsin\left(-\frac{3}{10}\right)\right]$    (b) $\cos\left(\arccos\frac{1}{2}\right)$

   (c) $\tan(\arctan 14)$

8  (a) $\sin\left(\sin^{-1}\frac{2}{3}\right)$    (b) $\cos\left[\cos^{-1}\left(-\frac{1}{5}\right)\right]$

   (c) $\tan[\tan^{-1}(-9)]$

9  (a) $\sin^{-1}\left(\sin\dfrac{\pi}{3}\right)$    (b) $\cos^{-1}\left[\cos\left(\dfrac{5\pi}{6}\right)\right]$

   (c) $\tan^{-1}\left[\tan\left(-\dfrac{\pi}{6}\right)\right]$

10  (a) $\arcsin\left[\sin\left(-\dfrac{\pi}{2}\right)\right]$    (b) $\arccos(\cos 0)$

    (c) $\arctan\left(\tan\dfrac{\pi}{4}\right)$

11  (a) $\arcsin\left(\sin\dfrac{5\pi}{4}\right)$    (b) $\arccos\left(\cos\dfrac{5\pi}{4}\right)$

    (c) $\arctan\left(\tan\dfrac{7\pi}{4}\right)$

12  (a) $\sin^{-1}\left(\sin\dfrac{2\pi}{3}\right)$    (b) $\cos^{-1}\left(\cos\dfrac{4\pi}{3}\right)$

    (c) $\tan^{-1}\left(\tan\dfrac{7\pi}{6}\right)$

13  (a) $\sin\left[\cos^{-1}\left(-\frac{1}{2}\right)\right]$    (b) $\cos(\tan^{-1} 1)$

    (c) $\tan[\sin^{-1}(-1)]$

14  (a) $\sin(\tan^{-1}\sqrt{3})$    (b) $\cos(\sin^{-1} 1)$

    (c) $\tan(\cos^{-1} 0)$

15  (a) $\cot\left(\sin^{-1}\frac{2}{3}\right)$    (b) $\sec\left[\tan^{-1}\left(-\frac{3}{5}\right)\right]$

    (c) $\csc\left[\cos^{-1}\left(-\frac{1}{4}\right)\right]$

16  (a) $\cot\left[\sin^{-1}\left(-\frac{2}{3}\right)\right]$    (b) $\sec\left(\tan^{-1}\frac{7}{4}\right)$

    (c) $\csc\left(\cos^{-1}\frac{1}{5}\right)$

17  (a) $\sin\left(\arcsin\frac{1}{2} + \arccos 0\right)$

    (b) $\cos\left[\arctan\left(-\frac{3}{4}\right) - \arcsin\frac{4}{5}\right]$

    (c) $\tan\left(\arctan\frac{4}{5} + \arccos\frac{8}{17}\right)$

**18** (a) $\sin\left[\sin^{-1}\frac{5}{13} - \cos^{-1}\left(-\frac{3}{5}\right)\right]$

(b) $\cos\left(\sin^{-1}\frac{4}{5} + \tan^{-1}\frac{3}{4}\right)$

(c) $\tan\left[\cos^{-1}\frac{1}{2} - \sin^{-1}\left(-\frac{1}{2}\right)\right]$

**19** (a) $\sin\left[2\arccos\left(-\frac{3}{5}\right)\right]$  (b) $\cos\left(2\sin^{-1}\frac{15}{17}\right)$

(c) $\tan\left(2\tan^{-1}\frac{3}{4}\right)$

**20** (a) $\sin\left(2\tan^{-1}\frac{5}{12}\right)$  (b) $\cos\left(2\arccos\frac{9}{41}\right)$

(c) $\tan\left[2\arcsin\left(-\frac{8}{17}\right)\right]$

**21** (a) $\sin\left[\frac{1}{2}\sin^{-1}\left(-\frac{7}{25}\right)\right]$  (b) $\cos\left(\frac{1}{2}\tan^{-1}\frac{8}{15}\right)$

(c) $\tan\left(\frac{1}{2}\cos^{-1}\frac{3}{5}\right)$

**22** (a) $\sin\left[\frac{1}{2}\cos^{-1}\left(-\frac{3}{5}\right)\right]$  (b) $\cos\left(\frac{1}{2}\sin^{-1}\frac{12}{13}\right)$

(c) $\tan\left(\frac{1}{2}\tan^{-1}\frac{40}{9}\right)$

**Exer. 23–30:** Write the expression as an algebraic expression in $x$ for $x > 0$.

**23** $\sin\left(\tan^{-1}x\right)$      **24** $\tan\left(\arccos x\right)$

**25** $\sec\left(\sin^{-1}\dfrac{x}{\sqrt{x^2+4}}\right)$  **26** $\cot\left(\sin^{-1}\dfrac{\sqrt{x^2-9}}{x}\right)$

**27** $\sin\left(2\sin^{-1}x\right)$      **28** $\cos\left(2\tan^{-1}x\right)$

**29** $\cos\left(\frac{1}{2}\arccos x\right)$      **30** $\tan\left(\dfrac{1}{2}\cos^{-1}\dfrac{1}{x}\right)$

**Exer. 31–32:** Complete the statements.

**31** (a) As $x \to -1^+$, $\sin^{-1}x \to$ _____

(b) As $x \to 1^-$, $\cos^{-1}x \to$ _____

(c) As $x \to \infty$, $\tan^{-1}x \to$ _____

**32** (a) As $x \to 1^-$, $\sin^{-1}x \to$ _____

(b) As $x \to -1^+$, $\cos^{-1}x \to$ _____

(c) As $x \to -\infty$, $\tan^{-1}x \to$ _____

**Exer. 33–42:** Sketch the graph of the equation.

**33** $y = \sin^{-1}2x$      **34** $y = \frac{1}{2}\sin^{-1}x$

**35** $y = \sin^{-1}(x+1)$

**36** $y = \sin^{-1}(x-2) + \dfrac{\pi}{2}$

**37** $y = \cos^{-1}\frac{1}{2}x$      **38** $y = 2\cos^{-1}x$

**39** $y = 2 + \tan^{-1}x$      **40** $y = \tan^{-1}2x$

**41** $y = \sin\left(\arccos x\right)$      **42** $y = \sin\left(\sin^{-1}x\right)$

**Exer. 43–46:** The given equation has the form $y = f(x)$. **(a)** Find the domain of $f$. **(b)** Find the range of $f$. **(c)** Solve for $x$ in terms of $y$.

**43** $y = \frac{1}{2}\sin^{-1}(x-3)$      **44** $y = 3\tan^{-1}(2x+1)$

**45** $y = 4\cos^{-1}\frac{2}{3}x$      **46** $y = 2\sin^{-1}(3x-4)$

**Exer. 47–50:** Solve the equation for $x$ in terms of $y$ if $x$ is restricted to the given interval.

**47** $y = -3 - \sin x$;      $\left[-\dfrac{\pi}{2}, \dfrac{\pi}{2}\right]$

**48** $y = 2 + 3\sin x$;      $\left[-\dfrac{\pi}{2}, \dfrac{\pi}{2}\right]$

**49** $y = 15 - 2\cos x$;   $[0, \pi]$

**50** $y = 6 - 3\cos x$;   $[0, \pi]$

**Exer. 51–52:** Solve the equation for $x$ in terms of $y$ if $0 < x < \pi$ and $0 < y < \pi$.

**51** $\dfrac{\sin x}{3} = \dfrac{\sin y}{4}$      **52** $\dfrac{4}{\sin x} = \dfrac{7}{\sin y}$

**Exer. 53–60:** Use inverse trigonometric functions to find the solutions of the equation that are in the given interval, and approximate the solutions to four decimal places.

**53** $\cos^2 x + 2\cos x - 1 = 0$;      $[0, 2\pi)$

**54** $\sin^2 x - \sin x - 1 = 0$;      $[0, 2\pi)$

**55** $2\tan^2 t + 9\tan t + 3 = 0$;      $\left(-\dfrac{\pi}{2}, \dfrac{\pi}{2}\right)$

**56** $3\sin^2 t + 7\sin t + 3 = 0$;      $\left[-\dfrac{\pi}{2}, \dfrac{\pi}{2}\right]$

**57** $15\cos^4 x - 14\cos^2 x + 3 = 0$;      $[0, \pi]$

**58** $3\tan^4 \theta - 19\tan^2 \theta + 2 = 0$;      $\left(-\dfrac{\pi}{2}, \dfrac{\pi}{2}\right)$

**59** $6\sin^3 \theta + 18\sin^2 \theta - 5\sin \theta - 15 = 0$;      $\left(-\dfrac{\pi}{2}, \dfrac{\pi}{2}\right)$

**60** $6\sin 2x - 8\cos x + 9\sin x - 6 = 0$;      $\left(-\dfrac{\pi}{2}, \dfrac{\pi}{2}\right)$

**Exer. 61–62:** If an earthquake has a total horizontal displacement of $S$ meters along its fault line, then the horizontal movement $M$ of a point on the surface of the earth $d$ kilometers from the fault line can be estimated using the formula

$$M = \frac{S}{2}\left(1 - \frac{2}{\pi}\tan^{-1}\frac{d}{D}\right),$$

where $D$ is the depth (in kilometers) below the surface of the focal point of the earthquake.

**61 Earthquake movement** For the San Francisco earthquake of 1906, $S$ was 4 meters and $D$ was 3.5 kilometers. Approximate $M$ for the stated values of $d$.

(a) 1 kilometer      (b) 4 kilometers

(c) 10 kilometers

**62 Earthquake movement** Approximate the depth $D$ of the focal point of an earthquake with $S = 3$ m if a point on the surface of the earth 5 kilometers from the fault line moved 0.6 meter horizontally.

**63 Tracking a sailboat** As shown in the figure, a sailboat is following a straight-line course $l$. The shortest distance from a tracking station $T$ to the course is $d$ miles. As the boat sails, the tracking station records its distance $k$ from $T$ and its direction $\theta$ with respect to $T$. Angle $\alpha$ specifies the direction of the sailboat.

(a) Express $\alpha$ in terms of $d$, $k$, and $\theta$.

(b) Estimate $\alpha$ to the nearest degree if $d = 50$ mi, $k = 210$ mi, and $\theta = 53.4°$.

**EXERCISE 63**

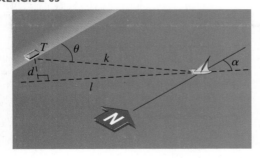

**64 Calculating viewing angles** An art critic whose eye level is 6 feet above the floor views a painting that is 10 feet in height and is mounted 4 feet above the floor, as shown in the figure.

(a) If the critic is standing $x$ feet from the wall, express the viewing angle $\theta$ in terms of $x$.

(b) Use the addition formula for the tangent to show that

$$\theta = \tan^{-1}\left(\frac{10x}{x^2 - 16}\right).$$

(c) For what value of $x$ is $\theta = 45°$?

**EXERCISE 64**

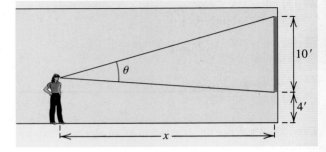

**Exer. 65–70:** Verify the identity.

**65** $\sin^{-1} x = \tan^{-1}\dfrac{x}{\sqrt{1 - x^2}}$

**66** $\arccos x + \arccos\sqrt{1 - x^2} = \dfrac{\pi}{2},\ 0 \le x \le 1$

**67** $\arcsin(-x) = -\arcsin x$

**68** $\arccos(-x) = \pi - \arccos x$

**69** $\arctan x + \arctan\dfrac{1}{x} = \dfrac{\pi}{2},\ x > 0$

**70** $2\cos^{-1} x = \cos^{-1}(2x^2 - 1),\ 0 \le x \le 1$

**Exer. 71–72:** Graph $f$, and determine its domain and range.

**71** $f(x) = 2\sin^{-1}(x - 1) + \cos^{-1}\frac{1}{2}x$

**72** $f(x) = \frac{1}{2}\tan^{-1}(1 - 2x) + 3\tan^{-1}\sqrt{x + 2}$

**Exer. 73–74:** Use a graph to estimate the solutions of the equation.

**73** $\sin^{-1} 2x = \tan^{-1}(1 - x)$

**74** $\cos^{-1}(x - \frac{1}{3}) = 2\sin^{-1}(\frac{1}{2} - x)$

**75 Designing a solar collector** In designing a collector for solar power, an important consideration is the amount of sunlight that is transmitted through the glass into the water being heated. If the angle of incidence $\theta$ of the sun's rays is measured from a line perpendicular to the surface of the glass, then the fraction $f(\theta)$ of sunlight reflected off the glass can be approximated by

$$f(\theta) = \frac{1}{2}\left(\frac{\sin^2 \alpha}{\sin^2 \beta} + \frac{\tan^2 \alpha}{\tan^2 \beta}\right),$$

where

$\alpha = \theta - \gamma$, $\beta = \theta + \gamma$, and $\gamma = \sin^{-1}(\sin \theta/1.52)$. Graph $f$ for $0 < \theta < \pi/2$, and estimate $\theta$ when $f(\theta) = 0.2$.

76 **Designing a solar collector**   The altitude of the sun is the angle $\phi$ that the sun's rays make with the horizon at a given time and place. Determining $\phi$ is important when tilting a solar collector to obtain maximum efficiency. On June 21 at a latitude of 51.7°N, the altitude of the sun can be approximated using the formula

$$\sin \phi = \sin 23.5° \sin 51.7° + \cos 23.5° \cos 51.7° \cos H,$$

where $H$ is called the hour angle, with $H = -\pi/2$ at 6 A.M., $H = 0$ at noon, and $H = \pi/2$ at 6 P.M.

(a) Solve the formula for $\phi$, and graph the resulting equation for $-\pi/2 \le H \le \pi/2$.

(b) Estimate the times when $\phi = 45°$.

# CHAPTER 6 REVIEW EXERCISES

**Exer. 1–22: Verify the identity.**

1  $(\cot^2 x + 1)(1 - \cos^2 x) = 1$

2  $\cos \theta + \sin \theta \tan \theta = \sec \theta$

3  $\dfrac{(\sec^2 \theta - 1)\cot \theta}{\tan \theta \sin \theta + \cos \theta} = \sin \theta$

4  $(\tan x + \cot x)^2 = \sec^2 x \csc^2 x$

5  $\dfrac{1}{1 + \sin t} = (\sec t - \tan t)\sec t$

6  $\dfrac{\sin(\alpha - \beta)}{\cos(\alpha + \beta)} = \dfrac{\tan \alpha - \tan \beta}{1 - \tan \alpha \tan \beta}$

7  $\tan 2u = \dfrac{2 \cot u}{\csc^2 u - 2}$       8  $\cos^2 \dfrac{v}{2} = \dfrac{1 + \sec v}{2 \sec v}$

9  $\dfrac{\tan^3 \phi - \cot^3 \phi}{\tan^2 \phi + \csc^2 \phi} = \tan \phi - \cot \phi$

10  $\dfrac{\sin u + \sin v}{\csc u + \csc v} = \dfrac{1 - \sin u \sin v}{-1 + \csc u \csc v}$

11  $\left(\dfrac{\sin^2 x}{\tan^4 x}\right)^3 \left(\dfrac{\csc^3 x}{\cot^6 x}\right)^2 = 1$

12  $\dfrac{\cos \gamma}{1 - \tan \gamma} + \dfrac{\sin \gamma}{1 - \cot \gamma} = \cos \gamma + \sin \gamma$

13  $\dfrac{\cos(-t)}{\sec(-t) + \tan(-t)} = 1 + \sin t$

14  $\dfrac{\cot(-t) + \csc(-t)}{\sin(-t)} = \dfrac{1}{1 - \cos t}$

15  $\sqrt{\dfrac{1 - \cos t}{1 + \cos t}} = \dfrac{1 - \cos t}{|\sin t|}$

16  $\sqrt{\dfrac{1 - \sin \theta}{1 + \sin \theta}} = \dfrac{|\cos \theta|}{1 + \sin \theta}$

17  $\cos\left(x - \dfrac{5\pi}{2}\right) = \sin x$

18  $\tan\left(x + \dfrac{3\pi}{4}\right) = \dfrac{\tan x - 1}{1 + \tan x}$

19  $\frac{1}{4}\sin 4\beta = \sin \beta \cos^3 \beta - \cos \beta \sin^3 \beta$

20  $\tan \frac{1}{2}\theta = \csc \theta - \cot \theta$

21  $\sin 8\theta = 8 \sin \theta \cos \theta\,(1 - 2\sin^2 \theta)(1 - 8 \sin^2 \theta \cos^2 \theta)$

22  $\arctan x = \dfrac{1}{2}\arctan \dfrac{2x}{1 - x^2}, \quad -1 < x < 1$

**Exer. 23–40: Find the solutions of the equation that are in the interval $[0, 2\pi)$.**

23  $2\cos^3 \theta - \cos \theta = 0$

24  $2\cos \alpha + \tan \alpha = \sec \alpha$

25  $\sin \theta = \tan \theta$       26  $\csc^5 \theta - 4\csc \theta = 0$

27  $2\cos^3 t + \cos^2 t - 2\cos t - 1 = 0$

28  $\cos x \cot^2 x = \cos x$

29  $\sin \beta + 2\cos^2 \beta = 1$

30  $\cos 2x + 3\cos x + 2 = 0$

31  $2\sec u \sin u + 2 = 4\sin u + \sec u$

32  $\tan 2x \cos 2x = \sin 2x$

33  $2 \cos 3x \cos 2x = 1 - 2 \sin 3x \sin 2x$

34  $\sin x \cos 2x + \cos x \sin 2x = 0$

35  $\cos \pi x + \sin \pi x = 0$    36  $\sin 2u = \sin u$

37  $2 \cos^2 \frac{1}{2}\theta - 3 \cos \theta = 0$

38  $\sec 2x \csc 2x = 2 \csc 2x$

39  $\sin 5x = \sin 3x$    40  $\cos 3x = -\cos 2x$

**Exer. 41–44: Find the exact value.**

41  $\cos 75°$    42  $\tan 285°$

43  $\sin 195°$    44  $\csc \dfrac{\pi}{8}$

**Exer. 45–56: If $\theta$ and $\phi$ are acute angles such that $\csc \theta = \frac{5}{3}$ and $\cos \phi = \frac{8}{17}$, find the exact value.**

45  $\sin (\theta + \phi)$    46  $\cos (\theta + \phi)$

47  $\tan (\phi + \theta)$    48  $\tan (\theta - \phi)$

49  $\sin (\phi - \theta)$    50  $\sin (\theta - \phi)$

51  $\sin 2\phi$    52  $\cos 2\phi$

53  $\tan 2\theta$    54  $\sin \frac{1}{2}\theta$

55  $\tan \frac{1}{2}\theta$    56  $\cos \frac{1}{2}\phi$

57  Express as a sum or difference:
(a) $\sin 7t \sin 4t$    (b) $\cos \frac{1}{4}u \cos (-\frac{1}{6}u)$
(c) $6 \cos 5x \sin 3x$    (d) $4 \sin 3\theta \cos 7\theta$

58  Express as a product:
(a) $\sin 8u + \sin 2u$    (b) $\cos 3\theta - \cos 8\theta$
(c) $\sin \frac{1}{4}t - \sin \frac{1}{5}t$    (d) $3 \cos 2x + 3 \cos 6x$

**Exer. 59–70: Find the exact value of the expression whenever it is defined.**

59  $\cos^{-1} \left(\dfrac{\sqrt{3}}{2}\right)$    60  $\arcsin \left(\dfrac{\sqrt{2}}{2}\right)$

61  $\arctan \sqrt{3}$    62  $\arccos \left(\tan \dfrac{3\pi}{4}\right)$

63  $\arcsin \left(\sin \dfrac{5\pi}{4}\right)$    64  $\cos^{-1} \left(\cos \dfrac{5\pi}{4}\right)$

65  $\sin \left[\arccos \left(-\dfrac{\sqrt{3}}{2}\right)\right]$    66  $\tan (\tan^{-1} 2)$

67  $\sec (\sin^{-1} \frac{3}{2})$    68  $\cos^{-1} (\sin 0)$

69  $\cos (\sin^{-1} \frac{15}{17} - \sin^{-1} \frac{8}{17})$

70  $\cos (2 \sin^{-1} \frac{4}{5})$

**Exer. 71–74: Sketch the graph of the equation.**

71  $y = \cos^{-1} 3x$    72  $y = 4 \sin^{-1} x$

73  $y = 1 - \sin^{-1} x$    74  $y = \sin (\frac{1}{2} \cos^{-1} x)$

75  Express $\cos (\alpha + \beta + \gamma)$ in terms of trigonometric functions of $\alpha$, $\beta$, and $\gamma$.

76  **Force of a foot**  When an individual is walking, the magnitude $F$ of the vertical force of one foot on the ground (see the figure) can be described by

$$F = A(\cos bt - a \cos 3bt),$$

where $t$ is time in seconds, $A > 0$, $b > 0$, and $0 < a < 1$.

(a) Show that $F = 0$ when $t = -\pi/(2b)$ and $t = \pi/(2b)$. (The time $t = -\pi/(2b)$ corresponds to the moment when the foot first touches the ground and the weight of the body is being supported by the other foot.)

(b) The maximum force occurs when

$$3a \sin 3bt = \sin bt.$$

If $a = \frac{1}{3}$, find the solutions of this equation for the interval $-\pi/(2b) < t < \pi/(2b)$.

(c) If $a = \frac{1}{3}$, express the maximum force in terms of $A$.

**EXERCISE 76**

77  Shown in the figure is a graph of the equation

$$y = \sin x - \frac{1}{2} \sin 2x + \frac{1}{3} \sin 3x.$$

The $x$-coordinates of the turning points are solutions of the equation $\cos x - \cos 2x + \cos 3x = 0$. Use a sum-to-product formula to find these $x$-coordinates.

EXERCISE 77

78 **Visual distinction**   The human eye can distinguish between two distant points $P$ and $Q$ provided the angle of
resolution $\theta$ is not too small. Suppose $P$ and $Q$ are $x$
units apart and are $d$ units from the eye, as illustrated in
the figure.

(a) Express $x$ in terms of $d$ and $\theta$.

(b) For a person with normal vision, the smallest distinguishable angle of resolution is about 0.0005 radian.
If a pen 6 inches long is viewed by such an individual at a distance of $d$ feet, for what values of $d$ will
the end points of the pens be distinguishable?

EXERCISE 78

79 **Satellites**   A satellite $S$ circles a planet at a distance $d$
miles from the planet's surface. The portion of the
planet's surface that is visible from the satellite is determined by the angle $\theta$ indicated in the figure.

(a) Assuming that the planet is spherical in shape, express $d$ in terms of $\theta$ and the radius $r$ of the planet.

(b) Approximate $\theta$ for a satellite 300 miles from the
surface of the earth, using $r = 4000$ mi.

EXERCISE 79

80 **Urban canyons**   Because of the tall buildings and relatively narrow streets in some inner cities, the amount of
sunlight illuminating these "canyons" is greatly reduced. If $h$ is the average height of the buildings and $w$
is the width of the street, the narrowness $N$ of the street
is defined by $N = h/w$. The angle $\theta$ of the horizon is
defined by $\tan \theta = N$. (The value $\theta = 63°$ may result in
an 85% loss of illumination.) Approximate the angle of
the horizon for the following values of $h$ and $w$.

(a) $h = 400$ ft,   $w = 80$ ft

(b) $h = 55$ m,   $w = 30$ m

*Trigonometric functions and vectors may be used to specify the position of parts of robotic arms.*

# CHAPTER 7

■ In the first two sections of this chapter we consider methods of solving oblique triangles using the law of sines and the law of cosines. We next introduce the trigonometric form for complex numbers and use it to find all $n$ solutions of equations of the form $w^n = z$, where $n$ is any positive integer and $w$ and $z$ are complex numbers. The last two sections contain an introduction to vectors, a topic that has many applications in engineering, the natural sciences, and advanced mathematics. ■

# APPLICATIONS

# OF

# TRIGONOMETRY

# 7.1    THE LAW OF SINES

FIGURE I

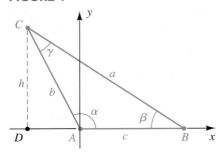

An **oblique triangle** is a triangle that does not contain a right angle. We shall use the letters $A$, $B$, $C$, $a$, $b$, $c$, $\alpha$, $\beta$, and $\gamma$ for parts of triangles as we did in Chapter 5. Given triangle $ABC$, let us place angle $\alpha$ in standard position so that $B$ is on the positive $x$-axis. The case for $\alpha$ obtuse is illustrated in Figure 1; however, the following discussion is also valid if $\alpha$ is acute.

Consider the line through $C$ parallel to the $y$-axis and intersecting the $x$-axis at point $D$. If we let $d(C, D) = h$, then the $y$-coordinate of $C$ is $h$. From the definitions of trigonometric functions of any angle,

$$\sin \alpha = \frac{h}{b}, \quad \text{so} \quad h = b \sin \alpha.$$

Referring to right triangle $BDC$, we see that

$$\sin \beta = \frac{h}{a}, \quad \text{so} \quad h = a \sin \beta.$$

Consequently,    $b \sin \alpha = a \sin \beta,$

which we may write as    $$\frac{\sin \alpha}{a} = \frac{\sin \beta}{b}.$$

If we place $\alpha$ in standard position with $C$ on the positive $x$-axis, then by the same reasoning,

$$\frac{\sin \alpha}{a} = \frac{\sin \gamma}{c}.$$

The last two equalities give us the following result.

**The Law of Sines**

> If $ABC$ is an oblique triangle labeled in the usual manner, then
>
> $$\frac{\sin \alpha}{a} = \frac{\sin \beta}{b} = \frac{\sin \gamma}{c}.$$

Note that the law of sines consists of the following three formulas:

(1) $\dfrac{\sin \alpha}{a} = \dfrac{\sin \beta}{b}$    (2) $\dfrac{\sin \alpha}{a} = \dfrac{\sin \gamma}{c}$    (3) $\dfrac{\sin \beta}{b} = \dfrac{\sin \gamma}{c}$

To apply any one of these formulas to a specific triangle, we must know the values of three of the four variables. If we substitute these three values into the appropriate formula, we can then solve for the value of the fourth variable. It follows that the law of sines can be used to find the remaining parts of an oblique triangle whenever we know either of the following (the three letters in parentheses are used to denote the known parts, with S representing a side and A an angle):

**(1)** two sides and an angle *opposite* one of them (SSA)

**(2)** two angles and any side (AAS or ASA)

In the next section we will discuss the law of cosines and show how it can be used to find the remaining parts of an oblique triangle when given the following:

**(1)** two sides and the angle *between* them (SAS)

**(2)** three sides (SSS)

The law of sines cannot be applied directly to the last two cases.
 The law of sines can also be written in the form

$$\frac{a}{\sin \alpha} = \frac{b}{\sin \beta} = \frac{c}{\sin \gamma}.$$

Instead of memorizing the three formulas associated with the law of sines, it may be more convenient to remember the following statement, which takes all of them into account.

**The Law of Sines
(General Form)**

> In any triangle, the ratio of the sine of an angle to the side opposite that angle is equal to the ratio of the sine of another angle to the side opposite that angle.

In examples and exercises involving triangles, we shall assume that known lengths of sides and angles have been obtained by measurement and hence are approximations to exact values. Unless directed otherwise, when finding parts of triangles we shall round off answers according to the following rule: *If known sides or angles are stated to a certain accuracy, then unknown sides or angles should be calculated to the same accuracy.* To illustrate, if known sides are stated to the nearest 0.1, then unknown sides should be calculated to the nearest 0.1. If known angles are stated to the nearest 10′, then unknown angles should be calculated to the nearest 10′. Similar remarks hold for accuracy to the nearest 0.01, 0.1°, and so on.

**EXAMPLE 1**    *Using the law of sines (ASA)*

Given triangle $ABC$ with $\alpha = 48°$, $\gamma = 57°$, and $b = 47$, approximate the remaining parts.

**FIGURE 2**

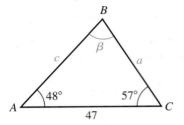

**SOLUTION**    The triangle is sketched in Figure 2. Since the sum of the angles of a triangle is 180°,

$$\beta = 180° - (57° + 48°) = 75°.$$

Since side $b$ and all three angles are known, we can find $a$ by using the formula in the law of sines that involves $a$, $\alpha$, $b$, and $\beta$:

$$\frac{a}{\sin \alpha} = \frac{b}{\sin \beta} \qquad \text{law of sines}$$

$$a = \frac{b \sin \alpha}{\sin \beta} \qquad \text{solve for } a$$

$$= \frac{47 \sin 48°}{\sin 75°} \qquad \text{substitute for } b, \alpha, \text{ and } \beta$$

$$\approx 36 \qquad \text{approximate to the nearest integer}$$

To find $c$, we merely replace $\dfrac{a}{\sin \alpha}$ with $\dfrac{c}{\sin \gamma}$ in the preceding solution for $a$, obtaining

$$c = \frac{b \sin \gamma}{\sin \beta} = \frac{47 \sin 57°}{\sin 75°} \approx 41.$$

**FIGURE 3**

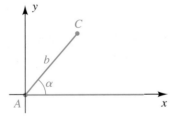

Data such as those in Example 1 lead to exactly one triangle $ABC$. However, if two sides and an angle *opposite* one of them are given, a unique triangle is not always determined. To illustrate, suppose that $a$ and $b$ are to be lengths of sides of triangle $ABC$ and that a given angle $\alpha$ is to be opposite the side of length $a$. Let us examine the case for $\alpha$ acute. Place $\alpha$ in standard position and consider the line segment $AC$ of length $b$ on the terminal side of $\alpha$, as shown in Figure 3. The third vertex, $B$, should be somewhere on the $x$-axis. Since the length $a$ of the side opposite $\alpha$ is given, we may find $B$ by striking off a circular arc of length $a$ with center at $C$. The four possible outcomes are illustrated in Figure 4 (without the coordinate axes).

**FIGURE 4**

(a)    (b)    (c)    (d)

The four possibilities in the figure may be described as follows:

**(a)** The arc does not intersect the *x*-axis and no triangle is formed.

**(b)** The arc is tangent to the *x*-axis and a right triangle is formed.

**(c)** The arc intersects the positive *x*-axis in two distinct points and two triangles are formed.

**(d)** The arc intersects both the positive and nonpositive parts of the *x*-axis and one triangle is formed.

The particular case that occurs in a given problem will become evident when the solution is attempted. For example, if we solve the equation

$$\frac{\sin \alpha}{a} = \frac{\sin \beta}{b}$$

and obtain $\sin \beta > 1$, then no triangle exists and we have case (a). If we obtain $\sin \beta = 1$, then $\beta = 90°$ and hence (b) occurs. If $\sin \beta < 1$, then there are two possible choices for the angle $\beta$. By checking both possibilities, we may determine whether (c) or (d) occurs.

If the measure of $\alpha$ is greater than $90°$, then a triangle exists if and only if $a > b$ (see Figure 5). Since we may have more than one possibility when two sides and an angle opposite one of them are given, this situation is sometimes called the **ambiguous case.**

**FIGURE 5**

**(a)** $a < b$

**(b)** $a > b$

### EXAMPLE 2 *Using the law of sines (SSA)*

Approximate the remaining parts of triangle *ABC* with $\alpha = 67°$, $a = 100$, and $c = 125$.

**SOLUTION** Since we know $\alpha$, $a$, and $c$, we can find $\gamma$ by using the formula in the law of sines that involves $a$, $\alpha$, $c$, and $\gamma$:

$$\frac{\sin \gamma}{c} = \frac{\sin \alpha}{a} \qquad \text{law of sines}$$

$$\sin \gamma = \frac{c \sin \alpha}{a} \qquad \text{solve for } \sin \gamma$$

$$= \frac{125 \sin 67°}{100} \qquad \text{substitute for } c, \alpha, \text{ and } a$$

$$\approx 1.1506 \qquad \text{approximate}$$

Since $\sin \gamma > 1$, no triangle can be constructed with the given parts.

### EXAMPLE 3 *Using the law of sines (SSA)*

Approximate the remaining parts of triangle *ABC* with $a = 12.4$, $b = 8.7$, and $\beta = 36.7°$.

**SOLUTION**    To find $\alpha$, we proceed as follows:

$$\frac{\sin \alpha}{a} = \frac{\sin \beta}{b} \qquad \text{law of sines}$$

$$\sin \alpha = \frac{a \sin \beta}{b} \qquad \text{solve for } \sin \alpha$$

$$= \frac{12.4 \sin 36.7°}{8.7} \qquad \text{substitute for } a, \beta, \text{ and } b$$

$$\approx 0.8518 \qquad \text{approximate}$$

There are two possible angles $\alpha$ between $0°$ and $180°$ such that $\sin \alpha$ is approximately 0.8518. The reference angle $\alpha_R$ is

$$\alpha_R \approx \sin^{-1}(0.8518) \approx 58.4°.$$

Consequently, the two possibilities for $\alpha$ are

$$\alpha_1 \approx 58.4° \quad \text{and} \quad \alpha_2 = 180° - \alpha_1 \approx 121.6°.$$

**FIGURE 6**

The angle $\alpha_1 \approx 58.4°$ gives us triangle $A_1BC$ in Figure 6, and the angle $\alpha_2 \approx 121.6°$ gives us triangle $A_2BC$.

If we let $\gamma_1$ and $\gamma_2$ denote the third angles of the triangles $A_1BC$ and $A_2BC$ corresponding to the angles $\alpha_1$ and $\alpha_2$, respectively, then

$$\gamma_1 = 180° - (\alpha_1 + \beta) \approx 180° - (58.4° + 36.7°) \approx 84.9°$$
$$\gamma_2 = 180° - (\alpha_2 + \beta) \approx 180° - (121.6° + 36.7°) \approx 21.7°.$$

If $c_1$ is the side opposite $\gamma_1$ in triangle $A_1BC$, then

$$\frac{c_1}{\sin \gamma_1} = \frac{a}{\sin \alpha_1} \qquad \text{law of sines}$$

$$c_1 = \frac{a \sin \gamma_1}{\sin \alpha_1} \qquad \text{solve for } c_1$$

$$\approx \frac{12.4 \sin 84.9°}{\sin 58.4°} \approx 14.5. \qquad \text{substitute and approximate}$$

Thus, the remaining parts of triangle $A_1BC$ are

$$\alpha_1 \approx 58.4°, \quad \gamma_1 \approx 84.9°, \quad \text{and} \quad c_1 \approx 14.5.$$

Similarly, if $c_2$ is the side opposite $\gamma_2$ in triangle $A_2BC$, then

$$c_2 = \frac{a \sin \gamma_2}{\sin \alpha_2} \approx \frac{12.4 \sin 21.7°}{\sin 121.6°} \approx 5.4,$$

and the remaining parts of triangle $A_2BC$ are

$$\alpha_2 \approx 121.6°, \quad \gamma_2 \approx 21.7°, \quad \text{and} \quad c_2 \approx 5.4.$$

### EXAMPLE 4     *Using an angle of elevation*

When the angle of elevation of the sun is 64°, a telephone pole that is tilted at an angle of 9° directly away from the sun casts a shadow 21 feet long on level ground. Approximate the length of the pole.

**SOLUTION**    The problem is illustrated in Figure 7 (not to scale). Triangle $ABC$ in Figure 8 also displays the given facts. Either drawing is sufficient for our purpose. Note that in Figure 8 we have calculated the following angles:

$$\beta = 90° - 9° = 81°$$
$$\gamma = 180° - (64° + 81°) = 35°$$

**FIGURE 7**

**FIGURE 8**

To find the length of the pole—that is, side $a$ of triangle $ABC$—we proceed as follows:

$$\frac{a}{\sin 64°} = \frac{21}{\sin 35°} \qquad \text{law of sines}$$

$$a = \frac{21 \sin 64°}{\sin 35°} \approx 33 \qquad \text{solve for } a \text{ and approximate}$$

Thus, the telephone pole is approximately 33 feet in length.

### EXAMPLE 5     *Using bearings*

A point $P$ on level ground is 3.0 kilometers due north of a point $Q$. A runner proceeds in the direction N25°E from $Q$ to a point $R$, and then from $R$ to $P$ in the direction S70°W. Approximate the distance run.

**FIGURE 9**

**FIGURE 10**

**SOLUTION**   The notation used to specify directions was introduced in Section 5.7. The arrows in Figure 9 show the path of the runner, together with a north-south (dashed) line from $R$ to another point $S$.

Since the lines through $PQ$ and $RS$ are parallel, it follows from geometry that the alternate interior angles $PQR$ and $QRS$ both have measure $25°$. Hence,

$$\angle PRQ = 70° - 25° = 45°.$$

These observations give us triangle $PQR$ in Figure 10 with

$$\angle QPR = 180° - (25° + 45°) = 110°.$$

We apply the law of sines twice:

$$\frac{q}{\sin 25°} = \frac{3.0}{\sin 45°} \quad \text{and} \quad \frac{p}{\sin 110°} = \frac{3.0}{\sin 45°}$$

Hence,

$$q = \frac{3.0 \sin 25°}{\sin 45°} \approx 1.8,$$

$$p = \frac{3.0 \sin 110°}{\sin 45°} \approx 4.0.$$

The distance run, $p + q$, is approximately $1.8 + 4.0 = 5.8$ km.

**EXAMPLE 6**   *Locating a school of fish*

A commercial fishing boat uses sonar equipment to detect a school of fish 2 miles east of the boat and traveling in the direction N51°W at a rate of 8 mi/hr (see Figure 11).

**FIGURE 11**

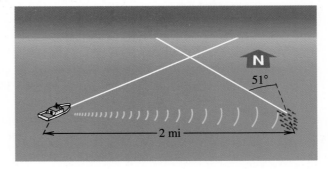

**(a)** If the boat travels at 20 mi/hr, approximate, to the nearest 0.1°, the direction it should head to intercept the school.

**(b)** Find, to the nearest minute, the time it will take the boat to reach the fish.

**FIGURE 12**

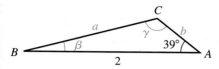

**SOLUTION**

**(a)** The problem is illustrated by the triangle in Figure 12, with the school of fish at $A$, the boat at $B$, and the point of interception at $C$. Note that $\alpha = 90° - 51° = 39°$. To obtain $\beta$, we begin as follows:

$$\frac{\sin \beta}{b} = \frac{\sin 39°}{a} \qquad \text{law of sines}$$

$$\sin \beta = \frac{b}{a} \sin 39° \qquad \text{solve for } \sin \beta$$

We next find $b/a$, letting $t$ denote the time at which the boat and fish meet at $C$:

$$a = 20t, \qquad b = 8t \qquad \text{(distance) = (rate)(time)}$$

$$\frac{b}{a} = \frac{8t}{20t} = \frac{2}{5} \qquad \text{divide } b \text{ by } a$$

$$\sin \beta = \tfrac{2}{5} \sin 39° \qquad \text{substitute for } b/a$$

$$\beta = \sin^{-1} \left( \tfrac{2}{5} \sin 39° \right) \approx 14.6° \qquad \text{approximate}$$

Since $90° - 14.6° = 75.4°$, the boat should travel in the (approximate) direction N75.4°E.

**(b)** Let us first find the distance $a$ from $B$ to $C$. We begin by noting that

$$\gamma \approx 180° - (39° + 14.6°) = 126.4°.$$

Next we have

$$\frac{a}{\sin \alpha} = \frac{c}{\sin \gamma} \qquad \text{law of sines}$$

$$a = \frac{c \sin \alpha}{\sin \gamma} \qquad \text{solve for } a$$

$$\approx \frac{2 \sin 39°}{\sin 126.4°} \approx 1.56 \text{ mi.} \qquad \text{substitute and approximate}$$

Using $a = 20t$, we find the time $t$ for the boat to reach $C$:

$$t = \frac{a}{20} \approx \frac{1.56}{20} \approx 0.08 \text{ hr} \approx 5 \text{ min}$$

## 7.1 EXERCISES

**Exer. 1–16:** Approximate the remaining parts of triangle *ABC*.

| | | |
|---|---|---|
| 1  $\alpha = 41°$, | $\gamma = 77°$, | $a = 10.5$ |
| 2  $\beta = 20°$, | $\gamma = 31°$, | $b = 210$ |

| | | |
|---|---|---|
| 3  $\alpha = 27°40'$, | $\beta = 52°10'$, | $a = 32.4$ |
| 4  $\beta = 50°50'$, | $\gamma = 70°30'$, | $c = 537$ |
| 5  $\alpha = 42°10'$, | $\gamma = 61°20'$, | $b = 19.7$ |
| 6  $\alpha = 103.45°$, | $\gamma = 27.19°$, | $b = 38.84$ |

7   $\gamma = 81°$,        $c = 11$,        $b = 12$

8   $\alpha = 32.32°$,     $c = 574.3$,     $a = 263.6$

9   $\gamma = 53°20'$,     $a = 140$,       $c = 115$

10  $\alpha = 27°30'$,     $c = 52.8$,      $a = 28.1$

11  $\gamma = 47.74°$,     $a = 131.08$,    $c = 97.84$

12  $\alpha = 42.17°$,     $a = 5.01$,      $b = 6.12$

13  $\alpha = 65°10'$,     $a = 21.3$,      $b = 18.9$

14  $\beta = 113°10'$,     $b = 248$,       $c = 195$

15  $\beta = 121.624°$,    $b = 0.283$,     $c = 0.178$

16  $\gamma = 73.01°$,     $a = 17.31$,     $c = 20.24$

17  **Surveying**   To find the distance between two points $A$ and $B$ that lie on opposite banks of a river, a surveyor lays off a line segment $AC$ of length 240 yards along one bank and determines that the measures of $\angle BAC$ and $\angle ACB$ are $63°20'$ and $54°10'$, respectively (see the figure). Approximate the distance between $A$ and $B$.

EXERCISE 17

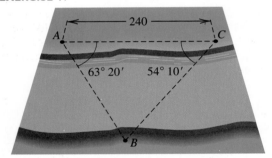

18  **Surveying**   To determine the distance between two points $A$ and $B$, a surveyor chooses a point $C$ that is 375 yards from $A$ and 530 yards from $B$. If $\angle BAC$ has measure $49°30'$, approximate the distance between $A$ and $B$.

19  **Cable car route**   As shown in the figure, a cable car carries passengers from a point $A$, which is 1.2 miles

EXERCISE 19

from a point $B$ at the base of a mountain, to a point $P$ at the top of the mountain. The angles of elevation of $P$ from $A$ and $B$ are $21°$ and $65°$, respectively.

(a) Approximate the distance between $A$ and $P$.

(b) Approximate the height of the mountain.

20  **Length of a shadow**   A straight road makes an angle of $15°$ with the horizontal. When the angle of elevation of the sun is $57°$, a vertical pole at the side of the road casts a shadow 75 feet long directly down the road, as shown in the figure. Approximate the length of the pole.

EXERCISE 20

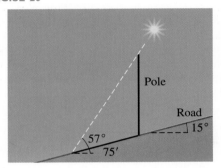

21  **Height of a hot-air balloon**   The angles of elevation of a balloon from two points $A$ and $B$ on level ground are $24°10'$ and $47°40'$, respectively. As shown in the figure, points $A$ and $B$ are 8.4 miles apart and the balloon is between the points, in the same vertical plane. Approximate the height of the balloon above the ground.

EXERCISE 21

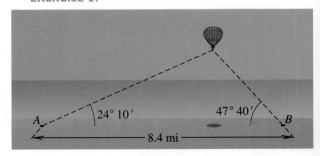

22  **Installing a solar panel**   Shown in the figure is a solar panel 10 feet in width, which is to be attached to a roof that makes an angle of $25°$ with the horizontal. Approximate the length $d$ of the brace that is needed for the panel to make an angle of $45°$ with the horizontal.

EXERCISE 22

23 **Distance to an airplane** A straight road makes an angle of 22° with the horizontal. From a certain point $P$ on the road, the angle of elevation of an airplane at point $A$ is 57°. At the same instant, from another point $Q$, 100 meters farther up the road, the angle of elevation is 63°. As indicated in the figure, the points $P$, $Q$, and $A$ lie in the same vertical plane. Approximate the distance from $P$ to the airplane.

EXERCISE 23

24 **Surveying** A surveyor notes that the direction from point $A$ to point $B$ is S63°W and the direction from $A$ to $C$ is S38°W. The distance from $A$ to $B$ is 239 yards, and the distance from $B$ to $C$ is 374 yards. Approximate the distance from $A$ to $C$.

25 **Sighting a forest fire** A forest ranger at an observation point $A$ sights a fire in the direction N27°10′E. Another ranger at an observation point $B$, 6.0 miles due east of $A$, sights the same fire at N52°40′W. Approximate the distance from each of the observation points to the fire.

26 **Leaning tower of Pisa** The leaning tower of Pisa was originally perpendicular to the ground and 179 feet tall. Because of sinking into the earth, it now leans at a certain angle $\theta$ from the perpendicular, as shown in the figure. When the top of the tower is viewed from a point 150 feet from the center of its base, the angle of elevation is 53.3°.

(a) Approximate the angle $\theta$.

(b) Approximate the distance $d$ that the center of the top of the tower has moved from the perpendicular.

EXERCISE 26

27 **Height of a cathedral** A cathedral is located on a hill, as shown in the figure. When the top of the spire is viewed from the base of the hill, the angle of elevation is 48°. When it is viewed at a distance of 200 feet from the base of the hill, the angle of elevation is 41°. The hill rises at an angle of 32°. Approximate the height of the cathedral.

EXERCISE 27

28 **Sighting from a helicopter** A helicopter hovers at an altitude that is 1000 feet above a mountain peak of altitude 5210 feet, as indicated in the figure on page 476. A second, taller peak is viewed from both the mountaintop and the helicopter. From the helicopter, the angle of depression is 43°, and from the mountaintop, the angle of elevation is 18°.

(a) Approximate the distance from peak to peak.

(b) Approximate the altitude of the taller peak.

**EXERCISE 28**

29 The volume $V$ of the right triangular prism shown in the figure is $\frac{1}{3}Bh$, where $B$ is the area of the base and $h$ is the height of the prism. Approximate

(a) $h$    (b) $V$

**EXERCISE 29**

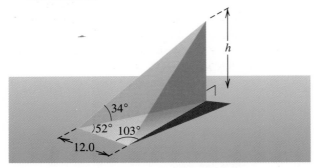

↖ 30 **Design for a jet fighter**  Shown in the figure is a plan for the top of a wing of a jet fighter.

(a) Approximate angle $\phi$.

(b) If the fuselage is 4.80 feet wide, approximate the wing span $CC'$.

(c) Approximate the area of triangle $ABC$.

**EXERCISE 30**

31 **Software for surveyors**  Computer software for surveyors makes use of coordinate systems to locate geographic positions. An offshore oil well at point $R$ in the figure is viewed from points $P$ and $Q$, and $\angle QPR$ and $\angle RQP$ are found to be $55°50'$ and $65°22'$, respectively. If points $P$ and $Q$ have coordinates (1487.7, 3452.8) and (3145.8, 5127.5), respectively, approximate the coordinates of $R$.

**EXERCISE 31**

32 **Mollweide's formula**  The following equation, called Mollweide's formula, is sometimes used to check solutions to triangles because it involves all the angles and sides:

$$\frac{a+b}{c} = \frac{\cos\frac{1}{2}(\alpha - \beta)}{\sin\frac{1}{2}\gamma}$$

(a) Use the law of sines to show that

$$\frac{a+b}{c} = \frac{\sin\alpha + \sin\beta}{\sin\gamma}.$$

(b) Use a sum-to-product formula and a double-angle formula to verify Mollweide's formula.

# 7.2 THE LAW OF COSINES

In the preceding section we stated that the law of sines cannot be applied directly to find the remaining parts of an oblique triangle when given either of the following two cases:

(1) two sides and the angle *between* them (SAS)

(2) three sides (SSS)

For these cases we may apply the *law of cosines,* which follows.

**The Law of Cosines**

If *ABC* is a triangle labeled in the usual manner, then

(1) $a^2 = b^2 + c^2 - 2bc \cos \alpha$

(2) $b^2 = a^2 + c^2 - 2ac \cos \beta$

(3) $c^2 = a^2 + b^2 - 2ab \cos \gamma$

**FIGURE 13**

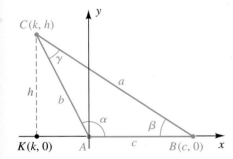

*PROOF* Let us prove the first formula. Given triangle *ABC*, place $\alpha$ in standard position, as illustrated in Figure 13. We have pictured $\alpha$ as obtuse; however, our discussion is also valid if $\alpha$ is acute. Consider the dashed line through *C*, parallel to the *y*-axis and intersecting the *x*-axis at the point $K(k, 0)$. If we let $d(C, K) = h$, then *C* has coordinates $(k, h)$. By the definition of trigonometric functions of any angle,

$$\cos \alpha = \frac{k}{b} \quad \text{and} \quad \sin \alpha = \frac{h}{b}.$$

Solving for *k* and *h* gives us

$$k = b \cos \alpha \quad \text{and} \quad h = b \sin \alpha.$$

Since the segment *AB* has length *c*, the coordinates of *B* are $(c, 0)$, and we obtain the following:

$$
\begin{aligned}
a^2 = [d(B, C)]^2 &= (k - c)^2 + (h - 0)^2 && \text{distance formula} \\
&= (b \cos \alpha - c)^2 + (b \sin \alpha)^2 && \text{substitute for } k \text{ and } h \\
&= b^2 \cos^2 \alpha - 2bc \cos \alpha + c^2 + b^2 \sin^2 \alpha && \text{square} \\
&= b^2(\cos^2 \alpha + \sin^2 \alpha) + c^2 - 2bc \cos \alpha && \text{factor the first and last terms} \\
&= b^2 + c^2 - 2bc \cos \alpha && \text{Pythagorean identity}
\end{aligned}
$$

Our result is the first formula stated in the law of cosines. The second and third formulas may be obtained by placing $\beta$ and $\gamma$, respectively, in standard position on a coordinate system. ∎

Note that if $\alpha = 90°$ in Figure 13, then $\cos \alpha = 0$ and the law of cosines reduces to $a^2 = b^2 + c^2$. This shows that the Pythagorean theorem is a special case of the law of cosines.

Instead of memorizing each of the three formulas of the law of cosines, it is more convenient to remember the following statement, which takes all of them into account.

**The Law of Cosines (General Form)**

> The square of the length of any side of a triangle equals the sum of the squares of the lengths of the other two sides minus twice the product of the lengths of the other two sides and the cosine of the angle between them.

Given two sides and the included angle of a triangle, we can use the law of cosines to find the third side. We may then use the law of sines to find another angle of the triangle. Whenever this procedure is followed, it is best to find the angle opposite the shortest side, since that angle is always acute. In this way, we avoid the possibility of obtaining two solutions when solving a trigonometric equation involving that angle, as illustrated in the following example. Of course, we can also use the law of cosines to find another angle.

**EXAMPLE 1    *Using the law of cosines (SAS)***

Approximate the remaining parts of triangle $ABC$ if $a = 5.0$, $c = 8.0$, and $\beta = 77°$.

*SOLUTION*    The triangle is sketched in Figure 14. Since $\beta$ is the angle *between* sides $a$ and $c$, we begin by approximating $b$ (the side opposite $\beta$) as follows:

$$b^2 = a^2 + c^2 - 2ac \cos \beta \qquad \text{law of cosines}$$
$$= (5.0)^2 + (8.0)^2 - 2(5.0)(8.0) \cos 77° \qquad \text{substitute for } a, c, \text{ and } \beta$$
$$= 89 - 80 \cos 77° \approx 71.0 \qquad \text{simplify and approximate}$$
$$b \approx \sqrt{71.0} \approx 8.4 \qquad \text{take the square root}$$

Let us find another angle of the triangle using the law of sines. In accordance with the remarks preceding this example, we shall apply the law of sines and find $\alpha$, since it is the angle opposite the shortest side $a$.

$$\frac{\sin \alpha}{a} = \frac{\sin \beta}{b} \qquad \text{law of sines}$$
$$\sin \alpha = \frac{a \sin \beta}{b} \qquad \text{solve for } \sin \alpha$$
$$= \frac{5.0 \sin 77°}{\sqrt{71.0}} \approx 0.5782 \qquad \text{substitute and approximate}$$

**FIGURE 14**

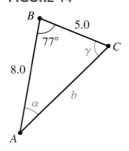

Since $\alpha$ is acute,

$$\alpha = \sin^{-1}(0.5782) \approx 35.3°.$$

Finally, since $\alpha + \beta + \gamma = 180°$, we have

$$\gamma = 180° - (\alpha + \beta) \approx 180° - (35.3° + 77°) = 67.7°.$$

Given the three sides of a triangle, we can use the law of cosines to find *any* of the three angles. We shall always find the largest angle first— that is, *the angle opposite the longest side*—since this practice will guarantee that the remaining angles are acute. We may then find another angle of the triangle by using either the law of sines or the law of cosines. Note that when an angle is found by means of the law of cosines, there is no ambiguous case, since we always obtain a unique angle between $0°$ and $180°$.

**EXAMPLE 2** *Using the law of cosines (SSS)*

If triangle $ABC$ has sides $a = 90$, $b = 70$, and $c = 40$, approximate angles $\alpha$, $\beta$, and $\gamma$ to the nearest degree.

**SOLUTION** In accordance with the remarks preceding this example, we first find the angle opposite the longest side $a$. Thus, we choose the form of the law of cosines that involves $\alpha$ and proceed as follows:

$$a^2 = b^2 + c^2 - 2bc \cos \alpha \qquad \text{law of cosines}$$

$$\cos \alpha = \frac{b^2 + c^2 - a^2}{2bc} \qquad \text{solve for } \cos \alpha$$

$$= \frac{70^2 + 40^2 - 90^2}{2(70)(40)} = -\frac{2}{7} \qquad \text{substitute and simplify}$$

$$\alpha = \cos^{-1}\left(-\tfrac{2}{7}\right) \approx 106.6° \approx 107° \qquad \text{approximate } \alpha$$

We may now use either the law of sines or the law of cosines to find $\beta$. Let us use the law of cosines in this case:

$$b^2 = a^2 + c^2 - 2ac \cos \beta \qquad \text{law of cosines}$$

$$\cos \beta = \frac{a^2 + c^2 - b^2}{2ac} \qquad \text{solve for } \cos \beta$$

$$= \frac{90^2 + 40^2 - 70^2}{2(90)(40)} = \frac{2}{3} \qquad \text{substitute and simplify}$$

$$\beta = \cos^{-1}\left(\tfrac{2}{3}\right) \approx 48.2° \approx 48° \qquad \text{approximate } \beta$$

Lastly, since $\alpha + \beta + \gamma = 180°$, we have

$$\gamma = 180° - (\alpha + \beta) \approx 180° - (107° + 48°) = 25°.$$

### EXAMPLE 3 *Approximating the diagonals of a parallelogram*

A parallelogram has sides of lengths 30 centimeters and 70 centimeters and one angle of measure 65°. Approximate the length of each diagonal to the nearest centimeter.

**SOLUTION**     The parallelogram $ABCD$ and its diagonals $AC$ and $BD$ are shown in Figure 15. Using triangle $ABC$ with $\angle ABC = 65°$, we may approximate $AC$ as follows:

$$(AC)^2 = (30)^2 + (70)^2 - 2(30)(70) \cos 65° \qquad \text{law of cosines}$$
$$\approx 900 + 4900 - 4200(0.4226) \approx 4025.1 \qquad \text{approximate}$$
$$AC \approx \sqrt{4025.1} \approx 63 \text{ cm} \qquad \text{take the square root}$$

Similarly, using triangle $BAD$ and $\angle BAD = 180° - 65° = 115°$, we may approximate $BD$ as follows:

$$(BD)^2 = (30)^2 + (70)^2 - 2(30)(70) \cos 115° \approx 7575.0 \qquad \text{law of cosines}$$
$$BD \approx \sqrt{7575.0} \approx 87 \text{ cm} \qquad \text{take the square root}$$

**FIGURE 15**

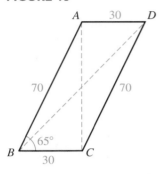

### EXAMPLE 4 *Finding the length of a cable*

A vertical pole 40 feet tall stands on a hillside that makes an angle of 17° with the horizontal. Approximate the minimal length of cable that will reach from the top of the pole to a point 72 feet downhill from the base of the pole.

**SOLUTION**     The sketch in Figure 16 displays the given data. We wish to find $AC$. Referring to the figure, we see that

$$\angle ABD = 90° - 17° = 73° \quad \text{and} \quad \angle ABC = 180° - 73° = 107°.$$

Using triangle $ABC$, we may approximate $AC$ as follows:

$$(AC)^2 = (72)^2 + (40)^2 - 2(72)(40) \cos 107° \approx 8468 \qquad \text{law of cosines}$$
$$AC \approx \sqrt{8468} \approx 92 \text{ ft} \qquad \text{take the square root}$$

**FIGURE 16**

The law of cosines can be used to derive a formula for the area of a triangle. Let us first prove a preliminary result.

Given triangle $ABC$, place angle $\alpha$ in standard position (see Figure 13). As shown in the proof of the law of cosines, the altitude $h$ from vertex $C$ is $h = b \sin \alpha$. Since the area $\mathcal{A}$ of the triangle is given by $\mathcal{A} = \frac{1}{2}ch$, we see that

$$\mathcal{A} = \tfrac{1}{2}bc \sin \alpha.$$

Our argument is independent of the specific angle that is placed in standard position. By taking $\beta$ and $\gamma$ in standard position, we obtain the formulas

$$\mathcal{A} = \tfrac{1}{2}ac \sin \beta \quad \text{and} \quad \mathcal{A} = \tfrac{1}{2}ab \sin \gamma.$$

All three formulas are covered in the following statement.

**Area of a Triangle**

> The area of a triangle equals one-half the product of the lengths of any two sides and the sine of the angle between them.

The next two examples illustrate uses of this result.

**EXAMPLE 5    *Approximating the area of a triangle***

Approximate the area of triangle $ABC$ with $a = 2.20$ cm, $b = 1.30$ cm, and $\gamma = 43.2°$.

**SOLUTION**    Since $\gamma$ is the angle between sides $a$ and $b$, we may use the preceding result directly, as follows:

$$\mathcal{A} = \tfrac{1}{2}ab \sin \gamma \qquad \text{area of a triangle formula}$$
$$= \tfrac{1}{2}(2.20)(1.30) \sin 43.2° \approx 0.98 \text{ cm}^2 \qquad \text{substitute and approximate}$$

**EXAMPLE 6    *Approximating the area of a triangle***

Approximate the area of triangle $ABC$ if $a = 5.0$ cm, $b = 3.0$ cm, and $\alpha = 37°$.

**SOLUTION**    To apply the formula for the area of a triangle, we must use the angle $\gamma$ between sides $a$ and $b$. Since we are given $a$, $b$, and $\alpha$, let us first find $\beta$ as follows:

$$\frac{\sin \beta}{b} = \frac{\sin \alpha}{a} \qquad \text{law of sines}$$

$$\sin \beta = \frac{b \sin \alpha}{a} \qquad \text{solve for } \sin \beta$$

$$= \frac{3.0 \sin 37°}{5.0} \qquad \text{substitute for } b,\ \alpha,\ \text{and } a$$

$$\beta_R = \sin^{-1}\left(\frac{3.0 \sin 37°}{5.0}\right) \approx 21° \qquad \text{reference angle for } \beta$$

$$\beta \approx 21° \quad \text{or} \quad \beta \approx 159° \qquad \beta_R \text{ or } 180° - \beta_R$$

*(continued)*

We reject $\beta \approx 159°$, because then $\alpha + \beta = 196° \geq 180°$. Hence, $\beta \approx 21°$ and

$$\gamma = 180° - (\alpha + \beta) \approx 180° - (37° + 21°) = 122°.$$

Finally, we approximate the area of the triangle as follows:

$$\mathcal{A} = \tfrac{1}{2}ab \sin \gamma \qquad \text{area of a triangle formula}$$
$$\approx \tfrac{1}{2}(5.0)(3.0) \sin 122° \approx 6.4 \text{ cm}^2 \qquad \text{substitute and approximate}$$

We shall use the preceding result for the area of a triangle to derive *Heron's formula,* which expresses the area of a triangle in terms of the lengths of its sides.

**Heron's Formula**

> The area $\mathcal{A}$ of a triangle with sides $a$, $b$, and $c$ is given by
> $$\mathcal{A} = \sqrt{s(s-a)(s-b)(s-c)},$$
> where $s$ is one-half the perimeter; that is, $s = \tfrac{1}{2}(a+b+c)$.

*PROOF*    The following equations are equivalent:

$$\mathcal{A} = \tfrac{1}{2}bc \sin \alpha$$
$$= \sqrt{\tfrac{1}{4}b^2c^2 \sin^2 \alpha}$$
$$= \sqrt{\tfrac{1}{4}b^2c^2(1 - \cos^2 \alpha)}$$
$$= \sqrt{\tfrac{1}{2}bc(1 + \cos \alpha) \cdot \tfrac{1}{2}bc(1 - \cos \alpha)}$$

We shall obtain Heron's formula by replacing the expressions under the radical sign by expressions involving only $a$, $b$, and $c$. We solve formula 1 of the law of cosines for $\cos \alpha$ and then substitute as follows:

$$\frac{1}{2}bc(1 + \cos \alpha) = \frac{1}{2}bc\left(1 + \frac{b^2 + c^2 - a^2}{2bc}\right)$$
$$= \frac{1}{2}bc\left(\frac{2bc + b^2 + c^2 - a^2}{2bc}\right)$$
$$= \frac{2bc + b^2 + c^2 - a^2}{4}$$
$$= \frac{(b+c)^2 - a^2}{4}$$
$$= \frac{(b+c) + a}{2} \cdot \frac{(b+c) - a}{2}$$

We use the same type of manipulations on the second expression under the radical sign:

$$\frac{1}{2}bc(1 - \cos \alpha) = \frac{a - b + c}{2} \cdot \frac{a + b + c}{2}$$

If we now substitute for the expressions under the radical sign, we obtain

$$\mathcal{A} = \sqrt{\frac{b + c + a}{2} \cdot \frac{b + c - a}{2} \cdot \frac{a - b + c}{2} \cdot \frac{a + b - c}{2}}.$$

Letting $s = \frac{1}{2}(a + b + c)$, we see that

$$s - a = \frac{b + c - a}{2}, \quad s - b = \frac{a - b + c}{2}, \quad s - c = \frac{a + b - c}{2}.$$

Substitution in the last formula for $\mathcal{A}$ gives us Heron's formula. ■

### EXAMPLE 7  Using Heron's formula

A triangular field has sides of lengths 125 yards, 160 yards, and 225 yards. Approximate the number of acres in the field. (One acre is equivalent to 4840 square yards.)

SOLUTION  We shall find the area of the field using Heron's formula with $a = 125$, $b = 160$, and $c = 225$:

$$s = \frac{1}{2}(125 + 160 + 225) = \frac{1}{2}(510) = 255$$
$$s - a = 255 - 125 = 130$$
$$s - b = 255 - 160 = 95$$
$$s - c = 255 - 225 = 30$$

Substituting in Heron's formula gives us

$$\mathcal{A} = \sqrt{(255)(130)(95)(30)} \approx 9720 \text{ yd}^2.$$

Since there are 4840 square yards in one acre, the number of acres is $\frac{9720}{4840}$, or approximately 2.

## 7.2 EXERCISES

Exer. 1–10: Approximate the remaining parts of triangle ABC.

1  $\alpha = 60°$, $b = 20$, $c = 30$

2  $\gamma = 45°$, $b = 10.0$, $a = 15.0$

3  $\beta = 150°$, $a = 150$, $c = 30$

4  $\beta = 73°50'$, $c = 14.0$, $a = 87.0$

5  $\gamma = 115°10'$, $a = 1.10$, $b = 2.10$

6  $\alpha = 23°40'$, $c = 4.30$, $b = 70.0$

7  $a = 2.0,$    $b = 3.0,$    $c = 4.0$

8  $a = 10,$    $b = 15,$    $c = 12$

9  $a = 25.0,$    $b = 80.0,$    $c = 60.0$

10  $a = 20.0,$    $b = 20.0,$    $c = 10.0$

11  **Dimensions of a triangular plot**  The angle at one corner of a triangular plot of ground is $73°40'$, and the sides that meet at this corner are 175 feet and 150 feet long. Approximate the length of the third side.

12  **Surveying**  To find the distance between two points $A$ and $B$, a surveyor chooses a point $C$ that is 420 yards from $A$ and 540 yards from $B$. If angle $ACB$ has measure $63°10'$, approximate the distance between $A$ and $B$.

13  **Distance between automobiles**  Two automobiles leave a city at the same time and travel along straight highways that differ in direction by $84°$. If their speeds are 60 mi/hr and 45 mi/hr, respectively, approximately how far apart are the cars at the end of 20 minutes?

14  **Angles of a triangular plot**  A triangular plot of land has sides of lengths 420 feet, 350 feet, and 180 feet. Approximate the smallest angle between the sides.

15  **Distance between ships**  A ship leaves port at 1:00 P.M. and travels S35°E at the rate of 24 mi/hr. Another ship leaves the same port at 1:30 P.M. and travels S20°W at 18 mi/hr. Approximately how far apart are the ships at 3:00 P.M.?

16  **Flight distance**  An airplane flies 165 miles from point $A$ in the direction $130°$ and then travels in the direction $245°$ for 80 miles. Approximately how far is the airplane from $A$?

17  **Jogger's course**  A jogger runs at a constant speed of one mile every 8 minutes in the direction S40°E for 20 minutes and then in the direction N20°E for the next 16 minutes. Approximate, to the nearest tenth of a mile, the distance from the end point to the starting point of the jogger's course.

18  **Surveying**  Two points $P$ and $Q$ on level ground are on opposite sides of a building. To find the distance between the points, a surveyor chooses a point $R$ that is 300 feet from $P$ and 438 feet from $Q$ and then determines that angle $PRQ$ has measure $37°40'$ (see the figure). Approximate the distance between $P$ and $Q$.

**EXERCISE 18**

19  **Motorboat's course**  A motorboat traveled along a triangular course having sides of lengths 2 kilometers, 4 kilometers, and 3 kilometers, respectively. The first side was traversed in the direction N20°W and the second in a direction S$\theta$°W, where $\theta°$ is the degree measure of an acute angle. Approximate, to the nearest minute, the direction in which the third side was traversed.

20  **Angle of a box**  The rectangular box shown in the figure has dimensions $8'' \times 6'' \times 4''$. Approximate the angle $\theta$ formed by a diagonal of the base and a diagonal of the $6'' \times 4''$ side.

**EXERCISE 20**

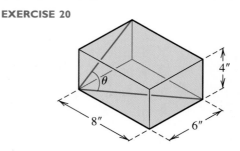

21  **Distances in a baseball diamond**  A baseball diamond has four bases (forming a square) that are 90 feet apart; the pitcher's mound is 60.5 feet from home plate. Approximate the distance from the pitcher's mound to each of the other three bases.

22  A rhombus has sides of length 100 centimeters, and the angle at one of the vertices is $70°$. Approximate the lengths of the diagonals to the nearest tenth of a centimeter.

23  **Reconnaissance**  A reconnaissance airplane $P$, flying at 10,000 feet above a point $R$ on the surface of the water, spots a submarine $S$ at an angle of depression of $37°$ and a tanker $T$ at an angle of depression of $21°$, as

shown in the figure. In addition, $\angle SPT$ is found to be 110°. Approximate the distance between the submarine and the tanker.

**EXERCISE 23**

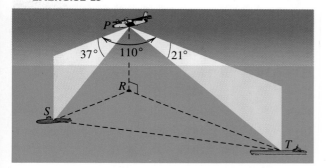

24 **Correcting a ship's course**   A cruise ship sets a course N47°E from an island to a port on the mainland, which is 150 miles away. After moving through strong currents, the ship is off course at a position $P$ that is N33°E and 80 miles from the island, as illustrated in the figure.

(a) Approximately how far is the ship from the port?

(b) In what direction should the ship head to correct its course?

**EXERCISE 24**

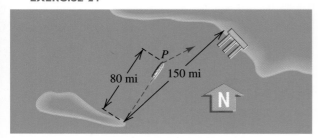

25 **Seismology**   Seismologists investigate the structure of the earth's interior by analyzing seismic waves caused by earthquakes. If the interior of the earth is assumed to be homogeneous, then these waves will travel in straight lines at a constant velocity $v$. The figure shows a cross-sectional view of the earth, with the epicenter at $E$ and an observation station at $S$. Use the law of cosines to show that the time $t$ for a wave to travel through the earth's interior from $E$ to $S$ is given by

$$t = \frac{2R}{v} \sin \frac{\theta}{2},$$

where $R$ is the radius of the earth and $\theta$ is the indicated angle with vertex at the center of the earth.

**EXERCISE 25**

26 **Calculating distances**   The distance across the river shown in the figure can be found without measuring angles. Two points $B$ and $C$ on the opposite shore are selected, and line segments $AB$ and $AC$ are extended as shown. Points $D$ and $E$ are chosen as indicated, and distances $BC$, $BD$, $BE$, $CD$, and $CE$ are then measured. Suppose that $BC = 184$ ft, $BD = 102$ ft, $BE = 218$ ft, $CD = 236$ ft, and $CE = 80$ ft.

(a) Approximate the distances $AB$ and $AC$.

(b) Approximate the shortest distance across the river from point $A$.

**EXERCISE 26**

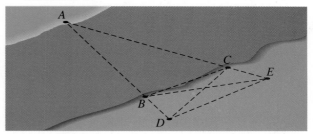

27 **Penrose tiles**   Penrose tiles are formed from a rhombus $ABCD$ having sides of length 1 and an interior angle of 72°. First a point $P$ is located that lies on the diagonal

**EXERCISE 27**

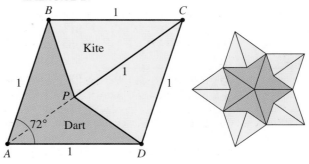

*AC* and is a distance 1 from vertex *C*, and then segments *PB* and *PD* are drawn to the other vertices of the diagonal, as shown in the figure. The two tiles formed are called a dart and a kite. Three-dimensional counterparts of these tiles have been applied in molecular chemistry.

(a) Find the degree measures of ∠*BPC*, ∠*APB*, and ∠*ABP*.

(b) Approximate, to the nearest 0.01, the length of segment *BP*.

(c) Approximate, to the nearest 0.01, the area of a kite and the area of a dart.

28 *Automotive design*  The rear hatchback door of an automobile is 42 inches long. A strut with a fully extended length of 24 inches is to be attached to the door and the body of the car so that when the door is opened completely, the strut is vertical and the rear clearance is 32 inches, as shown in the figure. Approximate the lengths of segments *TQ* and *TP*.

**EXERCISE 28**

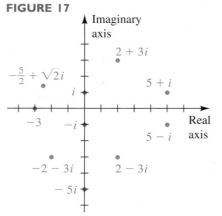

**Exer. 29–36: Approximate the area of triangle *ABC*.**

29 $\alpha = 60°$,     $b = 20$,        $c = 30$

30 $\gamma = 45°$,     $b = 10.0$,     $a = 15.0$

31 $\alpha = 40.3°$,   $\beta = 62.9°$,  $b = 5.63$

32 $\alpha = 35.7°$,   $\gamma = 105.2°$,  $b = 17.2$

33 $\alpha = 80.1°$,   $a = 8.0$,       $b = 3.4$

34 $\gamma = 32.1°$,   $a = 14.6$,     $c = 15.8$

35 $a = 25.0$,     $b = 80.0$,     $c = 60.0$

36 $a = 20.0$,     $b = 20.0$,     $c = 10.0$

**Exer. 37–38: A triangular field has sides of lengths *a*, *b*, and *c* (in yards). Approximate the number of acres in the field (1 acre = 4840 yd²).**

37 $a = 115$,     $b = 140$,       $c = 200$

38 $a = 320$,     $b = 350$,       $c = 500$

**Exer. 39–40: Approximate the area of a parallelogram that has sides of lengths *a* and *b* (in feet) if one angle at a vertex has measure *θ*.**

39 $a = 12.0$,     $b = 16.0$,     $\theta = 40°$

40 $a = 40.3$,     $b = 52.6$,     $\theta = 100°$

## 7.3  TRIGONOMETRIC FORM FOR COMPLEX NUMBERS

**FIGURE 17**

In Section 1.1 we represented real numbers by points on a coordinate line. We can obtain geometric representations for complex numbers by using points in a coordinate plane. Specifically, each complex number $a + bi$ determines a unique ordered pair $(a, b)$. The corresponding point $P(a, b)$ in a coordinate plane is the **geometric representation** of $a + bi$. To emphasize that we are assigning complex numbers to points in a plane, we may label the point $P(a, b)$ as $a + bi$. A coordinate plane with a complex number assigned to each point is referred to as a **complex plane** instead of an *xy*-plane. The *x*-axis is the **real axis** and the *y*-axis the **imaginary axis.** In Figure 17 we have represented several complex numbers geometrically. Note that to obtain the point corresponding to the conjugate $a - bi$ of any complex number $a + bi$, we simply reflect through the real axis.

The absolute value $|a|$ of a real number $a$ is the distance between the origin and the point on the *x*-axis that corresponds to $a$. Thus, it is natural to interpret the absolute value $|a + bi|$ of a complex number as the distance $\sqrt{a^2 + b^2}$ between the origin of a complex plane and the point $(a, b)$ that corresponds to $a + bi$.

**Definition of the Absolute Value of a Complex Number**

The **absolute value** $|a + bi|$ **of a complex number** $a + bi$ is
$$|a + bi| = \sqrt{a^2 + b^2}.$$

**EXAMPLE 1**   *Finding the absolute value of a complex number*

Find

**(a)** $|2 - 6i|$     **(b)** $|3i|$

**SOLUTION**   We use the previous definition:

**(a)**            $|2 - 6i| = \sqrt{2^2 + (-6)^2} = \sqrt{40} = 2\sqrt{10} \approx 6.3$

**(b)**            $|3i| = |0 + 3i| = \sqrt{0^2 + 3^2} = \sqrt{9} = 3$

---

The points corresponding to all complex numbers that have a fixed absolute value $k$ are on a circle of radius $k$ with center at the origin in the complex plane. For example, the points corresponding to the complex numbers $z$ with $|z| = 1$ are on a unit circle.

Let us consider a nonzero complex number $z = a + bi$ and its geometric representation $P(a, b)$, as illustrated in Figure 18. Let $\theta$ be any angle in standard position whose terminal side lies on the segment $OP$, and let $r = |z| = \sqrt{a^2 + b^2}$. Since $\cos \theta = a/r$ and $\sin \theta = b/r$, we see that $a = r \cos \theta$ and $b = r \sin \theta$. Substituting for $a$ and $b$ in $z = a + bi$, we obtain

$$z = a + bi = (r \cos \theta) + (r \sin \theta)i = r(\cos \theta + i \sin \theta).$$

This expression is called the **trigonometric form for the complex number** $a + bi$. A common abbreviation is

$$r(\cos \theta + i \sin \theta) = r \operatorname{cis} \theta.$$

The trigonometric form for $z = a + bi$ is not unique, since there are an unlimited number of different choices for the angle $\theta$. When the trigonometric form is used, the absolute value $r$ of $z$ is sometimes referred to as the **modulus** of $z$ and an angle $\theta$ associated with $z$ as an **argument** (or **amplitude**) of $z$.

We may summarize our discussion as follows.

**FIGURE 18**
$z = a + bi = r(\cos \theta + i \sin \theta)$

**Trigonometric Form for** $a + bi$

Let $z = a + bi$. If $r = |z| = \sqrt{a^2 + b^2}$ and if $\theta$ is an argument of $z$, then
$$z = r(\cos \theta + i \sin \theta) = r \operatorname{cis} \theta.$$

**EXAMPLE 2**    *Expressing a complex number in trigonometric form*

Express the complex number in trigonometric form with $0 \le \theta < 2\pi$:

**(a)** $-4 + 4i$    **(b)** $2\sqrt{3} - 2i$    **(c)** $2 + 7i$    **(d)** $-2 + 7i$

**SOLUTION**    We begin by representing each complex number geometrically and labeling its modulus $r$ and argument $\theta$, as in Figure 19.

**FIGURE 19**

**(a)**

**(b)**

**(c)**

**(d)**
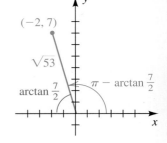

We next substitute for $r$ and $\theta$ in the trigonometric form:

**(a)**    $-4 + 4i = 4\sqrt{2}\left(\cos\dfrac{3\pi}{4} + i\sin\dfrac{3\pi}{4}\right) = 4\sqrt{2}\,\text{cis}\,\dfrac{3\pi}{4}$

**(b)**    $2\sqrt{3} - 2i = 4\left(\cos\dfrac{11\pi}{6} + i\sin\dfrac{11\pi}{6}\right) = 4\,\text{cis}\,\dfrac{11\pi}{6}$

**(c)** $2 + 7i = \sqrt{53}\left[\cos\left(\arctan\tfrac{7}{2}\right) + i\sin\left(\arctan\tfrac{7}{2}\right)\right] = \sqrt{53}\,\text{cis}\,\left(\arctan\tfrac{7}{2}\right)$

**(d)**    $-2 + 7i = \sqrt{53}\left[\cos\left(\pi - \arctan\tfrac{7}{2}\right) + i\sin\left(\pi - \arctan\tfrac{7}{2}\right)\right]$
$= \sqrt{53}\,\text{cis}\,\left(\pi - \arctan\tfrac{7}{2}\right)$

---

If we allow arbitrary values for $\theta$, there are many other trigonometric forms for the complex numbers in Example 2. Thus, for $-4 + 4i$ in part (a) we could use

$$\theta = \frac{3\pi}{4} + 2\pi n \quad \text{for any integer } n.$$

If, for example, we let $n = 1$ and $n = -1$, we obtain

$$-4 + 4i = 4\sqrt{2}\left(\cos\frac{11\pi}{4} + i\sin\frac{11\pi}{4}\right) = 4\sqrt{2}\,\text{cis}\,\frac{11\pi}{4}$$

$$-4 + 4i = 4\sqrt{2}\left[\cos\left(-\frac{5\pi}{4}\right) + i\sin\left(-\frac{5\pi}{4}\right)\right] = 4\sqrt{2}\,\text{cis}\,\left(-\frac{5\pi}{4}\right).$$

In general, *arguments for the same complex number always differ by a multiple of $2\pi$.*

If complex numbers are expressed in trigonometric form, then multiplication and division may be performed as indicated in the next theorem.

**Theorem on Products and Quotients of Complex Numbers**

If trigonometric forms for two complex numbers $z_1$ and $z_2$ are

$$z_1 = r_1(\cos \theta_1 + i \sin \theta_1) \quad \text{and} \quad z_2 = r_2(\cos \theta_2 + i \sin \theta_2),$$

then

**(1)** $z_1 z_2 = r_1 r_2 [\cos (\theta_1 + \theta_2) + i \sin (\theta_1 + \theta_2)]$

**(2)** $\dfrac{z_1}{z_2} = \dfrac{r_1}{r_2} [\cos (\theta_1 - \theta_2) + i \sin (\theta_1 - \theta_2)], \quad z_2 \neq 0$

**PROOF**    We shall prove (1) here and leave the proof of (2) as an exercise. Thus,

$$z_1 z_2 = r_1(\cos \theta_1 + i \sin \theta_1) \cdot r_2(\cos \theta_2 + i \sin \theta_2)$$
$$= r_1 r_2 [(\cos \theta_1 \cos \theta_2 - \sin \theta_1 \sin \theta_2)$$
$$+ i(\sin \theta_1 \cos \theta_2 + \cos \theta_1 \sin \theta_2)].$$

Applying the addition formulas for $\cos (\theta_1 + \theta_2)$ and $\sin (\theta_1 + \theta_2)$ gives us (1).    ∎

Part (1) of the preceding theorem states that *the modulus of a product of two complex numbers is the product of their moduli, and an argument is the sum of their arguments.* An analogous statement can be made for (2).

**EXAMPLE 3    Using trigonometric forms to find products and quotients**

If $z_1 = 2\sqrt{3} - 2i$ and $z_2 = -1 + \sqrt{3}i$, use trigonometric forms to find **(a)** $z_1 z_2$ and **(b)** $z_1/z_2$. Check by using algebraic methods.

**SOLUTION**    The complex number $2\sqrt{3} - 2i$ is represented geometrically in Figure 19(b). If we use $\theta = -\pi/6$ in the trigonometric form, then

$$z_1 = 2\sqrt{3} - 2i = 4\left[\cos\left(-\frac{\pi}{6}\right) + i \sin\left(-\frac{\pi}{6}\right)\right].$$

The complex number $z_2 = -1 + \sqrt{3}i$ is represented geometrically in Figure 20. A trigonometric form is

$$z_2 = -1 + \sqrt{3}i = 2\left(\cos \frac{2\pi}{3} + i \sin \frac{2\pi}{3}\right).$$

*(continued)*

FIGURE 20

(a) We apply (1) of the theorem on products and quotients of complex numbers:

$$z_1 z_2 = 4 \cdot 2\left[\cos\left(-\frac{\pi}{6} + \frac{2\pi}{3}\right) + i \sin\left(-\frac{\pi}{6} + \frac{2\pi}{6}\right)\right]$$

$$= 8\left(\cos\frac{\pi}{2} + i \sin\frac{\pi}{2}\right) = 8(0 + i) = 8i$$

Using algebraic methods to check our result, we have

$$z_1 z_2 = (2\sqrt{3} - 2i)(-1 + \sqrt{3}i)$$
$$= (-2\sqrt{3} + 2\sqrt{3}) + (2 + 6)i = 0 + 8i = 8i.$$

(b) We apply (2) of the theorem:

$$\frac{z_1}{z_2} = \frac{4}{2}\left[\cos\left(-\frac{\pi}{6} - \frac{2\pi}{3}\right) + i \sin\left(-\frac{\pi}{6} - \frac{2\pi}{3}\right)\right]$$

$$= 2\left[\cos\left(-\frac{5\pi}{6}\right) + i \sin\left(-\frac{5\pi}{6}\right)\right]$$

$$= 2\left[-\frac{\sqrt{3}}{2} + i\left(-\frac{1}{2}\right)\right] = -\sqrt{3} - i$$

Using algebraic methods, we obtain

$$\frac{z_1}{z_2} = \frac{2\sqrt{3} - 2i}{-1 + \sqrt{3}i} \cdot \frac{-1 - \sqrt{3}i}{-1 - \sqrt{3}i}$$
$$= \frac{(-2\sqrt{3} - 2\sqrt{3}) + (2 - 6)i}{4} = -\sqrt{3} - i.$$

## 7.3 EXERCISES

Exer. 1–10: Find the absolute value.

1   $|3 - 4i|$ 

2   $|5 + 8i|$

3   $|-6 - 7i|$ 

4   $|1 - i|$

5   $|8i|$ 

6   $|i^7|$

7   $|i^{500}|$ 

8   $|-15i|$

9   $|0|$ 

10   $|-15|$

Exer. 11–20: Represent the complex number geometrically.

11   $4 + 2i$ 

12   $-5 + 3i$

13   $3 - 5i$ 

14   $-2 - 6i$

15   $-(3 - 6i)$ 

16   $(1 + 2i)^2$

17   $2i(2 + 3i)$ 

18   $(-3i)(2 - i)$

19   $(1 + i)^2$ 

20   $4(-1 + 2i)$

Exer. 21–46: Express the complex number in trigonometric form with $0 \le \theta < 2\pi$.

21   $1 - i$ 

22   $\sqrt{3} + i$

23   $-4\sqrt{3} + 4i$ 

24   $-2 - 2i$

25   $2\sqrt{3} + 2i$ 

26   $3 - 3\sqrt{3}i$

27   $-4 - 4i$ 

28   $-10 + 10i$

29   $-20i$ 

30   $-6i$

31  12

32  15

33  −7

34  −5

35  $6i$

36  $4i$

37  $-5 - 5\sqrt{3}i$

38  $\sqrt{3} - i$

39  $2 + i$

40  $3 + 2i$

41  $-3 + i$

42  $-4 + 2i$

43  $-5 - 3i$

44  $-2 - 7i$

45  $4 - 3i$

46  $1 - 3i$

**Exer. 47–56: Express in the form $a + bi$, where $a$ and $b$ are real numbers.**

47  $4\left(\cos \dfrac{\pi}{4} + i \sin \dfrac{\pi}{4}\right)$

48  $8\left(\cos \dfrac{7\pi}{4} + i \sin \dfrac{7\pi}{4}\right)$

49  $6\left(\cos \dfrac{2\pi}{3} + i \sin \dfrac{2\pi}{3}\right)$

50  $12\left(\cos \dfrac{4\pi}{3} + i \sin \dfrac{4\pi}{3}\right)$

51  $5(\cos \pi + i \sin \pi)$

52  $3\left(\cos \dfrac{3\pi}{2} + i \sin \dfrac{3\pi}{2}\right)$

53  $\sqrt{34} \operatorname{cis} \left(\tan^{-1} \tfrac{3}{5}\right)$

54  $\sqrt{53} \operatorname{cis} \left[\tan^{-1} \left(-\tfrac{2}{7}\right)\right]$

55  $\sqrt{5} \operatorname{cis} \left[\tan^{-1} \left(-\tfrac{1}{2}\right)\right]$

56  $\sqrt{10} \operatorname{cis} \left(\tan^{-1} 3\right)$

**Exer. 57–64: Use trigonometric forms to find $z_1 z_2$ and $z_1/z_2$.**

57  $z_1 = -1 + i,$     $z_2 = 1 + i$

58  $z_1 = \sqrt{3} - i,$     $z_2 = -\sqrt{3} - i$

59  $z_1 = -2 - 2\sqrt{3}i,$   $z_2 = 5i$

60  $z_1 = -5 + 5i,$     $z_2 = -3i$

61  $z_1 = -10,$     $z_2 = -4$

62  $z_1 = 2i,$     $z_2 = -3i$

63  $z_1 = 4,$     $z_2 = 2 - i$

64  $z_1 = -3,$     $z_2 = 5 + 2i$

65  Prove (2) of the theorem on products and quotients of complex numbers.

66  (a) Extend (1) of the theorem on products and quotients of complex numbers to three complex numbers.

(b) Generalize (1) of the theorem to $n$ complex numbers.

## 7.4 DE MOIVRE'S THEOREM AND $n$TH ROOTS OF COMPLEX NUMBERS

If $z$ is a complex number and $n$ is a positive integer, then a complex number $w$ is an ***$n$th root*** of $z$ if $w^n = z$. We will show that every nonzero complex number has $n$ different $n$th roots. Since $\mathbb{R}$ is contained in $\mathbb{C}$, it will also follow that every nonzero real number has $n$ different $n$th (complex) roots. If $a$ is a positive real number and $n = 2$, then we already know that the roots are $\sqrt{a}$ and $-\sqrt{a}$.

If, in the theorem on products and quotients of complex numbers, we let both $z_1$ and $z_2$ equal the complex number $z = r(\cos \theta + i \sin \theta)$, we obtain

$$z^2 = r \cdot r[\cos (\theta + \theta) + i \sin (\theta + \theta)]$$
$$= r^2(\cos 2\theta + i \sin 2\theta).$$

Applying the same theorem to $z^2$ and $z$ gives us

$$z^2 \cdot z = (r^2 \cdot r)[\cos (2\theta + \theta) + i \sin (2\theta + \theta)],$$

or

$$z^3 = r^3(\cos 3\theta + i \sin 3\theta).$$

Applying the theorem to $z^3$ and $z$, we obtain

$$z^4 = r^4(\cos 4\theta + i \sin 4\theta).$$

In general, we have the following result, named after the French mathematician Abraham De Moivre (1667–1754).

**De Moivre's Theorem**

For every integer $n$,
$$[r(\cos\theta + i\sin\theta)]^n = r^n(\cos n\theta + i\sin n\theta).$$

### EXAMPLE I     *Using De Moivre's theorem*

Use De Moivre's theorem to change $(1 + i)^{20}$ to the form $a + bi$, where $a$ and $b$ are real numbers.

**FIGURE 21**

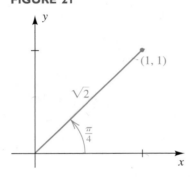

*SOLUTION*     It would be tedious to change $(1 + i)^{20}$ using algebraic methods. Let us therefore introduce a trigonometric form for $1 + i$. Referring to Figure 21, we see that

$$1 + i = \sqrt{2}\left(\cos\frac{\pi}{4} + i\sin\frac{\pi}{4}\right).$$

We now apply De Moivre's theorem:

$$(1 + i)^{20} = (2^{1/2})^{20}\left[\cos\left(20\cdot\frac{\pi}{4}\right) + i\sin\left(20\cdot\frac{\pi}{4}\right)\right]$$
$$= 2^{10}(\cos 5\pi + i\sin 5\pi) = 2^{10}(-1 + 0i) = -1024$$

The number $-1024$ is of the form $a + bi$ with $a = -1024$ and $b = 0$.

---

If a nonzero complex number $z$ has an $n$th root $w$, then $w^n = z$. If trigonometric forms for $w$ and $z$ are

$$w = s(\cos\alpha + i\sin\alpha) \quad\text{and}\quad z = r(\cos\theta + i\sin\theta),$$

then applying De Moivre's theorem to $w^n = z$ yields

$$s^n(\cos n\alpha + i\sin n\alpha) = r(\cos\theta + i\sin\theta).$$

If two complex numbers are equal, then so are their absolute values. Consequently, $s^n = r$, and since $s$ and $r$ are nonnegative, $s = \sqrt[n]{r}$. Substituting $s^n$ for $r$ in the last displayed equation and dividing both sides by $s^n$, we obtain

$$\cos n\alpha + i\sin n\alpha = \cos\theta + i\sin\theta.$$

Since the arguments of equal complex numbers differ by a multiple of $2\pi$, there is an integer $k$ such that $n\alpha = \theta + 2\pi k$. Dividing both sides of the last equation by $n$, we see that

$$\alpha = \frac{\theta + 2\pi k}{n} \quad\text{for some integer } k.$$

Substituting in the trigonometric form for $w$ gives us the formula

$$w = \sqrt[n]{r}\left[\cos\left(\frac{\theta + 2\pi k}{n}\right) + i\sin\left(\frac{\theta + 2\pi k}{n}\right)\right].$$

If we substitute $k = 0, 1, \ldots, n - 1$ successively, we obtain $n$ different $n$th roots of $z$. No other value of $k$ will produce a new $n$th root. For example, if $k = n$, we obtain the angle $(\theta + 2\pi n)/n$, or $(\theta/n) + 2\pi$, which gives us the same $n$th root as $k = 0$. Similarly, $k = n + 1$ yields the same $n$th root as $k = 1$, and so on. The same is true for negative values of $k$. We have proved the following theorem.

**Theorem on $n$th Roots**

> If $z = r(\cos\theta + i\sin\theta)$ is any nonzero complex number and if $n$ is any positive integer, then $z$ has exactly $n$ different $n$th roots $w_0, w_1, w_2, \ldots, w_{n-1}$. These roots, for $\theta$ in radians, are
>
> $$w_k = \sqrt[n]{r}\left[\cos\left(\frac{\theta + 2\pi k}{n}\right) + i\sin\left(\frac{\theta + 2\pi k}{n}\right)\right]$$
>
> or, equivalently, for $\theta$ in degrees,
>
> $$w_k = \sqrt[n]{r}\left[\cos\left(\frac{\theta + 360°k}{n}\right) + i\sin\left(\frac{\theta + 360°k}{n}\right)\right],$$
>
> where $k = 0, 1, \ldots, n - 1$.

The $n$th roots of $z$ in this theorem all have absolute value $\sqrt[n]{r}$, and hence their geometric representations lie on a circle of radius $\sqrt[n]{r}$ with center at $O$. Moreover, they are equispaced on this circle, since the difference in the arguments of successive $n$th roots is $2\pi/n$.

**EXAMPLE 2    Finding the fourth roots of a complex number**

Find the four fourth roots of $-8 - 8\sqrt{3}i$.

**SOLUTION**    The geometric representation of $-8 - 8\sqrt{3}i$ is shown in Figure 22. Introducing trigonometric form, we have

$$-8 - 8\sqrt{3}i = 16(\cos 240° + i\sin 240°).$$

Using the theorem on $n$th roots with $n = 4$ and noting that $\sqrt[4]{16} = 2$, we find that the fourth roots are

$$w_k = 2\left[\cos\left(\frac{240° + 360°k}{4}\right) + i\sin\left(\frac{240° + 360°k}{4}\right)\right]$$

for $k = 0, 1, 2, 3$. This formula may be written

$$w_k = 2[\cos(60° + 90°k) + i\sin(60° + 90°k)].$$

*(continued)*

**FIGURE 22**

Substituting 0, 1, 2, and 3 for $k$ in $(60° + 90°k)$ gives us the four fourth roots:

$$w_0 = 2(\cos 60° + i \sin 60°) = 1 + \sqrt{3}i$$
$$w_1 = 2(\cos 150° + i \sin 150°) = -\sqrt{3} + i$$
$$w_2 = 2(\cos 240° + i \sin 240°) = -1 - \sqrt{3}i$$
$$w_3 = 2(\cos 330° + i \sin 330°) = \sqrt{3} - i$$

**EXAMPLE 3    *Finding the sixth roots of a real number***

**(a)** Find the six sixth roots of $-1$.

**(b)** Represent the roots geometrically.

**SOLUTION**

**(a)** Writing $-1 = 1(\cos \pi + i \sin \pi)$ and using the theorem on $n$th roots with $n = 6$, we find that the sixth roots of $-1$ are given by

$$w_k = \cos \left( \frac{\pi + 2\pi k}{6} \right) + i \sin \left( \frac{\pi + 2\pi k}{6} \right)$$

for $k = 0, 1, 2, 3, 4, 5$. We can also write $w_k$ in the form

$$w_k = \cos \left( \frac{\pi}{6} + \frac{\pi}{3}k \right) + i \sin \left( \frac{\pi}{6} + \frac{\pi}{3}k \right).$$

Substituting 0, 1, 2, 3, 4, 5 for $k$, we obtain the six sixth roots of $-1$:

$$w_0 = \cos \frac{\pi}{6} + i \sin \frac{\pi}{6} = \frac{\sqrt{3}}{2} + \frac{1}{2}i$$

$$w_1 = \cos \frac{\pi}{2} + i \sin \frac{\pi}{2} = i$$

$$w_2 = \cos \frac{5\pi}{6} + i \sin \frac{5\pi}{6} = -\frac{\sqrt{3}}{2} + \frac{1}{2}i$$

$$w_3 = \cos \frac{7\pi}{6} + i \sin \frac{7\pi}{6} = -\frac{\sqrt{3}}{2} - \frac{1}{2}i$$

$$w_4 = \cos \frac{3\pi}{2} + i \sin \frac{3\pi}{2} = -i$$

$$w_5 = \cos \frac{11\pi}{6} + i \sin \frac{11\pi}{6} = \frac{\sqrt{3}}{2} - \frac{1}{2}i$$

**(b)** Since $|-1| = 1$, the points that represent the roots of $-1$ all lie on the unit circle shown in Figure 23. Moreover, they are equispaced on this circle.

**FIGURE 23**

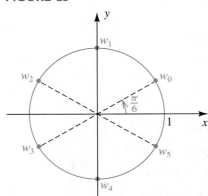

The special case in which $z = 1$ is of particular interest. The $n$ distinct $n$th roots of 1 are called the **$n$th roots of unity.** In particular, if $n = 3$, we call these roots the **cube roots of unity.**

**EXAMPLE 4    *Finding the cube roots of unity***

Find the three cube roots of unity.

**SOLUTION**    Writing $1 = \cos 0 + i \sin 0$ and using the theorem on $n$th roots with $n = 3$, we obtain

$$w_k = \cos \frac{2\pi k}{3} + i \sin \frac{2\pi k}{3}$$

for $k = 0, 1, 2$. Substituting for $k$ gives us the three roots:

$$w_0 = \cos 0 + i \sin 0 = 1$$
$$w_1 = \cos \frac{2\pi}{3} + i \sin \frac{2\pi}{3} = -\frac{1}{2} + \frac{\sqrt{3}}{2}i$$
$$w_2 = \cos \frac{4\pi}{3} + i \sin \frac{4\pi}{3} = -\frac{1}{2} - \frac{\sqrt{3}}{2}i$$

Note that finding the $n$th roots of a complex number $c$, as we did in Examples 2–4, is equivalent to finding all the solutions of the equation

$$x^n = c, \quad \text{or} \quad x^n - c = 0.$$

(See Exercises 23–30.)

## 7.4 EXERCISES

**Exer. 1–12:** Use De Moivre's theorem to change the given complex number to the form $a + bi$, where $a$ and $b$ are real numbers.

1  $(3 + 3i)^5$

2  $(1 + i)^{12}$

3  $(1 - i)^{10}$

4  $(-1 + i)^8$

5  $(1 - \sqrt{3}i)^3$

6  $(1 - \sqrt{3}i)^5$

7  $\left(-\dfrac{\sqrt{2}}{2} + \dfrac{\sqrt{2}}{2}i\right)^{15}$

8  $\left(\dfrac{\sqrt{2}}{2} + \dfrac{\sqrt{2}}{2}i\right)^{25}$

9  $\left(-\dfrac{\sqrt{3}}{2} - \dfrac{1}{2}i\right)^{20}$

10  $\left(-\dfrac{\sqrt{3}}{2} - \dfrac{1}{2}i\right)^{50}$

11  $(\sqrt{3} + i)^7$

12  $(-2 - 2i)^{10}$

13  Find the two square roots of $1 + \sqrt{3}i$.

14  Find the two square roots of $-9i$.

15  Find the four fourth roots of $-1 - \sqrt{3}i$.

16  Find the four fourth roots of $-8 + 8\sqrt{3}i$.

17  Find the three cube roots of $-27i$.

18  Find the three cube roots of $64i$.

**Exer. 19–22: Find the indicated roots, and represent them geometrically.**

19  The six sixth roots of unity

20  The eight eighth roots of unity

21  The five fifth roots of $1 + i$

22  The five fifth roots of $-\sqrt{3} - i$

**Exer. 23–30: Find the solutions of the equation.**

23  $x^4 - 16 = 0$

24  $x^6 - 64 = 0$

25  $x^6 + 64 = 0$

26  $x^5 + 1 = 0$

27  $x^3 + 8i = 0$

28  $x^3 - 64i = 0$

29  $x^5 - 243 = 0$

30  $x^4 + 81 = 0$

## 7.5    VECTORS

Quantities such as area, volume, length, temperature, and time have magnitude only and can be completely characterized by a single real number (with an appropriate unit of measurement such as in.$^2$, ft$^3$, cm, deg, or sec). A quantity of this type is a **scalar quantity,** and the corresponding real number is a **scalar.** A concept such as velocity or force has both magnitude and direction and is often represented by a **directed line segment**—that is, a line segment to which a direction has been assigned. Another name for a directed line segment is a **vector.**

As shown in Figure 24, we use $\overrightarrow{PQ}$ to denote the vector with **initial point** $P$ and **terminal point** $Q$, and we indicate the direction of the vector by placing the arrowhead at $Q$. The **magnitude** of $\overrightarrow{PQ}$ is the length of the segment $PQ$ and is denoted by $\|\overrightarrow{PQ}\|$. As in the figure, we use boldface letters such as **u** and **v** to denote vectors whose end points are not specified. In handwritten work, a notation such as $\vec{u}$ or $\vec{v}$ is often used.

Vectors that have the same magnitude and direction are said to be **equivalent.** In mathematics, a vector is determined only by its magnitude and direction, not by its location. Thus, we regard equivalent vectors, such as those in Figure 24, as **equal** and write

$$\mathbf{u} = \overrightarrow{PQ}, \quad \mathbf{v} = \overrightarrow{PQ}, \quad \text{and} \quad \mathbf{u} = \mathbf{v}.$$

Thus, *a vector may be translated from one location to another, provided neither the magnitude nor the direction is changed.*

We can represent many physical concepts by vectors. To illustrate, suppose an airplane is descending at a constant speed of 100 mi/hr and the line of flight makes an angle of 20° with the horizontal. Both of these facts are represented by the vector **v** of magnitude 100 in Figure 25. The vector **v** is a **velocity vector.**

A vector that represents a pull or push of some type is a **force vector.** The force exerted when a person holds a 5-pound weight is illustrated by

**FIGURE 24**  Equal vectors

**FIGURE 25**   Velocity vector

**FIGURE 26**   Force vector

the vector **F** of magnitude 5 in Figure 26. This force has the same magnitude as the force exerted on the weight by gravity, but it acts in the opposite direction. As a result, there is no movement upward or downward.

We sometimes use $\overrightarrow{AB}$ to represent the path of a point (or particle) as it moves along the line segment from $A$ to $B$. We then refer to $\overrightarrow{AB}$ as a **displacement** of the point (or particle). As in Figure 27, a displacement $\overrightarrow{AB}$ followed by a displacement $\overrightarrow{BC}$ leads to the same point as the single displacement $\overrightarrow{AC}$. By definition, the vector $\overrightarrow{AC}$ is the **sum** of $\overrightarrow{AB}$ and $\overrightarrow{BC}$, and we write

$$\overrightarrow{AC} = \overrightarrow{AB} + \overrightarrow{BC}.$$

Since vectors may be translated from one location to another, *any* two vectors may be added by placing the initial point of the second vector on the terminal point of the first and then drawing the line segment from the initial point of the first to the terminal point of the second, as in Figure 27.

Another way to find the sum is to choose vectors $\overrightarrow{PQ}$ and $\overrightarrow{PR}$ that are equal to $\overrightarrow{AB}$ and $\overrightarrow{BC}$, respectively, and have the same initial point $P$, as shown in Figure 28. If we construct parallelogram $RPQS$, then, since $\overrightarrow{PR} = \overrightarrow{QS}$, it follows that $\overrightarrow{PS} = \overrightarrow{PQ} + \overrightarrow{PR}$. If $\overrightarrow{PQ}$ and $\overrightarrow{PR}$ are two forces acting at $P$, then $\overrightarrow{PS}$ is the **resultant force**—that is, the single force that produces the same effect as the two combined forces.

If $c$ is a scalar and **v** is a vector, then $c\mathbf{v}$ is defined as a vector whose magnitude is $|c|$ times the magnitude $\|\mathbf{v}\|$ of **v** and whose direction is either the same as that of **v** (if $c > 0$) or opposite that of **v** (if $c < 0$). Illustrations are given in Figure 29. We refer to $c\mathbf{v}$ as a **scalar multiple** of **v**.

**FIGURE 27**   Sum of vectors

**FIGURE 28**   Resultant force

**FIGURE 29**   Scalar multiples

**FIGURE 30**

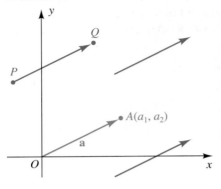

Throughout the remainder of this section we shall restrict our discussion to vectors that lie in an *xy*-plane. If $\overrightarrow{PQ}$ is such a vector, then, as indicated in Figure 30, there are many vectors that are equivalent to $\overrightarrow{PQ}$; however, there is exactly *one* equivalent vector $\mathbf{a} = \overrightarrow{OA}$ with initial point at the origin. In this sense, *each vector determines a unique ordered pair of real numbers,* the coordinates $(a_1, a_2)$ of the terminal point $A$. Conversely, every ordered pair $(a_1, a_2)$ determines the vector $\overrightarrow{OA}$, where $A$ has coordinates $(a_1, a_2)$. Thus, *there is a one-to-one correspondence between vectors in an xy-plane and ordered pairs of real numbers.* This correspondence allows us to interpret a vector as both a directed line segment *and* an ordered pair of real numbers. To avoid confusion with the notation for open intervals or points, we use the symbol $\langle a_1, a_2 \rangle$ for an ordered pair that represents a vector, and we denote it by a boldface letter—for example, $\mathbf{a} = \langle a_1, a_2 \rangle$. The numbers $a_1$ and $a_2$ are the **components** of the vector $\langle a_1, a_2 \rangle$. If $A$ is the point $(a_1, a_2)$, as in Figure 30, we call $\overrightarrow{OA}$ the **position vector** for $\langle a_1, a_2 \rangle$ or for the *point A*.

**FIGURE 31**    Magnitude $\|\mathbf{a}\|$

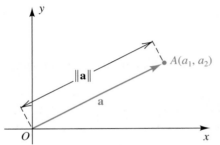

The preceding discussion shows that vectors have two different natures, one geometric and the other algebraic. Often we do not distinguish between the two. It should always be clear from our discussion whether we are referring to ordered pairs or directed line segments.

The **magnitude** $\|\mathbf{a}\|$ of the vector $\mathbf{a} = \langle a_1, a_2 \rangle$ is, by definition, the length of its position vector $OA$, as illustrated in Figure 31. The following definition states this fact another way.

**Definition of the Magnitude of a Vector**

The magnitude $\|\mathbf{a}\|$ of the vector $\mathbf{a} = \langle a_1, a_2 \rangle$ is

$$\|\mathbf{a}\| = \|\langle a_1, a_2 \rangle\| = \sqrt{a_1^2 + a_2^2}.$$

**FIGURE 32**

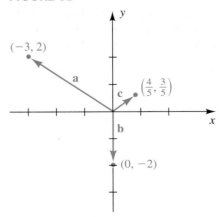

**EXAMPLE 1**    *Finding the magnitude of a vector*

Sketch the vectors for

$$\mathbf{a} = \langle -3, 2 \rangle, \quad \mathbf{b} = \langle 0, -2 \rangle, \quad \mathbf{c} = \langle \tfrac{4}{5}, \tfrac{3}{5} \rangle$$

on a coordinate plane, and find the magnitude of each vector.

**SOLUTION**    The vectors are sketched in Figure 32. By the definition of the magnitude of a vector,

$$\|\mathbf{a}\| = \|\langle -3, 2 \rangle\| = \sqrt{(-3)^2 + 2^2} = \sqrt{13}$$
$$\|\mathbf{b}\| = \|\langle 0, -2 \rangle\| = \sqrt{0^2 + (-2)^2} = \sqrt{4} = 2$$
$$\|\mathbf{c}\| = \|\langle \tfrac{4}{5}, \tfrac{3}{5} \rangle\| = \sqrt{(\tfrac{4}{5})^2 + (\tfrac{3}{5})^2} = \sqrt{\tfrac{25}{25}} = 1.$$

Consider the vectors $\overrightarrow{OA}$ and $\overrightarrow{OB}$ corresponding to $\mathbf{a} = \langle a_1, a_2 \rangle$ and $\mathbf{b} = \langle b_1, b_2 \rangle$, respectively, as illustrated in Figure 33. If $\overrightarrow{OC}$ corresponds to $\mathbf{c} = \langle a_1 + b_1, a_2 + b_2 \rangle$, we can show, using slopes, that $O, A, C,$ and $B$ are vertices of a parallelogram; that is,

$$\overrightarrow{OA} + \overrightarrow{OB} = \overrightarrow{OC}.$$

**FIGURE 33**

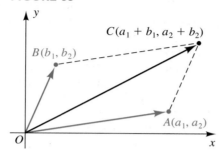

Expressing this equation in terms of ordered pairs leads to the following.

**Definition of Addition of Vectors**

$$\langle a_1, a_2 \rangle + \langle b_1, b_2 \rangle = \langle a_1 + b_1, a_2 + b_2 \rangle$$

Note that to add two vectors, we add corresponding components.

**ILLUSTRATION**

**Addition of Vectors**

■  $\langle 3, -4 \rangle + \langle 2, 7 \rangle = \langle 3 + 2, -4 + 7 \rangle = \langle 5, 3 \rangle$

■  $\langle 5, 1 \rangle + \langle -5, 1 \rangle = \langle 5 + (-5), 1 + 1 \rangle = \langle 0, 2 \rangle$

It can also be shown that if $c$ is a scalar and $\overrightarrow{OA}$ corresponds to $\mathbf{a} = \langle a_1, a_2 \rangle$, then the ordered pair determined by $c\overrightarrow{OA}$ is $(ca_1, ca_2)$, as illustrated in Figure 34 for $c > 1$. This leads to the next definition.

**FIGURE 34**

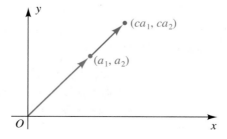

**Definition of a Scalar**
**Multiple of a Vector**

$$c\langle a_1, a_2 \rangle = \langle ca_1, ca_2 \rangle$$

Thus, to find a scalar multiple of a vector, we multiply each component by the scalar.

**ILLUSTRATION**

**Scalar Multiple of a Vector**

- $2\langle -3, 4 \rangle = \langle 2(-3), 2(4) \rangle = \langle -6, 8 \rangle$
- $-2\langle -3, 4 \rangle = \langle (-2)(-3), (-2)(4) \rangle = \langle 6, -8 \rangle$
- $1\langle 5, 2 \rangle = \langle 1 \cdot 5, 1 \cdot 2 \rangle = \langle 5, 2 \rangle$

**EXAMPLE 2     Finding a scalar multiple of a vector**

If $\mathbf{a} = \langle 2, 1 \rangle$, find $3\mathbf{a}$ and $-2\mathbf{a}$, and sketch all three vectors in a coordinate plane.

**SOLUTION**     Using the definition of scalar multiples of vectors, we find

$$3\mathbf{a} = 3\langle 2, 1 \rangle = \langle 3 \cdot 2, 3 \cdot 1 \rangle = \langle 6, 3 \rangle$$

$$-2\mathbf{a} = -2\langle 2, 1 \rangle = \langle (-2) \cdot 2, (-2) \cdot 1 \rangle = \langle -4, -2 \rangle.$$

The vectors are sketched in Figure 35.

**FIGURE 35**

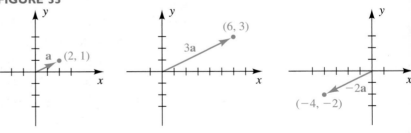

The **zero vector 0** and the **negative** $-\mathbf{a}$ of a vector $\mathbf{a} = \langle a_1, a_2 \rangle$ are defined as follows.

**Definition of 0 and −a**

$$\mathbf{0} = \langle 0, 0 \rangle \quad \text{and} \quad -\mathbf{a} = -\langle a_1, a_2 \rangle = \langle -a_1, -a_2 \rangle$$

### ILLUSTRATION

#### The Zero Vector and the Negative of a Vector

- $\langle 3, 5 \rangle + \mathbf{0} = \langle 3, 5 \rangle + \langle 0, 0 \rangle = \langle 3 + 0, 5 + 0 \rangle = \langle 3, 5 \rangle$
- $-\langle 3, -5 \rangle = \langle -3, -(-5) \rangle = \langle -3, 5 \rangle$
- $\langle 3, -5 \rangle + \langle -3, 5 \rangle = \langle 3 + (-3), -5 + 5 \rangle = \langle 0, 0 \rangle = \mathbf{0}$
- $0\langle 2, 3 \rangle = \langle 0 \cdot 2, 0 \cdot 3 \rangle = \langle 0, 0 \rangle = \mathbf{0}$
- $5 \cdot \mathbf{0} = 5\langle 0, 0 \rangle = \langle 5 \cdot 0, 5 \cdot 0 \rangle = \langle 0, 0 \rangle = \mathbf{0}$

We next state properties of addition and scalar multiples of vectors for any vectors **a**, **b** and scalars $c$, $d$. You should have little difficulty in remembering these properties, since they resemble familiar properties of real numbers.

**Properties of Addition and Scalar Multiples of Vectors**

| | |
|---|---|
| **(I)** $\mathbf{a} + \mathbf{b} = \mathbf{b} + \mathbf{a}$ | **(5)** $c(\mathbf{a} + \mathbf{b}) = c\mathbf{a} + c\mathbf{b}$ |
| **(2)** $\mathbf{a} + (\mathbf{b} + \mathbf{c}) = (\mathbf{a} + \mathbf{b}) + \mathbf{c}$ | **(6)** $(c + d)\mathbf{a} = c\mathbf{a} + d\mathbf{a}$ |
| **(3)** $\mathbf{a} + \mathbf{0} = \mathbf{a}$ | **(7)** $(cd)\mathbf{a} = c(d\mathbf{a}) = d(c\mathbf{a})$ |
| **(4)** $\mathbf{a} + (-\mathbf{a}) = \mathbf{0}$ | **(8)** $1\mathbf{a} = \mathbf{a}$ |
| | **(9)** $0\mathbf{a} = \mathbf{0} = c\mathbf{0}$ |

*PROOF*  Let $\mathbf{a} = \langle a_1, a_2 \rangle$ and $\mathbf{b} = \langle b_1, b_2 \rangle$. To prove property 1, we note that

$$\mathbf{a} + \mathbf{b} = \langle a_1 + b_1, a_2 + b_2 \rangle = \langle b_1 + a_1, b_2 + a_2 \rangle = \mathbf{b} + \mathbf{a}.$$

The proof of property 5 is as follows:

$$
\begin{aligned}
c(\mathbf{a} + \mathbf{b}) &= c\langle a_1 + b_1, a_2 + b_2 \rangle && \text{definition of addition}\\
&= \langle c(a_1 + b_1), c(a_2 + b_2) \rangle && \text{definition of scalar multiple}\\
&= \langle ca_1 + cb_1, ca_2 + cb_2 \rangle && \text{distributive property}\\
&= \langle ca_1, ca_2 \rangle + \langle cb_1, cb_2 \rangle && \text{definition of addition}\\
&= c\mathbf{a} + c\mathbf{b} && \text{definition of scalar multiple}
\end{aligned}
$$

Proofs of the remaining properties are similar.  ■

**Vector subtraction** (denoted by $-$) is defined by $\mathbf{a} - \mathbf{b} = \mathbf{a} + (-\mathbf{b})$. If we use the ordered pair notation for **a** and **b**, then $-\mathbf{b} = \langle -b_1, -b_2 \rangle$, and we obtain the following.

**Definition of Subtraction of Vectors**

$$\mathbf{a} - \mathbf{b} = \langle a_1, a_2 \rangle - \langle b_1, b_2 \rangle = \langle a_1 - b_1, a_2 - b_2 \rangle$$

Thus, to find $\mathbf{a} - \mathbf{b}$, we merely subtract the components of $\mathbf{b}$ from the corresponding components of $\mathbf{a}$.

**ILLUSTRATION**

**FIGURE 36**

**Subtraction of Vectors** If $\mathbf{a} = \langle 5, -4 \rangle$ and $\mathbf{b} = \langle -3, 2 \rangle$

■ $\mathbf{a} - \mathbf{b} = \langle 5, -4 \rangle - \langle -3, 2 \rangle$
$= \langle 5 - (-3), -4 - 2 \rangle = \langle 8, -6 \rangle$

■ $2\mathbf{a} - 3\mathbf{b} = 2\langle 5, -4 \rangle - 3\langle -3, 2 \rangle$
$= \langle 10, -8 \rangle - \langle -9, 6 \rangle = \langle 19, -14 \rangle$

If $\mathbf{a}$ and $\mathbf{b}$ are arbitrary vectors, then

$$\mathbf{b} + (\mathbf{a} - \mathbf{b}) = \mathbf{a};$$

that is, $\mathbf{a} - \mathbf{b}$ *is the vector that, when added to* $\mathbf{b}$, *gives us* $\mathbf{a}$. If we represent $\mathbf{a}$ and $\mathbf{b}$ by vectors $\overrightarrow{PQ}$ and $\overrightarrow{PR}$ *with the same initial point,* as in Figure 36, then $\overrightarrow{RQ}$ represents $\mathbf{a} - \mathbf{b}$.

The special vectors $\mathbf{i}$ and $\mathbf{j}$ are defined as follows.

**Definition of i and j**

$$\mathbf{i} = \langle 1, 0 \rangle, \quad \mathbf{j} = \langle 0, 1 \rangle$$

A **unit vector** is a vector of magnitude 1. The vectors $\mathbf{i}$ and $\mathbf{j}$ are unit vectors, as is the vector $\mathbf{c} = \langle \frac{4}{5}, \frac{3}{5} \rangle$ in Example 1.

The vectors $\mathbf{i}$ and $\mathbf{j}$ can be used to obtain an alternative way of denoting vectors. Specifically, if $\mathbf{a} = \langle a_1, a_2 \rangle$, then

$$\mathbf{a} = \langle a_1, 0 \rangle + \langle 0, a_2 \rangle = a_1 \langle 1, 0 \rangle + a_2 \langle 0, 1 \rangle.$$

This result gives us the following.

**i, j Form for Vectors**

$$\mathbf{a} = \langle a_1, a_2 \rangle = a_1 \mathbf{i} + a_2 \mathbf{j}$$

**ILLUSTRATION**

**i, j Form**

■ $\langle 5, 2 \rangle = 5\mathbf{i} + 2\mathbf{j}$

■ $\langle -3, 4 \rangle = -3\mathbf{i} + 4\mathbf{j}$

■ $\langle 0, -6 \rangle = 0\mathbf{i} + (-6)\mathbf{j} = -6\mathbf{j}$

Vectors corresponding to **i**, **j**, and **a** are illustrated in Figure 37. Since **i** and **j** are unit vectors, $a_1\mathbf{i}$ and $a_2\mathbf{j}$ may be represented by horizontal and vertical vectors of magnitudes $\|a_1\|$ and $\|a_2\|$, respectively, as illustrated in Figure 38. For this reason we call $a_1$ the **horizontal component** and $a_2$ the **vertical component** of the vector **a**.

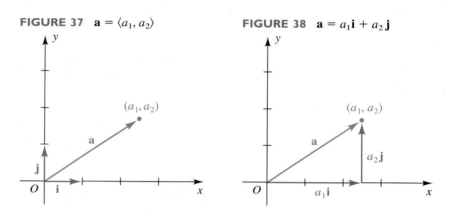

FIGURE 37 $\mathbf{a} = \langle a_1, a_2 \rangle$  FIGURE 38 $\mathbf{a} = a_1\mathbf{i} + a_2\mathbf{j}$

The vector sum $a_1\mathbf{i} + a_2\mathbf{j}$ is a **linear combination** of **i** and **j**. Rules for addition, subtraction, and multiplication by a scalar may be written as follows, with $\mathbf{b} = \langle b_1, b_2 \rangle = b_1\mathbf{i} + b_2\mathbf{j}$:

$$(a_1\mathbf{i} + a_2\mathbf{j}) + (b_1\mathbf{i} + b_2\mathbf{j}) = (a_1 + b_1)\mathbf{i} + (a_2 + b_2)\mathbf{j}$$

$$(a_1\mathbf{i} + a_2\mathbf{j}) - (b_1\mathbf{i} + b_2\mathbf{j}) = (a_1 - b_1)\mathbf{i} + (a_2 - b_2)\mathbf{j}$$

$$c(a_1\mathbf{i} + a_2\mathbf{j}) = (ca_1)\mathbf{i} + (ca_2)\mathbf{j}$$

These formulas show that we may regard linear combinations of **i** and **j** as algebraic sums.

**EXAMPLE 3** *Expressing a vector as a linear combination of* **i** *and* **j**

If $\mathbf{a} = 5\mathbf{i} + \mathbf{j}$ and $\mathbf{b} = 4\mathbf{i} - 7\mathbf{j}$, express $3\mathbf{a} - 2\mathbf{b}$ as a linear combination of **i** and **j**.

*SOLUTION*

$$3\mathbf{a} - 2\mathbf{b} = 3(5\mathbf{i} + \mathbf{j}) - 2(4\mathbf{i} - 7\mathbf{j})$$
$$= (15\mathbf{i} + 3\mathbf{j}) - (8\mathbf{i} - 14\mathbf{j})$$
$$= 7\mathbf{i} + 17\mathbf{j}$$

**FIGURE 39**

Let $\theta$ be an angle in standard position, measured from the positive $x$-axis to the vector $\mathbf{a} = \langle a_1, a_2 \rangle = a_1 \mathbf{i} + a_2 \mathbf{j}$, as illustrated in Figure 39. Since

$$\cos \theta = \frac{a_1}{\|\mathbf{a}\|} \quad \text{and} \quad \sin \theta = \frac{a_2}{\|\mathbf{a}\|},$$

we obtain the following formulas.

**Formulas for Horizontal and Vertical Components of $\mathbf{a} = \langle a_1, a_2 \rangle$**

> If the vector $\mathbf{a}$ and the angle $\theta$ are defined as above, then
> $$a_1 = \|\mathbf{a}\| \cos \theta \quad \text{and} \quad a_2 = \|\mathbf{a}\| \sin \theta.$$

Using these formulas, we have

$$\begin{aligned}
\mathbf{a} = \langle a_1, a_2 \rangle &= \langle \|\mathbf{a}\| \cos \theta, \|\mathbf{a}\| \sin \theta \rangle \\
&= \|\mathbf{a}\| \cos \theta \, \mathbf{i} + \|\mathbf{a}\| \sin \theta \, \mathbf{j} \\
&= \|\mathbf{a}\|(\cos \theta \, \mathbf{i} + \sin \theta \, \mathbf{j}).
\end{aligned}$$

This is analogous to the trigonometric form of a complex number.

**EXAMPLE 4**   *Expressing wind velocity as a vector*
If the wind is blowing at 12 mi/hr in the direction N40°W, express its velocity as a vector $\mathbf{v}$.

**FIGURE 40**

*SOLUTION*   The vector $\mathbf{v}$ is illustrated in Figure 40, where the angle $\theta = 90° + 40° = 130°$. Using the formulas for horizontal and vertical components with $\mathbf{v} = \langle v_1, v_2 \rangle$ gives us

$$v_1 = \|\mathbf{v}\| \cos \theta = 12 \cos 130°, \quad v_2 = \|\mathbf{v}\| \sin \theta = 12 \sin 130°.$$

Hence,
$$\begin{aligned}
\mathbf{v} = v_1 \mathbf{i} + v_2 \mathbf{j} \\
&= (12 \cos 130°)\mathbf{i} + (12 \sin 130°)\mathbf{j} \\
&\approx (-7.7)\mathbf{i} + (9.2)\mathbf{j}.
\end{aligned}$$

**EXAMPLE 5**   *Finding a resultant vector*
Two forces $\overrightarrow{PQ}$ and $\overrightarrow{PR}$ of magnitudes 5.0 kilograms and 8.0 kilograms, respectively, act at a point $P$. The direction of $\overrightarrow{PQ}$ is N20°E, and the direction of $\overrightarrow{PR}$ is N65°E. Approximate the magnitude and direction of the resultant vector $\overrightarrow{PS}$.

**FIGURE 41**

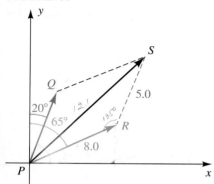

**SOLUTION**   The forces are represented geometrically in Figure 41. Note that the angles from the positive $x$-axis to $\overrightarrow{PQ}$ and $\overrightarrow{PR}$ have measures $70°$ and $25°$, respectively. Using the formulas for horizontal and vertical components, we obtain the following:

$$\overrightarrow{PQ} = (5 \cos 70°)\mathbf{i} + (5 \sin 70°)\mathbf{j}$$
$$\overrightarrow{PR} = (8 \cos 25°)\mathbf{i} + (8 \sin 25°)\mathbf{j}$$

Since $\overrightarrow{PS} = \overrightarrow{PQ} + \overrightarrow{PR}$,

$$\overrightarrow{PS} = (5 \cos 70° + 8 \cos 25°)\mathbf{i} + (5 \sin 70° + 8 \sin 25°)\mathbf{j}$$
$$\approx (9.0)\mathbf{i} + (8.1)\mathbf{j}.$$

Consequently,

$$\|\overrightarrow{PS}\| \approx \sqrt{(9.0)^2 + (8.1)^2} \approx 12.1.$$

We can also find $\|\overrightarrow{PS}\|$ by using the law of cosines (see Example 3 of Section 7.2). Since $\angle QPR = 45°$, it follows that $\angle PRS = 135°$, and hence

$$\|\overrightarrow{PS}\|^2 = (8.0)^2 + (5.0)^2 - 2(8.0)(5.0) \cos 135° \approx 145.6$$

and                                $$\|\overrightarrow{PS}\| \approx \sqrt{145.6} \approx 12.1.$$

If $\theta$ is the angle from the positive $x$-axis to the resultant $PS$, then using the (approximate) coordinates $(8.9606, 8.0794)$ of $S$, we obtain the following:

$$\tan \theta \approx \frac{8.0794}{8.9606} \approx 0.9017$$
$$\theta \approx \tan^{-1}(0.9017) \approx 42°$$

Hence, the direction of $\overrightarrow{PS}$ is approximately N48°E.

## 7.5 EXERCISES

**Exer. 1–6: Find a + b, a − b, 4a + 5b, and 4a − 5b.**

1  $\mathbf{a} = \langle 2, -3 \rangle,$     $\mathbf{b} = \langle 1, 4 \rangle$

2  $\mathbf{a} = \langle -2, 6 \rangle,$     $\mathbf{b} = \langle 2, 3 \rangle$

3  $\mathbf{a} = -\langle 7, -2 \rangle,$     $\mathbf{b} = 4\langle -2, 1 \rangle$

4  $\mathbf{a} = 2\langle 5, -4 \rangle,$     $\mathbf{b} = -\langle 6, 0 \rangle$

5  $\mathbf{a} = \mathbf{i} + 2\mathbf{j},$     $\mathbf{b} = 3\mathbf{i} - 5\mathbf{j}$

6  $\mathbf{a} = -3\mathbf{i} + \mathbf{j},$     $\mathbf{b} = -3\mathbf{i} + \mathbf{j}$

**Exer. 7–10: Sketch vectors corresponding to a, b, a + b, 2a, and −3b.**

7  $\mathbf{a} = 3\mathbf{i} + 2\mathbf{j},$     $\mathbf{b} = -\mathbf{i} + 5\mathbf{j}$

8  $\mathbf{a} = -5\mathbf{i} + 2\mathbf{j},$   $\mathbf{b} = \mathbf{i} - 3\mathbf{j}$

9  $\mathbf{a} = \langle -4, 6 \rangle,$     $\mathbf{b} = \langle -2, 3 \rangle$

10  $\mathbf{a} = \langle 2, 0 \rangle,$     $\mathbf{b} = \langle -2, 0 \rangle$

**Exer. 11–16:** Use components to express the sum or difference as a scalar multiple of one of the vectors a, b, c, d, e, or f shown in the figure.

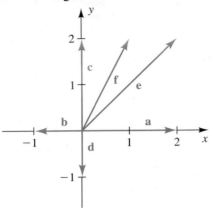

| 11 a + b | 12 c − d |
| 13 b + e | 14 f − b |
| 15 b + d | 16 e + c |

**Exer. 17–26:** If a = $\langle a_1, a_2 \rangle$, b = $\langle b_1, b_2 \rangle$, c = $\langle c_1, c_2 \rangle$, and m and n are real numbers, prove the stated property.

17 a + (b + c) = (a + b) + c

18 a + 0 = a

19 a + (−a) = 0          20 (m + n)a = ma + na

21 (mn)a = m(na) = n(ma)    22 1a = a

23 0a = 0 = m0             24 (−m)a = −ma

25 −(a + b) = −a − b       26 m(a − b) = ma − mb

27 If v = $\langle a, b \rangle$, prove that the magnitude of 2v is twice the magnitude of v.

28 If v = $\langle a, b \rangle$ and k is any real number, prove that the magnitude of kv is |k| times the magnitude of v.

**Exer. 29–36:** Find the magnitude of a and the smallest positive angle θ from the positive x-axis to the vector OP that corresponds to a.

| 29 a = $\langle 3, -3 \rangle$ | 30 a = $\langle -2, -2\sqrt{3} \rangle$ |
| 31 a = $\langle -5, 0 \rangle$ | 32 a = $\langle 0, 10 \rangle$ |
| 33 a = −4i + 5j | 34 a = 10i − 10j |
| 35 a = −18j | 36 a = 2i − 3j |

**Exer. 37–40:** The vectors a and b represent two forces acting at the same point, and θ is the smallest positive angle between a and b. Approximate the magnitude of the resultant force.

37 a = 40 lb,    b = 70 lb,    θ = 45°

38 a = 5.5 lb,   b = 6.2 lb,   θ = 60°

39 a = 2.0 kg,   b = 8.0 kg,   θ = 120°

40 a = 30 kg,    b = 50 kg,    θ = 150°

**Exer. 41–44:** The magnitudes and directions of two forces acting at a point P are given in (a) and (b). Approximate the magnitude and direction of the resultant vector.

| 41 (a) 90 kg,  N75°W | (b) 60 kg,  S5°E |
| 42 (a) 20 kg,  S17°W | (b) 50 kg,  N82°W |
| 43 (a) 6.0 lb, 110° | (b) 2.0 lb, 215° |
| 44 (a) 70 lb, 320° | (b) 40 lb, 30° |

**Exer. 45–48:** Approximate the horizontal and vertical components of the vector that is described.

45 A quarterback releases a football with a velocity of 50 ft/sec at an angle of 35° with the horizontal.

46 A child pulls a sled through the snow by exerting a force of 20 pounds at an angle of 40° with the horizontal.

47 The biceps muscle, in supporting the forearm and a weight held in the hand, exerts a force of 20 pounds. As shown in the figure, the muscle makes an angle of 108° with the forearm.

**EXERCISE 47**

48 A jet airplane approaches a runway at an angle of 7.5° with the horizontal, traveling at a velocity of 160 mi/hr.

Exer. 49–52: If forces $F_1, F_2, \ldots, F_n$ act at a point $P$, the net (or resultant) force $F$ is the sum $F_1 + F_2 + \cdots + F_n$. If $F = 0$, the forces are said to be in equilibrium. The given forces act at the origin $O$ of an $xy$-plane.

(a) Find the net force $F$.

(b) Find an additional force $G$ such that equilibrium occurs.

**49** $F_1 = \langle 4, 3 \rangle$, $\quad F_2 = \langle -2, -3 \rangle$, $\quad F_3 = \langle 5, 2 \rangle$

**50** $F_1 = \langle -3, -1 \rangle$, $\quad F_2 = \langle 0, -3 \rangle$, $\quad F_3 = \langle 3, 4 \rangle$

**51**

**52**

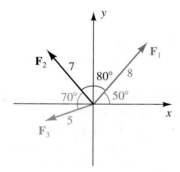

**53 Tugboat force** Two tugboats are towing a large ship into port, as shown in the figure. The larger tug exerts a force of 4000 pounds on its cable, and the smaller tug exerts a force of 3200 pounds on its cable. If the ship is to travel in a straight line from $A$ to $B$, approximate the

**EXERCISE 53**

angle $\theta$ that the larger tug must make with the line segment $AB$.

**54 Gravity simulation** Shown in the figure is a simple apparatus that may be used to simulate gravity conditions on other planets. A rope is attached to an astronaut who maneuvers on an inclined plane that makes an angle of $\theta$ degrees with the horizontal.

(a) If the astronaut weighs 160 pounds, find the $x$- and $y$-components of the downward force (see the figure for axes).

(b) The $y$-component in part (a) is the weight of the astronaut relative to the inclined plane. The astronaut would weigh 27 pounds on the moon and 60 pounds on Mars. Approximate the angles $\theta$ (to the nearest 0.01°) so that the inclined-plane apparatus will simulate walking on these surfaces.

**EXERCISE 54**

**55 Airplane course and ground speed** An airplane with an airspeed of 200 mi/hr is flying in the direction 50°,

**EXERCISE 55**

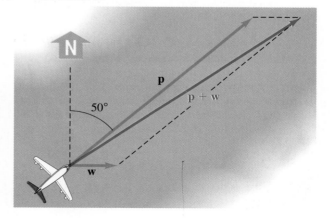

and a 40 mi/hr wind is blowing directly from the west. As shown in the figure, these facts may be represented by vectors **p** and **w** of magnitudes 200 and 40, respectively. The direction of the resultant **p** + **w** gives the true course of the airplane relative to the ground, and the magnitude $\|\mathbf{p} + \mathbf{w}\|$ is the ground speed of the airplane. Approximate the true course and ground speed.

**56 Airplane course and ground speed**  Refer to Exercise 55. An airplane is flying in the direction 140° with an airspeed of 500 mi/hr, and a 30 mi/hr wind is blowing in the direction 65°. Approximate the true course and ground speed of the airplane.

**57 Airplane course and ground speed**  An airplane pilot wishes to maintain a true course in the direction 250° with a ground speed of 400 mi/hr when the wind is blowing directly north at 50 mi/hr. Approximate the required airspeed and compass heading.

**58 Wind direction and speed**  An airplane is flying in the direction 20° with an airspeed of 300 mi/hr. Its ground speed and true course are 350 mi/hr and 30°, respectively. Approximate the direction and speed of the wind.

**59 Rowboat navigation**  The current in a river flows directly from the west at a rate of 1.5 ft/sec. A person who rows a boat at a rate of 4 ft/sec in still water wishes to row directly north across the river. Approximate, to the nearest degree, the direction in which the person should row.

**60 Motorboat navigation**  For a motorboat moving at a speed of 30 mi/hr to travel directly north across a river, it must aim at a point that has the bearing N15°E. If the current is flowing directly west, approximate the rate at which it flows.

**61 Flow of ground water**  Ground-water contaminants can enter a community's drinking water by migrating through porous rock into the aquifer. If underground water flows with a velocity $\mathbf{v}_1$ through an interface between one type of rock and a second type of rock, its velocity changes to $\mathbf{v}_2$, and both the direction and the speed of the flow can be obtained using the formula

$$\frac{\|\mathbf{v}_1\|}{\|\mathbf{v}_2\|} = \frac{\tan \theta_1}{\tan \theta_2},$$

where the angles $\theta_1$ and $\theta_2$ are as shown in the figure.

For sandstone, $\|\mathbf{v}_1\| = 8.2$ cm/day; for limestone, $\|\mathbf{v}_2\| = 3.8$ cm/day. If $\theta_1 = 30°$, approximate the vectors $\mathbf{v}_1$ and $\mathbf{v}_2$ in **i**, **j** form.

**EXERCISE 61**

**62 Flow of ground water**  Refer to Exercise 61. Contaminated ground water is flowing through silty sand with the direction of flow $\theta_1$ and speed (in cm/day) given by the vector $\mathbf{v}_1 = 20\mathbf{i} - 82\mathbf{j}$. When the flow enters a region of clean sand, its rate increases to 725 cm/day. Find the new direction of flow by approximating $\theta_2$.

**63 Robotic movement**  Vectors are useful for describing movement of robots.

(a) The robot's arm illustrated in the first figure can rotate at the joint connections $P$ and $Q$. The upper arm, represented by **a**, is 15 inches long, and the forearm (including the hand), represented by **b**, is 17 inches long. Approximate the coordinates of the point $R$ in the hand by using **a** + **b**.

**EXERCISE 63(a)**

(b) If the upper arm is rotated 85° and the forearm is rotated an additional 35°, as illustrated in the second figure, approximate the new coordinates of $R$ by using $\mathbf{c} + \mathbf{d}$.

EXERCISE 63(b)

**EXERCISE 64(a)**

(b) Suppose the robot's upper arm is rotated 75°, and then the forearm is rotated −80°, and finally the hand is rotated an additional 40°, as shown in the second figure. Approximate the new coordinates of $R$ by using $\mathbf{d} + \mathbf{e} + \mathbf{f}$.

EXERCISE 64(b)

64 **Robotic movement**   Refer to Exercise 63.

(a) Suppose the wrist joint of the robot's arm is allowed to rotate at the joint connection $S$ and the arm is located as shown in the first figure. The upper arm has a length of 15 inches; the forearm, without the hand, has a length of 10 inches; and the hand has a length of 7 inches. Approximate the coordinates of $R$ by using $\mathbf{a} + \mathbf{b} + \mathbf{c}$.

---

**7.6   THE DOT PRODUCT**

The *dot product* of two vectors has many applications. We begin with an algebraic definition.

**Definition of the Dot Product**

Let $\mathbf{a} = \langle a_1, a_2 \rangle = a_1 \mathbf{i} + a_2 \mathbf{j}$ and $\mathbf{b} = \langle b_1, b_2 \rangle = b_1 \mathbf{i} + b_2 \mathbf{j}$. The **dot product** $\mathbf{a} \cdot \mathbf{b}$ of $\mathbf{a}$ and $\mathbf{b}$ is

$$\mathbf{a} \cdot \mathbf{b} = \langle a_1, a_2 \rangle \cdot \langle b_1, b_2 \rangle = a_1 b_1 + a_2 b_2.$$

The symbol $\mathbf{a} \cdot \mathbf{b}$ is read "$\mathbf{a}$ dot $\mathbf{b}$." We also refer to the dot product as the **scalar product** or the **inner product.** Note that $\mathbf{a} \cdot \mathbf{b}$ *is a real number and not a vector,* as illustrated in the following example.

**EXAMPLE I     *Finding the dot product of two vectors***

Find $\mathbf{a} \cdot \mathbf{b}$.

**(a)** $\mathbf{a} = \langle -5, 3 \rangle, \quad \mathbf{b} = \langle 2, 6 \rangle$     **(b)** $\mathbf{a} = 4\mathbf{i} + 6\mathbf{j}, \quad \mathbf{b} = 3\mathbf{i} - 7\mathbf{j}$

**SOLUTION**

**(a)**     $\langle -5, 3 \rangle \cdot \langle 2, 6 \rangle = (-5)(2) + (3)(6) = -10 + 18 = 8$

**(b)**     $(4\mathbf{i} + 6\mathbf{j}) \cdot (3\mathbf{i} - 7\mathbf{j}) = (4)(3) + (6)(-7) = 12 - 42 = -30$

**Properties of the Dot Product**

If $\mathbf{a}$, $\mathbf{b}$, $\mathbf{c}$ are vectors and $c$ is a real number, then

**(1)** $\mathbf{a} \cdot \mathbf{a} = \|\mathbf{a}\|^2$

**(2)** $\mathbf{a} \cdot \mathbf{b} = \mathbf{b} \cdot \mathbf{a}$

**(3)** $\mathbf{a} \cdot (\mathbf{b} + \mathbf{c}) = \mathbf{a} \cdot \mathbf{b} + \mathbf{a} \cdot \mathbf{c}$

**(4)** $(c\mathbf{a}) \cdot \mathbf{b} = c(\mathbf{a} \cdot \mathbf{b}) = \mathbf{a} \cdot (c\mathbf{b})$

**(5)** $\mathbf{0} \cdot \mathbf{a} = 0$

*PROOF*     The proof of each property follows from the definition of the dot product and the properties of real numbers. Thus, if $\mathbf{a} = \langle a_1, a_2 \rangle$, $\mathbf{b} = \langle b_1, b_2 \rangle$, and $\mathbf{c} = \langle c_1, c_2 \rangle$, then

$$\mathbf{a} \cdot (\mathbf{b} + \mathbf{c}) = \langle a_1, a_2 \rangle \cdot \langle b_1 + c_1, b_2 + c_2 \rangle \qquad \text{definition of addition}$$

$$= a_1(b_1 + c_1) + a_2(b_2 + c_2) \qquad \text{definition of dot product}$$

$$= (a_1 b_1 + a_2 b_2) + (a_1 c_1 + a_2 c_2) \qquad \text{real number properties}$$

$$= \mathbf{a} \cdot \mathbf{b} + \mathbf{a} \cdot \mathbf{c}, \qquad \text{definition of dot product}$$

which proves property 3. The proofs of the remaining properties are left as exercises. ■

**FIGURE 42**

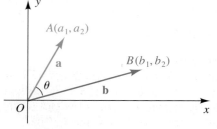

Any two nonzero vectors $\mathbf{a} = \langle a_1, a_2 \rangle$ and $\mathbf{b} = \langle b_1, b_2 \rangle$ may be represented in a coordinate plane by directed line segments from the origin $O$ to the points $A(a_1, a_2)$ and $B(b_1, b_2)$, respectively. The **angle $\theta$ between $\mathbf{a}$ and $\mathbf{b}$** is, by definition, $\angle AOB$ (see Figure 42). Note that $0 \leq \theta \leq \pi$ and that $\theta = 0$ if $\mathbf{a}$ and $\mathbf{b}$ have the same direction or $\theta = \pi$ if $\mathbf{a}$ and $\mathbf{b}$ have the opposite direction.

**Definition of Parallel and Orthogonal Vectors**

Let $\theta$ be the angle between two nonzero vectors **a** and **b**.

**(1)** **a** and **b** are **parallel** if $\theta = 0$ or $\theta = \pi$.

**(2)** **a** and **b** are **orthogonal** if $\theta = \dfrac{\pi}{2}$.

The vectors **a** and **b** in Figure 42 are parallel if and only if they lie on the same line that passes through the origin. In this case, $\mathbf{b} = c\mathbf{a}$ for some real number $c$. The vectors are orthogonal if and only if they lie on mutually perpendicular lines that pass through the origin. We assume that the zero vector **0** is parallel and orthogonal to *every* vector **a**.

The next theorem shows the close relationship between the angle between two vectors and their dot product.

**Theorem on the Dot Product**

If $\theta$ is the angle between two nonzero vectors **a** and **b**, then

$$\mathbf{a} \cdot \mathbf{b} = \|\mathbf{a}\|\,\|\mathbf{b}\|\cos\theta.$$

**PROOF**    If **a** and **b** are not parallel, we have a situation similar to that illustrated in Figure 42. We may then apply the law of cosines to triangle $AOB$. Since the lengths of the three sides of the triangle are $\|\mathbf{a}\|$, $\|\mathbf{b}\|$, and $d(A, B)$,

$$[d(A, B)]^2 = \|\mathbf{a}\|^2 + \|\mathbf{b}\|^2 - 2\|\mathbf{a}\|\,\|\mathbf{b}\|\cos\theta.$$

Using the distance formula and the definition of the magnitude of a vector, we obtain

$$(b_1 - a_1)^2 + (b_2 - a_2)^2 = (a_1^2 + a_2^2) + (b_1^2 + b_2^2) - 2\|\mathbf{a}\|\,\|\mathbf{b}\|\cos\theta,$$

which reduces to

$$-2a_1b_1 - 2a_2b_2 = -2\|\mathbf{a}\|\,\|\mathbf{b}\|\cos\theta.$$

Dividing both sides of the last equation by $-2$ gives us

$$a_1b_1 + a_2b_2 = \|\mathbf{a}\|\,\|\mathbf{b}\|\cos\theta,$$

which is equivalent to what we wished to prove.

If **a** and **b** are parallel, then either $\theta = 0$ or $\theta = \pi$, and therefore $\mathbf{b} = c\mathbf{a}$ for some real number $c$ with $c > 0$ if $\theta = 0$ and $c < 0$ if $\theta = \pi$. We can show, using properties of the dot product, that $\mathbf{a} \cdot (c\mathbf{a}) = \|\mathbf{a}\|\,\|c\mathbf{a}\|\cos\theta$, and hence the theorem is true for all nonzero vectors **a** and **b**. ■

**Theorem on the Cosine of the Angle Between Vectors**

If $\theta$ is the angle between two nonzero vectors **a** and **b**, then

$$\cos \theta = \frac{\mathbf{a} \cdot \mathbf{b}}{\|\mathbf{a}\| \|\mathbf{b}\|}.$$

**EXAMPLE 2    *Finding the angle between two vectors***

Find the angle between $\mathbf{a} = \langle 4, -3 \rangle$ and $\mathbf{b} = \langle 1, 2 \rangle$.

**FIGURE 43**

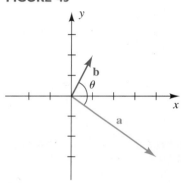

**SOLUTION**    The vectors are sketched in Figure 43. We apply the preceding theorem:

$$\cos \theta = \frac{\mathbf{a} \cdot \mathbf{b}}{\|\mathbf{a}\| \|\mathbf{b}\|} = \frac{(4)(1) + (-3)(2)}{\sqrt{16 + 9}\sqrt{1 + 4}} = \frac{-2}{5\sqrt{5}} = \frac{-2\sqrt{5}}{25}$$

Hence,

$$\theta = \arccos\left(\frac{-2\sqrt{5}}{25}\right) \approx 100.3°.$$

**EXAMPLE 3    *Showing that two vectors are parallel***

Let $\mathbf{a} = \frac{1}{2}\mathbf{i} - 3\mathbf{j}$ and $\mathbf{b} = -2\mathbf{i} + 12\mathbf{j}$.

**(a)** Show that **a** and **b** are parallel.

**(b)** Find the scalar $c$ such that $\mathbf{b} = c\mathbf{a}$.

**SOLUTION**

**(a)** By definition, the vectors **a** and **b** are parallel if and only if the angle $\theta$ between them is either $0$ or $\pi$. Since the cosine of this angle is

$$\cos \theta = \frac{\mathbf{a} \cdot \mathbf{b}}{\|\mathbf{a}\| \|\mathbf{b}\|} = \frac{\left(\frac{1}{2}\right)(-2) + (-3)(12)}{\sqrt{\frac{1}{4} + 9}\sqrt{4 + 144}} = \frac{-37}{37} = -1,$$

we see that

$$\theta = \arccos(-1) = \pi.$$

**(b)** Since **a** and **b** are parallel, there *is* a scalar $c$ such that $\mathbf{b} = c\mathbf{a}$; that is,

$$-2\mathbf{i} + 12\mathbf{j} = c(\tfrac{1}{2}\mathbf{i} - 3\mathbf{j}) = \tfrac{1}{2}c\mathbf{i} - 3c\mathbf{j}.$$

Equating **i** and **j** components gives us

$$-2 = \tfrac{1}{2}c \quad \text{and} \quad 12 = -3c.$$

Thus, $c = -4$; that is, $\mathbf{b} = -4\mathbf{a}$. Note that **a** and **b** have opposite directions, since $c < 0$.

Using the formula $\mathbf{a} \cdot \mathbf{b} = \|\mathbf{a}\| \|\mathbf{b}\| \cos \theta$, together with the fact that two vectors are orthogonal if and only if the angle between them is $\pi/2$ (or one of the vectors is $\mathbf{0}$), gives us the following result.

**Theorem on Orthogonal Vectors**

Two vectors $\mathbf{a}$ and $\mathbf{b}$ are orthogonal if and only if $\mathbf{a} \cdot \mathbf{b} = 0$.

**EXAMPLE 4   *Showing that two vectors are orthogonal***

Show that the pair of vectors is orthogonal:

**(a)** $\mathbf{i}, \mathbf{j}$     **(b)** $2\mathbf{i} + 3\mathbf{j}, 6\mathbf{i} - 4\mathbf{j}$

*SOLUTION*   We may use the theorem on orthogonal vectors to prove orthogonality by showing that the dot product of each pair is zero:

**(a)**          $\mathbf{i} \cdot \mathbf{j} = \langle 1, 0 \rangle \cdot \langle 0, 1 \rangle = (1)(0) + (0)(1) = 0 + 0 = 0$

**(b)**     $(2\mathbf{i} + 3\mathbf{j}) \cdot (6\mathbf{i} - 4\mathbf{j}) = (2)(6) + (3)(-4) = 12 - 12 = 0$

---

**Definition of comp$_\mathbf{b}$ a**

Let $\theta$ be the angle between two nonzero vectors $\mathbf{a}$ and $\mathbf{b}$. The **component of a along b**, denoted by comp$_\mathbf{b}$ **a**, is given by

$$\text{comp}_\mathbf{b}\, \mathbf{a} = \|\mathbf{a}\| \cos \theta.$$

The geometric significance of the preceding definition with $\theta$ acute or obtuse is illustrated in Figure 44, where the $x$- and $y$-axes are not shown.

**FIGURE 44**   comp$_\mathbf{b}$ **a** $= \|\mathbf{a}\| \cos \theta$

**(a)**

**(b)**

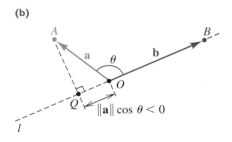

If angle $\theta$ is acute, then, as in Figure 44(a), we can form a right triangle by constructing a line segment $AQ$ perpendicular to the line $l$ through $O$ and $B$. Note that $\overrightarrow{OQ}$ has the same direction as $\overrightarrow{OB}$. Referring to (a) of the figure, we see that

$$\cos \theta = \frac{d(O, Q)}{\|\mathbf{a}\|} \quad \text{or, equivalently,} \quad \|\mathbf{a}\| \cos \theta = d(O, Q).$$

If $\theta$ is obtuse, then, as in Figure 44(b), we again construct $AQ$ perpendicular to $l$. In this case, the direction of $\overrightarrow{OQ}$ is opposite that of $\overrightarrow{OB}$, and since $\cos \theta$ is negative,

$$\cos \theta = \frac{-d(O, Q)}{\|\mathbf{a}\|} \quad \text{or, equivalently,} \quad \|\mathbf{a}\| \cos \theta = -d(O, Q).$$

special cases
for the component
of **a** along **b**

(1) If $\theta = \pi/2$, then **a** is orthogonal to **b** and $\text{comp}_\mathbf{b}\, \mathbf{a} = 0$.

(2) If $\theta = 0$, then **a** has the same direction as **b** and $\text{comp}_\mathbf{b}\, \mathbf{a} = \|\mathbf{a}\|$.

(3) If $\theta = \pi$, then **a** and **b** have opposite direction and $\text{comp}_\mathbf{b}\, \mathbf{a} = -\|\mathbf{a}\|$.

The preceding discussion shows that the component of **a** along **b** may be found by *projecting* the end point of **a** onto the line $l$ containing **b**. For this reason, $\|\mathbf{a}\| \cos \theta$ is sometimes called the **projection of a on b** and is denoted by $\text{proj}_\mathbf{b}\, \mathbf{a}$.

Formula for $\text{comp}_\mathbf{b}\, \mathbf{a}$

> If **a** and **b** are nonzero vectors, then
>
> $$\text{comp}_\mathbf{b}\, \mathbf{a} = \frac{\mathbf{a} \cdot \mathbf{b}}{\|\mathbf{b}\|}.$$

**PROOF**   If $\theta$ is the angle between **a** and **b**, then, from the theorem on the dot product,

$$\mathbf{a} \cdot \mathbf{b} = \|\mathbf{a}\| \|\mathbf{b}\| \cos \theta.$$

Dividing both sides of this equation by $\|\mathbf{b}\|$ gives us

$$\frac{\mathbf{a} \cdot \mathbf{b}}{\|\mathbf{b}\|} = \|\mathbf{a}\| \cos \theta = \text{comp}_\mathbf{b}\, \mathbf{a}. \quad \blacksquare$$

**EXAMPLE 5**   *Finding the components of one vector along another*

If $\mathbf{c} = 10\mathbf{i} + 4\mathbf{j}$ and $\mathbf{d} = 3\mathbf{i} - 2\mathbf{j}$, find $\text{comp}_\mathbf{d}\, \mathbf{c}$ and $\text{comp}_\mathbf{c}\, \mathbf{d}$, and illustrate these numbers graphically.

**SOLUTION**   The vectors **c** and **d** and the desired components are illustrated in Figure 45. We use the formula for $\text{comp}_\mathbf{b}\, \mathbf{a}$, as follows:

$$\text{comp}_\mathbf{d}\, \mathbf{c} = \frac{\mathbf{c} \cdot \mathbf{d}}{\|\mathbf{d}\|} = \frac{(10)(3) + (4)(-2)}{\sqrt{9 + 4}} = \frac{22}{\sqrt{13}} \approx 6.10$$

$$\text{comp}_\mathbf{c}\, \mathbf{d} = \frac{\mathbf{d} \cdot \mathbf{c}}{\|\mathbf{c}\|} = \frac{(3)(10) + (-2)(4)}{\sqrt{100 + 16}} = \frac{22}{\sqrt{116}} \approx 2.04$$

**FIGURE 45**

We shall conclude this section with a physical application of the dot product. First let us briefly discuss the scientific concept of *work*.

A **force** may be thought of as the physical entity that is used to describe a push or pull on an object. For example, a force is needed to push or pull an object along a horizontal plane, to lift an object off the ground, or to move a charged particle through an electromagnetic field. Forces are often measured in pounds. If an object weighs 10 pounds, then, by definition, the force required to lift it (or hold it off the ground) is 10 pounds. A force of this type is a **constant force,** since its magnitude does not change while it is applied to the given object.

If a constant force $F$ is applied to an object, moving it a distance $d$ in the direction of the force, then, by definition, the **work** $W$ done is

$$W = Fd.$$

If $F$ is measured in pounds and $d$ in feet, then the units for $W$ are foot-pounds (ft-lb). In the cgs system a **dyne** is used as the unit of force. If $F$ is expressed in dynes and $d$ in centimeters, then the unit for $W$ is the dyne-centimeter, or **erg.** In the mks system, the **newton** is used as the unit of force. If $F$ is in newtons and $d$ is in meters, then the unit for $W$ is the newton-meter, or **joule.**

### EXAMPLE 6    *Finding the work done by a constant force*

Find the work done in pushing an automobile along a level road from a point $A$ to another point $B$, 40 feet from $A$, while exerting a constant force of 90 pounds.

*SOLUTION*    The problem is illustrated in Figure 46, where we have pictured the road as part of a line $l$. Since the constant force is $F = 90$ lb and the distance the automobile moves is $d = 40$ ft, the work done is

$$W = (90)(40) = 3600 \text{ ft-lb}.$$

**FIGURE 46**

The formula $W = Fd$ is very restrictive, since it can be used only if the force is applied along the line of motion. More generally, suppose that a vector **a** represents a force and that its point of application moves along a vector **b**. This is illustrated in Figure 47, where the force **a** is used to pull an object along a level path from $O$ to $B$, and $\mathbf{b} = \overrightarrow{OB}$.

**FIGURE 47**

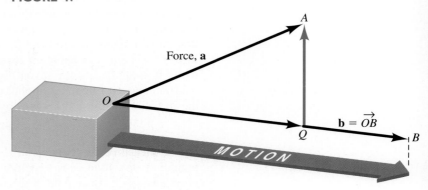

The vector **a** is the sum of the vectors $OQ$ and $QA$, where $\overrightarrow{QA}$ is orthogonal to **b**. Since $\overrightarrow{QA}$ does not contribute to the horizontal movement, we may assume that the motion from $O$ to $B$ is caused by $\overrightarrow{OQ}$ alone. Applying $W = Fd$, we know that the work is the product of $\|\overrightarrow{OQ}\|$ and $\|\mathbf{b}\|$. Since $\|\overrightarrow{OQ}\| = \text{comp}_\mathbf{b}\,\mathbf{a}$, we obtain

$$W = (\text{comp}_\mathbf{b}\,\mathbf{a})\|\mathbf{b}\| = (\|\mathbf{a}\|\cos\theta)\|\mathbf{b}\| = \mathbf{a}\cdot\mathbf{b},$$

where $\theta$ represents $\angle AOQ$. This leads to the following definition.

**Definition of Work**

The **work** $W$ done by a constant force **a** as its point of application moves along a vector **b** is $W = \mathbf{a}\cdot\mathbf{b}$.

**FIGURE 48**

**EXAMPLE 7    Finding the work done by a constant force**

The magnitude and direction of a constant force are given by $\mathbf{a} = 2\mathbf{i} + 5\mathbf{j}$. Find the work done if the point of application of the force moves from the origin to the point $P(4, 1)$.

**SOLUTION**    The force **a** and the vector $\mathbf{b} = \overrightarrow{OP}$ are sketched in Figure 48. Since $\mathbf{b} = \langle 4, 1 \rangle = 4\mathbf{i} + \mathbf{j}$, we have, from the preceding definition,

$$W = \mathbf{a}\cdot\mathbf{b} = (2\mathbf{i} + 5\mathbf{j})\cdot(4\mathbf{i} + \mathbf{j})$$
$$= (2)(4) + (5)(1) = 13.$$

If, for example, the unit of length is feet and the magnitude of the force is measured in pounds, then the work done is 13 ft-lb.

**EXAMPLE 8   *Finding the work done against gravity***

A small cart weighing 100 pounds is pushed up an incline that makes an angle of 30° with the horizontal, as shown in Figure 49. Find the work done against gravity in pushing the cart a distance of 80 feet.

**FIGURE 49**

**FIGURE 50**

**SOLUTION**    Let us introduce an *xy*-coordinate system, as shown in Figure 50. The vector $PQ$ represents the force of gravity acting vertically downward with a magnitude of 100 pounds. The corresponding vector **F** is $0\mathbf{i} - 100\mathbf{j}$. The point of application of this force moves along the vector $\overrightarrow{PR}$ of magnitude 80. If $\overrightarrow{PR}$ corresponds to $\mathbf{a} = a_1\mathbf{i} + a_2\mathbf{j}$, then, referring to triangle *PTR*, we see that

$$a_1 = 80 \cos 30° = 40\sqrt{3},$$
$$a_2 = 80 \sin 30° = 40,$$

and hence                    $\mathbf{a} = 40\sqrt{3}\mathbf{i} + 40\mathbf{j}.$

Applying the definition, we find that the work done *by* gravity is

$$\mathbf{F} \cdot \mathbf{a} = (0\mathbf{i} - 100\mathbf{j}) \cdot (40\sqrt{3}\mathbf{i} + 40\mathbf{j}) = 0 - 4000 = -4000 \text{ ft-lb.}$$

The work done *against* gravity is

$$-\mathbf{F} \cdot \mathbf{a} = 4000 \text{ ft-lb.}$$

## 7.6 EXERCISES

**Exer. 1–8:** Find (a) the dot product of the two vectors and (b) the angle between the two vectors.

1  $\langle -2, 5\rangle$,   $\langle 3, 6\rangle$

2  $\langle 4, -7\rangle$,   $\langle -2, 3\rangle$

3  $4\mathbf{i} - \mathbf{j}$,   $-3\mathbf{i} + 2\mathbf{j}$

4  $8\mathbf{i} - 3\mathbf{j}$,   $2\mathbf{i} - 7\mathbf{j}$

5  $9\mathbf{i}$,   $5\mathbf{i} + 4\mathbf{j}$

6  $6\mathbf{j}$,   $-4\mathbf{i}$

7  $\langle 10, 7\rangle$,   $\langle -2, -\frac{7}{5}\rangle$

8  $\langle -3, 6\rangle$,   $\langle -1, 2\rangle$

**Exer. 9–12:** Show that the vectors are orthogonal.

9  $\langle 4, -1\rangle$,   $\langle 2, 8\rangle$

10  $\langle 3, 6\rangle$,   $\langle 4, -2\rangle$

11  $-4\mathbf{j}$,   $-7\mathbf{i}$

12  $8\mathbf{i} - 4\mathbf{j}$,   $-6\mathbf{i} - 12\mathbf{j}$

**Exer. 13–16:** Show that the vectors are parallel, and determine whether they have the same or the opposite direction.

13 $\mathbf{a} = 3\mathbf{i} - 5\mathbf{j}$, $\quad \mathbf{b} = -\frac{12}{7}\mathbf{i} + \frac{20}{7}\mathbf{j}$

14 $\mathbf{a} = -\frac{5}{2}\mathbf{i} + 6\mathbf{j}$, $\quad \mathbf{b} = -10\mathbf{i} + 24\mathbf{j}$

15 $\mathbf{a} = \langle \frac{2}{3}, \frac{1}{2} \rangle$, $\qquad \mathbf{b} = \langle 8, 6 \rangle$

16 $\mathbf{a} = \langle 6, 18 \rangle$, $\qquad \mathbf{b} = \langle -4, -12 \rangle$

**Exer. 17–20:** Determine $c$ such that the two vectors are orthogonal.

17 $3\mathbf{i} - 2\mathbf{j}$, $\quad 4\mathbf{i} + 5c\mathbf{j}$ $\qquad$ 18 $4c\mathbf{i} + \mathbf{j}$, $\quad 9c\mathbf{i} - 25\mathbf{j}$

19 $9\mathbf{i} - 16c\mathbf{j}$, $\ \mathbf{i} + 4c\mathbf{j}$ $\qquad$ 20 $5c\mathbf{i} + 3\mathbf{j}$, $\quad 2\mathbf{i} + 7\mathbf{j}$

**Exer. 21–28:** Given that $\mathbf{a} = \langle 2, -3 \rangle$, $\mathbf{b} = \langle 3, 4 \rangle$, and $\mathbf{c} = \langle -1, 5 \rangle$, find the number.

21 (a) $\mathbf{a} \cdot (\mathbf{b} + \mathbf{c})$ $\qquad$ (b) $\mathbf{a} \cdot \mathbf{b} + \mathbf{a} \cdot \mathbf{c}$

22 (a) $\mathbf{b} \cdot (\mathbf{a} - \mathbf{c})$ $\qquad$ (b) $\mathbf{b} \cdot \mathbf{a} - \mathbf{b} \cdot \mathbf{c}$

23 $(2\mathbf{a} + \mathbf{b}) \cdot (3\mathbf{c})$ $\qquad$ 24 $(\mathbf{a} - \mathbf{b}) \cdot (\mathbf{b} + \mathbf{c})$

25 $\text{comp}_{\mathbf{c}}\, \mathbf{b}$ $\qquad$ 26 $\text{comp}_{\mathbf{b}}\, \mathbf{c}$

27 $\text{comp}_{\mathbf{b}}\, (\mathbf{a} + \mathbf{c})$ $\qquad$ 28 $\text{comp}_{\mathbf{c}}\, \mathbf{c}$

**Exer. 29–32:** If c represents a constant force, find the work done if the point of application of c moves along the line segment from $P$ to $Q$.

29 $\mathbf{c} = 3\mathbf{i} + 4\mathbf{j}$; $\qquad P(0, 0)$, $\qquad Q(5, -2)$

30 $\mathbf{c} = -10\mathbf{i} + 12\mathbf{j}$; $\quad P(0, 0)$, $\qquad Q(4, 7)$

31 $\mathbf{c} = 6\mathbf{i} + 4\mathbf{j}$; $\qquad P(2, -1)$, $\quad Q(4, 3)$
(*Hint:* Find a vector $\mathbf{b} = \langle b_1, b_2 \rangle$ such that $\mathbf{b} = \overrightarrow{PQ}$.)

32 $\mathbf{c} = -\mathbf{i} + 7\mathbf{j}$; $\qquad P(-2, 5)$, $\quad Q(6, 1)$

33 A constant force of magnitude 4 has the same direction as $\mathbf{j}$. Find the work done if its point of application moves from $P(0, 0)$ to $Q(8, 3)$.

34 A constant force of magnitude 10 has the same direction as $-\mathbf{i}$. Find the work done if its point of application moves from $P(0, 1)$ to $Q(1, 0)$.

**Exer. 35–40:** Prove the property if a and b are vectors and c is a real number.

35 $\mathbf{a} \cdot \mathbf{a} = \|\mathbf{a}\|^2$ $\qquad$ 36 $\mathbf{a} \cdot \mathbf{b} = \mathbf{b} \cdot \mathbf{a}$

37 $(c\mathbf{a}) \cdot \mathbf{b} = c(\mathbf{a} \cdot \mathbf{b})$ $\qquad$ 38 $c(\mathbf{a} \cdot \mathbf{b}) = \mathbf{a} \cdot (c\mathbf{b})$

39 $\mathbf{0} \cdot \mathbf{a} = 0$

40 $(\mathbf{a} + \mathbf{b}) \cdot (\mathbf{a} - \mathbf{b}) = \mathbf{a} \cdot \mathbf{a} - \mathbf{b} \cdot \mathbf{b}$

41 ***Pulling a wagon*** A child pulls a wagon along level ground by exerting a force of 20 pounds on a handle that makes an angle of 30° with the horizontal, as shown in the figure. Find the work done in pulling the wagon 100 feet.

**EXERCISE 41**

42 ***Pulling a wagon*** Refer to Exercise 41. Find the work done if the wagon is pulled, with the same force, 100 feet up an incline that makes an angle of 30° with the horizontal, as shown in the figure.

**EXERCISE 42**

43 ***The sun's rays*** The sun has a radius of 432,000 miles, and its center is 93,000,000 miles from the center of the earth. Let **v** and **w** be the vectors illustrated in the figure.

(a) Express **v** and **w** in **i**, **j** form.

(b) Approximate the angle between **v** and **w**.

**EXERCISE 43**

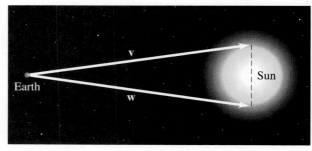

**44** *July sunlight*   The intensity $I$ of sunlight (in watts/m$^2$) can be calculated using the formula $I = ke^{-c/\sin \phi}$, where $k$ and $c$ are positive constants and $\phi$ is the angle between the sun's rays and the horizon. The amount of sunlight striking a vertical wall facing the sun is equal to the component of the sun's rays along the horizontal. If, during July, $\phi = 30°$, $k = 978$, and $c = 0.136$, approximate the total amount of sunlight striking a vertical wall that has an area of $160$ m$^2$.

**45** *Determining horsepower*   The amount of horsepower $P$ produced by an engine can be determined by using the formula $P = \frac{1}{550}(\mathbf{F} \cdot \mathbf{v})$, where $\mathbf{F}$ is the force (in pounds) exerted by the engine and $\mathbf{v}$ is the velocity (in ft/sec) of an object moved by the engine. An engine pulls with a force of 2200 pounds on a cable that makes an angle $\theta$

with the horizontal, moving a cart horizontally, as shown in the figure. Find the horsepower of the engine if the speed of the cart is 8 ft/sec when $\theta = 30°$.

EXERCISE 45

## CHAPTER 7 REVIEW EXERCISES

**Exer. 1–4: Find the exact values of the remaining parts of triangle $ABC$.**

**1** $\alpha = 60°$,   $b = 6$,      $c = 7$

**2** $\gamma = 30°$,   $a = 2\sqrt{3}$,   $c = 2$

**3** $\alpha = 60°$,   $\beta = 45°$,   $b = 100$

**4** $a = 2$,      $b = 3$,      $c = 4$

**Exer. 5–8: Approximate the remaining parts of triangle $ABC$.**

**5** $\beta = 67°$,      $\gamma = 75°$,   $b = 12$

**6** $\alpha = 23°30'$,   $c = 125$,   $a = 152$

**7** $\beta = 115°$,   $a = 4.6$,   $c = 7.3$

**8** $a = 37$,      $b = 55$,   $c = 43$

**Exer. 9–10: Approximate the area of triangle $ABC$ to the nearest 0.1 square unit.**

**9** $\alpha = 75°$,   $b = 20$,   $c = 30$

**10** $a = 4$,      $b = 7$,   $c = 10$

**Exer. 11–16: Express the complex number in trigonometric form with $0 \le \theta < 2\pi$.**

**11** $-10 + 10i$

**12** $2 - 2\sqrt{3}i$

**13** $-17$

**14** $-12i$

**15** $-5\sqrt{3} - 5i$

**16** $4 + 5i$

**Exer. 17–18: Express in the form $a + bi$, where $a$ and $b$ are real numbers.**

**17** $20\left(\cos \dfrac{11\pi}{6} + i \sin \dfrac{11\pi}{6}\right)$

**18** $13 \operatorname{cis} (\tan^{-1} \frac{5}{12})$

**Exer. 19–20: Use trigonometric forms to find $z_1 z_2$ and $z_1/z_2$.**

**19** $z_1 = -3\sqrt{3} - 3i$,   $z_2 = 2\sqrt{3} + 2i$

**20** $z_1 = 2\sqrt{2} + 2\sqrt{2}i$,   $z_2 = -1 - i$

**Exer. 21–24: Use De Moivre's theorem to change the given complex number to the form $a + bi$, where $a$ and $b$ are real numbers.**

**21** $(-\sqrt{3} + i)^9$

**22** $\left(\dfrac{\sqrt{2}}{2} - \dfrac{\sqrt{2}}{2}i\right)^{30}$

**23** $(3 - 3i)^5$

**24** $(2 + 2\sqrt{3}i)^{10}$

**25** Find the three cube roots of $-27$.

**26** Let $z = 1 - \sqrt{3}i$.

   (a) Find $z^{24}$.   (b) Find the three cube roots of $z$.

**27** Find the solutions of the equation $x^5 - 32 = 0$.

28  If $\mathbf{a} = \langle -4, 5\rangle$ and $\mathbf{b} = \langle 2, -8\rangle$, sketch vectors corresponding to

(a) $\mathbf{a} + \mathbf{b}$     (b) $\mathbf{a} - \mathbf{b}$     (c) $2\mathbf{a}$     (d) $-\frac{1}{2}\mathbf{b}$

29  If $\mathbf{a} = 2\mathbf{i} + 5\mathbf{j}$ and $\mathbf{b} = 4\mathbf{i} - \mathbf{j}$, find the vector or number corresponding to

(a) $4\mathbf{a} + \mathbf{b}$                (b) $2\mathbf{a} - 3\mathbf{b}$

(c) $\|\mathbf{a} - \mathbf{b}\|$                (d) $\|\mathbf{a}\| - \|\mathbf{b}\|$

30  If $\mathbf{a} = \langle a_1, a_2\rangle$, $\mathbf{r} = \langle x, y\rangle$, and $c > 0$, describe the set of all points $P(x, y)$ such that $\|\mathbf{r} - \mathbf{a}\| = c$.

31  If $\mathbf{a}$ and $\mathbf{b}$ are vectors with the same initial point and angle $\theta$ between them, prove that

$$\|\mathbf{a} - \mathbf{b}\|^2 = \|\mathbf{a}\|^2 + \|\mathbf{b}\|^2 - 2\|\mathbf{a}\|\,\|\mathbf{b}\|\cos\theta.$$

32  *A ship's course*   A ship is sailing at a speed of 14 mi/hr in the direction S50°E. Express its velocity $\mathbf{v}$ as a vector.

33  The magnitudes and directions of two forces are 72 kg, S60°E and 46 kg, N74°E, respectively. Approximate the magnitude and direction of the resultant force.

34  *Wind speed and direction*   An airplane is flying in the direction 80° with an airspeed of 400 mi/hr. Its ground speed and true course are 390 mi/hr and 90°, respectively. Approximate the direction and speed of the wind.

35  If $\mathbf{a} = \langle 2, -3\rangle$ and $\mathbf{b} = \langle -1, -4\rangle$, find each of the following:

(a) $\mathbf{a} \cdot \mathbf{b}$          (b) the angle between $\mathbf{a}$ and $\mathbf{b}$

(c) $\text{comp}_{\mathbf{a}}\,\mathbf{b}$

36  If $\mathbf{a} = 6\mathbf{i} - 2\mathbf{j}$ and $\mathbf{b} = \mathbf{i} + 3\mathbf{j}$, find each of the following:

(a) $(2\mathbf{a} - 3\mathbf{b}) \cdot \mathbf{a}$

(b) the angle between $\mathbf{a}$ and $\mathbf{a} + \mathbf{b}$

(c) $\text{comp}_{\mathbf{a}}\,(\mathbf{a} + \mathbf{b})$

37  A constant force has the magnitude and direction of $\mathbf{a} = 7\mathbf{i} + 4\mathbf{j}$. Find the work done when the point of application of $\mathbf{a}$ moves along the $x$-axis from $P(-5, 0)$ to $Q(3, 0)$.

38  *Skateboard racecourse*   A course for a skateboard race consists of a 200-meter downhill run and a 150-meter level portion. The angle of elevation of the starting point of the race from the finishing line is 27.4°. What angle does the hill make with the horizontal?

39  *Height of a skyscraper*   If a skyscraper is viewed from the top of a 50-foot building, the angle of elevation is 59°. If it is viewed from street level, the angle of elevation is 62° (see the figure).

(a) Use the law of sines to approximate the shortest distance between the tops of the two buildings.

(b) Approximate the height of the skyscraper.

**EXERCISE 39**

40  *Distances between cities*   The beach communities of San Clemente and Long Beach are 41 miles apart, along a fairly straight stretch of coastline. Shown in the figure is the triangle formed by the two cities and the town of Avalon at the southeast corner of Santa Catalina Island. Angles *ALS* and *ASL* are found to be 66.4° and 47.2°, respectively.

(a) Approximate the distance from Avalon to each of the two cities.

(b) Approximate the shortest distance from Avalon to the coast.

**EXERCISE 40**

41  *Surveying*   A surveyor wishes to find the distance between two inaccessible points $A$ and $B$. As shown in the figure, two points $C$ and $D$ are selected from which it is possible to view both $A$ and $B$. The distance $CD$ and the angles $ACD$, $ACB$, $BDC$, and $BDA$ are then measured. If $CD = 120$ ft, $\angle ACD = 115°$, $\angle ACB = 92°$, $\angle BDC = 125°$, and $\angle BDA = 100°$, approximate the distance $AB$.

**EXERCISE 41**

**42  Radio contact**  Two girls with two-way radios are at the intersection of two country roads that meet at a 105° angle (see the figure). One begins walking in a northerly direction along one road at the rate of 5 mi/hr; at the same time the other walks east along the other road at the same rate. If each radio has a range of 10 miles, how long will the girls maintain contact?

**EXERCISE 42**

**43  Robotic design**  Shown in the figure is a design for a robotic arm with two moving parts. The dimensions are chosen to emulate a human arm. The upper arm $AC$ and lower arm $CP$ rotate through angles $\theta_1$ and $\theta_2$, respectively, to hold an object at point $P(x, y)$.

**EXERCISE 43**

(a) Show that $\angle ACP = 180° - (\theta_2 - \theta_1)$.

(b) Find $d(A, P)$, and then use part (a) and the law of cosines to show that

$$1 + \cos(\theta_2 - \theta_1) = \frac{x^2 + (y - 26)^2}{578}.$$

(c) If $x = 25$, $y = 4$, and $\theta_1 = 135°$, approximate $\theta_2$.

**44  Rescue efforts**  A child is trapped 45 feet down an abandoned mine shaft that slants at an angle of 78° from the horizontal. A rescue tunnel is to be dug 50 feet from the shaft opening (see the figure).

(a) At what angle $\theta$ should the tunnel be dug?

(b) If the tunnel can be dug at a rate of 3 ft/hr, how many hours will it take to reach the child?

**EXERCISE 44**

**45  Design for a jet fighter**  Shown in the figure is a plan for the top of a wing of a jet fighter.

(a) Approximate angle $\phi$.

(b) Approximate the area of quadrilateral $ABCD$.

(c) If the fuselage is 5.8 feet wide, approximate the wing span $CC'$.

**EXERCISE 45**

*Systems of equations may be applied to finding points at which curved paths intersect.*

■ Applications of mathematics some-times require working simultaneously with more than one equation in several variables—that is, with a system of equations. In this chapter we develop methods for finding solutions that are common to all the equations in a sys-tem. Of particular importance are the techniques involving matrices, because they are well suited for computer pro-grams and can be readily applied to systems containing any number of linear equations in any number of variables. We shall also consider systems of ine-qualities and linear programming—topics that are of major importance in business applications and statistics. The final four sections provide an introduc-tion to the algebra of matrices and de-terminants. ■

# SYSTEMS OF
# EQUATIONS AND
# INEQUALITIES

## 8.1   SYSTEMS OF EQUATIONS

**FIGURE 1**

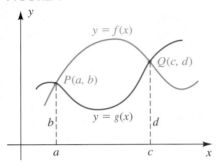

Consider the graphs of the two functions $f$ and $g$, illustrated in Figure 1. In applications it is often necessary to find points such as $P(a, b)$ and $Q(c, d)$ at which the graphs intersect. Since $P(a, b)$ is on each graph, the pair $(a, b)$ is a **solution** of *both* of the equations $y = f(x)$ and $y = g(x)$; that is,

$$b = f(a) \quad \text{and} \quad b = g(a).$$

We say that $(a, b)$ is a solution of the **system of equations** (or simply **system**)

$$\begin{cases} y = f(x) \\ y = g(x) \end{cases}$$

where the brace is used to indicate that the equations are to be treated simultaneously. Similarly, the pair $(c, d)$ is a solution of the system. To **solve** a system of equations means to find all the solutions.

As a special case, consider the system

$$\begin{cases} y = x^2 \\ y = 2x + 3 \end{cases}$$

**FIGURE 2**

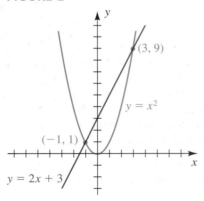

The graphs of the equations are the parabola and line sketched in Figure 2. The following table shows that the points $(-1, 1)$ and $(3, 9)$ are on both graphs.

| $(x, y)$ | $y = x^2$ | $y = 2x + 3$ |
|---|---|---|
| $(-1, 1)$ | $1 = (-1)^2$, or $1 = 1$ | $1 = 2(-1) + 3$, or $1 = 1$ |
| $(3, 9)$ | $9 = 3^2$, or $9 = 9$ | $9 = 2(3) + 3$, or $9 = 9$ |

Hence, $(-1, 1)$ and $(3, 9)$ are solutions of the system.

The preceding discussion does not give us a strategy for actually finding the solutions. The next two examples illustrate how to find the solutions of the system using only algebraic methods.

**EXAMPLE 1**   *Solving a system of two equations*

Solve the system

$$\begin{cases} y = x^2 \\ y = 2x + 3 \end{cases}$$

*SOLUTION*     If $(x, y)$ is a solution of the system, then the variable $y$ in the equation $y = 2x + 3$ must satisfy the condition $y = x^2$. Hence, we *substitute* $x^2$ for $y$ in $y = 2x + 3$:

$$x^2 = 2x + 3 \qquad \text{substitute } y = x^2 \text{ in } y = 2x + 3$$
$$x^2 - 2x - 3 = 0 \qquad \text{subtract } 2x + 3$$
$$(x + 1)(x - 3) = 0 \qquad \text{factor}$$
$$x + 1 = 0, \quad x - 3 = 0 \qquad \text{set each factor equal to 0}$$
$$x = -1, \qquad x = 3 \qquad \text{solve for } x$$

This gives us the $x$-values for the solutions $(x, y)$ of the system. To find the corresponding $y$-values, we may use either $y = x^2$ or $y = 2x + 3$. Using $y = x^2$, we find that

$$\text{if} \quad x = -1, \quad \text{then} \quad y = (-1)^2 = 1;$$

and

$$\text{if} \quad x = 3, \quad \text{then} \quad y = 3^2 = 9.$$

Hence, the solutions of the system are $(-1, 1)$ and $(3, 9)$.

We could also have found the solutions by substituting $y = 2x + 3$ in the *first* equation, $y = x^2$, obtaining

$$2x + 3 = x^2.$$

The remainder of the solution is the same.

---

Given the system in Example 1, we *could* have solved one of the equations for $x$ in terms of $y$ and then substituted in the other equation, obtaining an equation in $y$ alone. Solving the latter equation would give us the $y$-values for the solutions of the system. The $x$-values could then be found using one of the given equations. In general, we may use the following guidelines, where $u$ and $v$ denote any two variables (*possibly* $x$ and $y$). This technique is called the **method of substitution.**

**Guidelines for the Method of Substitution for Two Equations in Two Variables**

1   Solve one of the equations for one variable $u$ in terms of the other variable $v$.

2   Substitute the expression for $u$ found in guideline 1 in the other equation, obtaining an equation in $v$ alone.

3   Find the solutions of the equation in $v$ obtained in guideline 2.

4   Substitute the $v$-values found in guideline 3 in the equation of guideline 1 to find the corresponding $u$-values.

5   Check each pair $(u, v)$ found in guideline 4 in the given system.

**EXAMPLE 2     *Using the method of substitution***

Solve the following system and then sketch the graph of each equation, showing the points of intersection:

$$\begin{cases} x + y^2 = 6 \\ x + 2y = 3 \end{cases}$$

**SOLUTION**     We shall follow the guidelines with $u = x$ and $v = y$.

**Guideline 1**  Solve the second equation for $x$ in terms of $y$:

$$x = 3 - 2y$$

**Guideline 2**  Substitute the expression for $x$ found in guideline 1 in the first equation of the system:

$$(3 - 2y) + y^2 = 6 \quad \text{substitute } x = 3 - 2y \text{ in } x + y^2 = 6$$
$$y^2 - 2y - 3 = 0 \quad \text{simplify}$$

**Guideline 3**  Solve the equation in guideline 2 for $y$:

$$(y - 3)(y + 1) = 0 \qquad \text{factor } y^2 - 2y - 3$$
$$y - 3 = 0, \quad y + 1 = 0 \qquad \text{set each factor equal to 0}$$
$$y = 3, \qquad y = -1 \qquad \text{solve for } y$$

These are the only possible $y$-values for the solutions of the system.

**Guideline 4**  Use the equation $x = 3 - 2y$ from guideline 1 to find the corresponding $x$-values:

$$\text{if} \quad y = 3, \quad \text{then} \quad x = 3 - 2(3) = 3 - 6 = -3$$
$$\text{if} \quad y = -1, \quad \text{then} \quad x = 3 - 2(-1) = 3 + 2 = 5$$

Thus, possible solutions are $(-3, 3)$ and $(5, -1)$.

**Guideline 5**  Substituting $x = -3$ and $y = 3$ in $x + y^2 = 6$, the first equation of the system, yields $-3 + 9 = 6$, a true statement. Substituting $x = -3$ and $y = 3$ in $x + 2y = 3$, the second equation of the system, yields $-3 + 6 = 3$, also a true statement. Hence, $(-3, 3)$ is a solution of the system. In a similar manner, we may check that $(5, -1)$ is also a solution.

The graphs of the two equations (a parabola and a line) are sketched in Figure 3, showing the two points of intersection.

**FIGURE 3**

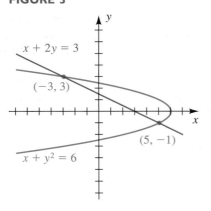

In future examples we will not list the specific guidelines that are used in finding solutions of systems.

In solving certain systems using the method of substitution, it is convenient to let $u$ or $v$ in the guidelines denote an *expression* involving another variable. This technique is illustrated in the next example with $u = x^2$.

**EXAMPLE 3    *Using the method of substitution***

Solve the following system and then sketch the graph of each equation, showing the points of intersection:

$$\begin{cases} x^2 + y^2 = 25 \\ x^2 + y \ = 19 \end{cases}$$

**SOLUTION**    We proceed as follows:

$$x^2 = 19 - y \qquad \text{solve } x^2 + y = 19 \text{ for } x^2$$
$$(19 - y) + y^2 = 25 \qquad \text{substitute } x^2 = 19 - y \text{ in } x^2 + y^2 = 25$$
$$y^2 - y - 6 = 0 \qquad \text{simplify}$$
$$(y - 3)(y + 2) = 0 \qquad \text{factor}$$
$$y - 3 = 0, \quad y + 2 = 0 \qquad \text{set each factor equal to } 0$$
$$y = 3, \qquad y = -2 \qquad \text{solve for } y$$

These are the only possible $y$-values for the solutions of the system. To find the corresponding $x$-values, we use $x^2 = 19 - y$:

$$\text{if} \quad y = 3, \quad \text{then} \quad x^2 = 19 - 3 = 16 \quad \text{and} \quad x = \pm 4$$
$$\text{if} \quad y = -2, \quad \text{then} \quad x^2 = 19 - (-2) = 21 \quad \text{and} \quad x = \pm\sqrt{21}$$

Thus, the only possible solutions of the systems are

$$(4, 3), \quad (-4, 3), \quad (\sqrt{21}, -2), \quad \text{and} \quad (-\sqrt{21}, -2).$$

We can check by substitution in the given equations that all four pairs are solutions.

The graph of $x^2 + y^2 = 25$ is a circle of radius 5 with center at the origin, and the graph of $y = 19 - x^2$ is a parabola with a vertical axis. The graphs are sketched in Figure 4. The points of intersection correspond to the solutions of the system.

There are, of course, other ways to find the solutions. We could solve the first equation for $x^2$, $x^2 = 25 - y^2$, and then substitute in the second, obtaining $25 - y^2 + y = 19$. Another method is to solve the second equation for $y$, $y = 19 - x^2$, and substitute in the first.

**FIGURE 4**

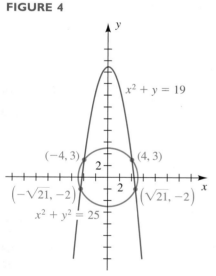

We can also consider equations in three variables $x$, $y$, and $z$, such as

$$x^2 y + xz + 3^y = 4z^3.$$

Such an equation has a **solution** $(a, b, c)$ if substitution of $a$, $b$, and $c$, for $x$, $y$, and $z$, respectively, yields a true statement. We refer to $(a, b, c)$ as an **ordered triple** of real numbers. Systems of equations are **equivalent systems** provided they have the same solutions. A system of equations in three variables and the solutions of the system are defined as in the two-variable case. Similarly, we can consider systems of *any* number of equations in *any* number of variables.

The method of substitution can be extended to these more complicated systems. For example, given three equations in three variables, suppose that it is possible to solve one of the equations for one variable in terms of the remaining two variables. By substituting that expression in each of the other equations, we obtain a system of two equations in two variables. The solutions of the two-variable system can then be used to find the solutions of the original system.

**EXAMPLE 4**    *Solving a system of three equations*

Solve the system

$$\begin{cases} x - y + z = 2 \\ xyz = 0 \\ 2y + z = 1 \end{cases}$$

**SOLUTION**    We proceed as follows:

$$z = 1 - 2y \quad \text{solve } 2y + z = 1 \text{ for } z$$

$$\begin{cases} x - y + (1 - 2y) = 2 \\ xy(1 - 2y) = 0 \end{cases} \quad \text{substitute } z = 1 - 2y \text{ in the first two equations}$$

$$\begin{cases} x - 3y - 1 = 0 \\ xy(1 - 2y) = 0 \end{cases} \quad \text{equivalent system}$$

We now find the solutions of the last system:

$$x = 3y + 1 \quad \text{solve } x - 3y - 1 = 0 \text{ for } x$$

$$(3y + 1)y(1 - 2y) = 0 \quad \begin{array}{l} \text{substitute } x = 3y + 1 \text{ in} \\ xy(1 - 2y) = 0 \end{array}$$

$$3y + 1 = 0, \quad y = 0, \quad 1 - 2y = 0 \quad \text{set each factor equal to 0}$$

$$y = -\tfrac{1}{3}, \quad y = 0, \quad y = \tfrac{1}{2} \quad \text{solve for } y$$

These are the only possible $y$-values for the solutions of the system.
To obtain the corresponding $x$-values, we substitute for $y$ in the equation $x = 3y + 1$, obtaining

$$x = 0, \quad x = 1, \quad \text{and} \quad x = \tfrac{5}{2}.$$

Using $z = 1 - 2y$ gives us the corresponding $z$-values

$$z = \tfrac{5}{3}, \quad z = 1, \quad \text{and} \quad z = 0.$$

Thus, the solutions $(x, y, z)$ of the original system must be among the ordered triples

$$\left(0, -\tfrac{1}{3}, \tfrac{5}{3}\right), \quad (1, 0, 1), \quad \text{and} \quad \left(\tfrac{5}{2}, \tfrac{1}{2}, 0\right).$$

Checking each shows that the three ordered triples are solutions of the system.

**EXAMPLE 5** *An application of a system of equations*

Is it possible to construct an aquarium with a glass top and two square ends that holds 16 ft$^3$ of water and requires 40 ft$^2$ of glass? (Disregard the thickness of the glass.)

*SOLUTION* We begin by sketching a typical aquarium and labeling it as in Figure 5, with $x$ and $y$ in feet. Referring to the figure and using formulas for volume and area, we see that

volume of the aquarium $= x^2y$      length $\times$ width $\times$ height

square feet of glass required $= 2x^2 + 4xy$   2 ends, 2 sides, top and bottom

**FIGURE 5**

Since the volume is to be 16 ft$^3$ and the area of the glass required is 40 ft$^2$, we obtain the following system of equations:

$$\begin{cases} x^2y = 16 \\ 2x^2 + 4xy = 40 \end{cases}$$

We find the solutions as follows:

$$y = \frac{16}{x^2} \quad \text{solve } x^2y = 16 \text{ for } y$$

$$2x^2 + 4x\left(\frac{16}{x^2}\right) = 40 \quad \text{substitute } y = 16/x^2 \text{ in } 2x^2 + 4xy = 40$$

$$x^2 + \frac{32}{x} = 20 \quad \text{cancel } x \text{ and divide by 2}$$

$$x^3 + 32 = 20x \quad \text{multiply by } x \ (x \neq 0)$$

$$x^3 - 20x + 32 = 0 \quad \text{subtract } 20x$$

We next look for rational solutions of the last equation. Dividing the polynomial $x^3 - 20x + 32$ synthetically by $x - 2$ gives us

$$\begin{array}{r|rrrr} 2 & 1 & 0 & -20 & 32 \\ & & 2 & 4 & -32 \\ \hline & 1 & 2 & -16 & 0 \end{array}$$

*(continued)*

Thus, one solution of $x^3 - 20x + 32 = 0$ is 2, and the remaining two solutions are zeros of the quotient $x^2 + 2x - 16$—that is, roots of the depressed equation

$$x^2 + 2x - 16 = 0.$$

By the quadratic formula,

$$x = \frac{-2 \pm \sqrt{2^2 - 4(1)(-16)}}{2} = \frac{-2 \pm 2\sqrt{17}}{2} = -1 \pm \sqrt{17}.$$

Since $x$ is positive, we may discard $x = -1 - \sqrt{17}$. Hence, the only possible values of $x$ are

$$x = 2 \quad \text{and} \quad x = -1 + \sqrt{17} \approx 3.12.$$

The corresponding $y$-values can be found by substituting for $x$ in the equation $y = 16/x^2$. Letting $x = 2$ gives us $y = \frac{16}{4} = 4$. Using these values, we obtain the dimensions 2 feet by 2 feet by 4 feet for an aquarium. Letting $x = -1 + \sqrt{17}$, we obtain $y = 16/(-1 + \sqrt{17})^2$, which simplifies to $y = \frac{1}{8}(9 + \sqrt{17}) \approx 1.64$. Thus, approximate dimensions for another aquarium are 3.12 feet by 3.12 feet by 1.64 feet.

## 8.1 EXERCISES

**Exer. 1–30: Use the method of substitution to solve the system.**

1. $\begin{cases} y = x^2 - 4 \\ y = 2x - 1 \end{cases}$

2. $\begin{cases} y = x^2 + 1 \\ x + y = 3 \end{cases}$

3. $\begin{cases} y^2 = 1 - x \\ x + 2y = 1 \end{cases}$

4. $\begin{cases} y^2 = x \\ x + 2y + 3 = 0 \end{cases}$

5. $\begin{cases} 2y = x^2 \\ y = 4x^3 \end{cases}$

6. $\begin{cases} x - y^3 = 1 \\ 2x = 9y^2 + 2 \end{cases}$

7. $\begin{cases} x + 2y = -1 \\ 2x - 3y = 12 \end{cases}$

8. $\begin{cases} 3x - 4y + 20 = 0 \\ 3x + 2y + 8 = 0 \end{cases}$

9. $\begin{cases} 2x - 3y = 1 \\ -6x + 9y = 4 \end{cases}$

10. $\begin{cases} 4x - 5y = 2 \\ 8x - 10y = -5 \end{cases}$

11. $\begin{cases} x + 3y = 5 \\ x^2 + y^2 = 25 \end{cases}$

12. $\begin{cases} 3x - 4y = 25 \\ x^2 + y^2 = 25 \end{cases}$

13. $\begin{cases} x^2 + y^2 = 8 \\ y - x = 4 \end{cases}$

14. $\begin{cases} x^2 + y^2 = 25 \\ 3x + 4y = -25 \end{cases}$

15. $\begin{cases} x^2 + y^2 = 9 \\ y - 3x = 2 \end{cases}$

16. $\begin{cases} x^2 + y^2 = 16 \\ y + 2x = -1 \end{cases}$

17. $\begin{cases} x^2 + y^2 = 16 \\ 2y - x = 4 \end{cases}$

18. $\begin{cases} x^2 + y^2 = 1 \\ y + 2x = -3 \end{cases}$

19. $\begin{cases} (x - 1)^2 + (y + 2)^2 = 10 \\ x + y = 1 \end{cases}$

20. $\begin{cases} xy = 2 \\ 3x - y + 5 = 0 \end{cases}$

21. $\begin{cases} y = 20/x^2 \\ y = 9 - x^2 \end{cases}$

22. $\begin{cases} x = y^2 - 4y + 5 \\ x - y = 1 \end{cases}$

23. $\begin{cases} y^2 - 4x^2 = 4 \\ 9y^2 + 16x^2 = 140 \end{cases}$

24. $\begin{cases} 25y^2 - 16x^2 = 400 \\ 9y^2 - 4x^2 = 36 \end{cases}$

25. $\begin{cases} x^2 - y^2 = 4 \\ x^2 + y^2 = 12 \end{cases}$

26. $\begin{cases} 6x^3 - y^3 = 1 \\ 3x^3 + 4y^3 = 5 \end{cases}$

27. $\begin{cases} x + 2y - z = -1 \\ 2x - y + z = 9 \\ x + 3y + 3z = 6 \end{cases}$

28. $\begin{cases} 2x - 3y - z^2 = 0 \\ x - y - z^2 = -1 \\ x^2 - xy = 0 \end{cases}$

$$29 \begin{cases} x^2 + z^2 = 5 \\ 2x + y = 1 \\ y + z = 1 \end{cases} \qquad 30 \begin{cases} x + 2z = 1 \\ 2y - z = 4 \\ xyz = 0 \end{cases}$$

*circuit*

31 The perimeter of a rectangle is 40 inches, and its area is 96 in.$^2$. Find its length and width.

32 Find the values of $b$ such that the system

$$\begin{cases} x^2 + y^2 = 4 \\ y = x + b \end{cases}$$

has

(a) one solution      (b) two solutions

(c) no solution

Interpret (a)–(c) graphically.

33 Is there a real number $x$ such that $x = 2^{-x}$? Decide by displaying graphically the system

$$\begin{cases} y = x \\ y = 2^{-x} \end{cases}$$

34 Is there a real number $x$ such that $x = \log x$? Decide by displaying graphically the system

$$\begin{cases} y = x \\ y = \log x \end{cases}$$

35 **Constructing tubing** Sections of cylindrical tubing are to be made from thin rectangular sheets that have an area of 200 in.$^2$ (see the figure). Is it possible to construct a tube that has a volume of 200 in.$^3$? If so, find the dimensions of the rectangular sheet.

**EXERCISE 35**

36 Shown in the figure is the graph of $y = x^2$ and a line of slope $m$ that passes through the point $(1, 1)$. Find the value of $m$ such that the line intersects the graph only at $(1, 1)$, and interpret graphically.

**EXERCISE 36**

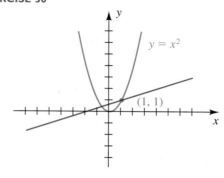

37 **Fish population** In fishery science, spawner-recruit functions are used to predict the number of adult fish $R$ in next year's breeding population from an estimate $S$ of the number of fish presently spawning.

(a) For a certain species of fish, $R = aS/(S + b)$. Estimate $a$ and $b$ from the data in the following table.

| Year | 1991 | 1992 | 1993 |
|---|---|---|---|
| **Number spawning** | 40,000 | 60,000 | 72,000 |

(b) Predict the breeding population for the year 1994.

38 **Fish population** Refer to Exercise 37. Ricker's spawner-recruit function is given by

$$R = aSe^{-bS}$$

for positive constants $a$ and $b$. This relationship predicts low recruitment from very high stocks and has been found to be appropriate for many species, such as arctic cod. Rework Exercise 37 using Ricker's spawner-recruit function.

39 **Competition for food** A *competition model* is a collection of equations that specifies how two or more species interact in competition for the food resources of an ecosystem. Let $x$ and $y$ denote the numbers (in hundreds) of two competing species, and suppose that the respective rates of growth $R_1$ and $R_2$ are given by

$$R_1 = 0.01x(50 - x - y),$$
$$R_2 = 0.02y(100 - y - 0.5x).$$

Determine the population levels $(x, y)$ at which both rates of growth are zero. (Such population levels are called *stationary points*.)

40 **Fencing a region** A rancher has 2420 feet of fence to enclose a rectangular region that lies along a straight

river. If no fence is used along the river (see the figure), is it possible to enclose 10 acres of land? Recall that 1 acre = 43,560 ft$^2$.

**EXERCISE 40**

**41  Constructing an aquarium**  Refer to Example 5. Is it possible to construct a small aquarium with an *open* top and two square ends that holds 2 ft$^3$ of water and requires 8 ft$^2$ of glass? If so, approximate the dimensions. (Disregard the thickness of the glass.)

**42  Isoperimetric problem**  The isoperimetric problem is to prove that of all plane geometric figures with the same perimeter (isoperimetric figures), the circle has the greatest area. Show that no rectangle has both the same area and the same perimeter as any circle.

**43  Moiré pattern**  A moiré pattern is formed when two geometrically regular patterns are superimposed. Shown in the figure is a pattern obtained from the family of circles $x^2 + y^2 = n^2$ and the family of horizontal lines $y = m$ for integers $m$ and $n$.

(a) Show that the points of intersection of the circle $x^2 + y^2 = n^2$ and the line $y = n - 1$ lie on a parabola.

(b) Work part (a) using the line $y = n - 2$.

**EXERCISE 43**

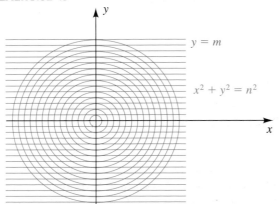

**44  Dimensions of a pill**  A spherical pill has diameter 1 centimeter. A second pill in the shape of a right circular cylinder is to be manufactured with the same volume and twice the surface area of the spherical pill.

(a) If $r$ is the radius and $h$ is the height of the cylindrical pill, show that $6r^2h = 1$ and $r^2 + rh = 1$. Conclude that $6r^3 - 6r + 1 = 0$.

(b) The positive solutions of $6r^3 - 6r + 1 = 0$ are approximately 0.172 and 0.903. Find the corresponding heights, and interpret these results.

**45  Hammer throw**  A hammer thrower is working on his form in a small practice area. The hammer spins, generating a circle with a radius of 5 feet, and when released, it hits a tall screen that is 50 feet from the center of the throwing area. Let coordinate axes be introduced as shown in the figure (not to scale).

(a) If the hammer is released at $(-4, -3)$ and travels in the tangent direction, where will it hit the screen?

(b) If the hammer is to hit at $(0, -50)$, where on the circle should it be released?

**EXERCISE 45**

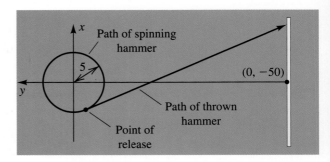

**46  Path of a tossed ball**  A person throws a ball from the edge of a hill, at an angle of 45° with the horizontal, as illustrated in the figure. The ball lands 50 feet down the hill, which has slope $-\frac{3}{4}$. Using calculus, it can be shown that the path of the ball is given by $y = ax^2 + x + c$ for some constants $a$ and $c$.

EXERCISE 46

(a) Disregarding the height of the person, find an equation for the path.

(b) What is the maximum height of the ball *off the ground?*

Exer. 47–50: Graph the two equations on the same coordinate plane and estimate the coordinates of the points of intersection.

47  $y = 5x^3 - 5x$;       $x^2 + y^2 = 4$

48  $9x^2 + y^2 = 9$;       $y = \tan^{-1} x$

49  $|x + \ln |x|| + y^2 = 0$;   $\dfrac{x^2}{4} + \dfrac{y^2}{2.25} = 1$

50  $y^3 - e^{x/2} = x$;       $y + 0.85x^2 = 2.1$

## 8.2  SYSTEMS OF LINEAR EQUATIONS IN TWO VARIABLES

An equation $ax + by = c$ (or, equivalently, $ax + by - c = 0$), with $a$ and $b$ not both zero, is a linear equation in two variables $x$ and $y$. Similarly, $ax + by + cz = d$ is a linear equation in three variables $x$, $y$, and $z$. We may also consider linear equations in four, five, or *any* number of variables. The most common systems of equations are those in which every equation is linear. In this section we shall consider only systems of two linear equations in two variables. Systems involving more than two variables are discussed in the next section.

Two systems of equations are equivalent if they have the same solutions. To find the solutions of a system, we may manipulate the equations until we obtain an equivalent system of simple equations for which the solutions can be found readily. Some manipulations (or *transformations*) that lead to equivalent systems are stated without proof in the next theorem.

**Theorem on Equivalent Systems**

Given a system of equations, an equivalent system results if

**(1)** two equations are interchanged.

**(2)** an equation is multiplied or divided by a nonzero constant.

**(3)** a constant multiple of one equation is added to another equation.

A *constant multiple* of an equation is obtained by multiplying *each* term of the equation by the same nonzero constant $k$. When applying part (3) of the theorem, we often use the phrase *add to one equation $k$ times* *any other equation.* To *add* two equations means to add corresponding sides of the equations.

The next example illustrates how the theorem on equivalent systems may be used to solve a system of linear equations.

### EXAMPLE 1    *Using the theorem on equivalent systems*

Solve the system

$$\begin{cases} x + 3y = -1 \\ 2x - y = 5 \end{cases}$$

**SOLUTION**    Using the preceding theorem gives us the following equivalent systems:

$$\begin{cases} x + 3y = -1 \\ 2x - y = 5 \end{cases} \quad \text{given}$$

$$\begin{cases} x + 3y = -1 \\ 6x - 3y = 15 \end{cases} \quad \text{multiply the second equation by 3}$$

$$\begin{cases} x + 3y = -1 \\ 7x = 14 \end{cases} \quad \text{add the first equation to the second}$$

We see from the last system that $7x = 14$, and hence $x = \frac{14}{7} = 2$. To find the corresponding $y$-value, we substitute 2 for $x$ in $x + 3y = -1$, obtaining $y = -1$. Thus, $(2, -1)$ is the only solution of the system.

There are many other ways to use the theorem on equivalent systems to find the solution. One such method is to proceed as follows:

$$\begin{cases} x + 3y = -1 \\ 2x - y = 5 \end{cases} \quad \text{given}$$

$$\begin{cases} -2x - 6y = 2 \\ 2x - y = 5 \end{cases} \quad \text{multiply the first equation by } -2$$

$$\begin{cases} -2x - 6y = 2 \\ - 7y = 7 \end{cases} \quad \text{add the first equation to the second}$$

We see from the last system that $-7y = 7$, or $y = -1$. To find the corresponding $x$-value, we could substitute $-1$ for $y$ in $x + 3y = -1$, obtaining $x = 2$. Hence, $(2, -1)$ is the solution.

The graphs of the two equations are lines that intersect at the point $(2, -1)$, as shown in Figure 6.

**FIGURE 6**

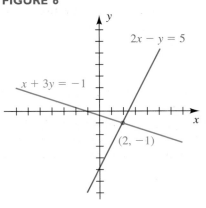

The technique used in Example 1 is called the **method of elimination**, since it involves the elimination of a variable from one of the equations.

8.2 SYSTEMS OF LINEAR EQUATIONS IN TWO VARIABLES     535

The method of elimination usually leads to solutions in fewer steps than does the method of substitution discussed in the preceding section.

**EXAMPLE 2**     *A system of linear equations with an infinite number of solutions*

Solve the system

$$\begin{cases} 3x + y = 6 \\ 6x + 2y = 12 \end{cases}$$

**FIGURE 7**

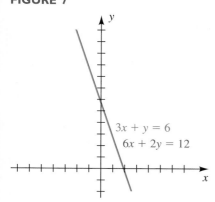

$3x + y = 6$
$6x + 2y = 12$

**SOLUTION**     Multiplying the second equation by $\frac{1}{2}$ gives us

$$\begin{cases} 3x + y = 6 \\ 3x + y = 6 \end{cases}$$

Thus, $(a, b)$ is a solution if and only if $3a + b = 6$—that is, $b = 6 - 3a$. It follows that the solutions consist of ordered pairs of the form $(a, 6 - 3a)$, where $a$ is any real number. If we wish to find particular solutions, we may substitute various values for $a$. A few solutions are $(0, 6)$, $(1, 3)$, $(3, -3)$, $(-2, 12)$, and $(\sqrt{2}, 6 - 3\sqrt{2})$.

The graph of each equation is the same line, as shown in Figure 7.

**EXAMPLE 3**     *A system of linear equations with no solutions*

Solve the system

$$\begin{cases} 3x + y = 6 \\ 6x + 2y = 20 \end{cases}$$

**FIGURE 8**

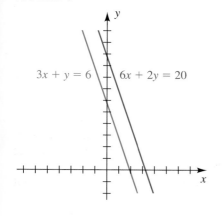

$3x + y = 6$     $6x + 2y = 20$

**SOLUTION**     If we add to the second equation $-2$ times the first equation, $-6x - 2y = -12$, we obtain the equivalent system

$$\begin{cases} 3x + y = 6 \\ 0 = 8 \end{cases}$$

The last equation can be written $0x + 0y = 8$, which is false for every ordered pair $(x, y)$. Thus, the system has no solution.

The graphs of the two equations in the given system are lines that have the same slope and hence are parallel (see Figure 8). The conclusion that the system has no solution corresponds to the fact that these lines do not intersect.

The preceding three examples illustrate typical outcomes of solving a system of two linear equations in two variables: there is either exactly one solution, an infinite number of solutions, or no solution. A system is **con-**

**sistent** if it has at least one solution. A system with an infinite number of solutions is **dependent and consistent.** A system is **inconsistent** if it has no solution.

Since the graph of any linear equation $ax + by = c$ is a line, *exactly one* of the three cases listed in the following table holds for any system of two such equations.

Characteristics of a System of Two
Linear Equations in Two Variables

| Graphs | Number of solutions | Classification |
|---|---|---|
| Nonparallel lines | One solution | Consistent system |
| Identical lines | Infinite number of solutions | Dependent and consistent system |
| Parallel lines | No solution | Inconsistent system |

In practice, there should be little difficulty determining which of the three cases occurs. The case of the unique solution will become apparent when suitable transformations are applied to the system, as illustrated in Example 1. The case of an infinite number of solutions is similar to that of Example 2, where one of the equations can be transformed into the other. The case of no solution is indicated by a contradiction, such as the statement $0 = 8$, which appeared in Example 3.

Certain applied problems can be solved by introducing systems of two linear equations, as illustrated in the next two examples.

**EXAMPLE 4**    *An application of a system of linear equations*

A produce company has a 100-acre farm on which it grows lettuce and cabbage. Each acre of cabbage requires 600 hours of labor, and each acre of lettuce needs 400 hours of labor. If 45,000 hours are available and if all land and labor resources are to be used, find the number of acres of each crop that should be planted.

*SOLUTION*    Let us introduce variables to denote the unknown quantities as follows:

$$x = \text{number of acres of cabbage}$$

$$y = \text{number of acres of lettuce}$$

Thus, the number of hours of labor required for each crop can be expressed as follows:

$$600x = \text{number of hours required for cabbage}$$

$$400y = \text{number of hours required for lettuce}$$

Using the facts that the total number of acres is 100 and the total number of hours available is 45,000 leads to the following system:

$$\begin{cases} x + y = 100 \\ 600x + 400y = 45{,}000 \end{cases}$$

We next use the method of elimination:

$$\begin{cases} x + y = 100 \\ 6x + 4y = 450 \end{cases} \qquad \text{divide the second equation by 100}$$

$$\begin{cases} -6x - 6y = -600 \\ 6x + 4y = 450 \end{cases} \qquad \text{multiply the first equation by } -6$$

$$\begin{cases} -6x - 6y = -600 \\ -2y = -150 \end{cases} \qquad \text{add the first equation to the second}$$

We see from the last equation that $-2y = -150$, or $y = 75$. Substituting 75 for $y$ in $x + y = 100$ gives us $x = 25$. Hence, the company should plant 25 acres of cabbage and 75 acres of lettuce.

---

**EXAMPLE 5** *Finding the speed of the current in a river*

A motorboat, operating at full throttle, made a trip 4 miles upstream (against a constant current) in 15 minutes. The return trip (with the same current and at full throttle) took 12 minutes. Find the speed of the current and the equivalent speed of the boat in still water.

*SOLUTION*    We begin by introducing variables to denote the unknown quantities. Thus, let

$$x = \text{speed of boat (in mi/hr)}$$

$$y = \text{speed of current (in mi/hr)}.$$

We plan to use the formula $d = rt$, where $d$ denotes the distance traveled, $r$ the rate, and $t$ the time. Since the current slows the boat as it travels upstream but adds to its speed as it travels downstream, we obtain

$$\text{upstream rate} = x - y \quad \text{(in mi/hr)}$$

$$\text{downstream rate} = x + y \quad \text{(in mi/hr)}.$$

The time (in hours) traveled in each direction is

$$\text{upstream time} = \tfrac{15}{60} = \tfrac{1}{4} \text{ hr}$$

$$\text{downstream time} = \tfrac{12}{60} = \tfrac{1}{5} \text{ hr}.$$

*(continued)*

The distance is 4 miles for each trip. Substituting in $d = rt$ gives us the system

$$\begin{cases} 4 = (x - y)\left(\tfrac{1}{4}\right) \\ 4 = (x + y)\left(\tfrac{1}{5}\right) \end{cases}$$

Applying the theorem on equivalent systems, we obtain

$$\begin{cases} x - y = 16 \\ x + y = 20 \end{cases}$$  multiply the first equation by 4 and the second by 5

$$\begin{cases} x - y = 16 \\ 2x \quad\;\; = 36 \end{cases}$$  add the first equation to the second

We see from the last equation that $2x = 36$, or $x = 18$. Substituting 18 for $x$ in $x + y = 20$ gives us $y = 2$. Hence, the speed of the boat in still water is 18 mi/hr, and the speed of the current is 2 mi/hr.

## 8.2 EXERCISES

**Exer. 1–20: Solve the system.**

1 $\begin{cases} 2x + 3y = 2 \\ x - 2y = 8 \end{cases}$

2 $\begin{cases} 4x + 5y = 13 \\ 3x + y = -4 \end{cases}$

3 $\begin{cases} 2x + 5y = 16 \\ 3x - 7y = 24 \end{cases}$

4 $\begin{cases} 7x - 8y = 9 \\ 4x + 3y = -10 \end{cases}$

5 $\begin{cases} 3r + 4s = 3 \\ r - 2s = -4 \end{cases}$

6 $\begin{cases} 9u + 2v = 0 \\ 3u - 5v = 17 \end{cases}$

7 $\begin{cases} 5x - 6y = 4 \\ 3x + 7y = 8 \end{cases}$

8 $\begin{cases} 2x + 8y = 7 \\ 3x - 5y = 4 \end{cases}$

9 $\begin{cases} \tfrac{1}{3}c + \tfrac{1}{2}d = 5 \\ c - \tfrac{2}{3}d = -1 \end{cases}$

10 $\begin{cases} \tfrac{1}{2}t - \tfrac{1}{3}v = \tfrac{3}{2} \\ \tfrac{2}{3}t + \tfrac{1}{4}v = \tfrac{5}{12} \end{cases}$

11 $\begin{cases} \sqrt{3}x - \sqrt{2}y = 2\sqrt{3} \\ 2\sqrt{2}x + \sqrt{3}y = \sqrt{2} \end{cases}$

12 $\begin{cases} 0.11x - 0.03y = 0.25 \\ 0.12x + 0.05y = 0.70 \end{cases}$

13 $\begin{cases} 2x - 3y = 5 \\ -6x + 9y = 12 \end{cases}$

14 $\begin{cases} 3p - q = 7 \\ -12p + 4q = 3 \end{cases}$

15 $\begin{cases} 3m - 4n = 2 \\ -6m + 8n = -4 \end{cases}$

16 $\begin{cases} x - 5y = 2 \\ 3x - 15y = 6 \end{cases}$

17 $\begin{cases} 2y - 5x = 0 \\ 3y + 4x = 0 \end{cases}$

18 $\begin{cases} 3x + 7y = 9 \\ y = 5 \end{cases}$

19 $\begin{cases} \dfrac{2}{x} + \dfrac{3}{y} = -2 \\ \dfrac{4}{x} - \dfrac{5}{y} = 1 \end{cases}$  $\left(Hint: \text{Let } u = \dfrac{1}{x} \text{ and } v = \dfrac{1}{y}.\right)$

20 $\begin{cases} \dfrac{3}{x-1} + \dfrac{4}{y+2} = 2 \\ \dfrac{6}{x-1} - \dfrac{7}{y+2} = -3 \end{cases}$

**Exer. 21–24: Solve the system for $a$ and $b$. (Hint: Treat terms such as $e^{3x}$, $\cos x$, and $\sin x$ as "constant coefficients.")**

21 $\begin{cases} ae^{3x} + be^{-3x} = 0 \\ a(3e^{3x}) + b(-3e^{-3x}) = e^{3x} \end{cases}$

22 $\begin{cases} ae^{-x} + be^{4x} = 0 \\ -ae^{-x} + b(4e^{4x}) = 2 \end{cases}$

23 $\begin{cases} a\cos x + b\sin x = 0 \\ -a\sin x + b\cos x = \tan x \end{cases}$

24 $\begin{cases} a\cos x + b\sin x = 0 \\ -a\sin x + b\cos x = \sin x \end{cases}$

25 **Ticket sales** The price of admission to a high school play was $1.50 for students and $2.25 for nonstudents. If 450 tickets were sold for a total of $777.75, how many of each kind were purchased?

26 **Air travel** An airline that flies from Los Angeles to Albuquerque with a stopover in Phoenix charges a fare of $45 to Phoenix and a fare of $60 from Los Angeles to Albuquerque. A total of 185 passengers boarded the plane in Los Angeles, and fares totaled $10,500. How many passengers got off the plane in Phoenix?

27 **Crayon dimensions** A crayon 8 centimeters in length and 1 centimeter in diameter will be made from 5 cm³ of colored wax. The crayon is to have the shape of a cylinder surmounted by a small conical tip (see the figure). Find the length $x$ of the cylinder and the height $y$ of the cone.

EXERCISE 27

28 **Rowing a boat** A man rows a boat 500 feet upstream against a constant current in 10 minutes. He then rows 300 feet downstream (with the same current) in 5 minutes. Find the speed of the current and the equivalent rate at which he can row in still water.

29 **Table top dimensions** A large table for a conference room is to be constructed in the shape of a rectangle with two semicircles at the ends (see the figure). The table is to have a perimeter of 40 feet, and the area of the rectangular portion is to be twice the sum of the areas of the two ends. Find the length $l$ and width $w$ of the rectangular portion.

EXERCISE 29

30 **Investment income** A woman has $15,000 to invest in two funds that pay simple interest at the rates of 6% and 8%. Interest on the 6% fund is tax-exempt; however, income tax must be paid on interest on the 8% fund. Being in a high tax bracket, the woman does not wish to invest the entire sum in the 8% account. Is there a way of investing the money so that she will receive $1000 in interest at the end of one year?

31 **Bobcat population** A bobcat population is classified by age into kittens (less than 1 year old) and adults (at least 1 year old). All adult females, including those born the prior year, have a litter each June, with an average litter size of 3 kittens. The springtime population of bobcats in a certain area is estimated to be 6000, and the male-female ratio is one. Estimate the number of adults and kittens in the population.

32 **Flow rates** A 300-gallon water storage tank is filled by a single inlet pipe, and two identical outlet pipes can be used to supply water to the surrounding fields (see the figure). It takes 5 hours to fill an empty tank when both outlet pipes are open. When one outlet pipe is closed, it takes 3 hours to fill the tank. Find the flow rates (in gallons per hour) in and out of the pipes.

EXERCISE 32

33 **Mixing a silver alloy** A silversmith has two alloys, one containing 35% silver and the other 60% silver. How much of each should be melted and combined to obtain 100 grams of an alloy containing 50% silver?

34 **Mixing nuts** A merchant wishes to mix peanuts costing $3 per pound with cashews costing $8 per pound to obtain 60 pounds of a mixture costing $5 per pound. How many pounds of each variety should be mixed?

35 **Air travel** An airplane, flying with a tail wind, travels 1200 miles in 2 hours. The return trip, against the wind, takes $2\frac{1}{2}$ hours. Find the cruising speed of the plane and

the speed of the wind (assume that both rates are constant).

**36 Filling orders** A stationery company sells two types of notepads to college bookstores, the first wholesaling for 50¢ and the second for 70¢. The company receives an order for 500 notepads, together with a check for $286. If the order fails to specify the number of each type, how should the company fill the order?

**†37 Acceleration** As a ball rolls down an inclined plane, its velocity $v(t)$ (in cm/sec) at time $t$ (in seconds) is given by $v(t) = v_0 + at$ for initial velocity $v_0$ and acceleration $a$ (in cm/sec$^2$). If $v(2) = 16$ and $v(5) = 25$, find $v_0$ and $a$.

**38 Vertical projection** If an object is projected vertically upward from an altitude of $s_0$ feet with an initial velocity of $v_0$ ft/sec, then its distance $s(t)$ above the ground after $t$ seconds is

$$s(t) = -16t^2 + v_0 t + s_0.$$

If $s(1) = 84$ and $s(2) = 116$, what are $v_0$ and $s_0$?

**†39 Planning production** A small furniture company manufacturers sofas and recliners. Each sofa requires 8 hours of labor and $60 in materials, while a recliner can be built for $35 in 6 hours. The company has 340 hours of labor available each week and can afford to buy $2250 worth of materials. How many recliners and sofas can be produced if all labor hours and all materials must be used?

**40 Livestock diet** A rancher is preparing an oat-cornmeal mixture for livestock. Each ounce of oats provides 4 grams of protein and 18 grams of carbohydrates, and an ounce of cornmeal provides 3 grams of protein and 24 grams of carbohydrates. How many ounces of each can be used to meet the nutritional goals of 200 grams of protein and 1320 grams of carbohydrates per feeding?

**41 Services swap** A plumber and an electrician are each doing repairs on their offices and agree to swap services. The number of hours spent on each of the projects is shown in the following table.

| | Plumber's office | Electrician's office |
|---|---|---|
| Plumber's hours | 6 | 4 |
| Electrician's hours | 5 | 6 |

They would prefer to call the matter even, but because of tax laws, they must charge for all work performed. They agree to select hourly wage rates so that the bill on

each project will match the income that each person would ordinarily receive for a comparable job.

(a) If $x$ and $y$ denote the hourly wages of the plumber and electrician, respectively, show that

$$6x + 5y = 10x \quad \text{and} \quad 4x + 6y = 11y.$$

Describe the solutions to this system.

(b) If the plumber ordinarily makes $20 per hour, what should the electrician charge?

**42** Find equations for the altitudes of the triangle with vertices $A(-3, 2)$, $B(5, 4)$, and $C(3, -8)$, and find the point at which the altitudes intersect.

**43 Warming trend in Paris** As a result of urbanization, the temperatures in Paris have increased. In 1891 the average daily minimum and maximum temperatures were 5.8 °C and 15.1 °C, respectively. Between 1891 and 1968, these average temperatures rose 0.019 °C/yr and 0.011 °C/yr, respectively. Assuming the increases were linear, find the year when the difference between the minimum and maximum temperatures was 9 °C, and determine the corresponding average maximum temperature.

**44 Long distance telephone rates** A telephone company charges customers a certain amount for the first minute of a long distance call and another amount for each additional minute. A customer makes two calls to the same city—a 36-minute call for $7.27 and a 13-minute call for $2.67.

(a) Determine the cost for the first minute and the cost for each additional minute.

(b) If there is a federal tax rate of 3.2% and a state tax rate of 7.2% on all long distance calls, find, to the nearest minute, the longest call to the same city whose cost will not exceed $5.00.

**45 Price and demand** Suppose consumers will buy 1,000,000 T-shirts if the selling price is $10, but for each $1 increase in price, they will buy 100,000 fewer T-shirts. Moreover, suppose vendors will order 2,000,000 T-shirts if the selling price is $15, and for every $1 increase in price, they will order an additional 150,000.

(a) Express the number $Q$ of T-shirts consumers will buy if the selling price is $p$ dollars.

(b) Express the number $K$ of T-shirts vendors will order if the selling price is $p$ dollars.

(c) Determine the market price—that is, the price when $Q = K$.

## 8.3 SYSTEMS OF LINEAR EQUATIONS IN MORE THAN TWO VARIABLES

For systems of linear equations containing more than two variables, we can use either the method of substitution explained in Section 8.1 or the method of elimination developed in Section 8.2. The method of elimination is the shorter and more straightforward technique for finding solutions. In addition, it leads to the matrix technique, discussed in this section.

**EXAMPLE 1** *Using the method of elimination to solve a system of linear equations*

Solve the system

$$\begin{cases} x - 2y + 3z = 4 \\ 2x + y - 4z = 3 \\ -3x + 4y - z = -2 \end{cases}$$

**SOLUTION**

$$\begin{cases} x - 2y + 3z = 4 \\ 5y - 10z = -5 \\ -3x + 4y - z = -2 \end{cases}$$ add $-2$ times the first equation to the second equation

$$\begin{cases} x - 2y + 3z = 4 \\ 5y - 10z = -5 \\ -2y + 8z = 10 \end{cases}$$ add 3 times the first equation to the third equation

$$\begin{cases} x - 2y + 3z = 4 \\ y - 2z = -1 \\ -2y + 8z = 10 \end{cases}$$ multiply the second equation by $\frac{1}{5}$

$$\begin{cases} x - 2y + 3z = 4 \\ y - 2z = -1 \\ 4z = 8 \end{cases}$$ add 2 times the second equation to the third equation

$$\begin{cases} x - 2y + 3z = 4 \\ y - 2z = -1 \\ z = 2 \end{cases}$$ multiply the third equation by $\frac{1}{4}$

The solutions of the last system are easy to find by **back substitution.** From the third equation, we see that $z = 2$. Substituting 2 for $z$ in the second equation, $y - 2z = -1$, we get $y = 3$. Finally, we find the $x$-value by substituting $y = 3$ and $z = 2$ in the first equation, $x - 2y + 3z = 4$, obtaining $x = 4$. Thus, there is one solution, $(4, 3, 2)$.

Any system of three linear equations in three variables has either a *unique solution,* an *infinite number of solutions,* or *no solution.* As for two equations in two variables, the terminology used to describe these systems is *consistent, dependent and consistent,* or *inconsistent,* respectively.

If we analyze the method of solution in Example 1, we see that the symbols used for the variables are immaterial. The *coefficients* of the variables are what we must consider. Thus, if different symbols such as $r$, $s$, and $t$ are used for the variables, we obtain the system

$$\begin{cases} r - 2s + 3t = \phantom{-}4 \\ 2r + \phantom{-}s - 4t = \phantom{-}3 \\ -3r + 4s - \phantom{-}t = -2 \end{cases}$$

The method of elimination can then proceed exactly as in the example. Since this is true, it is possible to simplify the process. Specifically, we introduce a scheme for keeping track of the coefficients in such a way that we do not have to write down the variables. Referring to the preceding system, we first check that variables appear in the same order in each equation and that terms not involving variables are to the right of the equal signs. We then list the numbers that are involved in the equations as follows:

$$\begin{bmatrix} 1 & -2 & 3 & 4 \\ 2 & 1 & -4 & 3 \\ -3 & 4 & -1 & -2 \end{bmatrix}$$

An array of numbers of this type is called a **matrix.** The **rows** of the matrix are the numbers that appear next to each other *horizontally:*

| 1 | −2 | 3 | 4 | first row, $R_1$ |
|---|----|---|---|------------------|
| 2 | 1 | −4 | 3 | second row, $R_2$ |
| −3 | 4 | −1 | −2 | third row, $R_3$ |

The **columns** of the matrix are the numbers that appear next to each other *vertically:*

| first column, $C_1$ | second column, $C_2$ | third column, $C_3$ | fourth column, $C_4$ |
|---------------------|----------------------|---------------------|----------------------|
| 1 | −2 | 3 | 4 |
| 2 | 1 | −4 | 3 |
| −3 | 4 | −1 | −2 |

The matrix obtained from a system of linear equations in the preceding manner is the **matrix of the system.** If we delete the last column of this matrix, the remaining array of numbers is the **coefficient matrix.** Since the matrix of the system can be obtained from the coefficient matrix by adjoining one column, we call it the **augmented coefficient matrix** or simply the **augmented matrix.** Later, when we use matrices to find the solutions of a system of linear equations, we shall introduce a vertical line segment in the augmented matrix to indicate where the equal signs would appear in the corresponding system of equations, as in the next illustration.

### ILLUSTRATION

### Coefficient Matrix and Augmented Matrix

| | coefficient | augmented |
|---|---|---|
| system | matrix | matrix |

$$\begin{cases} x - 2y + 3z = 4 \\ 2x + y - 4z = 3 \\ -3x + 4y - z = -2 \end{cases} \qquad \begin{bmatrix} 1 & -2 & 3 \\ 2 & 1 & -4 \\ -3 & 4 & -1 \end{bmatrix} \qquad \begin{bmatrix} 1 & -2 & 3 & 4 \\ 2 & 1 & -4 & 3 \\ -3 & 4 & -1 & -2 \end{bmatrix}$$

Before discussing a matrix method of solving a system of linear equations, let us state a general definition of a matrix. We shall use a **double subscript notation,** denoting the number that appears in row $i$ and column $j$ by $a_{ij}$. The **row subscript** of $a_{ij}$ is $i$, and the **column subscript** is $j$.

**Definition of a Matrix**

Let $m$ and $n$ be positive integers. An $m \times n$ **matrix** is an array of the following form, where each $a_{ij}$ is a real number:

$$\begin{bmatrix} a_{11} & a_{12} & a_{13} & \cdots & a_{1n} \\ a_{21} & a_{22} & a_{23} & \cdots & a_{2n} \\ a_{31} & a_{32} & a_{33} & \cdots & a_{3n} \\ \vdots & \vdots & \vdots & & \vdots \\ a_{m1} & a_{m2} & a_{m3} & \cdots & a_{mn} \end{bmatrix}$$

The notation $m \times n$ in the definition is read "$m$ by $n$." We often say that the matrix *is* $m \times n$ and call $m \times n$ the **size** of the matrix. It is possible to consider matrices in which the symbols $a_{ij}$ represent complex numbers, polynomials, or other mathematical objects; however, we shall not do so in this text. The rows and columns of a matrix are defined as before. Thus, the matrix in the definition has $m$ rows and $n$ columns. Note that $a_{23}$ is in row 2 and column 3 and $a_{32}$ is in row 3 and column 2. Each $a_{ij}$ is an **element of the matrix.** The elements $a_{11}, a_{22}, a_{33}, \ldots$ are the **main diagonal elements.** If $m = n$, the matrix is a **square matrix of order $n$.**

### ILLUSTRATION

#### $m \times n$ **Matrices**

$$
\underset{2 \times 3}{\blacksquare \begin{bmatrix} -5 & 3 & 1 \\ 7 & 0 & -2 \end{bmatrix}}
\qquad
\underset{2 \times 2}{\blacksquare \begin{bmatrix} 5 & -1 \\ 2 & 3 \end{bmatrix}}
\qquad
\underset{1 \times 3}{\blacksquare \begin{bmatrix} 3 & 1 & -2 \end{bmatrix}}
$$

$$
\underset{3 \times 2}{\blacksquare \begin{bmatrix} 2 & -1 \\ 0 & 1 \\ 8 & 3 \end{bmatrix}}
\qquad
\underset{3 \times 1}{\blacksquare \begin{bmatrix} -4 \\ 0 \\ 5 \end{bmatrix}}
$$

To find the solutions of a system of linear equations, we begin with the augmented matrix. If a variable does not appear in an equation, we assume that the coefficient is zero. We then work with the rows of the matrix *just as though they were equations.* The only items missing are the symbols for the variables, the addition signs used between terms, and the equal signs. We simply keep in mind that the numbers in the first column are the coefficients of the first variable, the numbers in the second column are the coefficients of the second variable, and so on. The rules for transforming a matrix are formulated so that they always produce a matrix of an equivalent system of equations.

The next theorem is a restatement, in terms of matrices, of the theorem on equivalent systems in Section 8.2. In part (2) of the theorem, the terminology *a row is multiplied by a nonzero constant* means that each element in the row is multiplied by the constant. To *add* two rows of a matrix, as in part (3), we add corresponding elements in each row.

**Matrix Row Transformation Theorem**

> Given a matrix of a system of linear equations, a matrix of an equivalent system results if
>
> **(1)** two rows are interchanged.
>
> **(2)** a row is multiplied by a nonzero constant.
>
> **(3)** a constant multiple of one row is added to another row.

We refer to 1–3 as the **elementary row transformations** of a matrix. If a matrix is obtained from another matrix by one or more elementary row transformations, the two matrices are said to be **equivalent** or, more precisely, **row equivalent.** We shall use the symbols in the following chart to denote elementary row transformations of a matrix, where the arrow $\rightarrow$ may be read "replaces." Thus, for the transformation $k\mathrm{R}_i \rightarrow \mathrm{R}_i$, the constant multiple $k\mathrm{R}_i$ *replaces* $\mathrm{R}_i$. Similarly, for $k\mathrm{R}_i + \mathrm{R}_j \rightarrow \mathrm{R}_j$, the sum $k\mathrm{R}_i + \mathrm{R}_j$ *replaces* $\mathrm{R}_j$. For convenience, we shall write $(-1)\mathrm{R}_i$ as $-\mathrm{R}_i$.

**Elementary Row Transformations of a Matrix**

| Symbol | Meaning |
|---|---|
| $R_i \leftrightarrow R_j$ | Interchange rows $R_i$ and $R_j$ |
| $kR_i \rightarrow R_i$ | Multiply row $R_i$ by $k$ |
| $kR_i + R_j \rightarrow R_j$ | Add $kR_i$ to row $R_j$ |

We shall next rework Example 1 using matrices. You should compare the two solutions, since analogous steps are employed in each case.

**EXAMPLE 2** *Using matrices to solve a system of linear equations*

Solve the system

$$\begin{cases} x - 2y + 3z = 4 \\ 2x + y - 4z = 3 \\ -3x + 4y - z = -2 \end{cases}$$

**SOLUTION** We begin with the matrix of the system—that is, with the augmented matrix:

$$\begin{bmatrix} 1 & -2 & 3 & | & 4 \\ 2 & 1 & -4 & | & 3 \\ -3 & 4 & -1 & | & -2 \end{bmatrix}$$

We next apply elementary row transformations to obtain another (simpler) matrix of an equivalent system of equations. These transformations correspond to the manipulations used for equations in Example 1. We will place appropriate symbols between equivalent matrices.

$$\begin{bmatrix} 1 & -2 & 3 & | & 4 \\ 2 & 1 & -4 & | & 3 \\ -3 & 4 & -1 & | & -2 \end{bmatrix} \begin{matrix} \\ -2R_1 + R_2 \rightarrow R_2 \\ 3R_1 + R_3 \rightarrow R_3 \end{matrix} \begin{bmatrix} 1 & -2 & 3 & | & 4 \\ 0 & 5 & -10 & | & -5 \\ 0 & -2 & 8 & | & 10 \end{bmatrix} \begin{matrix} \\ \text{add } -2R_1 \text{ to } R_2 \\ \text{add } 3R_1 \text{ to } R_3 \end{matrix}$$

$$\frac{1}{5}R_2 \rightarrow R_2 \begin{bmatrix} 1 & -2 & 3 & | & 4 \\ 0 & 1 & -2 & | & -1 \\ 0 & -2 & 8 & | & 10 \end{bmatrix} \text{multiply } R_2 \text{ by } \tfrac{1}{5}$$

$$2R_2 + R_3 \rightarrow R_3 \begin{bmatrix} 1 & -2 & 3 & | & 4 \\ 0 & 1 & -2 & | & -1 \\ 0 & 0 & 4 & | & 8 \end{bmatrix} \text{add } 2R_2 \text{ to } R_3$$

$$\frac{1}{4}R_3 \rightarrow R_3 \begin{bmatrix} 1 & -2 & 3 & | & 4 \\ 0 & 1 & -2 & | & -1 \\ 0 & 0 & 1 & | & 2 \end{bmatrix} \text{multiply } R_3 \text{ by } \tfrac{1}{4}$$

*(continued)*

We use the final matrix to return to the system of equations

$$\begin{cases} x - 2y + 3z = 4 \\ \phantom{x - 2}y - 2z = -1 \\ \phantom{x - 2y - 2}z = 2 \end{cases}$$

which is equivalent to the original system. The solution $x = 4$, $y = 3$, $z = 2$ may now be found by back substitution, as in Example 1.

The final matrix in the solution of Example 2 is in **echelon form.** In general, a matrix is in echelon form if it satisfies the following conditions.

**Echelon Form of a Matrix**

> **(1)** The first nonzero number in each row, reading from left to right, is 1.
>
> **(2)** The column containing the first nonzero number in any row is to the left of the column containing the first nonzero number in the row below.
>
> **(3)** Rows consisting entirely of zeros may appear at the bottom of the matrix.

The following is an illustration of a $6 \times 7$ matrix in echelon form. The symbols $a_{ij}$ represent real numbers.

**ILLUSTRATION**

**Echelon Form**

$$\begin{bmatrix} 1 & a_{12} & a_{13} & a_{14} & a_{15} & a_{16} & a_{17} \\ 0 & 1 & a_{23} & a_{24} & a_{25} & a_{26} & a_{27} \\ 0 & 0 & 0 & 1 & a_{35} & a_{36} & a_{37} \\ 0 & 0 & 0 & 0 & 0 & 1 & a_{47} \\ 0 & 0 & 0 & 0 & 0 & 0 & 0 \\ 0 & 0 & 0 & 0 & 0 & 0 & 0 \end{bmatrix}$$

The following guidelines may be used to find echelon forms.

**Guidelines for Finding the Echelon Form of a Matrix**

**1** Locate the *first* column that contains nonzero elements, and apply elementary row transformations to get the number 1 into the first row of that column.

**2** Apply elementary row transformations of the type $kR_1 + R_j \rightarrow R_j$ for $j > 1$ to get 0 underneath the 1 obtained in guideline 1 in each of the remaining rows.

**3** *Disregard the first row.* Locate the next column that contains nonzero elements, and apply elementary row transformations to get the number 1 into the *second* row of that column.

**4** Apply elementary row transformations of the type $kR_2 + R_j \rightarrow R_j$ for $j > 2$ to get 0 underneath the 1 obtained in guideline 3 in each of the remaining rows.

**5** *Disregard the first and second rows.* Locate the next column that contains nonzero elements, and repeat the procedure.

**6** Continue the process until the echelon form is reached.

Not all echelon forms contain rows consisting of only zeros (see Example 2).

We can use elementary row operations to transform the matrix of any system of linear equations to echelon form. The echelon form can then be used to produce a system of equations that is equivalent to the original system. The solutions of the given system may be found by back substitution. The next example illustrates this technique for a system of four linear equations.

**EXAMPLE 3**    *Using an echelon form to solve a system of linear equations*

Solve the system

$$\begin{cases} -2x + 3y + 4z \qquad\quad = -1 \\ \quad x \qquad - 2z + 2w = \quad 1 \\ \qquad\quad y + \ z - \ w = \quad 0 \\ \ 3x + \ y - 2z - \ w = \quad 3 \end{cases}$$

**SOLUTION**    We have arranged the equations so that the same variables appear in vertical columns. We begin with the augmented matrix and then obtain an echelon form as described in the guidelines.

*(continued)*

$$\begin{bmatrix} -2 & 3 & 4 & 0 & | & -1 \\ 1 & 0 & -2 & 2 & | & 1 \\ 0 & 1 & 1 & -1 & | & 0 \\ 3 & 1 & -2 & -1 & | & 3 \end{bmatrix} \xrightarrow{R_1 \leftrightarrow R_2} \begin{bmatrix} 1 & 0 & -2 & 2 & | & 1 \\ -2 & 3 & 4 & 0 & | & -1 \\ 0 & 1 & 1 & -1 & | & 0 \\ 3 & 1 & -2 & -1 & | & 3 \end{bmatrix}$$

$$\begin{array}{c} \\ 2R_1 + R_2 \to R_2 \\ \\ -3R_1 + R_4 \to R_4 \end{array} \begin{bmatrix} 1 & 0 & -2 & 2 & | & 1 \\ 0 & 3 & 0 & 4 & | & 1 \\ 0 & 1 & 1 & -1 & | & 0 \\ 0 & 1 & 4 & -7 & | & 0 \end{bmatrix}$$

$$R_2 \leftrightarrow R_3 \begin{bmatrix} 1 & 0 & -2 & 2 & | & 1 \\ 0 & 1 & 1 & -1 & | & 0 \\ 0 & 3 & 0 & 4 & | & 1 \\ 0 & 1 & 4 & -7 & | & 0 \end{bmatrix}$$

$$\begin{array}{c} \\ \\ -3R_2 + R_3 \to R_3 \\ -R_2 + R_4 \to R_4 \end{array} \begin{bmatrix} 1 & 0 & -2 & 2 & | & 1 \\ 0 & 1 & 1 & -1 & | & 0 \\ 0 & 0 & -3 & 7 & | & 1 \\ 0 & 0 & 3 & -6 & | & 0 \end{bmatrix}$$

$$\begin{array}{c} \\ \\ \\ R_3 + R_4 \to R_4 \end{array} \begin{bmatrix} 1 & 0 & -2 & 2 & | & 1 \\ 0 & 1 & 1 & -1 & | & 0 \\ 0 & 0 & -3 & 7 & | & 1 \\ 0 & 0 & 0 & 1 & | & 1 \end{bmatrix}$$

$$-\tfrac{1}{3}R_3 \to R_3 \begin{bmatrix} 1 & 0 & -2 & 2 & | & 1 \\ 0 & 1 & 1 & -1 & | & 0 \\ 0 & 0 & 1 & -\tfrac{7}{3} & | & -\tfrac{1}{3} \\ 0 & 0 & 0 & 1 & | & 1 \end{bmatrix}$$

The final matrix is in echelon form and corresponds to the following system of equations:

$$\begin{cases} x & - 2z + 2w = & 1 \\ & y + z - w = & 0 \\ & z - \tfrac{7}{3}w = & -\tfrac{1}{3} \\ & w = & 1 \end{cases}$$

We now use back substitution to find the solution. From the last equation we see that $w = 1$. Substituting in the third equation, $z - \tfrac{7}{3}w = -\tfrac{1}{3}$, we get

$$z - \tfrac{7}{3}(1) = -\tfrac{1}{3}, \quad \text{or} \quad z = \tfrac{6}{3} = 2.$$

Substituting $w = 1$ and $z = 2$ in the second equation, $y + z - w = 0$, we obtain

$$y + 2 - 1 = 0, \quad \text{or} \quad y = -1.$$

Finally, from the first equation, $x - 2z + 2w = 1$, we have

$$x - 2(2) + 2(1) = 1, \quad \text{or} \quad x = 3.$$

Hence, the system has one solution, $x = 3$, $y = -1$, $z = 2$, and $w = 1$.

After obtaining an echelon form, it is often convenient to apply additional elementary row operations of the type $kR_i + R_j \rightarrow R_j$ so that 0 also appears *above* the first 1 in each row. We refer to the resulting matrix as being in **reduced echelon form.** The following is an illustration of a $6 \times 7$ matrix in reduced echelon form. (Compare it with the echelon form on page 546.)

**ILLUSTRATION**

**Reduced Echelon Form**

$$\begin{bmatrix} 1 & 0 & a_{13} & 0 & a_{15} & 0 & a_{17} \\ 0 & 1 & a_{23} & 0 & a_{25} & 0 & a_{27} \\ 0 & 0 & 0 & 1 & a_{35} & 0 & a_{37} \\ 0 & 0 & 0 & 0 & 0 & 1 & a_{47} \\ 0 & 0 & 0 & 0 & 0 & 0 & 0 \\ 0 & 0 & 0 & 0 & 0 & 0 & 0 \end{bmatrix}$$

**EXAMPLE 4** *Using a reduced echelon form to solve a system of linear equations*

Solve the system in Example 3 using reduced echelon form.

*SOLUTION* We begin with the echelon form obtained in Example 3 and apply additional row operations as follows:

$$\begin{bmatrix} 1 & 0 & -2 & 2 & | & 1 \\ 0 & 1 & 1 & -1 & | & 0 \\ 0 & 0 & 1 & -\frac{7}{3} & | & -\frac{1}{3} \\ 0 & 0 & 0 & 1 & | & 1 \end{bmatrix} \begin{matrix} -2R_4 + R_1 \rightarrow R_1 \\ R_4 + R_2 \rightarrow R_2 \\ \frac{7}{3}R_4 + R_3 \rightarrow R_3 \end{matrix} \begin{bmatrix} 1 & 0 & -2 & 0 & | & -1 \\ 0 & 1 & 1 & 0 & | & 1 \\ 0 & 0 & 1 & 0 & | & 2 \\ 0 & 0 & 0 & 1 & | & 1 \end{bmatrix}$$

$$\begin{matrix} 2R_3 + R_1 \rightarrow R_1 \\ -R_3 + R_2 \rightarrow R_2 \end{matrix} \begin{bmatrix} 1 & 0 & 0 & 0 & | & 3 \\ 0 & 1 & 0 & 0 & | & -1 \\ 0 & 0 & 1 & 0 & | & 2 \\ 0 & 0 & 0 & 1 & | & 1 \end{bmatrix}$$

The system of equations corresponding to the reduced echelon form gives us the solution *without* using back substitution:

$$x = 3, \quad y = -1, \quad z = 2, \quad w = 1$$

Sometimes it is necessary to consider systems in which the number of equations is not the same as the number of variables. The same matrix techniques are applicable, as illustrated in the next example.

**EXAMPLE 5**    *Solving a system of two linear equations in three variables*

Solve the system

$$\begin{cases} 2x + 3y + 4z = 1 \\ 3x + 4y + 5z = 3 \end{cases}$$

*SOLUTION*    We shall begin with the augmented matrix and then find a reduced echelon form. There are many different ways of getting the number 1 into the first position of the first row. For example, the elementary row transformation $\frac{1}{2}R_1 \rightarrow R_1$ or $-\frac{1}{3}R_2 + R_1 \rightarrow R_1$ would accomplish this in one step. Another way, which does not involve fractions, is demonstrated in the following steps:

$$\begin{bmatrix} 2 & 3 & 4 & | & 1 \\ 3 & 4 & 5 & | & 3 \end{bmatrix} \xrightarrow{R_1 \leftrightarrow R_2} \begin{bmatrix} 3 & 4 & 5 & | & 3 \\ 2 & 3 & 4 & | & 1 \end{bmatrix}$$

$$\xrightarrow{-R_2 + R_1 \rightarrow R_1} \begin{bmatrix} 1 & 1 & 1 & | & 2 \\ 2 & 3 & 4 & | & 1 \end{bmatrix}$$

$$\xrightarrow{-2R_1 + R_2 \rightarrow R_2} \begin{bmatrix} 1 & 1 & 1 & | & 2 \\ 0 & 1 & 2 & | & -3 \end{bmatrix}$$

$$\xrightarrow{-R_2 + R_1 \rightarrow R_1} \begin{bmatrix} 1 & 0 & -1 & | & 5 \\ 0 & 1 & 2 & | & -3 \end{bmatrix}$$

The reduced echelon form is the matrix of the system

$$\begin{cases} x & - & z = & 5 \\ & y + 2z = & -3 \end{cases}$$

or, equivalently,

$$\begin{cases} x = & z + 5 \\ y = & -2z - 3 \end{cases}$$

There are an infinite number of solutions to this system; they can be found by assigning $z$ any value $c$ and then using the last two equations to express $x$ and $y$ in terms of $c$. This gives us

$$x = c + 5, \quad y = -2c - 3, \quad z = c.$$

Thus, the solutions of the system consist of all ordered triples of the form

$$(c + 5, -2c - 3, c)$$

for any real number $c$. The solutions may be checked by substituting $c + 5$ for $x$, $-2c - 3$ for $y$, and $c$ for $z$ in the two given equations.

We can obtain any number of solutions for the system by substituting specific real numbers for $c$. For example, if $c = 0$, we obtain $(5, -3, 0)$; if $c = 2$, we have $(7, -7, 2)$; and so on.

There are other ways to specify the general solution. For example, starting with $x = z + 5$ and $y = -2z - 3$, we could let $z = d - 5$ for any real number $d$. In this case,

$$x = z + 5 = (d - 5) + 5 = d$$
$$y = -2z - 3 = -2(d - 5) - 3 = -2d + 7,$$

and the solutions of the system have the form

$$(d, -2d + 7, d - 5).$$

These triples produce the same solutions as $(c + 5, -2c - 3, c)$. For example, if $d = 5$, we get $(5, -3, 0)$; if $d = 7$, we obtain $(7, -7, 2)$; and so on.

---

A system of linear equations is **homogeneous** if all the terms that do not contain variables—that is, the *constant terms*—are zero. A system of homogeneous equations always has the **trivial solution** obtained by substituting zero for each variable. Nontrivial solutions sometimes exist. The procedure for finding solutions is the same as that used for nonhomogeneous systems.

**EXAMPLE 6**    *Solving a homogeneous system of linear equations*

Solve the homogeneous system

$$\begin{cases} x - y + 4z = 0 \\ 2x + y - z = 0 \\ -x - y + 2z = 0 \end{cases}$$

*SOLUTION*    We begin with the augmented matrix and find a reduced echelon form:

$$\begin{bmatrix} 1 & -1 & 4 & | & 0 \\ 2 & 1 & -1 & | & 0 \\ -1 & -1 & 2 & | & 0 \end{bmatrix} \begin{matrix} \\ -2R_1 + R_2 \rightarrow R_2 \\ R_1 + R_3 \rightarrow R_3 \end{matrix} \begin{bmatrix} 1 & -1 & 4 & | & 0 \\ 0 & 3 & -9 & | & 0 \\ 0 & -2 & 6 & | & 0 \end{bmatrix}$$

$$\frac{1}{3}R_2 \rightarrow R_2 \begin{bmatrix} 1 & -1 & 4 & | & 0 \\ 0 & 1 & -3 & | & 0 \\ 0 & -2 & 6 & | & 0 \end{bmatrix}$$

$$\begin{matrix} R_2 + R_1 \rightarrow R_1 \\ \\ 2R_2 + R_3 \rightarrow R_3 \end{matrix} \begin{bmatrix} 1 & 0 & 1 & | & 0 \\ 0 & 1 & -3 & | & 0 \\ 0 & 0 & 0 & | & 0 \end{bmatrix}$$

*(continued)*

The reduced echelon form corresponds to the system

$$\begin{cases} x \quad + \quad z = 0 \\ \quad y - 3z = 0 \end{cases}$$

or, equivalently,

$$\begin{cases} x = -z \\ y = 3z \end{cases}$$

Assigning any value $c$ to $z$, we obtain $x = -c$ and $y = 3c$. The solutions consist of all ordered triples of the form $(-c, 3c, c)$ for any real number $c$.

**EXAMPLE 7**   **_A homogeneous system_**
**_with only the trivial solution_**

Solve the system

$$\begin{cases} x + y + z = 0 \\ x - y + z = 0 \\ x - y - z = 0 \end{cases}$$

_SOLUTION_    We begin with the augmented matrix and find a reduced echelon form:

$$\begin{bmatrix} 1 & 1 & 1 & | & 0 \\ 1 & -1 & 1 & | & 0 \\ 1 & -1 & -1 & | & 0 \end{bmatrix} \begin{matrix} \\ -R_1 + R_2 \to R_2 \\ -R_1 + R_3 \to R_3 \end{matrix} \begin{bmatrix} 1 & 1 & 1 & | & 0 \\ 0 & -2 & 0 & | & 0 \\ 0 & -2 & -2 & | & 0 \end{bmatrix}$$

$$-\tfrac{1}{2}R_2 \to R_2 \begin{bmatrix} 1 & 1 & 1 & | & 0 \\ 0 & 1 & 0 & | & 0 \\ 0 & -2 & -2 & | & 0 \end{bmatrix}$$

$$\begin{matrix} -R_2 + R_1 \to R_1 \\ \\ 2R_2 + R_3 \to R_3 \end{matrix} \begin{bmatrix} 1 & 0 & 1 & | & 0 \\ 0 & 1 & 0 & | & 0 \\ 0 & 0 & -2 & | & 0 \end{bmatrix}$$

$$-\tfrac{1}{2}R_3 \to R_3 \begin{bmatrix} 1 & 0 & 1 & | & 0 \\ 0 & 1 & 0 & | & 0 \\ 0 & 0 & 1 & | & 0 \end{bmatrix}$$

$$-R_3 + R_1 \to R_1 \begin{bmatrix} 1 & 0 & 0 & | & 0 \\ 0 & 1 & 0 & | & 0 \\ 0 & 0 & 1 & | & 0 \end{bmatrix}$$

The reduced echelon form is the matrix of the system

$$x = 0, \quad y = 0, \quad z = 0.$$

Thus, the only solution for the given system is the trivial one, $(0, 0, 0)$.

The next two examples illustrate applied problems.

**EXAMPLE 8**    *Using a system of equations*
*to determine maximum profit*

A manufacturer of electrical equipment has the following information about the weekly profit from the production and sale of a type of electric motor.

| Production level $x$ | 25 | 50 | 100 |
|---|---|---|---|
| Profit $P(x)$ (dollars) | 5250 | 7500 | 4500 |

**(a)** Determine $a$, $b$, and $c$ so that the graph of $P(x) = ax^2 + bx + c$ fits this information.

**(b)** According to the quadratic function $P$ in part (a), how many motors should be produced each week for maximum profit? What is the maximum weekly profit?

*SOLUTION*

**(a)** We see from the table that the graph of $P(x) = ax^2 + bx + c$ contains the points $(25, 5250)$, $(50, 7500)$, and $(100, 4500)$. This gives us the system of equations

$$\begin{cases} 5250 = 625a + 25b + c \\ 7500 = 2500a + 50b + c \\ 4500 = 10{,}000a + 100b + c \end{cases}$$

We can verify that the solution is $a = -2$, $b = 240$, $c = 500$.

**(b)** From part (a),

$$P(x) = -2x^2 + 240x + 500.$$

Since $a = -2 < 0$, the graph of the quadratic function $P$ is a parabola that opens downward. By the formula on page 149, the $x$-coordinate of the vertex (the highest point on the parabola) is

$$x = \frac{-b}{2a} = \frac{-240}{2(-2)} = \frac{-240}{-4} = 60.$$

Hence, for the maximum profit, the manufacturer should produce and sell 60 motors per week. The maximum weekly profit is

$$P(60) = -2(60)^2 + 240(60) + 500 = \$7700.$$

**EXAMPLE 9**    *Solving a mixture problem*

A merchant wishes to mix two grades of peanuts costing $3 and $4 per pound, respectively, with cashews costing $8 per pound, to obtain 140 pounds of a mixture costing $6 per pound. If the merchant also wants the amount of cheaper-grade peanuts to be twice that of the better-grade peanuts, how many pounds of each variety should be mixed?

*SOLUTION*    Let us introduce three variables as follows:

$$x = \text{number of pounds of peanuts at \$3 per pound}$$
$$y = \text{number of pounds of peanuts at \$4 per pound}$$
$$z = \text{number of pounds of cashews at \$8 per pound}$$

We refer to the statement of the problem and obtain the following system:

$$\begin{cases} x + y + z = 140 \\ 3x + 4y + 8z = 6(140) \\ x \qquad\quad = 2y \end{cases}$$

You may verify that the solution of this system is $x = 40$, $y = 20$, $z = 80$. Thus, the merchant should use 40 pounds of the $3 peanuts, 20 pounds of the $4 peanuts, and 80 pounds of cashews.

## 8.3 EXERCISES

**Exer. 1–26: Use matrices to solve the system.**

1. $\begin{cases} x - 2y - 3z = -1 \\ 2x + y + z = 6 \\ x + 3y - 2z = 13 \end{cases}$

2. $\begin{cases} x + 3y - z = -3 \\ 3x - y + 2z = 1 \\ 2x - y + z = -1 \end{cases}$

3. $\begin{cases} 5x + 2y - z = -7 \\ x - 2y + 2z = 0 \\ 3y + z = 17 \end{cases}$

4. $\begin{cases} 4x - y + 3z = 6 \\ -8x + 3y - 5z = -6 \\ 5x - 4y = -9 \end{cases}$

5. $\begin{cases} 2x + 6y - 4z = 1 \\ x + 3y - 2z = 4 \\ 2x + y - 3z = -7 \end{cases}$

6. $\begin{cases} x + 3y - 3z = -5 \\ 2x - y + z = -3 \\ -6x + 3y - 3z = 4 \end{cases}$

7. $\begin{cases} 2x - 3y + 2z = -3 \\ -3x + 2y + z = 1 \\ 4x + y - 3z = 4 \end{cases}$

8. $\begin{cases} 2x - 3y + z = 2 \\ 3x + 2y - z = -5 \\ 5x - 2y + z = 0 \end{cases}$

9. $\begin{cases} x + 3y + z = 0 \\ x + y - z = 0 \\ x - 2y - 4z = 0 \end{cases}$

10. $\begin{cases} 2x - y + z = 0 \\ x - y - 2z = 0 \\ 2x - 3y - z = 0 \end{cases}$

11. $\begin{cases} 2x + y + z = 0 \\ x - 2y - 2z = 0 \\ x + y + z = 0 \end{cases}$

12. $\begin{cases} x + y - 2z = 0 \\ x - y - 4z = 0 \\ y + z = 0 \end{cases}$

13. $\begin{cases} 3x - 2y + 5z = 7 \\ x + 4y - z = -2 \end{cases}$

14. $\begin{cases} 2x - y + 4z = 8 \\ -3x + y - 2z = 5 \end{cases}$

15. $\begin{cases} 4x - 2y + z = 5 \\ 3x + y - 4z = 0 \end{cases}$

16. $\begin{cases} 5x + 2y - z = 10 \\ y + z = -3 \end{cases}$

17. $\begin{cases} x + 2y - z - 3w = 2 \\ 3x + y - 2z - w = 6 \\ x + y + 3z - 2w = -3 \\ -2x - 2y + 3z + w = -9 \end{cases}$

18. $\begin{cases} x - 2y - 5z + w = -1 \\ 2x - y + z + w = 1 \\ 3x - 2y - 4z - 2w = 1 \\ x + y + 3z - 2w = 2 \end{cases}$

19 $\begin{cases} 2x - y - 2z + 2s - 5t = 2 \\ x + 3y - 2z + s - 2t = -5 \\ -x + 4y + 2z - 3s + 8t = -4 \\ 3x - 2y - 4z + s - 3t = -3 \\ 4x - 6y + z - 2s + t = 10 \end{cases}$

20 $\begin{cases} 3x + 2y + z + 3u + v + w = 1 \\ 2x + y - 2z + 3u - v + 4w = 6 \\ 6x + 3y + 4z - u + 2v + w = -6 \\ x + y + z + u - v - w = 8 \\ -2x - 2y + z - 3u + 2v - 3w = -10 \\ x - 3y + 2z + u + 3v + w = -1 \end{cases}$

21 $\begin{cases} 5x + 2z = 1 \\ y - 3z = 2 \\ 2x + y = 3 \end{cases}$   22 $\begin{cases} 2x - 3y = 12 \\ 3y + z = -2 \\ 5x - 3z = 3 \end{cases}$

23 $\begin{cases} 4x - 3y = 1 \\ 2x + y = -7 \\ -x + y = -1 \end{cases}$   24 $\begin{cases} 2x + 3y = -2 \\ x + y = 1 \\ x - 2y = 13 \end{cases}$

25 $\begin{cases} 2x + 3y = 5 \\ x - 3y = 4 \\ x + y = -2 \end{cases}$   26 $\begin{cases} 4x - y = 2 \\ 2x + 2y = 1 \\ 4x - 5y = 3 \end{cases}$

27 **Mixing acid solutions** Three solutions contain a certain acid. The first contains 10% acid, the second 30%, and the third 50%. A chemist wishes to use all three solutions to obtain a 50-liter mixture containing 32% acid. If the chemist wants to use twice as much of the 50% solution as of the 30% solution, how many liters of each solution should be used?

28 **Filling a pool** A swimming pool can be filled by three pipes, A, B, and C. Pipe A alone can fill the pool in 8 hours. If pipes A and C are used together, the pool can be filled in 6 hours. If B and C are used together, it takes 10 hours. How long does it take to fill the pool if all three pipes are used?

29 **Production capability** A company has three machines A, B, and C that are each capable of producing a certain item. However, because of a lack of skilled operators, only two of the machines can be used simultaneously. The following table indicates production over a three-day period using various combinations of the machines.

| Machines used | Hours used | Items produced |
|---|---|---|
| A and B | 6 | 4500 |
| A and C | 8 | 3600 |
| B and C | 7 | 4900 |

How long would it take each machine, if used alone, to produce 1000 items?

30 **Electrical resistance** In electrical circuits, the formula $1/R = (1/R_1) + (1/R_2)$ is used to find the total resistance $R$ if two resistors $R_1$ and $R_2$ are connected in parallel. Given three resistors A, B, and C, suppose that the total resistance is 48 ohms if A and B are connected in parallel, 80 ohms if B and C are connected in parallel, and 60 ohms if A and C are connected in parallel. Find the resistances of A, B, and C.

31 **Mixing fertilizers** A supplier of lawn products has three types of grass fertilizer $G_1$, $G_2$, and $G_3$, having nitrogen contents of 30%, 20%, and 15%, respectively. The supplier plans to mix them, obtaining 600 pounds of fertilizer with a 25% nitrogen content. The mixture is to contain 100 pounds more of type $G_3$ than of type $G_2$. How much of each type should be used?

32 **Particle acceleration** If a particle moves along a coordinate line with a constant acceleration $a$ (in cm/sec$^2$), then at time $t$ (in seconds) its distance $s(t)$ (in centimeters) from the origin is

$$s(t) = \tfrac{1}{2}at^2 + v_0t + s_0$$

for the velocity $v_0$ and distance $s_0$ from the origin at $t = 0$. If the distances of the particle from the origin at $t = \frac{1}{2}$, $t = 1$, and $t = \frac{3}{2}$ are 7, 11, and 17, respectively, find $a$, $v_0$, and $s_0$.

33 **Electrical currents** Shown in the figure is a schematic of an electrical circuit containing three resistors, a 6-volt battery, and a 12-volt battery. It can be shown,

EXERCISE 33

using Kirchhoff's laws, that the three currents $I_1$, $I_2$, and $I_3$ are solutions of the following system of equations:

$$\begin{cases} I_1 - I_2 + I_3 = 0 \\ R_1 I_1 + R_2 I_2 = 6 \\ R_2 I_2 + R_3 I_3 = 12 \end{cases}$$

Find the three currents if

(a) $R_1 = R_2 = R_3 = 3$ ohms

(b) $R_1 = 4$ ohms, $R_2 = 1$ ohm, and $R_3 = 4$ ohms

34 **Bird population**  A stable population of 35,000 birds lives on three islands. Each year 10% of the population on island A migrates to island B, 20% of the population on island B migrates to island C, and 5% of the population on island C migrates to island A. Find the number of birds on each island if the population count on each island does not vary from year to year.

35 **Blending coffees**  A shop specializes in preparing blends of gourmet coffees. From Columbian, Brazilian, and Kenyan coffees, the owner wishes to prepare 1-pound bags that will sell for $8.50. The cost per pound of these coffees is $10, $6, and $8, respectively. The amount of Columbian is to be three times the amount of Brazilian. Find the amount of each type of coffee in the blend.

36 **Cattle population**  A rancher has 750 head of cattle consisting of 400 adults (aged 2 or more years), 150 yearlings, and 200 calves. The following information is known about this particular species. Each spring an adult female gives birth to a single calf, and 75% of these calves will survive the first year. The yearly survival percentages for yearlings and adults are 80% and 90%, respectively. The male-female ratio is one in all age classes. Estimate the population of each age class

(a) next spring          (b) last spring

37 **Traffic flow**  Shown in the figure is a system of four one-way streets leading into the center of a city. The numbers in the figure denote the average number of vehicles per hour that travel in the directions shown. A total of 300 vehicles enter the area and 300 vehicles leave the area every hour. Signals at intersections A, B, C, and D are to be timed in order to avoid congestion, and this timing will determine traffic flow rates $x_1$, $x_2$, $x_3$, and $x_4$.

(a) If the number of vehicles entering an intersection per hour must equal the number leaving the inter-

section per hour, describe the traffic flow rates at each intersection with a system of equations.

(b) If the signal at intersection C is timed so that $x_3$ is equal to 100, find $x_1$, $x_2$, and $x_4$.

(c) Make use of the system in part (a) to explain why $75 \le x_3 \le 150$.

**EXERCISE 37**

38 If $f(x) = ax^3 + bx + c$, determine $a$, $b$, and $c$ such that the graph of $f$ passes through the points $P(-3, -12)$, $Q(-1, 22)$, and $R(2, 13)$.

39 **Air pollution**  Between 1850 and 1985 approximately 155 billion metric tons of carbon was added to the earth's atmosphere and the climate became about 0.5 °C warmer, an indication of the *greenhouse effect*. It is estimated that doubling the carbon dioxide ($CO_2$) in the atmosphere would result in an average global temperature increase of 4–5 °C. The future amount $A$ of $CO_2$ in the atmosphere in parts per million is sometimes estimated using the formula $A = a + ct + ke^{rt}$, where $a$, $c$, and $k$ are constants, $r$ is the percentage increase in the emission of $CO_2$, and $t$ is time in years, with $t = 0$ corresponding to 1990. Suppose it is estimated that in the year 2070, $A$ will be 800 if $r = 2.5\%$ and $A$ will be 560 if $r = 1.5\%$. If, in 1990, $A = 340$ and $r = 1\%$, find the year in which the amount of $CO_2$ in the atmosphere will have doubled.

40 **Air pollution**  Refer to Exercise 39. Suppose it is estimated that in the year 2030, $A$ will be 455 if $r = 2.0\%$ and $A$ will be 430 if $r = 1.5\%$. If, in 1990, $A = 340$ and $r = 2.5\%$, find the year in which the amount of $CO_2$ in the atmosphere will have doubled.

**Exer. 41–42: Find an equation of the parabola that has a vertical axis and passes through the given points.**

41  $P(2, 5)$,     $Q(-2, -3)$,   $R(1, 6)$

42  $P(3, -1)$,   $Q(1, -7)$,    $R(-2, 14)$

**Exer. 43–44:** Find an equation of the circle of the form $x^2 + y^2 + ax + by + c = 0$ that passes through the given points.

**43** $P(2, 1)$, $Q(-1, -4)$, $R(3, 0)$

**44** $P(-5, 5)$, $Q(-2, -4)$, $R(2, 4)$

**45** If $f(x) = ax^3 + bx^2 + cx + d$, find $a$, $b$, $c$, and $d$ if the graph of $f$ is to pass through $(-1, 2)$, $(0.5, 2)$, $(1, 3)$, and $(2, 4.5)$.

**46** If $f(x) = ax^4 + bx^3 + cx^2 + dx + e$, find $a$, $b$, $c$, $d$, and $e$ if the graph of $f$ is to pass through $(-2, 1.5)$, $(-1, -2)$, $(1, -3)$, $(2, -3.5)$, and $(3, -4.8)$.

## 8.4 PARTIAL FRACTIONS

In this section we show how systems of equations can be used to help decompose rational expressions into sums of simpler expressions. This technique is useful in advanced mathematics courses.

We may verify that

$$\frac{2}{x^2 - 1} = \frac{1}{x - 1} + \frac{-1}{x + 1}$$

by adding the fractions $1/(x - 1)$ and $-1/(x + 1)$ to obtain $2/(x^2 - 1)$. The expression on the right side of this equation is called the *partial fraction decomposition* of $2/(x^2 - 1)$.

It is theoretically possible to write *any* rational expression as a sum of rational expressions whose denominators involve powers of polynomials of degree not greater than two. Specifically, if $f(x)$ and $g(x)$ are polynomials *and the degree of $f(x)$ is less than the degree of $g(x)$*, it can be proved that

$$\frac{f(x)}{g(x)} = F_1 + F_2 + \cdots + F_r$$

such that each $F_k$ has one of the forms

$$\frac{A}{(px + q)^m} \quad \text{or} \quad \frac{Ax + B}{(ax^2 + bx + c)^n},$$

where $A$ and $B$ are real numbers, $m$ and $n$ are nonnegative integers, and the quadratic polynomial $ax^2 + bx + c$ is irreducible over $\mathbb{R}$ (that is, has no real zero). The sum $F_1 + F_2 + \cdots + F_r$ is the **partial fraction decomposition** of $f(x)/g(x)$, and each $F_k$ is a **partial fraction.**

For the partial fraction decomposition of $f(x)/g(x)$ to be found, *it is essential that $f(x)$ have lower degree than $g(x)$*. If this is not the case, we can use long division to obtain such an expression. For example, given

$$\frac{x^3 - 6x^2 + 5x - 3}{x^2 - 1},$$

we obtain

$$\frac{x^3 - 6x^2 + 5x - 3}{x^2 - 1} = x - 6 + \frac{6x - 9}{x^2 - 1}.$$

We then find the partial fraction decomposition of $(6x - 9)/(x^2 - 1)$. The following guidelines can be used to obtain decompositions.

**Guidelines for Finding Partial Fraction Decompositions of $f(x)/g(x)$**

1  If the degree of the numerator $f(x)$ is not lower than the degree of the denominator $g(x)$, use long division to obtain the proper form.

2  Factor the denominator $g(x)$ into a product of linear factors $px + q$ or irreducible quadratic factors $ax^2 + bx + c$, and collect repeated factors so that $g(x)$ is a product of *different* factors of the form $(px + q)^m$ or $(ax^2 + bx + c)^n$ for a nonnegative integer $m$ or $n$.

3  Apply the following rules to the factors found in guideline 2.
   **Rule A:** For each factor of the form $(px + q)^m$ with $m \geq 1$, the partial fraction decomposition contains a sum of $m$ partial fractions of the form

   $$\frac{A_1}{px + q} + \frac{A_2}{(px + q)^2} + \cdots + \frac{A_m}{(px + q)^m},$$

   where each numerator $A_k$ is a real number.
   **Rule B:** For each factor of the form $(ax^2 + bx + c)^n$ with $n \geq 1$ and $ax^2 + bx + c$ irreducible, the partial fraction decomposition contains a sum of $n$ partial fractions of the form

   $$\frac{A_1x + B_1}{ax^2 + bx + c} + \frac{A_2x + B_2}{(ax^2 + bx + c)^2} + \cdots + \frac{A_nx + B_n}{(ax^2 + bx + c)^n},$$

   where each $A_k$ and each $B_k$ is a real number.

4  Find the numbers $A_k$ and $B_k$ in guideline 3.

We shall apply the preceding guidelines in the following examples. For the sake of convenience, we will use the variables $A$, $B$, $C$, and so on, rather than the subscripted variables $A_k$ and $B_k$ given in the guidelines.

**EXAMPLE 1**    *A partial fraction decomposition in which each denominator is linear*

Find the partial fraction decomposition of

$$\frac{4x^2 + 13x - 9}{x^3 + 2x^2 - 3x}.$$

## SOLUTION

**Guideline 1** The degree of the numerator, 2, is less than the degree of the denominator 3, so long division is not required.

**Guideline 2** We factor the denominator:

$$x^3 + 2x^2 - 3x = x(x^2 + 2x - 3) = x(x + 3)(x - 1)$$

**Guideline 3** Each factor of the denominator has the form stated in Rule A, with $m = 1$. Thus, to the factor $x$ there corresponds a partial fraction of the form $A/x$. Similarly, to the factors $x + 3$ and $x - 1$ there correspond partial fractions of the form $B/(x + 3)$ and $C/(x - 1)$, respectively. The partial fraction decomposition has the form

$$\frac{4x^2 + 13x - 9}{x^3 + 2x^2 - 3x} = \frac{A}{x} + \frac{B}{x + 3} + \frac{C}{x - 1}.$$

**Guideline 4** We find the values of $A$, $B$, and $C$ in guideline 3. Multiplying both sides of the partial fraction decomposition by the least common denominator, $x(x + 3)(x - 1)$, gives us

$$\begin{aligned}
4x^2 + 13x - 9 &= A(x + 3)(x - 1) + Bx(x - 1) + Cx(x + 3) \\
&= A(x^2 + 2x - 3) + B(x^2 - x) + C(x^2 + 3x) \\
&= (A + B + C)x^2 + (2A - B + 3C)x - 3A.
\end{aligned}$$

Equating the coefficients of like powers of $x$ on each side of the last equation, we obtain the system of equations

$$\begin{cases}
A + B + C = 4 \\
2A - B + 3C = 13 \\
-3A = -9
\end{cases}$$

Using the methods of the preceding section yields the solution $A = 3$, $B = -1$, and $C = 2$. Hence, the partial fraction decomposition is

$$\frac{4x^2 + 13x - 9}{x(x + 3)(x - 1)} = \frac{3}{x} + \frac{-1}{x + 3} + \frac{2}{x - 1}.$$

There is an alternative way to find $A$, $B$, and $C$ if all factors of the denominator are linear and nonrepeated, as in this example. Instead of equating coefficients and using a system of equations, we begin with the equation

$$4x^2 + 13x - 9 = A(x + 3)(x - 1) + Bx(x - 1) + Cx(x + 3).$$

We next substitute values for $x$ that make the factors $x$, $x - 1$, and $x - 3$, zero. If we let $x = 0$ and simplify, we obtain

$$-9 = -3A, \quad \text{or} \quad A = 3.$$

Letting $x = 1$ in the equation leads to $8 = 4C$, or $C = 2$. Finally, if $x = -3$, then we have $-12 = 12B$, or $B = -1$.

**EXAMPLE 2** *A partial fraction decomposition containing a repeated linear factor*

Find the partial fraction decomposition of

$$\frac{x^2 + 10x - 36}{x(x - 3)^2}.$$

*SOLUTION*

**Guideline 1** The degree of the numerator, 2, is less than the degree of the denominator, 3, so long division is not required.

**Guideline 2** The denominator, $x(x - 3)^2$, is already in factored form.

**Guideline 3** By Rule A with $m = 1$, there is a partial fraction of the form $A/x$ corresponding to the factor $x$. Next, applying Rule A with $m = 2$, we find that the factor $(x - 3)^2$ determines a sum of two partial fractions of the form $B/(x - 3)$ and $C/(x - 3)^2$. Thus, the partial fraction decomposition has the form

$$\frac{x^2 + 10x - 36}{x(x - 3)^2} = \frac{A}{x} + \frac{B}{x - 3} + \frac{C}{(x - 3)^2}.$$

**Guideline 4** To find $A$, $B$, and $C$, we begin by multiplying both sides of the partial fraction decomposition in guideline 3 by the lcd, $x(x - 3)^2$:

$$x^2 + 10x - 36 = A(x - 3)^2 + Bx(x - 3) + Cx$$
$$= A(x^2 - 6x + 9) + B(x^2 - 3x) + Cx$$
$$= (A + B)x^2 + (-6A - 3B + C)x + 9A$$

We next equate the coefficients of like powers of $x$, obtaining the system

$$\begin{cases} A + B = 1 \\ -6A - 3B + C = 10 \\ 9A = -36 \end{cases}$$

This system of equations has the solution $A = -4$, $B = 5$, $C = 1$. The partial fraction decomposition is therefore

$$\frac{x^2 + 10x - 36}{x(x - 3)^2} = \frac{-4}{x} + \frac{5}{x - 3} + \frac{1}{(x - 3)^2}.$$

As in Example 1, we could also obtain $A$ and $C$ by beginning with the equation

$$x^2 + 10x - 36 = A(x - 3)^2 + Bx(x - 3) + Cx$$

and then substituting values for $x$ that make the factors, $x - 3$ and $x$, zero. Thus, letting $x = 3$, we obtain $3 = 3C$, or $C = 1$. Letting $x = 0$ gives us $-36 = 9A$, or $A = -4$. The value of $B$ may then be found by using one of the equations in the system.

**EXAMPLE 3** *A partial fraction decomposition containing an irreducible quadratic factor*

Find the partial fraction decomposition of

$$\frac{4x^3 - x^2 + 15x - 29}{2x^3 - x^2 + 8x - 4}.$$

**SOLUTION**

**Guideline 1** The degree of the numerator, 3, is *equal* to the degree of the denominator. Thus, long division is required, and we obtain

$$\frac{4x^3 - x^2 + 15x - 29}{2x^3 - x^2 + 8x - 4} = 2 + \frac{x^2 - x - 21}{2x^3 - x^2 + 8x - 4}.$$

**Guideline 2** The denominator may be factored by grouping, as follows:

$$2x^3 - x^2 + 8x - 4 = x^2(2x - 1) + 4(2x - 1) = (x^2 + 4)(2x - 1)$$

**Guideline 3** Applying Rule B to the irreducible quadratic factor $x^2 + 4$ in guideline 2, we see that one of the partial fractions has the form $(Ax + B)/(x^2 + 4)$. By Rule A, there is also a partial fraction $C/(2x - 1)$ corresponding to $2x - 1$. Consequently,

$$\frac{x^2 - x - 21}{2x^3 - x^2 + 8x - 4} = \frac{Ax + B}{x^2 + 4} + \frac{C}{2x - 1}.$$

**Guideline 4** Multiplying both sides of the partial fraction decomposition in guideline 3 by the lcd, $(x^2 + 4)(2x - 1)$, we obtain

$$x^2 - x - 21 = (Ax + B)(2x - 1) + C(x^2 + 4)$$
$$= 2Ax^2 - Ax + 2Bx - B + Cx^2 + 4C$$
$$= (2A + C)x^2 + (-A + 2B)x - B + 4C.$$

This leads to the system

$$\begin{cases} 2A \quad\quad + C = 1 \\ -A + 2B \quad\quad = -1 \\ \quad\quad - B + 4C = -21 \end{cases}$$

This system has the solution $A = 3$, $B = 1$, and $C = -5$. Thus, the partial fraction decomposition in guideline 3 is

$$\frac{x^2 - x - 21}{2x^3 - x^2 + 8x - 4} = \frac{3x + 1}{x^2 + 4} + \frac{-5}{2x - 1},$$

and therefore the decomposition of the given expression (see guideline 1) is

$$\frac{4x^3 - x^2 + 15x - 29}{2x^3 - x^2 + 8x - 4} = 2 + \frac{3x + 1}{x^2 + 4} + \frac{-5}{2x - 1}.$$

**EXAMPLE 4    *A partial fraction decomposition containing a repeated quadratic factor***

Find the partial fraction decomposition of

$$\frac{5x^3 - 3x^2 + 7x - 3}{(x^2 + 1)^2}.$$

**SOLUTION**

**Guideline 1**  The degree of the numerator, 3, is less than the degree of the denominator, 4, so long division is not required.

**Guideline 2**  The denominator, $(x^2 + 1)^2$, is already in factored form.

**Guideline 3**  We apply Rule B, with $n = 2$, to $(x^2 + 1)^2$ to obtain the partial fraction decomposition

$$\frac{5x^3 - 3x^2 + 7x - 3}{(x^2 + 1)^2} = \frac{Ax + B}{x^2 + 1} + \frac{Cx + D}{(x^2 + 1)^2}.$$

**Guideline 4**  Multiplying both sides of the decomposition in guideline 3 by $(x^2 + 1)^2$ gives us

$$5x^3 - 3x^2 + 7x - 3 = (Ax + B)(x^2 + 1) + Cx + D$$
$$= Ax^3 + Bx^2 + (A + C)x + (B + D).$$

Comparing the coefficients of $x^3$ and $x^2$, we obtain $A = 5$ and $B = -3$. From the coefficients of $x$, we see that $A + C = 7$. Thus, $C = 7 - A = 7 - 5 = 2$. Finally, comparing the constant terms gives us $B + D = -3$ and $D = -3 - B = -3 - (-3) = 0$. Therefore, the partial fraction decomposition is

$$\frac{5x^3 - 3x^2 + 7x - 3}{(x^2 + 1)^2} = \frac{5x - 3}{x^2 + 1} + \frac{2x}{(x^2 + 1)^2}.$$

## 8.4 EXERCISES

**Exer. 1–28: Find the partial fraction decomposition.**

1  $\dfrac{8x - 1}{(x - 2)(x + 3)}$

2  $\dfrac{x - 29}{(x - 4)(x + 1)}$

3  $\dfrac{x + 34}{x^2 - 4x - 12}$

4  $\dfrac{5x - 12}{x^2 - 4x}$

5  $\dfrac{4x^2 - 15x - 1}{(x - 1)(x + 2)(x - 3)}$

6  $\dfrac{x^2 + 19x + 20}{x(x + 2)(x - 5)}$

7  $\dfrac{4x^2 - 5x - 15}{x^3 - 4x^2 - 5x}$

8  $\dfrac{37 - 11x}{(x + 1)(x^2 - 5x + 6)}$

9  $\dfrac{2x + 3}{(x - 1)^2}$

10  $\dfrac{5x^2 - 4}{x^2(x + 2)}$

11  $\dfrac{19x^2 + 50x - 25}{3x^3 - 5x^2}$

12  $\dfrac{10 - x}{x^2 + 10x + 25}$

13  $\dfrac{x^2 - 6}{(x + 2)^2(2x - 1)}$

14  $\dfrac{2x^2 + x}{(x - 1)^2(x + 1)^2}$

15  $\dfrac{3x^3 + 11x^2 + 16x + 5}{x(x + 1)^3}$

16  $\dfrac{4x^3 + 3x^2 + 5x - 2}{x^3(x + 2)}$

17  $\dfrac{x^2 + x - 6}{(x^2 + 1)(x - 1)}$

18  $\dfrac{x^2 - x - 21}{(x^2 + 4)(2x - 1)}$

**19** $\dfrac{9x^2 - 3x + 8}{x^3 + 2x}$

**20** $\dfrac{2x^3 + 2x^2 + 4x - 3}{x^4 + x^2}$

**25** $\dfrac{3x^2 - 16}{x^2 - 4x}$

**26** $\dfrac{2x^2 + 7x}{x^2 + 6x + 9}$

**21** $\dfrac{4x^3 - x^2 + 4x + 2}{(x^2 + 1)^2}$

**22** $\dfrac{3x^3 + 13x - 1}{(x^2 + 4)^2}$

**27** $\dfrac{4x^3 + 4x^2 - 4x + 2}{2x^2 - x - 1}$

**23** $\dfrac{2x^4 - 2x^3 + 6x^2 - 5x + 1}{x^3 - x^2 + x - 1}$

**24** $\dfrac{x^3}{x^3 - 3x^2 + 9x - 27}$

**28** $\dfrac{x^5 - 5x^4 + 7x^3 - x^2 - 4x + 12}{x^3 - 3x^2}$

# 8.5 SYSTEMS OF INEQUALITIES

In Chapter 1 we restricted our discussion to inequalities in one variable. We shall now consider inequalities in two variables $x$ and $y$, such as those shown in the following illustration.

### ILLUSTRATION

**Inequalities in $x$ and $y$**

- $y^2 < x + 4$
- $3x - 4y > 12$
- $x^2 + y^2 \le 16$

A **solution** of an inequality in $x$ and $y$ is an ordered pair $(a, b)$ that yields a true statement if $a$ and $b$ are substituted for $x$ and $y$, respectively. To **solve** an inequality in $x$ and $y$ means to find all the solutions. The **graph** of such an inequality is the set of all points $(a, b)$ in an $xy$-plane that correspond to the solutions. Two inequalities are **equivalent** if they have the same solutions.

Given an inequality in $x$ and $y$, if we replace the inequality symbol with an equal sign, we obtain an equation whose graph usually separates the $xy$-plane into several regions. We shall consider only equations having the property that if $R$ is one such region and if a **test point** $(p, q)$ in $R$ yields a solution of the inequality, then *every* point in $R$ yields a solution. The following guidelines may then be used to sketch the graph of the inequality.

**Guidelines for Sketching the Graph of an Inequality in $x$ and $y$**

1. Replace the inequality symbol with an equal sign and graph the resulting equation. Use dashes if the inequality symbol is $<$ or $>$ to indicate that no point on the graph yields a solution. Use a solid line or curve for $\le$ or $\ge$ to indicate that solutions of the equation are also solutions of the inequality.

2. If $R$ is a region of the $xy$-plane determined by the graph in guideline 1 and if a test point $(p, q)$ in $R$ yields a solution of the inequality, then every point in $R$ yields a solution. Shade $R$ to indicate this fact. If $(p, q)$ is not a solution, then *no* point in $R$ yields a solution and $R$ is left unshaded.

The use of these guidelines is demonstrated in the next example.

**FIGURE 9**

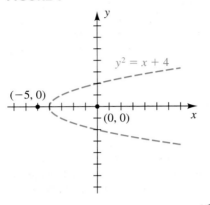

**EXAMPLE 1    *Sketching the graph of an inequality***

Find the solutions and sketch the graph of the inequality $y^2 < x + 4$.

**SOLUTION**

**Guideline 1** We replace $<$ with $=$, obtaining $y^2 = x + 4$. The graph of this equation is a parabola, symmetric with respect to the $x$-axis and having $x$-intercept $-4$ and $y$-intercepts $\pm 2$. Since the inequality symbol is $<$, we sketch the parabola using dashes, as in Figure 9.

**Guideline 2** The graph in guideline 1 separates the $xy$-plane into two regions, one to the *right* of the parabola and the other to the *left*. Let us choose test points $(0, 0)$ and $(-5, 0)$ in the regions (see Figure 9) and substitute for $x$ and $y$ in $y^2 < x + 4$ as follows:

**FIGURE 10**

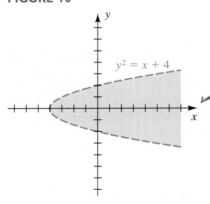

> ✓ ***Test point*** $(0, 0)$    LS: $0^2 = 0$
>
> RS: $0 + 4 = 4$

Since $0 < 4$ is a *true* statement, $(0, 0)$ *is* a solution of the inequality. Hence, *all* points to the right of the parabola yield solutions, so we shade this region, as in Figure 10.

> ✓ ***Test point*** $(-5, 0)$    LS: $0^2 = 0$
>
> RS: $-5 + 4 = -1$

Since $0 < -1$ is a *false* statement, $(-5, 0)$ *is not* a solution of the inequality. Hence, *no* point to the left of the parabola is a solution, and we leave that region unshaded.

---

**FIGURE 11**

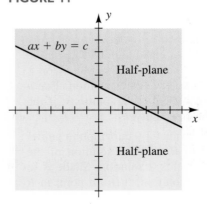

A **linear inequality** is an inequality that can be written in one of the following forms, where $a$, $b$, and $c$ are real numbers:

$$ax + by < c, \quad ax + by > c, \quad ax + by \le c, \quad ax + by \ge c$$

The line $ax + by = c$ separates the $xy$-plane into two **half-planes,** as illustrated in Figure 11. The solutions of the inequality consist of all points in *one* of these half-planes, where the line is included for $\le$ or $\ge$ and is not included for $<$ or $>$. *For a linear inequality, only one test point $(p, q)$ is required,* because if $(p, q)$ is a solution, then the half-plane with $(p, q)$ in it contains all the solutions, whereas if $(p, q)$ is *not* a solution, then the *other* half-plane contains the solutions.

**FIGURE 12**

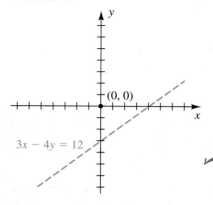

**EXAMPLE 2** *Sketching the graph of a linear inequality*

Sketch the graph of the inequality $3x - 4y > 12$.

**SOLUTION**  Replacing $>$ with $=$ gives us the line $3x - 4y = 12$, sketched with dashes in Figure 12. This line separates the $xy$-plane into two half-planes, one *above* the line and the other *below* the line. It is convenient to choose the test point $(0, 0)$ above the line and substitute in $3x - 4y > 12$, as follows:

**Test point $(0, 0)$**   LS: $3 \cdot 0 - 4 \cdot 0 = 0 - 0 = 0$

RS: $12$

Since $0 > 12$ is a false statement, $(0, 0)$ is not a solution. Thus, no point above the line is a solution, and the solutions of $3x - 4y > 12$ are given by the points in the half-plane *below* the line. The graph is sketched in Figure 13.

**FIGURE 13**

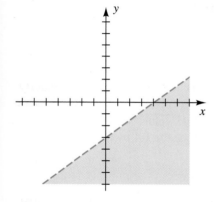

As we did with equations, we sometimes work simultaneously with several inequalities in two variables—that is, with a **system of inequalities**. The **solutions** of a system of inequalities are the solutions common to all inequalities in the system. The **graph** of a system of inequalities consists of the points corresponding to the solutions. The following examples illustrate a method for solving systems of inequalities.

**EXAMPLE 3** *Solving a system of linear inequalities*

Sketch the graph of the system

**FIGURE 14**

$$\begin{cases} x + y \le 4 \\ 2x - y \le 4 \end{cases}$$

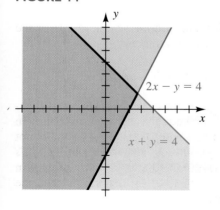

**SOLUTION**  We replace each $\le$ with $=$ and then sketch the resulting line, as shown in Figure 14. Using the test point $(0, 0)$, we see that the solutions of the system correspond to the points *below* (and on) the line $x + y = 4$ and *above* (and on) the line $2x - y = 4$. Shading these half-planes with different colors, as in Figure 14, we have as the graph of the system the points that are in *both* regions, indicated by the purple portion of the figure.

**FIGURE 15**

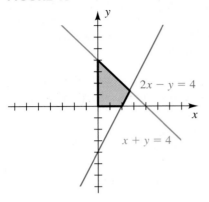

**EXAMPLE 4** *Solving a system of linear inequalities*

Sketch the graph of the system

$$\begin{cases} x + y \le 4 \\ 2x - y \le 4 \\ x \ge 0 \\ y \ge 0 \end{cases}$$

**SOLUTION** The first two inequalities are the same as those considered in Example 3, and hence the points on the graph of the system must lie within the purple region shown in Figure 14. In addition, the third and fourth inequalities in the system tell us that the points must lie in the first quadrant or on its boundaries. This gives us the region shown in Figure 15.

**EXAMPLE 5** *Solving a system of inequalities*
*containing absolute values*

Sketch the graph of the system

$$\begin{cases} |x| \le 2 \\ |y| > 1 \end{cases}$$

**FIGURE 16**

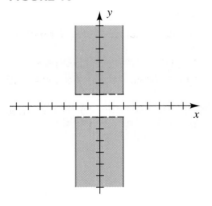

**SOLUTION** Using properties of absolute value, we see that $(x, y)$ is a solution of the system if and only if *both* of the following conditions are true:

**(1)** $-2 \le x \le 2$

**(2)** $y < -1$ or $y > 1$

Thus, a point $(x, y)$ on the graph of the system must lie between (or on) the vertical lines $x = \pm 2$ and also either below the horizontal line $y = -1$ or above the line $y = 1$. The graph is sketched in Figure 16.

**FIGURE 17**

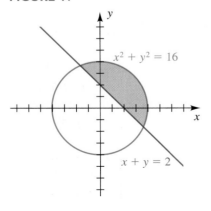

**EXAMPLE 6** *Solving a system of inequalities*

Sketch the graph of the system

$$\begin{cases} x^2 + y^2 \le 16 \\ x + y \ge 2 \end{cases}$$

**SOLUTION** The graphs of $x^2 + y^2 = 16$ and $x + y = 2$ are the circle and line shown in Figure 17. Using the test point $(0, 0)$, we see that the points that yield solutions of the system must lie inside (or on) the circle and also above (or on) the line. This gives us the region sketched in Figure 17.

**EXAMPLE 7    An application of a system of inequalities**

The manager of a baseball team wishes to buy bats and balls costing $12 and $3 each, respectively. At least five bats and ten balls are required, and the total cost is not to exceed $180. Find a system of inequalities that describes all possibilities, and sketch the graph.

**SOLUTION**    We begin by letting $x$ denote the number of bats and $y$ the number of balls. Since the cost of a bat is $12 and the cost of a ball is $3, we see that

$$12x = \text{cost of } x \text{ bats}$$

$$3y = \text{cost of } y \text{ balls.}$$

Since the total cost is not to exceed $180, we must have

$$12x + 3y \le 180$$

or, equivalently,                $$y \le -4x + 60.$$

Since at least five bats and ten balls are required, we also have

$$x \ge 5 \quad \text{and} \quad y \ge 10.$$

The graph of $y \le -4x + 60$ is the half-plane that lies *below* (or on) the line $y = -4x + 60$ shown in Figure 18.

The graph of $x \ge 5$ is the region to the right of (or on) the vertical line $x = 5$, and the graph of $y \ge 10$ is the region above (or on) the horizontal line $y = 10$.

The graph of the system—that is, the points common to the three half-planes—is the triangular region sketched in Figure 18.

**FIGURE 18**

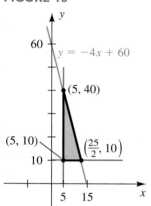

## 8.5 EXERCISES

**Exer. 1–10: Sketch the graph of the inequality.**

1  $3x - 2y < 6$

2  $4x + 3y < 12$

3  $2x + 3y \ge 2y + 1$

4  $2x - y > 3$

5  $y + 2 < x^2$

6  $y^2 - x \le 0$

7  $x^2 + 1 \le y$

8  $y - x^3 < 1$

9  $yx^2 \ge 1$

10  $x^2 + 4 \ge y$

**Exer. 11–26: Sketch the graph of the system of inequalities.**

11  $\begin{cases} 3x + y < 3 \\ 4 - y < 2x \end{cases}$

12  $\begin{cases} y + 2 < 2x \\ y - x > 4 \end{cases}$

13  $\begin{cases} y - x < 0 \\ 2x + 5y < 10 \end{cases}$

14  $\begin{cases} 2y - x \le 4 \\ 3y + 2x < 6 \end{cases}$

15  $\begin{cases} 3x + y \le 6 \\ y - 2x \ge 1 \\ x \ge -2 \\ y \le 4 \end{cases}$

16  $\begin{cases} 3x - 4y \ge 12 \\ x - 2y \le 2 \\ x \ge 9 \\ y \le 5 \end{cases}$

17  $\begin{cases} x + 2y \le 8 \\ 0 \le x \le 4 \\ 0 \le y \le 3 \end{cases}$

18  $\begin{cases} 2x + 3y \ge 6 \\ 0 \le x \le 5 \\ 0 \le y \le 4 \end{cases}$

19 $\begin{cases} |x| \geq 2 \\ |y| < 3 \end{cases}$

20 $\begin{cases} |x| \geq 4 \\ |y| \geq 3 \end{cases}$

21 $\begin{cases} |x + 2| \leq 1 \\ |y - 3| < 5 \end{cases}$

22 $\begin{cases} |x - 2| \leq 5 \\ |y - 4| > 2 \end{cases}$

23 $\begin{cases} x^2 + y^2 \leq 4 \\ x + y \geq 1 \end{cases}$

24 $\begin{cases} x^2 + y^2 > 1 \\ x^2 + y^2 < 4 \end{cases}$

25 $\begin{cases} x^2 \leq 1 - y \\ x \geq 1 + y \end{cases}$

26 $\begin{cases} x - y^2 < 0 \\ x + y^2 > 0 \end{cases}$

29

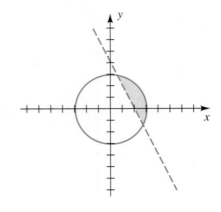

**Exer. 27–34: Find a system of inequalities whose graph is shown.**

27

30

28

31

**32**

**33**

**34**

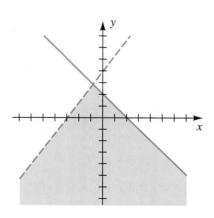

**35** *Inventory levels* A store sells two brands of television sets. Customer demand indicates that it is necessary to stock at least twice as many sets of brand A as of brand B. It is also necessary to have on hand at least 10 sets of brand B. There is room for not more than 100 sets in the store. Find and graph a system of inequalities that describes all possibilities for stocking the two brands.

**36** *Ticket prices* An auditorium contains 600 seats. For an upcoming event, tickets will be priced at $8.00 for some seats and $5.00 for others. At least 225 tickets are to be priced at $5.00, and total sales of at least $3000 are desired. Find and graph a system of inequalities that describes all possibilities for pricing the two types of tickets.

**37** *Investment strategy* A woman with $15,000 to invest decides to place at least $2000 in a high-risk, high-yield investment and at least three times that amount in a low-risk, low-yield investment. Find and graph a system of inequalities that describes all possibilities for placing the money in the two investments.

**38** *Inventory levels* The manager of a college bookstore stocks two types of notebooks, the first wholesaling for 55 cents and the second for 85 cents. The maximum amount to be spent is $600, and an inventory of at least 300 of the 85-cent variety and 400 of the 55-cent variety is desired. Find and graph a system of inequalities that describes all possibilities for stocking the two types of notebooks.

**39** *Dimensions of a can* An aerosol can is to be constructed in the shape of a circular cylinder with a small cone on the top. The total height of the can is to be no more than 9 inches, and the cylinder must contain at least 75% of the total volume. In addition, the height of the conical top must be at least 1 inch. Find and graph a system of inequalities that describes all possibilities for the relationship between the height $y$ of the cylinder and the height $x$ of the cone.

**40** *Dimensions of a window* A stained-glass window is to be constructed in the form of a rectangle surmounted by a semicircle (see the figure). The total height $h$ of the window can be no more than 6 feet, and the area of the rectangular part must be at least twice the area of the semicircle. In addition, the diameter $d$ of the semicircle must be at least 2 feet. Find and graph a system of inequalities that describes all possibilities for the base and height of the rectangular part.

**EXERCISE 40**

41 **Locating a power plant**  A nuclear power plant will be constructed to serve the power needs of cities A and B. City B is 100 miles due east of A. The state has promised that the plant will be at least 60 miles from each city. It is not possible, however, to locate the plant south of either city because of rough terrain, and the plant must be within 100 miles of both A and B. Assuming A is at the origin, find and graph a system of inequalities that describes all possible locations for the plant.

42 **Allocating space**  A man has a rectangular back yard that is 50 feet wide and 60 feet deep. He plans to construct a pool area and a patio area, as shown in the figure, where $y \geq 10$. He can spend at most $10,500 on the project. The patio area must be at least as large as the pool area. The pool area will cost $5 per square foot, and the patio will cost $3 per square foot. Find and graph a system of inequalities that describes all possibilities for the width of the patio and pool areas.

**EXERCISE 42**

**Exer. 43–44: Graph the inequality.**

43  $64y^3 - x^3 \leq e^{1-2x}$         44  $e^{5y} - e^{-x} \geq x^4$

**Exer. 45–48: Graph the system of inequalities.**

45  $\begin{cases} 5^{1-y} \geq x^4 + x^2 + 1 \\ x + 3y \geq x^{5/3} \end{cases}$     46  $\begin{cases} x^4 + y^5 < 2^x \\ \ln (x^2 + 1) < y^3 \end{cases}$

47  $\begin{cases} x^4 - 2x < 3y \\ x + 2y < x^3 - 5 \end{cases}$     48  $\begin{cases} e^x + x^2 \leq 2^{x+2y} \\ 2^{x+2y} \leq x^3 2^y \\ x > 0 \end{cases}$

## 8.6  LINEAR PROGRAMMING

**FIGURE 19**

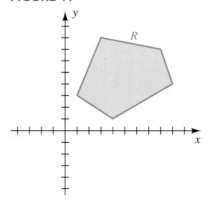

If a system of inequalities contains only linear inequalities of the form

$$ax + by \leq c \quad \text{or} \quad ax + by \geq c,$$

where $a$, $b$, and $c$ are real numbers, then the graph of the system may be a region $R$ in the $xy$-plane bounded by a polygon—possibly of the type illustrated in Figure 19 (for a specific illustration, see Example 4 and Figure 15 of Section 8.5). For problems in **linear programming,** we consider such systems together with an expression of the form

$$C = Ax + By + K,$$

where $A$, $B$, and $K$ are real numbers and $(x, y)$ is a point in $R$ (that is, a solution of the system). Since for each $(x, y)$ we obtain a specific value for $C$, we call $C$ a *function of two variables x and y.* In linear programming, $C$

is called an **objective function,** and the inequalities in the system are referred to as the **constraints** on $C$. The solutions of the system—that is, the pairs $(x, y)$ corresponding to the points in $R$—are called the **feasible solutions** for the problem.

In typical business applications, the value of $C$ may represent cost, profit, loss, or a physical resource, and the goal is to find a specific point $(x, y)$ in $R$ at which $C$ takes on its maximum or minimum value. The methods of linear programming greatly simplify the task of finding this point. Specifically, it can be shown that *the maximum and minimum values of C occur at a vertex of R.* This fact is used in the next example.

**EXAMPLE 1**   *Finding the maximum and minimum values of an objective function*

Find the maximum and minimum values of the objective function given by $C = 7x + 3y$ subject to the following constraints:

$$\begin{cases} x - 2y \geq -10 \\ 2x + \phantom{y} y \leq \phantom{1}10 \\ \phantom{2x + y} x \geq \phantom{1}0 \\ \phantom{2x + x} y \geq \phantom{1}0 \end{cases}$$

**FIGURE 20**

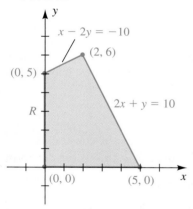

$y$

$x - 2y = -10$

$(2, 6)$

$(0, 5)$

$R$

$2x + y = 10$

$(0, 0)$        $(5, 0)$        $x$

**SOLUTION**    The graph of the system of inequalities determined by the constraints is the region $R$ bounded by the quadrilateral sketched in Figure 20. From the preceding discussion, the maximum and minimum values of $C$ must occur at a vertex of $R$. The values at the vertices are given in the following table.

| Vertex | Value of $C = 7x + 3y$ |
|--------|------------------------|
| $(0, 0)$ | $7(0) + 3(0) = 0$ |
| $(0, 5)$ | $7(0) + 3(5) = 15$ |
| $(5, 0)$ | $7(5) + 3(0) = 35$ |
| $(2, 6)$ | $7(2) + 3(6) = 32$ |

Hence, the minimum value $C = 0$ occurs if $x = 0$ and $y = 0$. The maximum value $C = 35$ occurs if $x = 5$ and $y = 0$.

In the preceding example, we say that the maximum value of $C$ on $R$ occurs at the vertex $(5, 0)$. To verify this fact, let us solve $C = 7x + 3y$ for $y$, obtaining

$$y = -\frac{7}{3}x + \frac{C}{3}.$$

For each $C$, the graph of this equation is a line of slope $-\frac{7}{3}$ and $y$-intercept $C/3$, as illustrated in Figure 21. To find the maximum value of $C$, we simply determine which of these lines that intersect the region has the largest $y$-intercept $C/3$. Referring to Figure 21, we see that the required line passes through $(5, 0)$. Similarly, for the minimum value of $C$, we determine the line $y = (-7/3)x + (C/3)$ that intersects the region and has the *smallest* $y$-intercept. This is the line through $(0, 0)$.

**FIGURE 21**

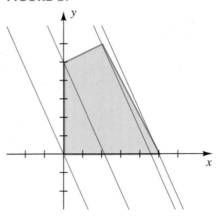

We shall call a problem that can be expressed in the form of Example 1 a **linear programming problem.** To solve such problems, we may use the following guidelines.

**Guidelines for Solving a Linear Programming Problem**

1   Sketch the region $R$ determined by the system of constraints.

2   Find the vertices of $R$.

3   Calculate the value of the objective function $C$ at each vertex of $R$.

4   Determine the maximum or minimum values of $C$ by using the vertices of $R$ in guideline 3 that yield the largest or smallest values, respectively.

In the next example we encounter a linear programming problem in which the minimum value of the objective function occurs at more than one point.

**EXAMPLE 2** *Solving a linear programming problem*

Find the minimum value of the objective function $C = 2x + 6y$ subject to the following constraints:

$$\begin{cases} 2x + 3y \geq 12 \\ x + 3y \geq 9 \\ x \geq 0 \\ y \geq 0 \end{cases}$$

**FIGURE 22**

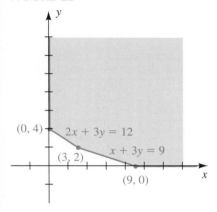

**SOLUTION** We shall follow the guidelines.

**Guideline 1** The graph of the system of inequalities determined by the constraints is the unbounded region $R$ sketched in Figure 22.

**Guideline 2** The vertices of $R$ are $(0, 4)$, $(3, 2)$, and $(9, 0)$, as shown in the figure.

**Guideline 3** The value of $C$ at each vertex of $R$ is given in the following table.

| Vertex | Value of $C = 2x + 6y$ |
|--------|------------------------|
| $(0, 4)$ | $2(0) + 6(4) = 24$ |
| $(3, 2)$ | $2(3) + 6(2) = 18$ |
| $(9, 0)$ | $2(9) + 6(0) = 18$ |

**Guideline 4** The table in guideline 3 shows that the minimum value of $C$, 18, occurs at *two* vertices, $(3, 2)$ and $(9, 0)$. Moreover, if $(x, y)$ is any point on the line segment joining these points, then $(x, y)$ is a solution of the equation $x + 3y = 9$, and hence

$$C = 2x + 6y = 2(x + 3y) = 2(9) = 18.$$

Thus, the minimum value $C = 18$ occurs at *every* point on this line segment.

---

In the next two examples we consider applications of linear programming. For such problems it is necessary to use given information and data to formulate the system of constraints and the objective function. Once this has been accomplished, we may apply the guidelines as we did in the solution to Example 2.

**EXAMPLE 3** *Maximizing profit*

A firm manufactures two products X and Y. For each product, it is necessary to use three different machines A, B, and C. To manufacture one unit

*(continued)*

of product X, machine A must be used for 3 hours, machine B for 1 hour, and machine C for 1 hour. To manufacture one unit of product Y requires 2 hours on A, 2 hours on B, and 1 hour on C. The profit on product X is $500 per unit, and the profit on product Y is $350 per unit. Machine A is available for a total of 24 hours per day; however, B can be used for only 16 hours and C for 9 hours. Assuming the machines are available when needed (subject to the noted total hour restrictions), determine the number of units of each product that should be manufactured each day in order to maximize the profit.

*SOLUTION*    The following table summarizes the data given in the statement of the problem.

| Machine | Hours required for one unit of X | Hours required for one unit of Y | Hours available |
|---------|:---:|:---:|:---:|
| A | 3 | 2 | 24 |
| B | 1 | 2 | 16 |
| C | 1 | 1 | 9 |

Let us introduce the following variables:

$$x = \text{number of units of X manufactured each day}$$

$$y = \text{number of units of Y manufactured each day}$$

Using the first row of the table, we note that each unit of X requires 3 hours on machine A, and hence $x$ units require $3x$ hours. Similarly, since each unit of Y requires 2 hours on A, $y$ units require $2y$ hours. Hence, the total number of hours per day that machine A must be used is $3x + 2y$. This, together with the fact that A can be used for at most 24 hours per day, gives us the first constraint in the following system of inequalities: $3x + 2y \leq 24$. The second and third constraints are obtained by using the same type of reasoning for rows 2 and 3 of the table. The last two constraints, $x \geq 0$ and $y \geq 0$, are true because $x$ and $y$ cannot be negative.

**FIGURE 23**

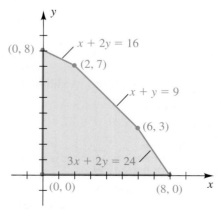

$$\begin{cases} 3x + 2y \leq 24 \\ x + 2y \leq 16 \\ x + \ y \leq \ 9 \\ x \geq \ 0 \\ y \geq \ 0 \end{cases}$$

The graph of this system is the region $R$ in Figure 23.

Since the product of each unit of product X yields a profit of $500 and each unit of product Y yields a profit of $350, the profit $P$ obtained by producing $x$ units of X together with $y$ units of Y is

$$P = 500x + 350y.$$

This is the objective function for the problem. The maximum value of $P$ must occur at one of the vertices of $R$ in Figure 23. The values of $P$ at these vertices are given in the following table.

| Vertex | Value of $P = 500x + 350y$ |
|--------|-----------------------------|
| $(0, 0)$ | $500(0) + 350(0) = 0$ |
| $(0, 8)$ | $500(0) + 350(8) = 2800$ |
| $(8, 0)$ | $500(8) + 350(0) = 4000$ |
| $(2, 7)$ | $500(2) + 350(7) = 3450$ |
| $(6, 3)$ | $500(6) + 350(3) = 4050$ |

We see from the table that a maximum profit of $4050 occurs for a daily production of 6 units of product X and 3 units of product Y.

Example 3 illustrates maximization of profit. The next example demonstrates how linear programming can be used to minimize the cost in a certain situation.

**EXAMPLE 4   *Minimizing cost***

A distributor of compact disk players has two warehouses $W_1$ and $W_2$. There are 80 units stored at $W_1$ and 70 units at $W_2$. Two customers A and B order 35 units and 60 units, respectively. The shipping cost from each warehouse to A and B is determined according to the following table. How should the order be filled to minimize the total shipping cost?

| Warehouse | Customer | Shipping cost per unit |
|-----------|----------|------------------------|
| $W_1$ | A | $ 8 |
| $W_1$ | B | 12 |
| $W_2$ | A | 10 |
| $W_2$ | B | 13 |

*SOLUTION*   Let us begin by introducing the following variables:

$$x = \text{number of units sent to A from } W_1$$
$$y = \text{number of units sent to B from } W_1$$

Since A ordered 35 units and B ordered 60 units, we must have

$$35 - x = \text{number of units sent to A from } W_2$$
$$60 - y = \text{number of units sent to B from } W_2.$$

*(continued)*

Our goal is to determine values for $x$ and $y$ that make the total shipping cost minimal.

The number of units shipped from $W_1$ cannot exceed 80, and the number shipped from $W_2$ cannot exceed 70. Expressing these facts in terms of inequalities gives us

$$\begin{cases} x + y \le 80 \\ (35 - x) + (60 - y) \le 70 \end{cases}$$

Simplifying, we obtain the first two constraints in the following system. The last two constraints are true because the largest values of $x$ and $y$ are 35 and 60, respectively.

$$\begin{cases} x + y \le 80 \\ x + y \ge 25 \\ 0 \le x \le 35 \\ 0 \le y \le 60 \end{cases}$$

The graph of this system is the region $R$ shown in Figure 24.

Let $C$ denote the total cost (in dollars) of shipping the disk players to customers A and B. We see from the table of shipping costs that the following are true:

cost of shipping 35 units to A $= 8x + 10(35 - x)$

cost of shipping 60 units to B $= 12y + 13(60 - y)$

Hence, the total cost is

$$C = 8x + 10(35 - x) + 12y + 13(60 - y).$$

Simplifying gives us the following objective function:

$$C = 1130 - 2x - y$$

To determine the minimum value of $C$ on $R$, we need check only the vertices shown in Figure 24, as in the following table.

**FIGURE 24**

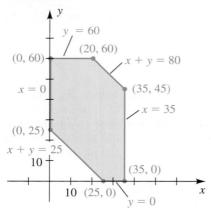

| Vertex | Value of $C = 1130 - 2x - y$ |
|--------|------------------------------|
| $(0, 25)$ | $1130 - 2(0) - 25 = 1105$ |
| $(0, 60)$ | $1130 - 2(0) - 60 = 1070$ |
| $(20, 60)$ | $1130 - 2(20) - 60 = 1030$ |
| $(35, 45)$ | $1130 - 2(35) - 45 = 1015$ |
| $(35, 0)$ | $1130 - 2(35) - 0 = 1060$ |
| $(25, 0)$ | $1130 - 2(25) - 0 = 1080$ |

We see from the table that the minimal shipping cost $1015 occurs if $x = 35$ and $y = 45$. This means that the distributor should ship all of the disk players to A from $W_1$ and none from $W_2$. In addition, the distributor should ship 45 units to B from $W_1$ and 15 units to B from $W_2$. (Note that the *maximum* shipping cost will occur if $x = 0$ and $y = 25$—that is, if all 35 units are shipped to A from $W_2$ and if B receives 25 units from $W_1$ and 35 units from $W_2$.)

---

The examples in this section are elementary linear programming problems in two variables that can be solved by basic methods. The much more complicated problems in many variables that occur in practice may be solved by employing matrix techniques that are adapted for solution by computers.

## 8.6 EXERCISES

**Exer. 1–2: Find the maximum and minimum values of the objective function $C$ on the region in the figure.**

1 $C = 3x + 2y + 5$

2 $C = 2x + 7y + 3$

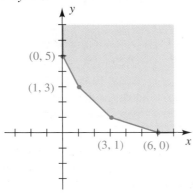

**Exer. 3–4: Sketch the region $R$ determined by the given constraints, and label its vertices. Find the maximum value of $C$ on $R$.**

3 $C = 3x + y$;      $x \geq 0$, $y \geq 0$,
   $3x - 4y \geq -12$,    $3x + 2y \leq 24$,    $3x - y \leq 15$

4 $C = 4x - 2y$;
   $x - 2y \geq -8$,    $7x - 2y \leq 28$,    $x + y \geq 4$

**Exer. 5–6: Sketch the region $R$ determined by the given constraints, and label its vertices. Find the minimum value of $C$ on $R$.**

5 $C = 3x + 6y$;    $x \geq 0$, $y \geq 0$,
   $2x + 3y \geq 12$,    $2x + 5y \geq 16$

6 $C = 6x + y$;    $y \geq 0$,
   $3x + y \geq 3$,    $x + 5y \leq 15$,    $2x + y \leq 12$

**Exer. 7–8: Sketch the region $R$ determined by the given constraints, and label its vertices. Describe the set of points for which $C$ is a maximum on $R$.**

7 $C = 2x + 4y$;    $x \geq 0$, $y \geq 0$,
   $x - 2y \geq -8$,    $\frac{1}{2}x + y \leq 6$,    $3x + 2y \leq 24$

8 $C = 6x + 3y$;    $x \geq 2$, $y \geq 1$,
   $2x + 3y \leq 19$,    $x + 0.5y \leq 6.5$

9 **Production scheduling** A manufacturer of tennis rackets makes a profit of $15 on each oversized racket and $8 on each standard racket. To meet dealer demand, daily production of standard rackets should be between 30 and 80, and production of oversized rackets should be between 10 and 30. To maintain high quality, the

total number of rackets produced should not exceed 80 per day. How many of each type should be manufactured daily to maximize the profit?

10  **Production scheduling**  A manufacturer of CB radios makes a profit of $25 on a deluxe model and $30 on a standard model. The company wishes to produce at least 80 deluxe models and at least 100 standard models per day. To maintain high quality, the daily production should not exceed 200 radios. How many of each type should be produced daily in order to maximize the profit?

11  **Minimizing cost**  Two substances S and T each contain two types of ingredients I and G. One pound of S contains 2 ounces of I and 4 ounces of G. One pound of T contains 2 ounces of I and 6 ounces of G. A manufacturer plans to combine quantities of the two substances to obtain a mixture that contains at least 9 ounces of I and 20 ounces of G. If the cost of S is $3.00 per pound and the cost of T is $4.00 per pound, how much of each substance should be used to keep the cost to a minimum?

12  **Maximizing gross profit**  A stationery company makes two types of notebooks: a deluxe notebook with subject dividers, which sells for $1.25, and a regular notebook, which sells for $0.90. The production cost is $1.00 for each deluxe notebook and $0.75 for each regular notebook. The company has the facilities to manufacture between 2000 and 3000 deluxe and between 3000 and 6000 regular notebooks, but not more than 7000 altogether. How many notebooks of each type should be manufactured to maximize the difference between the selling prices and the production cost?

13  **Minimizing shipping costs**  Refer to Example 4 of this section. If the shipping costs are $12 per unit from $W_1$ to A, $10 per unit from $W_2$ to A, $16 per unit from $W_1$ to B, and $12 per unit from $W_2$ to B, determine how the order should be filled to minimize shipping cost.

14  **Minimizing cost**  A coffee company purchases mixed lots of coffee beans and then grades them into premium, regular, and unusable beans. The company needs at least 280 tons of premium-grade and 200 tons of regular-grade coffee beans. The company can purchase ungraded coffee from two suppliers in any amount desired. Samples from the two suppliers contain the following percentages of premium, regular, and unusable beans:

| Supplier | Premium | Regular | Unusable |
| --- | --- | --- | --- |
| A | 20% | 50% | 30% |
| B | 40% | 20% | 40% |

If supplier A charges $125 per ton and B charges $200 per ton, how much should the company purchase from each supplier to fulfill its needs at minimum cost?

15  **Planning crop acreage**  A farmer, in the business of growing fodder for livestock, has 90 acres available for planting alfalfa and corn. The cost of seed per acre is $4 for alfalfa and $6 for corn. The total cost of labor will amount to $20 per acre for alfalfa and $10 per acre for corn. The expected income from alfalfa is $110 per acre, and from corn, $150 per acre. If the farmer does not wish to spend more than $480 for seed and $1400 for labor, how many acres of each crop should be planted to obtain the maximum profit?

16  **Machinery scheduling**  A small firm manufactures bookshelves and desks for microcomputers. For each product it is necessary to use a table saw and a power router. To manufacture each bookshelf, the saw must be used for $\frac{1}{2}$ hour and the router for 1 hour. A desk requires the use of each machine for 2 hours. The profits are $20 per bookshelf and $50 per desk. If the saw can be used 8 hours per day and the router for 12 hours per day, how many bookshelves and desks should be manufactured each day to maximize the profit?

17  **Minimizing a mixture's cost**  Three substances X, Y, and Z each contain four ingredients A, B, C, and D. The percentage of each ingredient and the cost in cents per ounce of each substance are given in the following table.

| Substance | Ingredients | | | | Cost per ounce |
| --- | --- | --- | --- | --- | --- |
| | A | B | C | D | |
| X | 20% | 10% | 25% | 45% | 25¢ |
| Y | 20% | 40% | 15% | 25% | 35¢ |
| Z | 10% | 20% | 25% | 45% | 50¢ |

If the cost is to be minimal, how many ounces of each substance should be combined to obtain a mixture of 20 ounces containing at least 14% A, 16% B, and 20% C? What combination would make the cost greatest?

**18 Maximizing profit** A man plans to operate a stand at a one-day fair at which he will sell bags of peanuts and bags of candy. He has $400 available to purchase his stock, which will cost 40¢ per bag of peanuts and 80¢ per bag of candy. He intends to sell the peanuts at $1.00 and the candy at $1.60 per bag. His stand can accommodate up to 500 bags of peanuts and 400 bags of candy. From past experience he knows that he will sell no more than a total of 700 bags. Find the number of bags of each that he should have available in order to maximize his profit. What is the maximum profit?

**19 Maximizing passenger capacity** A small community wishes to purchase used vans and small buses for its public transportation system. The community can spend no more than $100,000 for the vehicles and no more than $500 per month for maintenance. The vans sell for $10,000 each and average $100 per month in maintenance costs. The corresponding cost estimates for each bus are $20,000 and $75 per month. If each van can carry 15 passengers and each bus can accommodate 25 riders, determine the number of vans and buses that should be purchased to maximize the passenger capacity of the system.

**20 Minimizing fuel cost** Refer to Exercise 19. The monthly fuel cost (based on 5000 miles of service) is $550 for each van and $850 for each bus. Find the number of vans and buses that should be purchased to minimize the monthly fuel costs if the passenger capacity of the system must be at least 75.

**21 Stocking a fish farm** A fish farmer will purchase no more than 5000 young trout and bass from the hatchery and will feed them a special diet for the next year. The cost of food per fish will be $0.50 for trout and $0.75 for bass, and the total cost is not to exceed $3000. At the end of the year, a typical trout will weigh 3 pounds, and a bass will weigh 4 pounds. How many fish of each type should be stocked in the pond in order to maximize the total number of pounds of fish at the end of the year?

**22 Dietary planning** A hospital dietician wishes to prepare a corn-squash vegetable dish that will provide at least 3 grams of protein and cost no more than 36¢ per serving. An ounce of creamed corn provides $\frac{1}{2}$ gram of protein and costs 4¢. An ounce of squash supplies $\frac{1}{4}$ gram of protein and costs 3¢. For taste, there must be at least 2 ounces of corn and at least as much squash as corn. It is important to keep the total number of ounces in a serving as small as possible. Find the combination of corn and squash that will minimize the amount of ingredients used per serving.

**23 Planning storage units** A contractor has a large building that she wishes to convert into a series of rental storage spaces. She will construct basic 8 ft × 10 ft units and deluxe 12 ft × 10 ft units that contain extra shelves and a clothes closet. Market considerations dictate that there be at least twice as many basic units as deluxe units and that the basic units rent for $40 per month and the deluxe units for $75 per month. At most 7200 ft$^2$ is available for the storage spaces, and no more than $30,000 can be spent on construction. If each basic unit will cost $300 to make and each deluxe unit will cost $600, how many units of each type should be constructed to maximize monthly revenue?

**24 A moose's diet** A moose feeding primarily on tree leaves and aquatic plants is capable of digesting no more than 33 kilograms of these foods daily. Although the aquatic plants are lower in energy content, the animal must eat at least 17 kilograms to satisfy its sodium requirement. A kilogram of leaves provides four times as much energy as a kilogram of aquatic plants. Find the combination of foods that maximizes the daily energy intake.

## 8.7 THE ALGEBRA OF MATRICES

Matrices were introduced in Section 8.3 as an aid to finding solutions of systems of equations. In this section we discuss some of their properties. These properties are important in advanced fields of mathematics and in applications.

In the following definition, the symbol $(a_{ij})$ denotes an $m \times n$ matrix $A$ of the type displayed in the definition on page 543. We use similar notations for the matrices $B$ and $C$.

**Definition of Equality and Addition of Matrices**

Let $A = (a_{ij})$, $B = (b_{ij})$, and $C = (c_{ij})$ be $m \times n$ matrices.

(1) $A = B$ if and only if $a_{ij} = b_{ij}$ for every $i$ and $j$.

(2) $C = A + B$ if and only if $c_{ij} = a_{ij} + b_{ij}$ for every $i$ and $j$.

Note that two matrices are equal if and only if they have the same size and corresponding elements are equal.

**ILLUSTRATION**

**Equality of Matrices**

$$\begin{bmatrix} 1 & 0 & 5 \\ \sqrt[3]{8} & 3^2 & -2 \end{bmatrix} = \begin{bmatrix} (-1)^2 & 0 & \sqrt{25} \\ 2 & 9 & -2 \end{bmatrix}$$

Using the parentheses notation for matrices, we may write the definition of addition of two $m \times n$ matrices as

$$(a_{ij}) + (b_{ij}) = (a_{ij} + b_{ij}).$$

Thus, to add two matrices, we add the elements in corresponding positions in each matrix. *Two matrices can be added only if they have the same size.*

**ILLUSTRATION**

**Addition of Matrices**

$$\begin{bmatrix} 4 & -5 \\ 0 & 4 \\ -6 & 1 \end{bmatrix} + \begin{bmatrix} 3 & 2 \\ 7 & -4 \\ -2 & 1 \end{bmatrix} = \begin{bmatrix} 4+3 & -5+2 \\ 0+7 & 4+(-4) \\ -6+(-2) & 1+1 \end{bmatrix} = \begin{bmatrix} 7 & -3 \\ 7 & 0 \\ -8 & 2 \end{bmatrix}$$

$$\begin{bmatrix} 2 & 3 \\ -4 & 1 \end{bmatrix} + \begin{bmatrix} -2 & -3 \\ 4 & -1 \end{bmatrix} = \begin{bmatrix} 0 & 0 \\ 0 & 0 \end{bmatrix}$$

$$\begin{bmatrix} 1 & 3 & -2 \\ 0 & -5 & 4 \end{bmatrix} + \begin{bmatrix} 0 & 0 & 0 \\ 0 & 0 & 0 \end{bmatrix} = \begin{bmatrix} 1 & 3 & -2 \\ 0 & -5 & 4 \end{bmatrix}$$

The **$m \times n$ zero matrix,** denoted by $O$, is the matrix with $m$ rows and $n$ columns in which every element is 0.

### ILLUSTRATION

**Zero Matrices**

$$\blacksquare \begin{bmatrix} 0 & 0 \\ 0 & 0 \end{bmatrix} \qquad \blacksquare \begin{bmatrix} 0 & 0 \\ 0 & 0 \\ 0 & 0 \end{bmatrix} \qquad \blacksquare \begin{bmatrix} 0 & 0 & 0 & 0 \\ 0 & 0 & 0 & 0 \end{bmatrix}$$

The **additive inverse** $-A$ of the matrix $A = (a_{ij})$ is the matrix $(-a_{ij})$ obtained by changing the sign of each nonzero element of $A$.

### ILLUSTRATION

**Additive Inverse**

$$\blacksquare \quad -\begin{bmatrix} 2 & -3 & 4 \\ -1 & 0 & 5 \end{bmatrix} = \begin{bmatrix} -2 & 3 & -4 \\ 1 & 0 & -5 \end{bmatrix}$$

The proof of the next theorem follows from the definition of addition of matrices.

**Theorem on Matrix Properties**

If $A$, $B$, and $C$ are $m \times n$ matrices and if $O$ is the $m \times n$ zero matrix, then

(1) $A + B = B + A$

(2) $A + (B + C) = (A + B) + C$

(3) $A + O = A$

(4) $A + (-A) = O$

**Subtraction** of two $m \times n$ matrices is defined by

$$A - B = A + (-B).$$

Using the parentheses notation, we have

$$(a_{ij}) - (b_{ij}) = (a_{ij}) + (-b_{ij})$$
$$= (a_{ij} - b_{ij}).$$

Thus, to subtract two matrices, we subtract the elements in corresponding positions.

**ILLUSTRATION**

**Subtraction of Matrices**

$$\begin{bmatrix} 4 & -5 \\ 0 & 4 \\ -6 & 1 \end{bmatrix} - \begin{bmatrix} 3 & 2 \\ 7 & -4 \\ -2 & 1 \end{bmatrix} = \begin{bmatrix} 4-3 & -5-2 \\ 0-7 & 4-(-4) \\ -6-(-2) & 1-1 \end{bmatrix} = \begin{bmatrix} 1 & -7 \\ -7 & 8 \\ -4 & 0 \end{bmatrix}$$

**Definition of the Product of a Real Number and a Matrix**

The **product** of a real number $c$ and an $m \times n$ matrix $A = (a_{ij})$ is

$$cA = (ca_{ij}).$$

Note that to find $cA$, we multiply each element of $A$ by $c$.

**ILLUSTRATION**

**Product of a Real Number and a Matrix**

$$3\begin{bmatrix} 4 & -1 \\ 2 & 3 \end{bmatrix} = \begin{bmatrix} 3 \cdot 4 & 3 \cdot (-1) \\ 3 \cdot 2 & 3 \cdot 3 \end{bmatrix} = \begin{bmatrix} 12 & -3 \\ 6 & 9 \end{bmatrix}$$

We can prove the following.

**Theorem on Matrix Properties**

If $A$ and $B$ are $m \times n$ matrices and if $c$ and $d$ are real numbers, then

**(1)** $c(A + B) = cA + cB$

**(2)** $(c + d)A = cA + dA$

**(3)** $(cd)A = c(dA)$

The next definition, of the product $AB$ of two matrices, may seem unusual, but it has many uses in mathematics and applications. For multiplication, unlike addition, $A$ and $B$ may have different sizes; however, *the number of columns of A must be the same as the number of rows of B.* Thus, if $A$ is $m \times n$, then $B$ must be $n \times p$ for some $p$. As we shall see, the size of $AB$ is then $m \times p$. If $C = AB$, then a method for finding the element $c_{ij}$ in row $i$ and column $j$ of $C$ is given in the following guidelines.

**Guidelines for Finding $c_{ij}$ in the Product $C = AB$ If $A$ Is $m \times n$ and $B$ Is $n \times p$**

**1**   Single out the $i$th row $R_i$ of $A$ and the $j$th column $C_j$ of $B$:

$$\begin{bmatrix} a_{11} & a_{12} & \cdots & a_{1n} \\ \vdots & \vdots & & \vdots \\ a_{i1} & a_{i2} & \cdots & a_{in} \\ \vdots & \vdots & & \vdots \\ a_{m1} & a_{m2} & \cdots & a_{mn} \end{bmatrix} \begin{bmatrix} b_{11} & \cdots & b_{1j} & \cdots & b_{1p} \\ b_{21} & \cdots & b_{2j} & \cdots & b_{2p} \\ \vdots & & \vdots & & \vdots \\ b_{n1} & \cdots & b_{nj} & \cdots & b_{np} \end{bmatrix}$$

**2**   *Simultaneously* move to the right along $R_i$ and down $C_j$, multiplying pairs of elements, to obtain

$$a_{i1}b_{1j}, \ a_{i2}b_{2j}, \ a_{i3}b_{3j}, \ \ldots, \ a_{in}b_{nj}.$$

**3**   Add the products of the pairs in guideline 2 to obtain $c_{ij}$:

$$c_{ij} = a_{i1}b_{1j} + a_{i2}b_{2j} + a_{i3}b_{3j} + \cdots + a_{in}b_{nj}$$

Using the guidelines, we see that the element $c_{11}$ in the first row and the first column of $AB$ is

$$c_{11} = a_{11}b_{11} + a_{12}b_{21} + a_{13}b_{31} + \cdots + a_{1n}b_{n1}.$$

The element $c_{mp}$ in the last row and the last column of $AB$ is

$$c_{mp} = a_{m1}b_{1p} + a_{m2}b_{2p} + a_{m3}b_{3p} + \cdots + a_{mn}b_{np}.$$

The preceding discussion is summarized in the next definition.

**Definition of the Product of Two Matrices**

Let $A = (a_{ij})$ be an $m \times n$ matrix and let $B = (b_{ij})$ be an $n \times p$ matrix. The **product** $AB$ is the $m \times p$ matrix $C = (c_{ij})$ such that

$$c_{ij} = a_{i1}b_{1j} + a_{i2}b_{2j} + a_{i3}b_{3j} + \cdots + a_{in}b_{nj}$$

for $i = 1, 2, \ldots, m$ and $j = 1, 2, \ldots, p$.

The following diagram may help you remember the relationship between sizes of matrices when working with a product $AB$.

The next illustration contains some special cases.

**ILLUSTRATION**

### Sizes of Matrices in Products

| Size of $A$ | Size of $B$ | Size of $AB$ |
|---|---|---|
| $2 \times 3$ | $3 \times 5$ | $2 \times 5$ |
| $4 \times 2$ | $2 \times 3$ | $4 \times 3$ |
| $3 \times 1$ | $1 \times 3$ | $3 \times 3$ |
| $1 \times 3$ | $3 \times 1$ | $1 \times 1$ |
| $5 \times 3$ | $3 \times 5$ | $5 \times 5$ |
| $5 \times 3$ | $5 \times 3$ | $AB$ is not defined |

In the following example we find the product of two specific matrices.

**EXAMPLE 1**    *Finding the product of two matrices*

Find the product $AB$ if

$$A = \begin{bmatrix} 1 & 2 & -3 \\ 4 & 0 & -2 \end{bmatrix} \quad \text{and} \quad B = \begin{bmatrix} 5 & -4 & 2 & 0 \\ -1 & 6 & 3 & 1 \\ 7 & 0 & 5 & 8 \end{bmatrix}.$$

**SOLUTION**    The matrix $A$ is $2 \times 3$, and the matrix $B$ is $3 \times 4$. Hence, the product $C = AB$ is defined and is $2 \times 4$. We next use the guidelines to find the elements $c_{11}, c_{12}, \ldots$ of the product. For instance, to find the element $c_{23}$ we single out the second row $R_2$ of $A$ and the third column $C_3$ of $B$, as illustrated below, and then use guidelines 2 and 3 to obtain

$$c_{23} = 4 \cdot 2 + 0 \cdot 3 + (-2) \cdot 5 = -2.$$

$$\begin{bmatrix} 1 & 2 & -3 \\ 4 & 0 & -2 \end{bmatrix} \begin{bmatrix} 5 & -4 & 2 & 0 \\ -1 & 6 & 3 & 1 \\ 7 & 0 & 5 & 8 \end{bmatrix} = \begin{bmatrix} & & & \\ & & -2 & \end{bmatrix}$$

Similarly, to find the element $c_{12}$ in row 1 and column 2 of the product, we proceed as follows:

$$c_{12} = 1 \cdot (-4) + 2 \cdot 6 + (-3) \cdot 0 = 8$$

$$\begin{bmatrix} 1 & 2 & -3 \\ 4 & 0 & -2 \end{bmatrix} \begin{bmatrix} 5 & -4 & 2 & 0 \\ -1 & 6 & 3 & 1 \\ 7 & 0 & 5 & 8 \end{bmatrix} = \begin{bmatrix} & 8 & & \\ & & -2 & \end{bmatrix}$$

The remaining elements of the product are calculated as follows, where we have indicated the row of $A$ and the column of $B$ that are used when guideline 1 is applied.

| Row of $A$ | Column of $B$ | Element of $C$ |
|:---:|:---:|:---|
| $R_1$ | $C_1$ | $c_{11} = 1 \cdot 5 + 2 \cdot (-1) + (-3) \cdot 7 = -18$ |
| $R_1$ | $C_3$ | $c_{13} = 1 \cdot 2 + 2 \cdot 3 + (-3) \cdot 5 = -7$ |
| $R_1$ | $C_4$ | $c_{14} = 1 \cdot 0 + 2 \cdot 1 + (-3) \cdot 8 = -22$ |
| $R_2$ | $C_1$ | $c_{21} = 4 \cdot 5 + 0 \cdot (-1) + (-2) \cdot 7 = 6$ |
| $R_2$ | $C_2$ | $c_{22} = 4 \cdot (-4) + 0 \cdot 6 + (-2) \cdot 0 = -16$ |
| $R_2$ | $C_4$ | $c_{24} = 4 \cdot 0 + 0 \cdot 1 + (-2) \cdot 8 = -16$ |

Hence,

$$AB = \begin{bmatrix} 1 & 2 & -3 \\ 4 & 0 & -2 \end{bmatrix} \begin{bmatrix} 5 & -4 & 2 & 0 \\ -1 & 6 & 3 & 1 \\ 7 & 0 & 5 & 8 \end{bmatrix}$$

$$= \begin{bmatrix} -18 & 8 & -7 & -22 \\ 6 & -16 & -2 & -16 \end{bmatrix}.$$

A matrix is a **row matrix** if it has only one row. A **column matrix** has only one column. The following illustration contains some products involving row and column matrices. You should check each entry in the products.

### ILLUSTRATION

**Products Involving Row and Column Matrices**

- $\begin{bmatrix} -2 & 4 \\ 0 & -1 \\ 5 & 3 \end{bmatrix} \begin{bmatrix} -2 \\ 1 \end{bmatrix} = \begin{bmatrix} 8 \\ -1 \\ -7 \end{bmatrix}$
- $\begin{bmatrix} 3 & -1 & 2 \end{bmatrix} \begin{bmatrix} -2 & 4 \\ 0 & -1 \\ 5 & 3 \end{bmatrix} = \begin{bmatrix} 4 & 19 \end{bmatrix}$

- $\begin{bmatrix} -2 \\ 3 \end{bmatrix} \begin{bmatrix} 1 & 5 \end{bmatrix} = \begin{bmatrix} -2 & -10 \\ 3 & 15 \end{bmatrix}$
- $\begin{bmatrix} 1 & 5 \end{bmatrix} \begin{bmatrix} -2 \\ 3 \end{bmatrix} = \begin{bmatrix} 13 \end{bmatrix}$

The product operation for matrices is not commutative. For example, if $A$ is $2 \times 3$ and $B$ is $3 \times 4$, then $AB$ may be found, since the number of columns of $A$ is the same as the number of rows of $B$. However, $BA$ is undefined, since the number of columns of $B$ is different from the number of rows of $A$. Even if $AB$ and $BA$ are both defined, it is often true that these products are different. This is illustrated in the next example, along with the fact that the product of two nonzero matrices may equal a zero matrix.

**EXAMPLE 2**    *Matrix multiplication is not commutative*

If $A = \begin{bmatrix} 2 & 2 \\ -1 & -1 \end{bmatrix}$ and $B = \begin{bmatrix} 1 & 2 \\ 1 & 2 \end{bmatrix}$, show that $AB \neq BA$.

**SOLUTION**    Using the definition of the product of two matrices, we obtain the following:

$$AB = \begin{bmatrix} 2 & 2 \\ -1 & -1 \end{bmatrix}\begin{bmatrix} 1 & 2 \\ 1 & 2 \end{bmatrix} = \begin{bmatrix} 4 & 8 \\ -2 & -4 \end{bmatrix}$$

$$BA = \begin{bmatrix} 1 & 2 \\ 1 & 2 \end{bmatrix}\begin{bmatrix} 2 & 2 \\ -1 & -1 \end{bmatrix} = \begin{bmatrix} 0 & 0 \\ 0 & 0 \end{bmatrix}$$

Hence, $AB \neq BA$. Note that the last equality shows that *the product of two nonzero matrices can equal a zero matrix.*

---

Although matrix multiplication is not commutative, it is associative. Thus, if $A$ is $m \times n$, $B$ is $n \times p$, and $C$ is $p \times q$, then

$$A(BC) = (AB)C.$$

The distributive properties also hold if the matrices involved have the proper number of rows and columns. If $A_1$ and $A_2$ are $m \times n$ matrices and if $B_1$ and $B_2$ are $n \times p$ matrices, then

$$A_1(B_1 + B_2) = A_1B_1 + A_1B_2$$

$$(A_1 + A_2)B_1 = A_1B_1 + A_2B_1.$$

As a special case, if all matrices are square, of order $n$, then both the associative and distributive properties are true.

We conclude this section with an application of the product of two matrices.

**EXAMPLE 3**    *An application of a matrix product*

**(a)** Three investors $I_1$, $I_2$, and $I_3$ each own a certain number of shares of four stocks $S_1$, $S_2$, $S_3$, and $S_4$, according to matrix $A$. Matrix $B$ contains the present value $V$ of each share of each stock. Find $AB$ and interpret the meaning of its elements.

$$\text{investors} \begin{cases} I_1 \\ I_2 \\ I_3 \end{cases} \overbrace{\begin{bmatrix} \overset{S_1}{50} & \overset{S_2}{100} & \overset{S_3}{30} & \overset{S_4}{25} \\ 100 & 150 & 10 & 30 \\ 100 & 50 & 40 & 100 \end{bmatrix}}^{\text{number of shares of stock}} = A, \quad \text{stocks} \begin{cases} S_1 \\ S_2 \\ S_3 \\ S_4 \end{cases} \overbrace{\begin{bmatrix} 20.37 \\ 16.21 \\ 90.80 \\ 42.75 \end{bmatrix}}^{\overset{\text{share value}}{V}} = B$$

**(b)** Matrix $C$ contains the change in the value of each stock for the last week. Find $AC$ and interpret the meaning of its elements.

$$\text{stocks} \begin{cases} S_1 \\ S_2 \\ S_3 \\ S_4 \end{cases} \begin{bmatrix} +1.03 \\ -0.22 \\ -1.35 \\ +0.15 \end{bmatrix} = C$$

**SOLUTION**

**(a)** Since $A$ is a $3 \times 4$ matrix and $B$ is a $4 \times 1$ matrix, the product $AB$ is a $3 \times 1$ matrix.

$$AB = \begin{bmatrix} 50 & 100 & 30 & 25 \\ 100 & 150 & 10 & 30 \\ 100 & 50 & 40 & 100 \end{bmatrix} \begin{bmatrix} 20.37 \\ 16.21 \\ 90.80 \\ 42.75 \end{bmatrix} = \begin{bmatrix} 6432.25 \\ 6659.00 \\ 10{,}754.50 \end{bmatrix}$$

The first element in the product $AB$, 6432.25, was obtained from the computation

$$50(20.37) + 100(16.21) + 30(90.80) + 25(42.75)$$

and represents the total value that investor $I_1$ has in all four stocks. Similarly, the second and third elements represent the total value for investors $I_2$ and $I_3$, respectively.

**(b)**

$$AC = \begin{bmatrix} 50 & 100 & 30 & 25 \\ 100 & 150 & 10 & 30 \\ 100 & 50 & 40 & 100 \end{bmatrix} \begin{bmatrix} +1.03 \\ -0.22 \\ -1.35 \\ +0.15 \end{bmatrix} = \begin{bmatrix} -7.25 \\ 61.00 \\ 53.00 \end{bmatrix}$$

The first element in the product $AC$, $-7.25$, indicates that the total value that investor $I_1$ has in all four stocks went down $7.25 in the last week. The second and third elements indicate that the total value that investors $I_2$ and $I_3$ have in all four stocks went up $61.00 and $53.00, respectively.

## 8.7 EXERCISES

Exer. 1–8: Find, if possible, $A + B$, $A - B$, $2A$, and $-3B$.

**1** $A = \begin{bmatrix} 5 & -2 \\ 1 & 3 \end{bmatrix}$, $B = \begin{bmatrix} 4 & 1 \\ -3 & 2 \end{bmatrix}$

**2** $A = \begin{bmatrix} 3 & 0 \\ -1 & 2 \end{bmatrix}$, $B = \begin{bmatrix} 3 & -4 \\ 1 & 1 \end{bmatrix}$

**3** $A = \begin{bmatrix} 6 & -1 \\ 2 & 0 \\ -3 & 4 \end{bmatrix}$,     $B = \begin{bmatrix} 3 & 1 \\ -1 & 5 \\ 6 & 0 \end{bmatrix}$

**4** $A = \begin{bmatrix} 0 & -2 & 7 \\ 5 & 4 & -3 \end{bmatrix}$,     $B = \begin{bmatrix} 8 & 4 & 0 \\ 0 & 1 & 4 \end{bmatrix}$

**5** $A = [4 \quad -3 \quad 2]$,     $B = [7 \quad 0 \quad -5]$

**6** $A = \begin{bmatrix} 7 \\ -16 \end{bmatrix}$,     $B = \begin{bmatrix} -11 \\ 9 \end{bmatrix}$

**7** $A = \begin{bmatrix} 3 & -2 & 2 \\ 0 & 1 & -4 \\ -3 & 2 & -1 \end{bmatrix}$, $B = \begin{bmatrix} 4 & 0 \\ 2 & -1 \\ -1 & 3 \end{bmatrix}$

**8** $A = [2 \quad 1]$,     $B = [3 \quad -1 \quad 5]$

**Exer. 9–10: Find the given element of the matrix product $C = AB$ in the listed exercise.**

**9** $c_{21}$;   Exercise 15

**10** $c_{23}$;   Exercise 16

**Exer. 11–22: Find, if possible, $AB$ and $BA$.**

**11** $A = \begin{bmatrix} 2 & 6 \\ 3 & -4 \end{bmatrix}$,     $B = \begin{bmatrix} 5 & -2 \\ 1 & 7 \end{bmatrix}$

**12** $A = \begin{bmatrix} 4 & -2 \\ -2 & 1 \end{bmatrix}$,     $B = \begin{bmatrix} 2 & 1 \\ 4 & 2 \end{bmatrix}$

**13** $A = \begin{bmatrix} 3 & 0 & -1 \\ 0 & 4 & 2 \\ 5 & -3 & 1 \end{bmatrix}$, $B = \begin{bmatrix} 1 & -5 & 0 \\ 4 & 1 & -2 \\ 0 & -1 & 3 \end{bmatrix}$

**14** $A = \begin{bmatrix} 5 & 0 & 0 \\ 0 & -3 & 0 \\ 0 & 0 & 2 \end{bmatrix}$, $B = \begin{bmatrix} 3 & 0 & 0 \\ 0 & 4 & 0 \\ 0 & 0 & -2 \end{bmatrix}$

**15** $A = \begin{bmatrix} 4 & -3 & 1 \\ -5 & 2 & 2 \end{bmatrix}$, $B = \begin{bmatrix} 2 & 1 \\ 0 & 1 \\ -4 & 7 \end{bmatrix}$

**16** $A = \begin{bmatrix} 2 & 1 & -1 & 0 \\ 3 & -2 & 0 & 5 \\ -2 & 1 & 4 & 2 \end{bmatrix}$, $B = \begin{bmatrix} 5 & -3 & 1 \\ 1 & 2 & 0 \\ -1 & 0 & 4 \\ 0 & -2 & 3 \end{bmatrix}$

**17** $A = \begin{bmatrix} 1 & 2 & 3 \\ 4 & 5 & 6 \\ 7 & 8 & 9 \end{bmatrix}$, $B = \begin{bmatrix} 1 & 0 & 0 \\ 0 & 1 & 0 \\ 0 & 0 & 1 \end{bmatrix}$

**18** $A = \begin{bmatrix} 1 & 2 & 3 \\ 2 & 3 & 1 \\ 3 & 1 & 2 \end{bmatrix}$, $B = \begin{bmatrix} 2 & 0 & 0 \\ 0 & 2 & 0 \\ 0 & 0 & 2 \end{bmatrix}$

**19** $A = [-3 \quad 7 \quad 2]$,     $B = \begin{bmatrix} 1 \\ 4 \\ -5 \end{bmatrix}$

**20** $A = [4 \quad 8]$,     $B = \begin{bmatrix} -3 \\ 2 \end{bmatrix}$

**21** $A = \begin{bmatrix} 2 & 0 & 1 \\ -1 & 2 & 0 \end{bmatrix}$, $B = \begin{bmatrix} 1 & -1 & 2 \\ 3 & 1 & 0 \\ 0 & 2 & 1 \end{bmatrix}$

**22** $A = [3 \quad -1 \quad 4]$,     $B = \begin{bmatrix} -2 \\ 5 \end{bmatrix}$

**Exer. 23–26: Find $AB$.**

**23** $A = \begin{bmatrix} 4 & -2 \\ 0 & 3 \\ -7 & 5 \end{bmatrix}$,     $B = \begin{bmatrix} 3 \\ 4 \end{bmatrix}$

**24** $A = \begin{bmatrix} 4 \\ -3 \\ 2 \end{bmatrix}$,     $B = [5 \quad 1]$

**25** $A = \begin{bmatrix} 2 & 1 & 0 & -3 \\ -7 & 0 & -2 & 4 \end{bmatrix}$, $B = \begin{bmatrix} 4 & -2 & 0 \\ 1 & 1 & -2 \\ 0 & 0 & 5 \\ -3 & -1 & 0 \end{bmatrix}$

**26** $A = \begin{bmatrix} 1 & 2 & -3 \\ 4 & -5 & 6 \end{bmatrix}$, $B = \begin{bmatrix} 1 & -1 & 0 & 2 \\ -2 & 3 & 1 & 0 \\ 0 & 4 & 0 & -3 \end{bmatrix}$

**Exer. 27–30: Let**

$$A = \begin{bmatrix} 1 & 2 \\ 0 & -3 \end{bmatrix}, \quad B = \begin{bmatrix} 2 & -1 \\ 3 & 1 \end{bmatrix}, \quad C = \begin{bmatrix} 3 & 1 \\ -2 & 0 \end{bmatrix}.$$

**Verify the statement.**

**27** $(A + B)(A - B) \neq A^2 - B^2$,
    where $A^2 = AA$ and $B^2 = BB$.

28 $(A + B)(A + B) \neq A^2 + 2AB + B^2$

29 $A(B + C) = AB + AC$

30 $A(BC) = (AB)C$

**Exer. 31–34:** Verify the identity for

$$A = \begin{bmatrix} a & b \\ c & d \end{bmatrix}, \quad B = \begin{bmatrix} p & q \\ r & s \end{bmatrix}, \quad C = \begin{bmatrix} w & x \\ y & z \end{bmatrix}$$

and real numbers $m$ and $n$.

31 $m(A + B) = mA + mB$       32 $(m + n)A = mA + nA$

33 $A(B + C) = AB + AC$       34 $A(BC) = (AB)C$

35 *Value of inventory* A computer store stocks the following kinds of disks, each available in five colors: $5\frac{1}{4}''$ DS/DD (Double Sided/Double Density) priced at 33¢ each, $5\frac{1}{4}''$ DS/HD (High Density) priced at 42¢ each, and $3\frac{1}{2}''$ DS/DD priced at 51¢ each. The store's current inventory is as follows:

| Disk type | Colors | | | | |
|---|---|---|---|---|---|
| | Red | Green | Blue | Orange | Yellow |
| $5\frac{1}{4}''$ DS/DD | 400 | 400 | 300 | 250 | 100 |
| $5\frac{1}{4}''$ DS/HD | 550 | 450 | 500 | 200 | 100 |
| $3\frac{1}{2}''$ DS/DD | 500 | 500 | 600 | 300 | 200 |

(a) Organize these data into an inventory matrix $A$ and a price matrix $B$ so that the product $C = AB$ is defined.

(b) Find $C$.

(c) Interpret the meaning of element $c_{51}$ in $C$.

36 *Bird migration trends*   Refer to Exercise 34 of Section 8.3. Suppose the initial populations on islands A, B, and C are 12,000, 9000, and 14,000, respectively.

(a) Represent the initial populations with a $1 \times 3$ matrix $D$. Represent the proportions of the populations that migrate to each island with a $3 \times 3$ matrix $E$. (*Hint:* The first row of $E$ is 0.90, 0.10, and 0.00— indicating that 90% of the birds on A stay on A, 10% of the birds on A migrate to B, and no birds on A migrate to C.)

(b) Find the product $F = DE$, and interpret the meaning of the elements of $F$.

(c) Using a computational device, multiply $F$ times $E$, and continue to multiply the result by $E$ until a pattern becomes apparent. What is your conclusion? (Some specific keystrokes for the TI-81/82 graphics calculator are given in Appendix I.)

(d) Suppose the initial population matrix is $D = [34{,}000 \quad 500 \quad 500]$. Multiply $D$ times $E$, and continue to multiply the result by $E$ until a pattern becomes apparent. What is your conclusion?

## 8.8 THE INVERSE OF A MATRIX

Throughout this section and the next two sections we shall restrict our discussion to *square* matrices. The symbol $I_n$ will denote the square matrix of order $n$ that has 1 in each position on the main diagonal and 0 elsewhere. We call $I_n$ the **identity matrix of order $n$.**

### ILLUSTRATION

**Identity Matrices**

$$I_2 = \begin{bmatrix} 1 & 0 \\ 0 & 1 \end{bmatrix} \qquad I_3 = \begin{bmatrix} 1 & 0 & 0 \\ 0 & 1 & 0 \\ 0 & 0 & 1 \end{bmatrix}$$

We can show that if $A$ is any square matrix of order $n$, then

$$AI_n = A = I_nA.$$

### ILLUSTRATION

$AI_3 = A = I_3A$

$$\begin{bmatrix} a_{11} & a_{12} & a_{13} \\ a_{21} & a_{22} & a_{23} \\ a_{31} & a_{32} & a_{33} \end{bmatrix} \begin{bmatrix} 1 & 0 & 0 \\ 0 & 1 & 0 \\ 0 & 0 & 1 \end{bmatrix} = \begin{bmatrix} a_{11} & a_{12} & a_{13} \\ a_{21} & a_{22} & a_{23} \\ a_{31} & a_{32} & a_{33} \end{bmatrix}$$

$$= \begin{bmatrix} 1 & 0 & 0 \\ 0 & 1 & 0 \\ 0 & 0 & 1 \end{bmatrix} \begin{bmatrix} a_{11} & a_{12} & a_{13} \\ a_{21} & a_{22} & a_{23} \\ a_{31} & a_{32} & a_{33} \end{bmatrix}$$

Sometimes an $n \times n$ matrix $A$ has an **inverse**—a matrix $B$ such that $AB = I_n = BA$. If $A$ has an inverse, we denote it by $A^{-1}$ and write

$$AA^{-1} = I_n = A^{-1}A.$$

The symbol $A^{-1}$ is read "$A$ inverse." For matrices, the symbol $1/A$ does not represent the inverse $A^{-1}$.

If a square matrix $A$ has an inverse, we can calculate $A^{-1}$ using elementary row operations. If $A = (a_{ij})$ is $n \times n$, we begin with the $n \times 2n$ matrix

$$\left[ \begin{array}{cccc|cccc} a_{11} & a_{12} & \cdots & a_{1n} & 1 & 0 & \cdots & 0 \\ a_{21} & a_{22} & \cdots & a_{2n} & 0 & 1 & \cdots & 0 \\ \vdots & \vdots & & \vdots & \vdots & \vdots & & \vdots \\ a_{n1} & a_{n2} & \cdots & a_{nn} & 0 & 0 & \cdots & 1 \end{array} \right]$$

in which the $n \times n$ identity matrix $I_n$ appears to the right of the vertical rule. We next apply a succession of elementary row transformations, as we did in Section 8.3 to find reduced echelon forms, until we arrive at a matrix of the form

$$\left[ \begin{array}{cccc|cccc} 1 & 0 & \cdots & 0 & b_{11} & b_{12} & \cdots & b_{1n} \\ 0 & 1 & \cdots & 0 & b_{21} & b_{22} & \cdots & b_{2n} \\ \vdots & \vdots & & \vdots & \vdots & \vdots & & \vdots \\ 0 & 0 & \cdots & 1 & b_{n1} & b_{n2} & \cdots & b_{nn} \end{array} \right]$$

in which the identity matrix $I_n$ appears to the left of the vertical rule. It can be shown that the $n \times n$ matrix $(b_{ij})$ is the inverse $A^{-1}$.

**EXAMPLE 1**   *Finding the inverse of a 2 × 2 matrix*

Find $A^{-1}$ if $A = \begin{bmatrix} 3 & 5 \\ 1 & 4 \end{bmatrix}$.

*SOLUTION*   We begin with the matrix

$$\left[ \begin{array}{cc|cc} 3 & 5 & 1 & 0 \\ 1 & 4 & 0 & 1 \end{array} \right].$$

Next we perform elementary row transformations until the identity matrix $I_2$ appears on the left of the vertical rule, as follows:

$$\left[ \begin{array}{cc|cc} 3 & 5 & 1 & 0 \\ 1 & 4 & 0 & 1 \end{array} \right] \mathbf{R_1 \leftrightarrow R_2} \left[ \begin{array}{cc|cc} 1 & 4 & 0 & 1 \\ 3 & 5 & 1 & 0 \end{array} \right]$$

$$\mathbf{-3R_1 + R_2 \rightarrow R_2} \left[ \begin{array}{cc|cc} 1 & 4 & 0 & 1 \\ 0 & -7 & 1 & -3 \end{array} \right]$$

$$\mathbf{-\frac{1}{7}R_2 \rightarrow R_2} \left[ \begin{array}{cc|cc} 1 & 4 & 0 & 1 \\ 0 & 1 & -\frac{1}{7} & \frac{3}{7} \end{array} \right]$$

$$\mathbf{-4R_2 + R_1 \rightarrow R_1} \left[ \begin{array}{cc|cc} 1 & 0 & \frac{4}{7} & -\frac{5}{7} \\ 0 & 1 & -\frac{1}{7} & \frac{3}{7} \end{array} \right]$$

By the previous discussion,

$$A^{-1} = \begin{bmatrix} \frac{4}{7} & -\frac{5}{7} \\ -\frac{1}{7} & \frac{3}{7} \end{bmatrix} = \frac{1}{7} \begin{bmatrix} 4 & -5 \\ -1 & 3 \end{bmatrix}.$$

Let us verify that $AA^{-1} = I_2 = A^{-1}A$:

$$\begin{bmatrix} 3 & 5 \\ 1 & 4 \end{bmatrix} \begin{bmatrix} \frac{4}{7} & -\frac{5}{7} \\ -\frac{1}{7} & \frac{3}{7} \end{bmatrix} = \begin{bmatrix} 1 & 0 \\ 0 & 1 \end{bmatrix} = \begin{bmatrix} \frac{4}{7} & -\frac{5}{7} \\ -\frac{1}{7} & \frac{3}{7} \end{bmatrix} \begin{bmatrix} 3 & 5 \\ 1 & 4 \end{bmatrix}$$

**EXAMPLE 2**   *Finding the inverse of a 3 × 3 matrix*

Find $A^{-1}$ if $A = \begin{bmatrix} -1 & 3 & 1 \\ 2 & 5 & 0 \\ 3 & 1 & -2 \end{bmatrix}$.

*SOLUTION*

$$\begin{bmatrix} -1 & 3 & 1 & | & 1 & 0 & 0 \\ 2 & 5 & 0 & | & 0 & 1 & 0 \\ 3 & 1 & -2 & | & 0 & 0 & 1 \end{bmatrix} \xrightarrow{-R_1 \to R_1} \begin{bmatrix} 1 & -3 & -1 & | & -1 & 0 & 0 \\ 2 & 5 & 0 & | & 0 & 1 & 0 \\ 3 & 1 & -2 & | & 0 & 0 & 1 \end{bmatrix}$$

$$\xrightarrow[{-3R_1 + R_3 \to R_3}]{-2R_1 + R_2 \to R_2} \begin{bmatrix} 1 & -3 & -1 & | & -1 & 0 & 0 \\ 0 & 11 & 2 & | & 2 & 1 & 0 \\ 0 & 10 & 1 & | & 3 & 0 & 1 \end{bmatrix}$$

$$\xrightarrow{-R_3 + R_2 \to R_2} \begin{bmatrix} 1 & -3 & -1 & | & -1 & 0 & 0 \\ 0 & 1 & 1 & | & -1 & 1 & -1 \\ 0 & 10 & 1 & | & 3 & 0 & 1 \end{bmatrix}$$

$$\xrightarrow[{-10R_2 + R_3 \to R_3}]{3R_2 + R_1 \to R_1} \begin{bmatrix} 1 & 0 & 2 & | & -4 & 3 & -3 \\ 0 & 1 & 1 & | & -1 & 1 & -1 \\ 0 & 0 & -9 & | & 13 & -10 & 11 \end{bmatrix}$$

$$\xrightarrow{-\frac{1}{9}R_3 \to R_3} \begin{bmatrix} 1 & 0 & 2 & | & -4 & 3 & -3 \\ 0 & 1 & 1 & | & -1 & 1 & -1 \\ 0 & 0 & 1 & | & -\frac{13}{9} & \frac{10}{9} & -\frac{11}{9} \end{bmatrix}$$

$$\xrightarrow[{-R_3 + R_2 \to R_2}]{-2R_3 + R_1 \to R_1} \begin{bmatrix} 1 & 0 & 0 & | & -\frac{10}{9} & \frac{7}{9} & -\frac{5}{9} \\ 0 & 1 & 0 & | & \frac{4}{9} & -\frac{1}{9} & \frac{2}{9} \\ 0 & 0 & 1 & | & -\frac{13}{9} & \frac{10}{9} & -\frac{11}{9} \end{bmatrix}$$

Consequently,

$$A^{-1} = \begin{bmatrix} -\frac{10}{9} & \frac{7}{9} & -\frac{5}{9} \\ \frac{4}{9} & -\frac{1}{9} & \frac{2}{9} \\ -\frac{13}{9} & \frac{10}{9} & -\frac{11}{9} \end{bmatrix} = \frac{1}{9}\begin{bmatrix} -10 & 7 & -5 \\ 4 & -1 & 2 \\ -13 & 10 & -11 \end{bmatrix}.$$

You may verify that $AA^{-1} = I_3 = A^{-1}A$.

---

If the procedure used in Examples 1 and 2 does not lead to an identity matrix to the left of the vertical rule, then the matrix $A$ has no inverse.

We may apply inverses of matrices to solutions of systems of linear equations. Consider the case of two linear equations in two unknowns:

$$\begin{cases} a_{11}x + a_{12}y = k_1 \\ a_{21}x + a_{22}y = k_2 \end{cases}$$

This system can be expressed in terms of matrices as

$$\begin{bmatrix} a_{11}x + a_{12}y \\ a_{21}x + a_{22}y \end{bmatrix} = \begin{bmatrix} k_1 \\ k_2 \end{bmatrix}.$$

If we let

$$A = \begin{bmatrix} a_{11} & a_{12} \\ a_{21} & a_{22} \end{bmatrix}, \quad X = \begin{bmatrix} x \\ y \end{bmatrix}, \quad \text{and} \quad B = \begin{bmatrix} k_1 \\ k_2 \end{bmatrix},$$

then a *matrix form* for the system is

$$AX = B.$$

If $A^{-1}$ exists, then multiplying both sides of the last equation by $A^{-1}$ gives us $A^{-1}AX = A^{-1}B$. Since $A^{-1}A = I_2$ and $I_2X = X$, this leads to

$$X = A^{-1}B,$$

from which the solution $(x, y)$ may be found. This technique may be extended to systems of $n$ linear equations in $n$ unknowns.

**EXAMPLE 3**    *Solving a system of linear equations using an inverse of a matrix*

Solve the system of equations:

$$\begin{cases} -x + 3y + z = 1 \\ 2x + 5y \quad\quad = 3 \\ 3x + y - 2z = -2 \end{cases}$$

**SOLUTION**    If we let

$$A = \begin{bmatrix} -1 & 3 & 1 \\ 2 & 5 & 0 \\ 3 & 1 & -2 \end{bmatrix}, \quad X = \begin{bmatrix} x \\ y \\ z \end{bmatrix}, \quad \text{and} \quad B = \begin{bmatrix} 1 \\ 3 \\ -2 \end{bmatrix},$$

then $AX = B$. This implies that $X = A^{-1}B$. The matrix $A^{-1}$ was found in Example 2. Hence,

$$\begin{bmatrix} x \\ y \\ z \end{bmatrix} = \frac{1}{9}\begin{bmatrix} -10 & 7 & -5 \\ 4 & -1 & 2 \\ -13 & 10 & -11 \end{bmatrix}\begin{bmatrix} 1 \\ 3 \\ -2 \end{bmatrix} = \frac{1}{9}\begin{bmatrix} 21 \\ -3 \\ 39 \end{bmatrix} = \begin{bmatrix} \frac{7}{3} \\ -\frac{1}{3} \\ \frac{13}{3} \end{bmatrix}.$$

Thus, $x = \frac{7}{3}$, $y = -\frac{1}{3}$, $z = \frac{13}{3}$, and the ordered triple $\left(\frac{7}{3}, -\frac{1}{3}, \frac{13}{3}\right)$ is the solution of the given system.

The method of solution in Example 3 is beneficial only if $A^{-1}$ is known (or can be easily computed) or if many systems with the same coefficient matrix are to be considered. The preferred technique for solving an arbitrary system of linear equations is the matrix method discussed in Section 8.3.

There are other, more important uses for the inverse of a matrix; however, they occur in more advanced fields of mathematics and in applications of such fields.

## 8.8 EXERCISES

**Exer. 1–10: Find the inverse of the matrix, if it exists.**

1. $\begin{bmatrix} 2 & -4 \\ 1 & 3 \end{bmatrix}$

2. $\begin{bmatrix} 3 & 2 \\ 4 & 5 \end{bmatrix}$

3. $\begin{bmatrix} 2 & 4 \\ 4 & 8 \end{bmatrix}$

4. $\begin{bmatrix} 3 & -1 \\ 6 & -2 \end{bmatrix}$

5. $\begin{bmatrix} 3 & -1 & 0 \\ 2 & 2 & 0 \\ 0 & 0 & 4 \end{bmatrix}$

6. $\begin{bmatrix} 3 & 0 & 2 \\ 0 & 1 & 0 \\ -4 & 0 & 2 \end{bmatrix}$

7. $\begin{bmatrix} -2 & 2 & 3 \\ 1 & -1 & 0 \\ 0 & 1 & 4 \end{bmatrix}$

8. $\begin{bmatrix} 1 & 2 & 3 \\ -2 & 1 & 0 \\ 3 & -1 & 1 \end{bmatrix}$

9. $\begin{bmatrix} 2 & 0 & 0 \\ 0 & 4 & 0 \\ 0 & 0 & 6 \end{bmatrix}$

10. $\begin{bmatrix} 1 & 1 & 1 \\ 2 & 2 & 2 \\ 3 & 3 & 3 \end{bmatrix}$

11. State conditions on $a$ and $b$ that guarantee that the matrix $\begin{bmatrix} a & 0 \\ 0 & b \end{bmatrix}$ has an inverse, and find a formula for the inverse, if it exists.

12. If $abc \neq 0$, find the inverse of $\begin{bmatrix} a & 0 & 0 \\ 0 & b & 0 \\ 0 & 0 & c \end{bmatrix}$.

13. If $A = \begin{bmatrix} a_{11} & a_{12} & a_{13} \\ a_{21} & a_{22} & a_{23} \\ a_{31} & a_{32} & a_{33} \end{bmatrix}$, show that $AI_3 = A = I_3A$.

14. Show that $AI_4 = A = I_4A$ for every square matrix $A$ of order 4.

**Exer. 15–18: Solve the system using the method in Example 3. (Refer to Exercises 1–2 and 7–8.)**

15. $\begin{cases} 2x - 4y = c \\ x + 3y = d \end{cases}$

  (a) $\begin{bmatrix} c \\ d \end{bmatrix} = \begin{bmatrix} 3 \\ 1 \end{bmatrix}$     (b) $\begin{bmatrix} c \\ d \end{bmatrix} = \begin{bmatrix} -2 \\ 5 \end{bmatrix}$

16. $\begin{cases} 3x + 2y = c \\ 4x + 5y = d \end{cases}$

  (a) $\begin{bmatrix} c \\ d \end{bmatrix} = \begin{bmatrix} -1 \\ 1 \end{bmatrix}$     (b) $\begin{bmatrix} c \\ d \end{bmatrix} = \begin{bmatrix} 4 \\ 3 \end{bmatrix}$

17. $\begin{cases} -2x + 2y + 3z = c \\ x - y = d \\ y + 4z = e \end{cases}$

  (a) $\begin{bmatrix} c \\ d \\ e \end{bmatrix} = \begin{bmatrix} 1 \\ 3 \\ -2 \end{bmatrix}$     (b) $\begin{bmatrix} c \\ d \\ e \end{bmatrix} = \begin{bmatrix} -1 \\ 0 \\ 4 \end{bmatrix}$

18. $\begin{cases} x + 2y + 3z = c \\ -2x + y = d \\ 3x - y + z = e \end{cases}$

  (a) $\begin{bmatrix} c \\ d \\ e \end{bmatrix} = \begin{bmatrix} -1 \\ 4 \\ 2 \end{bmatrix}$     (b) $\begin{bmatrix} c \\ d \\ e \end{bmatrix} = \begin{bmatrix} -3 \\ -2 \\ 1 \end{bmatrix}$

**Exer. 19–20: (a) Express the system in the matrix form $AX = B$. (b) Approximate $A^{-1}$, using four-decimal-place accuracy for its elements. (c) Use $X = A^{-1}B$ to approximate the solution of the system to four-decimal-place accuracy.**

19. $\begin{cases} 3.1x + 6.7y - 8.7z = 1.5 \\ 4.1x - 5.1y + 0.2z = 2.1 \\ 0.6x + 1.1y - 7.4z = 3.9 \end{cases}$

  (*Note:* Some specific keystrokes for the TI-81/82 graphics calculator are given in Appendix I.)

20. $\begin{cases} 5.6x + 8.4y - 7.2z + 4.2w = 8.1 \\ 8.4x + 9.2y - 6.1z - 6.2w = 5.3 \\ -7.2x - 6.1y + 9.2z + 4.5w = 0.4 \\ 4.2x - 6.2y - 4.5z + 5.8w = 2.7 \end{cases}$

# 8.9 DETERMINANTS

Associated with each square matrix $A$ is a number called the **determinant of $A$**, denoted by $|A|$. This notation should not be confused with the symbol for the absolute value of a real number. To avoid any misunderstanding, the expression "det $A$" is sometimes used in place of $|A|$. We shall define $|A|$ by beginning with the case in which $A$ has order 1 and then increasing the order one at a time. As we shall see in Section 8.10, these definitions arise in a natural way when systems of linear equations are solved.

If $A$ is a square matrix of order 1, then $A$ has only one element. Thus, $A = [a_{11}]$ and we define $|A| = a_{11}$. If $A$ is a square matrix of order 2, then

$$A = \begin{bmatrix} a_{11} & a_{12} \\ a_{21} & a_{22} \end{bmatrix}$$

and the determinant of $A$ is defined by

$$|A| = a_{11}a_{22} - a_{21}a_{12}.$$

Another notation for $|A|$ is obtained by replacing the brackets used for $A$ with vertical bars, as follows.

**Definition of the Determinant of a $2 \times 2$ Matrix $A$**

$$|A| = \begin{vmatrix} a_{11} & a_{12} \\ a_{21} & a_{22} \end{vmatrix} = a_{11}a_{22} - a_{21}a_{12}$$

**EXAMPLE 1** *Finding the determinant of a $2 \times 2$ matrix*

Find $|A|$ if $A = \begin{bmatrix} 2 & -1 \\ 4 & -3 \end{bmatrix}$.

**SOLUTION** By definition,

$$|A| = \begin{vmatrix} 2 & -1 \\ 4 & -3 \end{vmatrix} = (2)(-3) - (4)(-1) = -6 + 4 = -2.$$

For square matrices of order $n > 1$, it is convenient to introduce the following terminology.

**Definition of Minors and Cofactors**

Let $A = (a_{ij})$ be a square matrix of order $n > 1$.

(1) The **minor** $M_{ij}$ of the element $a_{ij}$ is the determinant of the matrix of order $n - 1$ obtained by deleting row $i$ and column $j$.

(2) The **cofactor** $A_{ij}$ of the element $a_{ij}$ is $A_{ij} = (-1)^{i+j}M_{ij}$.

To determine the minor of an element, we delete the row and column in which the element appears and then find the determinant of the resulting square matrix. This process is demonstrated in the following illustration, where deletions of rows and columns in a $3 \times 3$ matrix are indicated with horizontal and vertical line segments, respectively.

## ILLUSTRATION

### Minors

**Matrix**

$$\begin{bmatrix} a_{11} & a_{12} & a_{13} \\ a_{21} & a_{22} & a_{23} \\ a_{31} & a_{32} & a_{33} \end{bmatrix}$$

**Minor**

$$M_{11} = \begin{vmatrix} a_{22} & a_{23} \\ a_{32} & a_{33} \end{vmatrix} = a_{22}a_{33} - a_{32}a_{23}$$

$$\begin{bmatrix} a_{11} & a_{12} & a_{13} \\ a_{21} & a_{22} & a_{23} \\ a_{31} & a_{32} & a_{33} \end{bmatrix}$$

$$M_{12} = \begin{vmatrix} a_{21} & a_{23} \\ a_{31} & a_{33} \end{vmatrix} = a_{21}a_{33} - a_{31}a_{23}$$

$$\begin{bmatrix} a_{11} & a_{12} & a_{13} \\ a_{21} & a_{22} & a_{23} \\ a_{31} & a_{32} & a_{33} \end{bmatrix}$$

$$M_{13} = \begin{vmatrix} a_{21} & a_{22} \\ a_{31} & a_{32} \end{vmatrix} = a_{21}a_{32} - a_{31}a_{22}$$

$$\begin{bmatrix} a_{11} & a_{12} & a_{13} \\ a_{21} & a_{22} & a_{23} \\ a_{31} & a_{32} & a_{33} \end{bmatrix}$$

$$M_{23} = \begin{vmatrix} a_{11} & a_{12} \\ a_{31} & a_{32} \end{vmatrix} = a_{11}a_{32} - a_{31}a_{12}$$

For the matrix in the preceding illustration, there are five other minors— $M_{21}$, $M_{22}$, $M_{31}$, $M_{32}$, and $M_{33}$—that can be obtained in similar fashion.

To obtain the cofactor of $a_{ij}$ of a square matrix $A = (a_{ij})$, we find the minor and multiply it by 1 or $-1$, depending on whether the sum of $i$ and $j$ is even or odd, respectively. Another way to remember the sign $(-1)^{i+j}$ associated with the cofactor $A_{ij}$ is to consider the following checkerboard style of plus and minus signs:

$$\begin{bmatrix} + & - & + & - & \cdots \\ - & + & - & + & \cdots \\ + & - & + & - & \cdots \\ - & + & - & + & \cdots \\ \vdots & \vdots & \vdots & \vdots & \end{bmatrix}$$

**EXAMPLE 2**     *Finding minors and cofactors*

If $A = \begin{bmatrix} 1 & -3 & 3 \\ 4 & 2 & 0 \\ -2 & -7 & 5 \end{bmatrix}$, find $M_{11}$, $M_{21}$, $M_{22}$, $A_{11}$, $A_{21}$, and $A_{22}$.

*SOLUTION*   Deleting appropriate rows and columns of $A$, we obtain

$$M_{11} = \begin{vmatrix} 2 & 0 \\ -7 & 5 \end{vmatrix} = (2)(5) - (-7)(0) = 10$$

$$M_{21} = \begin{vmatrix} -3 & 3 \\ -7 & 5 \end{vmatrix} = (-3)(5) - (-7)(3) = 6$$

$$M_{22} = \begin{vmatrix} 1 & 3 \\ -2 & 5 \end{vmatrix} = (1)(5) - (-2)(3) = 11.$$

To obtain the cofactors, we prefix the corresponding minors with the proper signs. Thus, using the definition of cofactor, we have

$$A_{11} = (-1)^{1+1}M_{11} = (1)(10) = 10$$

$$A_{21} = (-1)^{2+1}M_{21} = (-1)(6) = -6$$

$$A_{22} = (-1)^{2+2}M_{22} = (1)(11) = 11.$$

We can also use the checkerboard style of plus and minus signs to determine the proper signs.

The determinant $|A|$ of a square matrix of order 3 is defined as follows.

**Definition of the
Determinant of a
3 × 3 Matrix A**

$$|A| = \begin{vmatrix} a_{11} & a_{12} & a_{13} \\ a_{21} & a_{22} & a_{23} \\ a_{31} & a_{32} & a_{33} \end{vmatrix} = a_{11}A_{11} + a_{12}A_{12} + a_{13}A_{13}$$

Since $A_{11} = (-1)^{1+1}M_{11} = M_{11}$, $A_{12} = (-1)^{1+2}M_{12} = -M_{12}$, and $A_{13} = (-1)^{1+3}M_{13} = M_{13}$, the preceding definition may also be written

$$|A| = a_{11}M_{11} - a_{12}M_{12} + a_{13}M_{13}.$$

If we express $M_{11}$, $M_{12}$, and $M_{13}$ using elements of $A$ and rearrange terms, we obtain the following formula for $|A|$:

$$|A| = a_{11}a_{22}a_{33} - a_{11}a_{23}a_{32} - a_{12}a_{21}a_{33} + a_{12}a_{23}a_{31} + a_{13}a_{21}a_{32} - a_{13}a_{22}a_{31}$$

The definition of $|A|$ for a square matrix $A$ of order 3 displays a pattern of multiplying each element in row 1 by its cofactor and then adding to find $|A|$. This process is referred to as *expanding $|A|$ by the first row*. By actually carrying out the computations, we can show that $|A|$ can be expanded in similar fashion by using any row or column. As an illustration, the expansion by the second column is

$$|A| = a_{12}A_{12} + a_{22}A_{22} + a_{32}A_{32}$$

$$= a_{12}\left( - \begin{vmatrix} a_{21} & a_{23} \\ a_{31} & a_{33} \end{vmatrix} \right) + a_{22}\left( + \begin{vmatrix} a_{11} & a_{13} \\ a_{31} & a_{33} \end{vmatrix} \right) + a_{32}\left( - \begin{vmatrix} a_{11} & a_{13} \\ a_{21} & a_{23} \end{vmatrix} \right).$$

Applying the definition to the determinants in parentheses, multiplying as indicated, and rearranging the terms in the sum, we could arrive at the formula for $|A|$ in terms of the elements of $A$. Similarly, the expansion by the third row is

$$|A| = a_{31}A_{31} + a_{32}A_{32} + a_{33}A_{33}.$$

Once again we can show that this result agrees with previous expansions.

**EXAMPLE 3    *Finding the determinant of a $3 \times 3$ matrix***

Find $|A|$ if $A = \begin{bmatrix} -1 & 3 & 1 \\ 2 & 5 & 0 \\ 3 & 1 & -2 \end{bmatrix}$.

**SOLUTION**    Since the second row contains a zero, we shall expand $|A|$ by that row, because then we need to evaluate only two cofactors. Thus,

$$|A| = (2)A_{21} + (5)A_{22} + (0)A_{23}.$$

Using the definition of cofactors, we have

$$A_{21} = (-1)^{2+1}M_{21} = -\begin{vmatrix} 3 & 1 \\ 1 & -2 \end{vmatrix} = -[(3)(-2) - (1)(1)] = 7$$

$$A_{22} = (-1)^{2+2}M_{22} = \begin{vmatrix} -1 & 1 \\ 3 & -2 \end{vmatrix} = [(-1)(-2) - (3)(1)] = -1.$$

Consequently,

$$|A| = (2)(7) + (5)(-1) + (0)A_{23} = 14 - 5 + 0 = 9.$$

The following definition of the determinant of a matrix of arbitrary order $n$ is patterned after that used for the determinant of a matrix of order 3.

**Definition of the Determinant of an $n \times n$ Matrix $A$**

The **determinant $|A|$ of a matrix $A$ of order $n$** is the cofactor expansion by the first row:

$$|A| = a_{11}A_{11} + a_{12}A_{12} + \cdots + a_{1n}A_{1n}$$

In terms of minors,

$$|A| = a_{11}M_{11} - a_{12}M_{12} + \cdots + a_{1n}(-1)^{1+n}M_{1n}.$$

The number $|A|$ may be found by using *any* row or column, as stated in the following theorem.

**Theorem on Expansion of Determinants**

> If $A$ is a square matrix of order $n > 1$, then the determinant $|A|$ may be found by multiplying the elements of any row (or column) by their respective cofactors and adding the resulting products.

The proof of this theorem may be found in texts on linear algebra. The theorem is useful if many zeros appear in a row or column, as illustrated in the following example.

**EXAMPLE 4    *Finding the determinant of a 4 × 4 matrix***

Find $|A|$ if $A = \begin{bmatrix} 1 & 0 & 2 & 5 \\ -2 & 1 & 5 & 0 \\ 0 & 0 & -3 & 0 \\ 0 & -1 & 0 & 3 \end{bmatrix}$.

**SOLUTION**    Note that all but one of the elements in the third row are zero. Hence, if we expand $|A|$ by the third row, there will be at most one nonzero term. Specifically,

$$|A| = (0)A_{31} + (0)A_{32} + (-3)A_{33} + (0)A_{34} = -3A_{33}$$

with

$$A_{33} = (-1)^{3+3} \begin{vmatrix} 1 & 0 & 5 \\ -2 & 1 & 0 \\ 0 & -1 & 3 \end{vmatrix}.$$

We expand $A_{33}$ by column 1:

$$A_{33} = (1) \begin{vmatrix} 1 & 0 \\ -1 & 3 \end{vmatrix} + (-2)\left(- \begin{vmatrix} 0 & 5 \\ -1 & 3 \end{vmatrix}\right) + (0) \begin{vmatrix} 0 & 5 \\ 1 & 0 \end{vmatrix}$$

$$= (1)(3) + (-2)(-5) + (0)(-5) = 3 + 10 + 0 = 13$$

Thus,

$$|A| = -3A_{33} = (-3)(13) = -39.$$

In general, if all but one element in some row (or column) of $A$ are zero and if the determinant $|A|$ is expanded by that row (or column), then all terms drop out except the product of that element with its cofactor. If *all* elements in a row (or column) are zero, we have the following.

**Theorem on a Row of Zeros**

> If every element of a row (or column) of a square matrix $A$ is zero, then $|A| = 0$.

**PROOF**    If every element in a row (or column) of a matrix $A$ is zero, the expansion by that row (or column) is zero.  ▪

## 8.9 EXERCISES

Exer. 1–4: Find all the minors and cofactors of the elements in the matrix.

1   $\begin{bmatrix} 7 & -1 \\ 5 & 0 \end{bmatrix}$
2   $\begin{bmatrix} -6 & 4 \\ 3 & 2 \end{bmatrix}$

3   $\begin{bmatrix} 2 & 4 & -1 \\ 0 & 3 & 2 \\ -5 & 7 & 0 \end{bmatrix}$
4   $\begin{bmatrix} 5 & -2 & 1 \\ 4 & 7 & 0 \\ -3 & 4 & -1 \end{bmatrix}$

Exer. 5–8: Find the determinant of the matrix in the given exercise.

5  Exercise 1
6  Exercise 2

7  Exercise 3
8  Exercise 4

Exer. 9–20: Find the determinant of the matrix.

9   $\begin{bmatrix} -5 & 4 \\ -3 & 2 \end{bmatrix}$
10   $\begin{bmatrix} 6 & 4 \\ -3 & 2 \end{bmatrix}$

11   $\begin{bmatrix} a & -a \\ b & -b \end{bmatrix}$
12   $\begin{bmatrix} c & d \\ -d & c \end{bmatrix}$

13   $\begin{bmatrix} 3 & 1 & -2 \\ 4 & 2 & 5 \\ -6 & 3 & -1 \end{bmatrix}$
14   $\begin{bmatrix} 2 & -5 & 1 \\ -3 & 1 & 6 \\ 4 & -2 & 3 \end{bmatrix}$

15   $\begin{bmatrix} -5 & 4 & 1 \\ 3 & -2 & 7 \\ 2 & 0 & 6 \end{bmatrix}$
16   $\begin{bmatrix} 2 & 7 & -3 \\ 1 & 0 & 4 \\ 4 & -1 & -2 \end{bmatrix}$

17   $\begin{bmatrix} 3 & -1 & 2 & 0 \\ 4 & 0 & -3 & 5 \\ 0 & 6 & 0 & 0 \\ 1 & 3 & -4 & 2 \end{bmatrix}$
18   $\begin{bmatrix} 2 & 5 & 1 & 0 \\ -4 & 0 & -3 & 0 \\ 3 & -2 & 1 & 6 \\ -1 & 4 & 2 & 0 \end{bmatrix}$

19   $\begin{bmatrix} 0 & b & 0 & 0 \\ 0 & 0 & c & 0 \\ a & 0 & 0 & 0 \\ 0 & 0 & 0 & d \end{bmatrix}$
20   $\begin{bmatrix} a & u & v & w \\ 0 & b & x & y \\ 0 & 0 & c & z \\ 0 & 0 & 0 & d \end{bmatrix}$

Exer. 21–28: Verify the identity by expanding each determinant.

21   $\begin{vmatrix} a & b \\ c & d \end{vmatrix} = - \begin{vmatrix} c & d \\ a & b \end{vmatrix}$
22   $\begin{vmatrix} a & b \\ c & d \end{vmatrix} = - \begin{vmatrix} b & a \\ d & c \end{vmatrix}$

23   $\begin{vmatrix} a & kb \\ c & kd \end{vmatrix} = k \begin{vmatrix} a & b \\ c & d \end{vmatrix}$
24   $\begin{vmatrix} a & b \\ kc & kd \end{vmatrix} = k \begin{vmatrix} a & b \\ c & d \end{vmatrix}$

25   $\begin{vmatrix} a & b \\ c & d \end{vmatrix} = \begin{vmatrix} a & b \\ ka+c & kb+d \end{vmatrix}$

26   $\begin{vmatrix} a & b \\ c & d \end{vmatrix} = \begin{vmatrix} a & ka+b \\ c & kc+d \end{vmatrix}$

27   $\begin{vmatrix} a & b \\ c & d \end{vmatrix} + \begin{vmatrix} a & e \\ c & f \end{vmatrix} = \begin{vmatrix} a & b+e \\ c & d+f \end{vmatrix}$

28   $\begin{vmatrix} a & b \\ c & d \end{vmatrix} + \begin{vmatrix} a & b \\ e & f \end{vmatrix} = \begin{vmatrix} a & b \\ c+e & d+f \end{vmatrix}$

29  Let $A = (a_{ij})$ be a square matrix of order $n$ such that $a_{ij} = 0$ if $i < j$. Show that $|A| = a_{11}a_{22} \cdots a_{nn}$.

30  If $A = (a_{ij})$ is any $2 \times 2$ matrix such that $|A| \neq 0$, show that $A$ has an inverse, and find a general formula for $A^{-1}$.

Exer. 31–34: Let $I = I_2$ be the identity matrix of order 2, and let $f(x) = |A - xI|$. Find (a) the polynomial $f(x)$ and (b) the zeros of $f(x)$. (In the study of matrices, $f(x)$ is the *characteristic polynomial of A*, and the zeros of $f(x)$ are the *characteristic values (eigenvalues) of A*.)

31  $A = \begin{bmatrix} 1 & 2 \\ 3 & 2 \end{bmatrix}$
32  $A = \begin{bmatrix} 3 & 1 \\ 2 & 2 \end{bmatrix}$

33  $A = \begin{bmatrix} -3 & -2 \\ 2 & 2 \end{bmatrix}$
34  $A = \begin{bmatrix} 2 & -4 \\ -3 & 5 \end{bmatrix}$

Exer. 35–38: Let $I = I_3$ and let $f(x) = |A - xI|$. Find (a) the polynomial $f(x)$ and (b) the zeros of $f(x)$.

35  $A = \begin{bmatrix} 1 & 0 & 0 \\ 1 & 0 & -2 \\ -1 & 1 & -3 \end{bmatrix}$
36  $A = \begin{bmatrix} 2 & 1 & 0 \\ -1 & 0 & 0 \\ 1 & 3 & 2 \end{bmatrix}$

37  $A = \begin{bmatrix} 0 & 2 & -2 \\ -1 & 3 & 1 \\ -3 & 3 & 1 \end{bmatrix}$
38  $A = \begin{bmatrix} 3 & 2 & 2 \\ 1 & 0 & 2 \\ -1 & -1 & 0 \end{bmatrix}$

**Exer. 39–42:** Express the determinant in the form $ai + bj + ck$ for real numbers $a$, $b$, and $c$.

39
$$\begin{vmatrix} i & j & k \\ 2 & -1 & 6 \\ -3 & 5 & 1 \end{vmatrix}$$

40
$$\begin{vmatrix} i & j & k \\ 1 & -2 & 3 \\ 2 & 1 & -4 \end{vmatrix}$$

41
$$\begin{vmatrix} i & j & k \\ 5 & -6 & -1 \\ 3 & 0 & 1 \end{vmatrix}$$

42
$$\begin{vmatrix} i & j & k \\ 4 & -6 & 2 \\ -2 & 3 & -1 \end{vmatrix}$$

**Exer. 43–44:** Let $I = I_3$ and $f(x) = |A - xI|$. (a) Find the polynomial $f(x)$. (b) Graph $f$, and estimate the characteristic values of $A$.

43 $A = \begin{bmatrix} 1 & 0 & 1 \\ 0 & 2 & 1 \\ 1 & 1 & -2 \end{bmatrix}$

44 $A = \begin{bmatrix} 3 & -1 & -1 \\ -1 & 1 & 0 \\ -1 & 0 & -2 \end{bmatrix}$

## 8.10 PROPERTIES OF DETERMINANTS

Evaluating a determinant by using the expansion theorem stated in Section 8.9 is inefficient for matrices of high order. For example, if a determinant of a matrix of order 10 is expanded by any row, a sum of 10 terms is obtained, and each term contains the determinant of a matrix of order 9, which is a cofactor of the original matrix. If any of the latter determinants is expanded by a row (or column), a sum of 9 terms is obtained, each containing the determinant of a matrix of order 8. Hence, at this stage there are 90 determinants of matrices of order 8 to evaluate. The process could be continued until only determinants of matrices of order 2 remain. You may verify that there are 1,814,400 such matrices of order 2! Unless many elements of the original matrix are zero, it is an enormous task to carry out all of the computations.

In this section we discuss rules that simplify the process of evaluating determinants. The main use for these rules is to introduce zeros into the determinant. They may also be used to change the determinant to **echelon form**—that is, to a form in which the elements below the main diagonal elements are all zero (see Section 8.3). The transformations on rows stated in the next theorem are the same as the elementary row transformations of a matrix introduced in Section 8.3. However, for determinants we may also employ similar transformations on columns.

**Theorem on Row and Column Transformations of a Determinant**

Let $A$ be a square matrix of order $n$.

(1) If a matrix $B$ is obtained from $A$ by interchanging two rows (or columns), then $|B| = -|A|$.

(2) If $B$ is obtained from $A$ by multiplying every element of one row (or column) of $A$ by a real number $k$, then $|B| = k|A|$.

(3) If $B$ is obtained from $A$ by adding $k$ times any row (or column) of $A$ to another row (or column) for a real number $k$, then $|B| = |A|$—that is, the determinants of $B$ and $A$ are equal.

When using the theorem, we refer to the rows (or columns) of the *determinant* in the obvious way. For example, property 3 may be phrased as follows: *Adding k times any row (or column) to another row (or column) of a determinant does not affect the value of the determinant.*

Row transformations of determinants will be specified by means of the symbols $R_i \leftrightarrow R_j$, $kR_i \rightarrow R_i$, and $kR_i + R_j \rightarrow R_j$, which were introduced in Section 8.3. Analogous symbols are used for column transformations. For example, $kC_i + C_j \rightarrow C_j$ means "add $k$ times the $i$th column to the $j$th column."

Property 2 of the theorem on row and column transformations is useful for finding factors of determinants. To illustrate, for a determinant of a matrix of order 3, we have the following:

$$\begin{vmatrix} a_{11} & a_{12} & a_{13} \\ ka_{21} & ka_{22} & ka_{23} \\ a_{31} & a_{32} & a_{33} \end{vmatrix} = k \begin{vmatrix} a_{11} & a_{12} & a_{13} \\ a_{21} & a_{22} & a_{23} \\ a_{31} & a_{32} & a_{33} \end{vmatrix}$$

Similar formulas hold if $k$ is a common factor of the elements of any other row or column. When referring to this manipulation, we often use the phrase *k is a common factor of the row (or column).*

The following are illustrations of the preceding theorem, with the reason for each equality stated at the right.

**ILLUSTRATION**

**Transformations of Determinants**

$$\begin{vmatrix} 2 & 0 & 1 \\ 6 & 4 & 3 \\ 0 & 3 & 5 \end{vmatrix} = - \begin{vmatrix} 6 & 4 & 3 \\ 2 & 0 & 1 \\ 0 & 3 & 5 \end{vmatrix} \qquad R_1 \leftrightarrow R_2$$

$$\begin{vmatrix} 2 & 0 & 1 \\ 6 & 4 & 3 \\ 0 & 3 & 5 \end{vmatrix} = 2 \begin{vmatrix} 1 & 0 & 1 \\ 3 & 4 & 3 \\ 0 & 3 & 5 \end{vmatrix} \qquad \text{2 is a common factor of column 1}$$

$$\begin{vmatrix} 1 & -3 & 4 \\ 2 & -1 & 0 \\ 3 & 1 & 6 \end{vmatrix} = \begin{vmatrix} 1 & -3 & 4 \\ 0 & 5 & -8 \\ 3 & 1 & 6 \end{vmatrix} \qquad -2R_1 + R_2 \rightarrow R_2$$

$$\begin{vmatrix} 1 & -3 & 4 \\ 2 & -1 & 0 \\ 3 & 1 & 6 \end{vmatrix} = \begin{vmatrix} -5 & -3 & 4 \\ 0 & -1 & 0 \\ 5 & 1 & 6 \end{vmatrix} \qquad 2C_2 + C_1 \rightarrow C_1$$

**Theorem on Identical Rows**

If two rows (or columns) of a square matrix $A$ are identical, then $|A| = 0$.

*PROOF*    If $B$ is the matrix obtained from $A$ by interchanging the two identical rows (or columns), then $B$ and $A$ are the same and, consequently, $|B| = |A|$. However, by property 1 of the theorem on row and column transformations of a determinant, $|B| = -|A|$, and hence $-|A| = |A|$. Thus, $2|A| = 0$, and therefore $|A| = 0$.   ■

**EXAMPLE I**    *Using row and column transformations*

Find $|A|$ if $A = \begin{bmatrix} 0 & 5 & -1 & 6 \\ 1 & 0 & -2 & 3 \\ -3 & 2 & 0 & -5 \end{bmatrix}.$

*SOLUTION*    We plan to use property 3 of the theorem on row and column transformations to introduce many zeros in some row or column. It is convenient to work with an element of the matrix that equals 1, since this enables us to avoid the use of fractions. If 1 is not an element of the original matrix, it is always possible to introduce the number 1 by using property 2 or 3 of the theorem. In this example, 1 appears in row 3, and we proceed as follows, with the reason for each equality stated at the right.

$$\begin{vmatrix} 2 & 3 & 0 & 4 \\ 0 & 5 & -1 & 6 \\ 1 & 0 & -2 & 3 \\ -3 & 2 & 0 & -5 \end{vmatrix} = \begin{vmatrix} 0 & 3 & 4 & -2 \\ 0 & 5 & -1 & 6 \\ 1 & 0 & -2 & 3 \\ 0 & 2 & -6 & 4 \end{vmatrix} \qquad \begin{array}{l} -2R_3 + R_1 \to R_1 \\ \\ \\ 3R_3 + R_4 \to R_4 \end{array}$$

$$= (1) \begin{vmatrix} 3 & 4 & -2 \\ 5 & -1 & 6 \\ 2 & -6 & 4 \end{vmatrix} \qquad \begin{array}{l} \text{expand by the} \\ \text{first column} \end{array}$$

$$= \begin{vmatrix} 23 & 4 & 22 \\ 0 & -1 & 0 \\ -28 & -6 & -32 \end{vmatrix} \qquad \begin{array}{l} 5C_2 + C_1 \to C_1 \\ \\ 6C_2 + C_3 \to C_3 \end{array}$$

$$= (-1) \begin{vmatrix} 23 & 22 \\ -28 & -32 \end{vmatrix} \qquad \begin{array}{l} \text{expand by the} \\ \text{second row} \end{array}$$

$$= (-1)[(23)(-32) - (-28)(22)] \qquad \begin{array}{l} \text{definition of} \\ \text{determinant} \end{array}$$

$$= 120$$

**EXAMPLE 2**    *Removing common factors*
*from rows of a determinant*

Find $|A|$ if $A = \begin{bmatrix} 14 & -6 & 4 \\ 4 & -5 & 12 \\ -21 & 9 & -6 \end{bmatrix}.$

*SOLUTION*

$$|A| = 2 \begin{vmatrix} 7 & -3 & 2 \\ 4 & -5 & 12 \\ -21 & 9 & -6 \end{vmatrix} \qquad \text{2 is a common factor of row 1}$$

$$= (2)(-3) \begin{vmatrix} 7 & -3 & 2 \\ 4 & -5 & 12 \\ 7 & -3 & 2 \end{vmatrix} \qquad -3 \text{ is a common factor of row 3}$$

$$= 0 \qquad\qquad\qquad \text{two rows are identical}$$

**EXAMPLE 3    *Removing a common factor from a column***

Without expanding, show that $a - b$ is a factor of $\begin{vmatrix} 1 & 1 & 1 \\ a & b & c \\ a^2 & b^2 & c^2 \end{vmatrix}$.

*SOLUTION*

$$\begin{vmatrix} 1 & 1 & 1 \\ a & b & c \\ a^2 & b^2 & c^2 \end{vmatrix} = \begin{vmatrix} 0 & 1 & 1 \\ a - b & b & c \\ a^2 - b^2 & b^2 & c^2 \end{vmatrix} \qquad -C_2 + C_1 \to C_1$$

$$= (a - b) \begin{vmatrix} 0 & 1 & 1 \\ 1 & b & c \\ a + b & b^2 & c^2 \end{vmatrix} \qquad \begin{array}{l} a - b \text{ is a common factor} \\ \text{of column 1} \end{array}$$

Determinants arise in the study of solutions of systems of linear equations. To illustrate, let us consider two linear equations in two variables $x$ and $y$:

$$\begin{cases} a_{11}x + a_{12}y = k_1 \\ a_{21}x + a_{22}y = k_2 \end{cases}$$

where at least one nonzero coefficient appears in each equation. We may assume that $a_{11} \neq 0$, for otherwise $a_{12} \neq 0$, and we could then regard $y$ as the first variable instead of $x$. We shall use elementary row transformations to obtain the matrix of an equivalent system with $a_{21} = 0$ as follows:

$$\begin{bmatrix} a_{11} & a_{12} & \bigg| & k_1 \\ a_{21} & a_{22} & \bigg| & k_2 \end{bmatrix} \xrightarrow{-\frac{a_{21}}{a_{11}} R_1 + R_2 \to R_2} \begin{bmatrix} a_{11} & a_{12} & \bigg| & k_1 \\ 0 & a_{22} - \left(\dfrac{a_{21}a_{12}}{a_{11}}\right) & \bigg| & k_2 - \left(\dfrac{a_{21}k_1}{a_{11}}\right) \end{bmatrix}$$

$$\xrightarrow{a_{11}R_2 \to R_2} \begin{bmatrix} a_{11} & a_{12} & \bigg| & k_1 \\ 0 & (a_{11}a_{22} - a_{21}a_{12}) & \bigg| & (a_{11}k_2 - a_{21}k_1) \end{bmatrix}$$

Thus, the given system is equivalent to

$$\begin{cases} a_{11}x + a_{12}y = k_1 \\ (a_{11}a_{22} - a_{21}a_{12})y = a_{11}k_2 - a_{21}k_1 \end{cases}$$

which may also be written

$$\begin{cases} a_{11}x + a_{12}y = k_1 \\ \begin{vmatrix} a_{11} & a_{12} \\ a_{21} & a_{22} \end{vmatrix} y = \begin{vmatrix} a_{11} & k_1 \\ a_{21} & k_2 \end{vmatrix} \end{cases}$$

If $\begin{vmatrix} a_{11} & a_{12} \\ a_{21} & a_{22} \end{vmatrix} \neq 0$, we can solve the second equation for $y$, obtaining

$$y = \frac{\begin{vmatrix} a_{11} & k_1 \\ a_{21} & k_2 \end{vmatrix}}{\begin{vmatrix} a_{11} & a_{12} \\ a_{21} & a_{22} \end{vmatrix}}.$$

The corresponding value for $x$ may be found by substituting for $y$ in the first equation, which leads to

$$x = \frac{\begin{vmatrix} k_1 & a_{12} \\ k_2 & a_{22} \end{vmatrix}}{\begin{vmatrix} a_{11} & a_{12} \\ a_{21} & a_{22} \end{vmatrix}}.$$

This proves that *if the determinant of the coefficient matrix of a system of two linear equations in two variables is not zero, then the system has a unique solution.* The last two formulas for $x$ and $y$ as quotients of determinants constitute **Cramer's rule** for two variables.

There is an easy way to remember Cramer's rule. Let

$$D = \begin{bmatrix} a_{11} & a_{12} \\ a_{21} & a_{22} \end{bmatrix}$$

be the coefficient matrix of the system, and let $D_x$ denote the matrix obtained from $D$ by replacing the coefficients $a_{11}$, $a_{21}$ of $x$ by the numbers $k_1$, $k_2$, respectively. Similarly, let $D_y$ denote the matrix obtained from $D$ by replacing the coefficients $a_{12}$, $a_{22}$ of $y$ by the numbers $k_1$, $k_2$, respectively. Thus,

$$D_x = \begin{bmatrix} k_1 & a_{12} \\ k_2 & a_{22} \end{bmatrix}, \quad D_y = \begin{bmatrix} a_{11} & k_1 \\ a_{21} & k_2 \end{bmatrix}.$$

If $|D| \neq 0$, the solution $(x, y)$ is given by the following formulas.

**Cramer's Rule for Two Variables**

$$x = \frac{|D_x|}{|D|}, \quad y = \frac{|D_y|}{|D|}$$

**EXAMPLE 4**     *Using Cramer's rule to solve a system of two linear equations*

Use Cramer's rule to solve the system

$$\begin{cases} 2x - 3y = -4 \\ 5x + 7y = \phantom{-}1 \end{cases}$$

*SOLUTION*     The determinant of the coefficient matrix is

$$|D| = \begin{vmatrix} 2 & -3 \\ 5 & 7 \end{vmatrix} = 29.$$

Using the notation introduced previously, we have

$$|D_x| = \begin{vmatrix} -4 & -3 \\ 1 & 7 \end{vmatrix} = -25, \quad |D_y| = \begin{vmatrix} 2 & -4 \\ 5 & 1 \end{vmatrix} = 22.$$

Hence,     $$x = \frac{|D_x|}{|D|} = \frac{-25}{29}, \quad y = \frac{|D_y|}{|D|} = \frac{22}{29}.$$

Thus, the system has the unique solution $\left(-\frac{25}{29}, \frac{22}{29}\right)$.

---

Cramer's rule can be extended to systems of $n$ linear equations in $n$ variables $x_1, x_2, \ldots, x_n$, where each equation has the form

$$a_1 x_1 + a_2 x_2 + \cdots + a_n x_n = k.$$

To solve such a system, let $D$ denote the coefficient matrix and let $D_{x_i}$ denote the matrix obtained by replacing the coefficients of $x_i$ in $D$ by the numbers $k_1, \ldots, k_n$ that appear in the column to the right of the equal signs in the system. If $|D| \neq 0$, then the system has the following unique solution.

**Cramer's Rule (General Form)**

$$x_1 = \frac{|D_{x_1}|}{|D|}, \quad x_2 = \frac{|D_{x_2}|}{|D|}, \quad \ldots, \quad x_n = \frac{|D_{x_n}|}{|D|}$$

**EXAMPLE 5**     *Using Cramer's rule to solve a system of three linear equations*

Use Cramer's rule to solve the system

$$\begin{cases} x \phantom{- y} - 2z = 3 \\ \phantom{x} - y + 3z = 1 \\ 2x \phantom{- y} + 5z = 0 \end{cases}$$

**SOLUTION** We shall merely list the various determinants. You should check the results.

$$|D| = \begin{vmatrix} 1 & 0 & -2 \\ 0 & -1 & 3 \\ 2 & 0 & 5 \end{vmatrix} = -9, \quad |D_x| = \begin{vmatrix} 3 & 0 & -2 \\ 1 & -1 & 3 \\ 0 & 0 & 5 \end{vmatrix} = -15$$

$$|D_y| = \begin{vmatrix} 1 & 3 & -2 \\ 0 & 1 & 3 \\ 2 & 0 & 5 \end{vmatrix} = 27, \quad |D_z| = \begin{vmatrix} 1 & 0 & 3 \\ 0 & -1 & 1 \\ 2 & 0 & 0 \end{vmatrix} = 6$$

By Cramer's rule, the solution is

$$x = \frac{|D_x|}{|D|} = \frac{-15}{-9} = \frac{5}{3}, \quad y = \frac{|D_y|}{|D|} = \frac{27}{-9} = -3, \quad z = \frac{|D_z|}{|D|} = \frac{6}{-9} = -\frac{2}{3}.$$

Cramer's rule is an inefficient method to apply if the system has a large number of equations, since many determinants of matrices of high order must be evaluated. Note also that Cramer's rule cannot be used directly if $|D| = 0$ or if the number of equations is not the same as the number of variables. For numerical calculations, the matrix method is superior to Cramer's rule; however, the Cramer's rule formulation is theoretically useful.

## 8.10 EXERCISES

**Exer. 1–14: Without expanding, explain why the statement is true.**

**1** $\begin{vmatrix} 1 & 0 & 1 \\ 0 & 1 & 1 \\ 1 & 1 & 0 \end{vmatrix} = -\begin{vmatrix} 1 & 0 & 1 \\ 1 & 1 & 0 \\ 0 & 1 & 1 \end{vmatrix}$

**2** $\begin{vmatrix} 1 & 0 & 1 \\ 0 & 1 & 1 \\ 1 & 1 & 0 \end{vmatrix} = -\begin{vmatrix} 1 & 1 & 0 \\ 0 & 1 & 1 \\ 1 & 0 & 1 \end{vmatrix}$

**3** $\begin{vmatrix} 1 & 0 & 1 \\ 2 & 1 & 0 \\ 1 & 1 & 2 \end{vmatrix} = \begin{vmatrix} 1 & 0 & 1 \\ 2 & 1 & 0 \\ 0 & 1 & 1 \end{vmatrix}$

**4** $\begin{vmatrix} 1 & 1 & 2 \\ 1 & 0 & 1 \\ 2 & 1 & 1 \end{vmatrix} = \begin{vmatrix} 0 & 1 & 1 \\ 1 & 0 & 1 \\ 2 & 1 & 1 \end{vmatrix}$

**5** $\begin{vmatrix} 2 & 4 & 2 \\ 1 & 2 & 4 \\ 2 & 6 & 4 \end{vmatrix} = 4\begin{vmatrix} 1 & 2 & 1 \\ 1 & 2 & 4 \\ 1 & 3 & 2 \end{vmatrix}$

**6** $\begin{vmatrix} 2 & 1 & 6 \\ 4 & 3 & 3 \\ 2 & 1 & 3 \end{vmatrix} = 6\begin{vmatrix} 1 & 1 & 2 \\ 2 & 3 & 1 \\ 1 & 1 & 1 \end{vmatrix}$

**7** $\begin{vmatrix} 1 & -1 & 2 \\ 1 & 2 & -1 \\ 1 & -1 & 2 \end{vmatrix} = 0$

**8** $\begin{vmatrix} 1 & -1 & 1 \\ 0 & 1 & 0 \\ -1 & 1 & -1 \end{vmatrix} = 0$

**9** $\begin{vmatrix} 1 & 5 \\ -3 & 2 \end{vmatrix} = -\begin{vmatrix} 1 & 5 \\ 3 & -2 \end{vmatrix}$

**10** $\begin{vmatrix} 2 & -2 \\ 1 & 1 \end{vmatrix} = -\begin{vmatrix} -2 & 2 \\ 1 & 1 \end{vmatrix}$

**11** $\begin{vmatrix} 0 & 0 & 1 \\ 1 & 0 & 0 \\ 0 & 0 & 2 \end{vmatrix} = 0$

**12** $\begin{vmatrix} 1 & 0 & 1 \\ 0 & 0 & 0 \\ 1 & 1 & 0 \end{vmatrix} = 0$

13 $\begin{vmatrix} 1 & -1 & -2 \\ -1 & 2 & 1 \\ 0 & 1 & 1 \end{vmatrix} = \begin{vmatrix} 1 & -1 & 0 \\ -1 & 2 & -1 \\ 0 & 1 & 1 \end{vmatrix}$

14 $\begin{vmatrix} a & 0 & 0 \\ 0 & b & 0 \\ 0 & 0 & c \end{vmatrix} = -\begin{vmatrix} 0 & 0 & a \\ 0 & b & 0 \\ c & 0 & 0 \end{vmatrix}$

**Exer. 15–24:** Find the determinant of the matrix after introducing zeros, as in Example 1.

15 $\begin{bmatrix} 3 & 1 & 0 \\ -2 & 0 & 1 \\ 1 & 3 & -1 \end{bmatrix}$

16 $\begin{bmatrix} -3 & 0 & 4 \\ 1 & 2 & 0 \\ 4 & 1 & -1 \end{bmatrix}$

17 $\begin{bmatrix} 5 & 4 & 3 \\ -3 & 2 & 1 \\ 0 & 7 & -2 \end{bmatrix}$

18 $\begin{bmatrix} 0 & 2 & -6 \\ 5 & 1 & -3 \\ 6 & -2 & 5 \end{bmatrix}$

19 $\begin{bmatrix} 2 & 2 & -3 \\ 3 & 6 & 9 \\ -2 & 5 & 4 \end{bmatrix}$

20 $\begin{bmatrix} 3 & 8 & 5 \\ 5 & 3 & -6 \\ 2 & 4 & -2 \end{bmatrix}$

21 $\begin{bmatrix} 3 & 1 & -2 & 2 \\ 2 & 0 & 1 & 4 \\ 0 & 1 & 3 & 5 \\ -1 & 2 & 0 & -3 \end{bmatrix}$

22 $\begin{bmatrix} 3 & 2 & 0 & 4 \\ -2 & 0 & 5 & 0 \\ 4 & -3 & 1 & 6 \\ 2 & -1 & 2 & 0 \end{bmatrix}$

23 $\begin{bmatrix} 2 & -2 & 0 & 0 & -3 \\ 3 & 0 & 3 & 2 & -1 \\ 0 & 1 & -2 & 0 & 2 \\ -1 & 2 & 0 & 3 & 0 \\ 0 & 4 & 1 & 0 & 0 \end{bmatrix}$

24 $\begin{bmatrix} 2 & 0 & -1 & 0 & 2 \\ 1 & 3 & 0 & 0 & 1 \\ 0 & 4 & 3 & 0 & -1 \\ -1 & 2 & 0 & -2 & 0 \\ 0 & 1 & 5 & 0 & -4 \end{bmatrix}$

25 Show that
$$\begin{vmatrix} 1 & 1 & 1 \\ a & b & c \\ a^2 & b^2 & c^2 \end{vmatrix} = (a - b)(b - c)(c - a).$$
(*Hint:* See Example 3.)

26 Show that
$$\begin{vmatrix} 1 & 1 & 1 \\ a & b & c \\ a^3 & b^3 & c^3 \end{vmatrix} = (a - b)(b - c)(c - a)(a + b + c).$$

27 If
$$A = \begin{bmatrix} a_{11} & a_{12} & a_{13} & a_{14} \\ 0 & a_{22} & a_{23} & a_{24} \\ 0 & 0 & a_{33} & a_{34} \\ 0 & 0 & 0 & a_{44} \end{bmatrix},$$
show that $|A| = a_{11}a_{22}a_{33}a_{44}$.

28 If
$$A = \begin{bmatrix} a & b & 0 & 0 \\ c & d & 0 & 0 \\ 0 & 0 & e & f \\ 0 & 0 & g & h \end{bmatrix},$$
show that
$$|A| = \begin{vmatrix} a & b \\ c & d \end{vmatrix} \begin{vmatrix} e & f \\ g & h \end{vmatrix}.$$

29 If $A = (a_{ij})$ and $B = (b_{ij})$ are arbitrary square matrices of order 2, show that $|AB| = |A||B|$.

30 If $A = (a_{ij})$ is a square matrix of order $n$ and $k$ is any real number, show that $|kA| = k^n|A|$. (*Hint:* Use property 2 of the theorem on row and column transformations of a determinant.)

31 Use properties of determinants to show that the following is an equation of a line through the points $(x_1, y_1)$ and $(x_2, y_2)$:
$$\begin{vmatrix} x & y & 1 \\ x_1 & y_1 & 1 \\ x_2 & y_2 & 1 \end{vmatrix} = 0$$

32 Use properties of determinants to show that the following is an equation of a circle through three noncollinear points $(x_1, y_1)$, $(x_2, y_2)$, and $(x_3, y_3)$:
$$\begin{vmatrix} x^2 + y^2 & x & y & 1 \\ x_1^2 + y_1^2 & x_1 & y_1 & 1 \\ x_2^2 + y_2^2 & x_2 & y_2 & 1 \\ x_3^2 + y_3^2 & x_3 & y_3 & 1 \end{vmatrix} = 0$$

**Exer. 33–42:** Use Cramer's rule, whenever applicable, to solve the system.

33 $\begin{cases} 2x + 3y = 2 \\ x - 2y = 8 \end{cases}$

34 $\begin{cases} 4x + 5y = 13 \\ 3x + y = -4 \end{cases}$

35 $\begin{cases} 2x + 5y = 16 \\ 3x - 7y = 24 \end{cases}$

36 $\begin{cases} 7x - 8y = 9 \\ 4x + 3y = -10 \end{cases}$

37 $\begin{cases} 2x - 3y = 5 \\ -6x + 9y = 12 \end{cases}$

38 $\begin{cases} 3p - q = 7 \\ -12p + 4q = 3 \end{cases}$

$$39 \quad \begin{cases} x - 2y - 3z = -1 \\ 2x + y + z = 6 \\ x + 3y - 2z = 13 \end{cases}$$

$$40 \quad \begin{cases} x + 3y - z = -3 \\ 3x - y + 2z = 1 \\ 2x - y + z = -1 \end{cases}$$

$$41 \quad \begin{cases} 5x + 2y - z = -7 \\ x - 2y + 2z = 0 \\ 3y + z = 17 \end{cases}$$

$$42 \quad \begin{cases} 4x - y + 3z = 6 \\ -8x + 3y - 5z = -6 \\ 5x - 4y = -9 \end{cases}$$

## CHAPTER 8 REVIEW EXERCISES

**Exer. 1–16: Solve the system.**

$$1 \quad \begin{cases} 2x - 3y = 4 \\ 5x + 4y = 1 \end{cases}$$

$$2 \quad \begin{cases} x - 3y = 4 \\ -2x + 6y = 2 \end{cases}$$

$$3 \quad \begin{cases} y + 4 = x^2 \\ 2x + y = -1 \end{cases}$$

$$4 \quad \begin{cases} x^2 + y^2 = 25 \\ x - y = 7 \end{cases}$$

$$5 \quad \begin{cases} 9x^2 + 16y^2 = 140 \\ x^2 - 4y^2 = 4 \end{cases}$$

$$6 \quad \begin{cases} 2x = y^2 + 3z \\ x = y^2 + z - 1 \\ x^2 = xz \end{cases}$$

$$7 \quad \begin{cases} \dfrac{1}{x} + \dfrac{3}{y} = 7 \\ \dfrac{4}{x} - \dfrac{2}{y} = 1 \end{cases}$$

$$8 \quad \begin{cases} 2^x + 3^{y+1} = 10 \\ 2^{x+1} - 3^y = 5 \end{cases}$$

$$9 \quad \begin{cases} 3x + y - 2z = -1 \\ 2x - 3y + z = 4 \\ 4x + 5y - z = -2 \end{cases}$$

$$10 \quad \begin{cases} x + 3y = 0 \\ y - 5z = 3 \\ 2x + z = -1 \end{cases}$$

$$11 \quad \begin{cases} 4x - 3y - z = 0 \\ x - y - z = 0 \\ 3x - y + 3z = 0 \end{cases}$$

$$12 \quad \begin{cases} 2x + y - z = 0 \\ x - 2y + z = 0 \\ 3x + 3y + 2z = 0 \end{cases}$$

$$13 \quad \begin{cases} 4x + 2y - z = 1 \\ 3x + 2y + 4z = 2 \end{cases}$$

$$14 \quad \begin{cases} 2x + y = 6 \\ x - 3y = 17 \\ 3x + 2y = 7 \end{cases}$$

$$15 \quad \begin{cases} \dfrac{4}{x} + \dfrac{1}{y} + \dfrac{2}{z} = 4 \\ \dfrac{2}{x} + \dfrac{3}{y} - \dfrac{1}{z} = 1 \\ \dfrac{1}{x} + \dfrac{1}{y} + \dfrac{1}{z} = 4 \end{cases}$$

$$16 \quad \begin{cases} 2x - y + 3z - w = -3 \\ 3x + 2y - z + w = 13 \\ x - 3y + z - 2w = -4 \\ -x + y + 4z + 3w = 0 \end{cases}$$

**Exer. 17–20: Find the partial fraction decomposition.**

$$17 \quad \frac{4x^2 + 54x + 134}{(x + 3)(x^2 + 4x - 5)}$$

$$18 \quad \frac{2x^2 + 7x + 9}{x^2 + 2x + 1}$$

$$19 \quad \frac{x^2 + 14x - 13}{x^3 + 5x^2 + 4x + 20}$$

$$20 \quad \frac{x^3 + 2x^2 + 2x + 16}{x^4 + 7x^2 + 10}$$

**Exer. 21–24: Sketch the graph of the system.**

$$21 \quad \begin{cases} x^2 + y^2 < 16 \\ y - x^2 > 0 \end{cases}$$

$$22 \quad \begin{cases} y - x \le 0 \\ y + x \ge 2 \\ x \le 5 \end{cases}$$

$$23 \quad \begin{cases} x - 2y \le 2 \\ y - 3x \le 4 \\ 2x + y \le 4 \end{cases}$$

$$24 \quad \begin{cases} x^2 - y < 0 \\ y - 2x < 5 \\ xy < 0 \end{cases}$$

**Exer. 25–34: Express as a single matrix.**

$$25 \quad \begin{bmatrix} 2 & -1 & 0 \\ 3 & 0 & -2 \end{bmatrix} \begin{bmatrix} 2 & -1 & 3 \\ 0 & 3 & 0 \\ 1 & 4 & 2 \end{bmatrix}$$

$$26 \quad \begin{bmatrix} 4 & 2 \\ 5 & -3 \end{bmatrix} \begin{bmatrix} 3 \\ 7 \end{bmatrix}$$

$$27 \quad \begin{bmatrix} 2 & 0 \\ 1 & 4 \\ -2 & 3 \end{bmatrix} \begin{bmatrix} 0 & 2 & -3 \\ 4 & 5 & 1 \end{bmatrix}$$

$$28 \quad \begin{bmatrix} 0 & -2 & 3 \\ 4 & 1 & 2 \end{bmatrix} \begin{bmatrix} 2 & 0 \\ 3 & 8 \\ 2 & -7 \end{bmatrix}$$

$$29 \quad 2\begin{bmatrix} 0 & -1 & -4 \\ 3 & 2 & 1 \end{bmatrix} - 3\begin{bmatrix} 4 & -2 & 1 \\ 0 & 5 & -1 \end{bmatrix}$$

$$30 \quad \begin{bmatrix} 1 & 3 \\ 2 & 4 \end{bmatrix} \begin{bmatrix} a & 0 \\ 0 & a \end{bmatrix}$$

$$31 \quad \begin{bmatrix} a & 0 \\ 0 & b \end{bmatrix} \begin{bmatrix} 1 & 3 \\ 2 & 4 \end{bmatrix}$$

$$32 \quad \begin{bmatrix} 3 & 2 \\ 0 & 0 \end{bmatrix} \begin{bmatrix} -2 & 0 \\ 3 & 0 \end{bmatrix}$$

**33** $\begin{bmatrix} 1 & 2 \\ 3 & 4 \end{bmatrix} \left\{ \begin{bmatrix} 2 & -4 \\ 3 & 7 \end{bmatrix} + \begin{bmatrix} 1 & 5 \\ -2 & -3 \end{bmatrix} \right\}$

**34** $\begin{bmatrix} 3 & 2 & 5 \\ -3 & 4 & 7 \\ 6 & 5 & 1 \end{bmatrix} \begin{bmatrix} 3 & 2 & 5 \\ -3 & 4 & 7 \\ 6 & 5 & 1 \end{bmatrix}^{-1}$

**Exer. 35–38: Find the inverse of the matrix.**

**35** $\begin{bmatrix} 5 & -4 \\ -3 & 2 \end{bmatrix}$

**36** $\begin{bmatrix} 2 & -1 & 0 \\ 1 & 4 & 2 \\ 3 & -2 & 1 \end{bmatrix}$

**37** $\begin{bmatrix} 1 & 0 & 0 \\ 0 & 4 & 7 \\ 0 & 1 & 2 \end{bmatrix}$

**38** $\begin{bmatrix} 2 & 0 & 5 \\ 0 & 3 & -1 \\ 3 & 4 & 0 \end{bmatrix}$

**39** Use the result of Exercise 35 to solve the system
$$\begin{cases} 5x - 4y = 30 \\ -3x + 2y = -16 \end{cases}$$

**40** Use the result of Exercise 36 to solve the system
$$\begin{cases} 2x - y = -5 \\ x + 4y + 2z = 15 \\ 3x - 2y + z = -7 \end{cases}$$

**Exer. 41–50: Find the determinant of the matrix.**

**41** $[-6]$

**42** $\begin{bmatrix} 3 & 4 \\ -6 & -5 \end{bmatrix}$

**43** $\begin{bmatrix} 3 & -4 \\ 6 & 8 \end{bmatrix}$

**44** $\begin{bmatrix} 0 & 4 & -3 \\ 2 & 0 & 4 \\ -5 & 1 & 0 \end{bmatrix}$

**45** $\begin{bmatrix} 2 & -3 & 5 \\ -4 & 1 & 3 \\ 3 & 2 & -1 \end{bmatrix}$

**46** $\begin{bmatrix} 3 & 1 & -2 \\ -5 & 2 & -4 \\ 7 & 3 & -6 \end{bmatrix}$

**47** $\begin{bmatrix} 5 & 0 & 0 & 0 \\ 6 & -3 & 0 & 0 \\ 1 & 4 & -4 & 0 \\ 7 & 2 & 3 & 2 \end{bmatrix}$

**48** $\begin{bmatrix} 1 & 2 & 0 & 3 & 1 \\ -2 & -1 & 4 & 1 & 2 \\ 3 & 0 & -1 & 0 & -1 \\ 2 & -3 & 2 & -4 & 2 \\ -1 & 1 & 0 & 1 & 3 \end{bmatrix}$

**49** $\begin{bmatrix} 2 & 0 & 1 & 0 & -1 \\ 0 & 1 & 0 & 1 & 2 \\ 2 & -2 & 1 & -2 & 0 \\ 0 & 0 & -2 & 0 & 1 \\ 1 & -1 & 0 & -1 & 0 \end{bmatrix}$

**50** $\begin{bmatrix} 1 & 2 & 0 & 0 & 0 \\ 3 & 4 & 0 & 0 & 0 \\ 0 & 0 & 1 & 2 & 3 \\ 0 & 0 & 2 & -1 & 1 \\ 0 & 0 & 1 & 3 & -1 \end{bmatrix}$

**Exer. 51–52: Solve the equation $|A - xI| = 0$.**

**51** $A = \begin{bmatrix} 2 & 3 \\ 1 & -4 \end{bmatrix}$, $\quad I = I_2$

**52** $A = \begin{bmatrix} 2 & -1 & 3 \\ 0 & 4 & 0 \\ 1 & 0 & -2 \end{bmatrix}$, $\quad I = I_3$

**Exer. 53–54: Without expanding, explain why the statement is true.**

**53** $\begin{vmatrix} 2 & 4 & -6 \\ 1 & 4 & 3 \\ 2 & 2 & 0 \end{vmatrix} = 12 \begin{vmatrix} 1 & 1 & -1 \\ 1 & 2 & 1 \\ 2 & 1 & 0 \end{vmatrix}$

**54** $\begin{vmatrix} a & b & c \\ d & e & f \\ g & h & k \end{vmatrix} = \begin{vmatrix} d & e & f \\ g & h & k \\ a & b & c \end{vmatrix}$

**55** Find the determinant of the $n \times n$ matrix $(a_{ij})$ in which $a_{ij} = 0$ for $i \neq j$.

**56** Without expanding, show that
$$\begin{vmatrix} 1 & a & b+c \\ 1 & b & a+c \\ 1 & c & a+b \end{vmatrix} = 0.$$

**Exer. 57–58: Use Cramer's rule to solve the system.**

**57** $\begin{cases} 5x - 6y = 4 \\ 3x + 7y = 8 \end{cases}$

**58** $\begin{cases} 2x - 3y + 2z = -3 \\ -3x + 2y + z = 1 \\ 4x + y - 3z = 4 \end{cases}$

**59** *Watering a field* A rotating sprinkler head with a range of 50 feet is to be placed in the center of a rectangular field (see the figure). If the area of the field is 4000 ft$^2$ and the water is to just reach the corners, find the dimensions of the field.

**EXERCISE 59**

**60** Find equations of the two lines that are tangent to the circle $x^2 + y^2 = 1$ and pass through the point $(0, 3)$. (*Hint:* Let $y = mx + 3$ and determine conditions on $m$ that will ensure that the system has only one solution.)

**61** *Payroll accounting*    An accountant must pay taxes and payroll bonuses to employees from the company's profits of $50,000. The total tax is 40% of the amount left after bonuses are paid, and the total paid in bonuses is 10% of the amount left after taxes. Find the total tax and the total bonus amount.

**62** *Track dimensions*    A circular track is to have a 10-foot-wide running lane around the outside (see the figure). The inside distance around the track is to be 90% of the outside distance. Find the dimensions of the track.

**EXERCISE 62**

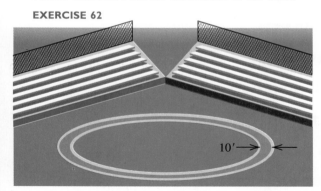

**63** *Flow rates*    Three inlet pipes A, B, and C can be used to fill a 1000-ft$^3$ water storage tank. When all three pipes are in operation, the tank can be filled in 10 hours. When only A and B are used, the time increases to 20 hours. With pipes A and C, the tank can be filled in 12.5 hours. Find the individual flow rates (in ft$^3$/hr) for each of the three pipes.

**64** *Warehouse shipping charges*    To fill an order for 150 office desks, a furniture distributor must ship the desks from two warehouses. The shipping cost per desk is $24 from the western warehouse and $35 from the eastern warehouse. If the total shipping charge is $4205, how many desks are shipped from each location?

**65** *Express mail rates*    An express-mail company charges $15 for overnight delivery of a letter, provided the dimensions of the standard envelope satisfy the following three conditions: (a) the length, the larger of the two dimensions, must be at most 12 inches; (b) the width must be at most 8 inches; (c) the width must be at least one-half the length. Find and graph a system of inequalities that describes all the possibilities for dimensions of a standard envelope.

**66** *Activities of a deer*    A deer spends the day in three basic activities: rest, searching for food, and grazing. At least 6 hours each day must be spent resting, and the number of hours spent searching for food will be at least two times the number of hours spent grazing. Using $x$ as the number of hours spent searching for food and $y$ as the number of hours spent grazing, find and graph the system of inequalities that describes the possible divisions of the day.

**67** *Production scheduling*    A company manufactures a power lawn mower and a power edger. These two products are of such high quality that the company can sell all the products it makes, but production capability is limited in the areas of machining, welding, and assembly. Each week, the company has 600 hours available for machining, 300 hours for welding, and 550 hours for assembly. The number of hours required for the production of a single item is shown in the following table. The profits from the sale of a mower and an edger are $100 and $80, respectively. How many mowers and edgers should be made each week to maximize the profit?

| Product | Machining | Welding | Assembly |
| --- | --- | --- | --- |
| Mower | 6 | 2 | 5 |
| Edger | 4 | 3 | 5 |

**68** *Maximizing investment income*    A retired couple wishes to invest $150,000, diversifying the investment in three areas: a high-risk stock that has an expected annual rate of return (or interest) of 15%, a low-risk stock that has an expected annual return of 10%, and government-issued bonds that pay annual interest of 8% and involve no risk. To protect the value of the investment, the couple wishes to place at least twice as much in the low-risk stock as in the high-risk stock and use the remainder to buy bonds. How should the money be invested to maximize the expected annual return?

*Sequences occur in high-density computer chips and in many other aspects of science and industry.*

■ *Sequences and summation notation, discussed in the first section, are very important in advanced mathematics and applications. Of special interest are arithmetic and geometric sequences, considered in Sections 9.2 and 9.3. We then discuss the method of mathematical induction, a process that is often used to prove that each statement in an infinite sequence of statements is true. As an application, we use it to prove the binomial theorem in Section 9.5. The last part of the chapter deals with counting processes that occur frequently in mathematics and everyday life. These include the concepts of permutations, combinations, and probability.*

■

# CHAPTER 9

# SEQUENCES, SERIES, AND PROBABILITY

## 9.1   INFINITE SEQUENCES AND SUMMATION NOTATION

An arbitrary *infinite sequence* may be denoted as follows.

**Infinite Sequence Notation**

$$a_1, a_2, a_3, \ldots, a_n, \ldots$$

For convenience, we often refer to infinite sequences as *sequences*. We may regard an infinite sequence as a collection of real numbers that is in one-to-one correspondence with the positive integers. Each number $a_k$ is a **term** of the sequence. The sequence is *ordered* in the sense that there is a **first term** $a_1$, a *second term* $a_2$, a *forty-fifth term* $a_{45}$, and, if $n$ denotes an arbitrary positive integer, an **$n$th term** $a_n$. Infinite sequences are often defined by stating a formula for the $n$th term.

Infinite sequences occur frequently in mathematics. For example, the sequence

$$0.6, 0.66, 0.666, 0.6666, 0.66666, \ldots$$

may be used to represent the rational number $\frac{2}{3}$. In this case the $n$th term gets closer and closer to $\frac{2}{3}$ as $n$ increases.

We may regard an infinite sequence as a function. Recall from Section 2.1 that a function $f$ is a correspondence that assigns to each number $x$ in the domain $D$ exactly one number $f(x)$ in the range $R$. If we restrict the domain to the positive integers 1, 2, 3, . . . , we obtain an infinite sequence, as in the following definition.

**Definition of Infinite Sequence**

An **infinite sequence** is a function whose domain is the set of positive integers.

In our work, the range of an infinite sequence will be a set of real numbers.

If a function $f$ is an infinite sequence, then to each positive integer $n$ there corresponds a real number $f(n)$. These numbers in the range of $f$ may be represented by writing

$$f(1), f(2), f(3), \ldots, f(n), \ldots.$$

To obtain the subscript form of a sequence, as shown at the beginning of this section, we let $a_n = f(n)$ for every positive integer $n$.

From the definition of equality of functions we see that a sequence

$$a_1, a_2, a_3, \ldots, a_n, \ldots$$

is **equal** to a sequence

$$b_1, b_2, b_3, \ldots, b_n, \ldots$$

if and only if $a_k = b_k$ for every positive integer $k$.

Another notation for a sequence with $n$th term $a_n$ is $\{a_n\}$. For example, the sequence $\{2^n\}$ has $n$th term $a_n = 2^n$. Using sequence notation, we write this sequence as follows:

$$2^1, 2^2, 2^3, \ldots, 2^n, \ldots$$

By definition, the sequence $\{2^n\}$ is the function $f$ with $f(n) = 2^n$ for every positive integer $n$.

### EXAMPLE I  *Finding terms of a sequence*

List the first four terms and the tenth term of each sequence:

**(a)** $\left\{\dfrac{n}{n+1}\right\}$  **(b)** $\{2 + (0.1)^n\}$  **(c)** $\left\{(-1)^{n+1}\dfrac{n^2}{3n-1}\right\}$  **(d)** $\{4\}$

**SOLUTION**  To find the first four terms, we substitute, successively, $n = 1, 2, 3$, and $4$ in the formula for $a_n$. The tenth term is found by substituting 10 for $n$. Doing this and simplifying gives us the following:

| Sequence | $n$th term $a_n$ | First four terms | Tenth term |
|---|---|---|---|
| **(a)** $\left\{\dfrac{n}{n+1}\right\}$ | $\dfrac{n}{n+1}$ | $\dfrac{1}{2}, \dfrac{2}{3}, \dfrac{3}{4}, \dfrac{4}{5}$ | $\dfrac{10}{11}$ |
| **(b)** $\{2 + (0.1)^n\}$ | $2 + (0.1)^n$ | $2.1, 2.01, 2.001, 2.0001$ | $2.0000000001$ |
| **(c)** $\left\{(-1)^{n+1}\dfrac{n^2}{3n-1}\right\}$ | $(-1)^{n+1}\dfrac{n^2}{3n-1}$ | $\dfrac{1}{2}, -\dfrac{4}{5}, \dfrac{9}{8}, -\dfrac{16}{11}$ | $-\dfrac{100}{29}$ |
| **(d)** $\{4\}$ | $4$ | $4, 4, 4, 4$ | $4$ |

For some sequences we state the first term $a_1$, together with a rule for obtaining any term $a_{k+1}$ from the preceding term $a_k$ whenever $k \geq 1$. We call such a statement a **recursive definition,** and the sequence is said to be defined **recursively.**

### EXAMPLE 2  *Finding terms of a recursively defined sequence*

Find the first four terms and the $n$th term of the infinite sequence defined recursively as follows:

$$a_1 = 3, \quad a_{k+1} = 2a_k \quad \text{for } k \geq 1$$

*SOLUTION*    The first four terms are

$$a_1 = 3$$
$$a_2 = 2a_1 = 2 \cdot 3 = 6$$
$$a_3 = 2a_2 = 2 \cdot 2 \cdot 3 = 2^2 \cdot 3 = 12$$
$$a_4 = 2a_3 = 2 \cdot 2 \cdot 2 \cdot 3 = 2^3 \cdot 3 = 24.$$

We have written the terms as products to gain some insight into the nature of the *n*th term. Continuing, we obtain $a_5 = 2^4 \cdot 3$, $a_6 = 2^5 \cdot 3$, and, in general,

$$a_n = 2^{n-1} \cdot 3$$

for every positive integer *n*.

---

If only the first few terms of an infinite sequence are known, then it is impossible to predict additional terms. For example, if we were given 3, 6, 9, . . . and asked to find the fourth term, we could not proceed without further information. The infinite sequence with *n*th term

$$a_n = 3n + (1 - n)^3(2 - n)^2(3 - n)$$

has for its first four terms 3, 6, 9, and 120. It is possible to describe sequences in which the first three terms are 3, 6, and 9 and the fourth term is *any* given number. This shows that when we work with an infinite sequence it is essential to have either specific information about the *n*th term or a general scheme for obtaining each term from the preceding one.

We sometimes need to find the sum of many terms of an infinite sequence. To express such sums easily, we use **summation notation.** Given an infinite sequence

$$a_1, a_2, a_3, \ldots, a_n, \ldots,$$

the symbol $\sum_{k=1}^{m} a_k$ represents the sum of the first *m* terms, as follows.

**Summation Notation**

$$\sum_{k=1}^{m} a_k = a_1 + a_2 + a_3 + \cdots + a_m$$

The Greek capital letter sigma, $\Sigma$, indicates a sum, and the symbol $a_k$ represents the *k*th term. The letter *k* is the **index of summation,** or the **summation variable,** and the numbers 1 and *m* indicate the smallest and largest values of the summation variable.

**EXAMPLE 3**    *Evaluating a sum*

Find the sum $\sum_{k=1}^{4} k^2(k - 3)$.

**SOLUTION**    In this case, $a_k = k^2(k-3)$. To find the sum, we merely substitute, in succession, the integers 1, 2, 3, and 4 for $k$ and add the resulting terms:

$$\sum_{k=1}^{4} k^2(k-3) = 1^2(1-3) + 2^2(2-3) + 3^2(3-3) + 4^2(4-3)$$

$$= (-2) + (-4) + 0 + 16 = 10$$

The letter we use for the summation variable is immaterial. To illustrate, if $j$ is the summation variable, then

$$\sum_{j=1}^{m} a_j = a_1 + a_2 + a_3 + \cdots + a_m,$$

which is the same sum as $\sum_{k=1}^{m} a_k$. As a specific example, the sum in Example 3 can be written

$$\sum_{j=1}^{4} j^2(j-3).$$

If $n$ is a positive integer, then the sum of the first $n$ terms of an infinite sequence will be denoted by $S_n$. For example, given the infinite sequence $a_1, a_2, a_3, \ldots, a_n, \ldots$,

$$S_1 = a_1$$
$$S_2 = a_1 + a_2$$
$$S_3 = a_1 + a_2 + a_3$$
$$S_4 = a_1 + a_2 + a_3 + a_4$$

and, in general,

$$S_n = \sum_{k=1}^{n} a_k = a_1 + a_2 + \cdots + a_n.$$

Note that we can also write

$$S_1 = a_1$$
$$S_2 = S_1 + a_2$$
$$S_3 = S_2 + a_3$$
$$S_4 = S_3 + a_4$$

and, for every $n > 1$,

$$S_n = S_{n-1} + a_n.$$

The real number $S_n$ is called the **$n$th partial sum** of the infinite sequence $a_1, a_2, a_3, \ldots, a_n, \ldots$, and the sequence

$$S_1, S_2, S_3, \ldots, S_n, \ldots$$

is a **sequence of partial sums.** Sequences of partial sums are important in the study of *infinite series,* a topic in calculus. We shall discuss some special types of infinite series in Section 9.3.

**EXAMPLE 4**    *Finding the terms of a sequence of partial sums*

Find the first four terms and the $n$th term of the sequence of partial sums associated with the sequence $1, 2, 3, \ldots, n, \ldots$ of positive integers.

**SOLUTION**    If we let $a_n = n$, then the first four terms of the sequence of partial sums are

$$S_1 = a_1 = 1$$
$$S_2 = S_1 + a_2 = 1 + 2 = 3$$
$$S_3 = S_2 + a_3 = 3 + 3 = 6$$
$$S_4 = S_3 + a_4 = 6 + 4 = 10.$$

The $n$th partial sum $S_n$ (that is, the sum of $1, 2, 3, \ldots, n$) can be written in either of the following forms:

$$S_n = 1 + 2 \quad\quad + 3 \quad\quad + \cdots + (n - 2) + (n - 1) + n$$
$$S_n = n + (n - 1) + (n - 2) + \cdots + 3 \quad\quad + 2 \quad\quad + 1$$

Adding corresponding terms on each side of these equations gives us

$$2S_n = \underbrace{(n + 1) + (n + 1) + (n + 1) + \cdots + (n + 1) + (n + 1) + (n + 1)}_{n \text{ times}}.$$

Since the expression $(n + 1)$ appears $n$ times on the right side of the last equation, we see that

$$2S_n = n(n + 1) \quad \text{or, equivalently,} \quad S_n = \frac{n(n + 1)}{2}.$$

If $a_k$ is the same for every positive integer $k$—say $a_k = c$ for a real number $c$—then

$$\sum_{k=1}^{n} a_k = a_1 + a_2 + a_3 + \cdots + a_n$$
$$= c + c + c + \cdots + c = nc.$$

We have proved property 1 of the following theorem.

**Theorem on the Sum of a Constant**

$$(1) \ \sum_{k=1}^{n} c = nc \qquad (2) \ \sum_{k=m}^{n} c = (n - m + 1)c$$

To prove property 2, we may write

$$\sum_{k=m}^{n} c = \sum_{k=1}^{n} c - \sum_{k=1}^{m-1} c \qquad \text{subtract the first } m - 1 \text{ terms from the sum of } n \text{ terms}$$

$$= nc - (m - 1)c \qquad \text{use property 1 for each sum}$$

$$= (n - m + 1)c \qquad \text{factor out } c$$

## ILLUSTRATION

### Sum of a Constant

■ $\displaystyle\sum_{k=1}^{4} 7 = 7 + 7 + 7 + 7 = 4 \cdot 7 = 28$

■ $\displaystyle\sum_{k=1}^{10} \pi = 10\pi$

■ $\displaystyle\sum_{k=3}^{8} 9 = (8 - 3 + 1)(9) = 6(9) = 54$

■ $\displaystyle\sum_{k=10}^{20} 5 = (20 - 10 + 1)(5) = 11(5) = 55$

As shown in property 2 of the preceding theorem, the domain of the summation variable does not have to begin at 1. For example,

$$\sum_{k=4}^{8} a_k = a_4 + a_5 + a_6 + a_7 + a_8.$$

As another variation, if the first term of an infinite sequence is $a_0$, as in

$$a_0, a_1, a_2, \ldots, a_n, \ldots,$$

then we may consider sums of the form

$$\sum_{k=0}^{n} a_k = a_0 + a_1 + a_2 + \cdots + a_n,$$

which is the sum of the first $n + 1$ terms of the sequence.

If the summation variable does not appear in the term $a_k$, then the *entire term* may be considered a constant. For example,

$$\sum_{j=1}^{n} a_k = n \cdot a_k,$$

since $j$ does not appear in the term $a_k$.

## EXAMPLE 5    *Evaluating a sum*

Find the sum $\displaystyle\sum_{k=0}^{3} \frac{2^k}{k+1}$.

## SOLUTION

$$\sum_{k=0}^{3} \frac{2^k}{k+1} = \frac{2^0}{0+1} + \frac{2^1}{1+1} + \frac{2^2}{2+1} + \frac{2^3}{3+1}$$

$$= 1 + 1 + \tfrac{4}{3} + 2 = \tfrac{16}{3}$$

Summation notation can be used to denote polynomials. Thus, if

$$f(x) = a_0 + a_1 x + a_2 x^2 + \cdots + a_n x^n,$$

then

$$f(x) = \sum_{k=0}^{n} a_k x^k.$$

The following theorem concerning sums has many uses.

**Theorem on Sums**

If $a_1, a_2, \ldots, a_n, \ldots$ and $b_1, b_2, \ldots, b_n, \ldots$ are infinite sequences, then for every positive integer $n$,

(1) $\displaystyle\sum_{k=1}^{n} (a_k + b_k) = \sum_{k=1}^{n} a_k + \sum_{k=1}^{n} b_k$

(2) $\displaystyle\sum_{k=1}^{n} (a_k - b_k) = \sum_{k=1}^{n} a_k - \sum_{k=1}^{n} b_k$

(3) $\displaystyle\sum_{k=1}^{n} ca_k = c\left(\sum_{k=1}^{n} a_k\right)$    for every real number $c$

**PROOF** To prove formula 1, we first write

$$\sum_{k=1}^{n} (a_k + b_k) = (a_1 + b_1) + (a_2 + b_2) + (a_3 + b_3) + \cdots + (a_n + b_n).$$

Using commutative and associative properties many times, we may rearrange the terms on the right-hand side to produce

$$\sum_{k=1}^{n} (a_k + b_k) = (a_1 + a_2 + a_3 + \cdots + a_n) + (b_1 + b_2 + b_3 + \cdots + b_n).$$

Expressing the right-hand side in summation notation gives us formula 1. For formula 3, we have

$$\sum_{k=1}^{n} (ca_k) = ca_1 + ca_2 + ca_3 + \cdots + ca_n$$

$$= c(a_1 + a_2 + a_3 + \cdots + a_n)$$

$$= c\left(\sum_{k=1}^{n} a_k\right).$$

The proof of formula 2 is left as an exercise. ∎

## 9.1 EXERCISES

**Exer. 1–16: Find the first four terms and the eighth term of the sequence.**

1 $\{12 - 3n\}$

2 $\left\{\dfrac{3}{5n - 2}\right\}$

3 $\left\{\dfrac{3n - 2}{n^2 + 1}\right\}$

4 $\left\{10 + \dfrac{1}{n}\right\}$

5 $\{9\}$

6 $\{\sqrt{2}\}$

7 $\{2 + (-0.1)^n\}$

8 $\{4 + (0.1)^n\}$

9 $\left\{(-1)^{n-1} \dfrac{n + 7}{2n}\right\}$

10 $\left\{(-1)^n \dfrac{6 - 2n}{\sqrt{n + 1}}\right\}$

11 $\{1 + (-1)^{n+1}\}$

12 $\{(-1)^{n+1} + (0.1)^{n-1}\}$

13 $\left\{\dfrac{2^n}{n^2 + 2}\right\}$

14 $\{(n - 1)(n - 2)(n - 3)\}$

15 $a_n$ is the number of decimal places in $(0.1)^n$.

16 $a_n$ is the number of positive integers less than $n^3$.

**Exer. 17–24: Find the first five terms of the recursively defined infinite sequence.**

17 $a_1 = 2, \qquad a_{k+1} = 3a_k - 5$

   (*Note:* Some specific keystrokes for the TI-81/82 graphics calculator are given in Appendix I.)

18 $a_1 = 5, \qquad a_{k+1} = 7 - 2a_k$

19 $a_1 = -3, \quad a_{k+1} = a_k^2$

20 $a_1 = 128, \quad a_{k+1} = \frac{1}{4}a_k$

21 $a_1 = 5, \qquad a_{k+1} = ka_k$

22 $a_1 = 3, \qquad a_{k+1} = 1/a_k$

23 $a_1 = 2, \qquad a_{k+1} = (a_k)^k$

24 $a_1 = 2, \qquad a_{k+1} = (a_k)^{1/k}$

**Exer. 25–28: Find the first four terms of the sequence of partial sums for the given sequence.**

25  $\{3 + \frac{1}{2}n\}$

26  $\{1/n^2\}$

27  $\{(-1)^n n^{-1/2}\}$

28  $\{(-1)^n 1/2^n\}$

**Exer. 29–44: Find the sum.**

29  $\sum_{k=1}^{5} (2k - 7)$

30  $\sum_{k=1}^{6} (10 - 3k)$

31  $\sum_{k=1}^{4} (k^2 - 5)$

32  $\sum_{k=1}^{10} [1 + (-1)^k]$

33  $\sum_{k=0}^{5} k(k - 2)$

34  $\sum_{k=0}^{4} (k - 1)(k - 3)$

35  $\sum_{k=3}^{6} \frac{k - 5}{k - 1}$

36  $\sum_{k=1}^{6} \frac{3}{k + 1}$

37  $\sum_{k=1}^{5} (-3)^{k-1}$

38  $\sum_{k=0}^{4} 3(2^k)$

39  $\sum_{k=1}^{100} 100$

40  $\sum_{k=1}^{1000} 5$

41  $\sum_{k=253}^{571} \frac{1}{3}$

42  $\sum_{k=137}^{428} 2.1$

43  $\sum_{j=1}^{7} \frac{1}{2}k^2$

44  $\sum_{k=0}^{5} (3j + 2)$

45  Prove formula 2 of the theorem on sums.

46  Extend formula 1 of the theorem on sums to

$$\sum_{k=1}^{n} (a_k + b_k + c_k).$$

47  Terms of the sequence defined recursively by $a_1 = 5$, $a_{k+1} = \sqrt{a_k}$ may be approximated with a calculator by entering 5 and pressing the $\boxed{\sqrt{x}}$ key repeatedly. Describe what happens to the terms of the sequence as $k$ increases.

48  Terms of the sequence defined recursively by $a_1 = 1$, $a_{k+1} = \cos a_k$ may be approximated with a calculator by entering 1 and pressing the $\boxed{\text{COS}}$ key (in radian mode) repeatedly. Describe what happens to the terms of the sequence as $k$ increases.

49  **Bode's sequence**  Bode's sequence, defined by
$$a_1 = 0.4, \quad a_k = 0.1(3 \cdot 2^{k-2} + 4) \quad \text{for } k \geq 2,$$
can be used to approximate distances of planets from the sun. These distances are measured in astronomical units, with 1 AU = 93,000,000 mi. For example, the third term corresponds to earth and the fifth term to the minor planet Ceres. Approximate the first five terms of the sequence.

50  Approximations to $\pi$ may be obtained from the sequence
$$x_1 = 3, \quad x_{k+1} = x_k - \tan x_k.$$
(a) Find the first five terms of this sequence.
(b) What happens to the terms of the sequence when $x_1 = 6$?

51  A test question lists the first four terms of a sequence as 2, 4, 6, and 8 and asks for the fifth term. Show that the fifth term can be any real number $a$ by finding the $n$th term of a sequence that has for its first five terms 2, 4, 6, 8, and $a$.

52  **Growth of bacteria**  The number of bacteria in a certain culture is initially 500, and the culture doubles in size every day.
(a) Find the number of bacteria present after one day, two days, and three days.
(b) Find a formula for the number of bacteria present after $n$ days.

53  **The Fibonacci sequence**  The Fibonacci sequence is defined recursively by
$$a_1 = 1, \quad a_2 = 1, \quad a_{k+1} = a_k + a_{k-1} \quad \text{for } k \geq 2.$$
(a) Find the first ten terms of the sequence.
(b) The terms of the sequence $r_k = a_{k+1}/a_k$ give progressively better approximations to $\tau$, the golden ratio. Approximate the first ten terms of this sequence.

54  **The Fibonacci sequence**  The Fibonacci sequence can be defined by the formula
$$a_n = \frac{1}{\sqrt{5}}\left(\frac{1 + \sqrt{5}}{2}\right)^n - \frac{1}{\sqrt{5}}\left(\frac{1 - \sqrt{5}}{2}\right)^n.$$
Find the first eight terms and show that they agree with those found using the definition in Exercise 53.

**Exer. 55–56:** Some calculators use an algorithm similar to the following to approximate $\sqrt{N}$ for a positive real number $N$: Let $x_1 = N/2$ and find successive approximations $x_2, x_3, \ldots$ by using

$$x_2 = \frac{1}{2}\left(x_1 + \frac{N}{x_1}\right), \quad x_3 = \frac{1}{2}\left(x_2 + \frac{N}{x_2}\right), \quad \ldots$$

until the desired accuracy is obtained. Use this method to approximate the radical to six-decimal-place accuracy.

**55** $\sqrt{5}$ (See Appendix I.)

**56** $\sqrt{18}$

**57** The equation $\frac{1}{3}\sqrt[3]{x} - x + 2 = 0$ has a root near 2. To approximate this root, rewrite the equation as $x = \frac{1}{3}\sqrt[3]{x} + 2$. Let $x_1 = 2$ and find successive approximations $x_2, x_3, \ldots$ by using the formulas

$$x_2 = \frac{1}{3}\sqrt[3]{x_1} + 2, \quad x_3 = \frac{1}{3}\sqrt[3]{x_2} + 2, \quad \ldots$$

until four-decimal-place accuracy is obtained. (See Appendix I.)

**58** The equation $2x + \dfrac{1}{x^4 + x + 2} = 0$ has a root near 0. Use a procedure similar to that in Exercise 57 to approximate this root to four-decimal-place accuracy.

**Exer. 59–60:** (a) Show that $f$ takes on both positive and negative values on the interval $[1, 2]$. (b) Use the method of Exercise 57, with $x_1 = 1.5$, to approximate a zero of $f$ to two-decimal-place accuracy.

**59** $f(x) = x - 2 + \log x$

**60** $f(x) = \log x - 10^{-x}$ (*Hint:* Solve for $x$ in $\log x$.)

**Exer. 61–63:** For the given $n$th term $a_n = f(n)$ of a sequence, use the graph of $y = f(x)$ on the interval $[1, 100]$ to verify that as $n$ increases without bound, $a_n$ approaches some real number $c$.

**61** $a_n = \left(1 + \dfrac{1}{n} + \dfrac{1}{2n^2}\right)^n$ **62** $a_n = n^{1/n}$

**63** $a_n = \left(\dfrac{1}{n}\right)^{1/n}$

# 9.2 ARITHMETIC SEQUENCES

In this section and the next we consider two special types of sequences: arithmetic and geometric. The first type may be defined as follows.

**Definition of Arithmetic Sequence**

A sequence $a_1, a_2, \ldots, a_n, \ldots$ is an **arithmetic sequence** if there is a real number $d$ such that for every positive integer $k$,

$$a_{k+1} = a_k + d.$$

The number $d = a_{k+1} - a_k$ is called the **common difference** of the sequence.

Note that the common difference $d$ is the difference of *any* two successive terms of an arithmetic sequence.

**EXAMPLE 1** *Showing that a sequence is arithmetic*

Show that the sequence

$$1, 4, 7, 10, \ldots, 3n - 2, \ldots$$

is arithmetic, and find the common difference.

SOLUTION    If $a_n = 3n - 2$, then for every positive integer $k$,

$$a_{k+1} - a_k = [3(k + 1) - 2] - (3k - 2)$$
$$= 3k + 3 - 2 - 3k + 2 = 3.$$

Hence, the given sequence is arithmetic with common difference 3.

Given an arithmetic sequence, we know that

$$a_{k+1} = a_k + d$$

for every positive integer $k$. This gives us a recursive formula for obtaining successive terms. Beginning with any real number $a_1$, we can obtain an arithmetic sequence with common difference $d$ simply by adding $d$ to $a_1$, then to $a_1 + d$, and so on, obtaining

$$a_1, a_1 + d, a_1 + 2d, a_1 + 3d, a_1 + 4d, \ldots .$$

The $n$th term $a_n$ of this sequence is given by the next formula.

**The $n$th Term of an Arithmetic Sequence**

$$a_n = a_1 + (n - 1)d$$

**EXAMPLE 2**    *Finding a specific term of an arithmetic sequence*
The first three terms of an arithmetic sequence are 20, 16.5, and 13. Find the fifteenth term.

SOLUTION    The common difference is

$$a_2 - a_1 = 16.5 - 20 = -3.5.$$

Substituting $n = 15$, $a_1 = 20$, and $d = -3.5$ in the formula for the $n$th term of an arithmetic sequence, $a_n = a_1 + (n - 1)d$, gives us

$$a_{15} = 20 + (15 - 1)(-3.5) = 20 - 49 = -29.$$

**EXAMPLE 3**    *Finding a specific term of an arithmetic sequence*
If the fourth term of an arithmetic sequence is 5 and the ninth term is 20, find the sixth term.

SOLUTION    We are given $a_4 = 5$ and $a_9 = 20$ and wish to find $a_6$. The following are equivalent systems of equations in the variables $a_1$ and $d$:

$$\begin{cases} a_4 = a_1 + (4 - 1)d & \text{let } n = 4 \text{ in } a_n = a_1 + (n - 1)d \\ a_9 = a_1 + (9 - 1)d & \text{let } n = 9 \text{ in } a_n = a_1 + (n - 1)d \end{cases}$$

$$\begin{cases} 5 = a_1 + 3d & a_4 = 5 \\ 20 = a_1 + 8d & a_9 = 20 \end{cases}$$

Subtracting the first equation of the system from the second gives us $15 = 5d$, or $d = 3$. Thus, $a_1 = 5 - 3d = -4$. Hence,

$$a_6 = a_1 + (6 - 1)d \qquad \text{let } n = 6 \text{ in } a_n = a_1 + (n - 1)d$$
$$= (-4) + (5)(3) = 11. \quad a_1 = -4 \text{ and } d = 3$$

The next theorem contains a formula for the $n$th partial sum $S_n$ of an arithmetic sequence.

**Theorem: Formulas for $S_n$**

> If $a_1, a_2, \ldots, a_n, \ldots$ is an arithmetic sequence with common difference $d$, then the $n$th partial sum $S_n$ is given by either
>
> $$S_n = \frac{n}{2}[2a_1 + (n - 1)d] \quad \text{or} \quad S_n = \frac{n}{2}(a_1 + a_n).$$

*PROOF*   We may write

$$S_n = a_1 + a_2 + a_3 + \cdots + a_n$$
$$= a_1 + (a_1 + d) + (a_1 + 2d) + \cdots + [a_1 + (n - 1)d].$$

Employing commutative and associative properties many times, we obtain

$$S_n = (a_1 + a_1 + a_1 + \cdots + a_1) + [d + 2d + \cdots + (n - 1)d],$$

with $a_1$ appearing $n$ times within the first pair of parentheses. Thus,

$$S_n = na_1 + d[1 + 2 + \cdots + (n - 1)].$$

The expression within brackets is the sum of the first $n - 1$ positive integers. Using the formula $S_n = n(n + 1)/2$ from Example 4 of Section 9.1, but with $n - 1$ in place of $n$, we have

$$1 + 2 + \cdots + (n - 1) = \frac{(n - 1)n}{2}.$$

Substituting in the last equation for $S_n$ and factoring out $n/2$ gives us

$$S_n = na_1 + d\frac{(n - 1)n}{2} = \frac{n}{2}[2a_1 + (n - 1)d].$$

Since $a_n = a_1 + (n - 1)d$, the last equation is equivalent to

$$S_n = \frac{n}{2}(a_1 + a_n). \quad \blacksquare$$

**EXAMPLE 4    *Finding a sum of even integers***

Find the sum of all the even integers from 2 through 100.

*SOLUTION*    This problem is equivalent to finding the sum of the first 50 terms of the arithmetic sequence

$$2, 4, 6, \ldots, 2n, \ldots.$$

Substituting $n = 50$, $a_1 = 2$, and $a_{50} = 100$ in $S_n = (n/2)(a_1 + a_n)$ produces

$$S_{50} = \tfrac{50}{2}(2 + 100) = 2550.$$

Alternatively, we may use $S_n = \dfrac{n}{2}[2a_1 + (n - 1)d]$ with $d = 2$:

$$S_{50} = \tfrac{50}{2}[2(2) + (50 - 1)2] = 25[4 + 98] = 2550$$

---

The **arithmetic mean** of two numbers $a$ and $b$ is defined as $(a + b)/2$. This is the **average** of $a$ and $b$. Note that

$$a, \frac{a + b}{2}, b$$

is a (finite) arithmetic sequence. This concept may be generalized as follows: If $c_1, c_2, \ldots, c_k$ are real numbers such that

$$a, c_1, c_2, \ldots, c_k, b$$

is a (finite) arithmetic sequence, then $c_1, c_2, \ldots, c_k$ are **$k$ arithmetic means** between the numbers $a$ and $b$. The process of determining these numbers is referred to as *inserting $k$ arithmetic means between $a$ and $b$*.

**EXAMPLE 5    *Inserting arithmetic means***

Insert three arithmetic means between 2 and 9.

*SOLUTION*    We wish to find three real numbers $c_1$, $c_2$, and $c_3$ such that the following is a (finite) arithmetic sequence:

$$2, c_1, c_2, c_3, 9$$

We may regard this sequence as an arithmetic sequence with first term $a_1 = 2$ and fifth term $a_5 = 9$. To find the common difference $d$, we proceed as follows:

| | |
|---|---|
| $a_5 = a_1 + (5 - 1)d$ | let $n = 5$ in $a_n = a_1 + (n - 1)d$ |
| $9 = 2 + 4d$ | $a_5 = 9$ and $a_1 = 2$ |
| $d = \tfrac{7}{4}$ | solve for $d$ |

Thus, the arithmetic means are

$$c_1 = a_1 + d = 2 + \tfrac{7}{4} = \tfrac{15}{4}$$

$$c_2 = c_1 + d = \tfrac{15}{4} + \tfrac{7}{4} = \tfrac{22}{4} = \tfrac{11}{2}$$

$$c_3 = c_2 + d = \tfrac{22}{4} + \tfrac{7}{4} = \tfrac{29}{4}.$$

**FIGURE 1**

$a_1 = 18$
$a_2$
$a_3$
$a_4$
$a_5$
$a_6$
$a_7$
$a_8$
$a_9 = 24$

### EXAMPLE 6 *An application of an arithmetic sequence*

A carpenter wishes to construct a ladder with nine rungs whose lengths decrease uniformly from 24 inches at the base to 18 inches at the top. Determine the lengths of the seven intermediate rungs.

**SOLUTION** The ladder is sketched in Figure 1. The lengths of the rungs are to form an arithmetic sequence $a_1, a_2, \ldots, a_9$ with $a_1 = 18$ and $a_9 = 24$. Hence, we need to insert seven arithmetic means between 18 and 24. Using $a_n = a_1 + (n - 1)d$ with $n = 9$, $a_1 = 18$, and $a_9 = 24$ gives us

$$24 = 18 + 8d, \quad \text{or} \quad 8d = 6.$$

Hence, $d = \tfrac{6}{8} = 0.75$ and the intermediate rungs have lengths (in inches)

$$18.75, \ 19.5, \ 20.25, \ 21, \ 21.75, \ 22.5, \ 23.25.$$

It is sometimes necessary to express a sum in terms of summation notation, as illustrated in the next example.

### EXAMPLE 7 *Expressing a sum in summation notation*

Express in terms of summation notation:

$$\tfrac{1}{4} + \tfrac{2}{9} + \tfrac{3}{14} + \tfrac{4}{19} + \tfrac{5}{24} + \tfrac{6}{29}$$

**SOLUTION** The six terms of the sum do not form an arithmetic sequence; however, the numerators and denominators of the fractions, *considered separately,* are each arithmetic. Specifically, we have the following:

> *Numerators:* 1, 2, 3, 4, 5, 6    common difference 1
>
> *Denominators:* 4, 9, 14, 19, 24, 29    common difference 5

Using the formula for the $n$th term of an arithmetic sequence twice, we obtain the following $n$th term for each sequence:

$$a_n = a_1 + (n - 1)d = 1 + (n - 1)1 = n$$

$$a_n = a_1 + (n - 1)d = 4 + (n - 1)5 = 5n - 1$$

*(continued)*

Hence, the $n$th term of the given sum is $n/(5n - 1)$, and we may write

$$\frac{1}{4} + \frac{2}{9} + \frac{3}{14} + \frac{4}{19} + \frac{5}{24} + \frac{6}{29} = \sum_{n=1}^{6} \frac{n}{5n - 1}.$$

## 9.2 EXERCISES

**Exer. 1–2: Show that the given sequence is arithmetic, and find the common difference.**

1  $-6, -2, 2, \ldots, 4n - 10, \ldots$

2  $53, 48, 43, \ldots, 58 - 5n, \ldots$

**Exer. 3–10: Find the fifth term, the tenth term, and the $n$th term of the arithmetic sequence.**

3  $2, 6, 10, 14, \ldots$           4  $16, 13, 10, 7, \ldots$

5  $3, 2.7, 2.4, 2.1, \ldots$

6  $-6, -4.5, -3, -1.5, \ldots$

7  $-7, -3.9, -0.8, 2.3, \ldots$

8  $x - 8, x - 3, x + 2, x + 7, \ldots$

9  $\ln 3, \ln 9, \ln 27, \ln 81, \ldots$

10  $\log 1000, \log 100, \log 10, \log 1, \ldots$

**Exer. 11–12: Find the common difference for the arithmetic sequence with the specified terms.**

11  $a_2 = 21, \quad a_6 = -11$           12  $a_4 = 14, \quad a_{11} = 35$

**Exer. 13–18: Find the specified term of the arithmetic sequence that has the two given terms.**

13  $a_{12}; \quad a_1 = 9.1, \qquad a_2 = 7.5$

14  $a_{11}; \quad a_1 = 2 + \sqrt{2}, \quad a_2 = 3$

15  $a_1; \quad a_6 = 2.7, \qquad a_7 = 5.2$

16  $a_1; \quad a_8 = 47, \qquad a_9 = 53$

17  $a_{15}; \quad a_3 = 7, \qquad a_{20} = 43$

18  $a_{10}; \quad a_2 = 1, \qquad a_{18} = 49$

**Exer. 19–22: Find the sum $S_n$ of the arithmetic sequence that satisfies the stated conditions.**

19  $a_1 = 40, \quad d = -3, \quad n = 30$

20  $a_1 = 5, \quad d = 0.1, \quad n = 40$

21  $a_1 = -9, \quad a_{10} = 15, \quad n = 10$

22  $a_7 = \frac{7}{3}, \quad d = -\frac{2}{3}, \quad n = 15$

**Exer. 23–26: Find the sum.**

23  $\displaystyle\sum_{k=1}^{20} (3k - 5)$           24  $\displaystyle\sum_{k=1}^{12} (7 - 4k)$

25  $\displaystyle\sum_{k=1}^{18} (\tfrac{1}{2}k + 7)$           26  $\displaystyle\sum_{k=1}^{10} (\tfrac{1}{4}k + 3)$

**Exer. 27–32: Express the sum in terms of summation notation. (Answers are not unique.)**

27  $1 + 3 + 5 + 7$           28  $2 + 4 + 6 + 8 + 10$

29  $1 + 3 + 5 + \cdots + 73$           30  $2 + 4 + 6 + \cdots + 150$

31  $\frac{3}{7} + \frac{6}{11} + \frac{9}{15} + \frac{12}{19} + \frac{15}{23} + \frac{18}{27}$

32  $\frac{5}{13} + \frac{10}{11} + \frac{15}{9} + \frac{20}{7}$

**Exer. 33–34: Find the number of terms in the arithmetic sequence with the given conditions.**

33  $a_1 = -2, \quad d = \tfrac{1}{4}, \quad S = 21$

34  $a_6 = -3, \quad d = 0.2, \quad S = -33$

35  Insert five arithmetic means between 2 and 10.

36  Insert three arithmetic means between 3 and $-5$.

37  (a) Find the number of integers between 32 and 395 that are divisible by 6.

   (b) Find their sum.

38  (a) Find the number of negative integers greater than $-500$ that are divisible by 33.

   (b) Find their sum.

39  **Log pile**  A pile of logs has 24 logs in the bottom layer, 23 in the second layer, 22 in the third, and so on. The top layer contains 10 logs. Find the total number of logs in the pile.

**40 Stadium seating** The first ten rows of seating in a certain section of a stadium have 30 seats, 32 seats, 34 seats, and so on. The eleventh through the twentieth rows each contain 50 seats. Find the total number of seats in the section.

**41 Constructing a grain bin** A grain bin is to be constructed in the shape of a frustum of a cone (see the figure). The bin is to be 10 feet tall with 11 metal rings positioned uniformly around it, from the 4-foot opening at the bottom to the 24-foot opening at the top. Find the total length of metal needed to make the rings.

EXERCISE 41

$a_n$

24'

10'

4'

$a_1$

**42 Coasting downhill** A bicycle rider coasts downhill, traveling 4 feet the first second. In each succeeding second, the rider travels 5 feet farther than in the preceding second. If the rider reaches the bottom of the hill in 11 seconds, find the total distance traveled.

**43 Prize money** A contest will have five cash prizes totaling $5000, and there will be a $100 difference between successive prizes. Find the first prize.

**44 Sales bonuses** A company is to distribute $46,000 in bonuses to its top ten salespeople. The tenth salesperson on the list will receive $1000, and the difference in bonus money between successively ranked salespeople is to be constant. Find the bonus for each salesperson.

**45 Distance an object falls** Assuming air resistance is negligible, a small object that is dropped from a hot air balloon falls 16 feet during the first second, 48 feet during the second second, 80 feet during the third second, 112 feet during the fourth second, and so on. Find an expression for the distance the object falls in $n$ seconds.

**46** If $f$ is a linear function, show that the sequence with $n$th term $a_n = f(n)$ is an arithmetic sequence.

**47 Genetic sequence** The sequence defined recursively by $x_{k+1} = x_k/(1 + x_k)$ occurs in genetics in the study of the elimination of a deficient gene from a population. Show that the sequence whose $n$th term is $1/x_n$ is arithmetic.

**48 Dimensions of a maze** Find the total length of the broken-line curve in the figure if the width of the maze formed by the curve is 16 inches and all halls in the maze have width 1 inch. What is the length if the width of the maze is 32 inches?

EXERCISE 48

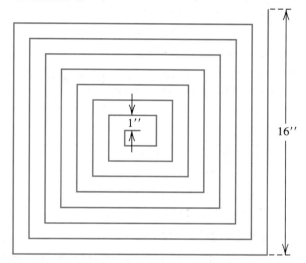

1''

16''

Exer. 49–50: Depreciation methods are sometimes used by businesses and individuals to estimate the value of an asset over a life span of $n$ years. In the sum-of-year's-digits method, for each year $k = 1, 2, 3, \ldots, n$, the value of an asset is decreased by the fraction $A_k = \dfrac{n - k + 1}{T_n}$ of its initial cost, where $T_n = 1 + 2 + 3 + \cdots + n$.

**49** (a) If $n = 8$, find $A_1, A_2, A_3, \ldots, A_8$.

  (b) Show that the sequence in (a) is arithmetic, and find $S_8$.

  (c) If the initial value of an asset is $1000, how much has been depreciated after 4 years?

**50** (a) If $n$ is any positive integer, find $A_1, A_2, A_3, \ldots, A_n$.

  (b) Show that the sequence in (a) is arithmetic, and find $S_n$.

## 9.3    GEOMETRIC SEQUENCES

The second special type of sequence that we will discuss—the geometric sequence—occurs frequently in applications.

**Definition of Geometric Sequence**

A sequence $a_1, a_2, \ldots, a_n, \ldots$ is a **geometric sequence** if $a_1 \neq 0$ and if there is a real number $r \neq 0$ such that for every positive integer $k$,

$$a_{k+1} = a_k r.$$

The number $r = \dfrac{a_{k+1}}{a_k}$ is called the **common ratio** of the sequence.

Note that the common ratio $r = a_{k+1}/a_k$ is the ratio of *any* two successive terms of a geometric sequence.

The formula $a_{k+1} = a_k r$ provides a recursive method for obtaining terms of a geometric sequence. Beginning with any nonzero real number $a_1$, we multiply by the number $r$ successively, obtaining

$$a_1, \ a_1 r, \ a_1 r^2, \ a_1 r^3, \ \ldots .$$

The $n$th term $a_n$ of this sequence is given by the next formula.

**Formula for the $n$th Term of a Geometric Sequence**

$$a_n = a_1 r^{n-1}$$

**EXAMPLE I**    *Finding terms of a geometric sequence*

A geometric sequence has first term 3 and common ratio $-\frac{1}{2}$. Find the first five terms and the tenth term.

**SOLUTION**    If we let $a_1 = 3$ and $r = -\frac{1}{2}$, then the first five terms are

$$3, \ -\tfrac{3}{2}, \ \tfrac{3}{4}, \ -\tfrac{3}{8}, \ \tfrac{3}{16}.$$

Using the formula $a_n = a_1 r^{n-1}$ with $n = 10$, we find that the tenth term is

$$a_{10} = a_1 r^9 = 3\left(-\tfrac{1}{2}\right)^9 = -\tfrac{3}{512}.$$

**EXAMPLE 2**    *Finding a specific term of a geometric sequence*

The third term of a geometric sequence is 5, and the sixth term is $-40$. Find the eighth term.

**SOLUTION** We are given $a_3 = 5$ and $a_6 = -40$ and wish to find $a_8$. The following are equivalent systems of equations in the variables $a_1$ and $r$:

$$\begin{cases} a_3 = a_1 r^{3-1} & \text{let } n = 3 \text{ in } a_n = a_1 r^{n-1} \\ a_6 = a_1 r^{6-1} & \text{let } n = 6 \text{ in } a_n = a_1 r^{n-1} \end{cases}$$

$$\begin{cases} 5 = a_1 r^2 & a_3 = 5 \\ -40 = a_1 r^5 & a_6 = -40 \end{cases}$$

Solving the first equation of the system for $a_1$ gives us $a_1 = 5/r^2$. Substituting this expression in the second equation yields

$$-40 = \frac{5}{r^2} \cdot r^5.$$

Simplifying, we get $r^3 = -8$, and hence $r = -2$. We next use $a_1 = 5/r^2$ to obtain

$$a_1 = \frac{5}{(-2)^2} = \frac{5}{4}.$$

Finally, using $a_n = a_1 r^{n-1}$ with $n = 8$ gives us

$$a_8 = a_1 r^7 = \left(\tfrac{5}{4}\right)(-2)^7 = -160.$$

---

The next theorem contains a formula for the $n$th partial sum $S_n$ of a geometric sequence.

**Theorem: Formula for $S_n$**

> The $n$th partial sum $S_n$ of a geometric sequence with first term $a_1$ and common ratio $r \neq 1$ is
>
> $$S_n = a_1 \frac{1 - r^n}{1 - r}.$$

**PROOF** By definition, the $n$th partial sum $S_n$ of a geometric sequence is

(1) $$S_n = a_1 + a_1 r + a_1 r^2 + \cdots + a_1 r^{n-2} + a_1 r^{n-1}.$$

Multiplying both sides of (1) by $r$, we obtain

(2) $$r S_n = a_1 r + a_1 r^2 + a_1 r^3 + \cdots + a_1 r^{n-1} + a_1 r^n.$$

If we subtract equation (2) from equation (1), all but two terms on the right-hand side cancel and we obtain the following:

$$S_n - r S_n = a_1 - a_1 r^n \qquad \text{subtract (2) from (1)}$$
$$S_n(1 - r) = a_1(1 - r^n) \qquad \text{factor both sides}$$
$$S_n = a_1 \frac{1 - r^n}{1 - r} \qquad \text{divide by } (1 - r) \quad \blacksquare$$

**EXAMPLE 3**    *Finding a sum of terms of a geometric sequence*

If $1, 0.3, 0.09, 0.027, \ldots$ is a geometric sequence, find the sum of the first five terms.

**SOLUTION**    If we let $a_1 = 1$, $r = 0.3$, and $n = 5$ in the formula for $S_n$ stated in the preceding theorem, we obtain

$$S_5 = (1)\frac{1 - (0.3)^5}{1 - 0.3} = 1.4251.$$

**EXAMPLE 4**    *The rapid growth of terms of a geometric sequence*

A man wishes to save money by setting aside 1 cent the first day, 2 cents the second day, 4 cents the third day, and so on.

**(a)** If he continues to double the amount set aside each day, how much must he set aside on the fifteenth day?

**(b)** Assuming he does not run out of money, what is the total amount saved at the end of 30 days?

**SOLUTION**

**(a)** The amount (in cents) set aside on successive days forms a geometric sequence

$$1, 2, 4, 8, \ldots,$$

with first term 1 and common ratio 2. We find the amount to be set aside on the fifteenth day by using $a_n = a_1 r^{n-1}$ with $a_1 = 1$ and $n = 15$:

$$a_{15} = a_1 r^{14} = 1 \cdot 2^{14} = 16{,}384$$

Thus, \$163.84 should be set aside on the fifteenth day.

**(b)** To find the total amount saved after 30 days, we use the formula for $S_n$ with $n = 30$, obtaining (in cents)

$$S_{30} = (1)\frac{1 - 2^{30}}{1 - 2} = 1{,}073{,}741{,}823.$$

Thus, the total amount saved is \$10,737,418.23.

In a manner analogous to our work with arithmetic sequences, if $a$ and $b$ are positive real numbers, then we call a positive number $c$ the **geometric mean** of $a$ and $b$ if $a, c, b$ is a geometric sequence. If the common ratio is $r$, then

$$r = \frac{c}{a} = \frac{b}{c}, \quad \text{or} \quad c^2 = ab.$$

Taking the square root of both sides of the last equation, we see that *the geometric mean of the positive numbers a and b is* $\sqrt{ab}$. As a generalization, $k$ positive real numbers $c_1, c_2, \ldots, c_k$ are $k$ **geometric means** between $a$ and $b$ if $a, c_1, c_2, \ldots, c_k, b$ is a geometric sequence. The process of determining these numbers is referred to as *inserting k geometric means between a and b.*

### ILLUSTRATION

### Geometric Means

| Numbers | Geometric mean |
|---|---|
| ■ 20, 45 | $\sqrt{20 \cdot 45} = \sqrt{900} = 30$ |
| ■ 3, 4 | $\sqrt{3 \cdot 4} = 2\sqrt{3} \approx 3.46$ |

Given the geometric series with first term $a_1$ and common ratio $r \neq 1$, we may write the formula for $S_n$ of the preceding theorem in the form

$$S_n = \frac{a_1}{1-r} - \frac{a_1}{1-r}r^n.$$

If $|r| < 1$, then $r^n$ *approaches 0 as n increases* without bound. Thus, $S_n$ approaches $a_1/(1-r)$ as $n$ increases without bound. Using the notation we developed for rational functions in Section 3.5, we have

$$S_n \to \frac{a_1}{1-r} \quad \text{as} \quad n \to \infty.$$

The number $a_1/(1-r)$ is called the *sum S* of the **infinite geometric series**

$$a_1 + a_1r + a_1r^2 + \cdots + a_1r^{n-1} + \cdots.$$

This gives us the next result.

**Theorem on the Sum of an Infinite Geometric Series**

If $|r| < 1$, then the infinite geometric series

$$a_1 + a_1r + a_1r^2 + \cdots + a_1r^{n-1} + \cdots$$

has the sum

$$S = \frac{a_1}{1-r}.$$

The preceding theorem implies that if we add more and more terms of the indicated infinite geometric series, the sums get closer and closer to $a_1/(1-r)$. The next example illustrates how the theorem can be used to show that every real number represented by a repeating decimal is rational.

**EXAMPLE 5**     *Expressing an infinite repeating decimal as a rational number*

Find a rational number that corresponds to the infinite repeating decimal $5.4\overline{27}$, where the block of digits underneath the bar is repeated indefinitely.

**SOLUTION**     From the decimal expression $5.4272727\ldots$, we obtain the infinite series

$$5.4 + 0.027 + 0.00027 + 0.0000027 + \cdots.$$

Beginning with the second term, $0.027$, the infinite series has the form given in the theorem on the sum of an infinite geometric series, with $a_1 = 0.027$ and $r = 0.01$. Hence, the sum $S$ of this infinite geometric series is

$$S = \frac{0.027}{1 - 0.01} = \frac{0.027}{0.99} = \frac{27}{990} = \frac{3}{110}.$$

Thus, the desired number is

$$5.4 + \tfrac{3}{110} = \tfrac{594}{110} + \tfrac{3}{110} = \tfrac{597}{110}.$$

A check by division shows that $\tfrac{597}{110}$ corresponds to $5.4\overline{27}$.

---

In general, given any infinite sequence $a_1, a_2, \ldots, a_n, \ldots$, the expression

$$a_1 + a_2 + \cdots + a_n + \cdots$$

is called an **infinite series** or simply a **series.** We denote this series by

$$\sum_{n=1}^{\infty} a_n.$$

Each number $a_k$ is a **term** of the series, and $a_n$ is the **nth term.** Since only *finite* sums may be added algebraically, it is necessary to define what is meant by an *infinite sum.* Consider the sequence of partial sums

$$S_1, S_2, \ldots, S_n, \ldots.$$

If there is a number $S$ such that $S_n \to S$ as $n \to \infty$, then, as in our discussion of infinite geometric series, $S$ is the **sum** of the infinite series and we write

$$S = a_1 + a_2 + \cdots + a_n + \cdots.$$

In the previous example we found that the infinite repeating decimal $5.4272727\ldots$ corresponds to the rational number $\tfrac{597}{110}$. Since $\tfrac{597}{110}$ is the sum of an infinite series determined by the decimal, we may write

$$\tfrac{597}{110} = 5.4 + 0.027 + 0.00027 + 0.0000027 + \cdots.$$

If the terms of an infinite sequence are alternately positive and negative, as in the expression

$$a_1 + (-a_2) + a_3 + (-a_4) + \cdots + [(-1)^{n+1}a_n] + \cdots$$

for positive real numbers $a_k$, then the expression is an **alternating infinite series,** and we write it in the form

$$a_1 - a_2 + a_3 - a_4 + \cdots + (-1)^{n+1}a_n + \cdots.$$

The most common types of alternating infinite series are infinite geometric series in which the common ratio $r$ is negative.

**EXAMPLE 6    *Finding the sum of an infinite geometric series***

Find the sum $S$ of the alternating infinite geometric series

$$\sum_{n=1}^{\infty} 3\left(-\tfrac{2}{3}\right)^{n-1} = 3 - 2 + \tfrac{4}{3} - \tfrac{8}{9} + \cdots + 3\left(-\tfrac{2}{3}\right)^{n-1} + \cdots.$$

**SOLUTION**    Using the formula for $S$ in the theorem on the sum of an infinite geometric series, with $a_1 = 3$ and $r = -\tfrac{2}{3}$, we obtain

$$S = \frac{a_1}{1 - r} = \frac{3}{1 - \left(-\tfrac{2}{3}\right)} = \frac{3}{\tfrac{5}{3}} = \frac{9}{5}.$$

**FIGURE 2**

**EXAMPLE 7    *An application of an infinite geometric series***

A rubber ball is dropped from a height of 10 meters. Suppose it rebounds one-half the distance after each fall, as illustrated by the arrows in Figure 2. Find the total distance the ball travels before coming to rest.

**SOLUTION**    If the ball always rebounds half the distance it falls, then, theoretically, it *never* comes to rest. However, the sum of the distances it travels downward and the sum of the distances it travels on the rebounds form two finite geometric series:

*Downward series:*   $10 + 5 + 2.5 + 1.25 + 0.625 + \cdots$

*Upward series:*          $5 + 2.5 + 1.25 + 0.625 + \cdots$

We assume that the total distance $S$ the ball travels can be found by adding the sums of these infinite series. This gives us

$$S = 10 + 2[5 + 2.5 + 1.25 + 0.625 + \cdots]$$
$$= 10 + 2\left[5 + 5\left(\tfrac{1}{2}\right) + 5\left(\tfrac{1}{2}\right)^2 + 5\left(\tfrac{1}{2}\right)^3 + \cdots\right].$$

Using the formula $S = a_1/(1 - r)$ with $a_1 = 5$ and $r = \tfrac{1}{2}$, we obtain

$$S = 10 + 2\left[\frac{5}{1 - \tfrac{1}{2}}\right] = 10 + 2(10) = 30 \text{ m}.$$

## 9.3 EXERCISES

**Exer. 1–2: Show that the given sequence is geometric, and find the common ratio.**

1  $5, -\frac{5}{4}, \frac{5}{16}, \ldots, 5\left(-\frac{1}{4}\right)^{n-1}, \ldots$

2  $\frac{1}{7}, \frac{3}{7}, \frac{9}{7}, \ldots, \frac{1}{7}(3)^{n-1}, \ldots$

**Exer. 3–14: Find the fifth term, the eighth term, and the $n$th term of the geometric sequence.**

3  $8, 4, 2, 1, \ldots$

4  $4, 1.2, 0.36, 0.108, \ldots$

5  $300, -30, 3, -0.3, \ldots$

6  $1, -\sqrt{3}, 3, -3\sqrt{3}, \ldots$

7  $5, 25, 125, 625, \ldots$

8  $2, 6, 18, 54, \ldots$

9  $4, -6, 9, -13.5, \ldots$

10  $162, -54, 18, -6, \ldots$

11  $1, -x^2, x^4, -x^6, \ldots$

12  $1, -\dfrac{x}{3}, \dfrac{x^2}{9}, -\dfrac{x^3}{27}, \ldots$

13  $2, 2^{x+1}, 2^{2x+1}, 2^{3x+1}, \ldots$

14  $10, 10^{2x-1}, 10^{4x-3}, 10^{6x-5}, \ldots$

**Exer. 15–16: Find all possible values of $r$ for a geometric sequence with the two given terms.**

15  $a_4 = 3, \quad a_6 = 9$

16  $a_3 = 4, \quad a_7 = \frac{1}{4}$

17  Find the sixth term of the geometric sequence whose first two terms are 4 and 6.

18  Find the seventh term of the geometric sequence whose second and third terms are 2 and $-\sqrt{2}$.

19  Given a geometric sequence such that $a_4 = 4$ and $a_7 = 12$, find $r$ and $a_{10}$.

20  Given a geometric sequence such that $a_2 = 3$ and $a_5 = -81$, find $r$ and $a_9$.

**Exer. 21–24: Find the sum.**

21  $\displaystyle\sum_{k=1}^{10} 3^k$

22  $\displaystyle\sum_{k=1}^{9} (-\sqrt{5})^k$

23  $\displaystyle\sum_{k=0}^{9} \left(-\tfrac{1}{2}\right)^{k+1}$

24  $\displaystyle\sum_{k=1}^{7} (3^{-k})$

**Exer. 25–28: Express the sum in terms of summation notation. (Answers are not unique.)**

25  $2 + 4 + 8 + 16 + 32 + 64 + 128$

26  $2 - 4 + 8 - 16 + 32 - 64$

27  $\frac{1}{4} - \frac{1}{12} + \frac{1}{36} - \frac{1}{108}$

28  $3 + \frac{3}{5} + \frac{3}{25} + \frac{3}{125} + \frac{3}{625}$

**Exer. 29–36: Find the sum of the infinite geometric series, if it exists.**

29  $1 - \frac{1}{2} + \frac{1}{4} - \frac{1}{8} + \cdots$

30  $2 + \frac{2}{3} + \frac{2}{9} + \frac{2}{27} + \cdots$

31  $1.5 + 0.015 + 0.00015 + \cdots$

32  $1 - 0.1 + 0.01 - 0.001 + \cdots$

33  $\sqrt{2} - 2 + \sqrt{8} - 4 + \cdots$

34  $1 + \frac{3}{2} + \frac{9}{4} + \frac{27}{8} + \cdots$

35  $256 + 192 + 144 + 108 + \cdots$

36  $250 - 100 + 40 - 16 + \cdots$

**Exer. 37–44: Find the rational number represented by the repeating decimal.**

37  $0.\overline{23}$

38  $0.0\overline{71}$

39  $2.4\overline{17}$

40  $10.\overline{5}$

41  $5.\overline{146}$

42  $3.2\overline{394}$

43  $1.6\overline{124}$

44  $123.61\overline{83}$

45  Find the geometric mean of 12 and 48.

46  Find the geometric mean of 20 and 25.

47  Insert two geometric means between 4 and 500.

48  Insert three geometric means between 2 and 512.

49  *Using a vacuum pump*  A vacuum pump removes one-half of the air in a container with each stroke. After 10 strokes, what percentage of the original amount of air remains in the container?

50  *Calculating depreciation*  The yearly depreciation of a certain machine is 25% of its value at the beginning of the year. If the original cost of the machine is $20,000, what is its value after 6 years?

51  *Growth of bacteria*  A certain culture initially contains 10,000 bacteria and increases by 20% every hour.

(a)  Find a formula for the number $N(t)$ of bacteria present after $t$ hours.

(b)  How many bacteria are in the culture at the end of 10 hours?

**52 Interest on savings** An amount of money $P$ is deposited in a savings account that pays interest at a rate of $r$ percent per year compounded quarterly; the principal and accumulated interest are left in the account. Find a formula for the total amount in the account after $n$ years.

**53 Rebounding ball** A rubber ball is dropped from a height of 60 feet. If it rebounds approximately two-thirds the distance after each fall, use an infinite geometric series to approximate the total distance the ball travels before coming to rest.

**54 Motion of a pendulum** The bob of a pendulum swings through an arc 24 centimeters long on its first swing. If each successive swing is approximately five-sixths the length of the preceding swing, use an infinite geometric series to approximate the total distance the bob travels before coming to rest.

**55 Multiplier effect** A manufacturing company that has just located in a small community will pay two million dollars per year in salaries. It has been estimated that 60% of these salaries will be spent in the local area, and 60% of the money spent will again change hands within the community. This process, called the *multiplier effect*, will be repeated ad infinitum. Find the total amount of local spending that will be generated by company salaries.

**56 Pest eradication** In a pest eradication program, $N$ sterilized male flies are released into the general population each day. It is estimated that 90% of these flies will survive a given day.

(a) Show that the number of sterilized flies in the population $n$ days after the program has begun is

$$N + (0.9)N + (0.9)^2N + \cdots + (0.9)^{n-1}N.$$

(b) If the *long-range* goal of the program is to keep 20,000 sterilized males in the population, how many flies should be released each day?

**57 Drug dosage** A certain drug has a half-life of about 2 hours in the bloodstream. The drug is formulated to be administered in doses of $D$ milligrams every 4 hours, but $D$ is yet to be determined.

(a) Show that the number of milligrams of drug in the bloodstream after the $n$th dose has been administered is

$$D + \tfrac{1}{4}D + \cdots + \left(\tfrac{1}{4}\right)^{n-1}D,$$

and that this sum is approximately $\tfrac{4}{3}D$ for large values of $n$.

(b) A level of more than 500 milligrams of the drug in the bloodstream is considered to be dangerous. Find the largest possible dose that can be given repeatedly over a long period of time.

**58 Genealogy** Shown in the figure is a family tree displaying the current generation (you) and 3 prior generations, with a total of 12 grandparents. If you were to trace your family history back 10 generations, how many grandparents would you find?

**EXERCISE 58**

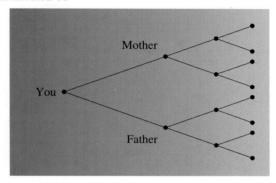

**59** The first figure shows some terms of a sequence of squares $S_1, S_2, \ldots, S_k, \ldots$. Let $a_k$, $A_k$, and $P_k$ denote the side, area, and perimeter, respectively, of the square $S_k$. The square $S_{k+1}$ is constructed from $S_k$ by connecting four points on $S_k$, with each point a distance of $\tfrac{1}{4}a_k$ from a vertex, as shown in the second figure.

(a) Find the relationship between $a_{k+1}$ and $a_k$.

(b) Find $a_n$, $A_n$, and $P_n$.

(c) Calculate $\displaystyle\sum_{n=1}^{\infty} P_n$.

**EXERCISE 59**

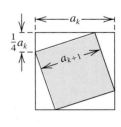

**60** The figure shows several terms of a sequence consisting of alternating circles and squares. Each circle is inscribed in a square, and each square (excluding the largest) is inscribed in a circle. Let $S_n$ denote the area of the $n$th square and $C_n$ the area of the $n$th circle.

   **(a)** Find the relationships between $S_n$ and $C_n$ and between $C_n$ and $S_{n+1}$.

   **(b)** What portion of the largest square is shaded in the figure?

EXERCISE 60

**61  Compound interest**  If a deposit of $100 is made on the first day of each month into an account that pays 6% interest compounded monthly, determine the amount in the account after 18 years.

**62  Compound interest**  Refer to Exercise 61. Show that if the monthly deposit is $P$ dollars and the rate is $r\%$ compounded monthly, then the amount $A$ in the account after $n$ months is given by

$$A = P\left(\frac{12}{r} + 1\right)\left[\left(1 + \frac{r}{12}\right)^n - 1\right].$$

**63  Compound interest**  Use Exercise 62 to find $A$ if the monthly deposit $P = \$100$, $r = 8\%$, and $n = 60$.

**64  Compound interest**  Refer to Exercise 62. If $r = 10\%$, approximately how many years are required to accumulate $100,000 if the monthly deposit $P$ is

   **(a)** $100           **(b)** $200

**Exer. 65–66:** The *double-declining balance method* is a method of depreciation in which, for each year $k = 1, 2, 3, \ldots, n$, the value of an asset is decreased by the fraction $A_k = \dfrac{2}{n}\left(1 - \dfrac{2}{n}\right)^{k-1}$ of its initial cost.

**65  (a)** If $n = 5$, find $A_1, A_2, \ldots, A_5$.

   **(b)** Show that the sequence in (a) is geometric, and find $S_5$.

   **(c)** If the initial value of an asset is $25,000, how much of its value has been depreciated after 2 years?

**66  (a)** If $n$ is any positive integer, find $A_1, A_2, \ldots, A_n$.

   **(b)** Show that the sequence in (a) is geometric, and find $S_n$.

## 9.4    MATHEMATICAL INDUCTION

If $n$ is a positive integer and we let $P_n$ denote the mathematical statement $(xy)^n = x^n y^n$, we obtain the following *infinite sequence of statements:*

Statement $P_1$:  $(xy)^1 = x^1 y^1$

Statement $P_2$:  $(xy)^2 = x^2 y^2$

Statement $P_3$:  $(xy)^3 = x^3 y^3$

$$\vdots \qquad\qquad \vdots$$

Statement $P_n$:  $(xy)^n = x^n y^n$

$$\vdots \qquad\qquad \vdots$$

It is easy to show that $P_1$, $P_2$, and $P_3$ are *true* statements. However, it is impossible to check the validity of $P_n$ for *every* positive integer $n$. Showing that $P_n$ is true for every $n$ requires the following principle.

**Principle of Mathematical Induction**

> If with each positive integer $n$ there is associated a statement $P_n$, then all the statements $P_n$ are true, provided the following two conditions are satisfied:
>
> **(1)** $P_1$ is true.
> **(2)** Whenever $k$ is a positive integer such that $P_k$ is true, then $P_{k+1}$ is also true.

To help understand this principle, consider an infinite sequence of statements labeled

$$P_1, P_2, P_3, \ldots, P_n, \ldots$$

that satisfy conditions (1) and (2). By (1), statement $P_1$ is true. Since condition (2) holds, whenever a statement $P_k$ is true the *next* statement $P_{k+1}$ is also true. Hence, since $P_1$ is true, $P_2$ is also true, by (2). However, if $P_2$ is true, then, by (2), we see that the next statement $P_3$ is true. Once again, if $P_3$ is true, then, by (2), $P_4$ is also true. If we continue in this manner, we can argue that if $n$ is any *particular* integer, then $P_n$ is true, since we can use condition (2) one step at a time, eventually reaching $P_n$. Although this type of reasoning does not actually *prove* the principle of mathematical induction, it certainly makes it plausible. The principle is proved in advanced algebra using postulates for the positive integers.

When applying the principle of mathematical induction, we always follow two steps.

**Steps in Applying the Principle of Mathematical Induction**

> **1** Show that $P_1$ is true.
> **2** *Assume* that $P_k$ is true, and then prove that $P_{k+1}$ is true.

Step 2 often causes confusion. Note that we do not *prove* that $P_k$ is true (except for $k = 1$). Instead, we show that *if* $P_k$ happens to be true, then the statement $P_{k+1}$ is also true. We refer to the assumption that $P_k$ is true as the **induction hypothesis.**

**EXAMPLE 1    *Using the principle of mathematical induction***

Use mathematical induction to prove that for every positive integer $n$, the sum of the first $n$ positive integers is

$$\frac{n(n+1)}{2}.$$

**SOLUTION**    If $n$ is any positive integer, let $P_n$ denote the statement

$$1 + 2 + 3 + \cdots + n = \frac{n(n+1)}{2}.$$

The following are some special cases of $P_n$:
  If $n = 1$, then $P_1$ is

$$1 = \frac{1(1+1)}{2}; \quad \text{that is,} \quad 1 = 1.$$

If $n = 2$, then $P_2$ is

$$1 + 2 = \frac{2(2+1)}{2}; \quad \text{that is,} \quad 3 = 3.$$

If $n = 3$, then $P_3$ is

$$1 + 2 + 3 = \frac{3(3+1)}{2}; \quad \text{that is,} \quad 6 = 6.$$

If $n = 4$, then $P_4$ is

$$1 + 2 + 3 + 4 = \frac{4(4+1)}{2}; \quad \text{that is,} \quad 10 = 10.$$

Although it is instructive to check the validity of $P_n$ for several values of $n$ as we have done, it is unnecessary to do so. We need only apply the two-step process outlined prior to this example. Thus, we proceed as follows:

**Step 1**  If we substitute $n = 1$ in $P_n$, then the left-hand side contains only the number 1 and the right-hand side is $\dfrac{1(1+1)}{2}$, which also equals 1. Hence, $P_1$ is true.

**Step 2**  Assume that $P_k$ is true. Thus, the induction hypothesis is

$$1 + 2 + 3 + \cdots + k = \frac{k(k+1)}{2}.$$

Our goal is to show that $P_{k+1}$ is true—that is,

$$1 + 2 + 3 + \cdots + (k+1) = \frac{(k+1)[(k+1)+1]}{2}.$$

We may prove that the last formula is true by rewriting the left side and using the induction hypothesis as follows:

$$1 + 2 + 3 + \cdots + (k + 1) = (1 + 2 + 3 + \cdots + k) + (k + 1) \qquad \text{group the first } k \text{ terms}$$

$$= \frac{k(k + 1)}{2} + (k + 1) \qquad \text{induction hypothesis}$$

$$= \frac{k(k + 1) + 2(k + 1)}{2} \qquad \text{add}$$

$$= \frac{(k + 1)(k + 2)}{2} \qquad \text{factor out } k + 1$$

$$= \frac{(k + 1)[(k + 1) + 1]}{2} \qquad \text{change form of } k + 2$$

This shows that $P_{k+1}$ is true, and therefore the proof by mathematical induction is complete.

---

**EXAMPLE 2** *Using the principle of mathematical induction*

Prove that for each positive integer $n$,

$$1^2 + 3^2 + \cdots + (2n - 1)^2 = \frac{n(2n - 1)(2n + 1)}{3}.$$

**SOLUTION** For each positive integer $n$, let $P_n$ denote the given statement. Note that this is a formula for the sum of the squares of the first $n$ odd positive integers. We again follow the two-step procedure.

**Step 1** Substituting 1 for $n$ in $P_n$, we obtain

$$1^2 = \frac{(1)(2 - 1)(2 + 1)}{3} = \frac{3}{3} = 1.$$

This shows that $P_1$ is true.

**Step 2** Assume that $P_k$ is true. Thus, the induction hypothesis is

$$1^2 + 3^2 + \cdots + (2k - 1)^2 = \frac{k(2k - 1)(2k + 1)}{3}.$$

We wish to show that $P_{k+1}$ is true—that is,

$$1^2 + 3^2 + \cdots + [2(k + 1) - 1]^2 = \frac{(k + 1)[2(k + 1) - 1][2(k + 1) + 1]}{3}.$$

This equation simplifies to

$$1^2 + 3^2 + \cdots + (2k + 1)^2 = \frac{(k + 1)(2k + 1)(2k + 3)}{3}.$$

*(continued)*

Remember that the next to last term on the left-hand side of the equation is $(2k - 1)^2$. In a manner similar to that used in the solution of Example 1, we may prove the formula for $P_{k+1}$ by rewriting the left side and using the induction hypothesis as follows:

$$1^2 + 3^2 + \cdots + (2k + 1)^2 = [1^2 + 3^2 + \cdots + (2k - 1)^2] + (2k + 1)^2 \quad \text{group the first } k \text{ terms}$$

$$= \frac{k(2k - 1)(2k + 1)}{3} + (2k + 1)^2 \quad \text{induction hypothesis}$$

$$= \frac{k(2k - 1)(2k + 1) + 3(2k + 1)^2}{3} \quad \text{add}$$

$$= \frac{(2k + 1)[k(2k - 1) + 3(2k + 1)]}{3} \quad \text{factor out } 2k + 1$$

$$= \frac{(2k + 1)(2k^2 + 5k + 3)}{3} \quad \text{simplify}$$

$$= \frac{(k + 1)(2k + 1)(2k + 3)}{3} \quad \text{factor and change order}$$

This shows that $P_{k+1}$ is true, and hence $P_n$ is true for every $n$.

---

**EXAMPLE 3**    *Using the principle of mathematical induction*
Prove that 2 is a factor of $n^2 + 5n$ for every positive integer $n$.

**SOLUTION**    For each positive integer $n$, let $P_n$ denote the following statement:

$$2 \text{ is a factor of } n^2 + 5n$$

We shall follow the two-step procedure.

**Step 1**  If $n = 1$, then

$$n^2 + 5n = 1^2 + 5 \cdot 1 = 6 = 2 \cdot 3.$$

Thus, 2 is a factor of $n^2 + 5n$ for $n = 1$; that is, $P_1$ is true.

**Step 2**  Assume that $P_k$ is true. Thus, the induction hypothesis is

$$2 \text{ is a factor of } k^2 + 5k$$

or, equivalently,                $k^2 + 5k = 2p$

for some integer $p$.

We wish to show that $P_{k+1}$ is true—that is,

$$2 \text{ is a factor of } (k + 1)^2 + 5(k + 1).$$

We may do this as follows:

$$(k + 1)^2 + 5(k + 1) = k^2 + 2k + 1 + 5k + 5 \qquad \text{multiply}$$
$$= (k^2 + 5k) + (2k + 6) \qquad \text{rearrange terms}$$
$$= 2p + 2(k + 3) \qquad \text{induction hypothesis and factor } 2k + 6$$
$$= 2(p + k + 3) \qquad \text{factor out 2}$$

Since 2 is a factor of the last expression, $P_{k+1}$ is true, and hence $P_n$ is true for every $n$.

---

Let $j$ be a positive integer, and suppose that with each integer $n \geq j$ there is associated a statement $P_n$. For example, if $j = 6$, then the statements are numbered $P_6, P_7, P_8, \ldots$. The principle of mathematical induction may be extended to cover this situation. To prove that the statements $P_n$ are true for $n \geq j$, we use the following two steps, in the same manner as we did for $n \geq 1$.

**Steps in Applying the Extended Principle of Mathematical Induction for $P_k$, $k \geq j$**

1  Show that $P_j$ is true.
2  Assume that $P_k$ is true with $k \geq j$, and prove that $P_{k+1}$ is true.

**EXAMPLE 4**  *Using the extended principle of mathematical induction*

Let $a$ be a nonzero real number such that $a > -1$. Prove that

$$(1 + a)^n > 1 + na$$

for every integer $n \geq 2$.

**SOLUTION**  For each positive integer $n$, let $P_n$ denote the inequality $(1 + a)^n > 1 + na$. Note that $P_1$ is *false*, since $(1 + a)^1 = 1 + (1)(a)$. However, we can show that $P_n$ is true for $n \geq 2$ by using the extended principle with $j = 2$.

**Step 1**  We first note that $(1 + a)^2 = 1 + 2a + a^2$. Since $a \neq 0$, we have $a^2 > 0$, and therefore $1 + 2a + a^2 > 1 + 2a$, or $(1 + a)^2 > 1 + 2a$. Hence, $P_2$ is true.

**Step 2**  Assume that $P_k$ is true. Thus, the induction hypothesis is

$$(1 + a)^k > 1 + ka.$$

We wish to show that $P_{k+1}$ is true—that is,

$$(1 + a)^{k+1} > 1 + (k + 1)a.$$

*(continued)*

To prove the last inequality, we first observe the following:

$$(1 + a)^{k+1} = (1 + a)^k(1 + a) \quad \text{law of exponents}$$
$$> (1 + ka)(1 + a) \quad \text{induction hypothesis and } 1 + a > 0$$

We next note that

$$(1 + ka)(1 + a) = 1 + ka + a + ka^2 \quad \text{multiply}$$
$$= 1 + (ka + a) + ka^2 \quad \text{group terms}$$
$$= 1 + (k + 1)a + ka^2 \quad \text{factor out } a$$
$$> 1 + (k + 1)a. \quad \text{since } ka^2 > 0$$

The last two inequalities give us

$$(1 + a)^{k+1} > 1 + (k + 1)a.$$

Thus, $P_{k+1}$ is true, and the proof by mathematical induction is complete.

## 9.4 EXERCISES

**Exer. 1–26: Prove that the statement is true for every positive integer $n$.**

1  $2 + 4 + 6 + \cdots + 2n = n(n + 1)$

2  $1 + 4 + 7 + \cdots + (3n - 2) = \dfrac{n(3n - 1)}{2}$

3  $1 + 3 + 5 + \cdots + (2n - 1) = n^2$

4  $3 + 9 + 15 + \cdots + (6n - 3) = 3n^2$

5  $2 + 7 + 12 + \cdots + (5n - 3) = \frac{1}{2}n(5n - 1)$

6  $2 + 6 + 18 + \cdots + 2 \cdot 3^{n-1} = 3^n - 1$

7  $1 + 2 \cdot 2 + 3 \cdot 2^2 + \cdots + n \cdot 2^{n-1} = 1 + (n - 1) \cdot 2^n$

8  $(-1)^1 + (-1)^2 + (-1)^3 + \cdots + (-1)^n = \dfrac{(-1)^n - 1}{2}$

9  $1^2 + 2^2 + 3^2 + \cdots + n^2 = \dfrac{n(n + 1)(2n + 1)}{6}$

10  $1^3 + 2^3 + 3^3 + \cdots + n^3 = \left[\dfrac{n(n + 1)}{2}\right]^2$

11  $\dfrac{1}{1 \cdot 2} + \dfrac{1}{2 \cdot 3} + \dfrac{1}{3 \cdot 4} + \cdots + \dfrac{1}{n(n + 1)} = \dfrac{n}{n + 1}$

12  $\dfrac{1}{1 \cdot 2 \cdot 3} + \dfrac{1}{2 \cdot 3 \cdot 4} + \dfrac{1}{3 \cdot 4 \cdot 5} + \cdots + \dfrac{1}{n(n + 1)(n + 2)} = \dfrac{n(n + 3)}{4(n + 1)(n + 2)}$

13  $3 + 3^2 + 3^3 + \cdots + 3^n = \frac{3}{2}(3^n - 1)$

14  $1^3 + 3^3 + 5^3 + \cdots + (2n - 1)^3 = n^2(2n^2 - 1)$

15  $n < 2^n$

16  $1 + 2n \leq 3^n$

17  $1 + 2 + 3 + \cdots + n < \frac{1}{8}(2n + 1)^2$

18  If $0 < a < b$, then $\left(\dfrac{a}{b}\right)^{n+1} < \left(\dfrac{a}{b}\right)^n$.

19  3 is a factor of $n^3 - n + 3$.

20  2 is a factor of $n^2 + n$.

21  4 is a factor of $5^n - 1$.

22  9 is a factor of $10^{n+1} + 3 \cdot 10^n + 5$.

23  If $a$ is greater than 1, then $a^n > 1$.

24  If $r \neq 1$, then
$$a + ar + ar^2 + \cdots + ar^{n-1} = \dfrac{a(1 - r^n)}{1 - r}.$$

**25** $a - b$ is a factor of $a^n - b^n$.
(*Hint:* $a^{k+1} - b^{k+1} = a^k(a - b) + (a^k - b^k)b$.)

**26** $a + b$ is a factor of $a^{2n-1} + b^{2n-1}$.

**Exer. 27–32:** Find the smallest positive integer $j$ for which the statement is true. Use the extended principle of mathematical induction to prove that the formula is true for every integer greater than $j$.

**27** $n + 12 \leq n^2$

**28** $n^2 + 18 \leq n^3$

**29** $5 + \log_2 n \leq n$

**30** $n^2 \leq 2^n$

**31** $2n + 2 \leq 2^n$

**32** $n \log_2 n + 20 \leq n^2$

**Exer. 33–36:** Express the sum in terms of $n$.

**33** $\displaystyle\sum_{k=1}^{n} (k^2 + 3k + 5)$

(*Hint:* Use the theorem on sums to write the sum as

$$\sum_{k=1}^{n} k^2 + 3 \sum_{k=1}^{n} k + \sum_{k=1}^{n} 5.$$

Next use Exercise 9, Example 4 of Section 9.1, and the theorem on the sum of a constant.)

**34** $\displaystyle\sum_{k=1}^{n} (3k^2 - 2k + 1)$

**35** $\displaystyle\sum_{k=1}^{n} (2k - 3)^2$

**36** $\displaystyle\sum_{k=1}^{n} (k^3 + 2k^2 - k + 4)$ (*Hint:* Use Exercise 10.)

**Exer. 37–38:** (a) Evaluate the given formula for the stated values of $n$, and solve the resulting system of equations for $a$, $b$, $c$, and $d$. (This method can sometimes be used to obtain formulas for sums.) (b) Compare the result in part (a) with the indicated exercise, and explain why this method does not prove that the formula is true for every $n$.

**37** $1^2 + 2^2 + 3^2 + \cdots + n^2 = an^3 + bn^2 + cn$;
$n = 1, 2, 3$ (Exercise 9)

**38** $1^3 + 2^3 + 3^3 + \cdots + n^3 = an^4 + bn^3 + cn^2 + dn$;
$n = 1, 2, 3, 4$ (Exercise 10)

**Exer. 39–40:** (a) Use the method of Exercises 37–38 to find a formula for the sum. (b) Verify that the formula found in part (a) is true for every $n$.

**39** $1^4 + 2^4 + 3^4 + \cdots + n^4$

**40** $2^3 + 4^3 + 6^3 + \cdots + (2n)^3$

**Exer. 41–44:** Prove that the statement is true for every positive integer $n$.

**41** $\sin(\theta + n\pi) = (-1)^n \sin\theta$

**42** $\cos(\theta + n\pi) = (-1)^n \cos\theta$

**43** Prove De Moivre's theorem:
$$[r(\cos\theta + i\sin\theta)]^n = r^n(\cos n\theta + i\sin n\theta)$$
for every positive integer $n$.

**44** Prove that for every positive integer $n \geq 3$, the sum of the interior angles of an $n$-sided polygon is given by $(n - 2) \cdot 180°$.

# 9.5 THE BINOMIAL THEOREM

A **binomial** is a sum $a + b$, where $a$ and $b$ represent numbers. If $n$ is a positive integer, then a general formula for *expanding* $(a + b)^n$ (that is, for expressing it as a sum) is given by the **binomial theorem.** In this section we shall use mathematical induction to establish this general formula. The following special cases can be obtained by multiplication:

$$(a + b)^2 = a^2 + 2ab + b^2$$
$$(a + b)^3 = a^3 + 3a^2b + 3ab^2 + b^3$$
$$(a + b)^4 = a^4 + 4a^3b + 6a^2b^2 + 4ab^3 + b^4$$
$$(a + b)^5 = a^5 + 5a^4b + 10a^3b^2 + 10a^2b^3 + 5ab^4 + b^5$$

These expansions of $(a + b)^n$ for $n = 2, 3, 4$, and 5 have the following properties:

**(1)** There are $n + 1$ terms, the first being $a^n$ and the last $b^n$.

**(2)** As we proceed from any term to the next, the power of $a$ decreases by 1 and the power of $b$ increases by 1. For each term, the sum of the exponents of $a$ and $b$ is $n$.

**(3)** Each term has the form $(c)a^{n-k}b^k$, where the coefficient $c$ is a real number and $k = 0, 1, 2, \ldots, n$.

**(4)** The following formula is true for each of the first $n$ terms of the expansion:

$$\frac{(\text{coefficient of term}) \cdot (\text{exponent of } a)}{\text{number of term}} = \text{coefficient of next term}$$

The following table illustrates property 4 for the expansion of $(a + b)^5$.

| Term | Number of term | Coefficient of term | Exponent of $a$ | Coefficient of next term |
|---|---|---|---|---|
| $a^5$ | 1 | 1 | 5 | $\dfrac{1 \cdot 5}{1} = 5$ |
| $5a^4b$ | 2 | 5 | 4 | $\dfrac{5 \cdot 4}{2} = 10$ |
| $10a^3b^2$ | 3 | 10 | 3 | $\dfrac{10 \cdot 3}{3} = 10$ |
| $10a^2b^3$ | 4 | 10 | 2 | $\dfrac{10 \cdot 2}{4} = 5$ |
| $5ab^4$ | 5 | 5 | 1 | $\dfrac{5 \cdot 1}{5} = 1$ |

Let us next consider $(a + b)^n$ for an arbitrary positive integer $n$. The first term is $a^n$, which has coefficient 1. If we assume that property 4 is true, we obtain the successive coefficients listed in the next table.

| Term | Number of term | Coefficient of term | Exponent of $a$ | Coefficient of next term |
|---|---|---|---|---|
| $a^n$ | 1 | 1 | $n$ | $\dfrac{1 \cdot n}{1} = n$ |
| $\dfrac{n}{1}a^{n-1}b$ | 2 | $\dfrac{n}{1}$ | $n - 1$ | $\dfrac{n(n - 1)}{2 \cdot 1}$ |
| $\dfrac{n(n - 1)}{2 \cdot 1}a^{n-2}b^2$ | 3 | $\dfrac{n(n - 1)}{2 \cdot 1}$ | $n - 2$ | $\dfrac{n(n - 1)(n - 2)}{3 \cdot 2 \cdot 1}$ |
| $\dfrac{n(n - 1)(n - 2)}{3 \cdot 2 \cdot 1}a^{n-3}b^3$ | 4 | $\dfrac{n(n - 1)(n - 2)}{3 \cdot 2 \cdot 1}$ | $n - 3$ | $\dfrac{n(n - 1)(n - 2)(n - 3)}{4 \cdot 3 \cdot 2 \cdot 1}$ |

The pattern that appears in the fifth column leads to the following formula for the coefficient of the general term.

**Coefficient of the (k + 1)st Term in the Expansion of $(a + b)^n$**

$$\frac{n \cdot (n - 1) \cdot (n - 2) \cdot (n - 3) \cdots \cdot (n - k + 1)}{k \cdot (k - 1) \cdots \cdot 3 \cdot 2 \cdot 1}, \quad k = 1, 2, \ldots, n$$

The $(k + 1)$st coefficient can be written in a compact form by using **factorial notation.** If $n$ is any nonnegative integer, then the symbol $n!$ (*n factorial*) is defined as follows.

**Definition of $n!$**

(1) $n! = n(n - 1)(n - 2) \cdots 1$  if  $n > 0$

(2) $0! = 1$

Thus, if $n > 0$, then $n!$ is the product of the first $n$ positive integers. The definition $0! = 1$ is used so that certain formulas involving factorials are true for all *nonnegative* integers.

**ILLUSTRATION**

**$n$ Factorial**

- $1! = 1$
- $2! = 2 \cdot 1 = 2$
- $3! = 3 \cdot 2 \cdot 1 = 6$
- $4! = 4 \cdot 3 \cdot 2 \cdot 1 = 24$

- $5! = 5 \cdot 4 \cdot 3 \cdot 2 \cdot 1 = 120$
- $6! = 6 \cdot 5 \cdot 4 \cdot 3 \cdot 2 \cdot 1 = 720$
- $7! = 7 \cdot 6 \cdot 5 \cdot 4 \cdot 3 \cdot 2 \cdot 1 = 5040$
- $8! = 8 \cdot 7 \cdot 6 \cdot 5 \cdot 4 \cdot 3 \cdot 2 \cdot 1 = 40,320$

Notice the rapid growth of $n!$ as $n$ increases. We can show with the aid of a calculator that

$$20! \approx 2.4 \times 10^{18} \quad \text{and} \quad 50! \approx 3.0 \times 10^{64}.$$

We sometimes wish to simplify quotients where both the numerator and the denominator contain factorials, as shown in the next illustration.

**ILLUSTRATION**

**Simplifying Quotients of Factorials**

- $\dfrac{7!}{5!} = \dfrac{7 \cdot 6 \cdot 5!}{5!} = 7 \cdot 6 = 42$

- $\dfrac{10!}{6!} = \dfrac{10 \cdot 9 \cdot 8 \cdot 7 \cdot 6!}{6!} = 10 \cdot 9 \cdot 8 \cdot 7 = 5040$

As in the preceding illustration, if $n$ and $k$ are positive integers and $k < n$, then

$$\frac{n!}{(n-k)!} = \frac{n \cdot (n-1) \cdot (n-2) \cdots (n-k+1) \cdot [(n-k)!]}{(n-k)!}$$

$$= n \cdot (n-1) \cdot (n-2) \cdots (n-k+1),$$

which is the numerator of the coefficient of the $(k+1)$st term of $(a+b)^n$. Dividing by the denominator $k!$ gives us the following alternative form for the $(k+1)$st coefficient:

$$\frac{n \cdot (n-1) \cdot (n-2) \cdots (n-k+1)}{k!} = \frac{n!}{k!(n-k)!}$$

These numbers are called **binomial coefficients** and are often denoted by either the symbol $\binom{n}{k}$ or the symbol $C(n, k)$. Thus, we have the following.

**Coefficient of the $(k+1)$st Term in the Expansion of $(a+b)^n$ (Alternative Form)**

$$\binom{n}{k} = C(n, k) = \frac{n!}{k!(n-k)!}, \quad k = 0, 1, 2, \ldots, n$$

The symbols $\binom{n}{k}$ and $C(n, k)$ are sometimes read "$n$ choose $k$."

**EXAMPLE 1    Evaluating $\binom{n}{k}$**

Find $\binom{5}{0}, \binom{5}{1}, \binom{5}{2}, \binom{5}{3}, \binom{5}{4},$ and $\binom{5}{5}$.

**SOLUTION**    These six numbers are the coefficients in the expansion of $(a+b)^5$, which we tabulated earlier in this section. By definition,

$$\binom{5}{0} = \frac{5!}{0!(5-0)!} = \frac{5!}{1 \cdot 5!} = 1$$

$$\binom{5}{1} = \frac{5!}{1!(5-1)!} = \frac{5!}{1 \cdot 4!} = \frac{5 \cdot 4!}{4!} = 5$$

$$\binom{5}{2} = \frac{5!}{2!(5-2)!} = \frac{5!}{2!3!} = \frac{5 \cdot 4 \cdot 3!}{2 \cdot 3!} = \frac{20}{2} = 10$$

$$\binom{5}{3} = \frac{5!}{3!(5-3)!} = \frac{5!}{3!2!} = \frac{5 \cdot 4 \cdot 3!}{3! \cdot 2} = \frac{20}{2} = 10$$

$$\binom{5}{4} = \frac{5!}{4!(5-4)!} = \frac{5!}{4!1!} = \frac{5 \cdot 4!}{4!} = 5$$

$$\binom{5}{5} = \frac{5!}{5!(5-5)!} = \frac{5!}{5!0!} = \frac{1}{0!} = \frac{1}{1} = 1.$$

The binomial theorem may be stated as follows.

## The Binomial Theorem

$$(a + b)^n = a^n + \binom{n}{1}a^{n-1}b + \binom{n}{2}a^{n-2}b^2 + \cdots + \binom{n}{k}a^{n-k}b^k + \cdots + \binom{n}{n-1}ab^{n-1} + b^n$$

Using summation notation, we may write the binomial theorem

$$(a + b)^n = \sum_{k=0}^{n} \binom{n}{k} a^{n-k}b^k.$$

An alternative statement of the binomial theorem is as follows. (A proof is given at the end of this section.)

## The Binomial Theorem (Alternative Form)

$$(a + b)^n = a^n + na^{n-1}b + \frac{n(n-1)}{2!}a^{n-2}b^2 + \cdots + \frac{n(n-1)(n-2)\cdots(n-k+1)}{k!}a^{n-k}b^k + \cdots + nab^{n-1} + b^n$$

The following examples may be solved either by using the general formulas for the binomial theorem or by repeated use of property 4 stated at the beginning of this section.

**EXAMPLE 2**   *Finding a binomial expansion*

Find the binomial expansion of $(2x + 3y^2)^4$.

**SOLUTION**   We use the binomial theorem with $a = 2x$, $b = 3y^2$, and $n = 4$:

$$(2x + 3y^2)^4 = (2x^4) + \binom{4}{1}(2x)^3(3y^2) + \binom{4}{2}(2x)^2(3y^2)^2 + \binom{4}{3}(2x)(3y^2)^3 + (3y^2)^4$$

$$= 16x^4 + 4(8x^3)(3y^2) + 6(4x^2)(9y^4) + 4(2x)(27y^6) + 81y^8$$

$$= 16x^4 + 96x^3y^2 + 216x^2y^4 + 216xy^6 + 81y^8$$

The next example illustrates that if either $a$ or $b$ is negative, then the terms of the expansion are alternately positive and negative.

**EXAMPLE 3**   *Finding a binomial expansion*

Expand $\left(\dfrac{1}{x} - 2\sqrt{x}\right)^5$.

**SOLUTION**    The binomial coefficients for $(a + b)^5$ were calculated in Example 1. Thus, if we let $a = 1/x$, $b = -2\sqrt{x}$, and $n = 5$ in the binomial theorem, we obtain

$$\left(\frac{1}{x} - 2\sqrt{x}\right)^5 = \left(\frac{1}{x}\right)^5 + 5\left(\frac{1}{x}\right)^4(-2\sqrt{x}) + 10\left(\frac{1}{x}\right)^3(-2\sqrt{x})^2$$

$$+ 10\left(\frac{1}{x}\right)^2(-2\sqrt{x})^3 + 5\left(\frac{1}{x}\right)(-2\sqrt{x})^4 + (-2\sqrt{x})^5,$$

which can be written as

$$\left(\frac{1}{x} - 2\sqrt{x}\right)^5 = \frac{1}{x^5} - \frac{10}{x^{7/2}} + \frac{40}{x^2} - \frac{80}{x^{1/2}} + 80x - 32x^{5/2}.$$

To find a specific term in the expansion of $(a + b)^n$, it is convenient to first find the exponent $k$ that is to be assigned to $b$. Notice that, by the binomial theorem, *the exponent of $b$ is always one less than the number of the term.* Once $k$ is found, we know that the exponent of $a$ is $n - k$ and the coefficient is $\binom{n}{k}$.

**EXAMPLE 4**    *Finding a specific term of a binomial expansion*
Find the fifth term in the expansion of $(x^3 + \sqrt{y})^{13}$.

**SOLUTION**    Let $a = x^3$ and $b = \sqrt{y}$. The exponent of $b$ in the fifth term is 4, and hence the exponent of $a$ is 9. From the discussion of the preceding paragraph we obtain

$$\binom{13}{4}(x^3)^9(\sqrt{y})^4 = \frac{13!}{4!(13 - 4)!}x^{27}y^2 = \frac{13 \cdot 12 \cdot 11 \cdot 10}{4!}x^{27}y^2 = 715x^{27}y^2.$$

**EXAMPLE 5**    *Finding a specific term of a binomial expansion*
Find the term involving $q^{10}$ in the binomial expansion of $(\frac{1}{3}p + q^2)^{12}$.

**SOLUTION**    From the statement of the binomial theorem with $a = \frac{1}{3}p$, $b = q^2$, and $n = 12$, each term in the expansion has the form

$$\binom{n}{k}a^{n-k}b^k = \binom{12}{k}\left(\frac{1}{3}p\right)^{12-k}(q^2)^k.$$

Since $(q^2)^k = q^{2k}$, we must let $k = 5$ to obtain the term involving $q^{10}$. Doing so gives us

$$\binom{12}{5}\left(\frac{1}{3}p\right)^{12-5}(q^2)^5 = \frac{12!}{5!(12-5)!}\left(\frac{1}{3}\right)^7 p^7 q^{10} = \frac{88}{243}p^7 q^{10}.$$

There is an interesting triangular array of numbers called **Pascal's triangle,** which can be used to obtain binomial coefficients. The numbers are arranged as follows:

```
                    1
                 1     1
              1     2     1
           1     3     3     1
        1     4     6     4     1
     1     5    10    10     5     1
  1     6    15    20    15     6     1
  .     .     .     .     .     .     .     .
     .     .     .     .     .     .     .
```

The numbers in the second row are the coefficients in the expansion of $(a + b)^1$; those in the third row are the coefficients determined by $(a + b)^2$; those in the fourth row are obtained from $(a + b)^3$; and so on. Each number in the array that is different from 1 can be found by adding the two numbers in the previous row that appear above and immediately to the left and right of the number, as illustrated in the solution of the next example.

**EXAMPLE 6**    *Using Pascal's triangle*

Find the eighth row of Pascal's triangle, and use it to expand $(a + b)^7$.

*SOLUTION*    Let us rewrite the seventh row and then use the process described above. In the following display the arrows indicate which two numbers in row seven are added to obtain the numbers in row eight.

```
     1     6     15     20     15     6     1
  1     7     21     35     35     21     7     1
```

The eighth row gives us the coefficients in the expansion of $(a + b)^7$:

$$(a + b)^7 = a^7 + 7a^6 b + 21a^5 b^2 + 35a^4 b^3 + 35a^3 b^4 + 21a^2 b^5 + 7ab^6 + b^7$$

Pascal's triangle is useful for expanding small powers of $a + b$; however, for expanding large powers or finding a specific term, as in Examples 4 and 5, the general formula given by the binomial theorem is more useful.

We shall conclude this section by giving a proof of the binomial theorem using mathematical induction.

**PROOF OF THE BINOMIAL THEOREM**    For each positive integer $n$, let $P_n$ denote the statement given in the alternative form of the binomial theorem.

**Step 1**    If $n = 1$, the statement reduces to $(a + b)^1 = a^1 + b^1$. Consequently, $P_1$ is true.

**Step 2**    Assume that $P_k$ is true. Thus, the induction hypothesis is

$$(a + b)^k = a^k + ka^{k-1}b + \frac{k(k-1)}{2!}a^{k-2}b^2 + \cdots + \frac{k(k-1)(k-2)\cdots(k-r+2)}{(r-1)!}a^{k-r+1}b^{r-1}$$

$$+ \frac{k(k-1)(k-2)\cdots(k-r+1)}{r!}a^{k-r}b^r + \cdots + kab^{k-1} + b^k.$$

We have shown both the $r$th and the $(r + 1)$st term in the expansion. To prove that $P_{k+1}$ is true, we first write

$$(a + b)^{k+1} = (a + b)^k(a + b).$$

Using the induction hypothesis to substitute for $(a + b)^k$ and then multiplying that expression by $a + b$, we obtain

$$(a + b)^{k+1} = \left[ a^{k+1} + ka^k b + \frac{k(k-1)}{2!}a^{k-1}b^2 + \cdots + \frac{k(k-1)\cdots(k-r+1)}{r!}a^{k-r+1}b^r + \cdots + ab^k \right]$$

$$+ \left[ a^k b + ka^{k-1}b^2 + \cdots + \frac{k(k-1)\cdots(k-r+2)}{(r-1)!}a^{k-r+1}b^r + \cdots + kab^k + b^{k+1} \right],$$

where the terms in the first pair of brackets result from multiplying the right side of the induction hypothesis by $a$ and the terms in the second pair of brackets result from multiplying by $b$. We next rearrange and combine terms:

$$(a + b)^{k+1} = a^{k+1} + (k + 1)a^k b + \left[ \frac{k(k-1)}{2!} + k \right]a^{k-1}b^2 + \cdots$$

$$+ \left[ \frac{k(k-1)\cdots(k-r+1)}{r!} + \frac{k(k-1)\cdots(k-r+2)}{(r-1)!} \right]a^{k-r+1}b^r$$

$$+ \cdots + (1 + k)ab^k + b^{k+1}$$

If the coefficients are simplified, we obtain statement $P_n$ with $k + 1$ substituted for $n$. Thus, $P_{k+1}$ is true, and therefore $P_n$ holds for every positive integer $n$, which completes the proof. ■

## 9.5 EXERCISES

Exer. 1–12: Evaluate the expression.

| | |
|---|---|
| 1  $2!6!$ | 2  $3!4!$ |
| 3  $7!0!$ | 4  $5!0!$ |
| 5  $\dfrac{8!}{5!}$ | 6  $\dfrac{6!}{3!}$ |
| 7  $\dbinom{5}{5}$ | 8  $\dbinom{7}{0}$ |
| 9  $\dbinom{7}{5}$ | 10  $\dbinom{8}{4}$ |
| 11  $\dbinom{13}{4}$ | 12  $\dbinom{52}{2}$ |

Exer. 13–14: Rewrite as an expression that does not contain factorials.

13  $\dfrac{(2n+2)!}{(2n)!}$

14  $\dfrac{(3n+1)!}{(3n-1)!}$

Exer. 15–28: Use the binomial theorem to expand and simplify.

| | |
|---|---|
| 15  $(4x-y)^3$ | 16  $(x^2+2y)^3$ |
| 17  $(a+b)^6$ | 18  $(a+b)^4$ |
| 19  $(a-b)^7$ | 20  $(a-b)^5$ |
| 21  $(3x-5y)^4$ | 22  $(2t-s)^5$ |
| 23  $(\tfrac{1}{3}x+y^2)^5$ | 24  $(\tfrac{1}{2}c+d^3)^4$ |
| 25  $\left(\dfrac{1}{x^2}+3x\right)^6$ | 26  $\left(\dfrac{1}{x^3}-2x\right)^5$ |
| 27  $\left(\sqrt{x}-\dfrac{1}{\sqrt{x}}\right)^5$ | 28  $\left(\sqrt{x}+\dfrac{1}{\sqrt{x}}\right)^5$ |

Exer. 29–44: Without expanding completely, find the indicated term(s) in the expansion of the expression.

29  $(3c^{2/5}+c^{4/5})^{25}$;  first three terms

30  $(x^3+5x^{-2})^{20}$;  first three terms

31  $(4b^{-1}-3b)^{15}$;  last three terms

32  $(s-2t^3)^{12}$;  last three terms

33  $\left(\dfrac{3}{c}+\dfrac{c^2}{4}\right)^7$;  sixth term

34  $(3a^2-\sqrt{b})^9$;  fifth term

35  $(\tfrac{1}{3}u+4v)^8$;  seventh term

36  $(3x^2-y^3)^{10}$;  fourth term

37  $(x^{1/2}+y^{1/2})^8$;  middle term

38  $(rs^2+t)^7$;  two middle terms

39  $(2y+x^2)^8$;  term that contains $x^{10}$

40  $(x^2-2y^3)^5$;  term that contains $y^6$

41  $(3b^3-2a^2)^4$;  term that contains $b^9$

42  $(\sqrt{c}+\sqrt{d})^8$;  term that contains $c^2$

43  $\left(3x-\dfrac{1}{4x}\right)^6$;  term that does not contain $x$

44  $(xy-3y^{-3})^8$;  term that does not contain $y$

45  Approximate $(1.2)^{10}$ by using the first three terms in the expansion of $(1+0.2)^{10}$, and compare your answer with that obtained using a calculator.

46  Approximate $(0.9)^4$ by using the first three terms in the expansion of $(1-0.1)^4$, and compare your answer with that obtained using a calculator.

Exer. 47–48: Simplify the expression using the binomial theorem.

47  $\dfrac{(x+h)^4-x^4}{h}$

48  $\dfrac{(x+h)^5-x^5}{h}$

49  Show that $\dbinom{n}{1}=\dbinom{n}{n-1}$ for $n\geq 1$.

50  Show that $\dbinom{n}{0}=\dbinom{n}{n}$ for $n\geq 0$.

# 9.6   PERMUTATIONS

**FIGURE 3**

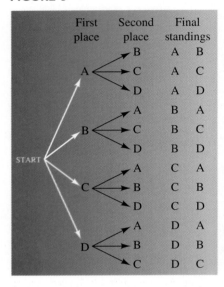

Suppose that four teams are involved in a tournament in which first, second, third, and fourth places will be determined. For identification purposes, we label the teams A, B, C, and D. Let us find the number of different ways that first and second place can be decided. It is convenient to use a **tree diagram** as in Figure 3. After the word START, the four possibilities for first place are listed. From each of these an arrow points to a possible second-place finisher. The final standings list the possible outcomes, from left to right. They are found by following the different paths (*branches* of the tree) that lead from the word START to the second-place team. The total number of outcomes is 12, which is the product of the number of choices (4) for first place and the number of choices (3) for second place (after first has been determined).

Let us now find the total number of ways that first, second, third, and fourth positions can be filled. To sketch a tree diagram, we may begin by drawing arrows from the word START to each possible first-place finisher A, B, C, or D. Next we draw arrows from those to possible second-place finishers, as was done in Figure 3. From each second-place position we then draw arrows indicating the possible third-place positions. Finally, we draw arrows to the fourth-place team. If we consider only the case in which team A finishes in first place, we have the diagram shown in Figure 4.

**FIGURE 4**

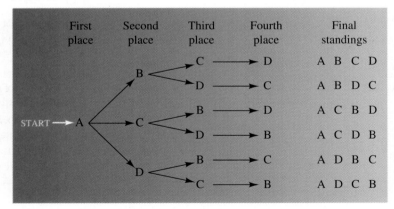

Note that there are six possible final standings in which team A occupies first place. In a complete tree diagram there would also be three other branches of this type corresponding to first-place finishes for B, C, and D.

A complete diagram would display the following 24 possibilities for the final standings:

ABCD, ABDC, ACBD, ACDB, ADBC, ADCB,

BACD, BADC, BCAD, BCDA, BDAC, BDCA,

CABD, CADB, CBAD, CBDA, CDAB, CDBA,

DABC, DACB, DBAC, DBCA, DCAB, DCBA.

Note that the number of possibilities (24) is the product of the number of ways (4) that first place may occur, the number of ways (3) that second place may occur (after first place has been determined), the number of possible outcomes (2) for third place (after the first two places have been decided), and the number of ways (1) that fourth place can occur (after the first three places have been taken).

The preceding discussion illustrates the following general rule, which we accept as a basic axiom of counting.

**Fundamental Counting Principle**

> Let $E_1, E_2, \ldots, E_k$ be a sequence of $k$ events. If, for each $i$, the event $E_i$ can occur in $m_i$ ways, then the total number of ways all the events may take place is the product $m_1 m_2 \cdots m_k$.

Returning to our first illustration, we let $E_1$ represent the determination of the first-place team, so that $m_1 = 4$. If $E_2$ denotes the determination of the second-place team, then $m_2 = 3$. Hence, the number of outcomes for the sequence $E_1, E_2$ is $4 \cdot 3 = 12$, which is the same as that found by means of the tree diagram. If we proceed to $E_3$, the determination of the third-place team, then $m_3 = 2$, and hence $m_1 m_2 m_3 = 24$. Finally, if $E_1, E_2$, and $E_3$ have occurred, there is only one possible outcome for $E_4$. Thus, $m_4 = 1$ and $m_1 m_2 m_3 m_4 = 24$.

Instead of teams, let us now regard $a$, $b$, $c$, and $d$ merely as symbols and consider the various *orderings*, or *arrangements*, that may be assigned to these symbols, taking them either two at a time, three at a time, or four at a time. By abstracting in this way we may apply our methods to other similar situations. The arrangements we have discussed are **arrangements without repetitions,** since a symbol may not be used more than once in an arrangement. In Example 1 we shall consider arrangements in which repetitions *are* allowed.

Previously we defined ordered pairs and ordered triples. Similarly, an *ordered 4-tuple* is a set containing four elements $x_1, x_2, x_3, x_4$ in which an ordering has been specified, so that one of the elements may be referred to as the *first element,* another as the *second element,* and so on. The symbol $(x_1, x_2, x_3, x_4)$ is used for the ordered 4-tuple having first element $x_1$, second element $x_2$, third element $x_3$, and fourth element $x_4$. In general, for any positive integer $r$, we speak of the **ordered $r$-tuple**

$$(x_1, x_2, \ldots, x_r)$$

as a set of elements in which $x_1$ is designated as the first element, $x_2$ as the second element, and so on.

### EXAMPLE 1    *Determining the number of r-tuples*

Using only the letters $a$, $b$, $c$, and $d$, determine how many of the following can be obtained:

**(a)** ordered triples     **(b)** ordered 4-tuples     **(c)** ordered $r$-tuples

### SOLUTION

**(a)** We must determine the number of symbols of the form $(x_1, x_2, x_3)$ that can be obtained using only the letters $a$, $b$, $c$, and $d$. This is not the same as listing first, second, and third place as in our previous illustration, since we have not ruled out the possibility of repetitions. For example, $(a, b, a)$, $(a, a, b)$, and $(a, a, a)$ are different ordered triples. If, for $i = 1, 2, 3$, we let $E_i$ represent the determination of $x_i$ in the ordered triple $(x_1, x_2, x_3)$, then, since repetitions are allowed, there are four possibilities—$a$, $b$, $c$, and $d$—for each of $E_1$, $E_2$, and $E_3$. Hence, by the fundamental counting principle, the total number of ordered triples is $4 \cdot 4 \cdot 4$, or 64.

**(b)** The number of possible ordered 4-tuples $(x_1, x_2, x_3, x_4)$ is $4 \cdot 4 \cdot 4 \cdot 4$, or 256.

**(c)** The number of ordered $r$-tuples is the product $4 \cdot 4 \cdot 4 \cdots\cdots 4$, with 4 appearing as a factor $r$ times. That product equals $4^r$.

### EXAMPLE 2    *Choosing class officers*

A class consists of 60 girls and 40 boys. In how many ways can a president, vice-president, treasurer, and secretary be chosen if the treasurer must be a girl, the secretary must be a boy, and a student may not hold more than one office?

### SOLUTION    If an event is specialized in some way (for example, the treasurer *must* be a girl), then that event should be performed before any nonspecialized events. Thus, we let $E_1$ represent the choice of treasurer and $E_2$ the choice of secretary. Next we let $E_3$ and $E_4$ denote the choices for president and vice-president, respectively. As in the fundamental counting principle, we let $m_i$ denote the number of different ways $E_i$ can occur for $i = 1, 2, 3$, and 4. It follows that $m_1 = 60$, $m_2 = 40$, $m_3 = 98$,

and $m_4 = 97$. By the fundamental counting principle, the total number of possibilities is

$$60 \cdot 40 \cdot 98 \cdot 97 = 22,814,400.$$

When working with sets, we are usually not concerned about the order or arrangement of the elements. In the remainder of this section, however, the arrangement of the elements will be our main concern.

**Definition of Permutation**

Let $S$ be a set of $n$ elements and let $1 \le r \le n$. A **permutation** of $r$ elements of $S$ is an arrangement, without repetitions, of $r$ elements.

We also use the phrase **permutation of $n$ elements taken $r$ at a time.** The symbol $P(n, r)$ will denote the number of different permutations of $r$ elements that can be obtained from a set containing $n$ elements. As a special case, $P(n, n)$ denotes the number of arrangements of $n$ elements of $S$—that is, the number of ways of arranging *all* the elements of $S$.

In our first discussion involving the four teams A, B, C, and D, we had $P(4, 2) = 12$, since there were 12 different ways of arranging the four teams in groups of two. We also showed that the number of ways to arrange all the elements A, B, C, and D is 24. In permutation notation we would write this result as $P(4, 4) = 24$.

The next theorem gives us a general formula for $P(n, r)$.

**Theorem on the Number of Different Permutations**

Let $S$ be a set of $n$ elements and let $1 \le r \le n$. The number of different permutations of $r$ elements of $S$ is

$$P(n, r) = n(n - 1)(n - 2) \cdots (n - r + 1).$$

*PROOF*    The problem of determining $P(n, r)$ is equivalent to determining the number of different $r$-tuples $(x_1, x_2, \ldots, x_r)$ such that each $x_i$ is an element of $S$ and no element of $S$ appears twice in the same $r$-tuple. We may find this number by means of the fundamental counting principle. For each $i = 1, 2, \ldots, r$, let $E_i$ represent the determination of the element $x_i$ and let $m_i$ be the number of different ways of choosing $x_i$. We wish to apply the sequence $E_1, E_2, \ldots, E_r$. We have $n$ possible choices for $x_1$, and consequently $m_1 = n$. Since repetitions are not allowed, we have $n - 1$ choices for $x_2$, so $m_2 = n - 1$. Continuing in this manner, we successively obtain $m_3 = n - 2$, $m_4 = n - 3$, and ultimately $m_r = n - (r - 1)$ or, equivalently, $m_r = n - r + 1$. Hence, using the fundamental counting principle, we obtain the formula for $P(n, r)$.  ■

Note that *the formula for $P(n, r)$ in the previous theorem contains exactly $r$ factors on the right-hand side*, as shown in the following illustration.

### ILLUSTRATION

**Number of Different Permutations**

- $P(n, 1) = n$
- $P(n, 2) = n(n - 1)$
- $P(n, 3) = n(n - 1)(n - 2)$
- $P(n, 4) = n(n - 1)(n - 2)(n - 3)$

**EXAMPLE 3**    *Evaluating $P(n, r)$*

Find $P(5, 2)$, $P(6, 4)$, and $P(5, 5)$.

**SOLUTION**    We use the formula for $P(n, r)$ in the preceding theorem:

$$5 - 2 + 1 = 4, \quad \text{so} \quad P(5, 2) = 5 \cdot 4 = 20$$
$$6 - 4 + 1 = 3, \quad \text{so} \quad P(6, 4) = 6 \cdot 5 \cdot 4 \cdot 3 = 360$$
$$5 - 5 + 1 = 1, \quad \text{so} \quad P(5, 5) = 5 \cdot 4 \cdot 3 \cdot 2 \cdot 1 = 120$$

**EXAMPLE 4**    *Arranging the batting order for a baseball team*

A baseball team consists of nine players. Find the number of ways of arranging the first four positions in the batting order if the pitcher is excluded.

**SOLUTION**    We wish to find the number of permutations of 8 objects taken 4 at a time. Using the formula for $P(n, r)$ with $n = 8$ and $r = 4$, we have $n - r + 1 = 5$, and it follows that

$$P(8, 4) = 8 \cdot 7 \cdot 6 \cdot 5 = 1680.$$

The next result gives us a form for $P(n, r)$ that involves the factorial symbol.

**Factorial Form for $P(n, r)$**

If $n$ is a positive integer and $1 \leq r \leq n$, then

$$P(n, r) = \frac{n!}{(n - r)!}.$$

**PROOF**   If we let $r = n$ in the formula for $P(n, r)$ in the theorem on permutations, we obtain the number of different arrangements of *all* the elements of a set consisting of $n$ elements. In this case,

$$n - r + 1 = n - n + 1 = 1$$

and        $$P(n, n) = n(n - 1)(n - 2) \cdots 3 \cdot 2 \cdot 1 = n!.$$

Consequently, $P(n, n)$ is the product of the first $n$ positive integers. This result is also given by the factorial form, for if $r = n$, then

$$P(n, n) = \frac{n!}{(n - n)!} = \frac{n!}{0!} = \frac{n!}{1} = n!.$$

If $1 \leq r < n$, then

$$\frac{n!}{(n - r)!} = \frac{n(n - 1)(n - 2) \cdots (n - r + 1) \cdot [(n - r)!]}{(n - r)!}$$

$$= n(n - 1)(n - 2) \cdots (n - r + 1).$$

This agrees with the formula for $P(n, r)$ in the theorem on permutations.

---

**EXAMPLE 5**   *Evaluating $P(n, r)$ using factorials*

Use the factorial form for $P(n, r)$ to find $P(5, 2)$, $P(6, 4)$, and $P(5, 5)$.

**SOLUTION**

$$P(5, 2) = \frac{5!}{(5 - 2)!} = \frac{5!}{3!} = \frac{5 \cdot 4 \cdot 3!}{3!} = 5 \cdot 4 = 20$$

$$P(6, 4) = \frac{6!}{(6 - 4)!} = \frac{6!}{2!} = \frac{6 \cdot 5 \cdot 4 \cdot 3 \cdot 2!}{2!} = 6 \cdot 5 \cdot 4 \cdot 3 = 360$$

$$P(5, 5) = \frac{5!}{(5 - 5)!} = \frac{5!}{0!} = \frac{5!}{1} = 5 \cdot 4 \cdot 3 \cdot 2 \cdot 1 = 120$$

---

## 9.6 EXERCISES

**Exer. 1–8: Find the number.**

1  $P(7, 3)$

2  $P(8, 5)$

3  $P(9, 6)$

4  $P(5, 3)$

5  $P(5, 5)$

6  $P(4, 4)$

7  $P(6, 1)$

8  $P(5, 1)$

9  How many three-digit numbers can be formed from the digits 1, 2, 3, 4, and 5 if repetitions

(a) are not allowed?        (b) are allowed?

10  Work Exercise 9 for four-digit numbers.

11  How many numbers can be formed from the digits 1, 2, 3, and 4 if repetitions are not allowed? (*Note:* 42 and 231 are examples of such numbers.)

12 Determine the number of positive integers less than 10,000 that can be formed from the digits 1, 2, 3, and 4 if repetitions are allowed.

13 *Basketball standings*   If eight basketball teams are in a tournament, find the number of different ways that first, second, and third place can be decided, assuming ties are not allowed.

14 *Basketball standings*   Work Exercise 13 for 12 teams.

15 *Wardrobe mix'n'match*   A girl has four skirts and six blouses. How many different skirt-blouse combinations can she wear?

16 *Wardrobe mix'n'match*   Refer to Exercise 15. If the girl also has three sweaters, how many different skirt-blouse-sweater combinations can she wear?

17 *License plate numbers*   In a certain state, automobile license plates start with one letter of the alphabet, followed by five digits (0, 1, 2, . . . , 9). Find how many different license plates are possible if

  (a) the first digit following the letter cannot be 0

  (b) the first letter cannot be O or I and the first digit cannot be 0

18 *Tossing dice*   Two dice are tossed, one after the other. In how many different ways can they fall? List the number of different ways the sum of the dots can equal

  (a) 3     (b) 5     (c) 7     (d) 9     (e) 11

19 *Seating arrangement*   A row of six seats in a classroom is to be filled by selecting individuals from a group of ten students.

  (a) In how many different ways can the seats be occupied?

  (b) If there are six boys and four girls in the group and if boys and girls are to be alternated, find the number of different seating arrangements.

20 *Scheduling courses*   A student in a certain college may take mathematics at 8, 10, 11, or 2 o'clock; English at 9, 10, 1, or 2; and history at 8, 11, 2, or 3. Find the number of different ways in which the student can schedule the three courses.

21 *True-or-false test*   In how many different ways can a test consisting of ten true-or-false questions be completed?

22 *Multiple-choice test*   A test consists of six multiple-choice questions, and there are five choices for each question. In how many different ways can the test be completed?

23 *Seating arrangement*   In how many different ways can eight people be seated in a row?

24 *Book arrangement*   In how many different ways can ten books be arranged on a shelf?

25 *Semaphore*   With six different flags, how many different signals can be sent by placing three flags, one above the other, on a flag pole?

26 *Selecting books*   In how many different ways can five books be selected from a 12-volume set of books?

27 *Radio call letters*   How many four-letter radio station call letters can be formed if the first letter must be K or W and repetitions

  (a) are not allowed?          (b) are allowed?

28 *Fraternity designations*   There are 24 letters in the Greek alphabet. How many fraternities may be specified by choosing three Greek letters if repetitions

  (a) are not allowed?          (b) are allowed?

29 *Phone numbers*   How many seven-digit phone numbers can be formed from the digits 0, 1, 2, 3, . . . , 9 if the first digit may not be 0?

30 *Baseball batting order*   After selecting nine players for a baseball game, the manager of the team arranges the batting order so that the pitcher bats last and the best hitter bats third. In how many different ways can the remainder of the batting order be arranged?

31 *ATM access code*   A customer remembers that 2, 4, 7, and 9 are the digits of a four-digit access code for an automatic bank-teller machine. Unfortunately, the customer has forgotten the order of the digits. Find the largest possible number of trials necessary to obtain the correct code.

32 *ATM access code*   Work Exercise 31 if the digits are 2, 4, and 7 and one of these digits is repeated in the four-digit code.

33 *Selecting theater seats*   Three married couples have purchased tickets for a play. Spouses are to be seated next to each other, and the six seats are in a row. In how many ways can the six people be seated?

34 *Horserace results*   Ten horses are entered in a race. If the possibility of a tie for any place is ignored, in how many ways can the first-, second-, and third-place winners be determined?

35 The commutative and associative laws of addition guarantee that the sum of integers 1 through 10 is indepen-

dent of the order in which the numbers are added. In how many different ways can these integers be summed?

**36 Shuffling cards**

(a) In how many ways can a standard deck of 52 cards be shuffled?

(b) In how many ways can the cards be shuffled so that the four aces appear on the top of the deck?

**37 Numerical palindromes** A palindrome is an integer, such as 45654, that reads the same backward and forward.

(a) How many five-digit palindromes are there?

(b) How many $n$-digit palindromes are there?

**38 Color arrangements** Each of the six squares shown in the figure is to be filled with any one of ten possible colors. How many ways are there of coloring the strip shown in the figure so that no two adjacent squares have the same color?

EXERCISE 38

**39** This exercise requires a graphing utility that can graph $x!$.

(a) Graph $y = \dfrac{x! e^x}{x^x \sqrt{2\pi x}}$ on $(0, 20]$, and estimate the horizontal asymptote.

(b) Use the graph in part (a) to find an approximation for $n!$ if $n$ is a large positive integer.

**40** (a) What happens if a calculator is used to find $P(150, 50)$? Explain.

(b) Approximate $r$ if $P(150, 50) = 10^r$ by using the following formula from advanced mathematics:

$$\log n! \approx \frac{n \ln n - n}{\ln 10}$$

## 9.7 DISTINGUISHABLE PERMUTATIONS AND COMBINATIONS

Certain problems involve finding different arrangements of objects, some of which are indistinguishable. For example, suppose we are given five disks of the same size, of which three are black, one is white, and one is red. Let us find the number of ways they can be arranged in a row so that different color arrangements are obtained. If the disks were all different colors, then the number of arrangements would be 5!, or 120. However, since some of the disks have the same appearance, we cannot obtain 120 different arrangements. To clarify this point, let us write

B  B  B  W  R

for the arrangement having black disks in the first three positions in the row, the white disk in the fourth position, and the red disk in the fifth position. The first three disks can be arranged in 3!, or 6, different ways, but these arrangements cannot be distinguished from one another because the first three disks look alike. We say that those 3! permutations are **nondistinguishable.** Similarly, given any other arrangement, say

B  R  B  W  B,

there are 3! different ways of arranging the three black disks, but again each such arrangement is nondistinguishable from the others. Let us call two arrangements of objects **distinguishable permutations** if one arrangement cannot be obtained from the other by rearranging like objects. Thus, B B B W R and B R B W B are distinguishable permutations of the five disks. Let $k$ denote the number of distinguishable permutations. Since to each such arrangement there correspond 3! *nondistinguishable* permutations, we must have $3!k = 5!$, the number of permutations of five *different* objects. Hence, $k = 5!/3! = 5 \cdot 4 = 20$. By the same type of reasoning we can obtain the following extension of this discussion.

**First Theorem on Distinguishable Permutations**

> If $r$ objects in a collection of $n$ objects are alike and if the remaining objects are different from each other and from the $r$ objects, then the number of distinguishable permutations of the $n$ objects is
>
> $$\frac{n!}{r!}.$$

We can generalize this theorem to the case in which there are several subcollections of nondistinguishable objects. For example, consider eight disks, of which four are black, three are white, and one is red. In this case, with each arrangement, such as

$$B \quad W \quad B \quad W \quad B \quad W \quad B \quad R,$$

there are 4! arrangements of the black disks and 3! arrangements of the white disks that have no effect on the color arrangement. Hence, 4!3! possible arrangements of the disks will not produce distinguishable permutations. If we let $k$ denote the number of *distinguishable* permutations, then $4!3!k = 8!$, since 8! is the number of permutations we would obtain if the disks were all different. Thus, the number of distinguishable permutations is

$$k = \frac{8!}{4!3!} = \frac{8 \cdot 7 \cdot 6 \cdot 5}{3!} \cdot \frac{4!}{4!} = 280.$$

The following general result can be proved.

**Second Theorem on Distinguishable Permutations**

> If, in a collection of $n$ objects, $n_1$ are alike, $n_2$ are alike of another kind, . . . , $n_k$ are alike of a further kind, and
>
> $$n = n_1 + n_2 + \cdots + n_k,$$
>
> then the number of distinguishable permutations of the $n$ objects is
>
> $$\frac{n!}{n_1!n_2!\cdots n_k!}.$$

**EXAMPLE I    *Finding the number of distinguishable permutations***

Find the number of distinguishable permutations of the letters in the word *Mississippi.*

**SOLUTION**    In this example we are given a collection of eleven objects in which four are of one kind (the letter *s*), four are of another kind (*i*), two are of a third kind (*p*), and one is of a fourth kind (*M*). Hence, by the preceding theorem, the number of distinguishable permutations is

$$\frac{11!}{4!4!2!1!} = 34{,}650.$$

When we work with permutations, our concern is with the orderings or arrangements of elements. Let us now ignore the order or arrangement of elements and consider the following question: Given a set containing $n$ distinct elements, in how many ways can a subset of $r$ elements be chosen with $r \leq n$? Before answering, let us state a definition.

**Definition of Combination**

Let $S$ be a set of $n$ elements and let $1 \leq r \leq n$. A **combination** of $r$ elements of $S$ is a subset of $S$ that contains $r$ distinct elements.

If $S$ contains $n$ elements, we also use the phrase **combination of $n$ elements taken $r$ at a time.** The symbol $C(n, r)$ will denote the number of combinations of $r$ elements that can be obtained from a set of $n$ elements.

**Theorem on the Number of Combinations**

The number of combinations of $r$ elements that can be obtained from a set of $n$ elements is

$$C(n, r) = \frac{n!}{(n - r)!r!}, \quad 1 \leq r \leq n.$$

**PROOF**    If $S$ contains $n$ elements, then, to find $C(n, r)$, we must find the total number of subsets of the form

$$\{x_1, x_2, \ldots, x_r\}$$

such that the $x_i$ are *different* elements of $S$. Since elements $x_1, x_2, \ldots, x_r$ can be arranged in $r!$ different ways, each such subset produces $r!$ different $r$-tuples. Thus, the total number of different $r$-tuples is $r!C(n, r)$. However, in the previous section we found that the total number of $r$-tuples is

$$P(n, r) = \frac{n!}{(n - r)!}.$$

Hence, $$r!C(n, r) = \frac{n!}{(n - r)!}.$$

Dividing both sides of the last equation by $r!$ gives us the formula for $C(n, r)$. ∎

The formula for $C(n, r)$ is identical to the formula for the binomial coefficient $\binom{n}{r}$ in Section 9.5.

### EXAMPLE 2    *Choosing a baseball squad*

A little league baseball squad has six outfielders, seven infielders, five pitchers, and two catchers. Each outfielder can play any of the three outfield positions, and each infielder can play any of the four infield positions. In how many ways can a team of nine players be chosen?

**SOLUTION**    The number of ways of choosing three outfielders from the six candidates is

$$C(6, 3) = \frac{6!}{(6 - 3)!3!} = \frac{6!}{3!3!} = \frac{6 \cdot 5 \cdot 4 \cdot 3!}{3 \cdot 2 \cdot 1 \cdot 3!} = 20.$$

The number of ways of choosing the four infielders is

$$C(7, 4) = \frac{7!}{(7 - 4)!4!} = \frac{7 \cdot 6 \cdot 5 \cdot 4!}{3!4!} = \frac{7 \cdot 6 \cdot 5}{3 \cdot 2 \cdot 1} = 35.$$

There are five ways of choosing a pitcher and two choices for the catcher. It follows from the fundamental counting principle that the total number of ways to choose a team is

$$20 \cdot 35 \cdot 5 \cdot 2 = 7000.$$

Note that if $r = n$, the formula for $C(n, r)$ becomes

$$C(n, n) = \frac{n!}{(n - n)!n!} = \frac{n!}{0!n!} = \frac{n!}{1 \cdot n!} = 1.$$

It is convenient to assign a meaning to $C(n, r)$ if $r = 0$. If the formula is to be true in this case, then we must have

$$C(n, 0) = \frac{n!}{(n - 0)!0!} = \frac{n!}{n!0!} = \frac{n!}{n! \cdot 1} = 1.$$

Hence, we *define* $C(n, 0) = 1$, which is the same as $C(n, n)$. Finally, for consistency, we also *define* $C(0, 0) = 1$. Thus, $C(n, r)$ has meaning for all nonnegative integers $n$ and $r$ with $r \leq n$.

**EXAMPLE 3** *Finding the number of subsets of a set*

Let $S$ be a set of $n$ elements. Find the number of distinct subsets of $S$.

**SOLUTION** Let $r$ be any nonnegative integer such that $r \le n$. From our previous work, the number of subsets of $S$ that consist of $r$ elements is $C(n, r)$, or $\binom{n}{r}$. Hence, to find the total number of subsets, we find the sum

$$\binom{n}{0} + \binom{n}{1} + \binom{n}{2} + \binom{n}{3} + \cdots + \binom{n}{n}.$$

This is precisely the binomial expansion of $(1 + 1)^n$. Thus, there are $2^n$ subsets of a set of $n$ elements. In particular, a set of 3 elements has $2^3$, or 8, different subsets. A set of 4 elements has $2^4$, or 16, subsets. A set of 10 elements has $2^{10}$, or 1024, subsets.

## 9.7 EXERCISES

**Exer. 1–8: Find the number.**

1  $C(7, 3)$

2  $C(8, 4)$

3  $C(9, 8)$

4  $C(6, 2)$

5  $C(n, n - 1)$

6  $C(n, 1)$

7  $C(7, 0)$

8  $C(5, 5)$

**Exer. 9–10: Find the number of possible color arrangements for the 12 given disks, arranged in a row.**

9  5 black, 3 red, 2 white, 2 green

10  3 black, 3 red, 3 white, 3 green

11  Find the number of distinguishable permutations of the letters in the word *bookkeeper*.

12  Find the number of distinguishable permutations of the letters in the word *moon*. List all the permutations.

13  *Choosing basketball teams* Ten people wish to play in a basketball game. In how many different ways can two teams of five players each be formed?

14  *Selecting test questions* A student may answer any six of ten questions on an examination.

(a) In how many ways can six questions be selected?

(b) How many selections are possible if the first two questions must be answered?

**Exer. 15–16: Consider any eight points such that no three are collinear.**

15  How many lines are determined?

16  How many triangles are determined?

17  *Book arrangement* A student has five mathematics books, four history books, and eight fiction books. In how many different ways can they be arranged on a shelf if books in the same category are kept next to one another?

18  *Selecting a basketball team* A basketball squad consists of twelve players.

(a) Disregarding positions, in how many ways can a team of five be selected?

(b) If the center of a team must be selected from two specific individuals on the squad and the other four members of the team from the remaining ten players, find the number of different teams possible.

19  *Selecting a football team* A football squad consists of three centers, ten linemen who can play either guard or tackle, three quarterbacks, six halfbacks, four ends, and four fullbacks. A team must have one center, two guards, two tackles, two ends, two halfbacks, a quarterback, and a fullback. In how many different ways can a team be selected from the squad?

**20 *Arranging keys on a ring*** In how many different ways can seven keys be arranged on a key ring if the keys can slide completely around the ring?

**21 *Committee selection*** A committee of 3 men and 2 women is to be chosen from a group of 12 men and 8 women. Determine the number of different ways of selecting the committee.

**22 *Birth order*** Let the letters G and B denote a girl birth and a boy birth, respectively. For a family of three boys and three girls, one possible birth order is G G G B B B. How many birth orders are possible for these six children?

**Exer. 23–24: Shown in the figure is a street map and a possible path from point *A* to point *B*. How many other possible paths are there from *A* to *B* if moves are restricted to the right or up? (*Hint:* If R denotes a move one unit right and U denotes a move one unit up, then the path in Exercise 23 can be specified by R U U R R U R.)**

**23**

**24**

**25 *Lotto selections*** To win a state lottery game, a player must correctly select six numbers from the numbers 1 through 49.

(a) Find the total number of selections possible.

(b) Work part (a) if a player selects only even numbers.

**26 *Office assignments*** A mathematics department has ten faculty members but only nine offices, so one office must be shared by two individuals. In how many different ways can the offices be assigned?

**27 *Tennis tournament*** In a round-robin tennis tournament, every player meets every other player exactly once. How many players can participate in a tournament of 45 matches?

**28 *True-or-false test*** A true-or-false test has 20 questions.

(a) In how many different ways can the test be completed?

(b) In how many different ways can a student answer 10 questions correctly?

**29 *Basketball championship series*** The winner of the seven-game NBA championship series is the team that wins four games. In how many different ways can the series be extended to seven games?

**30** A geometric design is determined by joining every pair of vertices of an octagon (see the figure).

(a) How many triangles in the design have their three vertices on the octagon?

(b) How many quadrilaterals in the design have their four vertices on the octagon?

**EXERCISE 30**

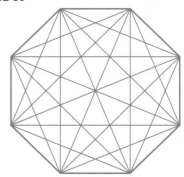

**31** *Ice cream selections* An ice cream parlor stocks 31 different flavors and advertises that it serves almost 4500 different triple scoop cones, with each scoop being a different flavor. How was this number obtained?

**32** *Choices of hamburger condiments* A fast food restaurant advertises that it offers any combination of 8 condiments on a hamburger, thus giving a customer 256 choices. How was this number obtained?

Exer. 33–34: **(a)** Calculate the sum $S_n$ for $n = 1, 2, 3,$ ..., 10, where if $n < r$, then $\binom{n}{r} = 0$. **(b)** Predict a general formula for $S_n$.

**33** $\binom{n}{1} + \binom{n}{3} + \binom{n}{5} + \binom{n}{7} + \cdots$

**34** $(1)\binom{n}{1} - (2)\binom{n}{2} + (3)\binom{n}{3} - (4)\binom{n}{4} + (5)\binom{n}{5} - \cdots$

## 9.8 PROBABILITY

If two dice are tossed, what are the chances of rolling a 7? If a person is dealt five cards from a standard deck of 52 playing cards, what is the likelihood of obtaining three aces? In the seventeenth century, similar questions about games of chance led to the study of *probability*. Since that time, the theory of probability has grown extensively. It is now used to predict outcomes of a large variety of situations that arise in the natural and social sciences.

Any chance process, such as flipping a coin, rolling a die, being dealt a card from a deck, determining if a manufactured item is defective, or finding the blood pressure of an individual, is an **experiment**. A result of an experiment is an **outcome**. We shall restrict our discussion to experiments for which outcomes are **equally likely**. This means, for example, that if a coin is flipped, we assume that the possibility of obtaining a head is the same as that of obtaining a tail. Similarly, if a die is tossed, we assume that the die is *fair*—that is, there is an equal chance of obtaining either a 1, 2, 3, 4, 5, or 6. The set $S$ of all possible outcomes of an experiment is the **sample space** of the experiment. Thus, if the experiment consists of flipping a coin and we let H or T denote the outcome of obtaining a head or tail, respectively, then the sample space $S$ may be denoted by

$$S = \{H, T\}.$$

If a fair die is tossed as an experiment, then the set $S$ of all possible outcomes (the sample space) is

$$S = \{1, 2, 3, 4, 5, 6\}.$$

The following definition expresses, in mathematical terms, the notion of obtaining *particular* outcomes of an experiment.

**Definition of Event**

Let $S$ be the sample space of an experiment. An **event** associated with the experiment is any subset $E$ of $S$.

Let us consider the experiment of tossing a single die, so that the sample space is $S = \{1, 2, 3, 4, 5, 6\}$. If $E = \{4\}$, then the event $E$ associated with the experiment consists of the outcome of obtaining a 4 on the toss. Different events may be associated with the same experiment. For example, if we let $E = \{1, 3, 5\}$, then this event consists of obtaining an odd number on a toss of the die.

As another illustration, suppose the experiment consists of flipping two coins, one after the other. If we let HH denote the outcome in which two heads appear, HT that of a head on the first coin and a tail on the second, and so on, then the sample space $S$ of the experiment may be denoted by

$$S = \{HH, HT, TH, TT\}.$$

If we let $\qquad\qquad\qquad E = \{HT, TH\},$

then the event $E$ consists of the appearance of a head on one of the coins and a tail on the other.

Next we shall define what is meant by the *probability* of an event. Throughout our discussion we will assume that the sample space $S$ of an experiment contains only a finite number of elements. If $E$ is an event, *the symbols $n(E)$ and $n(S)$ will denote the number of elements in E and S, respectively.*

**Definition of the Probability of an Event**

Let $S$ be the sample space of an experiment and $E$ an event. The **probability** $P(E)$ of $E$ is given by

$$P(E) = \frac{n(E)}{n(S)}.$$

Since $E$ is a subset of $S$, we see that

$$0 \le n(E) \le n(S).$$

Dividing by $n(S)$, we obtain

$$0 \le P(E) \le 1.$$

Note that $P(E) = 1$ if $E = S$, and $P(E) = 0$ if $E$ contains no elements.

The next example provides several illustrations of the preceding definition if $E$ contains exactly one element.

**EXAMPLE I**   *Finding the probability of an event*

**(a)** If a coin is flipped, find the probability that a head will turn up.

**(b)** If a fair die is tossed, find the probability of obtaining a 4.

**(c)** If two coins are flipped, find the probability that both coins turn up heads.

**SOLUTION**   For each experiment we shall list sets $S$ and $E$ and then use the definition of probability to find $P(E)$.

**(a)**
$$S = \{H, T\}, \quad E = \{H\}, \quad P(E) = \frac{n(E)}{n(S)} = \frac{1}{2}$$

**(b)**
$$S = \{1, 2, 3, 4, 5, 6\}, \quad E = \{4\}, \quad P(E) = \frac{n(E)}{n(S)} = \frac{1}{6}$$

**(c)**
$$S = \{HH, HT, TH, TT\}, \quad E = \{HH\}, \quad P(E) = \frac{n(E)}{n(S)} = \frac{1}{4}$$

In part (a) of Example 1 we found that the probability of obtaining a head on a flip of a coin is $\frac{1}{2}$. We take this to mean that if a coin is flipped many times, the number of times that a head turns up should be approximately one-half the total number of flips. Thus, for 100 flips, a head should turn up approximately 50 times. It is unlikely that this number will be *exactly* 50. A probability of $\frac{1}{2}$ implies that if we let the number of flips increase, then the number of times a head turns up *approaches* one-half the total number of flips. Similar remarks can be made for parts (b) and (c) of Example 1.

In the next two examples we consider experiments in which an event contains more than one element.

**EXAMPLE 2**   *Finding probabilities when two dice are tossed*

If two dice are tossed, what is the probability of rolling a sum of

**(a)** 7?      **(b)** 9?

**SOLUTION**   Let us refer to one die as *the first die* and the other as *the second die*. We shall use ordered pairs to represent outcomes as follows: (2, 4) denotes the outcome of obtaining a 2 on the first die and a 4 on the second; (5, 3) represents a 5 on the first die and a 3 on the second; and so on. Since there are six different possibilities for the first number of the ordered pair and, with each of these, six possibilities for the second number, the total number of ordered pairs is 36. Hence, if $S$ is the sample space, then $n(S) = 36$.

*(continued)*

(a) The event $E$ corresponding to rolling a sum of 7 is given by

$$E = \{(1, 6), (2, 5), (3, 4), (4, 3), (5, 2), (6, 1)\},$$

and consequently    $P(E) = \dfrac{n(E)}{n(S)} = \dfrac{6}{36} = \dfrac{1}{6}.$

(b) If $E$ is the event corresponding to rolling a sum of 9, then

$$E = \{(3, 6), (4, 5), (5, 4), (6, 3)\}$$

and    $P(E) = \dfrac{n(E)}{n(S)} = \dfrac{4}{36} = \dfrac{1}{9}.$

In the next example (and in the exercises), when it is stated that one or more cards are drawn from a deck, we mean that each card is removed from a standard 52-card deck and is *not* replaced before the next card is drawn.

### EXAMPLE 3    *Finding the probability of drawing a certain hand of cards*

Suppose five cards are drawn from a standard deck of 52 playing cards. Find the probability that all five cards are hearts.

**SOLUTION**    The sample space $S$ of the experiment is the set of all possible five-card hands that can be formed from the 52 cards in the deck. It follows from our work in the preceding section that $n(S) = C(52, 5)$.

Since there are 13 cards in the heart suit, the number of different ways of obtaining a hand that contains five hearts is $C(13, 5)$. Hence, if $E$ represents this event, then

$$P(E) = \frac{n(E)}{n(S)} = \frac{C(13, 5)}{C(52, 5)} = \frac{\dfrac{13!}{5!8!}}{\dfrac{52!}{5!47!}}.$$

We may show that

$$P(E) = \frac{1287}{2{,}598{,}960} \approx 0.0005 = \frac{5}{10{,}000} = \frac{1}{2000}.$$

This result implies that if the experiment is performed many times, a five-card heart hand should be drawn approximately once every 2000 times.

Suppose $S$ is the sample space of an experiment and $E_1$ and $E_2$ are two events associated with the experiment. If $E_1$ and $E_2$ have no elements in common, they are called *disjoint sets,* and we write $E_1 \cap E_2 = \varnothing$ (the *empty set*). In this case, if one event occurs, the other cannot occur; they are **mutually exclusive events.** Thus, if $E = E_1 \cup E_2$, then

$$n(E) = n(E_1 \cup E_2) = n(E_1) + n(E_2).$$

Hence,

$$P(E) = \frac{n(E_1) + n(E_2)}{n(S)} = \frac{n(E_1)}{n(S)} + \frac{n(E_2)}{n(S)},$$

or

$$P(E) = P(E_1) + P(E_2).$$

The probability of $E$ is therefore the sum of the probabilities of $E_1$ and $E_2$. We have proved the following.

**Theorem on Mutually Exclusive Events**

If $E_1$ and $E_2$ are mutually exclusive events and $E = E_1 \cup E_2$, then

$$P(E) = P(E_1 \cup E_2) = P(E_1) + P(E_2).$$

The preceding theorem can be extended to any number of events $E_1$, $E_2, \ldots, E_k$ that are mutually exclusive in the sense that if $i \neq j$, then $E_i \cap E_j = \varnothing$. The conclusion of the theorem is then

$$P(E) = P(E_1 \cup E_2 \cup \cdots \cup E_k) = P(E_1) + P(E_2) + \cdots + P(E_k).$$

**EXAMPLE 4** *Finding probabilities when two dice are tossed*

If two dice are tossed, find the probability of rolling a sum of either 7 or 9.

**SOLUTION** Let $E_1$ denote the event of rolling 7 and $E_2$ that of rolling 9. Since $E_1$ and $E_2$ cannot occur simultaneously, they are mutually exclusive events. We wish to find the probability of the event $E = E_1 \cup E_2$. From Example 2 we know that $P(E_1) = \frac{1}{6}$ and $P(E_2) = \frac{1}{9}$. Hence, by the last theorem,

$$P(E) = P(E_1) + P(E_2)$$
$$= \frac{1}{6} + \frac{1}{9} = \frac{5}{18} = 0.2\overline{7}.$$

If $E_1$ and $E_2$ are events that possibly have elements in common, then the following can be proved.

**Theorem on the Probability of the Occurrence of Either of Two Events**

If $E_1$ and $E_2$ are any two events, then
$$P(E_1 \cup E_2) = P(E_1) + P(E_2) - P(E_1 \cap E_2).$$

Note that if $E_1$ and $E_2$ are mutually exclusive, then $E_1 \cap E_2 = \varnothing$ and $P(E_1 \cap E_2) = 0$. Hence, the last theorem includes, as a special case, the theorem on mutually exclusive events.

**EXAMPLE 5**   *Finding the probability of selecting a certain card from a deck*

If a single card is selected from a 52-card deck, find the probability that the card is either a jack or a spade.

*SOLUTION*    Let $E_1$ denote the event that the card is a jack and $E_2$ the event that it is a spade. The events $E_1$ and $E_2$ are *not* mutually exclusive, since there is one card—the jack of spades—in both events, and hence $P(E_1 \cap E_2) = \frac{1}{52}$. By the preceding theorem, the probability that the card is either a jack or a spade is

$$P(E_1 \cup E_2) = P(E_1) + P(E_2) - P(E_1 \cap E_2)$$
$$= \tfrac{4}{52} + \tfrac{13}{52} - \tfrac{1}{52} = \tfrac{16}{52} = \tfrac{4}{13} \approx 0.31.$$

In solving probability problems, it is often helpful to categorize the outcomes of a sample space $S$ into an event $E$ and the set $E'$ of elements of $S$ that are not in $E$. We call $E'$ the **complement** of $E$. Note that
$$E \cup E' = S \quad \text{and} \quad n(E) + n(E') = n(S).$$
Dividing both sides of the last equation by $n(S)$ gives us
$$\frac{n(E)}{n(S)} + \frac{n(E')}{n(S)} = 1.$$
Hence,
$$P(E) + P(E') = 1, \quad \text{or} \quad P(E) = 1 - P(E').$$
We shall use the last formula in the next example.

**EXAMPLE 6**   *Finding the probability of drawing a certain hand of cards*

If 13 cards are drawn from a 52-card deck, what is the probability that at least 2 of the cards are hearts?

*SOLUTION* If $P(k)$ denotes the probability of getting $k$ hearts, then the probability of getting at least two hearts is

$$P(2) + P(3) + P(4) + \cdots + P(13).$$

Since the only remaining probabilities are $P(0)$ and $P(1)$, the desired probability is equal to

$$1 - [P(0) + P(1)].$$

To calculate $P(k)$ for any $k$, we may regard the deck as being split into two groups: hearts and non-hearts. For $P(0)$ we note that of the 13 hearts in the deck, we get none; and of the 39 non-hearts, we get 13. Since the number of ways to choose 13 cards from a 52-card deck is $C(52, 13)$, we see that

$$P(0) = \frac{n(0)}{n(S)} = \frac{C(13, 0) \cdot C(39, 13)}{C(52, 13)} \approx 0.0128.$$

The probability $P(1)$ corresponds to getting 1 of the hearts and 12 of the 39 non-hearts. Thus,

$$P(1) = \frac{n(1)}{n(S)} = \frac{C(13, 1) \cdot C(39, 12)}{C(52, 13)} \approx 0.0801.$$

Hence, the desired probability is

$$1 - [P(0) + P(1)] \approx 1 - [0.0128 + 0.0801] = 0.9071.$$

Two events $E_1$ and $E_2$ are said to be **independent events** if the occurrence of one does not influence the occurrence of the other. The following can be proved.

**Theorem on Independent Events**

If $E_1$ and $E_2$ are independent events, then

$$P(E_1 \cap E_2) = P(E_1) \cdot P(E_2).$$

In words, the theorem states that if $E_1$ and $E_2$ are independent events, the probability that *both* $E_1$ and $E_2$ occur simultaneously is the product of their probabilities. Note that if two events $E_1$ and $E_2$ are mutually exclusive, then $P(E_1 \cap E_2) = 0$ and they cannot be independent. (We assume that both $E_1$ and $E_2$ are not empty.)

**FIGURE 5**

**EXAMPLE 7**    *An application of probability*
*to an electrical system*

An electrical system has open-close switches $s_1$, $s_2$, and $s_3$, as shown in Figure 5. The switches operate independently of one another, and current will flow from $A$ to $B$ either if $s_1$ is closed or if *both* $s_2$ and $s_3$ are closed.

**(a)** If $S_k$ denotes the event that $s_k$ is closed, where $k = 1, 2, 3$, express, in terms of $P(S_1)$, $P(S_2)$, and $P(S_3)$, the probability $p$ that current will flow from $A$ to $B$.

**(b)** Find $p$ if $P(S_k) = \frac{1}{2}$ for each $k$.

*SOLUTION*

**(a)** The probability $p$ that either $S_1$ or *both* $S_2$ and $S_3$ occur is

$$p = P(S_1 \cup (S_2 \cap S_3)).$$

Using the theorem on the probability of the occurrence of either of two events $S_1$ or $S_2 \cap S_3$, we obtain

$$p = P(S_1) + P(S_2 \cap S_3) - P(S_1 \cap (S_2 \cap S_3)).$$

Applying the theorem on independent events twice gives us

$$p = P(S_1) + P(S_2) \cdot P(S_3) - P(S_1) \cdot P(S_2 \cap S_3).$$

Finally, using the theorem on independent events one more time, we see that

$$p = P(S_1) + P(S_2) \cdot P(S_3) - P(S_1) \cdot P(S_2) \cdot P(S_3).$$

**(b)** If $P(S_k) = \frac{1}{2}$ for each $k$, then from part (a), the probability that current will flow from $A$ to $B$ is

$$p = \tfrac{1}{2} + \tfrac{1}{2} \cdot \tfrac{1}{2} - \tfrac{1}{2} \cdot \tfrac{1}{2} \cdot \tfrac{1}{2} = \tfrac{5}{8} = 0.625.$$

**EXAMPLE 8**    *A continuation of Example 7*

Refer to Example 7. If the probability that $s_k$ is closed is the same for each $k$, determine $P(S_k)$ such that $p = 0.99$.

*SOLUTION*    Since the probability $P(S_k)$ is the same for each $k$, we let $P(S_k) = x$ for $k = 1, 2, 3$. Substituting in the formula for $p$ obtained in part (a) of Example 7, we obtain

$$p = x + x \cdot x - x \cdot x \cdot x = -x^3 + x^2 + x.$$

Letting $p = 0.99$ gives us the equation

$$-x^3 + x^2 + x = 0.99.$$

Graphing $y = -x^3 + x^2 + x - 0.99$ using a standard viewing rectangle, we see that there are three $x$-intercepts. The desired probability must lie between $x = 0$ and $x = 1$ and should be fairly close to 1. Using the view-

**FIGURE 6**   [0.8, 1] by [−0.01, 0.01]

ing rectangle dimensions [0.8, 1] by [−0.01, 0.01] with Xscl = 0.1 and Yscl = 0.01, we obtain a sketch similar to Figure 6. Using tracing and zoom-in features gives us $x \approx 0.93$. Hence, $P(S_k) \approx 0.93$.

Note that the probability that an individual switch is closed is *less* than the probability that current will flow through the system.

It is often of interest to know what amount of return we can expect on an investment in a game of chance. The following definition will help us answer questions that fall in this category.

**Definition of Expected Value**

Suppose a variable can have payoff amounts $a_1, a_2, \ldots, a_n$ with corresponding probabilities $p_1, p_2, \ldots, p_n$. The **expected value** $EV$ of the variable is given by

$$EV = a_1 p_1 + a_2 p_2 + \cdots + a_n p_n = \sum_{k=1}^{n} a_k p_k.$$

**EXAMPLE 9**   *Expected value of a single pull-tab*

States that run lotteries often offer games in which a certain number of pull-tabs are printed, some being redeemable for money and the rest worthless. Suppose that in a particular game there are 4000 pull-tabs, 432 of which are redeemable according to the following table.

| Number of pull-tabs | Value |
|:---:|:---:|
| 4 | $100 |
| 8 | 50 |
| 20 | 20 |
| 400 | 2 |

Find the expected value of a pull-tab that sells for $1.

**SOLUTION**   The payoff amounts $100, $50, $20, and $2 have probabilities $\frac{4}{4000}$, $\frac{8}{4000}$, $\frac{20}{4000}$, and $\frac{400}{4000}$, respectively. The remaining 3568 pull-tabs have a payoff amount of $0. By the preceding definition, the expected value of a single pull-tab is

$$EV = 100 \cdot \tfrac{4}{4000} + 50 \cdot \tfrac{8}{4000} + 20 \cdot \tfrac{20}{4000} + 2 \cdot \tfrac{400}{4000} + 0 \cdot \tfrac{3568}{4000}$$
$$= \tfrac{2000}{4000} = \$0.50.$$

Thus, after subtracting the $1 cost of the pull-tab, we can expect to *lose* $0.50 on each pull-tab we buy. Note that we cannot lose $0.50 on any

*(continued)*

individual pull-tab, but we can expect to lose this amount on each pull-tab in the long run. This game yields a terribly poor return for the buyer and a healthy profit for the seller.

The expected value of $0.50 obtained in Example 9 may be considered to be the amount we would expect to pay to play the game if the game were fair—that is, if we would not expect to win or lose any money after playing the game many times.

In this section we have merely introduced several basic concepts about probability. The interested person is referred to entire books and courses devoted to this branch of mathematics.

## 9.8 EXERCISES

1 Each suit in a 52-card deck is made up of an ace (A), nine numbered cards (2, 3, . . . , 10), and three face cards (J, Q, K). An experiment consists of drawing a single card from a 52-card deck followed by rolling a single die.

(a) Describe the sample space $S$ of the experiment, and find $n(S)$.

(b) Let $E_1$ be the event consisting of the outcomes in which a numbered card is drawn and the number of dots on the die is the same as the number on the card. Find $n(E_1)$, $n(E_1')$, and $P(E_1)$.

(c) Let $E_2$ be the event in which the card drawn is a face card, and let $E_3$ be the event in which the number of dots on the die is even. Are $E_2$ and $E_3$ mutually exclusive? Are they independent? Find $P(E_2)$, $P(E_3)$, $P(E_2 \cap E_3)$, and $P(E_2 \cup E_3)$.

(d) Are $E_1$ and $E_2$ mutually exclusive? Are they independent? Find $P(E_1 \cap E_2)$ and $P(E_1 \cup E_2)$.

2 An experiment consists of selecting a letter from the alphabet and one of the digits 0, 1, . . . , 9.

(a) Describe the sample space $S$ of the experiment, and find $n(S)$.

(b) Suppose the letters of the alphabet are assigned numbers as follows: $A = 1$, $B = 2$, . . . , $Z = 26$. Let $E_1$ be the event in which the units' digit of the number assigned to the letter of the alphabet is the same as the digit selected. Find $n(E_1)$, $n(E_1')$, and $P(E_1)$.

(c) Let $E_2$ be the event that the letter is one of the five vowels and $E_3$ the event that the digit is a prime number. Are $E_2$ and $E_3$ mutually exclusive? Are

they independent? Find $P(E_2)$, $P(E_3)$, $P(E_2 \cap E_3)$, and $P(E_2 \cup E_3)$.

(d) Let $E_4$ be the event that the numerical value of the letter is even. Are $E_2$ and $E_4$ mutually exclusive? Are they independent? Find $P(E_2 \cap E_4)$ and $P(E_2 \cup E_4)$.

**Exer. 3–4: A single card is drawn from a 52-card deck. Find the probability that the card is as specified.**

3 (a) a king

(b) a king or a queen

(c) a king, a queen, or a jack

4 (a) a heart

(b) a heart or a diamond

(c) a heart, a diamond, or a club

**Exer. 5–6: A single die is tossed. Find the probability that the die is as specified.**

5 (a) a 4       (b) a 6               (c) a 4 or a 6

6 (a) an even number       (b) a number divisible by 5

(c) an even number or a number divisible by 5

**Exer. 7–8: An urn contains five red balls, six green balls, and four white balls. If a single ball is drawn, find the probability that the ball is as specified.**

7 (a) red       (b) green       (c) red or white

8 (a) white       (b) green or white       (c) not green

**Exer. 9–10: Two dice are tossed. Find the probability that the sum is as specified.**

9 (a) 11       (b) 8               (c) 11 or 8

10 (a) greater than 9          (b) an odd number

**Exer. 11–12: Three dice are tossed. Find the probability of the specified event.**

11 A sum of 5

12 A 6 turns up on exactly one die

13 If three coins are flipped, find the probability that exactly two heads turn up.

14 If four coins are flipped, find the probability of obtaining two heads and two tails.

**Exer. 15–20: Suppose five cards are drawn from a 52-card deck. Find the probability of obtaining the indicated cards.**

15 Four of a kind (such as four aces or four kings)

16 Three aces and two kings

17 Four diamonds and one spade

18 Five face cards

19 A flush (five cards, all of the same suit)

20 A royal flush (an ace, king, queen, jack, and 10 of the same suit)

21 If a single die is tossed, find the probability of obtaining an odd number or a prime number.

22 A single card is drawn from a 52-card deck. Find the probability that the card is either red or a face card.

23 If the probability of a baseball player's getting a hit in one time at bat is 0.326, find the probability that the player gets no hits in 4 times at bat.

24 If the probability of a basketball player's making a free throw is 0.9, find the probability that the player makes at least 1 of 2 free throws.

**Exer. 25–26: The outcomes 1, 2, . . . , 6 of an experiment and their probabilities are listed in the table.**

| Outcome | 1 | 2 | 3 | 4 | 5 | 6 |
|---|---|---|---|---|---|---|
| Probability | 0.25 | 0.10 | 0.15 | 0.20 | 0.25 | 0.05 |

For the indicated events, find (a) $P(E_2)$, (b) $P(E_1 \cap E_2)$, (c) $P(E_1 \cup E_2)$, and (d) $P(E_2 \cup E_3')$.

25 $E_1 = \{1, 2\}$;          $E_2 = \{2, 3, 4\}$;     $E_3 = \{4, 6\}$

26 $E_1 = \{1, 2, 3, 6\}$;   $E_2 = \{3, 4\}$;     $E_3 = \{4, 5, 6\}$

**Exer. 27–28: A box contains 10 red chips, 20 blue chips, and 30 green chips. If 5 chips are drawn from the box, find the probability of drawing the indicated chips.**

27 (a) all blue

   (b) at least 1 green

   (c) at most 1 red

28 (a) exactly 4 green

   (b) at least 2 red

   (c) at most 2 blue

29 **True-or-false test**   A true-or-false test consists of eight questions. If a student guesses the answer for each question, find the probability that

   (a) eight answers are correct

   (b) seven answers are correct and one is incorrect

   (c) six answers are correct and two are incorrect

   (d) at least six answers are correct

30 **Committee selection**   A 6-member committee is to be chosen by drawing names of individuals from a hat. If the hat contains the names of 8 men and 14 women, find the probability that the committee will consist of 3 men and 3 women.

**Exer. 31–32: Five cards are drawn from a 52-card deck. Find the probability of the specified event.**

31 Obtaining at least one ace

32 Obtaining at least one heart

33 **Tossing dice**   If two dice are tossed, find the probability that the sum is greater than 5.

34 **Tossing dice**   If three dice are tossed, find the probability that the sum is less than 16.

35 **Family make up**   Assuming that girl-boy births are equally probable, find the probability that a family with five children has

   (a) all boys          (b) at least one girl

36 **Slot machine**   A standard slot machine contains three reels, and each reel contains 20 symbols. If the first reel has five bells, the middle reel four bells, and the last reel two bells, find the probability of obtaining three bells in a row.

37 **ESP experiment**   In a simple experiment designed to test ESP, four cards (jack, queen, king, and ace) are shuffled and then placed face down on a table. The subject then attempts to identify each of the four cards, giving a different name to each of the cards. If the indi-

vidual is guessing, find the probability of correctly identifying

(a) all four cards     (b) exactly two of the four cards

**38  Tossing dice**  Three dice are tossed.

  (a) Find the probability that all dice show the same number of dots.

  (b) Find the probability that the numbers of dots on the dice are all different.

  (c) Work parts (a) and (b) for *n* dice.

**39  Trick dice**  For a normal die, the sum of the dots on opposite faces is 7. Shown in the figure is a pair of trick dice in which the *same* number of dots appears on opposite faces. Find the probability of rolling a sum of

(a) 7          (b) 8

**EXERCISE 39**

**40  Carnival game**  In a common carnival game, three balls are rolled down an incline into slots numbered 1 through 9, as shown in the figure. Because the slots are so narrow, players have no control over where the balls collect. A prize is given if the sum of the three numbers is less than 7. Find the probability of winning a prize.

**EXERCISE 40**

**41  Probability demonstration**  Shown in the figure is a small version of a probability demonstration device. A small ball is dropped into the top of the maze and tumbles to the bottom. Each time the ball strikes an obstacle, there is a 50% chance that the ball will move to the left. Find the probability that the ball ends up in the slot

(a) on the far left          (b) in the middle

**EXERCISE 41**

**42  Roulette**  In the American version of roulette, a ball is spun around a wheel and has an equal chance of landing in any one of 38 slots numbered 0, 00, 1, 2, . . . , 36. Shown in the figure is a standard betting layout for roulette, where the color of the oval corresponds to the color of the slot on the wheel. Find the probability that the ball lands

(a) in a black slot

(b) in a black slot twice in succession

**EXERCISE 42**

**43  Selecting lottery numbers**  In one version of a popular lottery game, a player selects six of the numbers from 1 to 54. The agency in charge of the lottery also selects six numbers. What is the probability that the player will match the six numbers if two 50¢ tickets are purchased? (This jackpot is worth at least $2 million in prize money and grows according to the number of tickets sold.)

**44  Lottery**  Refer to Exercise 43. The player can win about $1000 for matching five of the six numbers and about $40 for matching four of the six numbers. Find the probability that the player will win some amount of prize money on the purchase of one ticket.

**45  Playing craps**  In the game of *craps,* there are two ways a player can win a *pass line* bet. The player wins immediately if two dice are rolled and their sum is 7 or 11. If their sum is 4, 5, 6, 8, 9, or 10, the player can still win a pass line bet if this same number (called the *point*) is rolled again before a 7 is rolled. Find the probability that the player wins

(a) a pass line bet on the first roll

(b) a pass line bet with a 4 on the first roll

(c) on any pass line bet

**46 Crapless craps** Refer to Exercise 45. In the game of craps, a player loses a pass line bet if a sum of 2, 3, or 12 is obtained on the first roll (referred to as "craps"). In another version of the game, called *crapless craps,* the player does not lose by rolling craps and does not win by rolling an 11 on the first roll. Instead, the player wins if the first roll is a 7 or if the point (2–12, excluding 7) is repeated before a 7 is rolled. Find the probability that the player wins on a pass line bet in crapless craps.

**Exer. 47–48:** Refer to Examples 7 and 8. **(a)** Find $p$ for the electrical system shown in the figure if $P(S_k) = 0.9$ for each $k$. **(b)** Use a graph to estimate $P(S_k)$ if $p = 0.99$.

**47**

**48**

**49 Birthday probability**

(a) Show that the probability $p$ that $n$ people all have different birthdays is given by

$$p = \frac{365!}{365^n(365 - n)!}.$$

(b) If a room contains 32 people, approximate the probability that two or more people have the *same* birthday. (First approximate $\ln p$ by using the following formula from advanced mathematics:

$$\ln n! \approx n \ln n - n.)$$

**50 Birthday probability** Refer to Exercise 49. Find the smallest number of people in a room such that the probability that everyone has a different birthday is less than $\frac{1}{2}$.

**51 A bet in craps** Refer to Exercise 45. A player receives $2 for winning a $1 pass line bet. Approximate the expected value of a $1 bet.

**52 A bet in roulette** Refer to Exercise 42. If a player bets $1 that the ball will land in a black slot, he or she will receive $2 if it does. Approximate the expected value of a $1 bet.

**53 Contest prize winnings** A contest offers the following cash prizes.

| Number of prizes | 1 | 10 | 100 | 1000 |
|---|---|---|---|---|
| Prize values | $1,000,000 | $100,000 | $10,000 | $1000 |

If the sponsor expects 20 million contestants, find the expected value for a single contestant.

**54 Tournament prize winnings** A bowling tournament is handicapped so that all 80 bowlers are equally matched. The tournament prizes are listed in the table.

| Place | 1st | 2nd | 3rd | 4th | 5th–10th |
|---|---|---|---|---|---|
| Prize | $1000 | $500 | $300 | $200 | $100 |

Find the expected winnings for one contestant.

# CHAPTER 9 REVIEW EXERCISES

**Exer. 1–4:** Find the first four terms and the seventh term of the sequence that has the given $n$th term.

**1** $\left\{\dfrac{5n}{3 - 2n^2}\right\}$

**2** $\{(-1)^{n+1} - (0.1)^n\}$

**3** $\{1 + (-\frac{1}{2})^{n-1}\}$

**4** $\left\{\dfrac{2^n}{(n + 1)(n + 2)(n + 3)}\right\}$

**Exer. 5–8:** Find the first five terms of the recursively defined infinite sequence.

**5** $a_1 = 10,\quad a_{k+1} = 1 + (1/a_k)$

**6** $a_1 = 2,\quad a_{k+1} = a_k!$

**7** $a_1 = 9,\quad a_{k+1} = \sqrt{a_k}$

**8** $a_1 = 1,\quad a_{k+1} = (1 + a_k)^{-1}$

Exer. 9–12: Evaluate.

**9** $\sum_{k=1}^{5} (k^2 + 4)$

**10** $\sum_{k=2}^{6} \dfrac{2k - 8}{k - 1}$

**11** $\sum_{k=7}^{100} 10$

**12** $\sum_{k=1}^{4} (2^k - 10)$

Exer. 13–24: Express the sum in terms of summation notation. (Answers are not unique.)

**13** $3 + 6 + 9 + 12 + 15$

**14** $4 + 2 + 1 + \frac{1}{2} + \frac{1}{4} + \frac{1}{8}$

**15** $\dfrac{1}{1 \cdot 2} + \dfrac{1}{2 \cdot 3} + \dfrac{1}{3 \cdot 4} + \cdots + \dfrac{1}{99 \cdot 100}$

**16** $\dfrac{1}{1 \cdot 2 \cdot 3} + \dfrac{1}{2 \cdot 3 \cdot 4} + \dfrac{1}{3 \cdot 4 \cdot 5} + \cdots + \dfrac{1}{98 \cdot 99 \cdot 100}$

**17** $\frac{1}{2} + \frac{2}{5} + \frac{3}{8} + \frac{4}{11}$

**18** $\frac{1}{4} + \frac{2}{9} + \frac{3}{14} + \frac{4}{19}$

**19** $100 - 95 + 90 - 85 + 80$

**20** $1 - \frac{1}{2} + \frac{1}{3} - \frac{1}{4} + \frac{1}{5} - \frac{1}{6} + \frac{1}{7}$

**21** $a_0 + a_1 x^4 + a_2 x^8 + \cdots + a_{25} x^{100}$

**22** $a_0 + a_1 x^3 + a_2 x^6 + \cdots + a_{20} x^{60}$

**23** $1 - \dfrac{x^2}{2} + \dfrac{x^4}{4} - \dfrac{x^6}{6} + \cdots + (-1)^n \dfrac{x^{2n}}{2n}$

**24** $1 + x + \dfrac{x^2}{2} + \dfrac{x^3}{3} + \cdots + \dfrac{x^n}{n}$

**25** Find the tenth term and the sum of the first ten terms of the arithmetic sequence whose first two terms are $4 + \sqrt{3}$ and 3.

**26** Find the sum of the first eight terms of the arithmetic sequence in which the fourth term is 9 and the common difference is $-5$.

**27** The fifth and thirteenth terms of an arithmetic sequence are 5 and 77, respectively. Find the first term and the tenth term.

**28** Insert four arithmetic means between 20 and $-10$.

**29** Find the tenth term of the geometric sequence whose first two terms are $\frac{1}{8}$ and $\frac{1}{4}$.

**30** If a geometric sequence has 3 and $-0.3$ as its third and fourth terms, find the eighth term.

**31** Find the geometric mean of 4 and 8.

**32** In a certain geometric sequence, the eighth term is 100 and the common ratio is $-\frac{3}{2}$. Find the first term.

**33** Given an arithmetic sequence such that $S_{12} = 402$ and $a_{12} = 50$, find $a_1$ and $d$.

**34** Given a geometric sequence such that $a_5 = \frac{1}{16}$ and $r = \frac{3}{2}$, find $a_1$ and $S_5$.

Exer. 35–38: Evaluate.

**35** $\sum_{k=1}^{15} (5k - 2)$

**36** $\sum_{k=1}^{10} (6 - \frac{1}{2}k)$

**37** $\sum_{k=1}^{10} (2^k - \frac{1}{2})$

**38** $\sum_{k=1}^{8} (\frac{1}{2} - 2^k)$

**39** Find the sum of the infinite geometric series
$$1 - \tfrac{2}{5} + \tfrac{4}{25} - \tfrac{8}{125} + \cdots.$$

**40** Find the rational number whose decimal representation is 6.274.

Exer. 41–45: Prove that the statement is true for every positive integer $n$.

**41** $2 + 5 + 8 + \cdots + (3n - 1) = \dfrac{n(3n + 1)}{2}$

**42** $2^2 + 4^2 + 6^2 + \cdots + (2n)^2 = \dfrac{2n(2n + 1)(n + 1)}{3}$

**43** $\dfrac{1}{1 \cdot 3} + \dfrac{1}{3 \cdot 5} + \dfrac{1}{5 \cdot 7} + \cdots + \dfrac{1}{(2n - 1)(2n + 1)} = \dfrac{n}{2n + 1}$

**44** $1 \cdot 2 + 2 \cdot 3 + 3 \cdot 4 + \cdots + n(n + 1) = \dfrac{n(n + 1)(n + 2)}{3}$

**45** 3 is a factor of $n^3 + 2n$.

**46** Prove that $n^2 + 3 < 2^n$ for every positive integer $n \geq 5$.

Exer. 47–48: Find the smallest positive integer $j$ for which the statement is true. Use the extended principle of mathematical induction to prove that the formula is true for every integer greater than $j$.

**47** $2^n \leq n!$

**48** $10^n \leq n^n$

Exer. 49–50: Use the binomial theorem to expand and simplify the expression.

**49** $(x^2 - 3y)^6$

**50** $(2a + b^3)^4$

Exer. 51–54: Without expanding completely, find the indicated term(s) in the expansion of the expression.

**51** $(a^{2/5} + 2a^{-3/5})^{20}$;   first three terms

**52** $(b^3 - \frac{1}{2}c^2)^9$;         sixth term

**53** $(4a^2 - b)^7$;        term that contains $a^{10}$

**54** $(2c^3 + 5c^{-2})^{10}$;   term that does not contain $c$

**55** *Building blocks*   Ten-foot lengths of $2 \times 2$ lumber are to be cut into five pieces to form children's building blocks; the lengths of the five blocks are to form an arithmetic sequence.

(a) Show that the difference $d$ in lengths must be less than 1 foot.

(b) If the smallest block is to have a length of 6 inches, find the lengths of the other four pieces.

**56  Constructing a ladder**  A ladder is to be constructed with 16 rungs whose lengths decrease uniformly from 20 inches at the base to 16 inches at the top. Find the total length of material needed for the rungs.

**57**  Shown in the first figure is a broken-line curve obtained by taking two adjacent sides of a square, each of length $s_n$, decreasing the length of the side by a factor $f$ with $0 < f < 1$, and forming two sides of a smaller square of length $s_{n+1} = f \cdot s_n$. The process is then repeated ad infinitum. If $s_1 = 1$ in the second figure, express the length of the resulting (infinite) broken-line curve in terms of $f$.

**EXERCISE 57**

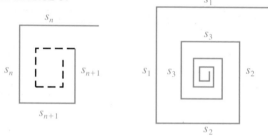

**58  Rebounding ball**  When a ball is dropped from a height of $h$ feet, it reaches the ground in $\sqrt{h}/4$ seconds. The ball rebounds to a height of $d$ feet in $\sqrt{d}/4$ seconds. If a rubber ball is dropped from a height of 10 feet and rebounds to three-fourths of its height after each fall, how many seconds elapse before the ball comes to rest?

**59  Selecting cards**

(a) In how many ways can 13 cards be selected from a deck of 52 cards?

(b) In how many ways can 13 cards be selected to obtain five spades, three hearts, three clubs, and two diamonds?

**60**  How many four-digit numbers can be formed from the digits 1, 2, 3, 4, 5, and 6 if repetitions

(a) are not allowed?        (b) are allowed?

**61  Selecting test questions**

(a) If a student must answer 8 of 12 questions on an examination, how many different selections of questions are possible?

(b) How many selections are possible if the first three questions must be answered?

**62  Color arrangements**  If six black, five red, four white, and two green disks are to be arranged in a row, what is the number of possible color arrangements?

**63  Coin toss**  Find the probability that the coins will match if

(a) two boys each toss a coin

(b) three boys each toss a coin

**64  Dealing cards**  If 4 cards are dealt from a 52-card deck, find the probability that

(a) all four cards will be the same color

(b) the cards dealt will alternate red-black-red-black

**65  Raffle probabilities**  If 1000 tickets are sold for a raffle, find the probability of winning if an individual purchases

(a) 1 ticket        (b) 10 tickets        (c) 50 tickets

**66  Coin toss**  If four coins are flipped, find the probability of obtaining one head and three tails.

**67  True-or-false quiz**  A quiz consists of six true-or-false questions; at least four correct answers are required for a passing grade. If a student guesses at each answer, what is the probability of

(a) passing?                (b) failing?

**68  Die and card probabilities**  If a single die is tossed and then a card is drawn from a 52-card deck, what is the probability of obtaining

(a) a 6 on the die and the king of hearts?

(b) a 6 on the die or the king of hearts?

**69  Population demographics**  In a town of 5000 people, 1000 are over 60 years old and 2000 are female. It is known that 40% of the females are over 60. What is the probability that a randomly chosen individual from the town is either female or over 60?

**70  Backgammon moves**  In the game of backgammon, players are allowed to move their counters the same number of spaces as the sum of the dots on two dice. However, if a double is rolled (that is, both dice show the same number of dots), then players may move their counters twice the sum of the dots. What is the probability that a player will be able to move his or her counters at least 10 spaces on a given roll?

**71  Games in a series**  Two equally matched baseball teams are playing a series of games. The first team to win four games wins the series. Find the expected number of games in the series.

*Conic sections are employed in the design of
a large variety of architectural structures.*

■ Plane geometry includes the study of figures—such as lines, circles, and triangles—that lie in a plane. Theorems are proved by reasoning deductively from certain postulates. In analytic geometry, plane geometric figures are investigated by introducing coordinate systems and then using equations and formulas. If the study of analytic geometry were to be summarized by means of one statement, perhaps the following would be appropriate: Given an equation, find its graph, and conversely, given a graph, find its equation. In this chapter we shall apply coordinate methods to several basic plane figures. ■

# CHAPTER 10

# TOPICS FROM ANALYTIC GEOMETRY

## 10.1  PARABOLAS

The *conic sections*, also called *conics*, can be obtained by intersecting a double-napped right circular cone with a plane. By varying the position of the plane, we obtain a *circle*, an *ellipse*, a *parabola*, or a *hyperbola*, as illustrated in Figure 1.

**FIGURE 1**

(a) Circle      (b) Ellipse      (c) Parabola      (d) Hyperbola

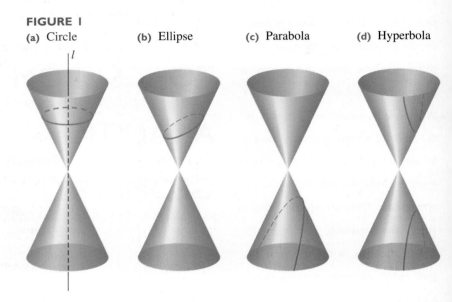

*Degenerate conics* are obtained if the plane intersects the cone in only one point or along either one or two lines that lie on the cone. Conic sections were studied extensively by the ancient Greeks, who discovered properties that enable us to state their definitions in terms of points and lines, as we do in our discussion.

From our work in Section 2.3, if $a \neq 0$, the graph of $y = ax^2 + bx + c$ is a *parabola* with a vertical axis. We shall next state a general definition of a parabola and derive equations for parabolas that have either a vertical axis or a horizontal axis.

**Definition of a Parabola**

A **parabola** is the set of all points in a plane equidistant from a fixed point $F$ (the **focus**) and a fixed line $l$ (the **directrix**) that lie in the plane.

**FIGURE 2**

**FIGURE 3**

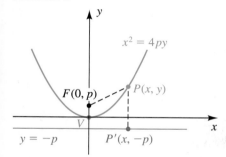

We shall assume that $F$ is not on $l$, for this would result in a line. If $P$ is a point in the plane and $P'$ is the point on $l$ determined by a line through $P$ that is perpendicular to $l$ (see Figure 2), then, by the preceding definition, $P$ is on the parabola if and only if the distances $d(P, F)$ and $d(P, P')$ are equal. The **axis** of the parabola is the line through $F$ that is perpendicular to the directrix. The **vertex** of the parabola is the point $V$ on the axis halfway from $F$ to $l$. The vertex is the point on the parabola that is closest to the directrix.

To obtain a simple equation for a parabola, place the $y$-axis along the axis of the parabola, with the origin at the vertex $V$, as shown in Figure 3. In this case, the focus $F$ has coordinates $(0, p)$ for some real number $p \neq 0$, and the equation of the directrix is $y = -p$. (The figure shows the case $p > 0$.) By the distance formula, a point $P(x, y)$ is on the parabola if and only if $d(P, F) = d(P, P')$—that is, if

$$\sqrt{(x - 0)^2 + (y - p)^2} = \sqrt{(x - x)^2 + (y + p)^2}.$$

We square both sides and simplify:

$$x^2 + (y - p)^2 = (y + p)^2$$
$$x^2 + y^2 - 2py + p^2 = y^2 + 2py + p^2$$
$$x^2 = 4py$$

An equivalent equation for the parabola is

$$y = \frac{1}{4p}x^2.$$

We have shown that the coordinates of every point $(x, y)$ on the parabola satisfy $x^2 = 4py$. Conversely, if $(x, y)$ is a solution of $x^2 = 4py$, then by reversing the previous steps we see that the point $(x, y)$ is on the parabola. If $p > 0$, the parabola opens upward, as in Figure 3. If $p < 0$, the parabola opens downward. The graph is symmetric with respect to the $y$-axis, since substitution of $-x$ for $x$ does not change the equation $x^2 = 4py$.

If we interchange the roles of $x$ and $y$, we obtain

$$y^2 = 4px \quad \text{or, equivalently,} \quad x = \frac{1}{4p}y^2.$$

This is an equation of a parabola with vertex at the origin, focus $F(p, 0)$, and opening right if $p > 0$ or left if $p < 0$. The equation of the directrix is $x = -p$.

For convenience we often refer to "the parabola $x^2 = 4py$ (or $y^2 = 4px$)" instead of "the parabola with equation $x^2 = 4py$ (or $y^2 = 4px$)." The next chart summarizes our discussion.

**Parabolas with Vertex** $V(0, 0)$

| Equation | Graph for $p > 0$ | Graph for $p < 0$ |
|---|---|---|
| $x^2 = 4py$ <br><br> or $y = \dfrac{1}{4p}x^2$ | | |
| $y^2 = 4px$ <br><br> or $x = \dfrac{1}{4p}y^2$ | | |

We see from the chart that for any nonzero real number $a$, the graph of the **standard equation** $y = ax^2$ or $x = ay^2$ is a parabola with vertex $V(0, 0)$. Moreover, $a = 1/(4p)$ or, equivalently, $p = 1/(4a)$, where $|p|$ is the distance between the focus $F$ and vertex $V$. To find the directrix $l$, recall that $l$ is also a distance $|p|$ from $V$.

**EXAMPLE 1**     *Finding the focus and directrix of a parabola*

Find the focus and directrix of the parabola $y = -\frac{1}{6}x^2$, and sketch its graph.

**SOLUTION**     The equation has the form $y = ax^2$ with $a = -\frac{1}{6}$. As in the preceding chart, $a = 1/(4p)$, and hence

$$p = \frac{1}{4a} = \frac{1}{4\left(-\frac{1}{6}\right)} = -\frac{3}{2}.$$

**FIGURE 4**

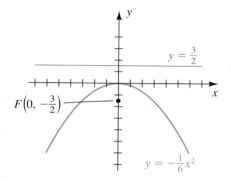

Thus, the parabola opens downward and has focus $F\left(0, -\frac{3}{2}\right)$, as illustrated in Figure 4. The directrix is the horizontal line $y = \frac{3}{2}$, which is a distance $\frac{3}{2}$ above $V$, as shown in the figure.

**EXAMPLE 2** *Finding an equation of a parabola satisfying prescribed conditions*

**(a)** Find an equation of a parabola that has vertex at the origin, opens right, and passes through the point $P(7, -3)$.

**(b)** Find the focus of the parabola.

*SOLUTION*

**(a)** The parabola is sketched in Figure 5. As in the chart, an equation of the parabola has the form $x = ay^2$ for some number $a$. If $P(7, -3)$ is on the graph, then

$$7 = a(-3)^2, \quad \text{or} \quad a = \tfrac{7}{9}.$$

Hence, an equation for the parabola is $x = \frac{7}{9}y^2$.

**(b)** The focus is a distance $p$ to the right of the vertex, where

$$p = \frac{1}{4a} = \frac{1}{4\left(\frac{7}{9}\right)} = \frac{9}{28}.$$

Thus, the focus has coordinates $\left(\frac{9}{28}, 0\right)$.

**FIGURE 5**

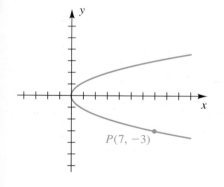

To extend our discussion to a parabola with vertex *not* at the origin, we use a **translation of axes,** as illustrated in Figure 6, where the $x$- and

**FIGURE 6**

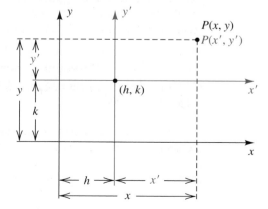

$y$-axes are shifted to positions—denoted by $x'$ and $y'$—that are parallel to their original positions. Every point $P$ in the plane then has two different ordered-pair representations: $P(x, y)$ in the $xy$-system and $P(x', y')$ in the $x'y'$-system. If the origin of the new $x'y'$-system has coordinates $(h, k)$ in the $xy$-plane, as illustrated in Figure 6, we see that

$$x = x' + h \quad \text{and} \quad y = y' + k.$$

These formulas are true for all values of $h$ and $k$. Equivalent formulas are

$$x' = x - h \quad \text{and} \quad y' = y - k.$$

This gives us the following.

**Translation of Axes Formulas**

If $(x, y)$ are the coordinates of a point $P$ in an $xy$-plane and if $(x', y')$ are the coordinates of $P$ in an $x'y'$-plane with origin at the point $(h, k)$ of the $xy$-plane, then

(1) $x = x' + h, \quad y = y' + k$

(2) $x' = x - h, \quad y' = y - k$

To illustrate the translation of axes formulas, let us consider

$$(x')^2 = 4py',$$

which we know is an equation of a parabola with vertex at the origin $O'$ of the $x'y'$-plane. Using formula 2, we see that

$$(x - h)^2 = 4p(y - k)$$

is an equation of the same parabola in the $xy$-plane with vertex $V(h, k)$. The focus is $F(h, k + p)$, and the directrix is $y = k - p$, as shown in the first row of the next chart. Squaring the left side of $(x - h)^2 = 4p(y - k)$ and simplifying leads to an equation of the form

$$y = ax^2 + bx + c,$$

where $a$, $b$, and $c$ are real numbers. Conversely, if $a \neq 0$, then the graph of $y = ax^2 + bx + c$ is a parabola with a vertical axis. As with $y = ax^2$, we can show that $a = 1/(4p)$ or, equivalently, $p = 1/(4a)$.

Similarly, if we begin with $(y')^2 = 4px'$ and use the translation of axes formulas, we obtain

$$(y - k)^2 = 4p(x - h),$$

which is an equation of a parabola in the $xy$-plane with vertex $V(h, k)$, opening right if $p > 0$ or left if $p < 0$, as shown in the second row of the following chart (where, for convenience, we have taken $V(h, k)$ in the first quadrant). This equation may be written in the form $x = ay^2 + by + c$, where $a = 1/(4p)$ or, equivalently, $p = 1/(4a)$.

**Parabolas with Vertex $V(h, k)$**

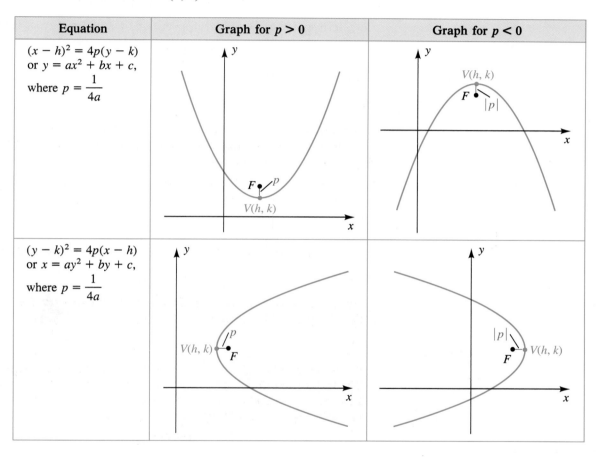

| Equation | Graph for $p > 0$ | Graph for $p < 0$ |
|---|---|---|
| $(x - h)^2 = 4p(y - k)$ or $y = ax^2 + bx + c$, where $p = \dfrac{1}{4a}$ | | |
| $(y - k)^2 = 4p(x - h)$ or $x = ay^2 + by + c$, where $p = \dfrac{1}{4a}$ | | |

**EXAMPLE 3**  *Sketching a parabola with a horizontal axis*

Discuss and sketch the graph of $2x = y^2 + 8y + 22$.

**SOLUTION**  The equation can be written in the form shown in the second row, $x = ay^2 + by + c$, so we see from the previous chart that the graph is a parabola with a horizontal axis. We first write the given equation as

$$y^2 + 8y = 2x - 22$$

**FIGURE 7**

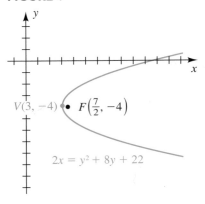

$2x = y^2 + 8y + 22$

and then complete the square on the left by adding 16 to both sides:

$$y^2 + 8y + 16 = 2x - 6$$
$$(y + 4)^2 = 2(x - 3)$$

Referring to the last chart, we see that $h = 3$, $k = -4$, and $4p = 2$ or, equivalently, $p = \frac{1}{2}$. This gives us the following:

The vertex $V(h, k)$ is $V(3, -4)$.

The focus is $F(h + p, k) = F(3 + \frac{1}{2}, -4)$, or $F(\frac{7}{2}, -4)$.

The directrix is $x = h - p = 3 - \frac{1}{2}$, or $x = \frac{5}{2}$.

The parabola is sketched in Figure 7.

**FIGURE 8**

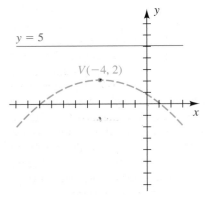

**EXAMPLE 4**    *Finding an equation of a parabola given its vertex and directrix*

A parabola has vertex $V(-4, 2)$ and directrix $y = 5$. Express the equation of the parabola in the form $y = ax^2 + bx + c$.

**SOLUTION**    The vertex and directrix are shown in Figure 8. The dashes indicate a possible position for the parabola. The last chart shows that an equation of the parabola is

$$(x - h)^2 = 4p(y - k),$$

with $h = -4$ and $k = 2$ and with $p$ equal to *negative* 3, since $V$ is 3 units *below* the directrix. This gives us

$$(x + 4)^2 = -12(y - 2).$$

The last equation can be expressed in the form $y = ax^2 + bx + c$, as follows:

$$x^2 + 8x + 16 = -12y + 24$$
$$12y = -x^2 - 8x + 8$$
$$y = -\tfrac{1}{12}x^2 - \tfrac{2}{3}x + \tfrac{2}{3}$$

**FIGURE 9**

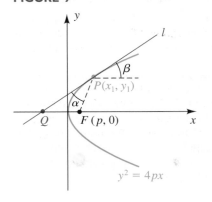

$y^2 = 4px$

An important property is associated with a tangent line to a parabola. (A *tangent line* to a parabola is a line that has exactly one point in common with the parabola but does not cut through the parabola.) Suppose $l$ is the tangent line at a point $P(x_1, y_1)$ on the graph of $y^2 = 4px$, and let $F$ be the focus. As in Figure 9, let $\alpha$ denote the angle between $l$ and the line segment $FP$, and let $\beta$ denote the angle between $l$ and the indicated horizontal

**FIGURE 10**

(a) Searchlight mirror

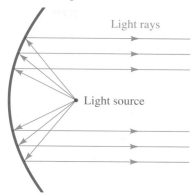

Light rays

Light source

(b) Telescope mirror

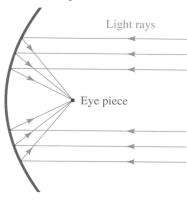

Light rays

Eye piece

**FIGURE 11**

2′  12′

half-line with end point $P$. It can be shown that $\alpha = \beta$. This *reflective property* has many applications. For example, the shape of the mirror in a searchlight is obtained by revolving a parabola about its axis. The resulting three-dimensional surface is said to be *generated* by the parabola and is called a **paraboloid.** The **focus** of the paraboloid is the same as the focus of the generating parabola. If a light source is placed at $F$, then, by a law of physics (the angle of reflection equals the angle of incidence), a beam of light will be reflected along a line parallel to the axis (see Figure 10(a)). The same principle is employed in the construction of mirrors for telescopes or solar ovens—a beam of light coming toward the parabolic mirror and parallel to the axis will be reflected into the focus (see Figure 10(b)). Antennas for radar systems, radio telescopes, and field microphones used at football games also make use of this property.

**FIGURE 12**

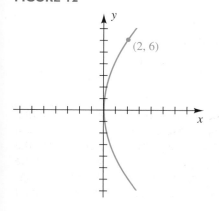

(2, 6)

**EXAMPLE 5**   *Locating the focus of a satellite TV antenna*

The interior of a satellite TV antenna is a dish having the shape of a (finite) paraboloid that has diameter 12 feet and is 2 feet deep, as shown in Figure 11. Find the distance from the center of the dish to the focus.

*SOLUTION*   The generating parabola is sketched on an $xy$-plane in Figure 12, where we have taken the vertex of the parabola at the origin and the axis along the $x$-axis. An equation of the parabola is $y^2 = 4px$, where $p$ is the required distance from the center of the dish to the focus. Since the point $(2, 6)$ is on the parabola, we obtain

$$6^2 = 4p \cdot 2, \quad \text{or} \quad p = \tfrac{36}{8} = 4.5 \text{ ft.}$$

In the next example we use a graphing utility to sketch the graph of a parabola with a horizontal axis.

**EXAMPLE 6    *Graphing half-parabolas***

Graph $x = y^2 + 2y - 4$.

**SOLUTION**    Some specific keystrokes for the TI-81/82 graphics calculator are given in Appendix I. The graph is a parabola with a horizontal axis. We begin by solving the equivalent equation

$$y^2 + 2y - 4 - x = 0$$

for $y$ in terms of $x$ by using the quadratic formula with $a = 1$, $b = 2$, and $c = -4 - x$:

$$y = \frac{-2 \pm \sqrt{2^2 - 4(1)(-4 - x)}}{2(1)} \qquad \text{quadratic formula}$$

$$= \frac{-2 \pm \sqrt{20 + 4x}}{2} \qquad \text{simplify}$$

$$= -1 \pm \sqrt{x + 5} \qquad \text{simplify}$$

**FIGURE 13**   $[-15, 15]$ by $[-10, 10]$

The last equation, $y = -1 \pm \sqrt{x + 5}$, represents the top half of the parabola ($y = -1 + \sqrt{x + 5}$) and the bottom half ($y = -1 - \sqrt{x + 5}$). Note that $y = -1$ is the axis of the parabola.

Next, we make the assignments

$$Y_1 = \sqrt{x + 5}, \quad Y_2 = -1 + Y_1, \quad \text{and} \quad Y_3 = -1 - Y_1.$$

We select the functions $Y_2$ and $Y_3$ to be graphed with a standard viewing rectangle, and we obtain a display similar to Figure 13.

## *10.1 EXERCISES*

**Exer. 1–12: Find the vertex, focus, and directrix of the parabola. Sketch its graph, showing the focus and the directrix.**

1  $8y = x^2$

2  $20x = y^2$

3  $2y^2 = -3x$

4  $x^2 = -3y$

5  $(x + 2)^2 = -8(y - 1)$

6  $(x - 3)^2 = \frac{1}{2}(y + 1)$

7  $(y - 2)^2 = \frac{1}{4}(x - 3)$

8  $(y + 1)^2 = -12(x + 2)$

9  $y = x^2 - 4x + 2$

10  $y^2 + 14y + 4x + 45 = 0$

11  $x^2 + 20y = 10$

12  $y^2 - 4y - 2x - 4 = 0$

**Exer. 13–16: Find an equation for the parabola shown in the figure.**

13

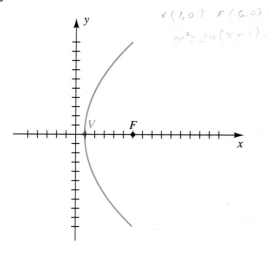

*handwritten:* $V(1,0)$  $F(6,0)$
$y^2 = 20(x-1)$

16

$P$

$V(3, -2)$

14

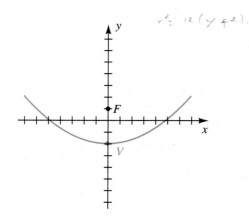

*handwritten:* $x^2 = 12(y+2)$

**Exer. 17–28: Find an equation of the parabola that satisfies the given conditions.**

17  Focus $F(2, 0)$,     directrix $x = -2$   *handwritten:* $y^2 = 8x$

18  Focus $F(0, -4)$,    directrix $y = 4$

19  Focus $F(6, 4)$,     directrix $y = -2$   *handwritten:* $(x-6)^2 = 12(y-1)$

20  Focus $F(-3, -2)$,   directrix $y = 1$

21  Vertex $V(3, -5)$,   directrix $x = 2$

22  Vertex $V(-2, 3)$,   directrix $y = 5$

23  Vertex $V(-1, 0)$,   focus $F(-4, 0)$

24  Vertex $V(1, -2)$,   focus $F(1, 0)$

25  Vertex at the origin, symmetric to the $y$-axis, and passing through the point $(2, -3)$

26  Vertex at the origin, symmetric to the $y$-axis, and passing through the point $(6, 3)$

27  Vertex $V(-3, 5)$, axis parallel to the $x$-axis, and passing through the point $(5, 9)$

28  Vertex $V(3, -2)$, axis parallel to the $x$-axis, and $y$-intercept 1

15

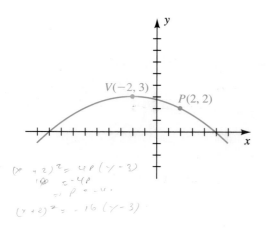

$V(-2, 3)$   $P(2, 2)$

*handwritten:*
$(x+2)^2 = 4p(y-3)$
$16 = -4p$
$p = -4$
$(x+2)^2 = -16(y-3)$

**Exer. 29–32: Find an equation for the set of points in an $xy$-plane that are equidistant from the point $P$ and the line $l$.**

29  $P(0, 5)$;     $l: y = -3$       30  $P(7, 0)$;     $l: x = 1$

31  $P(-6, 3)$;    $l: x = -2$       32  $P(5, -2)$;    $l: y = 4$

**Exer. 33–36: Find an equation for the indicated half of the parabola.**

33 Lower half of $(y + 1)^2 = x + 3$

34 Upper half of $(y - 2)^2 = x - 4$

35 Right half of $(x + 1)^2 = y - 4$

36 Left half of $(x + 3)^2 = y + 2$

**Exer. 37–38: Find an equation for the parabola that has a vertical axis and passes through the given points.**

37 $P(2, 5),$    $Q(-2, -3),$    $R(1, 6)$

38 $P(3, -1),$    $Q(1, -7),$    $R(-2, 14)$

**Exer. 39–40: Find an equation for the parabola that has a horizontal axis and passes through the given points.**

39 $P(-1, 1),$    $Q(11, -2),$    $R(5, -1)$

40 $P(2, 1),$    $Q(6, 2),$    $R(12, -1)$

41 *Telescope mirror*   A mirror for a reflecting telescope has the shape of a (finite) paraboloid of diameter 8 inches and depth 1 inch. How far from the center of the mirror will the incoming light collect?

EXERCISE 41

42 *Antenna dish*   A satellite antenna dish has the shape of a paraboloid that is 10 feet across at the open end and is 3 feet deep. At what distance from the center of the dish should the receiver be placed to receive the greatest intensity of sound waves?

43 *Searchlight reflector*   A searchlight reflector has the shape of a paraboloid with the light source at the focus. If the reflector is 3 feet across at the opening and 1 foot deep, where is the focus?

44 *Flashlight mirror*   A flashlight mirror has the shape of a paraboloid of diameter 4 inches and depth $\frac{3}{4}$ inch, as shown in the figure. Where should the bulb be placed so that the emitted light rays are parallel to the axis of the paraboloid?

EXERCISE 44

45 *Receiving dish*   A sound receiving dish used at outdoor sporting events is constructed in the shape of a paraboloid with its focus 5 inches from the vertex. Determine the width of the dish if the depth is to be 2 feet.

46 *Receiving dish*   Work Exercise 45 if the receiver is 9 inches from the vertex.

47 *Parabolic reflector*
  (a) The focal length of the (finite) paraboloid in the figure is the distance $p$ between its vertex and focus. Express $p$ in terms of $r$ and $h$.
  (b) A reflector is to be constructed with a focal length of 10 feet and a depth of 5 feet. Find the radius of the reflector.

EXERCISE 47

48 *Confocal parabolas*   The parabola $y^2 = 4p(x + p)$ has its focus at the origin and axis along the x-axis. By as-

EXERCISE 48

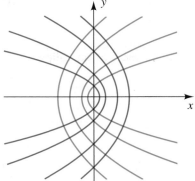

signing different values to $p$, we obtain a family of confocal parabolas, as shown in the figure. Such families occur in the study of electricity and magnetism. Show that there are exactly two parabolas in the family that pass through a given point $P(x_1, y_1)$ if $y_1 \neq 0$.

49  **Jodrell Bank radio telescope**   A radio telescope has the shape of a paraboloid of revolution with focal length $p$ and diameter of base $2a$. From calculus, the surface area $S$ available for collecting radio waves is

$$S = \frac{8\pi p^2}{3}\left[\left(1 + \frac{a^2}{4p^2}\right)^{3/2} - 1\right].$$

One of the largest radio telescopes, located in Jodrell Bank, Cheshire, England, has diameter 250 feet and focal length 50 feet. Approximate $S$ to the nearest square foot.

50  **Focal length and angular velocity**   A cylindrical container, partially filled with mercury, is rotated about its axis so that the angular speed of each cross section is $\omega$ radians/second (see Example 6 of Section 5.1). From physics, the function $f$, whose graph generates the inside surface of the mercury (see the figure), is given by

$$f(x) = \tfrac{1}{64}\omega^2 x^2 + k,$$

where $k$ is a constant. Determine the angular velocity $\omega$ that will result in a focal length of 2 feet.

EXERCISE 50

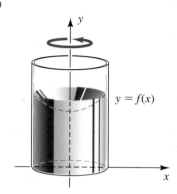

Exer. 51–52: Graph the equation.

51  $x = -y^2 + 2y + 5$          52  $x = 2y^2 + 3y - 7$

Exer. 53–54: Graph the parabolas on the same coordinate plane, and estimate the points of intersection.

53  $y = x^2 - 2.1x - 1$;          $x = y^2 + 1$

54  $y = -2.1x^2 + 0.1x + 1.2$;   $x = 0.6y^2 + 1.7y - 1.1$

# 10.2   ELLIPSES

An ellipse may be defined as follows. (*Foci* is the plural of *focus*.)

**Definition of an Ellipse**

An **ellipse** is the set of all points in a plane, the sum of whose distances from two fixed points (the **foci**) in the plane is a positive constant.

We can construct an ellipse on paper as follows: Insert two pushpins in the paper at any points $F$ and $F'$, and fasten the ends of a piece of string to the pins. After looping the string around a pencil and drawing it tight, as at

FIGURE 14

FIGURE 15

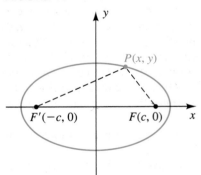

point $P$ in Figure 14, move the pencil, keeping the string tight. The sum of the distances $d(F, P)$ and $d(F', P)$ is the length of the string and hence is constant; thus, the pencil will trace out an ellipse with foci at $F$ and $F'$. The midpoint of the segment $F'F$ is called the **center** of the ellipse. By changing the positions of $F$ and $F'$ while keeping the length of the string fixed, we can vary the shape of the ellipse considerably. If $F$ and $F'$ are far apart so that $d(F, F')$ is almost the same as the length of the string, the ellipse is flat. If $d(F, F')$ is close to zero, the ellipse is almost circular. If $F = F'$, we obtain a circle with center $F$.

To obtain a simple equation for an ellipse, choose the $x$-axis as the line through the two foci $F$ and $F'$, with the center of the ellipse at the origin. If $F$ has coordinates $(c, 0)$ with $c > 0$, then, as in Figure 15, $F'$ has coordinates $(-c, 0)$. Hence, the distance between $F$ and $F'$ is $2c$. The constant sum of the distances of $P$ from $F$ and $F'$ will be denoted by $2a$. To obtain points that are not on the $x$-axis, we must have $2a > 2c$—that is, $a > c$. By definition, $P(x, y)$ is on the ellipse if and only if the following equivalent equations are true:

$$d(P, F) + d(P, F') = 2a$$
$$\sqrt{(x - c)^2 + (y - 0)^2} + \sqrt{(x + c)^2 + (y - 0)^2} = 2a$$
$$\sqrt{(x - c)^2 + y^2} = 2a - \sqrt{(x + c)^2 + y^2}$$

Squaring both sides of the last equation gives us

$$x^2 - 2cx + c^2 + y^2 = 4a^2 - 4a\sqrt{(x + c)^2 + y^2} + x^2 + 2cx + c^2 + y^2,$$

or

$$a\sqrt{(x + c)^2 + y^2} = a^2 + cx.$$

Squaring both sides again yields

$$a^2(x^2 + 2cx + c^2 + y^2) = a^4 + 2a^2cx + c^2x^2,$$

or

$$x^2(a^2 - c^2) + a^2y^2 = a^2(a^2 - c^2).$$

Dividing both sides by $a^2(a^2 - c^2)$, we obtain

$$\frac{x^2}{a^2} + \frac{y^2}{a^2 - c^2} = 1.$$

Recalling that $a > c$ and therefore $a^2 - c^2 > 0$, we let

$$b = \sqrt{a^2 - c^2}, \quad \text{or} \quad b^2 = a^2 - c^2.$$

This gives us

$$\frac{x^2}{a^2} + \frac{y^2}{b^2} = 1.$$

Since $c > 0$ and $b^2 = a^2 - c^2$, it follows that $a^2 > b^2$ and hence $a > b$.

We have shown that the coordinates of every point $(x, y)$ on the ellipse in Figure 16 satisfy the equation $(x^2/a^2) + (y^2/b^2) = 1$. Conversely, if $(x, y)$ is a solution of this equation, then by reversing the preceding steps we see that the point $(x, y)$ is on the ellipse.

**FIGURE 16**

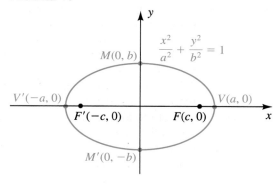

We may find the $x$-intercepts of the ellipse by letting $y = 0$ in the equation. Doing so gives us $x^2/a^2 = 1$, or $x^2 = a^2$. Consequently, the $x$-intercepts are $a$ and $-a$. The corresponding points $V(a, 0)$ and $V'(-a, 0)$ on the graph are called the **vertices** of the ellipse (see Figure 16). The line segment $V'V$ is called the **major axis**. Similarly, letting $x = 0$ in the equation, we obtain $y^2/b^2 = 1$, or $y^2 = b^2$. Hence, the $y$-intercepts are $b$ and $-b$. The segment between $M'(0, -b)$ and $M(0, b)$ is called the **minor axis** of the ellipse. The major axis is always longer than the minor axis, since $a > b$.

Applying tests for symmetry, we see that the ellipse is symmetric with respect to the $x$-axis, the $y$-axis, and the origin.

Similarly, if we take the foci on the $y$-axis, we obtain the equation

$$\frac{x^2}{b^2} + \frac{y^2}{a^2} = 1.$$

In this case the vertices of the ellipse are $(0, \pm a)$, and the end points of the minor axis are $(\pm b, 0)$, as shown in Figure 17.

The preceding discussion may be summarized as follows.

**FIGURE 17**

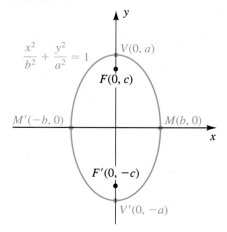

**Standard Equations of an Ellipse with Center at the Origin**

The graph of

$$\frac{x^2}{a^2} + \frac{y^2}{b^2} = 1 \quad \text{or} \quad \frac{x^2}{b^2} + \frac{y^2}{a^2} = 1,$$

where $a > b > 0$, is an ellipse with center at the origin. The length of the major axis is $2a$, and the length of the minor axis is $2b$. The foci are a distance $c$ from the origin, where $c^2 = a^2 - b^2$.

We have shown that an equation of an ellipse with center at the origin and foci on a coordinate axis can always be written in the form

$$\frac{x^2}{p} + \frac{y^2}{q} = 1, \quad \text{or} \quad qx^2 + py^2 = pq,$$

with $p$ and $q$ positive and $p \neq q$. If $p > q$, the major axis is on the $x$-axis; if $p < q$, the major axis is on the $y$-axis. It is unnecessary to memorize these facts, because in any given problem the major axis can be determined by examining the $x$- and $y$-intercepts.

**EXAMPLE I     _Sketching an ellipse with center at the origin_**

Sketch the graph of $2x^2 + 9y^2 = 18$, and find the foci.

**FIGURE 18**

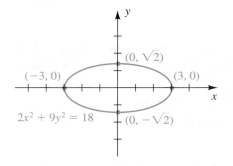

**SOLUTION**     The graph is an ellipse with center at the origin and foci on a coordinate axis. To find the $x$-intercepts, we let $y = 0$, obtaining

$$2x^2 = 18, \quad \text{or} \quad x = \pm 3.$$

To find the $y$-intercepts, we let $x = 0$, obtaining

$$9y^2 = 18, \quad \text{or} \quad y = \pm\sqrt{2}.$$

We are now able to sketch the ellipse (Figure 18). Since $\sqrt{2} \approx 1.4 < 3$, the major axis is on the $x$-axis.

To find the foci, we let $a = 3$, $b = \sqrt{2}$, and we calculate

$$c^2 = a^2 - b^2 = 3^2 - (\sqrt{2})^2 = 7.$$

Thus, $c = \sqrt{7}$, and the foci are $(\pm\sqrt{7}, 0)$.

**EXAMPLE 2     _Sketching an ellipse with center at the origin_**

Sketch the graph of $9x^2 + 4y^2 = 25$, and find the foci.

**FIGURE 19**

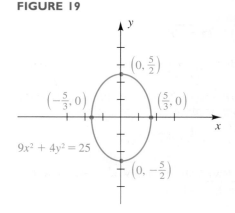

**SOLUTION**     As in Example 1, the graph is an ellipse with center at the origin and foci on a coordinate axis. To find the $x$-intercepts, we let $y = 0$, obtaining

$$9x^2 = 25, \quad \text{or} \quad x = \pm\tfrac{5}{3}.$$

To find the $y$-intercepts, we let $x = 0$, obtaining

$$4y^2 = 25, \quad \text{or} \quad y = \pm\tfrac{5}{2}.$$

These calculations give us the sketch in Figure 19. Since $\tfrac{5}{3} < \tfrac{5}{2}$, the major axis is on the $y$-axis.

To find the foci, we let $a = \tfrac{5}{2}$, $b = \tfrac{5}{3}$, and we calculate

$$c^2 = a^2 - b^2 = \left(\tfrac{5}{2}\right)^2 - \left(\tfrac{5}{3}\right)^2 = \tfrac{125}{36}.$$

Thus, $c = \sqrt{125/36} = 5\sqrt{5}/6 \approx 1.86$, and the foci are $(0, \pm 5\sqrt{5}/6)$.

**EXAMPLE 3**    *Finding an equation of an ellipse given its vertices and foci*

Find an equation of the ellipse with vertices $(\pm 4, 0)$ and foci $(\pm 2, 0)$.

**SOLUTION**    Since the foci are on the $x$-axis and are equidistant from the origin, the major axis is on the $x$-axis and the ellipse has center $(0, 0)$. Thus, a general equation of an ellipse is

$$\frac{x^2}{a^2} + \frac{y^2}{b^2} = 1.$$

Since the vertices are $(\pm 4, 0)$, we conclude that $a = 4$. Since the foci are $(\pm 2, 0)$, we have $c = 2$. Hence,

$$b^2 = a^2 - c^2 = 4^2 - 2^2 = 12,$$

and an equation of the ellipse is

$$\frac{x^2}{16} + \frac{y^2}{12} = 1.$$

In certain applications it is necessary to work with only one-half of an ellipse. The next example indicates how to find equations in such cases.

**EXAMPLE 4**    *Finding equations for half-ellipses*

Find equations for the upper half, lower half, left half, and right half of the ellipse $9x^2 + 4y^2 = 25$.

**SOLUTION**    The graph of the entire ellipse was sketched in Figure 19. To find equations for the upper and lower halves, we solve for $y$ in terms of $x$, as follows:

$$9x^2 + 4y^2 = 25 \qquad\qquad \text{given}$$

$$y^2 = \frac{25 - 9x^2}{4} \qquad\qquad \text{solve for } y^2$$

$$y = \pm\sqrt{\frac{25 - 9x^2}{4}} = \pm\frac{1}{2}\sqrt{25 - 9x^2} \qquad \text{take the square root}$$

Since $\sqrt{25 - 9x^2} \geq 0$, it follows that equations for the upper and lower halves are $y = \frac{1}{2}\sqrt{25 - 9x^2}$ and $y = -\frac{1}{2}\sqrt{25 - 9x^2}$, respectively.

To find equations for the left and right halves, we use a procedure similar to that above and solve for $x$ in terms of $y$, obtaining

$$x = \pm\sqrt{\frac{25 - 4y^2}{9}} = \pm\frac{1}{3}\sqrt{25 - 4y^2}.$$

The left half of the ellipse has the equation $x = -\frac{1}{3}\sqrt{25 - 4y^2}$, and the right half is given by $x = \frac{1}{3}\sqrt{25 - 4y^2}$.

We can use the translation of axes formulas from the last section to extend our work to an ellipse with center at any point $C(h, k)$ in the $xy$-plane. For example, since the graph of

$$\frac{(x')^2}{a^2} + \frac{(y')^2}{b^2} = 1$$

is an ellipse with center at $O'$ in an $x'y'$-plane (see Figure 20), its equation relative to the $xy$-coordinate system is

$$\frac{(x - h)^2}{a^2} + \frac{(y - k)^2}{b^2} = 1.$$

**FIGURE 20**

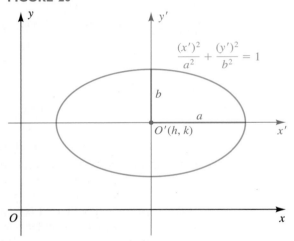

Squaring terms in the last equation and simplifying gives us an equation of the form

$$Ax^2 + Cy^2 + Dx + Ey + F = 0,$$

where the coefficients are real numbers and both $A$ and $C$ are positive. Conversely, if we start with such an equation, then by completing squares we can obtain a form that displays the center of the ellipse and the lengths of the major and minor axes. This technique is illustrated in the next example.

**EXAMPLE 5**   *Sketching an ellipse with center $(h, k)$*

Discuss and sketch the graph of the equation

$$16x^2 + 9y^2 + 64x - 18y - 71 = 0.$$

**SOLUTION**    We begin by grouping the terms containing $x$ and those containing $y$:

$$(16x^2 + 64x) + (9y^2 - 18y) = 71$$

Next, we factor out the coefficients of $x^2$ and $y^2$ as follows:

$$16(x^2 + 4x + \underline{\quad}) + 9(y^2 - 2y + \underline{\quad}) = 71$$

We now complete the squares for the expressions within parentheses:

$$16(x^2 + 4x + 4) + 9(y^2 - 2y + 1) = 71 + 64 + 9$$

By adding 4 to the expression within the first parentheses we have added 64 to the left side of the equation, and hence we must compensate by adding 64 to the right side. Similarly, by adding 1 to the expression within the second parentheses we have added 9 to the left side, and consequently we must also add 9 to the right side. The last equation may be written

$$16(x + 2)^2 + 9(y - 1)^2 = 144.$$

Dividing by 144 to obtain 1 on the right side gives us

$$\frac{(x + 2)^2}{9} + \frac{(y - 1)^2}{16} = 1,$$

which is of the form

$$\frac{(x')^2}{9} + \frac{(y')^2}{16} = 1,$$

with $x' = x + 2$ and $y' = y - 1$. This corresponds to letting $h = -2$ and $k = 1$ in the translation of axes formulas.

The graph of the equation $((x')^2/9) + ((y')^2/16) = 1$ is an ellipse with center at the origin $O'$ in the $x'y'$-plane and major axis on the $y'$-axis. It follows that the graph of the given equation is an ellipse with center $C(-2, 1)$ in the $xy$-plane and major axis on the vertical line $x = -2$. Using $a = 4$ and $b = 3$ gives us the ellipse in Figure 21.

**FIGURE 21**

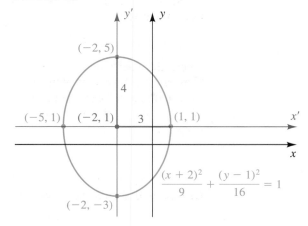

*(continued)*

To find the foci, we first calculate

$$c^2 = a^2 - b^2 = 4^2 - 3^2 = 7.$$

The distance from the center of the ellipse to the foci is $c = \sqrt{7}$. Since the center is $(-2, 1)$, the foci are $(-2, 1 \pm \sqrt{7})$.

---

Graphics calculators and computer programs are often unable to plot the graphs of an equation of the form

$$Ax^2 + Cy^2 + Dx + Ey + F = 0,$$

such as that considered in the last example. In most cases we must first solve the equation for $y$ in terms of $x$ and then plot the two resulting functions, as illustrated in the next example.

### EXAMPLE 6    *Graphing half-ellipses*

Use a graphing utility to sketch the graph of

$$3x^2 + 4y^2 + 12x - 8y + 9 = 0.$$

SOLUTION    The equation may be regarded as a quadratic equation in $y$ of the form $ay^2 + by + c = 0$ by rearranging terms as follows:

$$4y^2 - 8y + (3x^2 + 12x + 9) = 0$$

Applying the quadratic formula to the previous equation with $a = 4$, $b = -8$, and $c = 3x^2 + 12x + 9$ gives us

$$y = \frac{-(-8) \pm \sqrt{(-8)^2 - 4(4)(3x^2 + 12x + 9)}}{2(4)}$$

$$= \frac{8 \pm \sqrt{64 - 16(3x^2 + 12x + 9)}}{8}$$

$$= 1 \pm \tfrac{1}{8}\sqrt{64 - 16(3x^2 + 12x + 9)}.$$

**FIGURE 22**   $[-6, 6]$ by $[-4, 4]$

Note that we did not completely simplify the radicand, since we will be using a graphing utility.

As in Example 6 of the previous section, we now make the assignments

$$Y_1 = \tfrac{1}{8}\sqrt{64 - 16(3x^2 + 12x + 9)}, \quad Y_2 = 1 + Y_1, \quad \text{and} \quad Y_3 = 1 - Y_1.$$

Finally, we select $Y_2$ and $Y_3$ to be graphed, choose the viewing rectangle $[-6, 6]$ by $[-4, 4]$, and obtain a display similar to Figure 22.

---

Ellipses can be very flat or almost circular. To obtain information about the *roundness* of an ellipse, we sometimes use the term *eccentricity*, which is defined as follows, with $a$, $b$, and $c$ having the same meanings as before.

**Definition of Eccentricity**

The **eccentricity** $e$ of an ellipse is

$$e = \frac{c}{a} = \frac{\sqrt{a^2 - b^2}}{a}.$$

Consider the ellipse $(x^2/a^2) + (y^2/b^2) = 1$, and suppose that the length $2a$ of the major axis is fixed and the length $2b$ of the minor axis is variable (note that $0 < b < a$). Since $b^2$ is positive, $a^2 - b^2 < a^2$ and hence $\sqrt{a^2 - b^2} < a$. Dividing both sides of the last inequality by $a$ gives us $\sqrt{a^2 - b^2}/a < 1$, or $0 < e < 1$. If $b$ is close to 0, then $\sqrt{a^2 - b^2} \approx a$, $e \approx 1$, and the ellipse is very flat. This case is illustrated in Figure 23(a), with $a = 2$, $b = 0.3$, and $e \approx 0.99$. If $b$ is close to $a$, then $\sqrt{a^2 - b^2} \approx 0$, $e \approx 0$, and the ellipse is almost circular. This case is illustrated in Figure 23(b), with $a = 2$, $b = 1.9999$, and $e \approx 0.01$.

**FIGURE 23**
(a) Eccentricity almost 1        (b) Eccentricity almost 0

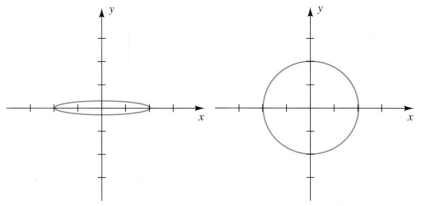

After many years of analyzing an enormous amount of empirical data, the German astronomer Johannes Kepler (1571–1630) formulated three laws that describe the motion of planets about the sun. Kepler's first law states that the orbit of each planet in the solar system is an ellipse with the sun at one focus. Most of these orbits are almost circular, so their corresponding eccentricities are close to 0. To illustrate, for Earth, $e \approx 0.017$; for Mars, $e \approx 0.093$; and for Uranus, $e \approx 0.046$. The orbits of Mercury and Pluto are less circular, with eccentricities of 0.206 and 0.249, respectively.

Many comets have elliptical orbits with the sun at a focus. In this case the eccentricity $e$ is close to 1, and the ellipse is very flat. In the next example we use the **astronomical unit** (AU)—that is, the average distance from the earth to the sun—to specify large distances (1 AU ≈ 93,000,000 mi).

### EXAMPLE 7 *Approximating a distance in an elliptical path*

Halley's comet has an elliptical orbit with eccentricity $e = 0.967$. The closest that Halley's comet comes to the sun is 0.587 AU. Approximate the maximum distance of the comet from the sun, to the nearest 0.1 AU.

**FIGURE 24** Halley's comet

*SOLUTION* Figure 24 illustrates the orbit of the comet, where $c$ is the distance from the center of the ellipse to a focus (the sun) and $2a$ is the length of the major axis.

Since $a - c$ is the minimum distance between the sun and the comet, we have (in AU)

$$a - c = 0.587, \quad \text{or} \quad a = c + 0.587.$$

Since $e = c/a = 0.967$, we obtain the following:

$$
\begin{aligned}
c &= 0.967a && \text{multiply by } a \\
&= 0.967(c + 0.587) && \text{substitute for } a \\
&\approx 0.967c + 0.568 && \text{multiply} \\
c - 0.967c &\approx 0.568 && \text{subtract } 0.967c \\
c &\approx \frac{0.568}{0.033} \approx 17.2 && \text{solve for } c
\end{aligned}
$$

Since $a = c + 0.587$, we obtain

$$a \approx 17.2 + 0.587 \approx 17.8,$$

and the maximum distance between the sun and the comet is

$$a + c \approx 17.8 + 17.2 = 35.0 \text{ AU.}$$

**FIGURE 25**

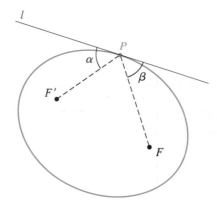

An ellipse has a *reflective property* analogous to that of the parabola discussed at the end of the previous section. To illustrate, let $l$ denote the tangent line at a point $P$ on an ellipse with foci $F$ and $F'$, as shown in Figure 25. If $\alpha$ is the acute angle between $F'P$ and $l$ and if $\beta$ is the acute angle between $FP$ and $l$, it can be shown that $\alpha = \beta$. Thus, if a ray of light or sound emanates from one focus, it is reflected to the other focus. This property is used in the design of certain types of optical equipment.

## FIGURE 26

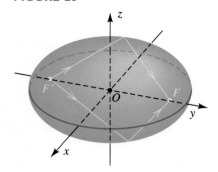

If the ellipse with center $O$ and foci $F'$ and $F$ on the x-axis is revolved about the x-axis, as illustrated in Figure 26, we obtain a three-dimensional surface called an **ellipsoid**. The upper half or lower half is a **hemi-ellipsoid**, as is the right half or left half. Sound waves or other impulses that are emitted from the focus $F'$ will be reflected off the ellipsoid into the focus $F$. This property is used in the design of *whispering galleries*—structures with ellipsoidal ceilings, in which a person who whispers at one focus can be heard at the other focus. Examples of whispering galleries may be found in the Rotunda of the Capitol Building in Washington, D.C., and in the Mormon Tabernacle in Salt Lake City.

The reflective property of ellipsoids (and hemi-ellipsoids) is used in modern medicine in a device called a *lithotripter*, which disintegrates kidney stones by means of high-energy underwater shock waves. After taking extremely accurate measurements, the operator positions the patient so that the stone is at a focus. Ultra–high frequency shock waves are then produced at the other focus, and reflected waves break up the kidney stone. Recovery time with this technique is usually 3–4 days, instead of the 2–3 weeks with conventional surgery. Moreover, the mortality rate is less than 0.01%, as compared to 2–3% for traditional surgery.

## 10.2 EXERCISES

**Exer. 1–14: Find the vertices and foci of the ellipse. Sketch its graph, showing the foci.**

1 $\dfrac{x^2}{9} + \dfrac{y^2}{4} = 1$

2 $\dfrac{x^2}{25} + \dfrac{y^2}{16} = 1$

3 $\dfrac{x^2}{15} + \dfrac{y^2}{16} = 1$

4 $\dfrac{x^2}{45} + \dfrac{y^2}{49} = 1$

5 $4x^2 + y^2 = 16$

6 $y^2 + 9x^2 = 9$

7 $4x^2 + 25y^2 = 1$

8 $10y^2 + x^2 = 5$

9 $\dfrac{(x - 3)^2}{16} + \dfrac{(y + 4)^2}{9} = 1$

10 $\dfrac{(x + 2)^2}{25} + \dfrac{(y - 3)^2}{4} = 1$

11 $4x^2 + 9y^2 - 32x - 36y + 64 = 0$

12 $x^2 + 2y^2 + 2x - 20y + 43 = 0$

13 $25x^2 + 4y^2 - 250x - 16y + 541 = 0$

14 $4x^2 + y^2 = 2y$

**Exer. 15–18: Find an equation for the ellipse shown in the figure.**

15

16

17

18

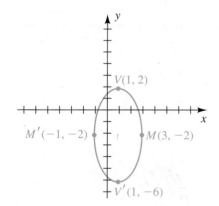

**Exer. 19–30: Find an equation for the ellipse that has its center at the origin and satisfies the given conditions.**

19  Vertices $V(\pm 8, 0)$,              foci $F(\pm 5, 0)$

20  Vertices $V(0, \pm 7)$,              foci $F(0, \pm 2)$

21  Vertices $V(0, \pm 5)$,              minor axis of length 3

22  Foci $F(\pm 3, 0)$,              minor axis of length 2

23  Vertices $V(0, \pm 6)$,              passing through $(3, 2)$

24  Passing through $(2, 3)$ and $(6, 1)$

25  Eccentricity $\frac{3}{4}$,              vertices $V(0, \pm 4)$

26  Eccentricity $\frac{1}{2}$,              vertices on the $x$-axis,
passing through $(1, 3)$

27  $x$-intercepts $\pm 2$,              $y$-intercepts $\pm \frac{1}{3}$

28  $x$-intercepts $\pm \frac{1}{2}$,              $y$-intercepts $\pm 4$

29  Horizontal major axis of length 8, minor axis of length 5

30  Vertical major axis of length 7, minor axis of length 6

**Exer. 31–32: Find the points of intersection of the graphs of the equations. Sketch both graphs on the same coordinate plane, and show the points of intersection.**

31  $\begin{cases} x^2 + 4y^2 = 20 \\ x + 2y = 6 \end{cases}$          32  $\begin{cases} x^2 + 4y^2 = 36 \\ x^2 + y^2 = 12 \end{cases}$

**Exer. 33–36: Find an equation for the set of points in an $xy$-plane such that the sum of the distances from $F$ and $F'$ is $k$.**

33  $F(3, 0)$,      $F'(-3, 0)$;      $k = 10$

34  $F(12, 0)$,    $F'(-12, 0)$;    $k = 26$

35  $F(0, 15)$,    $F'(0, -15)$;    $k = 34$

36  $F(0, 8)$,      $F'(0, -8)$;      $k = 20$

**Exer. 37–44: Determine whether the graph of the equation is the upper, lower, left, or right half of an ellipse, and find an equation for the ellipse.**

37  $y = 11\sqrt{1 - \dfrac{x^2}{49}}$          38  $y = -6\sqrt{1 - \dfrac{x^2}{25}}$

39  $x = -\frac{1}{3}\sqrt{9 - y^2}$          40  $x = \frac{4}{5}\sqrt{25 - y^2}$

41  $x = 1 + 2\sqrt{1 - \dfrac{(y + 2)^2}{9}}$

42  $x = -2 - 5\sqrt{1 - \dfrac{(y - 1)^2}{16}}$

43  $y = 2 - 7\sqrt{1 - \dfrac{(x + 1)^2}{9}}$

44  $y = -1 + \sqrt{1 - \dfrac{(x - 3)^2}{16}}$

**45** *Dimensions of an arch*  An arch of a bridge is semi-elliptical, with major axis horizontal. The base of the arch is 30 feet across, and the highest part of the arch is 10 feet above the horizontal roadway, as shown in the figure. Find the height of the arch 6 feet from the center of the base.

**EXERCISE 45**

**46** *Designing a bridge*  A bridge is to be constructed across a river that is 200 feet wide. The arch of the bridge is to be semielliptical and must be constructed so that a ship less than 50 feet wide and 30 feet high can pass safely through the arch, as shown in the figure.

(a) Find an equation for the arch.

(b) Approximate the height of the arch in the middle of the bridge.

**EXERCISE 46**

**47** *Earth's orbit*  Assume that the length of the major axis of the earth's orbit is 186,000,000 miles and that the eccentricity is 0.017. Approximate, to the nearest 1000 miles, the maximum and minimum distances between the earth and the sun.

**48** *Mercury's orbit*  The planet Mercury travels in an elliptical orbit that has eccentricity 0.206 and major axis of length 0.774 AU. Find the maximum and minimum distances between Mercury and the sun.

**49** *Elliptical reflector*  The basic shape of an elliptical reflector is a hemi-ellipsoid of height $h$ and diameter $k$, as shown in the figure. Waves emitted from focus $F$ will reflect off the surface into focus $F'$.

(a) Express the distances $d(V, F)$ and $d(V, F')$ in terms of $h$ and $k$.

(b) An elliptical reflector of height 17 centimeters is to be constructed so that waves emitted from $F$ are reflected to a point $F'$ that is 32 centimeters from $V$. Find the diameter of the reflector and the location of $F$.

**EXERCISE 49**

**50** *Lithotripter operation*  A lithotripter of height 15 centimeters and diameter 18 centimeters is to be constructed as shown in the figure. High-energy underwater shock waves will be emitted from the focus $F$ that is closest to the vertex $V$.

(a) Find the distance from $V$ to $F$.

(b) How far from $V$ (in the vertical direction) should a kidney stone be located?

**EXERCISE 50**

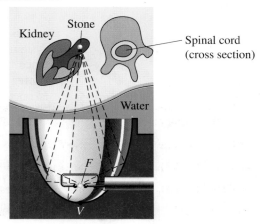

51 **Whispering gallery**  The ceiling of a whispering gallery has the shape of the hemi-ellipsoid shown in Figure 26, with the highest point of the ceiling 15 feet above the elliptical floor and the vertices of the floor 50 feet apart. If two people are standing at the foci $F'$ and $F$, how far from the vertices are their feet?

52 An ellipse has a vertex at the origin and foci $F_1(p, 0)$ and $F_2(p + 2c, 0)$, as shown in the figure. If the focus at

EXERCISE 52

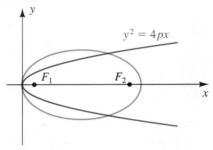

$F_1$ is fixed and $(x, y)$ is on the ellipse, show that $y^2$ approaches $4px$ as $c \rightarrow \infty$. (Thus, as $c \rightarrow \infty$, the ellipse takes on the shape of a parabola.)

**Exer. 53–56: Graph the ellipses on the same coordinate plane, and estimate their points of intersection.**

53 $\dfrac{x^2}{2.9} + \dfrac{y^2}{2.1} = 1$;    $\dfrac{x^2}{4.3} + \dfrac{(y - 2.1)^2}{4.9} = 1$

54 $\dfrac{x^2}{3.9} + \dfrac{y^2}{2.4} = 1$;    $\dfrac{(x + 1.9)^2}{4.1} + \dfrac{y^2}{2.5} = 1$

55 $\dfrac{(x + 0.1)^2}{1.7} + \dfrac{y^2}{0.9} = 1$;    $\dfrac{x^2}{0.9} + \dfrac{(y - 0.25)^2}{1.8} = 1$

56 $\dfrac{x^2}{3.1} + \dfrac{(y - 0.2)^2}{2.8} = 1$;    $\dfrac{(x + 0.23)^2}{1.8} + \dfrac{y^2}{4.2} = 1$

## 10.3    HYPERBOLAS

The definition of a hyperbola is similar to that of an ellipse. The only change is that instead of using the *sum* of distances from two fixed points, we use the *difference*.

**Definition of a Hyperbola**

A **hyperbola** is the set of all points in a plane, the difference of whose distances from two fixed points (the **foci**) in the plane is a positive constant.

FIGURE 27

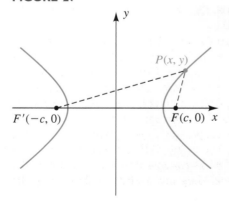

To find a simple equation for a hyperbola, we choose a coordinate system with foci at $F(c, 0)$ and $F'(-c, 0)$ and denote the (constant) distance by $2a$. The midpoint of the segment $F'F$ (the origin) is called the **center** of the hyperbola. Referring to Figure 27, we see that a point $P(x, y)$ is on the hyperbola if and only if either one of the following is true:

(1) $d(P, F') - d(P, F) = 2a$    or    (2) $d(P, F) - d(P, F') = 2a$

If $P$ is not on the $x$-axis, then from Figure 27 we see that

$$d(P, F) < d(F', F) + d(P, F'),$$

because the length of one side of a triangle is always less than the sum of the lengths of the other two sides. Similarly,

$$d(P, F') < d(F', F) + d(P, F).$$

Equivalent forms for the previous two inequalities are

$$d(P, F) - d(P, F') < d(F', F) \quad \text{and} \quad d(P, F') - d(P, F) < d(F', F).$$

Since the differences on the left sides of these inequalities both equal $2a$ and since $d(F', F) = 2c$, the last two inequalities imply that $2a < 2c$, or $a < c$. (Recall that for ellipses, we had $a > c$.)

Next, equations (1) and (2) may be replaced by the single equation

$$|d(P, F) - d(P, F')| = 2a.$$

Using the distance formula to find $d(P, F)$ and $d(P, F')$, we obtain an equation of the hyperbola:

$$\left| \sqrt{(x - c)^2 + (y - 0)^2} - \sqrt{(x + c)^2 + (y - 0)^2} \right| = 2a$$

Employing the type of simplification procedure that we used to derive an equation for an ellipse, we can rewrite the preceding equation as

$$\frac{x^2}{a^2} - \frac{y^2}{c^2 - a^2} = 1.$$

Finally, if we let

$$b^2 = c^2 - a^2 \quad \text{with} \quad b > 0$$

in the preceding equation, we obtain

$$\frac{x^2}{a^2} - \frac{y^2}{b^2} = 1.$$

We have shown that the coordinates of every point $(x, y)$ on the hyperbola in Figure 27 satisfy the equation $(x^2/a^2) - (y^2/b^2) = 1$. Conversely, if $(x, y)$ is a solution of this equation, then by reversing steps we see that the point $(x, y)$ is on the hyperbola.

Applying tests for symmetry, we see that the hyperbola is symmetric with respect to both axes and the origin. We may find the $x$-intercepts of the hyperbola by letting $y = 0$ in the equation. Doing so gives us $x^2/a^2 = 1$, or $x^2 = a^2$, and consequently the $x$-intercepts are $a$ and $-a$. The corresponding points $V(a, 0)$ and $V'(-a, 0)$ on the graph are called the **vertices** of the hyperbola (see Figure 28). The line segment $V'V$ is called the **transverse axis.** The graph has no $y$-intercept, since the equation $-y^2/b^2 = 1$ has the *complex* solutions $y = \pm bi$. The points $W(0, b)$ and $W'(0, -b)$ are end points of the **conjugate axis** $W'W$. The points $W$ and $W'$ are not on the hyperbola; however, as we shall see, they are useful for describing the graph.

**FIGURE 28**

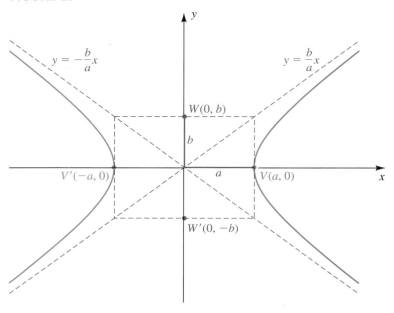

Solving the equation $(x^2/a^2) - (y^2/b^2) = 1$ for $y$ gives us

$$y = \pm \frac{b}{a} \sqrt{x^2 - a^2}.$$

If $x^2 - a^2 < 0$ or, equivalently, $-a < x < a$, then there are no points $(x, y)$ on the graph. There *are* points $P(x, y)$ on the graph if $x \geq a$ or $x \leq -a$.

It can be shown that *the lines $y = \pm(b/a)x$ are asymptotes for the hyperbola.* These asymptotes serve as excellent guides for sketching the graph. A convenient way to sketch the asymptotes is to first plot the vertices $V(a, 0)$, $V'(-a, 0)$ and the points $W(0, b)$, $W'(0, -b)$ (see Figure 28). If vertical and horizontal lines are drawn through these end points of the transverse and conjugate axes, respectively, then the diagonals of the resulting rectangle have slopes $b/a$ and $-b/a$. Hence, by extending these diagonals we obtain the asymptotes $y = \pm(b/a)x$. The hyperbola is then sketched as in Figure 28, using the asymptotes as guides. The two parts that make up the hyperbola are called the **right branch** and the **left branch** of the hyperbola.

Similarly, if we take the foci on the $y$-axis, we obtain the equation

$$\frac{y^2}{a^2} - \frac{x^2}{b^2} = 1.$$

In this case the vertices of the hyperbola are $(0, \pm a)$ and the end points of the conjugate axis are $(\pm b, 0)$, as shown in Figure 29. The asymptotes are $y = \pm(a/b)x$ (*not* $y = \pm(b/a)x$, as in the previous case), and we now refer to the **upper branch** and **lower branch** of the hyperbola.

**FIGURE 29**

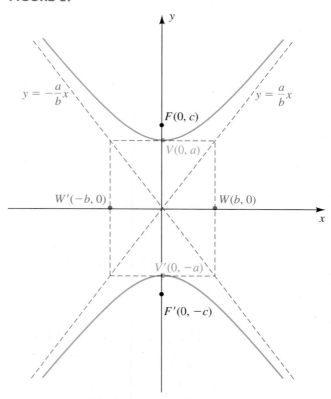

The preceding discussion may be summarized as follows.

**Standard Equations of a Hyperbola with Center at the Origin**

The graph of

$$\frac{x^2}{a^2} - \frac{y^2}{b^2} = 1 \quad \text{or} \quad \frac{y^2}{a^2} - \frac{x^2}{b^2} = 1$$

is a hyperbola with center at the origin. The length of the transverse axis is $2a$, and the length of the conjugate axis is $2b$. The foci are a distance $c$ from the origin, where $c^2 = a^2 + b^2$.

We have shown that an equation of a hyperbola with center at the origin and foci on a coordinate axis can always be written in the form

$$\frac{x^2}{p} + \frac{y^2}{q} = 1, \quad \text{or} \quad qx^2 + py^2 = pq,$$

*where p and q have opposite signs.* The vertices are on the x-axis if $p$ is positive or on the y-axis if $q$ is positive.

**EXAMPLE I** *Sketching a hyperbola with center at the origin*

Sketch the graph of $9x^2 - 4y^2 = 36$. Find the foci and equations of the asymptotes.

**SOLUTION**    From the remarks preceding this example, the graph is a hyperbola with center at the origin. To express the given equation in a standard form, we divide both sides by 36 and simplify, obtaining

$$\frac{x^2}{4} - \frac{y^2}{9} = 1.$$

**FIGURE 30**

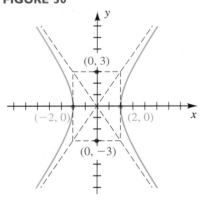

Comparing $(x^2/4) - (y^2/9) = 1$ to $(x^2/a^2) - (y^2/b^2) = 1$, we see that $a^2 = 4$ and $b^2 = 9$; that is, $a = 2$ and $b = 3$. The hyperbola has its vertices on the x-axis, since there are x-intercepts and no y-intercepts. The vertices $(\pm 2, 0)$ and the end points $(0, \pm 3)$ of the conjugate axis determine a rectangle whose diagonals (extended) give us the asymptotes. The graph of the equation is sketched in Figure 30.

To find the foci, we calculate

$$c^2 = a^2 + b^2 = 4 + 9 = 13.$$

Thus, $c = \sqrt{13}$, and the foci are $(\pm\sqrt{13}, 0)$.

The equations of the asymptotes, $y = \pm\frac{3}{2}x$, can be found by referring to the graph or to the equations $y = \pm(b/a)x$.

The preceding example indicates that for hyperbolas it is not always true that $a > b$, as is the case for ellipses. In fact, we may have $a < b$, $a > b$, or $a = b$.

**EXAMPLE 2** *Sketching a hyperbola with center at the origin*

Sketch the graph of $4y^2 - 2x^2 = 1$. Find the foci and equations of the asymptotes.

**SOLUTION**    To express the given equation in a standard form, we write

$$\frac{y^2}{\frac{1}{4}} - \frac{x^2}{\frac{1}{2}} = 1.$$

**FIGURE 31**

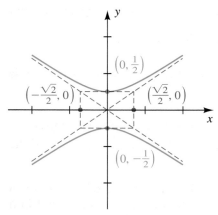

Thus,

$$a^2 = \tfrac{1}{4}, \quad b^2 = \tfrac{1}{2}, \quad c^2 = a^2 + b^2 = \tfrac{3}{4},$$

and consequently

$$a = \frac{1}{2}, \quad b = \frac{1}{\sqrt{2}}, \quad c = \frac{\sqrt{3}}{2}.$$

The hyperbola has its vertices on the $y$-axis, since there are $y$-intercepts and no $x$-intercepts. The vertices are $(0, \pm\tfrac{1}{2})$, the end points of the conjugate axes are $(\pm 1/\sqrt{2}, 0)$, and the foci are $(0, \pm\sqrt{3}/2)$. The graph is sketched in Figure 31.

To find the equations of the asymptotes, we refer to the figure or use $y = \pm(a/b)x$, obtaining $y = \pm(\sqrt{2}/2)x$.

**EXAMPLE 3**   *Finding an equation of a hyperbola satisfying prescribed conditions*

A hyperbola has vertices $(\pm 3, 0)$ and passes through the point $P(5, 2)$. Find its equation, foci, and asymptotes.

**FIGURE 32**

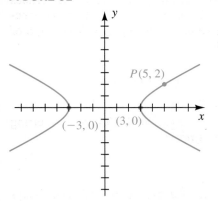

**SOLUTION**   We begin by sketching a hyperbola with vertices $(\pm 3, 0)$ that passes through the point $P(5, 2)$, as in Figure 32.

An equation of the hyperbola has the form

$$\frac{x^2}{3^2} - \frac{y^2}{b^2} = 1.$$

Since $P(5, 2)$ is on the hyperbola, the $x$- and $y$-coordinates satisfy this equation; that is,

$$\frac{5^2}{3^2} - \frac{2^2}{b^2} = 1.$$

Solving for $b^2$ gives us $b^2 = \tfrac{9}{4}$, and hence an equation for the hyperbola is

$$\frac{x^2}{9} - \frac{y^2}{\frac{9}{4}} = 1$$

or, equivalently,        $x^2 - 4y^2 = 9.$

To find the foci, we first calculate

$$c^2 = a^2 + b^2 = 9 + \tfrac{9}{4} = \tfrac{45}{4}.$$

Hence, $c = \sqrt{\tfrac{45}{4}} = \tfrac{3}{2}\sqrt{5} \approx 3.35$, and the foci are $\left(\pm\tfrac{3}{2}\sqrt{5}, 0\right)$.

The general equations of the asymptotes are $y = \pm(b/a)x$. Substituting $a = 3$ and $b = \tfrac{3}{2}$ gives us $y = \pm\tfrac{1}{2}x$.

The next example indicates how to find equations for certain parts of a hyperbola.

### EXAMPLE 4    *Finding equations of portions of a hyperbola*

The hyperbola $9x^2 - 4y^2 = 36$ was discussed in Example 1. Solve the equation as indicated, and describe the resulting graph.

**(a)** For $x$ in terms of $y$      **(b)** For $y$ in terms of $x$

### SOLUTION

**(a)** We solve for $x$ in terms of $y$ as follows:

$$9x^2 - 4y^2 = 36 \qquad \text{given}$$

$$x^2 = \frac{36 + 4y^2}{9} \qquad \text{solve for } x^2$$

$$x = \pm \tfrac{2}{3}\sqrt{9 + y^2} \qquad \text{factor out 4 and take the square root}$$

The graph of $x = \tfrac{2}{3}\sqrt{9 + y^2}$ is the right branch of the hyperbola shown in Figure 30, and the graph of $x = -\tfrac{2}{3}\sqrt{9 + y^2}$ is the left branch.

**(b)** We solve for $y$ in terms of $x$ as follows:

$$9x^2 - 4y^2 = 36 \qquad \text{given}$$

$$4y^2 = 9x^2 - 36 \qquad \text{add } 4y^2 \text{ and subtract 36}$$

$$y^2 = \frac{9(x^2 - 4)}{4} \qquad \text{factor out 9 and divide by 4}$$

$$y = \pm \tfrac{3}{2}\sqrt{x^2 - 4} \qquad \text{take the square root}$$

The graph of $y = \tfrac{3}{2}\sqrt{x^2 - 4}$ is the upper half of the right and left branches, and the graph of $y = -\tfrac{3}{2}\sqrt{x^2 - 4}$ is the lower half of these branches.

---

As was the case for ellipses, we may use the translation of axes formulas to generalize our work. The following example illustrates this technique.

### EXAMPLE 5    *Sketching a hyperbola with center $(h, k)$*

Discuss and sketch the graph of the equation

$$9x^2 - 4y^2 - 54x - 16y + 29 = 0.$$

**SOLUTION** We arrange our work using a procedure similar to that used for ellipses in Example 5 of the previous section:

$$(9x^2 - 54x) + (-4y^2 - 16y) = -29 \qquad \text{group terms}$$
$$9(x^2 - 6x + \underline{\quad}) - 4(y^2 + 4y + \underline{\quad}) = -29 \qquad \text{factor out 9 and } -4$$
$$9(x^2 - 6x + 9) - 4(y^2 + 4y + 4) = -29 + 81 - 16 \qquad \text{complete the squares}$$
$$9(x - 3)^2 - 4(y + 2)^2 = 36 \qquad \text{factor}$$
$$\frac{(x - 3)^2}{4} - \frac{(y + 2)^2}{9} = 1 \qquad \text{divide by 36}$$

The last equation has the form

$$\frac{(x')^2}{4} - \frac{(y')^2}{9} = 1,$$

**FIGURE 33**

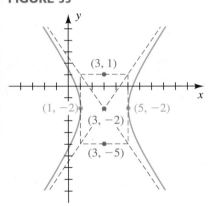

with $x' = x - 3$ and $y' = y + 2$. Thus, we translate the $x$- and $y$-axes to the new origin $C(3, -2)$. The graph is a hyperbola with vertices on the $x'$-axis (the line $y = -2$) and

$$a^2 = 4, \quad b^2 = 9, \quad c^2 = a^2 + b^2 = 13.$$

Hence,

$$a = 2, \quad b = 3, \quad c = \sqrt{13}.$$

As illustrated in Figure 33, the vertices are $(3 \pm 2, -2)$—that is, $(5, -2)$ and $(1, -2)$. The end points of the conjugate axis are $(3, -2 \pm 3)$—that is, $(3, 1)$ and $(3, -5)$. The foci are $(3 \pm \sqrt{13}, -2)$, and equations of the asymptotes are

$$y + 2 = \pm \tfrac{3}{2}(x - 3).$$

---

The results of Sections 10.1 through 10.3 indicate that the graph of every equation of the form

$$Ax^2 + Cy^2 + Dx + Ey + F = 0$$

is a conic, except for certain degenerate cases in which a point, one or two lines, or no graph is obtained. Although we have considered only special examples, our methods can be applied to any such equation. If $A$ and $C$ are equal and not 0, then the graph, when it exists, is a circle or, in exceptional cases, a point. If $A$ and $C$ are unequal but have the same sign, then, by completing squares and properly translating axes, we obtain an equation whose graph, when it exists, is an ellipse (or a point). If $A$ and $C$ have opposite signs, an equation of a hyperbola is obtained or possibly, in the degenerate case, two intersecting straight lines. If either $A$ or $C$ (but not both) is 0, the graph is a parabola or, in certain cases, a pair of parallel lines.

We shall conclude this section with an application involving hyperbolas.

### EXAMPLE 6    *Locating a ship*

Coast Guard station A is 200 miles directly east of another station B. A ship is sailing on a line parallel to and 50 miles north of the line through A and B. Radio signals are sent out from A and B at the rate of 980 ft/$\mu$sec (microsecond). If, at 1:00 P.M., the signal from B reaches the ship 400 $\mu$sec after the signal from A, locate the position of the ship at that time.

*SOLUTION*    Let us introduce a coordinate system, as shown in Figure 34(a), with the stations at points A and B on the x-axis and the ship at P on the line $y = 50$. Since at 1:00 P.M. it takes 400 $\mu$sec longer for the signal to arrive from B than from A, the difference $d_1 - d_2$ in the indicated distances at that time is

$$d_1 - d_2 = (980)(400) = 392{,}000 \text{ ft.}$$

**FIGURE 34**

(a)                                                    (b)

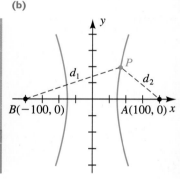

Dividing by 5280 (ft/mi) gives us

$$d_1 - d_2 = \frac{392{,}000}{5280} = 74.\overline{24} \text{ mi.}$$

At 1:00 P.M., point P is on the right branch of a hyperbola whose equation is $(x^2/a^2) - (y^2/b^2) = 1$ (see Figure 34(b)), consisting of all points whose difference in distances from the foci B and A is $d_1 - d_2$. In our derivation of the equation $(x^2/a^2) - (y^2/b^2) = 1$, we let $d_1 - d_2 = 2a$; it follows that in the present situation

$$a = \frac{74.\overline{24}}{2} = 37.\overline{12} \quad \text{and} \quad a^2 \approx 1378.$$

Since the distance $c$ from the origin to either focus is 100,

$$b^2 = c^2 - a^2 \approx 10{,}000 - 1378, \quad \text{or} \quad b^2 \approx 8622.$$

Hence, an (approximate) equation for the hyperbola that has foci $A$ and $B$ and passes through $P$ is

$$\frac{x^2}{1378} - \frac{y^2}{8622} = 1.$$

FIGURE 35

If we let $y = 50$ (the $y$-coordinate of $P$), we obtain

$$\frac{x^2}{1378} - \frac{2500}{8622} = 1.$$

Solving for $x$ gives us $x \approx 42.16$. Rounding off to the nearest mile, we find that the coordinates of $P$ are approximately $(42, 50)$.

An extension of the method used in Example 6 is the basis for the navigational system LORAN (for Long Range Navigation). This system involves two pairs of radio transmitters, such as those located at $T$, $T'$ and $S$, $S'$ in Figure 35. Suppose that signals sent out by the transmitters at $T$ and $T'$ reach a radio receiver in a ship located at some point $P$. The difference in the times of arrival of the signals can be used to determine the difference in the distances of $P$ from $T$ and $T'$. Thus, $P$ lies on one branch of a hyperbola with foci at $T$ and $T'$. Repeating this process for the other pair of transmitters, we see that $P$ also lies on one branch of a hyperbola with foci at $S$ and $S'$. The intersection of these two branches determines the position of $P$.

FIGURE 36

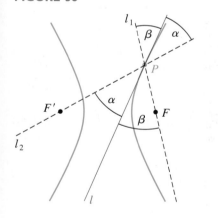

A hyperbola has a *reflective property* analogous to that of the ellipse discussed at the end of the previous section. To illustrate, let $l$ denote the tangent line at a point $P$ on a hyperbola with foci $F$ and $F'$, as shown in Figure 36. If $\alpha$ is the acute angle between $F'P$ and $l$ and if $\beta$ is the acute angle between $FP$ and $l$, it can be shown that $\alpha = \beta$. If a ray of light is directed along the line $l_1$ toward $F$, it will be reflected back at $P$ along the line $l_2$ toward $F'$. This property is used in the design of telescopes of the Cassegrain type (see Exercise 52).

## 10.3 EXERCISES

Exer. 1–16: Find the vertices, the foci, and the equations of the asymptotes of the hyperbola. Sketch its graph, showing the asymptotes and the foci.

1 $\dfrac{x^2}{9} - \dfrac{y^2}{4} = 1$

2 $\dfrac{y^2}{49} - \dfrac{x^2}{16} = 1$

3 $\dfrac{y^2}{9} - \dfrac{x^2}{4} = 1$

4 $\dfrac{x^2}{49} - \dfrac{y^2}{16} = 1$

5 $x^2 - \dfrac{y^2}{24} = 1$

6 $y^2 - \dfrac{x^2}{15} = 1$

**7** $y^2 - 4x^2 = 16$          **8** $x^2 - 2y^2 = 8$

**9** $16x^2 - 36y^2 = 1$          **10** $y^2 - 16x^2 = 1$

**11** $\dfrac{(y + 2)^2}{9} - \dfrac{(x + 2)^2}{4} = 1$

**12** $\dfrac{(x - 3)^2}{25} - \dfrac{(y - 1)^2}{4} = 1$

**13** $144x^2 - 25y^2 + 864x - 100y - 2404 = 0$

**14** $y^2 - 4x^2 - 12y - 16x + 16 = 0$

**15** $4y^2 - x^2 + 40y - 4x + 60 = 0$

**16** $25x^2 - 9y^2 + 100x - 54y + 10 = 0$

**Exer. 17–20:** Find an equation for the hyperbola shown in the figure.

**17**

**18**

**19**

**20**

**Exer. 21–32:** Find an equation for the hyperbola that has its center at the origin and satisfies the given conditions.

**21** Foci $F(0, \pm 4)$,          vertices $V(0, \pm 1)$

**22** Foci $F(\pm 8, 0)$,          vertices $V(\pm 5, 0)$

**23** Foci $F(\pm 5, 0)$,          vertices $V(\pm 3, 0)$

**24** Foci $F(0, \pm 3)$,          vertices $V(0, \pm 2)$

**25** Foci $F(0, \pm 5)$,          conjugate axis of length 4

**26** Vertices $V(\pm 4, 0)$,   passing through $(8, 2)$

**27** Vertices $V(\pm 3, 0)$,   asymptotes $y = \pm 2x$

**28** Foci $F(0, \pm 10)$,        asymptotes $y = \pm\frac{1}{3}x$

**29** $x$-intercepts $\pm 5$,       asymptotes $y = \pm 2x$

**30** $y$-intercepts $\pm 2$,       asymptotes $y = \pm\frac{1}{4}x$

**31** Vertical transverse axis of length 10, conjugate axis of length 14

**32** Horizontal transverse axis of length 6, conjugate axis of length 2

**Exer. 33–34: Find the points of intersection of the graphs of the equations. Sketch both graphs on the same coordinate plane, and show the points of intersection.**

**33** $\begin{cases} y^2 - 4x^2 = 16 \\ y - x = 4 \end{cases}$

**34** $\begin{cases} x^2 - y^2 = 4 \\ y^2 - 3x = 0 \end{cases}$

**Exer. 35–38: Find an equation for the set of points in an $xy$-plane such that the difference of the distances from $F$ and $F'$ is $k$.**

**35** $F(13, 0)$, $F'(-13, 0)$; $k = 24$

**36** $F(5, 0)$, $F'(-5, 0)$; $k = 8$

**37** $F(0, 10)$, $F'(0, -10)$; $k = 16$

**38** $F(0, 17)$, $F'(0, -17)$; $k = 30$

**Exer. 39–46: Describe the part of a hyperbola given by the equation.**

**39** $x = \frac{5}{4}\sqrt{y^2 + 16}$

**40** $x = -\frac{5}{4}\sqrt{y^2 + 16}$

**41** $y = \frac{3}{7}\sqrt{x^2 + 49}$

**42** $y = -\frac{3}{7}\sqrt{x^2 + 49}$

**43** $y = -\frac{9}{4}\sqrt{x^2 - 16}$

**44** $y = \frac{9}{4}\sqrt{x^2 - 16}$

**45** $x = -\frac{2}{3}\sqrt{y^2 - 36}$

**46** $x = \frac{2}{3}\sqrt{y^2 - 36}$

**47** The graphs of the equations

$$\frac{x^2}{a^2} - \frac{y^2}{b^2} = 1 \quad \text{and} \quad \frac{x^2}{a^2} - \frac{y^2}{b^2} = -1$$

are called *conjugate hyperbolas*. Sketch the graphs of both equations on the same coordinate plane, with $a = 5$ and $b = 3$, and describe the relationship between the two graphs.

**48** Find an equation of the hyperbola with foci $(h \pm c, k)$ and vertices $(h \pm a, k)$, where

$$0 < a < c \quad \text{and} \quad c^2 = a^2 + b^2.$$

**49** *Alpha particles* In 1911, the physicist Ernest Rutherford (1871–1937) discovered that if alpha particles are shot toward the nucleus of an atom, they are eventually repulsed away from the nucleus along hyperbolic paths. The figure illustrates the path of a particle that starts toward the origin along the line $y = \frac{1}{2}x$ and comes within 3 units of the nucleus. Find an equation of the path.

**EXERCISE 49**

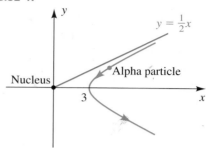

**50** *Airplane maneuver* An airplane is flying along the hyperbolic path illustrated in the figure. If an equation of the path is $2y^2 - x^2 = 8$, determine how close the airplane comes to a town located at $(3, 0)$. (*Hint:* Let $S$ denote the square of the distance from a point $(x, y)$ on the path to $(3, 0)$, and find the minimum value of $S$.)

**EXERCISE 50**

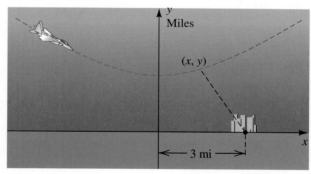

**51** *Locating a ship* A ship is traveling a course that is 100 miles from, and parallel to, a straight shoreline. The ship sends out a distress signal that is received by two Coast Guard stations A and B, located 200 miles apart, as shown in the figure. By measuring the difference in signal reception times, it is determined that the ship is 160 miles closer to B than to A. Where is the ship?

**EXERCISE 51**

**52 *Design of a telescope*** The Cassegrain telescope design (dating back to 1672) makes use of the reflective properties of both the parabola and the hyperbola. Shown in the figure is a (split) parabolic mirror, with focus at $F_1$ and axis along the line $l$, and a second hyperbolic mirror, with one focus also at $F_1$ and transverse axis along $l$. Where do incoming light waves parallel to the common axis finally collect?

**EXERCISE 52**

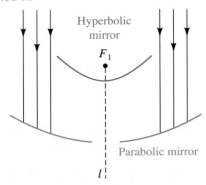

Hyperbolic mirror

$F_1$

Parabolic mirror

$l$

**Exer. 53–54: Graph the hyperbolas on the same coordinate plane, and estimate their first-quadrant point of intersection.**

**53** $\dfrac{(y - 0.1)^2}{1.6} - \dfrac{(x + 0.2)^2}{0.5} = 1;$

$\dfrac{(y - 0.5)^2}{2.7} - \dfrac{(x - 0.1)^2}{5.3} = 1$

**54** $\dfrac{(x - 0.1)^2}{0.12} - \dfrac{y^2}{0.1} = 1;$ $\dfrac{x^2}{0.9} - \dfrac{(y - 0.3)^2}{2.1} = 1$

**Exer. 55–56: Graph the hyperbolas on the same coordinate plane, and determine the number of points of intersection.**

**55** $\dfrac{(x - 0.3)^2}{1.3} - \dfrac{y^2}{2.7} = 1;$ $\dfrac{y^2}{2.8} - \dfrac{(x - 0.2)^2}{1.2} = 1$

**56** $\dfrac{(x + 0.2)^2}{1.75} - \dfrac{(y - 0.5)^2}{1.6} = 1;$

$\dfrac{(x - 0.6)^2}{2.2} - \dfrac{(y + 0.4)^2}{2.35} = 1$

## 10.4 PLANE CURVES AND PARAMETRIC EQUATIONS

If $f$ is a function, the graph of the equation $y = f(x)$ is often called a *plane curve*. However, this definition is restrictive, because it excludes many useful graphs. The following definition is more general.

**Definition of Plane Curve**

A **plane curve** is a set $C$ of ordered pairs $(f(t), g(t))$, where $f$ and $g$ are functions defined on an interval $I$.

For simplicity, we often refer to a plane curve as a **curve.** The **graph** of $C$ in the preceding definition consists of all points $P(t) = (f(t), g(t))$ in an $xy$-plane, for $t$ in $I$. We shall use the term *curve* interchangeably with *graph of a curve*. We sometimes regard the point $P(t)$ as tracing the curve $C$ as $t$ varies through the interval $I$.

The graphs of several curves are sketched in Figure 37, where $I$ is a closed interval $[a, b]$. In (a) of the figure, $P(a) \neq P(b)$, and $P(a)$ and $P(b)$ are called the **end points** of $C$. The curve in (a) intersects itself; that is, two different values of $t$ produce the same point. If $P(a) = P(b)$, as in

**FIGURE 37**

**(a)** Curve

**(b)** Closed curve

**(c)** Simple closed curve

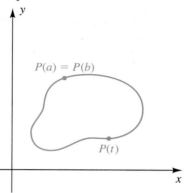

Figure 37(b), then $C$ is a **closed curve.** If $P(a) = P(b)$ and $C$ does not intersect itself at any other point, as in Figure 37(c), then $C$ is a **simple closed curve.**

A convenient way to represent curves is given in the next definition.

---

**Definition of Parametric Equations**

Let $C$ be the curve consisting of all ordered pairs $(f(t), g(t))$, where $f$ and $g$ are defined on an interval $I$. The equations

$$x = f(t), \quad y = g(t),$$

for $t$ in $I$, are **parametric equations** for $C$ with **parameter** $t$.

---

The curve $C$ in this definition is referred to as a **parametrized curve,** and the parametric equations are a **parametrization** for $C$. We often use the notation

$$x = f(t), \quad y = g(t); \quad t \text{ in } I$$

to indicate the domain $I$ of $f$ and $g$. Sometimes it may be possible to eliminate the parameter and obtain a familiar equation in $x$ and $y$ for $C$. In simple cases we may sketch a graph of a parametrized curve by plotting points and connecting them in order of increasing $t$, as illustrated in the next example.

**EXAMPLE 1    Sketching the graph of a parametrized curve**
Sketch the graph of the curve $C$ that has the parametrization

$$x = 2t, \quad y = t^2 - 1; \quad -1 \le t \le 2.$$

**SOLUTION**    We use the parametric equations to tabulate coordinates of points $P(x, y)$ on $C$, as follows.

| $t$ | $-1$ | $-\frac{1}{2}$ | $0$ | $\frac{1}{2}$ | $1$ | $\frac{3}{2}$ | $2$ |
|---|---|---|---|---|---|---|---|
| $x$ | $-2$ | $-1$ | $0$ | $1$ | $2$ | $3$ | $4$ |
| $y$ | $0$ | $-\frac{3}{4}$ | $-1$ | $-\frac{3}{4}$ | $0$ | $\frac{5}{4}$ | $3$ |

**FIGURE 38**

$x = 2t, \; y = t^2 - 1; \; -1 \leq t \leq 2$

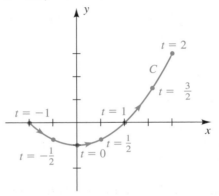

Plotting points leads to the sketch in Figure 38. The arrowheads on the graph indicate the direction in which $P(x, y)$ traces the curve as $t$ *increases* from $-1$ to $2$.

We may obtain a clearer description of the graph by eliminating the parameter. Solving the first parametric equation for $t$, we obtain $t = \frac{1}{2}x$. Substituting this expression for $t$ in the second equation gives us

$$y = (\tfrac{1}{2}x)^2 - 1.$$

The graph of this equation in $x$ and $y$ is a parabola symmetric with respect to the $y$-axis with vertex $(0, -1)$. However, since $x = 2t$ and $-1 \leq t \leq 2$, we see that $-2 \leq x \leq 4$ for points $(x, y)$ on $C$, and hence $C$ is that part of the parabola between the points $(-2, 0)$ and $(4, 3)$ shown in Figure 38.

As indicated by the arrowheads in Figure 38, the point $P(x, y)$ traces the curve $C$ from *left to right* as $t$ increases. The parametric equations

$$x = -2t, \quad y = t^2 - 1; \quad -2 \leq t \leq 1$$

give us the same graph; however, as $t$ increases, $P(x, y)$ traces the curve from *right to left*. For other parametrizations, the point $P(x, y)$ may oscillate back and forth as $t$ increases.

The **orientation** of a parametrized curve $C$ is the direction determined by *increasing* values of the parameter. We often indicate an orientation by placing arrowheads on $C$, as in Figure 38. If $P(x, y)$ moves back and forth as $t$ increases, we may place arrows *alongside* of $C$.

As we have observed, a curve may have different orientations, depending on the parametrization. To illustrate, the curve $C$ in Example 1 is given parametrically by any of the following:

$$x = 2t, \quad y = t^2 - 1; \qquad -1 \leq t \leq 2$$
$$x = t, \quad y = \tfrac{1}{4}t^2 - 1; \qquad -2 \leq t \leq 4$$
$$x = t^3, \quad y = \tfrac{1}{4}t^6 - 1; \quad \sqrt[3]{-2} \leq t \leq \sqrt[3]{4}$$

The next example demonstrates that it is sometimes useful to eliminate the parameter *before* plotting points.

**EXAMPLE 2** *Describing the motion of a point*

A point moves in a plane such that its position $P(x, y)$ at time $t$ is given by

$$x = a \cos t, \quad y = a \sin t; \quad t \text{ in } \mathbb{R},$$

where $a > 0$. Describe the motion of the point.

**SOLUTION** We may eliminate the parameter by rewriting the parametric equations as

$$\frac{x}{a} = \cos t, \quad \frac{y}{a} = \sin t$$

and using the identity $\cos^2 t + \sin^2 t = 1$ to obtain

$$\left(\frac{x}{a}\right)^2 + \left(\frac{y}{a}\right)^2 = 1,$$

or

$$x^2 + y^2 = a^2.$$

This shows that the point $P(x, y)$ moves on the circle $C$ of radius $a$ with center at the origin (see Figure 39). The point is at $A(a, 0)$ when $t = 0$, at $(0, a)$ when $t = \pi/2$, at $(-a, 0)$ when $t = \pi$, at $(0, -a)$ when $t = 3\pi/2$, and back at $A(a, 0)$ when $t = 2\pi$. Thus, $P$ moves around $C$ in a counterclockwise direction, making one revolution every $2\pi$ units of time. The orientation of $C$ is indicated by the arrowheads in the figure.

Note that in this example we may interpret $t$ geometrically as the radian measure of the angle generated by the line segment $OP$.

**FIGURE 39**

$x = a \cos t, \; y = a \sin t; \; t \text{ in } \mathbb{R}$

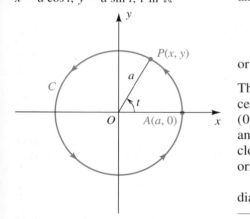

**EXAMPLE 3** *Sketching the graph of a parametrized curve*

Sketch the graph of the curve $C$ that has the parametrization

$$x = -2 + t^2, \quad y = 1 + 2t^2; \quad t \text{ in } \mathbb{R},$$

and indicate the orientation.

**SOLUTION** To eliminate the parameter, we use the first equation to obtain $t^2 = x + 2$ and then substitute for $t^2$ in the second equation. Thus,

$$y = 1 + 2(x + 2).$$

*(continued)*

The graph of the last equation is the line of slope 2 through the point $(-2, 1)$, as indicated by the dashes in Figure 40(a). However, since $t^2 \geq 0$, we see from the parametric equations for $C$ that

$$x = -2 + t^2 \geq -2 \quad \text{and} \quad y = 1 + 2t^2 \geq 1.$$

Thus, the graph of $C$ is that part of the line to the right of $(-2, 1)$ (the point corresponding to $t = 0$), as shown in Figure 40(b). The orientation is indicated by the arrows alongside of $C$. As $t$ increases in the interval $(-\infty, 0]$, $P(x, y)$ moves down the curve toward the point $(-2, 1)$. As $t$ increases in $[0, \infty)$, $P(x, y)$ moves up the curve away from $(-2, 1)$.

**FIGURE 40**

(a)                                    (b)

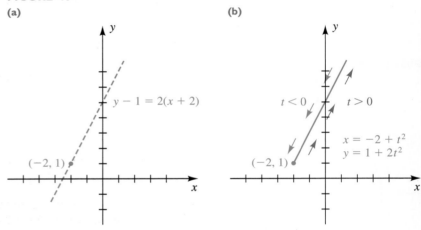

If a curve $C$ is described by an equation $y = f(x)$ for some function $f$, then an easy way to obtain parametric equations for $C$ is to let

$$x = t, \quad y = f(t),$$

where $t$ is in the domain of $f$. For example, if $y = x^3$, then parametric equations are

$$x = t, \quad y = t^3; \quad t \text{ in } \mathbb{R}.$$

We can use many different substitutions for $x$, provided that as $t$ varies through some interval, $x$ takes on every value in the domain of $f$. Thus, the graph of $y = x^3$ is also given by

$$x = t^{1/3}, \quad y = t; \quad t \text{ in } \mathbb{R}.$$

Note, however, that the parametric equations

$$x = \sin t, \quad y = \sin^3 t; \quad t \text{ in } \mathbb{R}$$

give only that part of the graph of $y = x^3$ between the points $(-1, -1)$ and $(1, 1)$.

**EXAMPLE 4** *Finding parametric equations for a line*

Find three parametrizations for the line of slope $m$ through the point $(x_1, y_1)$.

**SOLUTION** By the point-slope form, an equation for the line is

$$y - y_1 = m(x - x_1).$$

If we let $x = t$, then $y - y_1 = m(t - x_1)$ and we obtain the parametrization

$$x = t, \quad y = y_1 + m(t - x_1); \quad t \text{ in } \mathbb{R}.$$

We obtain another parametrization for the line if we let $x - x_1 = t$. In this case $y - y_1 = mt$, and we have

$$x = x_1 + t, \quad y = y_1 + mt; \quad t \text{ in } \mathbb{R}.$$

As a third illustration, if we let $x - x_1 = \tan t$, then

$$x = x_1 + \tan t, \quad y = y_1 + m \tan t; \quad -\frac{\pi}{2} < t < \frac{\pi}{2}.$$

There are many other parametrizations for the line.

---

Parametric equations of the form

$$x = a \sin \omega_1 t, \quad y = b \cos \omega_2 t; \quad t \geq 0,$$

where $a$, $b$, $\omega_1$, and $\omega_2$ are constants, occur in electrical theory. The variables $x$ and $y$ usually represent voltages or currents at time $t$. The resulting curve is often difficult to sketch; however, using an oscilloscope and imposing voltages or currents on the input terminals, we can represent the graph, a **Lissajous figure,** on the screen of the oscilloscope. Graphing utilities are very helpful in obtaining these complicated graphs.

**EXAMPLE 5** *Graphing a Lissajous figure*

Sketch the graph of the Lissajous figure that has the parametrization

$$x = \sin 2t, \quad y = \cos t; \quad 0 \leq t \leq 2\pi.$$

Determine the values of $t$ that correspond to the curve in each quadrant.

**SOLUTION** Some specific keystrokes for the TI-81/82 graphics calculator are given in Appendix I. We first need to set our graphing utility in a parametric mode. Next we make the assignments

$$X_{1T} = \sin 2t \quad \text{and} \quad Y_{1T} = \cos t,$$

where the subscript 1T on X and Y indicates that $X_{1T}$ and $Y_{1T}$ represent the first *pair* of parametric equations.

*(continued)*

When graphing parametric equations, we need to assign minimum (Tmin) and maximum (Tmax) values to the parameter $t$, in addition to viewing rectangle dimensions. We also need to select an increment, or step value (Tstep), for $t$. A typical value for Tstep is $\pi/30 \approx 0.105$. If a smaller value of Tstep is chosen, the accuracy of the sketch is increased, but so is the amount of time needed to sketch the graph.

For this example, we use Tmin = 0, Tmax = 6.3 (use $2\pi$ if it is available), and Tstep = 0.1. Since $x$ and $y$ are between $-1$ and 1, we will assign $-1$ to Ymin and 1 to Ymax. To maintain our 3:2 screen proportion, we select $-1.5$ for Xmin and 1.5 for Xmax, and then we graph $X_{1T}$ and $Y_{1T}$ to obtain the Lissajous figure in Figure 41.

Referring to the parametric equations, we see that as $t$ increases from 0 to $\pi/2$, the point $P(x, y)$ starts at (0, 1) and traces the part of the curve in quadrant I (in a generally clockwise direction). As $t$ increases from $\pi/2$ to $\pi$, $P(x, y)$ traces the part in quadrant III (in a counterclockwise direction). For $\pi < t < 3\pi/2$, we obtain the part in quadrant IV; and $3\pi/2 < t < 2\pi$ gives us the part in quadrant II.

**FIGURE 41**   $[-1.5, 1.5]$ by $[-1, 1]$

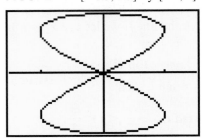

## EXAMPLE 6   *Finding parametric equations for a cycloid*

The curve traced by a fixed point $P$ on the circumference of a circle as the circle rolls along a line in a plane is called a **cycloid.** Find parametric equations for a cycloid.

*SOLUTION*    Suppose the circle has radius $a$ and that it rolls along (and above) the $x$-axis in the positive direction. If one position of $P$ is the origin, then Figure 42 displays part of the curve and a possible position of the circle. The v-shaped part of the curve at $x = 2\pi a$ is called a **cusp.**

**FIGURE 42**

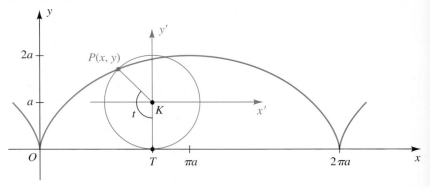

Let $K$ denote the center of the circle and $T$ the point of tangency with the $x$-axis. We introduce, as a parameter $t$, the radian measure of angle *TKP*. The distance the circle has rolled is $d(O, T) = at$. Consequently, the

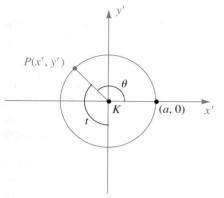

**FIGURE 43**

coordinates of $K$ are $(at, a)$. If we consider an $x'y'$-coordinate system with origin at $K(at, a)$ and if $P(x', y')$ denotes the point $P$ relative to this system, then, by the translation of axes formulas with $h = at$ and $k = a$,

$$x = at + x', \quad y = a + y'.$$

If, as in Figure 43, $\theta$ denotes an angle in standard position on the $x'y'$-plane, then $\theta = (3\pi/2) - t$. Hence,

$$x' = a \cos \theta = a \cos \left( \frac{3\pi}{2} - t \right) = -a \sin t$$

$$y' = a \sin \theta = a \sin \left( \frac{3\pi}{2} - t \right) = -a \cos t,$$

and substitution in $x = at + x'$, $y = a + y'$ gives us parametric equations for the cycloid:

$$x = a(t - \sin t), \quad y = a(1 - \cos t); \quad t \text{ in } \mathbb{R}$$

If $a < 0$, then the graph of $x = a(t - \sin t)$, $y = a(1 - \cos t)$ is the inverted cycloid that results if the circle of Example 6 rolls *below* the $x$-axis. This curve has a number of important physical properties. To illustrate, suppose a thin wire passes through two fixed points $A$ and $B$, as shown in Figure 44, and that the shape of the wire can be changed by bending it in any manner. Suppose further that a bead is allowed to slide along the wire and the only force acting on the bead is gravity. We now ask which of all the possible paths will allow the bead to slide from $A$ to $B$ in the least amount of time. It is natural to believe that the desired path is the straight line segment from $A$ to $B$; however, this is not the correct answer. The path that requires the least time coincides with the graph of an inverted cycloid with $A$ at the origin. Because the velocity of the bead increases more rapidly along the cycloid than along the line through $A$ and $B$, the bead reaches $B$ more rapidly, even though the distance is greater.

**FIGURE 44**

There is another interesting property of this **curve of least descent.** Suppose that $A$ is the origin and $B$ is the point with $x$-coordinate $\pi |a|$— that is, the lowest point on the cycloid in the first arc to the right of $A$. If the bead is released at *any* point between $A$ and $B$, it can be shown that the time required for it to reach $B$ is always the same.

Variations of the cycloid occur in applications. For example, if a motorcycle wheel rolls along a straight road, then the curve traced by a fixed point on one of the spokes is a cycloidlike curve. In this case the curve does not have cusps, nor does it intersect the road (the $x$-axis) as does the graph of a cycloid. If the wheel of a train rolls along a railroad track, then the curve traced by a fixed point on the circumference of the wheel (which extends below the track) contains loops at regular intervals. Other cycloids are defined in Exercises 33 and 34.

## 10.4 EXERCISES

**Exer. 1–22:** Find an equation in $x$ and $y$ whose graph contains the points on the curve $C$. Sketch the graph of $C$ and indicate the orientation.

1. $x = t - 2$, $\quad y = 2t + 3$; $\quad 0 \le t \le 5$

2. $x = 1 - 2t$, $\quad y = 1 + t$; $\quad -1 \le t \le 4$

3. $x = t^2 + 1$, $\quad y = t^2 - 1$; $\quad -2 \le t \le 2$

4. $x = t^3 + 1$, $\quad y = t^3 - 1$; $\quad -2 \le t \le 2$

5. $x = 4t^2 - 5$, $\quad y = 2t + 3$; $\quad t$ in $\mathbb{R}$

6. $x = t^3$, $\quad y = t^2$; $\quad t$ in $\mathbb{R}$

7. $x = e^t$, $\quad y = e^{-2t}$; $\quad t$ in $\mathbb{R}$

8. $x = \sqrt{t}$, $\quad y = 3t + 4$; $\quad t \ge 0$

9. $x = 2 \sin t$, $\quad y = 3 \cos t$; $\quad 0 \le t \le 2\pi$

10. $x = \cos t - 2$, $\quad y = \sin t + 3$; $\quad 0 \le t \le 2\pi$

11. $x = \sec t$, $\quad y = \tan t$; $\quad -\pi/2 < t < \pi/2$

12. $x = \cos 2t$, $\quad y = \sin t$; $\quad -\pi \le t \le \pi$

13. $x = t^2$, $\quad y = 2 \ln t$; $\quad t > 0$

14. $x = \cos^3 t$, $\quad y = \sin^3 t$; $\quad 0 \le t \le 2\pi$

15. $x = \sin t$, $\quad y = \csc t$; $\quad 0 < t \le \pi/2$

16. $x = e^t$, $\quad y = e^{-t}$; $\quad t$ in $\mathbb{R}$

17. $x = t$, $\quad y = \sqrt{t^2 - 1}$; $\quad |t| \ge 1$

18. $x = -2\sqrt{1 - t^2}$, $y = t$; $\quad |t| \le 1$

19. $x = t$, $\quad y = \sqrt{t^2 - 2t + 1}$; $\quad 0 \le t \le 4$

20. $x = 2t$, $\quad y = 8t^3$; $\quad -1 \le t \le 1$

21. $x = (t + 1)^3$, $\quad y = (t + 2)^2$; $\quad 0 \le t \le 2$

22. $x = \tan t$, $\quad y = 1$; $\quad -\pi/2 < t < \pi/2$

**Exer. 23–24:** Curves $C_1$, $C_2$, $C_3$, and $C_4$ are given parametrically, for $t$ in $\mathbb{R}$. Sketch their graphs and indicate orientations.

23. $C_1$: $x = t^2$, $\quad y = t$
    $C_2$: $x = t^4$, $\quad y = t^2$
    $C_3$: $x = \sin^2 t$, $\quad y = \sin t$
    $C_4$: $x = e^{2t}$, $\quad y = -e^t$

24. $C_1$: $x = t$, $\quad y = 1 - t$
    $C_2$: $x = 1 - t^2$, $\quad y = t^2$
    $C_3$: $x = \cos^2 t$, $\quad y = \sin^2 t$
    $C_4$: $x = \ln t - t$, $\quad y = 1 + t - \ln t$; $\quad t > 0$

**Exer. 25–26:** The parametric equations specify the position of a moving point $P(x, y)$ at time $t$. Sketch the graph and indicate the motion of $P$ as $t$ increases.

25. (a) $x = \cos t$, $\quad y = \sin t$; $\quad 0 \le t \le \pi$
    (b) $x = \sin t$, $\quad y = \cos t$; $\quad 0 \le t \le \pi$
    (c) $x = t$, $\quad y = \sqrt{1 - t^2}$; $\quad -1 \le t \le 1$

26. (a) $x = t^2$, $\quad y = 1 - t^2$; $\quad 0 \le t \le 1$
    (b) $x = 1 - \ln t$, $y = \ln t$; $\quad 1 \le t \le e$
    (c) $x = \cos^2 t$, $\quad y = \sin^2 t$; $\quad 0 \le t \le 2\pi$

27. Show that
$$x = a \cos t + h, \quad y = b \sin t + k; \quad 0 \le t \le 2\pi$$
are parametric equations of an ellipse with center $(h, k)$ and axes of lengths $2a$ and $2b$.

28. Show that
$$x = a \sec t + h, \quad y = b \tan t + k;$$
$$-\pi/2 < t < 3\pi/2 \text{ and } t \ne \pi/2$$
are parametric equations of a hyperbola with center $(h, k)$, transverse axis of length $2a$, and conjugate axis of length $2b$. Determine the values of $t$ for each branch.

**Exer. 29–30:** (a) Find three parametrizations that give the same graph as the given equation. (b) Find three parametrizations that give only a portion of the graph of the given equation.

29. $y = x^2$

30. $y = \ln x$

31. Refer to Example 5.

(a) Describe the Lissajous figure given by $f(t) = a \sin \omega t$ and $g(t) = b \cos \omega t$ for $t \ge 0$ and $a \ne b$.

(b) Suppose $f(t) = a \sin \omega_1 t$ and $g(t) = b \sin \omega_2 t$, where $\omega_1$ and $\omega_2$ are positive rational numbers, and write $\omega_2/\omega_1$ as $m/n$ for positive integers $m$ and $n$. Show that if $p = 2\pi n/\omega_1$, then both $f(t + p) = f(t)$ and $g(t + p) = g(t)$. Conclude that the curve retraces itself every $p$ units of time.

**32** Shown in the figure is the Lissajous figure given by

$$x = 2 \sin 3t, \quad y = 3 \sin 1.5t; \quad t \geq 0.$$

Find the period of the figure—that is, the length of the smallest $t$-interval that traces the curve.

EXERCISE 32

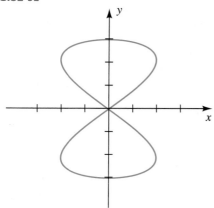

**33** A circle $C$ of radius $b$ rolls on the outside of the circle $x^2 + y^2 = a^2$, and $b < a$. Let $P$ be a fixed point on $C$, and let the initial position of $P$ be $A(a, 0)$, as shown in the figure. If the parameter $t$ is the angle from the positive $x$-axis to the line segment from $O$ to the center of $C$, show that parametric equations for the curve traced by $P$ (an *epicycloid*) are

$$x = (a + b) \cos t - b \cos \left( \frac{a + b}{b} t \right)$$

$$y = (a + b) \sin t - b \sin \left( \frac{a + b}{b} t \right);$$

$$0 \leq t \leq 2\pi.$$

EXERCISE 33

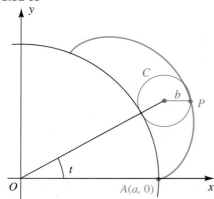

**34** If the circle $C$ of Exercise 33 rolls on the inside of the second circle (see the figure), then the curve traced by $P$ is a *hypocycloid*.

**(a)** Show that parametric equations for this curve are

$$x = (a - b) \cos t + b \cos \left( \frac{a - b}{b} t \right)$$

$$y = (a - b) \sin t - b \sin \left( \frac{a - b}{b} t \right);$$

$$0 \leq t \leq 2\pi.$$

**(b)** If $b = \frac{1}{4}a$, show that $x = a \cos^3 t$, $y = a \sin^3 t$ and sketch the graph.

EXERCISE 34

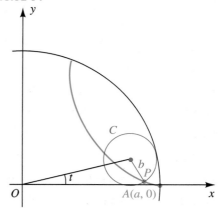

**35** If $b = \frac{1}{3}a$ in Exercise 33, find parametric equations for the epicycloid and sketch the graph.

**36** The radius of circle $B$ is one-third that of circle $A$. How many revolutions will circle $B$ make as it rolls around circle $A$ until it reaches its starting point? (*Hint:* Use Exercise 35.)

**Exer. 37–40: Graph the curve.**

**37** $x = 3 \sin^5 t, \qquad y = 3 \cos^5 t; \qquad 0 \leq t \leq 2\pi$

**38** $x = 8 \cos t - 2 \cos 4t, \quad y = 8 \sin t - 2 \sin 4t;$
$$0 \leq t \leq 2\pi$$

**39** $x = 3t - 2 \sin t, \qquad y = 3 - 2 \cos t; \qquad -8 \leq t \leq 8$

**40** $x = 2t - 3 \sin t, \qquad y = 2 - 3 \cos t; \qquad -8 \leq t \leq 8$

**Exer. 41–44:** Graph the given curves on the same coordinate plane, and describe the shape of the resulting figure.

**41** $C_1$: $x = 2 \sin 3t$,     $y = 3 \cos 2t$;     $-\pi/2 \le t \le \pi/2$
$C_2$: $x = \frac{1}{4} \cos t + \frac{3}{4}$, $y = \frac{1}{4} \sin t + \frac{3}{2}$;     $0 \le t \le 2\pi$
$C_3$: $x = \frac{1}{4} \cos t - \frac{3}{4}$, $y = \frac{1}{4} \sin t + \frac{3}{2}$;     $0 \le t \le 2\pi$
$C_4$: $x = \frac{3}{4} \cos t$,     $y = \frac{1}{4} \sin t$;     $0 \le t \le 2\pi$
$C_5$: $x = \frac{1}{4} \cos t$,     $y = \frac{1}{8} \sin t + \frac{3}{4}$;     $\pi \le t \le 2\pi$

**42** $C_1$: $x = \frac{3}{2} \cos t + 1$, $y = \sin t - 1$;     $-\pi/2 \le t \le \pi/2$
$C_2$: $x = \frac{3}{2} \cos t + 1$, $y = \sin t + 1$;     $-\pi/2 \le t \le \pi/2$
$C_3$: $x = 1$,     $y = 2 \tan t$;     $-\pi/4 \le t \le \pi/4$

**43** $C_1$: $x = \tan t$,     $y = 3 \tan t$;     $0 \le t \le \pi/4$
$C_2$: $x = 1 + \tan t$, $y = 3 - 3 \tan t$;     $0 \le t \le \pi/4$
$C_3$: $x = \frac{1}{2} + \tan t$, $y = \frac{3}{2}$;     $0 \le t \le \pi/4$

**44** $C_1$: $x = 1 + \cos t$, $y = 1 + \sin t$;     $\pi/3 \le t \le 2\pi$
$C_2$: $x = 1 + \tan t$, $y = 1$;     $0 \le t \le \pi/4$

# 10.5   POLAR COORDINATES

**FIGURE 45**

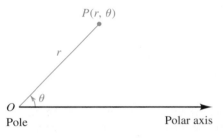

In a rectangular coordinate system, the ordered pair $(a, b)$ denotes the point whose directed distances from the $x$- and $y$-axes are $b$ and $a$, respectively. Another method for representing points is to use *polar coordinates*. We begin with a fixed point $O$ (the **origin**, or **pole**) and a directed half-line (the **polar axis**) with end point $O$. Next we consider any point $P$ in the plane different from $O$. If, as illustrated in Figure 45, $r = d(O, P)$ and $\theta$ denotes the measure of any angle determined by the polar axis and $OP$, then $r$ and $\theta$ are **polar coordinates** of $P$ and the symbols $(r, \theta)$ or $P(r, \theta)$ are used to denote $P$. As usual, $\theta$ is considered positive if the angle is generated by a counterclockwise rotation of the polar axis and negative if the rotation is clockwise. Either radian or degree measure may be used for $\theta$.

The polar coordinates of a point are not unique. For example, $(3, \pi/4)$, $(3, 9\pi/4)$, and $(3, -7\pi/4)$ all represent the same point (see Figure 46). We shall also allow $r$ to be negative. In this case, instead of measuring $|r|$ units along the terminal side of the angle $\theta$, we measure along the half-line with end point $O$ that has direction *opposite* that of the terminal side. The points corresponding to the pairs $(-3, 5\pi/4)$ and $(-3, -3\pi/4)$ are also plotted in Figure 46.

**FIGURE 46**

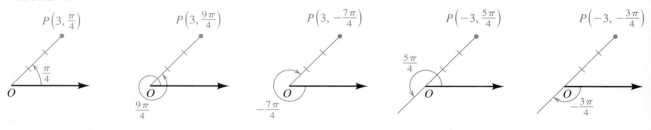

We agree that the pole $O$ has polar coordinates $(0, \theta)$ for *any* $\theta$. An assignment of ordered pairs of the form $(r, \theta)$ to points in a plane is a **polar coordinate system,** and the plane is an $r\theta$-**plane.**

**FIGURE 47**

**FIGURE 48**

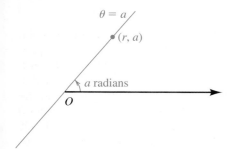

A **polar equation** is an equation in $r$ and $\theta$. A **solution** of a polar equation is an ordered pair $(a, b)$ that leads to equality if $a$ is substituted for $r$ and $b$ for $\theta$. The **graph** of a polar equation is the set of all points (in an $r\theta$-plane) that correspond to the solutions.

The simplest polar equations are $r = a$ and $\theta = a$, where $a$ is a nonzero real number. Since the solutions of the polar equation $r = a$ are of the form $(a, \theta)$ for *any* angle $\theta$, it follows that the graph is a circle of radius $|a|$ with center at the pole. A graph for $a > 0$ is sketched in Figure 47. The same graph is obtained for $r = -a$.

The solutions of the polar equation $\theta = a$ are of the form $(r, a)$ for *any* real number $r$. Since the (angle) coordinate $a$ is constant, the graph is a line through the origin, as illustrated in Figure 48 for the case $0 < a < \pi/2$.

In the following examples we obtain the graphs of polar equations by plotting points. As you proceed through this section, you should try to recognize forms of polar equations so that you will be able to sketch their graphs by plotting few, if any, points.

**EXAMPLE 1**  *Sketching the graph of a polar equation*

Sketch the graph of the polar equation $r = 4 \sin \theta$.

**SOLUTION**  The following table displays some solutions of the equation. We have included a third row in the table that contains one-decimal-place approximations to $r$.

| $\theta$ | $0$ | $\dfrac{\pi}{6}$ | $\dfrac{\pi}{4}$ | $\dfrac{\pi}{3}$ | $\dfrac{\pi}{2}$ | $\dfrac{2\pi}{3}$ | $\dfrac{3\pi}{4}$ | $\dfrac{5\pi}{6}$ | $\pi$ |
|---|---|---|---|---|---|---|---|---|---|
| $r$ | $0$ | $2$ | $2\sqrt{2}$ | $2\sqrt{3}$ | $4$ | $2\sqrt{3}$ | $2\sqrt{2}$ | $2$ | $0$ |
| $r$ (approx.) | $0$ | $2$ | $2.8$ | $3.4$ | $4$ | $3.4$ | $2.8$ | $2$ | $0$ |

**FIGURE 49**

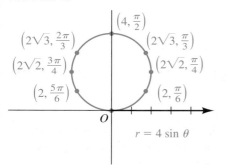

$r = 4 \sin \theta$

The points in an $r\theta$-plane that correspond to the pairs in the table appear to lie on a circle of radius 2, and we draw the graph accordingly (see Figure 49). As an aid to plotting points, we have extended the polar axis in the negative direction and introduced a vertical line through the pole.

The proof that the graph of $r = 4 \sin \theta$ is a circle is given in Example 8. Additional points obtained by letting $\theta$ vary from $\pi$ to $2\pi$ lie on the same circle. For example, the solution $(-2, 7\pi/6)$ gives us the same point as $(2, \pi/6)$; the point corresponding to $(-2\sqrt{2}, 5\pi/4)$ is the same as that obtained from $(2\sqrt{2}, \pi/4)$; and so on. If we let $\theta$ increase through all real numbers, we obtain the same points again and again because of the periodicity of the sine function.

**EXAMPLE 2    *Sketching the graph of a polar equation***
Sketch the graph of the polar equation $r = 2 + 2 \cos \theta$.

SOLUTION    Since the cosine function decreases from 1 to $-1$ as $\theta$ varies from 0 to $\pi$, it follows that $r$ decreases from 4 to 0 in this $\theta$-interval. The following table exhibits some solutions of $r = 2 + 2 \cos \theta$, together with one-decimal-place approximations to $r$.

| $\theta$ | $0$ | $\dfrac{\pi}{6}$ | $\dfrac{\pi}{4}$ | $\dfrac{\pi}{3}$ | $\dfrac{\pi}{2}$ | $\dfrac{2\pi}{3}$ | $\dfrac{3\pi}{4}$ | $\dfrac{5\pi}{6}$ | $\pi$ |
|---|---|---|---|---|---|---|---|---|---|
| $r$ | $4$ | $2 + \sqrt{3}$ | $2 + \sqrt{2}$ | $3$ | $2$ | $1$ | $2 - \sqrt{2}$ | $2 - \sqrt{3}$ | $0$ |
| $r$ (approx.) | $4$ | $3.7$ | $3.4$ | $3$ | $2$ | $1$ | $0.6$ | $0.3$ | $0$ |

Plotting points in an $r\theta$-plane leads to the upper half of the graph sketched in Figure 50. (We have used polar coordinate graph paper, which displays lines through $O$ at various angles and concentric circles with centers at the pole.)

**FIGURE 50**

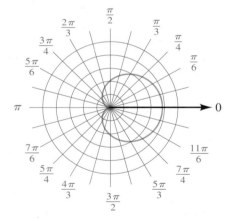

If $\theta$ increases from $\pi$ to $2\pi$, then $\cos \theta$ increases from $-1$ to $1$, and consequently $r$ increases from 0 to 4. Plotting points for $\pi \le \theta \le 2\pi$ gives us the lower half of the graph.

The same graph may be obtained by taking other intervals of length $2\pi$ for $\theta$.

The heart-shaped graph in Example 2 is a **cardioid.** In general, the graph of any of the following polar equations, with $a \neq 0$, is a cardioid:

$$r = a(1 + \cos \theta) \qquad r = a(1 + \sin \theta)$$
$$r = a(1 - \cos \theta) \qquad r = a(1 - \sin \theta)$$

If $a$ and $b$ are not zero, then the graphs of the following polar equations are **limaçons:**

$$r = a + b \cos \theta \qquad r = a + b \sin \theta$$

Note that the special limaçons in which $|a| = |b|$ are cardioids. Some limaçons contain a loop, as shown in the next example.

**EXAMPLE 3**   *Sketching the graph of a polar equation*

Sketch the graph of the polar equation $r = 2 + 4 \cos \theta$.

**SOLUTION**   Coordinates of some points in an $r\theta$-plane that correspond to $0 \le \theta \le \pi$ are listed in the following table.

| $\theta$ | 0 | $\dfrac{\pi}{6}$ | $\dfrac{\pi}{4}$ | $\dfrac{\pi}{3}$ | $\dfrac{\pi}{2}$ | $\dfrac{2\pi}{3}$ | $\dfrac{3\pi}{4}$ | $\dfrac{5\pi}{6}$ | $\pi$ |
|---|---|---|---|---|---|---|---|---|---|
| $r$ | 6 | $2 + 2\sqrt{3}$ | $2 + 2\sqrt{2}$ | 4 | 2 | 0 | $2 - 2\sqrt{2}$ | $2 - 2\sqrt{3}$ | $-2$ |
| $r$ (approx.) | 6 | 5.4 | 4.8 | 4 | 2 | 0 | $-0.8$ | $-1.4$ | $-2$ |

Note that $r = 0$ at $\theta = 2\pi/3$. The values of $r$ are negative if $2\pi/3 < \theta \le \pi$, and this leads to the lower half of the small loop in Figure 51. Letting $\theta$ range from $\pi$ to $2\pi$ gives us the upper half of the small loop and the lower half of the large loop.

**FIGURE 51**

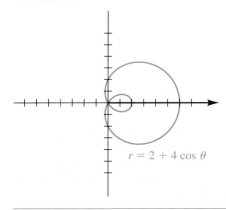

$r = 2 + 4 \cos \theta$

**EXAMPLE 4**    *Sketching the graph of a polar equation*

Sketch the graph of the polar equation $r = a \sin 2\theta$ for $a > 0$.

**SOLUTION**    Instead of tabulating solutions, let us reason as follows. If $\theta$ increases from 0 to $\pi/4$, then $2\theta$ varies from 0 to $\pi/2$, and hence $\sin 2\theta$ increases from 0 to 1. It follows that $r$ increases from 0 to $a$ in the $\theta$-interval $[0, \pi/4]$. If we next let $\theta$ increase from $\pi/4$ to $\pi/2$, then $2\theta$ changes from $\pi/2$ to $\pi$, and hence $\sin 2\theta$ decreases from 1 to 0. Thus, $r$ decreases from $a$ to 0 in the $\theta$-interval $[\pi/4, \pi/2]$. The corresponding points on the graph constitute the first-quadrant loop illustrated in Figure 52. Note that the point $P(r, \theta)$ traces the loop in a *counterclockwise* direction (indicated by the arrows) as $\theta$ increases from 0 to $\pi/2$.

If $\pi/2 \le \theta \le \pi$, then $\pi \le 2\theta \le 2\pi$, and therefore $r = a \sin 2\theta \le 0$. Thus, *if* $\pi/2 < \theta < \pi$, *then r is negative and the points $P(r, \theta)$ are in the fourth quadrant.* If $\theta$ increases from $\pi/2$ to $\pi$, then we can show, by plotting points, that $P(r, \theta)$ traces (in a counterclockwise direction) the loop shown in the fourth quadrant.

Similarly, for $\pi \le \theta \le 3\pi/2$ we get the loop in the third quadrant, and for $3\pi/2 \le \theta \le 2\pi$ we get the loop in the second quadrant. Both loops are traced in a counterclockwise direction as $\theta$ increases. You should verify these facts by plotting some points with, say, $a = 1$. In Figure 52 we have plotted only those points on the graph that correspond to the largest numerical values of $r$.

**FIGURE 52**

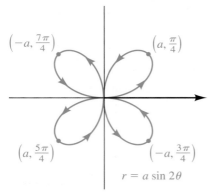

$r = a \sin 2\theta$

The graph in Example 4 is a **four-leafed rose.** In general, a polar equation of the form

$$r = a \sin n\theta \quad \text{or} \quad r = a \cos n\theta$$

for any positive integer $n$ greater than 1 and any nonzero real number $a$ has a graph that consists of a number of loops through the origin. If $n$ is even, there are $2n$ loops, and if $n$ is odd, there are $n$ loops.

The graph of the polar equation $r = a\theta$ for any nonzero real number $a$ is a **spiral of Archimedes.** The case $a = 1$ is considered in the next example.

**EXAMPLE 5**    *Sketching the graph of a spiral of Archimedes*

Sketch the graph of the polar equation $r = \theta$ for $\theta \ge 0$.

**SOLUTION**    The graph consists of all points that have polar coordinates of the form $(c, c)$ for every real number $c \ge 0$. Thus, the graph contains the points $(0, 0)$, $(\pi/2, \pi/2)$, $(\pi, \pi)$, and so on. As $\theta$ increases, $r$ increases at the same rate, and the spiral winds around the origin in a counterclockwise direction, intersecting the polar axis at $0, 2\pi, 4\pi, \ldots$, as illustrated in Figure 53.

**FIGURE 53**

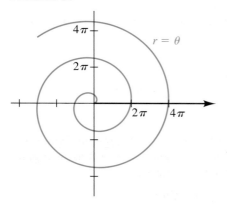

$r = \theta$

If $\theta$ is allowed to be negative, then as $\theta$ decreases through negative values, the resulting spiral winds around the origin and is the symmetric image, with respect to the vertical axis, of the curve sketched in Figure 53.

Let us next superimpose an $xy$-plane on an $r\theta$-plane so that the positive $x$-axis coincides with the polar axis. Any point $P$ in the plane may then be assigned rectangular coordinates $(x, y)$ or polar coordinates $(r, \theta)$. If $r > 0$, we have a situation similar to that illustrated in Figure 54(a); if $r < 0$, we have that shown in (b) of the figure. In Figure 54(b), for later purposes, we have also plotted the point $P'$, having polar coordinates $(|r|, \theta)$ and rectangular coordinates $(-x, -y)$.

**FIGURE 54**

(a) $r > 0$                                                      (b) $r < 0$

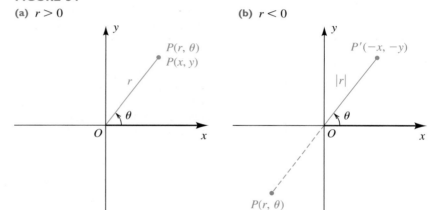

The following result specifies relationships between $(x, y)$ and $(r, \theta)$, where it is assumed that the positive $x$-axis coincides with the polar axis.

**Relationships Between Rectangular and Polar Coordinates**

The rectangular coordinates $(x, y)$ and polar coordinates $(r, \theta)$ of a point $P$ are related as follows:

**(1)** $x = r \cos \theta, \quad y = r \sin \theta$

**(2)** $r^2 = x^2 + y^2, \quad \tan \theta = \dfrac{y}{x}$ if $x \neq 0$

**PROOF**    Although we have pictured $\theta$ as an acute angle in Figure 54, the discussion that follows is valid for all angles. If $r > 0$ as in Figure 54(a), then $\cos \theta = x/r$ and $\sin \theta = y/r$, and hence

$$x = r \cos \theta, \quad y = r \sin \theta.$$

If $r < 0$, then $|r| = -r$, and from Figure 54(b) we see that

$$\cos \theta = \frac{-x}{|r|} = \frac{-x}{-r} = \frac{x}{r}, \quad \sin \theta = \frac{-y}{|r|} = \frac{-y}{-r} = \frac{y}{r}.$$

Multiplication by $r$ gives us relationship (1), and therefore these formulas hold if $r$ is either positive or negative. If $r = 0$, then the point is the pole and we again see that the formulas in (1) are true.

The formulas in (2) follow readily from Figure 54. ■

We may use the preceding result to change from one system of coordinates to the other.

**FIGURE 55**

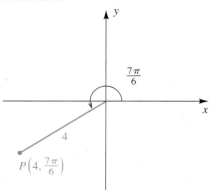

### EXAMPLE 6    *Changing polar coordinates to rectangular coordinates*

If $(4, 7\pi/6)$ are polar coordinates of a point $P$, find the rectangular coordinates of $P$.

*SOLUTION*    The point $P$ is plotted in Figure 55. Substituting $r = 4$ and $\theta = 7\pi/6$ in formula (1) of the preceding result, we obtain the following:

$$x = r \cos \theta = 4 \cos (7\pi/6) = 4(-\sqrt{3}/2) = -2\sqrt{3}$$

$$y = r \sin \theta = 4 \sin (7\pi/6) = 4(-1/2) = -2$$

Hence, the rectangular coordinates of $P$ are $(-2\sqrt{3}, -2)$.

### EXAMPLE 7    *Changing rectangular coordinates to polar coordinates*

If $(-1, \sqrt{3})$ are rectangular coordinates of a point $P$, find three different pairs of polar coordinates $(r, \theta)$ for $P$.

*SOLUTION*    Three possibilities for $\theta$ are illustrated in Figure 56(a)–(c). Using $x = -1$ and $y = \sqrt{3}$ in (2) of the previous result, we obtain

$$r^2 = x^2 + y^2 = (-1)^2 + (\sqrt{3})^2 = 4,$$

and since $r$ is positive in Figure 56(a), $r = 2$. Using

$$\tan \theta = \frac{y}{x} = \frac{\sqrt{3}}{-1} = -\sqrt{3},$$

we see that the reference angle for $\theta$ is $\theta_R = \pi/3$, and hence

$$\theta = \pi - \frac{\pi}{3} = \frac{2\pi}{3}.$$

Thus, $(2, 2\pi/3)$ is one pair of polar coordinates for $P$.

**FIGURE 56**

**(a)**

**(b)**

**(c)**

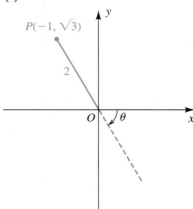

Referring to Figure 56(b) and the values obtained for (a), we obtain

$$r = 2 \quad \text{and} \quad \theta = \frac{2\pi}{3} + 2\pi = \frac{8\pi}{3}.$$

Hence, $(2, 8\pi/3)$ is another pair of polar coordinates for $P$.

In Figure 56(c), $\theta = -\pi/3$. In this case we use $r = -2$ to obtain $(-2, -\pi/3)$ as a third pair of polar coordinates for $P$.

We may use the relationships between rectangular and polar coordinates to transform a polar equation to an equation in $x$ and $y$, and vice versa. This procedure is illustrated in the next three examples.

**EXAMPLE 8** *Changing a polar equation to an equation in x and y*

Find an equation in $x$ and $y$ that has the same graph as the polar equation $r = a \sin \theta$, with $a \neq 0$. Sketch the graph.

*SOLUTION* From (1) of the preceding result, a relationship between $\sin \theta$ and $y$ is given by $y = r \sin \theta$. To introduce the expression $r \sin \theta$ into the equation $r = a \sin \theta$, we multiply both sides by $r$, obtaining

$$r^2 = ar \sin \theta.$$

*(continued)*

**FIGURE 57**

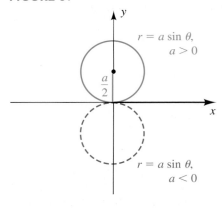

Next, using $r^2 = x^2 + y^2$ and $y = r \sin \theta$, we have

$$x^2 + y^2 = ay,$$

or

$$x^2 + y^2 - ay = 0.$$

Completing the square in $y$ gives us

$$x^2 + y^2 - ay + \left( \frac{a}{2} \right)^2 = \left( \frac{a}{2} \right)^2,$$

or

$$x^2 + \left( y - \frac{a}{2} \right)^2 = \left( \frac{a}{2} \right)^2.$$

In the $xy$-plane, the graph of the last equation is a circle with center $(0, a/2)$ and radius $|a|/2$, as illustrated in Figure 57 for the case $a > 0$ (the solid circle) and $a < 0$ (the dashed circle).

**FIGURE 58**

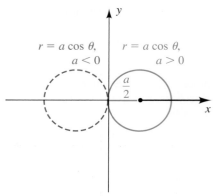

Using the same method as in the preceding example, we can show that the graph of $r = a \cos \theta$, with $a \neq 0$, is a circle of radius $a/2$ of the type illustrated in Figure 58.

**EXAMPLE 9**    *Changing an equation in x and y to a polar equation*

Find a polar equation for the hyperbola $x^2 - y^2 = 16$.

**SOLUTION**    Using the formulas $x = r \cos \theta$ and $y = r \sin \theta$, we obtain the following polar equations:

$$(r \cos \theta)^2 - (r \sin \theta)^2 = 16 \qquad \text{given}$$
$$r^2 \cos^2 \theta - r^2 \sin^2 \theta = 16 \qquad \text{square the terms}$$
$$r^2(\cos^2 \theta - \sin^2 \theta) = 16 \qquad \text{factor out } r^2$$
$$r^2 \cos 2\theta = 16 \qquad \text{double-angle formula}$$
$$r^2 = \frac{16}{\cos 2\theta} \qquad \text{divide by } \cos 2\theta$$
$$r^2 = 16 \sec 2\theta \qquad \text{reciprocal identity}$$

The division by $\cos 2\theta$ is allowable because $\cos 2\theta \neq 0$. (Note that if $\cos 2\theta = 0$, then $r^2 \cos 2\theta \neq 16$.)

**EXAMPLE 10**    *Finding a polar equation of a line*

Find a polar equation of an arbitrary line.

*SOLUTION*    Every line in an *xy*-coordinate plane is the graph of a linear equation of the form $ax + by = c$. Using the formulas $x = r \cos \theta$ and $y = r \sin \theta$ gives us the following equivalent polar equations:

$$ar \cos \theta + br \sin \theta = c \quad \text{substitute for } x \text{ and } y$$

$$r(a \cos \theta + b \sin \theta) = c \quad \text{factor out } r$$

If $a \cos \theta + b \sin \theta \neq 0$, the last equation may be written as follows:

$$r = \frac{c}{a \cos \theta + b \sin \theta}$$

If we superimpose an *xy*-plane on an *r*θ-plane, then the graph of a polar equation may be symmetric with respect to the *x*-axis (the polar axis), the *y*-axis (the line $\theta = \pi/2$), or the origin (the pole). Some typical symmetries are illustrated in Figure 59. The next result summarizes these symmetries.

**FIGURE 59**   Symmetries of graphs of polar equations

**(a)** Polar axis

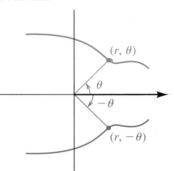

**(b)** Line $\theta = \pi/2$

**(c)** Pole

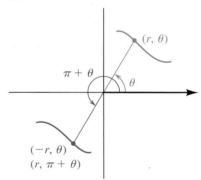

**Tests for Symmetry**

(**1**) The graph of $r = f(\theta)$ is symmetric with respect to the polar axis if substitution of $-\theta$ for $\theta$ leads to an equivalent equation.

(**2**) The graph of $r = f(\theta)$ is symmetric with respect to the vertical line $\theta = \pi/2$ if substitution of either (a) $\pi - \theta$ for $\theta$ or (b) $-r$ for $r$ and $-\theta$ for $\theta$ leads to an equivalent equation.

(**3**) The graph of $r = f(\theta)$ is symmetric with respect to the pole if substitution of either (a) $\pi + \theta$ for $\theta$ or (b) $-r$ for $r$ leads to an equivalent equation.

To illustrate, since cos $(-\theta) = \cos \theta$, the graph of the polar equation $r = 2 + 4 \cos \theta$ in Example 3 is symmetric with respect to the polar axis, by test (1). Since sin $(\pi - \theta) = \sin \theta$, the graph in Example 1 is symmetric with respect to the line $\theta = \pi/2$, by test (2). The graph of the four-leafed rose in Example 4 is symmetric with respect to the polar axis, the line $\theta = \pi/2$, and the pole. Other tests for symmetry may be stated; however, those we have listed are among the easiest to apply.

Unlike the graph of an equation in $x$ and $y$, the graph of a polar equation $r = f(\theta)$ can be symmetric with respect to the polar axis, the line $\theta = \pi/2$, or the pole *without* satisfying one of the preceding tests for symmetry. This is true because of the many different ways of specifying a point in polar coordinates.

Another difference between rectangular and polar coordinate systems is that the points of intersection of two graphs cannot always be found by solving the polar equations simultaneously. To illustrate, from Example 1, the graph of $r = 4 \sin \theta$ is a circle of diameter 4 with center at $(2, \pi/2)$ (see Figure 60). Similarly, the graph of $r = 4 \cos \theta$ is a circle of diameter 4 with center at $(2, 0)$ on the polar axis. Referring to Figure 60, we see that the coordinates of the point of intersection $P(2\sqrt{2}, \pi/4)$ in quadrant I satisfy both equations; however, the origin $O$, which is on each circle, *cannot* be found by solving the equations simultaneously. Thus, in searching for points of intersection of polar graphs, it is sometimes necessary to refer to the graphs themselves, *in addition* to solving the two equations simultaneously. An alternative method is to use different (equivalent) equations for the graphs.

**FIGURE 60**

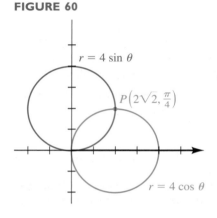

## 10.5 EXERCISES

**Exer. 1–34: Sketch the graph of the polar equation.**

1 $r = 5$

2 $r = -2$

3 $\theta = -\pi/6$

4 $\theta = \pi/4$

5 $r = 3 \cos \theta$

6 $r = -2 \sin \theta$

7 $r = 4 \cos \theta + 2 \sin \theta$

8 $r = 6 \cos \theta - 2 \sin \theta$

9 $r = 4(1 - \sin \theta)$

10 $r = 3(1 + \cos \theta)$

11 $r = -6(1 + \cos \theta)$

12 $r = 2(1 + \sin \theta)$

13 $r = 2 + 4 \sin \theta$

14 $r = 1 + 2 \cos \theta$

15 $r = \sqrt{3} - 2 \sin \theta$

16 $r = 2\sqrt{3} - 4 \cos \theta$

17 $r = 2 - \cos \theta$

18 $r = 5 + 3 \sin \theta$

19 $r = 4 \csc \theta$

20 $r = -3 \sec \theta$

21 $r = 8 \cos 3\theta$

22 $r = 2 \sin 4\theta$

23 $r = 3 \sin 2\theta$

24 $r = 8 \cos 5\theta$

25 $r^2 = 4 \cos 2\theta$  (lemniscate)

26 $r^2 = -16 \sin 2\theta$

27 $r = 2^\theta, \theta \geq 0$  (spiral)

28 $r = e^{2\theta}, \theta \geq 0$  (logarithmic spiral)

29 $r = 2\theta, \theta \geq 0$

30 $r\theta = 1, \theta > 0$  (spiral)

31 $r = 6 \sin^2 (\theta/2)$

32 $r = -4 \cos^2 (\theta/2)$

33 $r = 2 + 2 \sec \theta$  (conchoid)

34 $r = 1 - \csc \theta$

**Exer. 35–40: Change the polar coordinates to rectangular coordinates.**

35 (a) $(3, \pi/4)$  (b) $(-1, 2\pi/3)$

36 (a) $(5, 5\pi/6)$       (b) $(-6, 7\pi/3)$

37 (a) $(8, -2\pi/3)$      (b) $(-3, 5\pi/3)$

38 (a) $(4, -\pi/4)$      (b) $(-2, 7\pi/6)$

39 $(6, \arctan \frac{3}{4})$      40 $(10, \arccos (-\frac{1}{3}))$

**Exer. 41–44: Change the rectangular coordinates to polar coordinates with $r > 0$ and $0 \le \theta \le 2\pi$.**

41 (a) $(-1, 1)$      (b) $(-2\sqrt{3}, -2)$

42 (a) $(3\sqrt{3}, 3)$      (b) $(2, -2)$

43 (a) $(7, -7\sqrt{3})$      (b) $(5, 5)$

44 (a) $(-2\sqrt{2}, -2\sqrt{2})$      (b) $(-4, 4\sqrt{3})$

45 Which polar coordinates represent the same point as $(3, \pi/3)$?

     (a) $(3, 7\pi/3)$    (b) $(3, -\pi/3)$    (c) $(-3, 4\pi/3)$

     (d) $(3, -2\pi/3)$    (e) $(-3, -2\pi/3)$    (f) $(-3, -\pi/3)$

46 Which polar coordinates represent the same point as $(4, -\pi/2)$?

     (a) $(4, 5\pi/2)$    (b) $(4, 7\pi/2)$    (c) $(-4, -\pi/2)$

     (d) $(4, -5\pi/2)$    (e) $(-4, -3\pi/2)$    (f) $(-4, \pi/2)$

**Exer. 47–56: Find a polar equation that has the same graph as the equation in $x$ and $y$.**

47 $x = -3$      48 $y = 2$

49 $x^2 + y^2 = 16$      50 $x^2 = 8y$

51 $2y = -x$      52 $y = 6x$

53 $y^2 - x^2 = 4$      54 $xy = 8$

55 $(x - 1)^2 + y^2 = 1$

56 $(x + 2)^2 + (y - 3)^2 = 13$

**Exer. 57–74: Find an equation in $x$ and $y$ that has the same graph as the polar equation. Use it to help sketch the graph in an $r\theta$-plane.**

57 $r \cos \theta = 5$      58 $r \sin \theta = -2$

59 $r - 6 \sin \theta = 0$      60 $r = 2$

61 $\theta = \pi/4$      62 $r = 4 \sec \theta$

63 $r^2(4 \sin^2 \theta - 9 \cos^2 \theta) = 36$

64 $r^2(\cos^2 \theta + 4 \sin^2 \theta) = 16$

65 $r^2 \cos 2\theta = 1$      66 $r^2 \sin 2\theta = 4$

67 $r(\sin \theta - 2 \cos \theta) = 6$

68 $r(3 \cos \theta - 4 \sin \theta) = 12$

69 $r(\sin \theta + r \cos^2 \theta) = 1$    70 $r(r \sin^2 \theta - \cos \theta) = 3$

71 $r = 8 \sin \theta - 2 \cos \theta$    72 $r = 2 \cos \theta - 4 \sin \theta$

73 $r = \tan \theta$      74 $r = 6 \cot \theta$

75 If $P_1(r_1, \theta_1)$ and $P_2(r_2, \theta_2)$ are points in an $r\theta$-plane, use the law of cosines to prove that
$$[d(P_1, P_2)]^2 = r_1^2 + r_2^2 - 2r_1r_2 \cos (\theta_2 - \theta_1).$$

76 Prove that the graph of each polar equation is a circle, and find its center and radius.

     (a) $r = a \sin \theta, a \neq 0$

     (b) $r = b \cos \theta, b \neq 0$

     (c) $r = a \sin \theta + b \cos \theta, a \neq 0$ and $b \neq 0$

**Exer. 77–78: Graph the polar equation for the indicated values of $\theta$, and use the graph to determine symmetries.**

77 $r = 2 \sin^2 \theta \tan^2 \theta; \quad -\pi/3 \le \theta \le \pi/3$

     (*Note:* Some specific keystrokes for the TI-81/82 graphics calculator are given in Appendix I.)

78 $r = \dfrac{4}{1 + \sin^2 \theta}; \quad 0 \le \theta \le 2\pi$

**Exer. 79–80: Graph the polar equations on the same coordinate plane, and estimate the points of intersection of the graphs.**

79 $r = 8 \cos 3\theta, \quad\quad r = 4 - 2.5 \cos \theta$

80 $r = 2 \sin^2 \theta, \quad\quad r = \frac{3}{4}(\theta + \cos^2 \theta)$

---

# 10.6   POLAR EQUATIONS OF CONICS

The following theorem combines the definitions of parabola, ellipse, and hyperbola into a unified description of the conic sections. The constant $e$ in the statement of the theorem is the **eccentricity** of the conic. The point

$F$ is a **focus** of the conic, and the line $l$ is a **directrix.** Possible positions of $F$ and $l$ are illustrated in Figure 61.

**FIGURE 61**

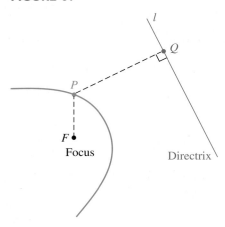

Focus

Directrix

**Theorem on Conics**

Let $F$ be a fixed point and $l$ a fixed line in a plane. The set of all points $P$ in the plane, such that the ratio $d(P, F)/d(P, Q)$ is a positive constant $e$ with $d(P, Q)$ the distance from $P$ to $l$, is a conic section. The conic is a parabola if $e = 1$, an ellipse if $0 < e < 1$, and a hyperbola if $e > 1$.

*PROOF*   If $e = 1$, then $d(P, F) = d(P, Q)$ and, by definition, the resulting conic is a parabola with focus $F$ and directrix $l$.

Suppose next that $0 < e < 1$. It is convenient to introduce a polar coordinate system in the plane with $F$ as the pole and $l$ perpendicular to the polar axis at the point $D(d, 0)$, with $d > 0$, as illustrated in Figure 62. If $P(r, \theta)$ is a point in the plane such that $d(P, F)/d(P, Q) = e < 1$, then $P$ lies to the left of $l$. Let $C$ be the projection of $P$ on the polar axis. Since

$$d(P, F) = r \quad \text{and} \quad d(P, Q) = d - r \cos \theta,$$

it follows that $P$ satisfies the condition in the theorem if and only if the following are true:

$$\frac{r}{d - r \cos \theta} = e$$

$$r = de - er \cos \theta$$

$$r(1 + e \cos \theta) = de$$

$$r = \frac{de}{1 + e \cos \theta}$$

**FIGURE 62**

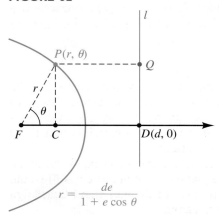

$$r = \frac{de}{1 + e \cos \theta}$$

The same equations are obtained if $e = 1$; however, there is no point $(r, \theta)$ on the graph if $1 + \cos \theta = 0$.

An equation in $x$ and $y$ corresponding to $r = de - er \cos \theta$ is

$$\sqrt{x^2 + y^2} = de - ex.$$

Squaring both sides and rearranging terms leads to

$$(1 - e^2)x^2 + 2de^2x + y^2 = d^2e^2.$$

Completing the square and simplifying, we obtain

$$\left(x + \frac{de^2}{1 - e^2}\right)^2 + \frac{y^2}{1 - e^2} = \frac{d^2e^2}{(1 - e^2)^2}.$$

Finally, dividing both sides by $d^2e^2/(1 - e^2)^2$ gives us an equation of the form

$$\frac{(x - h)^2}{a^2} + \frac{y^2}{b^2} = 1,$$

with $h = -de^2/(1 - e^2)$. Consequently, the graph is an ellipse with center at the point $(h, 0)$ on the $x$-axis and with

$$a^2 = \frac{d^2e^2}{(1 - e^2)^2} \quad \text{and} \quad b^2 = \frac{d^2e^2}{1 - e^2}.$$

Since

$$c^2 = a^2 - b^2 = \frac{d^2e^4}{(1 - e^2)^2},$$

we obtain $c = de^2/(1 - e^2)$, and hence $|h| = c$. This proves that $F$ is a focus of the ellipse. It also follows that $e = c/a$. A similar proof may be given for the case $e > 1$. ∎

We also can show that every conic that is not degenerate may be described by means of the statement in the theorem on conics. This gives us a formulation of conic sections that is equivalent to the one used previously. Since the theorem includes all three types of conics, it is sometimes regarded as a definition for the conic sections.

If we had chosen the focus $F$ to the *right* of the directrix, as illustrated in Figure 63 (with $d > 0$), then the equation $r = de/(1 - e \cos \theta)$ would have resulted. (Note the minus sign in place of the plus sign.) Other sign changes occur if $d$ is allowed to be negative.

**FIGURE 63**

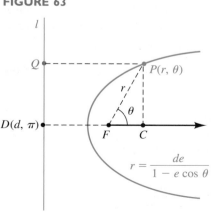

$$r = \frac{de}{1 - e \cos \theta}$$

If we had taken *l parallel* to the polar axis through one of the points $(d, \pi/2)$ or $(d, 3\pi/2)$, as illustrated in Figure 64, then the corresponding equations would have contained $\sin \theta$ instead of $\cos \theta$.

**FIGURE 64**

(a)

(b)

$$r = \frac{de}{1 + e \sin \theta}$$

$$r = \frac{de}{1 - e \sin \theta}$$

The following theorem summarizes our discussion.

**Theorem on Polar Equations of Conics**

A polar equation that has one of the four forms

$$r = \frac{de}{1 \pm e \cos \theta} \quad \text{or} \quad r = \frac{de}{1 \pm e \sin \theta}$$

is a conic section. The conic is a parabola if $e = 1$, an ellipse if $0 < e < 1$, or a hyperbola if $e > 1$.

**EXAMPLE 1**   *Sketching the graph of a polar equation of an ellipse*

Describe and sketch the graph of the polar equation

$$r = \frac{10}{3 + 2 \cos \theta}.$$

**SOLUTION**   We first divide the numerator and denominator of the fraction by 3 to obtain the constant term 1 in the denominator:

$$r = \frac{\frac{10}{3}}{1 + \frac{2}{3} \cos \theta}$$

This equation has one of the forms in the preceding theorem, with $e = \frac{2}{3}$. Thus, the graph is an ellipse with focus $F$ at the pole and major axis along the polar axis. We find the end points of the major axis by letting $\theta = 0$ and $\theta = \pi$. This gives us the points $V(2, 0)$ and $V'(10, \pi)$. Hence,

$$2a = d(V', V) = 12, \quad \text{or} \quad a = 6.$$

**FIGURE 65**

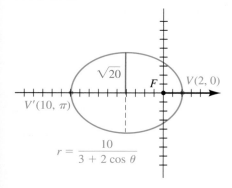

$$r = \frac{10}{3 + 2\cos\theta}$$

The center of the ellipse is the midpoint $(4, \pi)$ of the segment $V'V$. Using the fact that $e = c/a$, we obtain

$$c = ae = 6\left(\tfrac{2}{3}\right) = 4.$$

Hence,

$$b^2 = a^2 - c^2 = 36 - 16 = 20.$$

Thus, $b = \sqrt{20}$. The graph is sketched in Figure 65. For reference, we have superimposed an $xy$-coordinate system on the polar system.

**EXAMPLE 2**    *Sketching the graph of a polar*
*equation of a hyperbola*

Describe and sketch the graph of the polar equation

$$r = \frac{10}{2 + 3\sin\theta}.$$

*SOLUTION*    To express the equation in the proper form, we divide the numerator and denominator of the fraction by 2:

$$r = \frac{5}{1 + \tfrac{3}{2}\sin\theta}$$

Thus, $e = \tfrac{3}{2}$, and, by the theorem on polar equations of conics, the graph is a hyperbola with a focus at the pole. The expression $\sin\theta$ tells us that the transverse axis of the hyperbola is perpendicular to the polar axis. To find the vertices, we let $\theta = \pi/2$ and $\theta = 3\pi/2$ in the given equation. This gives us the points $V(2, \pi/2)$ and $V'(-10, 3\pi/2)$. Hence,

$$2a = d(V, V') = 8, \quad \text{or} \quad a = 4.$$

The points $(5, 0)$ and $(5, \pi)$ on the graph can be used to sketch the lower branch of the hyperbola. The upper branch is obtained by symmetry, as illustrated in Figure 66. If we desire more accuracy or additional information, we calculate

$$c = ae = 4\left(\tfrac{3}{2}\right) = 6$$

and

$$b^2 = c^2 - a^2 = 36 - 16 = 20.$$

Asymptotes may then be constructed in the usual way.

**FIGURE 66**

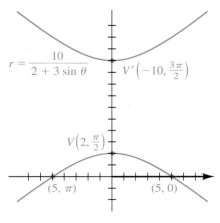

$$r = \frac{10}{2 + 3\sin\theta}$$

**EXAMPLE 3**    *Sketching the graph of a polar equation of a parabola*

Sketch the graph of the polar equation

$$r = \frac{15}{4 - 4\cos\theta}.$$

**SOLUTION**    To obtain the proper form, we divide the numerator and denominator by 4:

$$r = \frac{\frac{15}{4}}{1 - \cos\theta}$$

Consequently, $e = 1$, and, by the theorem on polar equations of conics, the graph is a parabola with focus at the pole. We may obtain a sketch by plotting the points that correspond to the quadrantal angles indicated in the following table.

**FIGURE 67**

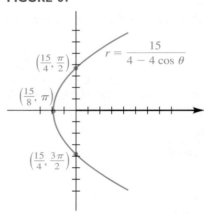

| $\theta$ | 0 | $\dfrac{\pi}{2}$ | $\pi$ | $\dfrac{3\pi}{2}$ |
|---|---|---|---|---|
| $r$ | undefined | $\frac{15}{4}$ | $\frac{15}{8}$ | $\frac{15}{4}$ |

Note that there is no point on the graph corresponding to $\theta = 0$, since the denominator $1 - \cos\theta$ is 0 for that value. Plotting the three points and using the fact that the graph is a parabola with focus at the pole gives us the sketch in Figure 67.

If we desire only a rough sketch of a conic, then the technique employed in Example 3 is recommended. To use this method, we plot (if possible) points corresponding to $\theta = 0$, $\pi/2$, $\pi$, and $3\pi/2$. These points, together with the type of conic (obtained from the value of the eccentricity $e$), readily lead to the sketch.

**EXAMPLE 4**    *Expressing a polar equation of a conic in terms of x and y*

Find an equation in $x$ and $y$ that has the same graph as the polar equation

$$r = \frac{15}{4 - 4\sin\theta}.$$

**SOLUTION**   We multiply both sides of the polar equation by the lcd, $4 - 4 \sin \theta$, and then use relationships between rectangular and polar coordinates as follows:

$$r(4 - 4 \sin \theta) = 15$$
$$4r = 15 + 4r \sin \theta$$
$$4(\pm\sqrt{x^2 + y^2}) = 15 + 4y$$

Squaring both sides and simplifying yields

$$16(x^2 + y^2) = 225 + 120y + 16y^2$$
$$16x^2 = 120y + 225.$$

**EXAMPLE 5**   *Finding a polar equation of a conic satisfying prescribed conditions*

Find a polar equation of the conic with a focus at the pole, eccentricity $e = \frac{1}{2}$, and directrix $r = -3 \sec \theta$.

**SOLUTION**   The equation $r = -3 \sec \theta$ of the directrix may be written $r \cos \theta = -3$, which is equivalent to $x = -3$ in a rectangular coordinate system. This gives us the situation illustrated in Figure 63, with $d = 3$. Hence, a polar equation has the form

$$r = \frac{de}{1 - e \cos \theta}.$$

We now substitute $d = 3$ and $e = \frac{1}{2}$:

$$r = \frac{3\left(\frac{1}{2}\right)}{1 - \frac{1}{2} \cos \theta}$$

$$r = \frac{3}{2 - \cos \theta}$$

## 10.6 EXERCISES

Exer. 1–12: Find the eccentricity and classify the conic. Sketch the graph and label the vertices.

1  $r = \dfrac{12}{6 + 2 \sin \theta}$

2  $r = \dfrac{12}{6 - 2 \sin \theta}$

3  $r = \dfrac{12}{2 - 6 \cos \theta}$

4  $r = \dfrac{12}{2 + 6 \cos \theta}$

5  $r = \dfrac{3}{2 + 2 \cos \theta}$

6  $r = \dfrac{3}{2 - 2 \sin \theta}$

7  $r = \dfrac{4}{\cos \theta - 2}$

8  $r = \dfrac{4 \sec \theta}{2 \sec \theta - 1}$

9  $r = \dfrac{6 \csc \theta}{2 \csc \theta + 3}$

10  $r = \dfrac{8 \csc \theta}{2 \csc \theta - 5}$

**11** $r = \dfrac{4 \csc \theta}{1 + \csc \theta}$

**12** $r = \csc \theta (\csc \theta - \cot \theta)$

**Exer. 13–24:** Find equations in $x$ and $y$ for the polar equations in Exercises 1–12.

**Exer. 25–32:** Find a polar equation of the conic with focus at the pole that has the given eccentricity and equation of directrix.

**25** $e = \frac{1}{3}$,  $r = 2 \sec \theta$       **26** $e = 1$,  $r \cos \theta = 5$

**27** $e = \frac{4}{3}$,  $r \cos \theta = -3$       **28** $e = 3$,  $r = -4 \sec \theta$

**29** $e = 1$,  $r \sin \theta = -2$       **30** $e = 4$,  $r = -3 \csc \theta$

**31** $e = \frac{2}{5}$,  $r = 4 \csc \theta$       **32** $e = \frac{3}{4}$,  $r \sin \theta = 5$

**Exer. 33–34:** Find a polar equation of the parabola with focus at the pole and the given vertex.

**33** $V\left(4, \dfrac{\pi}{2}\right)$       **34** $V(5, 0)$

**Exer. 35–36:** An ellipse has a focus at the pole with the given center $C$ and vertex $V$. Find **(a)** the eccentricity and **(b)** a polar equation for the ellipse.

**35** $C\left(3, \dfrac{\pi}{2}\right)$,  $V\left(1, \dfrac{3\pi}{2}\right)$       **36** $C(2, \pi)$,  $V(1, 0)$

**37 Kepler's first law**   Kepler's first law asserts that planets travel in elliptical orbits with the sun at one focus. To find an equation of an orbit, place the pole $O$ at the

center of the sun and the polar axis along the major axis of the ellipse (see the figure).

**(a)** Show that an equation of the orbit is

$$r = \frac{(1 - e^2)a}{1 - e \cos \theta},$$

where $e$ is the eccentricity and $2a$ is the length of the major axis.

**(b)** The perihelion distance $r_{\text{per}}$ and aphelion distance $r_{\text{aph}}$ are defined as the minimum and maximum distances, respectively, of a planet from the sun. Show that $r_{\text{per}} = a(1 - e)$ and $r_{\text{aph}} = a(1 + e)$.

**EXERCISE 37**

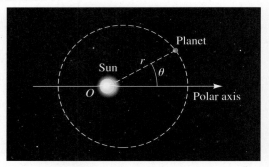

**38 Kepler's first law**   Refer to Exercise 37. The planet Pluto travels in an elliptical orbit of eccentricity 0.249. If the perihelion distance is 29.62 AU, find a polar equation for the orbit and estimate the aphelion distance.

# CHAPTER 10 REVIEW EXERCISES

**Exer. 1–16:** Find the vertices and foci of the conic, and sketch its graph.

**1** $y^2 = 64x$

**2** $y = 8x^2 + 32x + 33$

**3** $9y^2 = 144 - 16x^2$

**4** $9y^2 = 144 + 16x^2$

**5** $x^2 - y^2 - 4 = 0$

**6** $25x^2 + 36y^2 = 1$

**7** $25y = 100 - x^2$

**8** $3x^2 + 4y^2 - 18x + 8y + 19 = 0$

**9** $x^2 - 9y^2 + 8x + 90y - 210 = 0$

**10** $x = 2y^2 + 8y + 3$

**11** $4x^2 + 9y^2 + 24x - 36y + 36 = 0$

**12** $4x^2 - y^2 - 40x - 8y + 88 = 0$

**13** $y^2 - 8x + 8y + 32 = 0$

**14** $4x^2 + y^2 - 24x + 4y + 36 = 0$

**15** $x^2 - 9y^2 + 8x + 7 = 0$

**16** $y^2 - 2x^2 + 6y + 8x - 3 = 0$

**Exer. 17–18:** Find the standard equation of a parabola with a vertical axis that satisfies the given conditions.

**17** $x$-intercepts $-10$ and $-4$, $y$-intercept $80$

**18** $x$-intercepts $-11$ and $3$, passing through $(2, 39)$

**Exer. 19–28: Find an equation for the conic that satisfies the given conditions.**

19 Hyperbola, with vertices $V(0, \pm 7)$ and end points of conjugate axis $(\pm 3, 0)$

20 Parabola, with focus $F(-4, 0)$ and directrix $x = 4$

21 Parabola, with focus $F(0, -10)$ and directrix $y = 10$

22 Parabola, with vertex at the origin, symmetric to the $x$-axis, and passing through the point $(5, -1)$

23 Ellipse, with vertices $V(0, \pm 10)$ and foci $F(0, \pm 5)$

24 Hyperbola, with foci $F(\pm 10, 0)$ and vertices $V(\pm 5, 0)$

25 Hyperbola, with vertices $V(0, \pm 6)$ and asymptotes $y = \pm 9x$

26 Ellipse, with foci $F(\pm 2, 0)$ and passing through the point $(2, \sqrt{2})$

27 Ellipse, with eccentricity $\frac{2}{3}$ and end points of minor axis $(\pm 5, 0)$

28 Ellipse, with eccentricity $\frac{3}{4}$ and foci $F(\pm 12, 0)$

29 (a) Determine $A$ so that the point $(2, -3)$ is on the conic $Ax^2 + 2y^2 = 4$.

   (b) Is the conic an ellipse or a hyperbola?

30 If a square with sides parallel to the coordinate axes is inscribed in the ellipse $(x^2/a^2) + (y^2/b^2) = 1$, express the area $A$ of the square in terms of $a$ and $b$.

31 Find the standard equation of the circle that has center at the focus of the parabola $y = \frac{1}{8}x^2$ and passes through the the origin.

32 A point $P(x, y)$ is the same distance from $(4, 0)$ as it is from the circle $x^2 + y^2 = 4$, as illustrated in the figure. Show that the collection of all such points forms a branch of a hyperbola, and sketch its graph.

**EXERCISE 32**

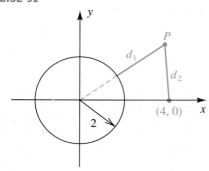

**Exer. 33–37: Find an equation in $x$ and $y$ whose graph contains the points on the curve $C$. Sketch the graph of $C$, and indicate the orientation.**

33  $x = 3 + 4t,$       $y = t - 1;$       $-2 \le t \le 2$

34  $x = \sqrt{-t},$      $y = t^2 - 4;$      $t \le 0$

35  $x = \cos^2 t - 2,$   $y = \sin t + 1;$   $0 \le t \le 2\pi$

36  $x = \sqrt{t},$       $y = 2^{-t};$       $t \ge 0$

37  $x = \dfrac{1}{t} + 1,$   $y = \dfrac{2}{t} - t;$   $0 < t \le 4$

38 Curves $C_1, C_2, C_3,$ and $C_4$ are given parametrically for $t$ in $\mathbb{R}$. Sketch their graphs, and discuss their similarities and differences.

$C_1$: $x = t,$            $y = \sqrt{16 - t^2}$
$C_2$: $x = -\sqrt{16 - t},$  $y = -\sqrt{t}$
$C_3$: $x = 4 \cos t,$       $y = 4 \sin t$
$C_4$: $x = e^t,$           $y = -\sqrt{16 - e^{2t}}$

39 Change $(5, 7\pi/4)$ to rectangular coordinates.

40 Change $(2\sqrt{3}, -2)$ to polar coordinates with $r > 0$ and $0 \le \theta < 2\pi$.

**Exer. 41–52: Sketch the graph of the polar equation.**

41  $r = -4 \sin \theta$

42  $r = 8 \sec \theta$

43  $r = 3 \sin 5\theta$

44  $r = 6 - 3 \cos \theta$

45  $r = 3 - 3 \sin \theta$

46  $r = 2 + 4 \cos \theta$

47  $r^2 = 9 \sin 2\theta$

48  $2r = \theta$

49  $r = \dfrac{8}{1 - 3 \sin \theta}$

50  $r = 6 - r \cos \theta$

51  $r = \dfrac{6}{3 + 2 \cos \theta}$

52  $r = \dfrac{-6 \csc \theta}{1 - 2 \csc \theta}$

**Exer. 53–56: Find a polar equation that has the same graph as the equation in $x$ and $y$.**

53  $y^2 = 4x$

54  $x^2 + y^2 - 3x + 4y = 0$

55  $2x - 3y = 8$

56  $x^2 + y^2 = 2xy$

**Exer. 57–62: Find an equation in $x$ and $y$ that has the same graph as the polar equation.**

57  $r^2 = \tan \theta$

58  $r = 2 \cos \theta + 3 \sin \theta$

59  $r^2 = 4 \sin 2\theta$

60  $\theta = \sqrt{3}$

61  $r = 5 \sec \theta + 3r \sec \theta$

62  $r^2 \sin \theta = 6 \csc \theta + r \cot \theta$

# APPENDIXES

GRAPHING WITH A TI GRAPHICS CALCULATOR

This appendix provides basic instructions for graphing on a TI-81 or TI-82 graphics calculator, as well as using some advanced features. Most of the examples in this appendix are taken from examples or exercises in the text, but the solutions presented are not the same as the solutions that appear in the text. The solutions here are intended to highlight TI-81/82 features and illustrate the specific keystrokes, whereas the solutions in the text are of a more general nature. For example, the reasons for selecting a particular viewing rectangle are discussed more in the text than in this appendix.

Working through this appendix is an excellent way to familiarize yourself with both the TI-81/82 and how a graphics calculator is used in this text—thereby reducing the time needed for in-class instruction concerning the graphics calculator. The *Guidebook* that comes with the calculator has a section entitled ''Getting Started.'' Consulting this section first will help you become familiar with the keyboard and with calculating arithmetic expressions.

We begin by providing every keystroke needed to obtain the screens shown in the figures (the screens are from the TI-81—the TI-82 screens are similar). As we progress, we will assume that *all* keystroke sequences are not needed. When different keystrokes are needed to perform an operation on the TI-82, we have noted the difference near the first occurrence. We conclude the appendix with a brief discussion of additional features found on the TI-82 that are not on the TI-81.

**EXAMPLE I**   *Sketching two simple graphs*

Sketch the graphs of $y = x$ and $y = x^2$ on the same screen.

**FIGURE I**

(a) Assignments

(b) Standard viewing rectangle

$[-10, 10]$ by $[-10, 10]$

(c) "Square" standard viewing rectangle

$[-15, 15]$ by $[-10, 10]$

(d) Viewing rectangle information

**SOLUTION**   After pressing the $\boxed{\text{ON}}$ key, we make the assignments

$$Y_1 = x \quad \text{and} \quad Y_2 = x^2$$

using the following keystroke sequence:

*Make Y assignments*

$Y_1$:   $\boxed{Y=}$ $\boxed{\text{CLEAR}}$ $\boxed{\text{X|T}}$ $\boxed{\text{ENTER}}$
$Y_2$:   $\boxed{\text{CLEAR}}$ $\boxed{\text{X|T}}$ $\boxed{x^2}$ $\boxed{\text{ENTER}}$

*TI-82 Note*   The $\boxed{\text{X|T}}$ key is labeled $\boxed{\text{X,T,}\theta}$.

Figure 1(a) shows the current screen. Next we choose a standard viewing rectangle by pressing $\boxed{\text{ZOOM}}$ $\boxed{6}$, and we obtain Figure 1(b). Note that the graph of $y = x$ does not appear to bisect the first quadrant. We can remedy this situation by pressing $\boxed{\text{ZOOM}}$ $\boxed{5}$, which gives us a screen that provides a true proportion; this screen is shown in Figure 1(c). By pressing the $\boxed{\text{RANGE}}$ key, we display the current viewing rectangle dimensions, as shown in Figure 1(d). The terms Xmin, Xmax, etc., were discussed on page 87.

*TI-82 Note*   The $\boxed{\text{RANGE}}$ key is labeled $\boxed{\text{WINDOW}}$.

We often need to determine the coordinates of a point to some specified accuracy. The next example illustrates a technique for doing so.

### EXAMPLE 2   *Practicing the tracing and zoom-in features*

Familiarize yourself with the tracing and zoom-in features by finding the
points of intersection of the graphs in Example 1.

**FIGURE 2**

(a) After pressing

(b) After a zoom-in

$[-15, 15]$ by $[-10, 10]$                    $[-3.6, 3.7]$ by $[-2.3, 2.7]$

**SOLUTION**   Press GRAPH to get the last graphs drawn on the screen.
Now press TRACE . The current screen is shown in Figure 2(a). Note the
blinking cursor on your screen and its $x$- and $y$-coordinates at the bottom
of the screen. Now press ZOOM 2 ENTER to *zoom in* on the position of
the cursor (see Figure 2(b)). The actual viewing rectangle dimensions of
Figure 2(b) depend on the original position of the cursor and the values of
the ZOOM factors (see choice 4 under the ZOOM menu).

*TI-82 Note*   The ZOOM factors are choice 4 under the MEMORY submenu of the ZOOM menu.

We see that $(0, 0)$ is a likely candidate for a point of intersection, and
we will try to "zero in" on the other point of intersection. Press TRACE
again, press the ▷ key several times to get to about $x = 0.5$, and then
press the cursor keys ▽ and △ , noticing that the cursor jumps between
the graphs of $Y_1$ and $Y_2$. Now press the ▷ key to get closer to the point of
intersection. To get as close to the point as desired, repeat the process of
alternately pressing the key sequence ZOOM 2 ENTER and the TRACE
with the ◁ and ▷ cursor keys. This process is referred to in the text as
*zooming-in* and *tracing*. The other point of intersection is actually $(1, 1)$.

We will now examine specific keystrokes and screens for some of the
examples in the text. The next example is Example 9 in Section 1.7.

**EXAMPLE 3**    *Estimating points of intersection of graphs*

Estimate the points of intersection of the graphs of $y = 2x - 1$ and $y = x^2 - 3$.

**FIGURE 3**

(a) Assignments

(b) Standard viewing rectangle

$[-15, 15]$ by $[-10, 10]$

(c) Selecting box corners

$[-15, 15]$ by $[-10, 10]$

(d) Enhanced view

$[1.7, 4.3]$ by $[2.7, 5.2]$

**SOLUTION**

*Make Y assignments*

$Y_1$:    $\boxed{Y=}$ $\boxed{\text{CLEAR}}$ $2$ $\boxed{\text{X|T}}$ $\boxed{-}$ $1$ $\boxed{\triangledown}$

$Y_2$:    $\boxed{\text{CLEAR}}$ $\boxed{\text{X|T}}$ $\boxed{x^2}$ $\boxed{-}$ $3$ (see Figure 3(a))

Note that the ''2'' entered in the $Y_1$ assignment represents data entry as opposed to menu choice $\boxed{2}$.

*Select VR (viewing rectangle)*

$\boxed{\text{ZOOM}}$ $\boxed{6}$ $\boxed{\text{ZOOM}}$ $\boxed{5}$ (see Figure 3(b))

We will now show another approach to zooming in on a region—using a *box*.

*Select box corners*

$\boxed{\text{ZOOM}}$ $\boxed{1}$ $\boxed{\triangle}$ (8 times) $\boxed{\triangleright}$ (5 times) $\boxed{\text{ENTER}}$ $\boxed{\triangleright}$ (8 times) $\boxed{\triangle}$ (8 times)

(see Figure 3(c)) $\boxed{\text{ENTER}}$ (see Figure 3(d))

You may now trace and zoom as in Example 2 or by using the box method. From this point on, we will simply say ''trace and zoom'' when trying to find specific points—you select the method.

The approximate coordinates for the point of intersection in the first quadrant are (2.7, 4.5). To determine coordinates for the point of intersection in the third quadrant, press ⎡ZOOM⎤⎡6⎤, and then trace and zoom to obtain the approximate coordinates (−0.7, −2.5).

---

The next example (Example 10 in Section 1.7) demonstrates how to graph more than four functions on the TI-81 by using the DRAW menu. TI-82 users can simply use $Y_5$, but will find the following use of DRAW informative.

### EXAMPLE 4    *Estimating points of intersection of graphs*

Estimate the points of intersection of the circles with equations $x^2 + y^2 = 25$ and $x^2 + y^2 - 4y = 12$.

**FIGURE 4**

(a) Assignments

(b) Graph of $Y_1$, $Y_2$, and $Y_4$

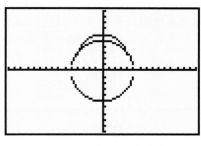

$[-15, 15]$ by $[-10, 10]$

(c) Adding $Y_5$

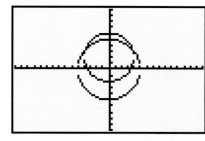

$[-15, 15]$ by $[-10, 10]$

**SOLUTION**    Solving the equation $x^2 + y^2 = 25$ for $y$ in terms of $x$ gives us $y = \pm\sqrt{25 - x^2}$. We assign $\sqrt{25 - x^2}$ to $Y_1$ and $-Y_1$ to $Y_2$.

*Make Y assignments*

$Y_1$:    ⎡Y =⎤ ⎡CLEAR⎤ ⎡2nd⎤ ⎡√⎤ ⎡(⎤ 25 ⎡−⎤ ⎡X|T⎤ ⎡$x^2$⎤ ⎡)⎤ ⎡▽⎤

$Y_2$:    ⎡CLEAR⎤ ⎡(−)⎤ ⎡2nd⎤ ⎡Y-VARS⎤ ⎡1⎤ ⎡▽⎤

**TI-82 Note**    When accessing the Y-values, you need to first select the "Function" submenu (choice 1) under the Y-VARS menu.

Note the difference between the subtraction key ⎡−⎤ and the negation key ⎡(−)⎤. The subtraction key is used for operations such as $3 - 4$, whereas the negation key is used to enter an expression such as $-2$. In the remaining examples, we will not mention the use of the negation key for data entry. Solving $x^2 + y^2 - 4y = 12$ for $y$ in terms of $x$ gives us $y = 2 \pm \sqrt{16 - x^2}$. We assign $\sqrt{16 - x^2}$ to $Y_3$, $2 + Y_3$ to $Y_4$, and $2 - Y_3$ to $Y_5$.

$Y_3$:    ⎡2nd⎤ ⎡√⎤ ⎡(⎤ 16 ⎡−⎤ ⎡X|T⎤ ⎡$x^2$⎤ ⎡)⎤ ⎡▽⎤

$Y_4$:    2 ⎡+⎤ ⎡2nd⎤ ⎡Y-VARS⎤ ⎡3⎤ ⎡ENTER⎤

*Turn off $Y_3$*    ⎡△⎤ ⎡◁⎤ ⎡ENTER⎤

*(continued)*

*TI-82 Note*    The $\boxed{\triangle}$ key needs to be pressed twice to correctly position the cursor.

Note that the "=" sign by $Y_3$ is not shaded (see Figure 4(a)).

*Graph* $Y_1$, $Y_2$, *and* $Y_4$    $\boxed{\text{ZOOM}}$ $\boxed{6}$ (oval shape) $\boxed{\text{ZOOM}}$ $\boxed{5}$ (see Figure 4(b))

*Add* $Y_5$    $\boxed{\text{2nd}}$ $\boxed{\text{DRAW}}$ $\boxed{6}$ $2$ $\boxed{-}$ $\boxed{\text{2nd}}$ $\boxed{\text{Y-VARS}}$ $\boxed{3}$ $\boxed{\text{ENTER}}$ (see Figure 4(c))

Now trace and zoom in on the point of intersection in the first quadrant. The graph of $Y_5$ is lost, but it is not needed to find this point. We estimate the point's coordinates to be $(3.8, 3.25)$. Since both circles are symmetric with respect to the $y$-axis, the other point of intersection is approximately $(-3.8, 3.25)$.

---

We proceed to parts (a) and (b) of Example 9 in Section 2.1.

**EXAMPLE 5**    *Analyzing the graph of a function*

Let $f(x) = x^{2/3} - 3$. Sketch the graph of $f$ and find $f(-2)$.

**FIGURE 5**

(a) Assignment

(b) Graph of $f$

$[-15, 15]$ by $[-10, 10]$

(c) Finding the function value

**SOLUTION**    The assignment of X^(2/3) − 3 to $Y_1$ produces only the portion of the graph with $x \geq 0$. Three ways to represent the function and obtain the entire graph of $f$ are $\sqrt[3]{X^2} - 3$, $(X^\wedge(1/3))^\wedge 2 - 3$, and $(X^\wedge 2)^\wedge(1/3) - 3$. We will use the first representation to demonstrate use of the special cube root key. As a general rule, it is best to represent $x^{m/n}$ as $(X^\wedge(1/n))^\wedge m$. Be sure to clear $Y_2$, $Y_3$, and $Y_4$ after entering $Y_1$.

*Make Y assignment*    $Y_1$:  $\boxed{Y =}$ $\boxed{\text{CLEAR}}$ $\boxed{\text{MATH}}$ $\boxed{4}$ $\boxed{\text{X|T}}$ $\boxed{x^2}$ $\boxed{-}$ $3$ (see Figure 5(a))

*Select VR*    $\boxed{\text{ZOOM}}$ $\boxed{6}$ $\boxed{\text{ZOOM}}$ $\boxed{5}$ (see Figure 5(b))

We will find $f(-2)$ by storing $-2$ in memory location X and then querying the value of $Y_1$.

*Store $-2$ in X*      | 2nd | | QUIT | | CLEAR | $-2$ | STO▷ | | X (or X|T) | | ENTER |

*Query* $Y_1$      | 2nd | | Y-VARS | | 1 | | ENTER |

Figure 5(c) shows the current screen and indicates that $f(-2) \approx -1.41$. To obtain two digits after the decimal point, press the following key sequence:

*Change precision*      | MODE | | ▽ | | ▷ | | ▷ | | ▷ | | ENTER | | 2nd | | QUIT |

We can now recall the last operation by pressing | 2nd | | ENTRY | , and then we press | ENTER | to obtain the value $-1.41$ (also shown in Figure 5(c)). Change your precision back to "Float" before proceeding.

---

The next example (Example 8 in Section 2.4) shows how to examine a function of a function—that is, a composite function.

**EXAMPLE 6**   *Graphically analyzing a composite function*

Let $f(x) = x^2 - 16$ and $g(x) = \sqrt{x}$. Sketch $y = (f \circ g)(x)$, and use the graph to find the domain of $f \circ g$. Find $f(g(3))$.

**FIGURE 6**

(a) Assignments

(b) Standard viewing rectangle

$[-15, 15]$ by $[-10, 10]$

(c) Changing RANGE values

```
RANGE
Xmin=-10
Xmax=50
Xscl=5
Ymin=-20
Ymax=20
Yscl=5
Xres=1
```

(d) Graph of $y = (f \circ g)(x)$

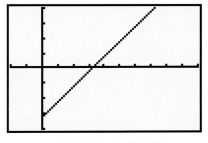

$[-10, 50]$ by $[-20, 20]$

*Make Y assignments*

**SOLUTION**      We begin by making the assignments $Y_1 = \sqrt{x}$ and $Y_2 = (Y_1)^2 - 16$.

$Y_1$: $\boxed{Y =}$ $\boxed{\text{CLEAR}}$ $\boxed{\text{2nd}}$ $\boxed{\sqrt{\phantom{x}}}$ $\boxed{\text{X|T}}$ $\boxed{\text{ENTER}}$

$Y_2$: $\boxed{\text{2nd}}$ $\boxed{\text{Y-VARS}}$ $\boxed{1}$ $\boxed{x^2}$ $\boxed{-}$ 16 $\boxed{\text{ENTER}}$ (see Figure 6(a))

Since we want only the graph of $Y_2$, we turn off $Y_1$ before graphing.

*Turn off $Y_1$*

$\boxed{\triangle}$ $\boxed{\triangle}$ $\boxed{\triangleleft}$ $\boxed{\text{ENTER}}$ (done in Figure 6(a))

After pressing $\boxed{\text{ZOOM}}$ $\boxed{5}$ to obtain Figure 6(b), we change to a different viewing rectangle to see a more complete picture of the graph of $Y_2$.

*Change VR*

$\boxed{\text{RANGE}}$ $-10$ $\boxed{\triangledown}$ $50$ $\boxed{\triangledown}$ $5$ $\boxed{\triangledown}$ $-20$ $\boxed{\triangledown}$ $20$ $\boxed{\triangledown}$ $5$ (see Figure 6(c))

**TI-82 Note**      After pressing $\boxed{\text{WINDOW}}$, you need to press $\boxed{\triangledown}$ to correctly position the cursor.

Now press $\boxed{\text{GRAPH}}$ to obtain Figure 6(d). We see that the graph is a half-line with end point $(0, -16)$, indicating that the domain of $Y_2$ is all $x \geq 0$. Note that the viewing rectangle selected gives us a $60 \times 40$ graph and maintains a true proportion since the line appears to be at a 45° angle from the horizontal (a slope of 1). It is fairly easy to see that the $x$-intercept is between 15 and 20 (it is 16), as shown in Figure 6(d).

To find $f(g(3))$, we store 3 in X and then query $Y_2$, obtaining $-13$.

*Find $f(g(3))$*

$\boxed{\text{2nd}}$ $\boxed{\text{QUIT}}$ 3 $\boxed{\text{STO}\triangleright}$ $\boxed{\text{X|T}}$ $\boxed{\text{ENTER}}$

$\boxed{\text{2nd}}$ $\boxed{\text{Y-VARS}}$ $\boxed{2}$ $\boxed{\text{ENTER}}$

The next example (Example 8 in Section 3.3) involves a polynomial that has both small and large roots, making the selection of a viewing rectangle rather difficult. Since you should be familiar by now with assigning simple expressions to Y-values and selecting various viewing rectangles, we omit some of the keystrokes.

**EXAMPLE 7**      *Exploring the graph of a polynomial*

Find the zeros of $f(x) = x^3 - 1000x^2 - x + 1000$.

**FIGURE 7**

(a) Small view of $f$                          (b) Larger view of $f$

                          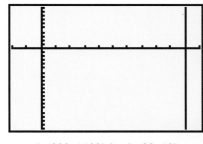

$[-15, 15]$ by $[-10, 10]$                    $[-200, 1100]$ by $[-20, 10]$

(c) Assignments

(d) Compressed graph of $f$

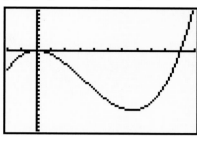

$[-200, 1100]$ by $[-20, 10]$

**SOLUTION**   We assign $x^3 - 1000x^2 - x + 1000$ to $Y_1$ and use a standard viewing rectangle to obtain Figure 7(a). Be sure to clear the other Y-values.

*Make Y assignment*   $Y_1$:   $\boxed{Y=}$ $\boxed{\text{CLEAR}}$ $\boxed{\text{X|T}}$ $\boxed{\text{MATH}}$ $\boxed{3}$ $\boxed{-}$ $1000$ $\boxed{\text{X|T}}$ $\boxed{x^2}$ $\boxed{-}$ $\boxed{\text{X|T}}$ $\boxed{+}$ $1000$
$\boxed{\text{ENTER}}$

*Select VR*   $\boxed{\text{ZOOM}}$ $\boxed{6}$ $\boxed{\text{ZOOM}}$ $\boxed{5}$ (see Figure 7(a))

As explained on page 210, we determine that the roots are $\pm 1$ and 1000. After some experimentation, we select the viewing rectangle $[-200, 1100]$ by $[-20, 10]$ with Xscl = 100 and Yscl = 1. The graph in Figure 7(b) clearly shows the root at $x = 1000$, but obscures the roots at $x = \pm 1$.

We will try to find a multiplier that will compress the graph of $Y_1$ so that the turning points are visible. Again, after some trial and error, we arrive at $10^{-7}$ as an acceptable multiplier of $Y_1$. The keystroke sequence for $Y_2$ is as follows:

*Make Y assignment*   $Y_2$:   $\boxed{Y=}$ $\boxed{\triangledown}$ $\boxed{\triangledown}$ $10$ $\boxed{\wedge}$ $\boxed{(}$ $\boxed{-7}$ $\boxed{)}$ $\boxed{\text{2nd}}$ $\boxed{\text{Y-VARS}}$ $\boxed{1}$
*Turn off* $Y_1$   $\boxed{\triangle}$ $\boxed{\triangle}$ $\boxed{\triangleleft}$ $\boxed{\text{ENTER}}$ (see Figure 7(c))

Now graph $Y_2$ to obtain Figure 7(d). Experiment by changing the exponent on the multiplying factor. Keep in mind that Figure 7(d) is the graph of $f$ with a vertical compression factor of $10^7$ (ten million!).

---

The next example (Example 8 in Section 3.5) illustrates a method for graphing rational functions and finding equations of the vertical asymptotes.

**EXAMPLE 8**   *Sketching the graph of a rational function*
Sketch the graph of

$$f(x) = \frac{x^2 - x}{9x^3 - 9x^2 - 22x + 8}$$

and find equations of the vertical asymptotes.

**FIGURE 8**

(a) Assignments

(b) Connected mode view of *f*

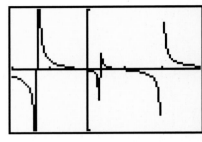

$[-2, 3]$ by $[-1, 1]$

(c) Dot mode view of *f*

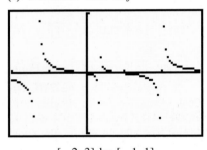

$[-2, 3]$ by $[-1, 1]$

(d) Graph of the denominator

$[-2, 3]$ by $[-1, 1]$

**SOLUTION**     We begin by making the assignments

$$Y_1 = x^2 - x, \quad Y_2 = 9x^3 - 9x^2 - 22x + 8, \quad \text{and} \quad Y_3 = Y_1/Y_2.$$

*Make Y assignments*

$Y_1$:   $\boxed{Y =}$ $\boxed{\text{CLEAR}}$ $\boxed{\text{X}|\text{T}}$ $\boxed{x^2}$ $\boxed{-}$ $\boxed{\text{X}|\text{T}}$ $\boxed{\text{ENTER}}$

$Y_2$:   $\boxed{\text{CLEAR}}$ $\boxed{9}$ $\boxed{\text{X}|\text{T}}$ $\boxed{\text{MATH}}$ $\boxed{3}$ $\boxed{-}$ $\boxed{9}$ $\boxed{\text{X}|\text{T}}$ $\boxed{x^2}$ $\boxed{-}$ $\boxed{22}$ $\boxed{\text{X}|\text{T}}$ $\boxed{+}$ $\boxed{8}$ $\boxed{\text{ENTER}}$

$Y_3$:   $\boxed{\text{CLEAR}}$ $\boxed{\text{2nd}}$ $\boxed{\text{Y-VARS}}$ $\boxed{1}$ $\boxed{\div}$ $\boxed{\text{2nd}}$ $\boxed{\text{Y-VARS}}$ $\boxed{2}$

Now turn off $Y_1$ and $Y_2$ (see Figure 8(a)) and graph $Y_3$ using $\boxed{\text{ZOOM}}$ $\boxed{6}$. This graph gives us virtually no indication of the true shape of *f*. Change the viewing rectangle to $[-2, 3]$ by $[-1, 1]$ and regraph $Y_3$ to obtain Figure 8(b). Notice the near vertical lines on the screen. These lines result from the calculator's trying to draw a line from one pixel on the screen to another. To eliminate the near vertical lines, we will change to *dot mode*.

*Select "Dot"*

$\boxed{\text{MODE}}$ $\boxed{\nabla}$ (4 times) $\boxed{\triangleright}$ $\boxed{\text{ENTER}}$

Press $\boxed{\text{GRAPH}}$ to redraw $Y_3$. Figure 8(c) gives us a reasonable view of the graph of *f* and indicates that there are three vertical asymptotes. Since there is no point to zoom in on in order to find an equation of the vertical asymptotes, we will abandon the graph of $Y_3$ and examine the graph of $Y_2$ (the denominator)—looking for its zeros. Graphing only $Y_2$ (turn off $Y_3$)

with the same viewing rectangle, but using *connected mode*, gives us Figure 8(d).

By tracing and zooming in on the zeros in Figure 8(d), we obtain the approximate values $-1.33$, $0.33$, and $2.00$. As discussed on page 235, the rational root theorem indicates that the zeros may be $-\frac{4}{3}$, $\frac{1}{3}$, and 2. By storing each of these values in X and querying $Y_2$ (obtaining 0 each time), we are reasonably sure that these are the zeros of $Y_2$. Since none of these zeros are zeros of the numerator, we have the following equations of the vertical asymptotes for $f$:

$$x = -\tfrac{4}{3}, \quad x = \tfrac{1}{3}, \quad \text{and} \quad x = 2$$

---

The TI-81/82 has an interesting process for handling any statements involving the relational operators ($=, \neq, >, \geq, <, \leq$), which are found by pressing $\boxed{\text{2nd}}$ $\boxed{\text{TEST}}$. To illustrate this process, we examine the statements

$$5 > 3, \quad 3 = 2, \quad \text{and} \quad 2 \leq 2.$$

These statements and their results are shown in Figure 9. The result of 1 in the second line of Figure 9 indicates that the statement $5 > 3$ is true, whereas the 0 in the fourth line indicates that the statement $3 = 2$ is false. We can also compare function values—say, $Y_1 > Y_2$—and then graph the result. We will demonstrate this procedure in the next two examples.

**FIGURE 9**

The next example (Example 6 in Section 4.2) involves the graph of a function in which the natural exponential function plays a key role.

**EXAMPLE 9**  *Sketching a Gompertz growth curve*

Sketch the graph of the **Gompertz growth function** $G$ given by

$$G(t) = ke^{(-Ae^{-Bt})}$$

on the interval $[0, 5]$ for $k = 1.1$, $A = 3.2$, and $B = 1.1$, and estimate the time $t$ (to three decimal places) at which $G(t) = 1$.

**FIGURE 10**

(a) Assignment

(b) Graph of $G$

[0, 5] by [0, 2]

(c) Graph of the line $y = 1$ if $Y_1 > 1$

[0, 5] by [0, 2]

(d) Trace to the line $y = 1$

[0, 5] by [0, 2]

**SOLUTION**    We begin by making the assignment (using $x$ for $t$)

$$Y_1 = 1.1e^{(-3.2e^{-1.1t})}.$$

*Make Y assignment*    $Y_1$:   $\boxed{Y =}$ $\boxed{\text{CLEAR}}$ 1.1 $\boxed{\text{2nd}}$ $\boxed{e^x}$ $\boxed{(}$ $-3.2$ $\boxed{\text{2nd}}$ $\boxed{e^x}$ $\boxed{(}$ $-1.1$ $\boxed{\text{X|T}}$ $\boxed{)}$ $\boxed{)}$
$\boxed{\text{ENTER}}$ (see Figure 10(a))

From the discussion on page 260, we know that the maximum value of $G$ is 1.1, so we choose the viewing rectangle [0, 5] by [0, 2]. Graphing $G$ gives us Figure 10(b). The end point values of the graph are approximately (0, 0.045) and (5, 1.086).

To determine the time when $y = G(t) = 1$, we will use the procedure discussed before this example. We first change to *dot mode* and then assign $Y_1 > 1$ to $Y_2$.

*Make Y assignment*    $Y_2$:   $\boxed{Y =}$ $\boxed{\triangledown}$ $\boxed{\triangledown}$ $\boxed{\text{2nd}}$ $\boxed{\text{Y-VARS}}$ $\boxed{1}$ $\boxed{\text{2nd}}$ $\boxed{\text{TEST}}$ $\boxed{3}$ 1

Thus, when $Y_1 \le 1$, the graph of $Y_2$ will be the horizontal line $y = 0$, and when $Y_1 > 1$, the graph of $Y_2$ will be the horizontal line $y = 1$. Turning off $Y_1$ and graphing $Y_2$ gives us Figure 10(c). Press $\boxed{\text{TRACE}}$ and note that the displayed $y$-coordinate is 0. Press $\boxed{\triangleright}$ successively until the $y$-coordinate becomes 1 (see Figure 10(d)). Now press $\boxed{\text{ZOOM}}$ $\boxed{2}$ $\boxed{\text{ENTER}}$ and repeat the trace and zoom procedure until you obtain the desired three-decimal-place accuracy for $x = t \approx 3.194$.

The following example (Exercise 71 in Section 4.3) is a good illustration of the power of a graphing utility, since it is impossible to find the exact solution using only algebraic methods.

**EXAMPLE 10**  *Approximating a solution to an inequality*

Graph $f(x) = 2.2 \log (x + 2)$ and $g(x) = \ln x$, and estimate the solution of the inequality $f(x) \geq g(x)$.

**FIGURE 11**

(a) Assignments

(b) Graph of $f$ and $g$

$[-2, 16]$ by $[-4, 8]$

(c) Graph of $Y_3 = Y_1 \geq Y_2$

$[-2, 16]$ by $[-4, 8]$

(d) Graph of $Y_3 = (Y_1 \geq Y_2) + 1$

$[-2, 16]$ by $[-4, 8]$

**SOLUTION**  We begin by making the assignments

$$Y_1 = 2.2 \log (x + 2) \quad \text{and} \quad Y_2 = \ln x.$$

*Make Y assignments*

$Y_1$:  $\boxed{Y =}\,\boxed{\text{CLEAR}}\,2.2\,\boxed{\text{LOG}}\,\boxed{(}\,\boxed{\text{X|T}}\,\boxed{+}\,2\,\boxed{)}\,\boxed{\text{ENTER}}$

$Y_2$:  $\boxed{\text{CLEAR}}\,\boxed{\text{LN}}\,\boxed{\text{X|T}}\,\boxed{\text{ENTER}}$ (see Figure 11(a))

Since the domain of $f$ is $(-2, \infty)$ and the domain of $g$ is $(0, \infty)$, we select Xmin $= -2$. Choosing the viewing rectangle $[-2, 16]$ by $[-4, 8]$ and graphing $Y_1$ and $Y_2$ in *connected mode* gives us Figure 11(b).

It is difficult to see whether the graph of $f$ intersects the graph of $g$ in Figure 11(b). We will again apply the procedure that precedes Example 9. Turn off $Y_1$ and $Y_2$ and assign $Y_1 \geq Y_2$ to $Y_3$.

*Make Y assignment*

$Y_3$:  $\boxed{\text{2nd}}\,\boxed{\text{Y-VARS}}\,\boxed{1}\,\boxed{\text{2nd}}\,\boxed{\text{TEST}}\,\boxed{4}\,\boxed{\text{2nd}}\,\boxed{\text{Y-VARS}}\,\boxed{2}$

*(continued)*

Now graph $Y_3$ in *dot mode* to obtain Figure 11(c). It is now easy to determine the interval on which $f(x) \geq g(x)$—the interval corresponds to the line segment shown in Figure 11(c). By tracing to and then zooming in on the right end point of the line segment of $y = 1$ in Figure 11(c), we determine this interval to be (0, 14.90).

As a variation of the assignment for $Y_3$, we could assign the expression $(Y_1 \geq Y_2) + 1$ to $Y_3$ and obtain the graph in Figure 11(d). The advantage of using this assignment is that the graph of $Y_3$ is always visible since it never coincides with the $x$-axis.

Another (more common) approach to solving an inequality of the form $f(x) \geq g(x)$ (as in the last example) is to examine the equivalent inequality

$$f(x) - g(x) \geq 0$$

by assigning $f(x) - g(x)$ to $Y_1$ and then determine all values of $x$ that correspond to nonnegative values of $Y_1$.

The next example (Example 8 in Section 2.2) shows how we can utilize the true-or-false output concept demonstrated in the last two examples in graphing a piecewise-defined function.

**EXAMPLE 11**    *Sketching the graph of a piecewise-defined function*

Sketch the graph of the function $f$ if

$$f(x) = \begin{cases} 2x + 3 & \text{if } x < 0 \\ x^2 & \text{if } 0 \leq x < 2 \\ 1 & \text{if } x \geq 2 \end{cases}$$

**FIGURE 12**

(a) Assignment

(b) Graph of $f$ in connected mode

$[-6, 6]$ by $[-3, 5]$

(c) Graph of $f$ in dot mode

$[-6, 6]$ by $[-3, 5]$

**SOLUTION**    We begin by making the assignment

$$Y_1 = (2x + 3)(x < 0) + x^2(0 \le x)(x < 2) + 1(x \ge 2).$$

$\underbrace{\phantom{(2x+3)(x<0)}}_{\text{first piece}}$  $\underbrace{\phantom{x^2(0 \le x)(x < 2)}}_{\text{second piece}}$  $\underbrace{\phantom{1(x \ge 2)}}_{\text{third piece}}$

*Make Y assignment*    $Y_1$:    $\boxed{Y=}$ $\boxed{\text{CLEAR}}$ $\boxed{(}$ $2$ $\boxed{\text{X}|\text{T}}$ $\boxed{+}$ $3$ $\boxed{)}$ $\boxed{(}$ $\boxed{\text{X}|\text{T}}$ $\boxed{\text{2nd}}$ $\boxed{\text{TEST}}$ $5$ $0$ $\boxed{)}$ $\boxed{+}$
$\boxed{\text{X}|\text{T}}$ $\boxed{x^2}$ $\boxed{(}$ $0$ $\boxed{\text{2nd}}$ $\boxed{\text{TEST}}$ $6$ $\boxed{\text{X}|\text{T}}$ $\boxed{)}$ $\boxed{(}$ $\boxed{\text{X}|\text{T}}$ $\boxed{\text{2nd}}$ $\boxed{\text{TEST}}$ $5$ $2$ $\boxed{)}$ $\boxed{+}$
$1$ $\boxed{(}$ $\boxed{\text{X}|\text{T}}$ $\boxed{\text{2nd}}$ $\boxed{\text{TEST}}$ $4$ $2$ $\boxed{)}$ (see Figure 12(a))

As the variable $x$ takes on values from Xmin to Xmax, the inequality $x < 0$ in the first piece will have a value of either 1 (if $x < 0$) or 0 (if $x \ge 0$). This value is multiplied by the value of $2x + 3$ and assigned to $Y_1$. In the second piece, note that *both* $0 \le x$ and $x < 2$ must be true for the value of $x^2$ to be assigned to $Y_1$. The general idea is that each piece is "on" only when $x$ takes on the associated domain values.

Graphing the function using the viewing rectangle dimensions $[-6, 6]$ by $[-3, 5]$ exhibits the most important features of the graph (see Figure 12(b)). Since Figure 12(b) was drawn in *connected mode*, the calculator included lines between the end points of the pieces. To eliminate these lines, we can redraw the graph in *dot mode*, as shown in Figure 12(c). Note that the graphics calculator makes no distinction between including and excluding an end point (some software packages do).

An alternative method for representing the function $f$ is to assign each piece to a Y-value as follows:

$$Y_1 = (2x + 3)(x < 0), \quad Y_2 = x^2(0 \le x)(x < 2), \quad Y_3 = 1(x \ge 2)$$

Then graph $Y_4 = Y_1 + Y_2 + Y_3$ to obtain the graph of $f$ (be sure to turn off $Y_1$, $Y_2$, and $Y_3$).

---

The result discussed in the next example (Example 6 in Section 5.3) plays an important role in advanced mathematics.

**EXAMPLE 12**    *Sketching the graph of $f(x) = (\sin x)/x$*

If $f(x) = (\sin x)/x$, sketch the graph of $f$ on $[-\pi, \pi]$, and investigate the behavior of $f(x)$ as $x \to 0^-$ and as $x \to 0^+$.

**FIGURE 13**

(a) Assignment

(b) Graph of $f$

$[-3.15, 3.15]$ by $[-2.1, 2.1]$

(c) Zoom-in at the point $(0, 1)$

$[-0.001, 0.001]$ by $[0.999, 1.001]$

**SOLUTION** Note that $f$ is undefined at $x = 0$, because substitution of 0 for $x$ yields the meaningless expression 0/0. We begin by assigning $(\sin x)/x$ to $Y_1$ (the parentheses are optional). Be sure to have your calculator in "Rad"ian mode.

*Make Y assignment*  $Y_1$: $\boxed{Y=}$ $\boxed{\text{CLEAR}}$ $\boxed{(}$ $\boxed{\text{SIN}}$ $\boxed{\text{X}|\text{T}}$ $\boxed{)}$ $\boxed{\div}$ $\boxed{\text{X}|\text{T}}$ (see Figure 13(a))

Since $\pi \approx 3.1416$, we use the viewing rectangle $[-3.15, 3.15]$ by $[-2.1, 2.1]$. Selecting *connected mode* and graphing $Y_1$ gives us Figure 13(b). To get a better view of the graph of $f$ when $x = 0$, we choose the viewing rectangle $[-0.001, 0.001]$ by $[0.999, 1.001]$ to obtain Figure 13(c). Use of tracing and zoom-in suggests that

$$\text{as } x \to 0^-, \quad f(x) \to 1 \quad \text{and as} \quad x \to 0^+, \quad f(x) \to 1.$$

There is a hole in the graph at the point $(0, 1)$; however, most graphing utilities are not capable of showing this fact.

Our graphical technique does not *prove* that $f(x) \to 1$ as $x \to 0$, but it does make it appear highly probable. A rigorous proof, based on the definition of $\sin x$ and geometric considerations, can be found in calculus texts.

---

The next example (Example 6 in Section 5.6) discusses *simultaneous* plotting of functions—an alternative method for graphing several functions (as opposed to *sequential* plotting).

**EXAMPLE 13**  *Sketching the graph of a sum of two trigonometric functions*

Sketch the graphs of $y = \cos x$, $y = \sin x$, and $y = \cos x + \sin x$ on the same coordinate plane for $0 \le x \le 3\pi$.

**FIGURE 14**

(a) Assignments

(b) Graph of $Y_1$, $Y_2$, and $Y_3$

$[0, 9.42]$ by $[-3.14, 3.14]$

(c) Enhanced graph of $Y_1$, $Y_2$, and $Y_3$          (d) Graphing in "Simul" mode

          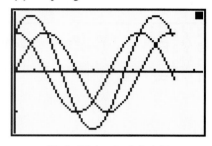

$[0, 9.42]$ by $[-1.5, 1.5]$                    $[0, 9.42]$ by $[-1.5, 1.5]$

**SOLUTION**   We make the following assignments:

$$Y_1 = \cos x, \quad Y_2 = \sin x, \quad \text{and} \quad Y_3 = Y_1 + Y_2$$

*Make Y assignments*       $Y_1$:  [Y =] [CLEAR] [COS] [X|T] [ENTER]

$Y_2$:  [SIN] [X|T] [ENTER]

$Y_3$:  [2nd] [Y-VARS] [1] [+] [2nd] [Y-VARS] [2]  (see Figure 14(a))

Since $3\pi \approx 9.42$ and we desire a $3:2$ horizontal:vertical screen proportion, we choose the viewing rectangle $[0, 9.42]$ by $[-3.14, 3.14]$, obtaining Figure 14(b). The clarity of the graph can be enhanced by changing the viewing rectangle to $[0, 9.42]$ by $[-1.5, 1.5]$, as in Figure 14(c).

Figure 14(c) was obtained using sequential plotting—that is, the graphs of $Y_1$, $Y_2$, and $Y_3$ were plotted in their respective order. To change from the sequential mode to a simultaneous mode in which the three graphs are plotted at the same time, use the following key sequence:

*Select "Simul"*       [MODE] [▽ (5 times)] [▷] [ENTER]

Now press [GRAPH] and you will see $Y_1$, $Y_2$, and $Y_3$ plotted at the same time—that is, in the simultaneous mode (see Figure 14(d)). Select "Sequence" mode before proceeding.

---

A topic that you may study in calculus is the relationship of a curve and its associated **circle of curvature,** or **osculating circle.** For now, we may think of the circle of curvature of a curve as the circle that best coincides with the curve at a point that is on the curve.

**EXAMPLE 14**   *Finding points of intersection of two nearly identical graphs*

For the graph of $y = \sec x$ and the point $P(0, 1)$, the equation of the circle of curvature is $x^2 + (y - 2)^2 = 1$. Determine the portion of the interval $(-1.57, 1.57)$ on which the graph of $y = \sec x$ is below the graph of the lower half of its circle of curvature.

**FIGURE 15**

(a) Assignments

(b) Graph of $Y_1$ and $Y_2$

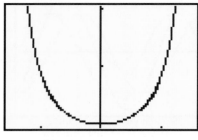

[−1.57, 1.57] by [0.95, 3.05]

(c) Graph of $Y_3 = (Y_1 < Y_2) + 1$

[0, 1] by [0.5, 2.5]

**SOLUTION** Solving the equation $x^2 + (y - 2)^2 = 1$ for $y$ gives us $y = 2 \pm \sqrt{1 - x^2}$, where the plus sign corresponds to the upper half of the circle and the minus sign corresponds to the lower half of the circle. Hence, we make the assignments

$$Y_1 = 1/\cos x \quad \text{and} \quad Y_2 = 2 - \sqrt{1 - x^2}.$$

*Make Y assignments*

$Y_1$: $\boxed{Y =}$ $\boxed{\text{CLEAR}}$ $\boxed{1}$ $\boxed{\div}$ $\boxed{\text{COS}}$ $\boxed{X|T}$ $\boxed{\text{ENTER}}$

$Y_2$: $\boxed{\text{CLEAR}}$ $\boxed{2}$ $\boxed{-}$ $\boxed{\sqrt{}}$ $\boxed{(}$ $\boxed{1}$ $\boxed{-}$ $\boxed{X|T}$ $\boxed{x^2}$ $\boxed{)}$ $\boxed{\text{ENTER}}$

(see Figure 15(a))

To maintain a 3:2 proportion on the screen, we select the viewing rectangle $[-1.57, 1.57]$ by $[0.95, 3.05]$ ($\frac{2}{3} \cdot 1.57 \approx 1.05$ and the $y$-coordinate of the center of the circle of curvature is 2). Pressing $\boxed{\text{GRAPH}}$ gives us Figure 15(b), and we see that the two graphs nearly coincide.

Using a trace and zoom routine near $P$ yields two nearly identical and nearly horizontal lines. Once again, we use the procedure described before Example 9 to help us determine the points of intersection of $Y_1$ and $Y_2$, and we assign $(Y_1 < Y_2) + 1$ to $Y_3$. Since the domain of $Y_2$ is $[-1, 1]$, the graphs of $Y_1$ and $Y_2$ are symmetric with respect to the $y$-axis, and we desire to see only the $y$-values 1 and 2, we choose the viewing rectangle $[0, 1]$ by $[0.5, 2.5]$. Switching to *dot mode*, turning off $Y_1$ and $Y_2$, and graphing only $Y_3$ gives us Figure 15(c).

Figure 15(c) indicates that the graph of $y = \sec x$ is *above* the graph of the semicircle on the interval $(0, 0.977)$ and *below* it on $(0.977, 1]$. Thus, the graph of $y = \sec x$ is lower than the graph of $y = 2 - \sqrt{1 - x^2}$ on $[-1, -0.977) \cup (0.977, 1]$.

In the next example (Example 11 in Section 6.6), we show how a graphing utility can be extremely helpful in determining whether an equation involving inverse trigonometric functions is an identity and, if the equation is not an identity, in finding solutions of the equation.

**EXAMPLE 15    *Investigating an equation***

Determine whether the equation

$$\arctan x = \frac{\arcsin x}{\arccos x}$$

is an identity. If it is not an identity, then approximate the values of $x$ for which the equation is true—that is, solve the equation.

**FIGURE 16**

(a) Assignments

(b) Selecting RANGE values

(c) Graph of $Y_1$ and $Y_2$

$[-1, 1]$ by $[-1.57, 1.57]$

*SOLUTION*    We begin by making the assignments

$$Y_1 = \tan^{-1} x \quad \text{and} \quad Y_2 = \sin^{-1} x/\cos^{-1} x.$$

*Make Y assignments*

$Y_1$: $\boxed{Y=}$ $\boxed{\text{CLEAR}}$ $\boxed{\text{2nd}}$ $\boxed{\text{TAN}^{-1}}$ $\boxed{\text{X|T}}$ $\boxed{\text{ENTER}}$

$Y_2$: $\boxed{\text{CLEAR}}$ $\boxed{\text{2nd}}$ $\boxed{\text{SIN}^{-1}}$ $\boxed{\text{X|T}}$ $\boxed{\div}$ $\boxed{\text{2nd}}$ $\boxed{\text{COS}^{-1}}$ $\boxed{\text{X|T}}$ $\boxed{\text{ENTER}}$ (see Figure 16(a))

Since the domain of $\sin^{-1}$ and $\cos^{-1}$ is $[-1, 1]$ and the range of $\tan^{-1}$ is $(-\pi/2, \pi/2)$, we choose the viewing rectangle dimensions $[-1, 1]$ by $[-1.57, 1.57]$ with Xscl = 0.1 and Yscl = 0.2.

*Select VR*

$\boxed{\text{RANGE}}$ $-1$ $\boxed{\triangledown}$ $1$ $\boxed{\triangledown}$ $.1$ $\boxed{\triangledown}$ $-1.57$ $\boxed{\triangledown}$ $1.57$ $\boxed{\triangledown}$ $.2$ (see Figure 16(b))

Graphing $Y_1$ and $Y_2$ in the *connected mode* gives us Figure 16(c). Since the graphs representing $Y_1$ and $Y_2$ are not the same, we know that *the given equation is not an identity.* However, because the graphs intersect at least twice, we know that the equation has at least two solutions. A little exploration shows that there are exactly two points of intersection.

It appears that $x = 0$ is a solution, and a quick check in the given equation verifies that this is true ($\arctan 0 = 0$ and $(\arcsin 0)/\arccos 0 = 0/(\pi/2) = 0$). To estimate the point of intersection in the first quadrant, we use the tracing and zoom-in features in determine that the point has the approximate coordinates $(0.450, 0.423)$. Hence,

$$x = 0 \quad \text{and} \quad x \approx 0.450$$

are the values of $x$ for which the given equation is true.

The next example (Exercise 36 in Section 8.7) illustrates how to handle matrix multiplication on the TI-81/82.

### EXAMPLE 16    *Investigating long-term trends with matrix multiplication*

Read Exercise 34 in Section 8.3 and Exercise 36 in Section 8.7. Solve part (c) of Exercise 36. (*Note:* It is not important that you understand how these matrices were formed if you are just working through this appendix. The point here is to show the solution on a TI-81/82.)

**FIGURE 17**

(a) Matrix [A]    (b) Matrix [B]    (c) Result of [A] × [B]

```
[A]*[B]→[A]
[ 11500 8400 15…
```

**SOLUTION**    Make the matrix assignments

$$[A] = [12{,}000 \quad 9000 \quad 14{,}000] \quad \text{and} \quad [B] = \begin{bmatrix} 0.90 & 0.10 & 0.00 \\ 0.00 & 0.80 & 0.20 \\ 0.05 & 0.00 & 0.95 \end{bmatrix}.$$

*Make matrix assignments*    [A]: [MATRX] [▷] [1] 1 [ENTER] 3 [ENTER] 12000 [ENTER] 9000
[ENTER] 14000 [ENTER] (see Figure 17(a))

**TI-82 Note**    After pressing [MATRX], you need to press [▷] twice to get to the EDIT submenu.

[B]: [MATRX] [▷] [2] 3 [ENTER] 3 [ENTER] .9 [ENTER] .1 [ENTER] 0
[ENTER] 0 [ENTER] .8 [ENTER] .2 [ENTER] .05 [ENTER] 0 [ENTER]
.95 [ENTER] (see Figure 17(b))

We want to multiply [A] times [B] and then multiply that result by [B]. To perform this operation, use the following keystroke sequence:

*Perform matrix multiplication*    [2nd] [QUIT] [CLEAR] [2nd] [[A]] [×] [2nd] [[B]] [STO▷] [2nd] [[A]] [ENTER]

**TI-82 Note**    Use [MATRX] [1] for [2nd] [[A]] and [MATRX] [2] for [2nd] [[B]].

The current screen is shown in Figure 17(c). The matrix in the second line represents the product of [A] and [B], and this product is now stored in [A]. (To see the third element, press [▷] three times, and then press [◁] three times to return to the screen in Figure 17(c).)

To repeat this multiplication, we need only press ENTER . This next matrix represents the product of the matrix in line 2 and [B]. Note that we could also press 2nd ENTRY ENTER to perform the multiplication along with recalling the operation. After pressing the ENTER key about 50 times (you may want to change your precision to 0 to keep the lines shorter), you see that the matrix stabilizes to [10,000 5000 20,000]. These numbers represent the bird populations on the islands after a long period of time.

We now examine how to solve a system of equations on the TI-81/82. The system given is the system in Exercise 19 in Section 8.8.

**EXAMPLE 17**    *Solving a system of equations*
Solve the system of equations

$$\begin{cases} 3.1x + 6.7y - 8.7z = 1.5 \\ 4.1x - 5.1y + 0.2z = 2.1 \\ 0.6x + 1.1y - 7.4z = 3.9 \end{cases}$$

**FIGURE 18**
(a) Matrix [A]

(b) Matrix [B]

(c) $[A]^{-1} \times [B]$

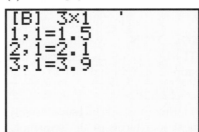

**SOLUTION**    We make the assignments

$$[A] = \begin{bmatrix} 3.1 & 6.7 & -8.7 \\ 4.1 & -5.1 & 0.2 \\ 0.6 & 1.1 & -7.4 \end{bmatrix} \quad \text{and} \quad [B] = \begin{bmatrix} 1.5 \\ 2.1 \\ 3.9 \end{bmatrix}.$$

*Make matrix assignments*    [A]:  MATRX ▷ 1 3 ENTER 3 ENTER 3.1 ENTER 6.7 ENTER
−8.7 ENTER 4.1 ENTER −5.1 ENTER .2 ENTER .6 ENTER
1.1 ENTER −7.4 ENTER (see Figure 18(a))

[B]:  MATRX ▷ 2 3 ENTER 1 ENTER 1.5 ENTER 2.1 ENTER 3.9
ENTER (see Figure 18(b))

*(continued)*

*Find the inverse of a matrix*

We can solve the system by finding the product $[A]^{-1}[B]$.

$\boxed{2\text{nd}}\;\boxed{\text{QUIT}}\;\boxed{\text{CLEAR}}\;\boxed{2\text{nd}}\;\boxed{[A]}\;\boxed{x^{-1}}\;\boxed{\times}\;\boxed{2\text{nd}}\;\boxed{[B]}\;\boxed{\text{ENTER}}$

Figure 18(c) indicates that the solution of the system is

$$x \approx -0.1081, \quad y \approx -0.5227, \quad \text{and} \quad z \approx -0.6135.$$

Also shown in Figure 18(c) is the display for finding the determinant of $[A]$.

*Find the determinant of a matrix*

$\boxed{\text{MATRX}}\;\boxed{5}\;\boxed{2\text{nd}}\;\boxed{[A]}\;\boxed{\text{ENTER}}$

*TI-82 Note*   To find the determinant, use $\boxed{\text{MATRX}}\;\boxed{\triangleright}\;\boxed{1}\;\boxed{\text{MATRX}}\;\boxed{1}$ $\boxed{\text{ENTER}}$

We next present two methods for finding terms of a recursively defined sequence on the TI-81/82. The next example uses the sequence in Exercise 17 in Section 9.1.

**FIGURE 19**

*Find the terms of a sequence*

**EXAMPLE 18**   *Finding terms of a recursively defined sequence*

Find the first five terms of the infinite sequence defined recursively as follows:

$$a_1 = 2, \quad a_{k+1} = 3a_k - 5$$

**SOLUTION**   We first present a method that relies on the fact that the TI-81/82 stores its last result in a memory location called ANS. We will store 2 in ANS and then assign our next entry to be 3 times ANS minus 5.

$\boxed{\text{CLEAR}}\;2\;\boxed{\text{ENTER}}\;3\;\boxed{2\text{nd}}\;\boxed{\text{ANS}}\;\boxed{-}\;5\;\boxed{\text{ENTER}}$

Pressing $\boxed{\text{ENTER}}$ successively gives us the terms of the sequence. The next four terms of the sequence, as shown in Figure 19, are

$$a_2 = 1, \quad a_3 = -2, \quad a_4 = -11, \quad \text{and} \quad a_5 = -38.$$

**FIGURE 20**

```
5/2
                    2.5
0.5(Ans+5/Ans)
                    2.25
              2.236111111
              2.236067978
              2.236067978
```

We can solve Exercise 55 in Section 9.1 in a manner similar to that used in Example 18. After reading the directions on page 623, enter 5/2 and then execute 0.5(ANS + 5/ANS) repeatedly to obtain an approximation of $\sqrt{5}$ (see Figure 20).

A second method for finding terms of a recursively defined sequence is to store a computed value in a memory location other than ANS. This method is similar to the first method, but frees up the ANS location. The next example is Exercise 57 in Section 9.1.

**FIGURE 21**

*Find the terms of a sequence*

**FIGURE 22**

*Compute a permutation*

*Compute a combination*

**EXAMPLE 19**   *Finding terms of a recursively defined sequence*

The equation $\frac{1}{3}\sqrt[3]{x} - x + 2 = 0$ has a root near 2. To approximate this root, rewrite the equation as $x = \frac{1}{3}\sqrt[3]{x} + 2$ (of the form $x = f(x)$). Let $x_1 = 2$ and find successive approximations $x_2, x_3, \ldots$ by using the formulas

$$x_2 = \tfrac{1}{3}\sqrt[3]{x_1} + 2, \quad x_3 = \tfrac{1}{3}\sqrt[3]{x_2} + 2, \quad \ldots$$

until four-decimal-place accuracy is obtained.

**SOLUTION**   We store the value of $x_1$, 2, in memory location X. We next store $\frac{1}{3}\sqrt[3]{X} + 2$ in X. The calculator will first use 2 for X to evaluate the cube root and then store the value of the entire expression in X.

$\boxed{\text{CLEAR}}\, 2\, \boxed{\text{STO}\triangleright}\, \boxed{\text{X|T}}\, \boxed{\text{ENTER}}\, \boxed{(}\, 1\, \boxed{\div}\, 3\, \boxed{)}\, \boxed{\text{MATH}}\, \boxed{4}\, \boxed{\text{X|T}}\, \boxed{+}\, 2$
$\boxed{\text{STO}\triangleright}\, \boxed{\text{X|T}}\, \boxed{\text{ENTER}}$

Pressing $\boxed{\text{ENTER}}$ gives us successively better approximations to the root of the equation, as indicated in Figure 21.

It is fairly easy to generate arithmetic and geometric sequences using the TI-81/82. For example, to generate the arithmetic sequence with first term 4 and common difference 7, we need only enter 4 and then execute the instruction ANS + 7 successively, as shown in Figure 22.

Another feature of the TI-81/82 is the ease of computation of permutations and combinations. For example, to compute $P(52, 5)$ and $C(52, 5)$, use the following keystrokes:

$\boxed{\text{CLEAR}}\, 52\, \boxed{\text{MATH}}\, \boxed{\triangleleft}\, \boxed{2}\, 5\, \boxed{\text{ENTER}}$
$52\, \boxed{\text{MATH}}\, \boxed{\triangleleft}\, \boxed{3}\, 5\, \boxed{\text{ENTER}}$

The results are shown in Figure 23.

**FIGURE 23**

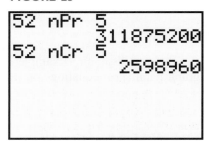

In the next example (Example 6 in Section 10.1) we demonstrate how to use the TI-81/82 to sketch the graph of a parabola that has a horizontal axis.

**EXAMPLE 20    *Graphing half-parabolas***

Sketch the graph of $x = y^2 + 2y - 4$.

**FIGURE 24**

(a) Assignments                    (b) Graph of the parabola

$[-15, 15]$ by $[-10, 10]$

**SOLUTION**     We begin by solving the equivalent equation

$$y^2 + 2y - 4 - x = 0$$

for $y$ in terms of $x$ by using the quadratic formula with $a = 1$, $b = 2$, and $c = -4 - x$, obtaining $y = -1 \pm \sqrt{x + 5}$. The last equation represents the top and bottom halves of the parabola in a function form.

Next, we make the assignments

$$Y_1 = \sqrt{x + 5}, \quad Y_2 = -1 + Y_1, \quad \text{and} \quad Y_3 = -1 - Y_1.$$

*Make Y assignments*

$Y_1$:    $\boxed{Y=}$ $\boxed{\text{CLEAR}}$ $\boxed{\text{2nd}}$ $\boxed{\sqrt{\ }}$ $\boxed{(}$ $\boxed{X|T}$ $\boxed{+}$ $\boxed{5}$ $\boxed{)}$ $\boxed{\text{ENTER}}$

$Y_2$:    $\boxed{\text{CLEAR}}$ $-1$ $\boxed{+}$ $\boxed{\text{2nd}}$ $\boxed{\text{Y-VARS}}$ $\boxed{1}$ $\boxed{\text{ENTER}}$

$Y_3$:    $\boxed{\text{CLEAR}}$ $-1$ $\boxed{-}$ $\boxed{\text{2nd}}$ $\boxed{\text{Y-VARS}}$ $\boxed{1}$ (see Figure 24(a))

Turn off $Y_1$, choose the viewing rectangle dimensions $[-15, 15]$ by $[-10, 10]$, and graph $Y_2$ and $Y_3$ to obtain Figure 24(b).

---

The next example (Example 5 in Section 10.4) demonstrates how to graph parametric equations on the TI-81/82.

**EXAMPLE 21    *Graphing a set of parametric equations***

Sketch the graph of the Lissajous figure that has the parametrization

$$x = \sin 2t, \quad y = \cos t; \quad 0 \le t \le 2\pi.$$

**FIGURE 25**

(a) Changing to "Param" mode

(b) Parametric assignments

(c) Parametric RANGE values

(d) Lissajous figure

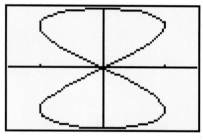

$[-1.5, 1.5]$ by $[-1, 1]$

*SOLUTION*   We first need to set the TI-81/82 in a parametric mode. The "Param" or "Par" option is found under MODE next to "Function" or "Func."

*Select "Param"*    $\boxed{\text{MODE}}$ $\boxed{\triangledown}$ (3 times) $\boxed{\triangleright}$ $\boxed{\text{ENTER}}$ (see Figure 25(a))

Next we make the assignments

$$X_{1T} = \sin 2t \quad \text{and} \quad Y_{1T} = \cos t,$$

where the subscript "1T" on X and Y indicates that $X_{1T}$ and $Y_{1T}$ represent the first *pair* of parametric equations.

*Make parametric assignments*   $X_{1T}$:   $\boxed{\text{Y}=}$ $\boxed{\text{CLEAR}}$ $\boxed{\text{SIN}}$ 2 $\boxed{\text{X|T}}$ $\boxed{\text{ENTER}}$

$Y_{1T}$:   $\boxed{\text{CLEAR}}$ $\boxed{\text{COS}}$ $\boxed{\text{X|T}}$ $\boxed{\text{ENTER}}$ (see Figure 25(b))

When graphing parametric equations, we need to assign minimum (Tmin) and maximum (Tmax) values to the parameter $t$, in addition to viewing rectangle dimensions. We also need to select an increment, or step value (Tstep), for $t$. A typical value for Tstep is $\pi/30 \approx 0.105$. If a smaller value of Tstep is chosen, the accuracy of the sketch is increased, but so is the amount of time needed to sketch the graph.

*(continued)*

For this example, we use Tmin = 0, Tmax = 6.3, and Tstep = 0.1. Since $x$ and $y$ are between $-1$ and 1, we will assign $-1$ to Ymin and 1 to Ymax. To maintain our $3:2$ screen proportion, we select $-1.5$ for Xmin and 1.5 for Xmax. The specific keystrokes are as follows:

*Select RANGE values*

RANGE 0 ▽ 6.3 ▽ .1 ▽ −1.5 ▽ 1.5 ▽ 1 ▽ −1 ▽ 1 ▽ 1 (see Figure 25(c))

Pressing GRAPH gives us the Lissajous figure in Figure 25(d). Pressing TRACE reveals the current values of X, Y, and T. Experiment with the ◁ and ▷ cursor keys to familiarize yourself with the orientation of the curve.

Using parametric equations on a graphics calculator can greatly simplify the representation of equations that are not in function form. For example, to graph the parabola in Example 20, we can make the assignments

$$X_{1T} = t^2 + 2t - 4 \quad \text{and} \quad Y_{1T} = t$$

in "Param" mode. Next, since the axis of symmetry is $y = -1$ and since $y = t$, we set Tmin and Ymin to $-5$, Tmax and Ymax to 3, Xmin to $-6$, Xmax to 6, and Tstep to 0.1.

Graphing gives us a figure similar to Figure 24(b). To graph selected portions of the parabola, we can vary the values of Tmin and Tmax. Try a variation of this graph using Tmin = $-4$ and Tmax = 2.

The next example (Exercise 77 in Section 10.5) demonstrates how to graph a polar equation on the TI-81.

**EXAMPLE 22** *Graphing a polar equation*

Graph $r = 2 \sin^2 \theta \tan^2 \theta$ for $-\pi/3 \le \theta \le \pi/3$.

**FIGURE 26**

(a) Polar assignments

(b) Polar RANGE values

(c) Graph of $r = 2 \sin^2 \theta \tan^2 \theta$          (d) Graph of $r = 6 \sin 2\theta$

 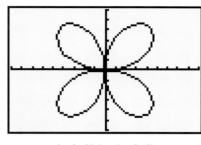

$[-9, 9]$ by $[-6, 6]$                 $[-9, 9]$ by $[-6, 6]$

**SOLUTION**    As with parametric equations, we select the ''Param'' mode to graph polar equations. Also, switch ''Rect'' to ''Polar'' under the MODE menu. Since $\cos \theta = x/r$ and $\sin \theta = y/r$, we make use of the equations

$$x = f(t) \cos t \quad \text{and} \quad y = f(t) \sin t$$

to sketch graphs of the form $r = f(t)$ (or $r = f(\theta)$). To avoid repetitive typing, we make the assignments

$$X_{1T} = X_{2T} \cos T \quad \text{and} \quad Y_{1T} = X_{2T} \sin T.$$

Now we need only assign the function, $f(t)$, to $X_{2T}$. For this example, assign $2(\sin T)^2 (\tan T)^2$ to $X_{2T}$.

*TI-82 Note*    The solution is much simpler on the TI-82 than on the TI-81. Merely select ''Pol'' under the MODE menu and assign the function to $r_1$ under the Y = menu.

*Make polar assignments*    $X_{1T}$: [Y =] [CLEAR] [2nd] [Y-VARS] [7] [COS] [X|T] [ENTER]

$Y_{1T}$: [CLEAR] [2nd] [Y-VARS] [7] [SIN] [X|T] [ENTER]

$X_{2T}$: 2 [(] [SIN] [X|T] [)] [$x^2$] [(] [TAN] [X|T] [)] [$x^2$] (see Figure 26(a))

Since $\pi/3 \approx 1.05$, we will use $-1.05$ for Tmin and $1.05$ for Tmax.

*Select RANGE values*    [RANGE] $-1.05$ [▽] $1.05$ [▽] $.105$ [▽] $-9$ [▽] $9$ [▽] $1$ [▽] $-6$ [▽] $6$ [▽] $1$ (see Figure 26(b))

Only $X_{1T}$ and $Y_{1T}$ are turned on, since there is not a paired Y assignment for $X_{2T}$. Press [GRAPH] to obtain Figure 26(c). Pressing [TRACE] displays the values of T, R, and $\theta$ (same as T).

You may want to practice with another figure, such as a four-leafed rose. If so, go back and assign $6 \sin 2T$ to $X_{2T}$, $0$ to Tmin, and $6.3$ to Tmax (see Figure 26(d)). You will see how easily polar equations can be graphed with this setup.

The TI-82 has many features not available on the TI-81. Some of these features dramatically change the way you will approach the solution to a problem, whereas other features provide extra capabilities, such as the ability to split the screen and store and recall graphs. We will briefly summarize a few of these features.

■ **Creating tables.** The TI-82 has the ability to create a table of values for an independent variable and various function values. Under $\boxed{\text{TblSet}}$, you can set the first table value and the increment value for the independent variable. The values can be seen by pressing $\boxed{\text{2nd}}$ $\boxed{\text{TABLE}}$. Use the cursor keys to scroll through the table values. This feature is particularly helpful for comparing several function values using the same value of the independent variable.

■ **Locating extrema.** The TI-82 has two procedures, called fMin and fMax, that can be used to find the low and high points, respectively, of a graph of a function. For example, graph $Y_1 = x^3 - x$ with a standard viewing rectangle. We see that there is a high point to the left of the $y$-axis (between $x = -1$ and $x = 0$) and a low point to the right of the $y$-axis (between $x = 0$ and $x = 1$). To find the high point, enter fMax$(Y_1,X,-1,0)$. The specific keystrokes are as follows:

$\boxed{\text{2nd}}$ $\boxed{\text{QUIT}}$ $\boxed{\text{MATH}}$ $\boxed{7}$ $\boxed{\text{2nd}}$ $\boxed{\text{Y-VARS}}$ $\boxed{1}$ $\boxed{1}$ $\boxed{,}$ $\boxed{\text{X,T,}\theta}$ $\boxed{,}$ $-1$ $\boxed{,}$ $\boxed{0}$ $\boxed{)}$ $\boxed{\text{ENTER}}$

The result that is returned is the $x$-value of the high point—in this case, $x \approx -0.577$. By querying $Y_1$, we find that the associated $y$-value is approximately 0.385. We can find the coordinates of the low point in a similar fashion using fMin$(Y_1,X,0,1)$.

■ **Solving equations.** The TI-82 has a versatile *solve* procedure that can be used for solving equations, finding $x$-intercepts, and finding points of intersection of graphs. The parameters we need to provide to the solve procedure are an expression, the name of the independent variable, and an approximate value of a zero of the expression. As an example, assign $x^3 - 3x + 1$ to $Y_1$ and graph it. Notice that there are three zeros of $Y_1$. To approximate the zero between $x = 1$ and $x = 2$, we can enter

$\boxed{\text{2nd}}$ $\boxed{\text{QUIT}}$ $\boxed{\text{MATH}}$ $\boxed{0}$ $\boxed{\text{2nd}}$ $\boxed{\text{Y-VARS}}$ $\boxed{1}$ $\boxed{1}$ $\boxed{,}$ $\boxed{\text{X,T,}\theta}$ $\boxed{,}$ $1.5$ $\boxed{)}$ $\boxed{\text{ENTER}}$,

which returns a value of $x \approx 1.532$ for the zero. Also available on the TI-82 are a special root finding routine and a routine for finding points of intersection of graphs.

■ **Listing values.** A single value entry can often be replaced by a list of values on the TI-82. For example, entering log {1,10,100} returns {0 1 2}, indicating three results:

$$\log 1 = 0, \quad \log 10 = 1, \quad \text{and} \quad \log 100 = 2$$

Another example of the use of lists is to assign $\{2,4,8\}\sin X$ to $Y_1$ and then graph $Y_1$. The result is the screen display with graphs of $y = 2\sin x$, $y = 4\sin x$, and $y = 8\sin x$.

There are other features available on the TI-82 that are extremely useful for studying calculus, but these are beyond the scope of this text.

(Graphs of trigonometric functions, inverse trigonometric functions, and conics appear on other endpapers in this text.)

Horizontal line; constant function

Vertical line

Identity function

Absolute value function

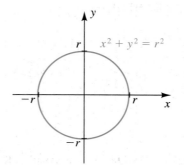

Circle with center $(0, 0)$ and radius $r$

Semicircles

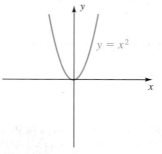

Parabola with vertical axis; squaring function

Parabola with horizontal axis

Square root function

Cube root function

A graph with a cusp
at the origin

Cubing function

Greatest integer function

Reciprocal function

A rational function

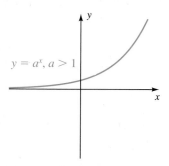

Exponential growth
function (includes natural
exponential function)

Exponential decay
function

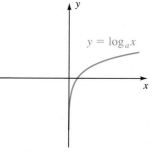

Logarithmic function
(includes common and
natural logarithmic
functions)

The graph of $y = f(x)$ is shown in black in each figure.
The domain of $f$ is $[-1, 3]$ and the range of $f$ is $[-4, 3]$.

$$y = g(x) = f(x) + 3$$

The graph of $f$ is shifted vertically upward 3 units.
Domain of $g$: $[-1, 3]$      Range of $g$: $[-1, 6]$

$$y = h(x) = f(x) - 4$$

The graph of $f$ is shifted vertically downward 4 units.
Domain of $h$: $[-1, 3]$      Range of $h$: $[-8, -1]$

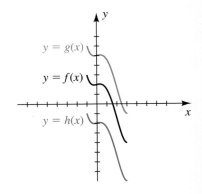

$$y = g(x) = f(x - 3)$$

The graph of $f$ is shifted horizontally to the right 3 units.
Domain of $g$: $[2, 6]$      Range of $g$: $[-4, 3]$

$$y = h(x) = f(x + 6)$$

The graph of $f$ is shifted horizontally to the left 6 units.
Domain of $h$: $[-7, -3]$      Range of $h$: $[-4, 3]$

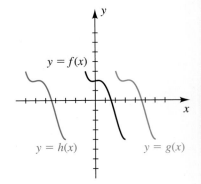

$$y = g(x) = 2f(x) \quad [2 > 1]$$

The graph of $f$ is stretched vertically by a factor of 2.
Domain of $g$: $[-1, 3]$      Range of $g$: $[-8, 6]$

$$y = h(x) = \tfrac{1}{2}f(x) \quad [\tfrac{1}{2} < 1]$$

The graph of $f$ is compressed vertically by a factor of 2.
Domain of $h$: $[-1, 3]$      Range of $h$: $[-2, \tfrac{3}{2}]$

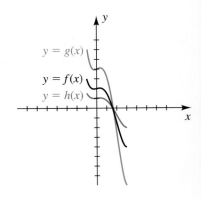

$$y = g(x) = f(2x) \quad [2 > 1]$$

The graph of $f$ is compressed horizontally by a factor of 2.
Domain of $g$: $\left[-\frac{1}{2}, \frac{3}{2}\right]$    Range of $g$: $[-4, 3]$

$$y = h(x) = f(\tfrac{1}{2}x) \quad [\tfrac{1}{2} < 1]$$

The graph of $f$ is stretched horizontally by a factor of 2.
Domain of $h$: $[-2, 6]$    Range of $h$: $[-4, 3]$

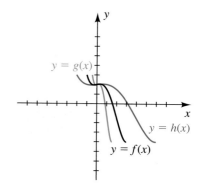

$$y = g(x) = -f(x)$$

The graph of $f$ is reflected through the $x$-axis.
Domain of $g$: $[-1, 3]$    Range of $g$: $[-3, 4]$

$$y = h(x) = f(-x)$$

The graph of $f$ is reflected through the $y$-axis.
Domain of $h$: $[-3, 1]$    Range of $h$: $[-4, 3]$

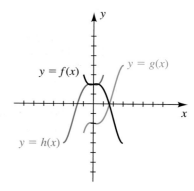

$$y = g(x) = |f(x)|$$

Reflect points on $f$ with negative $y$-values through the $x$-axis.
Domain of $g$: $[-1, 3]$    Range of $g$: $[0, 4]$

$$y = h(x) = f(|x|)$$

Reflect points on $f$ with positive $x$-values through the $y$-axis.
Domain of $h$: $[-3, 3]$    Range of $h$: $[-4, 3]$ at most.
In this case, the range
is a subset of $[-4, 3]$.

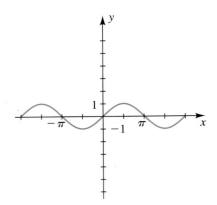

$y = \sin x$

Domain: $\mathbb{R}$

Range: $[-1, 1]$

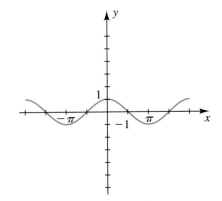

$y = \cos x$

Domain: $\mathbb{R}$

Range: $[-1, 1]$

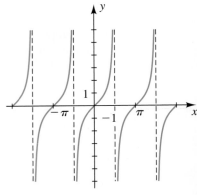

$y = \tan x$

Domain: $x \neq \dfrac{\pi}{2} + \pi n$

Range: $\mathbb{R}$

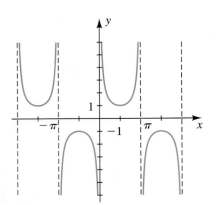

$y = \csc x$

Domain: $x \neq \pi n$

Range: $(-\infty, -1] \cup [1, \infty)$

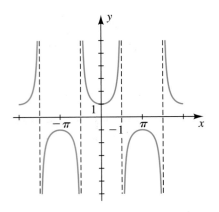

$y = \sec x$

Domain: $x \neq \dfrac{\pi}{2} + \pi n$

Range: $(-\infty, -1] \cup [1, \infty)$

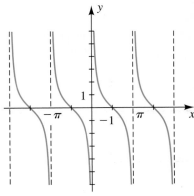

$y = \cot x$

Domain: $x \neq \pi n$

Range: $\mathbb{R}$

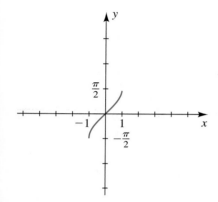

$y = \sin^{-1} x$
Domain: $[-1, 1]$

Range: $\left[ -\dfrac{\pi}{2}, \dfrac{\pi}{2} \right]$

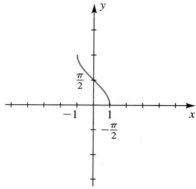

$y = \cos^{-1} x$
Domain: $[-1, 1]$

Range: $[0, \pi]$

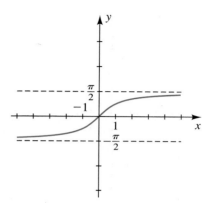

$y = \tan^{-1} x$
Domain: $\mathbb{R}$

Range: $\left( -\dfrac{\pi}{2}, \dfrac{\pi}{2} \right)$

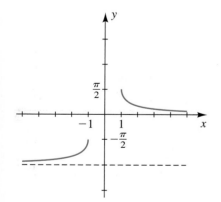

$y = \csc^{-1} x$
Domain: $(-\infty, -1] \cup [1, \infty)$

Range: $\left( -\pi, -\dfrac{\pi}{2} \right] \cup \left( 0, \dfrac{\pi}{2} \right]$

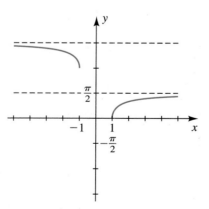

$y = \sec^{-1} x$
Domain: $(-\infty, -1] \cup [1, \infty)$

Range: $\left[ 0, \dfrac{\pi}{2} \right) \cup \left[ \pi, \dfrac{3\pi}{2} \right)$

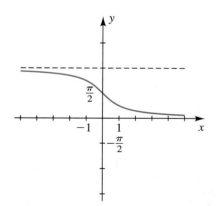

$y = \cot^{-1} x$
Domain: $\mathbb{R}$

Range: $(0, \pi)$

# ANSWERS TO SELECTED EXERCISES

A *Student's Solutions Manual* to accompany this textbook is available from your college bookstore. The guide contains detailed solutions to approximately one-third of the exercises, as well as strategies for solving other exercises in the text.

## CHAPTER I

### EXERCISES 1.1

1 (a) Negative  (b) Positive  (c) Negative
  (d) Positive
3 (a) $<$  (b) $>$  (c) $=$
5 (a) $>$  (b) $>$  (c) $>$
7 (a) $x < 0$  (b) $y \geq 0$  (c) $q \leq \pi$  (d) $2 < d < 4$
  (e) $t \geq 5$  (f) $-z \leq 3$  (g) $\frac{p}{q} \leq 7$  (h) $\frac{1}{w} \geq 9$
  (i) $|x| > 7$
9 (a) 5  (b) 3  (c) 11
11 (a) $-15$  (b) $-3$  (c) 11
13 (a) $4 - \pi$  (b) $4 - \pi$  (c) $1.5 - \sqrt{2}$
15 (a) 4  (b) 12  (c) 12  (d) 8
17 (a) 10  (b) 9  (c) 9  (d) 19
19 $|7 - x| < 5$  21 $|-3 - x| \geq 8$  23 $|x - 4| \leq 3$
25 $-x - 3$  27 $2 - x$  29 $b - a$  31 $x^2 + 4$
33 $\neq$  35 $=$  37 $\neq$  39 $=$
41 Construct a right triangle with sides of lengths $\sqrt{2}$ and 1. The hypotenuse will have length $\sqrt{3}$. Next construct a right triangle with sides of lengths $\sqrt{3}$ and $\sqrt{2}$. The hypotenuse will have length $\sqrt{5}$.
43 The large rectangle has area $a(b + c)$. The sum of the areas of the two small rectangles is $ab + ac$.
45 (a) $4.27 \times 10^5$  (b) $9.8 \times 10^{-8}$  (c) $8.1 \times 10^8$
47 (a) 830,000  (b) 0.000 000 000 002 9
  (c) 563,000,000

49 $1.7 \times 10^{-24}$  51 $5.87 \times 10^{12}$
53 $1.678 \times 10^{-24}$ g  55 $4.1472 \times 10^6$ frames
57 (a) 125  (b) 21

### EXERCISES 1.2

1 $\dfrac{16}{81}$  3 $\dfrac{9}{8}$  5 $\dfrac{-47}{3}$  7 $\dfrac{1}{8}$  9 $\dfrac{1}{25}$  11 $8x^9$
13 $\dfrac{9}{2}$  15 $\dfrac{4}{xy}$  17 $\dfrac{9y^6}{x^8}$  19 $\dfrac{81}{64}y^6$  21 $\dfrac{s^6}{4r^8}$
23 $\dfrac{20y}{x^3}$  25 $9x^{10}y^{14}$  27 $24x^{3/2}$  29 $\dfrac{1}{9a^4}$
31 $\dfrac{8}{x^{1/2}}$  33 $4x^2y^4$  35 $\dfrac{3}{x^3y^2}$  37 1  39 $x^{3/4}$
41 $(a + b)^{2/3}$  43 $(x^2 + y^2)^{1/2}$
45 (a) $4x\sqrt{x}$  (b) $8x\sqrt{x}$
47 (a) $8 - \sqrt[3]{y}$  (b) $\sqrt[3]{8 - y}$  49 9  51 $-2\sqrt[5]{2}$
53 $\dfrac{1}{2}\sqrt[3]{4}$  55 $\dfrac{3y^3}{x^2}$  57 $\dfrac{2a^2}{b}$  59 $\dfrac{1}{2y^2}\sqrt{6xy}$
61 $\dfrac{x}{3}\sqrt[4]{15x^2y^3}$  63 $\dfrac{3x^5}{y^2}$  65 $5x^2y^5\sqrt{2}$
67 $|x^3|y^2$  69 $x^2|y - 1|^3$  71 $\neq$; $(a^r)^2 = a^{2r} \neq a^{(r^2)}$
73 $\neq$; $(ab)^{xy} = a^{xy}b^{xy} \neq a^xb^y$
75 $=$; $\sqrt[n]{\dfrac{1}{c}} = \left(\dfrac{1}{c}\right)^{1/n} = \dfrac{1^{1/n}}{c^{1/n}} = \dfrac{1}{\sqrt[n]{c}}$  77 \$232,825.78
79 2.82 m  81 The 120-kg lifter

### EXERCISES 1.3

1 $6u^2 - 13u - 12$  3 $4y^2 - 5x$  5 $4x^2 - 9y^2$
7 $9x^2 + 12xy + 4y^2$  9 $x - y$
11 $x^3 - 6x^2y + 12xy^2 - 8y^3$  13 $(8x + 3)(x - 7)$
15 Irreducible  17 $(2x - 5)^2$
19 $x^2(x + 2)(x - 2)$  21 $(4x - y^2)(16x^2 + 4xy^2 + y^4)$

**23** $(4x + 3)(16x^2 - 12x + 9)$

**25** $3(x + 3)(x - 3)(x + 1)$

**27** $(a + b)(a - b)(a^2 - ab + b^2)(a^2 + ab + b^2)$

**29** $(x + 2 + 3y)(x + 2 - 3y)$    **31** $\dfrac{y + 5}{y^2 + 5y + 25}$

**33** $\dfrac{x}{x - 1}$    **35** $\dfrac{6s - 7}{(3s + 1)^2}$    **37** $\dfrac{5x^2 + 2}{x^3}$

**39** $\dfrac{4(2t + 5)}{t + 2}$    **41** $\dfrac{2(2x + 3)}{3x - 4}$    **43** $\dfrac{2x - 1}{x}$

**45** $\dfrac{11u^2 + 18u + 5}{u(3u + 1)}$    **47** $-\dfrac{x + 5}{(x + 2)^2}$    **49** $a + b$

**51** $x + y$    **53** $\dfrac{rs}{r^2 - s^2}$    **55** $2x + h - 3$

**57** $-\dfrac{3}{(x - 1)(a - 1)}$    **59** $-\dfrac{3x^2 + 3xh + h^2}{x^3(x + h)^3}$

**61** $\dfrac{t + 10\sqrt{t} + 25}{t - 25}$    **63** $\dfrac{\sqrt[3]{a^2} + \sqrt[3]{ab} + \sqrt[3]{b^2}}{a - b}$

**65** $\dfrac{1}{(a + b)(\sqrt{a} + \sqrt{b})}$    **67** $\dfrac{2}{\sqrt{2(x + h) + 1} + \sqrt{2x + 1}}$

**69** $4x^{4/3} - x^{1/3} + 5x^{-2/3}$    **71** $x^{-1} + 4x^{-3} + 4x^{-5}$

**73** $\dfrac{1 + x^5}{x^3}$    **75** $\dfrac{1 - x^2}{x^{1/2}}$

**77** $(3x + 2)^3(36x^2 - 37x + 6)$

**79** $\dfrac{(2x + 1)^2(8x^2 + x - 24)}{(x^2 - 4)^{1/2}}$    **81** $\dfrac{(3x + 1)^5(39x - 89)}{(2x - 5)^{1/2}}$

**83** $\dfrac{27x^2 - 24x + 2}{(6x + 1)^4}$    **85** $\dfrac{4x(1 - x^2)}{(x^2 + 2)^4}$    **87** $\dfrac{x^2 + 12}{(x^2 + 4)^{4/3}}$

**89** $\dfrac{6(3 - 2x)}{(4x^2 + 9)^{3/2}}$

**91** Area of I is $(x - y)x$, area of II is $(x - y)y$, and
$A = x^2 - y^2 = (x - y)x + (x - y)y = (x - y)(x + y)$.

## EXERCISES 1.4

**1** 1    **3** $-\dfrac{24}{29}$    **5** $-\dfrac{3}{61}$

**7** All real numbers except $\pm 2$    **9** No solution

**11** $-\dfrac{2}{3}, \dfrac{1}{5}$    **13** $-\dfrac{1}{2}$    **15** $\pm\dfrac{3}{5}$    **17** $3 \pm \sqrt{17}$

**19** $-2 \pm \sqrt{2}$    **21** $-\dfrac{2}{3}, 2$    **23** No solution

**25** $\pm\dfrac{2}{3}, 2$    **27** 0, 25    **29** 6    **31** 9

**33** $\pm\dfrac{1}{10}\sqrt{70 \pm 10\sqrt{29}}$    **35** $\pm 2, \pm 3$    **37** $\dfrac{8}{27}, -8$

**39 (a)** 8    **(b)** $\pm 8$    **(c)** No real solutions
**(d)** 625    **(e)** No real solutions

**41 (a)** $x = \dfrac{y \pm \sqrt{2y^2 - 1}}{2}$    **(b)** $y = -2x \pm \sqrt{8x^2 + 1}$

**43 (a)** $0; -4{,}500{,}000$    **(b)** $2.13 \times 10^{-7}$

**45** $r = \dfrac{A - P}{Pt}$    **47** $q = \dfrac{p(1 - S)}{S(1 - p)}$    **49** $q = \dfrac{fp}{p - f}$

**51** $v = \sqrt{\dfrac{2K}{m}}$    **53** $r = \dfrac{-\pi h + \sqrt{\pi^2 h^2 + 2\pi A}}{2\pi}$

**55** $C = 2\sqrt{R^2 - d^2}$    **57** 180 mo (or 15 yr)

**59 (a)** After 64 sec    **(b)** 96 m and 128 m, respectively

**61** 1237.5 ft    **63** 7 ft

**65 (a)** 206.25 ft    **(b)** 40 mi/hr

**67 (a)** $d = 100\sqrt{20t^2 + 4t + 1}$    **(b)** 3:30 P.M.

**69** \$200 or \$175

**71 (a)** 1 cm    **(b)** capsule volume $\approx$ 0.360 cm$^3$, tablet volume $\approx$ 0.393 cm$^3$    **73** \$4.00

**75** There are two possible routes corresponding to
$x \approx 0.6743$ mi and $x \approx 2.2887$ mi.

## EXERCISES 1.5

**1** $2 + 4i$    **3** $18 - 3i$    **5** $41 - 11i$    **7** $17 - i$

**9** $21 - 20i$    **11** $-24 - 7i$    **13** 25    **15** $-i$

**17** $i$    **19** $\dfrac{3}{10} - \dfrac{3}{5}i$    **21** $\dfrac{1}{2} - i$    **23** $\dfrac{34}{53} + \dfrac{40}{53}i$

**25** $\dfrac{2}{5} + \dfrac{4}{5}i$    **27** $-142 - 65i$    **29** $-2 - 14i$

**31** $-\dfrac{44}{113} + \dfrac{95}{113}i$    **33** $\dfrac{21}{2}i$    **35** $x = 4, y = -16$

**37** $x = 1, y = 3$    **39** $3 \pm 2i$    **41** $-2 \pm 3i$

**43** $\dfrac{5}{2} \pm \dfrac{1}{2}\sqrt{55}i$    **45** $-\dfrac{1}{8} \pm \dfrac{1}{8}\sqrt{47}i$

**47** $-5, \dfrac{5}{2} \pm \dfrac{5}{2}\sqrt{3}i$    **49** $\pm 4, \pm 4i$    **51** $\pm 2i, \pm\dfrac{3}{2}i$

**53** $0, -\dfrac{3}{2} \pm \dfrac{1}{2}\sqrt{7}i$

**55** If $w = c + di$, then $\overline{z + w} = \overline{(a + bi) + (c + di)}$
$= \overline{(a + c) + (b + d)i} = (a + c) - (b + d)i$
$= (a - bi) + (c - di) = \bar{z} + \bar{w}$.

**57** $\overline{z \cdot w} = \overline{(a + bi) \cdot (c + di)} = \overline{(ac - bd) + (ad + bc)i}$
$= (ac - bd) - (ad + bc)i = ac - adi - bd - bci$
$= a(c - di) - bi(c - di) = (a - bi) \cdot (c - di)$
$= \bar{z} \cdot \bar{w}$

**59** If $\bar{z} = z$, then $a - bi = a + bi$ and hence $-bi = bi$, or
$2bi = 0$. Thus, $b = 0$ and $z = a$ is real. Conversely, if
$z$ is real, then $b = 0$ and hence $\bar{z} = \overline{a + 0i} = a - 0i = a + 0i = z$.

## EXERCISES 1.6

**1** $(-\infty, -2)$

**3** $(-5, -3]$

**5** $-5 < x \le 8$    **7** $(12, \infty)$    **9** $[9, 19)$

**11** $\left(-\dfrac{2}{3}, \infty\right)$    **13** $\left(\dfrac{4}{3}, \infty\right)$

**15** All real numbers except 1    **17** $(-3.01, -2.99)$

**19** $\left(-\infty, \dfrac{2}{3}\right] \cup [4, \infty)$    **21** $(-\infty, \infty)$

**23** $(-\infty, 3) \cup (3, \infty)$    **25** $(-4, 4)$    **27** $\left(-\dfrac{1}{3}, \dfrac{1}{2}\right)$

**29** $(-2, 3)$    **31** $\left(-\infty, -\dfrac{5}{2}\right] \cup [1, \infty)$    **33** $\left(-\dfrac{3}{5}, \dfrac{3}{5}\right)$

**35** $(-\infty, -2) \cup (-2, -1) \cup \{0\}$    **37** $(-2, 0) \cup (0, 1]$
**39** $(-2, 2] \cup (5, \infty)$    **41** $(-\infty, -3) \cup (0, 3)$

**43** $\left(\dfrac{3}{2}, \dfrac{7}{3}\right)$    **45** $(-\infty, -1) \cup \left(2, \dfrac{7}{2}\right]$

**47** $\left(1, \dfrac{5}{3}\right) \cup [2, 5]$    **49** $(-1, 0) \cup (1, \infty)$

**51 (a)** $-8, -2$    **(b)** $-8 < x < -2$
  **(c)** $(-\infty, -8) \cup (-2, \infty)$    **53** $|w - 148| \le 2$

**55** $4 \le p < 6$    **57** $6\dfrac{2}{3}$ yr    **59** $\dfrac{1}{2}$ sec

**61** $0 \le v < 30$

## EXERCISES 1.7

**1 (a)** The line parallel to the $y$-axis that intersects the
    $x$-axis at $(-2, 0)$
  **(b)** The line parallel to the $x$-axis that intersects the
    $y$-axis is $(0, 3)$
  **(c)** All points to the right of and on the $y$-axis
  **(d)** All points in quadrants I and III
  **(e)** All points below the $x$-axis
  **(f)** All points on the $y$-axis

**3 (a)** $\sqrt{29}$    **(b)** $\left(5, -\dfrac{1}{2}\right)$

**5** $d(A, C)^2 = d(A, B)^2 + d(B, C)^2$; area $= 28$
**7** $d(A, B) = d(B, C) = d(C, D) = d(D, A)$ and
    $d(A, C)^2 = d(A, B)^2 + d(B, C)^2$
**9** $(13, -28)$    **11** $d(A, C) = d(B, C) = \sqrt{145}$
**13** $(0, 3 + \sqrt{11}), (0, 3 - \sqrt{11})$

**15** $a < \dfrac{2}{5}$ or $a > 4$

**17**

**19**

**21**

**23**

**25**

**27**

**29**

**31**

**33**

**35**

**37 (a)** 21, 23    **(b)** 25, 27    **(c)** 29

**39**

**41**

**43**

**45**

**47**

**49**

**51** $(x - 2)^2 + (y + 3)^2 = 25$    **53** $\left(x - \dfrac{1}{4}\right)^2 + y^2 = 5$

**55** $(x + 4)^2 + (y - 6)^2 = 41$
**57** $(x + 3)^2 + (y - 6)^2 = 9$
**59** $(x + 4)^2 + (y - 4)^2 = 16$
**61** $(x - 1)^2 + (y - 2)^2 = 34$    **63** $C(2, -3); r = 7$

**65** $C(0, -2); r = 11$    **67** $C(3, -1); r = \dfrac{1}{2}\sqrt{70}$

**69** $C(-2, 1); r = 0$ (a point)
**71** Not a circle, since $r^2$ cannot equal $-2$
**73** $y = \sqrt{36 - x^2}; y = -\sqrt{36 - x^2};$
     $x = \sqrt{36 - y^2}; x = -\sqrt{36 - y^2}$
**75** $y = -1 + \sqrt{49 - (x - 2)^2};$
     $y = -1 - \sqrt{49 - (x - 2)^2};$
     $x = 2 + \sqrt{49 - (y + 1)^2}; x = 2 - \sqrt{49 - (y + 1)^2}$

**77 (a)** Inside    **(b)** On    **(c)** Outside
**79 (a)** 2    **(b)** $3 \pm \sqrt{5}$    **81** $(x + 2)^2 + (y - 3)^2 = 25$
**83** $\sqrt{5}$    **85** $[-15, -3) \cup (2, 15]$
**87** $(-1, 0) \cup (0, 1)$

**89**

$-1.2, 0.5, 1.6$

$[-6, 6]$ by $[-4, 4]$

**91**

$(0.6, 0.8), (-0.6, -0.8)$

$[-3, 3]$ by $[-2, 2]$

**93**

$(0.999, 0.968),$
$(0.251, 0.032)$

$[-3, 3]$ by $[-2, 2]$

**EXERCISES 1.8**

**1** $m = -\dfrac{3}{4}$    **3** $m = 0$

**5** $m$ is undefined

**7** The slopes of opposite sides are equal.
**9** The slopes of opposite sides are equal, and the slopes of two adjacent sides are negative reciprocals.
**11** $(-12, 0)$

**13**

**15**

**17**

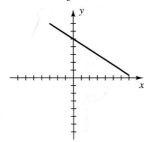

**19 (a)** $x = 5$    **(b)** $y = -2$    **21** $4x + y = 17$
**23** $3x + y = 12$    **25** $11x + 7y = 9$
**27** $5x - 2y = 18$    **29** $5x + 2y = 29$
**31** $y = \frac{3}{4}x - 3$    **33** $y = -\frac{1}{3}x + \frac{11}{3}$
**35** $5x - 7y = -15$    **37** $y = -x$
**39** $m = -\frac{2}{3}$, $b = 5$    **41** $m = \frac{4}{3}$, $b = -3$

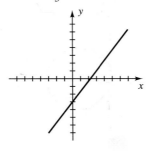

**43 (a)** $y = 3$    **(b)** $y = -\frac{1}{2}x$    **(c)** $y = -\frac{3}{2}x + 1$
**(d)** $y + 2 = -(x - 3)$

**45** $\dfrac{x}{3/2} + \dfrac{y}{-3} = 1$    **47** $(x - 3)^2 + (y + 2)^2 = 49$

**49** Approximately 23 weeks
**51 (a)** 25.2 tons    **(b)** As large as 3.4 tons
**53 (a)** $y = \dfrac{5}{14}x$    **(b)** 58

**55 (a)** $W = \dfrac{20}{3}t + 10$    **(d)**
**(b)** 50 lb
**(c)** 9 yrs

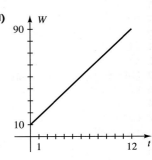

**57** $H = -\dfrac{8}{3}T + \dfrac{7520}{3}$

**59 (a)** $T = 0.032t + 13.5$    **(b)** 16.22 °C
**61 (a)** $E = 0.55R + 3600$    **(b)** $P = 0.45R - 3600$
**(c)** $8000
**63 (a)** Yes: the creature at $x = 3$    **(b)** No
**65** 34.95 mi/hr
**67** $a = 0.321$; $b = -0.9425$

**69** $(-19, 13)$

$[-30, 3]$ by $[-2, 20]$

**71** $(-0.8, -0.6)$, $(4.8, -3.4)$, $(2, 5)$; right isosceles triangle

$[-15, 15]$ by $[-10, 10]$

## CHAPTER 1 REVIEW EXERCISES

**1** $-x - 3$    **2** $-(x-2)(x-3)$    **3** $\dfrac{b^3}{a^8}$    **4** $-\dfrac{p^8}{2q}$

**5** $\dfrac{x^3z}{y^{10}}$    **6** $\dfrac{16x^2}{z^4y^6}$    **7** $\dfrac{b^6}{a^2}$    **8** $\dfrac{y-x^2}{x^2y}$

**9** $2xyz\sqrt[3]{x^2z}$    **10** $2ab\sqrt{ac}$    **11** $\dfrac{1-\sqrt{t}}{t}$

**12** $c^2d^4$    **13** $\dfrac{2x}{y^2}$    **14** $\dfrac{x+6\sqrt{x}+9}{9-x}$

**15** $x^4 + x^3 - x^2 + x - 2$    **16** $-x^2 + 18x + 7$

**17** $6a^2 + 11ab - 35b^2$    **18** $16r^4 - 24r^2s + 9s^2$

**19** $169a^4 - 16b^2$    **20** $8a^3 + 12a^2b + 6ab^2 + b^3$

**21** $81x^4 - 72x^2y^2 + 16y^4$

**22** $a^2 + b^2 + c^2 + d^2 + 2(ab + ac + ad + bc + bd + cd)$

**23** $(14x + 9)(2x - 1)$    **24** $(4a^2 + 3b^2)^2$

**25** $8(x + 2y)(x^2 - 2xy + 4y^2)$

**26** $u^3v(v - u)(v^2 + uv + u^2)$

**27** $(p^4 + q^4)(p^2 + q^2)(p + q)(p - q)$

**28** $x^2(x - 4)^2$    **29** $(x - 7 + 7y)(x - 7 - 7y)$

**30** $(x - 2)(x + 2)^2(x^2 - 2x + 4)$    **31** $\dfrac{27}{(4x - 5)(10x + 1)}$

**32** $\dfrac{5x^2 - 6x - 20}{x(x + 2)^2}$    **33** $\dfrac{x^3 + 1}{x^2 + 1}$    **34** $\dfrac{ab}{a + b}$

**35** $\dfrac{1}{x + 3}$    **36** $\dfrac{2(5x^2 + x + 4)}{(6x + 1)^{2/3}(4 - x^2)^2}$    **37** $-\dfrac{5}{6}$

**38** $-4, \dfrac{3}{2}$    **39** $-\dfrac{2}{3} \pm \dfrac{1}{3}\sqrt{19}$    **40** $\pm\dfrac{5}{2}, \pm\sqrt{2}$

**41** $\pm\dfrac{1}{2}\sqrt{7}, -\dfrac{2}{5}$    **42** $-\dfrac{3}{2}, 2$    **43** $-5, 4$

**44** $\dfrac{1}{4}, \dfrac{1}{9}$    **45** $2$    **46** $5$    **47** $\left(\dfrac{2}{3}, \infty\right)$

**48** $\left(-\dfrac{11}{4}, \dfrac{9}{4}\right)$    **49** $\left(-\infty, -\dfrac{3}{10}\right)$    **50** $\left(-7, \dfrac{7}{2}\right)$

**51** $(-\infty, 1) \cup (5, \infty)$    **52** $\left(-\infty, \dfrac{11}{3}\right] \cup [7, \infty)$

**53** $\left(-\infty, -\dfrac{3}{2}\right) \cup \left(\dfrac{2}{5}, \infty\right)$    **54** $[-2, 5]$

**55** $(-\infty, -2) \cup \{0\} \cup [3, \infty)$

**56** $(-3, -1) \cup (-1, 2]$    **57** $\left(-\infty, -\dfrac{3}{2}\right) \cup (2, 9)$

**58** $(-\infty, -5) \cup [-1, 5)$    **59** $(1, \infty)$

**60** $(0, 1) \cup (2, 3)$    **61** $R = \sqrt[4]{\dfrac{8FVL}{\pi P}}$

**62** $r = \dfrac{-\pi hR + \sqrt{12\pi hV - 3\pi^2h^2R^2}}{2\pi h}$    **63** $-55 + 48i$

**64** $\dfrac{9}{85} + \dfrac{2}{85}i$    **65** $-\dfrac{9}{53} - \dfrac{48}{53}i$    **66** $-2 - 5i$

**67** $d(A, B)^2 + d(A, C)^2 = d(B, C)^2$; area $= 10$

**68** **(a)** $\sqrt{265}$    **(b)** $\left(-\dfrac{13}{2}, 1\right)$    **(c)** $(-11, -23)$

**69** $(0, 1), (0, 11)$    **70** $-2 < a < 1$

**71** $(x - 7)^2 + (y + 4)^2 = 149$

**72** $(x - 3)^2 + (y + 2)^2 = 169$

**73** $x = -2 - \sqrt{9 - y^2}$

**74** The slope of $AD$ and $BC$ is $\dfrac{2}{3}$.

**75** **(a)** $18x + 6y = 7$    **(b)** $2x - 6y = 3$

**76** $(x + 5)^2 + (y + 1)^2 = 81$    **77** $x + y = -3$

**78** $5x - y = 23$    **79** $2x - 3y = 5$

**80** $C(0,6); r = \sqrt{5}$

**81**

**82**

**83**

**84**

**85**

**86**

**87**

**88**

**89** $2x - 5y = 10$ **90** $(x + 2)^2 + (y - 1)^2 = 25$
**91** 11.055% **92** 6 oz of vegetables and 4 oz of meat
**93** 80 gal of 20% solution and 40 gal of 50% solution
**94** 1 hr 40 min
**95** (a) $d = \sqrt{2900t^2 - 200t + 4}$

(b) $t = \dfrac{5 + 2\sqrt{19{,}603}}{145} \approx 1.97$, or approximately

11:58 A.M.

**96** There are two arrangements: 40 ft × 25 ft and
50 ft × 20 ft.
**97** (a) $2\sqrt{2}$ ft (b) 2 ft **98** Over \$100,000
**99** $T > 279.57$ K **100** 36 to 38 trees/acre
**101** (a) 259 ft (b) 2002
**102** (a) $V = 6000t + 89{,}000$ (b) $2\dfrac{1}{3}$

# CHAPTER 2

## EXERCISES 2.1

**1** $-6, -4, -24$ **3** $-12, -22, -36$
**5** (a) $5a - 2$ (b) $-5a - 2$ (c) $-5a + 2$
(d) $5a + 5h - 2$ (e) $5a + 5h - 4$ (f) 5
**7** (a) $a^2 - a + 3$ (b) $a^2 + a + 3$ (c) $-a^2 + a - 3$
(d) $a^2 + 2ah + h^2 - a - h + 3$
(e) $a^2 + h^2 - a - h + 6$ (f) $2a + h - 1$
**9** $2x + h$ **11** $\dfrac{1}{\sqrt{x - 3} + \sqrt{a - 3}}$
**13** (a) $\dfrac{4}{a^2}$ (b) $\dfrac{1}{4a^2}$ (c) $4a$ (d) $2a$
**15** (a) $\dfrac{2a}{a^2 + 1}$ (b) $\dfrac{a^2 + 1}{2a}$ (c) $\dfrac{2\sqrt{a}}{a + 1}$ (d) $\dfrac{\sqrt{2a^3 + 2a}}{a^2 + 1}$
**17** (a) $[-3, 4]$ (b) $[-2, 2]$ (c) 0 (d) $-1, \dfrac{1}{2}, 2$
(e) $\left(-1, \dfrac{1}{2}\right) \cup (2, 4]$
**19** $\left[-\dfrac{7}{2}, \infty\right)$ **21** $[-3, 3]$
**23** All real numbers except $-2$, 0, and 2
**25** $\left[\dfrac{3}{2}, 4\right) \cup (4, \infty)$ **27** $(2, \infty)$ **29** $[-2, 2]$

**31** (a)
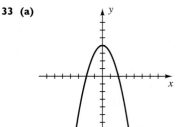
(b) $D = (-\infty, \infty)$,
$R = (-\infty, \infty)$
(c) Increasing on
$(-\infty, \infty)$

**33** (a)
(b) $D = (-\infty, \infty)$,
$R = (-\infty, 4]$
(c) Increasing on
$(-\infty, 0]$,
decreasing on
$[0, \infty)$

**35** (a)
(b) $D = [-4, \infty)$,
$R = [0, \infty)$
(c) Increasing on
$[-4, \infty)$

**37** (a)
(b) $D = (-\infty, \infty)$,
$R = \{-2\}$
(c) Constant on
$(-\infty, \infty)$

**39** (a)

(b) $D = [-6, 6]$,
$R = [-6, 0]$
(c) Decreasing on
$[-6, 0]$,
increasing on
$[0, 6]$

**41** $f(x) = \dfrac{1}{6}x + \dfrac{3}{2}$ **43** Yes **45** No **47** Yes

**49** No **51** No **53** $V = 4x(15 - x)(10 - x)$

**55** (a) $y = \dfrac{500}{x}$ (b) $C = 300x + \dfrac{100{,}000}{x} - 600$

**57** $S(h) = 6h - 50$

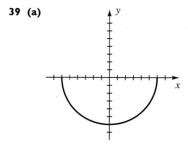

**59 (a)** $y = 2.5t + 33$

**(b)**

(7, 50.5)   (10, 58)

(6, 48)

10

1

t

The yearly increase in height

**(c)** 58 in.

**61** $d = 2\sqrt{t^2 + 2500}$

**63 (a)** $y = \sqrt{h^2 + 2hr}$   **(b)** 1280.6 mi

**65** $d = \sqrt{90,400 + x^2}$   **67** 20 lb   **69** $\dfrac{50}{9}$ ohms

**71** $d$ is multiplied by 9.

**73 (a)**

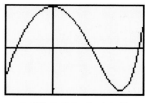

$[-2, 2]$ by $[-2, 2]$

**(b)** $[-0.75, 0.75]$
**(c)** Decreasing on $[-2, -0.55]$ and on $[0.55, 2]$, increasing on $[-0.55, 0.55]$

**75 (a)**

$[-0.7, 1.4]$ by $[-1.1, 1]$

**(b)** $[-1.03, 1]$
**(c)** Increasing on $[-0.7, 0]$ and on $[1.06, 1.4]$, decreasing on $[0, 1.06]$

**77 (a)** 8   **(b)** $\pm 8$   **(c)** No real solutions   **(d)** 625
**(e)** No real solutions

## EXERCISES 2.2

**1** Odd   **3** Even   **5** Neither
**7** Even   **9** Odd

**11**

**13**

**15**

**17**

**19**

**21**

**23**

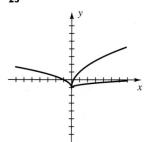

**25** $(7, -3)$

**27 (a)**

**(b)**

**(c)**

**(d)**

**(k)**

**(l)**

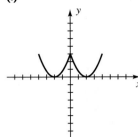

**29 (a)** $y = f(x + 9) + 1$   **(b)** $y = -f(x)$
   **(c)** $y = -f(x + 7) - 1$
**31 (a)** $y = f(x + 4)$   **(b)** $y = f(x) + 1$   **(c)** $y = f(-x)$

**(e)**

**(f)**

**33**

**35**

**(g)**

**(h)**

**37**

**(i)**

**(j)**

**39 (a)**

**(b)**

**(c)**

**(d)**

**47**

**(e)**

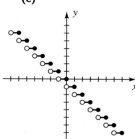

**41** If $x > 0$, two different points on the graph have $x$-coordinate $x$.

**43**

**45**

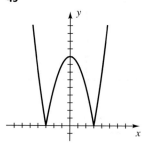

**49 (a)** $D = [-2, 6]$, $R = [-16, 8]$
**(b)** $D = [-4, 12]$, $R = [-4, 8]$
**(c)** $D = [1, 9]$, $R = [-3, 9]$
**(d)** $D = [-4, 4]$, $R = [-7, 5]$
**(e)** $D = [-6, 2]$, $R = [-4, 8]$
**(f)** $D = [-2, 6]$, $R = [-8, 4]$
**(g)** $D = [-6, 6]$, $R = [-4, 8]$
**(h)** $D = [-2, 6]$, $R = [0, 8]$

**51** $T(x) = \begin{cases} 0.15x & \text{if } x \le 20{,}000 \\ 0.20x - 1000 & \text{if } x > 20{,}000 \end{cases}$

**53** $\begin{cases} 1.20x & \text{if } 0 \le x \le 10{,}000 \\ 1.50x - 3000 & \text{if } 10{,}000 < x \le 15{,}000 \\ 1.80x - 7500 & \text{if } x > 15{,}000 \end{cases}$

**55** $(-3.12, 22.00)$

**57** $(-\infty, -3) \cup (-3, 1.87) \cup (4.13, \infty)$

*EXERCISES 2.3*

**1** $y = a(x + 3)^2 + 1$     **3** $y = ax^2 - 3$
**5** $f(x) = -(x + 2)^2 - 4$     **7** $f(x) = 2(x - 3)^2 + 4$
**9** $f(x) = -3(x + 1)^2 - 2$

**11** $f(x) = -\dfrac{3}{4}(x - 6)^2 - 7$

**13 (a)** Min: $f(2) = -4$     **(c)**
**(b)** $0, 4$

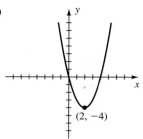

$(2, -4)$

**15 (a)** Max: $f\left(\frac{11}{24}\right) = \frac{841}{48}$  **(c)**

**(b)** $-\frac{3}{4}, \frac{5}{3}$

$\left(\frac{11}{24}, \frac{841}{48}\right)$

**17 (a)** Min: $f\left(-\frac{4}{3}\right) = 0$

**(b)** $-\frac{4}{3}$

**(c)**

$\left(-\frac{4}{3}, 0\right)$

**19 (a)** Min: $f(-2) = 5$
**(b)** None
**(c)**

$(-2, 5)$

**21 (a)** Max: $f(5) = 7$   **(b)** $5 \pm \frac{1}{2}\sqrt{14} \approx 6.87, 3.13$

**(c)**

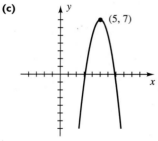
$(5, 7)$

**23** $y = \frac{1}{8}(x - 4)^2 - 1$   **25** $y = -\frac{4}{9}(x + 2)^2 + 4$

**27** $y = 3(x - 0)^2 - 2$   **29** $y = -\frac{5}{9}(x - 3)^2 + 5$

**31** $y = -\frac{1}{4}(x - 1)^2 + 4$   **33** 6.125   **35** 24.72 km

**37** 10.5 lb   **39 (a)** 424 ft   **(b)** 100 ft
**41** 20 and 20

**43 (a)** $y = 250 - \frac{3}{4}x$   **(b)** $A = x\left(250 - \frac{3}{4}x\right)$

**(c)** $166\frac{2}{3}$ ft by 125 ft

**45** $y = -\frac{4}{27}\left(x - \frac{9}{2}\right)^2 + 3$

**47 (a)** $y = \frac{1}{500}x^2 + 10$   **(b)** 282 ft   **49** 2 ft

**51** 500 pairs
**53 (a)** $R(x) = 500x(30 - x)$

**(b)** \$15

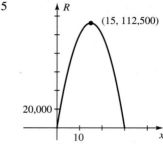
(15, 112,500)
20,000
10

**55**

$[-3, 3]$ by $[-2, 2]$

$(-0.57, 0.64)$,
$(0.02, -0.27)$,
$(0.81, -0.41)$

**57**

$[-8, 4]$ by $[-1, 7]$

Smaller values of $a$ result in a wider parabola; larger values of $a$ result in a narrower parabola.

*EXERCISES 2.4*

**1 (a)** 15   **(b)** $-3$   **(c)** 54   **(d)** $\frac{2}{3}$

**3 (a)** $3x^2 + 1$; $3 - x^2$; $2x^4 + 3x^2 - 2$; $\frac{x^2 + 2}{2x^2 - 1}$

**(b)** $\mathbb{R}$   **(c)** All real numbers except $\pm\frac{1}{2}\sqrt{2}$

**5 (a)** $2\sqrt{x + 5}$; 0; $x + 5$; 1   **(b)** $[-5, \infty)$
**(c)** $(-5, \infty)$

**7 (a)** $\frac{3x^2 + 6x}{(x - 4)(x + 5)}$; $\frac{x^2 + 14x}{(x - 4)(x + 5)}$; $\frac{2x^2}{(x - 4)(x + 5)}$;

$\frac{2(x + 5)}{x - 4}$

**(b)** All real numbers except $-5$ and 4
**(c)** All real numbers except $-5$, 0, and 4

**9 (a)** $-2x^2 - 1$   **(b)** $-4x^2 + 4x - 1$   **(c)** $4x - 3$
**(d)** $-x^4$

**11 (a)** $6x + 9$   **(b)** $6x - 8$   **(c)** $-3$   **(d)** $10$

**13 (a)** $75x^2 + 4$   **(b)** $15x^2 + 20$   **(c)** $304$   **(d)** $155$

**15 (a)** $8x^2 - 2x - 5$   **(b)** $4x^2 + 6x - 9$   **(c)** $31$
**(d)** $45$

**17 (a)** $8x^3 - 20x$   **(b)** $128x^3 - 20x$   **(c)** $-24$
**(d)** $3396$

**19 (a)** $7$   **(b)** $-7$   **(c)** $7$   **(d)** $-7$

**21 (a)** $x + 2 - 3\sqrt{x + 2};\ [-2, \infty)$
**(b)** $\sqrt{x^2 - 3x + 2};\ (-\infty, 1] \cup [2, \infty)$

**23 (a)** $3x - 4;\ [0, \infty)$
**(b)** $\sqrt{3x^2 - 12};\ (-\infty, -2] \cup [2, \infty)$

**25 (a)** $\sqrt{\sqrt{x + 5} - 2};\ [-1, \infty)$
**(b)** $\sqrt{\sqrt{x - 2} + 5};\ [2, \infty)$

**27 (a)** $\sqrt{3 - \sqrt{x^2 - 16}};\ [-5, -4] \cup [4, 5]$
**(b)** $\sqrt{-x - 13};\ (-\infty, -13]$

**29 (a)** $x;\ \mathbb{R}$   **(b)** $x;\ \mathbb{R}$

**31 (a)** $\dfrac{1}{x^6}$; all nonzero real numbers
**(b)** $\dfrac{1}{x^6}$; all nonzero real numbers

**33 (a)** $\dfrac{1}{5 - x}$; all real numbers except 4 and 5
**(b)** $\dfrac{-2x + 5}{-3x + 7}$; all real numbers except 2 and $\dfrac{7}{3}$

**35** $-3 \pm \sqrt{2}$   **37 (a)** $5$   **(b)** $6$   **(c)** $6$   **(d)** $5$

**39** $20\sqrt{x^2 + 1}$   **41** Odd   **43** $31.26$

**45** $A = 36\pi t^2$   **47** $r = 9\sqrt[3]{t}$

**49** $h = 5\sqrt{t^2 + 8t}$   **51** $d = \sqrt{90,400 + (500 + 150t)^2}$

Exer. 53–60: Answers are not unique.

**53** $u = x^2 + 3x,\ y = u^{1/3}$   **55** $u = x - 3,\ y = u^{-4}$
**57** $u = x^4 - 2x^2 + 5,\ y = u^5$

**59** $u = \sqrt{x + 4},\ y = \dfrac{u - 2}{u + 2}$   **61** $5 \times 10^{-13}$

**63 (a)** $Y_1 = x$, graph $Y_3 = -2Y_2$

$[-12, 12]$ by $[-16, 8]$

**(b)** $Y_1 = 0.5x$, graph $Y_2$

$[-12, 12]$ by $[-16, 8]$

**(c)** $Y_1 = x - 3$, graph $Y_3 = Y_2 + 1$

$[-12, 12]$ by $[-6, 10]$

**(d)** $Y_1 = x + 2$, graph $Y_3 = Y_2 - 3$

$[-12, 12]$ by $[-8, 8]$

**(e)** $Y_1 = -x$, graph $Y_2$

$[-12, 12]$ by $[-8, 8]$

**(f)** $Y_1 = x$, graph $Y_3 = -Y_2$

$[-12, 12]$ by $[-8, 8]$

**(g)** $Y_1 = $ abs $x$, graph $Y_2$

[−12, 12] by [−8, 8]

**(h)** $Y_1 = x$, graph $Y_3 = $ abs $Y_2$

[−2, 6] by [0, 8]

## EXERCISES 2.5

**1** Yes    **3** No    **5** Yes    **7** No    **9** No
**11** Yes

Exer. 13–16: Show that $f(g(x)) = x = g(f(x))$.

**13**               **15**

**17** $f^{-1}(x) = \dfrac{x-5}{3}$    **19** $f^{-1}(x) = \dfrac{2x+1}{3x}$

**21** $f^{-1}(x) = \dfrac{5x+2}{2x-3}$    **23** $f^{-1}(x) = -\sqrt{\dfrac{2-x}{3}}$

**25** $f^{-1}(x) = \sqrt[3]{\dfrac{x+5}{2}}$    **27** $f^{-1}(x) = 3 - x^2$, $x \geq 0$

**29** $f^{-1}(x) = (x-1)^3$    **31** $f^{-1}(x) = x$

**33** $f^{-1}(x) = -\sqrt{x+4}$

**35 (a)**

**(b)** $D = [-1, 2]$;
$R = \left[\dfrac{1}{2}, 4\right]$

**(c)** $D_1 = \left[\dfrac{1}{2}, 4\right]$;
$R_1 = [-1, 2]$

**37 (a)**

**(b)** $D = [-3, 3]$;
$R = [-2, 2]$
**(c)** $D_1 = [-2, 2]$;
$R_1 = [-3, 3]$

**39 (a)** Since $f$ is one-to-one, an inverse exists;
$$f^{-1}(x) = \dfrac{x-b}{a}$$
**(b)** No; not one-to-one
**41 (c)** The graph of $f$ is symmetric about the line $y = x$.
Thus, $f(x) = f^{-1}(x)$.
**43** Yes

**45**

**(a)** [−0.27, 1.22]
**(b)** [−0.20, 3.31];
[−0.27, 1.22]

[−1, 2] by [−1, 4]

**47** $x = \dfrac{0.4996 + \sqrt{(-0.4996)^2 - 4(0.0833)(3.5491 - D)}}{2(0.0833)}$

## CHAPTER 2 REVIEW EXERCISES

**1 (a)** $\dfrac{1}{2}$    **(b)** $-\dfrac{1}{\sqrt{2}}$    **(c)** 0    **(d)** $-\dfrac{x}{\sqrt{3-x}}$

**(e)** $-\dfrac{x}{\sqrt{x+3}}$   **(f)** $\dfrac{x^2}{\sqrt{x^2+3}}$   **(g)** $\dfrac{x^2}{x+3}$

**2** Positive    **3** Positive

**4 (a)** $\left[\dfrac{4}{3}, \infty\right)$; $[0, \infty)$

**(b)** All real numbers except $-3$; $(0, \infty)$

**5** $-2a - h + 1$     **6** $-\dfrac{1}{(a + h + 2)(a + 2)}$

**7** $f(x) = \dfrac{5}{2}x - \dfrac{1}{2}$

**8 (a)** Odd     **(b)** Neither     **(c)** Even

**9**

**10**

**11 (a)**

**(b)** $D = \mathbb{R}$; $R = \mathbb{R}$
**(c)** Decreasing on $(-\infty, \infty)$

**12 (a)**

**(b)** $D = \mathbb{R}$;
$R = \{1000\}$
**(c)** Constant on $(-\infty, \infty)$

**13 (a)**

**(b)** $D = \mathbb{R}$;
$R = [0, \infty)$
**(c)** Decreasing on $(-\infty, -3]$, increasing on $[-3, \infty)$

**14 (a)**

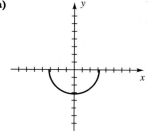

**(b)** $D = [-\sqrt{10}, \sqrt{10}]$;
$R = [-\sqrt{10}, 0]$
**(c)** Decreasing on $[-\sqrt{10}, 0]$, increasing on $[0, \sqrt{10}]$

**15 (a)**

**(b)** $D = [-1, \infty)$;
$R = (-\infty, 1]$
**(c)** Decreasing on $[-1, \infty)$

**16 (a)**

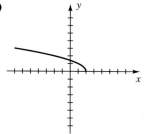

**(b)** $D = (-\infty, 2]$;
$R = [0, \infty)$
**(c)** Decreasing on $(-\infty, 2]$

**17 (a)**

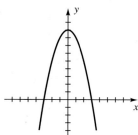

**(b)** $D = \mathbb{R}$;
$R = (-\infty, 9]$
**(c)** Increasing on $(-\infty, 0]$, decreasing on $[0, \infty)$

**18 (a)**

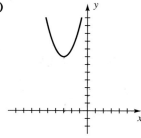

**(b)** $D = \mathbb{R}$;
$R = [7, \infty)$
**(c)** Decreasing on $(-\infty, -3]$, increasing on $[-3, \infty)$

**19 (a)**

**(b)** $D = \mathbb{R}$;
$R = [0, \infty)$

**(c)** Decreasing on $(-\infty, 0]$, increasing on $[0, 2]$, constant on $[2, \infty)$

**(e)**

**(f)**

**20 (a)**

**22 (a)**   **(b)**

**(b)** $D = \mathbb{R}$; $R = \{\ldots, -3, -1, 1, 3, \ldots\}$
**(c)** Constant on $[n, n + 1)$, where $n$ is any integer

**21 (a)**   **(b)**

**(c)**   **(d)**

**(c)**   **(d)**

**(e)**   **(f)**

(g)

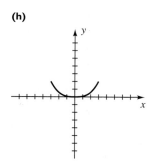

(h)

**23** $y = \frac{1}{2}(x - 2)^2 - 4$     **24** $y = -|x - 2| - 1$

**25** Min: $f(-3) = 4$     **26** Max: $f(5) = -7$
**27** Max: $f(-1) = -37$     **28** Min: $f(4) = -108$

**29** $f(x) = -2(x - 3)^2 + 4$     **30** $y = \frac{3}{2}(x - 3)^2 - 2$

**31 (a)** $[0, 2]$   **(b)** $(0, 2]$     **32 (a)** $-1$   **(b)** $\sqrt{13}$
**33 (a)** $18x^2 + 9x - 1$     **(b)** $6x^2 - 15x + 5$

**34 (a)** $\sqrt{\dfrac{3 + 2x^2}{x^2}}$     **(b)** $\dfrac{1}{3x + 2}$

**35 (a)** $\sqrt{28 - x}$; $[3, 28]$     **(b)** $\sqrt{\sqrt{25 - x^2} - 3}$; $[-4, 4]$

**36 (a)** $\dfrac{1}{x + 3}$; all real numbers except $-3$ and $0$

**(b)** $\dfrac{6x + 4}{x}$; all real numbers except $-\dfrac{2}{3}$ and $0$

**37** $u = x^2 - 5x$, $y = \sqrt[3]{u}$     **38** Yes

**39 (a)** $f^{-1}(x) = \dfrac{10 - x}{15}$     **40 (a)** $f^{-1}(x) = -\sqrt{\dfrac{9 - x}{2}}$

**(b)**

**(b)**

**41 (a)** 2   **(b)** 4   **(c)** 2   **(d)** 2   **(e)** $x > 2$
**42 (a)** 5   **(b)** 7   **(c)** 4
   **(d)** Not enough information is given.

**43 (a)** $V = 6000t + 89,000$   **(b)** $2\dfrac{1}{3}$

**44 (a)** $F = \dfrac{9}{5}C + 32$   **(b)** $1.8\ °F$

**45 (a)** $y = -\dfrac{4}{5}x + 20$   **(b)** $V = 4x\left(-\dfrac{4}{5}x + 20\right)$

**46** $C = \dfrac{3\pi(r^3 + 16)}{10r}$

**47 (a)** $V = 10t$
   **(b)** $V = 200h^2$ for $0 \le h \le 6$;
      $V = 7200 + 3200(h - 6)$ for $6 < h \le 9$
   **(c)** $h = \sqrt{\dfrac{t}{20}}$ for $0 \le t \le 720$; $h = 6 + \dfrac{t - 720}{320}$ for
      $720 < t \le 1680$

**48 (a)** $r = \dfrac{1}{2}x$   **(b)** $y = \dfrac{5}{4\pi} - \dfrac{1}{48}x^3$

**49 (a)** $y = \dfrac{bh}{a - b}$   **(b)** $V = \dfrac{1}{3}\pi h(a^2 + ab + b^2)$

**(c)** $\dfrac{200}{7\pi} \approx 9.1$ ft

**50** Increases 250%     **51** 375     **52** 10,125 watts

**53** $\dfrac{18}{13}$ hr after 1:00 P.M., or about 2:23 P.M.

**54** Radius of semicircle is $\dfrac{1}{8\pi}$ mi; length of rectangle is $\dfrac{1}{8}$ mi.

**55 (a)** 1 sec   **(b)** 4 ft
   **(c)** On the moon, 6 sec and 24 ft
**56 (a)** $(87.5, 17.5)$   **(b)** 30.625 units

# CHAPTER 3

*EXERCISES 3.1*

**1 (a)**

**(b)**

**3 (a)**

**(b)**

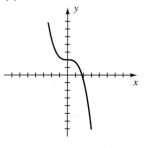

**19** $f(x) > 0$ if $-2 < x < 3$ or $x > 4$, $f(x) < 0$ if $x < -2$ or $3 < x < 4$

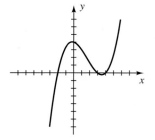

**21** $f(x) > 0$ if $x > 2$, $f(x) < 0$ if $x < -2$ or $|x| < 2$

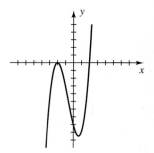

**5** $f(3) = -2 < 0$, $f(4) = 10 > 0$

**7** $f(2) = 5 > 0$, $f(3) = -5 < 0$

**9** $f\left(-\dfrac{1}{2}\right) = \dfrac{19}{32} > 0$, $f(-1) = -1 < 0$

**11** $f(x) > 0$ if $x > 2$, $f(x) < 0$ if $x < 2$

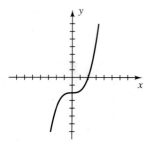

**13** $f(x) > 0$ if $|x| < 2$, $f(x) < 0$ if $|x| > 2$

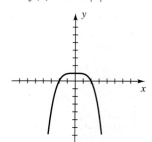

**23** $f(x) > 0$ if $|x| > 2$ or $|x| < \sqrt{2}$, $f(x) < 0$ if $\sqrt{2} < |x| < 2$

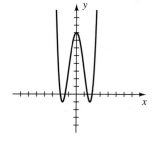

**25** $f(x) > 0$ if $|x| > 2$, $f(x) < 0$ if $|x| < 2$, $x \neq 0$, $x \neq 1$

**27** If $n$ is even, then $(-x)^n = x^n$ and hence $f(-x) = f(x)$. Thus, $f$ is an even function.

**29** $-\dfrac{4}{3}$

**31** $P(x) > 0$ on $\left(-\dfrac{1}{5}\sqrt{15}, 0\right)$ and $\left(\dfrac{1}{5}\sqrt{15}, \infty\right)$; $P(x) < 0$ on $\left(-\infty, -\dfrac{1}{5}\sqrt{15}\right)$ and $\left(0, \dfrac{1}{5}\sqrt{15}\right)$

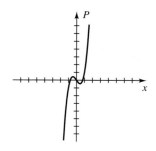

**15** $f(x) > 0$ if $|x| > 2$, $f(x) < 0$ if $0 < |x| < 2$

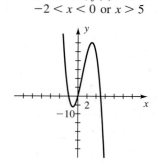

**17** $f(x) > 0$ if $x < -2$ or $0 < x < 5$, $f(x) < 0$ if $-2 < x < 0$ or $x > 5$

**33** **(b)** $V(x) > 0$ on $(0, 10)$ and $(15, \infty)$; allowable values for $x$ are in $(0, 10)$.

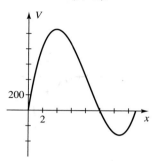

**35** **(a)** $T > 0$ for $0 < t < 12$; $T < 0$ for $12 < t < 24$

**(b)**

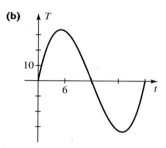

**(c)** $T(6) = 32.4 > 32$, $T(7) = 29.75 < 32$

**37** **(a)** $N(t) > 0$ for $0 < t < 5$

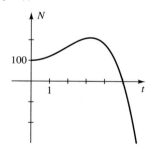

**(b)** The population becomes extinct after 5 years.

**39**

$[-4.5, 4.5]$ by $[-3, 3]$

$-1.89, 0.49, 1.20$

**41**

$[-4.5, 4.5]$ by $[-3, 3]$

$-1.88, 0.35,$ and $1.53$

**43**

$[-4.5, 4.5]$ by $[-3, 3]$

$(0.56, \infty)$

**45**

$[-4.5, 4.5]$ by $[-3, 3]$

$(-1.10, \infty)$

**47**

$[-4.5, 4.5]$ by $[-2, 4]$

$(-1.29, -0.77),$
$(0.085, 2.66),$
$(1.36, -0.42)$

## EXERCISES 3.2

**1** $2x^2 - x + 3$; $4x - 3$    **3** $\frac{3}{2}x$; $\frac{1}{2}x - 4$

**5** $0$; $7x + 2$    **7** $\frac{9}{2}$; $\frac{53}{2}$    **9** $26$    **11** $7$

**13** $f(-3) = 0$    **15** $f(-2) = 0$    **17** $x^3 - 3x^2 - 10x$

**19** $x^4 - 2x^3 - 9x^2 + 2x + 8$    **21** $2x^2 + x + 6$; $7$

**23** $x^2 - 3x + 1$; $-8$

**25** $3x^4 - 6x^3 + 12x^2 - 18x + 36$; $-65$

**27** $4x^3 + 2x^2 - 4x - 2$; $0$    **29** $73$    **31** $-0.0444$

**33** $8 + 7\sqrt{3}$    **35** $f(-2) = 0$    **37** $f\left(\frac{1}{2}\right) = 0$

**39** $3, 5$    **41** $f(c) > 0$    **43** $-14$

**45** If $f(x) = x^n - y^n$ and $n$ is even, then $f(-y) = 0$.
**47** (a) $V = \pi x^2(6 - x)$

(b) $\left(\frac{1}{2}(5 + \sqrt{45}), \frac{1}{2}(7 - \sqrt{45})\right)$

**49** (a) $A = 8x - 2x^3$    (b) $\sqrt{13} - 1 \approx 2.61$
**51** $-9.55$    **53** $-0.75, 1.96$

## EXERCISES 3.3

**1** $-4x^3 + 16x^2 - 4x - 24$    **3** $3x^3 + 3x^2 - 36x$
**5** $-2x^3 + 6x^2 - 8x + 24$
**7** $x^4 + 2x^3 - 23x^2 - 24x + 144$

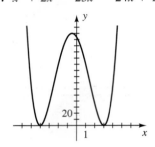

**9** $3x^6 - 27x^5 + 81x^4 - 81x^3$

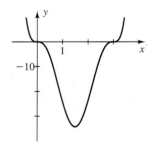

**11** $f(x) = \frac{7}{9}(x + 1)\left(x - \frac{3}{2}\right)(x - 3)$

**13** $f(x) = -1(x - 1)^2(x - 3)$

**15** $-\frac{2}{3}$ (multiplicity 1); 0 (multiplicity 2);

$\frac{5}{2}$ (multiplicity 3)

**17** $-\frac{3}{2}$ (multiplicity 2); 0 (multiplicity 3)

**19** $-4$ (multiplicity 3); $-3$ (multiplicity 2);
3 (multiplicity 5)

**21** $\pm 4i, \pm 3$ (each of multiplicity 1)
**23** $f(x) = (x + 3)^2(x + 2)(x - 1)$
**25** $f(x) = (x - 1)^5(x + 1)$

Exer. 27–34: The types of possible solutions are listed in the order positive, negative, nonreal complex.

**27** 3, 0, 0 or 1, 0, 2    **29** 0, 1, 2
**31** 2, 2, 0; 2, 0, 2; 0, 2, 2; 0, 0, 4
**33** 2, 3, 0; 2, 1, 2; 0, 3, 2; 0, 1, 4
**35** Upper 5, lower $-2$    **37** Upper 2, lower $-2$
**39** Upper 3, lower $-3$

**41** $f(x) = -\frac{1}{4}(x + 1)^2(x - 1)(x - 2)^3$

**43** (a) $f(x) = a(x + 3)^3(x + 1)(x - 2)^2$    (b) 108
**45** No    **47** Yes: $1.5(x - 2)(x - 5.2)(x - 10.1)$

**49** $f(t) = \frac{5}{3528}t(t - 5)(t - 19)(t - 24)$

**51**

As the multiplicity increases, the graph becomes more horizontal at (0.5, 0).

$[-3, 3]$ by $[-2, 2]$

**53**

$-1.2$ (multiplicity 2); 1.1 (multiplicity 1)

$[-3, 3]$ by $[-3, 1]$

**55** 2007 (when $t \approx 27.1$)

## EXERCISES 3.4

**1** $x^2 - 6x + 13$    **3** $(x - 2)(x^2 + 4x + 29)$
**5** $x(x + 1)(x^2 - 6x + 10)$
**7** $(x^2 - 8x + 25)(x^2 + 4x + 5)$
**9** $x(x^2 + 4)(x^2 - 2x + 2)$

Exer. 11–14: Show that none of the possible rational roots listed satisfy the equation.

**11** $\pm 1, \pm 2, \pm 3, \pm 6$    **13** $\pm 1, \pm 2$    **15** $-2, -1, 4$

**17** $-3, 2, \frac{5}{2}$    **19** $-7, \pm\sqrt{2}, 4$

**21** $-3, -\frac{2}{3}, 0$ (multiplicity 2), $\frac{1}{2}$

**23** $-\frac{3}{4}, -\frac{3}{4} \pm \frac{3}{4}\sqrt{7}i$

**25** $f(x) = (3x + 2)(2x - 1)(x - 1)^2(x - 2)$

**27** No. If $i$ is a root, then $-i$ is also a root. Hence, the polynomial would have factors $x - 1$, $x + 1$, $x - i$, $x + i$ and therefore would be of degree greater than 3.

**29** Since $n$ is odd and nonreal complex zeros occur in conjugate pairs for polynomials with real coefficients, there must be at least one real zero.

**31 (a)** The two boxes correspond to $x = 5$ and
  $x = 5(2 - \sqrt{2})$.
  **(b)** The box corresponding to $x = 5$

**33 (c)** In feet: 5, 12, and 13     **35 (b)** 4 ft

**37** None     **39** $-1.2, 0.8, -\dfrac{1}{2} \pm \dfrac{\sqrt{3}}{2}i$     **41** 10,200 m

## EXERCISES 3.5

**1 (a)**

**(b)** $D$ = all nonzero real numbers; $R = D$
**(c)** Decreasing on $(-\infty, 0)$ and on $(0, \infty)$

**3**

**5**

**7**

**9**

**11**

**13**

**15**

**17**

**19**

**21**

**23**

**25** $y = x - 2$

**27** $y = -\dfrac{1}{2}x$

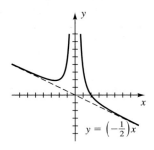

$y = \left(-\dfrac{1}{2}\right)x$

**29** $f(x) = \dfrac{2x - 3}{x + 1}$ for $x \neq -2$

$(-2, 7)$

**31** $f(x) = \dfrac{-1}{x + 1}$ for $x \neq 1$

$\left(1, -\dfrac{1}{2}\right)$

**33** $f(x) = x - 1$ for $x \neq -2$

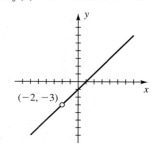

$(-2, -3)$

**35** $f(x) = \dfrac{x + 2}{x + 1}$ for $x \neq -2$

$(-2, 0)$

**37** $f(x) = \dfrac{3 - x}{x - 4}$

**39** $f(x) = \dfrac{6x^2 - 6x - 12}{x^3 - 7x + 6}$

**41** **(a)** $h = \dfrac{16}{(r + 0.5)^2} - 1$     **(b)** $V(r) = \pi r^2 h$
  **(c)** Exclude $r \leq 0$ and $r \geq 3.5$.

**43** **(a)** $V(t) = 50 + 5t,\ A(t) = 0.5t$     **(b)** $\dfrac{t}{10t + 100}$
  **(c)** As $t \to \infty$, $c(t) \to 0.1$ lb of salt per gal.

**45** **(a)** $0 < S < 4000$     **(b)** 4500     **(c)** 2000
  **(d)** A 125% increase in $S$ produces only a 12.5%
  increase in $R$.

**47** None          **49** $x = 0.999$

$[-9, 3]$ by $[-9, 3]$

$[0.7, 1.3]$ by $[0.8, 1.2]$

**51** **(a)** The graph of $g$ is the horizontal line $y = 1$ with
  holes at $x = 0,\ \pm 1,\ \pm 2,\ \pm 3$.
  **(b)** The graph of $h$ is the graph of $p$ with holes at
  $x = 0,\ \pm 1,\ \pm 2,\ \pm 3$.

## CHAPTER 3 REVIEW EXERCISES

**1** $f(x) > 0$ if $x > -2$,
$f(x) < 0$ if $x < -2$

**2** $f(x) > 0$ if $x < -\sqrt[6]{32}$
or $x > \sqrt[6]{32}$, $f(x) < 0$ if
$-\sqrt[6]{32} < x < \sqrt[6]{32}$

**3** $f(x) > 0$ if $-2 < x < 1$
or $1 < x < 3$, $f(x) < 0$
if $x < -2$ or $x > 3$

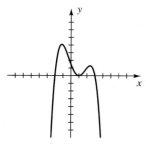

**4** $f(x) > 0$ if $-1 < x < 0$
or $0 < x < 2$, $f(x) < 0$
if $x < -1$ or $x > 2$

**5** $f(x) > 0$ if $-4 < x < 0$
or $x > 2$, $f(x) < 0$ if
$x < -4$ or $0 < x < 2$

**6** $f(x) > 0$ if $-4 < x < -2$,
$0 < x < 2$, or $x > 4$,
$f(x) < 0$ if $x < -4$,
$-2 < x < 0$, or
$2 < x < 4$

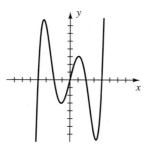

**7** $f(0) = -9 < 100$ and $f(10) = 561 > 100$. By the intermediate value theorem for polynomial functions, $f$ takes on every value between $-9$ and $561$. Hence, there is at least one real number $a$ in $[0, 10]$ such that $f(a) = 100$.

**8** Let $f(x) = x^5 - 3x^4 - 2x^3 - x + 1$. $f(0) = 1 > 0$ and $f(1) = -4 < 0$. By the intermediate value theorem for polynomial functions, $f$ takes on every value between $-4$ and $1$. Hence, there is at least one real number $a$ in $[0, 1]$ such that $f(a) = 0$.

**9** $3x^2 + 2$; $-21x^2 + 5x - 9$    **10** $4x - 1$; $2x - 1$

**11** $-132$    **12** $f(3) = 0$

**13** $6x^4 - 12x^3 + 24x^2 - 52x + 104$; $-200$

**14** $2x^2 + (5 + 2\sqrt{2})x + (2 + 5\sqrt{2})$; $11 + 2\sqrt{2}$

**15** $\dfrac{2}{41}(x^2 + 6x + 34)(x + 1)$

**16** $\dfrac{1}{4}x(x^2 - 2x + 2)(x - 3)$

**17** $x^7 + 6x^6 + 9x^5$

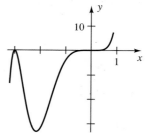

**18** $(x - 2)^3(x + 3)(x - 1)$

**19** 1 (multiplicity 5); $-3$ (multiplicity 1)

**20** $0$, $\pm i$ (all have multiplicity 2)

**21 (a)** Either 3 positive and 1 negative, or 1 positive, 1 negative, and 2 nonreal complex
**(b)** Upper bound 3; lower bound $-1$

**22 (a)** Either 2 positive and 3 negative; 2 positive, 1 negative, and 2 nonreal complex; 3 negative and 2 nonreal complex; or 1 negative and 4 nonreal complex
**(b)** Upper bound 2; lower bound $-3$

**23** Since there are only even powers,
$7x^6 + 2x^4 + 3x^2 + 10 \geq 10$ for every real number $x$.

**24** $-3, -2, -2 \pm i$    **25** $-\dfrac{1}{2}, \dfrac{1}{4}, \dfrac{3}{2}$    **26** $\pm\sqrt{6}, \pm 1$

**27**        **28**

**29**

**30**

**31**

**32**

**33**

**34**

**35**

**36**

**37 (a)** $\dfrac{1}{15,000}$

**(b)** $y \approx 0.9754 < 1$ if $x = 6.1$, and $y \approx 1.0006 > 1$ if $x = 6.2$

**38 (a)** $V = \dfrac{1}{4\pi}x(l^2 - x^2)$

**(b)** If $x > 0$, $V > 0$ when $0 < x < l$.

**39** $t = 4$ (10:00 A.M.) and $t = 16 - 4\sqrt{6} \approx 6.2020$ (12:12 P.M.)

**40** $\sqrt{5} < t < 4$

**41 (a)** $R = k$

**(b)** $k$ is the maximum rate at which the liver can remove alcohol from the bloodstream.

**42 (a)** $C(100) = \$2,000,000.00$ and $C(90) \approx \$163,636.36$

**(b)**

# CHAPTER 4

*EXERCISES 4.1*

**1** 5    **3** $-1, 3$    **5** $-\dfrac{4}{99}$    **7** $\dfrac{18}{5}$

**9 (a)**

**(b)**

**(c)**

**(d)**

**(e)**

**(f)**

**15**

**17**

**(g)**

**(h)**

**19**

**(i)**

**(j)**

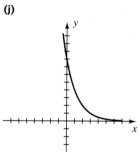

**21 (a)** 90    **(b)** 59    **(c)** 35
**23 (a)** 1039; 3118; 5400    **25 (a)** 50 mg; 25 mg;
   **(b)**    $\dfrac{25}{2}\sqrt{2} \approx 17.7$ mg
   **(b)**

**11**

**13**

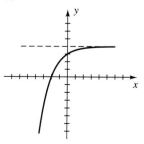

**27** $-\dfrac{1}{1600}$

**29 (a)** \$1010.00    **(b)** \$1020.10    **(c)** \$1061.52
   **(d)** \$1126.83

**31 (a)** \$7800    **(b)** \$4790    **(c)** \$2942

**33 (a)** \$1061.36    **(b)** \$1126.49    **(c)** \$1346.86
   **(d)** \$1814.02

**35 (a)** Examine the pattern formed by the value $y$ in the
   year $n$.
   **(b)** Solve $s = (1 - a)^T y_0$ for $a$.

**37 (a)** \$925.75    **(b)** \$243,270    **39** \$6346.40

**41** 6.58 yr

**43**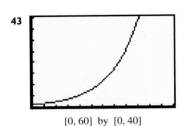

[0, 60] by [0, 40]

(a) 26.13    (b) 8.50

**45**

[−10.5, 10.5] by [−7, 7]

− 1.02, 2.14, 3.62

**47**

[−3, 3] by [−2, 2]

(a) Not one-to-one
(b) 0

**49**

[−4, 1] by [−2, 3]

(a) Increasing: [−3.37, −1.19] ∪ [0.52, 1];
decreasing: [−4, −3.37] ∪ [−1.19, 0.52]
(b) [−1.79, 1.94]

**51**

[0, 7.5] by [0, 5]

The maximum number
of sales approaches $k$.

**53**

[0, 40] by [0, 200000]

After approximately
32.8 yr

## EXERCISES 4.2

**1** (a)     (b)

**3** (a)     (b)

**5** $1510.59    **7** $13,806.92    **9** 13%    **11** 3, 4

**13** −1    **15** $-\dfrac{3}{4}$, 0    **17** $\dfrac{4}{(e^x + e^{-x})^2}$    **19** 27.43 g

**21** 261.1 million    **23** 13.5%    **25** 5412

**27** 7.44 in.    **29** 75.77 cm; 15.98 cm/yr

**31** $6.82 per hr    **33** (a) 7.19%    (b) 7.25%

**35**

[0, 60] by [0, 40]

(a) 29.96    (b) 8.15

**37 (a)**

[−7.5, 7.5] by [−5, 5]

**(b)**

**39 (a)**

[−4.5, 4.5] by [−3, 3]

**(b)**

**41**

[−3, 11] by [−10, 80]

−1.04, 2.11

**43**

[0, 4.5] by [0, 3]

$f(x)$ is closer to $e^x$ if $x \approx 0$; $g(x)$ is closer to $e^x$ if $x \approx 1$.

**45**

[−2, 2.5] by [−1, 2]

0.11, 0.79, 1.13

**47**

[0, 200] by [0, 8]

$y \approx 2.71 \approx e$

**49** 0.567

**51**

[−5.5, 5] by [−2, 5]

Increasing on $[-1, \infty)$; decreasing on $(-\infty, -1]$

**53 (a)** As $h$ increases, $C$ decreases.
**(b)** As $y$ increases, $C$ decreases.

## EXERCISES 4.3

**1 (a)** $\log_4 64 = 3$   **(b)** $\log_4 \dfrac{1}{64} = -3$   **(c)** $\log_t s = r$
**(d)** $\log_3 (4 - t) = x$   **(e)** $\log_5 \dfrac{a + b}{a} = 7t$
**(f)** $\log_{0.7} (5.3) = t$

**3 (a)** $2^5 = 32$   **(b)** $3^{-5} = \dfrac{1}{243}$   **(c)** $t^p = r$
**(d)** $3^5 = (x + 2)$   **(e)** $2^{3x+4} = m$   **(f)** $b^{3/2} = 512$

**5** $t = 3 \log_a \dfrac{5}{2}$   **7** $t = \dfrac{1}{C} \log_a \left(\dfrac{A - D}{B}\right)$

**9 (a)** $\log 100{,}000 = 5$   **(b)** $\log 0.001 = -3$
**(c)** $\log (y + 1) = x$   **(d)** $\ln p = 7$
**(e)** $\ln (3 - x) = 2t$

**11 (a)** $10^{50} = x$   **(b)** $10^{20t} = x$   **(c)** $e^{0.1} = x$
**(d)** $e^{4+3x} = w$   **(e)** $e^{1/6} = z - 2$

**13 (a)** 0   **(b)** 1   **(c)** Not possible   **(d)** 2   **(e)** 8
**(f)** 3   **(g)** −2

**15 (a)** 3   **(b)** 5   **(c)** 2   **(d)** −4   **(e)** 2
**(f)** −3   **(g)** $3e^2$

**17** 4   **19** No solution   **21** −1, −2   **23** 13

**25** 27   **27** $\pm\dfrac{1}{e}$   **29** 3

**31 (a)**

**(b)**

**(i)**

**(j)**

**(c)**

**(d)**

**(k)**

**(e)**

**(f)**

**33**

**35**

**(g)**

**(h)**

**37** $f(x) = \log_2 x$     **39** $f(x) = \log_2 x - 1$
**41** $f(x) = \log_2 (-x)$     **43** $f(x) = 2 \log_2 x$
**45 (a)** 4240    **(b)** 8.85    **(c)** 0.0237    **(d)** 9.97
     **(e)** 1.05    **(f)** 0.202

**47** $t = -1600 \log_2 \left( \dfrac{q}{q_0} \right)$     **49** $t = -\dfrac{L}{R} \ln \left( \dfrac{I}{20} \right)$

**51 (a)** 2    **(b)** 4    **(c)** 5
**53 (a)** 10    **(b)** 30    **(c)** 40    **55** In the year 2079
**57 (a)** $W = 2.4e^{1.84h}$    **(b)** 37.92 kg
**59 (a)** 10,007 ft    **(b)** 18,004 ft
**61 (a)** 305.9 kg    **(b)** (1) 20 yr (2) 19.8 yr
**63** 10.1 mi    **65** $2^{1/8} \approx 1.09$
**67 (a)** Pedestrians have faster average walking speeds in
     large cities.
    **(b)** 570,000

**69** 1.763    **71** (0, 14.90)

[−2, 16] by [−4, 8]

## EXERCISES 4.4

**I** **(a)** $\log_4 x + \log_4 z$    **(b)** $\log_4 y - \log_4 x$

**(c)** $\frac{1}{3} \log_4 z$

**3** $3 \log_a x + \log_a w - 2 \log_a y - 4 \log_a z$

**5** $\frac{1}{3} \log z - \log x - \frac{1}{2} \log y$

**7** $\frac{7}{4} \ln x - \frac{5}{4} \ln y - \frac{1}{4} \ln z$

**9** **(a)** $\log_3 (5xy)$    **(b)** $\log_3 \frac{2z}{x}$    **(c)** $\log_3 y^5$

**11** $\log_a \frac{x^2 \sqrt[3]{x - 2}}{(2x + 3)^5}$    **13** $\log \frac{y^{13/3}}{x^2}$    **15** $\ln x$    **17** $\frac{7}{2}$

**19** $5\sqrt{5}$    **21** No solution    **23** $-7$    **25** 1

**27** $-2$    **29** $\frac{-1 + \sqrt{65}}{2}$    **31** $-1 + \sqrt{1 + e}$

**33**

**35**

**37**

**39**

**41**

**43**

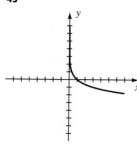

**45** $f(x) = \log_2 x^2$    **47** $f(x) = \log_2 (8x)$    **49** $y = \frac{b}{x^k}$

**51**

**53** **(a)** 0    **(b)** $R(2x) = R(x) + a \log 2$    **55** 0.29 cm

**57**

[0, 6] by [−1, 3]

$(0, 1.01] \cup [2.4, \infty)$

**59**

[0, 8] by [−1.67, 3.67]

1.41, 6.59

**61**

[0.2, 16] by [−4.77, 5.77]

**(a)** Increasing on [0.2, 0.63] and [6.87, 16]; decreasing on [0.63, 6.87]

**(b)** 4.61; −3.31

## EXERCISES 4.5

**1** $\dfrac{\log 8}{\log 5} \approx 1.29$    **3** $4 - \dfrac{\log 5}{\log 3} \approx 2.54$    **5** $1.1133$

**7** $-0.7325$    **9** $2$    **11** $\dfrac{\log (2/81)}{\log 24} \approx -1.16$

**13** $\dfrac{\log (8/25)}{\log (4/5)} \approx 5.11$    **15** $-3$    **17** $5$

**19** $\dfrac{2}{3}\sqrt{\dfrac{101}{11}} \approx 2.02$    **21** $1, 2$

**23** $\dfrac{\log (4 + \sqrt{19})}{\log 4} \approx 1.53$    **25** $1$ or $100$    **27** $10^{100}$

**29** $10{,}000$    **31** $x = \log (y \pm \sqrt{y^2 - 1})$

**33** $x = \dfrac{1}{2}\log\left(\dfrac{1+y}{1-y}\right)$    **35** $x = \ln (y + \sqrt{y^2 + 1})$

**37** $x = \dfrac{1}{2}\ln\left(\dfrac{y+1}{y-1}\right)$

**39** $y$-intercept $= \log_2 3$
$\approx 1.5850$

**41** $x$-intercept $= \log_4 3$
$\approx 0.7925$

**43 (a)** $2.2$    **(b)** $5$    **(c)** $8.3$

**45** Basic if pH $> 7$, acidic if pH $< 7$

**47** $11.58$ yr $\approx 11$ yr $7$ mo    **49** $86.4$ m

**51 (a)** *Hint:* Take the natural logarithm of both sides first.

   **(b)** Note that $f(e) = \dfrac{1}{e}$. Any horizontal line $y = k$, with $0 < k < \dfrac{1}{e}$, will intersect the graph at points $\left(x_1, \dfrac{\ln x_1}{x_1}\right)$ and $\left(x_2, \dfrac{\ln x_2}{x_2}\right)$, where $1 < x_1 < e$ and $x_2 > e$.

**53 (a)**      **(b)** $6.58$ min

**55 (a)** $t = \dfrac{\log (F/F_0)}{\log (1 - m)}$    **(b)** After $13{,}863$ generations

**57 (a)** $4.28$ ft    **(b)** $24.8$ yr    **59** $\dfrac{\ln (25/6)}{\ln (200/35)} \approx 0.82$

**61** The suspicion is correct.

**63** The suspicion is incorrect.

**65** None

$[-1, 17]$ by $[-1, 11]$

**67** $1.37, 9.94$

$[-1, 17]$ by $[-1, 11]$

**69**      $(-\infty, -0.32) \cup (1.52, 6.84)$

$[-5, 10]$ by $[-8, 2]$

## CHAPTER 4 REVIEW EXERCISES

**3**

**4**

**11**

**12**

**5**

**6**

**13**

**14**

**7**

**8**

**15**

**16**

**17 (a)** $-4$  **(b)** $0$  **(c)** $1$  **(d)** $4$  **(e)** $6$  **(f)** $8$
**(g)** $\dfrac{1}{2}$

**18 (a)** $\dfrac{1}{3}$  **(b)** $0$  **(c)** $1$  **(d)** $5$  **(e)** $1$  **(f)** $25$
**(g)** $\dfrac{1}{3}$

**19** $0$   **20** $9$   **21** $9$   **22** $\dfrac{33}{47}$   **23** $1$   **24** $99$

**25** $5 - \dfrac{\log 6}{\log 2}$   **26** $\pm\sqrt{\dfrac{\log 7}{\log 3}}$   **27** $\dfrac{\log (3/8)}{\log (32/9)}$   **28** $1$

**29** $\dfrac{1}{4}$, $1$, $4$   **30** No solution   **31** $\sqrt{5}$   **32** $2$

**33** $0$, $\pm 1$   **34** $\ln 2$   **35 (a)** $-3$, $2$   **(b)** $2$

**36 (a)** $8$   **(b)** $\pm 4$   **37** $4 \log x + \dfrac{2}{3}\log y - \dfrac{1}{3}\log z$

**9**

**10**

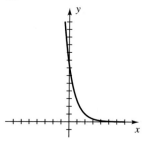

**38** $-\log(xy^2)$    **39** $x = \log\left(\dfrac{1 \pm \sqrt{1-4y^2}}{2y}\right)$

**40** If $y < 0$, then $x = \log\left(\dfrac{1 - \sqrt{1+4y^2}}{2y}\right)$.

If $y > 0$, then $x = \log\left(\dfrac{1 + \sqrt{1+4y^2}}{2y}\right)$.

**41** (a) 1.89   (b) 78.3   (c) 0.472
**42** (a) 0.924   (b) 0.00375   (c) 6.05
**43** (a) 2000
   (b) $2000(3^{1/6}) \approx 2401$; $2000(3^{1/2}) \approx 3464$; 6000
**44** \$1125.51
**45** (a)     (b) 8 days

**46** $N = 1000\left(\dfrac{3}{5}\right)^{t/3}$

**47** (a) After 11.39 yr   (b) 6.30 yr
**48** $t = (\ln 100)\dfrac{L}{R} \approx 4.6\dfrac{L}{R}$
**49** (a) $I = I_0 10^{\alpha/10}$
   (b) Examine $I(\alpha + 1)$, where $I(\alpha)$ is the intensity corresponding to $\alpha$ decibels.
**50** $t = -\dfrac{1}{k}\ln\left(\dfrac{a-L}{ab}\right)$   **51** $A = 10^{(R+5.1)/2.3} - 3000$
**52** $\dfrac{A_1}{A_2} = \dfrac{10^{(R+5.1)/2.3} - 3000}{10^{(R+7.5)/2.3} - 34{,}000}$   **53** 26,615.9 mi²
**54** $h = \dfrac{\ln(29/p)}{0.000034}$   **55** $v = a\ln\left(\dfrac{m_1 + m_2}{m_1}\right)$
**56** (a) $n = 10^{7.7-0.9R}$   (b) 12,589; 1585; 200
**57** (a) $E = 10^{11.4+1.5R}$   (b) $10^{24}$ ergs   **58** 110 days
**59** 86.8 cm; 9.715 cm/yr   **60** $t = -\dfrac{L}{R}\ln\left(\dfrac{V-RI}{V}\right)$
**61** (a) 26,749 yr   (b) 30%   **62** 16.91 yr

## CHAPTER 5

### EXERCISES 5.1

Exer. 1–4: The answers are not unique.

**1** (a) 480°, 840°, −240°, −600°
   (b) 495°, 855°, −225°, −585°
   (c) 330°, 690°, −390°, −750°

**3** (a) 260°, 980°, −100°, −460°
   (b) $\dfrac{17\pi}{6}, \dfrac{29\pi}{6}, -\dfrac{7\pi}{6}, -\dfrac{19\pi}{6}$
   (c) $\dfrac{7\pi}{4}, \dfrac{15\pi}{4}, -\dfrac{9\pi}{4}, -\dfrac{17\pi}{4}$
**5** (a) 84°42′26″   (b) 57.5°
**7** (a) 131°8′23″   (b) 43.58°
**9** (a) $\dfrac{5\pi}{6}$   (b) $-\dfrac{\pi}{3}$   (c) $\dfrac{5\pi}{4}$
**11** (a) $\dfrac{5\pi}{2}$   (b) $\dfrac{2\pi}{5}$   (c) $\dfrac{5\pi}{9}$
**13** (a) 120°   (b) 330°   (c) 135°
**15** (a) −630°   (b) 1260°   (c) 20°
**17** 114°35′30″   **19** 286°28′44″   **21** 37.6833°
**23** 115.4408°   **25** 63°10′8″   **27** 310°37′17″
**29** 2.5 cm
**31** (a) $2\pi \approx 6.28$ cm   (b) $8\pi \approx 25.13$ cm²
**33** (a) 1.75; $\dfrac{315}{\pi} \approx 100.27°$   (b) 14 cm²
**35** (a) $\dfrac{20\pi}{9} \approx 6.98$ m   (b) $\dfrac{80\pi}{9} \approx 27.93$ m²
**37** In miles: (a) 4189   (b) 3142   (c) 2094
   (d) 698   (e) 70
**39** $\dfrac{1}{8}$ radian $\approx 7°10′$   **41** $7.29 \times 10^{-5}$
**43** (a) $80\pi$   (b) $\dfrac{100\pi}{3} \approx 104.72$
**45** (a) $\dfrac{200\pi}{3}, 90\pi$   (b) $\dfrac{100\pi}{3}, \dfrac{105\pi}{4}$
**47** (a) $\dfrac{21\pi}{8} \approx 8.25$ ft   (b) $\dfrac{2}{3}d$
**49** Large   **51** 192.08

### EXERCISES 5.2

*Note:* Answers are in the order *sin, cos, tan, cot, sec, csc* for any exercises that require the values of the six trigonometric functions.

**1** $\dfrac{4}{5}, \dfrac{3}{5}, \dfrac{4}{3}, \dfrac{3}{4}, \dfrac{5}{3}, \dfrac{5}{4}$

**3** $\dfrac{2}{5}, \dfrac{\sqrt{21}}{5}, \dfrac{2}{\sqrt{21}}, \dfrac{\sqrt{21}}{2}, \dfrac{5}{\sqrt{21}}, \dfrac{5}{2}$

**5** $\dfrac{a}{\sqrt{a^2+b^2}}, \dfrac{b}{\sqrt{a^2+b^2}}, \dfrac{a}{b}, \dfrac{b}{a}, \dfrac{\sqrt{a^2+b^2}}{b}, \dfrac{\sqrt{a^2+b^2}}{a}$

**7** $\dfrac{b}{c}, \dfrac{\sqrt{c^2-b^2}}{c}, \dfrac{b}{\sqrt{c^2-b^2}}, \dfrac{\sqrt{c^2-b^2}}{b}, \dfrac{c}{\sqrt{c^2-b^2}}, \dfrac{c}{b}$

**9** $x = 8$; $y = 4\sqrt{3}$   **11** $x = 7\sqrt{2}$; $y = 7$
**13** $x = 4\sqrt{3}$; $y = 4$   **15** $\dfrac{3}{5}, \dfrac{4}{5}, \dfrac{3}{4}, \dfrac{4}{3}, \dfrac{5}{4}, \dfrac{5}{3}$

**17** $\dfrac{5}{13}, \dfrac{12}{13}, \dfrac{5}{12}, \dfrac{12}{5}, \dfrac{13}{12}, \dfrac{13}{5}$

**19** $\dfrac{\sqrt{11}}{6}, \dfrac{5}{6}, \dfrac{\sqrt{11}}{5}, \dfrac{5}{\sqrt{11}}, \dfrac{6}{5}, \dfrac{6}{\sqrt{11}}$

**21** $200\sqrt{3} \approx 346.4$ ft    **23 (a)** $-1$    **(b)** $-4$

**25 (a)** 5    **(b)** 5    **27** $1 - \sin\theta\cos\theta$

**29** $\sin\theta$    **31** $\cot\theta = \dfrac{\sqrt{1 - \sin^2\theta}}{\sin\theta}$

**33** $\sec\theta = \dfrac{1}{\sqrt{1 - \sin^2\theta}}$    **35** $\sin\theta = \dfrac{\sqrt{\sec^2\theta - 1}}{\sec\theta}$

Exer. 37–58: Typical verifications are given.

**37** $\cos\theta\sec\theta = \cos\theta(1/\cos\theta) = 1$

**39** $\sin\theta\sec\theta = \sin\theta(1/\cos\theta) = \sin\theta/\cos\theta = \tan\theta$

**41** $\dfrac{\csc\theta}{\sec\theta} = \dfrac{1/\sin\theta}{1/\cos\theta} = \dfrac{\cos\theta}{\sin\theta} = \cot\theta$

**43** $(1 + \cos\theta)(1 - \cos\theta) = 1 - \cos^2\theta = \sin^2\theta$

**45** $\cos^2\theta(\sec^2\theta - 1) = \cos^2\theta(\tan^2\theta) = \cos^2\theta \cdot \dfrac{\sin^2\theta}{\cos^2\theta}$
$$= \sin^2\theta$$

**47** $\dfrac{\sin\theta}{\csc\theta} + \dfrac{\cos\theta}{\sec\theta} = \dfrac{\sin\theta}{1/\sin\theta} + \dfrac{\cos\theta}{1/\cos\theta} = \sin^2\theta + \cos^2\theta = 1$

**49** $(1 + \sin\theta)(1 - \sin\theta) = 1 - \sin^2\theta = \cos^2\theta = \dfrac{1}{\sec^2\theta}$

**51** $\sec\theta - \cos\theta = \dfrac{1}{\cos\theta} - \cos\theta = \dfrac{1 - \cos^2\theta}{\cos\theta} = \dfrac{\sin^2\theta}{\cos\theta}$
$$= \dfrac{\sin\theta}{\cos\theta} \cdot \sin\theta = \tan\theta\sin\theta$$

**53** $(\cot\theta + \csc\theta)(\tan\theta - \sin\theta)$
$$= \cot\theta\tan\theta - \cot\theta\sin\theta + \csc\theta\tan\theta - \csc\theta\sin\theta$$
$$= \dfrac{1}{\tan\theta}\tan\theta - \dfrac{\cos\theta}{\sin\theta}\sin\theta + \dfrac{1}{\sin\theta}\dfrac{\sin\theta}{\cos\theta} - \dfrac{1}{\sin\theta}\sin\theta$$
$$= 1 - \cos\theta + \dfrac{1}{\cos\theta} - 1 = -\cos\theta + \sec\theta$$
$$= \sec\theta - \cos\theta$$

**55** $\sec^2\theta\csc^2\theta = (1 + \tan^2\theta)(1 + \cot^2\theta)$
$$= 1 + \tan^2\theta + \cot^2\theta + 1$$
$$= \sec^2\theta + \csc^2\theta$$

**57** $\log\csc\theta = \log\left(\dfrac{1}{\sin\theta}\right) = \log 1 - \log\sin\theta$
$$= 0 - \log\sin\theta = -\log\sin\theta$$

**59** $-\dfrac{3}{5}, \dfrac{4}{5}, -\dfrac{3}{4}, -\dfrac{4}{3}, \dfrac{5}{4}, -\dfrac{5}{3}$

**61** $-\dfrac{5}{\sqrt{29}}, -\dfrac{2}{\sqrt{29}}, \dfrac{5}{2}, \dfrac{2}{5}, -\dfrac{\sqrt{29}}{2}, -\dfrac{\sqrt{29}}{5}$

**63** $\dfrac{4}{\sqrt{17}}, -\dfrac{1}{\sqrt{17}}, -4, -\dfrac{1}{4}, -\sqrt{17}, \dfrac{\sqrt{17}}{4}$

**65** $\dfrac{4}{5}, \dfrac{3}{5}, \dfrac{4}{3}, \dfrac{3}{4}, \dfrac{5}{3}, \dfrac{5}{4}$

**67** $-\dfrac{7}{\sqrt{53}}, -\dfrac{2}{\sqrt{53}}, \dfrac{7}{2}, \dfrac{2}{7}, -\dfrac{\sqrt{53}}{2}, -\dfrac{\sqrt{53}}{7}$

*Note:* U denotes *undefined.*

**69 (a)** 1, 0, U, 0, U, 1    **(b)** 0, 1, 0, U, 1, U
   **(c)** $-1$, 0, U, 0, U, $-1$    **(d)** 0, $-1$, 0, U, $-1$, U

**71 (a)** IV    **(b)** III    **(c)** II    **(d)** III

**73** $\dfrac{3}{5}, -\dfrac{4}{5}, -\dfrac{3}{4}, -\dfrac{4}{3}, -\dfrac{5}{4}, \dfrac{5}{3}$

**75** $-\dfrac{5}{13}, \dfrac{12}{13}, -\dfrac{5}{12}, -\dfrac{12}{5}, \dfrac{13}{12}, -\dfrac{13}{5}$

**77** $-\dfrac{\sqrt{8}}{3}, -\dfrac{1}{3}, \sqrt{8}, \dfrac{1}{\sqrt{8}}, -3, -\dfrac{3}{\sqrt{8}}$

**79** $\dfrac{\sqrt{15}}{4}, -\dfrac{1}{4}, -\sqrt{15}, -\dfrac{1}{\sqrt{15}}, -4, \dfrac{4}{\sqrt{15}}$

**81** $-\tan\theta$    **83** $\sec\theta$    **85** $-\sin\dfrac{\theta}{2}$

## EXERCISES 5.3

**1** $\dfrac{8}{17}, -\dfrac{15}{17}, -\dfrac{8}{15}, -\dfrac{15}{8}, -\dfrac{17}{15}, \dfrac{17}{8}$

**3** $-\dfrac{7}{25}, \dfrac{24}{25}, -\dfrac{7}{24}, -\dfrac{24}{7}, \dfrac{25}{24}, -\dfrac{25}{7}$

**5 (a)** $\left(-\dfrac{3}{5}, -\dfrac{4}{5}\right)$    **(b)** $\left(-\dfrac{3}{5}, -\dfrac{4}{5}\right)$    **(c)** $\left(\dfrac{3}{5}, -\dfrac{4}{5}\right)$
   **(d)** $\left(-\dfrac{3}{5}, \dfrac{4}{5}\right)$

**7 (a)** $\left(\dfrac{12}{13}, \dfrac{5}{13}\right)$    **(b)** $\left(\dfrac{12}{13}, \dfrac{5}{13}\right)$    **(c)** $\left(-\dfrac{12}{13}, \dfrac{5}{13}\right)$
   **(d)** $\left(\dfrac{12}{13}, -\dfrac{5}{13}\right)$

**9 (a)** $(1, 0)$; 0, 1, 0, U, 1, U
   **(b)** $(-1, 0)$; 0, $-1$, 0, U, $-1$, U

**11 (a)** $(0, -1)$; $-1$, 0, U, 0, U, $-1$
   **(b)** $(0, 1)$; 1, 0, U, 0, U, 1

**13 (a)** $\left(\dfrac{\sqrt{2}}{2}, \dfrac{\sqrt{2}}{2}\right)$; $\dfrac{\sqrt{2}}{2}, \dfrac{\sqrt{2}}{2}$, 1, 1, $\sqrt{2}, \sqrt{2}$
   **(b)** $\left(-\dfrac{\sqrt{2}}{2}, \dfrac{\sqrt{2}}{2}\right)$; $\dfrac{\sqrt{2}}{2}, -\dfrac{\sqrt{2}}{2}, -1, -1, -\sqrt{2}, \sqrt{2}$

**15 (a)** $\left(-\dfrac{\sqrt{2}}{2}, -\dfrac{\sqrt{2}}{2}\right)$; $-\dfrac{\sqrt{2}}{2}, -\dfrac{\sqrt{2}}{2}$, 1, 1, $-\sqrt{2}, -\sqrt{2}$
   **(b)** $\left(\dfrac{\sqrt{2}}{2}, -\dfrac{\sqrt{2}}{2}\right)$; $-\dfrac{\sqrt{2}}{2}, \dfrac{\sqrt{2}}{2}, -1, -1, \sqrt{2}, -\sqrt{2}$

**17 (a)** $-1$    **(b)** $-\dfrac{\sqrt{2}}{2}$    **(c)** $-1$

**19 (a)** 1    **(b)** $-1$    **(c)** 1

Exer. 21–26: Typical verifications are given.

**21** $\sin(-x)\sec(-x) = (-\sin x)\sec x = (-\sin x)(1/\cos x)$
$$= -\tan x$$

**23** $\dfrac{\cot(-x)}{\csc(-x)} = \dfrac{-\cot x}{-\csc x} = \dfrac{\cos x/\sin x}{1/\sin x} = \cos x$

**25** $\dfrac{1}{\cos(-x)} - \tan(-x)\sin(-x) = \dfrac{1}{\cos x} - (-\tan x)(-\sin x)$
$$= \dfrac{1}{\cos x} - \dfrac{\sin x}{\cos x}\sin x$$
$$= \dfrac{1-\sin^2 x}{\cos x} = \dfrac{\cos^2 x}{\cos x} = \cos x$$

**27** (a) $0$   (b) $-1$    **29** (a) $\dfrac{\sqrt{2}}{2}$   (b) $-1$

**31** (a) $1$   (b) $-\infty$    **33** (a) $-1$   (b) $\infty$

**35** (a) $\infty$   (b) $\sqrt{2}$    **37** (a) $-\infty$   (b) $1$

**39** $\dfrac{3\pi}{2}, \dfrac{7\pi}{2}$    **41** $\dfrac{\pi}{6}, \dfrac{5\pi}{6}, \dfrac{13\pi}{6}, \dfrac{17\pi}{6}$    **43** $0, 2\pi, 4\pi$

**45** $\dfrac{\pi}{4}, \dfrac{7\pi}{4}, \dfrac{9\pi}{4}, \dfrac{15\pi}{4}$    **47** $\dfrac{\pi}{4}, \dfrac{5\pi}{4}$    **49** $0, \pi$

**51** (a) $-\dfrac{11\pi}{6}, -\dfrac{7\pi}{6}, \dfrac{\pi}{6}, \dfrac{5\pi}{6}$

(b) $-\dfrac{11\pi}{6} < x < -\dfrac{7\pi}{6}$ and $\dfrac{\pi}{6} < x < \dfrac{5\pi}{6}$

(c) $-2\pi \le x < -\dfrac{11\pi}{6}, \ -\dfrac{7\pi}{6} < x < \dfrac{\pi}{6}$, and

$\dfrac{5\pi}{6} < x \le 2\pi$

**53** (a) $-\dfrac{4\pi}{3}, -\dfrac{2\pi}{3}, \dfrac{2\pi}{3}, \dfrac{4\pi}{3}$

(b) $-2\pi \le x < -\dfrac{4\pi}{3}, \ -\dfrac{2\pi}{3} < x < \dfrac{2\pi}{3}$, and

$\dfrac{4\pi}{3} < x \le 2\pi$

(c) $-\dfrac{4\pi}{3} < x < -\dfrac{2\pi}{3}$ and $\dfrac{2\pi}{3} < x < \dfrac{4\pi}{3}$

**55**

**57**

**59**

**61**

**63** (a) $\left[-2\pi, -\dfrac{3\pi}{2}\right), \left(-\dfrac{3\pi}{2}, -\pi\right], \left[0, \dfrac{\pi}{2}\right), \left(\dfrac{\pi}{2}, \pi\right]$

(b) $\left[-\pi, -\dfrac{\pi}{2}\right), \left(-\dfrac{\pi}{2}, 0\right], \left[\pi, \dfrac{3\pi}{2}\right), \left(\dfrac{3\pi}{2}, 2\pi\right]$

**65** (a) The tangent function increases on *all* intervals on
which it is defined. Between $-2\pi$ and $2\pi$, these
intervals are $\left[-2\pi, -\dfrac{3\pi}{2}\right), \left(-\dfrac{3\pi}{2}, -\dfrac{\pi}{2}\right),$
$\left(-\dfrac{\pi}{2}, \dfrac{\pi}{2}\right), \left(\dfrac{\pi}{2}, \dfrac{3\pi}{2}\right),$ and $\left(\dfrac{3\pi}{2}, 2\pi\right].$

(b) The tangent function is *never* decreasing on any
interval for which it is defined.

**69** (a) $-0.8$   (b) $-0.9$   (c) $0.5, 2.6$

**71** (a) $-0.7$   (b) $0.4$   (c) $2.2, 4.1$

**73** $\pm 0.72, \pm 1.62, \pm 2.61,$    **75** $(\pm 2.03, 1.82);$
$\pm 2.98$               $(\pm 4.91, -4.81)$

$[-3.14, 3.14]$ by $[-2.09, 2.09]$     $[-6.28, 6.28]$ by $[-5.19, 3.19]$

**77** $0$    **79** $1$    **81** $1$

## EXERCISES 5.4

**1** (a) $60°$   (b) $20°$   (c) $22°$   (d) $60°$

**3** (a) $\dfrac{\pi}{4}$   (b) $\dfrac{\pi}{3}$   (c) $\dfrac{\pi}{6}$   (d) $\dfrac{\pi}{4}$

**5** (a) $\pi - 3 \approx 8.1°$   (b) $\pi - 2 \approx 65.4°$
(c) $2\pi - 5.5 \approx 44.9°$   (d) $32\pi - 100 \approx 30.4°$

**7** (a) $\dfrac{\sqrt{3}}{2}$   (b) $\dfrac{\sqrt{2}}{2}$    **9** (a) $-\dfrac{\sqrt{3}}{2}$   (b) $\dfrac{1}{2}$

**11** (a) $-\dfrac{\sqrt{3}}{3}$   (b) $-\sqrt{3}$    **13** (a) $-\dfrac{\sqrt{3}}{3}$   (b) $\sqrt{3}$

**15 (a)** $-2$  **(b)** $\dfrac{2}{\sqrt{3}}$   **17 (a)** $-\dfrac{2}{\sqrt{3}}$   **(b)** 2

**19 (a)** 0.958  **(b)** 0.778  **21 (a)** 0.387  **(b)** 0.472
**23 (a)** 2.650  **(b)** 3.179  **25 (a)** 30.46°  **(b)** 30°27′
**27 (a)** 74.88°  **(b)** 74°53′
**29 (a)** 24.94°  **(b)** 24°57′
**31 (a)** 76.38°  **(b)** 76°23′
**33 (a)** 0.9899  **(b)** $-0.1097$  **(c)** $-0.1425$
  **(d)** 0.7907  **(e)** $-11.2493$  **(f)** 1.3677
**35 (a)** 214.3°, 325.7°  **(b)** 41.5°, 318.5°
  **(c)** 70.3°, 250.3°  **(d)** 133.8°, 313.8°
  **(e)** 153.6°, 206.4°  **(f)** 42.3°, 137.7°
**37 (a)** 0.43, 2.71  **(b)** 1.69, 4.59  **(c)** 1.87, 5.01
  **(d)** 0.36, 3.50  **(e)** 0.96, 5.32  **(f)** 3.35, 6.07
**39** 0.28 cm
**41 (a)** The maximum occurs when the sun is rising in
    the east.
  **(b)** $\dfrac{\sqrt{2}}{4} \approx 35\%$
**43** $(9, 9\sqrt{3})$
**45 (a)** $x \approx -0.4161$, $y \approx 0.9093$
  **(b)** $x \approx -0.8838$, $y \approx -0.4678$

**(e)** 2, $8\pi$

**(f)** $\dfrac{1}{2}$, $\dfrac{\pi}{2}$

**(g)** 4, $2\pi$

**(h)** 1, $\dfrac{\pi}{2}$

## EXERCISES 5.5

**1 (a)** 4, $2\pi$

**(b)** 1, $\dfrac{\pi}{2}$

**3 (a)** 3, $2\pi$

**(b)** 1, $\dfrac{2\pi}{3}$

**(c)** $\dfrac{1}{4}$, $2\pi$

**(d)** 1, $8\pi$

**(c)** $\dfrac{1}{3}$, $2\pi$

**(d)** 1, $6\pi$

**(e)** $2, 6\pi$

**(f)** $\dfrac{1}{2}, \dfrac{2\pi}{3}$

**13** $1, \pi, \dfrac{\pi}{2}$

**15** $1, \dfrac{2\pi}{3}, -\dfrac{\pi}{3}$

**(g)** $3, 2\pi$

**(h)** $1, \dfrac{2\pi}{3}$

**17** $2, \dfrac{2\pi}{3}, \dfrac{\pi}{3}$

**19** $1, 4\pi, \dfrac{2\pi}{3}$

**5** $1, 2\pi, \dfrac{\pi}{2}$

**7** $3, 2\pi, -\dfrac{\pi}{6}$

**21** $6, 2, 0$

**23** $2, 4, 0$

**9** $1, 2\pi, -\dfrac{\pi}{2}$

**11** $4, 2\pi, \dfrac{\pi}{4}$

**25** $\dfrac{1}{2}, 1, 0$

**27** $5, \dfrac{2\pi}{3}, \dfrac{\pi}{6}$

**29** $3, 4\pi, \dfrac{\pi}{2}$

**31** $5, 6\pi, -\dfrac{\pi}{2}$

**33** $3, 2, -4$

**35** $\sqrt{2}, 4, \dfrac{1}{2}$

**37** $2, \pi, \dfrac{\pi}{2}$

**39** $5, \pi, -\pi$

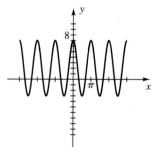

**41** **(a)** $4, 2\pi, -\pi$   **(b)** $y = 4 \sin (x + \pi)$

**43** **(a)** $2, 4, -3$   **(b)** $y = 2 \sin \left(\dfrac{\pi}{2}x + \dfrac{3\pi}{2}\right)$

**45** $4\pi$   **47** $a = 8, b = 4\pi$

**49**

**51**

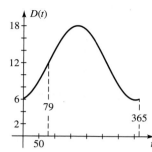

**53** **(a)** $f(t) = 10 \sin \left[\dfrac{\pi}{12}(t - 10)\right] + 0$ with $a = 10$,

$b = \dfrac{\pi}{12}, c = -\dfrac{5\pi}{6}, d = 0$

**(b)**

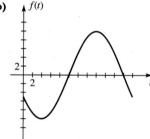

**55** **(a)** $f(t) = 10 \sin \left[\dfrac{\pi}{12}(t - 9)\right] + 20$ with $a = 10$,

$b = \dfrac{\pi}{12}, c = -\dfrac{3\pi}{4}, d = 20$

**(b)**

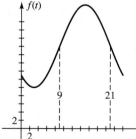

**57** As $x \to 0^-$ or as $x \to 0^+$, $y$ oscillates between $-1$ and $1$ and does not approach a unique value.

$[-2, 2]$ by $[-1.33, 1.33]$

**59** As $x \to 0^-$ or as $x \to 0^+$, $y$ appears to approach 2.

$[-2, 2]$ by $[-0.33, 2.33]$

**61** $y = 4$

$[-20, 20]$ by $[-1, 5]$

**63** $[-\pi, -1.63] \cup$
$[-0.45, 0.61] \cup$
$[1.49, 2.42]$

$[-3.14, 3.14]$ by $[-2.09, 2.09]$

*EXERCISES 5.6*

**1** $\pi$

**3** $\pi$

**5** $2\pi$

**7** $2\pi$

**9** $\pi$

**11** $\dfrac{\pi}{2}$

**13** $4\pi$

**15** $\dfrac{\pi}{2}$

**17** $2\pi$

**19** $\pi$

**21** $\dfrac{\pi}{2}$

**23** $3\pi$

**25** $\dfrac{\pi}{2}$

**27** $2\pi$

**41** $\pi$

**43** $6\pi$

**29** $2\pi$

**31** $\pi$

**45** $\pi$

**47** $4\pi$

**33** $6\pi$

**35** $\pi$

**49** 2

**51** 1

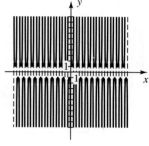

**53** $y = -\cot\left(x + \dfrac{\pi}{2}\right)$

**37** $4\pi$

**39** $2\pi$

**55**

**57**

**59**

**61**

**63**

**65**

**67** $e^{-x/4}$

$[-6.28, 6.28]$ by $[-4.19, 4.19]$

**69** $(-2.76, 3.09)$;
$(1.23, -3.68)$

$[-3.14, 3.14]$ by $[-4, 4]$

**71** $[-0.70, 0.12]$

$[-2, 2]$ by $[-1.33, 1.33]$

**73** $[-\pi, -1.31] \cup$
$[0.11, 0.95] \cup [2.39, \pi]$

$[-3.14, 3.14]$ by $[-2.09, 2.09]$

**75** **(a)** $A_0 e^{-\alpha z}$    **(b)** $\dfrac{\alpha}{k} z_0$    **(c)** $\dfrac{\ln 2}{\alpha}$

---

**EXERCISES 5.7**

**1** $\beta = 60°$, $a = \dfrac{20}{3}\sqrt{3}$, $c = \dfrac{40}{3}\sqrt{3}$

**3** $\alpha = 45°$, $a = b = 15\sqrt{2}$
**5** $\alpha = \beta = 45°$, $c = 5\sqrt{2}$
**7** $\alpha = 60°$, $\beta = 30°$, $a = 15$
**9** $\beta = 53°$, $a \approx 18$, $c \approx 30$
**11** $\alpha = 18°9'$, $a \approx 78.7$, $c \approx 252.6$
**13** $\alpha \approx 29°$, $\beta \approx 61°$, $c \approx 51$
**15** $\alpha \approx 69°$, $\beta \approx 21°$, $a \approx 5.4$    **17** $b = c \cos \alpha$
**19** $a = b \cot \beta$    **21** $c = a \csc \alpha$
**23** $b = \sqrt{c^2 - a^2}$
**25** $250\sqrt{3} + 4 \approx 437$ ft    **27** 28,800 ft    **29** 160 m
**31** 9659 ft    **33** **(a)** 58 ft    **(b)** 27 ft    **35** 51°20'
**37** 16.3°    **39** 1,459,379 ft$^2$    **41** 21.8°    **43** 20.2 m
**45** 29.7 km    **47** 3944 mi    **49** 126 mi/hr
**51** **(a)** 45%
     **(b)** Each satellite has a signal range of more than
     120°.

**53** $h = d \sin \alpha + c$    **55** $h = \dfrac{d}{\cot \alpha - \cot \beta}$

**57** $h = d(\tan \beta - \tan \alpha)$
**59** N70°E; N40°W; S15°W; S25°E
**61** **(a)** 55 mi    **(b)** S63°E    **63** 324 mi

---

**EXERCISES 5.8**

**1** **(a)** $\omega = 200\pi$ radians/min
     **(b)** In cm: $x = 20 \cos 200\pi t$, $y = 20 \sin 200\pi t$

**3** Amplitude, 10 cm; period, $\dfrac{1}{3}$ sec; frequency,
3 osc/sec. The point is at the origin at $t = 0$. It
moves upward with decreasing speed, reaching the
point with coordinate 10 at $t = \dfrac{1}{12}$. It then reverses
direction and moves downward, gaining speed until it
reaches the origin at $t = \dfrac{1}{6}$. It continues downward
with decreasing speed, reaching the point with
coordinate $-10$ at $t = \dfrac{1}{4}$. It then reverses direction
and moves upward with increasing speed, returning to
the origin at $t = \dfrac{1}{3}$.

**5** Amplitude, 4 cm; period, $\dfrac{4}{3}$ sec; frequency, $\dfrac{3}{4}$ osc/sec.
The motion is similar to that in Exercise 3; however,
the point starts 4 units above the origin and moves

downward, reaching the origin at $t = \frac{1}{3}$ and the point with coordinate $-4$ at $t = \frac{2}{3}$. It then reverses direction and moves upward, reaching the origin at $t = 1$ and its initial point at $t = \frac{4}{3}$.

**7** $d = 5 \cos \frac{2\pi}{3} t$

**9** Current lags emf by $\frac{1}{1440}$ sec.

**11 (a)** $\frac{\pi}{15}$ radian/sec

   **(b)** $h(t) = 60 - 50 \cos \frac{\pi}{15} t = 60 + 50 \sin \left( \frac{\pi}{15} t - \frac{\pi}{2} \right)$

**13 (a)** 8 ft; 12 min    **(b)** 105 km/hr

**15** $y = (4.5) \sin \left[ \frac{\pi}{6} (t - 11) \right] + 7.5$

## CHAPTER 5 REVIEW EXERCISES

**1** $\frac{11\pi}{6}, \frac{9\pi}{4}, -\frac{5\pi}{6}, \frac{4\pi}{3}, \frac{\pi}{5}$

**2** $810°, -120°, 315°, 900°, 36°$

**3 (a)** 0.1    **(b)** 0.2 $m^2$

**4 (a)** $\frac{35\pi}{12}$ cm    **(b)** $\frac{175\pi}{16}$ $cm^2$

**5** $x = 6\sqrt{3};\ y = 3\sqrt{3}$    **6** $x = \frac{7}{2}\sqrt{2};\ y = \frac{7}{2}\sqrt{2}$

**7** $\tan \theta = \sqrt{\sec^2 \theta - 1}$    **8** $\cot \theta = \sqrt{\csc^2 \theta - 1}$

Exer. 9–18: Typical verifications are given.

**9** $\sin \theta\ (\csc \theta - \sin \theta) = \sin \theta \csc \theta - \sin^2 \theta$
$$= 1 - \sin^2 \theta = \cos^2 \theta$$

**10** $\cos \theta\ (\tan \theta + \cot \theta) = \cos \theta \cdot \dfrac{\sin \theta}{\cos \theta} + \cos \theta \cdot \dfrac{\cos \theta}{\sin \theta}$
$$= \sin \theta + \frac{\cos^2 \theta}{\sin \theta} = \frac{\sin^2 \theta + \cos^2 \theta}{\sin \theta}$$
$$= \frac{1}{\sin \theta} = \csc \theta$$

**11** $(\cos^2 \theta - 1)(\tan^2 \theta + 1) = (\cos^2 \theta - 1)(\sec^2 \theta)$
$$= \cos^2 \theta \sec^2 \theta - \sec^2 \theta$$
$$= 1 - \sec^2 \theta$$

**12** $\dfrac{\sec \theta - \cos \theta}{\tan \theta} = \dfrac{\dfrac{1}{\cos \theta} - \cos \theta}{\dfrac{\sin \theta}{\cos \theta}} = \dfrac{\dfrac{1 - \cos^2 \theta}{\cos \theta}}{\dfrac{\sin \theta}{\cos \theta}} = \dfrac{\dfrac{\sin^2 \theta}{\cos \theta}}{\dfrac{\sin \theta}{\cos \theta}}$

$$= \dfrac{\dfrac{\sin \theta}{\cos \theta}}{\dfrac{1}{\cos \theta}} = \frac{\tan \theta}{\sec \theta}$$

**13** $\dfrac{1 + \tan^2 \theta}{\tan^2 \theta} = \dfrac{1}{\tan^2 \theta} + \dfrac{\tan^2 \theta}{\tan^2 \theta} = \cot^2 \theta + 1 = \csc^2 \theta$

**14** $\dfrac{\sec \theta + \csc \theta}{\sec \theta - \csc \theta} = \dfrac{\dfrac{1}{\cos \theta} + \dfrac{1}{\sin \theta}}{\dfrac{1}{\cos \theta} - \dfrac{1}{\sin \theta}} = \dfrac{\dfrac{\sin \theta + \cos \theta}{\cos \theta \sin \theta}}{\dfrac{\sin \theta - \cos \theta}{\cos \theta \sin \theta}}$

$$= \frac{\sin \theta + \cos \theta}{\sin \theta - \cos \theta}$$

**15** $\dfrac{\cot \theta - 1}{1 - \tan \theta} = \dfrac{\dfrac{\cos \theta}{\sin \theta} - 1}{1 - \dfrac{\sin \theta}{\cos \theta}} = \dfrac{\dfrac{\cos \theta - \sin \theta}{\sin \theta}}{\dfrac{\cos \theta - \sin \theta}{\cos \theta}}$

$$= \frac{(\cos \theta - \sin \theta)\cos \theta}{(\cos \theta - \sin \theta)\sin \theta} = \frac{\cos \theta}{\sin \theta} = \cot \theta$$

**16** $\dfrac{1 + \sec \theta}{\tan \theta + \sin \theta} = \dfrac{1 + \dfrac{1}{\cos \theta}}{\dfrac{\sin \theta}{\cos \theta} + \sin \theta} = \dfrac{\dfrac{\cos \theta + 1}{\cos \theta}}{\dfrac{\sin \theta + \sin \theta \cos \theta}{\cos \theta}} = \dfrac{\dfrac{\cos \theta + 1}{\cos \theta}}{\dfrac{\sin \theta(1 + \cos \theta)}{\cos \theta}}$

$$= \frac{1}{\sin \theta} = \csc \theta$$

**17** $\dfrac{\tan(-\theta) + \cot(-\theta)}{\tan \theta} = \dfrac{-\tan \theta - \cot \theta}{\tan \theta} = -\dfrac{\tan \theta}{\tan \theta} - \dfrac{\cot \theta}{\tan \theta}$
$$= -1 - \cot^2 \theta = -(1 + \cot^2 \theta)$$
$$= -\csc^2 \theta$$

**18** $-\dfrac{1}{\csc(-\theta)} - \dfrac{\cot(-\theta)}{\sec(-\theta)} = -\dfrac{1}{-\csc \theta} - \dfrac{-\cot \theta}{\sec \theta}$
$$= \sin \theta + \frac{\cos \theta/\sin \theta}{1/\cos \theta}$$
$$= \sin \theta + \frac{\cos^2 \theta}{\sin \theta} = \frac{\sin^2 \theta + \cos^2 \theta}{\sin \theta}$$
$$= \frac{1}{\sin \theta} = \csc \theta$$

**19** $\dfrac{\sqrt{33}}{7}, \dfrac{4}{7}, \dfrac{\sqrt{33}}{4}, \dfrac{4}{\sqrt{33}}, \dfrac{7}{4}, \dfrac{7}{\sqrt{33}}$

**20 (a)** $-\dfrac{4}{5}, \dfrac{3}{5}, -\dfrac{4}{3}, -\dfrac{3}{4}, \dfrac{5}{3}, -\dfrac{5}{4}$

**(b)** $\dfrac{2}{\sqrt{13}}, -\dfrac{3}{\sqrt{13}}, -\dfrac{2}{3}, -\dfrac{3}{2}, -\dfrac{\sqrt{13}}{3}, \dfrac{\sqrt{13}}{2}$

**(c)** $-1, 0, U, 0, U, -1$

**21 (a)** II   **(b)** III   **(c)** IV

**22 (a)** $-\dfrac{4}{5}, \dfrac{3}{5}, -\dfrac{4}{3}, -\dfrac{3}{4}, \dfrac{5}{3}, -\dfrac{5}{4}$

**(b)** $\dfrac{2}{\sqrt{13}}, -\dfrac{3}{\sqrt{13}}, -\dfrac{2}{3}, -\dfrac{3}{2}, -\dfrac{\sqrt{13}}{3}, \dfrac{\sqrt{13}}{2}$

**23** $(-1, 0); (0, -1); (0, 1); \left(-\dfrac{\sqrt{2}}{2}, -\dfrac{\sqrt{2}}{2}\right); (1, 0);$
$\left(\dfrac{\sqrt{3}}{2}, \dfrac{1}{2}\right)$

**24** $\left(\dfrac{3}{5}, \dfrac{4}{5}\right); \left(\dfrac{3}{5}, \dfrac{4}{5}\right); \left(-\dfrac{3}{4}, \dfrac{4}{5}\right); \left(-\dfrac{3}{5}, \dfrac{4}{5}\right)$

**25 (a)** $\dfrac{\pi}{4}, \dfrac{\pi}{6}, \dfrac{\pi}{8}$   **(b)** $65°, 43°, 8°$

**26 (a)** $1, 0, U, 0, U, 1$

**(b)** $\dfrac{\sqrt{2}}{2}, -\dfrac{\sqrt{2}}{2}, -1, -1, -\sqrt{2}, \sqrt{2}$

**(c)** $0, 1, 0, U, 1, U$

**(d)** $-\dfrac{1}{2}, \dfrac{\sqrt{3}}{2}, -\dfrac{\sqrt{3}}{3}, -\sqrt{3}, \dfrac{2}{\sqrt{3}}, -2$

**27 (a)** $-\dfrac{\sqrt{2}}{2}$   **(b)** $-\dfrac{\sqrt{3}}{3}$   **(c)** $-\dfrac{1}{2}$   **(d)** $-2$

**(e)** $-1$   **(f)** $-\dfrac{2}{\sqrt{3}}$

**28** $310.5°$

**29** $5, 2\pi$          **30** $\dfrac{2}{3}, 2\pi$

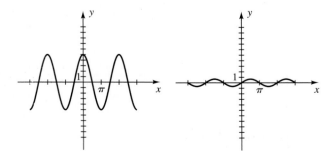

**31** $\dfrac{1}{3}, \dfrac{2\pi}{3}$          **32** $\dfrac{1}{2}, 6\pi$

**33** $3, 4\pi$          **34** $4, \pi$

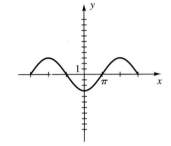

**35** $2, 2$          **36** $4, 4$

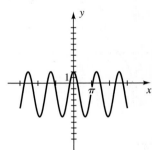

**37 (a)** $1.43, 2$   **(b)** $y = 1.43 \sin \pi x$

**38 (a)** $3.27, 3\pi$   **(b)** $y = -3.27 \sin \dfrac{2}{3}x$

**39 (a)** $3, \dfrac{4\pi}{3}$   **(b)** $y = -3 \cos \dfrac{3}{2}x$

**40 (a)** $2, \dfrac{\pi}{2}$   **(b)** $y = 2 \cos \dfrac{\pi}{2}x$

**41**

**42**

**49**

**50**

**43**

**44**

**51**

**52**

**53** $\alpha = 30°$, $a \approx 23$, $c \approx 46$

**54** $\beta = 35°20'$, $a \approx 310$, $c \approx 380$

**55** $\alpha \approx 68°$, $\beta \approx 22°$, $c \approx 67$

**56** $\alpha \approx 13°$, $\beta \approx 77°$, $b = 40$

**57 (a)** $\dfrac{109\pi}{6}$   **(b)** 440.2   **58** 1048 ft   **59** 52°

**60** Approximately 67,900,000 mi

**61** $\dfrac{6\pi}{5}$ radians $= 216°$   **62** 250 ft

**63 (a)** 231.0 ft   **(b)** 434.5   **64 (b)** 2 mi

**65 (a)** $T = h + d(\cos\alpha\tan\theta - \sin\alpha)$   **(b)** 22.54 ft

**66 (a)** $\dfrac{25}{3}\sqrt{3} \approx 14.43$ ft-candles   **(b)** 37.47°

**67 (b)** 4.69   **68 (a)** 74.05 in.   **(b)** 24.75 in.

**69 (a)** $S = 4a^2 \sin\theta$   **(b)** $V = \dfrac{4}{3}a^3 \sin^2\theta\cos\theta$

**70 (a)** $h = R\sec\dfrac{s}{R} - R$   **(b)** $h \approx 1650$ ft

**71**

**45**

**46**

**47**

**48**

**72** $y = 98.6 + (0.3) \sin\left(\dfrac{\pi}{12}t - \dfrac{11\pi}{12}\right)$

**73 (a)** 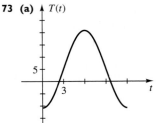　　**(b)** 20.8°C on July 1

**74 (a)** 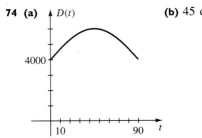　　**(b)** 45 days into summer

**75 (a)** The cork is in simple harmonic motion.
**(b)** $1 \le t \le 2$
**76 (a)** $P = (\cos \omega t, \sin \omega t)$　　**(b)** $x = 2 + \cos \omega t$
**(c)** $Q$ is in simple harmonic motion on the interval $[1, 3]$.

# CHAPTER 6

## EXERCISES 6.1

Exer. 1–50: Typical verifications are given for Exercises 1, 5, 9, . . . , 49.

**1** $\csc \theta - \sin \theta = \dfrac{1}{\sin \theta} - \sin \theta = \dfrac{1 - \sin^2 \theta}{\sin \theta} = \dfrac{\cos^2 \theta}{\sin \theta}$
$\qquad = \dfrac{\cos \theta}{\sin \theta} \cos \theta = \cot \theta \cos \theta$

**5** $\dfrac{\csc^2 \theta}{1 + \tan^2 \theta} = \dfrac{\csc^2 \theta}{\sec^2 \theta} = \dfrac{1/\sin^2 \theta}{1/\cos^2 \theta} = \dfrac{\cos^2 \theta}{\sin^2 \theta}$
$\qquad = \left(\dfrac{\cos \theta}{\sin \theta}\right)^2 = \cot^2 \theta$

**9** $\dfrac{1}{1 - \cos \gamma} + \dfrac{1}{1 + \cos \gamma} = \dfrac{1 + \cos \gamma + 1 - \cos \gamma}{1 - \cos^2 \gamma} = \dfrac{2}{\sin^2 \gamma}$
$\qquad = 2 \csc^2 \gamma$
**13** $\csc^4 t - \cot^4 t = (\csc^2 t + \cot^2 t)(\csc^2 t - \cot^2 t)$
$\qquad = (\csc^2 t + \cot^2 t)(1) = \csc^2 t + \cot^2 t$

**17** $\dfrac{\tan^2 x}{\sec x + 1} = \dfrac{\sec^2 x - 1}{\sec x + 1} = \dfrac{(\sec x + 1)(\sec x - 1)}{\sec x + 1}$
$\qquad = \sec x - 1 = \dfrac{1}{\cos x} - 1 = \dfrac{1 - \cos x}{\cos x}$
**21** $\sin^4 r - \cos^4 r = (\sin^2 r - \cos^2 r)(\sin^2 r + \cos^2 r)$
$\qquad = (\sin^2 r - \cos^2 r)(1) = \sin^2 r - \cos^2 r$
**25** $(\sec t + \tan t)^2 = \left(\dfrac{1}{\cos t} + \dfrac{\sin t}{\cos t}\right)^2 = \left(\dfrac{1 + \sin t}{\cos t}\right)^2$
$\qquad = \dfrac{(1 + \sin t)^2}{\cos^2 t} = \dfrac{(1 + \sin t)^2}{1 - \sin^2 t}$
$\qquad = \dfrac{(1 + \sin t)^2}{(1 + \sin t)(1 - \sin t)} = \dfrac{1 + \sin t}{1 - \sin t}$
**29** $\dfrac{1 + \csc \beta}{\cot \beta + \cos \beta} = \dfrac{1 + \dfrac{1}{\sin \beta}}{\dfrac{\cos \beta}{\sin \beta} + \cos \beta} = \dfrac{\dfrac{\sin \beta + 1}{\sin \beta}}{\dfrac{\cos \beta + \cos \beta \sin \beta}{\sin \beta}}$
$\qquad = \dfrac{\sin \beta + 1}{\cos \beta(1 + \sin \beta)} = \dfrac{1}{\cos \beta} = \sec \beta$
**33** $\text{RS} = \dfrac{\tan \alpha + \tan \beta}{1 - \tan \alpha \tan \beta} = \dfrac{\dfrac{\sin \alpha}{\cos \alpha} + \dfrac{\sin \beta}{\cos \beta}}{1 - \dfrac{\sin \alpha}{\cos \alpha} \cdot \dfrac{\sin \beta}{\cos \beta}}$
$\qquad = \dfrac{\dfrac{\sin \alpha \cos \beta + \cos \alpha \sin \beta}{\cos \alpha \cos \beta}}{\dfrac{\cos \alpha \cos \beta - \sin \alpha \sin \beta}{\cos \alpha \cos \beta}} = \dfrac{\sin \alpha \cos \beta + \cos \alpha \sin \beta}{\cos \alpha \cos \beta - \sin \alpha \sin \beta}$
$\qquad = \text{LS}$
**37** $\dfrac{1}{\tan \beta + \cot \beta} = \dfrac{1}{\dfrac{\sin \beta}{\cos \beta} + \dfrac{\cos \beta}{\sin \beta}} = \dfrac{1}{\dfrac{\sin^2 \beta + \cos^2 \beta}{\cos \beta \sin \beta}}$
$\qquad = \sin \beta \cos \beta$
**41** $\text{RS} = \sec^4 \phi - 4 \tan^2 \phi = (\sec^2 \phi)^2 - 4 \tan^2 \phi$
$\qquad = (1 + \tan^2 \phi)^2 - 4 \tan^2 \phi$
$\qquad = 1 + 2 \tan^2 \phi + \tan^4 \phi - 4 \tan^2 \phi$
$\qquad = 1 - 2 \tan^2 \phi + \tan^4 \phi$
$\qquad = (1 - \tan^2 \phi)^2 = \text{LS}$
**45** $\log 10^{\tan t} = \log_{10} 10^{\tan t} = \tan t$, since $\log_a a^x = x$.
**49** $\ln |\sec \theta + \tan \theta| = \ln \left|\dfrac{(\sec \theta + \tan \theta)(\sec \theta - \tan \theta)}{\sec \theta - \tan \theta}\right|$
$\qquad = \ln \left|\dfrac{\sec^2 \theta - \tan^2 \theta}{\sec \theta - \tan \theta}\right| = \ln \left|\dfrac{1}{\sec \theta - \tan \theta}\right|$
$\qquad = \ln |1| - \ln |\sec \theta - \tan \theta|$
$\qquad = -\ln |\sec \theta - \tan \theta|$

Exer. 51–62: A typical value of $t$ or $\theta$ and the resulting nonequality are given.

**51** $\pi,\ -1 \ne 1$　　**53** $\dfrac{3\pi}{2},\ 1 \ne -1$　　**55** $\dfrac{\pi}{4},\ 2 \ne 1$

**57** $\pi, -1 \neq 1$   **59** $\dfrac{\pi}{4}, \cos \sqrt{2} \neq 1$   **61** $\pi, -5 \neq 0$

**63** $a^3 \cos^3 \theta$   **65** $a \tan \theta \sin \theta$   **67** $a \sec \theta$

**69** $\dfrac{1}{a^2} \cos^2 \theta$   **71** $a \tan \theta$   **73** $a^4 \sec^3 \theta \tan \theta$

**75** The graph of $f$ appears to be that of $y = g(x) = -1$.

$$\dfrac{\sin^2 x - \sin^4 x}{(1 - \sec^2 x)\cos^4 x} = \dfrac{\sin^2 x(1 - \sin^2 x)}{-\tan^2 x \cos^4 x}$$

$$= \dfrac{\sin^2 x \cos^2 x}{-(\sin^2 x/\cos^2 x)\cos^4 x} = \dfrac{\sin^2 x \cos^2 x}{-\sin^2 x \cos^2 x}$$

$$= -1$$

**77** The graph of $f$ appears to be that of $y = g(x) = \cos x$.

$\sec x(\sin x \cos x + \cos^2 x) - \sin x$

$\qquad = \sec x \cos x(\sin x + \cos x) - \sin x$

$\qquad = (\sin x + \cos x) - \sin x = \cos x$

## EXERCISES 6.2

Exer. 1–34: $n$ denotes any integer.

**1** $\dfrac{5\pi}{4} + 2\pi n, \dfrac{7\pi}{4} + 2\pi n$   **3** $\dfrac{\pi}{3} + \pi n$

**5** $\dfrac{\pi}{3} + 2\pi n, \dfrac{5\pi}{3} + 2\pi n$   **7** No solution, since $\dfrac{\pi}{2} > 1$.

**9** All $\theta$ except $\theta = \dfrac{\pi}{2} + \pi n$

**11** $\dfrac{\pi}{12} + \pi n, \dfrac{11\pi}{12} + \pi n$   **13** $\dfrac{\pi}{2} + 3\pi n$

**15** $-\dfrac{\pi}{12} + 2\pi n, \dfrac{7\pi}{12} + 2\pi n$

**17** $\dfrac{\pi}{4} + \pi n, \dfrac{7\pi}{12} + \pi n$   **19** $\dfrac{2\pi}{3} + 2\pi n, \dfrac{4\pi}{3} + 2\pi n$

**21** $\dfrac{\pi}{4} + \dfrac{\pi}{2}n$   **23** $2\pi n, \dfrac{3\pi}{2} + 2\pi n$

**25** $\dfrac{\pi}{3} + \pi n, \dfrac{2\pi}{3} + \pi n$   **27** $\dfrac{4\pi}{3} + 2\pi n, \dfrac{5\pi}{3} + 2\pi n$

**29** $\dfrac{\pi}{6} + \pi n, \dfrac{5\pi}{6} + \pi n$   **31** $\dfrac{7\pi}{6} + 2\pi n, \dfrac{11\pi}{6} + 2\pi n$

**33** $\dfrac{\pi}{12} + \pi n, \dfrac{5\pi}{12} + \pi n$   **35** $e^{(\pi/2)+\pi n}$

**37** $\dfrac{3\pi}{8}, \dfrac{7\pi}{8}, \dfrac{11\pi}{8}, \dfrac{15\pi}{8}$   **39** $\dfrac{\pi}{3}, \dfrac{2\pi}{3}, \dfrac{4\pi}{3}, \dfrac{5\pi}{3}$

**41** $\dfrac{\pi}{6}, \dfrac{5\pi}{6}, \dfrac{3\pi}{2}$   **43** $0, \pi, \dfrac{\pi}{4}, \dfrac{3\pi}{4}, \dfrac{5\pi}{4}, \dfrac{7\pi}{4}$

**45** $\dfrac{\pi}{2}, \dfrac{3\pi}{2}, \dfrac{2\pi}{3}, \dfrac{4\pi}{3}$   **47** No solution   **49** $\dfrac{11\pi}{6}, \dfrac{\pi}{2}$

**51** $0, \dfrac{\pi}{2}$   **53** $\dfrac{\pi}{4}, \dfrac{5\pi}{4}$

**55** All $\alpha$ in $[0, 2\pi)$ except $0, \dfrac{\pi}{2}, \pi$, and $\dfrac{3\pi}{2}$

**57** $\dfrac{\pi}{2}, \dfrac{3\pi}{2}, \dfrac{7\pi}{6}, \dfrac{11\pi}{6}$   **59** $\dfrac{3\pi}{4}, \dfrac{7\pi}{4}$

**61** $15°30', 164°30'$   **63** $135°, 315°, 116°30', 296°30'$

**65** $41°50', 138°10', 194°30', 345°30'$   **67** 10

**69** $t \approx 3.50$ and $t \approx 8.50$   **71** (a) 3.29   (b) 4

**73** (a)

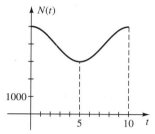

(b) $0 \le t < \dfrac{5}{3}$ and $\dfrac{25}{3} < t \le 10$

**75** $A\left(-\dfrac{4\pi}{3}, -\dfrac{2\pi}{3} + \dfrac{1}{2}\sqrt{3}\right), B\left(-\dfrac{2\pi}{3}, -\dfrac{\pi}{3} - \dfrac{1}{2}\sqrt{3}\right),$

$\qquad C\left(\dfrac{2\pi}{3}, \dfrac{\pi}{3} + \dfrac{1}{2}\sqrt{3}\right), D\left(\dfrac{4\pi}{3}, \dfrac{2\pi}{3} - \dfrac{1}{2}\sqrt{3}\right)$

**77** $\dfrac{7}{360}$   **79** $[0, 1.27] \cup [5.02, 2\pi]$

**81** $(0.39, 1.96) \cup (2.36, 3.53) \cup (5.11, 5.50)$

**83** $-1.48, 1.08$   **85** $\pm 1.00$

  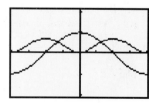

$[-3.14, 3.14]$ by $[-2.09, 2.09]$   $[-3.14, 3.14]$ by $[-2.09, 2.09]$

**87** $\pm 0.64, \pm 2.42$   **89** (a) $37.6°$   (b) $52.5°$

$[-3.14, 3.14]$ by $[-2.09, 2.09]$

## EXERCISES 6.3

**1** (a) $\cos 43°23'$   (b) $\sin 16°48'$   (c) $\cot \dfrac{\pi}{3}$

(d) $\csc 72.72°$

**3 (a)** $\sin\dfrac{3\pi}{20}$    **(b)** $\cos\left(\dfrac{2\pi-1}{4}\right)$    **(c)** $\cot\left(\dfrac{\pi-2}{2}\right)$

**(d)** $\sec\left(\dfrac{\pi}{2}-0.53\right)$

**5 (a)** $\dfrac{\sqrt{2}+\sqrt{3}}{2}$    **(b)** $\dfrac{\sqrt{6}-\sqrt{2}}{4}$

**7 (a)** $\sqrt{3}+1$    **(b)** $-2-\sqrt{3}$

**9 (a)** $\dfrac{\sqrt{2}-1}{2}$    **(b)** $\dfrac{\sqrt{6}+\sqrt{2}}{4}$

**11** $\cos 25°$    **13** $\sin(-5°)$    **15** $\sin(-5)$

**17 (a)** $\dfrac{77}{85}$    **(b)** $\dfrac{36}{85}$    **(c)** I

**19 (a)** $-\dfrac{24}{25}$    **(b)** $-\dfrac{24}{7}$    **(c)** IV

**21 (a)** $\dfrac{3\sqrt{21}-8}{25}\approx 0.23$    **(b)** $\dfrac{4\sqrt{21}+6}{25}\approx 0.97$    **(c)** I

**23** $\sin(\theta+\pi)=\sin\theta\cos\pi+\cos\theta\sin\pi$
$=\sin\theta(-1)+\cos\theta(0)=-\sin\theta$

**25** $\sin\left(x-\dfrac{5\pi}{2}\right)=\sin x\cos\dfrac{5\pi}{2}-\cos x\sin\dfrac{5\pi}{2}=-\cos x$

**27** $\cos(\theta-\pi)=\cos\theta\cos\pi+\sin\theta\sin\pi=-\cos\theta$

**29** $\cos\left(x+\dfrac{3\pi}{2}\right)=\cos x\cos\dfrac{3\pi}{2}-\sin x\sin\dfrac{3\pi}{2}=\sin x$

**31** $\tan\left(x-\dfrac{\pi}{2}\right)=\dfrac{\sin\left(x-\dfrac{\pi}{2}\right)}{\cos\left(x-\dfrac{\pi}{2}\right)}=\dfrac{\sin x\cos\dfrac{\pi}{2}-\cos x\sin\dfrac{\pi}{2}}{\cos x\cos\dfrac{\pi}{2}+\sin x\sin\dfrac{\pi}{2}}$

$=\dfrac{-\cos x}{\sin x}=-\cot x$

**33** $\tan\left(\theta+\dfrac{\pi}{2}\right)=\cot\left[\dfrac{\pi}{2}-\left(\theta+\dfrac{\pi}{2}\right)\right]$
$=\cot(-\theta)=-\cot\theta$

**35** $\sin\left(\theta+\dfrac{\pi}{4}\right)=\sin\theta\cos\dfrac{\pi}{4}+\cos\theta\sin\dfrac{\pi}{4}$

$=\dfrac{\sqrt{2}}{2}\sin\theta+\dfrac{\sqrt{2}}{2}\cos\theta$

$=\dfrac{\sqrt{2}}{2}(\sin\theta+\cos\theta)$

**37** $\tan\left(u+\dfrac{\pi}{4}\right)=\dfrac{\tan u+\tan\dfrac{\pi}{4}}{1-\tan u\tan\dfrac{\pi}{4}}=\dfrac{1+\tan u}{1-\tan u}$

**39** $\cos(u+v)+\cos(u-v)$
$=(\cos u\cos v-\sin u\sin v)+(\cos u\cos v+\sin u\sin v)$
$=2\cos u\cos v$

**41** $\sin(u+v)\cdot\sin(u-v)$
$=(\sin u\cos v+\cos u\sin v)\cdot(\sin u\cos v-\cos u\sin v)$
$=\sin^2 u\cos^2 v-\cos^2 u\sin^2 v$
$=\sin^2 u(1-\sin^2 v)-(1-\sin^2 u)\sin^2 v$
$=\sin^2 u-\sin^2 u\sin^2 v-\sin^2 v+\sin^2 u\sin^2 v$
$=\sin^2 u-\sin^2 v$

**43** $\dfrac{1}{\cot\alpha-\cot\beta}=\dfrac{1}{\dfrac{\cos\alpha}{\sin\alpha}-\dfrac{\cos\beta}{\sin\beta}}=\dfrac{1}{\dfrac{\cos\alpha\sin\beta-\cos\beta\sin\alpha}{\sin\alpha\sin\beta}}$

$=\dfrac{\sin\alpha\sin\beta}{\sin(\beta-\alpha)}$

**45** $\sin(u+v+w)=\sin[(u+v)+w]$
$=\sin(u+v)\cos w+$
$\cos(u+v)\sin w$
$=(\sin u\cos v+\cos u\sin v)\cos w+$
$(\cos u\cos v-\sin u\sin v)\sin w$
$=\sin u\cos v\cos w+\cos u\sin v\cos w+$
$\cos u\cos v\sin w-\sin u\sin v\sin w$

**47** $\cot(u+v)=\dfrac{\cos(u+v)}{\sin(u+v)}$

$=\dfrac{(\cos u\cos v-\sin u\sin v)(1/\sin u\sin v)}{(\sin u\cos v+\cos u\sin v)(1/\sin u\sin v)}$

$=\dfrac{\cot u\cot v-1}{\cot v+\cot u}$

**49** $\sin(u-v)=\sin[u+(-v)]$
$=\sin u\cos(-v)+\cos u\sin(-v)$
$=\sin u\cos v-\cos u\sin v$

**51** $\dfrac{f(x+h)-f(x)}{h}=\dfrac{\cos(x+h)-\cos x}{h}$

$=\dfrac{\cos x\cos h-\sin x\sin h-\cos x}{h}$

$=\dfrac{\cos x\cos h-\cos x}{h}-\dfrac{\sin x\sin h}{h}$

$=\cos x\left(\dfrac{\cos h-1}{h}\right)-\sin x\left(\dfrac{\sin h}{h}\right)$

**53** $0,\ \dfrac{\pi}{3},\ \dfrac{2\pi}{3}$    **55** $\dfrac{\pi}{6},\ \dfrac{\pi}{2},\ \dfrac{5\pi}{6}$

**57** $\dfrac{\pi}{12},\ \dfrac{5\pi}{12};\ \dfrac{3\pi}{4}$ is extraneous

**59 (a)** $f(x)=2\cos\left(2x-\dfrac{\pi}{6}\right)$    **(c)**

**(b)** $2,\ \pi,\ \dfrac{\pi}{12}$

**61 (a)** $f(x) = 2\sqrt{2} \cos\left(3x + \dfrac{\pi}{4}\right)$

**(b)** $2\sqrt{2}, \dfrac{2\pi}{3}, -\dfrac{\pi}{12}$

**(c)**

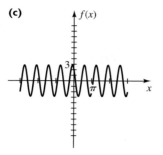

**63** $y = 10\sqrt{41} \cos\left(60\pi t - \tan^{-1}\dfrac{5}{4}\right)$

$\approx 10\sqrt{41} \cos(60\pi t - 0.8961)$

**65 (a)** $y = \sqrt{13} \cos(t - C)$ with $\tan C = \dfrac{3}{2}$; $\sqrt{13}, 2\pi$

**(b)** $t = C + \dfrac{\pi}{2} + \pi n \approx 2.55 + \pi n$ for every non-negative integer $n$

**67 (a)** $p(t) = A \sin \omega t + B \sin(\omega t + \tau)$
$= A \sin \omega t + B(\sin \omega t \cos \tau + \cos \omega t \sin \tau)$
$= (B \sin \tau) \cos \omega t + (A + B \cos \tau) \sin \omega t$
$= a \cos \omega t + b \sin \omega t$
with $a = B \sin \tau$ and $b = A + B \cos \tau$

**(b)** $C^2 = (B \sin \tau)^2 + (A + B \cos \tau)^2$
$= B^2 \sin^2 \tau + A^2 + 2AB \cos \tau + B^2 \cos^2 \tau$
$= A^2 + B^2(\sin^2 \tau + \cos^2 \tau) + 2AB \cos \tau$
$= A^2 + B^2 + 2AB \cos \tau$

**69 (a)** $C^2 = A^2 + B^2 + 2AB \cos \tau \le A^2 + B^2 + 2AB$,
since $\cos \tau \le 1$ and $A > 0$, $B > 0$. Thus,
$C^2 \le (A + B)^2$, and hence $C \le A + B$.
**(b)** $0, 2\pi$     **(c)** $\cos \tau > -B/(2A)$

**71** $(-2.97, -2.69)$,
$(-1.00, -0.37)$,
$(0.17, 0.46)$,
$(2.14, 2.77)$

$[-3.14, 3.14]$ by $[-5, 5]$

## EXERCISES 6.4

**1** $\dfrac{24}{25}, -\dfrac{7}{25}, -\dfrac{24}{7}$     **3** $-\dfrac{4}{9}\sqrt{2}, -\dfrac{7}{9}, \dfrac{4}{7}\sqrt{2}$

**5** $\dfrac{1}{10}\sqrt{10}, \dfrac{3}{10}\sqrt{10}, \dfrac{1}{3}$

**7** $-\dfrac{1}{2}\sqrt{2 + \sqrt{2}}, \dfrac{1}{2}\sqrt{2 - \sqrt{2}}, -\sqrt{2} - 1$

**9 (a)** $\dfrac{1}{2}\sqrt{2 - \sqrt{2}}$     **(b)** $\dfrac{1}{2}\sqrt{2 - \sqrt{3}}$     **(c)** $\sqrt{2} + 1$

**11** $\sin 10\theta = \sin(2 \cdot 5\theta) = 2 \sin 5\theta \cos 5\theta$

**13** $4 \sin\dfrac{x}{2} \cos\dfrac{x}{2} = 2 \cdot 2 \sin\dfrac{x}{2} \cos\dfrac{x}{2} = 2 \sin\left(2 \cdot \dfrac{x}{2}\right)$
$= 2 \sin x$

**15** $(\sin t + \cos t)^2 = \sin^2 t + 2 \sin t \cos t + \cos^2 t$
$= 1 + \sin 2t$

**17** $\sin 3u = \sin(2u + u) = \sin 2u \cos u + \cos 2u \sin u$
$= (2 \sin u \cos u) \cos u + (1 - 2 \sin^2 u) \sin u$
$= 2 \sin u \cos^2 u + \sin u - 2 \sin^3 u$
$= 2 \sin u(1 - \sin^2 u) + \sin u - 2 \sin^3 u$
$= 2 \sin u - 2 \sin^3 u + \sin u - 2 \sin^3 u$
$= 3 \sin u - 4 \sin^3 u = \sin u(3 - 4 \sin^2 u)$

**19** $\cos 4\theta = \cos(2 \cdot 2\theta) = 2 \cos^2 2\theta - 1$
$= 2(2 \cos^2 \theta - 1)^2 - 1$
$= 2(4 \cos^4 \theta - 4 \cos^2 \theta + 1) - 1$
$= 8 \cos^4 \theta - 8 \cos^2 \theta + 1$

**21** $\sin^4 t = (\sin^2 t)^2 = \left(\dfrac{1 - \cos 2t}{2}\right)^2$
$= \dfrac{1}{4}(1 - 2 \cos 2t + \cos^2 2t)$
$= \dfrac{1}{4} - \dfrac{1}{2} \cos 2t + \dfrac{1}{4}\left(\dfrac{1 + \cos 4t}{2}\right)$
$= \dfrac{1}{4} - \dfrac{1}{2} \cos 2t + \dfrac{1}{8} + \dfrac{1}{8} \cos 4t$
$= \dfrac{3}{8} - \dfrac{1}{2} \cos 2t + \dfrac{1}{8} \cos 4t$

**23** $\sec 2\theta = \dfrac{1}{\cos 2\theta} = \dfrac{1}{2 \cos^2 \theta - 1} = \dfrac{1}{2\left(\dfrac{1}{\sec^2 \theta}\right) - 1}$
$= \dfrac{1}{\dfrac{2 - \sec^2 \theta}{\sec^2 \theta}} = \dfrac{\sec^2 \theta}{2 - \sec^2 \theta}$

**25** $2 \sin^2 2t + \cos 4t = 2 \sin^2 2t + \cos(2 \cdot 2t)$
$= 2 \sin^2 2t + (1 - 2 \sin^2 2t) = 1$

**27** $\tan 3u = \tan(2u + u) = \dfrac{\tan 2u + \tan u}{1 - \tan 2u \tan u}$
$= \dfrac{\dfrac{2 \tan u}{1 - \tan^2 u} + \tan u}{1 - \dfrac{2 \tan u}{1 - \tan^2 u} \cdot \tan u} = \dfrac{\dfrac{2 \tan u + \tan u - \tan^3 u}{1 - \tan^2 u}}{\dfrac{1 - \tan^2 u - 2 \tan^2 u}{1 - \tan^2 u}}$
$= \dfrac{3 \tan u - \tan^3 u}{1 - 3 \tan^2 u} = \dfrac{\tan u(3 - \tan^2 u)}{1 - 3 \tan^2 u}$

**29** $\dfrac{3}{8} + \dfrac{1}{2} \cos \theta + \dfrac{1}{8} \cos 2\theta$

**31** $\dfrac{3}{8} - \dfrac{1}{2} \cos 4x + \dfrac{1}{8} \cos 8x$     **33** $0, \pi, \dfrac{2\pi}{3}, \dfrac{4\pi}{3}$

**35** $\dfrac{\pi}{3}, \dfrac{5\pi}{3}, \pi$    **37** $0, \pi$    **39** $0, \dfrac{\pi}{3}, \dfrac{5\pi}{3}$

**43 (a)** 1.20, 5.09

**(b)** $P\!\left(\dfrac{2\pi}{3}, -1.5\right)$, $Q(\pi, -1)$, $R\!\left(\dfrac{4\pi}{3}, -1.5\right)$

**45 (a)** $-\dfrac{3\pi}{2}, -\dfrac{\pi}{2}, \dfrac{\pi}{2}, \dfrac{3\pi}{2}$

**(b)** $0, \pm\pi, \pm 2\pi, \pm\dfrac{\pi}{4}, \pm\dfrac{3\pi}{4}, \pm\dfrac{5\pi}{4}, \pm\dfrac{7\pi}{4}$

**47 (b)** Yes, point $B$ is 25 miles from $A$.

**49 (a)** $V = \dfrac{5}{2}\sin\theta$    **(b)** 53.13°    **51 (b)** 12.43 mm

**53** The graph of $f$ appears to be that of $y = g(x) = \tan x$.

$$\frac{\sin 2x + \sin x}{\cos 2x + \cos x + 1} = \frac{2\sin x\cos x + \sin x}{(2\cos^2 x - 1) + \cos x + 1}$$

$$= \frac{\sin x(2\cos x + 1)}{\cos x(2\cos x + 1)} = \frac{\sin x}{\cos x} = \tan x$$

## EXERCISES 6.5

**1** $\dfrac{1}{2}\cos 4t - \dfrac{1}{2}\cos 10t$    **3** $\dfrac{1}{2}\cos 2u + \dfrac{1}{2}\cos 10u$

**5** $\sin 12\theta + \sin 6\theta$    **7** $\dfrac{3}{2}\sin 3x + \dfrac{3}{2}\sin x$

**9** $2\sin 4\theta \cos 2\theta$    **11** $-2\sin 4x \sin x$

**13** $-2\cos 5t \sin 2t$    **15** $2\cos\dfrac{3}{2}x \cos\dfrac{1}{2}x$

**17** $\dfrac{\sin 4t + \sin 6t}{\cos 4t - \cos 6t} = \dfrac{2\sin 5t \cos t}{2\sin 5t \sin t} = \cot t$

**19** $\dfrac{\sin u + \sin v}{\cos u + \cos v} = \dfrac{2\sin\frac{1}{2}(u+v)\cos\frac{1}{2}(u-v)}{2\cos\frac{1}{2}(u+v)\cos\frac{1}{2}(u-v)} = \tan\dfrac{1}{2}(u+v)$

**21** $\dfrac{\sin u - \sin v}{\sin u + \sin v} = \dfrac{2\cos\frac{1}{2}(u+v)\sin\frac{1}{2}(u-v)}{2\sin\frac{1}{2}(u+v)\cos\frac{1}{2}(u-v)}$

$$= \cot\frac{1}{2}(u+v)\tan\frac{1}{2}(u-v)$$

$$= \frac{\tan\frac{1}{2}(u-v)}{\tan\frac{1}{2}(u+v)}$$

**23** $4\cos x\cos 2x\sin 3x = 2\cos 2x\,(2\sin 3x\cos x)$

$= 2\cos 2x\,(\sin 4x + \sin 2x)$

$= (2\cos 2x\sin 4x) + (2\cos 2x\sin 2x)$

$= [\sin 6x - \sin(-2x)] + (\sin 4x - \sin 0)$

$= \sin 2x + \sin 4x + \sin 6x$

**25** $\dfrac{1}{2}\sin\left[(a+b)x\right] + \dfrac{1}{2}\sin\left[(a-b)x\right]$    **27** $\dfrac{\pi}{4}n$

**29** $\dfrac{\pi}{2}n$    **31** $\dfrac{\pi}{2} + \pi n, \dfrac{\pi}{12} + \dfrac{\pi}{2}n, \dfrac{5\pi}{12} + \dfrac{\pi}{2}n$

**33** $\dfrac{\pi}{7} + \dfrac{2\pi}{7}n, \dfrac{2\pi}{3}n$    **35** $\dfrac{\pi}{4}, \dfrac{3\pi}{4}, \dfrac{5\pi}{4}, \dfrac{7\pi}{4}, \dfrac{\pi}{2}, \dfrac{3\pi}{2}$

**37** $0, \pm\pi, \pm 2\pi, \pm\dfrac{\pi}{4}, \pm\dfrac{3\pi}{4}, \pm\dfrac{5\pi}{4}, \pm\dfrac{7\pi}{4}$

**39** $f(x) = \dfrac{1}{2}\sin\dfrac{\pi n}{l}(x+kt) + \dfrac{1}{2}\sin\dfrac{\pi n}{l}(x-kt)$

**41 (a)** $0, \pm 1.05, \pm 1.57, \pm 2.09, \pm 3.14$

**(b)** $0, \pm\dfrac{\pi}{3}, \pm\dfrac{\pi}{2}, \pm\dfrac{2\pi}{3}, \pm\pi$

$[-3.14, 3.14]$ by $[-2.09, 2.09]$

**43** The graph of $f$ appears to be that of $y = g(x) = \tan 2x$.

$$\frac{\sin x + \sin 2x + \sin 3x}{\cos x + \cos 2x + \cos 3x} = \frac{\sin 2x + (\sin 3x + \sin x)}{\cos 2x + (\cos 3x + \cos x)}$$

$$= \frac{\sin 2x + 2\sin 2x\cos x}{\cos 2x + 2\cos 2x\cos x}$$

$$= \frac{\sin 2x(1 + 2\cos x)}{\cos 2x(1 + 2\cos x)} = \frac{\sin 2x}{\cos 2x}$$

$$= \tan 2x$$

## EXERCISES 6.6

**1 (a)** $-\dfrac{\pi}{4}$    **(b)** $\dfrac{2\pi}{3}$    **(c)** $-\dfrac{\pi}{3}$

**3 (a)** $\dfrac{\pi}{3}$    **(b)** $\dfrac{\pi}{4}$    **(c)** $\dfrac{\pi}{6}$

**5 (a)** Not defined    **(b)** Not defined    **(c)** $\dfrac{\pi}{4}$

**7 (a)** $-\dfrac{3}{10}$    **(b)** $\dfrac{1}{2}$    **(c)** 14

**9 (a)** $\dfrac{\pi}{3}$    **(b)** $\dfrac{5\pi}{6}$    **(c)** $-\dfrac{\pi}{6}$

**11 (a)** $-\dfrac{\pi}{4}$    **(b)** $\dfrac{3\pi}{4}$    **(c)** $-\dfrac{\pi}{4}$

**13 (a)** $\dfrac{\sqrt{3}}{2}$    **(b)** $\dfrac{\sqrt{2}}{2}$    **(c)** Not defined

**15** (a) $\dfrac{\sqrt{5}}{2}$  (b) $\dfrac{\sqrt{34}}{5}$  (c) $\dfrac{4}{\sqrt{15}}$

**17** (a) $\dfrac{\sqrt{3}}{2}$  (b) $0$  (c) $-\dfrac{77}{36}$

**19** (a) $-\dfrac{24}{25}$  (b) $-\dfrac{161}{289}$  (c) $\dfrac{24}{7}$

**21** (a) $-\dfrac{1}{10}\sqrt{2}$  (b) $\dfrac{4}{17}\sqrt{17}$  (c) $\dfrac{1}{2}$

**23** $\dfrac{x}{\sqrt{x^2+1}}$  **25** $\dfrac{\sqrt{x^2+4}}{2}$  **27** $2x\sqrt{1-x^2}$

**29** $\sqrt{\dfrac{1+x}{2}}$  **31** (a) $-\dfrac{\pi}{2}$  (b) $0$  (c) $\dfrac{\pi}{2}$

**33**

**35**

**37**

**39**

**41**

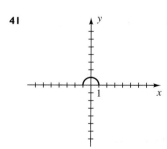

**43** (a) $2 \le x \le 4$
(b) $-\dfrac{\pi}{4} \le y \le \dfrac{\pi}{4}$
(c) $x = \sin 2y + 3$

**45** (a) $-\dfrac{3}{2} \le x \le \dfrac{3}{2}$  (b) $0 \le y \le 4\pi$

(c) $x = \dfrac{3}{2}\cos\dfrac{1}{4}y$

**47** $x = \sin^{-1}(-y-3)$    **49** $x = \cos^{-1}\left[\dfrac{1}{2}(15-y)\right]$

**51** $x = x_R$ or $x = \pi - x_R$, where $x_R = \sin^{-1}\left(\dfrac{3}{4}\sin y\right)$

**53** $\cos^{-1}(-1+\sqrt{2}) \approx 1.1437$,
$2\pi - \cos^{-1}(-1+\sqrt{2}) \approx 5.1395$

**55** $\tan^{-1}\dfrac{1}{4}(-9+\sqrt{57}) \approx -0.3478$,

$\tan^{-1}\dfrac{1}{4}(-9-\sqrt{57}) \approx -1.3337$

**57** $\cos^{-1}\dfrac{1}{5}\sqrt{15} \approx 0.6847$, $\cos^{-1}\left(-\dfrac{1}{5}\sqrt{15}\right) \approx 2.4569$,

$\cos^{-1}\dfrac{1}{3}\sqrt{3} \approx 0.9553$, $\cos^{-1}\left(-\dfrac{1}{3}\sqrt{3}\right) \approx 2.1863$

**59** $\sin^{-1}\left(\pm\dfrac{1}{6}\sqrt{30}\right) \approx \pm 1.1503$

**61** (a) $1.65$ m  (b) $0.92$ m  (c) $0.43$ m

**63** (a) $\alpha = \theta - \sin^{-1}\dfrac{d}{k}$  (b) $40°$

**65** Let $\alpha = \sin^{-1}x$ and $\beta = \tan^{-1}\dfrac{x}{\sqrt{1+x^2}}$ with $-\dfrac{\pi}{2} <$
$\alpha < \dfrac{\pi}{2}$ and $-\dfrac{\pi}{2} < \beta < \dfrac{\pi}{2}$. Thus, $\sin\alpha = x$ and
$\sin\beta = x$. Since the sine function is one-to-one on
$\left(-\dfrac{\pi}{2}, \dfrac{\pi}{2}\right)$, we have $\alpha = \beta$.

**67** Let $\alpha = \arcsin(-x)$ and $\beta = \arcsin x$ with $-\dfrac{\pi}{2} \le$
$\alpha \le \dfrac{\pi}{2}$ and $-\dfrac{\pi}{2} \le \beta \le \dfrac{\pi}{2}$. Thus, $\sin\alpha = -x$ and
$\sin\beta = x$. Consequently, $\sin\alpha = -\sin\beta = \sin(-\beta)$.
Since the sine function is one-to-one on $\left[-\dfrac{\pi}{2}, \dfrac{\pi}{2}\right]$,
we have $\alpha = -\beta$.

**69** Let $\alpha = \arctan x$ and $\beta = \arctan(1/x)$. Since $x > 0$,
we have $0 < \alpha < \dfrac{\pi}{2}$ and $0 < \beta < \dfrac{\pi}{2}$, and hence $0 <$
$\alpha + \beta < \pi$. Thus,
$\tan(\alpha+\beta) = \dfrac{\tan\alpha + \tan\beta}{1 - \tan\alpha\tan\beta} = \dfrac{x + (1/x)}{1 - x\cdot(1/x)} = \dfrac{x+(1/x)}{0}$.
Since the denominator is 0, $\tan(\alpha+\beta)$ is undefined
and hence $\alpha + \beta = \dfrac{\pi}{2}$.

**71** Domain: $[0, 2]$;

range: $\left[-\dfrac{\pi}{2}, \pi\right]$

$[-3, 6]$ by $[-2, 4]$

**73** 0.29

$[-3, 3]$ by $[-2, 2]$

**75** $\theta \approx 1.25 \approx 72°$

$[0, 1.57]$ by $[0, 1.05]$

## CHAPTER 6 REVIEW EXERCISES

**1** $(\cot^2 x + 1)(1 - \cos^2 x) = (\csc^2 x)(\sin^2 x) = 1$

**2** $\cos \theta + \sin \theta \tan \theta = \cos \theta + \sin \theta \cdot \dfrac{\sin \theta}{\cos \theta}$

$= \dfrac{\cos^2 \theta + \sin^2 \theta}{\cos \theta} = \dfrac{1}{\cos \theta} = \sec \theta$

**3** $\dfrac{(\sec^2 \theta - 1) \cot \theta}{\tan \theta \sin \theta + \cos \theta} = \dfrac{(\tan^2 \theta) \cot \theta}{\dfrac{\sin \theta}{\cos \theta} \cdot \sin \theta + \cos \theta}$

$= \dfrac{\tan \theta}{\dfrac{\sin^2 \theta + \cos^2 \theta}{\cos \theta}} = \dfrac{\sin \theta/\cos \theta}{1/\cos \theta} = \sin \theta$

**4** $(\tan x + \cot x)^2 = \left(\dfrac{\sin x}{\cos x} + \dfrac{\cos x}{\sin x}\right)^2 = \left(\dfrac{\sin^2 x + \cos^2 x}{\cos x \sin x}\right)^2$

$= \dfrac{1}{\cos^2 x \sin^2 x} = \sec^2 x \csc^2 x$

**5** $\dfrac{1}{1 + \sin t} \cdot \dfrac{1 - \sin t}{1 - \sin t} = \dfrac{1 - \sin t}{1 - \sin^2 t} = \dfrac{1 - \sin t}{\cos^2 t}$

$= \dfrac{1 - \sin t}{\cos t} \cdot \dfrac{1}{\cos t}$

$= \left(\dfrac{1}{\cos t} - \dfrac{\sin t}{\cos t}\right) \cdot \sec t$

$= (\sec t - \tan t) \sec t$

**6** $\dfrac{\sin (\alpha - \beta)}{\cos (\alpha + \beta)} = \dfrac{(\sin \alpha \cos \beta - \cos \alpha \sin \beta)/\cos \alpha \cos \beta}{(\cos \alpha \cos \beta - \sin \alpha \sin \beta)/\cos \alpha \cos \beta}$

$= \dfrac{\tan \alpha - \tan \beta}{1 - \tan \alpha \tan \beta}$

**7** $\tan 2u = \dfrac{2 \tan u}{1 - \tan^2 u} = \dfrac{2 \cdot \dfrac{1}{\cot u}}{1 - \dfrac{1}{\cot^2 u}} = \dfrac{\dfrac{2}{\cot u}}{\dfrac{\cot^2 u - 1}{\cot^2 u}}$

$= \dfrac{2 \cot u}{\cot^2 u - 1} = \dfrac{2 \cot u}{(\csc^2 u - 1) - 1} = \dfrac{2 \cot u}{\csc^2 u - 2}$

**8** $\cos^2 \dfrac{v}{2} = \dfrac{1 + \cos v}{2} = \dfrac{1 + \dfrac{1}{\sec v}}{2} = \dfrac{\dfrac{\sec v + 1}{\sec v}}{2}$

$= \dfrac{1 + \sec v}{2 \sec v}$

**9** $\dfrac{\tan^3 \phi - \cot^3 \phi}{\tan^2 \phi + \csc^2 \phi} = \dfrac{(\tan \phi - \cot \phi)[(\tan^2 \phi + \tan \phi \cot \phi + \cot^2 \phi)]}{[\tan^2 \phi + (1 + \cot^2 \phi)]}$

$= \tan \phi - \cot \phi$

**10** LS $= \dfrac{\sin u + \sin v}{\csc u + \csc v} = \dfrac{\sin u + \sin v}{\dfrac{1}{\sin u} + \dfrac{1}{\sin v}} = \dfrac{\sin u + \sin v}{\dfrac{\sin v + \sin u}{\sin u \sin v}}$

$= \sin u \sin v$

RS $= \dfrac{1 - \sin u \sin v}{-1 + \csc u \csc v} = \dfrac{1 - \sin u \sin v}{-1 + \dfrac{1}{\sin u \sin v}}$

$= \dfrac{1 - \sin u \sin v}{\dfrac{1 - \sin u \sin v}{\sin u \sin v}} = \sin u \sin v$

Since the LS and RS equal the same expression and the steps are reversible, the identity is verified.

**11** $\left(\dfrac{\sin^2 x}{\tan^4 x}\right)^3 \left(\dfrac{\csc^3 x}{\cot^6 x}\right)^2 = \left(\dfrac{\sin^6 x}{\tan^{12} x}\right)\left(\dfrac{\csc^6 x}{\cot^{12} x}\right) = \dfrac{(\sin x \csc x)^6}{(\tan x \cot x)^{12}}$

$= \dfrac{(1)^6}{(1)^{12}} = 1$

**12** $\dfrac{\cos \gamma}{1 - \tan \gamma} + \dfrac{\sin \gamma}{1 - \cot \gamma} = \dfrac{\cos \gamma}{\dfrac{\cos \gamma - \sin \gamma}{\cos \gamma}} + \dfrac{\sin \gamma}{\dfrac{\sin \gamma - \cos \gamma}{\sin \gamma}}$

$= \dfrac{\cos^2 \gamma}{\cos \gamma - \sin \gamma} + \dfrac{\sin^2 \gamma}{\sin \gamma - \cos \gamma}$

$= \dfrac{\cos^2 \gamma - \sin^2 \gamma}{\cos \gamma - \sin \gamma}$

$= \dfrac{(\cos \gamma + \sin \gamma)(\cos \gamma - \sin \gamma)}{\cos \gamma - \sin \gamma}$

$= \cos \gamma + \sin \gamma$

**13** $\dfrac{\cos (-t)}{\sec (-t) + \tan (-t)} = \dfrac{\cos t}{\sec t - \tan t} = \dfrac{\cos t}{\dfrac{1}{\cos t} - \dfrac{\sin t}{\cos t}}$

$= \dfrac{\cos t}{\dfrac{1 - \sin t}{\cos t}} = \dfrac{\cos^2 t}{1 - \sin t} = \dfrac{1 - \sin^2 t}{1 - \sin t}$

$= \dfrac{(1 - \sin t)(1 + \sin t)}{1 - \sin t} = 1 + \sin t$

**14**
$$\frac{\cot(-t) + \csc(-t)}{\sin(-t)} = \frac{-\cot t - \csc t}{-\sin t} = \frac{\dfrac{\cos t}{\sin t} + \dfrac{1}{\sin t}}{\sin t}$$
$$= \frac{\cos t + 1}{\sin^2 t} = \frac{\cos t + 1}{1 - \cos^2 t}$$
$$= \frac{\cos t + 1}{(1 - \cos t)(1 + \cos t)} = \frac{1}{1 - \cos t}$$

**15**
$$\sqrt{\frac{1 - \cos t}{1 + \cos t}} = \sqrt{\frac{(1 - \cos t)}{(1 + \cos t)} \cdot \frac{(1 - \cos t)}{(1 - \cos t)}} = \sqrt{\frac{(1 - \cos t)^2}{1 - \cos^2 t}}$$
$$= \sqrt{\frac{(1 - \cos t)^2}{\sin^2 t}} = \frac{|1 - \cos t|}{|\sin t|} = \frac{1 - \cos t}{|\sin t|},$$
since $(1 - \cos t) \geq 0$.

**16**
$$\sqrt{\frac{1 - \sin \theta}{1 + \sin \theta}} = \sqrt{\frac{(1 - \sin \theta)}{(1 + \sin \theta)} \cdot \frac{(1 + \sin \theta)}{(1 + \sin \theta)}} = \sqrt{\frac{1 - \sin^2 \theta}{(1 + \sin \theta)^2}}$$
$$= \sqrt{\frac{\cos^2 \theta}{(1 + \sin \theta)^2}} = \frac{|\cos \theta|}{|1 + \sin \theta|} = \frac{|\cos \theta|}{1 + \sin \theta},$$
since $(1 + \sin \theta) \geq 0$.

**17** $\cos\left(x - \dfrac{5\pi}{2}\right) = \cos x \cos \dfrac{5\pi}{2} + \sin x \sin \dfrac{5\pi}{2} = \sin x$

**18** $\tan\left(x + \dfrac{3\pi}{4}\right) = \dfrac{\tan x + \tan \dfrac{3\pi}{4}}{1 - \tan x \tan \dfrac{3\pi}{4}} = \dfrac{\tan x - 1}{1 + \tan x}$

**19** $\dfrac{1}{4} \sin 4\beta = \dfrac{1}{4} \sin(2 \cdot 2\beta) = \dfrac{1}{4}(2 \sin 2\beta \cos 2\beta)$
$$= \dfrac{1}{2}(2 \sin \beta \cos \beta)(\cos^2 \beta - \sin^2 \beta)$$
$$= \sin \beta \cos^3 \beta - \cos \beta \sin^3 \beta$$

**20** $\tan \dfrac{1}{2}\theta = \dfrac{1 - \cos \theta}{\sin \theta} = \dfrac{1}{\sin \theta} - \dfrac{\cos \theta}{\sin \theta} = \csc \theta - \cot \theta$

**21** $\sin 8\theta = 2 \sin 4\theta \cos 4\theta$
$$= 2(2 \sin 2\theta \cos 2\theta)(1 - 2 \sin^2 2\theta)$$
$$= 8 \sin \theta \cos \theta\,(1 - 2 \sin^2 \theta)[1 - 2(2 \sin \theta \cos \theta)^2]$$
$$= 8 \sin \theta \cos \theta(1 - 2 \sin^2 \theta)(1 - 8 \sin^2 \theta \cos^2 \theta)$$

**22** Let $\alpha = \arctan x$ and $\beta = \arctan \dfrac{2x}{1 - x^2}$. Because
$-1 < x < 1,\ -\dfrac{\pi}{4} < \alpha < \dfrac{\pi}{4}$. Thus, $\tan \alpha = x$ and
$$\tan \beta = \frac{2x}{1 - x^2} = \frac{2 \tan \alpha}{1 - \tan^2 \alpha} = \tan 2\alpha.$$ Since the
tangent function is one-to-one on $\left(-\dfrac{\pi}{2}, \dfrac{\pi}{2}\right)$, we have
$\beta = 2\alpha$ or, equivalently, $\alpha = \dfrac{1}{2}\beta$.

**23** $\dfrac{\pi}{2}, \dfrac{3\pi}{2}, \dfrac{\pi}{4}, \dfrac{7\pi}{4}, \dfrac{3\pi}{4}, \dfrac{5\pi}{4}$    **24** $\dfrac{7\pi}{6}, \dfrac{11\pi}{6}$    **25** $0, \pi$

**26** $\dfrac{\pi}{4}, \dfrac{3\pi}{4}, \dfrac{5\pi}{4}, \dfrac{7\pi}{4}$    **27** $0, \pi, \dfrac{2\pi}{3}, \dfrac{4\pi}{3}$

**28** $\dfrac{\pi}{2}, \dfrac{3\pi}{2}, \dfrac{\pi}{4}, \dfrac{5\pi}{4}, \dfrac{3\pi}{4}, \dfrac{7\pi}{4}$    **29** $\dfrac{7\pi}{6}, \dfrac{11\pi}{6}, \dfrac{\pi}{2}$

**30** $\dfrac{2\pi}{3}, \dfrac{4\pi}{3}, \pi$    **31** $\dfrac{\pi}{6}, \dfrac{5\pi}{6}, \dfrac{\pi}{3}, \dfrac{5\pi}{3}$

**32** All $x$ in $[0, 2\pi)$ except $\dfrac{\pi}{4}, \dfrac{3\pi}{4}, \dfrac{5\pi}{4}, \dfrac{7\pi}{4}$

**33** $\dfrac{\pi}{3}, \dfrac{5\pi}{3}$    **34** $0, \dfrac{\pi}{3}, \dfrac{2\pi}{3}, \pi, \dfrac{4\pi}{3}, \dfrac{5\pi}{3}$

**35** $\dfrac{3}{4}, \dfrac{7}{4}, \dfrac{11}{4}, \dfrac{15}{4}, \dfrac{19}{4}, \dfrac{23}{4}$    **36** $0, \pi, \dfrac{\pi}{3}, \dfrac{5\pi}{3}$

**37** $\dfrac{\pi}{3}, \dfrac{5\pi}{3}$    **38** $\dfrac{\pi}{6}, \dfrac{5\pi}{6}, \dfrac{7\pi}{6}, \dfrac{11\pi}{6}$

**39** $0, \dfrac{\pi}{8}, \dfrac{3\pi}{8}, \dfrac{5\pi}{8}, \dfrac{7\pi}{8}, \pi, \dfrac{9\pi}{8}, \dfrac{11\pi}{8}, \dfrac{13\pi}{8}, \dfrac{15\pi}{8}$

**40** $\dfrac{\pi}{5}, \dfrac{3\pi}{5}, \pi, \dfrac{7\pi}{5}, \dfrac{9\pi}{5}$    **41** $\dfrac{\sqrt{6} - \sqrt{2}}{4}$

**42** $-2 - \sqrt{3}$    **43** $\dfrac{\sqrt{2} - \sqrt{6}}{4}$    **44** $\dfrac{2}{\sqrt{2 - \sqrt{2}}}$

**45** $\dfrac{84}{85}$    **46** $-\dfrac{13}{85}$    **47** $-\dfrac{84}{13}$    **48** $-\dfrac{36}{77}$

**49** $\dfrac{36}{85}$    **50** $-\dfrac{36}{85}$    **51** $\dfrac{240}{289}$    **52** $-\dfrac{161}{289}$

**53** $\dfrac{24}{7}$    **54** $\dfrac{1}{10}\sqrt{10}$    **55** $\dfrac{1}{3}$    **56** $\dfrac{5}{34}\sqrt{34}$

**57** (a) $\dfrac{1}{2}\cos 3t - \dfrac{1}{2}\cos 11t$

   (b) $\dfrac{1}{2}\cos \dfrac{1}{12}u + \dfrac{1}{2}\cos \dfrac{5}{12}u$

   (c) $3 \sin 8x - 3 \sin 2x$    (d) $2 \sin 10\theta - 2 \sin 4\theta$

**58** (a) $2 \sin 5u \cos 3u$    (b) $2 \sin \dfrac{11}{2}\theta \sin \dfrac{5}{2}\theta$

   (c) $2 \cos \dfrac{9}{40}t \sin \dfrac{1}{40}t$    (d) $6 \cos 4x \cos 2x$

**59** $\dfrac{\pi}{6}$    **60** $\dfrac{\pi}{4}$    **61** $\dfrac{\pi}{3}$    **62** $\pi$    **63** $-\dfrac{\pi}{4}$

**64** $\dfrac{3\pi}{4}$    **65** $\dfrac{1}{2}$    **66** $2$    **67** Not defined    **68** $\dfrac{\pi}{2}$

**69** $\dfrac{240}{289}$    **70** $-\dfrac{7}{25}$

**71**             **72**

**73**           **74**

**75** $\cos (\alpha + \beta + \gamma) = \cos [(\alpha + \beta) + \gamma]$
$= \cos (\alpha + \beta) \cos \gamma - \sin (\alpha + \beta) \sin \gamma$
$= (\cos \alpha \cos \beta - \sin \alpha \sin \beta) \cos \gamma -$
$(\sin \alpha \cos \beta + \cos \alpha \sin \beta) \sin \gamma$
$= \cos \alpha \cos \beta \cos \gamma - \sin \alpha \sin \beta \cos \gamma -$
$\sin \alpha \cos \beta \sin \gamma - \cos \alpha \sin \beta \sin \gamma$

**76** **(b)** $t = 0, \pm \dfrac{\pi}{4b}$    **(c)** $\dfrac{2}{3}\sqrt{2}\, A$

**77** $\pm \dfrac{\pi}{4}, \pm \dfrac{3\pi}{4}, \pm \dfrac{5\pi}{4}, \pm \dfrac{7\pi}{4}, \pm \dfrac{\pi}{3}, \pm \dfrac{5\pi}{3}$

**78** **(a)** $x = 2d \tan \dfrac{1}{2}\theta$    **(b)** $d \le 1000$ ft

**79** **(a)** $d = r\left(\sec \dfrac{1}{2}\theta - 1\right)$    **(b)** $43°$

**80** **(a)** $78.7°$    **(b)** $61.4°$

# CHAPTER 7

## EXERCISES 7.1

**1** $\beta = 62°, b \approx 14.1, c \approx 15.6$
**3** $\gamma = 100°10', b \approx 55.1, c \approx 68.7$
**5** $\beta = 76°30', a \approx 13.6, c \approx 17.8$
**7** No triangle exists.
**9** $\alpha \approx 77°30', \beta \approx 49°10', b \approx 108; \alpha \approx 102°30',$
$\beta \approx 24°10', b \approx 59$
**11** $\alpha \approx 82.54°, \beta \approx 49.72°, b \approx 100.85; \alpha \approx 97.46°,$
$\beta \approx 34.80°, b \approx 75.45$
**13** $\beta \approx 53°40', \gamma \approx 61°10', c \approx 20.6$
**15** $\alpha \approx 25.993°, \gamma \approx 32.383°, a \approx 0.146$    **17** 219 yd
**19** **(a)** 1.6 mi    **(b)** 0.6 mi    **21** 2.7 mi    **23** 628 m
**25** 3.7 mi from $A$ and 5.4 mi from $B$    **27** 350 ft
**29** **(a)** 18.7    **(b)** 814    **31** (3949.9, 2994.2)

## EXERCISES 7.2

**1** $a \approx 26, \beta \approx 41°, \gamma \approx 79°$
**3** $b \approx 180, \alpha \approx 25°, \gamma \approx 5°$
**5** $c \approx 2.75, \alpha \approx 21°10', \beta \approx 43°40'$
**7** $\alpha \approx 29°, \beta \approx 47°, \gamma \approx 104°$
**9** $\alpha \approx 12°30', \beta \approx 136°30', \gamma \approx 31°00'$    **11** 196 ft
**13** 24 mi    **15** 39 mi    **17** 2.3 mi
**19** N55°31'E
**21** 63.7 ft from first and third base; 66.8 ft from second base
**23** 37,039 ft $\approx$ 7 mi
**25** *Hint:* Use the formula $\sin \dfrac{\theta}{2} = \sqrt{\dfrac{1 - \cos \theta}{2}}$.
**27** **(a)** $72°, 108°, 36°$    **(b)** 0.62    **(c)** 0.59, 0.36

Exer. 29–36: The answer is in square units.

**29** 260    **31** 11.21    **33** 13.1    **35** 517.0
**37** 1.62 acres    **39** 123.4 ft$^2$

## EXERCISES 7.3

**1** 5    **3** $\sqrt{85}$    **5** 8    **7** 1    **9** 0

*Note:* Point $P$ is the point corresponding to the geometric representation.

**11** $P(4, 2)$    **13** $P(3, -5)$    **15** $P(-3, 6)$
**17** $P(-6, 4)$    **19** $P(0, 2)$

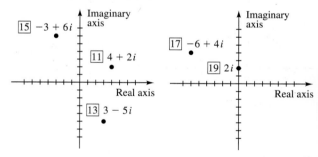

**21** $\sqrt{2}\,\text{cis}\,\dfrac{7\pi}{4}$    **23** $8\,\text{cis}\,\dfrac{5\pi}{6}$    **25** $4\,\text{cis}\,\dfrac{\pi}{6}$
**27** $4\sqrt{2}\,\text{cis}\,\dfrac{5\pi}{4}$    **29** $20\,\text{cis}\,\dfrac{3\pi}{2}$    **31** $12\,\text{cis}\,0$
**33** $7\,\text{cis}\,\pi$    **35** $6\,\text{cis}\,\dfrac{\pi}{2}$    **37** $10\,\text{cis}\,\dfrac{4\pi}{3}$
**39** $\sqrt{5}\,\text{cis}\,\left(\tan^{-1}\dfrac{1}{2}\right)$    **41** $\sqrt{10}\,\text{cis}\,\left[\tan^{-1}\left(-\dfrac{1}{3}\right) + \pi\right]$
**43** $\sqrt{34}\,\text{cis}\,\left(\tan^{-1}\dfrac{3}{5} + \pi\right)$

**45** $5 \operatorname{cis}\left[\tan^{-1}\left(-\frac{3}{4}\right) + 2\pi\right]$

**47** $2\sqrt{2} + 2\sqrt{2}i$    **49** $-3 + 3\sqrt{3}i$    **51** $-5$

**53** $5 + 3i$    **55** $2 - i$    **57** $-2, i$

**59** $10\sqrt{3} - 10i, -\frac{2}{5}\sqrt{3} + \frac{2}{5}i$    **61** $40, \frac{5}{2}$

**63** $8 - 4i, \frac{8}{5} + \frac{4}{5}i$

## EXERCISES 7.4

**1** $-972 - 972i$    **3** $-32i$    **5** $-8$

**7** $-\frac{1}{2}\sqrt{2} - \frac{1}{2}\sqrt{2}i$    **9** $-\frac{1}{2} - \frac{1}{2}\sqrt{3}i$

**11** $-64\sqrt{3} - 64i$    **13** $\pm\left(\frac{1}{2}\sqrt{6} + \frac{1}{2}\sqrt{2}i\right)$

**15** $\pm\left(\frac{\sqrt[4]{2}}{2} + \frac{\sqrt[4]{18}}{2}i\right), \pm\left(\frac{\sqrt[4]{18}}{2} - \frac{\sqrt[4]{2}}{2}i\right)$

**17** $3i, \pm\frac{3}{2}\sqrt{3} - \frac{3}{2}i$

**19** $\pm 1, \frac{1}{2} \pm \frac{1}{2}\sqrt{3}i,$    **21** $\sqrt[10]{2} \operatorname{cis}\theta$ with $\theta = 9°,$

$-\frac{1}{2} \pm \frac{1}{2}\sqrt{3}i$    $81°, 153°, 225°, 297°$

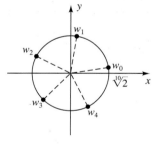

**23** $\pm 2, \pm 2i$    **25** $\pm 2i, \sqrt{3} \pm i, -\sqrt{3} \pm i$

**27** $2i, \pm\sqrt{3} - i$

**29** $3 \operatorname{cis}\theta$ with $\theta = 0°, 72°, 144°, 216°, 288°$

## EXERCISES 7.5

**1** $\langle 3, 1\rangle, \langle 1, -7\rangle, \langle 13, 8\rangle, \langle 3, -32\rangle$

**3** $\langle -15, 6\rangle, \langle 1, -2\rangle, \langle -68, 28\rangle, \langle 12, -12\rangle$

**5** $4\mathbf{i} - 3\mathbf{j}, -2\mathbf{i} + 7\mathbf{j}, 19\mathbf{i} - 17\mathbf{j}, -11\mathbf{i} + 33\mathbf{j}$

**7** Terminal points are $(3, 2), (-1, 5), (2, 7),$ $(6, 4), (3, -15).$

**9** Terminal points are $(-4, 6), (-2, 3),$ $(-6, 9), (-8, 12),$ $(6, -9).$

**11** $-\mathbf{b}$    **13** $\mathbf{f}$    **15** $-\frac{1}{2}\mathbf{e}$

**17** $\mathbf{a} + (\mathbf{b} + \mathbf{c}) = \langle a_1, a_2\rangle + (\langle b_1, b_2\rangle + \langle c_1, c_2\rangle)$
$= \langle a_1, a_2\rangle + \langle b_1 + c_1, b_2 + c_2\rangle$
$= \langle a_1 + b_1 + c_1, a_2 + b_2 + c_2\rangle$
$= \langle a_1 + b_1, a_2 + b_2\rangle + \langle c_1, c_2\rangle$
$= (\langle a_1, a_2\rangle + \langle b_1, b_2\rangle) + \langle c_1, c_2\rangle$
$= (\mathbf{a} + \mathbf{b}) + \mathbf{c}$

**19** $\mathbf{a} + (-\mathbf{a}) = \langle a_1, a_2\rangle + (-\langle a_1, a_2\rangle)$
$= \langle a_1, a_2\rangle + \langle -a_1, -a_2\rangle$
$= \langle a_1 - a_1, a_2 - a_2\rangle$
$= \langle 0, 0\rangle = \mathbf{0}$

**21** $(mn)\mathbf{a} = (mn)\langle a_1, a_2\rangle$
$= \langle (mn)a_1, (mn)a_2\rangle$
$= \langle mna_1, mna_2\rangle$
$= m\langle na_1, na_2\rangle$   or   $n\langle ma_1, ma_2\rangle$
$= m(n\langle a_1, a_2\rangle)$   or   $n(m\langle a_1, a_2\rangle)$
$= m(n\mathbf{a})$   or   $n(m\mathbf{a})$

**23** $0\mathbf{a} = 0\langle a_1, a_2\rangle = \langle 0a_1, 0a_2\rangle = \langle 0, 0\rangle = \mathbf{0}.$
Also, $m\mathbf{0} = m\langle 0, 0\rangle = \langle m0, m0\rangle = \langle 0, 0\rangle = \mathbf{0}.$

**25** $-(\mathbf{a} + \mathbf{b}) = -(\langle a_1, a_2\rangle + \langle b_1, b_2\rangle)$
$= -(\langle a_1 + b_1, a_2 + b_2\rangle)$
$= \langle -(a_1 + b_1), -(a_2 + b_2)\rangle$
$= \langle -a_1 - b_1, -a_2 - b_2\rangle$
$= \langle -a_1, -a_2\rangle + \langle -b_1, -b_2\rangle$
$= -\mathbf{a} + (-\mathbf{b}) = -\mathbf{a} - \mathbf{b}$

**27** $\|2\mathbf{v}\| = \|2\langle a, b\rangle\| = \|\langle 2a, 2b\rangle\| = \sqrt{(2a)^2 + (2b)^2}$
$= \sqrt{4a^2 + 4b^2} = 2\sqrt{a^2 + b^2} = 2\|\langle a, b\rangle\|$
$= 2\|\mathbf{v}\|$

**29** $3\sqrt{2}; \frac{7\pi}{4}$    **31** $5; \pi$    **33** $\sqrt{41}; \tan^{-1}\left(-\frac{5}{4}\right) + \pi$

**35** $18; \frac{3\pi}{2}$    **37** $102$ lb    **39** $7.2$ kg

**41** $89$ kg; S66°W    **43** $5.8$ lb; $129°$    **45** $40.96; 28.68$

**47** $-6.18$; $19.02$
**49** (a) $\mathbf{F} = \langle 7, 2 \rangle$ (b) $\mathbf{G} = -\mathbf{F} = \langle -7, -2 \rangle$
**51** (a) $\mathbf{F} \approx \langle -5.86, 1.13 \rangle$
(b) $\mathbf{G} = -\mathbf{F} \approx \langle 5.86, -1.13 \rangle$
**53** $\sin^{-1}(0.4) \approx 23.6°$ **55** $56°$; $232$ mi/hr
**57** $420$ mi/hr; $244°$ **59** N22°W
**61** $\mathbf{v}_1 \approx 4.1\mathbf{i} - 7.10\mathbf{j}$; $\mathbf{v}_2 \approx 0.98\mathbf{i} - 3.67\mathbf{j}$
**63** (a) $(24.51, 20.57)$ (b) $(-24.57, 18.10)$

## EXERCISES 7.6

**1** (a) $24$ (b) $\cos^{-1}\left(\dfrac{24}{\sqrt{29}\sqrt{45}}\right) \approx 48°22'$

**3** (a) $-14$ (b) $\cos^{-1}\left(\dfrac{-14}{\sqrt{17}\sqrt{13}}\right) \approx 160°21'$

**5** (a) $45$ (b) $\cos^{-1}\left(\dfrac{45}{\sqrt{81}\sqrt{41}}\right) \approx 38°40'$

**7** (a) $-\dfrac{149}{5}$ (b) $\cos^{-1}\left(\dfrac{-149/5}{\sqrt{149}\sqrt{149/25}}\right) = 180°$

**9** $\langle 4, -1 \rangle \cdot \langle 2, 8 \rangle = 0$ **11** $(-4\mathbf{j}) \cdot (-7\mathbf{i}) = 0$

**13** Opposite **15** Same **17** $\dfrac{6}{5}$ **19** $\pm\dfrac{3}{8}$

**21** (a) $-23$ (b) $-23$ **23** $-51$
**25** $17/\sqrt{26} \approx 3.33$ **27** $2.2$ **29** $7$ **31** $28$ **33** $12$
**35** $\mathbf{a} \cdot \mathbf{a} = \langle a_1, a_2 \rangle \cdot \langle a_1, a_2 \rangle = a_1^2 + a_2^2 = (\sqrt{a_1^2 + a_2^2})^2 = \|\mathbf{a}\|^2$
**37** $(c\mathbf{a}) \cdot \mathbf{b} = (c\langle a_1, a_2 \rangle) \cdot \langle b_1, b_2 \rangle$
$= \langle ca_1, ca_2 \rangle \cdot \langle b_1, b_2 \rangle$
$= ca_1b_1 + ca_2b_2$
$= c(a_1b_1 + a_2b_2) = c(\mathbf{a} \cdot \mathbf{b})$
**39** $\mathbf{0} \cdot \mathbf{a} = \langle 0, 0 \rangle \cdot \langle a_1, a_2 \rangle = 0(a_1) + 0(a_2) = 0 + 0 = 0$
**41** $1000\sqrt{3} \approx 1732$ ft-lb
**43** (a) $\mathbf{v} = (93 \times 10^6)\mathbf{i} + (0.432 \times 10^6)\mathbf{j}$;
$\mathbf{w} = (93 \times 10^6)\mathbf{i} - (0.432 \times 10^6)\mathbf{j}$
(b) $0.53°$
**45** $16\sqrt{3} \approx 27.7$ horsepower

## CHAPTER 7 REVIEW EXERCISES

**1** $a = \sqrt{43}$, $\beta = \cos^{-1}\left(\dfrac{4}{43}\sqrt{43}\right)$, $\gamma = \cos^{-1}\left(\dfrac{5}{86}\sqrt{43}\right)$
**2** $\alpha = 60°$, $\beta = 90°$, $b = 4$; $\alpha = 120°$, $\beta = 30°$, $b = 2$
**3** $\gamma = 75°$, $a = 50\sqrt{6}$, $c = 50(1 + \sqrt{3})$
**4** $\alpha = \cos^{-1}\left(\dfrac{7}{8}\right)$, $\beta = \cos^{-1}\left(\dfrac{11}{16}\right)$, $\gamma = \cos^{-1}\left(-\dfrac{1}{4}\right)$
**5** $\alpha = 38°$, $a \approx 8.0$, $c \approx 13$
**6** $\gamma \approx 19°10'$, $\beta \approx 137°20'$, $b \approx 258$
**7** $\alpha \approx 24°$, $\gamma \approx 41°$, $b \approx 10.1$
**8** $\alpha \approx 42°$, $\beta \approx 87°$, $\gamma \approx 51°$ **9** $290$ **10** $10.9$
**11** $10\sqrt{2}$ cis $\dfrac{3\pi}{4}$ **12** $4$ cis $\dfrac{5\pi}{3}$ **13** $17$ cis $\pi$

**14** $12$ cis $\dfrac{3\pi}{2}$ **15** $10$ cis $\dfrac{7\pi}{6}$
**16** $\sqrt{41}$ cis $\left(\tan^{-1}\dfrac{5}{4}\right)$ **17** $10\sqrt{3} - 10i$
**18** $12 + 5i$ **19** $-12 - 12\sqrt{3}\,i$, $-\dfrac{3}{2}$
**20** $-4\sqrt{2}\,i$, $-2\sqrt{2}$ **21** $-512i$ **22** $i$
**23** $-972 + 972i$ **24** $-2^{19} - 2^{19}\sqrt{3}\,i$
**25** $-3$, $\dfrac{3}{2} \pm \dfrac{3}{2}\sqrt{3}\,i$
**26** (a) $2^{24}$ (b) $\sqrt[3]{2}$ cis $\theta$ with $\theta = 100°, 220°, 340°$
**27** $2$ cis $\theta$ with $\theta = 0°, 72°, 144°, 216°, 288°$
**28** Terminal points are
$(-2, -3)$, $(-6, 13)$,
$(-8, 10)$, $(-1, 4)$.

**29** (a) $12\mathbf{i} + 19\mathbf{j}$ (b) $-8\mathbf{i} + 13\mathbf{j}$ (c) $\sqrt{40} \approx 6.32$
(d) $\sqrt{29} - \sqrt{17} \approx 1.26$
**30** Circle with center $(a_1, a_2)$ and radius $c$
**31** The vectors $\mathbf{a}$, $\mathbf{b}$, and $\mathbf{a} - \mathbf{b}$ form a triangle with the vector $\mathbf{a} - \mathbf{b}$ opposite angle $\theta$. The conclusion is a direct application of the law of cosines with sides $\|\mathbf{a}\|$, $\|\mathbf{b}\|$, and $\|\mathbf{a} - \mathbf{b}\|$.
**32** $\langle 14 \cos 40°, -14 \sin 40° \rangle$ **33** $109$; S78°E
**34** $183°$; $70$ mi/hr

**35** (a) $10$ (b) $\cos^{-1}\left(\dfrac{10}{\sqrt{13}\sqrt{17}}\right) \approx 47°44'$ (c) $\dfrac{10}{\sqrt{13}}$

**36** (a) $80$ (b) $\cos^{-1}\left(\dfrac{40}{\sqrt{40}\sqrt{50}}\right) \approx 26°34'$ (c) $\sqrt{40}$

**37** $56$ **38** $47.6°$ **39** (a) $449$ ft (b) $434$ ft
**40** (a) $33$ mi, $41$ mi (b) $30$ mi **41** $204$
**42** $1$ hour and $16$ minutes **43** (c) $158°$
**44** (a) $47°$ (b) $20$
**45** (a) $72°$ (b) $181.6$ ft$^2$ (c) $37.6$ ft

## CHAPTER 8

## EXERCISES 8.1

**1** $(3, 5)$, $(-1, -3)$ **3** $(1, 0)$, $(-3, 2)$
**5** $(0, 0)$, $\left(\dfrac{1}{8}, \dfrac{1}{128}\right)$ **7** $(3, -2)$ **9** No solution

**11** $(-4, 3), (5, 0)$    **13** $(-2, 2)$

**15** $\left(-\dfrac{3}{5} + \dfrac{1}{10}\sqrt{86}, \dfrac{1}{5} + \dfrac{3}{10}\sqrt{86}\right),$

$\left(-\dfrac{3}{5} - \dfrac{1}{10}\sqrt{86}, \dfrac{1}{5} - \dfrac{3}{10}\sqrt{86}\right)$

**17** $(-4, 0), \left(\dfrac{12}{5}, \dfrac{16}{5}\right)$    **19** $(0, 1), (4, -3)$

**21** $(\pm 2, 5), (\pm\sqrt{5}, 4)$

**23** $(\sqrt{2}, \pm 2\sqrt{3}), (-\sqrt{2}, \pm 2\sqrt{3})$

**25** $(2\sqrt{2}, \pm 2), (-2\sqrt{2}, \pm 2)$    **27** $(3, -1, 2)$

**29** $(1, -1, 2), (-1, 3, -2)$

**31** 12 in. × 8 in.

**33** Yes; a solution occurs between 0 and 1.

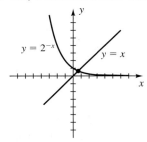

**35** Yes; $r = 2$ in., $h = \dfrac{50}{\pi}$ in.

**37** **(a)** $a = 120{,}000, b = 40{,}000$    **(b)** 77,143

**39** $(0, 0), (0, 100), (50, 0)$; the fourth solution $(-100, 150)$ is not meaningful.

**41** Yes; 1 ft × 1 ft × 2 ft or

$\dfrac{\sqrt{13} - 1}{2}$ ft × $\dfrac{\sqrt{13} - 1}{2}$ ft × $\dfrac{8}{(\sqrt{13} - 1)^2}$ ft

$\approx 1.30$ ft × 1.30 ft × 1.18 ft

**43** The points are on the parabola **(a)** $y = \dfrac{1}{2}x^2 - \dfrac{1}{2}$ and

**(b)** $y = \dfrac{1}{4}x^2 - 1$.

**45** **(a)** $(31.25, -50)$

**(b)** $\left(-\dfrac{3}{2}\sqrt{11}, -\dfrac{1}{2}\right) \approx (-4.975, -0.5)$

**47** $(\mp 0.56, \pm 1.92), (\mp 0.63, \pm 1.90),$
$(\pm 1.14, \pm 1.65)$

$[-6, 6]$ by $[-4, 4]$

**49** $(-1.44, \pm 1.04), (-0.12, \pm 1.50),$
$(0.10, \pm 1.50), (1.22, \pm 1.19)$

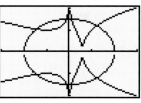

$[-3, 3]$ by $[-2, 2]$

*EXERCISES 8.2*

**1** $(4, -2)$    **3** $(8, 0)$    **5** $\left(-1, \dfrac{3}{2}\right)$    **7** $\left(\dfrac{76}{53}, \dfrac{28}{53}\right)$

**9** $\left(\dfrac{51}{13}, \dfrac{96}{13}\right)$    **11** $\left(\dfrac{8}{7}, -\dfrac{3}{7}\sqrt{6}\right)$    **13** No solution

**15** All ordered pairs $(m, n)$ such that $3m - 4n = 2$

**17** $(0, 0)$    **19** $\left(-\dfrac{22}{7}, -\dfrac{11}{5}\right)$

**21** $a = \dfrac{1}{6}, b = -\dfrac{1}{6}e^{6x}$

**23** $a = \cos x - \sec x, b = \sin x$

**25** 313 students, 137 nonstudents

**27** $x = \left(\dfrac{30}{\pi}\right) - 4 \approx 5.55$ cm, $y = 12 - \left(\dfrac{30}{\pi}\right) \approx 2.45$ cm

**29** $l = 10$ ft, $w = \dfrac{20}{\pi}$ ft    **31** 2400 adults, 3600 kittens

**33** 40 g of 35% alloy, 60 g of 60% alloy

**35** 540 mi/hr, 60 mi/hr    **37** $v_0 = 10, a = 3$

**39** 20 sofas, 30 recliners

**41** **(a)** $\left(c, \dfrac{4}{5}c\right)$ for an arbitrary $c > 0$    **(b)** \$16 per hour

**43** 1928; 15.5°C

**45** **(a)** $Q = -100{,}000p + 2{,}000{,}000$
**(b)** $K = 150{,}000p - 250{,}000$    **(c)** \$9.00

*EXERCISES 8.3*

**1** $(2, 3, -1)$    **3** $(-2, 4, 5)$    **5** No solution

**7** $\left(\dfrac{2}{3}, \dfrac{31}{21}, \dfrac{1}{21}\right)$

Exer. 9–16: There are other forms for the answers; $c$ is any real number.

**9** $(2c, -c, c)$    **11** $(0, -c, c)$

**13** $\left(\dfrac{12}{7} - \dfrac{9}{7}c, \dfrac{4}{7}c - \dfrac{13}{14}, c\right)$

**15** $\left(\dfrac{7}{10}c + \dfrac{1}{2}, \dfrac{19}{10}c - \dfrac{3}{2}, c\right)$    **17** $(1, 3, -1, 2)$

**19** $(2, -1, 3, 4, 1)$    **21** $\left(\dfrac{1}{11}, \dfrac{31}{11}, \dfrac{3}{11}\right)$

**23** $(-2, -3)$    **25** No solution

**27** 17 of 10%, 11 of 30%, 22 of 50%

**29** 4 hr for A, 2 hr for B, 5 hr for C

**31** 380 lb of $G_1$, 60 lb of $G_2$, 160 lb of $G_3$

**33** (a) $I_1 = 0, I_2 = 2, I_3 = 2$    (b) $I_1 = \dfrac{3}{4}, I_2 = 3, I_3 = \dfrac{9}{4}$

**35** $\dfrac{3}{8}$ lb Columbian, $\dfrac{1}{8}$ lb Brazilian, $\dfrac{1}{2}$ lb Kenyan

**37** (a) A: $x_1 + x_4 = 75$, B: $x_1 + x_2 = 150$,
   C: $x_2 + x_3 = 225$, D: $x_3 + x_4 = 150$
   (b) $x_1 = 25, x_2 = 125, x_4 = 50$
   (c) $x_3 = 150 - x_4 \le 150$;
      $x_3 = 225 - x_2 = 225 - (150 - x_1) = 75 + x_1 \ge 75$

**39** 2134    **41** $y = -x^2 + 2x + 5$

**43** $x^2 + y^2 - x + 3y - 6 = 0$

**45** $a = -\dfrac{4}{9}, b = \dfrac{11}{9}, c = \dfrac{17}{18}, d = \dfrac{23}{18}$

## EXERCISES 8.4

**1** $\dfrac{3}{x - 2} + \dfrac{5}{x + 3}$    **3** $\dfrac{5}{x - 6} - \dfrac{4}{x + 2}$

**5** $\dfrac{2}{x - 1} + \dfrac{3}{x + 2} - \dfrac{1}{x - 3}$    **7** $\dfrac{3}{x} + \dfrac{2}{x - 5} - \dfrac{1}{x + 1}$

**9** $\dfrac{2}{x - 1} + \dfrac{5}{(x - 1)^2}$    **11** $-\dfrac{7}{x} + \dfrac{5}{x^2} + \dfrac{40}{3x - 5}$

**13** $\dfrac{24/25}{x + 2} + \dfrac{2/5}{(x + 2)^2} - \dfrac{23/25}{2x - 1}$

**15** $\dfrac{5}{x} - \dfrac{2}{x + 1} + \dfrac{3}{(x + 1)^3}$    **17** $-\dfrac{2}{x - 1} + \dfrac{3x + 4}{x^2 + 1}$

**19** $\dfrac{4}{x} + \dfrac{5x - 3}{x^2 + 2}$    **21** $\dfrac{4x - 1}{x^2 + 1} + \dfrac{3}{(x^2 + 1)^2}$

**23** $2x + \dfrac{1}{x - 1} + \dfrac{3x}{x^2 + 1}$    **25** $3 + \dfrac{4}{x} + \dfrac{8}{x - 4}$

**27** $2x + 3 + \dfrac{2}{x - 1} - \dfrac{3}{2x + 1}$

## EXERCISES 8.5

**1**

**3**

**5**

**7**

**9**

**11**

**13**

**15**

**17**

**19**

**21**

**23**

**25**

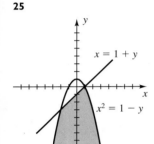

**27** $0 \le x < 3$, $y < -x + 4$, $y \ge x - 4$

**29** $x^2 + y^2 \le 9$, $y > -2x + 4$

**31** $y < x$, $y \le -x + 4$, $(x - 2)^2 + (y - 2)^2 \le 8$

**33** $y > \dfrac{1}{8}x + \dfrac{1}{2}$, $y \le x + 4$, $y \le -\dfrac{3}{4}x + 4$

**35** If $x$ and $y$ denote the number of sets of brand A and brand B, respectively, then a system is $x \ge 20$, $y \ge 10$, $x \ge 2y$, $x + y \le 100$.

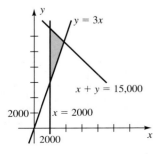

**37** If $x$ and $y$ denote the amount placed in the high-risk and low-risk investment, respectively, then a system is $x \ge 2000$, $y \ge 3x$, $x + y \le 15{,}000$.

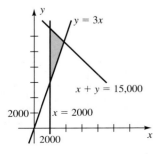

**39** $x + y \le 9$, $y \ge x$, $x \ge 1$

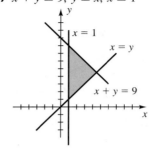

**41** If the plant is located at $(x, y)$, then a system is $(60)^2 \le x^2 + y^2 \le (100)^2$, $(60)^2 \le (x - 100)^2 + y^2 \le (100)^2$, $y \ge 0$.

**43**

**45**

$[-3.5, 4]$ by $[-1, 4]$          $[-1.5, 1.5]$ by $[-1, 1]$

**47** There is no solution.

$[-4.5, 4.5]$ by $[-3, 3]$

## EXERCISES 8.6

**1** Maximum of 27 at $(6, 2)$; minimum of 9 at $(0, 2)$
**3** Maximum of 21 at $(6, 3)$

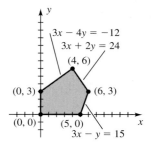

**5** Minimum of 21 at $(3, 2)$

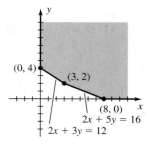

**7** $C$ has the maximum value 24 for any point on the line segment from $(2, 5)$ to $(6, 3)$.

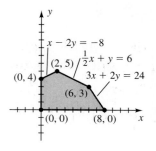

**9** 50 standard and 30 oversized
**11** 3.5 lb of S and 1 lb of T
**13** Send 25 from $W_1$ to A and 0 from $W_1$ to B.
Send 10 from $W_2$ to A and 60 from $W_2$ to B.
**15** None of alfalfa and 80 acres of corn
**17** Minimum cost: 16 oz X, 4 oz Y, 0 oz Z;
maximum cost: 0 oz X, 8 oz Y, 12 oz Z
**19** 2 vans and 4 buses    **21** 3000 trout and 2000 bass
**23** 60 small units and 20 deluxe units

## EXERCISES 8.7

**1** $\begin{bmatrix} 9 & -1 \\ -2 & 5 \end{bmatrix}, \begin{bmatrix} 1 & -3 \\ 4 & 1 \end{bmatrix}, \begin{bmatrix} 10 & -4 \\ 2 & 6 \end{bmatrix}, \begin{bmatrix} -12 & -3 \\ 9 & -6 \end{bmatrix}$

**3** $\begin{bmatrix} 9 & 0 \\ 1 & 5 \\ 3 & 4 \end{bmatrix}, \begin{bmatrix} 3 & -2 \\ 3 & -5 \\ -9 & 4 \end{bmatrix}, \begin{bmatrix} 12 & -2 \\ 4 & 0 \\ -6 & 8 \end{bmatrix}, \begin{bmatrix} -9 & -3 \\ 3 & -15 \\ -18 & 0 \end{bmatrix}$

**5** $[11 \quad -3 \quad -3], [-3 \quad -3 \quad 7],$
$[8 \quad -6 \quad 4], [-21 \quad 0 \quad 15]$

**7** Not possible, not possible,
$\begin{bmatrix} 6 & -4 & 4 \\ 0 & 2 & -8 \\ -6 & 4 & -2 \end{bmatrix}, \begin{bmatrix} -12 & 0 \\ -6 & 3 \\ 3 & -9 \end{bmatrix}$

**9** $-18$

**11** $\begin{bmatrix} 16 & 38 \\ 11 & -34 \end{bmatrix}, \begin{bmatrix} 4 & 38 \\ 23 & -22 \end{bmatrix}$

**13** $\begin{bmatrix} 3 & -14 & -3 \\ 16 & 2 & -2 \\ -7 & -29 & 9 \end{bmatrix}, \begin{bmatrix} 3 & -20 & -11 \\ 2 & 10 & -4 \\ 15 & -13 & 1 \end{bmatrix}$

**15** $\begin{bmatrix} 4 & 8 \\ -18 & 11 \end{bmatrix}, \begin{bmatrix} 3 & -4 & 4 \\ -5 & 2 & 2 \\ -51 & 26 & 10 \end{bmatrix}$

**17** $\begin{bmatrix} 1 & 2 & 3 \\ 4 & 5 & 6 \\ 7 & 8 & 9 \end{bmatrix}, \begin{bmatrix} 1 & 2 & 3 \\ 4 & 5 & 6 \\ 7 & 8 & 9 \end{bmatrix}$

**19** $[15], \begin{bmatrix} -3 & 7 & 2 \\ -12 & 28 & 8 \\ 15 & -35 & -10 \end{bmatrix}$

**21** $\begin{bmatrix} 2 & 0 & 5 \\ 5 & 3 & -2 \end{bmatrix}$, not possible    **23** $\begin{bmatrix} 4 \\ 12 \\ -1 \end{bmatrix}$

**25** $\begin{bmatrix} 18 & 0 & -2 \\ -40 & 10 & -10 \end{bmatrix}$

**35 (a)** $A = \begin{bmatrix} 400 & 550 & 500 \\ 400 & 450 & 500 \\ 300 & 500 & 600 \\ 250 & 200 & 300 \\ 100 & 100 & 200 \end{bmatrix}, B = \begin{bmatrix} \$0.33 \\ \$0.42 \\ \$0.51 \end{bmatrix}$ **(b)** $\begin{bmatrix} \$618.00 \\ \$576.00 \\ \$615.00 \\ \$319.50 \\ \$177.00 \end{bmatrix}$

**(c)** The $177.00 represents the amount the store would receive if all the yellow disks were sold.

## EXERCISES 8.8

**1** $\dfrac{1}{10}\begin{bmatrix} 3 & 4 \\ -1 & 2 \end{bmatrix}$     **3** Does not exist

**5** $\dfrac{1}{8}\begin{bmatrix} 2 & 1 & 0 \\ -2 & 3 & 0 \\ 0 & 0 & 2 \end{bmatrix}$     **7** $\dfrac{1}{3}\begin{bmatrix} -4 & -5 & 3 \\ -4 & -8 & 3 \\ 1 & 2 & 0 \end{bmatrix}$

**9** $\begin{bmatrix} \dfrac{1}{2} & 0 & 0 \\ 0 & \dfrac{1}{4} & 0 \\ 0 & 0 & \dfrac{1}{6} \end{bmatrix}$     **11** $ab \neq 0$; $\begin{bmatrix} \dfrac{1}{a} & 0 \\ 0 & \dfrac{1}{b} \end{bmatrix}$

**15** (a) $\left(\dfrac{13}{10}, -\dfrac{1}{10}\right)$     (b) $\left(\dfrac{7}{5}, \dfrac{6}{5}\right)$

**17** (a) $\left(-\dfrac{25}{3}, -\dfrac{34}{3}, \dfrac{7}{3}\right)$     (b) $\left(\dfrac{16}{3}, \dfrac{16}{3}, -\dfrac{1}{3}\right)$

**19** (a) $\begin{bmatrix} 3.1 & 6.7 & -8.7 \\ 4.1 & -5.1 & 0.2 \\ 0.6 & 1.1 & -7.4 \end{bmatrix}\begin{bmatrix} x \\ y \\ z \end{bmatrix} = \begin{bmatrix} 1.5 \\ 2.1 \\ 3.9 \end{bmatrix}$

(b) $\begin{bmatrix} 0.1474 & 0.1572 & -0.1691 \\ 0.1197 & -0.0696 & -0.1426 \\ 0.0297 & 0.0024 & -0.1700 \end{bmatrix}$

(c) $x \approx -0.1081$, $y \approx -0.5227$, $z \approx -0.6135$

## EXERCISES 8.9

**1** $M_{11} = 0 = A_{11}$; $M_{12} = 5$; $A_{12} = -5$;
$M_{21} = -1$; $A_{21} = 1$; $M_{22} = 7 = A_{22}$

**3** $M_{11} = -14 = A_{11}$; $M_{12} = 10$; $A_{12} = -10$;
$M_{13} = 15 = A_{13}$; $M_{21} = 7$; $A_{21} = -7$;
$M_{22} = -5 = A_{22}$; $M_{23} = 34$; $A_{23} = -34$;
$M_{31} = 11 = A_{31}$; $M_{32} = 4$; $A_{32} = -4$;
$M_{33} = 6 = A_{33}$

**5** 5     **7** −83     **9** 2     **11** 0

**13** −125     **15** 48     **17** −216     **19** $abcd$

**31** (a) $x^2 - 3x - 4$     (b) −1, 4

**33** (a) $x^2 + x - 2$     (b) −2, 1

**35** (a) $-x^3 - 2x^2 + x + 2$     (b) −2, −1, 1

**37** (a) $-x^3 + 4x^2 + 4x - 16$     (b) −2, 2, 4
**39** $-31i - 20j + 7k$     **41** $-6i - 8j + 18k$
**43** (a) $-x^3 + x^2 + 6x - 7$     (b) −2.51, 1.22, 2.29

$[-10, 11]$ by $[-12, 2]$

## EXERCISES 8.10

**1** $R_2 \leftrightarrow R_3$     **3** $-R_1 + R_3 \to R_3$
**5** 2 is a common factor of $R_1$ and $R_3$
**7** $R_1$ and $R_3$ are identical
**9** −1 is a common factor of $R_2$
**11** Every number in $C_2$ is 0     **13** $2C_1 + C_3 \to C_3$
**15** −10     **17** −142     **19** −183     **21** 44     **23** 359
**33** $(4, -2)$     **35** $(8, 0)$
**37** $|D| = 0$ so Cramer's rule cannot be used.
**39** $(2, 3, -1)$     **41** $(-2, 4, 5)$

## CHAPTER 8 REVIEW EXERCISES

**1** $\left(\dfrac{19}{23}, -\dfrac{18}{23}\right)$     **2** No solution     **3** $(-3, 5), (1, -3)$
**4** $(4, -3), (3, -4)$     **5** $(2\sqrt{3}, \pm\sqrt{2}), (-2\sqrt{3}, \pm\sqrt{2})$
**6** $(-1, \pm 1, -1), \left(0, \pm\dfrac{1}{2}\sqrt{6}, -\dfrac{1}{2}\right)$     **7** $\left(\dfrac{14}{17}, \dfrac{14}{27}\right)$
**8** $\left(\log_2 \dfrac{25}{7}, \log_3 \dfrac{15}{7}\right)$     **9** $\left(\dfrac{6}{11}, -\dfrac{7}{11}, 1\right)$
**10** $\left(-\dfrac{6}{29}, \dfrac{2}{29}, -\dfrac{17}{29}\right)$
**11** $(-2c, -3c, c)$ for any real number $c$     **12** $(0, 0, 0)$
**13** $\left(5c - 1, -\dfrac{19}{2}c + \dfrac{5}{2}, c\right)$ for any real number $c$
**14** $(5, -4)$     **15** $\left(-1, \dfrac{1}{2}, \dfrac{1}{3}\right)$     **16** $(3, -1, -2, 4)$
**17** $\dfrac{8}{x-1} - \dfrac{3}{x+5} - \dfrac{1}{x+3}$     **18** $2 + \dfrac{3}{x+1} + \dfrac{4}{(x+1)^2}$
**19** $-\dfrac{2}{x+5} + \dfrac{3x-1}{x^2+4}$     **20** $\dfrac{4}{x^2+2} + \dfrac{x-2}{x^2+5}$

**21**

**22**

**23**

**24**

**25** $\begin{bmatrix} 4 & -5 & 6 \\ 4 & -11 & 5 \end{bmatrix}$   **26** $\begin{bmatrix} 26 \\ -6 \end{bmatrix}$

**27** $\begin{bmatrix} 0 & 4 & -6 \\ 16 & 22 & 1 \\ 12 & 11 & 9 \end{bmatrix}$   **28** $\begin{bmatrix} 0 & -37 \\ 15 & -6 \end{bmatrix}$

**29** $\begin{bmatrix} -12 & 4 & -11 \\ 6 & -11 & 5 \end{bmatrix}$   **30** $\begin{bmatrix} a & 3a \\ 2a & 4a \end{bmatrix}$

**31** $\begin{bmatrix} a & 3a \\ 2b & 4b \end{bmatrix}$   **32** $\begin{bmatrix} 0 & 0 \\ 0 & 0 \end{bmatrix}$

**33** $\begin{bmatrix} 5 & 9 \\ 13 & 19 \end{bmatrix}$   **34** $\begin{bmatrix} 1 & 0 & 0 \\ 0 & 1 & 0 \\ 0 & 0 & 1 \end{bmatrix}$

**35** $-\dfrac{1}{2}\begin{bmatrix} 2 & 4 \\ 3 & 5 \end{bmatrix}$   **36** $\dfrac{1}{11}\begin{bmatrix} 8 & 1 & -2 \\ 5 & 2 & -4 \\ -14 & 1 & 9 \end{bmatrix}$

**37** $\begin{bmatrix} 1 & 0 & 0 \\ 0 & 2 & -7 \\ 0 & -1 & 4 \end{bmatrix}$   **38** $\dfrac{1}{37}\begin{bmatrix} -4 & -20 & 15 \\ 3 & 15 & -2 \\ 9 & 8 & -6 \end{bmatrix}$

**39** $(2, -5)$   **40** $(-1, 3, 2)$   **41** $-6$   **42** 9
**43** 48   **44** $-86$   **45** $-84$   **46** 0   **47** 120
**48** $-76$   **49** 0   **50** $-50$   **51** $-1 \pm 2\sqrt{3}$
**52** $4, \pm\sqrt{7}$
**53** 2 is a common factor of $R_1$, 2 is a common factor of $C_2$, and 3 is a common factor of $C_3$.

**54** Interchange $R_1$ with $R_2$ and then $R_2$ with $R_3$ to obtain the determinant on the right. The effect is to multiply by $-1$ twice.

**55** $a_{11}a_{22}a_{33}\cdots a_{nn}$   **57** $\left(\dfrac{76}{53}, \dfrac{28}{53}\right)$   **58** $\left(\dfrac{2}{3}, \dfrac{31}{21}, \dfrac{1}{21}\right)$

**59** $40\sqrt{5}$ ft $\times 20\sqrt{5}$ ft   **60** $y = \pm 2\sqrt{2}x + 3$
**61** Tax = \$18,750; bonus = \$3,125
**62** Inside radius = 90 ft, outside radius = 100 ft
**63** In ft$^3$/hr: A, 30; B, 20; C, 50
**64** Western 95, eastern 55

**65** If $x$ and $y$ denote the length and width, respectively, then a system is $x \le 12$, $y \le 8$, $y \ge \dfrac{1}{2}x$.

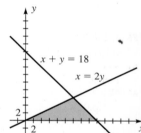

**66** $x + y \le 18$, $x \ge 2y$, $x \ge 0$, $y \ge 0$

**67** 80 mowers and 30 edgers
**68** High-risk \$50,000; low-risk \$100,000; bonds \$0

# CHAPTER 9

## EXERCISES 9.1

**1** 9, 6, 3, 0; $-12$   **3** $\dfrac{1}{2}, \dfrac{4}{5}, \dfrac{7}{10}, \dfrac{10}{17}; \dfrac{22}{65}$

**5** 9, 9, 9, 9; 9
**7** 1.9, 2.01, 1.999, 2.0001; 2.00000001

**9** $4, -\dfrac{9}{4}, \dfrac{5}{3}, -\dfrac{11}{8}; -\dfrac{15}{16}$   **11** 2, 0, 2, 0; 0

**13** $\dfrac{2}{3}, \dfrac{2}{3}, \dfrac{8}{11}, \dfrac{8}{9}; \dfrac{128}{33}$   **15** 1, 2, 3, 4; 8

**17** 2, 1, $-2$, $-11$, $-38$   **19** $-3, 3^2, 3^4, 3^8, 3^{16}$
**21** 5, 5, 10, 30, 120   **23** 2, 2, 4, $4^3$, $4^{12}$

**25** $\dfrac{7}{2}, \dfrac{15}{2}, 12, 17$

**27** $-1, -1 + \dfrac{1}{\sqrt{2}}, -1 + \dfrac{1}{\sqrt{2}} - \dfrac{1}{\sqrt{3}}, -\dfrac{1}{2} + \dfrac{1}{\sqrt{2}} - \dfrac{1}{\sqrt{3}}$

**29** $-5$    **31** $10$    **33** $25$    **35** $-\dfrac{17}{15}$    **37** $61$

**39** $10,000$    **41** $\dfrac{319}{3}$    **43** $\dfrac{7}{2}k^2$

**45** $\displaystyle\sum_{k=1}^{n} (a_k - b_k)$

$$= (a_1 - b_1) + (a_2 - b_2) + \cdots + (a_n - b_n)$$
$$= (a_1 + a_2 + \cdots + a_n) + (-b_1 - b_2 - \cdots - b_n)$$
$$= (a_1 + a_2 + \cdots + a_n) - (b_1 + b_2 + \cdots + b_n)$$
$$= \sum_{k=1}^{n} a_k - \sum_{k=1}^{n} b_k$$

**47** As $k$ increases, the terms approach 1.
**49** 0.4, 0.7, 1, 1.6, 2.8
**51** $a_n = 2n + \dfrac{1}{24}(n-1)(n-2)(n-3)(n-4)(a-10)$

(The answer is not unique.)
**53** (a) 1, 1, 2, 3, 5, 8, 13, 21, 34, 55
   (b) 1, 2, 1.5, 1.$\overline{6}$, 1.6, 1.625, 1.6153846, 1.6190476, 1.6176471, 1.6181818
**55** 2.236068    **57** 2.4493
**59** (a) $f(1) = -1 < 0, f(2) \approx 0.30 > 0$    (b) 1.76
**61** The graph has a horizontal asymptote $y \approx 2.718 \approx e$. Hence, $a_n$ approaches $e$.
**63** The graph has a horizontal asymptote $y \approx 1$. Hence, $a_n$ approaches 1.

## EXERCISES 9.2

**1** Show that $a_{k+1} - a_k = 4$.    **3** 18; 38; $4n - 2$
**5** 1.8; 0.3; $3.3 - 0.3n$    **7** 5.4; 20.9; $(3.1)n - (10.1)$
**9** $\ln 3^5$; $\ln 3^{10}$; $\ln 3^n$    **11** $-8$    **13** $-8.5$
**15** $-9.8$    **17** $\dfrac{551}{17}$    **19** $-105$    **21** 30    **23** 530

**25** $\dfrac{423}{2}$    **27** $\displaystyle\sum_{n=1}^{4} (2n - 1)$    **29** $\displaystyle\sum_{n=1}^{37} (2n - 1)$

**31** $\displaystyle\sum_{n=1}^{6} \dfrac{3n}{4n + 3}$    **33** 24    **35** $\dfrac{10}{3}, \dfrac{14}{3}, 6, \dfrac{22}{3}, \dfrac{26}{3}$

**37** (a) 60    (b) 12,780    **39** 255    **41** $154\pi$ ft
**43** $1200    **45** $16n^2$
**47** Show that the $(n + 1)$st term is 1 greater than the $n$th term.

**49** (a) $\dfrac{8}{36}, \dfrac{7}{36}, \dfrac{6}{36}, \ldots, \dfrac{1}{36}$    (b) $d = -\dfrac{1}{36}$; 1

   (c) $722.22

## EXERCISES 9.3

**1** Show that $\dfrac{a_{k+1}}{a_k} = -\dfrac{1}{4}$.    **3** $\dfrac{1}{2}$; $\dfrac{1}{16}$; $8\left(\dfrac{1}{2}\right)^{n-1} = 2^{4-n}$

**5** 0.03; $-0.00003$; $300(-0.1)^{n-1}$
**7** 3125; 390,625; $5^n$
**9** 20.25; $-68.34375$; $4(-1.5)^{n-1}$
**11** $x^8$; $-x^{14}$; $(-1)^{n-1}x^{2n-2}$
**13** $2^{4x+1}$; $2^{7x+1}$; $2^{(n-1)x+1}$    **15** $\pm\sqrt{3}$    **17** $\dfrac{243}{8}$

**19** $\sqrt[3]{3}$; 36    **21** 88,572    **23** $-\dfrac{1023}{3072}$    **25** $\displaystyle\sum_{n=1}^{7} 2^n$

**27** $\displaystyle\sum_{n=1}^{4} (-1)^{n+1} \dfrac{1}{4}\left(\dfrac{1}{3}\right)^{n-1}$    **29** $\dfrac{2}{3}$    **31** $\dfrac{50}{33}$

**33** Since $|r| = \sqrt{2} > 1$, the sum does not exist.

**35** 1024    **37** $\dfrac{23}{99}$    **39** $\dfrac{2393}{990}$    **41** $\dfrac{5141}{999}$

**43** $\dfrac{16,123}{9999}$    **45** 24    **47** 4, 20, 100, 500

**49** $\dfrac{25}{256}\% \approx 0.1\%$

**51** (a) $N(t) = 10,000(1.2)^t$    (b) 61,917    **53** 300 ft
**55** $3,000,000    **57** (b) 375 mg
**59** (a) $a_{k+1} = \dfrac{1}{4}\sqrt{10}a_k$

   (b) $a_n = \left(\dfrac{1}{4}\sqrt{10}\right)^{n-1} a_1, A_n = \left(\dfrac{5}{8}\right)^{n-1} A_1$,

   $$P_n = \left(\dfrac{1}{4}\sqrt{10}\right)^{n-1} P_1$$

   (c) $\dfrac{16a_1}{4 - \sqrt{10}}$

**61** $38,929.00    **63** $7396.67

**65** (a) $\dfrac{2}{5}, \dfrac{6}{25}, \dfrac{18}{125}, \dfrac{54}{625}, \dfrac{162}{3125}$

   (b) $r = \dfrac{3}{5}$; $\dfrac{2882}{3125} = 0.92224$    (c) $16,000

## EXERCISES 9.4

Exer. 1–32: A typical proof is given for Exercises 1, 5, 9, . . . , 29.

**1** (1) $P_1$ is true, since $2(1) = 1(1 + 1) = 2$.
   (2) Assume $P_k$ is true:
   $2 + 4 + 6 + \cdots + 2k = k(k + 1)$. Hence,
   $2 + 4 + 6 + \cdots + 2k + 2(k + 1)$
   $\qquad\qquad = k(k + 1) + 2(k + 1)$
   $\qquad\qquad = (k + 1)(k + 2)$
   $\qquad\qquad = (k + 1)(k + 1 + 1)$.
   Thus, $P_{k+1}$ is true, and the proof is complete.

**5** (1) $P_1$ is true, since $5(1) - 3 = \frac{1}{2}(1)[5(1) - 1] = 2$.

(2) Assume $P_k$ is true:

$$2 + 7 + 12 + \cdots + (5k - 3) = \frac{1}{2}k(5k - 1). \text{ Hence,}$$

$$2 + 7 + 12 + \cdots + (5k - 3) + 5(k + 1) - 3$$
$$= \frac{1}{2}k(5k - 1) + 5(k + 1) - 3$$
$$= \frac{5}{2}k^2 + \frac{9}{2}k + 2$$
$$= \frac{1}{2}(5k^2 + 9k + 4)$$
$$= \frac{1}{2}(k + 1)(5k + 4)$$
$$= \frac{1}{2}(k + 1)[5(k + 1) - 1].$$

Thus, $P_{k+1}$ is true, and the proof is complete.

**9** (1) $P_1$ is true, since $(1)^1 = \dfrac{1(1 + 1)[2(1) + 1]}{6} = 1$.

(2) Assume $P_k$ is true:

$$1^2 + 2^2 + 3^2 + \cdots + k^2 = \frac{k(k + 1)(2k + 1)}{6}. \text{ Hence,}$$

$$1^2 + 2^2 + 3^2 + \cdots + k^2 + (k + 1)^2$$
$$= \frac{k(k + 1)(2k + 1)}{6} + (k + 1)^2$$
$$= (k + 1)\left[\frac{k(2k + 1)}{6} + \frac{6(k + 1)}{6}\right]$$
$$= \frac{(k + 1)(2k^2 + 7k + 6)}{6}$$
$$= \frac{(k + 1)(k + 2)(2k + 3)}{6}.$$

Thus, $P_{k+1}$ is true, and the proof is complete.

**13** (1) $P_1$ is true, since $3^1 = \frac{3}{2}(3^1 - 1) = 3$.

(2) Assume $P_k$ is true:

$$3 + 3^2 + 3^3 + \cdots + 3^k = \frac{3}{2}(3^k - 1). \text{ Hence,}$$

$$3 + 3^2 + 3^3 + \cdots + 3^k + 3^{k+1} = \frac{3}{2}(3^k - 1) + 3^{k+1}$$
$$= \frac{3}{2} \cdot 3^k - \frac{3}{2} + 3 \cdot 3^k$$
$$= \frac{9}{2} \cdot 3^k - \frac{3}{2}$$
$$= \frac{3}{2}(3 \cdot 3^k - 1)$$
$$= \frac{3}{2}(3^{k+1} - 1).$$

Thus, $P_{k+1}$ is true, and the proof is complete.

**17** (1) $P_1$ is true, since $1 < \frac{1}{8}[2(1) + 1]^2 = \frac{9}{8}$.

(2) Assume $P_k$ is true:

$$1 + 2 + 3 + \cdots + k < \frac{1}{8}(2k + 1)^2. \text{ Hence,}$$

$$1 + 2 + 3 + \cdots + k + (k + 1) < \frac{1}{8}(2k + 1)^2 + (k +$$
$$= \frac{1}{2}k^2 + \frac{3}{2}k + \frac{9}{8}$$
$$= \frac{1}{8}(4k^2 + 12k + 9)$$
$$= \frac{1}{8}(2k + 3)^2$$
$$= \frac{1}{8}[2(k + 1) + 1]^2.$$

Thus, $P_{k+1}$ is true, and the proof is complete.

**21** (1) For $n = 1$, $5^n - 1 = 4$ and 4 is a factor of 4.

(2) Assume 4 is a factor of $5^k - 1$. The $(k + 1)$st term is

$$5^{k+1} - 1 = 5 \cdot 5^k - 1$$
$$= 5 \cdot 5^k - 5 + 4$$
$$= 5(5^k - 1) + 4.$$

By the induction hypothesis, 4 is a factor of $5^k - 1$ and 4 is a factor of 4, so 4 is a factor of the $(k + 1)$st term. Thus, $P_{k+1}$ is true, and the proof is complete.

**25** (1) For $n = 1$, $a - b$ is a factor of $a^1 - b^1$.

(2) Assume $a - b$ is a factor of $a^k - b^k$. Following the hint for the $(k + 1)$st term,

$$a^{k+1} - b^{k+1} = a^k \cdot a - b \cdot a^k + b \cdot a^k - b^k \cdot b$$
$$= a^k(a - b) + (a^k - b^k)b.$$

Since $(a - b)$ is a factor of $a^k(a - b)$ and since by the induction hypothesis $a - b$ is a factor of $(a^k - b^k)$, it follows that $a - b$ is a factor of the $(k + 1)$st term. Thus, $P_{k+1}$ is true, and the proof is complete.

**29** (1) $P_8$ is true, since $5 + \log_2 8 \le 8$.

(2) Assume $P_k$ is true: $5 + \log_2 k \le k$. Hence,

$$5 + \log_2(k + 1) < 5 + \log_2(k + k)$$
$$= 5 + \log_2 2k$$
$$= 5 + \log_2 2 + \log_2 k$$
$$= (5 + \log_2 k) + 1$$
$$\le k + 1.$$

Thus, $P_{k+1}$ is true, and the proof is complete.

**33** $\dfrac{n^3 + 6n^2 + 20n}{3}$    **35** $\dfrac{4n^3 - 12n^2 + 11n}{3}$

**37** (a) $a + b + c = 1$, $8a + 4b + 2c = 5$,

$27a + 9b + 3c = 14$; $a = \frac{1}{3}$, $b = \frac{1}{2}$, $c = \frac{1}{6}$

(b) The method used in part (a) shows that the formula is true for only $n = 1, 2, 3$.

**39 (a)** $\dfrac{1}{5}n^5 + \dfrac{1}{2}n^4 + \dfrac{1}{3}n^3 - \dfrac{1}{30}n$

**(b)** Use mathematical induction.

**41** (1) For $n = 1$,

$\sin(\theta + 1\pi)$

$= \sin\theta\cos\pi + \cos\theta\sin\pi$

$= -\sin\theta = (-1)^1\sin\theta.$

(2) Assume $P_k$ is true: $\sin(\theta + k\pi) = (-1)^k\sin\theta$.

Hence,

$\sin[\theta + (k+1)\pi]$

$= \sin[(\theta + k\pi) + \pi]$

$= \sin(\theta + k\pi)\cos\pi + \cos(\theta + k\pi)\sin\pi$

$= [(-1)^k\sin\theta]\cdot(-1) + \cos(\theta + k\pi)\cdot(0)$

$= (-1)^{k+1}\sin\theta.$

Thus, $P_{k+1}$ is true, and the proof is complete.

**43** (1) For $n = 1$,

$[r(\cos\theta + i\sin\theta)]^1 = r^1[\cos(1\theta) + i\sin(1\theta)].$

(2) Assume $P_k$ is true:

$[r(\cos\theta + i\sin\theta)]^k = r^k(\cos k\theta + i\sin k\theta)$. Hence,

$[r(\cos\theta + i\sin\theta)]^{k+1}$

$= [r(\cos\theta + i\sin\theta)]^k[r(\cos\theta + i\sin\theta)]$

$= r^k[\cos k\theta + i\sin k\theta][r(\cos\theta + i\sin\theta)]$

$= r^{k+1}[(\cos k\theta\cos\theta - \sin k\theta\sin\theta)$

$\qquad + i(\sin k\theta\cos\theta + \cos k\theta\sin\theta)]$

$= r^{k+1}[\cos(k+1)\theta + i\sin(k+1)\theta].$

Thus, $P_{k+1}$ is true, and the proof is complete.

**49** $\dbinom{n}{1} = \dfrac{n!}{(n-1)!\,1!} = n$ and

$\dbinom{n}{n-1} = \dfrac{n!}{[n-(n-1)]!(n-1)!}$

$= \dfrac{n!}{1!(n-1)!} = n$

## EXERCISES 9.6

**1** 210     **3** 60,480     **5** 120     **7** 6

**9 (a)** 60     **(b)** 125     **11** 64     **13** $P(8, 3) = 336$

**15** 24     **17 (a)** 2,340,000     **(b)** 2,160,000

**19 (a)** 151,200     **(b)** 5760     **21** 1024

**23** $P(8, 8) = 40,320$     **25** $P(6, 3) = 120$

**27 (a)** 27,600     **(b)** 35,152     **29** 9,000,000

**31** $P(4, 4) = 24$     **33** $3! \cdot 2^3 = 48$

**35** $P(10, 10) = 3,628,800$

**37 (a)** 900

**(b)** If $n$ is even, $9 \cdot 10^{(n/2)-1}$; if $n$ is odd, $9 \cdot 10^{(n-1)/2}$.

**39**

**(a)** $y = 1$

**(b)** $n! \approx \dfrac{n^n\sqrt{2\pi n}}{e^n}$

## EXERCISES 9.5

**1** 1440     **3** 5040     **5** 336     **7** 1     **9** 21

**11** 715     **13** $(2n+2)(2n+1)$

**15** $64x^3 - 48x^2y + 12xy^2 - y^3$

**17** $a^6 + 6a^5b + 15a^4b^2 + 20a^3b^3 + 15a^2b^4 + 6ab^5 + b^6$

**19** $a^7 - 7a^6b + 21a^5b^2 - 35a^4b^3 + 35a^3b^4 - 21a^2b^5$
$\qquad\qquad + 7ab^6 - b^7$

**21** $81x^4 - 540x^3y + 1350x^2y^2 - 1500xy^3 + 625y^4$

**23** $\dfrac{1}{243}x^5 + \dfrac{5}{81}x^4y^2 + \dfrac{10}{27}x^3y^4 + \dfrac{10}{9}x^2y^6 + \dfrac{5}{3}xy^8 + y^{10}$

**25** $x^{-12} + 18x^{-9} + 135x^{-6} + 540x^{-3} + 1215$
$\qquad\qquad + 1458x^3 + 729x^6$

**27** $x^{5/2} - 5x^{3/2} + 10x^{1/2} - 10x^{-1/2} + 5x^{-3/2} - x^{-5/2}$

**29** $3^{25}c^{10} + 25 \cdot 3^{24}c^{52/5} + 300 \cdot 3^{23}c^{54/5}$

**31** $-1680 \cdot 3^{13}b^{11} + 60 \cdot 3^{14}b^{13} - 3^{15}b^{15}$     **33** $\dfrac{189}{1024}c^8$

**35** $\dfrac{114,688}{9}u^2v^6$     **37** $70x^2y^2$     **39** $448y^3x^{10}$

**41** $-216b^9a^2$     **43** $-\dfrac{135}{16}$     **45** 4.8, 6.19

**47** $4x^3 + 6x^2h + 4xh^2 + h^3$

## EXERCISES 9.7

**1** 35     **3** 9     **5** $n$     **7** 1     **9** $\dfrac{12!}{5!3!2!2!} = 166,320$

**11** $\dfrac{10!}{3!2!2!1!1!1!} = 151,200$     **13** $C(10, 5) = 252$

**15** $C(8, 2) = 28$     **17** $(5! \cdot 4! \cdot 8!) \cdot 3! = 696,729,600$

**19** $3 \cdot C(10, 2) \cdot C(8, 2) \cdot C(4, 2) \cdot C(6, 2) \cdot 3 \cdot 4$
$\qquad\qquad = 4,082,400$

**21** $C(12, 3) \cdot C(8, 2) = 6160$     **23** $C(8, 3) = 56$

**25 (a)** $C(49, 6) = 13,983,816$     **(b)** $C(24, 6) = 134,596$

**27** $C(n, 2) = 45$ and hence $n = 10$     **29** $C(6, 3) = 20$

**31** By finding $C(31, 3) = 4495$

**33 (a)** 1, 2, 4, 8, 16, 32, 64, 128, 256, 512

**(b)** $S_n = 2^{n-1}$

## EXERCISES 9.8

**1 (a)** A representative outcome is (nine of clubs, 3);
312    **(b)** 20; 292; $\dfrac{20}{312}$

**(c)** No; yes; $\dfrac{72}{312}$; $\dfrac{156}{312}$; $\dfrac{36}{312}$; $\dfrac{192}{312}$

**(d)** Yes; no; 0; $\dfrac{92}{312}$

**3 (a)** $\dfrac{4}{52}$   **(b)** $\dfrac{8}{52}$   **(c)** $\dfrac{12}{52}$

**5 (a)** $\dfrac{1}{6}$   **(b)** $\dfrac{1}{6}$   **(c)** $\dfrac{2}{6}$

**7 (a)** $\dfrac{5}{15}$   **(b)** $\dfrac{6}{15}$   **(c)** $\dfrac{9}{15}$

**9 (a)** $\dfrac{2}{36}$   **(b)** $\dfrac{5}{36}$   **(c)** $\dfrac{7}{36}$   **11** $\dfrac{6}{216}$   **13** $\dfrac{3}{8}$

**15** $\dfrac{48 \cdot 13}{C(52, 5)} \approx 0.00024$   **17** $\dfrac{C(13, 4) \cdot C(13, 1)}{C(52, 5)} \approx 0.00358$

**19** $\dfrac{C(13, 5) \cdot 4}{C(52, 5)} \approx 0.00198$   **21** $\dfrac{4}{6}$

**23** $(0.674)^4 \approx 0.2064$

**25 (a)** 0.45   **(b)** 0.10   **(c)** 0.70   **(d)** 0.95

**27 (a)** $\dfrac{C(20, 5) \cdot C(40, 0)}{C(60, 5)} \approx 0.0028$

**(b)** $1 - \dfrac{C(30, 0) \cdot C(30, 5)}{C(60, 5)} \approx 0.9739$

**(c)** $\dfrac{C(10, 0) \cdot C(50, 5)}{C(60, 5)} + \dfrac{C(10, 1) \cdot C(50, 4)}{C(60, 5)} \approx 0.8096$

**29 (a)** $\dfrac{C(8, 8)}{2^8} \approx 0.00391$   **(b)** $\dfrac{C(8, 7)}{2^8} = 0.03125$

**(c)** $\dfrac{C(8, 6)}{2^8} = 0.109375$

**(d)** $\dfrac{C(8, 6) + C(8, 7) + C(8, 8)}{2^8} \approx 0.14453$

**31** $1 - \dfrac{C(48, 5)}{C(52, 5)} \approx 0.34116$   **33** $1 - \dfrac{10}{36} = \dfrac{26}{36}$

**35 (a)** $\dfrac{1}{32}$   **(b)** $1 - \dfrac{1}{32} = \dfrac{31}{32}$

**37 (a)** $\dfrac{C(4, 4)}{4!} = \dfrac{1}{24}$   **(b)** $\dfrac{C(4, 2)}{4!} = \dfrac{1}{4}$

**39 (a)** 0   **(b)** $\dfrac{1}{9}$   **41 (a)** $\dfrac{1}{16}$   **(b)** $\dfrac{C(4, 2)}{2^4} = \dfrac{6}{16}$

**43** $\dfrac{2}{25,827,165}$ (about 1 chance in 13 million)

**45 (a)** $\dfrac{8}{36}$   **(b)** $\dfrac{1}{36}$   **(c)** $\dfrac{244}{495} \approx 0.4929$

**47 (a)** 0.9639   **(b)** 0.95   **49 (b)** 0.76

**51** $0.99   **53** $0.20

## CHAPTER 9 REVIEW EXERCISES

**1** $5, -2, -1, -\dfrac{20}{29}; -\dfrac{7}{19}$

**2** $0.9, -1.01, 0.999, -1.0001; 0.9999999$

**3** $2, \dfrac{1}{2}, \dfrac{5}{4}, \dfrac{7}{8}; \dfrac{65}{64}$   **4** $\dfrac{1}{12}, \dfrac{1}{15}, \dfrac{1}{15}, \dfrac{8}{105}; \dfrac{8}{45}$

**5** $10, \dfrac{11}{10}, \dfrac{21}{11}, \dfrac{32}{21}, \dfrac{53}{32}$   **6** $2, 2, 2, 2, 2$

**7** $9, 3, \sqrt{3}, \sqrt[4]{3}, \sqrt[8]{3}$   **8** $1, \dfrac{1}{2}, \dfrac{2}{3}, \dfrac{3}{5}, \dfrac{5}{8}$

**9** 75   **10** $-\dfrac{37}{10}$   **11** 940   **12** $-10$   **13** $\displaystyle\sum_{n=1}^{5} 3n$

**14** $\displaystyle\sum_{n=1}^{6} 2^{3-n}$   **15** $\displaystyle\sum_{n=1}^{99} \dfrac{1}{n(n + 1)}$   **16** $\displaystyle\sum_{n=1}^{98} \dfrac{1}{n(n + 1)(n + 2)}$

**17** $\displaystyle\sum_{n=1}^{4} \dfrac{n}{3n - 1}$   **18** $\displaystyle\sum_{n=1}^{4} \dfrac{n}{5n - 1}$

**19** $\displaystyle\sum_{n=1}^{5} (-1)^{n+1}(105 - 5n)$   **20** $\displaystyle\sum_{n=1}^{7} (-1)^{n-1}\dfrac{1}{n}$

**21** $\displaystyle\sum_{n=0}^{25} a_n x^{4n}$   **22** $\displaystyle\sum_{n=0}^{20} a_n x^{3n}$   **23** $1 + \displaystyle\sum_{k=1}^{n} (-1)^k \dfrac{x^{2k}}{2k}$

**24** $1 + \displaystyle\sum_{k=1}^{n} \dfrac{x^k}{k}$   **25** $-5 - 8\sqrt{3}; -5 - 35\sqrt{3}$

**26** 52   **27** $-31; 50$   **28** $20, 14, 8, 2, -4, -10$

**29** 64   **30** $-0.00003$   **31** $4\sqrt{2}$   **32** $-\dfrac{12,800}{2187}$

**33** $17; 3$   **34** $\dfrac{1}{81}; \dfrac{211}{1296}$   **35** 570   **36** 32.5

**37** 2041   **38** $-506$   **39** $\dfrac{5}{7}$   **40** $\dfrac{6268}{999}$

**41** (1) $P_1$ is true, since $3(1) - 1 = \dfrac{1[3(1) + 1]}{2} = 2.$

(2) Assume $P_k$ is true:

$2 + 5 + 8 + \cdots + (3k - 1) = \dfrac{k(3k + 1)}{2}$. Hence,

$2 + 5 + 8 + \cdots + (3k - 1) + 3(k + 1) - 1$

$= \dfrac{k(3k + 1)}{2} + 3(k + 1) - 1$

$= \dfrac{3k^2 + k + 6k + 4}{2}$

$= \dfrac{3k^2 + 7k + 4}{2}$

$= \dfrac{(k + 1)(3k + 4)}{2}$

$= \dfrac{(k + 1)[3(k + 1) + 1]}{2}.$

Thus, $P_{k+1}$ is true, and the proof is complete.

**42** (1) $P_1$ is true, since $[2(1)]^2 = \dfrac{[2(1)][2(1) + 1][1 + 1]}{3} = 4.$

(2) Assume $P_k$ is true:

$$2^2 + 4^2 + 6^2 + \cdots + (2k)^2 = \dfrac{(2k)(2k + 1)(k + 1)}{3}.$$

Hence,

$$2^2 + 4^2 + 6^2 + \cdots + (2k)^2 + [2(k + 1)]^2$$
$$= \dfrac{(2k)(2k + 1)(k + 1)}{3} + [2(k + 1)]^2$$
$$= (k + 1)\left(\dfrac{4k^2 + 2k}{3} + \dfrac{12(k + 1)}{3}\right)$$
$$= \dfrac{(k + 1)(4k^2 + 14k + 12)}{3}$$
$$= \dfrac{2(k + 1)(2k + 3)(k + 2)}{3}.$$

Thus, $P_{k+1}$ is true, and the proof is complete.

**43** (1) $P_1$ is true, since $\dfrac{1}{[2(1) - 1][2(1) + 1]} = \dfrac{1}{2(1) + 1} = \dfrac{1}{3}.$

(2) Assume $P_k$ is true:

$$\dfrac{1}{1 \cdot 3} + \dfrac{1}{3 \cdot 5} + \dfrac{1}{5 \cdot 7} + \cdots + \dfrac{1}{(2k - 1)(2k + 1)} = \dfrac{k}{2k + 1}.$$

Hence,

$$\dfrac{1}{1 \cdot 3} + \dfrac{1}{3 \cdot 5} + \dfrac{1}{5 \cdot 7} + \cdots + \dfrac{1}{(2k - 1)(2k + 1)}$$
$$+ \dfrac{1}{(2k + 1)(2k + 3)} = \dfrac{k}{2k + 1} + \dfrac{1}{(2k + 1)(2k + 3)}$$
$$= \dfrac{k(2k + 3) + 1}{(2k + 1)(2k + 3)}$$
$$= \dfrac{2k^2 + 3k + 1}{(2k + 1)(2k + 3)}$$
$$= \dfrac{(2k + 1)(k + 1)}{(2k + 1)(2k + 3)}$$
$$= \dfrac{k + 1}{2(k + 1) + 1}.$$

Thus, $P_{k+1}$ is true, and the proof is complete.

**44** (1) $P_1$ is true, since $1(1 + 1) = \dfrac{(1)(1 + 1)(1 + 2)}{3} = 2.$

(2) Assume $P_k$ is true:

$$1 \cdot 2 + 2 \cdot 3 + 3 \cdot 4 + \cdots + k(k + 1)$$
$$= \dfrac{k(k + 1)(k + 2)}{3}.$$

Hence,

$$1 \cdot 2 + 2 \cdot 3 + 3 \cdot 4 + \cdots + k(k + 1) + (k + 1)(k + 2)$$
$$= \dfrac{k(k + 1)(k + 2)}{3} + (k + 1)(k + 2)$$
$$= (k + 1)(k + 2)\left(\dfrac{k}{3} + 1\right)$$
$$= \dfrac{(k + 1)(k + 2)(k + 3)}{3}.$$

Thus, $P_{k+1}$ is true, and the proof is complete.

**45** (1) For $n = 1$, $n^3 + 2n = 3$ and 3 is a factor of 3.

(2) Assume 3 is a factor of $k^3 + 2k$. The $(k + 1)$st term is

$$(k + 1)^3 + 2(k + 1) = k^3 + 3k^2 + 5k + 3$$
$$= (k^3 + 2k) + (3k^2 + 3k + 3)$$
$$= (k^3 + 2k) + 3(k^2 + k + 1).$$

By the induction hypothesis, 3 is a factor of $k^3 + 2k$ and 3 is a factor of $3(k^2 + k + 1)$, so 3 is a factor of the $(k + 1)$st term. Thus, $P_{k+1}$ is true, and the proof is complete.

**46** (1) $P_5$ is true, since $5^2 + 3 < 2^5.$

(2) Assume $P_k$ is true: $k^2 + 3 < 2^k$. Hence,

$$(k + 1)^2 + 3 = k^2 + 2k + 4$$
$$= (k^2 + 3) + (k + 1)$$
$$< 2^k + (k + 1)$$
$$< 2^k + 2^k$$
$$= 2 \cdot 2^k = 2^{k+1}$$

Thus, $P_{k+1}$ is true, and the proof is complete.

**47** (1) $P_4$ is true, since $2^4 \le 4!.$

(2) Assume $P_k$ is true: $2^k \le k!$. Hence,

$$2^{k+1} = 2 \cdot 2^k \le 2 \cdot k! < (k + 1) \cdot k! = (k + 1)!.$$

Thus, $P_{k+1}$ is true, and the proof is complete.

**48** (1) $P_{10}$ is true, since $10^{10} \le 10^{10}.$

(2) Assume $P_k$ is true: $10^k \le k^k$. Hence,

$$10^{k+1} = 10 \cdot 10^k \le 10 \cdot k^k < (k + 1) \cdot k^k$$
$$< (k + 1) \cdot (k + 1)^k = (k + 1)^{k+1}.$$

Thus, $P_{k+1}$ is true, and the proof is complete.

**49** $x^{12} - 18x^{10}y + 135x^8y^2 - 540x^6y^3 + 1215x^4y^4$
$\qquad\qquad\qquad\qquad - 1458x^2y^5 + 729y^6$

**50** $16a^4 + 32a^3b^3 + 24a^2b^6 + 8ab^9 + b^{12}$

**51** $a^8 + 40a^7 + 760a^6$  **52** $-\dfrac{63}{16}b^{12}c^{10}$

**53** $21{,}504a^{10}b^2$  **54** $52{,}500{,}000$

**55** (a) $d = 1 - \dfrac{1}{2}a_1$  (b) In ft: $1\dfrac{1}{4}, 2, 2\dfrac{3}{4}, 3\dfrac{1}{2}$

**56** 24 ft  **57** $\dfrac{2}{1 - f}$  **58** $\dfrac{7}{4}\sqrt{10} + \sqrt{30} \approx 11.01$

**59** (a) $P(52, 13) \approx 3.954 \times 10^{21}$

(b) $P(13, 5) \cdot P(13, 3) \cdot P(13, 3) \cdot P(13, 2)$
$\qquad\qquad\qquad\qquad \approx 7.094 \times 10^{13}$

**60** (a) $P(6, 4) = 360$  (b) $6^4 = 1296$

**61** (a) $C(12, 8) = 495$  (b) $C(9, 5) = 126$

**62** $\dfrac{17!}{6!5!4!2!} = 85{,}765{,}680$  **63** (a) $\dfrac{2}{4}$  (b) $\dfrac{2}{8}$

**64** (a) $\dfrac{P(26, 4) \cdot 2}{P(52, 4)} \approx 0.1104$  (b) $\dfrac{26^2 \cdot 25^2}{P(52, 4)} \approx 0.0650$

**65** (a) $\dfrac{1}{1000}$  (b) $\dfrac{10}{1000}$  (c) $\dfrac{50}{1000}$  **66** $\dfrac{C(4, 1)}{2^4} = \dfrac{4}{16}$

**67** (a) $\dfrac{C(6, 4) + C(6, 5) + C(6, 6)}{2^6} = \dfrac{22}{64}$  (b) $1 - \dfrac{22}{64} = \dfrac{42}{64}$

**68 (a)** $\dfrac{1}{312}$  **(b)** $\dfrac{57}{312}$  **69** 0.44  **70** $\dfrac{8}{36}$

**71** 5.8125

# CHAPTER 10

## EXERCISES 10.1

**1** $V(0, 0)$; $F(0, 2)$; $y = -2$  **3** $V(0, 0)$; $F\left(-\dfrac{3}{8}, 0\right)$;

$x = \dfrac{3}{8}$

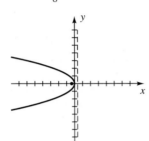

**5** $V(-2, 1)$; $F(-2, -1)$;  **7** $V(3, 2)$; $F\left(\dfrac{49}{16}, 2\right)$;

$y = 3$  $x = \dfrac{47}{16}$

**9** $V(2, -2)$; $F\left(2, -\dfrac{7}{4}\right)$;  **11** $V\left(0, \dfrac{1}{2}\right)$; $F\left(0, -\dfrac{9}{2}\right)$;

$y = -\dfrac{9}{4}$  $y = \dfrac{11}{2}$

**13** $y^2 = 20(x - 1)$  **15** $(x + 2)^2 = -16(y - 3)$
**17** $y^2 = 8x$  **19** $(x - 6)^2 = 12(y - 1)$
**21** $(y + 5)^2 = 4(x - 3)$  **23** $y^2 = -12(x + 1)$
**25** $3x^2 = -4y$  **27** $(y - 5)^2 = 2(x + 3)$
**29** $x^2 = 16(y - 1)$  **31** $(y - 3)^2 = -8(x + 4)$
**33** $y = -\sqrt{x + 3} - 1$  **35** $x = \sqrt{y - 4} - 1$
**37** $y = -x^2 + 2x + 5$  **39** $x = y^2 - 3y + 1$

**41** 4 in.  **43** $\dfrac{9}{16}$ ft from the center of the paraboloid

**45** $2\sqrt{480} \approx 43.82$ in.

**47 (a)** $p = \dfrac{r^2}{4h}$  **(b)** $10\sqrt{2}$ ft  **49** 64,968 ft²

**51**  **53** $(2.08, -1.04)$,
  $(2.92, 1.38)$

$[-11, 10]$ by $[-7, 7]$

$[-2, 4]$ by $[-3, 3]$

## EXERCISES 10.2

**1** $V(\pm 3, 0)$; $F(\pm\sqrt{5}, 0)$  **3** $V(0, \pm 4)$; $F(0, \pm 1)$

**5** $V(0, \pm 4)$; $F(0, \pm 2\sqrt{3})$  **7** $V\left(\pm\dfrac{1}{2}, 0\right)$;

  $F\left(\pm\dfrac{1}{10}\sqrt{21}, 0\right)$

**9** $V(3 \pm 4, -4)$;
$F(3 \pm \sqrt{7}, -4)$

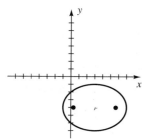

**11** $V(4 \pm 3, 2)$;
$F(4 \pm \sqrt{5}, 2)$

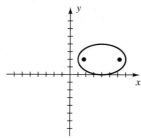

**43** Lower half of $\dfrac{(x+1)^2}{9} + \dfrac{(y-2)^2}{49} = 1$

**45** $\sqrt{84} \approx 9.2$ ft      **47** 94,581,000; 91,419,000

**49** (a) $d = h - \sqrt{h^2 - \dfrac{1}{4}k^2}$; $d' = h + \sqrt{h^2 - \dfrac{1}{4}k^2}$

(b) 16 cm; 2 cm from $V$

**51** 5 ft

**53** $(\pm 1.540, 0.618)$

**55** $(-0.88, 0.76)$,
$(-0.48, -0.91)$,
$(0.58, -0.81)$,
$(0.92, 0.59)$

**13** $V(5, 2 \pm 5)$;
$F(5, 2 \pm \sqrt{21})$

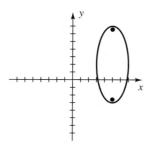

$[-6, 6]$ by $[-2, 6]$

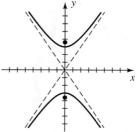

$[-3, 3]$ by $[-2, 2]$

**15** $\dfrac{x^2}{4} + \dfrac{y^2}{36} = 1$     **17** $\dfrac{(x+2)^2}{25} + \dfrac{(y-1)^2}{4} = 1$

**19** $\dfrac{x^2}{64} + \dfrac{y^2}{39} = 1$     **21** $\dfrac{4x^2}{9} + \dfrac{y^2}{25} = 1$

**23** $\dfrac{8x^2}{81} + \dfrac{y^2}{36} = 1$     **25** $\dfrac{x^2}{7} + \dfrac{y^2}{16} = 1$

**27** $\dfrac{x^2}{4} + 9y^2 = 1$     **29** $\dfrac{x^2}{16} + \dfrac{4y^2}{25} = 1$

**31** $(2, 2)$, $(4, 1)$

**EXERCISES 10.3**

**1** $V(\pm 3, 0)$; $F(\pm \sqrt{13}, 0)$;
$y = \pm \dfrac{2}{3}x$

**3** $V(0, \pm 3)$; $F(0, \pm \sqrt{13})$;
$y = \pm \dfrac{3}{2}x$

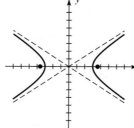

**33** $\dfrac{x^2}{25} + \dfrac{y^2}{16} = 1$     **35** $\dfrac{x^2}{64} + \dfrac{y^2}{289} = 1$

**37** Upper half of $\dfrac{x^2}{49} + \dfrac{y^2}{121} = 1$

**39** Left half of $x^2 + \dfrac{y^2}{9} = 1$

**41** Right half of $\dfrac{(x-1)^2}{4} + \dfrac{(y+2)^2}{9} = 1$

**5** $V(\pm 1, 0)$; $F(\pm 5, 0)$;
$y = \pm \sqrt{24}x$

**7** $V(0, \pm 4)$; $F(0, \pm 2\sqrt{5})$;
$y = \pm 2x$

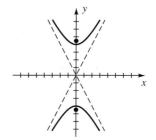

**9** $V\left(\pm\dfrac{1}{4}, 0\right)$;

$F\left(\pm\dfrac{1}{12}\sqrt{13}, 0\right)$;

$y = \pm\dfrac{2}{3}x$

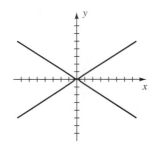

**11** $V(-2, -2 \pm 3)$;

$F(-2, -2 \pm \sqrt{13})$;

$(y + 2) = \pm\dfrac{3}{2}(x + 2)$

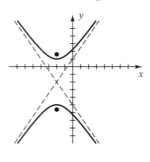

**13** $V(-3 \pm 5, -2)$;

$F(-3 \pm 13, -2)$;

$(y + 2) = \pm\dfrac{12}{5}(x + 3)$

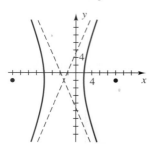

**15** $V(-2, -5 \pm 3)$;

$F(-2, -5 \pm 3\sqrt{5})$;

$(y + 5) = \pm\dfrac{1}{2}(x + 2)$

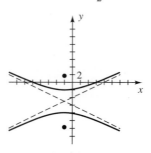

**17** $\dfrac{x^2}{9} - \dfrac{y^2}{16} = 1$    **19** $(y + 3)^2 - \dfrac{(x + 2)^2}{3} = 1$

**21** $y^2 - \dfrac{x^2}{15} = 1$    **23** $\dfrac{x^2}{9} - \dfrac{y^2}{16} = 1$    **25** $\dfrac{y^2}{21} - \dfrac{x^2}{4} = 1$

**27** $\dfrac{x^2}{9} - \dfrac{y^2}{36} = 1$    **29** $\dfrac{x^2}{25} - \dfrac{y^2}{100} = 1$

**31** $\dfrac{y^2}{25} - \dfrac{x^2}{49} = 1$

**33** $(0, 4), \left(\dfrac{8}{3}, \dfrac{20}{3}\right)$

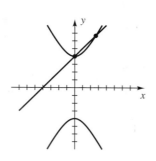

**35** $\dfrac{x^2}{144} - \dfrac{y^2}{25} = 1$    **37** $\dfrac{y^2}{64} - \dfrac{x^2}{36} = 1$

**39** Right branch of $\dfrac{x^2}{25} - \dfrac{y^2}{16} = 1$

**41** Upper branch of $\dfrac{y^2}{9} - \dfrac{x^2}{49} = 1$

**43** Lower halves of the branches of $\dfrac{x^2}{16} - \dfrac{y^2}{81} = 1$

**45** Left halves of the branches of $\dfrac{y^2}{36} - \dfrac{x^2}{16} = 1$

**47** The graphs have the same asymptotes.

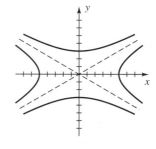

**49** $x = \sqrt{9 + 4y^2}$

**51** If a coordinate system similar to that in Example 6 is introduced, then the ship's coordinates are

$\left(\dfrac{80}{3}\sqrt{34}, 100\right) \approx (155.5, 100)$.

**53** $(0.741, 2.206)$      **55** None

$[-15, 15]$ by $[-10, 10]$      $[-15, 15]$ by $[-10, 10]$

**EXERCISES 10.4**

**1** $y = 2x + 7$          **3** $y = x - 2$

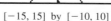

**5** $(y - 3)^2 = x + 5$

**7** $y = 1/x^2$

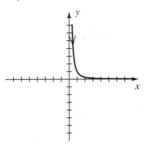

**21** $y = (x^{1/3} + 1)^2$

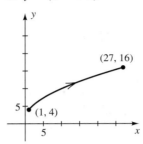

(27, 16)

(1, 4)

**9** $\dfrac{x^2}{4} + \dfrac{y^2}{9} = 1$

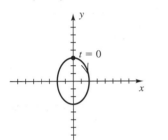

$t = 0$

**11** $x^2 - y^2 = 1$

**23** $C_1$

$C_2$

$t = 0$

**13** $y = \ln x$

**15** $y = 1/x$

$C_3$

$t = \dfrac{\pi}{2} + 2\pi n$

$t = \dfrac{3\pi}{2} + 2\pi n$

$C_4$

**17** $y = \sqrt{x^2 - 1}$

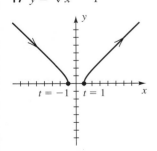

$t = -1$   $t = 1$

**19** $y = |x - 1|$

**25 (a)**

**(b)**

**(c)**

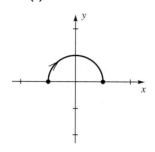

**29** Answers are not unique.

    **(a)** (1) $x = t$,         $y = t^2$;             $t \in \mathbb{R}$

        (2) $x = \tan t$,    $y = \tan^2 t$;      $-\dfrac{\pi}{2} < t < \dfrac{\pi}{2}$

        (3) $x = t^3$,        $y = t^6$;           $t \in \mathbb{R}$

    **(b)** (1) $x = e^t$,        $y = e^{2t}$;        $t \in \mathbb{R}$ (only gives $x > 0$)

        (2) $x = \sin t$,    $y = \sin^2 t$;     $t \in \mathbb{R}$ (only gives $-1 \le x \le 1$)

        (3) $x = \tan^{-1} t$,   $y = (\tan^{-1} t)^2$;   $t \in \mathbb{R}$ $\left(\text{only gives } -\dfrac{\pi}{2} < x < \dfrac{\pi}{2}\right)$

**31 (a)** The figure is an ellipse with center $(0, 0)$ and axes of lengths $2a$ and $2b$.

**35** $x = 4b \cos t - b \cos 4t$,
    $y = 4b \sin t - b \sin 4t$

**37**

$[-6, 6]$ by $[-4, 4]$

**39**

$[-30, 30]$ by $[-20, 20]$

**41** A mask with a mouth, nose, and eyes

**43** The letter A

**EXERCISES 10.5**

**1**

**3**

**5**

**7**

**9**

**11**

**13**

**15**

**29**

**31**

**33**

**17**

**19**

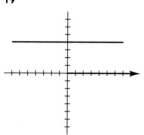

**35** (a) $\left(\frac{3}{2}\sqrt{2}, \frac{3}{2}\sqrt{2}\right)$    (b) $\left(\frac{1}{2}, -\frac{1}{2}\sqrt{3}\right)$

**21**

**23**

**37** (a) $(-4, -4\sqrt{3})$    (b) $\left(-\frac{3}{2}, \frac{3}{2}\sqrt{3}\right)$

**39** $\left(\frac{24}{5}, \frac{18}{5}\right)$    **41** (a) $\left(\sqrt{2}, \frac{3\pi}{4}\right)$    (b) $\left(4, \frac{7\pi}{6}\right)$

**43** (a) $\left(14, \frac{5\pi}{3}\right)$    (b) $\left(5\sqrt{2}, \frac{\pi}{4}\right)$    **45** (a), (c), (e)

**47** $r = -3 \sec \theta$    **49** $r = 4$    **51** $\theta = \tan^{-1}\left(-\frac{1}{2}\right)$

**53** $r^2 = -4 \sec 2\theta$    **55** $r = 2 \cos \theta$

**57** $x = 5$      **59** $x^2 + (y - 3)^2 = 9$

**25**

**27**

**61** $y = x$

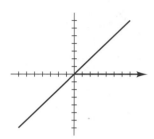

**63** $\dfrac{y^2}{9} - \dfrac{x^2}{4} = 1$

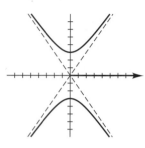

**75** Let $P_1(r_1, \theta_1)$ and $P_2(r_2, \theta_2)$ be points in an $r\theta$-plane. Let $a = r_1$, $b = r_2$, $c = d(P_1, P_2)$, and $\gamma = \theta_2 - \theta_1$. Substituting into the law of cosines, $c^2 = a^2 + b^2 - 2ab \cos \gamma$, gives us the formula.

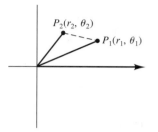

**65** $x^2 - y^2 = 1$

**67** $y - 2x = 6$

**77** Symmetric with respect to the polar axis

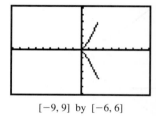

$[-9, 9]$ by $[-6, 6]$

**69** $y = -x^2 + 1$

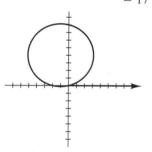

**71** $(x + 1)^2 + (y - 4)^2 = 17$

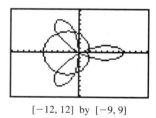

**79** The approximate polar coordinates are $(1.75, \pm 0.45)$, $(4.49, \pm 1.77)$, and $(5.76, \pm 2.35)$.

$[-12, 12]$ by $[-9, 9]$

## EXERCISES 10.6

**1** $\dfrac{1}{3}$, ellipse

**3** 3, hyperbola

**73** $y^2 = \dfrac{x^4}{1 - x^2}$

**5** 1, parabola

**7** $\frac{1}{2}$, ellipse

**3** $V(0, \pm 4)$; $F(0, \pm \sqrt{7})$

**4** $V(0, \pm 4)$; $F(0, \pm 5)$

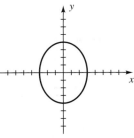

**9** $\frac{3}{2}$, hyperbola

**11** 1, parabola

**5** $V(\pm 2, 0)$; $F(\pm 2\sqrt{2}, 0)$

**6** $V\left(\pm \frac{1}{5}, 0\right)$;
$F\left(\pm \frac{1}{30}\sqrt{11}, 0\right)$

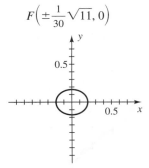

**13** $9x^2 + 8y^2 + 12y - 36 = 0$
**15** $8x^2 - y^2 + 36x + 36 = 0$
**17** $4y^2 + 12x - 9 = 0$    **19** $3x^2 + 4y^2 + 8x - 16 = 0$
**21** $4x^2 - 5y^2 + 36y - 36 = 0$; $x \neq \pm 3$

**23** $x^2 + 8y - 16 = 0$; $x \neq \pm 4$    **25** $r = \dfrac{2}{3 + \cos \theta}$

**27** $r = \dfrac{12}{3 - 4\cos \theta}$    **29** $r = \dfrac{2}{1 - \sin \theta}$

**31** $r = \dfrac{8}{5 + 2\sin \theta}$    **33** $r = \dfrac{8}{1 + \sin \theta}$

**35** (a) $\dfrac{3}{4}$    (b) $r = \dfrac{7}{4 - 3\sin \theta}$

**7** $V(0, 4)$; $F\left(0, -\frac{9}{4}\right)$

**8** $V(3 \pm 2, -1)$;
$F(3 \pm 1, -1)$

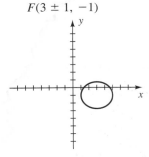

*CHAPTER 10 REVIEW EXERCISES*

**1** $V(0, 0)$; $F(16, 0)$    **2** $V(-2, 1)$; $F\left(-2, \frac{33}{32}\right)$

**9** $V(-4 \pm 1, 5)$;
$F\left(-4 \pm \frac{1}{3}\sqrt{10}, 5\right)$

**10** $V(-5, -2)$;
$F\left(-\frac{39}{8}, -2\right)$

**11** $V(-3 \pm 3, 2)$;
$F(-3 \pm \sqrt{5}, 2)$

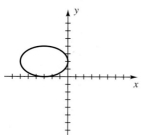

**12** $V(5, -4 \pm 2)$;
$F(5, -4 \pm \sqrt{5})$

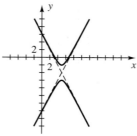

**29 (a)** $-\dfrac{7}{2}$    **(b)** Hyperbola    **30** $A = \dfrac{4a^2b^2}{a^2 + b^2}$

**31** $x^2 + (y - 2)^2 = 4$

**32** $\dfrac{(x - 2)^2}{1} - \dfrac{y^2}{3} = 1$,
$x \geq 3$, or
$x = 2 + \sqrt{1 + \dfrac{y^2}{3}}$

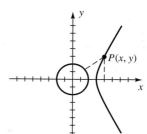

**13** $V(2, -4)$; $F(4, -4)$

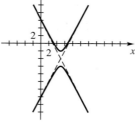

**14** $V(3, -2 \pm 2)$;
$F(3, -2 \pm \sqrt{3})$

**33** $x = 4y + 7$

**34** $y = x^4 - 4$

**15** $V(-4 \pm 3, 0)$;
$F(-4 \pm \sqrt{10}, 0)$

**16** $V(2, -3 \pm 2)$;
$F(2, -3 \pm \sqrt{6})$

**35** $(y - 1)^2 = -(x + 1)$

**36** $y = 2^{-x^2}$

**17** $y = 2(x + 7)^2 - 18$    **18** $y = -3(x + 4)^2 + 147$

**19** $\dfrac{y^2}{49} - \dfrac{x^2}{9} = 1$    **20** $y^2 = -16x$    **21** $x^2 = -40y$

**22** $x = 5y^2$    **23** $\dfrac{x^2}{75} + \dfrac{y^2}{100} = 1$    **24** $\dfrac{x^2}{25} - \dfrac{y^2}{75} = 1$

**25** $\dfrac{y^2}{36} - \dfrac{x^2}{\frac{4}{9}} = 1$    **26** $\dfrac{x^2}{8} + \dfrac{y^2}{4} = 1$

**27** $\dfrac{x^2}{25} + \dfrac{y^2}{45} = 1$    **28** $\dfrac{x^2}{256} + \dfrac{y^2}{112} = 1$

**37** $y = \dfrac{2x^2 - 4x + 1}{x - 1}$

$\left(\dfrac{5}{4}, -\dfrac{7}{2}\right)$

**38** $C_1$

$C_2$

$C_3$

$C_4$

**39** $\left(\dfrac{5}{2}\sqrt{2},\, -\dfrac{5}{2}\sqrt{2}\right)$   **40** $\left(4,\, \dfrac{11\pi}{6}\right)$

**41**

**42**

**43**

**44**

**45**

**46**

**47**

**48**

**49**

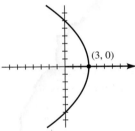

$\left(2, \dfrac{3\pi}{2}\right)$

$\left(-4, \dfrac{\pi}{2}\right)$

**50**

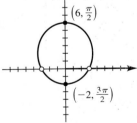

$(3, 0)$

**51** $\dfrac{2}{3}$, ellipse      **52** $\dfrac{1}{2}$, ellipse

$(6, \pi)$      $\left(\dfrac{6}{5}, 0\right)$

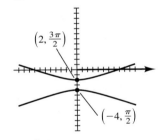

$\left(6, \dfrac{\pi}{2}\right)$

$\left(-2, \dfrac{3\pi}{2}\right)$

**53** $r = 4 \cot \theta \csc \theta$   **54** $r = 3\cos\theta - 4\sin\theta$
**55** $r(2\cos\theta - 3\sin\theta) = 8$   **56** $\theta = \dfrac{\pi}{4}$
**57** $x^3 + xy^2 = y$   **58** $x^2 + y^2 = 2x + 3y$
**59** $(x^2 + y^2)^2 = 8xy$   **60** $y = (\tan\sqrt{3})\,x$
**61** $8x^2 + 9y^2 + 10x - 25 = 0$   **62** $y^2 = 6 + x$

# INDEX